Mechanisms and Machine Science

Volume 113

This book series establishes a well-defined forum for monographs, edited Books, and proceedings on mechanical engineering with particular emphasis on MMS (Mechanism and Machine Science). The final goal is the publication of research that shows the development of mechanical engineering and particularly MMS in all technical aspects, even in very recent assessments. Published works share an approach by which technical details and formulation are discussed, and discuss modern formalisms with the aim to circulate research and technical achievements for use in professional, research, academic, and teaching activities.

This technical approach is an essential characteristic of the series. By discussing technical details and formulations in terms of modern formalisms, the possibility is created not only to show technical developments but also to explain achievements for technical teaching and research activity today and for the future.

The book series is intended to collect technical views on developments of the broad field of MMS in a unique frame that can be seen in its totality as an Encyclopaedia of MMS but with the additional purpose of archiving and teaching MMS achievements. Therefore, the book series will be of use not only for researchers and teachers in Mechanical Engineering but also for professionals and students for their formation and future work.

The series is promoted under the auspices of International Federation for the Promotion of Mechanism and Machine Science (IFToMM).

Prospective authors and editors can contact Mr. Pierpaolo Riva (publishing editor, Springer) at: pierpaolo.riva@springer.com

Indexed by SCOPUS and Google Scholar.

More information about this series at https://link.springer.com/bookseries/8779

Nguyen Van Khang · Nguyen Quang Hoang ·
Marco Ceccarelli
Editors

Advances in Asian Mechanism and Machine Science

Proceedings of IFToMM Asian MMS 2021

Springer

Editors
Nguyen Van Khang
Department of Applied Mechanics
Hanoi University of Science and Technology
Hanoi, Vietnam

Nguyen Quang Hoang
Department of Applied Mechanics
Hanoi University of Science and Technology
Hanoi, Vietnam

Marco Ceccarelli
Department of Industrial Engineering
University of Rome Tor Vergata
Rome, Italy

ISSN 2211-0984 ISSN 2211-0992 (electronic)
Mechanisms and Machine Science
ISBN 978-3-030-91891-0 ISBN 978-3-030-91892-7 (eBook)
https://doi.org/10.1007/978-3-030-91892-7

This Springer imprint is published by the registered company Springer Nature Switzerland AG
The registered company address is: Gewerbestrasse 11, 6330 Cham, Switzerland

Preface

The first conference on Asian Mechanism and Machine Science, in short Asian MMS, started in Taipei in 2010 as an initiative of IFToMM, the International Federation for the Promotion of Mechanism and Machine Science (MMS), as a specific forum for Asiatic communities to promote better relations and disseminations of MMS activities in Asia. Then following events were held successfully in Tokyo in 2012, in Tianjin in 2014, in Guangzhou in 2016, and in Bengaluru in 2018. This year the Asian MMS is organized in Vietnam, a very active IFToMM member organization. Despite the critical situation due the COVID-19 pandemic, the conference has attracted a large number of researchers coming mainly but not only from Asia in a wide range of topics coming mainly but not only from Asia, within the spirit of collaboration of the IFToMM mission. This sixth event of Asian MMS is organized in Hanoi during 15–18 December 2021, with a programme for online participation and paper presentations due to the COVID-19 pandemic limitations thanks to the great effort of local organizers.

The Asian MMS, although primarily intended for Asian countries, serves as a global platform for the participants to exchange ideas and present their research in the several fields of MMS in order to exchange and share new and innovative ideas.

The papers in this proceedings volume were accepted after a peer-review process, and then they are presented online at the relevant session of the conference which covers different topics on Mechanism Design and Theory, Machine and Robot Design, Gearing and Transmissions, Actuators and Sensors, Dynamics and Control of Multibody Systems, Vibration Techniques, Biomechanics, Micro and Nano Systems, and Mechatronics.

We have received 126 papers, of which 93 full papers were accepted after the review for presenting and being included in this proceedings volume together with four keynote contributions. The majority of the papers were from Vietnam, but submissions came also from other IFToMM communities such as China-Beijing, China-Taipei, India, Japan, Russia, Kazakhstan, Korea, Turkey, and even with collaboration from non-Asiatic countries such as Italy, Spain, Germany, France, Canada, USA, Mexico, and Brazil.

We would like to express our sincere gratitude to the reviewers, who contributed to the review process with their experience and expertise in due time with a speedy but rigorous review process. The authors of the papers are also acknowledged for having finalized the papers submission after careful revision according to the review comments. We believe that the published papers can be of interest and stimulus for readers for their future activity also with the aim to contribute with their results to the next events of the Asian MMS.

We would like to thank all the team members of the organizing committee, who helped with the conference organization and preparation of these proceedings. We would like to send also our appreciation to the Springer-Nature team for their support and patience in publishing this book in time for Asian MMS 2021 conference.

September 2021

Nguyen Van Khang
Marco Ceccarelli
Nguyen Quang Hoang

Organization

Programme Chairs

Nguyen Van Khang	Hanoi University of Science and Technology, Vietnam
Marco Ceccarelli	University of Rome Tor Vergata, Italy
Nguyen Quang Hoang	Hanoi University of Science and Technology, Vietnam

Programme Committee

Nguyen Van Khang	Hanoi University of Science and Technology, Vietnam
Marco Ceccarelli	University of Rome Tor Vergata, Italy

Contents

Machine and Robot Design

Flexible Robots and Mechanisms

Dynamics and Control of Machines and Robots

Artificial Intelligence and Robots

Actuators and Sensors Involving Mechanics

Biomechanics and Medical/Healthcare Devices

Micro and Nano Mechanisms and Systems

Engineering Vibrations and Nonlinear Dynamics

Keynotes

Past Achievements and Future Challenges of Mechanism Design for Robotics

Marco Ceccarelli(✉)

LARM2: Laboratory of Robot Mechatronics, Department of Industrial Engineering, University of Rome Tor Vergata, 00133 Rome, Italy
marco.ceccarelli@uniroma2.it

Abstract. Challenges for Robotics and Mechatronics can be considered from several viewpoints in technical, social, and financial ones as due to new designs and applications. In this keynote paper new horizons are discussed in terms of Innovation issues coming from Mechanism Design looking at an inspiration from past achievements. The attention is focused on challenging aspects that are related to the mechanical structure of a modern system as for the structure and operation when considering assigned tasks either in substituting or helping human operators. The keynote speech presents aspects emphasizing the role of mechanism design in system developments form the past and for the future as based on the fact that the action in performing tasks, either in coordination or not with human operators, is of mechanical nature due to motion and force transmission goals of the operation. The challenges of mechanism design are presented both in terms of technical solutions and community activity, since each of them depends, impacts, and generates each other. Examples of past and current solutions are presented to show how a mechanism design can be determinant for a design of robotic systems with novel successful achievements.

Keywords: Robotics · Design · Innovation · Mechanism design · History of MMS

1 A Summary

This keynote addresses the following questions:

- Everywhere is asked for Innovation in technological developments.
- But what does innovation mean in robot design? what is the role of mechanism design?
- What is the community identity and role in those challenges in technological developments in mechanism design for robots?

Today, Innovation is understood as a multidisciplinary activity to produce technological developments for practical implementations with benefits both of their producers and users within society improvements.

Figure 1 summarizes the concept of innovation as produced and exploited by individual actors within multidisciplinary frames with several different areas. The success

N. V. Khang et al. (Eds.): ASIAN MMS 2021, MMS 113, pp. 3–9, 2022.
https://doi.org/10.1007/978-3-030-91892-7_1

of innovation requires that all these aspects will be properly developed and concur to a final implementation of an innovative product/idea in the usage by a large community or public. However, it is to note the innovation is strongly based on technical ideas/achievements, but its success is obtained thanks other non-technical factors, as pointed out in [1–3].

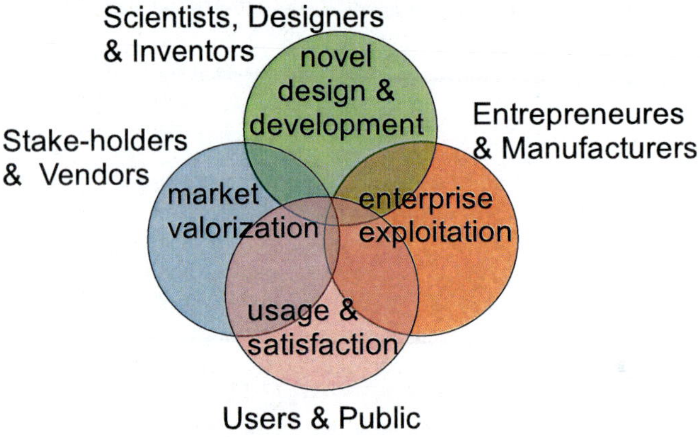

Fig. 1. A scheme for actors and areas in Innovation activity

Activity in Robotics requires a multidisciplinary cooperation even at technological levels so that it is oriented to multidisciplinary work that has facilitated and still stimulates innovation with all the above aspects.

Mechanism design is an important step in design procedures for new robot systems when considering the mechanical nature of the structure and task to be achieved referring to the following aspects:

- Mechanisms will be always an essential part of systems that help or substitute human beings in their actions and tasks
- New continuous needs require to update problems and solutions since Society continuously evolves with new and updated needs and requirements; thus, even mechanism designs are asked to be updated.

Figure 2 summarizes the main concepts and activities of a design process of robot systems with a central role of mechanism design giving a key role to the mechanical design matching operation and design issues also from the multidisciplinary (mechatronics) robot features.

The technical aspects in the first step can be considered completed with the definition of the intellectual achievements to be protected either with patents or publications to perform a technological transfer as well as dissemination and cultural assessments with new formation outcomes. Those aspects can be also worked out during the next phase of production exploit with involvement of entrepreneurs and manufacturing frames. The innovation character is definitively identified when the market valorization of the well-directed service solution is achieved successfully with large/proper sales of the innovative product after or because the users' acceptance and satisfaction for a public fruition. This will give a measure also of the impact of the proposed innovation both from social and financial viewpoints with expected long duration.

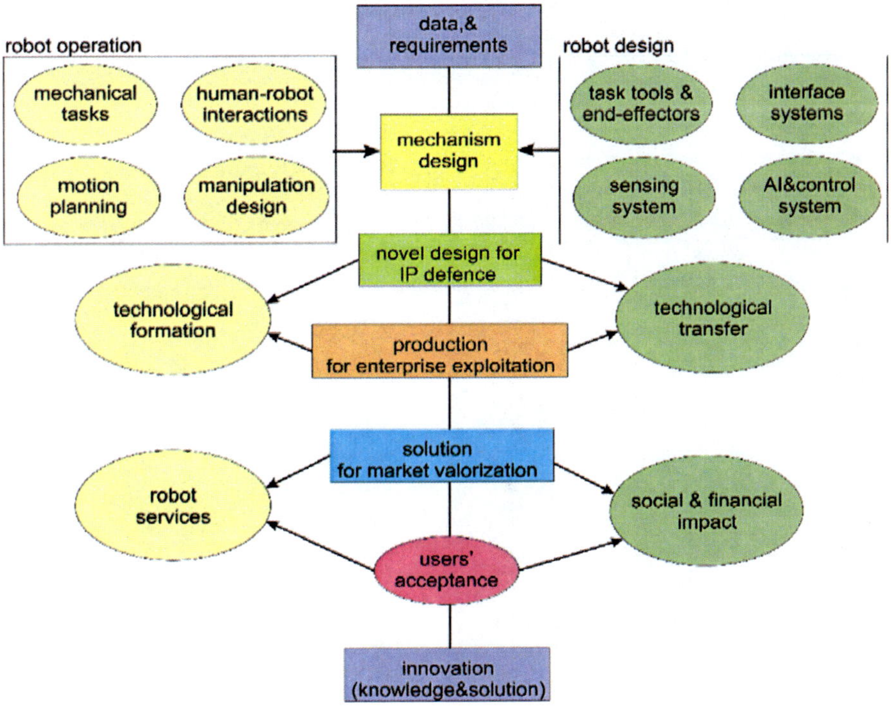

Fig. 2. A scheme of the design activity for robots as centered on Mechanism Design

Examples of Innovation in Robotics are discussed in the keynote to clarify the challenges and results as due to Mechanism Design for robots, both in new robot systems and community developments.

Figure 3 shows examples of humanoid robots as based on mechanism design: in Fig. 3 a) is the pioneer first humanoid Wabot-1 that was presented by prof. I. Kato at the first Romansy in 1972, [4], and in Fig. 3 b) is the LARMbot humanoid design by the author's team as based on several parallel mechanisms. [5]. One can note that both are conceived with an evident mechanism structure with essential mechanical components, although the integration with sensors, actuators, control units and artificial intelligence is essential.

Figure 4 shows a solution of L-CADELv2, a service robotic system for motion assistance of upper limbs, [6], that has been developed at LARM2 in Rome. The design was conceived with a modular design as in Fig. 4 a) with modules and equipment for a robotic service functioning with proper shape and size to be acceptable from the potential users who can be identified in physiotherapists, nurses, and ultimately patients, including elderly people for exercising). A lab prototype is shown in Fig. 4b) whose innovation of the design consists mainly in adjusting a cable driven parallel mechanisms for motion assistance with user-oriented features in conform and operation.

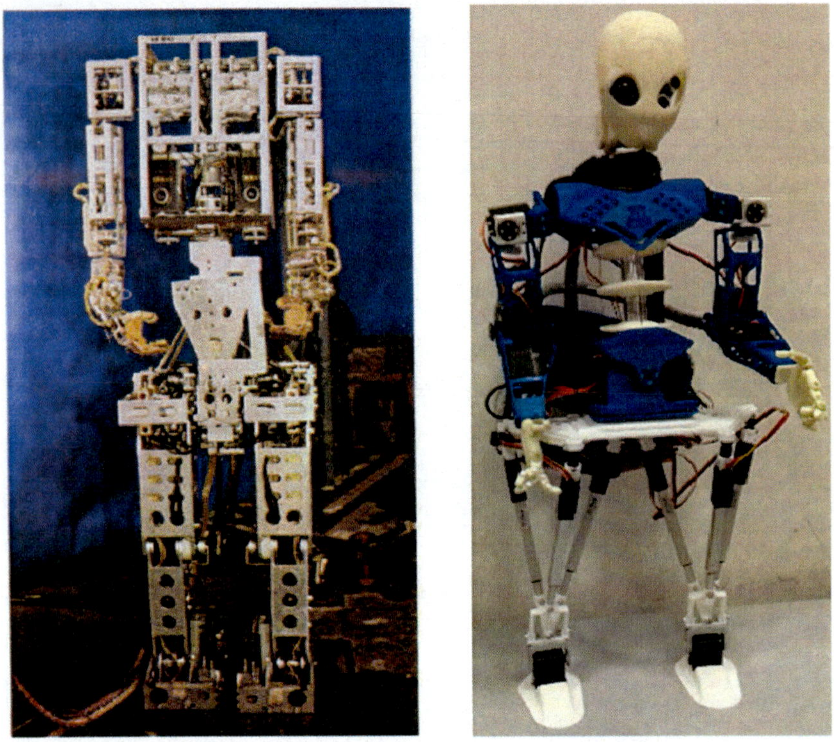

a) b)

Fig. 3. Examples of mechanism design in humanoid robots: a) Wabot 1 humanoid in 1973, [4]; b) LARM bot design in 2017, [5]

An example of innovation from community viewpoints is the start of the Romansy conference series in 1972, [4] Fig. 5 a), as the first pioneering conference forum on Robotics to discuss and share the problems and solution for robot evolution and application. Figure 5b), [7], shows the proceedings page of the MEDER 2021 conference series that is specifically started in 2009 on Mechanism Design for Robots to emphasize the role of the topic in robot developments.

Innovations is produced by inventors coming from a community that is produced by new figures of scientists, professionals, and producers. Significance of innovation is determined by the corresponding community and particularly significant can be considered the history and role of IFToMM in Robotics and MMS at large [3, 8, 9].

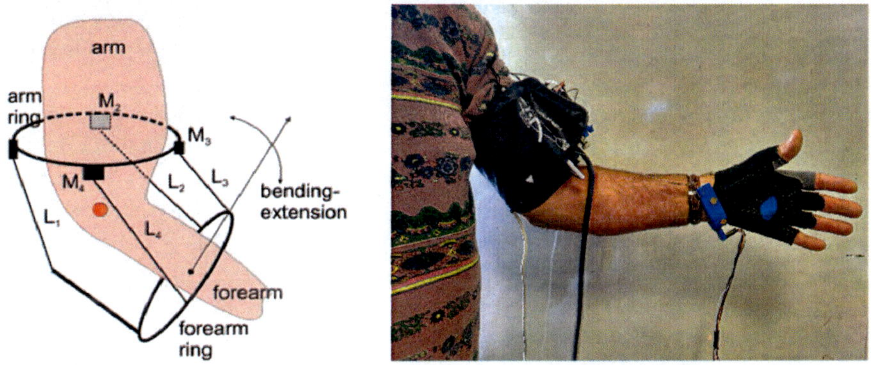

a) b)

Fig. 4. The L-CADELv2 as designed at LARM2 in Rome, [6]: a) the conceptual design; b) a prototype

a) b)

Fig. 5. Title page of proceedings of: a) Romansy in 1973, [4]; b) MEDER in 2021, [7].

The structure of IFToMM is summarized in Fig. 6 a) in terms of IFToMM bodies that are established in the IFToMM constitution (www.iftomm.net). IFToMM mission is indicated in the article 1 of the constitution as finalized to provide leadership for cooperation and development of modern results in Mechanism and Machine Sciences at worldwide levels. The mission statement indicates engineering aspects but also the practical scope of implementation of the activity of the community in developments with technological transfers for the benefit of the society with clear aims toward innovation.

The bodies of IFToMM can be described as in Fig. 6 a) as General Assembly (GA): it is the supreme body of the Federation; Executive Council; Committees of the General Assembly (GACs); Technical Committees (TCs); and Permanent Commissions (PCs). The structure and goals for PCs and TCs are aimed at running activities in their specific fields of interest by attracting researchers and practitioners, including young individuals, and by promoting and developing the field as outlined in Fig. 6 b).

Since its foundation in 1969 IFToMM activity has grown in many aspects, as for example concerning the number of Member Organizations (from the 13 founder ones to the current 44), the size and scale of conference events (on specific topics, at national and international levels, in addition to the MMS World Congress), and the number and focus of technical committees working on specific discipline areas of MMS, [9]. All the IFToMM activities are directed to innovation in MMS technical fields with a clear impact ultimately to the society welfare. IFToMM itself can be considered an innovation result when it is recognized as a product of new attention to a community working for technological developments of MMS.

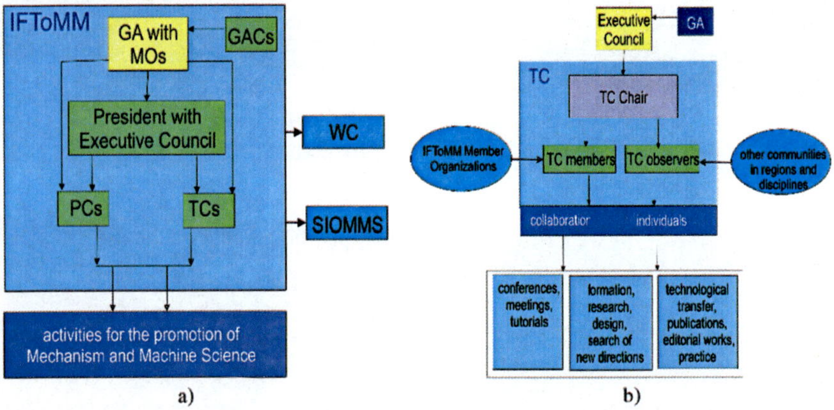

Fig. 6. Structure and activity of IFToMM: a) the overall scheme; b) a scheme for TCs

Significant examples of contributions of IFToMM with innovation contents can be summarized both in community aggregation and identification for new research subjects of large interest, new forums and publication frames, and new formation characters in research and profession. Specifically, in Robotics, the TC of Robotics and Mechatronics is the innovative community result that since the early days of IFToMM searches for new horizons in the field with main attention to mechanical implications of the mechatronic design and operation of robots.

References

1. Ceccarelli, M.: Innovation challenges for mechanism design. Mech. Mach. Theory **125**, 94–100 (2018). https://doi.org/10.1016/j.mechmachtheory.2017.11.026

2. Ceccarelli, M.: Innovation of MMS with inspiration from the past. Int. J. Appl. Mech. Eng. **21**(3), IX–XXII (2016)
3. Ceccarelli, M.: Figures and achievements in MMS as landmarks in history of MMS for inspiration of IFToMM activity. Mech. Mach. Theory **105**, 529–539 (2016). https://doi.org/10.1016/j.mechmachtheory.2016.07.012
4. Kobrinski, I., et al. (eds.): First CISM-IFToMM ROMANSY (Udine 5–8 September 1973). Springer, Wien (1974)
5. Ceccarelli, M., Cafolla, D., Russo, M., Carbone, G.: LARMBot humanoid design towards a prototype. MOJ Int. J. Appl. Bionics Biomech. **1**(2) (2017). https://doi.org/10.15406/mojabb.2017.01.00008
6. Ceccarelli, M., et al.: Design and experimental characterization of L-CADEL v2, an assistive device for elbow motion. Sensors (2021, in print)
7. Zeghloul, S., et al. (eds.): Mechanism Design for Robotics- MEDER 2021, Mechanisms and Machine Science, vol. 103. Springer, Dordrecht (2021)
8. Ceccarelli, M.: A short account of History of IFToMM and its role in MMS. Mech. Mach. Theory **89**, 75–91 (2015)
9. González-Cruz, C.A., Ceccarelli, M., Alimehmeti, M., Jáuregui-Correa, J.C.: Design and experience of a test-bed for gearboxes. In: Uhl, T. (ed.) Advances in Mechanism and Machine Science: IFToMM WC 2019. Mechanisms and Machine Science, vol. 73, pp. 967–976. Springer, Cham (2019). https://doi.org/10.1007/978-3-030-20131-9_96

On the Drive System of Robots Using a Differential Mechanism

Yusuke Sugahara(✉)

Tokyo Institute of Technology, Tokyo, Japan
sugahara.y.ab@m.titech.ac.jp

Abstract. Robots have generally many degrees of freedom and require many actuators to actuate. Designers have often wondered if it would be possible to assign actuators not to each joint but for each desired motion appropriately, and in such cases, the differential mechanism has attracted attention. Furthermore, environmental adaptability can also be achieved by utilizing its characteristics as a multiport system. In this paper, some examples of the use of differential mechanisms in robot mechanisms are introduced, and their aims and achievements are discussed.

Keywords: Differential mechanism · Differential gear · Robot mechanism

1 Drive System of Robots

Robots have generally many degrees of freedom and require many actuators to actuate. In a typical robot design, the drive mechanism for each joint axis is provided independently, and one actuator is provided for each Fig. 1(a). Sometimes, however, there are parts where it is not desirable to place actuators, or there are circumstances where it is not desirable to have a dedicated actuator to drive an axis. What would designers think in such cases?

The first idea is not to assign an actuator to each joint of the robot, but to make the appropriate actuators contribute cooperatively to the desired motion. In this case, a multiple-input-multiple-output drive mechanism, which is used to drive multiple axes in a coordinated manner, is the design candidate Fig. 1(b).

The second idea is to use fewer actuators to drive the system than the desired number of degrees of freedom. In this case, since the system is underactuated, nonlinear control is often used, but the output can be also stabilized by using the characteristics of the environment as a feedback system, as described later Fig. 1(c).

This paper tries to classify the differential mechanisms that sometimes attract attention in the above-mentioned design from its function, and then introduces application examples in robot mechanisms which the author has developed so far, and discuss the aims and results.

N. V. Khang et al. (Eds.): ASIAN MMS 2021, MMS 113, pp. 10–21, 2022.
https://doi.org/10.1007/978-3-030-91892-7_2

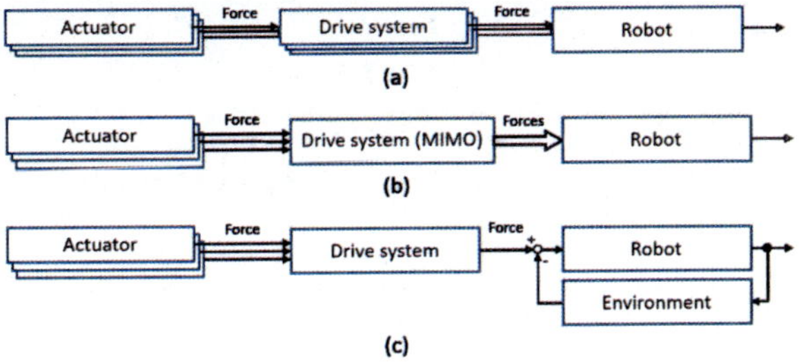

Fig. 1. Drive system of robots.

2 Robots and Differential Mechanisms

An example of a differential mechanism, the bevel-gear differential, is shown in Fig. 2. The mechanism has two degrees of freedom, so that when rotation is applied to two of the three axes, the motion of the other one is determined by it. This allows us to combine two inputs to produce one input, and also allows us to have one input produce two outputs [1]. This system can be described, according to Takahashi [2] and Hirose [3], as a *three-port system* in which the dynamic quantities from the three ports are always in balance. Furthermore, Svoboda [4] introduces it as an *Additive cell,* a kind of elementary computing mechanism, and it can "establish linear relations between mechanical motions of the cell, usually shaft rotations or slide displacements."

Here, there is a general relationship between the forces and displacements at the three ports [3], as follows:

$$a_1x_1 + a_2x_2 + a_3x_3 = 0 \tag{1}$$

$$f_1/a_1 = f_2/a_2 = f_3/a_3 \tag{2}$$

where x_i is a displacement, f_i is a force, and a_i is a constant, which is a kind of reduction ratio. x_i is called an *across variable* because it changes when it crosses the three-port system, and f_i is called a *through variable* because it does not change [2].

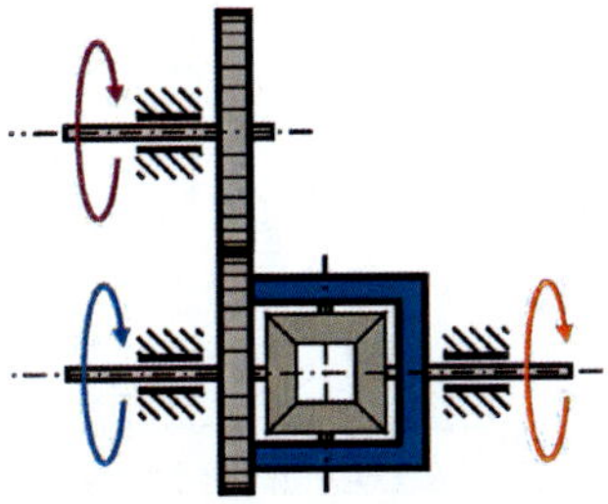

Fig. 2. Bevel-gear differential.

We can think of Eq. (1) as representing the displacement relationship and Eq. (2) as representing the force relationship. Therefore, let us try to classify the use of differential mechanisms in robotics according to this as follows:

Type 1 Using two displacements as input and the remaining one as output (using the relationship in Eq. (1)).
- a Driving of output by multiple power sources (2 inputs and 1 output).
- b Driving of outputs by multiple power sources (2 inputs and 2 outputs).

Type 2 Constraining the motion (using the relationship in Eq. (1)).

Type 3 Realizing the desired force relationship by balancing the three forces (using the relationship in Eq. (2)).
- a Adaptation to environment.
- b Indirect actuation force control using force relationships.

2.1 Type 1: Using Two Displacements as Input and the Remaining One as Output

Using the relationship in Eq. (1), the output of the remaining port can be controlled by controlling two of the three ports. There is also a method to combine two sets of this differential mechanism to make a two-input, two-output system.

1-a Driving of Output by Multiple Power Sources (2 Inputs and 1 Output). In some cranes and hoists, to achieve a redundant system for safety, the power of two motors is interfered by a differential mechanism [1]. This design allows the actuation to continue even if one of the motors fails.

The famous Toyota Hybrid System [5] is a mechanism that uses a differential mechanism to combine the output of the engine, the input power to the generator and the driving power of the vehicle. In this mechanism, the output of the engine, the input power to the generator, and the power of the motor directly connected to the output shaft can be controlled, so that the driving power of the vehicle, which is the output of this system, can be controlled, while simultaneously realizing "drive by motor only", "drive by engine and power generation", "drive by both engine and motor", and "deceleration by regeneration."

Furthermore, Wu *et al.* [6] proposed a power-assisted drive system for the walking chair, in which the input power by human power and the assist power by a motor are used as the input of a differential mechanism, and the power of one axis is the output.

1-b Driving of Outputs by Multiple Power Sources (2 Inputs and 2 Outputs). There are some research examples of combining two of the differential mechanisms to form a two-input, two-output system. This method is mainly used as an implementation of the concept of "Coupled Drive" [7] proposed by Hirose *et al.*

The coupled drive is a method of driving by coupling the output of multiple actuators, instead of using a dedicated actuator for each output degree of freedom, to reduce the overall weight of the system by using multiple actuators to output motion that requires large power.

In Hirose *et al.*'s "FDM coupled drive mechanism" [8], the housings of the two drive motors are rotatable, forming a kind of differential mechanism by itself. By combining two of these, when the two motors rotate in opposite directions, only one of the output shafts rotates, and when they rotate in the same direction, the other shaft rotates.

Haraguchi *et al.*'s "coupled differential gear mechanism" [9] similarly consists of two sets of differential gear mechanisms, where the torque and angular velocity of one output shaft is the sum of the two inputs, while the other output shaft is the difference of the two inputs.

As can be seen from these examples, this mechanism can also be considered as a coordinate transformation mechanism from the actuator displacement vector to the joint displacement vector.

The author has also proposed a similar concept of a mechanism to drive joints and wheels, which is introduced in Sect. 3.

2.2 Type 2: Constraining the Motion

This method similarly uses the relationship in Eq. (1), but without specifying the input-output assignments of the three displacements, using the characteristic that the three displacements are constrained to satisfy this equation.

The PAS-Arm developed by Higuchi [10] uses a differential mechanism combined with a continuous variable transmission, called a "linear combination mechanism," to constrain each joint angle of a passive three-degree-of-freedom spatial manipulator as shown in Eq. (1), thereby successfully constraining its end effector to a virtual spatial guiding surface.

2.3 Type 3: Realizing the Desired Force Relationship by Balancing the Three Forces

By using the relationship in Eq. (2), the desired force relationship can be obtained. This includes methods for equalizing the multiple forces of the mechanism acting on the environment and methods for precisely controlling the output force.

3-A Adaptation to Environment. In this method, power is input to only one of the three ports. Therefore in this case, the differential mechanism alone does not determine the motion. In this approach, the behavior can be often stabilized by constructing a static feedback system by utilizing interaction with the environment.

The most famous example may be the one used for automobiles to transmit power to the drive wheels [1]. Here, the torque from the engine is the input and the torque of the left and right driveshafts is the output, and the differential mechanism works to equalize the forces from the road on the left and right drive wheels.

The rocker-bogie mechanism [11] used in the exploration rover, although of simpler construction, also makes use of this relationship. Here, the gravitational force of the vehicle on the bogie link is the input and the reaction forces of the two wheels from the ground are the output, and the mechanism works so that the ratio of the forces is the desired value. The same effect can be observed for the rocker links. A similar principle has been used in the ankles of walking robots [12], and the author has also used it in a vertical transportation mechanism [13].

The connected differential mechanism [3], which is an extension and connecting of this principle, has been applied to grippers that can grasp flexible objects and foot mechanisms [14].

3-B Indirect Actuation Force Control Using Force Relationships. If the force in one of the three ports is controllable, the forces in the other two ports are also controllable. A clutch mechanism that uses this relationship to precisely control the transmission torque is introduced in Sect. 4.

3 Example 1: A Stair-Climbing Wheelchair with Active Wheeled Four-Bar Linkages [15,16]

TBW-1 Matsushima is a prototype stair climbing wheelchair developed by the author. The aim was to develop a prototype wheelchair that would allow wheelchair users to pass through fragile Japanese historic buildings, without causing larger damage than caused by the passage of walking people. Compared with the crawler-type rough terrain vehicles, the main characteristic of this prototype is that it does not apply concentrated load to the stair nosing and its ground pressure is low.

Its photograph and outline are shown in Fig. 3. The prototype consists of a commercial wheelchair, a leg mechanism with hip and ankle joints, which are active degrees of freedom in the direction of the pitch axis, respectively, and active wheeled four-bar linkage mechanisms connected to each side of the ankle joint. This four-bar linkage mechanism has a wheel at each vertex, and it can transform and drive each wheel as described below, and it basically performs stair climbing by repeatedly transforming into parallelogram mode and straight-line mode as shown in Fig. 4. As four or more wheels are constantly on the step while ascending and descending stairs, low ground pressure and a wide support polygon can be ensured in principle. Furthermore, by using rubber tires on half of the eight wheels and omnidirectional wheels on the other half, the prototype can turn about a point on the line connecting the ground contact points of the left and right rubber tires, not only while running on the ground, but also while ascending and descending stairs.

The transformation modes of this four-bar mechanism are shown in Fig. 5. Two adjacent vertices are active joints, which can be driven to transform from a

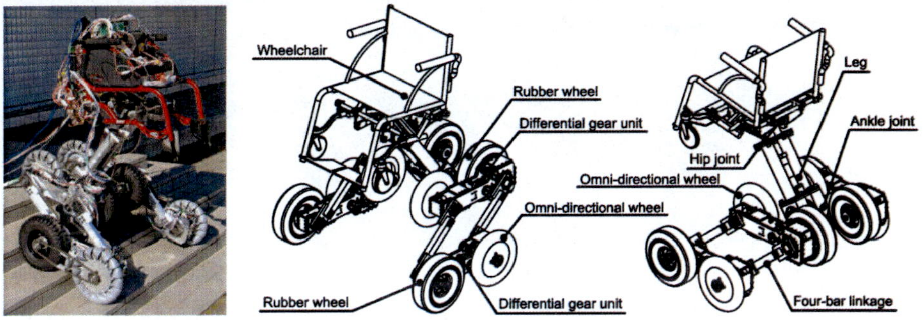

Fig. 3. TBW-1 Matsushima [16].

Fig. 4. Motion sequence of descending stairs [16].

parallelogram mode (Fig. 5(a)) to a straight line mode (Fig. 5(b)). All four links are of the same length, which allows the mechanism to transform to dog-leg mode (Fig. 5(c)) by controlling the joints appropriately in straight line mode. By using this dog-leg mode at the beginning of descending stairs and at the end of ascending stairs as shown in Fig. 4, static stability can be maintained at all moments of ascending and descending stairs.

Simply speaking, if the transformation to dog-leg mode is taken into account, three actuators, one for driving the wheels and two for transformation, are required to drive this four-bar linkage mechanism. However, as the time spent climbing stairs is not expected to be a large proportion of the total time spent using this wheelchair, it is not a good design to have an actuator that only operates when climbing stairs. Therefore in this study, the author proposed a mechanism in which the coupled drive by two actuators was used for driving joints and wheels, respectively.

The structure of the active wheeled four-bar linkage mechanism is shown in Fig. 6. The motors are arranged on two parallel links and the output of each motor is divided by a differential gear into two motions: the rotation angle of the wheels and the angular displacement of the two adjacent joints, i.e. the transformation of the four-bar mechanism. The displacements of the two adjacent

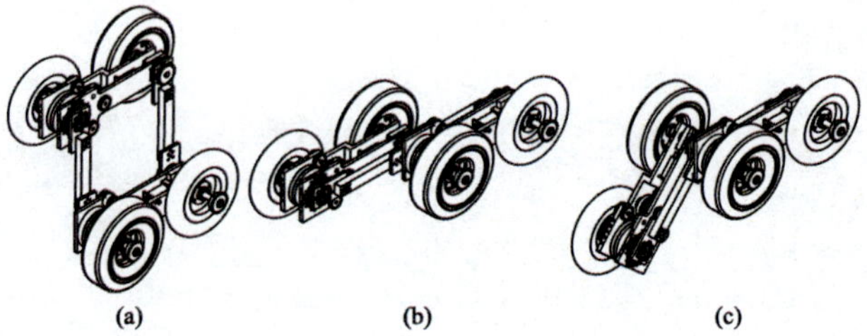

Fig. 5. Transformation of the four-bar linkage module. (a): Parallelogram mode, (b): Straight mode, (c): Dogleg mode [16].

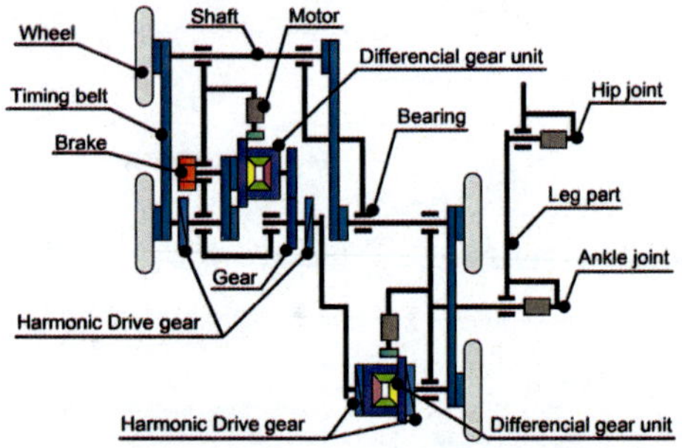

Fig. 6. Structure of the wheeled four-bar linkage mechanism [16].

joints are geometrically related and the four wheels are connected by a timing belt so that the rotation angles of all the wheels are equal.

Here, note that the two output shafts of one differential gear are connected to the wheel and the joint via a timing belt and spur gears respectively, so that the directions of rotation are reversed from each other. With this configuration, when the motors are rotated by the same angle in the same direction, the joints are driven and the four-bar mechanism is transformed, and conversely, when the motors are rotated by the same angle in different directions, the joints do not move and the wheels rotate. Since the two adjacent joints of the four-bar mechanism are the driving joints, the transformation is also possible in two different dog-leg modes.

For the kinematics and statics of this mechanism, the references [15,16] can be consulted.

4 Example 2: Regenerative Clutch [17,18]

Today, many so-called human-powered machines, such as bicycles, wheelchairs and manual hoists, that do not have motors and engines but are powered by the operator's power from handles or pedals, have been still widely used. Since these cannot be controlled by motor control, there are few examples of research on their robotization. If it is possible to control the human-powered machines

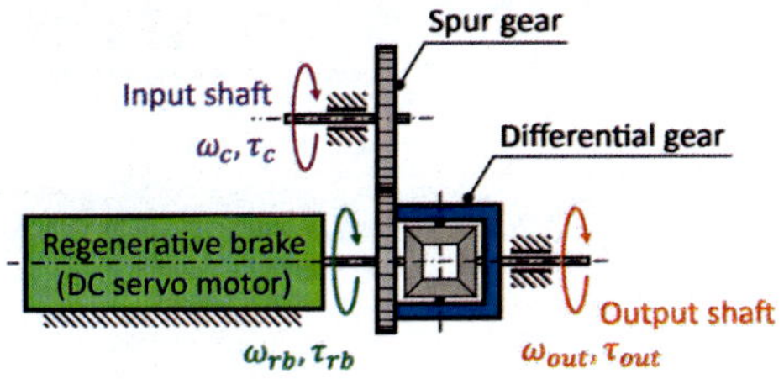

Fig. 7. Regenerative clutch.

(while they are still driven by human power), it should be possible to realize automation and multi-functionalization through robotic technology, while maintaining the advantages of human-powered machines, i.e. intrinsic safety, environmental friendliness and health enhancement ability. Based on this concept of *human-powered robotics*, the author has studied the methods to achieve motion control and robotization by controlling the internal power flow in systems driven by power applied by the operator from handles and pedals.

Here, commercially available powder clutches were initially used to control the power flow. However, powder clutches have problems of poor controllability, large heat generation, and low energy efficiency. Therefore, the author has focused on the regenerative braking function of the motor, which can be expected to have as good response as that of the motor and can also recover energy, and have proposed a servo clutch system using a regenerative brake and a differential gear mechanism.

The proposed *regenerative clutch* is a clutch system that utilizes the force relationship of the differential mechanism to control the transmission torque by controlling the braking torque with a regenerative brake [17]. The structure is shown in Fig. 7. The input shaft is connected to the carrier of the differential gear, one of the two shafts connected to the bevel gears is connected to the output shaft and the other to the shaft of the regenerative brake. By using the relation in Eq. (2), when the regenerative brake works, the same torque as the regenerative braking torque is generated on the output shaft. By applying power to the input shaft and controlling the regenerative brake, a kind of clutch mechanism that can indirectly control the transmitted torque to the output shaft can be realized.

An example of the step response of the prototype regenerative clutch is shown in Fig. 8. It can be seen that the transmitted torque rises in about 10 ms. Compared with a time constant of more than 100 ms for a commertialized powder clutch with a similar torque capacity, the regenerative clutch is confirmed to have a superior response.

Furthermore, by using two sets of these regenerative clutches for forward and reverse torque control respectively, a mechanism for servo control of joint axes for robots can be realized. The photograph and the structure of THR75, the prototype of the developed single-DOF joint mechanism, are shown in Fig. 9.

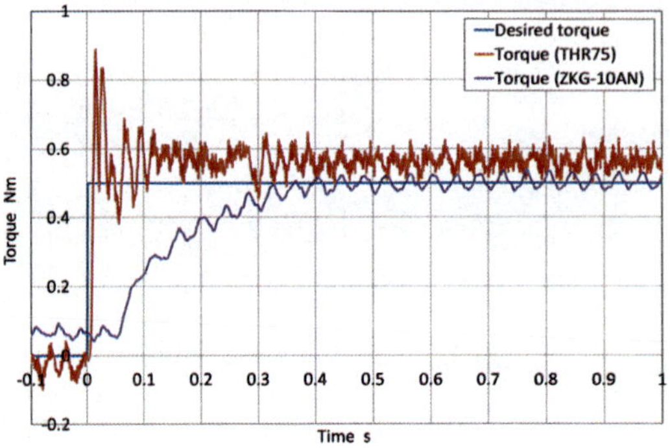

Fig. 8. Time response of the regenerative clutch [17]. $\tau_{ref} = 0.5$ Nm.

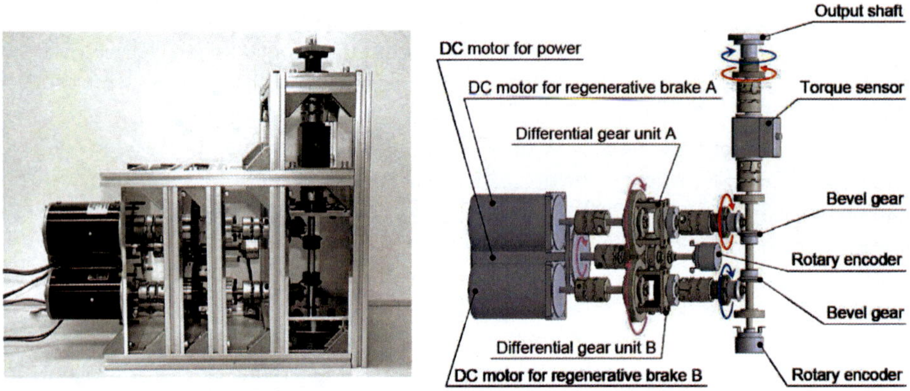

Fig. 9. THR75 [17].

The joint angle control experiment by PD control was carried out using this system, and it was confirmed that it had good control performance [17].

Also, by replacing one of these two sets of regenerative clutches with a constant-load spring, a structure that enables two degrees of freedom joint control can be realized with a total of two regenerative clutches. Figure 10 shows prototype THR78D, a multi-DOF human-powered robot that uses this principle to drive a five-bar mechanism. This prototype realized 2-DOF trajectory control driven by power physically supplied by operator [18].

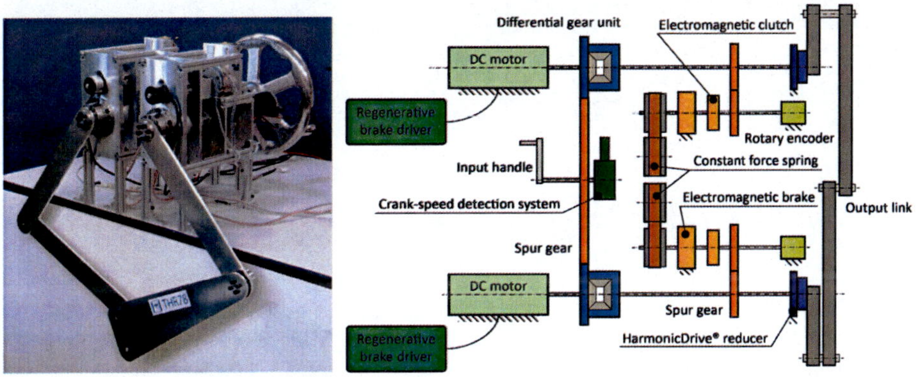

Fig. 10. THR78D [18].

5 Conclusions

This paper attempted to broadly classify the functions of differential mechanisms used in robots.

Regarding the use of the displacement relationship expressed by Eq. (1), it is interesting that the seemingly complex functions of coordinate transformation of joint displacements and motion constraint by a functional relationship can be realized by using only the mechanism.

On the other hand, Takaki [19] pointed out that an environmentally adaptive mechanism can be realized by daring to use an indeterminate chain whose motion is indeterminate. The differential mechanism we have focused on in this paper is a mechanism with exactly this property. According to the force relationship expressed in Eq. (2), if we consider one of the three ports as an input and another as an output, the output can be changed according to the stimulus from the environment input to the last port. This principle is utilized by the rocker-bogie mechanism for ground adaptation, and by the regenerative clutch to construct the control system.

However, it should be noted that the system itself is not necessarily stable, and must be stabilized by some other method for practical use. For example, in the rocker-bogie mechanism, stability is ensured by having a structure in which all wheels are forced to the ground by gravity. Regenerative clutches have the same characteristics and cannot generate torque in the reverse direction by themselves. For this reason, for servo control of the joint, it is necessary to use two sets, or to constantly apply a force in the opposite direction using a spring. It is important to design a system that can be stabilized by utilizing the characteristics of the system, including the environment.

As described above, by effectively using the differential mechanism, we can design robots that mechanically realize advanced functions. The differential mechanism may be said to be some kind of intelligent mechanism.

Acknowledgment. This work was partially supported by JKA and its promotional funds from the KEIRIN RACE, and by JSPS Grants-in-Aid for Scientific Research (KAKENHI), Grant Numbers 20K04376, 17K06254, 23760213 and 20760157.

References

1. Erdman, A.G., Sandor, G.N., Kota, S.: Mechanism Design. Analysis and Synthesis, vol. I, 4th edn. Prentice-Hall, Hoboken (2001)
2. Takahashi, Y.: Systems and Control. Iwanami Shoten, Tokyo (1960)
3. Hirose, S.: Connected differential mechanism and its applications. In: Pham, D.T., Heginbotham, W.B. (eds.) Robot Grippers. IFS Publications, Bedford (1986)
4. Svoboda, A.: Computing Mechanisms and Linkages. McGraw-Hill, New York (1948)
5. Abe, S., Kotani, T., Ibaraki, R., Tojima, K., Shamoto, S., Sakai, A.: A development of Toyota hybrid system. Toyota Tech. Rev. **47**(2), (1997)
6. Wu, Y., Nakamura, H., Takeda, Y., Higuchi, M., Sugimoto, K.: Development of a power assist system of a walking chair based on human arm characteristics. J. Adv. Mech. Des. Syst. Manuf. **1**(1), 141–154 (2007)
7. Hirose, S., Arikawa, K.: Coupled and decoupled actuation of robotic mechanisms. Adv. Robot. **15**(2), 125–138 (2001)
8. Hirose, S., Nishihara, T.: Development of powerful manipulator "Raiden I" (Basic design used FDM and FDM-CD). In: Proceedings of the 13th Annual Conference of the Robotics Society of Japan (1995). (in Japanese)
9. Haraguchi, R., Osuka, K., Sugie, T.: Development of output coupling mechanisms for mechanical systems. J. Robot. Mechatron. **15**(4), 432–441 (2003)
10. Higuchi, M.: Development of a human symbiotic assist arm "PAS-Arm" (basic concept and mechanism). J. Robot. Mechatron. **24**(3), 458–463 (2003)
11. Hacot, H., Dubowsky, S., Bidaud, P.: Analysis and simulation of a rocker-bogie exploration rover. In: Proceedings of the 12th CISM-IFToMM Symposium on Theory and Practice of Robots and Maniputators (ROMANSY12). Springer (1998)
12. Hirose, S., Yoneda, K., Tsukagoshi, H.: TITAN VII: quadruped walking and manipulating robot on a steep slope. In: Proceedings of the IEEE International Conference on Robotics and Automation (1997)
13. Ishii, T., Sugahara, Y., Matsuura, D., Takeda, Y., Matsumoto, T., Nakazawa, D.: Development of self-propelled vertical transportation mechanism that grips and runs on rail. Trans. JSME **86**(890), 20–00141 (2020). (in Japanese)
14. Ogata, M., Hirose, S.: Study on ankle mechanism for walking robots - fundamental considerations on its functions and morphology. In: Proceedings of the 9th Robotics Symposia (2004)
15. Sugahara, Y., Yonezawa, N., Kosuge, K.: A novel stair-climbing wheelchair with variable configuration four-bar linkage —mechanism design and kinematics—. In: Parenti Castelli, V., Schiehlen, W. (eds.) ROMANSY 18 Robot Design, Dynamics and Control. CICMS, vol. 524, pp. 167–174. Springer, Vienna (2010). https://doi.org/10.1007/978-3-7091-0277-0_19
16. Sugahara, Y., Yonezawa, N., Kosuge, K.: A novel stair-climbing wheelchair with transformable wheeled four-bar linkages. In: Proceedings of the 2010 IEEE/RSJ International Conference on Intelligent Robots and Systems (IROS 2010), pp. 3333–3339 (2010)

17. Sugahara, Y., Kikui, K., Endo, M., Okamoto, J., Matsuura, D., Takeda, Y.: A human-powered joint drive mechanism using regenerative clutches. In: Proceedings of the 2017 IEEE/RSJ International Conference on Intelligent Robots and Systems (IROS 2017), pp. 6337–6342 (2017)
18. Sugahara, Y., Tsukamoto, K., Endo, M., Okamoto, J., Matsuura, D., Takeda, Y.: A multi-DOF human-powered robot using regenerative clutches and constant-force springs. In: Proceedings of the 2019 IEEE/RSJ International Conference on Intelligent Robots and Systems (IROS 2019), pp. 4593–4599 (2019)
19. Takaki, T.: Open intelligence and mechanical mechanisms. J. Robot. Soc. Jpn. **36**(9), 605–608 (2018). (in Japanese)

Rational Design of a Micro-positioner with Elastic Hinges

Victor A. Glazunov, Alexey V. Orlov, and Pavel A. Skvortsov(✉)

Mechanical Engineering Research Institute of the Russian Academy of Sciences, Maly'j Khariton'evskij pereulok 4, Moscow, Russia

Abstract. The paper presents the results of numerical simulation and experimental verification of device for micropositioning objects. This device was developed by the staff of the IMASH RAN and based on piezo motors and flexure hinges. As a result of numerical simulation, the maximum equivalent stresses in flexible hinges were determined. The platform displacements obtained during numerical simulation are compared with the results of experimental verification. The obtained characteristics show good convergence of the results of numerical modeling and experimental research.

Keywords: Micropositioning · Flexure hinges · Piezo motors

1 Introduction

1.1 Literature Review

Today, the issue of creating devices for micropositioning of objects in the submicron range is quite actual. These micro-positioning devices have found application in such areas as microelectronics, astronomy, medical technology, bio-nanotechnology, machine tool construction, as well as in the creation of semiconductor devices and various microscopes. In the current difficult political and economic environment, the development of native micro-positioning systems is an important science task.

Currently, there are various approaches to the creation of micro-positioning devices. So, for example, the known design of a magnetostrictive positioner [1], which allows for ultra-precise positioning, however, such designs are usually performed in two stages with a limited range of displacements at the accurate stage. There are designs that combine a magnetostrictor and a hydraulic transducer [2], however, such devices do not provide high positioning accuracy and do not allow movement over a significant distance, providing high linear resolution and large force.

Specialists created structures based on rectangular piezobimorphs, but they were not widely used due to insufficient rigidity, and the designs of positioners based on circular piezobimorphs did not gain mass distribution due to their low rigidity [3, 4].

Known designs of thermosensitive micropositioners which are made by using composite materials and materials with shape memory [5]. The disadvantage of this design is its low efficiency due to small bending deformations and a small maximum deflection,

N. V. Khang et al. (Eds.): ASIAN MMS 2021, MMS 113, pp. 22–30, 2022.
https://doi.org/10.1007/978-3-030-91892-7_3

which does not allow to design a device with small dimensions for a given required deflection.

Known designs of micropositioners, in which the movement of the movable element of the metrological platform is carried out in a non-contact manner. This is achieved through the use of magnetic inserts made of magnetized ferromagnetic material [6]. The disadvantages of this design are the small value of the stroke, the complexity of the design, low metrological reliability of measurements and its inconsistency with the high resolution of the coordinate sensors, insufficient mechanical rigidity of the system, as well as the limited speed of the system's response to the deviation of the coordinates of the executive body from the nominal value.

The known design of a micropositioner based on the phenomenon of superconductor levitation in a magnetic field. This design uses Nd-Fe-b alloy permanent film magnets, which have high remanent magnetization. The main disadvantages of this design are the complexity and high cost of serial production, as well as efficiency only at temperatures from 4,2 K to 80 K [7].

1.2 Object of Study

Currently, the most widely used micropositioning devices are made by using precision linear motors or piezoceramic actuators. The operation of such devices is based on the reverse piezoelectric effect. The advantages of these devices include the stability of properties over time and over a wide temperature range, high Curie temperature, high mechanical strength, and relatively simple manufacturing technology. In such positioning devices, a system consisting of flexure hinges is created to transmit motion to the output link of the mechanism. Various sensors are used to monitor displacements and compensate for external environmental influences. The displacements of the output link can be recorded, for example, by using an interferometer. This design minimizes wear, excludes friction, backlash and noise and excludes the need for lubrication. At present, the most promising design of a manipulator for submicron displacements is a design which was created in the laboratory of the IMASH RAN [8, 9]. The kinematic diagram of the manipulator is shown in Fig. 1.

Currently, the issue of calculating and designing hinges, which will allow large displacements, while maintaining a sufficient margin of safety, remains relevant. This issue is considered in detail in works [10–12]. The calculation of these mechanisms can be based on the theory of bending of beams, according to which it is easy to determine the stresses and strains in a beam of the corresponding cross-section. However, the dependences given in the classical theories of bending work well only for small displacements, when the principle of invariability of the initial dimensions is valid. To calculate elastic hinges with a specific geometry in Fig. 2, even the smallest deformations must be taken into account in devices with submicron precision. This can be realized using numerical modeling.

1.3 Goal of the Work

The goal of this work is the strength analysis of the mechanism for submicron displacements, as well as experimental verification of the obtained results of numerical modeling and the removal of characteristics during the operation of the device.

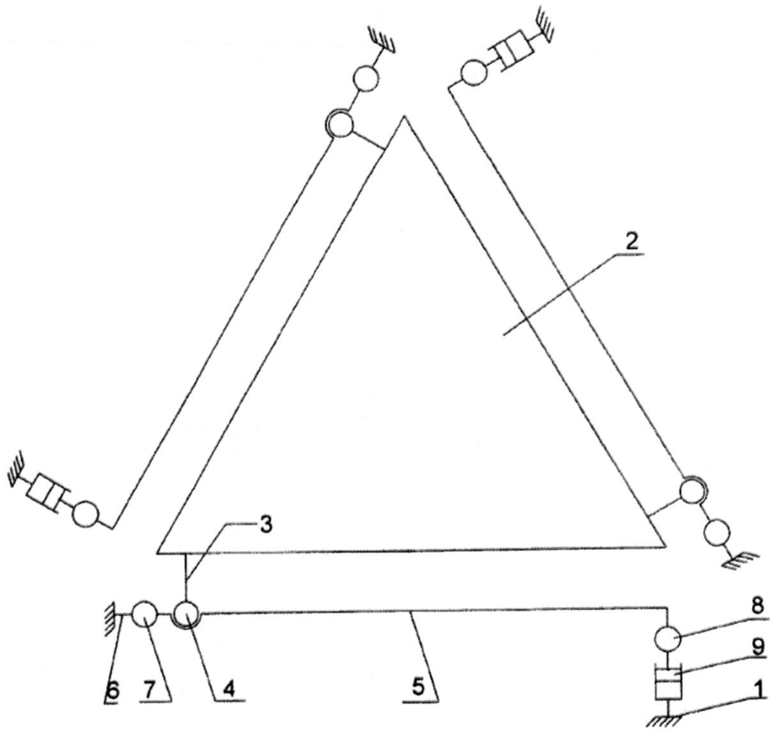

Fig. 1. Kinematic diagram of the manipulator for submicron movements, 1 - output link, 2 - movable platform, 3 - support links, 4.8 - kinematic pairs, 5 - levers, 6 - termination, 7 - rotary hinge, 9 - linear piezo motor

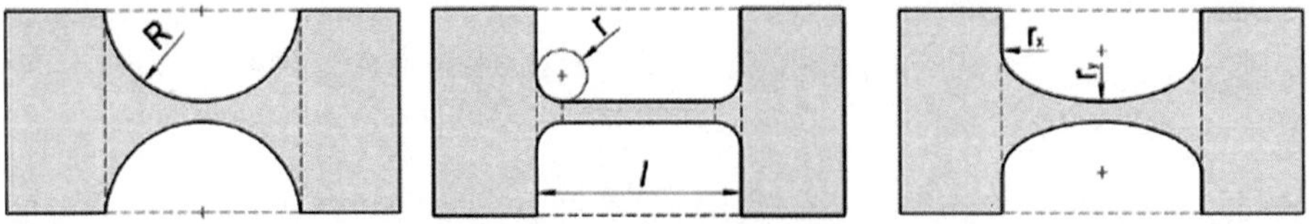

Fig. 2. Various versions of flexible hinges

2 Research of Device for Micropositioning

2.1 Numerical Simulation

The main purpose of the numerical simulation in this work is to determine the maximum equivalent stresses in the micropositioner structure, as well as displacements in the direction of the axis of the piezoelectric motor. Numerical modeling was carried out in the SolidWorks software package. The properties of beryllium bronze were used as the mechanical properties of the material with a modulus of elasticity E = 130 GPa, Poisson's ratio $\mu = 0.33$, ultimate strength $\sigma v = 500$ MPa [13].

The displacement created by one piezo motor was used as the load. The design diagram of the structure and the finite element model are shown in Fig. 3, 4.

In the course of the calculation, the value of the maximum equivalent stress was determined, which was 23.6 MPa, see Fig. 5, which indicates a large margin of safety and the possibility of using these hinges for large displacements.

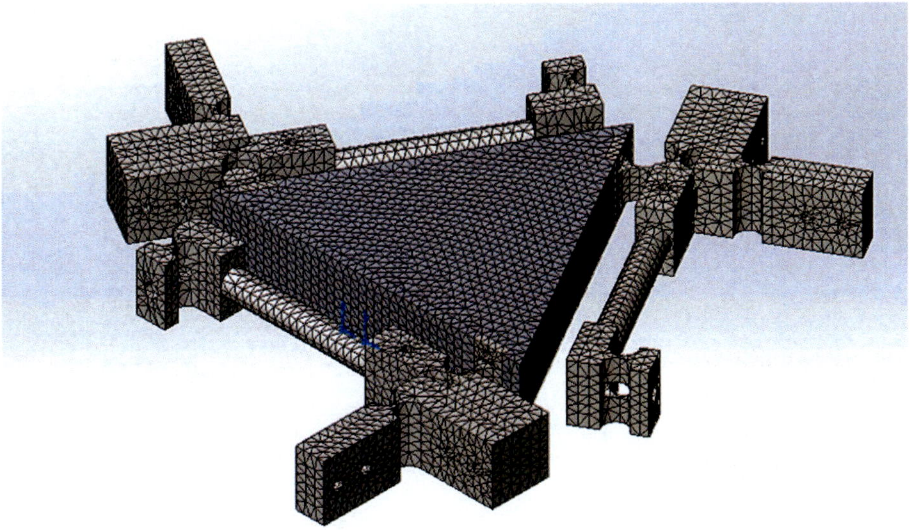

Fig. 3. Finite element model of the calculated micropositioner containing 67602 elements

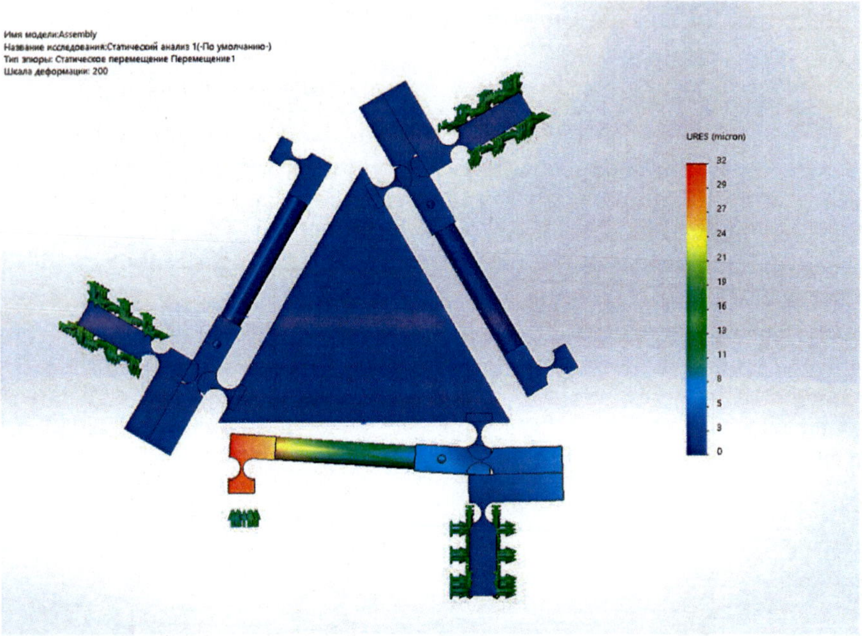

Fig. 4. Design scheme of the structure

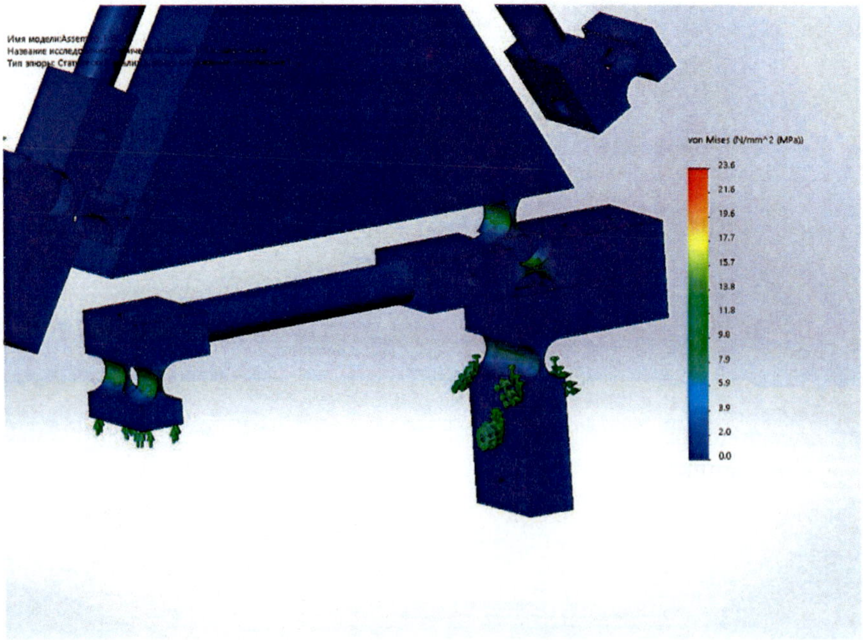

Fig. 5. Maximum equivalent stresses arising in the structure

3 Experimental Study

Checking the adequacy of the model and obtaining accurate characteristics of the platform movement were carried out on an experimental setup, see Fig. 6.

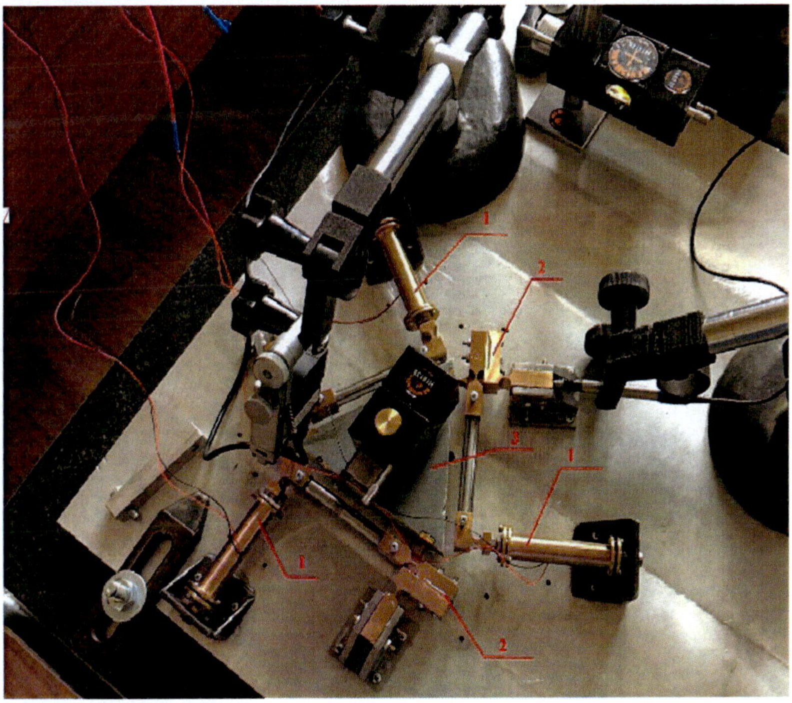

Fig. 6. Photo of the experimental setup, 1 - piezo motors, 2 - flexible hinges, 3 - movable platform

The platform was moved by one piezo motor. In the course of the experiment, a Reneshaw Michelson interferometer was used to measure the linear displacement of the platform, in the center of which an opaque mirror of the interferometer was fixed. The movement was carried out along the axis of the active piezo motor. The movement of the input link of the lever was recorded by an inductive displacement sensor. The obtained characteristics are shown in Fig. 7, 8 and 9. When comparing the displacement of the center of the platform, obtained as a result of numerical simulation with the results during the experiment, it was noted that the results have good convergence.

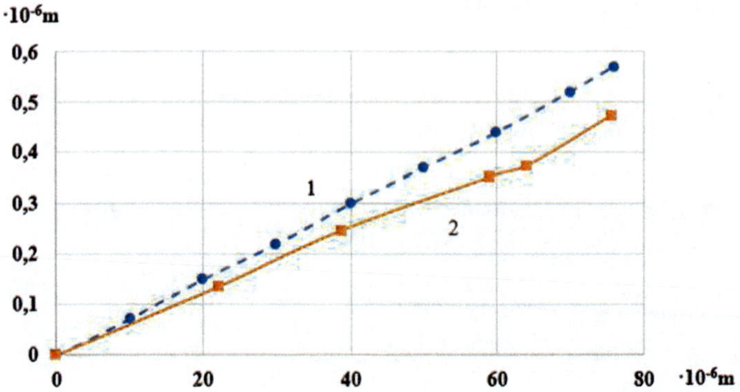

Fig. 7. Dependence of the displacement of the center of the platform on the displacement of the input link, 1 - the characteristic obtained during numerical simulation in SolidWorks, 2 - the characteristic obtained during the experiment

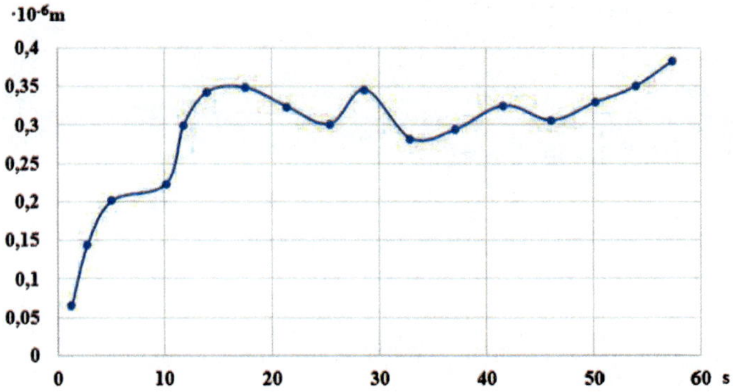

Fig. 8. Dependence of the axial displacement of the center of the platform on time when voltage of 100 V is applied to the motor

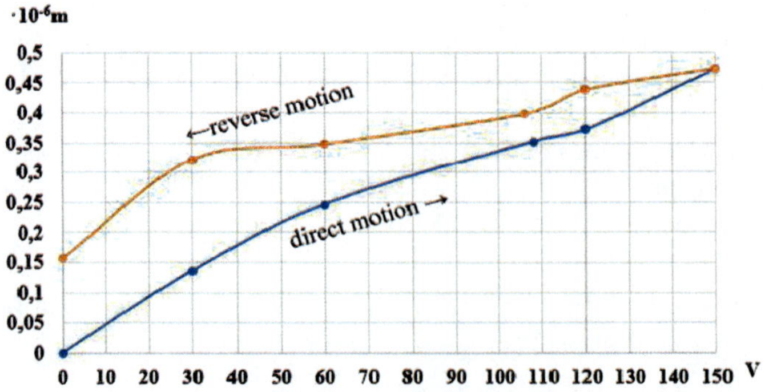

Fig. 9. Diagram of displacements of the center of the platform depending on the voltage on the motor during the forward and reverse motion

4 Conclusions

The paper describes the design of a device that is best able to solve the problem of submicron positioning of objects. In the course of numerical simulation, the maximum equivalent stresses in the structure were determined. The adequacy of the model was verified during the experiment. Also, in the course of the experiment, the dependence of the displacement of the center of the mobile platform on the displacement on the input link, the dependence of the displacement of the center of the mobile platform on time when a constant voltage of 100 V was applied, as well as the dependence of the displacement of the center of the platform on the voltage on the motor during forward and reverse strokes were plotted. A promising direction of research in this area is the analysis and synthesis of rational designs of flexible joints that allow large displacements, while having a sufficient margin of safety to increase the working area of the mechanism, as well as work to compensate and eliminate external environmental influences.

References

1. Aganin, V.A., et al.: Ustroistvo dlya tochnogo posistionirovaniya [Precise positioning device]. Patent RF, no. 2475354 (2012)
2. Rakhovsky, V.I.: Shirokodiapazonnoe ustroistvo dlya tochnogo positsionirovaniya [Broad range device for precise positioning]. Patent RF, no. 2279755 C2 (2006)
3. Bykov, V.A., et al.: Positsioner trechkoordinatni [Three-coordinate positioner]. Patent RF, no. 2297078 C1 (2007)
4. Sokolov, D.: At the sources of nanotechnology. Nanoindustriya **8**(46), 64–69 (2013)
5. Afonina, V.S., et al.: Mikromekhanicheskoe ustrojstvo, sposob ego izgotovleniya i sistema manipulirovaniya mikro- i nanoobèktami [Micromechanical device, method for its manufacture and system for manipulation of micro-and nanoobjects]. Patent RF no. 2458002 C2 (2012)
6. Indukaev, K.V., et al.: Mnogokoordinatnaya metrologicheskaya platforma [Multiaxis metrological platform]. Patent RF no. 2365953 C1 (2009)
7. Parilov, A.A.: Mikropozitsioner na osnove vysokoenergichnykh postoyanny'kh plenochny'kh magnitov [Micropositioner based on high-energy permanent film magnets]. Patent RF no. 26282 U1 (2002)
8. Orlov, A.V., et al.: Manipulyator dlya submikronny'kh peremeshhenij [Manipulator for submicron displacements]. Patent RF no. 2679260 C1 (2019)
9. Glazunov, V.A., Kheilo, S.V. (eds.): Mekhanizmy' perspektivny'kh robototekhnicheskikh sistem [Mechanisms of advanced robotic systems], 296 p. Technosphera Publication, Moscow (2020)
10. Zhu, B., Zhang, X., Liu, M., et al.: Topological and shape optimization of flexure hinges for designing compliant mechanisms using the level set method. Chin. J. Mech. Eng. **32**(13) (2019). https://doi.org/10.1186/s10033-019-0332-z. Accessed 20 July 2021
11. Lin, R., Zhang, X., Long, X., Fatikow, S.: Hybrid flexure hinges. Rev. Sci. Instrum. **84**, 085004 (2013). https://doi.org/10.1063/1.4818522. https://aip.scitation.org/doi/pdf/10.1063/1.4818522. Accessed 20 July 2021

12. Linß, S., Henning, S., Zentner, L.: Modeling and design of flexure hinge-based compliant mechanisms. Kinematics. IntechOpen (2019). https://doi.org/10.5772/intechopen.85224. https://www.researchgate.net/publication/332290865_Modeling_and_Design_of_Flexure_Hinge-Based_Compliant_Mechanisms. Accessed 20 July 2021
13. Babichev, A.P., Babushkina, N.A., Bratkovskij, A.M.: Fizicheskie velichiny': spravochnik [Physical quantities: a reference book]. Grigorèv, I.S., Mejlikhov, E.Z. (eds.), 1232 p. Energoatomizdat, Moscow (1991)

Vibration Suppression of Beam Structures by Multiple Dynamic Vibration Absorbers

Nguyen Phong Dien(✉), Nguyen Van Khang, Nguyen Thi Van Huong, and Hoang Trung Nghia

Hanoi University of Science and Technology, Hanoi, Vietnam
dien.nguyenphong@hust.edu.vn

Abstract. Vibration absorbers are frequently used to suppress the excessive vibrations in structural systems. In this paper, an imposing nodes technique is applied for vibration suppression of Euler-Bernoulli beams subjected to forced harmonic excitations by means of multiple dynamic vibration absorbers. A procedure based on Taguchi method is proposed to determine the optimum absorber parameters to suppress the vibration amplitude of the beams. Numerical tests are performed to show the effectiveness of the proposed procedure.

Keywords: Beam structures · Dynamic vibration absorber · Taguchi method · Harmonic excitations · Passive vibration control

1 Introduction

Beams are conventional constructions such as house beams, suspended cables in suspension structures, air traffic control towers, wind turbine columns, etc. In the course of the work, these structures are exposed to the effects of wind exploitation and operation. Under the influence of changing external frequencies, beam structures appear to be subjected to forced vibration. Since the frequency of the external force changes over a wide band, there is the possibility of a resonance which can cause structural damage. Therefore, the reduction of the amplitude range of the structure at resonant frequencies is a necessary task. To determine the resonance frequencies and to investigate the behavior of the system under the action of external forces, the natural frequencies of the structure must first be evaluated [1–4]. Vibration amplitudes at different points in the structure excited by the external force in a wide frequency band are then calculated.

The dynamic vibration absorbers are usually used to reduce the vibrations of the beams. Optimal parameter design for dynamic vibration absorber installed in beam structures becomes an interesting problem in recent years. It is well known that Taguchi method [5–7] for the product design process may be divided into three stages: system design, parameter design, and tolerance design. Taguchi method of parameter design is successfully applied to many mechanical systems: an acoustic muffler, a gear/pinion system, a spring, an electro-hydraulic servo system, a dynamic vibration absorber. In each system, the design parameters to be optimized are identified, along with the desired

N. V. Khang et al. (Eds.): ASIAN MMS 2021, MMS 113, pp. 31–45, 2022.
https://doi.org/10.1007/978-3-030-91892-7_4

response [8–18]. The problem of optimal installation positions of dynamic vibration absorbers has not been studied much [19, 20].

In this paper, a model of the beam subjected to harmonic forces is presented, in which a system of translational dynamic vibration absorbers is attached. The effectiveness of vibration reduction by dynamic vibration absorbers is then investigated. The optimization procedure for position parameters of the attached absorbers in a beam subjected to a uniform harmonic force is also presented.

2 Mathematical Model

Let us consider the model of a Euler-Bernoulli beam of the length L and the flexural rigidity EI, which is attached with a number of multiple dynmic vibration absorbers at positions x = η_j (j = 1, …, n_a) as shown in Fig. 1. For simplicity, it is assumed that the considered beam is homogeneous with an uniform cross section, where w is the dynamic deflection, u_j is vertical coordinate of j-absorber, k_j, d_j, m_j are stiffness, damping coefficients and mass of j-absorber, respectively, and $p(x, t)$ is the distributed force.

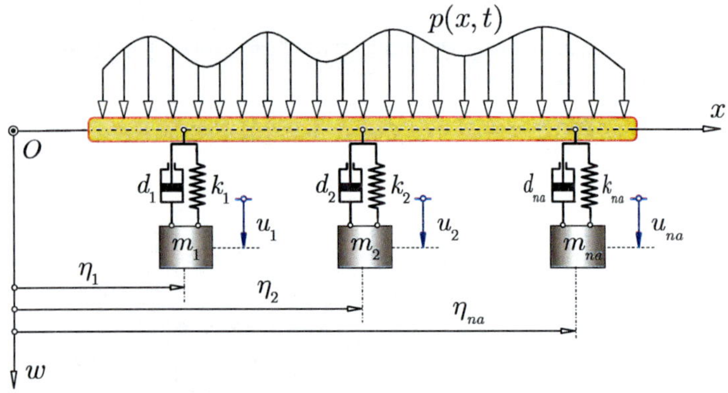

Fig. 1. Model of a beam with TMDs

Using the method of substructures, the system is now divided into $n_a + 1$ substructures, namely, the beam structure and n_a absorbers (Fig. 2).

Reaction forces have the following form

$$F_j(t) = k_j(u_j - \mathrm{w}_{\eta j}) + d_j(\dot{u}_j - \dot{\mathrm{w}}_{\eta j}) \quad j = (1, 2, \ldots, na) \tag{1}$$

where

$$\mathrm{w}_{\eta j} = \mathrm{w}(\eta_j, t), \quad \dot{\mathrm{w}}_{\eta j} = \frac{\partial \mathrm{w}(\eta_j, t)}{\partial t}.$$

Using Newton's second law, the equation describing the vibration of j-absorber can be expressed in the form

$$\mathrm{m}_j\ddot{u}_j = -F_j \tag{2}$$

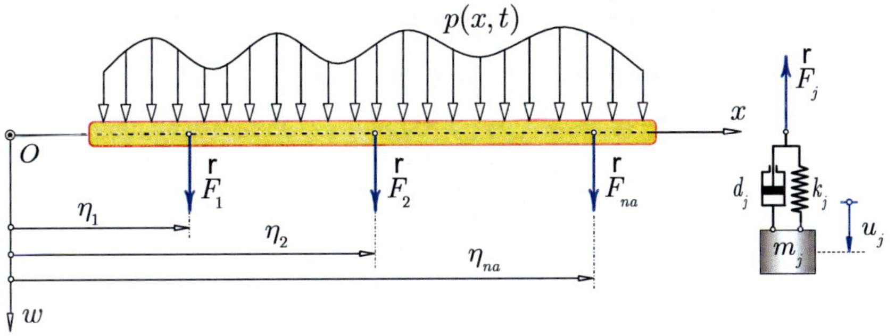

Fig. 2. Substructures: the beam and absorber

Substitution of Eq. (2) into Eq. (1) yields the vibration equation of j-absorber

$$m_j\ddot{u}_j + d_j\dot{u}_j + k_j u_j = k_j w_{hj} + d_j \dot{w}_{hj},\ \ j = 1, 2, ..., n_a. \tag{3}$$

Applying the basic principles of dynamics, the equation that describes transverse vibration of beam including internal friction is [1, 2]

$$\mu(\frac{\partial^2 w}{\partial t^2} + c^{(e)}\frac{\partial w}{\partial t}) + EI\left[\frac{\partial^4 w}{\partial x^4} + c^{(i)}\frac{\partial^5 w}{\partial x^4 \partial t}\right] = p(x, t) + \sum_{j=1}^{n_a} F_j \delta(x - \eta_j) \tag{4}$$

In Eq. (4), μ denotes mass per length unit, $c^{(e)}$, $c^{(i)}$ are damping coefficient and internal friction coefficient per length unit of beam, respectively, and the Delta-Dirac function $\delta(x - \eta_j)$ is defined by

$$\delta(x - \eta_j) = \begin{cases} 1 \ when \ \ x = \eta_j \\ 0 \ when \ \ x \neq \eta_j \end{cases} \tag{5}$$

The vibration equations according to Eqs. (3) and (4) are a mixed set of ordinary and partial differential equations. Four boundary conditions, two at $x = 0$ and two at $x = L$, and the initial conditions must be specified to find the solution of this set.

Using Ritz-Galerkin method, the solution of Eqs. (3) and (4) can be found in the form

$$w(x, t) = \sum_{r=1}^{n_b} X_r(x) q_r(t) \tag{6}$$

where $X_r(x)$ denotes the mode shape of the beam and $q_r(t)$ is the generalized displacement to be determined. Substituting Eq. (6) into Eqs. (4) and (3), we find

$$\ddot{q}_k(t) + (c^{(e)} + c^{(i)}\omega_k^2)\dot{q}_k(t) + \omega_k^2 q_k(t) = \frac{\int_0^l p(x, t) X_k(x) dx}{\mu \int_0^l X_k^2(x) dx} + \frac{\sum_{j=1}^{n_a} \int_0^l F_j(t) X_k(x) \delta(x - \eta_j) dx}{\mu \int_0^l X_k^2(x) dx},\ \ k = 1, ..., n_b. \tag{7}$$

$$m_j\ddot{u}_j(t) + d_j\dot{u}_j(t) + k_j u_j(t) - d_j\sum_{r=1}^{n_b} X_r(\eta_j)\dot{q}_r(t) - \mathrm{k}_j\sum_{r=1}^{n_b} X_r(\eta_j)q_r(t) = m_j g,\ \ j = 1, 2, \ldots, n_a. \tag{8}$$

where ω_k is eigenfrequency of the beam [1, 2]. Using the notations

$$2\delta_k = (c^{(e)} + c^{(i)}\omega_k^2),\ \ D_k = \mu\int_0^l X_k^2(x)dx = const \tag{9}$$

$$\alpha_k(t) = \int_0^l p(x,t)X_k(x)dx,\ \ h_k(t) = \frac{\alpha_k(t)}{D_k}, \tag{10}$$

it follows from Eq. (7) that

$$\ddot{q}_k(t) + 2\delta_k\dot{q}_k(t) + \omega_k^2 q_k(t) = h_k(t) + \frac{1}{D_k}\sum_{j=1}^{n_a}\int_0^l F_j(t)X_k(x)\delta(x-\eta_j)dx \tag{11}$$

Substitution of Eq. (6) into Eq. (1) yields

$$\begin{aligned} F_j(t) &= d_j\left[\dot{u}_j(t) - \dot{\mathrm{w}}_{\eta j}\right] + k_j\left[u_j(t) - \mathrm{w}_{\eta j}\right] \\ &= d_j\dot{u}_j(t) + k_j u_j(t) - d_j\sum_{r=1}^{n_b} X_r(\eta_j)\dot{q}_r(t) - k_j\sum_{r=1}^{n_b} X_r(\eta_j)q_r(t) \end{aligned} \tag{12}$$

in which $d_j, k_j, X_r(\eta_j)$ are the known constants. According to the property of the Delta-Dirac function we have

$$\sum_{j=1}^{n_a}\int_0^l F_j(t)X_k(x)\delta(x-\eta_j)dx = \sum_{j=1}^{n_a} F_j(t)X_k(\eta_j) \tag{13}$$

Substitution of Eq. (12) into Eq. (13) one obtains

$$\begin{aligned} \sum_{j=1}^{n_a} F_j(t)X_k(\eta_j) &= \sum_{j=1}^{n_a} X_k(\eta_j)\left[d_j\dot{u}_j(t) + k_j u_j(t)\right] \\ &- \sum_{j=1}^{n_a} X_k(\eta_j)\left\{\sum_{r=1}^{n_b}\left[d_j\dot{q}_r(t) + k_j q_r(t)\right]X_r(\eta_j)\right\} \end{aligned} \tag{14}$$

Substitution of Eq. (14) into Eq. (11) yields

$$\begin{aligned} \ddot{q}_k(t) + 2\delta_k\dot{q}_k(t) + \omega_k^2 q_k(t) &= h_k(t) \\ &+ \frac{1}{D_k}\sum_{j=1}^{n_a} X_k(\eta_j)\left[d_j\dot{u}_j(t) + k_j u_j(t)\right] \end{aligned}$$

$$-\frac{1}{D_k}\sum_{j=1}^{n_a} X_k(\eta_j)\left\{\sum_{r=1}^{n_b}\left[d_j\dot{q}_r(t)+k_jq_r(t)\right]X_r(\eta_j)\right\} \tag{15}$$

It follows from Eq. (8) that

$$\sum_{r=1}^{n_b}\left[d_j\dot{q}_r(t)+k_jq_r(t)\right]X_r(\eta_j)=m_j\ddot{u}_j(t)+d_j\dot{u}_j(t)+k_ju_j(t) \tag{16}$$

By substituting Eq. (16) into Eq. (15) leads to

$$\ddot{q}_k(t)+2\delta_k\dot{q}_k(t)+\omega_k^2q_k(t)=h_k(t)+\frac{1}{D_k}\sum_{j=1}^{n_a}X_k(\eta_j)\left[d_j\dot{u}_j(t)+k_ju_j(t)\right]$$

$$-\frac{1}{D_k}\sum_{j=1}^{n_a}X_k(\eta_j)\left\{m_j\ddot{u}_j(t)+d_j\dot{u}_j(t)+k_ju_j(t)\right\},\ k=1,2,\ldots n_b \tag{17}$$

Equations (17) and (8) consist of a system of $n = n_a + n_b$ ordinary differential equations that describes the vibration of the beam with dynamic vibration absorbers.

We consider now the vibration of the beams under harmonic excitation $p(t) = p_0 \sin \Omega t$. According to Eqs. (9) and (10) we have

$$\alpha_k(t)=[p_0\int_0^l X_k(x)dx]\sin\Omega t,\ \ h_k(t)=\frac{p_0\int_0^l X_k(x)dx}{D_k}\sin\Omega t$$

It follows that

$$h_k(t)=\hat{h}_k\sin\Omega t;\qquad \hat{h}_k=\frac{p_0\int_0^l X_k(x)dx}{D_k} \tag{18}$$

In this case, Eq. (17) has the following form

$$\ddot{q}_k(t)+2\delta_k\dot{q}_k(t)+\omega_k^2q_k(t)=\hat{h}_k\sin\Omega t-\frac{1}{D_k}\sum_{j=1}^{n_a}m_jX_k(\eta_j)\ddot{u}_j \tag{19}$$

Using the notations $\frac{d_j}{m_j} = 2\delta_{jc}$, $\frac{k_j}{m_j} = \omega_{jc}^2$, it follows from Eq. (8) that

$$\ddot{u}_j(t)+2\delta_{jc}\dot{u}_j(t)+\omega_{jc}^2u_j(t)=2\delta_{jc}\sum_{r=1}^{n_b}X_r(\eta_j)\dot{q}_r(t)+\omega_{jc}^2\sum_{r=1}^{n_b}X_r(\eta_j)q_r(t) \tag{20}$$

Equations (19) and (20) are a system of $n = n_a + n_b$ differential equations describing the transverse vibration of beam with a lot of dynamic vibration absorbers under the harmonic distributed force, in which $X_k(x)$ is the eigenfunction of beam. The concrete form of $X_k(x)$ depends on the boundary conditions of beam.

3 The Complex Frequency Response Function

In this section, we consider cases that often occur in structural, when the excitation frequency Ω is approximately equal to the fundamental frequency ω_1 of the beam ($\Omega \approx \omega_1$). It follows from Eq. (19) and Eq. (20) that

$$\ddot{q}_1(t) + 2\delta_1\dot{q}_1(t) + \omega_1^2 q_1(t) = \hat{h}_1 \sin \Omega t - \frac{1}{D_1}\sum_{j=1}^{n_a} m_j X_k(\eta_j)\ddot{u}_j, \tag{21}$$

$$\ddot{u}_j(t) + 2\delta_{jc}\dot{u}_j(t) + \omega_{jc}^2 u_j(t) = 2\delta_{jc}X_1(\eta_j)\dot{q}_1(t) + \omega_{jc}^2 X_1(\eta_j)q_1(t) \tag{22}$$

We use the following notations

$$x_s = q_1, \quad \omega_s = \omega_1, \quad \delta_s = \delta_1, \quad \hat{h}_s = \hat{h}_1 \tag{23}$$

Equation (21) can be written in the following form

$$\ddot{x}_s(t) + 2\delta_s\dot{x}_s(t) + \omega_s^2 x_s(t) = \hat{h}_1 \sin \Omega t - \frac{1}{D_1}\sum_{j=1}^{n_a} m_j X_1(\eta_j)\ddot{u}_j \tag{24}$$

The solution of Eqs. (2) and (24) can now be found using the method of frequency response function. We note that $\cos \Omega t = \mathrm{Re} e^{i\Omega t}$, $\sin \Omega t = \mathrm{Im} e^{i\Omega t}$, Eq. (24) can thus be written as follows

$$\ddot{x}_s(t) + 2\delta_s\dot{x}_s(t) + \omega_s^2 x_s(t) = \hat{h}_s e^{i\Omega t} - \frac{1}{D_1}\sum_{j=1}^{n_a} m_j X_1(\eta_j)\ddot{u}_j \tag{25}$$

We find the solutions of Eqs. (22) and (25) in the form

$$x_s(t) = H_s e^{i\Omega t}, \quad u_j(t) = H_{jc} e^{i\Omega t} \tag{26}$$

Substitution of Eq. (26) into Eqs. (25) and (22) lead to the system of linear algebraic equations

$$\left[\omega_s^2 - \Omega^2 + i2\delta_s\Omega\right]H_s = \hat{h}_s + \frac{\Omega^2}{D_1}\sum_{j=1}^{n_a} m_j X_1(\eta_j)H_{jc}, \tag{27}$$

$$\left[\omega_{jc}^2 - \Omega^2 + 2i\delta_{jc}\Omega\right]H_{jc} = \left[\omega_{jc}^2 + 2i\delta_{jc}\Omega\right]X_1(\eta_j)H_s \tag{28}$$

From Eq. (28) one has

$$H_{jc} = \frac{\left[\omega_{jc}^2 + i2\delta_{jc}\Omega\right]}{\left[\omega_{jc}^2 - \Omega^2 + i2\delta_{jc}\Omega\right]}X_1(\eta_j)H_s \tag{29}$$

Substitution of Eq. (29) into Eq. (27) yields

$$\left[\omega_s^2 - \Omega^2 + i2\delta_s\Omega\right]H_s = \hat{h}_s + H_s\frac{\Omega^2}{D_1}\sum_{j=1}^{n_a}\left(m_jX_1^2(\eta_j)\frac{\left[\omega_{jc}^2 + i2\delta_{jc}\Omega\right]}{\left[\omega_{jc}^2 - \Omega^2 + i2\delta_{jc}\Omega\right]}\right) \tag{30}$$

By introducing

$$a_{11} = \sum_{j=1}^{n_a} m_jX_1^2(\eta_j)\frac{\left[\omega_{jc}^2 + i2\delta_{jc}\Omega\right]}{\left[\omega_{jc}^2 - \Omega^2 + i2\delta_{jc}\Omega\right]}, \tag{31}$$

Equation (30) can be written in the following form

$$\left(\left[\omega_s^2 - \Omega^2 + i2\delta_s\Omega\right] - \frac{\Omega^2}{D_k}a_{11}\right)H_s = \hat{h}_s \tag{32}$$

From Eq. (32) one has

$$H_s = \frac{\hat{h}_s}{\left[\omega_s^2 - \Omega^2 + i2\delta_s\Omega\right] - \frac{\Omega^2}{D_1}a_{11}} \tag{33}$$

in which

$$\begin{aligned} a_{11} &= \sum_{j=1}^{n_a} m_jX_1^2(\eta_j)\frac{\left[\omega_{jc}^2 + i2\delta_{jc}\Omega\right]}{\left[\omega_{jc}^2 - \Omega^2 + i2\delta_{jc}\Omega\right]} \\ &= \sum_{j=1}^{n_a} m_jX_1^2(\eta_j)\frac{\left[\omega_{jc}^2 + i2\delta_{jc}\Omega\right]\left[\omega_{jc}^2 - \Omega^2 - i2\delta_{jc}\Omega\right]}{\left[\left(\omega_{jc}^2 - \Omega^2\right)^2 + 4\delta_{jc}^2\Omega^2\right]} \\ &= \sum_{j=1}^{n_a} m_jX_1^2(\eta_j)\frac{\left[\omega_{jc}^2\left(\omega_{jc}^2 - \Omega^2\right) + 4\delta_{jc}^2\Omega^2 - 2i\delta_{jc}\Omega^3\right]}{\left[\left(\omega_{jc}^2 - \Omega^2\right)^2 + 4\delta_{jc}^2\Omega^2\right]} \end{aligned} \tag{34}$$

The denominator in Eq. (33) takes the form

$$\left[\omega_s^2 - \Omega^2 + i2\delta_1\Omega\right] + \frac{\Omega^2}{D_1}a_{11} = \left[\omega_s^2 - \Omega^2 + i2\delta_s\Omega\right]$$
$$+ \frac{\Omega^2}{D_1}\sum_{j=1}^{n_a} m_j X_1^2(\eta_j)\frac{\left[\omega_{jc}^2\left(\omega_{jc}^2 - \Omega^2\right) + 4\delta_{jc}^2\Omega^2\right]}{\left[\left(\omega_{jc}^2 - \Omega^2\right)^2 + 4\delta_{jc}^2\Omega^2\right]}$$
$$- i\sum_{j=1}^{n_a} m_j X_1^2(\eta_j)\frac{2\Omega^3\delta_{jc}}{\left[\left(\omega_{jc}^2 - \Omega^2\right)^2 + 4\delta_{jc}^2\Omega^2\right]}$$

By introducing the following notations

$$a = \left[\omega_s^2 - \Omega^2\right] - \frac{\Omega^2}{D_1}\sum_{j=1}^{n_a} m_j X_1^2(\eta_j)\frac{\left[\omega_{jc}^2\left(\omega_{jc}^2 - \Omega^2\right) + 4\delta_{jc}^2\Omega^2\right]}{\left[\left(\omega_{jc}^2 - \Omega^2\right)^2 + 4\delta_{jc}^2\Omega^2\right]}, \tag{35}$$

$$b = 2\delta_s\Omega + \frac{\Omega^2}{D_1}\sum_{j=1}^{n_a} m_j X_1^2(\eta_j)\frac{2\Omega^3\delta_{jc}}{\left[\left(\omega_{jc}^2 - \Omega^2\right)^2 + 4\delta_{jc}^2\Omega^2\right]}, \tag{36}$$

function H_s can be written in the form

$$H_s = \frac{\hat{h}_s}{a + ib} = \frac{\hat{h}_s(a - ib)}{a^2 + b^2} \tag{37}$$

The modulus of the complex frequency response function H_s can now be calculated by the following formula

$$H = \left|\tilde{H}_s\right| = \frac{\hat{h}_s}{\sqrt{a^2 + b^2}} \tag{38}$$

The optimum problem is stated as follows: Find the parameters of dynamic vibration absorbers m_j, c_j, k_j for $j = 1, 2, \ldots, n_a$ which minimize the objective function according to Eq. (38).

4 Determination of Mass Parameters of Dynamic Absorbers

To determine the mass parameters of the vibration absorbers installed on the beam, we consider the bending vibrations of an unclamped beam with five vibration absorbers at the installation locations η_i as shown in Fig. 3.

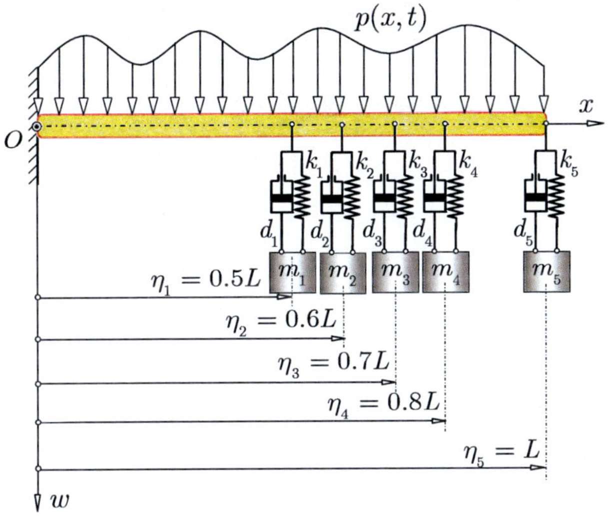

Fig. 3. A clamped -free Beam with five vibration absorbers

The model is a specific case though, but the calculation method is relatively general. The parameters of the beam are listed in Table 1. From the given parameters, it results $\omega_1 = 110.3005\ s^{-1}$. We consider the case of the distributed load $p = p_0 \sin \Omega t$, where $p_0 = 15$ N/m and $\Omega = \omega_1$.

Table 1. Beam parameters

Parameters	Value	Unit
EI	3.06×10^7	Nm^2
L	10	m
μ	24.5	kg/m
$c^{(e)}$	0.4	1/s
$c^{(i)}$	0.0001	s/m

In order to optimize the selection of the mass parameters of vibration absorbers, a procedure is proposed here to determine the masses of vibration absorbers based on the mode shapes of the beam. The mode shape of the beam and the mode function values at the installation positions are shown in Fig. 4 [3, 4].

The mass of the vibration absorbers is calculated according to the following expression

$$m_j = \alpha a_j \text{ for } j = 1, 2, \ldots, 5.$$

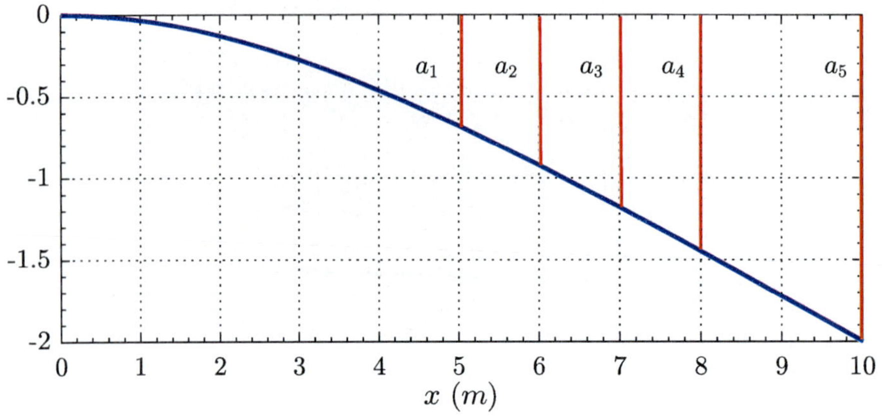

Fig. 4. First mode shapes of the beam and the positions of the vibration absorbers

where α is scaling factor, a_j denotes the value of the first mode shape at position η_j. According to [4] we have

$$a_1 = \|X_1(\eta_1)\| = 0.6790, \;\; a_2 = \|X_1(\eta_2)\| = 0.9223$$
$$a_3 = \|X_1(\eta_3)\| = 1.1818, \;\; a_4 = \|X_1(\eta_4)\| = 1.4510$$
$$a_5 = \|X_1(\eta_5)\| = 2.0000$$

According to the experience of the designers, it is possible to choose the total mass of the vibration absorbers about 1% of the mass of the carrier. So we have the formula

$$m_1 + m_2 + m_3 + m_4 + m_5 = \alpha(a_1 + a_2 + a_3 + a_4 + a_5) = \frac{m_b}{100}$$

It follows that

$$\alpha = \frac{m_b}{100(a_1 + a_2 + a_3 + a_4 + a_5)} = 0.393$$

Hence, the masses of vibration absorbers can then be determined as

$$m_1 = 0.2669, \;\; m_2 = 0.3625, \;\; m_3 = 0.4644,$$
$$m_4 = 0.5702, \;\; m_5 = 0.7860 \;\; (kg).$$

5 Determination of the Optimal Parameters of the Absorbers

In [10–14], Nguyen Van Khang and his colleagues have proposed a convenient algorithm to determine the optimal parameters of vibration dampers installed on the main system based on Taguchi method. Based on this, the optimal parameters d_i and k_i of 5 different TMDs at $\Omega = \omega_1$ as mentioned above can be determined and given in Table 2.

Table 2. The optimal parameters of the TMDs

	m (kg)	d (Ns/m)	k (N/m)
TMD1	0.2669	0.625	414.5
TMD2	0.3625	1.000	536.6
TMD3	0.4644	1.500	723.0
TMD4	0.5702	2.500	900.0
TMD5	0.7860	3.000	1225.4

When the excitation frequency Ω changes in a neighborhood with the first-order specific frequency ω_1, the optimum parameters of vibration absorbers with weighting factors are given in Table 3 and the corresponding frequency response functions shown in Fig. 5.

Table 3. The optimal parameters of the TMDs using the weighting factors

	m (kg)	d (Ns/m)	k (N/m)
TMD1	0.2669	1.6257	386.6
TMD2	0.3625	1.7942	523.2
TMD3	0.4644	3.2942	670.1
TMD4	0.5702	2.5696	793.5
TMD5	0.7860	4.6654	1165.4

From Fig. 5, it can be seen that the vibration of the beam at position x = L using 5 vibration absorbers mounted at L/2, 6L/10, 7L/10, 8L/10 and 10L/10 is 2.665 (mm), compared to before damper mounting is 44.01 (mm). Thus beam fluctuations are significantly reduced by about 94%.

The above results are based on the selection parameters using vibration modes. Now we choose the masses of vibration absorbers so that the total mass is equal to 1% of the mass of the beam and divide them evenly into the mass of the 5 vibration absorbers. Then, Taguchi method is applied to determine the parameters for hardness and viscosity resistance. The calculation results are shown in Fig. 6.

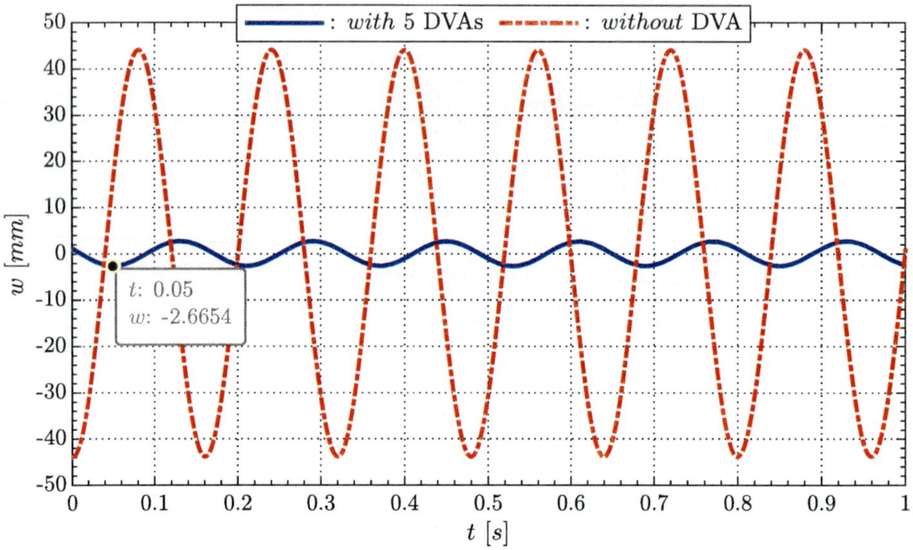

Fig. 5. The time response of a beam at $x = L$ with 5 vibration absorbers

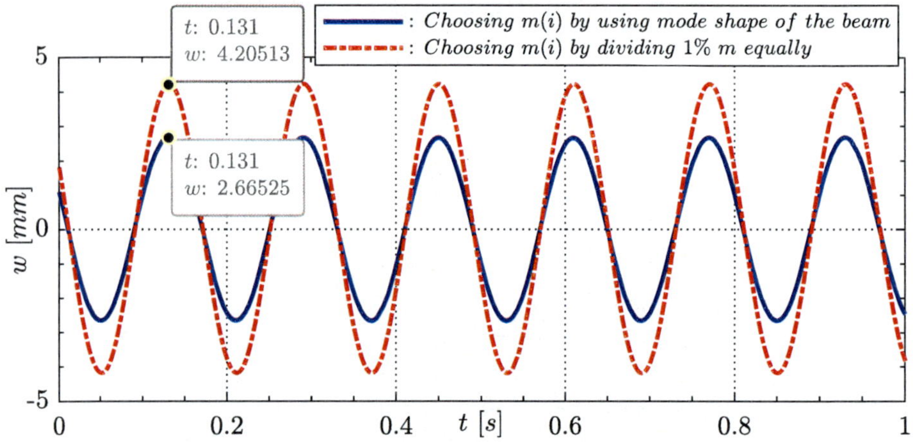

Fig. 6. Time response of beam at position $x = L$

6 Optimal Installation Positions of Dynamic Vibration Absorbers

Using the optimal parameters of vibration absorbers in Table 3, the vibration efficiency of installation positions of the vibration absorbers is now investigated. Three options for installing the vibration absorbers are shown in Figs. 7, 8 and 9.

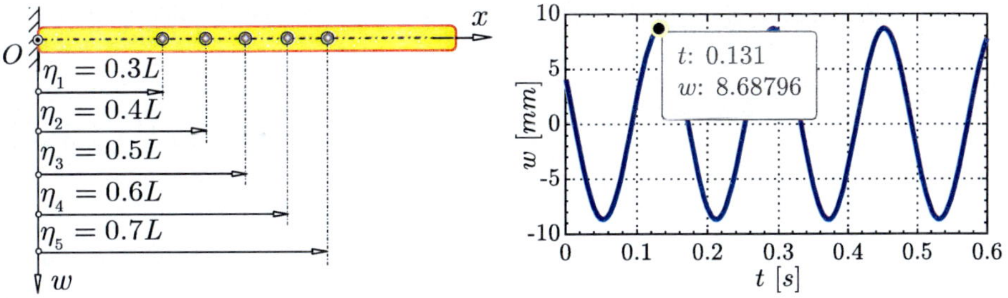

Fig. 7. Installation positions of vibration absorbers at *3L/10, 4L/10, 5L/10, 6L/10, 7L/10*

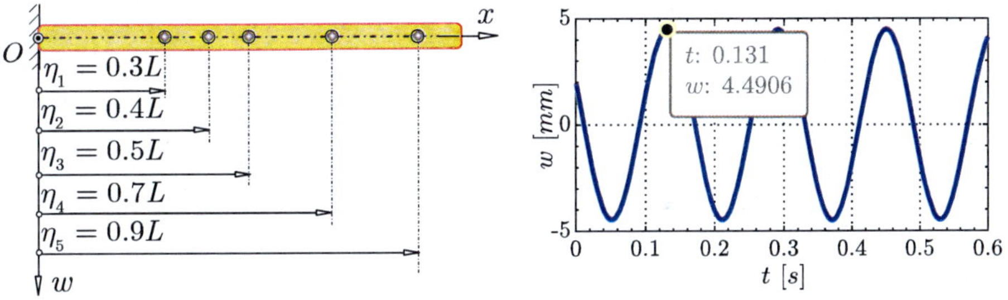

Fig. 8. Installation positions of vibration absorbers at 3/*10, 4L/10, 5L/10, 7L/10, 9L/10*

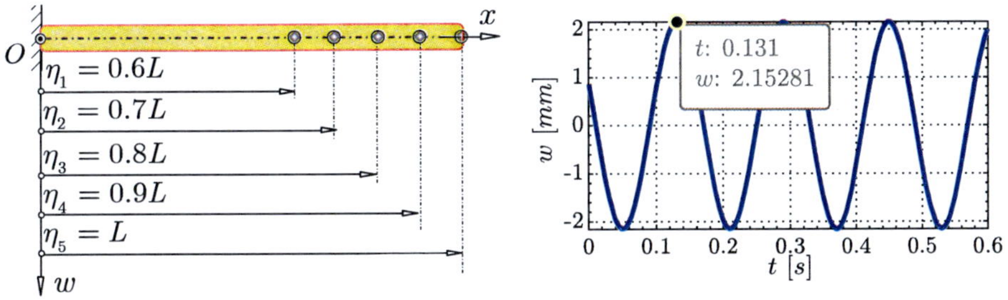

Fig. 9. Installation positions of vibration absorbers at *6L/10, 7L/10, 8L/10, 9L/10, 10L/10*

From the above calculation results, the results of vibration reduction according to the positions of vibration absorbers are summarized in Table 4.

From these numerical simulation results, it can be seen that: if the positions of the vibration absorbers are installed in the area of large mode amplitudes, the effect of vibration reduction is great, and the amplitude of vibration of the beam becomes small.

Table 4. Vibration amplitude and vibration reduction effect by various installation locations

Installation positions of vibration absorbers	5L/10 6L/10 7L/10 8L/10 10L/10	5L/10 6L/10 7L/10 9L/10 10L/10	3L/10 4L/10 5L/10 6L/10 7L/10	3L/10 4L/10 5L/10 7L/10 9L/10	6L/10 7L/10 8L/10 9L/10 10L/10	3L/10 5L/10 7L/10 9L/10 10L/10
w (mm)	2.665	2.46	8.698	4.491	2.153	2.622
Effect (%)	93.94	94.41	80.24	89.80	95.11	94.04

7 Conclusion

Determination of the optimal locations of dynamic vibration absorbers for the vibration control of beams is an important problem in practical application, but there is little research concerning this problem. In this paper, we propose an approach to determine the optimal mass parameters of dynamic vibration absorbers using the mode of beam transverse vibration. The determination of the optimal parameters of viscous elements and spring elements based on an algorithm derived from the experimental method by Taguchi. If the dynamic vibration absorbers are installed closer to the position at large amplitudes in vibration mode, the effect of reducing vibration is quite large and the amplitude of the bending vibrations becomes smaller. The numerical simulation results show the optimal results of the proposed method. It can be shown that the vibration damping effect is approx. 90%.

Acknowledgments. This paper was completed with the financial support of the Vietnam National Foundation for Science and Technology Development (NAFOSTED) under grant number 107.04–2020.28.

References

1. Meirovitch, L.: Dynamics and Control of Structures. Wiley, New York (1990)
2. Inman, D.J.: Vibration with Control. Wiley, Chichester (2006)
3. Kelly, S.G.: Fundamentals of Mechanical Vibrations. McGraw-Hill, Singapore (1993)
4. Rao, S.S.: Vibration of Continuous Systems. Wiley, Hoboken (2007)
5. Roy, R.K.: A Primer on the Taguchi Method. Society of Manufacturing Engineers, USA (2010)
6. Taguchi, G., Chowdhury, S., Wu, Y.: Taguchi's Quality Engineering Handbook. Wiley, Hoboken (2005)
7. Zambanini, R.A.: The application of Taguchi's method of parameter design to the design of mechanical systems. Master thesis, Lehigh University (1992)
8. Jacquot, R.G.: Optimal dynamic vibration absorbers for general beam systems. J. Sound Vib. **60**(4), 535–542 (1978)
9. Ozguven, H.N., Candir, B.: Suppressing the first and second resonances of beams by dynamic vibration absorbers. J. Sound Vib. **111**(3), 377–390 (1986)

10. Lin, Y.H., Cho, C.H.: Vibration suppression of beam structures transversed by multiple moving loads using a damped absorber. J. Mar. Sci. Technol. **1**, 39–48 (1993)
11. Chtiba, M.O., Chouba, S., El-Borgi, S., Nayfeh, A.H.: Confinement of vibrations in flexible structures using supplementary absorbers: dynamic optimization. J. Vib. Control **16**(3), 357–376 (2010)
12. Carpineto, N., Lacarbonara, W., Vestroni, F.: Mitigation of pedestrian-induced vibrations in suspension footbridges via multiple tuned mass dampers. J. Vib. Control **16**(5), 749–776 (2010)
13. Khang, N.V., Phuc, V.D., Huong, N.T.V., Duong, D.T.: Optimal control of transverse vibration of Euler-Bernoulli beam with multiple dynamic vibration absorbers using Taguchi's method. Vietnam J. Mech. **40**(3), 1–19 (2018)
14. Khang, N.V., Phuc, V.D., Duong, D.T., Huong, N.T.V.: A procedure for optimal design of a dynamic vibration absorber installed in the damped primary system based on Taguchi's method. Vietnam J. Sci. Technol. **55**(5), 649–661 (2018)
15. Khang, N.V., Duong, D.T., Huong, N.T.V., Dinh, N.D.T.T., Phuc, V.D.: Optimal control of vibration by multiple tuned liquid dampers using Taguchi method. J. Mech. Sci. Technol. **33**(4), 1563–1572 (2019)
16. Phuc, V.D.: Optimal control of vibration by combining many dynamic vibration absorbers. Ph.D. thesis in Vietnamese, Hanoi University of Science and Technology (2019)
17. Khang, N.V., Dien, N.P., Phuc, V.D., Duong, D.T., Huong, N.T.V.: Optimal positions of dynamic absorbers in transverse vibration control of beams. In: Proceedings of the National Conference on Dynamics and Control, pp 244–249. Natural Science and Technology Publishing House, Danang (2019)
18. Liu, K., Coppola, G.: Optimal design of damped dynamic vibration absorber for damped primary systems. Trans. Can. Soc. Mech. Eng. **34**(1), 119–135 (2010)
19. Latas, W.: Optimal positions of tunable translational and rotational dynamic absorbers in global vibration control in beams. J. Theor. Appl. Mech. **53**(2), 467–476 (2015)
20. Patil, S.S., Awasare, P.J.: Vibration suppression along the segment of beams by imposing nodes using multiple vibrations absorbers. J. Vibr. Anal. Measur. Control **4**(1), 29–39 (2016)

Mechanism Design and Theory

MechAnalyzer for Teaching of Kinematics of Linkage Mechanisms Through Simulations and Historical Context

Nilabro Saha[1], Rajeevlochana G. Chittawadigi[2(✉)], and Subir Kumar Saha[3]

[1] Department of Mechanical Engineering, National Institute of Technology Durgapur, Durgapur, India

[2] Department of Mechanical Engineering, Amrita School of Engineering, Bengaluru, Amrita Vishwa Vidyapeetham, Bengaluru, India
rg_chittawadigi@blr.amrita.edu

[3] Department of Mechanical Engineering, Indian Institute of Technology Delhi, New Delhi, India

Abstract. The superficial coverage of mechanism theory in undergraduate courses, and the difficulty to efficiently convey insights into mechanism kinematics through blackboard teaching alone, have made the deep knowledge of mechanisms and their variety rare among engineers. The adoption of 3D simulation software with specific focus towards kinematics and a rich catalogue of mechanisms, can make it easier for students to learn and appreciate, and for teachers to teach, mechanism theory. MechAnalyzer, in the development of which the second and third authors have been involved, is one such program. Due to its specific application as a learning tool, MechAnalyzer has a significantly lower barrier to entry compared to CAD/CAE tools and other software. In its multiple modules, it also aggregates features that may be available in many different software, that are directly related to an undergraduate Theory of Machines course. Many of its modules have been reported in the literature elsewhere. In this paper, the addition of new mechanisms such as six-bar linkages, exact straight-line mechanisms and walking mechanisms have been reported. A new Mechanism Explorer module has been introduced for curious users to further explore mechanism theory. A practical application of MechAnalyzer in the classroom teaching of mechanism science has been demonstrated. A kinematic formulation that makes it simpler to deploy new mechanisms in the software has also been presented.

Keywords: Mechanisms · Kinematics · Simulation · Education · MechAnalyzer

1 Introduction

The subject of kinematics of mechanisms is mainly taught today as a preface to dynamics in undergraduate courses. Although most contemporary engineers trained in the last quarter century have an understanding of microprocessors, control theory and mechatronics, a vast majority of them lack a deep knowledge of either kinematics or the vast variety

N. V. Khang et al. (Eds.): ASIAN MMS 2021, MMS 113, pp. 49–59, 2022.
https://doi.org/10.1007/978-3-030-91892-7_5

of mechanisms, which represents a certain "lost" body of knowledge in kinematics of machinery that was familiar to earlier generations (Moon, 2003 [1]).

Linkage mechanisms such as the slider-crank mechanism and the four-bar mechanism, however, are still ubiquitous in the machine world, in the IC engines, in lifting mechanisms for the hood of a car, etc. Therefore, a sound knowledge of linkage mechanisms, their kinematics and synthesis, is an important tool to a mechanical engineer. But usual blackboard teaching in a classroom setting prevents the teacher from efficiently conveying insights into the motion of mechanism to the students. This can be assisted/enhanced by the use of computer simulations. A plethora of other software have been developed with specific focus on mechanism theory including, but not limited to, commercial programs like *Working Model 2D*, *Ch Mechanism Toolkit*, *SAM*, and free programs such as *GIM* and *Mechanism Developer (MechDev)*.

MechAnalyzer (hereon referred to as MA), developed by the second and third authors as a Windows desktop application using Visual C# and OpenTK, is one such program. MA, with its in-built catalogue of mechanisms and unique feature set, is valuable not only as a teaching aid, but also in its ability to kindle a student's interest by allowing them the freedom to explore different intriguing and historically important linkage mechanisms. Further reading sections and the Mechanism Explorer module within the software allow a curious user to independently explore the history, applications, and a plethora of curated links related to each mechanism. Furthermore, MA has a significantly flatter learning curve compared to CAD/CAE programs and other programs mentioned above. A user can easily change the parameters of a model and immediately simulate it with the changes made. Version 6 (available to download for free at http://www.roboanalyzer.com/mechanalyzer.html) comes with many user interface improvements. New enhancements to the MA Core Module and newly added linkage mechanisms have been presented in this paper.

2 Mechanisms in MechAnalyzer

The catalogue of mechanisms in MA had been prepared by surveying the curricula of undergraduate Theory of Machines courses in Mechanical Engineering. Table 1 below lists the expanded catalogue of mechanisms currently available in the new version of MA.

MA includes not only the 3D models of the mechanisms, but also their animation, coupler curves, and kinematic analyses. Enhancements in MA regarding modules related to gears and cams been reported earlier by Dikshithaa et al. [2]. Further recent enhancements by the authors of this paper on new gear mechanisms in MA and improvements to the existing modules are also made available [3]. The linkage mechanisms newly added to MA (marked with a * in Table 1) have been reported in the following subsections.

2.1 Straight Line Mechanisms in MA

An area of "lost" kinematic knowledge (Moon, 2003 [1]) is the straight-line mechanisms. Their history can be traced back to the machine age in the 18th and 19th centuries, in

Table 1. Catalogue of mechanisms available in the latest version of MA

Lower pair mechanisms		Higher pair mechanisms
Four-bar mechanism	Six-bar linkage*	Simple gear train
Slider-crank mechanism	Peaucellier's inversor*	Compound gear train
Quick-return mechanisms	Hart's inversor*	Planetary gear train*
Steering mechanism	Jansen's linkage*	Standard open differential*
Pantograph mechanism	Ghassaei's linkage*	Cam and follower mechanism
	Klann's linkage*	

applications requiring transformation of a straight-line motion into circular motion, such as in the steam engine (Dijksman, 2004 [4]). Approximate straight-line linkages like Watt, Chebyshev, Robert, Evans, and Hoeken types that are obtained as inversions of the four-bar linkage (Tesar, 1965 [5]), have been part of MA since Version 3 (Lokesh, 2015 [6]) and are shown in Fig. 1 below. The dashed coupler curves in MA have been inspired by the similar representation in the Hrones and Nelson atlas of fourbar coupler curves, allowing users to infer at a glance, the instantaneous velocity of the coupler point, with longer dashes implying greater velocity.

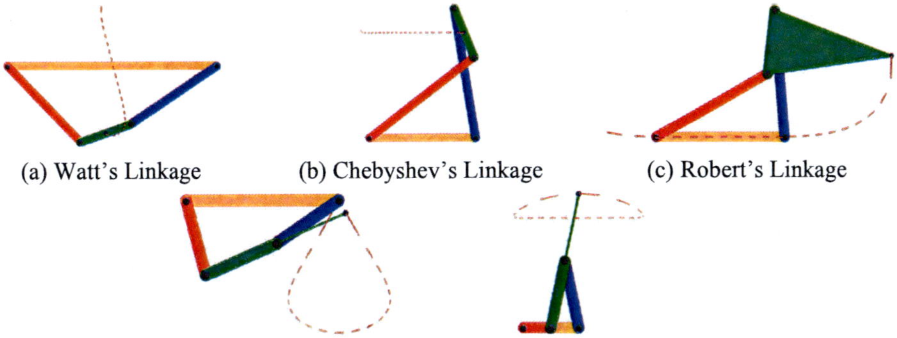

(a) Watt's Linkage (b) Chebyshev's Linkage (c) Robert's Linkage

(d) Evans' Linkage (e) Hoeken's Linkage

Fig. 1. Approximate straight-line mechanisms in MA

The variety of straight-line mechanisms available in MA has been expanded with the inclusion of exact straight-line mechanisms. The main user interface of MA is shown in Fig. 2a. The Fig. 2b shows Hart's straight-line mechanism of the first kind, presented in 1876 (Hart, 1876 [7]), technically known as the contraparallelogram chain of Hart (Dijksman, 1971 [8]). Figure 2c, on the other hand, shows the Peaucellier's inversor, which the French engineer Peaucellier had invented as early as 1864 (Peaucellier, 1864 [9]). His publication, however, had fallen into oblivion. So much so, that when six years later, in 1870, Peaucellier's linkage had been independently derived by Lippman Lipkin of St. Petersburg, the discovery had been attributed to him (Nolle, 1974 [10]). Both the

Hart's and Peaucellier's inversors have been shown to be cognates of the lesser known quadruplanar inversor of Sylvester and Kempe, and therefore of each other (Dijksman, 1971 [8]).

The Peaucellier linkage has been used in recent years to prove theorems about workspace topology in robotics. It is also said to have been used as a part of a blowing engine to ventilate the House of Commons in 1877 (Moon, 2003 [1]).

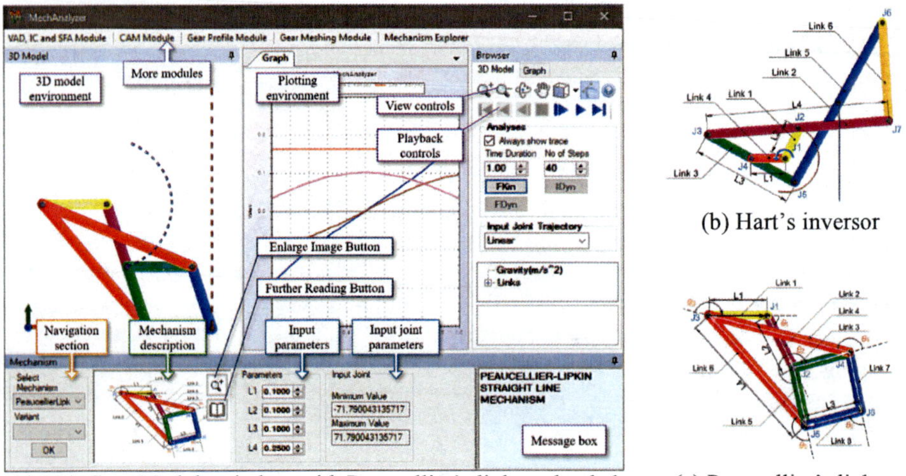

(a) MA Core Module window with Peaucellier's linkage loaded (b) Hart's inversor (c) Peaucellier's linkage

Fig. 2. MA user interface and exact straight-line mechanisms

2.2 Walking Mechanisms

In navigating rough terrain, walking mechanisms exhibit greater versatility and maneuverability over their wheeled counterparts, which can only travel over prepared land (Sheba, 2015 [11]). An obvious advantage of walking machines is the ability to avoid inconsistencies of the terrain, i.e., stepping over its local maxima and minima, resulting in less loss of energy during locomotion. Detailed investigation into the nature, mechanics, applications, advantages and disadvantages of walking machines can be found in (Shigley, 1960 [12]) and (Ghassaei, 2011 [13]).

As narrated in (Chen and McCarthy, 2021 [14]), Theodorus Jansen had designed an 8-bar walking mechanism which he had used in several of his walking machines (kinetic sculptures) since 1991. Figure 3 shows the Jansen's linkage which functions by producing a foot trajectory that has a straight-line motion along its interaction with the ground (called the stride phase). The link lengths that Jansen had originally used can be found in [13]. In fact, the need to make improvements to Jansen's linkage in terms of design simplicity, symmetry of locus about the vertical axis, and straightness and constant velocity of stride were the motivations for Amanda Ghassaei to design her optimized crank-based leg mechanism, the Ghassaei's linkage (Fig. 4).

Named after its inventor Joseph Klann, the Klann linkage simulates the gait of legged animals. Whereas Jansen's linkage required 8 links per leg, Klann's linkage (Fig. 5) requires only 6; and whereas 3 Jansen legs are necessary to produce smooth walking motion, the same can be achieved using only 2 Klann legs. Klann, in his patents [15, 16] has provided means to calculate the final link lengths of his linkage based on 6 independent input parameters. This means an infinite number of Klann linkages can be generated, each suited to some specific application. A straightforward procedure of designing a Klann linkage can be found in (Shannon, 2011 [17]).

The *Parameters* panel in MA for each mechanism has been presented in Fig. 3b, Fig. 4b and Fig. 5b showing the link lengths which are scaled values of the original link lengths proposed by the respective inventors.

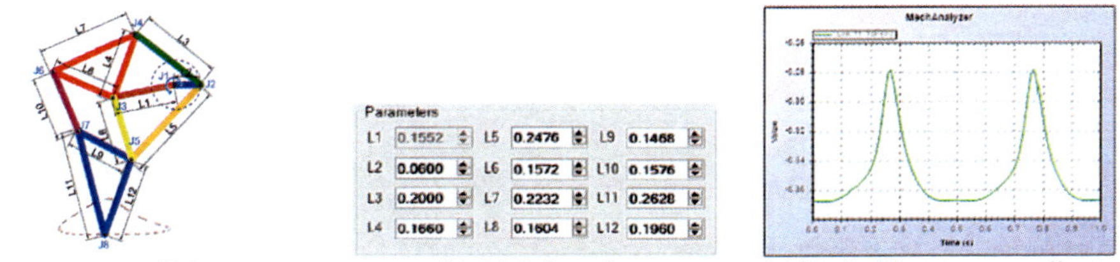

(a) Jansen's linkage (b) Link lengths (c) End effector y-coordinate

Fig. 3. Jansen's linkage in MA and end-effector position for two input crank rotations

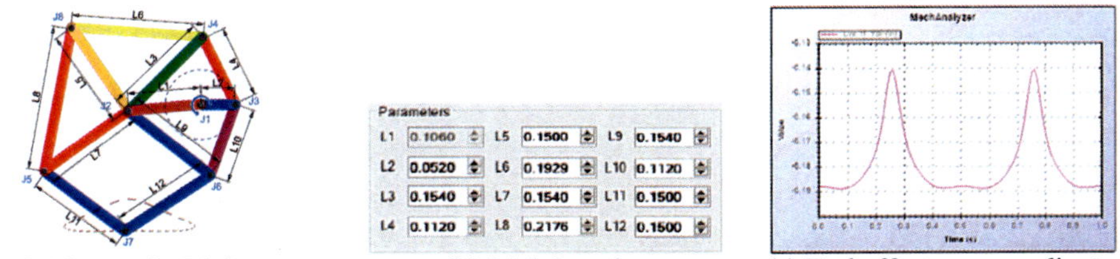

(a) Ghassaei's Linkage (b) Link lengths (c) End effector y-coordinate

Fig. 4. Ghassaei's linkage in MA and end-effector position for two input crank rotations

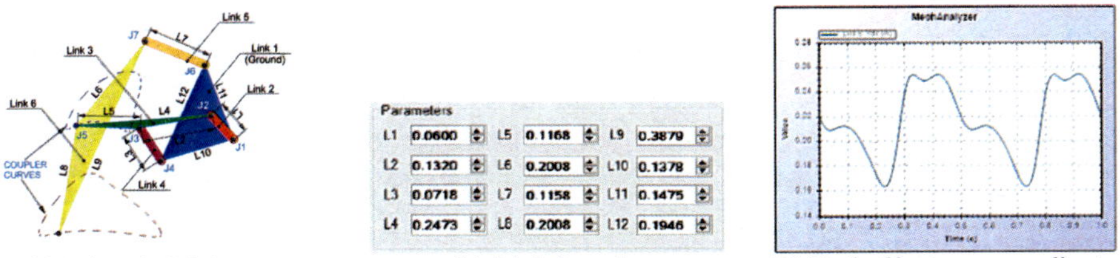

(a) Klann's Linkage (b) Link lengths (c) End effector y-coordinate

Fig. 5. Klann's linkage in MA and end-effector position for two input crank rotations

2.3 Six-Bar Linkages, Their Inversions and Variants

Six-bar linkages with one degree of freedom have two variants – the Watt's form and the Stephenson's form (Dijksman, 1971 [18]). Five different mechanisms are derivable by kinematic inversion of the Watt and Stephenson 6-bar chains, namely the Watt-I & II, and Stephenson-I, II & III linkages. Out of these, only three – Watt-I, and Stephenson-I & II – produce six-bar coupler curves, which have been thoroughly investigated by Primrose et al. [19, 20], and have been found to be of degrees 14, 14 and 16 respectively. The rest of the six-bar linkages generate four-bar coupler curves.

As shown in Fig. 6 below, MA includes the Watt-I & II, and Stephenson-I & III mechanisms. However, the Stephenson-II mechanism has been excluded since the kinematic formulation (described later in the paper) used for the forward kinematics would require solving a system of six second-degree algebraic equations.

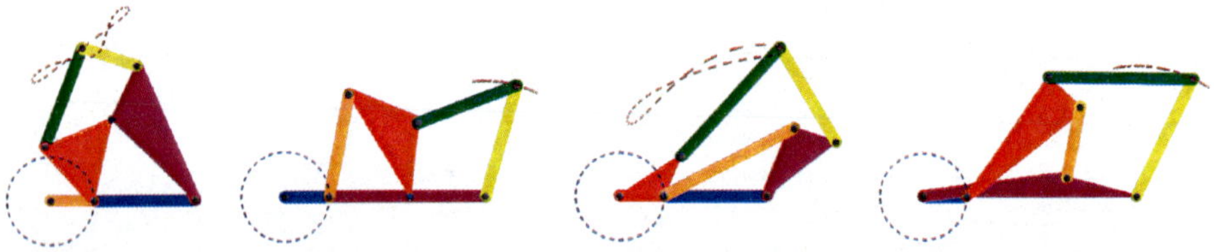

(a) Six-bar Watt-I chain (b) Six-bar Watt-II chain (c) Six-bar Stephenson-I chain (d) Six-bar Stephenson-III chain

Fig. 6. Variants of the six-bar chain and their coupler curves in MA

3 Mechanism Explorer

It is the conviction of the authors that interest in mechanism theory can only be kindled by placing it in the context of its historical development. However, without the right guidance and proper resources readily available, doing so is difficult both for the student as well as the teacher. With the goal of making such resources available to the user, each mechanism in MA now features a further reading button (below the enlarge image button, refer Fig. 2) that presents to the user a window with brief information about the mechanism along with a curated list of resources for further exploration. These resources and more have been also aggregated and presented as a new MA module, the 'Mechanism Explorer' module whose screenshot in Fig. 7 shows the page on the important developers of mechanism science. This represents our first steps towards extensive documentation of mechanism science within MA, the breadth of which is expected to grow rapidly with the further development of the software.

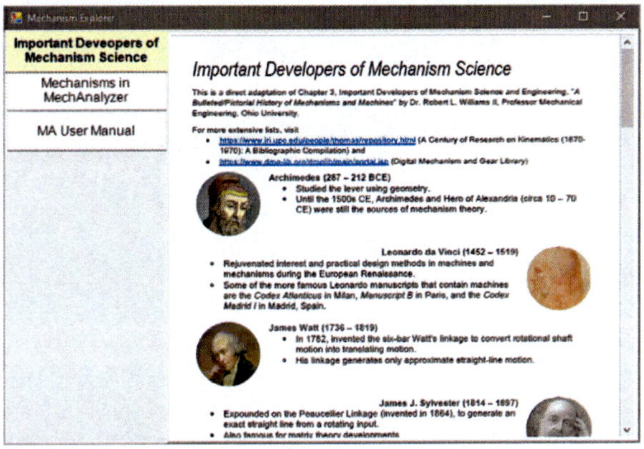

Fig. 7. MA Mechanism Explorer module

4 Using MA in the Classroom: Reconstruction of a Page from the Hrones and Nelson Atlas of Fourbar Coupler Curves

An introduction to mechanism synthesis in the classroom requires familiarization of students with the different kinds of coupler curves that can be generated by a fourbar linkage. This typically involves introducing students to the Hrones and Nelson atlas of fourbar coupler curves. It is the belief of the authors that such instruction can be made more fruitful by letting the students reconstruct a page from the atlas. This should practically demonstrate to students how to read the atlas; the scale invariance of the shape of the coupler curves; how the coupler curves evolve when the coupler point is varied along a straight line; how double points and other features of the curves emerge in the process, etc. Such goals can be achieved with minimal technological requirements by using MA along with Microsoft PowerPoint, LibreOffice Draw or other similar software. Figure 8 below shows a page from the fourbar atlas and its reconstruction using MA and LibreOffice Draw.

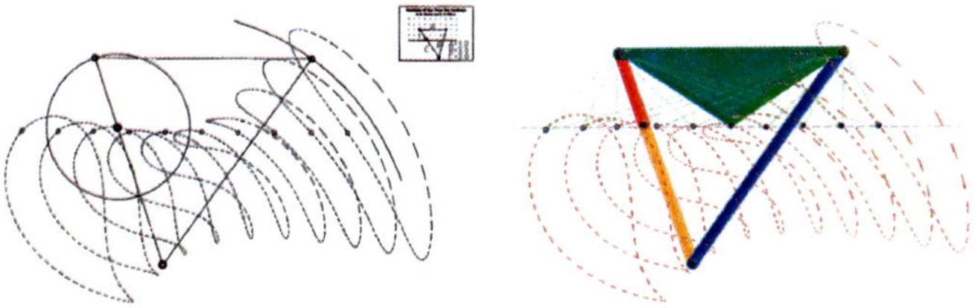

Fig. 8. A page from Hrones and Nelson's fourbar atlas and its reconstruction.

5 Kinematic Formulation

In a kinematic chain composed only of revolute pairs, the Cartesian position of each joint can be found as an *intersection of circles* with centers at the adjacent joints and

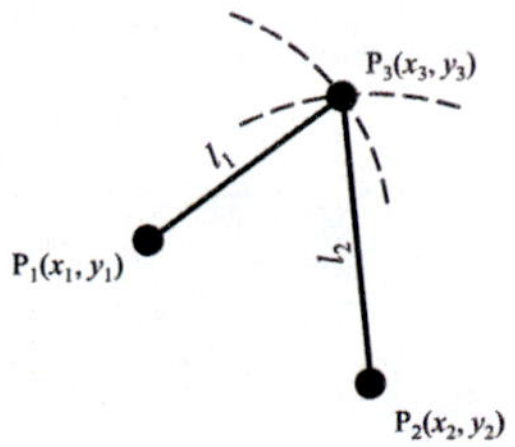

Fig. 9. Position analysis of joint 3, given positions of joints 1 and 2

radii equal to lengths of the adjacent links. Figure 9 above shows a part of such a chain. With the position of joints 1 and 2 known, the position of joint 3 can be deduced as the intersection of the circles $\mathbf{p}_{31}^{\mathrm{T}}\mathbf{p}_{31} = l_1^2$ and $\mathbf{p}_{32}^{\mathrm{T}}\mathbf{p}_{32} = l_2^2$, where $\mathbf{p}_i = (x_i, y_i)^{\mathrm{T}}$, $\mathbf{p}_{ij} = \left(x_{ij}, y_{ij}\right)^{\mathrm{T}} = \left(x_i - x_j, y_i - y_j\right)^{\mathrm{T}}$. This gives the possible positions of joint 3 as

$$\mathbf{p}_3 = \begin{pmatrix} x_3 \\ y_3 \end{pmatrix} = \frac{1}{2}\begin{pmatrix} x_1 + x_2 \\ y_1 + y_2 \end{pmatrix} + a\begin{pmatrix} x_2 - x_1 \\ y_2 - y_1 \end{pmatrix} \pm b\begin{pmatrix} y_2 - y_1 \\ x_1 - x_2 \end{pmatrix} \quad (1)$$

where $a = \frac{l_1^2 - l_2^2}{2R^2}$, $b = \sqrt{2\frac{l_1^2 + l_2^2}{R^2} - \frac{(l_1^2 - l_2^2)^2}{R^4} - 1}$, and $R = \sqrt{x_{21}^2 + y_{21}^2}$. The choice of the sign in Eq. (1) determines the branch of the solution. Given the position of joint 3, its linear velocity and linear acceleration can be derived by taking time derivatives on both sides of equations of the circles. This yields

$$\begin{pmatrix} \mathbf{p}_{31}^{\mathrm{T}} \\ \mathbf{p}_{32}^{\mathrm{T}} \end{pmatrix}\dot{\mathbf{p}}_3 = [\mathbf{P}]\dot{\mathbf{p}}_3 = \begin{pmatrix} \mathbf{p}_{31}^{\mathrm{T}}\dot{\mathbf{p}}_1 \\ \mathbf{p}_{32}^{\mathrm{T}}\dot{\mathbf{p}}_2 \end{pmatrix}, \quad \text{and } [\mathbf{P}]\ddot{\mathbf{p}}_3 = \begin{pmatrix} \mathbf{p}_{31}^{\mathrm{T}}\ddot{\mathbf{p}}_1 - \dot{\mathbf{p}}_{31}^{\mathrm{T}}\dot{\mathbf{p}}_{31} \\ \mathbf{p}_{32}^{\mathrm{T}}\ddot{\mathbf{p}}_2 - \dot{\mathbf{p}}_{32}^{\mathrm{T}}\dot{\mathbf{p}}_{32} \end{pmatrix}.$$

From these, $\dot{\mathbf{p}}_3$ and $\ddot{\mathbf{p}}_3$ can be easily obtained by inverting the 2×2 matrix $[\mathbf{P}]$, given $\dot{\mathbf{p}}_1, \ddot{\mathbf{p}}_1, \dot{\mathbf{p}}_2, \ddot{\mathbf{p}}_2$. This method of kinematic analysis had been outlined in (Hampali, 2015 [21]), an earlier work by the second and third authors of this paper. It has been further developed in this paper, and a similar approach may be used to obtain the kinematics of a sliding link. The advantages that this method affords over loop closure equations are: a comprehensive linear algebraic treatment; and easy modularization and automation to compute kinematics of more complex mechanisms. The following subsection illustrates this kinematic formulation as applied to the Peaucellier-Lipkin exact straight-line mechanism in MA.

5.1 Illustration: Forward Kinematic Analysis of Peaucellier-Lipkin Linkage

The links, joints, and joint angles are named as in Fig. 10a, same as within MA. Here, Link 1 is kept fixed. The input is provided to joint 1, making θ_1 the input joint angle. Therefore, the position vectors $\mathbf{p}_1 = (0, 0)^{\mathrm{T}}$, $\mathbf{p}_3 = (-l_1, 0)^{\mathrm{T}}$, and $\mathbf{p}_2 = (l_2 \cos\theta_1, l_2 \sin\theta_1)^{\mathrm{T}}$. Vectors $\mathbf{p}_4$ and $\mathbf{p}_5$ are the negative and positive branches respectively of the intersection of the circles with centers at $\mathbf{p}_3$ and $\mathbf{p}_2$ and radii l_4 and l_3 respectively. Subsequently, $\mathbf{p}_6$ is calculated as the negative branch of the intersection of circles $\mathbf{p}_{64}^{\mathrm{T}}\mathbf{p}_{64} = l_3^2$ and

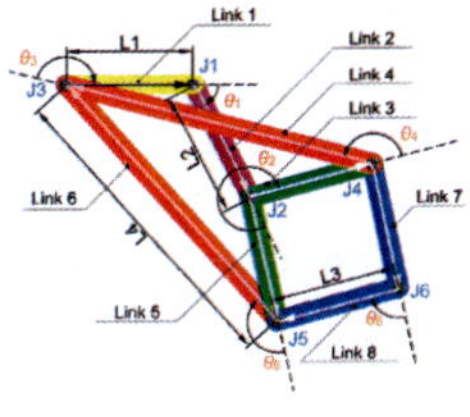

(a) Peaucellier's inversor

(b) Straight-line motion

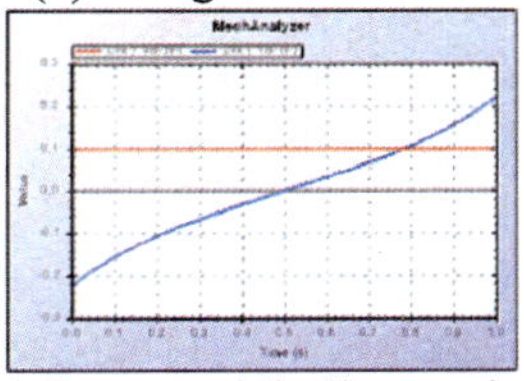

(c) Exact straight-line motion of end effector (plot in MA)

(d) Circular motion

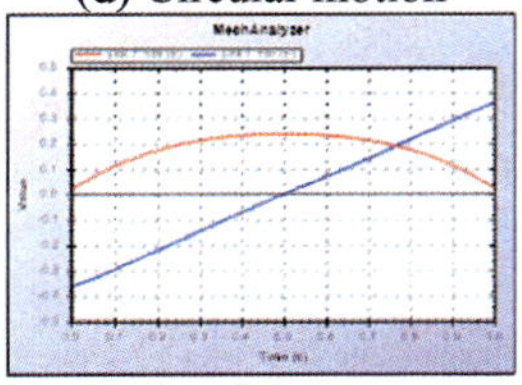

(e) Circular motion of end effector (plot in MA)

Fig. 10. Peaucellier's inversor in MA with linear input joint motion

$\mathbf{p}_{65}^T\mathbf{p}_{65} = l_3^2$. The linear velocity and acceleration vectors of joints 4, 5 and 6 are computed from

$$\begin{pmatrix} \mathbf{P}_{432} & 0 & 0 \\ 0 & \mathbf{P}_{532} & 0 \\ 0 & 0 & \mathbf{P}_{645} \end{pmatrix} \begin{pmatrix} \dot{\mathbf{p}}_4 \\ \dot{\mathbf{p}}_5 \\ \dot{\mathbf{p}}_6 \end{pmatrix} = \begin{pmatrix} \mathbf{u}_{432} \\ \mathbf{u}_{532} \\ \mathbf{u}_{645} \end{pmatrix}, \text{ and } \begin{pmatrix} \mathbf{P}_{432} & 0 & 0 \\ 0 & \mathbf{P}_{532} & 0 \\ 0 & 0 & \mathbf{P}_{645} \end{pmatrix} \begin{pmatrix} \ddot{\mathbf{p}}_4 \\ \ddot{\mathbf{p}}_5 \\ \ddot{\mathbf{p}}_6 \end{pmatrix} = \begin{pmatrix} \mathbf{v}_{432} \\ \mathbf{v}_{532} \\ \mathbf{v}_{645} \end{pmatrix},$$

where $(\mathbf{P}_{432})_{2\times 2} = \begin{pmatrix} \mathbf{p}_{43}^T \\ \mathbf{p}_{42}^T \end{pmatrix}$, $\mathbf{u}_{432} = \begin{pmatrix} \mathbf{p}_{43}^T\dot{\mathbf{p}}_3 \\ \mathbf{p}_{42}^T\dot{\mathbf{p}}_2 \end{pmatrix}$, $\mathbf{v}_{432} = \begin{pmatrix} \mathbf{p}_{43}^T\ddot{\mathbf{p}}_3 - \dot{\mathbf{p}}_{43}^T\dot{\mathbf{p}}_{43} \\ \mathbf{p}_{42}^T\ddot{\mathbf{p}}_2 - \dot{\mathbf{p}}_{42}^T\dot{\mathbf{p}}_{42} \end{pmatrix}$ and so on. Note that $\dot{\mathbf{p}}_3 = \ddot{\mathbf{p}}_3 = (0, 0)^T$.

From geometric considerations, it is known that the inverse of a circle is a straight line if the center of inversion lies on the circle. Otherwise, the inverse is a circle of a different diameter. These two cases correspond to $l_1 = l_2$ and $l_1 \neq l_2$ respectively, in our model. Correspondingly, Fig. 10b and c show the Peaucellier's inversor in MA tracing a straight line whereas Fig. 10d and e show it tracing a circular arc. Similar end-effector curves and plots can also be demonstrated for the Hart's inversor in MA.

6 Discussions and Conclusions

Two immediate mechanisms that are listed for inclusion in MA are Hart's inversor of the second kind and the quadruplanar inversor of Sylvester and Kempe. The inclusion of the six-bar Stephenson-II mechanism would make MA's catalogue more complete. However, within the kinematic formulation presented here, its position analysis requires the solution of six second-degree algebraic equations with six unknowns. An attempt may be made to numerically solve the system. The multi-dimensional Newton-Raphson method may be a plausible technique, which has been outlined in [22]. The geared five-bar linkage, straight line mechanisms based on the slider-crank chain, and other unique linkages may also be considered for inclusion in the future.

Earlier papers on MA have elucidated how MA can be used in a holistic study of Theory of Machines. In this paper, MA's application in learning kinematics of mechanisms has been highlighted. How MA may spark in students an interest in and an appreciation of the historical development of mechanisms and machines has also been explored. Immense scope for further work remains, including cataloging of more mechanisms. In conclusion, it should be reiterated that MA is not useful just for its simulation abilities (there are other, arguably better software for that purpose), but also for its contribution as a teaching and learning tool. It is hoped that a skilled teacher and an inquisitive student can draw far more value from MechAnalyzer than has been described in this and earlier papers.

References

1. Moon, F.C.: Franz Reuleaux: contributions to 19th century kinematics and theory of machines. Appl. Mech. Rev. **56**(2) (2003)
2. Dikshithaa, R., Jain, S., Swaminathan, J., Chittawadigi, R.G., Saha, S.K.: MechAnalyzer: software to teach kinematics concepts related to cams, gears, and instantaneous center. In: Sen, D., Mohan, S., Ananthasuresh, G.K. (eds.) Mechanism and Machine Science. LNME, pp. 135–149. Springer, Singapore (2021). https://doi.org/10.1007/978-981-15-4477-4_10
3. Saha, N., Chittawadigi, R.G., Saha, S.K.: MechAnalyzer: gear meshing visualization for effective teaching and learning. In: 5th International and 20th National Conference on Machines and Mechanisms (iNaCoMM 2021). PDPM Indian Institute of Information Technology, Design and Manufacturing, Jabalpur (PDPM IIITDMJ), India (2021)
4. Dijksman, E.: On the history of focal mechanisms and their derivatives. In: Ceccarelli, M. (ed.) International Symposium on History of Machines and Mechanisms, pp. 303–314. Springer, Dordrecht (2004). https://doi.org/10.1007/1-4020-2204-2_24
5. Tesar, D., Vidosic, J.P.: Analysis of approximate four-bar straight-line mechanisms. J. Eng. Ind. **87**(3), 291 (1965)
6. Lokesh, R., Chittawadigi, R.G., Saha, S.K.: MechAnalyzer: 3D simulation software to teach kinematics of machines. In: 2nd International and 17th National Conference on Machines and Mechanisms (iNaCoMM2015) (2015)
7. Hart, H.: On some cases of parallel motion. Proc. Lond. Math. Soc. **s1-8**(1), 286–291 (1876)
8. Dijksman, E.A.: A strong relationship between new and old inversion mechanisms. J. Eng. Ind. **93**(1), 334 (1971)
9. Peaucellier, A.: Note sur une question de geometrie de compas, Nouvelle Annales de Mathématique, Sér. II, Tome III, p. 344 (1864), and Sér. II, Tome XII, pp. 71–78 (1873)
10. Nolle, H.: Linkage coupler curve synthesis: a historical review—I. Developments up to 1875. Mech. Mach. Theory **9**(2), 147–168 (1974)
11. Sheba, J.K., Elara, M.R., Martínez-García, E., Tan-Phuc, L.: Synthesizing reconfigurable foot traces using a Klann mechanism. Robotica **35**(01), 189–205 (2015)
12. Shigley, J.E.: The Mechanics of Walking Vehicles (1960)
13. Ghassaei, A., Choi, P.P., Whitaker, D.: The Design and Optimization of a Crank-Based Leg Mechanism. Pomona, USA (2011)
14. Chen, K., McCarthy, J.M.: Kinematic synthesis of a modified Jansen leg mechanism. In: Lenarčič, J., Siciliano, B. (eds.) ARK 2020. SPAR, vol. 15, pp. 242–249. Springer, Cham (2021). https://doi.org/10.1007/978-3-030-50975-0_30
15. Klann, J.C.: Walking Device. U.S. Patent No. 6,260,862, 17 July 2001
16. Klann, J.C.: Walking Device. U.S. Patent No. 6,478,314, 12 November 2002

17. Shannon, T.: Development of a Museum-Quality Display of Mechanisms, Diss. Massachusetts Institute of Technology (2011)
18. Dijksman, E.A.: Six-bar cognates of Watt's form. J. Eng. Ind. Trans. ASME **93**(February), 183–190 (1971)
19. Primrose, E.J.F., Freudenstein, F., Roth, B.: Six-bar motion I. The Watt mechanism. Arch. Ration. Mech. Anal. **24**(1), 22–41 (1967)
20. Primrose, E.J.F., Freudenstein, F., Roth, B.: Six-bar motion II. The Stephenson-1 and Stephenson-2 mechanisms. Arch. Ration. Mech. Anal. **24**(1), 42–72 (1967)
21. Hampali, S., Chittawadigi, R.G., Saha, S.K.: MechAnalyzer: 3D model based mechanism learning software. In: Proceedings of the 14th IFToMM World Congress (2015)
22. Norton, R.L.: Design of Machinery: An Introduction to the Synthesis and Analysis of Mechanisms and Machines. McGraw-Hill Education (2020)

Kirigami Tessellation Based on the Two-Fold Symmetric Bricard 6R Linkage and Spherical 4R Linkage

Weiwei Lin[1], Fufu Yang[1,2](✉), and Jun Zhang[1,2]

[1] School of Mechanical Engineering and Automation, Fuzhou University, Fuzhou 350100, China
yangfufu@fzu.edu.cn

[2] Fujian Provincial Collaborative Innovation Center of High-End Equipment Manufacturing, Fuzhou 350100, China

Abstract. The paper presents a kirigami tessellation based on two-fold symmetric Bricard 6R linkage and spherical 4R linkage. The mobility and kinematics are analysed by the screw theory. The tessellation can be folded to cylindrical-like structure and to a dense polyhedral-like tubular structure via the interference between the sheets. The folding property of the tessellation is only related with two kinds of dihedral angles in the structures. Hence, it is potential to form variable sandwich layers which work with one DOF.

Keywords: Kirigami · Tessellation · Two-fold symmetric Bricard 6R linkage · Thin-walled tubes

1 Introduction

Origami, a traditional art originating in China and developing in Japan, has been used in a variety of applications, such as solar arrays [1, 2], robots [3, 4], honeycomb cores [5, 6] and metamaterials [7]. By regularly removing some parts of the paper surface, krigami releases some constraints to produce new forms and characteristics of motion, and is an important branch of origami.

Therefore, it has attracted some scholars and engineers to study, e.g. a tilted surface with flexible kirigami to replace the optical tracking [2] and a bionic creeping robot mimicking the crawling of snakes [4]. Meanwhile, kirigami has been applied to mechanical metamaterials [7] and honeycomb sandwich [6]. Fang [8] and Wang [9] used the self-locking of the origami structure to get large and complex assorted 3D-shapes kirigami. These 3D shape structures formed by kirigami have excellent mechanical properties.

In the application, structures with creases, which are able to conduct the deformation, are commonly used as one type of energy-absorbing structures. Song [10] takes origami patterns into thin-walled structures and finds that origami-patterned tubes are more suitable as energy-absorbing devices. Later, Ma [11, 12] and Ming [13] find tubes showing deformation in diamond mode have better mechanical properties than conventional origami patterns. However, the complex patterns make processing more difficult.

N. V. Khang et al. (Eds.): ASIAN MMS 2021, MMS 113, pp. 60–68, 2022.
https://doi.org/10.1007/978-3-030-91892-7_6

The kirigami structure, releasing some constraints, can get the thin-walled tubes in complex pattern. By the skeletonized pattern, the kirigami tessellation we propose can create new dihedrons and become a thin-walled tuber in diamond mode when it folds.

In this paper, inspired from an overconstrained linkage, we get a kirigami pattern and create a new tessellation based on it. The pattern, double-sym kiri, is introduced and the combination of two double-sym kiris in two directions are analysed in Sect. 2. The tessellation and its performance are shown in Sect. 3. Conclusion in Sect. 4 ends this paper.

2 Double-Sym Kiri

Pellegrino et al. [14] presented a model with both line-symmetric and plane-symmetric, which was called two-fold symmetric 6R linkage studied in detail by Chen and You [15]. In this linkage, some particular twist angles, $\pi/2$ and $3\pi/2$, were adopted, and we can get the generalized form of two-fold symmetric Bricard 6R linkage by releasing the constraints, namely the geometric conditions are:

$$\alpha_{12} = \alpha_{45} = \alpha, \quad \alpha_{34} = \alpha_{61} = \beta = 2\pi - \alpha \tag{1a}$$

$$\alpha_{23} = -\alpha_{56} = \gamma, \tag{1b}$$

$$a_{12} = a_{23} = a_{34} = a_{45} = a_{56} = a_{61} = a, \tag{1c}$$

$$R_2 = R_5 = -r, \quad R_3 = R_6 = r, \tag{1d}$$

where α_{ij} represents the twist angles between axes Z_i ($i = 1, 2, \ldots, 6$) and Z_j ($j = 1, 2, \ldots, 6$), a_{ij} represents the link length between axes Z_i and Z_j, and R_i is the common normal distance from axis X_i to X_{i+1}. The linkage is shown in Fig. 1(b), and hence has two symmetric planes 1 and 2.

Here, we are going to use the generalised two-fold symmetric Bricard 6R linkage to construct a kirigami pattern. To make all revolute joint axes (creases) are coplanar at the deployed configuration, all link lengths are zero, namely $a_{ij} = 0$, and the twist angles must satisfy $4\alpha + 2\gamma = 2\pi$, namely $\gamma = \pi - 2\alpha$. Therefore, the pattern is called double-sym kiri, γ is the design parameter, r is to determine the size of pattern. Here, the outer contour is limited as a square to facilitate the tessellation, see Fig. 1(a), where the solid line indicates the mountain creases and the dashed line indicates the valley creases.

In Fig. 1 (a), coordinate system O-$X_0Y_0Z_0$ is built, where the midpoint of B_1B_6 is chosen as the origin O, X_0-axis passes through point A_1, Y_0-axis directs to C_1, Z_0-axis determined by the right-hand rule, and A_i, B_i ($i = 1, 2, 3, 4$) and C_j ($j = 1, 2, 3, 4, 5, 6$) are endpoints.

The kinematic equations of two-fold symmetric Bricard 6R linkage are [16]:

$$\theta_1 = \theta_5, \theta_2 = \theta_4, \theta_3 = \theta_6, \theta_2 = \theta_1 + \pi \tag{2a}$$

$$\theta_3 = 2\arctan\frac{2\cos\frac{\gamma}{2}}{\sin\gamma\tan\frac{\theta_1}{2}}. \tag{2b}$$

Therefore, there are two kinds of kinematic angles for this linkage. To describe the motion more intuitively, dihedral angles are employed, $\varphi_{\rm m}$ presenting the one for the mountain creases and $\varphi_{\rm v}$ for valley creases. According to the definitions of dihedral angles and kinematic angles,

$$\varphi_{\rm v} = \theta_3, \varphi_{\rm m} = \pi + \theta_1 = \theta_4, \tag{3}$$

Equation (2b) becomes

$$\varphi_{\rm m} = \pi - 2\arctan\frac{2\cos\frac{\gamma}{2}}{\sin\gamma\tan\frac{\varphi_{\rm v}}{2}}, \tag{4}$$

and the curves of them for some γs are shown in Fig. 2. Here, $\gamma < \pi$ to ensure the double -sym kiri created. As γ increasing, $\varphi_{\rm m}$ gradually linearly correlated with $\varphi_{\rm v}$.

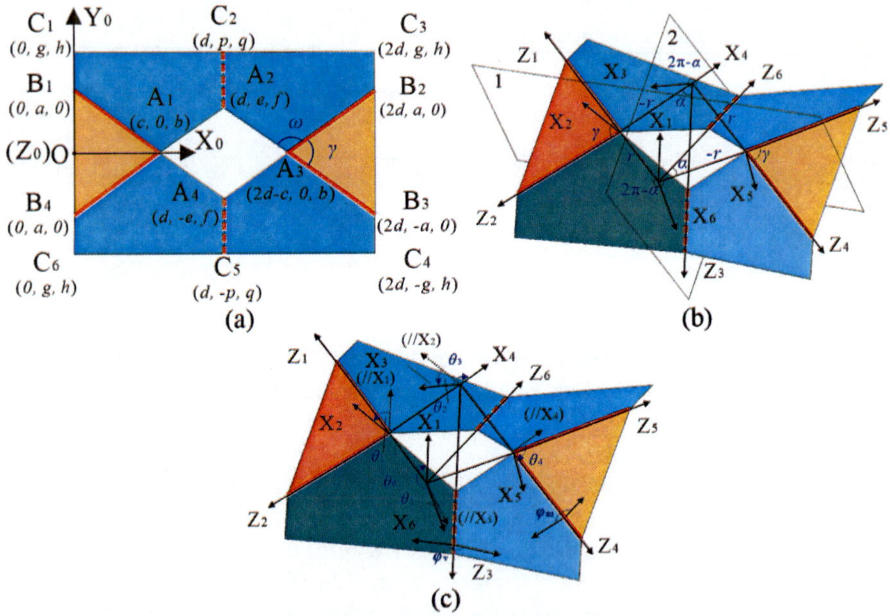

Fig. 1. The geometric conditions of double-sym kiri

3 Tessellation

Since the outer contour is rectangular, it is possible tessellated along X axis and Y axis, which are denoted as horizontal and vertical directions, respectively. In the vertical direction, two double-sym kiris are combined together [17], see Fig. 3 (a). The center vertex is connected by four creases which are perpendicular to each other. According to [18], the spherical 4R linkage can contains two planar motion paths, which are the

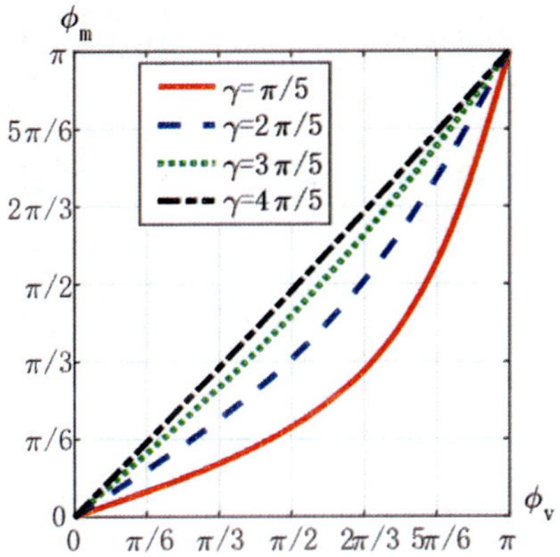

Fig. 2. Kinematic path of two kiri patterns in the horizontal direction

ways we are interested in. Therefore, each pair of two adjacent quadrilaterals in vertical direction is merged as one panel, see Fig. 3 (b).

Since the two-fold symmetric Bricard 6R linkage has one DOF,when the common revolute joint is chosen as the input, two 6R linkages are both determined in the same kinetic characteristics, hence the two-loop assembly has one DOF.

In the horizontal direction, see Fig. 4, two double-sym kiri patterns are connected by two identical spherical 4R linkages, namely Miura-oris. The dihedral angles in the spherical 4R linkage satisfy the equation: $\varphi_{s1} = \varphi_{s3} = \varphi_m$, $\varphi_{s2} = \varphi_{s4} = \varphi_s$.

The kinematic equation of spherical 4R linkage is [19]

$$U \sin\varphi_s + V \cos\varphi_s + W = 0, \tag{5}$$

where:
$U = -\sin\alpha_{23}\sin\alpha_{41}\sin\varphi_m$,
$V = \cos\alpha_{12}\sin\alpha_{23}\sin\alpha_{41}\cos\varphi_m + \sin\alpha_{12}\sin\alpha_{23}\cos\alpha_{41}$,
$W = \sin\alpha_{12}\cos\alpha_{23}\sin\alpha_{41}\cos\varphi_m - \cos\alpha_{12}\cos\alpha_{23}\cos\alpha_{41} + \cos\alpha_{34}$.
Substituting Eq. (4) to Eq. (5),

$$\frac{4\cos\frac{\varphi_v}{2}\sin\frac{\gamma}{2}(\cos\gamma + 1)\left[\cos\left(\frac{\varphi_v}{2} + \pi - \varphi_s\right) + \cos\frac{\varphi_v}{2}\right]}{\cos\varphi_v + \cos\gamma(\cos\varphi_v - 1) + 3} = 0. \tag{6}$$

According to Eq. (6), we can get $\varphi_s = \varphi_v$ or $\varphi_s = 0$. As the dihedral angles φ_s and φ_v are both variables, during the folding φ_v must always be equal to φ_s. Therefore, there are two kinds of dihedral angles in the two double-sym kiri patterns in horizontal direction.

For the two double-sym kiri patterns in horizontal direction, there are four kinematic loops and the structure consists of 12 links connected by 15 joints. The topological constrain graph is shown in Fig. 5, in which the vertices of polygon represent the rigid links and the edges of polygon represent the joints.

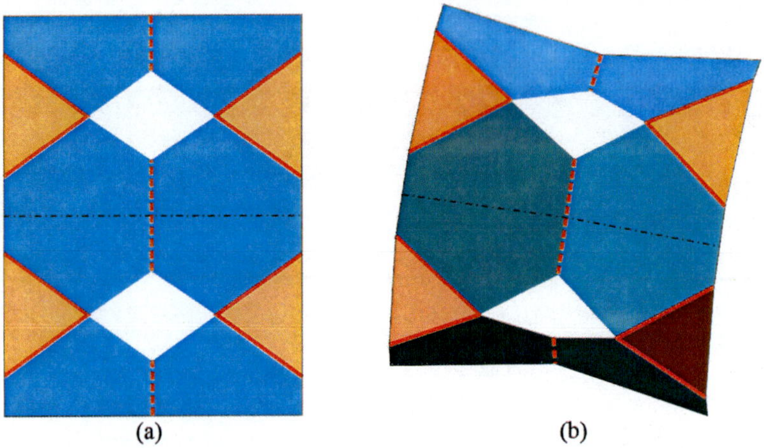

Fig. 3. Two double-sym kiri patterns in the vertical direction

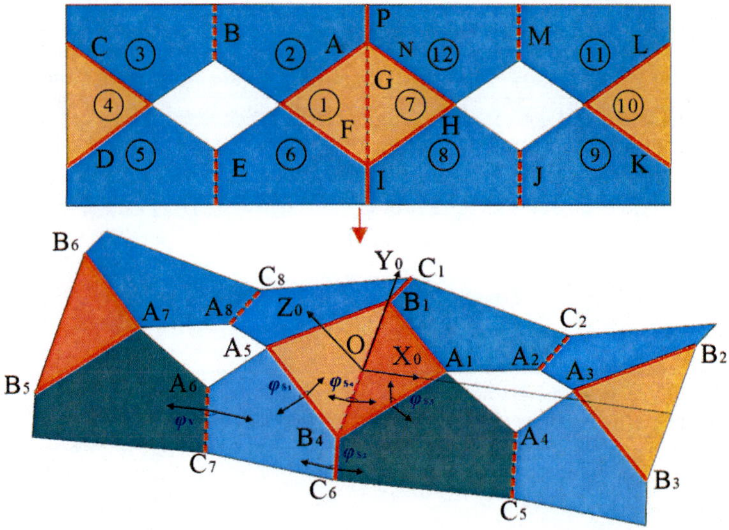

Fig. 4. Two double-sym kiri patterns in the horizontal direction

The coordinate system is established in the right kiri as in Sect. 2, see Fig. 4. Then the coordinates of joint endpoints A_i, B_i, C_j are listed in the Table 1. Then, screws of all creases are obtained, see Table 2.

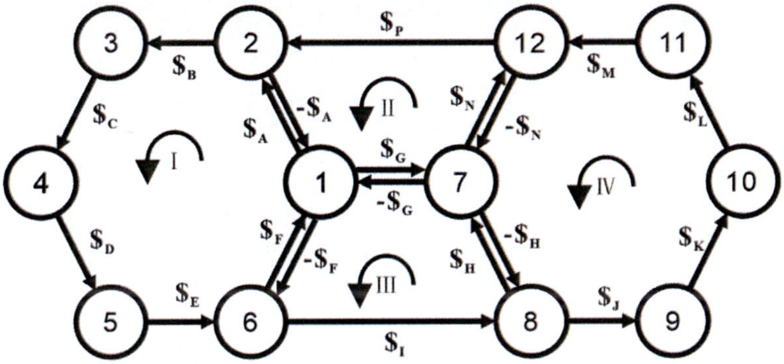

Fig. 5. The topological constrain graph of two double-sym kiris in the horizontal direction

Table 1. Coordinates of endpoints

Parameter	A1	A2	A3	A4	A5
Coordinates	$(c, 0, 0)$	(d, e, f)	$(2d{-}c, 0, 0)$	$(d, -e, f)$	$(-c, 0, 0)$
Parameter	A7	A7	A8	B1	B2
Coordinate	$(-d, -e, f)$	$(-2d + c, 0, 0)$	$(-d, e, f)$	$(0, a, b)$	$(2d, 1, b)$
Parameter	B3	B4	B5	B6	C1
Coordinates	$(2d, -1, b)$	$(0, -a, b)$	$(-2d, -1, b)$	$(-2d, 1, b)$	$(0, g, h)$
Parameter	C2	C5	C6	C7	C8
Coordinate	(d, p, q)	$(d, -p, q)$	$(0, -g, h)$	$(-d, -p, q)$	$(-d, p, q)$

According to the kinematics of multi-loop linkages [20], the constraint equations are

$$\begin{aligned}
&\$_{\mathbf{A}}\omega_{\mathrm{A}} + \$_{\mathbf{B}}\omega_{\mathrm{B}} + \$_{\mathbf{C}}\omega_{\mathrm{C}} + \$_{\mathbf{D}}\omega_{\mathrm{D}} + \$_{\mathbf{E}}\omega_{\mathrm{E}} + \$_{\mathbf{F}}\omega_{\mathrm{F}} = 0\\
&-\$_{\mathbf{A}}\omega_{\mathrm{A}} + \$_{\mathbf{G}}\omega_{\mathrm{G}} + \$_{\mathbf{N}}\omega_{\mathrm{N}} + \$_{\mathbf{P}}\omega_{\mathrm{P}} = 0\\
&-\$_{\mathbf{F}}\omega_{\mathrm{F}} - \$_{\mathbf{G}}\omega_{\mathrm{G}} + \$_{\mathbf{H}}\omega_{\mathrm{H}} + \$_{\mathbf{I}}\omega_{\mathrm{I}} = 0\\
&-\$_{\mathbf{H}}\omega_{\mathrm{H}} + \$_{\mathbf{J}}\omega_{\mathrm{J}} + \$_{\mathbf{K}}\omega_{\mathrm{K}} + \$_{\mathbf{L}}\omega_{\mathrm{L}} + \$_{\mathbf{M}}\omega_{\mathrm{M}} - \$_{\mathbf{N}}\omega_{\mathrm{N}} = 0
\end{aligned} \tag{7}$$

which can be written as the matrix form

$$\mathbf{M\omega} = 0, \tag{8}$$

$\boldsymbol{\omega}$ is the collection of velocities with a vector

$$\boldsymbol{\omega} = [\omega_{\mathrm{A}}, \omega_{\mathrm{B}}, \omega_{\mathrm{C}}, \omega_{\mathrm{D}}, \omega_{\mathrm{E}}, \omega_{\mathrm{F}}, \omega_{\mathrm{G}}, \omega_{\mathrm{H}}, \omega_{\mathrm{I}}, \omega_{\mathrm{J}}, \omega_{\mathrm{K}}, \omega_{\mathrm{L}}, \omega_{\mathrm{M}}, \omega_{\mathrm{N}}, \omega_{\mathrm{P}}]^{\mathrm{T}} \tag{9}$$

$\mathbf{M}$ is the coefficient matrix

$$\mathbf{M} = \left\{\begin{matrix}
\$_{\mathbf{A}} & \$_{\mathbf{B}} & \$_{\mathbf{C}} & \$_{\mathbf{D}} & \$_{\mathbf{E}} & \$_{\mathbf{F}} & \mathbf{0} & \mathbf{0} & \mathbf{0} & \mathbf{0} & \mathbf{0} & \mathbf{0} & \mathbf{0} & \mathbf{0} & \mathbf{0}\\
-\$_{\mathbf{A}} & \mathbf{0} & \mathbf{0} & \mathbf{0} & \mathbf{0} & \mathbf{0} & \$_{\mathbf{G}} & \mathbf{0} & \mathbf{0} & \mathbf{0} & \mathbf{0} & \mathbf{0} & \mathbf{0} & \$_{\mathbf{N}} & \$_{\mathbf{P}}\\
\mathbf{0} & \mathbf{0} & \mathbf{0} & \mathbf{0} & \mathbf{0} & -\$_{\mathbf{F}} & -\$_{\mathbf{G}} & \$_{\mathbf{H}} & \$_{\mathbf{I}} & \mathbf{0} & \mathbf{0} & \mathbf{0} & \mathbf{0} & \mathbf{0} & \mathbf{0}\\
\mathbf{0} & \mathbf{0} & \mathbf{0} & \mathbf{0} & \mathbf{0} & \mathbf{0} & \mathbf{0} & -\$_{\mathbf{H}} & \mathbf{0} & \$_{\mathbf{J}} & \$_{\mathbf{K}} & \$_{\mathbf{L}} & \$_{\mathbf{M}} & -\$_{\mathbf{N}} & \mathbf{0}
\end{matrix}\right\} \tag{10}$$

Table 2. Coordinates of screws

Screw $\$_i$	Direction Vector $\mathbf{S}_i$	Line distance $\mathbf{S}_{\mathbf{d}i}$
$\$_{\mathbf{A}}$	$[-c, -a, -b]$	$\left[c^3 - c, ac^2, bc^2\right]$
$\$_{\mathbf{B}}$	$[0, p-e, q-f]$	$\left[-d, p-(e-p)\left(pe-p^2+qf-q^2\right), q-(f-q)\left(pe-p^2+qf-q^2\right)\right]$
$\$_{\mathbf{C}}$	$[-c, a, b]$	$\left[c\left(b^2+a^2+2cd\right)-2d, a-a\left(b^2+a^2+2cd\right), b-b\left(b^2+a^2+2cd\right)\right]$
$\$_{\mathbf{D}}$	$[-c, -a, b]$	$\left[c\left(b^2+a^2+2cd\right)-2d, -a+a\left(b^2+a^2+2cd\right), b-b\left(b^2+a^2+2cd\right)\right]$
$\$_{\mathbf{E}}$	$[0, e-p, q-f]$	$\left[-d, -p+(e-p)\left(pe-p^2+qf-q^2\right), q-(f-q)\left(pe-p^2+qf-q^2\right)\right]$
$\$_{\mathbf{F}}$	$[-c, a, -b]$	$\left[c^3 - c, -ac^2, bc^2\right]$
$\$_{\mathbf{G}}$	$[0, -2a, 0]$	$\left[0, 4a^3 - a, b\right]$
$\$_{\mathbf{H}}$	$[-c, -a, b]$	$\left[c\left(a^2+b^2\right), a\left(a^2+b^2\right)-a, b-b\left(a^2+b^2\right)\right]$
$\$_{\mathbf{I}}$	$[0, a-g, h-b]$	$\left[0, (a-g)\left(ag-g^2+bh-h^2\right)-g, h-(b-h)\left(ag-g^2+bh-h^2\right)\right]$
$\$_{\mathbf{J}}$	$[0, e-p, q-f]$	$\left[d, -p+(e-p)\left(pe-p^2+qf-q^2\right), q-(f-q)\left(pe-p^2+qf-q^2\right)\right]$
$\$_{\mathbf{K}}$	$[c, -a, b]$	$\left[-c\left(b^2+a^2+2cd\right)+2d, -a+a\left(b^2+a^2+2cd\right), b-b\left(b^2+a^2+2cd\right)\right]$
$\$_{\mathbf{L}}$	$[c, a, b]$	$\left[-c\left(b^2+a^2-2cd\right)+2d, a-a\left(b^2+a^2+2cd\right), b-b\left(b^2+a^2+2cd\right)\right]$
$\$_{\mathbf{M}}$	$[0, p-e, q-f]$	$\left[d, p-(e-p)\left(pe-p^2+qf-q^2\right), q-(f-q)\left(pe-p^2+qf-q^2\right)\right]$
$\$_{\mathbf{N}}$	$[-c, a, b]$	$\left[c\left(a^2+b^2\right), a-a\left(a^2+b^2\right), b-b\left(a^2+b^2\right)\right]$
$\$_{\mathbf{P}}$	$[0, g-a, h-b]$	$\left[0, g-(a-g)\left(ag-g^2+bh-h^2\right), h-(b-h)\left(ag-g^2+bh-h^2\right)\right]$

where $\mathbf{O} = [0, 0, 0, 0, 0, 0]^{\mathrm{T}}$, $\$_i = [\mathbf{S}_i, \mathbf{S}_{\mathbf{d}i} \times \mathbf{S}_i]^{\mathrm{T}}$ (i = A, B, C…). The null space matrix $\mathbf{nM}$ can be obtained by Matlab

$$\mathbf{nM} = Q\left\{1, \frac{2b}{f-q}, -1, -1, \frac{2b}{f-q}, 1, \frac{a(h-b)+bg}{ah}, 1, \frac{2b}{h}, \frac{2b}{f-q}, -1, -1, -\frac{2b}{f-q}, 1, \frac{2b}{h}\right\}^{\mathrm{T}} \tag{11}$$

where $Q = \frac{h}{2b}$. There is one independent column, so the combination of the two double-sym kiris in the horizontal direction is one DOF.

Considering the above two directions, the tessellation based on two-fold symmetric Bricard 6R linkage and spherical 4R linkage can be constructed, see Fig. 6 (a). We can find that the kirigami presents a cylindrical-like structure during the folding from a flat state. It will generate physical interference when the skeletonized diamond shape disappears. This interference makes it possible to fold the pattern to a compact tube. The folding performance can be expressed by the equation from cubic geometric relationships

$$\cos\varepsilon = \frac{\cos\gamma - \cos^2\omega}{\sin^2\omega}\cos^2\left(\omega + \frac{\gamma}{2}\right) + \sin^2\left(\omega + \frac{\gamma}{2}\right), \tag{12}$$

where, ε denotes the angle of the two valley creases in the 6R mechanism when interference is generated. ω is an inner angle of the hexagonal sheet, which determines the skeletonized diamond.

To obtain a dense tubular structure, ε has to be set as the interior corner of a regular polygon and the number of columns of hexagonal sheets should be the number of sides of the polygon. Such as, when $\gamma = 60°$ and $\omega = 120°$, ε equals to 60° which is the interior corner of a regular triangle and requires three columns of hexagonal sheets, as Fig. 6 (a) shown. When $\gamma = 90°$, $\omega = 112.5°$, ε equals to 90° which is the interior corner of a square and requires four columns of hexagonal sheets, as Fig. 6 (b) shown.

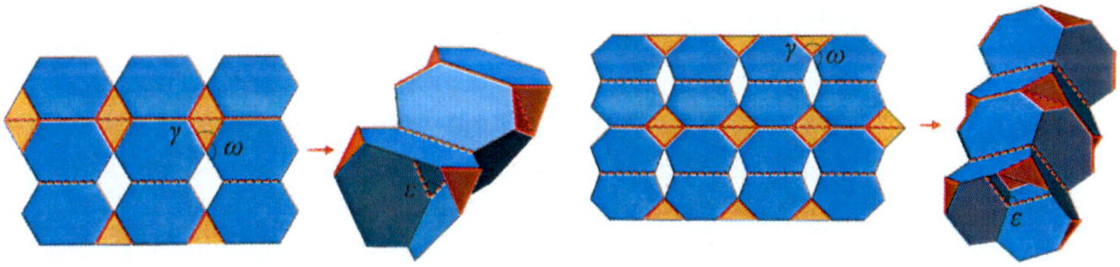

(a) (b)

Fig. 6. The compact tube. (a) Trigonometry-like tube. (b) Quadrilateral-like tube.

4 Conclusion

This paper presents a double-sym kiri pattern based on two-fold symmetric Bricard 6R linkage, and a tessellation is constructed with this kirigami and spherical 4R linkage. Kinematic analysis shows that the tessellation is with one DOF, and it can be folded as a cylindrical-like structure when the hollowed diamonds disappear. The angular relationship is determined by an equation that allows the formation of a dense polyhedron-like tubular structure via the interference between the sheets.

In the future, the derived patterns in general combinations based on two-fold symmetric Bricard 6R linkage and spherical 4R linkage will be studied and the mechanical analysis of the thin-walled tube will also be carried out.

Acknowledgment. The authors wish to appreciate the financial supports from the National Natural Science Foundation of China (Project No. 5190s5101), the Natural Science Foundation of Fujian Province, China (Project No. 2019J01209), and the TJU-FZU Independent Innovation Fund (Project No. TF-1901).

References

1. Koryo, M.: Method of packaging and deployment of large membranes in space. Inst. Space Astronaut. Sci. Rep. 1 (1985)
2. Lamoureux, A., Lee, K., Shlian, M., Forrest, S.R., Shtein, M.: Dynamic Kirigami structures for integrated solar tracking. Nat. Commun. **6**, 1–6 (2015)
3. Lee, D.Y., Kim, S.R., Kim, J.S., Park, J.J., Cho, K.J.: Origami wheel transformer: a variable-diameter wheel drive robot using an origami structure. Soft Robot. **4**, 163–180 (2017)

4. Rafsanjani, A., Zhang, Y., Liu, B., Rubinstein, S.M., Bertoldi, K.: Kirigami skins make a simple soft actuator crawl. Sci. Robot. **3** (2018)
5. Hou, Y., Neville, R., Scarpa, F., Remillat, C., Gu, B., Ruzzene, M.: Graded conventional-auxetic Kirigami sandwich structures: Flatwise compression and edgewise loading. Compos. Part B: Eng. **59**, 33–42 (2014). https://doi.org/10.1016/j.compositesb.2013.10.084
6. Neville, R.M., Monti, A., Hazra, K., Scarpa, F., Remillat, C., Farrow, I.R.: Transverse stiffness and strength of Kirigami zero-ν PEEK honeycombs. Compos. Struct. **114**, 30–40 (2014). https://doi.org/10.1016/j.compstruct.2014.04.001
7. Eidini, M., Paulino, G.H.: Unraveling metamaterial properties in zigzag-base folded sheets. Sci. Adv. **1**, e1500224 (2015)
8. Fang, H., Chu, S.A., Xia, Y., Wang, K.: Programmable self-locking origami mechanical metamaterials. Adv. Mater. **30**, 1706311 (2018)
9. Wang, X., Guest, S.D., Kamien, R.D.: Keeping it together: interleaved Kirigami extension assembly. Phys. Rev. X **10**, 11013 (2020)
10. Song, J., Chen, Y., Lu, G.: Axial crushing of thin-walled structures with origami patterns. Thin-Walled Struct. **54**, 65–71 (2012)
11. Zhou, C., Wang, B., Ma, J., You, Z.: Dynamic axial crushing of origami crash boxes. Int. J. Mech. Sci. **118**, 1–2 (2016)
12. Ma, J., You, Z.: Energy absorption of thin-walled beams with a pre-folded origami pattern. Thin-Walled Struct. **73**, 198–206 (2013)
13. Ming, S., et al.: The energy absorption of long origami-ending tubes with geometrical imperfections. Thin-Walled Struct. **161**, 107415 (2021)
14. Pellegrino, S., Green, C., Guest, S.D., Watt, A.: SAR advanced deployable structure. University of Cambridge Technical report (2000)
15. Chen, Y., You, Z.: Two-fold symmetrical 6R foldable frame and its bifurcations. Int. J. Solids Struct. **46**, 4504–4514 (2009)
16. Yang, F.F., Chen, K.J.: General kinematics of twofold-symmetric Bricard 6R linkage. J. Tianjin Univ. Accepted. (in Chinese)
17. Song, C.Y.: Kinematic study of overconstrained linkages and design of reconfigurable mechanisms. Ph.D. Nanyang Technological University (2013)
18. Chiang, C.H.: Kinematics of Spherical Mechanisms. Cambridge University Press, New York (1988)
19. Cervantes-Sánchez, J.J., Medellín-Castillo, H.I.: A robust classification scheme for spherical 4R linkages. Mech. Mach. Theory **37**(10), 1145–1163 (2002)
20. Wohlhart, K.: Screw spaces and connectivities in multiloop linkages. In: Lenarčič, J., Galletti, C. (eds.) On Advances in Robot Kinematics, pp. 97–106. Springer, Dordrecht (2004). https://doi.org/10.1007/978-1-4020-2249-4_11

Path Synthesis Method for Self-alignment Knee Exoskeleton

Rui Wu, Ruiqin Li(✉), Hailong Liang, and Fengping Ning

North University of China, Taiyuan 030051, China

Abstract. Axial misalignment of the exoskeleton through a purely rolling connection is unavoidable during movement, as the knee joint consists of a mixture of sliding and rolling movements. Axial misalignment can affect transmission efficiency and even create potential safety hazards. Various methods have been trying to address axial misalignment, most of which increase the complexity of the knee exoskeleton structure. In this paper, the path of the knee joint is obtained by extracting the coordinates of the knee joint motion marker points during the walking of an adult male. A power function is fitted to the path point to obtain a seventh-order polynomial function. The points on the function are taken as the target points of synthesis, and the knee path is synthesized to obtain the planar four-bar mechanism. The error between the path of the planar four-bar mechanism and the seventh polynomial function is analyzed. The results show that the planar four-bar mechanism can track the partial curve of the polynomial well. Therefore, the planar four-bar mechanism as a rotating connection can mitigate the axial misalignment.

Keywords: Knee exoskeleton · Self-alignment · Path synthesis

1 Introduction

The exoskeleton robot is a wearable robot that can help patients recover and improve endurance in healthy people [1–3]. Although Exoskeletons have made significant progress over the years, there are still many serious challenges [4, 5]. Axis misalignment of the exoskeleton is a problem that cannot be ignored [6–8]. The knee joint contains two kinds of movement, namely rolling and sliding, which causes the joint axis position of the knee to float. The fixed axis of the exoskeleton will be dislocated from the axis of the floating knee joint. Axis misalignment not only affects the exoskeleton and boosts efficiency, but also poses a safety risk [9, 10]. Axis dislocation usually has two reasons, one is the mismatch between the fixed mechanism size and the human body size, the other is the mismatch between the fixed rotation axis of the exoskeleton and the floating joint axis of the human body.

Many experts have conducted extensive research on the first part of the mechanism with the axis misalignment of the human body. The dominating measure is to increase the degree of freedom of the knee, making the joint of the exoskeleton adjustable. Stienen et al. [11] decoupled the motion of the knee joint, transforming the knee joint

N. V. Khang et al. (Eds.): ASIAN MMS 2021, MMS 113, pp. 69–79, 2022.
https://doi.org/10.1007/978-3-030-91892-7_7

motion into a composite motion of translation and rotation. These structures are feasible for rehabilitation exoskeletons, which may be too complex for enhanced exoskeletons [12]. Cempini et al. [13] designed a self-aligning mechanism based on kinematic static analysis. Schorsch et al. [3] also proposed a decoupling method to obtain an exoskeleton elbow mechanism with planar adjustable power. The adjustable exoskeleton of the knee joint can be adjusted for different wearers so that the exoskeleton joint in the initial position coincides with the axis of the human joint. But another joint dislocation was not addressed. That's the displacement caused by the float in the center of the joint. This kind of problem is more difficult to solve. Because the rotary joints in use today are axis fixed. The simple mechanism synthesis with the variable axis is a new problem for mechanism science with the development of science and technology. Niu [14] proposed a five-bar parallel mechanism with the ability to automatically adjust the center of rotation. Singh's [15] work on synthesizing exoskeleton mechanisms based on the path points of the hip joint in a person's natural gait has given us more insight. Sarkisian [16] demonstrates a lightweight and compact self-aligning mechanism that enhances the self-aligning mechanism while improving sitting and standing comfort. Various methods have been reported to solve the axis misalignment problem, but all of them increase the complexity of the knee joint structure. The planar four-bar mechanism is one of the simplest single-degree-of-freedom mechanisms. It can be applied to the design of the exoskeleton knee joint and other joints with the variable axis with appropriate size and structure. Seth et al. [17] conducted musculoskeletal dynamics studies, which provided data to support further studies of the knee joint path. In terms of linkage synthesis, Bai and Wu et al. [18–20] investigated the synthesis of a four-bar mechanism with coupler curves using different methods. Wu et al. [21] further investigated the mixed path and motion synthesis problem based on a conic filtering algorithm. Kafash and Nahvi [22] proposed a circular proximity function so that the synthesis variables do not increase with the number of target points, which reduces the computational cost of synthesis. These studies have laid the theoretical foundation for further research on self-aligned synthesis mechanisms [23]. The path is modeled based on the extraction of the actual position of the joint and the path mapping. The path synthesis into theory is applied to synthesize the parameterized conditions. And the error analysis of the trajectories of the synthesized obtained four-bar mechanism is carried out.

In this paper, a simple exoskeleton knee joint is obtained based on path synthesis. The path points of the knee joint during human movement were projected, and a polynomial was used to fit the projected points. Polynomials build exact points of exact synthesis based on equal step size. The knee joint structure of the exoskeleton was synthesis with the exact points obtained.

2 Methods of Path Modeling

To effectively restrain the dislocation between the exoskeleton knee joint and the human knee joint, the motion path of the human knee joint was analyzed. Due to the floating of the knee joint, it is difficult to design the exoskeleton knee joint. The knee joint is general the floating center joint that is rolling and sliding combined motion. We wanted to find a single degree of freedom mechanism that was simple enough to accommodate the motion of the knee joint. The synthesis problem of the knee joint is transformed into a simple mechanism to meet the requirements of axial floating. By analyzing the motion path of the knee joint, the four-bar mechanism can track the motion path of the knee joint well.

We extracted movement data from Seth et al.'s [17] musculoskeletal dynamics study during human lower limb movement. Assuming that the movement data of the left and right lower limbs were the same, we selected the movement data of the right lower limbs for the study. To analyze the slip and rotation of the knee axis, the actual motion path was projected. The points of motion states and marked points are shown in Fig. 1 below. The marking point position output selects the gait cycle. The path of the knee joint is determined by the mutual position of Mark points.

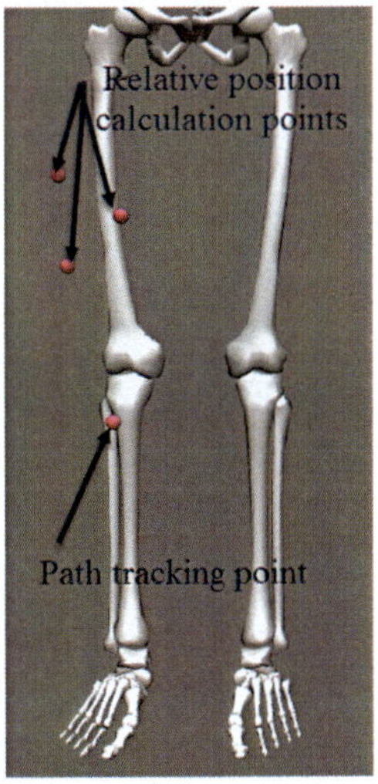

Fig. 1. The distribution position of lower limb movement track points

The center of rotation floats and changes with the movement of the knee joint. This also explains the internal forces that occur even after the axis is aligned. The knee joint in achieving the walking function is a single degree of freedom mechanism. The joint attachment mark points were obtained by calculating the data from Seth et al.'s [17] model, and a two-dimensional point cloud was obtained by projecting the path. The polynomial was used to fit the point, the curves are shown in Fig. 2, and the specific expression of the polynomial was obtained.

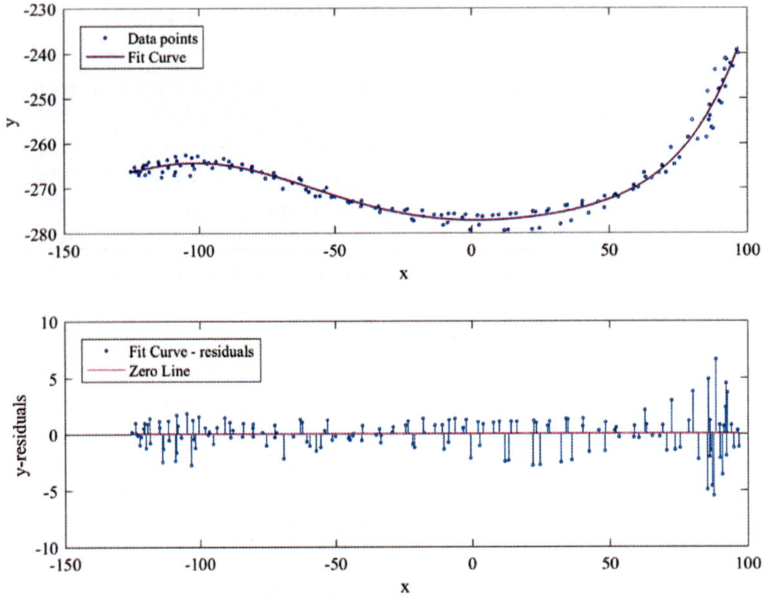

Fig. 2. The path fitting curve and the corresponding residuals

The specific expression was obtained by using a seventh-order polynomial for the fit. The residuals of all the fitted points were calculated based on the function. The fitted curves and residuals are shown in Fig. 2.

$$y = a_1x^7 + a_2x^6 + a_3x^5 + a_4x^4 + a_5x^3 + a_6x^2 + a_7x + a_8 \tag{1}$$

where

$a_1 = -1.328\text{e}{-14}$, $a_2 = 9.552\text{e}{-12}$, $a_3 = 2.429\text{e}{-09}$, $a_4 = -1.274\text{e}{-08}$, $a_5 = -6.696\text{e}{-06}$, $a_6 = 0.002003$, $a_7 = -0.006969$, $a_8 = -277.2$.

3 Precise Point

It can see that the point span range is about 220 mm. 10 mm, 6 mm, 4 mm, 2 mm isometric steps are selected to increase the accuracy of path synthesis, and the value of the polynomial is calculated. A total of 31 path points are obtained. The positions of the selected points are shown in Fig. 3, and the obtained path point data are shown in Table 1 below.

Table 1. List of target points

1	−125.30	−266.46	12	−15.30	−276.60	23	74.70	−262.59
2	−115.30	−265.21	13	−5.30	−277.11	24	78.70	−259.74
3	−105.30	−264.52	14	4.70	−277.19	25	82.70	−256.36
4	−95.30	−264.59	15	14.70	−276.89	26	86.70	−252.36
5	−85.30	−265.40	16	24.70	−276.23	27	88.70	−250.11
6	−75.30	−266.83	17	34.70	−275.19	28	90.70	−247.66
7	−65.30	−268.65	18	44.70	−273.65	29	92.70	−245.00
8	−55.30	−270.64	19	50.70	−272.40	30	94.70	−242.13
9	−45.30	−272.58	20	56.70	−270.79	31	96.70	−239.01
10	−35.30	−274.30	21	62.70	−268.73			
11	−25.30	−275.66	22	68.70	−266.05			

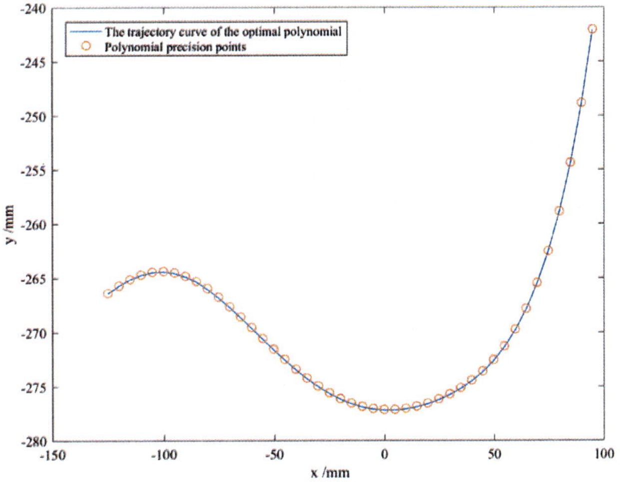

Fig. 3. The distribution of target points in the curve of path synthesis

4 Synthesis Four bar Mechanism

As shown in Fig. 4, a four-bar linkage is uniquely determined by nine parameters, i.e., r_1, r_2, r_3, r_4, r_5, β, x_A, y_A, α. Here, A(x_A, y_A) is the coordinate values of point A in the {O-XY}. r_1, r_2, r_3, r_4, r_5, are the link lengths. α is the angle between line AD and the horizontal axis. β is the angle between line *BC* and line *BM*. θ_2 is the rotation angle between line *AD* and line *AB*. θ_3 is the rotation angle between line *AD* and line *BC*. θ_4 is the rotation angle between line *AD* and line *CD*.

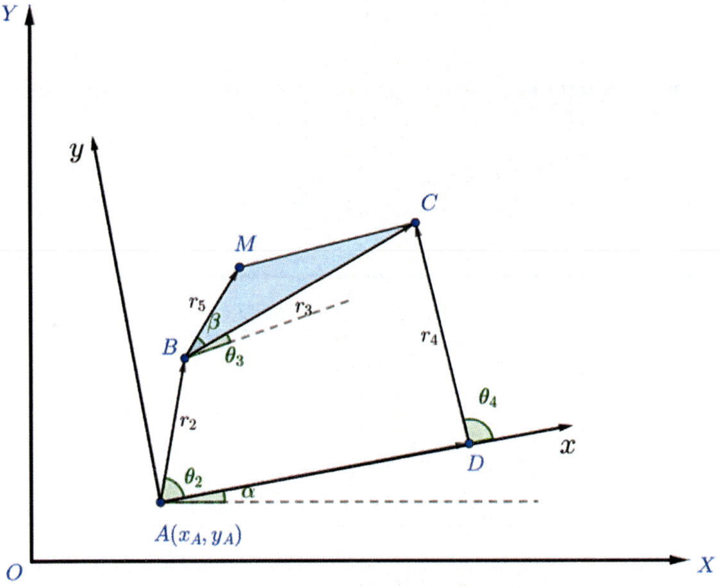

Fig. 4. Typical revolute four-bar linkages

The position equation of point M has been introduced many times in the literature [24]. For the completeness of this article, we will briefly explain the derivation method. Vector loop equation of four-bar linkage can be written as

$$\mathrm{r}_2 + \mathrm{r}_3 - \mathrm{r}_4 - \mathrm{r}_1 = 0 \tag{2}$$

$$\theta_3 = 2\arctan\left[\frac{m \pm \sqrt{l^2 + m^2 - n^2}}{l - n}\right] \tag{3}$$

where
$l = 2r_2r_3\cos\theta_2 - 2r_1r_3$, $m = 2r_2r_3\sin\theta_2$, $n = r_1^2 + r_2^2 + r_3^2 - r_4^2 - 2r_1r_2cos\theta_2$.
The position of point M can be expressed as

$$\begin{bmatrix} X_M \\ Y_M \end{bmatrix} = \begin{bmatrix} x_A \\ y_A \end{bmatrix} + \begin{bmatrix} cos\alpha & -sin\alpha \\ sin\alpha & cos\alpha \end{bmatrix}\begin{bmatrix} x_M \\ y_M \end{bmatrix} \tag{4}$$

where
$x_M = r_2cos\theta_2 + r_5cos\beta cos\theta_3 - r_5sin\beta sin\theta_3$,
$y_M = r_2sin\theta_2 + r_5cos\beta sin\theta_3 - r_5sin\beta cos\theta_3$.

4.1 Variable

The path synthesis of a four-bar mechanism has nine variables(r_1, r_2, r_3, r_4, r_5, β, x_A, y_A, α). If the error is calculated using a general objective function, the difficulty of calculation increases as the number of points given increases. Kafash and Nahvi [22] proposed to use the circular proximity function to divide the design variables into four optimization variables(x_A, y_A, r_3, β) and five design variables($r_1, r_2, r_4, r_5, \alpha$), and the target does not increase with the increase of target points.

4.2 Objective Function

The target function is defined as

$$f_{obj} = CPF_{dimensionless} + h_1 M_1 + h_2 M_2 \tag{5}$$

The values of M_1 and M_2 are defined as a constant of 1000. For detailed information about the $CPF_{dimensionless}$ function, please refer to the literature [22].

Where

$$h_1 = \begin{cases} 1 \; s + l > p + q \\ 0 \; s + l < p + q \end{cases} \tag{6}$$

$$h_2 = \begin{cases} 1 \; \text{r\_2} \neq \text{s} \\ 0 \; \text{r\_2} = \text{s} \end{cases} \tag{7}$$

$$s = max\{r_1, r_2, r_3, r_4\} \tag{8}$$

$$l = min\{r_1, r_2, r_3, r_4\} \tag{9}$$

$$\{p, q\} = \{r_1, r_2, r_3, r_4\} - \{s, l\} \tag{10}$$

The values of variables r_1 and r_4, as well as α, can be calculated using the following formula.

$$r_1 = \sqrt{(x_D - x_A)^2 + (y_D - y_A)^2} \tag{11}$$

$$r_4 = \frac{1}{n}\sqrt{\left(x_C^i - x_D\right)^2 + \left(y_C^i - y_D\right)^2} \tag{12}$$

$$\alpha = \arctan\left(\frac{y_D - y_A}{x_D - x_A}\right) \tag{13}$$

where

$$\begin{bmatrix} x_C^i \\ y_C^i \end{bmatrix} = \begin{bmatrix} x_A \\ y_A \end{bmatrix} + r_2 \begin{bmatrix} cos\theta_{AB}^i \\ sin\theta_{AB}^i \end{bmatrix} + r_3 \begin{bmatrix} cos\theta_{BC}^i \\ sin\theta_{BC}^i \end{bmatrix} \tag{14}$$

$$\theta_{AB}^i = arctan\left(\frac{Y_d^i - y_A}{X_d^i - x_A}\right) \pm arccos\left(\frac{r_2^2 + \left(X_d^i - x_A\right)^2 + \left(Y_d^i - y_A\right)^2 - r_5^2}{2r_2\sqrt{\left(X_d^i - x_A\right)^2 + \left(Y_d^i - y_A\right)^2}}\right) \tag{15}$$

$$\theta_{BM}^i = arctan\left(\frac{Y_d^i - y_A - r_2 sin\theta_{AB}^i}{X_d^i - x_A - r_2 cos\theta_{AB}^i}\right) \tag{16}$$

$$\theta_{BC}^i = \theta_{BM}^i - \beta \tag{17}$$

The calculation methods of r_2 and r_5 can be found in the literature [25]. For detailed information about the x_D, y_D, please refer to the literature [22]. $X_d{}^i$ and $Y_d{}^i$ are defined as the precision point in Table 1.

5 Results and Discussion

The example shows how to extract the path and solve the parameters of the planar four-bar mechanism. Table 1 is a classical synthesis problem (31 points) synthesizing a new coupler curve tracking a set of points. The exact points after discretization of the knee joint path. The results obtained using synthesis are shown in Table 2, and the plane tracking results are shown in Fig. 5.

Table 2. Link parameters for tracking knee joint motion

No	Error	r1	r2	r3	r4	r5	β	xA	yA	α
1	2.86E−03	265.85	51.16	198.69	135.99	483.61	−0.43	−278.58	137.92	−1.01
2	8.89E−05	913.68	99.01	295.70	987.02	335.60	0.81	−291.10	−435.23	1.44
3	3.35E−04	284.09	81.99	299.63	351.35	375.45	−1.35	−317.36	−44.57	−0.55
4	9.63E−03	205.60	62.38	223.86	91.77	344.72	−0.55	−238.01	−7.60	0.04

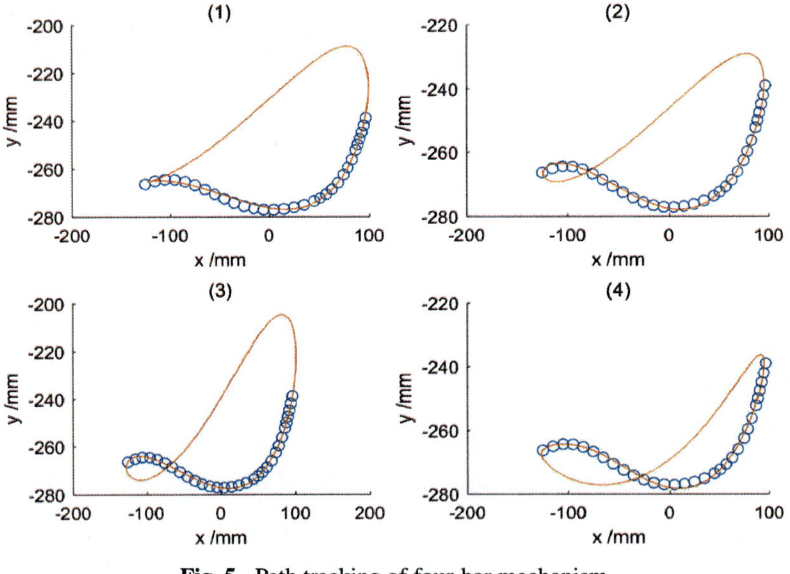

Fig. 5. Path tracking of four-bar mechanism

Four track access graphs are given in Fig. 5. In terms of tracking error, the error in Fig. 5(2) is the smallest. From the specific task of getting used to walking, the path diagram generated by the mechanism in Fig. 5(4) is more suitable for path point cloud because its path is within the range of motion. Both Fig. 5(1) and 5(3) have more room to choose from the perspective of adapting to multiple tasks because the range of motion is wider. Based on the data in Fig. 5(4), a 3D model of the knee joint connected by a four-bar mechanism is presented as shown in Fig. 6.

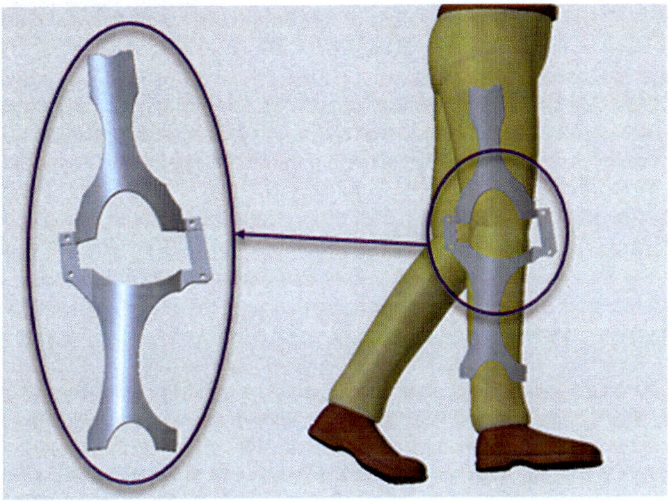

Fig. 6. Self-aligned 3D model of knee joint

Through the above analysis, the path generated by the four-bar mechanism as the joint of the knee joint can well track the path of the human knee joint. The design of the four-bar linkage minimizes the generation of internal forces and can track knee movement well. It is proved that this method can provide effective theoretical support for knee joint synthesis.

6 Conclusions

A method of path synthesis for a self-aligned self-aligned knee exoskeleton is proposed. Firstly, the axis of the knee joint is picked up and projected to form discrete points. A four-bar mechanism is obtained by synthesizing the path points according to the synthesis method. The path of the four-bar linkage has the same trend as that of the knee joint. The path error of the four-bar mechanism is further analyzed, and it is found that the discrete points can be tracked well to minimize the internal force generation.

The main contribution of this work is the proposal of a knee exoskeleton structure with a variable axis. In addition, with the self-adjusting mechanism proposed in [16], the mechanism does not require additional control to achieve an automatic change of the center of rotation. In addition, it is a relatively simple mechanism when compared with the five-bar mechanism proposed in [14].

The structure is suitable for exoskeletal knee joints, while the synthesis method applies to structural synthesis problems with variable axes. In addition, the proposed method can also be extended to the structural synthesis of the elbow joint, which will be further investigated in the future.

References

1. Zanotto, D., Akiyama, Y., Stegall, P., et al.: Knee joint misalignment in exoskeletons for the lower extremities: effects on user's gait. IEEE Trans. Rob. **31**(4), 978–987 (2015)

2. Christensen, S., Bai, S.: Kinematic analysis and design of a novel shoulder exoskeleton using a double parallelogram linkage. J. Mech. Robot. **10**(4), 1–10 (2018)
3. Schorsch, J.F., Keemink, A.Q.L., Stienen, A.H.A., et al.: A novel self-aligning mechanism to decouple force and torques for a planar exoskeleton joint. Mech. Sci. **5**(2), 29–35 (2014)
4. Shen, Z., Allison, G., Cui, L.: An integrated type and dimensional synthesis method to design one degree-of-freedom planar linkages with only revolute joints for exoskeletons. J. Mech. Des. **140**(9), 1–12 (2018)
5. D'elia, N., Vanetti, F., Cempini, M., et al.: Physical human-robot interaction of an active pelvis orthosis: toward ergonomic assessment of wearable robots. J. Neuroeng. Rehabil. **14**(1), 1–14 (2017)
6. Zhou, L., Li, Y., Bai, S.: A human-centered design optimization approach for robotic exoskeletons through biomechanical simulation. Robot. Auton. Syst. **91**(1), 337–347 (2017)
7. Junius, K., Degelaen, M., Lefeber, N., et al.: Bilateral, misalignment-compensating, full-DOF hip exoskeleton: design and kinematic validation. Appl. Bionics Biomech. **2017**, 1–14 (2017)
8. Li, J., Zhang, Z., Tao, C., et al.: A number synthesis method of the self-adapting upper-limb rehabilitation exoskeletons. Int. J. Adv. Rob. Syst. **14**(3), 1–14 (2017)
9. Chen, B., Zi, B., Wang, Z.Y., et al.: Knee exoskeletons for gait rehabilitation and human performance augmentation: a state-of-the-art. Mech. Mach. Theory **134**, 499–511 (2019)
10. Nuckols, R.W., Takahashi, K.Z., Farris, D.J., et al.: Mechanics of walking and running up and downhill: a joint-level perspective to guide design of lower-limb exoskeletons. PLoS ONE **15**(8), 1–20 (2020)
11. Stienen, A.H.A., Hekman, E.E.G., Van Der Helm, F.C.T., et al.: Self-Aligning exoskeleton axes through decoupling of joint rotations and translations. IEEE Trans. Rob. **25**(3), 628–633 (2009)
12. Li, J., Cao, Q., Zhang, C., et al.: Position solution of a novel four-DOFs self-aligning exoskeleton mechanism for upper limb rehabilitation. Mech. Mach. Theory **141**, 14–39 (2019)
13. Cempini, M., De Rossi, S.M.M., Lenzi, T., et al.: Self-alignment mechanisms for assistive wearable robots: a kinetostatic compatibility method. IEEE Trans. Rob. **29**(1), 236–250 (2013)
14. Niu, Y., Song, Z., Dai, J.: Kinematic analysis and optimization of a planar parallel compliant mechanism for self-alignment knee exoskeleton. Mech. Sci. **9**(2), 405–416 (2018)
15. Singh, R., Chaudhary, H., Singh, A.K.: A novel gait-inspired four-bar lower limb exoskeleton to guide the walking movement. J. Mech. Med. Biol. **19**(4), 1–24 (2019)
16. Sarkisian, S.V., Ishmael, M.K., Lenzi, T.: Self-aligning mechanism improves comfort and performance with a powered knee exoskeleton. IEEE Trans. Neural Syst. Rehabil. Eng. **29**(1), 629–640 (2021)
17. Seth, A., Hicks, J.L., Uchida, T.K., et al.: OpenSim: simulating musculoskeletal dynamics and neuromuscular control to study human and animal movement. PLoS Comput. Biol. **14**(7), 1–20 (2018)
18. Bai, S.P., Angeles, J.: Coupler-curve synthesis of four-bar linkages via a novel formulation. Mech. Mach. Theory **94**, 177–187 (2015)
19. Wu, R., Li, R., Bai, S.: A fully analytical method for coupler-curve synthesis of planar four-bar linkages. Mech. Mach. Theory **155**, 1–9 (2021)
20. Bai, S.: A note on the univariate nonic derived from the coupler curve of four-bar linkages. Mech. Mach. Theory **162**, 1–10 (2021)
21. Wu, R., Li, R., Liang, H., et al.: Mixed synthesis method of motion and path of planar four-bar linkages. Mech. Sci. **12**(1), 443–449 (2021)
22. Kafash, S.H., Nahvi, A.: Optimal synthesis of four-bar path generator linkages using circular proximity function. Mech. Mach. Theory **115**, 18–34 (2017)
23. Varedi-Koulaei, S.M., Rezagholizadeh, H.: Synthesis of the four-bar linkage as path generation by choosing the shape of the connecting rod. Proc. Inst. Mech. Eng. Part C-J. Mech. Eng. Sci. **234**(13), 2643–2652 (2020)

24. Acharyya, S.K., Mandal, M.: Performance of EAs for four-bar linkage synthesis. Mech. Mach. Theory **44**(9), 1784–1794 (2009)
25. Zhou, H., Cheung, E.H.M.: Optimal synthesis of crank–rocker linkages for path generation using the orientation structural error of the fixed link. Mech. Mach. Theory **36**(8), 973–982 (2001)

One-DOF Origami Boxes with Rigid and Flat Foldability

Yuanqing Gu[1,2] and Yan Chen[1,2](✉)

[1] School of Mechanical Engineering, Tianjin University, Tianjin 300072, China
yan_chen@tju.edu.cn

[2] Key Laboratory of Mechanism Theory and Equipment Design of Ministry of Education, Tianjin University, Tianjin 300072, China

Abstract. Rigid origami has great potential in engineering applications to deal with the folding of rigid sheets. Especially in the packaging and transportation industry, origami-inspired box-shaped foldable structures are more advantageous due to the large deployable ratio and low cost. In this paper, a novel crease pattern based on waterbomb pattern is proposed to generate a rigid origami box with flat foldability and a single degree of freedom (DOF). Based on the kinematic equivalence between the rigid origami pattern and the assembly of spherical linkages, kinematic properties of the corresponding integrated mechanism are investigated and formulated. By utilizing the vertex-splitting technique and introducing creases on each vertical facet of an origami box, more box-shaped structures with various geometry and shapes are constructed. In addition, the prototypes of the proposed origami boxes are fabricated to verifying their motion properties. The newly found method can be readily extended to the construction of foldable polygonal prism structures for various engineering applications.

Keywords: Rigid origami · Crease pattern · Origami boxes

1 Introduction

Origami-inspired foldable structures have been adopted in wide areas including aerospace devices [1], robotics [2], mechanical metamaterials [3], deployable origami tents [4], and so on. Particularly, rigid origami has a better advantage in engineering to deal with the folding of rigid sheets, i.e., its facets revolve around predetermined folds without any deformation during the continuous folding process [5]. From kinematics point of view, the rigid sheets and creases of rigid origami can be regarded as rigid links and revolute joints [6], respectively, e.g. four creases intersect at one point in space can be modeled as a spherical 4*R* (S4*R*) linkage. Therefore, a rigid origami pattern is equivalent to a network of spherical linkages, in which Miura-ori pattern and waterbomb pattern are frequently mentioned and investigated [7]. Apart from rigid foldability, flat foldability of one origami pattern is another important property that needs to be considered to acquire the large deployable ratio and low transportation cost [8], which refers to the capability to fold an origami pattern compactly into a flat overlapped sheet.

N. V. Khang et al. (Eds.): ASIAN MMS 2021, MMS 113, pp. 80–88, 2022.
https://doi.org/10.1007/978-3-030-91892-7_8

Folding rigid origami boxes or cartons is challenging and practical in both theoretical investigation and engineering application. Yet, there is little work on how to fold boxes and cartons rigidly and compactly into a flat overlapped configuration. Balkcom et al. [9] presented two rigid folding schemes for cubical bags but with specific height conditions. The first practical solution for rigid tall shopping bags was proposed by Wu and You [10], which can lead to direct applications in the packaging industry. On the other hand, kinematics analysis and stiffness characteristics have been demonstrated for folding the patterned sheets into box-shaped cartons of complex geometry and shapes [11, 12]. Furthermore, origami-inspired mechanisms related to three-dimensional crease patterns have been developed to provide more design inspiration [13, 14].

The crease patterns that enable origami cubes rigid and flat foldability with mobility one have been proposed in our previous work [15]. In order to enhance the engineering applications of origami structures such as foldable boxes, cartons and containers, this paper for the first time presents an approach to construct the novel crease patterns for origami box-shaped structures with various geometry.

The layout of the paper is as follows. Section 2 proposes a novel origami box based on waterbomb pattern. The variants of origami boxes are presented in Sect. 3, and the conclusion and further discussion are given in Sect. 4.

2 An Origami Box with the Novel Crease Pattern

The well-known crease pattern of a waterbomb origami in Fig. 1(a) is obtained by tesselating six-crease origami vertices, where the dashed lines stand for mountain creases and the solid lines denote valley creases. Here, we take a part of simplified crease pattern, highlighted in red line area in Fig. 1(a), as the planer net of a box with a square base (Fig. 1(b)), a crease pattern of the novel origami box is constructed in Fig. 1(c).

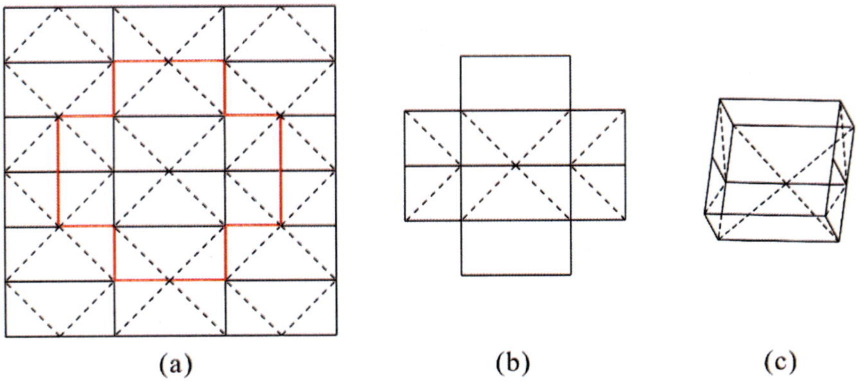

Fig. 1. (a) Crease pattern of a waterbomb origami. (b) The planer net of (c) a novel origami box with a square base.

On the base of this origami box, see Fig. 2, vertices A, B, C and D with identical geometry can be modelled as one-DOF S4*R* linkages, respectively. The identical six-crease vertices E and F can be modelled as S6*R* linkages, so do vertex G. Although vertices E, F, and G are all derived from waterbomb pattern, the initial configuration of vertices E and F are not flat in the fully deployable state of the box.

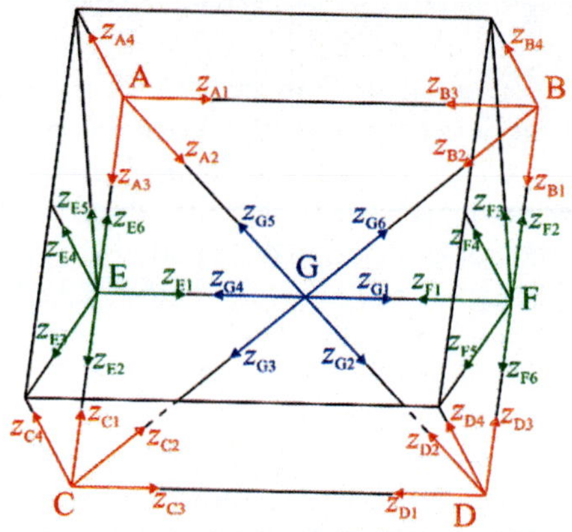

Fig. 2. The 4*R*-4*R*-4*R*-4*R*-6*R*-6*R*-6*R* integrated mechanism.

Nonetheless, the 4*R*-4*R*-4*R*-4*R*-6*R*-6*R*-6*R* (from vertices A to G) integrated mechanism is formed on the bottom of this origami box as shown in Fig. 2. Kinematic analysis of this integrated mechanism is carried out based on the matrix method with the D-H notations [16]. With this method, Fig. 3(a) shows the D-H parameters of a portion of a general spherical linkage, whose rotating axes intersect at one point in space. For a single closed loop linkage, the closure equation can be found in [16] as

$$\mathbf{Q}_{21}\mathbf{Q}_{32}\mathbf{Q}_{43}\ldots\mathbf{Q}_{i(i-1)}\mathbf{Q}_{1i} = \mathbf{I}. \tag{1}$$

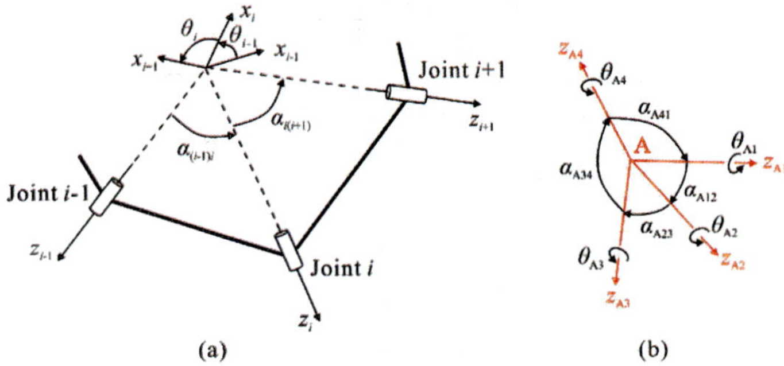

Fig. 3. The D-H notations of (a) a portion of a spherical linkage and (b) the S4*R* linkage A.

Take the S4*R* linkage A in Fig. 3(b) as an example, the geometrical parameters of the S4*R* linkages are identified as

$$\alpha_{i12} = \alpha_{i23} = \pi/4, \alpha_{i34} = \alpha_{i41} = \pi/2, \tag{2}$$

in which i = A, B, C, D..

For S6*R* linkages E and F, there exists

$$\alpha_{i12} = \alpha_{i61} = \pi/2, \alpha_{i23} = \alpha_{i34} = \alpha_{i45} = \alpha_{i56} = \pi/4\ (i = \text{E, F}), \tag{3}$$

and for S6*R* linkage G,

$$\alpha_{\text{G23}} = \alpha_{\text{G56}} = \pi/2, \alpha_{\text{G12}} = \alpha_{\text{G34}} = \alpha_{\text{G45}} = \alpha_{\text{G61}} = \pi/4. \tag{4}$$

Kinematics of each spherical linkage can be deduced through the above closure equation and geometric parameters. Firstly, the joint angle relationships of the S4*R* linkages can be derived as

$$\tan\frac{\theta_{i2}}{2} = \left(\sqrt{2}\tan\theta_{i1}\right)^{-1}, \theta_{i3} = \theta_{i1}, \tan\frac{\theta_{i4}}{2} = (\sin\theta_{i1})^{-1}\ (i = \text{A, B, C, D}). \tag{5}$$

As linkages A and B share a common revolute joint, so do linkages C and D, there exit further relations that

$$\theta_{\text{A1}} = \theta_{\text{B3}}, \theta_{\text{C3}} = \theta_{\text{D1}} \tag{6a}$$

$$\theta_{\text{A}j} = \theta_{\text{B}j},\quad \theta_{\text{C}j} = \theta_{\text{D}j}\ (j = 1,\ 2,\ 3,\ 4). \tag{6b}$$

Subsequently, $\theta_{\text{A2}}, \theta_{\text{B2}}, \theta_{\text{C2}}$ and θ_{D2} can be obtained as inputs of linkage G, so we can obtain the following relationships as

$$\theta_{\text{G2}} = \theta_{\text{D2}}, \theta_{\text{G3}} = \theta_{\text{C2}}, \theta_{\text{G5}} = \theta_{\text{A2}}, \theta_{\text{G6}} = \theta_{\text{B2}}, \tag{7a}$$

$$\tan\frac{\theta_{\text{G1}}}{2} = \tan\frac{\theta_{\text{G4}}}{2} = -\frac{\sqrt{2}}{2}\tan\frac{\theta_{\text{G2}}}{2}, \theta_{\text{G2}} = \theta_{\text{G3}} = \theta_{\text{G5}} = \theta_{\text{G6}}, \tag{7b}$$

which further leads to

$$\theta_{\text{A}j} = \theta_{\text{B}j} = \theta_{\text{C}j} = \theta_{\text{D}j}\ (j = 1,\ 2,\ 3,\ 4). \tag{8}$$

Similarly, due to the relations at common revolute joints of linkages E and F,

$$\theta_{\text{E1}} = \theta_{\text{G4}}, \theta_{\text{E2}} = \theta_{\text{C1}}, \theta_{\text{E6}} = \theta_{\text{A3}}, \tag{9a}$$

$$\theta_{\text{F1}} = \theta_{\text{G1}}, \theta_{\text{F2}} = \theta_{\text{B1}}, \theta_{\text{F6}} = \theta_{\text{D3}}, \tag{9b}$$

we can also derive that

$$\cos\theta_{\text{E4}} = (\cos\theta_{\text{E1}} - 1)\cos^2\theta_{\text{E2}} - 2\sin\theta_{\text{E1}}\sin\theta_{\text{E2}} + 1, \tag{10a}$$

$$\tan\frac{\theta_{E3}}{2}=\tan\frac{\theta_{E5}}{2}=\frac{-\sqrt{2}\sin\theta_{E4}+\sqrt{\begin{array}{l}2\sin^2\theta_{E4}+(\cos\theta_{E4}+1)^2\\-(2\cos\theta_{E1}+1-2\sin\theta_{E1}\sin\theta_{E2}-\cos\theta_{E4})^2\end{array}}}{2(\cos\theta_{E1}+1-\sin\theta_{E1}\sin\theta_{E2})},\tag{10b}$$

and the following relations can be given by combining Eqs. (7a, 7b–10a, 10b) as

$$\theta_{Ej}=\theta_{Fj}\ (j=1,2,\ldots,6).\tag{11}$$

From the above analysis, it can be seen that θ_{A1} is the only input of this integrated mechanism and the rest of joint angles can be determined by θ_{A1}. Hence, the proposed origami box has mobility one. Furthermore, dihedral angles are preferred to directly represent the folding motion, which can be expressed with respect to the joint angles as

$$\varphi_{i1}=\pi+\theta_{i1},\ \varphi_{i2}=\pi-\theta_{i2},\ \varphi_{i3}=\pi+\theta_{i3},\ \varphi_{i4}=\pi+\theta_{i4}\ i=\text{A, B, C, D}.\tag{12a}$$

$$\varphi_{G1}=\pi+\theta_{G1},\ \varphi_{G2}=\pi-\theta_{G2},\ \varphi_{G3}=\pi-\theta_{G3},\ \varphi_{G4}=\pi+\theta_{G4},\ \varphi_{G5}=\pi-\theta_{G5},\ \varphi_{G6}=\pi-\theta_{G6},\tag{12b}$$

$$\varphi_{E1}=\pi+\theta_{E1},\ \varphi_{E2}=\pi+\theta_{E2},\ \varphi_{E3}=\pi-\theta_{E3},\ \varphi_{E4}=\pi+\theta_{E4},\ \varphi_{E5}=\pi-\theta_{E5},\ \varphi_{E6}=\pi+\theta_{E6},\tag{12c}$$

$$\varphi_{F1}=\pi+\theta_{F1},\ \varphi_{F2}=\pi+\theta_{F2},\ \varphi_{F3}=\pi-\theta_{F3},\ \varphi_{F4}=\pi+\theta_{F4},\ \varphi_{F5}=\pi-\theta_{F5},\ \varphi_{F6}=\pi+\theta_{F6}.\tag{12d}$$

Hence, the kinematic relations of dihedral angles in this one-DOF origami box are presented as following equations. Given φ_{A1} as an input dihedral angle, the input-output curves of dihedral angles in this origami box are obtained and illustrated in Fig. 4. A physical prototype with flat foldability has been fabricated to demonstrate and verify the symmetric motion as shown in Fig. 5.

$$\begin{array}{l}\tan\frac{\varphi_{i2}}{2}=\sqrt{2}\tan\varphi_{i1},\theta_{i3}=\theta_{i1},\tan\frac{\varphi_{i4}}{2}=\sin\varphi_{i1}\ (i=\text{A, B, C, D}),\\ \varphi_{Aj}=\varphi_{Bj}=\varphi_{Cj}=\varphi_{Dj}\ (j=1,\ 2,\ 3,\ 4),\\ \varphi_{G2}=\varphi_{G3}=\varphi_{G5}=\varphi_{G6}=\varphi_{A2},\tan\frac{\varphi_{G1}}{2}=\tan\frac{\varphi_{G4}}{2}=\sqrt{2}\tan\frac{\varphi_{G2}}{2},\\ \varphi_{E1}=\varphi_{G4},\varphi_{E2}=\varphi_{E6}=\varphi_{A3},\\ \cos\varphi_{E4}=(\cos\varphi_{E1}+1)\cos^2\varphi_{E2}+2\sin\varphi_{E1}\sin\varphi_{E2}-1,\\ \tan\frac{\varphi_{E3}}{2}=\tan\frac{\varphi_{E5}}{2}=\frac{2(1-\cos\varphi_{E1}-\sin\varphi_{E1}\sin\varphi_{E2})}{\sqrt{2}\sin\varphi_{E4}+\sqrt{\begin{array}{l}2\sin^2\varphi_{E4}+(1-\cos\varphi_{E4})^2\\-(1-2\cos\varphi_{E1}-2\sin\varphi_{E1}\sin\varphi_{E2}+\cos\varphi_{E4})^2\end{array}}},\\ \varphi_{Ej}=\varphi_{Fj}\ (j=1,\ 2,\ldots,6).\end{array}\tag{13}$$

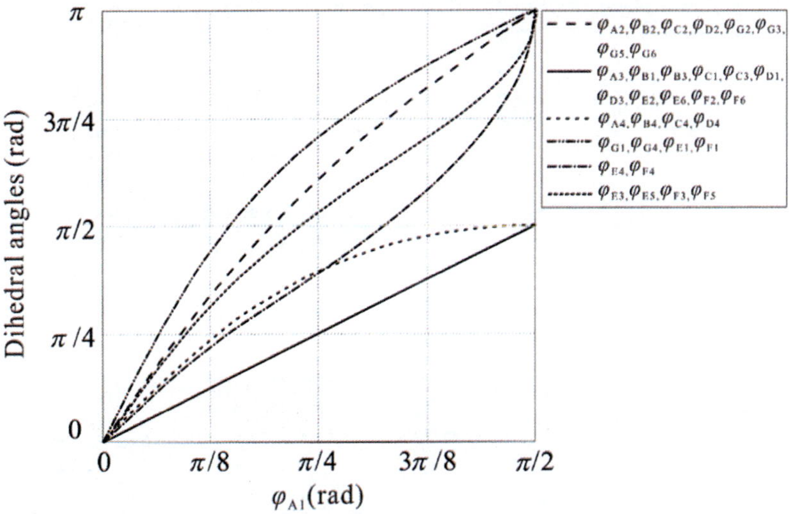

Fig. 4. Input-output curves of dihedral angles in the proposed origami box.

Fig. 5. Phototype of the origami box and its folding process.

3 The Variants of Origami Boxes

For the proposed crease pattern of origami box with a square base in Fig. 2, vertex-splitting technique [17] is adopted to construct the one-DOF and flat foldable origami boxes with distinct geometry. As shown in Fig. 6, by splitting six-crease vertex G, two four-crease vertices G_1 and G_2 connected by a common crease line (revolute joint) are obtained to form the rectangle base of the cuboid box. Kinematic equivalence and symmetric folding of the assembly of those two four-crease vertices have been revealed in [17], so one-DOF rigid and flat foldability is preserved in this cuboid box. In addition, by adjusting the geometric parameters of six-crease vertices E and F, the extended origami box with a symmetric hexagon base (as an example) has been identified and constructed in Fig. 7, which demonstrates the flexibility of our design approach in this paper. Noted that the purpose of vertex-splitting technique used in origami boxes is to extend the geometry of the flat foldable boxes for wider applications.

Further, inspired by the proposed design method for origami cube in our previous work [15], we aim to construct various polygonal prism structures by introducing creases

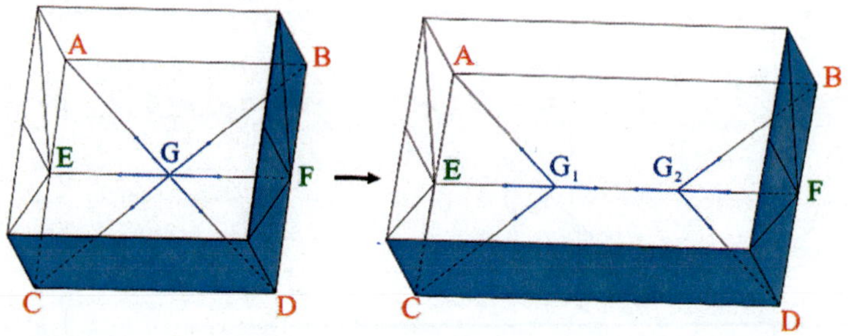

Fig. 6. The origami cuboid box with vertex-splitting technique.

Fig.7. The extended origami box with a symmetric hexagon base.

on each vertical facet of an origami box respectively. The crease patterns of pentagonal and hexagonal origami prism structures are shown in Fig. 8, in which the geometry conditions of flat foldability at each origami vertex should always be considered. The same approach described in Sect. 2 can also be utilized to analyse the kinematics in those origami prism structures.

For example, compared with the box in Fig. 2, four-crease vertices A and C have identical geometry in this hexagonal origami prism structure, so do vertices B and D. Here, four-crease vertex G becomes one S4*R* linkage, where each one is one-DOF. As sharing common creases, AB, BG, GD and DC, the motion of the assembly of those

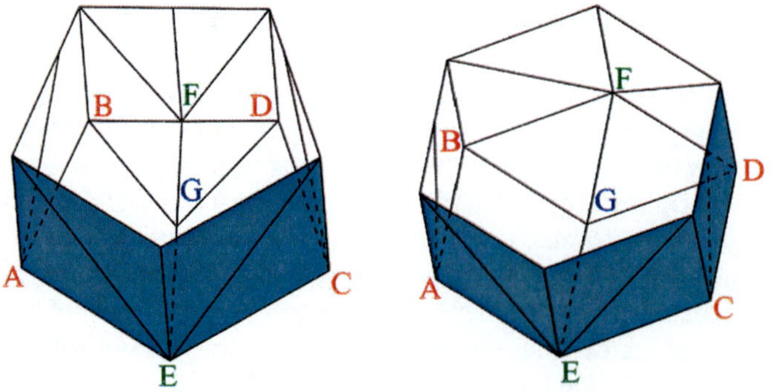

Fig. 8. The crease patterns of pentagonal and hexagonal origami prism structures.

Fig. 9. Folding process of hexagonal origami prism structures.

five S4*R* linkages can be obtained and identified if only one kinematic input is given to it. Meanwhile, the motion of three-DOF S6*R* linkage at vertex E is determined by the three inputs from vertices A, G and C, respectively. For vertex F, the motion is also defined from vertices B, G and D. Hence, the hexagonal origami prism structure has mobility one. The folding process of the physical model is shown in Fig. 9, which presents the plane-symmetric folding performance of this origami prism structure due to the symmetric layout of all the creases.

4 Conclusions and Discussion

In this paper, a novel origami box with one-DOF rigid and flat foldability has been proposed based on waterbomb pattern. Kinematics of the corresponding integrated mechanism has been investigated based on the kinematic equivalence between rigid origami pattern and the assembly of spherical linkages. Furthermore, more origami box-shaped structures with various geometry and shapes are developed by utilizing vertex-splitting technique and introducing additional creases. Physical models have been fabricated to reveal the folding process and validate the proposed approach.

Due to the folding motion derives from the motion of the integrated mechanism at the bottom of each origami box, the height of foldable structures can be selected without physical interference. In addition to folding the origami boxes, the new method proposed in this paper can be readily extended to construct deployable polyhedral structures for wider engineering applications including temporary aerospace residences, small satellites, containers, etc.

References

1. Morgan, J., Magleby, S.P., Howell, L.L.: An approach to designing origami-adapted aerospace mechanisms. J. Mech. Des. **138**(5), 052301 (2016)
2. Felton, S., Tolley, M., Demaine, E., Rus, D., Wood, R.: A method for building self-folding machines. Science **345**(6197), 644–646 (2014)
3. Fang, H., Chu, S.C.A., Xia, Y., Wang, K.W.: Programmable self-locking origami mechanical metamaterials. Adv. Mater. **30**, 1706311 (2018)
4. Liu, S., Lv, W., Chen, Y., Lu, G.: Deployable prismatic structures with rigid origami patterns. J. Mech. Robot. **8**, 031002 (2016)

5. Wang, K., Chen, Y.: Folding a patterned cylinder by rigid origami. In: Wang-Iverson, P., Lang, R., Yim, M. (eds.) Origami 5, pp. 265–276. AK Peters/CRC Press, New York (2011)
6. Dai, J.S., Rees Jones, J.: Mobility in metamorphic mechanisms of foldable/erectable kinds. J. Mech. Des. **121**, 375–382 (1999)
7. Feng, H., Ma, J., Yan, C., You, Z.: Twist of tubular mechanical metamaterials based on waterbomb origami. Sci. Rep. **8**(1), 9522 (2018)
8. Tachi, T.: Freeform variations of origami. J. Geom. Graph. **14**, 203–215 (2010)
9. Balkcom, D.J., Demaine, E.D., Demaine, M.L., Ochsendorf, J.A., You, Z.: Folding paper shopping bags. In: Origami4. Proceedings of the 4th International Meeting of Origami Science, Math, and Education, pp. 315–334 (2009)
10. Wu, W., You, Z.: A solution for folding rigid tall shopping bags. Proc. R. Soc. A **467**, 2561–2574 (2011)
11. Dai, J.S., Rees Jones, J.: Kinematics and mobility analysis of carton folds in packing manipulation. Proc. Inst. Mech. Eng. Part C: J. Mech. Eng. Sci. **216**(C10), 959–970 (2003)
12. Dai, J.S., Cannella, F.: Stiffness characteristics of carton folds for packaging. J. Mech. Des. Trans. **130**, 022305 (2008)
13. Wei, G., Dai, J.S.: Origami-inspired integrated planar-spherical overconstrained mechanisms. J. Mech. Des. **136**(5), 051003 (2014)
14. Chen, Y., Lv, W., Peng, R., Wei, G.: Mobile assemblies of four-spherical-4R-integrated linkages and the associated four-crease-integrated rigid origami patterns. Mech. Mach. Theory **142**, 103613 (2019)
15. Gu, Y., Chen, Y.: Origami cubes with one-DOF rigid and flat foldability. Int. J. Solids Struct. **207**, 250–261 (2020)
16. Denavit, J., Hartenberg, R.S.: A kinematic notation for lower-pair mechanisms based on matrices. Trans. ASME J. Appl. Mech. **22**, 215–221 (1955)
17. Zhang, X., Chen, Y.: Vertex-splitting on a diamond origami pattern. J. Mech. Robot. **11**(3), 031014 (2019)

Generalized Kinematics Error Prediction of CNC Milling Machines by Using Simulation Method

Hung Q. Nguyen[1,2](✉), Toan B. Pham[1,2], Luan C. Vuong[1,2], Duong N. Nguyen[1], and Nhi K. Ngo[1,2]

[1] Laboratory of Applied Mechanics (LAM), Ho Chi Minh City University of Technology (HCMUT), 268 Ly Thuong Kiet Street, District 10, Ho Chi Minh City, Vietnam
nqhung1706@hcmut.edu.vn

[2] Vietnam National University Ho Chi Minh City, Linh Trung Ward, Thu Duc District, Ho Chi Minh City, Vietnam

Abstract. Nowadays, precision mechanical products manufacture has become a necessity in daily life. Consequently, high accuracy CNC machines in modern manufacturing systems are always required. Geometric (kinematic) errors, cutting tool dimensions, and thermal errors generated during machining are three sources of error that affect the accuracy of CNC milling machines. Many studies have been presented on CNC milling machine geometric (kinematic) error prediction to increase product accuracy. In this paper, the authors proposed a method with the primary purpose of using kinematic equations to predict generalized kinematic errors of CNC milling machines. Therefore, it can be applied to solve the requirements of designing high-precision CNC milling machines with long-term working ability. The kinematic modeling for the multi-body system of the CNC machine system is built based on the kinetic theory of Dennavit - Hartenberg (D-H). Using the Dennavit - Hartenberg (D-H) parameters to set the transfer matrices that define the position of the tooltip and workpiece. Then, by considering additional error parameters, the authors develop a method to determine the error based on the forward position kinematics equations for CNC milling machines. In this study, the authors used the Monte Carlo method to calculate and simulate the maximum generalizes kinematic error from the kinetic equations. In addition, the position errors for each stage of the CNC milling machine have been determined by experimental measurement using precision measuring instruments. Finally, the error results from the survey are compared with each other to decide on the method's accuracy and effectiveness.

Keywords: CNC milling machine · Kinematic error · D-H kinematic equation · Monte Carlo Simulation method

1 Introduction

One of the requirements for creating a high-precision CNC machine is to minimize machining errors. Genetically, different error sources are basically categorized into three

N. V. Khang et al. (Eds.): ASIAN MMS 2021, MMS 113, pp. 89–100, 2022.
https://doi.org/10.1007/978-3-030-91892-7_9

subcategories: geometric (kinematic) errors, dynamics errors, and thermal errors [1]. The accuracy of machine tools is often affected by error sources such as geometric (kinematic) errors of machine components, assembly errors of machine parts, deformation under heavy working conditions of the machine. Because of these errors, the relative position between tool and workpiece deviate from their original states. This increases the inaccuracy of the machine.

Geometric (kinematic) errors of CNC machines are considered the most critical factors are causing errors in machining processes [2]. These errors come from the machine components, such as specification error in the accuracy of linear guide components, tolerance of the assembly parts, the rotational error, the rigidity of parts. Geometric (kinematic) errors account for closely 70% of the total error in machined tools [3]. Therefore, they contribute mainly to the inaccuracy of a machine tool. In addition, the imperfections in the geometric dimensions of machine parts can cause erroneous motion, called kinematic errors [4]. Reducing kinematic errors is very important for simultaneous multi-axis movement during machining because they determine the accuracy. Currently, there have been many studies evaluating kinematic errors based on a homogeneous transform matrix [5]. The authors have built D-H kinematic equations to determine the relationship of position between the moving links and the fixed coordinate system. From these results, it is possible to develop a method to evaluate the position error of the tooltip relative to the workpiece and the fixed coordinate system.

In this paper, the generalized kinematic error approach is used to predict the general geometric (kinematic) errors of the CNC milling machine. The accuracy of the method is verified by using computational simulations combined with experimental measurements. First, the forward kinematic equation of the CNC machine is built by using D-H parameters which contain information about the position of the machine coordinate systems. The relationship between the local coordinates of the moving stage and the fixed global coordinate system is shown through a homogeneous transformation matrix. For each stage in the CNC milling machine, there are many corresponding kinematic errors, leading to the general kinematic error of the position of the tooltip. Therefore, the author has combined the ideal transformation matrices with the matrices containing kinematic error parameters, thereby becoming a generalized error matrix. Then, the Monte Carlo simulation method with the input error boundaries is used to estimate the generalized kinematic error of the CNC machine. The simulation error results are verified through the experimental measurement results on the CNC machine. Finally, a discussion is presented to evaluate the effectiveness and scope of this survey method.

2 Material and Methods

2.1 The Denavit - Hartenberg Kinematics

The coordinate frame B_i is fixed to the link (i) and the coordinate frame B_{i-1} is fixed to the link (i − 1). Based on the D-H convention, the transformation matrix ${}^{i-1}T_i$ to transform coordinate frames B_i to B_{i-1} is presented [6].

$$
{}^{i-1}T_i = \begin{bmatrix} \cos\theta_i & -\sin\theta_i\cos\alpha_i & \sin\theta_i\sin\alpha_i & a_i\cos\theta_i \\ \sin\theta_i & \cos\theta_i\cos\alpha_i & -\cos\theta_i\sin\alpha_i & a_i\sin\theta_i \\ 0 & \sin\alpha_i & \cos\theta_i & d_i \\ 0 & 0 & 0 & 1 \end{bmatrix} \tag{1}
$$

The proposed kinematic diagram modeling for the 4-axis CNC milling machine (Fig. 1a) with three translation axes in 3 directions X Y Z and the rotation axis (A-axis).

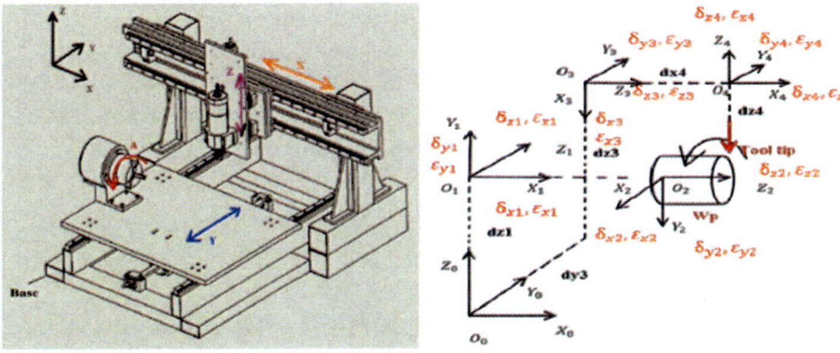

(a) (b)

Fig. 1. CNC milling machine: The topological structure (a) and D-H diagram with kinematic error (b)

2.2 The Geometric Kinematics Error of CNC Machine

As shown in Fig. 1a, the machine tool is divided into two systems. One is a translation system, which includes the X and Z-axis, and the other is a workpiece system that includes Y-axis and rotation A-axis. This system has a significant error budget and is difficult to determine (Table 1). Reducing this error is the primary purpose of this paper. The kinematic error of the multi-axis machine tool is presented in [7, 8]. The series of kinematic errors (Fig. 1b) included the coordinate systems of each kinematic unit. This series clearly and intuitively reflect the relationship between the movement of each axis and the effect on machine tool accuracy. The translation and workpiece systems are independent of each other. Therefore, in the process of determining error for the machine, if the influence of the rotation axis error is considered, the angular error value can be converted directly to the distance error or vice versa through the transformation matrix. In this section, the boundaries for angular and distance errors are calculated using a Monte Carlo simulation [9].

Monte Carlo Simulation Method Implementation Process:

1. Setting all errors and accuracy of CNC machine.
2. Configuration of the distance and angular error range for each axis.
3. Randomly generate distance and angular errors for each axis

Table 1. The error budgets of kinematic error.

Error symbol	Definition
δx	Average distance error along the X direction
δy	Average distance error along the Y direction
δz	Average distance error along the Z direction
εx	Average angle error of rotating around the X-axis
εy	Average angle error of rotating around the Y-axis
εz	Average angle error of rotating around the Z-axis

4. Construct kinematic model includes angular and distance errors for CNC machines.
5. Calculate the tooltip position concerning the workpiece from kinematic modeling.
6. Calculate the tooltip position concerning the workpiece by setting all error terms the zero.
7. Subtract tooltip position with error budgets from zero error terms tooltip position.
8. Calculate the total error of the machine.
9. Repeat the process using code from 1–100000 times.
10. The calculated results are presented by histogram plot with frequency and position error.

2.3 Theoretical of Kinematic Error

Kinematic errors are divided into two types: line (distance) errors and angular errors. Absolute vector survey in which the position error of the extrapolated coordinates is expressed as:

$$\Delta\vec{r} = \sum_{i=1}^{n} \frac{\partial \vec{r}_i}{\partial q_i} \Delta q_i \tag{2}$$

In there: Δq_i is the error of the point q_i at O_n (O is the origin, O_n at the n link). The absolute value of the distance error of the n link is expressed as:

$$|\Delta\vec{r}| = \sqrt{(\Delta x_n)^2 + (\Delta y_n)^2 + (\Delta z_n)^2} \tag{3}$$

Inside: Δx_n, Δy_n, Δz_n is the corresponding error of the coordinates at O_n when surveyed in the reference system, so the position error is a function of the Descartes coordinates of the point and is defined by:

$$\Delta r_n = \iiint\limits_{xyz} \Delta r_n(x, y, z).\frac{dxdydz}{V} \tag{4}$$

If we are only interested in the mean error of a finite number of points, we can use the concept of position arithmetic mean error in terms of:

$$\Delta r_{cp} = \sum_{i=1}^{n} \left(\frac{\Delta r_i}{n}\right) = \sum_{i=1}^{n} \left(\frac{\sqrt{(\Delta x_n)^2 + (\Delta y_n)^2 + (\Delta z_n)^2}}{n}\right) \tag{5}$$

2.4 Experimental Measurement Setting for the CNC Machine

There are many additional tools to serve the needs of precision measurement and alignment for multi-axis CNC milling machines. Research team using Ditron digital linear scale systems to measure the accuracy of CNC milling machines. Ditron digital Scales are used as a stand-alone measuring system with a digital display for three output axis, which can be convenient for installation. Along with high operability and accuracy, resolution of systems can be customized from 0.001–0.005 mm. The measurement systems can be set up as shown in Fig. 2.

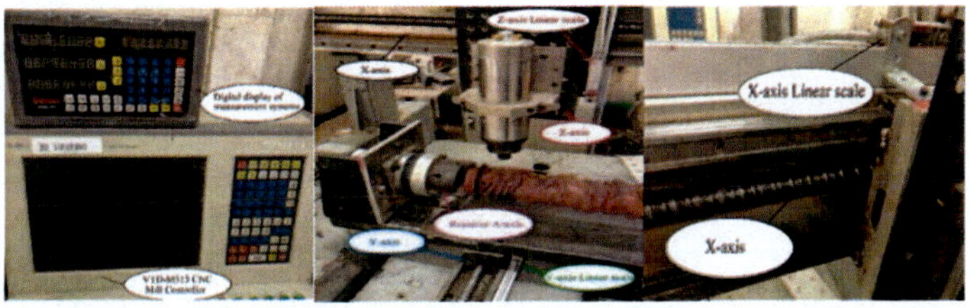

Fig. 2. Ditron Digital linear scale setup on CNC machine.

3 Results and Discussion

3.1 The Dennavit – Hartenberg Forward Kinematic

From the D-H kinetic diagram (Fig. 1.b), we can set up the Dennavit - Hartenberg parameter (Table 2) with the following parameters: θ_i, α_i, a_i, d_i.

Table 2. D-H parameter table of CNC milling machine:

Link	θ	αi	Ai (m)	Di (m)
Workpiece relative to the origin				
0	0	0	0.3	0.45
1	0	90	$dx_{01} + \mathrm{Y}$	$dx_{01}, dy_{01}, dz_{01}$
2	θ_2	90	0	$dx_{12}, dy_{12}, dz_{12}$
Tooltip relative to the origin				
3	0	90	$dx_{03} + \mathrm{X}$	$dx_{03}, dy_{03}, dz_{03}$
4	0	90	$\mathrm{Z} + dz_{34}$	$dx_{34}, dy_{34}, dz_{34}$

Calculate the homogeneous transform matrix of the tooltip relative to the workpiece.

$$\text{Including links } 1,\ 2 \text{ and workpiece: } {}^{\mathrm{wp}}T_0 = {}^{1}T_0.{}^{2}T_1.{}^{\mathrm{wp}}T_2 \quad (6)$$

With: $dx_{01} = 0\,\text{m}$, $dy_{01} = 0.3\,\text{m}$, $dz_{01} = 0.45\,\text{m}$; $dx_{12} = 0\,\text{m}$, $dy_{12} = 0\,\text{m}$, $dz_{12} = 0\,\text{m}$ are the generical dimension values of Links 1 and 2, r is the turning radius of the point of operation on the workpiece. Values of Y and θ_2 are the movement variable of link 1 translating in the Y direction and link two rotating around the X-axis (A-axis). Equation (6) becomes:

$$ {}^{wp}T_0 = \begin{bmatrix} Cos\theta_2 & 0 & Sin\theta_2 & 0 \\ \text{Sin}\theta_2 & 0 & -Cos\theta_2 & Y + r\cos\theta_2 + 0.3 \\ 0 & 0 & 1 & 0.45 + r\sin\theta_2 \\ 0 & 0 & 0 & 1 \end{bmatrix} \quad (7) $$

$$ \text{Includes links 3, 4, and tooltip: } {}^{Tool}T_0 = {}^{3}T_0.{}^{4}T_3.{}^{Tool}T_4 \quad (8) $$

$$ {}^{Tool}T_0 = \begin{bmatrix} 1\,0\,0\,X \\ 0\,1\,0\,0.6 \\ 0\,0\,1\,0.6 + Z \\ 0\,0\,0\,1 \end{bmatrix} \quad (9) $$

With: $dx_{03} = 0\,\text{m}$, $dy_{03} = 0.6\,\text{m}$, $\text{dz}_{03} = 0.6\,\text{m}$; $\text{dx}_{34} = 0\,\text{m}$, $\text{dy}_{34} = 0\,\text{m}$, $\text{dz}_{34} = 0\,\text{m}$ are the values of the generical dimensions of links 3 and 4. The X and Z values are the motion variables of link three translating in the X direction and link four translating in the Z direction. Using the coordinate transformation relationship between the milling machine motion axes, X, Y, and Z are used to represent the position coordinates of the tooltip $p = (X, Y, Z, 0)^T$ along with the three directions in the workpiece coordinate system (WCS), and I, J, and K are used to represent direction vectors $v = (I, J, K, 0)^T$. The following formula represents the relative coordinate matrix of the tooltip relative to the workpiece:

$$ p_1 = \left[{}^{Tool}T_0\right]^{-1}.{}^{\text{wp}}T_0.p_2 \quad (10) $$

$$ v_1 = \left[{}^{wp}T_0\right].v_2. \quad (11) $$

Position expression $p = (\text{X}, \text{Y}, \text{Z}, 1)^T$ of the tooltip at the initial position $p_1 = (0, 0, 0, 1)^T$ and direction vector $v_1 = (0, 0, 1, 0)^T$. The following formula represents the relative coordinate matrix of the tooltip relative to the workpiece:

$$ p = \left[{}^{Tool}T_0\right]^{-1}.{}^{\text{wp}}T_0.p_1 \quad (12) $$

$$ v = \left[{}^{wp}T_0\right].v_1 \quad (13) $$

For position $p_0 = (X_0, Y_0, Z_0, 1)^T$ and posture $v_0 = (I_0, J_0, K_0, 1)^T$, we obtained Eqs. (12) and (13) as:

$$\begin{Bmatrix} X_0 = -X \\ Y_0 = -0.3 + Y + r.\cos(\theta_2) \\ Z_0 = -0.15 + r.\sin(\theta_2) - Z \\ I_0 = \sin(\theta_2) \\ J_0 = -\cos(\theta_2) \\ K_0 = 1 \end{Bmatrix} \tag{14}$$

Equation (14) indicates that the position and posture of the instrument are in the ideal state, or there is no error in the established WCS. In special situations, X_O, Y_O, and Z_O are used to represent the actual position coordinates of the tooltip along with the three directions in WCS, and I_O, J_O, and K_O are used to represent the vectors. The realistic posture of the tool is shown in Fig. 3. Due to the kinematic error of the machine system, error transformation matrices E_{YA} and E_{XZ} were brought into the solution of the transformation matrix of actual position and posture of the tool in WCS, as stated in Eqs. (7) and (9) becomes:

$$\begin{aligned} {}^{Tool}T_0' &= {}^{Tool}T_0.E_{XZ} \\ {}^{wp}T_0' &= {}^{wp}T_0.E_{YA} \end{aligned} \tag{15}$$

$$\textit{With}: \quad E_{XZ} = \begin{bmatrix} 1 & -\varepsilon_z & \varepsilon_y & \delta_x \\ \varepsilon_z & 1 & -\varepsilon_x & \delta_y \\ \varepsilon_y & \varepsilon_x & 1 & \delta_z \\ 0 & 0 & 0 & 1 \end{bmatrix}; E_{YA} = \begin{bmatrix} 1 & -\varepsilon_z & \varepsilon_y & \delta_x \\ \varepsilon_z & 1 & 0 & \delta_y \\ \varepsilon_y & \varepsilon_x & 1 & \delta_z \\ 0 & 0 & 0 & 1 \end{bmatrix} \tag{16}$$

The transformation matrix included in Eqs. (15) and (16) is indicated as:

$${}^{wp}T_0' = \begin{bmatrix} C\theta_2 + S\theta_2\varepsilon_y & -C\theta_2\varepsilon_y + S\theta_2\varepsilon_x & C\theta_2\varepsilon_y + S\theta_2 & C\theta_2\delta_x + S\theta_2\delta_z \\ S\theta_2 - C\theta_2\varepsilon_y & -S\theta_2\varepsilon_Z - C\theta_2\varepsilon_x & S\theta_2\varepsilon_y - C\theta_2 & S\theta_2\delta_x + 0.3 - C\theta_2\delta_z + Y + r\cos\theta_2 \\ \varepsilon_y & \varepsilon_x & 1 & \delta_z + 0.45 + r\sin\theta_2 \\ 0 & 0 & 0 & 1 \end{bmatrix} \tag{17}$$

$${}^{Tool}T_0' = \begin{bmatrix} 1 & -\varepsilon_z & \varepsilon_y & \delta_x + X \\ \varepsilon_z & 1 & -\varepsilon_x & \delta_y + 0.6 \\ \varepsilon_y & \varepsilon_x & 1 & \delta_z + 0.6 + Z \\ 0 & 0 & 0 & 1 \end{bmatrix} \tag{18}$$

For position $p_0' = (X_0', Y_0', Z_0', 1)^T$ and posture $v_0' = (I_0', J_0', K_0', 1)^T$ we obtained Eqs. (17) and (18) as:

$$\begin{Bmatrix} X_0' = (c\theta_2 - s\theta_2 + 1)\delta_x + (s\theta_2 + c\theta_2\varepsilon_z + \varepsilon_y)\delta_z - (Y + rc\theta_2 + 0.3)\varepsilon_z + \varepsilon_y(0.45 + rs\theta_2) - X \\ Y_0 = (c\theta_2\varepsilon_z + s\theta_2)\delta_x + (s\theta_2\varepsilon_z - c\theta_2 - \varepsilon_x)\delta_z + \delta_y - 0.3 + Y + rc\theta_2 - \varepsilon_x(0.45 + rs\theta_2) \\ Z_0 = (c\theta_2\varepsilon_y + s\theta_2\varepsilon_x)\delta_x + (s\theta_2\varepsilon_y - c\theta_2\varepsilon_x + 2)\delta_z + (Y + rc\theta_2 + 0.3)\varepsilon_x + rs\theta_2 - 0.15 - Z \\ I_0 = s\theta_2 + c\theta_2\varepsilon_y \\ J_0 = s\theta_2\varepsilon_y - c\theta_2 \\ K_0 = 1 \end{Bmatrix} \tag{19}$$

3.2 Monte Carlo Simulation Method and Results

By setting all error terms as 0, the ideal position of the tool contact point when starting position can be calculated. Finally, the error vector (r) can be calculated as:

$$r_0 = p_0 - p_0^{'} \tag{20}$$

Components of error vectors are:

$$r = [\Delta x, \Delta y, \Delta z]^T \tag{21}$$

Where Δx, Δy and Δz are total errors in X-, Y- and Z directions, respectively. Position error metric can be calculated as:

$$\Delta r = \sqrt{(\Delta x^2 + \Delta y^2 + \Delta z^2)} \tag{22}$$

Based on the manufacturer's specifications, the parts used to assemble the CNC milling machine include: [10] THK Linear motion guide (Fig. 3), [11] THK Ball Screw (Fig. 4). We can determine the error boundaries range for the machine.

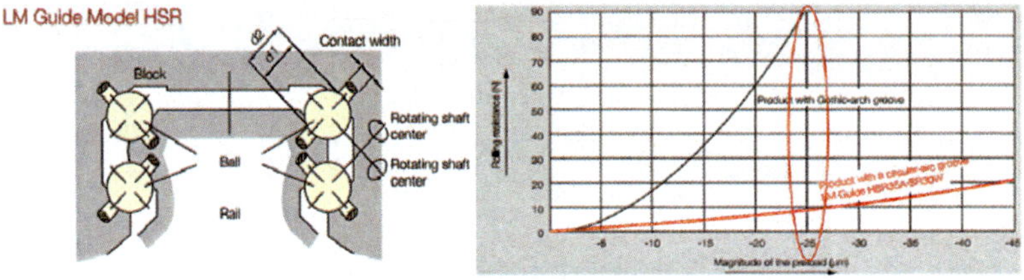

Fig. 3. THK Linear motion guide accuracy error. [10]

Unit:μm

Accuracy Grades		C0		C1		C2		C3		C5		Cp3		Cp5	
Normal travel ℓ_u[mm]		e_p	V_{up}	e_p	V_{up}	e_p	V_{up}	e_p	V_{up}	e_p	V_{up}	e_p	V_{up}	e_p	V_{up}
Above	Or less														
–	100	3	3	3.5	5	5	7	8	8	18	18	12	12	23	23
100	200	3.5	3	4.5	5	7	7	10	8	20	18	12	12	23	23
200	315	4	3.5	6	5	8	7	12	8	23	18	12	12	23	23
315	400	5	3.5	7	5	9	7	13	10	25	20	13	12	25	25
400	500	6	4	8	5	10	7	15	10	27	20	15	13	27	26
500	630	6	4	9	6	11	8	16	12	30	23	16	14	32	29
630	800	7	5	10	7	13	9	18	13	35	25	18	16	36	31
800	1000	8	6	11	8	15	10	21	15	40	27	21	17	40	34

Fig. 4. Tolerance on specified travel $\pm e_p$ and permissible travel variation V_{up} in relation to the nominal travel l_u for positioning THK ball screws. [11]

Error boundaries for linear motion can be set as:

$$0\,\mu\text{m} \le \delta x, \delta y, \delta z \le 25\,\mu\text{m} \tag{23}$$

Because the actual measurement systems can only determine the distance error for three translation axes, error boundaries for the rotary axis can be set as:

$$\varepsilon x, \varepsilon y, \varepsilon z = 0\,\text{arcmin} \tag{24}$$

Solving Eq. (10)–(24) with Monte Carlo Simulation method 100000 times. The following result histogram plot is obtained frequency with position error metric (μm), which is shown in Fig. 5

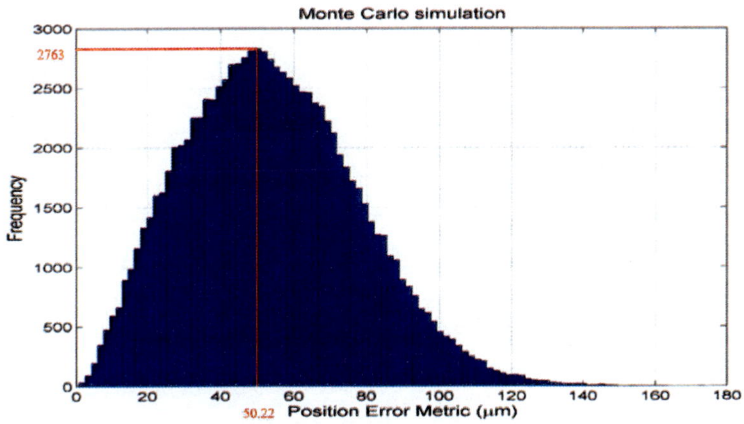

Fig. 5. Monte Carlo simulation results

The results (Fig. 5) show that the maximum position error (or the maximum generalized kinematics error) of the CNC milling machine is 50.22 µm with the highest occurrence frequency of 2763 times.

3.3 Measurement Results of CNC Machine by Using Ditron Digital Linear Scale

The test method for linear translation motion axis X, Y and Z is tested through three experimental measurements. The testing orbits of the tooltip are circular and half-circular in the XY, YZ and ZX plane, respectively, with a radius is 50 mm. The measuring results of the test are recorded in Table 3 and the error results are recorded in Table 4.

The diagram of the measuring results is plotted on the 2D plane, as shown in Fig. 6.

The error results from Table 4 shows that the average error of the 3 translation axes X, Y, Z is respectively: $\overline{\Delta X} = 0.028$ mm, $\overline{\Delta Y} = 0.022$ mm, $\overline{\Delta Z} = 0.024$ mm. Based on the theoretical of kinematic error, following Eq. (2)–(5), the experiment maximum position error (or the experiment generalized kinematics error) is $\overline{\Delta r_{cp}} = 0.043\text{ mm} \approx 43\,\mu\text{m}$.

Table 3. Measuring results on X-Y, X-Z, Y-Z plane:

Plane Angle (s°)	X-Y		X-Z		Y-Z	
	X_1(mm)	Y_1(mm)	X_2(mm)	Z_1 (mm)	Y_2(mm)	Z_2(mm)
0	50.000	0.000	50.000	0.000	49.989	−0.032
10	49.268	8.656	49.222	8.667	49.222	8.661
20	46.962	17.112	46.965	17.123	46.966	17.124
…						
180	−50.028	0.015	−50.023	0.017	−50.026	0.036
−10	−49.205	−8.657	−49.221	−8.661	49.258	−8.662
−20	−47.017	−17.129	−46.997	−17.117	47.013	−17.073
…						
−160	46.955	−17.084				
−170	49.206	−8.664	49.253	−8.697	49.218	−8.356
−180	49.986	−0.027	49.988	0.019	49.976	0.027

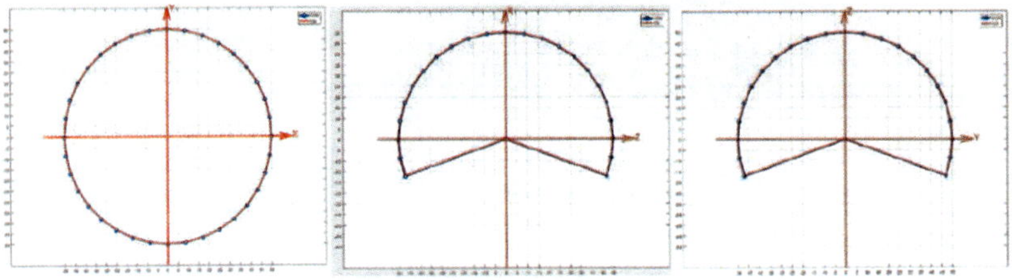

Fig. 6. The measuring results plot on X-Y, X-Z, Y-Z planes.

Table 4. The error results on X-Y, X-Z, Y-Z axis:

Error Angle	ΔX_1 (mm)	ΔY_1 (mm)	ΔX_2 (mm)	ΔZ_1 (mm)	ΔY_2 (mm)	ΔZ_2 (mm)
0	0.023	0.019	0.019	0.021	0.021	0.027
10	0.028	0.026	0.018	0.015	0.018	0.021
20	0.023	0.011	0.020	0.022	0.019	0.023
…						
180	0.028	0.015	0.023	0.017	0.026	0.023
−10	0.035	0.026	0.019	0.021	0.018	0.020
−20	0.032	0.028	0.012	0.016	0.028	0.028
…						
−160	0.030	0.017				
−170	0.034	0.018	0.013	0.014	0.022	0.027
−180	0.021	0.027	0.012	0.019	0.024	0.027

4 Conclusion

This paper presents an approach to predict the generalized kinematics error for CNC milling machines. The method has been presented step by step to calculate and simulate the kinematic error compared with the experimental measurement results. In this paper, the authors have successfully calculated the transformation matrix and the generalized kinematics error matrix for CNC milling machines. Then apply the Monte Carlo simulation method to simulate the result of the most significant generalized kinematics error. The calculation and simulation result shows that the maximum error is about 50.22 μm with the highest occurrence frequency of 2763 times. The result of a maximum experimental error determined by using measurement methods with digital linear scale applied on the CNC machine is about 43 μm. By comparing the results of simulation methods and experimental measurements, the authors may have concluded as follows:

- The experimental measurement error is smaller than the simulated calculation error, which is approximately equal to 7.22 μm. This difference is quite small, acceptable and can be explained by considering the actual operating conditions of the machine. So that, based on the Monte Carlo calculation and simulation method applying the generalized kinematics error equations, it is possible to predict the most significant error of the CNC milling machine.
- In this paper, experimental measurement of angular error is not performed because of certain limitations and depends on the measuring device. However, from the above results, the prediction ability of the proposed method is still guaranteed with the necessary accuracy.
- The proposed method has the primary purpose of predicting the maximum generalized kinematics error for CNC milling machines. That depends greatly on the health and working condition of the machine. Therefore, it is necessary to have a deeper study to indicate the effective working area to improve the accuracy of the machine.

Acknowledgement. This research is funded by Vietnam National University Ho Chi Minh City (VNU-HCM) under grant number C2020-20-12. We acknowledge the support of time and facilities from Ho Chi Minh City University of Technology (HCMUT), VNU - HCM for this study.

References

1. Florussen, G.H.J.: Accuracy analysis of multi-axis machines by 3D length measurements. Ph.D. thesis, Eindhoven University of Technology, Eindhoven, Netherlands (2002)
2. Sharif Uddin, M., Ibaraki, S., Matsubara, A., Matsushita, T.: Prediction and compensation of machining geometric errors of five-axis machining centers with kinematic errors. Precis. Eng. **33**(2), 194–201 (2009)
3. Liu, H., Li, B., Wang, X., Tan, G.: Characteristics of and measurement methods for geometric errors in CNC machine tools. Int. J. Adv. Manuf. Technol. **54**, 195–201 (2011)
4. Schwenke, H., Knapp, W., Haitjema, H., Weckenmann, A., Schmitt, R., Delbressine, F.: Geometric error measurement and compensation of machines - an update. CIRP Ann. Manuf. Technol. **57**, 660–675 (2008)

5. Lamikiz, A., López de Lacalle, L.N., Ocerin, O., Díez, D., Maidagan, E.: The Denavit and Hartenberg approach applied to evaluate the consequences in the tool tip position of geometrical errors in five-axis milling centres. Int. J. Adv. Manuf. Technol. **37**, 22–139 (2008)
6. Jazar, R.N.: Theory of Applied Robotics Kinematics, Dynamics, and Control, 2nd edn. Springer, Heidelberg (2010). https://doi.org/10.1007/978-1-4419-1750-8
7. Jiang, X.: Characterizing geometric errors in rotary axes of 5-axis machine tools. Ph.D. thesis, The University of Birmingham, Birmingham, UK (2014)
8. Eman, K.F., Wu, B.T., DeVries, M.F.: A generalized geometric error model for multi-axis machines. CIRP Ann. Manuf. Technol. **36**(1), 2530256 (1987)
9. Yilmaz, Y.E.: Design of a high precision hybrid am machine. MSc thesis, Middle East Technical University, Ankara, Turkey (2019)
10. THK Linear Motion Guide. https://www.thk.com/?q=us/node/5197. Accessed 10 June 2021
11. THK DIN Standard Compliant Ball Screw. https://www.thk.com/?q=us/node/5212. Accessed 10 June 2021

Mobile Assemblies of Bricard Linkages Inspired from Waterbomb Thick-Panel Origami

Xiao Zhang(✉) and Yan Chen

School of Mechanical Engineering, Tianjin University, Tianjin 300072, China
zhangxiaofun@tju.edu.cn

Abstract. Mobile assemblies of spatial linkages and thick-panel origami patterns have been used to design deployable structures for aerospace, such as solar arrays and antennas. The kinematic equivalence between a mobile assembly and thick-panel form has been used to design new of them based on the known assemblies and origami patterns. Here, waterbomb origami pattern is taken as the research object, and two types of plane-symmetric Bricard linkages are derived from the **D** and **W** thick-panel origami vertices, where linkages and thick-panel vertices are kinematically equivalent. Furthermore, new mobile assemblies of plane-symmetric Bricard linkages are constructed by using the transition method based on the multi-vertex waterbomb thick-panel origami pattern, and a special type mobile assembly with two motion paths is derived under the specific geometric conditions. A prototype is manufactured and two motion paths are verified according to the deployment sequences. This work proposes a bridge between waterbomb thick-panel origami and the assembly of plane-symmetric Bricard linkages and shows the possibility of design morphing assembly of linkages.

Keywords: Waterbomb origami · Bricard linkage · Mobile assembly

1 Introduction

Mobile assemblies of linkages are networks of unit linkages [1], which have great potential for engineering applications, such as shelters [2, 3], antennas [4, 5] and backbone for solar panels [6]. Based on typical unit linkages, such as Bennett linkages, Myard linkages and Bricard linkages, multiple mobile assemblies are constructed according to the tessellation method [7] and movable connection method [8]. These assemblies have different deployed configurations, such as planes [9], arches [7], parabolic cylindrical surfaces [10] and polyhedrons [11–13]. However, there is a limited number of mobile assemblies based on Bricard linkages due to the complex motion.

On the other hand, rigid origami patterns are generally kinematically equivalent to mobile assemblies of spherical linkages [14] and they have been used to be transited to new mobile assemblies of spatial linkages by regarding the thick-panel forms to bridges [1, 15]. These thick-panel forms are constructed based on the offset-crease method where four-crease, five-crease and six-crease vertices are kinematically equivalent to Bennett linkage, Myard linkage, Bricard linkage, respectively [16], which indicates a

N. V. Khang et al. (Eds.): ASIAN MMS 2021, MMS 113, pp. 101–108, 2022.
https://doi.org/10.1007/978-3-030-91892-7_10

multi-vertex thick-panel form is kinematically equivalent to multi-linkage assembly. By now, four-crease origami patterns and diamond origami patterns have been used to design mobile assemblies of Bennett linkages and mobile assemblies of plane-symmetric Bricard linkages, correspondingly [1, 15]. As waterbomb is a typical origami pattern consists of two types of six-crease origami vertices, it indicates the thick-panel form can be used to design a new assembly of Bricard linkages [17]. In addition, the geometric conditions about two motion paths of the waterbomb thick-panel origami [17] show the potential to construct a mobile assembly to realise morphing.

This paper aims at constructing mobile assemblies of plane-symmetric Bricard linkages from waterbomb thick-panel origami. The layout of the paper is listed as follows. Section 2 proposes two types of plane-symmetric Bricard linkages from vertices of waterbomb thick-panel origami. In Sect. 3, an assembly of plane-symmetric Bricard linkages with two paths is transited from multi-vertex waterbomb thick-panel origami. Conclusions are drawn with further discussion in Sect. 4.

2 Bricard Linkages Derived from Two Types of Six-Crease Vertices

A waterbomb origami pattern consists of two types of six-crease vertices, **D** type and **W** type. A crease pattern with eight **D** type vertices and ten **W** type vertices is shown in Fig. 1(a) where solid lines are mountain creases and dashed lines are valley creases. Its motion sequences under plane-symmetric conditions are shown in Fig. 1(b).

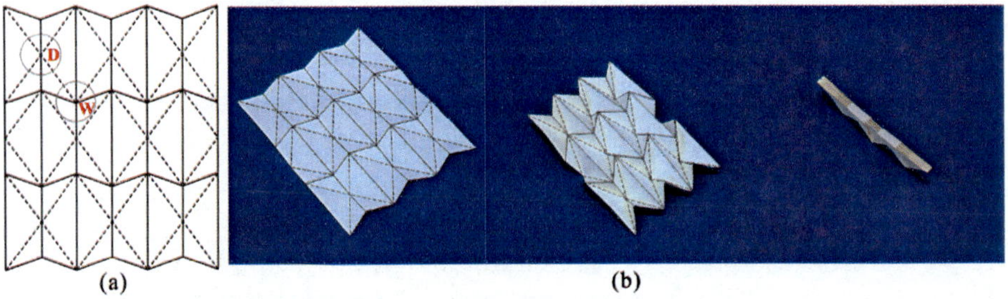

Fig. 1. A waterbomb origami. (a) The crease pattern with **D** and **W** types of vertices. (b) Motion sequences under plane-symmetric conditions of each vertex.

Figure 2(a) shows an enlarged vertex **D** with axes marked along the creases by z_i ($i = 1, 2, \ldots, 6$) and sector angles noted by $\alpha_{12} = \alpha_{34} = \alpha_{45} = \alpha_{61} = \alpha$ and $\alpha_{23} = \alpha_{56} = \pi - 2\alpha$ ($0 < \alpha \leq \pi/2$) to satisfy the flat foldability. Under the plane-symmetric condition, its thick-panel form can be constructed based on the offset-crease method, as shown in Fig. 2(b), which is corresponding to a plane-symmetric Bricard linkage [16]. To achieve the compact folding without interference, thicknesses of six panels satisfy

$$t_{12}^{D} = t_{23}^{D} = t_{34}^{D} = t_{45}^{D} = t_{56}^{D} = t_{61}^{D} = (2 + \mu)a, \tag{1}$$

where $\mu = \cos(\alpha + \beta) \sin\alpha / \sin\beta$ and another solution is $\mu = 1$ when $\alpha = \beta$ under compatibility conditions for waterbomb thick-panel form [17].

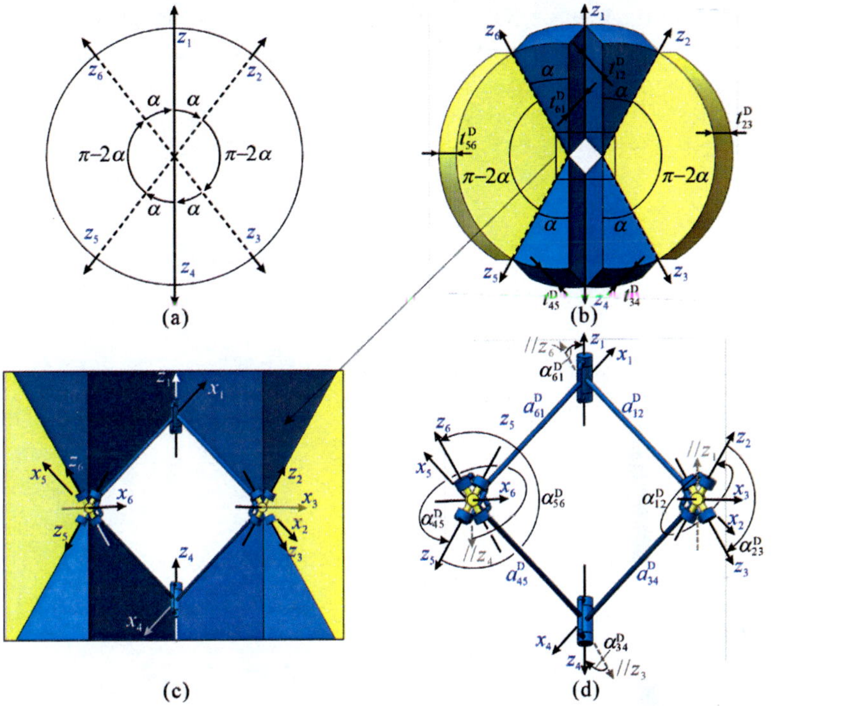

Fig. 2. The **D** type vertex and its corresponding Bricard linkage. (a) The crease pattern. (b) The corresponding thick-panel form and (c) the enlarged vertex. (d) The corresponding Bricard linkages with D-H notations.

The coordinate systems according to the Denavit and Hartenberg (D-H) notation [18] are marked in Fig. 2(c). By replacing the thick panels with links and connecting the adjacent revolute joints along the shortest distance, the corresponding plane-symmetric Bricard linkage is constructed in Fig. 2(d) with marked link lengths and twists. Hence, the single-vertex thick-panel origami in Fig. 2(c) and the special plane-symmetric Bricard linkage in Fig. 2(d) are kinematically equivalent. The relationships between twists of the linkage and sector angles of the origami vertex can be obtained as

$$\alpha_{12}^{\mathbf{D}} = -\alpha_{61}^{\mathbf{D}} = -\alpha,\ \ \alpha_{23}^{\mathbf{D}} = -\alpha_{56}^{\mathbf{D}} = \pi - 2\alpha,\ \ \alpha_{34}^{\mathbf{D}} = -\alpha_{45}^{\mathbf{D}} = \alpha, \tag{2}$$

and that between thicknesses of panels and lengths of links are

$$a_{12}^{\mathbf{D}} = a_{34}^{\mathbf{D}} = a_{45}^{\mathbf{D}} = a_{61}^{\mathbf{D}} = t_{12}^{\mathbf{D}} = (2+\mu)a,\ \ a_{23}^{\mathbf{D}} = a_{56}^{\mathbf{D}} = 0. \tag{3}$$

For the vertex **W** of waterbomb pattern in Fig. 3(a), six axes z_i ($i = 1, 2, \ldots, 6$) are marked along the creases with sector angles $\alpha_{12} = \alpha_{61} = \pi - \alpha - \beta$, $\alpha_{23} = \alpha_{56} = \beta$ and $\alpha_{34} = \alpha_{45} = \alpha$, where $0 < \alpha, \beta \leq \pi/2$. Similar to the **D** vertex, its thick-panel form can be constructed by offsetting hinges, as shown in Fig. 3(b). To obtain the compact folding property and flat unfolded profile, the thickness of panels satisfies

$$t_{12}^{\mathbf{W}} = t_{61}^{\mathbf{W}} = (1+\mu)a,\ t_{23}^{\mathbf{W}} = t_{56}^{\mathbf{W}} = a,\ t_{34}^{\mathbf{W}} = t_{45}^{\mathbf{W}} = \mu a,\ t_{12}' = t_{61}' = (2+\mu)a. \tag{4}$$

The enlarged thick-panel vertex in Fig. 3(c) shows the construction of the corresponding plane-symmetric Bricard linkage. Similar to the **D** vertex, the theoretical links connected to hinges are along the direction of the common normal of the adjacent axes, and the kinematic equivalence between single-vertex thick-panel origami in Fig. 3(c) and the plane-symmetric Bricard linkage in Fig. 3 (d) can be easily obtained.

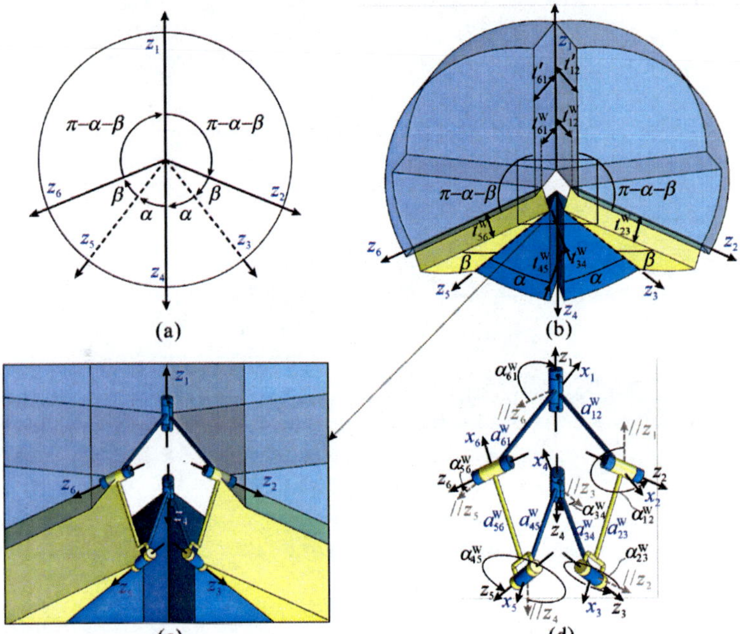

Fig. 3. The **W** type vertex and its corresponding Bricard linkage. (a) The crease pattern. (b) The corresponding thick-panel form and (c) the enlarged vertex. (d) The corresponding Bricard linkages with D-H notations.

According to the notations of the origami vertex and the linkage, relationships between twists of the linkage and sector angles of the origami vertex are

$$\alpha_{12}^{\mathbf{W}} = -\alpha_{61}^{\mathbf{W}} = \alpha + \beta - \pi, \quad \alpha_{23}^{\mathbf{W}} = -\alpha_{56}^{\mathbf{W}} = -\beta, \quad \alpha_{34}^{\mathbf{W}} = -\alpha_{45}^{\mathbf{W}} = \alpha. \tag{5}$$

To ensure the flat foldability of **W** thick-panel vertex, relationships between the thickness of panels and lengths of links satisfy

$$a_{12}^{\mathbf{W}} = a_{61}^{\mathbf{W}} = t_{12}^{\mathbf{W}} = t_{61}^{\mathbf{W}} = (1+\mu)a, \; a_{23}^{\mathbf{W}} = a_{56}^{\mathbf{W}} = t_{23}^{\mathbf{W}} = t_{56}^{\mathbf{W}} = a, \; a_{34}^{\mathbf{W}} = a_{45}^{\mathbf{W}} = t_{34}^{\mathbf{W}} = t_{45}^{\mathbf{W}} = \mu a. \tag{6}$$

3 Mobile Assemblies of Bricard Linkages from Multi-vertex Waterbomb Thick-Panel Origami

Figure 4(a) is the representative portion of a waterbomb pattern consisting of four vertices A, B, C, D with creases noted by a_i, b_i, c_i, d_i ($i = 1, 2,\ldots, 6$) and sector angles marked by α, β, $\pi - 2\alpha$, $\gamma = \pi - \alpha - \beta$. Figure 4(b) shows the corresponding thick-panel form which indicates an assembly of two types of plane-symmetric Bricard linkages. For panel P_1, it is connected to three linkages A, B, D with links a_{34}, b_{12}, d_{23}, as shown in Fig. 4(c). The shared joints between linkages A and D, linkages B and D, linkages A and B, i.e., a_3/d_3, b_2/d_2 and a_4/b_1, are combined into one. Then a closed loop is constructed based on the links and joints from a_3/d_3 to a_4/b_1 via a_{34}, through b_{12} to b_2/d_2, and then back to a_3/d_3 by d_{23}. Along the thickness direction of the panel P_1, the order of joints, a_4/b_1 in the bottom, b_2/d_2 in the middle and a_3/d_3 on the top are determined, respectively. The link lengths satisfy

$$a_{34}^{A} = b_{12}^{B} + d_{23}^{D} = (2 + \mu)a. \tag{7}$$

Similarly, the order of joints in the panel P_2 can be obtained as d_1/c_1, b_2/d_2 and b_3/c_6. When transit panels to be links, the arrangement of joints on a link can be determined by the order of joints in the panels. Hence, a mobile assembly of plane-symmetric Bricard linkages corresponding to the waterbomb thick-panel origami is constructed, as shown in Fig. 4(d). From Eqs. (2), (3), (5), (6) and (7), construction conditions of the mobile assembly of plane-symmetric Bricard linkages can be derived, as

$$\begin{aligned} &\alpha_{12}^{A} = \alpha_{12}^{C} = -\alpha,\ \alpha_{23}^{A} = \alpha_{23}^{C} = \pi - 2\alpha,\ \alpha_{34}^{A} = \alpha_{34}^{C} = \alpha, \\ &\alpha_{12}^{B} = \alpha_{12}^{D} = \alpha + \beta - \pi,\ \alpha_{23}^{B} = \alpha_{23}^{D} = -\beta,\ \alpha_{34}^{B} = \alpha_{34}^{D} = \alpha, \\ &a_{34}^{A} = b_{12}^{B} + d_{23}^{D},\ c_{61}^{C} = b_{23}^{B} + d_{12}^{D},\ k_{34}^{K} = k_{23}^{K} + k_{12}^{K}, \end{aligned} \tag{8}$$

where k and K represent *a*, *b*, *c*, *d* and A, B, C, D or **D** and **W**, respectively.

The angles α_{45}, α_{56}, α_{61} and link lengths k_{45}, k_{56}, k_{61} can be easily obtained based on Eqs. (2), (3), (5), (6) and (7). When $\mu = 1, \alpha + \beta < \pi/2, \alpha = \beta$, there are always two smooth folding paths [17]. Here, we arbitrarily choose a pattern with six vertices (Fig. 5) whose parameters satisfy $\mu = 1$ and $\alpha = \beta = 40°$ in Eq. (9). Two motion sequences of the prototype are shown in Fig. 5. Here, Fig. 5(a) shows the planar motion path of thick-panel origami from the flat configuration to the closed pack configuration, and the mobile assembly corresponds to the path from a bundle to a strip via the intermediate configuration. Similarly, Fig. 5(b) shows the arch motion path of thick-panel origami and the corresponding motion of mobile assembly of Bricard linkages. The curves of relationships of kinematic variables φ_i and ϕ_i ($i = 1, 2$) can be derived from Eqs. (10) and (11) [17], as shown in Fig. 5(c). The deployed configuration i and folded configuration ii of thick-panel origami are with $\varphi_i = \phi_i = 180°$ and $\varphi_i = \phi_i = 0$, respectively. Here, φ_i and ϕ_i are dihedral angles corresponding to crease i of thick-panel origami vertices **D** and **W**, respectively.

$$\begin{aligned} &\alpha_{12}^{A} = \alpha_{12}^{C} = \alpha_{12}^{E} = -\alpha = -40°,\ \alpha_{23}^{A} = \alpha_{23}^{C} = \alpha_{23}^{E} = 100°,\ \alpha_{34}^{A} = \alpha_{34}^{C} = \alpha_{34}^{E} = \alpha = 40°, \\ &\alpha_{12}^{B} = \alpha_{12}^{D} = \alpha_{12}^{F} = -20°,\ \alpha_{23}^{B} = \alpha_{23}^{D} = \alpha_{23}^{F} = -\beta = -40°,\ \alpha_{34}^{B} = \alpha_{34}^{D} = \alpha_{34}^{F} = \alpha = 40°, \end{aligned}$$

$$
\begin{aligned}
&a_{34}^{\mathrm{A}} = b_{12}^{\mathrm{B}} + d_{56}^{\mathrm{D}} = 3a,\ c_{61}^{\mathrm{C}} = b_{23}^{\mathrm{B}} + d_{45}^{\mathrm{D}} = 3a,\\
&a_{12}^{\mathbf{D}} = c_{12}^{\mathbf{D}} = e_{12}^{\mathbf{D}} = 3a,\ a_{23}^{\mathbf{D}} = c_{23}^{\mathbf{D}} = e_{23}^{\mathbf{D}} = 0,\ a_{34}^{\mathbf{D}} = c_{34}^{\mathbf{D}} = e_{34}^{\mathbf{D}} = 3a,\\
&b_{12}^{\mathbf{W}} = d_{12}^{\mathbf{W}} = f_{12}^{\mathbf{W}} = 3a,\ b_{23}^{\mathbf{W}} = d_{23}^{\mathbf{W}} = f_{23}^{\mathbf{W}} = 0,\ b_{34}^{\mathbf{W}} = d_{34}^{\mathbf{W}} = f_{34}^{\mathbf{W}} = 3a.
\end{aligned}
\tag{9}
$$

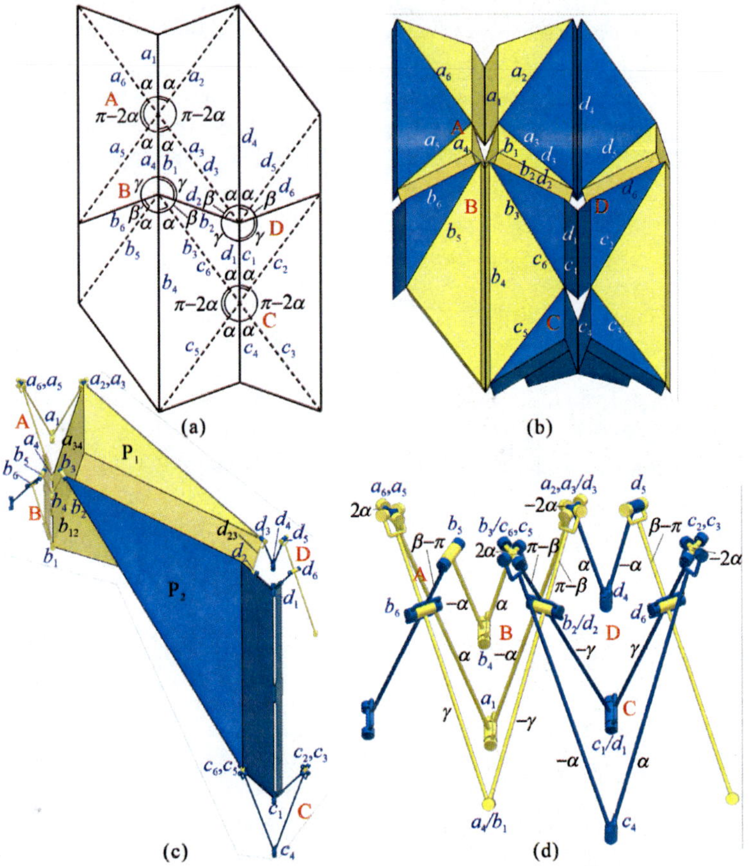

Fig. 4. The transition of multi-vertex waterbomb pattern. (a) Crease pattern with four vertices, where $\gamma = \pi - \alpha - \beta$. (b) Its thick-panel form. (c) The enlarged view of the two shared panels with four attached plane-symmetric Bricard linkages and (d) the corresponding mobile assembly with four Bricard linkages.

$$
\tan\frac{\varphi_1}{2} = \frac{1}{\cos\alpha}\tan\frac{\varphi_2}{2},\ \varphi_4 = \varphi_1,\ \varphi_2 = \varphi_3 = \varphi_4 = \varphi_5 = \varphi_6. \tag{10}
$$

$$
\tan\frac{\phi_1}{2} = \frac{1}{\cos\alpha}\tan\frac{\phi_3}{2},\ \tan\frac{\phi_2}{2} = \frac{\cos(\alpha+\beta)}{\cos\alpha}\tan\frac{\phi_3}{2},\ \phi_4 = \phi_1,\ \phi_5 = \phi_3,\ \phi_6 = \phi_2. \tag{11a}
$$

$$\begin{aligned}
\tan\frac{\phi_1}{2} &= \frac{\tan(\phi_3/2)\left(\sin^2(\alpha+\beta)\tan^2(\phi_3/2)+2\left(\sin^2\beta+\sin^2\alpha\right)\right)}{\sin(\alpha+\beta)(\sin\beta+\cos(\alpha+\beta)\sin\alpha)\tan^2(\phi_3/2)+4\sin\alpha\sin\beta\cos\beta},\\
\tan\frac{\phi_2}{2} &= \frac{\sin(\alpha+\beta)}{2\sin\alpha}\tan\frac{\phi_3}{2},\ \phi_5=\phi_3, \phi_6=\phi_2,\\
\tan\frac{\phi_4}{2} &= \frac{\tan(\phi_3/2)\left(4\sin\alpha\sin^2(\alpha+\beta)\tan^2(\phi_3/2)-8\sin\alpha\left(2\sin^2\beta-\sin^2(\alpha+\beta)\right)\right)}{(\cos(3\alpha+\beta)-8\cos(\alpha+\beta)+7\cos(\alpha-\beta))\sin(\alpha+\beta)\tan^2(\phi_3/2)+8\sin^2\alpha\sin2\beta},
\end{aligned} \tag{11b}$$

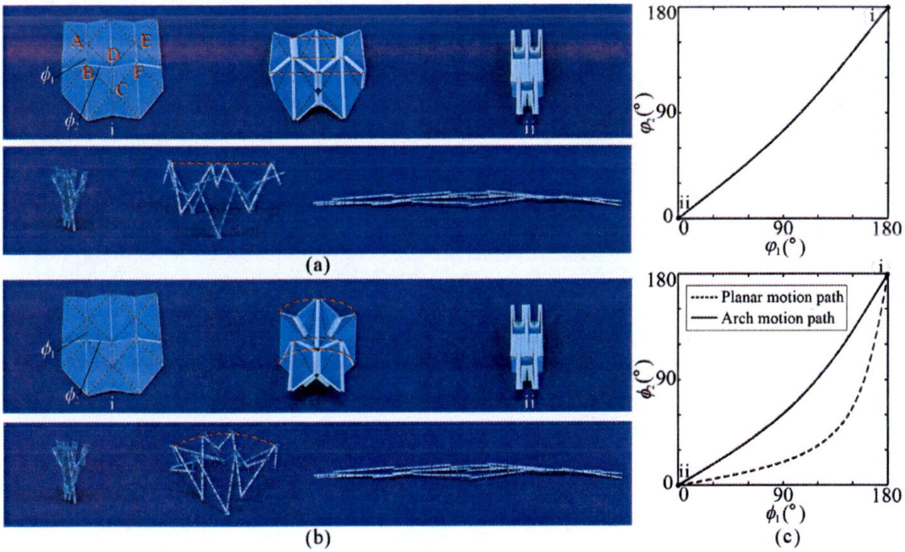

Fig. 5. Two paths of motion sequences of a waterbomb thick-panel origami and its corresponding mobile assembly of plane-symmetric Bricard linkages with $\alpha = \beta = 40°$. (a) The planar motion sequences. (b) The arch motion sequences. (c) The curves of φ_2 vs. φ_1 and ϕ_2 vs. ϕ_1.

4 Conclusions

In this paper, new mobile assemblies consisting of two types of plane-symmetric Bricard linkages have been constructed from waterbomb thick-panel origami based on their kinematic equivalence. According to the geometric conditions in the reference [17], waterbomb thick-panel origami with two smooth paths is constructed. By transiting the thick-panel form to a mobile assembly, the morphing mobile assembly is created, which has the planar deploying movement and the arch deploying movement. The two paths can be changed at the position where mobile assembly is in the fully folded configuration with all links in a line.

As the **D** type vertex is a line- and plane-symmetric origami vertex, we can conjuncture that this vertex can generate thick-panel origami kinematically equivalent to a line-symmetric Bricard linkage. The waterbomb thick-panel pattern with the **D** type vertex and plane-symmetric **W** type vertex could be transited to an assembly mixed with line-symmetric Bricard linkages and plane-symmetric Bricard linkages. This study also can be extended to the thick-panel forms from other six-crease rigid origami patterns, such as the Resch pattern. Furthermore, new mobile assemblies of spatial linkages can be obtained by our proposed approach.

References

1. Zhang, X., Chen, Y.: Mobile assemblies of Bennett linkages from four-crease origami patterns. Proc. Roy. Soc. A-Math. Phys. Eng. Sci. **474**, 20170621 (2018)
2. De Temmerman, N., Mollaert, M., Mele, V., et al.: Design and analysis of a foldable mobile shelter system. Int. J. Space Struct. **22**, 161–168 (2007)
3. Alegria Mira, L., Thrall, A.P., De Temmerman, N.: Deployable scissor arch for transitional shelters. Automat. Constr. **43**, 123–131 (2014)
4. You, Z.: Deployable structure of curved profile for space antennas. J. Aerosp. Eng. **13**, 139–143 (2000)
5. Cui, J., Huang, H., Li, B., Deng, Z.: A novel surface deployable antenna structure based on special form of Bricard linkages. In: Dai, S.J., Zoppi, M., Kong, X. (eds.) Advances in Reconfigurable Mechanisms and Robots I, pp. 783–792. Springer, London (2012)
6. Xu, Y., Lin, Q., Wang, X., Li, L., Cong, Q., Pan, B.: Mechanism design and dynamic analysis of a large-scale spatial deployable structure for space mission. In: 7th International Conference on Electronics and Information Engineering, p. 1032226. SPIE, Nanjing (2017)
7. You, Z., Chen, Y.: Motion Structures: Deployable Structural Assemblies of Mechanisms. Spon Press, New York (2014)
8. Huang, H., Deng, Z., Li, B.: Mobile assemblies of large deployable mechanisms. J. Space Eng. **5**, 1–14 (2012)
9. Ma, B., Huang, H.: Large deployable networks constructed by interconnected Bricard linkages. Adv. Mater. Res. **338**, 723–726 (2011)
10. Song, X., Deng, Z., Guo, H., Liu, R., Li, L., Liu, R.: Networking of Bennett linkages and its application on deployable parabolic cylindrical antenna. Mech. Mach. Theory **109**, 95–125 (2017)
11. Yang, F., You, Z., Chen, Y.: Mobile assembly of two Bennett linkages and its application to transformation between cuboctahedron and octahedron. Mech. Mach. Theory. **145**, 103698 (2020)
12. Xiu, H., Wang, K., Wei, G., Ren, L., Dai, J.: A Sarrus-like overconstrained eight-bar linkage and its associated Fulleroid-like platonic deployable mechanisms. Proc. Inst. Mech. Eng. C-J. Mech. Eng. Sci. **234**, 241–262 (2020)
13. Chen, Y., Yang, F., You, Z.: Transformation of polyhedrons. Int. J. Solids. Struct. **138**, 193–204 (2018)
14. Dai, J.S., Rees Jones, J.: Mobility in metamorphic mechanisms of foldable/erectable kinds. J. Mech. Design **121**, 375–382 (1999)
15. Zhang, X., Chen, Y.: The diamond thick-panel origami and the corresponding mobile assemblies of plane-symmetric Bricard linkages. Mech. Mach. Theory **130**, 585–604 (2018)
16. Chen, Y., Peng, R., You, Z.: Origami of thick panels. Science **349**, 396–400 (2015)
17. Chen, Y., Feng, H., Ma, J., Peng, R., You, Z.: Symmetric waterbomb origami. Proc. Roy. Soc. A-Math. Phys. Eng. Sci. **472**, 20150846 (2016)
18. Denavit, J., Hartenberg, R.S.: A kinematic notation for lower-pair mechanisms based on matrices. Trans. ASME J. Appl. Mech. **22**, 215–221 (1955)

Global Path and Local Motion Planning for Humanoid Climbing Robot Using Kinect Sensor

Dung Anh Nguyen[1](✉) and Akira Shimada[2]

[1] Hanoi University of Science and Technology, 1 Dai Co Viet, Hanoi, Vietnam
dung.nguyenanh@hust.edu.vn
[2] Shibaura Institute of Technology, 3-9-14 Shibaura, Minato-ku 108-8548, Tokyo, Japan
ashimada@sic.shibaura-it.ac.jp

Abstract. This paper presents a study on improving the adaptability of humanoid climbing robots in vertical environments. Although the robot is intended to be used for rescue and load-bearing in mountains or other hazardous areas in the future, this paper focuses on a path planning and local planning algorithm for humanoid robot wall climbing as the initial phase of our development. The first step is to acquire a depth map to extract accurate climbing holds on the vertical wall. Secondly, we propose a global planning algorithm for the humanoid robot using data from Kinect. During climbing, the humanoid robot utilizes the local planning algorithm, based on quasi-static equilibrium, to adjust its body posture to remain in an equilibrium state. Finally, all algorithms are evaluated with a simple, practical example for a humanoid climbing robot system, and its effectiveness is demonstrated experimentally in a real environment.

Keywords: Humanoid robot · Climbing · Motion planning · Kinect

1 Introduction

Recently, a number of scientists have developed climbing robots consisting of mechanics, electronics, cybernetics, control, and computer engineering. The conventional robots are specialized and equipped with sliding segments (Cepolina et al. [1]; Zhu et al. [2]), wheels (Longo and Muscato [3]; Sanchez et al. [4], or multi-leg systems (B. H. Wilcox et al. [5]; Daltorio et al. [6] to climb up walls or rampant terrain. There are two major concerns: a robot's locomotion and adhesion methods. In respect to the adhesion method [7–10], we have four kinds of climbing robots: magnetic, vacuum, or suction cups, gripping to the surface, and propulsion (Behnam Miripour [11]); Consequently, they are limited to environments consisting of glass, metal or other smooth surfaces.

To the best of our knowledge, Lemur IIb [12] represents the first attempt to build a free climbing robot. It is a planar four-limbed climbing robot created by NASA's Jet Propulsion Laboratory. Tenzing [13], another free climbing robot, was built at Dartmouth College. Like Lemur IIb, it is a four-limbed planar robot with two revolute joints in each

N. V. Khang et al. (Eds.): ASIAN MMS 2021, MMS 113, pp. 109–122, 2022.
https://doi.org/10.1007/978-3-030-91892-7_11

limb. Several studies researching four-limbed robot researches have been conducted done at Stanford by Tim Bretl [14–16], Kris Hauser [17, 18], and Ruixiang Zhang [19].

To climb a climbing wall, a humanoid robot typically starts by observing at the entire wall to create the map of the major terrain features, such as climbing hold positions, as well as climbing hold shapes. Based on this map, the best route is planned. This route only gives height level direction since the local motion planning for each fundamental movement is still lacking. To start the algorithm of trajectory planning for free climbing, the humanoid robot also acquires detail local planning around its current position. With the local motion planning, the robot identifies candidate holds that can be reached in order to move up. Therefore, a general scenario of climbing consists of the following steps:

- Step 1: Global sensing.
- Step 2: Global Path Planning.
- Step 3: Local motion planning (detailed planning for series of moves 4-3-4 contact-holds).
- Step 4: Execution of the moves.

Based on the general scenario, our project is divided into three technical parts: observation, planning, and motion control. Here we focus on the first two parts. In the observation part, we use a Kinect [21] for global sensing (step 1) to get information about the climbing wall, such as all the climbing holds positions and the climbing wall's dimensions. After acquiring the current environment, the path planning (step 2) begins the special motion planning algorithm named "Right Hand Search Algorithm" (RHSA) [20], which is utilized to identify the best route to climb up the wall. This algorithm can be used to construct the trajectory that connects the start position to the goal position. While following the route, the robot uses the local motion planning (step 3) to identify target holds and adjusts its center of mass to perform short-term goals.

2 Precondition and Preliminary Work

Hardware Structure and Assumptions on the Research
We selected the KHR-3HV manufactured by Kondo [23] as a prototype to develop a humanoid climbing robot. The humanoid robot has 22 degrees of freedom (DOF). The humanoid robot does not have any special mechanism to grasp holds, but angled brackets are used as hands.

This paper addresses the problem of global motion planning and local planning of a humanoid climbing robot in to "free climbing" on vertical surfaces. However, it is not easy to develop a humanoid climbing robot in natural environments. Therefore, this paper focus on an artificial climbing wall.

Climbing Wall and Climbing Wall Configuration. The robot's workspace is a straight vertical wall. The wall is made from wood with multi-holes (shown in Fig. 1). The holds are made of metallic rings, which the climbing hands can hook into, and the feet can be placed on, shown in Fig. 2.

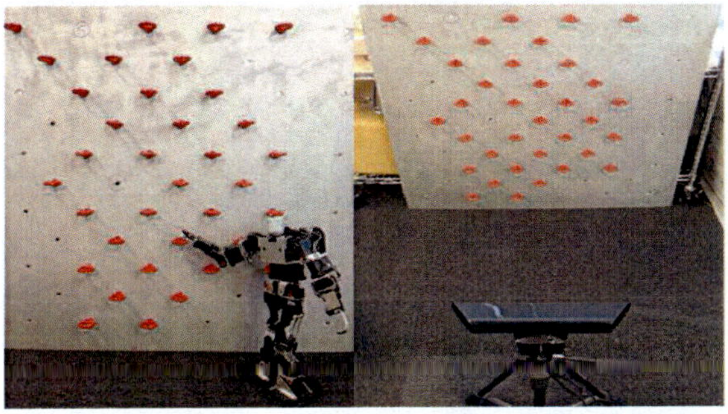

Fig. 1. Climbing wall and system location

Fig. 2. Climbing hold hand and foot

Equilibrium. The robot must apply contact forces with hands and legs at holds that compensate for gravity to remain balanced. For instance, two hands and left leg are at holds while the right hand is moving. In this case, the robot performs a 3-contact-holds posture to remain balance. If the robot keeps balance by four holds, the posture is called 4-contact-holds.

Motion and Transition. In this research, we assumed the robot's motion to be quasi-static [24]. A motion humanoid climbing robot consists of a sequence of 3- contact-holds and 4-contact-holds postures. If 3-contact- holds and 4-contact-holds posture share three holds, they are adjacent, and transition exits between them.

3 Observation Part

The overview of our proposed systems is shown in Fig. 3. Kinect receives all of the information of the climbing wall, and then it finds the locations of all the climbing holds. From the size of the robot, the controller creates the working area. Using this working area and applying the right-hand search algorithm, the best route can be found accordingly.

3.1 Pre-processing of Kinect Depth Data

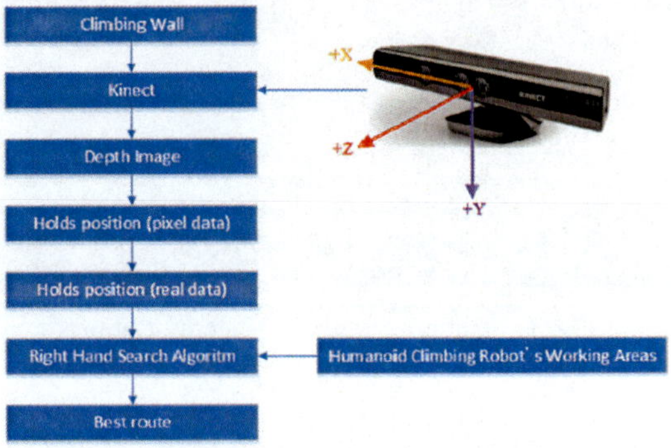

Fig. 3. Overview proposed systems

The depth value represents the vertical distance between the target point and the XOY plane of the Kinect coordinates, which is defined as a right-handed coordinate with the Kinect's facing direction as the positive Z direction and Kinect's right-side direction as the positive X direction (Fig. 3).

In OpenNI [22], the Kinect depth map is saved as a one-dimension array. Like the color image, the pixel at the top left corner of the depth value starts the array. When the resolution is 640X480, the positions of the pixel in the array can be illustrated as Table 1.

Table 1. The storing rule of Kinect depth values

1(1,1)	2(1,2)	…	639(1,639)	640(1,640)
	642(2,2)	…	1279(2,639)	640(2,640)
…	…	…	…	…
306561(480,1)	306562(480,2)	…	307199(480,639)	307200(480,640)

3.2 Searching Climbing Holds Position (Pixel Data)

To find out the climbing hold positions, the controller uses the different values of the holds and climbing wall since the climbing hold's depth value is smaller than the depth value of the wall. Assuming the climbing holds are the same size and the square shape of the climbing wall, we divided the wall into 11 rows and 11 columns for a total of 121 areas (Fig. 4(a)), for example. Using the simple algorithm shown below, the controller checks all areas to identify which areas have "hold" and which do not.

Each area is defined as an area with a "hold" if this area contains pixels whose depth value is smaller than the climbing wall's standard depth value. Any pixels with

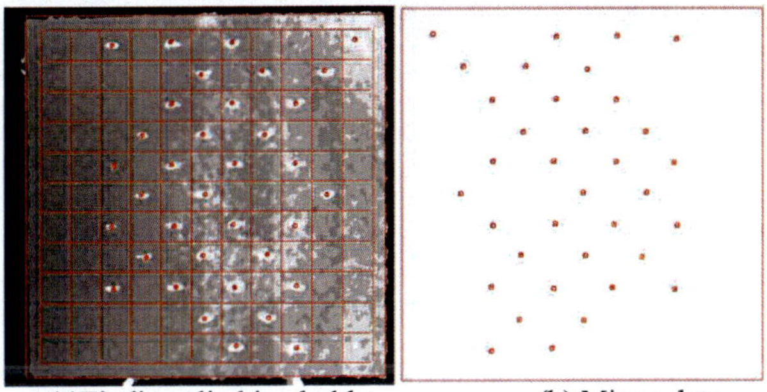

(a)Finding climbing holds (b) Mirror data

Fig. 4. Pixel data

the smallest depth value are defined as the pixels linked to a climbing hold (x, y depth value).

Since the depth values of the Kinect are opposite from the real dimensions. The pixel data is converted to the new pixel. Each climbing hold position (X, Y - pixel) is converted as Eq. (1).

$$\begin{cases} Y_{new} = Y \\ X_{new} = 640 - X \end{cases} \quad (1)$$

The result of every climbing hold position is shown in Fig. 4(b), for example.

The controller should rebuild the climbing wall information with the climbing hold positions (real dimension). From the Kinect data, the controller recognizes the four corners position (A, B, C, and D) of the climbing wall (see Fig. 5(a)) and also the real dimension of the climbing wall (a-mm). Then, the controller defines point A as a new origin; H denotes the climbing hold (with x_H, y_H pixel), point H's position can be converted to real dimensions using the following Eqs. (2).

$$\begin{cases} x = (x_H - x_A) * a/(x_B - x_A) \\ y = (y_H - y_A) * a/(y_B - y_A) \end{cases} \quad (2)$$

In the same manner, the controller converts all the climbing hold positions from pixel data to real data. To use the set of the climbing holds in the Right-Hand Search Algorithm, we define the set of all climbing holds as H. The order of climbing holds in the set H regulatory compliance: A hold that has a smaller distance (compared with the left top point) has a smaller index in the set H (see in Fig. 5(b)).

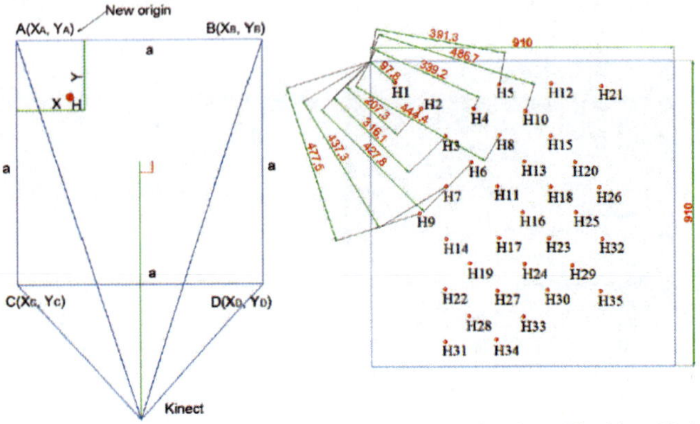

(a) Geometric relationship (b) Organization Climbing Holds

Fig. 5. Convert and label climbing holds

4 Global Path Planning

4.1 Defining Robot "Reachable Area"

Figure 6 describes the reachable area, reference point for scan. The reachable area for the hands and feet that is obtained depends on the size of the robot. The reachable area can be moved along with the robot's motion on the wall. Then the robot finds "right-hand reachable holds" under the robot's configurations. When the robot's right hand hooks into RHRH, the robot can find at least one hold within the left-hand reachable area and find a hold for each leg within the reachable area.

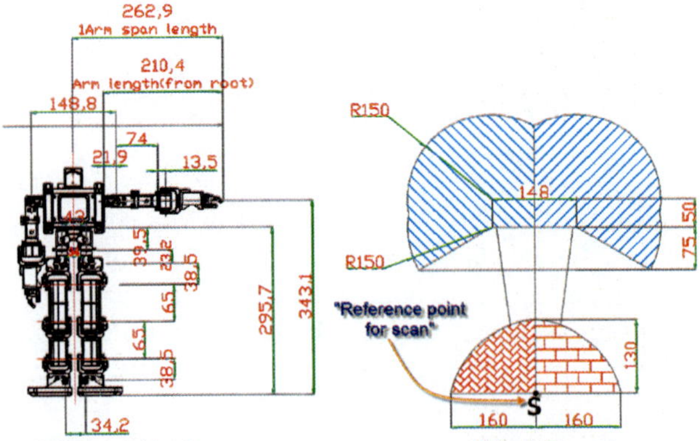

(a) Dimension of the robot (b) Reachable area

Fig. 6. Definition of "reachable area" and "reference point for scan"-S

4.2 Scanning Phase

As the first step of the presented right-hand search algorithm, RHSA, the robot needs to find a "right-hand reachable hold" by scanning the wall through a reference scanning point, S, shown in Fig. 6(a). First, depending on the robot's size illustrated in Fig. 6(b), the robot controller automatically finds the "reachable area" from S.

The controller uses the set of grid points O to sample the scan point S. The set O is generated as follows: The controller generates vertical and horizontal parallel lines through AB and AC (see in Fig. 5) with an equal offset distance. Then, we have O as the set of intersection points for those horizontal and vertical lines.

Scanning Valid Right-Hand Holds

One "right hand reachable holds (RHRH)" can be found for the reference scanning position S. The robot can find, at least one hold located in the left-hand reachable area and find one hold for each leg reachable area. The point S is sampled from the set O (defined in Sect. 4.2); (similar to sampling algorithm in [25]). The controller can find the robot's reachable area on the climbing wall when the corresponding S is found. The RHRH is the hold existing in the reachable space of the right hand, and simultaneously at least one group consisting of three holds in the reachable spaces of the left hand, right foot, and left foot. We call this group is the "family hold group". Note that one RHRH has multiple "family hold groups".

Table 2. 3 first right-hand reachable holds for S (910,270 mm)

V[1]	V[2]	V[3]	V[4]
H14, V1	H14, V2	H14, V3	H14, V4
H31, H34	H31, H34	H31, H34	H31, H34

An example is introduced: Select a scan point "S" shown in Fig. 7. The controller finds out one RHRH V1 as its family hold group, H14; H31; H34, which are left hand hold, left foot and right foot holds, respectively.

A set, V, is defined as a set of all the right-hand reachable holds, and RHRHs are stored as the array of V sets with an order of priority: left to right, bottom to top. For instance, RHRH V1 is stored in V[1] which contains one element [H14,V1,H31,H34]. If one V[i] set has n elements, there are n "family hold group" with respect to the RHRH. Now, by scanning the point successively, S from (910,0), (910,10), (910,910),..., (0,0), (0-910) (sampling from the set O), the controller finds every RHRHs. For the point S from (910,0) to (910,910), the controller finds out 4 RHRHs (Table 2). The controller adds these 4 RHRHs to the set of all candidates of start holds, which is denoted by St

St = *V1, V2, V3, V4* (see in Fig. 7).

Temporarily: *V* = *V1, V2, V3, V4*

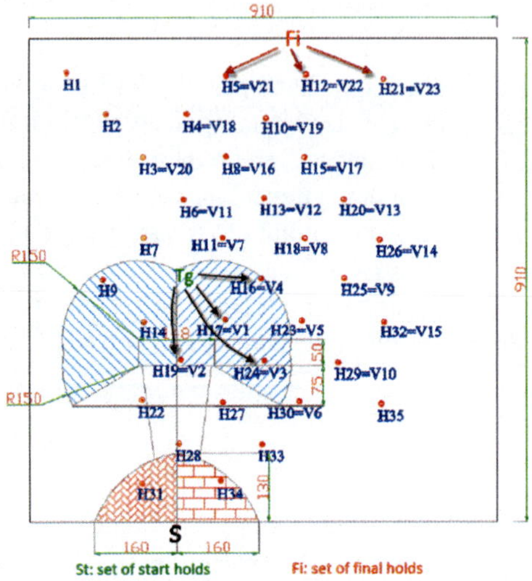

Fig. 7. Scan example using "Right-hand reachable holds."

Algorithm 1 Scanning "Right Hand Reachable Hold"

Generate initial V:
V←∅
O:= the set of all area addresses of the wall;
while (O is not empty) **do**
  S ← a chosen vertex from the set O;
  O ← O ∩{S}
  if (RHRH in right-hand reachable area) and (at least one hold located in the left hand, left leg, and right leg reachable areas) then Adds new RHRH to set *V;*
  end if
  Update RHRH's "family hold groups";
end while

The controller continues to choose point S from the set O. We remember that the "family hold groups" of old RHRHs maybe be updated related to a new point S, and a new RHRH could be found. For example, when point S is at the address (910, 290), the controller updates the information on the RHRHs (Fig. 8). Finally, we have set V. $V = V1, V2, V3, \ldots, V23$ (see in Fig. 7).

Like "start hold," we define the "final hold." The "final hold" belongs to the RHRH that is located in the highest row, as shown in Fig. 7. The controller adds these "final hold" to the set of all final holds (denoted by Fi). Completing the scanning process, we have the result:

$St = V1, V2, V3, V4$
$Fi = V21, V22, V23$
$V = V1, V2, V3, \ldots, V23$

V[1] (When V1 is hold H17) (red is new update VRHP)
$\begin{bmatrix} H9 & V1 \\ H31 & H34 \end{bmatrix} \begin{bmatrix} H14 & V1 \\ H31 & H34 \end{bmatrix}, \begin{bmatrix} H19 & V1 \\ H31 & H34 \end{bmatrix}$
V[2] (When V2 is hold H19)
$\begin{bmatrix} H9 & V2 \\ H31 & H34 \end{bmatrix}, \begin{bmatrix} H14 & V2 \\ H31 & H34 \end{bmatrix},$
V[3] (When V3 is hold H24) red is new update VRHP
$\begin{bmatrix} H9 & V3 \\ H31 & H34 \end{bmatrix}, \begin{bmatrix} H14 & V3 \\ H31 & H34 \end{bmatrix}, \begin{bmatrix} H19 & V3 \\ H31 & H34 \end{bmatrix}$
V[4] (When V4 is hold H16 (new VRHP))
$\begin{bmatrix} H9 & V4 \\ H31 & H34 \end{bmatrix} \begin{bmatrix} H14 & V4 \\ H31 & H34 \end{bmatrix}$

Fig. 8. The updating information of RHRHs when point S at (910,290)

Cost Evaluations and Building the Graph of "Right Hand. Reachable Holds".

Algorithm 2 Scanning "Right Hand Reachable Hold"

Input: RHRHs in array
Output: Cost between "adjacent holds."

```
Define RHRH;
Define the similar operator between two families hold groups: family hold groups are similar if the position of both legs and the left hand are similar;
Define cost: number of the family hold groups between two RHRHs;
cost:=0;
for (i from 0 to the number of family hold group of k^th RHRH -1), do
    for (j from 0 to the number of family hold group of h^th RHRH -1) do
        if (i^th and j^th family hold group of k^th RHRH and h^th RHRH are similar) then Set new RHRH to set V;
        end if
    end for
end for
```

In this subsection, we define "adjacent hold" and evaluate the cost between RHRHs; in addition, the controller builds the graph of RHRHs based on these costs. An RHRH[i] is called "adjacent hold" of RHRH[j] if RHRH[i] and RHRH[j] have at least one common family hold group. The cost between two RHRHs is the number of similar "family hold groups." When moving(climbing) between two adjacent holds having the largest cost, the robot has many options to utilize the most advantageous "family hold group" to minimize energy, avoid obstacles and singularity, or optimize the moving distance of the desired limb. Therefore, it is obviously beneficial when the robot can climb between adjacent holds with the largest costs. Now, we set up the algorithm to determine the cost between two RHRHs (see in Algorithm 2).

First, we define each VRHP as a structure containing elements such as V[3] = [[H9,V3;H31,H34], [H14,V3;H31,H34], [H19,V3;H31,H34], [H14,V3;H31,H28], [H7,V3;H31,H34], [H7,V3;H31,H28]].

V[3] is the third element in the *V[]* set (in the above section), while V3 is the valid right-hand hold (the hold H24).

Second, we define the "similar operator" between two elements. Elements A and B are similar to each other if A[1,1], A[2,1] and A[2,2] are equal to B[1,1], B[2,1] and B[2,2] respectively. Using the "for" loop to compare elements between two structures, we increase the cost by one if a similar operator occurs. In this way, after comparing all RHRHs with each other, the controller has a matrix containing the cost between every RHRHs.

The controller can create the graph (Ma-graph) between the "right-hand reachable holds" from this cost result. Obviously, climbing robots can move easily between two RHRHs have the large cost (more common "family holds group"), so the goal is to find a path of maximum total cost from "start hold" to "final hold."

4.3 The Query Phase

By using the way where the total cost is maximal, the robot can climb easily. First, the controller builds a new graph in which the complement costs with the maximum value (in the Ma-graph) substitute the old values being the costs between RHRHs, but the infinity values are kept to identify that robot cannot move between RHRHs which have a zero cost, and we have a new graph in Fig. 9.

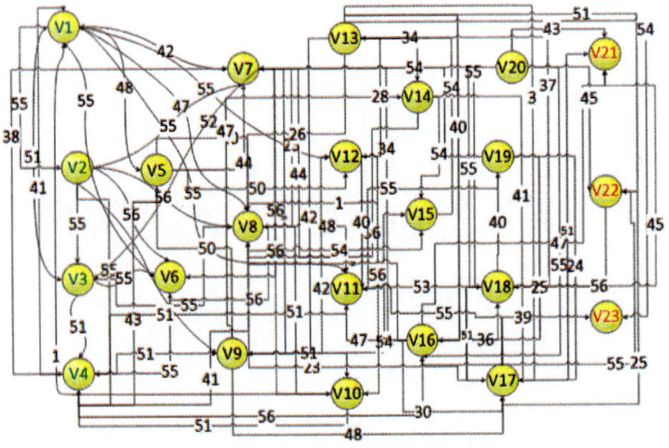

Fig. 9. The new graph RHRHs

With new graph, the easiest route to climb is the path of maximum total cost from "start hold" to "final hold." The Dijkstra algorithm [26] is used to solve this problem.

5 Local Motion Planning

In local motion planning, we assume that the short-term goal is to move the robot's right hand, currently at a red hold, to a new blue hold (see in Fig. 10(a)). To break the contact, the robot must first adjust its posture to redistribute the contact forces over the other three limbs so that the contact force applied on the red hold by the right hand becomes 0. See Fig. 10(a). At the time t_0 where each of the robot limbs is positioned at the four climbing holds. We call this set a 4-contact-holds. The robot breaks the contact at the red hold at time t_1 and moves at a 3-contact-holds (set of 3 contacted holds). The robot's right-hand reaches the blue hold (local target hold) at the time t_2 (see in Fig. 10(b)).

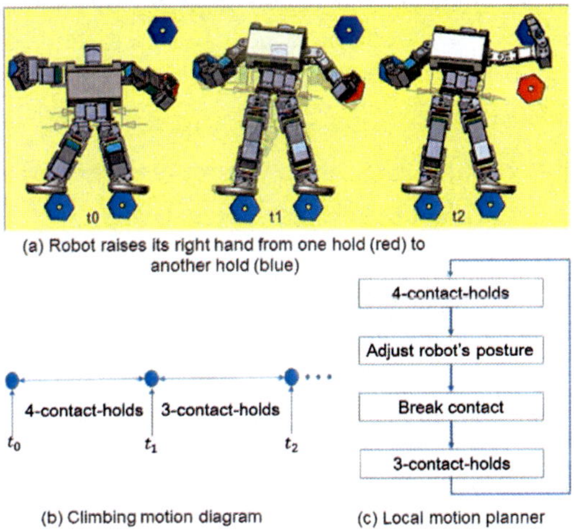

Fig. 10. Detail local motion planning

To continue climbing, the robot will then have to break contact at one of the other three limbs and then reach a new hold with the corresponding limb, etc...(see in Fig. 10(c). So overall, the entire climbing motion consists of successive moves. By applying this local planning, the humanoid robot can calculate the individual joint angles for each of the robot's joints. Figure 11 is an example of applying local motion planning to determine all joint angles. At the posture 0 (at the time t_0), the robot keeps balance on the wall by 4-contact-holds. From the posture 0 to posture 1, the robot adjusts the its body (at the time t_1). Therefore, at the posture 3, the robot raises its right hand to reach a new climbing hold (at the time t_2). In the same manner, from the posture 4 to 7, the robot raises its left hand to reach a new climbing hold.

6 Experiment with an Example

We implemented our global path and local motion planning on the KHR-3HV robot. Using the planner, the robot can climb up freely along the designed route. Of course, free climbing also requires various sensor feedback (vision, force sensor...) to make the climbing more precise and robust. However, for our example, an open-loop position controller is implemented without sensor feedback on the robot to execute the trajectory at a very slow velocity to reduce positioning errors and slipping risks.

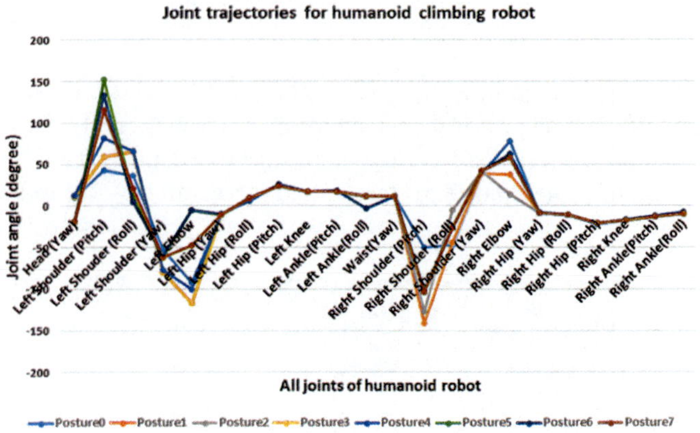

Fig. 11. Joint trajectories example

In this example, the climbing wall configuration is shown in Fig. 1. The path of right-hand [V1]-[V7]-[V16]-[V19]-[V22] was found on the current climbing wall configuration.

The humanoid robot starts climbing the wall with "the start hold" V[1]. The "family holds group" of V[1], which the robot uses to keep balance in this step, is $\begin{bmatrix} \text{H14} & \text{V1} \\ \text{H31} & \text{H34} \end{bmatrix}$ (show in Fig. 12 (24[s]).

Lessons from experiments, KHR-3HV is well designed and robust, capable of precise joint angle control. However, in this experiment, several limitations remain. For example, the robot is controlled by joint-angle control, achieving the desired posture but not always desired force.

Fig. 12. Snapshots of the humanoid climbing robot along the best route.

7 Conclusion

This paper describes the trajectory for an autonomously operating humanoid robot to climb up a wall. The Kinect sensor is used to detect the climbing hold positions, and we addressed the problem of motion planning for humanoid climbing robots. Then the "Right Hand Search Algorithm" is developed to determine the route to climb up the wall. This algorithm compared these paths to find out the best route based on the ease of the climbing. Then, local motion planning is proposed to support the robot in each rudimentary movement while the humanoid robot is following the global path. Finally, experiments demonstrate that our humanoid robot can reliably climb the rudimentary climbing wall.

References

1. Cepolina, F., Zoppi, M., Zurlo, G., Molfino, R.: A robotic cleaning agency. In: Proceedings of IAS2004 8th Conference on Intelligent Autonomous Systems, Amsterdam, The Netherlands, pp. 1153 1161 (2004)
2. Zhu, J., Sun, D., Tso, S.-K.: Application of a service climbing robot with motion planning and visual sensing. J. Robot. Syst. **20**(4), 189–199 (2004)
3. Longo, D., Muscato, G.: A modular approach for the design of the alicia3 climbing robot for industrial inspection. Ind. Robot: Int. J. **31**(2), 148158 (2004)
4. Snchez, J., Vzquez, F., Paz, E.: Machine vision guidance system for a modular climbing robot used in shipbuilding. In: Tokhi, M.O., Virk, G.S., Hossain, M.A. (eds.) Climbing and Walking Robots, pp. 893–900. Springer, Heidelberg (2006). https://doi.org/10.1007/3-540-26415-9_107
5. Wilcox, B.H., et al.: ATHLETE: a cargo handling and manipulation robot for the moon. J. Field Robot. **24**(5), 421–434 (2007)
6. Daltorio, K.A., Wei, T.E., Gorb, S.N., Ritzmann, R.E., Quinn, R.D.: Passive foot design and contact area analysis for climbing mini-whegs. In: Proceedings of the 2007 IEEE International Conference on Robotics and Automation, Roma, Italy, pp. 1274–1279 (2007)
7. Bach, F.W., Haferkamp, H., Lindemaier, J., Rachkov, M.: Underwater climbing robot for contact arc metal drilling and cutting. In: Proceedings of IEEE International Conference on Industrial Electronics, Control and Instrumentation, Taipei, Taiwan, pp. 1560–1565 (1996)
8. Briones, L., Bustamante, P., Serna, M.A.: Wallclimbing robot for inspection in nuclear power plants. In: Proceedings of IEEE International Conference Robotics and Automation, San Diego, CA, pp. 1409–1414 (1994)
9. Chen, I.-M., Yeo, S.: Locomotion of a two-dimensional walking-climbing robot using a closed-loop mechanism: from gait generation to navigation. Int. J. Robot. Res. **22**(1), 21–40 (2003)
10. Kang, T., Kim, H., Son, T., Choi, H.: Design of quadruped walking and climbing robot. In: Proceedings of IEEE/RSJ International Conference on Intelligent Robots and Systems, Las Vegas, NV (20030
11. Miripour, B.: Climbing and Walking Robots ISBN 978-953-307–030-8, Published by INTECH (2010)
12. Kennedy, B., Okon, A., Aghazarian, H., et al.: Lemur IIb: a robotic system for steep terrain access. Ind. Robot: Int. J. **33**(4), 265–269 (2006)
13. Linder, S.P., Wei, E., Clay, A.: Robotic rock climbing using computer vision and force feedback. In: Proceedings of IEEE International Conference on Robotics and Automation, pp. 4685–4690 (2005)

14. Bretl, T., Rock, S., Latombe, Jean-Claude., Kennedy, B., Aghazarian, H.: Free-climbing with a multi-use robot. In: Ang, M.H., Khatib, O. (eds.) Experimental Robotics IX. STAR, vol. 21, pp. 449–458. Springer, Heidelberg (2006). https://doi.org/10.1007/11552246_43
15. Bretl, T.: Multistep motion planning: application to freeclimbing robots. Ph.D Diss., Dept. of Aero Astronautics, Stanford University, Stanford, CA, June 2005
16. Bretl, T.: Motion planning of multi-limbed robots subject to equilibrium constraints: the free-climbing robot problem. Int. J. Robot. Res. **25**(4), 317–342 (2006)
17. Hauser, K.: Motion planning for legged and humanoid robots. Ph.D. Diss., Dept. of Computer Science, Stanford University, Stanford, CA, September 2008
18. Hauser, K., Latombe, J.C.: Multimodal motion planning in non-expansive spaces. Int. J. Robot. Res. **29**(7), 897–915 (2010)
19. Zhang, R.: Design and implementation of an autonomous climbing robot. Ph.D. Diss., Dept. of Computer Science, Stanford University, Stanford, CA, January 2014
20. Dung, N.A., Shimada, A.: A path-planning algorithm for humanoid climbing robot using kinect sensor. In: Proceedings of IEEE International Conference on Society of Instrument and Control EngineersFrBT2.5, Japan, September 2014
21. Kinect Sensor. http://www.microsoft.com/en-us/kinectforwindows/. Accessed 2014
22. OpenNI. http://www.openni.org. Accessed 2014
23. JKondo Kagaku co.,ltd. http://www.kondo-robot.com/. Accessed 2013
24. Goodwine, B., Burdick, J.: Motion planning for kinematic stratified systems with application to quasi-static legged locomotion and finger gaiting. IEEE Trans. Autom. Contr. **18**(2), 209–222 (2002)
25. Choset, H., Lynch, K.M., Hutchinson, S., Kantor, G.A., Burgard, W., Kavraki, L.E.:Sebastian Thrun Principles of Robot Motion: Theory, Algorithms, and Implementations, A Bradford Book, May 2005
26. Dijkstra, E.W.: A note on two problems in connexion with graphs. Numerische Mathematik **1**(1), 269–271 (1959). https://doi.org/10.1007/BF01386390

An Optimization Method for Dynamics of Non-holonomic Systems

Do Dang Khoa[1(✉)], Tran Si Kien[2], Phan Dang Phong[2], and Do Sanh[1]

[1] Hanoi University of Science and Technology, N.1 Dai Co Viet Road, Hanoi, Vietnam
khoa.dodang@hust.edu.vn

[2] National Research Institute of Mechanical Engineering, N.4 Pham Van Dong, Hanoi, Vietnam

Abstract. This paper proposes an optimization method for motion of non-holonomic mechanical systems. In such system, the number of degrees of freedom is less than the number of independent generalized coordinates, so the number of independent velocities (also the number of independent accelerations) is less than the number of independent coordinates. This makes it difficult to determine the boundary conditions for the system of motion equations. In this paper, the principle of compatibility and the method of transformation matrix are used to directly build the equations of motion of the constrained dynamics system in the form of ordinary differential equations (ODEs) avoiding the use of Lagrange multipliers and then the Pontryagin's maximum principle is used to solve the optimization of dynamics problem. A selection of boundary condition for Hamiltonian costate variables is also proposed. For illustration, the optimal control for the steering motion of an automobile while satisfying the non-holonomic constraints is carried out.

Keywords: Pontryagin's maximum principle · Principle of compatibility · Method of transformation matrix · Non-holonomic constraints

1 Introduction

Optimal motion of dynamics systems is usually solved by direct and indirect approaches. While direct approaches often lead to time-consuming computation due to optimization of the large number of parameters, indirect methods based on Pontryagin's maximum principle is proposed as a good candidate [1]. When dealing optimal problems with non-holonomic constraints, there exist two approaches. The first one is to lump the constraints into the conjugate system of differential equations as in [2]. The second one is to embed the constraints into the motion equations of the constrained dynamics system. The general method to study the motion of such systems is the Lagrange multiplier method that supplements Lagrange multipliers, thus increasing the complexity in deriving the motion equations of the system and more difficult to implement the Pontryagin's maximum principle to optimize the dynamics problems [3]. In this paper, a method to derive ODEs of motion of constrained dynamics system using the principle of compatibility is applied [4–6]. This method is similarly to the one using independent kinematic variations in [7]. The advantage of this method is to build a complete set of ODEs of generalized

N. V. Khang et al. (Eds.): ASIAN MMS 2021, MMS 113, pp. 123–133, 2022.
https://doi.org/10.1007/978-3-030-91892-7_12

variables of the constrained dynamics system that makes it easier to convert to the set of Hamiltonian conjugate variables for optimal problem implementation. An illustrative example to find steering torque of the automobile while keeping its minimum power and satisfying the non-holonomic constraints are simulated using the proposed methodology.

2 The Basis of the Method

2.1 Step 1: Equations of Motion of Constrained Dynamics Systems

The mechanical system's position represented by n generalized coordinates $q_i(i = \overline{1, n})$ and n generalized forces and subjected to s constraints in the following form:

$$f_\alpha(q_1, .., q_n, \dot{q}_1, .., \dot{q}_n) = 0 \quad ; \quad \alpha = \overline{1, s} \tag{1}$$

These constraints are non-integrable and are named as non-holonomic constraints. To derive the system equations of motion, we use the method of transformation matrix [8–10]. First, to determine the inertial matrix **A**, which is the symmetric, non-singular matrix of size n and the kinetic energy T of the mechanical system is written in the form:

$$T = 0.5\,\dot{\mathbf{q}}^T\mathbf{A}\dot{\mathbf{q}} \tag{2}$$

Where:

$$\dot{\mathbf{q}} = \left[\dot{q}_1, .., \dot{q}_n\right]^T \tag{3}$$

is a (n × 1) matrix of generalized velocities (symbol T in the upper right corner of the character indicating matrix transposition). From now on, the matrices are written in bold letters and the vector is considered as the column matrix. The equations of motion of mechanical system are written in the form:

$$\mathbf{DA\ddot{q}} = \mathbf{DQ} + \mathbf{DQ}^{qt} \tag{4}$$

where:

$$\mathbf{Q} = \left[Q_1 \,..\, Q_n\right]^T \tag{5}$$

is the matrix of generalized forces of the active forces, and

$$\mathbf{Q}^{qt} = \left[Q_1^{qt} \,..\, Q_n^{qt}\right]^T \tag{6}$$

is the matrix of generalized forces of the inertial forces whose components includes the generalized velocities, they are determined from matrix ∂_ιA, a matrix of size n in the form of

$$\partial_i\mathbf{A} = \begin{bmatrix} \frac{\partial a_{11}}{\partial q_i} & \frac{\partial a_{12}}{\partial q_i} & . & \frac{\partial a_{1n}}{\partial q_i} \\ \frac{\partial a_{12}}{\partial q_i} & \frac{\partial a_{22}}{\partial q_i} & . & \frac{\partial a_{2n}}{\partial q_i} \\ . & . & . & . \\ \frac{\partial a_{1n}}{\partial q_i} & \frac{\partial a_{2n}}{\partial q_i} & . & \frac{\partial a_{nn}}{\partial q_i} \end{bmatrix} \; : \quad i = \overline{1, n} \tag{7}$$

$$\mathbf{Q}^{qt} = \mathbf{Q}^{qt1} - \mathbf{Q}^{qt2} \tag{8}$$

$$\mathbf{Q}^{qt1} = \left[Q_1^{qt1} \; Q_2^{qt1} .. Q_n^{qt1} \right]^T; \quad Q_i^{qt1} = 0.5\dot{\mathbf{q}}^T \partial_i \mathbf{A}\dot{\mathbf{q}} \;\; ; \; i = \overline{1, n} \; ; \;\; \mathbf{Q}^{qt2} = \sum_{i=1}^{n} \partial_i \mathbf{A}(\dot{q}_i \dot{\mathbf{q}}) \tag{9}$$

Matrix **D** obtained by representing all the generalized accelerations in terms of independent generalized accelerations from the constraint equations. The constraint equations are written in the form:

$$\sum_{i=1}^{n} \frac{\partial f_\alpha}{\partial \dot{q}_i} \ddot{q}_i + \sum_{i,j=1}^{n} \frac{\partial^2 f_\alpha}{\partial q_i \partial q_j} = 0; \qquad \alpha = \overline{1, s} \tag{10}$$

The dependent generalized accelerations $\ddot{q}_\alpha$ ($\alpha = \overline{1, s}$; $s = n - k$) are calculated according to the independent generalized accelerations $\ddot{q}_\sigma$ ($\sigma = \overline{1, k}$; $k = n - s$)

$$\ddot{q}_\alpha = \sum_{\sigma=1}^{k=n-s} d_{\alpha\sigma} \ddot{q}_\sigma + ...; \;\; \alpha = \overline{1, s} \tag{11}$$

The unwritten terms do not contain the generalized accelerations. From now on, generalized accelerations are represented by independent generalized accelerations using matrix **D**. Matrix **D** has the following form:

$$\mathbf{D} = \begin{bmatrix} \mathbf{D}(k) \\ \mathbf{D}(s) \end{bmatrix} \tag{12}$$

Matrix **D**(k) is a (k × k) identity matrix, and matrix **D**(s) of size (s × k) has the following form:

$$\mathbf{D}(s) = \begin{bmatrix} d_{1,1} & .. & d_{1,k} \\ d_{2,1} & .. & d_{2,k} \\ . & .. & . \\ d_{s,1} & .. & d_{k+s,k} \end{bmatrix} \tag{13}$$

Where the elements of the matrix **D**(s) are the coefficients in Eq. (11). From matrix Eq. (4), we obtain k equations and s equations from (10) (constraint equations written in form of generalized accelerations), n generalized accelerations $\ddot{q}_i(i = \overline{1, n})$ will be calculated as a function of generalized coordinates and generalized velocities [11]:

$$\ddot{q}_i = f_i(q_1, q_2, \ldots, q_n, \dot{q}_1, \dot{q}_2, \ldots, \dot{q}_n) \; ; \;\; (i = \overline{1, n}) \tag{14}$$

Then, the conjugate variables will be set:

$$(q_1, q_2, \ldots, q_n, q_{n+1}, q_{n+2}, \ldots, q_{2n}, p_1, p_2, \ldots, p_{2n}) \tag{15}$$

where $\{q_1 \ldots q_n\}$ are converted from the generalized coordinates, $\{q_{n+1}, q_{n+2}, .., q_{2n}\}$ is the generalized velocity $q_{n+1} \equiv \dot{q}_1$; $q_{n+2} \equiv \dot{q}_2, .., q_{2n} \equiv \dot{q}_n$, variables $p_j(j = \overline{1, 2n})$ are determined from the Hamilton function.

Note that if the coefficients of the inertial matrix are constant, the components of inertial generalized forces are zero ($\mathbf{Q}^{qt} = 0$). In this case, the equations describing the motion of the system are the constraint Eqs. (10) and:

$$\mathbf{DA\ddot{q}} = \mathbf{DQ} \tag{16}$$

2.2 Step 2: Optimization Problem

Based on the Pontryagin's maximum principle [11], the motion of a mechanical system needs to be represented by first order equations with conjugate variables $\{q_i, p_i, z\}$; $(i = \overline{1, 2n})$, in which variable z determines the requirements of the optimization problem, these factors (control parameters) can affect motion characteristics to minimize (or maximize) a functional, called the objective functional

$$I = \min \int_0^{t_f} f_0 dt \tag{17}$$

f_0 is a function of control parameters u_i $(i = 1, 2, ..)$ and conjugate variables, t_f - operating time of the mechanical system.

To apply the Pontryagin's maximum principle to the mechanical system, it is necessary to write the equations of motion in terms of the conjugate variables by converting Lagrange variables to conjugate variables and determining control variables.

The conjugate variables p_i; $(i = \overline{1, 2n})$ are determined from Hamilton function H:

$$H = -f_0 + \sum_{i=1}^{2n} p_i f_i; \tag{18}$$

where:

$$f_i \equiv q_{n+i} (i = \overline{1, n}); \;\; f_{n+s} = \ddot{q}_s; \;\; (s = \overline{1, n}) \tag{19}$$

For the motion of the mechanical system to meet the requirements set by the Pontryagin's Principle, the control variables are found from the extreme conditions of the Hamilton function H. In particular, optimal controls (denoted by u_{ropt}) can be found from the following condition.

$$\frac{\partial H}{\partial u_r} = 0; \;\; r = 1, 2, \ldots \tag{20}$$

The quantities $\ddot{q}_s$; $(s = \overline{1, n})$ are calculated from Eq. (14). The system of conjugate equations is written in the following form:

$$\frac{dq_i(t)}{dt} = f_i (i = \overline{1, 2n});$$
$$\frac{dp_i(t)}{dt} = -\frac{\partial H}{\partial q_i}; \;\; (i = \overline{1, 2n});$$

$$\frac{dz(t)}{dt} = f_0 \quad (21)$$

Solving this system of equations by replacing the control variables obtained from (20) to the right side of the Eqs. (21) with the following boundary conditions:

$$\{q_i(0)\} \; và \; \{p_i(t_f)\}; \; (i = \overline{1, 2n}) \quad (22)$$

will give the results that meet the requirements of the optimal control problem.

The challenge is to determine the terminal boundary condition of the variable $p_i(t_f)$; $(i = \overline{1, 2n})$ when the motion of the mechanical system is subjected to a constraint (the holonomic constraints and the program constraints) and for the initial boundary condition $q_i(0)$, $(i = \overline{1, 2n})$ in the case of a non-holonomic system.

3 Illustrative Examples

Consider the steering motion of an automobile. The automobile is modeled in three main parts: the rear axle, the front axle, and the vehicle body. Neglecting the mass of the front, rear axles and car wheels, the mass of the vehicle body is m, the body center of mass lies on the centerline AB, the points A and B are located at the distances a and b from the center of mass C, respectively. The body moves in the horizontal plane, the center axis AB is inclined to the horizontal axis at an angle of φ, the inertia moment of the automobile about the center of mass C of the car is J_1. At the instance of time t, the car is moving such that the center axis AB is inclined to the x-axis at an angle of φ, the front wheels is steering around the vertical z axis by an angle of θ. The force F acts along the center axis AB. The inertial moment and steering torque of the steering wheel are J_2 and $M = M_0 - k\dot{\theta}$, respectively. Due to the non-holonomic constraints on the wheels, there exist reaction force $\mathbf{R_B}$ and $\mathbf{R_A}$ (combined from $\mathbf{R_{A1}}$ and $\mathbf{R_{A2}}$) as in Fig. 1. The challenge is to find minimum power of the steering torque while maintaining the non-holonomic constraints.

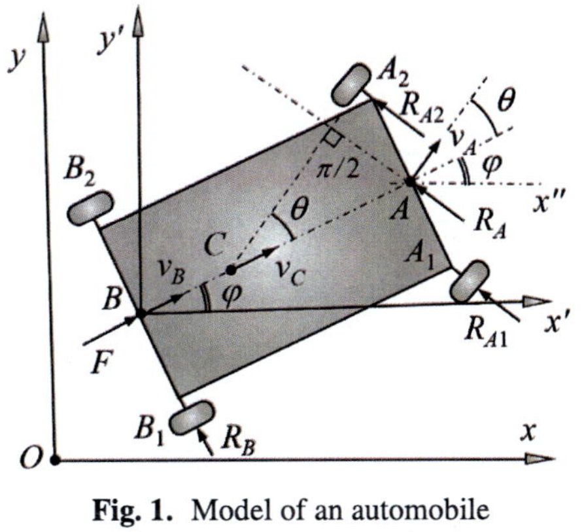

Fig. 1. Model of an automobile

Step 1. Select the generalized coordinates as coordinates of the mass center x_C, y_C, the inclined angle of the vehicle body φ, and the angle of steering wheel θ. The torque applied to the steering wheel is M. Select M_0 as the control parameter u. The kinetic energy expression is in the following form:

$$T = 0.5\,[m(\dot{x}_C^2 + \dot{y}_C^2) + J_1\dot{\varphi}^2 + J_2(\dot{\varphi} + \dot{\theta})^2] \tag{23}$$

The inertial matrix is a constant matrix in the form of:

$$\mathbf{A} = \begin{bmatrix} m & 0 & 0 & 0 \\ 0 & m & 0 & 0 \\ 0 & 0 & J_1 + J_2 & J_2 \\ 0 & 0 & 0 & J_2 \end{bmatrix} \tag{24}$$

Since the inertial matrix is a constant matrix, so the components of generalized forces of the inertial forces are zero, the generalized forces respect to the generalized coordinates $x_C, y_C, \varphi, \theta$ of the force F, and motor torque M are:

$$Q_1 = F\cos\varphi;\ Q_2 = F\sin\varphi;\ Q_3 = 0;\ Q_4 = u - k\dot{\theta} \tag{25}$$

When steering the wheels, the direction of the automobile changes an angle of θ. The rear wheel does not slip in a perpendicular direction to the center axis AB upon condition that the velocity of point B is along the center axis AB. Thus:

$$\dot{y}_B = \dot{x}_B \tan\varphi \tag{26}$$

Since the constraints in A and B consider as ideal constraints, the velocity of A must be perpendicular to the constraint force $\mathbf{R}_A$, i.e. meet the following condition:

$$\dot{y}_A = \dot{x}_A \tan(\varphi + \theta) \tag{27}$$

Thus, the following equations are obtained:

$$\begin{aligned} x_A &= x_C + a\cos\varphi;\quad y_A = y_C + a\sin\varphi; \\ x_B &= x_C - b\cos\varphi;\ y_B = y_C - b\sin\varphi \end{aligned} \tag{28}$$

Differentiate the above equations:

$$\begin{aligned} \dot{x}_A &= \dot{x}_C - a\sin\varphi\dot{\varphi};\quad \dot{y}_A = \dot{y}_C + a\cos\varphi\dot{\varphi}; \\ \dot{x}_B &= \dot{x}_C + b\sin\varphi\dot{\varphi};\quad \dot{y}_B = \dot{y}_C - b\cos\varphi\dot{\varphi}; \end{aligned} \tag{29}$$

Substitute Eq. (29) into Eqs. (26) and (27):

$$\begin{aligned} &\dot{x}_C\sin\varphi - \dot{y}_C\cos\varphi + b\dot{\varphi} = 0 \\ &\dot{x}_C\sin(\varphi + \theta) - \dot{y}_C\cos(\varphi + \theta) - a\cos\theta\dot{\theta} = 0 \end{aligned} \tag{30}$$

These are the motion constraints of the car which cannot be integral to obtain a system containing only generalized coordinates, thus it is not possible to remove the generalized velocities from the constraint equations (non-holonomic constraints). To derive the motion equations of the automobile, we will use Eq. (16) and Eq. (24) of matrix **A** and the generalized forces (25).

Differentiate Eq. (30), we have:

$$\begin{aligned}&\ddot{x}_C \sin\varphi - \ddot{y}_C \cos\varphi + b\ddot{\varphi} + \cos\varphi \dot{x}_C\dot{\varphi} + \sin\varphi \dot{y}_C\dot{\varphi} = 0;\\ &\ddot{x}_C \sin(\varphi+\theta) - \ddot{y}_C \cos(\varphi+\theta) - a\cos\theta\ddot{\varphi}\\ &+\cos(\varphi+\theta)(\dot{\varphi}+\dot{\theta})\dot{x}_C + \sin(\varphi+\theta)(\dot{\varphi}+\dot{\theta})\dot{y}_C + a\sin\theta\dot{\varphi}\dot{\theta} = 0;\end{aligned} \tag{31}$$

From there, matrix **D** of size (2 × 4) represents the dependent generalized accelerations $\ddot{x}_C, \ddot{y}_C$ respect to the independent generalized accelerations $\ddot{\varphi}, \ddot{\theta}$.

$$\begin{bmatrix} d_{13}\; d_{23}\; d_{33}\; d_{43} \\ d_{14}\; d_{24}\; d_{34}\; d_{44} \end{bmatrix} \tag{32}$$

Where:

$$\begin{aligned}d_{13} &= \frac{a\cos\varphi\cos\theta + b\cos(\varphi+\theta)}{\sin\theta};\, d_{23} = \frac{a\sin\varphi\cos\theta + b\cos(\varphi+\theta)}{\sin\theta};\, d_{33} = 1;\, d_{43} = 0;\\ d_{14} &= 0;\, d_{24} = 0;\, d_{34} = 0;\, d_{44} = 1;\end{aligned} \tag{33}$$

The motion equations of the automobile include two equations obtained from (16):

$$\begin{aligned}&d_{13}(m\ddot{x}_C - F\cos\varphi) + d_{23}(m\ddot{y}_C - F\sin\varphi) = 0;\\ &J_2\ddot{\theta} - M + k\dot{\theta} = 0;\end{aligned} \tag{34}$$

and two equations in the acceleration form (31). The result is a system of 4 motion equations of the automobile:

$$\begin{aligned}Eq1 &\equiv m(a\cos\varphi\cos\theta + b\cos(\varphi+\theta)\ddot{x}_C + m(a\sin\varphi\cos\theta + b\sin((\varphi+\theta)\ddot{y}_C\\ &\quad +(J_1+J_2)\sin\theta\ddot{\varphi} + J_2\sin\theta\ddot{\theta} - Fa + b)\cos\theta = 0;\\ Eq2 &\equiv J_2\ddot{\theta} - u + k\dot{\theta};\\ Eq3 &\equiv \ddot{x}_C\sin\varphi - \ddot{y}_C\cos\varphi + b\ddot{\varphi} + \cos\varphi\dot{x}_C\dot{\varphi} + \sin\varphi\dot{y}_C\dot{\varphi} = 0;\\ Eq4 &\equiv \ddot{x}_C\sin(\varphi+\theta) - \ddot{y}_C cos(\varphi+\theta) - a\cos\theta\ddot{\varphi}\\ &\quad +\cos(\varphi+\theta)(\dot{\varphi}+\dot{\theta})\dot{x}_C + \sin(\varphi+\theta)(\dot{\varphi}+\dot{\theta})\dot{y}_C + a\sin\theta\dot{\varphi}\dot{\theta} = 0;\end{aligned} \tag{35}$$

From the system of Eqs. (35), we can calculate the generalized accelerations $\ddot{x}_C, \ddot{y}_C, \ddot{\varphi}, \ddot{\theta}$ as a function of the generalized coordinates and the generalized velocities:

$$\begin{aligned}&\ddot{x}_C = \ddot{x}_C(x_C, y_C, \varphi, \theta, \dot{x}_C, \dot{y}_C, \dot{\varphi}, \dot{\theta});\; \ddot{y}_C = \ddot{y}_C(x_C, y_C, \varphi, \theta, \dot{x}_C, \dot{y}_C, \dot{\varphi}, \dot{\theta});\\ &\ddot{\varphi} = \ddot{\varphi}(x_C, y_C, \varphi, \theta, \dot{x}_C, \dot{y}_C, \dot{\varphi}, \dot{\theta});\; \ddot{\theta} = \ddot{\theta}(x_C, y_C, \varphi, \theta, \dot{x}_C, \dot{y}_C, \dot{\varphi}, \dot{\theta});\end{aligned} \tag{36}$$

Step 2. Optimal control problem:

The problem of interest is to determine torque $M = M_0 - k\dot{\theta}$ for minimum torque power when the automobile is in motion. Select control parameter is $u = M_0$, the objective functional has the following form:

$$I = \int_0^{tf} f_0 dt \to \min \tag{37}$$

Where function f_0 has the form of:

$$\begin{aligned} f_0 = 0.5\{[(u - k\dot{\theta})\,\dot{\theta}]^2 + [\dot{x}_C \sin(\varphi) - \dot{y}_C \cos(\varphi) + b\dot{\varphi}]^2 \\ +[\dot{x}_C \sin(\varphi + \theta) - \dot{y}_C \cos(\varphi + \theta) - a\cos(\theta)\dot{\theta}]^2\} \end{aligned} \tag{38}$$

The objective functional is supposed to minimize the power of steering torque and the non-holonomic constraints together. To solve this problem, we can use the Pontryagin's maximum principle [11]. Set the conjugate variables $\{q_i, p_i, z\}$; $(i = \overline{1, 8})$ where variables q_i $(i = \overline{1, 8})$ is converted from the following Lagrange variables:

$$q_1 = x_C, q_2 = y_C, q_3 = \varphi, q_4 = \theta, q_5 = \dot{x}_C, q_6 = \dot{y}_C, q_7 = \dot{\varphi}, q_8 = \dot{\theta} \tag{39}$$

and:

$$z = \int_0^t f_0 dt \tag{40}$$

The variables $p_i (i = \overline{1, 8})$ are determined from the function Hamilton H:

$$H = -f_0 + \sum_{i=1}^{8} f_i p_i \tag{41}$$

Where:

$$f_1 = \dot{x}_C; f_2 = \dot{y}_C, f_3 = \dot{\varphi}, f_4 = \dot{\theta}, f_5 = \ddot{x}_C, f_6 = \ddot{y}_C, f_7 = \ddot{\varphi}, f_8 = \ddot{\theta} \tag{42}$$

The quantities $\ddot{x}_C, \ddot{y}_C, \ddot{\varphi}, \ddot{\theta}$ are calculated from (35) where all quantities are expressed by Hamilton conjugate variables.

The control parameter u selected from the minimum condition of the objective functional (37) can be calculated from [12]:

$$\frac{\partial H}{\partial u} = 0 \tag{43}$$

Control parameter u calculated from (43) is called optimal control parameter (u_{opt}).

The automobile motion corresponding to the minimum of the objective functional (37) will be the root of the system of equations:

$$\frac{dq_i}{dt} = f_i; \quad (i = \overline{1, 8}); \ \frac{dp_i}{dt} = -\frac{\partial H}{\partial q_i}; \ (i = \overline{1, 8}); \ \frac{dz}{dt} = f_0 \tag{44}$$

after replacing control parameter $u = u_{opt}$ to the right sides of the system of Eqs. (44) and solving with the following boundary conditions

$$\left\{\begin{array}{l} q_1(0) = 0.5;\ q_2(0) = 0.375;\ q_3(0) = \arctan(0.75);\ q_4(0) = 0;\ q_5(0) = 4; \\ q_6(0) = 3;\ q_7(0) = 0;\ q_8(0) = 0.001;\ p_1(t_f) = h_{1f};\ p_2(t_f) = h_{2f}; \\ p_3(t_f) = h_{3f}, p_4(t_f) = h_{4f}, p_5(t_f) = 0, p_5(t_f) = 0, p_6(t_f) = 0, p_7(t_f) = 0; \\ p_8(t_f) = 0;\ z(0) = 0, G_f = 0 \end{array}\right\}$$

The boundary conditions respect to variables $p_i(i = \overline{1,8})$ where $p_i(t_f) = 0;\ (i = \overline{5,8})$ and $p_i(t_f);\ (i = \overline{1,4})$ is calculated by using constraint equations to form the function G

$$G = G_1 + \lambda G_2 \tag{45}$$

The quantities $h_i(i = \overline{1,4})$ and G_f will be calculated as follows:

$$G_f = G(t_f);\quad h_i = \frac{\partial G}{\partial q_i};\quad (i = \overline{1,4})\quad h_{if} = h_i(t_f);\quad i = \overline{1,4} \tag{46}$$

Where:

$$\begin{aligned} G_1 &\equiv \dot{x}_C \sin\varphi - \dot{y}_C \cos\varphi + b\dot{\varphi} = 0 \\ G_2 &\equiv \dot{x}_C \sin(\varphi + \theta) - \dot{y}_C \cos(\varphi + \theta) - a\cos\theta\dot{\theta} = 0 \end{aligned} \tag{47}$$

The numerical simulation of the problem is carried out based on the following data:

$m = 25.10^3$(kg); $J_1 = 45.10^3$(kgm^2); $J_2 = 10.10^3$(kgm^2); $a = 2.25$(m); $b = 0.5$(m); $F = 150.10^3$(N); $k = 15$ (Nms/rad); $t_f = 1.5$(s)

Simulation Result (Figs. 2, 3, 5, 6, 7 and 8):

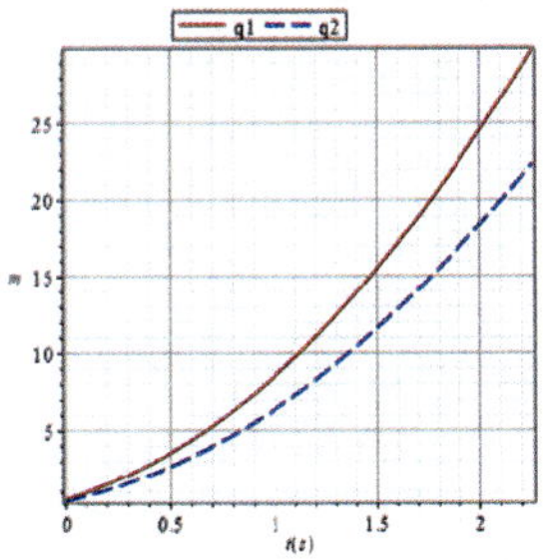

Fig. 2. Generalized coordinates q1, and q2

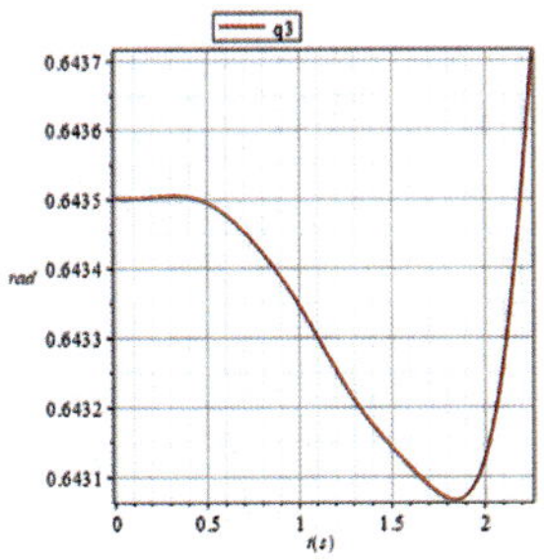

Fig. 3. Generalized coordinate q3

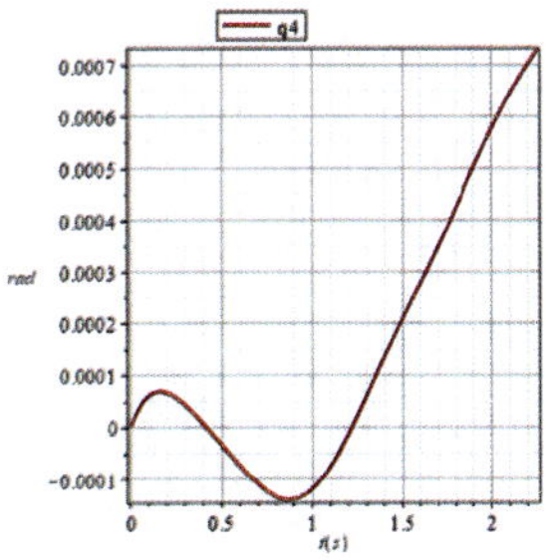

Fig. 4. Generalized coordinate q4

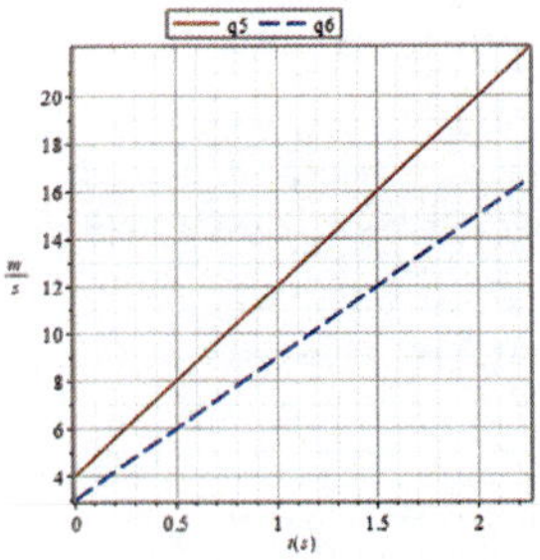

Fig. 5. Generalized velocities q5, and q6

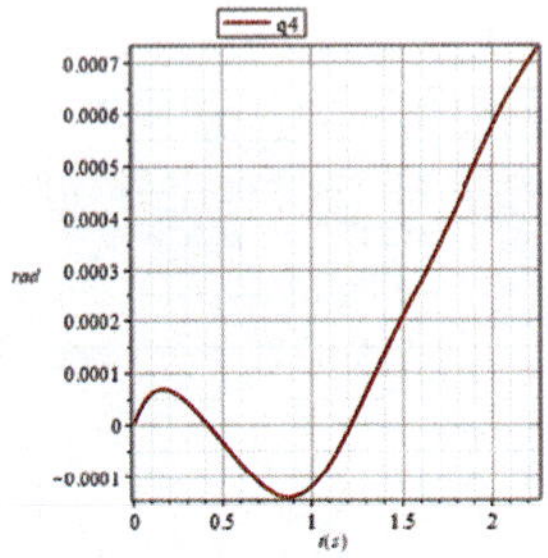

Fig. 6. Generalized coordinate q4

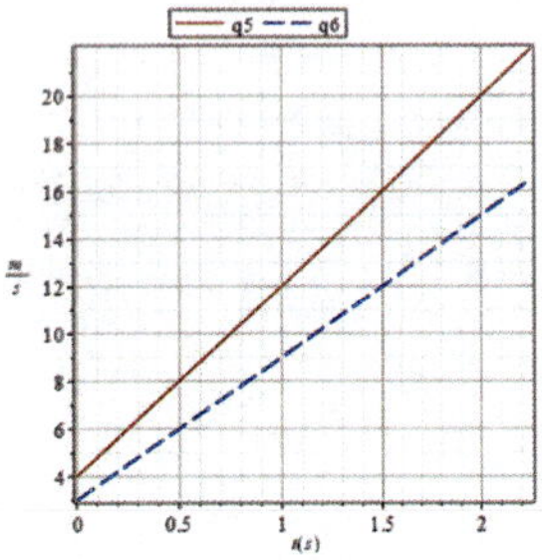

Fig. 7. Generalized velocities q5, and q6

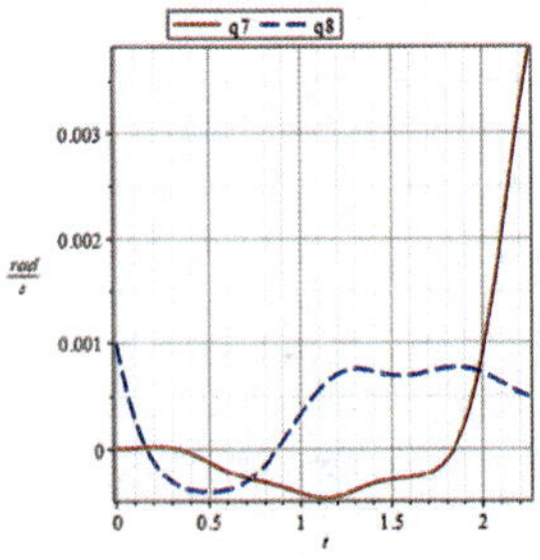

Fig. 8. Generalized velocities q7, and q8

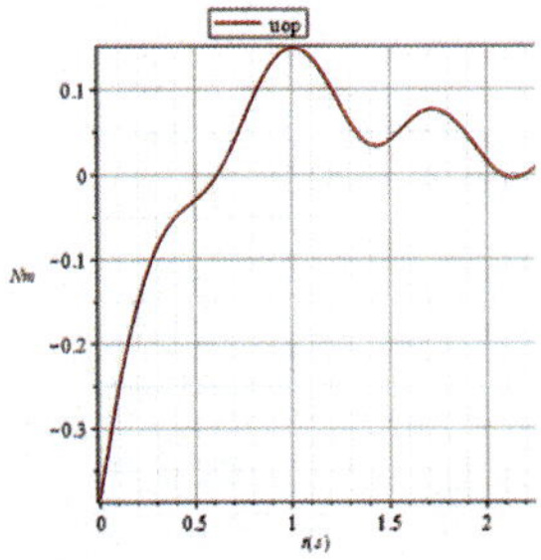

Fig. 9. Optimal control u(t)

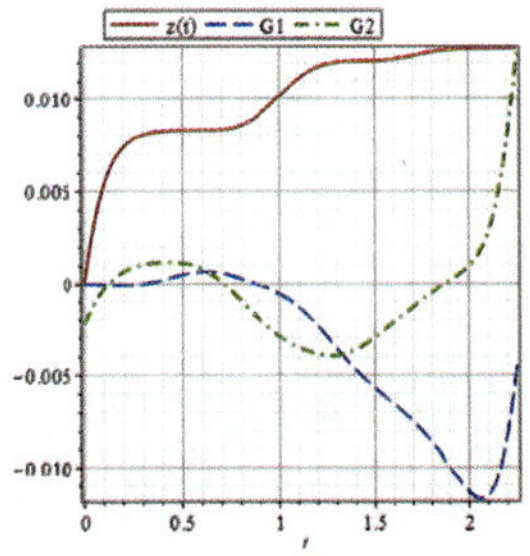

Fig. 10. Cost and constraints functions

From the simulation results, it is obviously seen that the generalized coordinates and velocities satisfy the non-holonomic constraints G1 and G2 while keeping low power of steering torque u(t) as shown in Fig. 10. The optimal control u(t) is also bounded and converges to zero for the simulation time (Fig. 9). The only disadvantage is the steering angle q_4 is just varying around a small value (Fig. 4) which might be due to the dynamic parameters such as thrusting force F kept constant. However, the proposed method is already proved to be applicable for optimal control problems of mechanical system of non-holonomic constraints.

4 Conclusion

The paper proposed a method to study the optimal control of non-holonomic linear mechanical systems. As is known, it will be difficult to use the Pontryagin's principle if mechanical system is subjected to position constraints (e.g. mechanical systems with holonomic constraints, program constraints) when determining boundary conditions for the conjugate variables at the end of the trajectory $p_j(t_f)$; $(j = \overline{1, 2n})$. In case of non-holonomic systems, it is also difficult to determine the initial condition for Lagrange variables $q_{i,}(0), \dot{q}_i(0)$; $(i = \overline{1, n})$ because of non-integrable constraints. To overcome the above difficulties, the ordinary equations of motion in matrix form are built by the principle of compatibility and the method of transformation matrix to avoid using

Lagrange multipliers. For illustration, the optimal control of the steering motion of an automobile is studied. The simulation results prove the credibility of the proposed method.

References

1. Chettibi, T., Lehtihet, H.E., Haddad, M., Hanchi, S.: Minimum cost trajectory planning for industrial robots. Eur. J. Mech. A/Solids **23**, 703–715 (2004)
2. Jeremic, B., Radulovic, R., Obradovic, A.: Analysis of the brachistochronic motion of a variable mass nonholonomic mechanical system. Theoret. Appl. Mech. **43**(1), 19–32 (2016)
3. Unda, J., Garcia de Jalon, J., Losantos, F., Emparantza, R.: A comparative study on some different formulations of the dynamic equations of constrained mechanical system. In: ASME Mechanisms Conference, Columbus, OH (1986)
4. Sanh, D.: On the principle of compatibility and equations of motion of constrained. In: Mechanical Systems ZAMM, Berlin, vol. 60, pp. 210–212 (1980)
5. Sanh, D.: On the motion of controlled mechanical systems. In: Advances in Mechanics, Varsawa, t.7, V. 2, pp. 3–24 (1984)
6. Sanh, D.: On the motion of constrained mechanisms, Thesis of Doctorate in Science, Hanoi University of Science and Technology, Vietnam (1984)
7. Kovecses, J., Piedboeuf, J., Lange, C.: Dynamics modeling and simulation of constrained robotic systems. IEEE/ASME Trans. Mechatron. **8**(2), 165–177 (2003)
8. Le, N.N., Phong, D.V., Sanh, D.: On numerical methods for constrained mechanical systems. In: Iutam Symposium on Recent Developments in Nonlinear Oscillations of Mechanical Systems. IUTAM Kluwer Academic Publishers (1999)
9. Do, S., Do, K.D.: Method of transmission matrix applying for investigation of motion of planar mechanism. Mach. Dyn. Res. **34**(40), 5–22 (2010)
10. Sanh, D., Phong, D.V., Khoa, D.D., Duc, T.: A method for solving the motion of constrained systems. In: Proceedings of the 16th Asian Pacific Vibration Conference, Hanoi, Vietnam, pp. 803–811 (2015)
11. Pontryagin, L.S., et al.: The mathematical theory of optimal processes. In: Triogoff, K. (ed.) Interscience Publishers, John Wiley and Sons. Inc, New York (1962)
12. Lewis, F.L., Vrabie, D., Syrmos, V.L.: Optimal Control, 3rd edn. Wiley, Hoboken (2012)

Design and Implementation of a Door Latch Unlocking Mechanism

Che-Min Lin[1], Yu-Hsun Chen[1(✉)], and Kuan-Lun Hsu[2(✉)]

[1] Department of Mechanical Engineering, National Taiwan University of Science and Technology, Taipei, Taiwan
{B10703218,yhchen}@mail.ntust.edu.tw
[2] Department of Mechanical Engineering, National Taiwan University, Taipei, Taiwan
kuanlunhsu@ntu.edu.tw

Abstract. In recent years, a variety of identification systems and unlocking methods have been developed for electronic locks. However, a signal failure or power outage may lead to the malfunction of the electronic lock. Therefore, this research aims at developing a device that can process electronic signals while retaining the physical features for locking/unlocking function. Firstly, this work surveys commercially used unlocking methods and characterizes suitable motion requirements for unlocking mechanisms. Next, a mechanism with variable topology composed of gears, linkage, and a friction ring is provided. By adjusting the number of the grippers and the location of the latches, the presented mechanism can be compatible with different door locking systems. The presented device can be installed on traditional keyed door locks and integrated with electronic locks. The flexibility of the door latch unlocking mechanism is verified theoretically and digitally. Besides, the prototyping of the presented design is implemented through a 3D printing process.

Keywords: Mechanism with variable topology · Kinematic analysis · Computer-aid simulation

1 Introduction

A lock is a device for protecting personal property and safety by restricting people's access to a specific space [1]. The traditional locks with keys are easy to operate and highly reliable, although they cannot work if the keys aren't well preserved by the owner. In recent years, with the development of technology, many convenient unlocking methods without keys are invented, such as entering a password, remote control through a smartphone, or biometric identification such as fingerprints, iris, and facial features. These innovative devices are generally composed of a sensor, a controlling system, and a motor. However, once the power supply or the control server fails, the lock function will subsequently become invalid. Therefore, this research proposes an accessory unlocking mechanism that can be unlocked through both the original key and extra modern methods. It can be installed on most types of conventional doors, such as the doorknob locks and

N. V. Khang et al. (Eds.): ASIAN MMS 2021, MMS 113, pp. 134–141, 2022.
https://doi.org/10.1007/978-3-030-91892-7_13

handlesets shown in Fig. 1. This innovative unlocking mechanism is powered by a motor, and a linkage is driven to rotate the original door lock and then unlatch it. In addition, the presented mechanism also preserves the keyed entry knob in case of an unexpected signal failure or a power outage. The locks can still be unlatched by a key without any electric supplementary.

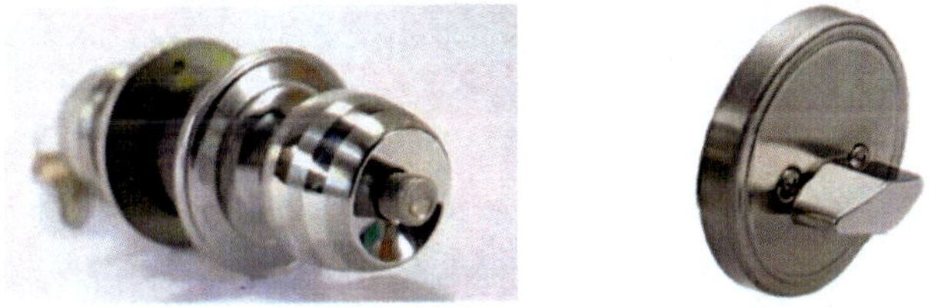

(a) Doorknob locks (b) Handlesets

Fig. 1. Two common types of door lock [2]

2 Door Latch Unlocking Mechanism

Based on a preliminary literature survey, we conclude that the unlocking mechanism is required to achieve both gripping and rotational motions so as to fit in the motion requirements of commercially used door latch mechanisms, including knob locks, cam locks, deadbolt lock types, and mortise locks.

The metamorphic mechanism is defined by three characteristics in multi-stage, multi-topological structure, and multi-degree of freedom [3–4]. Attempting to meet the aforementioned specifications, we propose a design solution as presented in Fig. 2. The motions produced by the unlocking mechanism can be separated into two stages, with different topological structures. At the first motion stage, the gripping motion is achieved by the four-bar linkage, as the mechanism sketch in Fig. 3(a). A friction ring serves as a circular frame (link 1, K_F). The pinion is driven by a motor, and the gear shield (link 2, K_P) can be equivalent to a sliding block traveling along a curved guide (link 1). A connecting link (link 3, K_L) is adjacent to the gear shield (link 2), and a gripper (link 4, K_G) is adjacent to the connecting link (link 3) and the ring (link 1) respectively.

Once the gripper clamps the knob, the driving force transmitted by the gear shield (link 2) is greater than the friction between the frame and the ring (link 1). Therefore, the relative movements within the four links (1, 2, 3, 4) are restricted, which is equivalent to one link (link 2') that rotates around the axis of the knob, as Fig. 3(b) shows. Consequently, at the second motion stage, a continual rotation of the gear shield can result in the rotation of the knob and the unlocking function. The topology structures [3] of the two stages can be expressed as the matrixes (1).

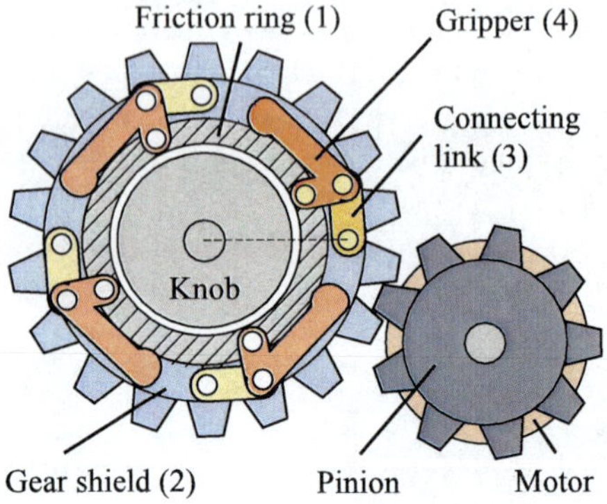

Fig. 2. Design concept

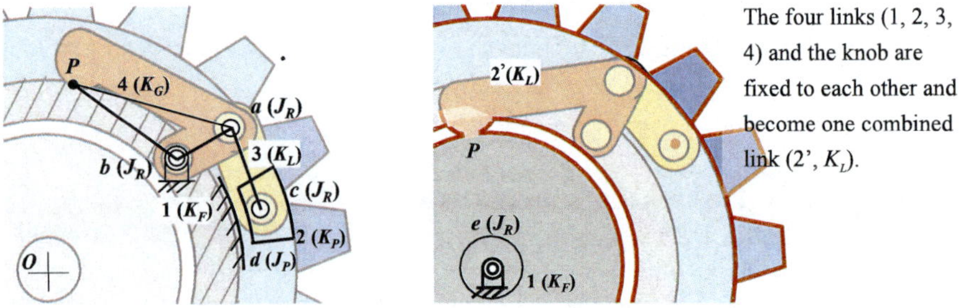

(a) The four-bar linkage at the 1st stage (b) The equivalent mechanism at the 2nd stage

Fig. 3. Mechanism sketch

$$\begin{bmatrix} K_F & P & 0 & R \\ d & K_P & R & 0 \\ 0 & c & K_L & R \\ b & 0 & a & K_G \end{bmatrix};\begin{bmatrix} K_F & R \\ e & K_L \end{bmatrix} \tag{1}$$

3 Kinematic and Dynamic Performance

The motion in the first stage is analyzed based on the vector loop method. As Fig. 4(a) shows, the position analysis of the gripper can be expressed as:

$$r_1\begin{bmatrix} \cos\theta_1 \\ \sin\theta_1 \end{bmatrix} + r_4\begin{bmatrix} \cos\theta_4 \\ \sin\theta_4 \end{bmatrix} = r_2\begin{bmatrix} \cos\theta_2 \\ \sin\theta_2 \end{bmatrix} + r_3\begin{bmatrix} \cos\theta_3 \\ \sin\theta_3 \end{bmatrix} \tag{2}$$

Then, by combining coefficient θ_3, parameter θ_4 can be solved as Eq. (3), where $\mathrm{A} = 2r_1r_4\cos\theta_1 - 2r_2r_4\cos\theta_2$, $\mathrm{B} = 2r_1r_4\sin\theta_1 - 2r_2r_4\sin\theta_2$, and $\mathrm{C} = r_1^2 + r_2^2 + r_4^2 - r_3^2 - 2r_1r_2\cos(\theta_1 - \theta_2)$. Therefore, parameter θ_3 can be solved as Eq. (4).

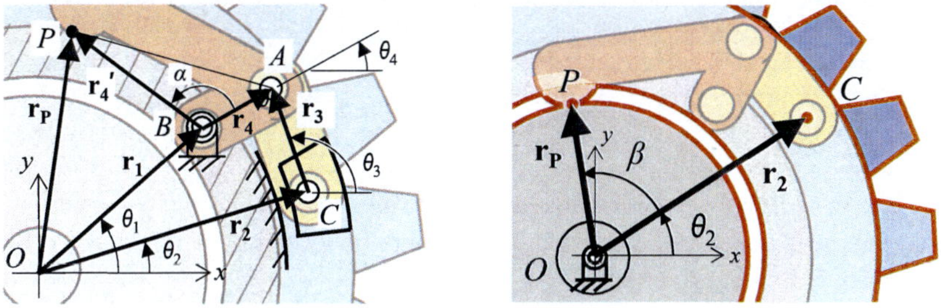

(a) The 1st stage (b) The 2nd stage

Fig. 4. Vector loop

$$\theta_4 = 2tan^{-1}(\frac{-B \pm \sqrt{B^2 - C^2 + A^2}}{C - A}) \tag{3}$$

$$\theta_3 = tan^{-1}\left(\frac{r_2 sin\theta_2 + r_1 sin\theta_1 - r_4 sin\theta_4}{r_2 cos\theta_2 + r_1 cos\theta_1 - r_4 cos\theta_4}\right) \tag{4}$$

Once θ_4 is determined, the gripper trajectory can be calculated by the position of point P, as expression of Eq. (5).

$$\mathbf{r_P} = \mathbf{r}_1 + \mathbf{r}_4$$

$$\begin{bmatrix} X_P \\ Y_P \end{bmatrix} = r_1 \begin{bmatrix} \cos\theta_1 \\ \sin\theta_1 \end{bmatrix} + r_4' \begin{bmatrix} \cos(\theta_4 + \alpha) \\ \sin(\theta_4 + \alpha) \end{bmatrix} \tag{5}$$

At the 2nd stage, Fig. 4(b), the gripper clamps the knob and rotates around the axis of the knob continuously. The trajectory of point P can be analyzed as Eqs. (6), where r_P and β are constant which can be determined by the final position of the gripper at the 1st stage.

$$\begin{bmatrix} X_P' \\ Y_P' \end{bmatrix} = r_P \begin{bmatrix} \cos(\theta_2 + \beta) \\ \sin(\theta_2 + \beta) \end{bmatrix} \tag{6}$$

The velocity can be analyzed as Eqs. (7). Since link 1 doesn't move in the first stage, angular velocity ω_1 equals 0. The angular velocity of the output link can be solved as Eq. (8).

$$r_1\omega_1 \begin{bmatrix} -\sin\theta_1 \\ \cos\theta_1 \end{bmatrix} + r_4\omega_4 \begin{bmatrix} -\sin\theta_4 \\ \cos\theta_4 \end{bmatrix} = r_2\omega_2 \begin{bmatrix} -\sin\theta_2 \\ \cos\theta_2 \end{bmatrix} + r_3\omega_3 \begin{bmatrix} -\sin\theta_3 \\ \cos\theta_3 \end{bmatrix} \tag{7}$$

$$\omega_4 = \omega_2 \cdot \frac{-r_2 r_3 sin\theta_2 cos\theta_3 + r_2 r_3 sin\theta_3 cos\theta_2}{-r_3 r_4 sin\theta_3 cos\theta_4 + r_3 r_4 sin\theta_4 cos\theta_3} \tag{8}$$

At the first motion stage, the forces applied to the mechanism can be analyzed as Fig. 5(a), where $\mathbf{T_2}$ is the driving torque transmitted by the gears, $\mathbf{T_4}$ is the output torque of the gripper around the pivot point B, $\mathbf{R_B}$ and $\mathbf{R_C}$ are the supporting forces from the fixed links. All of the links rotate slowly, so it can be analyzed as a static problem. According to energy method, output torque $\mathbf{T_4}$ is expressed as Eq. (9), assuming that the weight of the links can be neglected. supporting forces $\mathbf{R_B}$ and $\mathbf{R_C}$ do not generate work or energy, because point B doesn't move and the direction of $\mathbf{R_C}$ is perpendicular to the movement direction of point C.

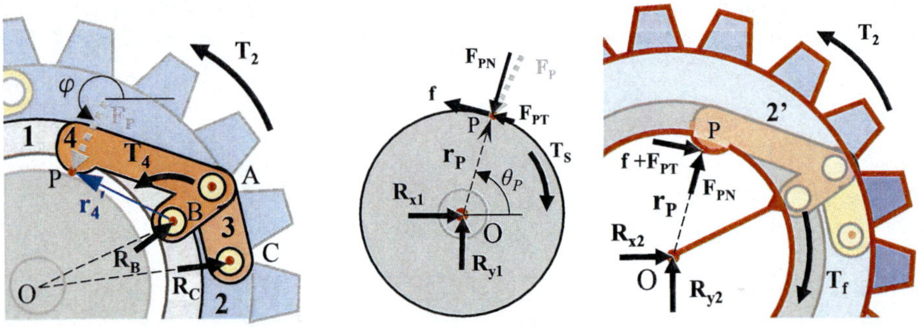

(a) Mechanism at the 1st stage (b1) Knob at the 2nd stage (b2) Gripper at the 2nd stage

Fig. 5. Force analysis

$$\mathbf{T}_4 \cdot \boldsymbol{\omega}_4 = \mathbf{T}_2 \cdot \boldsymbol{\omega}_2 \tag{9}$$

Therefore, the clamping force at point P is determined as Eq. (10). The direction of $\mathbf{F_P}$ is perpendicular to vector $\mathbf{r_4}$', where the angle $\varphi = \theta_4 + \alpha + \pi/2$.

$$\mathbf{F_P} = \frac{\mathbf{T}_4}{\mathbf{r}_4{}'} \tag{10}$$

At the second motion stage, the force analysis is presented in Fig. 5(b$_1$)–(b$_2$). $\mathbf{F_{PN}}$ and $\mathbf{F_{PT}}$ are the projections of the force $\mathbf{F_P}$ in the tangent and normal directions, $\mathbf{f}$ is the friction force between the gripper and the knob, as presented in Eq. (11), where μ_P is the friction coefficient between the two surfaces.

$$\mathbf{f} = \mathbf{F_{PN}} \cdot \mu_P \tag{11}$$

Moreover, $\mathbf{R_{x1}}$, $\mathbf{R_{y1}}$, $\mathbf{R_{x2}}$ and $\mathbf{R_{y2}}$ are the supporting forces from the fixed link. $\mathbf{T_S}$ is the resistance force of the inner spring of the door handle, so the torque to turn the door handle must be greater than $\mathbf{T_S}$, inequality (12). $\mathbf{T_f}$ is the friction torque between the friction ring and the frame. At the 1st stage, the friction $\mathbf{T_f}$ must be large enough to keep the friction ring stationary. At the 2nd stage, the knob can be rotated to unlock the door, so the friction torque $\mathbf{T_f}$ must satisfy the inequality (13).

$$r_P(f + F_{PT}) > T_S \tag{12}$$

$$T_f < T_2 - r_P(f + F_{PT}) \tag{13}$$

For the unlocking mechanism with variable topology proposed in this study, the key component is the friction ring that cannot be turned at the 1st motion stage but is rotated at the 2nd stage. This device is designed with considerations of the parameters related to the friction torque $\mathbf{T_f}$, including the friction coefficient of the material, dimensions, and the assembly preload of the friction ring.

4 Numerical Results

Through the dimensional design, the parameters of the vectors in Fig. 4(a) are shown in Table 1. For unlocking a doorknob, the friction ring is set at the position $\theta_1 = 30$ deg at the 1st stage. The input motion θ_2 is from 10° to 15° at the 1st stage and from 15° to 30° at the 2nd stage. As the result, the trajectory of point P of the unlock mechanism is generated through Eqs. (2)–(6) and a simulation by the software SolidWorks, Fig. 6. The motion path starts at the upper right position, travels along a curve to clamp, and then moves along a short curve around the knob to rotate and unlock it.

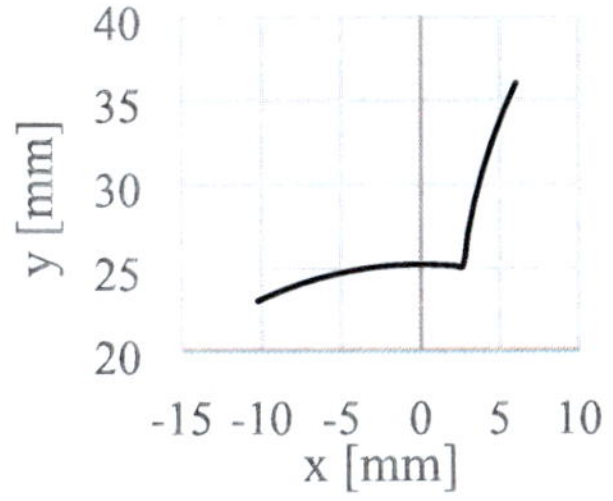

(a) Theoretical result

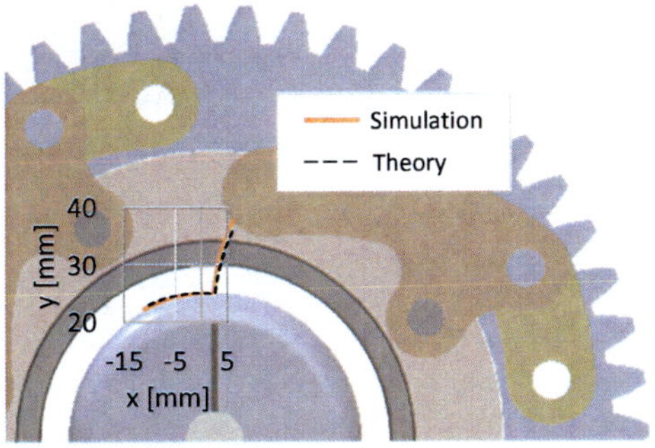

(b) Comparison of theoretical and simulated results

Fig. 6. Trajectory of point P

Table 1. Dimensional parameters

Symbol	r_1 [mm]	r_2 [mm]	r_3 [mm]	r_4 [mm]	r_4' [mm]	α [deg]
Value	41.56	58	20	18.78	33.6	128.3

A DC motor is used as a power source that provides the input speed of 2 rpm and input torque of 15 Nm. The number of teeth of the pinion and gear is 21 and 51, respectively. With the transmission of gears, the angular velocity and torque at gear shied are $\omega_2 =$ 4.94 deg/s and $T_2 = 36.43$ Nm. At the final position of the 1st stage ($\theta_2 = 15°$), the angular velocity of the gripper is determined $\omega_4 = 16.80$ deg/s through Eq. (8), and the clamping force at point P can be solved through Eqs. (9)–(10). A rubber sheet with friction coefficient of 0.5 is used for the gripper, so the friction is solved through Eq. (12). As the force parameters listed in Table 2, friction torque T_f must be less than 32.2640 Nm based on Eq. (13). Therefore, a rubber ring with a friction coefficient of 0.3 is used to make the friction ring, with a radius of 60 mm and an assembly preload.

Table 2. Force analysis

Symbol	T_2 [Nm]	F_P [N]	F_{PN} [N]	F_{PT} [N]	f [N]	T_f [Nm]
Value	36.43	318.89	318.80	7.18	159.40	32.26

For the handleset with a knob in Fig. 1(b), it also can be unlocked by the proposed mechanism. The angular position is set $\theta_1 = 75°$ and $\theta_2 = 56.8°$ at the initial position (Fig. 7(a)), $\theta_2 = 66.67°$ when it clamps (Fig. 7(b)), and θ_1 increases to 69° with the rotation of the gear shield to unlock the handleset (Fig. 7(c)). A prototype is made by 3D printing, and the feasibility is verified by the testing with a doorknob, as Fig. 8 shows.

(a) Initial position (b) Clamping position (c) Unlock the handleset

Fig. 7. 3D model and simulation for handleset

(a) Initial position (b) Clamping position (c) Unlock the doorknob

Fig. 8. Prototype

5 Conclusions

This research provides an accessory unlocking mechanism that can be integrated with many types of conventional doors. The presented mechanism with variable topology can retain the physical features for locking/unlocking function to avoid any signal failure or power outage of electronic locks. The flexibility of the design concept is verified theoretically, digitally, and experimentally. In the future, other potential design concepts addressed in this work will be further investigated. In addition, a face recognition and a remote control can be incorporated into this mechanism for the sake of convenience.

Acknowledgment. The authors are grateful to the Ministry of Science and Technology (Taipei, Taiwan) under Grant MOST 108–2218-E-011–031-MY2 for the support of this work.

References

1. Shi, K., Hsiao, K.H., Zhao, Y., Huang, C.F., Xiong, W.Y.: Structural analysis of ancient Chinese wooden locks. Mech. Mach. Theory **146**, 103741 (2020)
2. Master Lock official website. https://www.masterlock.eu/. Accessed 19 Apr 2021
3. Yan, H.S., Kuo, C.H.: Topological representations and characteristics of variable kinematic joints. ASME J. Mech. Design **128**(2), 384–391 (2006)
4. Dai, J.S., Rees Jones, J.: Matrix representation of topological changes in metamorphic mechanisms. ASME J. Mech. Design **127**(4), 837–840 (2005)

Gearing and Transmissions

An Expression Method of Kinematic and Structure Diagrams for Planetary Gear Systems

Vu Le Huy[1,2](✉) and Do Duc Nam[3]

[1] Faculty of Mechanical Engineering and Mechatronics, PHENIKAA University, Hanoi 12116, Vietnam
huy.vule@phenikaa-uni.edu.vn

[2] A&A Green Phoenix Group JSC, PHENIKAA Research and Technology Institute (PRATI), No.167 Hoang Ngan, Trung Hoa, Cau Giay, Hanoi 11313, Vietnam

[3] National Institute of Patent and Technology Exploitation, No. 39, Tran Hung Dao, Hoan Kiem, Hanoi, Vietnam

Abstract. Transmission system is the important part of every machine in industry, where planetary gear sets are used for the requirements of large speed ratio as well as small size. For those requirements, planetary gear systems, which are combined from at least the two planetary gear sets, have been being used popular. However, many expression methods to present the kinematics and structure of planetary gear systems are still being studied. This paper proposes a method of demonstration and notation for kinematic and structure diagrams with the advantages in structure analysis and kinematic calculation of the complex planetary gear systems.

Keywords: Structure · Kinematic · Diagram · Planetary · Gear System

1 Introduction

Machines, especially in industry, generally have power system containing power source and reducers such as belt, chain drives as well as gear box to obtain the large output torque. It is known that the output torque is proportional to the speed ratio which is the ratio of the input shaft speed to the output shaft speed, but it is inversely proportional to the size. Therefore, one requirement of the power systems is to have large ratio but with small size such as robotics, aerospace, aircraft, agriculture, drilling, medical applications, etc. Planetary gear transmission is one of the drives to be suitable for that requirement. One basic planetary mechanism sometimes does not satisfy the demand of transmission leading to the planetary gear drives joined together to become the systems which having larger ratio ranges, difference of kinematics, efficiency, and degree of freedom. The comparison and finding out one optimal combination are not simple tasks, so it needs to carefully explore joining methods, connection analyzing and relationship between mechanisms in the system. Besides, determination of the arrangement law, symbols and the organization of these structures is also considered.

N. V. Khang et al. (Eds.): ASIAN MMS 2021, MMS 113, pp. 145–156, 2022.
https://doi.org/10.1007/978-3-030-91892-7_14

There are many studies proposing the expression and analysis method of kinematic and structure diagrams of planetary gear systems. Structure diagram of planetary system is the simplest kind to present the system used to study the links between the planetary gear sets, while kinematics diagram presents fully and detailly the system in motion of the links. In kinematics diagram, the links (the sun gear, the ring gear, and the arm), the joining methods and the engaging of gears can be seen. From one structure diagram, many kinematics diagrams can be established. Early, the planetary gear mechanism as a part of gear chains with other mechanisms in the study of synthesis of kinematic structures [1], where the network concepts and combinatorial analysis were used to develop a method for kinematic structural classification and enumeration of the mechanisms. There method presented the structure diagrams with each element as a polygon, whose vertices represent kinematic pairs, while the gear pairs and turning-pair vertices are represented by the solid dots and hollow dots, respectively. This method was used to present the kinematic structures in the mechanism synthesis problem; however, it is difficult to understand the links between the mechanisms, especially with the planetary gear sets. It was used similarly for kinematic synthesis of cam-controlled planetary gear trains [2] but stopped at a single planetary gear set. Later, the books [3, 4] presented the structure chains of one planetary mechanism by one circle with the three basic links notated by the lines with the letters. However, there was no presentation method for the systems with multiple planetary gear sets. This method was used in the study of Gao [5] but the end of each link presented by a filled circle to perform kinematic analysis based on topology, which was called as topological graph. The topological graphs spread out on a large space with many nodes even for single planetary gear set. Recently, kinematic relationship of planetary systems was analyzed via block diagram by employing control techniques [6, 7]. The power-oriented graph, which is a graphical model technique, was used for the dynamic modeling of planetary gear sets, nevertheless, it is too complicated to present the structure diagrams. In the study of Wang [8], a hypergraph-based method was used to present the power flow and efficiency analysis of planetary transmission, but it was applied only for one single mechanism and a double-stage herringbone planetary gear transmission system. However, the methods for kinematic and structure diagrams in those studies were presented for the systems with one or two planetary gear sets, but there is no conventional method for the combination of more the planetary gear sets. This paper proposes a method of demonstration and notation for kinematic and structure diagrams by using the idea with the method for structure chains in the books [3, 4] but a different arrangement to easy present and understand the diagrams. The presented method here is applicable in structure and kinematic analysis for the complex planetary gear systems even with more degrees of freedom.

2 Demonstration of Planetary Gear Set and Systems

2.1 Planetary Gear Sets

Basically, a planetary gear set is expressed by structure sketch as shown in Fig. 1(a) by a circle and three lines, where the three links, i.e., the sun gear, the ring gear and the arm, are shown by the three lines and notated as k_1, k_2, k_3. The planetary gear set with the two outputs, i.e., the number of degrees of freedom (DOF) w is 2, is called as differential mechanism, and its symbol is v. If the DOF of the set is 1, the planetary gear set is called as planetary mechanism and symbolled by h. The symbol v or h is put in the circle of the structure sketches, and therefore this method is named as vh method.

When the links in a structure sketch are indicated as the input, the output and the fixed link, then structure diagram is obtained as Fig. 1(b). In this study, the input or active link of a mechanism is notated as A, the output or passive link of a mechanism is notated as B, and the fixed link is notated as F. When F is closed, the planetary gear set has only one transmission ($w = 1$) which is the planetary mechanism, and its structure diagram is presented in Fig. 1(c). From a structure diagram, some kinematic diagrams could be estimated corresponding to the links indicated exactly to be the sun gear (notated by the number 1), the ring gear (notated by the number 3) and the arm (notated by the number 0). A kinematic diagram of the planetary mechanism with the structure diagram in Fig. 1(c) is illustrated in Fig. 1(d), where the links, the joining methods and also the engaging of gears could be seen. In the kinematic diagrams, the planetary gears are numbered as 2. The number of kinematic diagrams obtained from a structure diagram of the planetary mechanism is 6 as maximum.

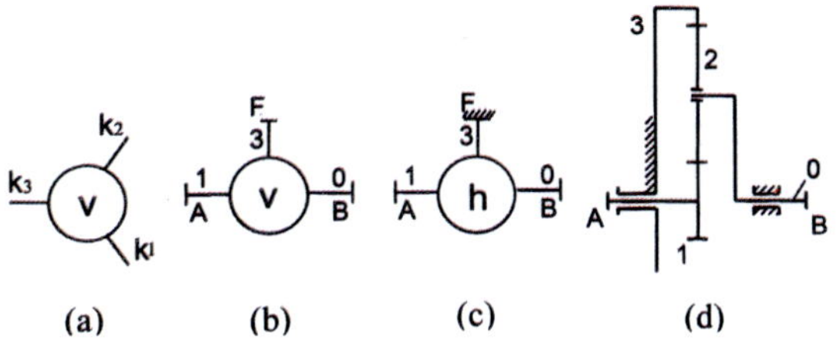

Fig. 1. Planetary gear set expressed with (a) Structure sketch, (b, c) Structure diagrams and (d) Kinematics diagram.

2.2 Planetary Gear Systems

By joining differential mechanisms, a differential system could be obtained with DOF $w > 1$. In this system, each differential mechanism is notated as v_i, where i is ordinal number of mechanisms in the system and has the value from 1 to the number of mechanisms n_M.

Structure sketch of a planetary system is joined from two or more mechanisms as shown in Fig. 2. Differential mechanism is placed in each top of basic geometric figures such as the lozenge for the system with 2 mechanisms in Fig. 2(a) and the hexagon for the system with 3 mechanisms in Fig. 2(b). For the cases of the systems with over than 3 mechanisms, their structure sketches are expressed by connecting the basic geometries. Figure 2(c) shows an example of a structure sketch of a system combined from 4 differential mechanisms. The mechanisms and their links in structure sketches are numbered following ordinal in the rules from bottom to top, and from left to right. By using the proposed method, the structure sketches of the systems with DOF of 2 were established as shown in Table 1, where the number of differential mechanisms in the systems is from 1 to 4.

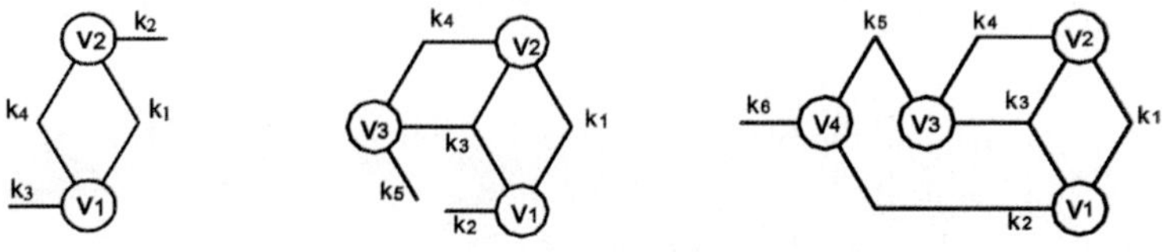

(a) (b) (c)

Fig. 2. Structure sketch of the planetary gear systems with (a) two, (b) three and (c) four differential mechanisms.

Table 1. Structure sketches of 2 DOF systems.

No.	Structure sketch	No.	Structure sketch	No.	Structure sketch
1	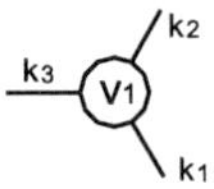	7	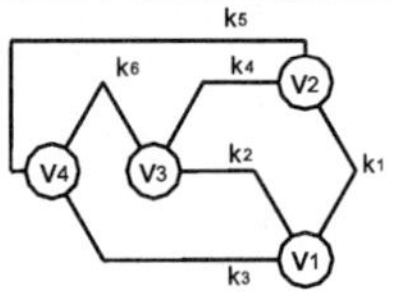	13	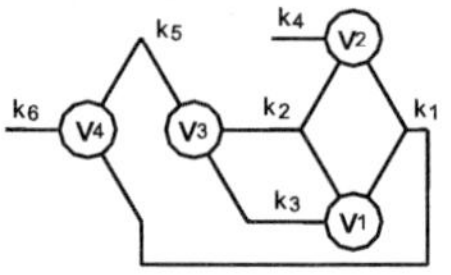
2	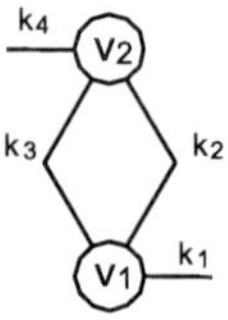	8	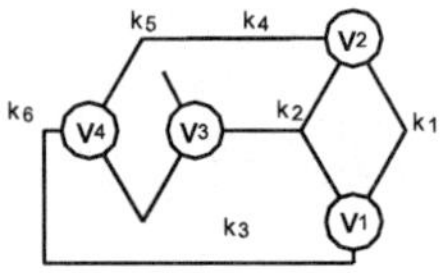	14	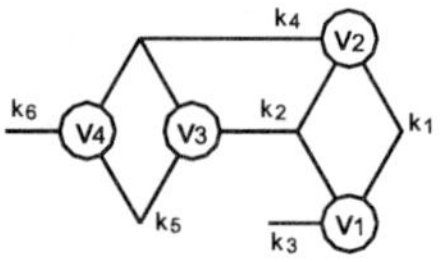
3	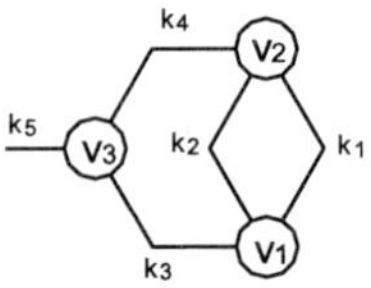	9	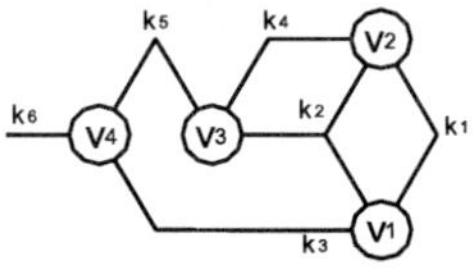	15	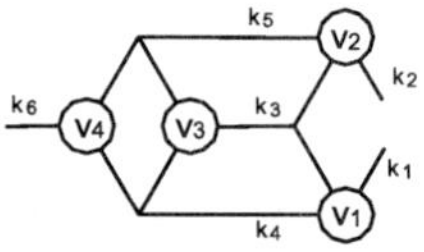
4	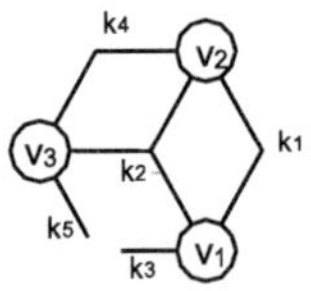	10	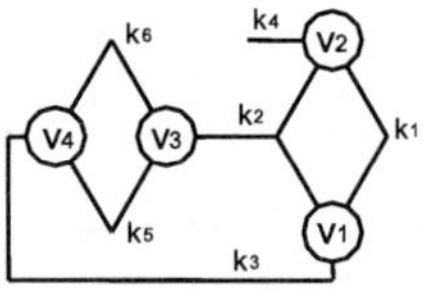	16	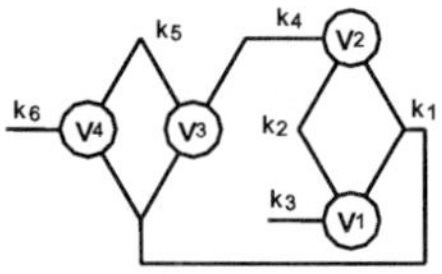
5	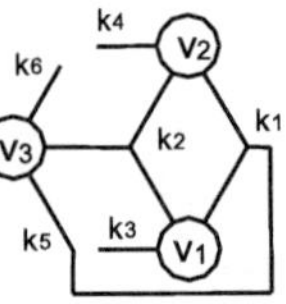	11	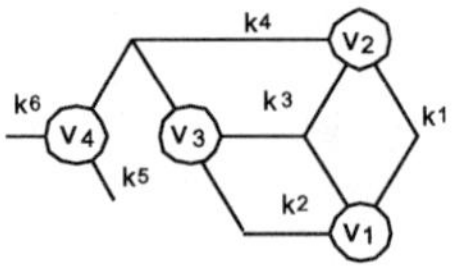	17	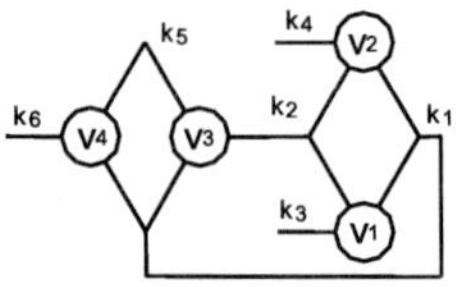
6	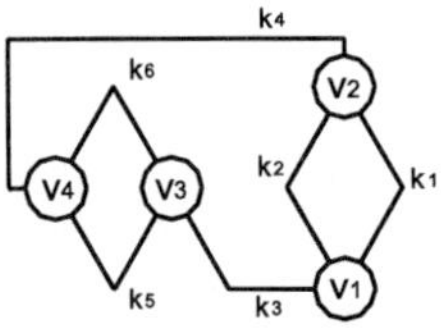	12	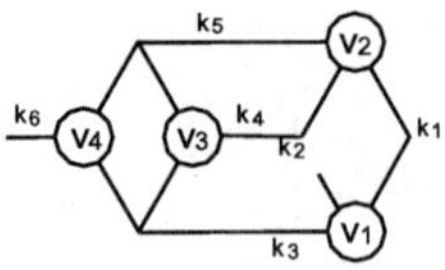	18	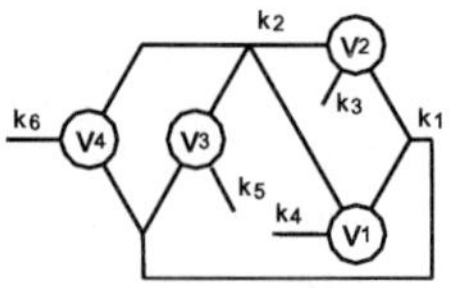

Structure diagrams of the systems are also expressed similarly as the structure sketches however it is needed to indicate the input, the output, and the fixed links. Besides, in order to reduce the number of DOF of the systems, the diagrams are also need to express control elements to constrain the DOF. Control elements used here are notated as f_i for brake to fix the links, l_i for clutch to make a rigid coupling between the links. By using the control elements in a structure diagram, the possible number of induced transmissions with DOF of 1 is calculated following combination theory as

$$n \leq C(w-1, c) = \frac{(w-1)!}{c!(w-c-1)!} \tag{1}$$

where c is the number of the control elements. The structure sketch with two differential mechanisms, which is numbered as 2 in Table 1, is used here for example. Without any control element, DOF of the system is 2. By applying 2 control elements to be brakes to the structure sketch, the possible number of transmissions is calculated as 4 corresponding to the 4 structure diagrams shown in Table 2. When fixing one single link or one joint, they became 1 DOF structure diagram, which is called pure planetary system (PPS) or close differential system (CDS). A PPS is the system with only the planetary mechanisms, while a CDS is the system combined by at least one differential mechanism with other planetary or differential mechanisms.

Table 2. Structure diagrams with 2 planetary gear sets and 2 brakes (f_1, f_2).

No.	Structure diagram	No.	Structure diagram
1	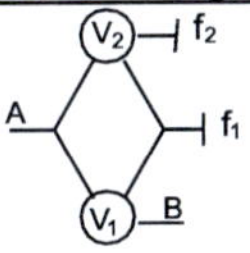	3	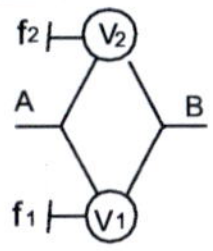
2	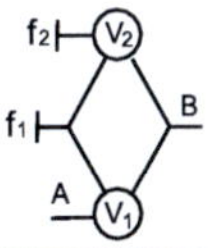	4	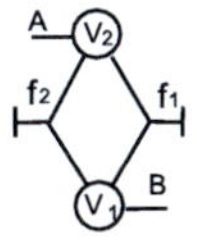

As mentioned above, from one structure diagram, there should be some kinematic diagrams estimated by converting the order of the links in combination. The number of kinematic diagrams obtained from a structure diagram could be calculated as

$$K = K_0^{n_M} \tag{2}$$

where n_M is the number of planetary mechanisms in the system, K_0 is the number of kinematic diagrams obtained from a structure diagram of the planetary mechanism. It is known as mentioned above that $K_0 = 6$.

2.3 Structure Analysis

From the above method of expression and organization of structure sketches, diagrams and kinematics diagrams, there are some comments obtained as follows.

- The new expression method surmounts the overlapping or intersection of the links in the joints as well as avoids using the hidden links through the mechanisms. This method shows more clearly the number of links in the joints, which makes advantage in the establishment and analyzing method of the system even the number of mechanisms and DOF changed.
- The expression of the diagrams here clearly shows the properties of joints and the relationships between the mechanisms in the system through the joints and the number of the basic links n_0. Since the number of DOF of the system is formulated as

$$w = n_0 - n_M \tag{3}$$

  the diagrams which do not ensure the number of DOF of the system should be rejected.
- The method arranges the sketches from the single links, which have not been joined to other links, to double joints (sketches 2, 3 in Table 1), 3 joints (sketches 4, 5, 6, 7, 8, 9, 10… in Table 1) and 4 joints (sketches 16, 17, 18… in Table 1) corresponding to the establishment and organization in succession depending on the increment of the number of mechanisms and complication of joints, the repeated and symmetric cases could be rejected.
- Total numbers of structure sketches with 1, 2, 3 and 4 mechanisms are different corresponding to DOF, i.e., the number of sketches with 2, 3 and 4 of DOF is 18, 20 and 13, respectively. The difference of these numbers expresses the property of joints, the relationship between DOF, the number of mechanisms n_M and the basic links n_0 (include single and joined links).

3 Notation of Planetary Gear Systems

3.1 Notation of the Structure Diagrams

In order to call briefly the structure diagrams, they are notated as presented here. The structure diagram of a planetary system is notated in general form as

$$\mathrm{A\underline{B}/T/\underline{A}B} \tag{4}$$

Where A and B are the h or v mechanisms having one basic link joined to the input and the output, respectively; $\underline{A}$ and $\underline{B}$ are the h or v mechanisms which have the basic links joining simultaneously to the output and the input of the system, but they were mentioned by A and B; T denotes the h or v mechanisms which have the basic links joining only internal joints. If there is not any mechanism with the basic links joining only internal joints, the structure diagram of the planetary system is notated as

$$A\underline{B}/\underline{A}B \tag{5}$$

It is noted that the total number of the symbols h and v in A, B and T should equal to the number of the mechanisms in the system. Besides, the total numbers of the symbols h and v in $A\underline{B}$ and $\underline{A}B$ should equal to the numbers of the basic links joining the input and the output, respectively.

For the case of the 1 DOF systems (PPS and CDS) with up to 3 mechanisms, the number of the structure diagrams could be estimated to be 23 as shown in Table 3. The notations of the 23 structure diagrams are also presented. For example, the notation of the structure diagram numbered as 23 is v$\underline{\text{v}}$/h/$\underline{\text{v}}$v, where there is a symbol v, $\underline{\text{v}}$, h, $\underline{\text{v}}$ and v corresponding to A, $\underline{B}$, T, $\underline{A}$ and B in Eq. (4), respectively. From the notation, the number of the mechanisms n_{M}, the numbers of the links joined to the input n_{I} and the output n_{O} is calculated as

$$n_{\text{M}} = n_{\text{A}} + n_{\text{T}} + n_{\text{B}} = 1 + 1 + 1 = 3 \tag{6}$$

$$n_{\text{I}} = n_{\text{A}} + n_{\underline{\text{B}}} = 1 + 1 = 2 \tag{7}$$

$$n_{\text{O}} = n_{\underline{\text{A}}} + n_{\text{B}} = 1 + 1 = 2 \tag{8}$$

where n_{A}, n_{T} and n_{B} is the number of the symbols v and h in A, T and B; $n_{\underline{\text{A}}}$ and $n_{\underline{\text{B}}}$ is the number of the symbols $\underline{\text{v}}$ and $\underline{\text{h}}$ in $\underline{A}$ and $\underline{B}$, respectively. It means that the number of differential and planetary mechanisms, the number of links which join to the input and the output as well as the sum of the fixed links could be determined from the notation of the structure diagrams. The mechanisms in a system are organized and orderly numbered following one unification principle making advantage for checking and forming the formulas which relate to the structure diagrams.

3.2 Notation of the Kinematics Diagrams

As mentioned above, kinematic diagram shows the links, the joining methods and also the engaging of gears in the system. From a structure diagram, some kinematic diagrams could be estimated. Therefore, the notation of a kinematics diagram should be based on the notation of the structure diagram which the kinematics diagram is derived. The kinematics diagram of a planetary system is notated in general form as

$$\left(A\underline{B}/T/\underline{A}B\right)^{\text{f}}_{\text{a/t/b}} \tag{9}$$

where the part in the parentheses is the notation of its structure diagram. The letters f, a, t and b should be the numbers indicating the elements in a planetary gear set where 0,

Table 3. Structure diagrams of 1 DOF systems with 1, 2 and 3 mechanisams.

No.	Notation	Structure diagram	No.	Notation	Structure diagram
1	h	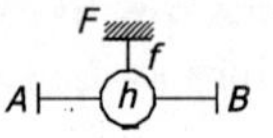	13	$\mathrm{h\underline{v}/h/v}$	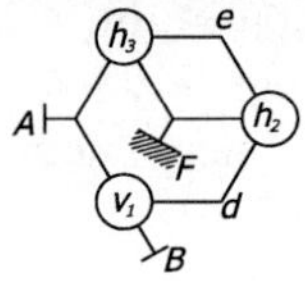
2	h/h	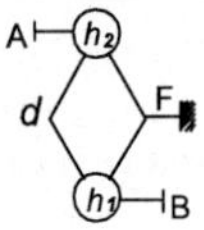	14	v/vh	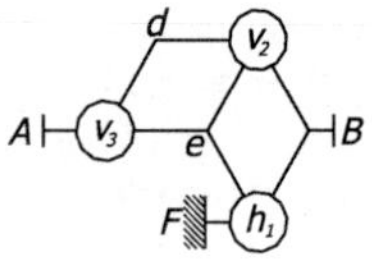
3	$\mathrm{v/\underline{v}h}$	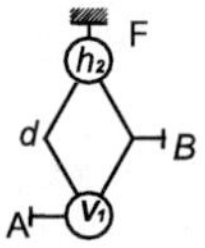	15	hv/v	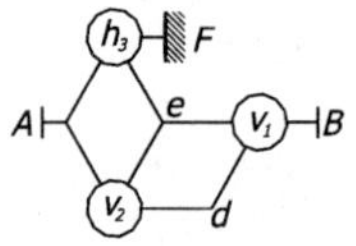
4	$\mathrm{h\underline{v}/v}$	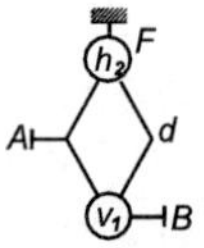	16	v/h/v	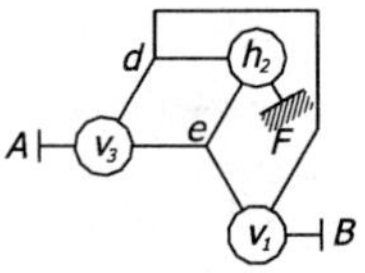
5	h/h/h	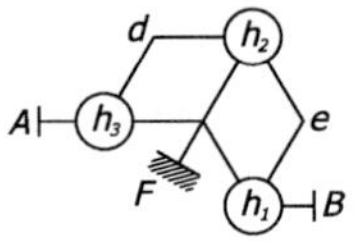	17	$\mathrm{v/\underline{v}vh}$	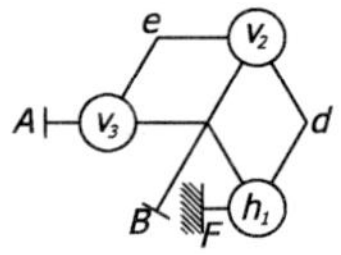
6	v/h/h	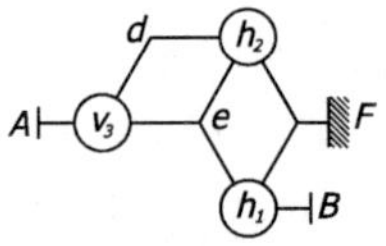	18	$\mathrm{hv\underline{v}/v}$	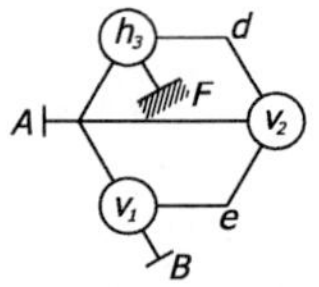
7	h/h/v	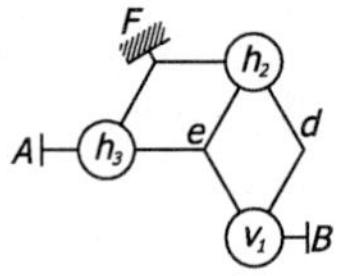	19	$\mathrm{vv/\underline{v}h}$	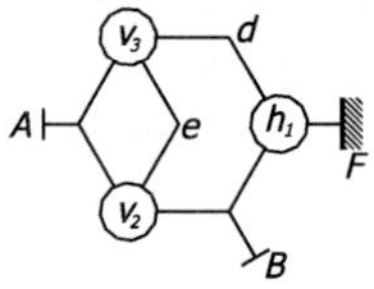
8	h/vh	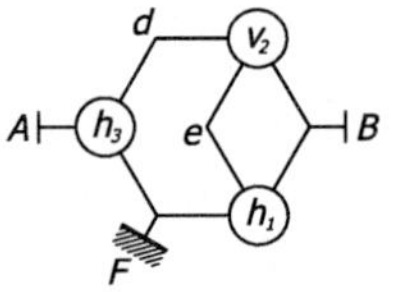	20	$\mathrm{h\underline{v}/vv}$	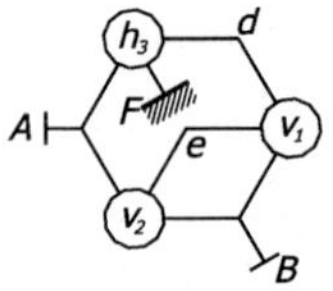
9	hv/h	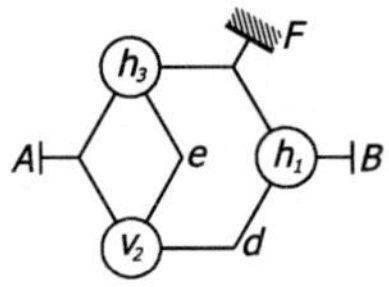	21	$\mathrm{v/h/\underline{v}v}$	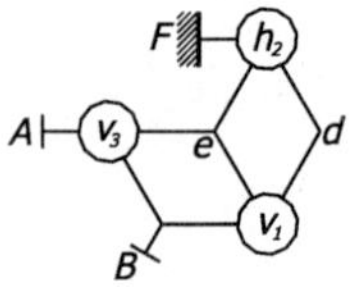

(continued)

Table 3. (*continued*)

10	v/hh	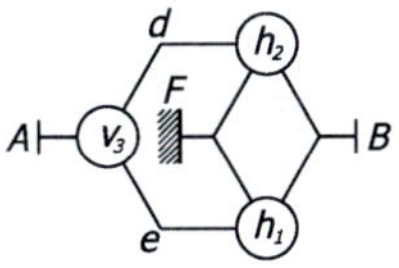	22	vv/h/v	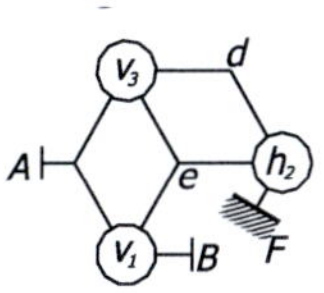
11	hh/v	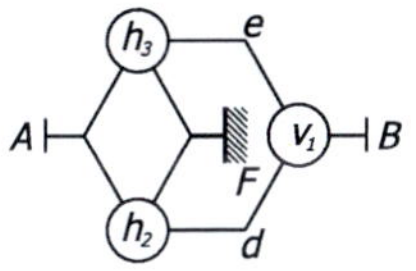	23	vv/h/vv	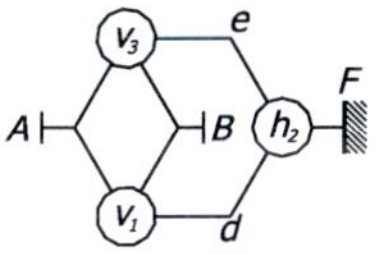
12	v/h/vh	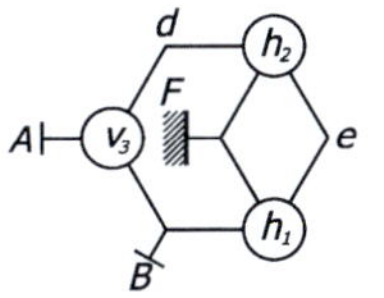			

1, 2 and 3 are corresponding to the arm, the sun gear, the planetary gears and the ring gear. The links joining the input or the output are put at the position of the letters a or b, respectively. The internal joints with the links, which need to clearly determine the connection of the links inside the system, are put at the position of the letter t. The fixed links of the h mechanisms have the indexed number put at the position of the letter f.

For the case of 1 DOF systems with 1, 2 and 3 mechanisms, the 23 notations and kinematic diagrams obtained from the 23 structure diagrams in Table 3 are presented in Table 4. There are three general rules taken out as follows:

- The notation of a kinematic diagram should correspond to the structure diagram. If the order of the active, intermediary, and passive mechanisms is in succession from left to right, then the notation of the kinematic diagram should be also in the same order of the mechanisms.
- The notation should be as simple as possible but must avoid confusion. Therefore, it is necessary to notate the active, passive, and fixed links. On the contrary, only the links participating in an internal joint of the system with three mechanisms should be indicated in the notation, and the other internal joint could be inferred. The order number of the mechanisms might be indicated as the subscript if it is necessary.
- The notation should be consistent, i.e., the position of the active, passive, and fixed links must be the same to the structure and kinematic diagrams.

Table 4. Kinematic diagrams of 1 DOF systems with 1, 2 and 3 mechanisms.

No.	Notation	Kinematic diagrams	No.	Notation	Kinematic diagrams
1	$(\text{h})^{3}_{1/0}$	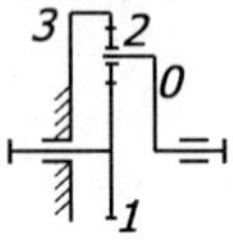	13	$(\text{h}\underline{\text{v}}/\text{h}/\text{v})^{33}_{11/01/3}$	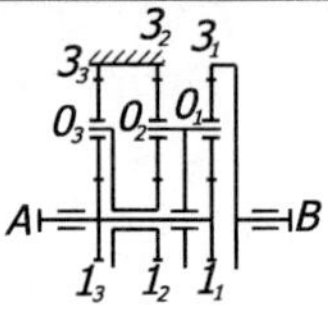
2	$(\text{h}/\text{h})^{30}_{1/3}$	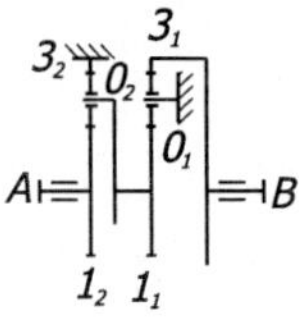	14	$(\text{v}/\text{vh})^{0}_{1/301/33}$	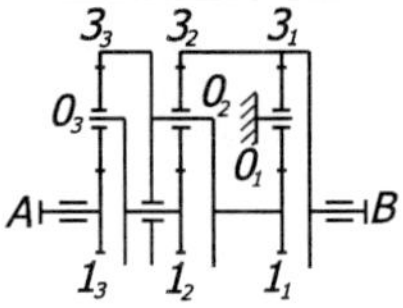
3	$(\text{v}/\underline{\text{v}}\text{h})^{0}_{1/33}$	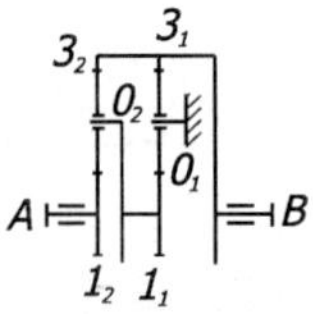	15	$(\text{hv}/\text{v})^{3}_{11/030/3}$	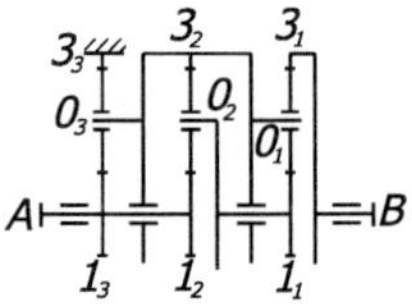
4	$(\text{h}\underline{\text{v}}/\text{v})^{3}_{13/0}$	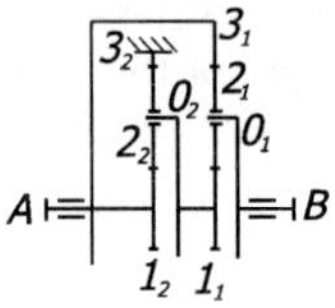	16	$(\text{v}/\text{h}/\text{v})^{3}_{1/301/3}$	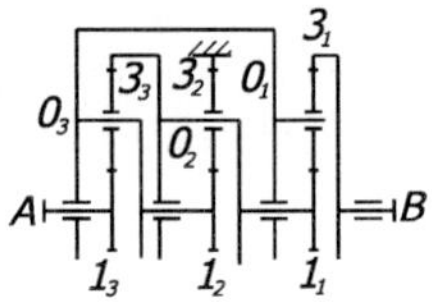
5	$(\text{h}/\text{h}/\text{h})^{333}_{1/0}$	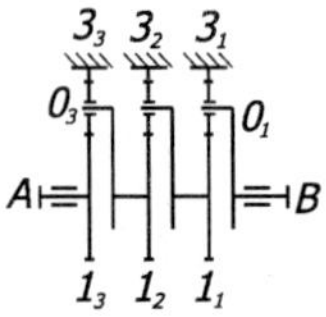	17	$(\text{v}/\underline{\text{v}}\text{vh})^{0}_{1/01/333}$	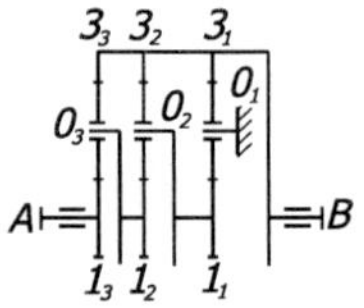
6	$(\text{v}/\text{h}/\text{h})^{33}_{1/011/0}$	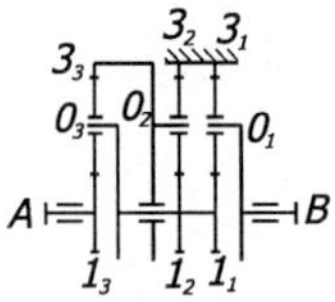	18	$(\text{hv}\underline{\text{v}}/\text{v})^{3}_{111/03/3}$	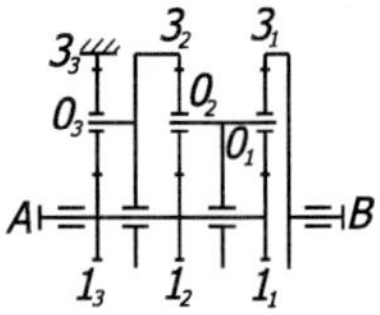
7	$(\text{h}/\text{h}/\text{v})^{33}_{1/010/3}$	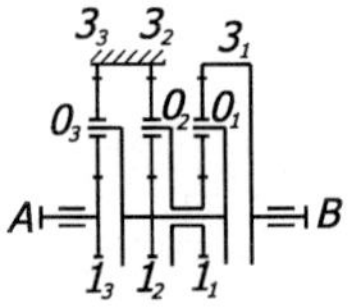	19	$(\text{vv}/\underline{\text{v}}\text{h})^{1}_{11/30/30}$	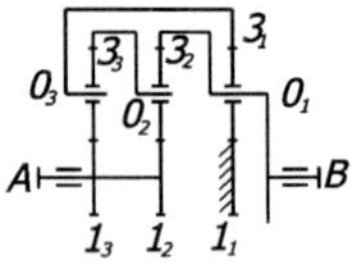
8	$(\text{h}/\text{vh})^{3}_{1/01/30}$	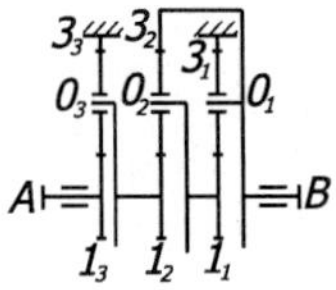	20	$(\text{h}\underline{\text{v}}/\text{vv})^{3}_{11/0_3 3_1/30}$	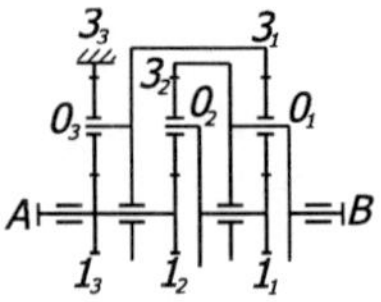

(continued)

Table 4. (*continued*)

9	$(hv/h)^{30}_{10/01/3}$	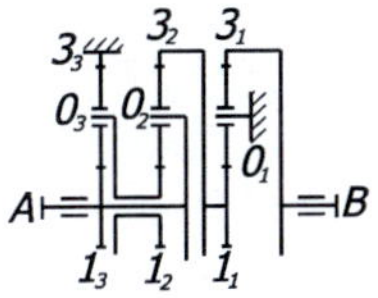	21	$(v/h/\underline{v}v)^{3}_{1/011/33}$	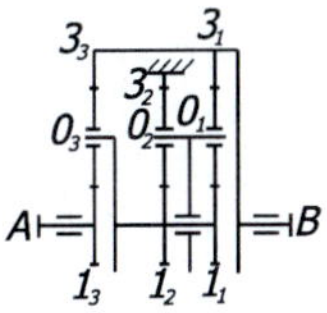
10	$(v/hh)^{30}_{1/01/31}$	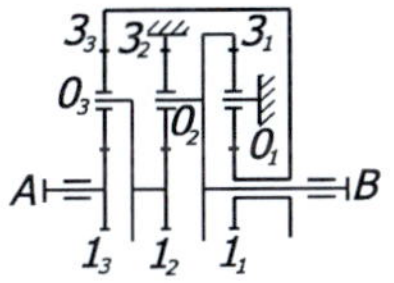	22	$(v\underline{v}/h/v)^{0}_{11/330/3}$	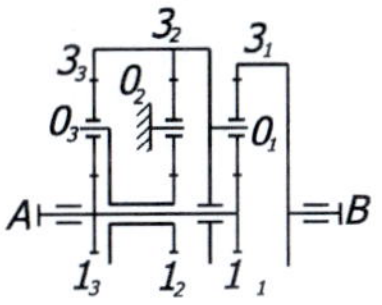
11	$(hh/v)^{03}_{11/3_3 0_1/3}$	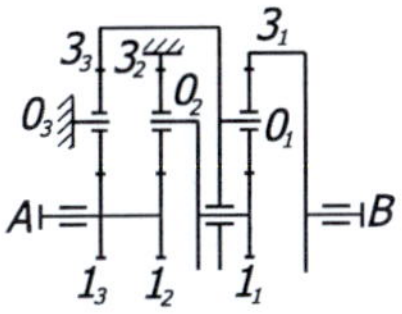	23	$(v\underline{v}/h/\underline{v}v)^{0}_{11/03/33}$	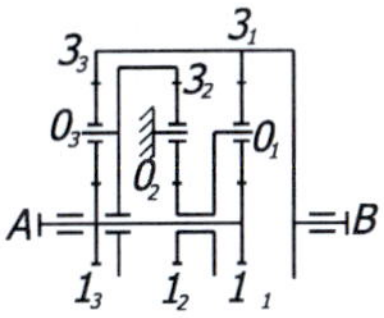
12	$(v/h/\underline{v}h)^{33}_{1/01/30}$	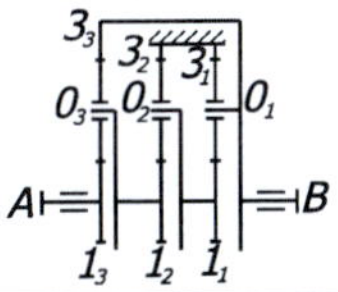			

4 Conclusion

This paper introduced a new method to demonstrate and notate the structure and kinematic diagrams of the complex planetary gear systems, which is named as vh method. It was implemented for the notation and diagrams of 1 DOF systems with 1, 2 and 3 mechanisms. This method overcomes the weak points of the previous methods, and helps to estimate, arrange as well as analyze the structure and kinematic diagrams with more advantages. The presented method is also applicable in structure and kinematic analysis for the complex planetary gear systems even with more degrees of freedom.

References

1. Buchsbaum, F., Freudenstein, F.: Synthesis of kinematic structure of geared kinematic chains and other mechanisms. J. Mech. **5**, 357–392 (1970)
2. Hsieh, W.H.: Kinematic synthesis of cam-controlled planetary gear trains. Mech. Mach. Theory **44**, 873–895 (2009)
3. Johannes, V.: Getriebetechnik Umlaufrädergetriebe. VEB Verlag Technik, Berlin (1978)
4. Beiz, W., Küttner, K.H.: DUBBEL - Taschenbuch für den Maschinenbau. Springer, Berlin (1995). https://doi.org/10.1007/978-3-662-54805-9
5. Gao, M.F., Hu, J.B.: Kinematic analysis of planetary gear trains based on topology. J. Mech. Design **140**(1), 012302 (2017)

6. Tsai, M.C., Huang, C.C., Lin, B.J.: Kinematic analysis of planetary gear systems using block diagrams. J. Mech. Des. **132**, 065001–065011 (2010)
7. Zanasi, R., Tebaldi, D.: Modeling of complex planetary gear sets using power-oriented graphs. IEEE Trans. Veh. Technol. **69**(12), 14470–14483 (2020)
8. Wang, Y., Yang, W., Tang, X., Lin, X., He, Z.: Power flow and efficiency analysis of high-speed heavy load herringbone planetary transmission using a hypergraph-based method. Appl. Sci. **10**(17), 5849 (2020)

Dynamic Response of Gear Transmission System under Barrel Rolling Flight Environment

Hao Cheng, Jing Wei(✉), Aiqiang Zhang, and Bin Peng

State Key Laboratory of Mechanical Transmissions, Chongqing University, Chongqing 400044, China

Abstract. This paper presents a dynamic modeling method of a gear transmission system considering the barrel roll flight environment. The kinematic equations of gear transmission system component are derived under non-inertial frame. The dynamics model of the gear rotor transmission system with time-varying positions was established using Lagrange energy method, and derived the gear transmission system dynamic contact stress calculation model combined with Hertz formula. The influence laws of barrel rolling acceleration and radius parameters on vibration response, axis trajectory, bearing support force and dynamic contact stress were solved and studied. The results show that considering maneuvering flight conditions, the frequency spectrum of vibration response appears time-varying rotation excitation frequency. The trajectory of the axis deviates with the barrel rolling parameters in different directions, and the influence of the driven gear with larger mass is more significant. The airframe flight will also affect the value and direction of the bearing's supporting force vector, and the dynamic contact stress increases with the amplification of barrel radius. This paper provides a theoretical basis for the dynamic stress calculation and reliability design of gear transmission in time-varying positions.

Keywords: Barrel roll maneuver flight · Gear transmission · Dynamic contact stress

1 Introduction

Maneuvering flight refers to the process in which an aircraft constantly changes its motion state (including velocity, acceleration, angular velocity, angular acceleration, etc.) in space. It is a unique operating environment of the gear transmission system of an aircraft and an important index to evaluate its performance [1]. In 1998, in the flight plan of "Black Hawk" helicopter in the United States, the engine rotor response under nearly 68 flight conditions was studied. It was found that 10 of them would increase the rotor load, and 2 of them had a significant effect on the response [2]. In this case, many scholars conducted relevant studies on the rotor dynamic characteristics in flight environment.

Sakata et al. [3] studied the dynamic characteristics of jet engine high-speed rotor system when the airframe was pitching and turning, and compared them with the experimental results. Duchemin et al. [4] used Lagrange energy method to derive the motion

N. V. Khang et al. (Eds.): ASIAN MMS 2021, MMS 113, pp. 157–167, 2022.
https://doi.org/10.1007/978-3-030-91892-7_15

equation of the rotor system under the condition of foundation motion, and studied the stability of the rotor system under sinusoidal rotation of the foundation, and verified it through experiments. Dakel et al. [5] established geometrically finite element models of rotors with 6 deterministic motions, and studied the steady-state dynamic characteristics of rotors with different configurations (symmetric and asymmetric) under the condition of supporting motion. Soni et al. [6] studied the vibration response of the magnetic levitation rotor system under the condition of pitch, roll and yaw of the foundation, and proved the superiority of the proposed new control method in the stability control of the magnetic levitation rotor system under the condition of foundation motion by comparing the system stability under the condition of foundation motion. However, there are few studies on the dynamics characteristics of the gear transmission system under flight environment, especially the study on dynamic contact stress, which have important guiding significance for the fatigue life calculation and reliability design of the system.

2 Dynamics Modeling of a Maneuvering Flight Gear Rotor

2.1 Gear Rotor Model

Taking the inner rotor gear pair of a certain type of coaxial rigid helicopter as the research object (Fig. 1), a 25-node model of the Timoshenko beam element of the shaft system is established, with Young's modulus $E = 2.1 \times 10^{11}$ Pa, Poisson's ratio $v = 0.316$, and density $\rho = 7860$ kg/m^3. and the main parameters are shown in Table 1.

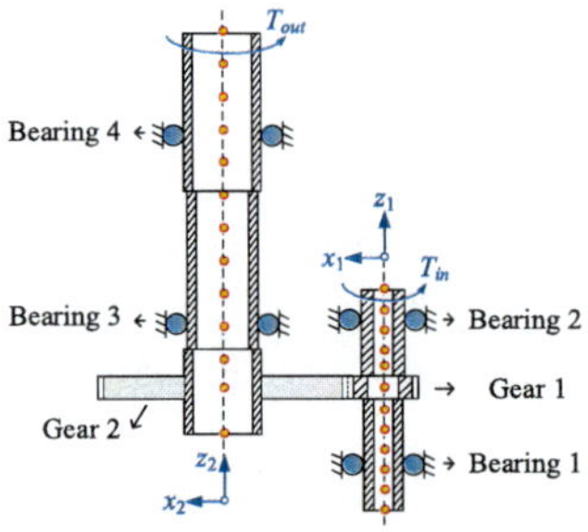

Fig. 1. Gear rotor model

2.2 Description of Space Motion

As a typical maneuvering, barrel roll refers to an effective technique for attack and defense in which the aircraft ensures its flight direction by drawing a spiral (Fig. 2).

Table 1. Gear system parameters

Number of tooth	z_g/z_p	23/63
Modulus	m	4.41 mm
Central distance	a	182.32 mm
Pressure angle	a_n	25°
Helical angle	β	10°
Stiffness of bearing	k_g / k_p	5.28e8/4.60e8 N/m
Torque	T	19851 N·m
Rated revolution	n	2198 r/min

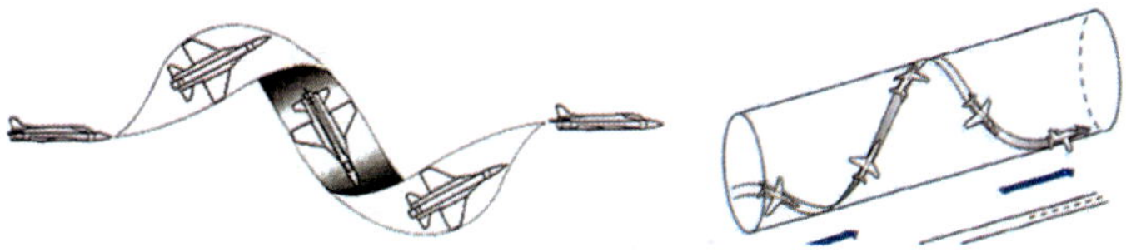

Fig. 2. Barrel roll movement

The gear rotor dynamics model established by *O-XYZ* in the fixed coordinate system on the ground is shown in Fig. 3(a). The body moving coordinate system O_m-$X_mY_mZ_m$ was established to represent the barrel rolling posture, O_m to represent the center of gravity of the aircraft, and the flight direction of the aircraft was the same as the horizontal axis *OX*. The gear rotor coordinate system o_c-$x_cy_cz_c$ represents the vibration of each node *c*, and o_c represents the center of mass of the gear rotor. The servo coordinate system O_m-$X_mY_mZ_m$ and the gear rotor coordinate system o_c-$x_cy_cz_c$ are the rotation coordinate systems, which rotate along their x_m and z_c axes respectively, as shown in Eq. (1).

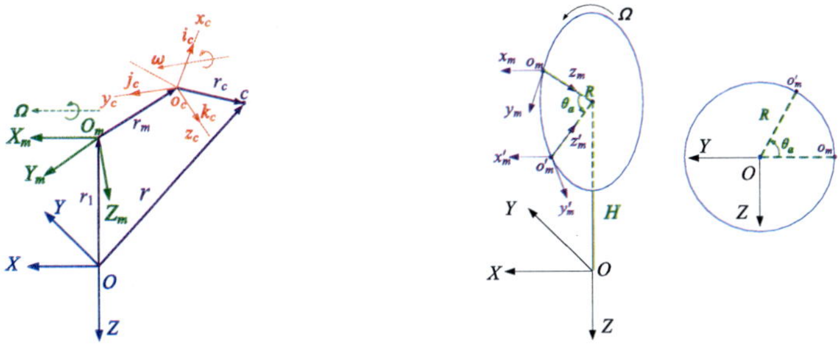

(a) Generalized coordinate system (b) Motion parameters of barrel roll

Fig. 3. System coordinate system setting

$$\begin{bmatrix} \boldsymbol{i}_m \\ \boldsymbol{j}_m \\ \boldsymbol{k}_m \end{bmatrix} = \begin{bmatrix} 1 & 0 & 0 \\ 0 & \cos(\theta_a) & -\sin(\theta_a) \\ 0 & \sin(\theta_a) & \cos(\theta_a) \end{bmatrix} \begin{bmatrix} \boldsymbol{i} \\ \boldsymbol{j} \\ \boldsymbol{k} \end{bmatrix}$$
$$\begin{bmatrix} \boldsymbol{i}_c \\ \boldsymbol{j}_c \\ \boldsymbol{k}_c \end{bmatrix} = \begin{bmatrix} \cos(\theta_c) & -\sin(\theta_c) & 0 \\ \sin(\theta_c) & \cos(\theta_c) & 0 \\ 0 & 0 & 1 \end{bmatrix} \begin{bmatrix} \boldsymbol{i}_m \\ \boldsymbol{j}_m \\ \boldsymbol{k}_m \end{bmatrix} \tag{1}$$

In the generalized coordinate system, the absolute motion of any node is described. $\boldsymbol{r}_1$ and $\boldsymbol{r}$ respectively represent the absolute radial vectors of points O_m and c in the coordinate system *OXYZ* [7], $\boldsymbol{r}_{\mathrm{m}}$ represents the offset of the gear rotor's centroid and the body's centroid, and $\boldsymbol{r}_{\mathrm{c}}$ represents the vibration displacement of each node, as shown in Eq. (2). The origin position of the body coordinate system and barrel rolling motion parameters are shown in Fig. 3(b). The parameters, such as barrel rolling radius $\boldsymbol{R}$, angular velocity $\boldsymbol{\Omega}$, angular acceleration $\dot{\boldsymbol{\Omega}}$, and the body's own angular velocity $\boldsymbol{\omega}_c$, are shown in Eq. (3).

$$\begin{cases} \boldsymbol{r} = \boldsymbol{r}_1 + \boldsymbol{r}_m + \boldsymbol{r}_{\mathrm{c}} \\ \boldsymbol{r}_1 = X \cdot \boldsymbol{i} + Y \cdot \boldsymbol{j} + Z \cdot \boldsymbol{k} \\ \boldsymbol{r}_m = x_m \cdot \boldsymbol{i}_m + y_m \cdot \boldsymbol{j}_m + z_m \cdot \boldsymbol{k}_m \\ \boldsymbol{r}_{\mathrm{c}} = x_{\mathrm{c}} \cdot \boldsymbol{i}_{\mathrm{c}} + y_{\mathrm{c}} \cdot \boldsymbol{j}_{\mathrm{c}} + z_{\mathrm{c}} \cdot \boldsymbol{k}_{\mathrm{c}} \end{cases} \tag{2}$$

$$\begin{cases} \boldsymbol{\Omega} = \Omega \cdot i_1, \dot{\boldsymbol{\Omega}} = \boldsymbol{0} \\ \omega = \boldsymbol{\omega_c} \cdot \boldsymbol{k}_c, \dot{\omega} = \dot{\omega}_c \cdot \boldsymbol{k}_c \\ \ddot{X} = a_e, \ddot{Y} = \boldsymbol{\Omega}^2 \cdot \boldsymbol{R} \cdot \sin(\theta_a), \ddot{Z} = -\boldsymbol{\Omega}^2 \cdot \boldsymbol{R} \cdot cos(\theta_a) \\ x_m = l_x, y_m = l_y, z_z = l_z \end{cases} \tag{3}$$

In Eq. (3), is the rotational angular displacement of the barrel rolling axis of the body $\boldsymbol{\theta_a}$, and $\boldsymbol{\theta_c}$ is the rotational angular displacement of the body itself. The derivative of $\boldsymbol{R}$ with respect to time solve the absolute velocity vector $\boldsymbol{v}_c$ at the point c in the inertial coordinate system *OXYZ*. The rotational angular velocity of the moving coordinate system O_m-$X_mY_mZ_m$ is $\boldsymbol{\Omega}$. According to Poisson's formula, the derivative of the basis vector with respect to time can be obtained as shown in Eq. (4). Then the absolute velocity/acceleration vector $\boldsymbol{v}_c/\boldsymbol{a}$ is shown in Eq. (5) and (6) respectively.

$$\frac{d\boldsymbol{i}}{dt} = \boldsymbol{\Omega} \times \boldsymbol{i}, \frac{d\boldsymbol{j}}{dt} = \boldsymbol{\Omega} \times \boldsymbol{j}, \frac{d\boldsymbol{k}}{dt} = \boldsymbol{\Omega} \times \boldsymbol{k} \tag{4}$$

$$\boldsymbol{v} = \ddot{X} \cdot \boldsymbol{i} + \ddot{Y} \cdot \boldsymbol{j} + \ddot{Z} \cdot \boldsymbol{k} + \boldsymbol{\Omega} \times \boldsymbol{r}_m + \boldsymbol{v}_{\mathrm{c}} + (\boldsymbol{\omega} + \boldsymbol{\Omega}) \times \boldsymbol{r}_{\mathrm{c}} \tag{5}$$

$$\begin{aligned} \boldsymbol{a} =& \ddot{X} \cdot \boldsymbol{i} + \ddot{Y} \cdot \boldsymbol{j} + \ddot{Z} \cdot \boldsymbol{k} + \ddot{x}_{\mathrm{c}} \cdot \boldsymbol{i}_{\mathrm{c}} + \ddot{y}_{\mathrm{c}} \cdot \boldsymbol{j}_{\mathrm{c}} + \ddot{z}_{\mathrm{c}} \cdot \boldsymbol{k}_{\mathrm{c}} + \dot{\boldsymbol{\Omega}} \times \boldsymbol{r}_m + \boldsymbol{\Omega} \times (\boldsymbol{\Omega} \times \boldsymbol{r}_m) \\ &+ \dot{\boldsymbol{\Omega}} \times \boldsymbol{r}_{\mathrm{c}} + 2 \cdot \boldsymbol{\Omega} \times \boldsymbol{v}_{\mathrm{c}} + \dot{\boldsymbol{\omega}} \times \boldsymbol{r}_{\mathrm{c}} + \boldsymbol{\omega} \times (\boldsymbol{\omega} \times \boldsymbol{r}_{\mathrm{c}}) + 2 \cdot \boldsymbol{\omega} \times \boldsymbol{v}_{\mathrm{c}} \\ &+ \boldsymbol{\omega} \times (\boldsymbol{\Omega} \times \boldsymbol{r}_{\mathrm{c}}) + \boldsymbol{\Omega} \times (\boldsymbol{\omega} \times \boldsymbol{r}_{\mathrm{c}}) + \boldsymbol{\Omega} \times (\boldsymbol{\Omega} \times \boldsymbol{r}_{\mathrm{c}}) \end{aligned} \tag{6}$$

2.3 Non-inertial System Dynamics Modeling

The differential equation of gear rotor motion under barrel rolling motion was derived from Lagrange equation, which is applicable to generalized coordinates, and its form is shown in Eq. (7), where $\boldsymbol{F}_{qi}$ is non-conservative force.

$$\frac{d}{dt}\left(\frac{\partial \boldsymbol{T}}{\partial \dot{\boldsymbol{q}}_i}\right) - \frac{\partial \boldsymbol{T}}{\partial \boldsymbol{q}_i} + \frac{\partial \boldsymbol{U}}{\partial \boldsymbol{q}_i} = \boldsymbol{F}_{qi} \tag{7}$$

First, define the coordinate system needed to describe the motion of the gear wheel body, and calculate the vector representing the rotation between them. Then, define the kinetic energy and strain energy of the shafting element.

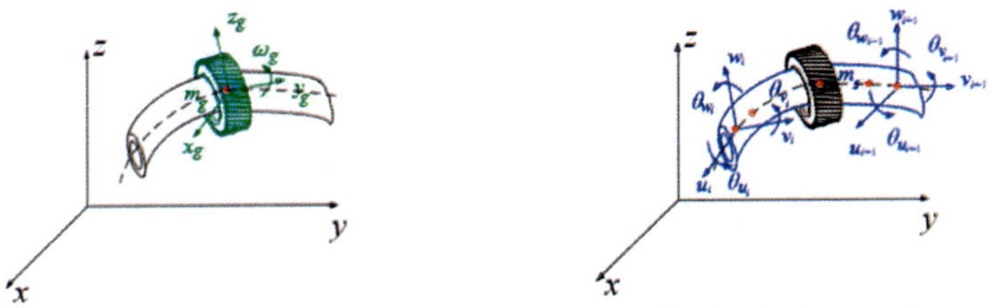

(a) Gear (b) Shafting beam unit

Fig. 4. Gear and shaft coordinate system

The body coordinate system of the gear wheel is shown in Fig. 4(a). The kinetic energy and potential energy of the wheel can be expressed by its absolute velocity as shown in Eq. (8, 9) respectively, where $\boldsymbol{m_g}$ and $\mathbf{I_g}$ are the mass and moment of inertia of the gear rotor; meanwhile, $\boldsymbol{v_g}$ and $\boldsymbol{\omega_g}$ are the absolute and angular velocities of the gear rotor in a non-inertial frame. $\boldsymbol{y_g}$ is the absolute displacement vector of the gear rotor in a non-inertial frame.

$$\boldsymbol{T}_g = \frac{1}{2} \cdot m_g \cdot (\boldsymbol{v_g})^T \cdot \boldsymbol{v_g} + \frac{1}{2} \cdot (\boldsymbol{\omega_g})^T \boldsymbol{I_g} \cdot \boldsymbol{\omega_g} \tag{8}$$

$$\boldsymbol{U}_g = m_g \cdot g \cdot \boldsymbol{y_g} \tag{9}$$

A 6-DOF Timoshenko beam element model is established considering the bending and torsional axis pendulum deformation, as shown in Fig. 4(b). The kinetic energy and potential energy are shown in Eqs. (10, 11), where $\boldsymbol{k_s}$ is the translational elastic stiffness, $\boldsymbol{K}_\theta$ is the torsional elastic stiffness.

$$\boldsymbol{T}_s = \frac{1}{2}[(\boldsymbol{v}_s)^T \cdot \boldsymbol{M}_s \cdot \boldsymbol{v_s} + (\boldsymbol{\omega}_s)^T \cdot \boldsymbol{I}_s \cdot \boldsymbol{\omega_s} + I_{sp} \cdot (\Omega_s - 2 \cdot \Omega_s \cdot \theta_x \cdot \dot{\theta}_y)] \tag{10}$$

$$\boldsymbol{V}_s = \frac{1}{2} \cdot \left[\boldsymbol{k_s} \cdot \boldsymbol{y}_s^2 + \boldsymbol{K}_\theta \cdot \boldsymbol{\theta}_s^2\right] \tag{11}$$

The kinetic energy and potential energy of the gear and shaft system were substituted into the Lagrange Eq. (7) to obtain the corresponding differential equations of motion, which were finally assembled into the differential equations of motion of the system, as shown in Eq. (12). It is assumed that the gear rotor is fixed on the shafting and the velocity vector is equal to the node of the shafting.

$$(\boldsymbol{m_g} + \boldsymbol{M_s})\ddot{\boldsymbol{X}} + [(\boldsymbol{\Omega} + \boldsymbol{\omega})\boldsymbol{G_s} + \boldsymbol{C}]\dot{\tilde{\boldsymbol{X}}} + (\boldsymbol{K_s} + \boldsymbol{\Omega\omega K_s^{\Omega\omega}} + \boldsymbol{\omega^2 K_s^{\omega 2}})\tilde{\boldsymbol{X}} = \boldsymbol{F} + \boldsymbol{F_G} + \boldsymbol{F_F} \tag{12}$$

Where $\boldsymbol{X} = \{u, v, w, \theta_w, \theta_v, \theta_w\}^{\mathrm{T}}$, $\boldsymbol{\Omega}$ and $\boldsymbol{\omega}$ are the body motion and angular velocity of shaft system; $\boldsymbol{G_s}$ is the gyro matrix; $\boldsymbol{C}$ is meshing damping; $\boldsymbol{K}, \boldsymbol{K_s}^{\Omega\omega}, \boldsymbol{K_s}^{\omega 2}$ are respectively the stiffness matrix of shaft system and the stiffness effect matrix of the attachment caused by foundation motion. $\boldsymbol{F}$ is the generalized force vector, including the input and output torque $\boldsymbol{T}$ of the transmission system, and the gear tooth meshing force $\boldsymbol{F_m}$. $\boldsymbol{F_G}$ represents gravity, $\boldsymbol{F_F}$ represents inertia force.

3 Dynamic Behaviors Analysis in Non-inertial System

By using the Precise Integration Method (PIM) to solve the motion differential equation, the vibration characteristics of the system were analyzed, and the simulation results were summarized.

3.1 Vibration Characteristics

When the barrel roll angular velocity of the airframe $\Omega = 4$ rad·s^{-1}, the influence of the acceleration $\boldsymbol{a_e}$ of the airframe on the displacement response of the driving and driven gear nodes is shown in Fig. 5. When $\boldsymbol{a_e} = 0\ g$, the vibration displacement is a high-frequency vibration dominated by frequency f_{m}. When the airframe is moving, the gravity of component becomes a time-varying excitation, i.e., $\boldsymbol{G}_i = -m_i \cdot g \cdot [\sin(\Omega \cdot t) \cdot \boldsymbol{i} + \cos(\Omega \cdot t) \cdot \boldsymbol{j}]$. Particularly when $\boldsymbol{a_e} = 20\ g$, the inertial force and gravity have a great influence on the vibration displacement of the massive gear 2, and f_Ω even becomes the dominant excitation frequency.

The influence of the radius $\boldsymbol{R}$ on the displacement response of the gear nodes is shown in Fig. 6. When $R = 0\ m$, the vibration displacement is dominated by f_m. When the body rotates, the centrifugal force along the local coordinate axis of each component becomes a time-varying term, that is, $\boldsymbol{F_F} = -m_i \cdot \omega^2 \cdot \boldsymbol{R} \cdot [\sin(\Omega \cdot t) \cdot \boldsymbol{i} + \cos(\Omega \cdot t) \cdot \boldsymbol{j}]$, the basic rotation frequency f_Ω is one of the main excitation frequencies.

3.2 Axis Trajectory

The displacements of gear nodes under the barrel rolling motion accelerations $\boldsymbol{a_e}$ were obtained, as shown in Fig. 7. The gear offset in the $\boldsymbol{x}$ direction has a great correlation with $\boldsymbol{a_e}$, but the direction is just opposite to $\boldsymbol{a_e}$, which conforms to D'Alembert's principle. Due to the large mass and inertia, the deviation of gear 2 is obviously larger than that of gear 1, and the deviation of $\boldsymbol{y}$ direction is about 700 μm when $\boldsymbol{a_e}$ is −40g.

Displacement responses of gear nodes under different radius $\boldsymbol{R}$ under barrel rolling motion were obtained, as shown in Fig. 8. The gear offset in the $\boldsymbol{y}$ direction is highly correlated with the drum radius $\boldsymbol{R}$. The deviation of gear 2 is obviously larger than that of gear 1, especially $\boldsymbol{y}$ direction which is about 1700 μm when $\boldsymbol{R}$ is 80 m.

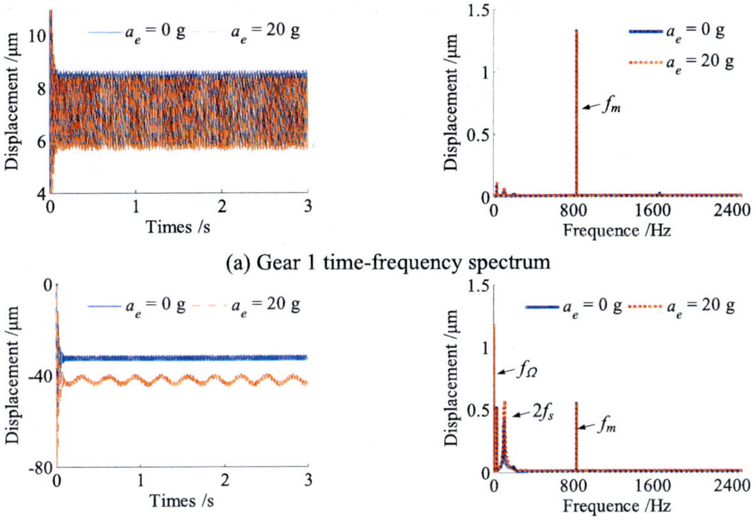

(a) Gear 1 time-frequency spectrum

(b) Gear 2 time-frequency spectrum

Fig. 5. The effect of $\boldsymbol{a_e}$ on vibration

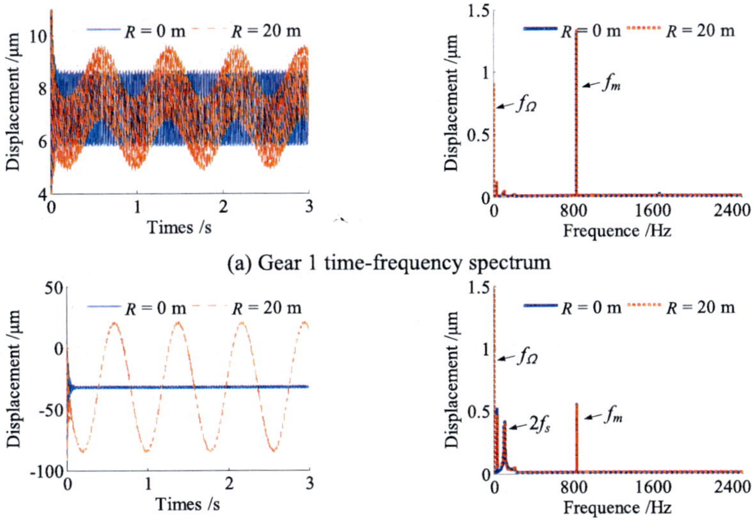

(a) Gear 1 time-frequency spectrum

(b) Gear 2 time-frequency spectrum

Fig. 6. The effect of $\boldsymbol{R}$ on vibration

3.3 Bearing Support Force

The changes of shafting offset will inevitably lead to the change of radial supporting force as shown in Fig. 9. And the influence law of acceleration $\boldsymbol{a_e}$ under barrel roll maneuvering on radial supporting force is shown in Fig. 10. For the gear 1, the bearing

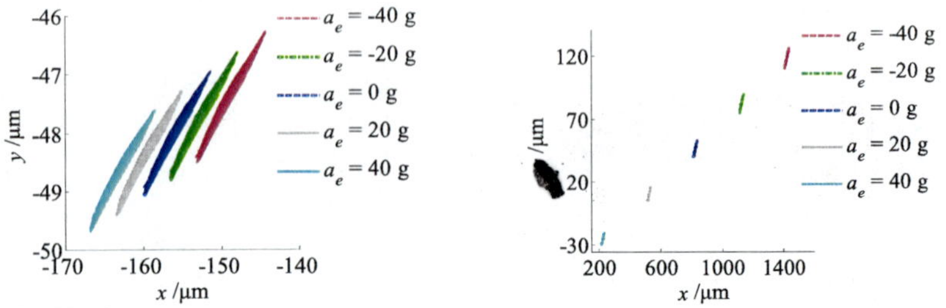

(a) Gear 1 centerline orbit (b) Gear 2 centerline orbit

Fig. 7. The effect of ae on centerline orbit

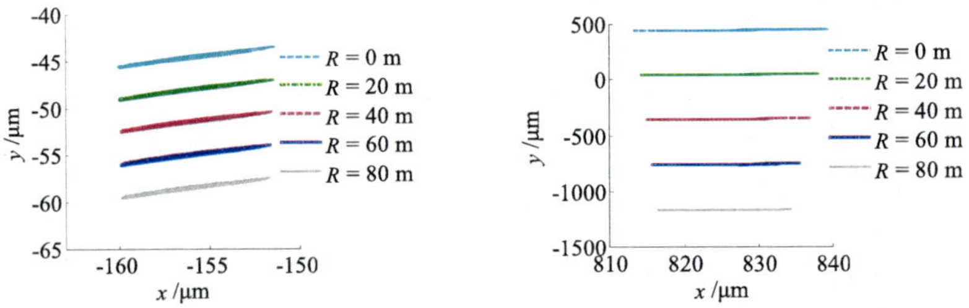

(a) Gear 1 centerline orbit (b) Gear 2 centerline orbit

Fig. 8. The effect of ***R*** on centerline orbit

force $\boldsymbol{F}_{\mathrm{b}1}$ increases from 5.05 kN ($\boldsymbol{a_e} = 0$ g) to 5.26 kN ($\boldsymbol{a_e} = 40$ g) and 5.24 kN ($\boldsymbol{a_e} = -40$ g) respectively, but decreases at $\boldsymbol{a_e} = 20$ g, and the bearing force Angle $\boldsymbol{\theta}_{\mathrm{b}1}$ changes from 24.4° to 25.7°. That is, the supporting force vector $\boldsymbol{F}_{\mathrm{b}1}$ is always in the first quadrant. When $\boldsymbol{a_e} = -40$ g, the supporting force $\boldsymbol{F}_{\mathrm{b}3}$ at bearing 3 increased significantly to 8.32 kN, which was 63.8% higher than that at $\boldsymbol{a_e} = 0$ g (5.08 kN). However, when $\boldsymbol{a_e} = 40$ g, the supporting force decreased to 2.12 kN. Bearing force Angle $\boldsymbol{\theta}_{\mathrm{b}3}$ changes from 18.7° to 50.3°, still in the first quadrant.

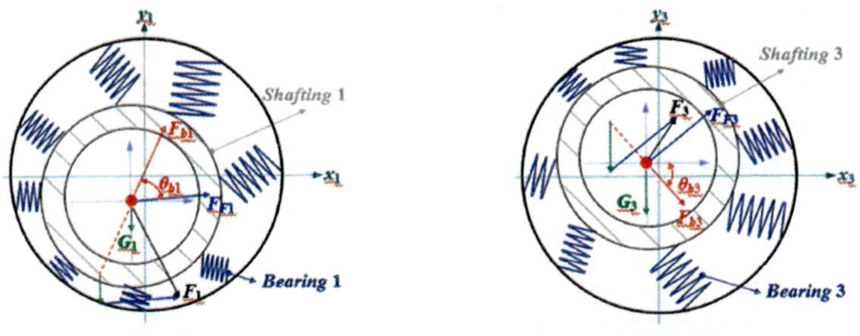

(a) Shafting 1 (b) Shafting 3

Fig. 9. Force balance relationship of shafting

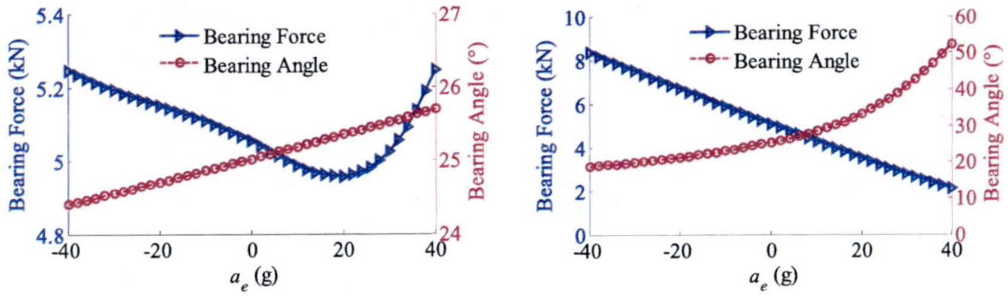

(a) Bearing 1 (b) Bearing 3

Fig. 10. The effect of $\boldsymbol{a_e}$ on bearing force

The influence rule of radius $\boldsymbol{R}$ on radial supporting force is shown in Fig. 11. For the gear 1, the supporting force $\boldsymbol{F}_{\mathrm{b1}}$ at bearing 1 decreases from 5.05 kN ($\boldsymbol{R} = 0$ m) to 4.88 kN ($\boldsymbol{R} = 60$ m) and then rises to 5.06 kN ($\boldsymbol{R} = 80$ m), and the bearing force Angle $\boldsymbol{\theta}_{\mathrm{b1}}$ changes from 25.0° to 18.5°, that is, the supporting force vector $\boldsymbol{F}_{\mathrm{b1}}$ is always in the first quadrant. When $\boldsymbol{R} > 20$ m (4.60 kN), the bearing force $\boldsymbol{F}_{\mathrm{b3}}$ increases significantly to 9.11 kN, which is 79.3% higher than that of $\boldsymbol{R} = 0$ (5.08 kN). Meanwhile, the bearing force Angle $\boldsymbol{\theta}_{\mathrm{b3}}$ changes from 25.3° to −59.73°. The results show that the supporting force vector $\boldsymbol{F}_{\mathrm{b3}}$ varies widely in the first and fourth quadrants.

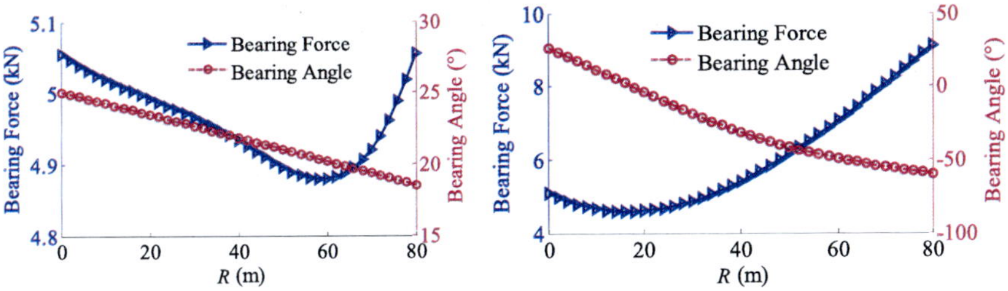

(a) Gear 1 centerline orbit (b) Gear 2 centerline orbit

Fig. 11. The effect of $\boldsymbol{R}$ on bearing force

3.4 Contact Dynamic Stress

The contact dynamic stress of gear pair is the direct reflection of the carrying capacity of gear, and it is also the direct cause of fatigue failure such as tooth fracture, tooth surface pitting and tooth surface peeling. Since additional effects under barrel rolling motion of the body will increase the force of the gear, which brings a major threat to the safety and reliability of the aircraft, it is necessary to analyze the change rule of dynamic stress of the gear rotor under barrel rolling motion. The calculation formula of contact dynamic stress of the gear rotor is shown in Eq. (13).

$$\boldsymbol{\sigma}_{\boldsymbol{Hkj}} = \sqrt{\frac{\boldsymbol{F}_{\boldsymbol{mj}} \cos \alpha_j}{\pi W \cos \alpha} \cdot \frac{1/\rho_{kj}}{\frac{1-v_1^2}{E_1} + \frac{1-v_2^2}{E_2}}} \tag{13}$$

σ_{Hkj} and ρ_{kj} represent the contact dynamic stress and integrated curvature of tooth to tooth at meshing position j and k respectively [8]. By solving Eq. (12), we can obtain the meshing force F_{mj} at any meshing position j, ρ_{kj}, v and E represent the main curvature radius, Poisson's ratio and elastic modulus of the gear. And the contact dynamic stress is greatly affected by the curvature radius of meshing point. The meshing force and dynamic stress of the gear pair when the inertial action is not taken into account in the external non-inertial condition are shown in Fig. 12.

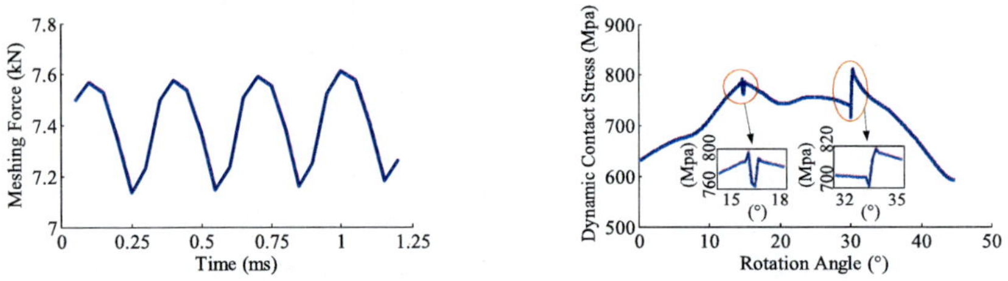

(a) Gear pair meshing force (b) Dynamic contact stress

Fig. 12. Gear pair dynamic contact stress

The influence of a_e and R parameters on the contact dynamic stress σ_H of the gear pair during barrel rolling of the body was studied respectively. The RMS value variation trend of contact dynamic stress σ_H is shown in Fig. 13. σ_H increases approximately non-linearly with a_e and R, but it is more sensitive to the radius R. The contact dynamic stress increases significantly at a radius of 80 m, about 4%. Therefore, increasing the barrel radius intensifies the stress between the gear pairs, and the additional effect of airframe maneuver flight should be considered in the design.

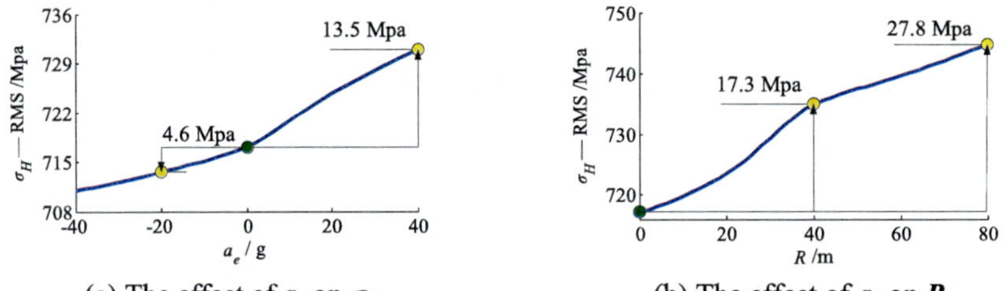

(a) The effect of a_e on σ_H (b) The effect of a_e on R

Fig. 13. Gear pair dynamic contact stress

4 Conclusion

The kinematic equation of gear transmission system component are derived under barrel rolling flight environment. The dynamic behaviors of the gear system under different non-inertial parameters were studied with dynamic mesh relationships. The main conclusions are shown as follows:

(1) The inertial force and gravity have a great influence on the vibration displacement behaviors of the gear transmissions. With the rotational motion, the rotational frequency f_{Ω} even becomes the dominant excitation frequency.
(2) Under the different parameters of the barrel rolling maneuvering, the trajectory of the axle center deviates in different directions, and the influence of the gear with larger mass is more significant, which leads to affect both value and direction of the bearing's supporting force vector.
(3) Combined with Hertz formula, the gear transmission system dynamic contact stress(σ_H) calculation model was derived. The $\boldsymbol{\sigma}_{\boldsymbol{H}}$ increases with the increase of barrel radius. And with increase of positive acceleration, the $\boldsymbol{\sigma}_{\boldsymbol{H}}$ increases.

Acknowledgments. The author is grateful for the support of the National Natural Science Foundation of China (grant number 51775058), the National Key R&D Program of China (No. 2020YFB2008101) and Innovation Group Science Foundation of Chongqing Natural Science Foundation (No. cstc2019jcyj-cxttX0003).

References

1. Jie, G., Peishen, Z., Zhenghong, G.: Advanced Flight Dynamics. National Defense Industry Press, Beijing (2004)
2. Kufeld, R., Bousman, W.: High load conditions measured on a UH-60A in maneuvering flight. J. Am. Helicopter Soc. **43**(03), 202–211 (1995)
3. Okamoto, S., Sakata, M., Kimura, K., et al.: Vibration analysis of a high speed and light weight rotor system subjected to a pitching or turning motion. J. Sound Vib. **184**(05), 871–885 (1995)
4. Duchemin, M., Berlioz, A., Ferraris, G.: Dynamic behavior and stability of a rotor under base excitation. J. Vib. Acoust. **128**(05), 576–585 (2006)
5. Dakel, M., Baguet, S., Dufour, R.: Steady-state dynamic behavior of an on-board rotor under combined base motions. J. Vib. Control **20**(15), 2254–2287 (2004)
6. Soni, T., Dutt, J., Das, A.: Parametric stability analysis of active magnetic bearing-supported rotor system with a novel control law subject to periodic base motion. IEEE Trans. Indu. Electron. **67**(02), 1160–1170 (2019)
7. Zhang, A., Wei, J., Shi, L., et al.: Modeling and dynamic response of parallel shaft gear transmission in non-inertial system. Nonlinear Dyn. **98**(2), 997–1017 (2019)
8. Zhang, C.: Mechanical Principle and Mechanical Design, 3rd edn. China Machine Press, Beijing (2018)

Online Intelligent Gear-Shift Decision of Vehicle Considering Driving Intention Using Moving Horizon Strategy

Jihao Feng, Datong Qin(✉), Kang Wang, and Yonggang Liu

State Key Laboratory of Mechanical Transmission,
Chonging University, Chongqing 400044, China
dtqin@cqu.edu.cn

Abstract. Gear-shift decision-making is the key to affect the power, economy and comfort of the vehicle. At present, gear-shift decision-making mostly relies on manual calibration, and it is impossible to make reasonable decisions based on real-time changes in driving intentions. The degree of intelligence needs to be improved. This paper proposes an online gear-shift decision-making method considering the driving intention. Based on the theoretical framework of model predictive control, in the predicting horizon, the multi-layer perception machine is used to predict the short-term vehicle states, and K-means clustering and deep belief network method are used to classify and recognize driving intentions. A gear-shift decision-making optimization model considering the fuel consumption and the driving intention is built in the predicting horizon, and the optimal gear-shift decision-making sequence is solved by the dynamic programming algorithm. The simulation and experiments show that the proposed online gear-shift decision-making method can improve the vehicle's adaptability to driving behavior while saving fuel, so as to improve the intelligence of the gear-shift decision-making of the vehicle.

Keywords: Online intelligent gear-shift decision · Driving intention · Deep learning · Model predictive control

1 Introduction

For the vehicle equipped with staged transmission (DCT/AMT/AT), the gear-shift position decision not only affects the fuel economy, but also affects the driving performance of the vehicle. Improving the adaptability of the gear-shift decision to the driver's driving demand while meeting the lower fuel consumption has important practical engineering significance [1, 2].

Modern vehicles mainly use two/three parameter shift curves [3, 4], and it can be divided into economic-type and power-type. But it not consider the complex changes in the driving intention and the driving environment during the driving process, which has certain limitations and is prone to unexpected gearshift [5, 6]. The gear-shift decision-making methods based on intelligent control are generally divided into intelligent correction type [7, 8] and intelligent learning type [9, 10]. The gear-shift decision-making

N. V. Khang et al. (Eds.): ASIAN MMS 2021, MMS 113, pp. 168–178, 2022.
https://doi.org/10.1007/978-3-030-91892-7_16

methods based on intelligent control mainly relies on the operation data of excellent drivers, the self-learning ability of fuzzy rules is weak, and the randomness of the neural network is strong, whose rationality directly affects the shift quality.

The gear-shift decision-making method based on online optimization generally takes the minimum fuel consumption in a short period of time as the optimization goal, and obtain the optimal gear-shift decision online based on real-time driving data [11, 12]. The real-time optimization method can overcome some shortcomings of the above two/three parameters and intelligent method, however, the above real-time method fails to take the driver's driving behavior into account in the optimal gear-shift decision-making process.

Therefore, in this paper, an online gear-shift decision method considering driving intention is proposed. Based on the MPC framework, multi-layer perceptron (MLP) is used for short-term prediction of the vehicle states, and clustering and deep belief network are used for classification and recognition of the driving intention. The optimal gear position sequence in the prediction horizon is solved in real time based on dynamic programming algorithm. The proposed online intelligent gear-shift decision-making control process is shown in Fig. 1.

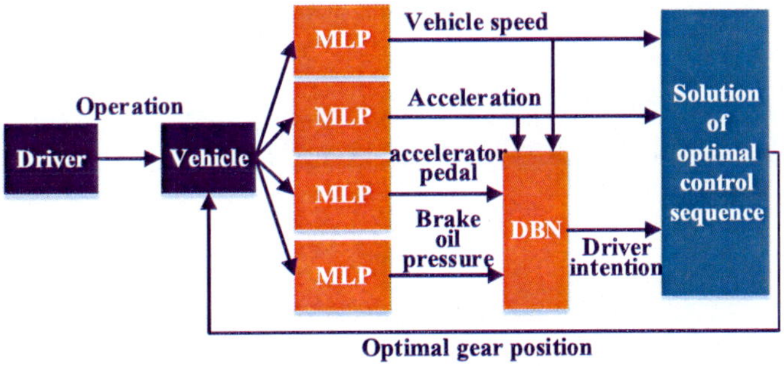

Fig. 1. The proposed real-time intelligent gear-shift decision-making control process.

2 Acquisition and Preprocessing of the Naturalistic Driving Data

In order to analyze and model driver's driving intention, a 7-speed DCT SUV is used for data acquisition test. Based on the CAN bus, the driving data is read out using OBD interface, and the data acquisition device is the vehicle recorder. We use sensors equipped with production vehicles, including rotational speed sensor, pressure sensor, throttle opening sensor and acceleration sensor. The collected signals include vehicle speed, acceleration, brake oil pressure, throttle opening, gear position, etc. The sampling frequency is 100 Hz. The experimental route is shown in Fig. 2. The road scene involves urban central road, suburban roads, mountain roads and urban high-speed roads.

Normally, the signal collected from the equipment has noise. Wavelet analysis finds the best approximation to the original signal through the signal transformation in time and frequency domain to distinguish the original signal from the noise signal. However, the wavelet denoising method needs to determine the wavelet basis function and the number of decomposition levels, so a data denoising method based on composite evaluation index is proposed. Root mean square error, signal-to-noise ratio and smoothness degree are selected as evaluation indexes. After comparative analysis of the three indexes, Sym8 wavelet basis function, and *heusure* threshold estimating method is selected, and the threshold function is soft threshold. And, the weight of root mean square error and smoothness degree are fused based on the variation coefficient, and to determine the wavelet decomposition level, and the final decomposition level is 5. The denoising effect is shown in Fig. 3.

Fig. 2. Data acquisition route and process.

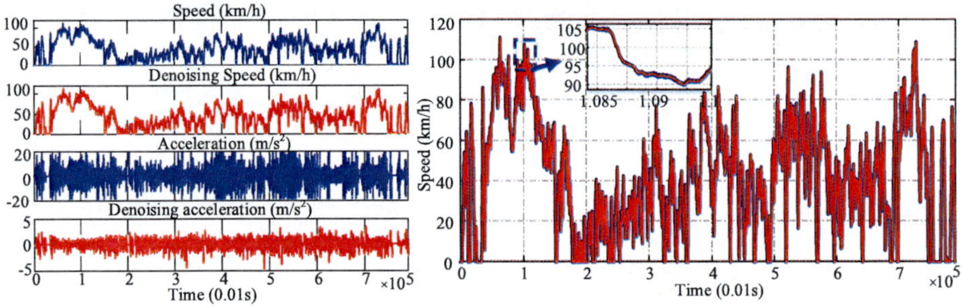

Fig. 3. Wavelet denoising effect based on Composite Index.

3 Vehicle States Prediction and Driving Intention Recognition

3.1 Driving Intention Classification and Recognition

Classification Method of Driving Intention Based on Clustering. For the automatic transmission system of the vehicle, effective driving intention recognition can help to develop advanced shift strategy, so as to improve the personalization of gear-shift decision and meet the driving demand of drivers. In this paper, the driving intention is divided into five categories, namely, rapid acceleration, acceleration, cruise, deceleration and rapid deceleration. The accelerator pedal and brake pedal directly reflect the driver's driving intention, while speed and acceleration indirectly reflect the driver's driving intention. Therefore, the driving behavior data is classified based on the characteristics of vehicle speed, acceleration, accelerator opening and braking oil pressure. In the previous research, the driver's driving intention data is usually manually calibrated, however, the manual calibration is subjective and workload is heavy. The K-means algorithm is an unsupervised learning method, which can effectively dig out the potential rules from a large number of data. Therefore, based on the selected classification features, this paper uses the K-means algorithm to cluster the driving behavior data, and divides the driving behavior data into five categories: rapid acceleration-1, acceleration-2, cruise-3, deceleration-4, rapid deceleration-5.

The flowchart of K-means algorithm is shown in Fig. 4. First, given a number of categories k (k > 0), then calculate the distance from the sample point to k clustering centers. The nearest category is the category to which the sample belongs. With the increase of the number of sample points, the cluster center points are continuously updated until all the data are divided into k categories to meet the principle of high similarity between the same category and low similarity between different categories. The similarity and distance are both defined by Euclidean distance:

$$d_{xy} = \sqrt{\sum_{i=1}^{n} (x_i - y_i)^2} \tag{1}$$

where, x_i represents the ith acceleration sampling data, y_i represents the cluster center value, and n represents the number of characteristic parameters of the cluster center. Driving intention classification results is shown in Fig. 4.

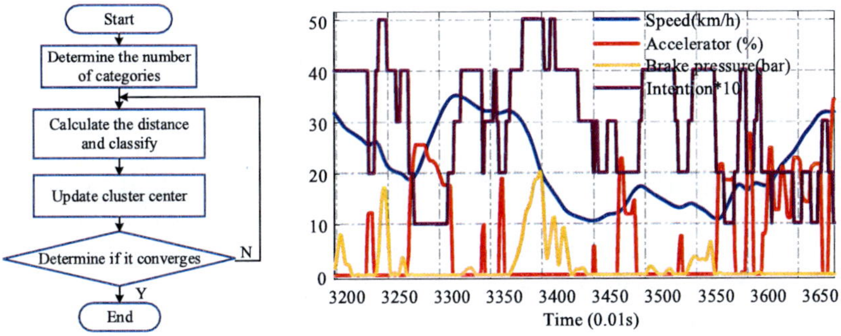

Fig. 4. Driving intention classification results based on K-means clustering method.

Driving Intention Recognition Model Based on Deep Belief Network (DBN). The DBN is a probability generation model, which is based on multiple restricted Boltzmann machines (RBMs) stacked. In the training process, the method of layer by layer training is used, which solves the problem that traditional neural network is not suitable for multi-layer network training. The whole DBN training is divided into two parts: forward stacked RBM learning from low level to high level and backward fine-tuning learning from high level to low level.

(1) **Forward stacked RBM learning**
In essence, pre training is the process of initializing network parameters, which is carried out by layer by layer unsupervised feature optimization algorithm. The network parameters to be initialized are the connection weights between layers and the bias values of neurons in each layer. Taking one layer RBM as an example (Fig. 5), RBM includes a visible layer v and a hidden layer h. There is no connection between each layer unit, and there is full connection between layers. Suppose that there are n visible elements in layer v and m hidden elements in layer h. Then, the energy of RBM as a system is defined as:

$$E(v, h|\theta) = -\sum_{i=1}^{n} a_i v_i - \sum_{j=1}^{m} b_j h_j - \sum_{i=1}^{n}\sum_{j=1}^{m} v_i W_{ij} h_j \tag{2}$$

where: v_i is the state of the ith visible unit; h_j is the state of the jth hidden cell; $\theta = \{W_{ij}, a_i, b_j\}$ are the parameters of RBM. W_{ij} is the connection weight between visible unit i and hidden unit j, a_i is the offset of visible unit i, and b_j is the offset of hidden unit j. Based on the energy function, the joint probability distribution of (v, h) is obtained:

$$P(v, h|\theta) = \frac{e^{-E(v,h|\theta)}}{Z(\theta)} \tag{3}$$

where $Z(\theta) = \sum_{v,h} e^{(v,h|\theta)}$ is the normalized factor, i.e. partition function. Then the marginal distribution (also called likelihood function) of joint probability distribution $P(v, h|\theta)$ can be expressed as:

$$P(v|\theta) = \frac{1}{Z(\theta)} \sum_{h} e^{-E(v,h|\theta)} \tag{4}$$

The task of learning RBM is to find out the parameter θ to fit the given training data. By maximizing the log likelihood function of RBM on the training set, a contrast difference algorithm is proposed in [13] to learn the parameters. Gibbs sampling is used to determine the state of the hidden unit and the visible unit, and the gradient rise method is used to maximize the likelihood function, so as to obtain the maximum likelihood function θ.

(2) **Network parameter tuning.**
After the completion of the pre-training, each layer of RBM can get the initialized parameters, which constitutes the preliminary framework of DBN. Next, the DBN needs to be optimized to further optimize the parameters of each layer of the network, so as to make the discrimination performance of the network better. The optimization process is a supervised learning process, that is, the labeled data is used for training, and the back propagation algorithm is used to fine tune the network parameters, so as to achieve the global optimization of the network. The weights and bias values of traditional BP neural network are all random initial trial, and then BP algorithm is used to optimize the weights and bias values until convergence. In the process of optimizing DBN, the weights of each layer of BP neural network can

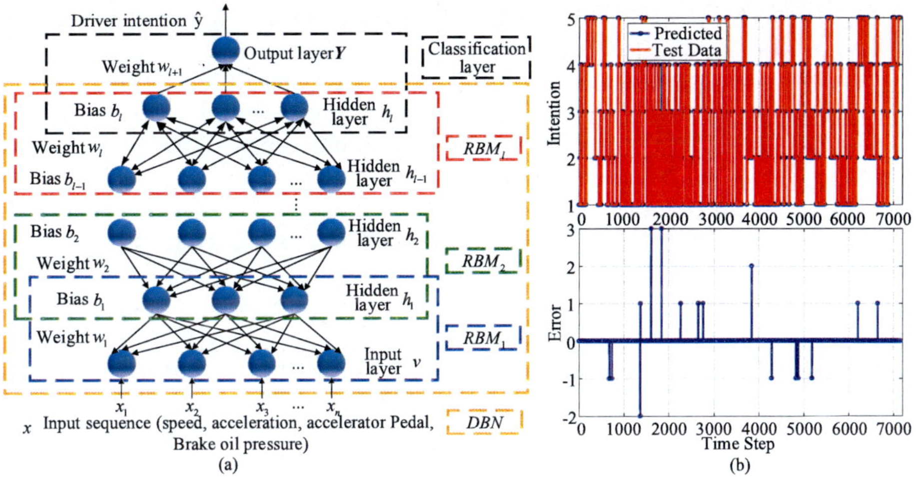

Fig. 5. (a) DBN network structure (b) Driving intention recognition results.

be initialized by the weights of DBN, instead of using the random initial values to initialize the network, and the DBN is expanded into BP neural network, Finally, BP algorithm is used to fine tune the parameters of the whole network, so that the classification performance of the network is better.

Based on the above driving intention classification data, the DBN network is used to classify the driving intention. The results are shown in Fig. 5. It can be seen that the accuracy of driving intention classification is high, with the accuracy of 99.4%.

3.2 Short Term Vehicle States Prediction Based on MLP

Based on the model predictive control theory, this paper makes an online optimization decision on the vehicle gear-shift position. In order to realize the online gear-shift position decision under the MPC framework, and to consider the driving intention in the real-time optimization problem, it is necessary to make short-term prediction of the vehicle speed, acceleration, accelerator opening and braking oil pressure. The prediction data is input into the DBN network to realize the prediction and recognition of the driving intention in the prediction horizon. The predicted driving intention is taken into account in the optimal control problem in the prediction horizon, so as to obtain the gear-shift position which can better meet the driver's demand.

For the short-term prediction of the vehicle states, we use the MLP model. Each state has an MLP for prediction. According to the characteristics of different predicted states, the topology of the four MLPs is different. Among them:

(1) Vehicle speed, accelerator pedal opening and braking oil pressure prediction: the prediction horizon is 10s (one data point per second). The input features of MLP are vehicle speed, acceleration, accelerator opening and braking oil pressure. Each feature is 10s, and the total input feature is 40s series data. The MLP model have

one hidden layer, and the number of nodes is 80 vehicle and accelerator prediction and 100 in braking oil pressure prediction.

(2) Acceleration prediction: the prediction horizon is 10s, and the input characteristics of MLP for acceleration prediction are the same as those for the vehicle speed prediction. The MLP has two hidden layers, the number of nodes in hidden layer 1 is 70, and the number of nodes in hidden layer 2 is 10.

The MLP model is trained based on several standard working conditions and experimental data, and the short-term prediction of four vehicle states is carried out, and the driving intention is recognized by the DBN, so as to realize the prediction of driving intention. The driving intention prediction information is provided to the optimization decision module, and the real-time gear-shift point considering the driving intention can be obtained after optimization calculation.

4 On Line Gear-Shift Points Optimization Considering Driving Intention

In this section, the real-time optimization strategy of the optimal gear-shift point is introduced. Based on the MPC framework, the dynamic programming algorithm is used to obtain the optimal decision variables in the prediction horizon.

The dynamic programming is used to optimize the gear-shift points in the prediction horizon. The gear-shift points in the prediction horizon are divided into N stages, and the step length Δt of each stage is 1s. When the shift value is taken as the state variable $x(k) = i(k)$ and the shift command is taken as the control variable $u(k) = u_g(k)$, the system state transition equation is as follows:

$$x(k+1) = x(k) + u(k) \tag{5}$$

In order to take fuel consumption and driving intention into account, four sub-objective functions are considered in the stage index function of DP:

(1) Fuel consumption: Based on the engine bench test data and vehicle dynamics equation (Eq. 6), the current fuel consumption value can be determined by engine speed and torque interpolation, then the fuel consumption(ml) in stage K is Q(k).

$$\frac{T_e i_g i_0 \eta_T}{r} = mgf + mg\theta + \frac{C_D A v^2}{21.15} + \sigma m\dot{v} \tag{6}$$

where T_e is the torque of the engine, i_g is the gear ratio, i_0 is the gear ratio of the main reducer, η_T is the transmission efficiency, m is the vehicle's mass, g is the acceleration owing to gravity, f is the rolling resistance coefficient, θ is the slope angle, C_D is the air drag coefficient, A is the vehicle frontal area (m^2), v is the vehicle speed (km/h), r is the wheel radius, and σ is the conversion coefficient of the rotating mass.

(2) Reserve power: the driver needs more power when accelerating, and the vehicle needs more reserve power at this time. Therefore, when there is an intention to

accelerate, using the reciprocal of the reserve power P_b as the index as the sub objective function, Δt is sample time, the reserve power index RE is:

$$RE(k) = \Delta t / P_b(k) \tag{7}$$

(3) Brake assist index: when there is deceleration intention, the engine can be used to brake to prevent the brake from heat fading, reduce the driver's working intensity of stepping on the brake pedal, and reduce the vehicle speed more quickly. Brake assist index BA can use the reciprocal of engine anti drag torque T_R as the index value of sub objective function:

$$BA(k) = \Delta t / T_R \tag{8}$$

(4) Number of shifts: the increase of shift times will significantly reduce the driving comfort, at the same time, it will lead to the clutch overheating and wear failure, so the shift times should be minimized. The shift frequency index Cg is:

$$Cg(k) = \mu |x(k+1) - x(k)| \tag{9}$$

where μ is the penalty factor for gearshift. In conclusion, the total objective function of stage k is as follows:

$$Z(x(k), u(k)) = Cg(k) + Q(k) + RE(k) + BA(k) \tag{10}$$

In the prediction horizon, the objective function recursive equation of dynamic programming is as follows:

$$\begin{cases} J_k(x(k)) = 0 \, , k = N \\ J_k^*(x(k)) = \min[Z(x(k), u(k)) + J_{k+1}^*(x(k+1))] \, , 0 \le k \le N-1 \end{cases} \tag{11}$$

where, the control constraints in the k phase are as follows:

$$u_g(k) = \begin{cases} [-1, 0] \, , i = 7 \\ [1, 0], \; i = 1 \\ [-1, 0, 1], \; others \end{cases} \tag{12}$$

The vehicle states constraints are as follows:

$$\begin{cases} 1 \le i(k) \le 7 \\ T_{e\min} \le T_e(k) \le T_{e\max} \\ n_{e\min} \le n_e(k) \le n_{e\max} \end{cases} \tag{13}$$

When the vehicle is running, the data of vehicle speed, acceleration, accelerator pedal opening and braking oil pressure are collected in real time. Four MLP models are used to predict the vehicle speed, acceleration, accelerator pedal and brake oil pressure, which are input into the DBN network to predict the driving intention. Finally, the vehicle speed, acceleration and driving intention information are input into the dynamic programming algorithm to calculate the gear in the prediction horizon in real time, and the first control variable is taken as the current control variable. In the next time step, continue the above steps, rolling optimization, so as to achieve real-time optimal gear-shift decision-making control.

5 Simulation Verification

In this section, the proposed method is verified based on the world light vehicle test cycle (WLTC) standard driving cycle. We compare the proposed strategy with DP, two parameter power-type and economic-type strategy. The simulation results are shown in Fig. 6. Figure 6 (a) shows the prediction results of vehicle speed, Fig. 6 (b) shows the prediction results of acceleration, Fig. 6 (c) shows the prediction results of accelerator pedal, and Fig. 6 (d) shows the prediction results of brake oil pressure. It can be seen from Fig. 6 (a)–(d) that the MLP model has a good prediction of the short-term states of the vehicle, in each prediction horizon (10s), the prediction results will gradually become worse, but also within the acceptable range. Figure 6 (e) shows the result of driving intention recognition. It can be seen that the driver's driving intention is accurately recognized, which provides driving behavior information for gear-shift decision-making.

Figure 6 (f) is the simulation result of power-type and economic-type strategy, and Fig. 6 (g) is the result of DP and the proposed strategy. Table 1 shows the performance indexes of four gear decision methods, including fuel consumption, Reserve power, Brake assist torque and number of shifts. It can be seen that the power-type fuel consumption is the highest, and the proposed strategy is equivalent to that of DP and economic strategy. The Reserve power and Brake assist torque indexes of the proposed and DP strategy are lower than that of the power-type, but far higher than the economic ones. For the index of number of shifts, the proposed strategy is far less than economic and power type. The reduction of number of shifts can help reduce unnecessary friction and heat generation of clutch, and help to reduce the failure rate of clutch. The proposed strategy is close to the indexes of DP, which also shows the effectiveness of the proposed online optimization strategy. In conclusion, the proposed strategy can meet the adaptability of driving behavior, improve the vehicle power, reduce the number of gear changes and significantly improve the driving quality under the condition of low fuel consumption.

Table 1. Performance index of each method under WLTC

	Power	Economic	DP	Proposed
Fuel consumption (L/100 km)	9.56	7.70	7.71	7.73
Reserve power (kW)	80648.78	31089.58	44189.83	43125.59
Brake assist torque (Nm)	6912.56	3708.132	5696.52	5642.12
Number of shifts	188	344	80	90

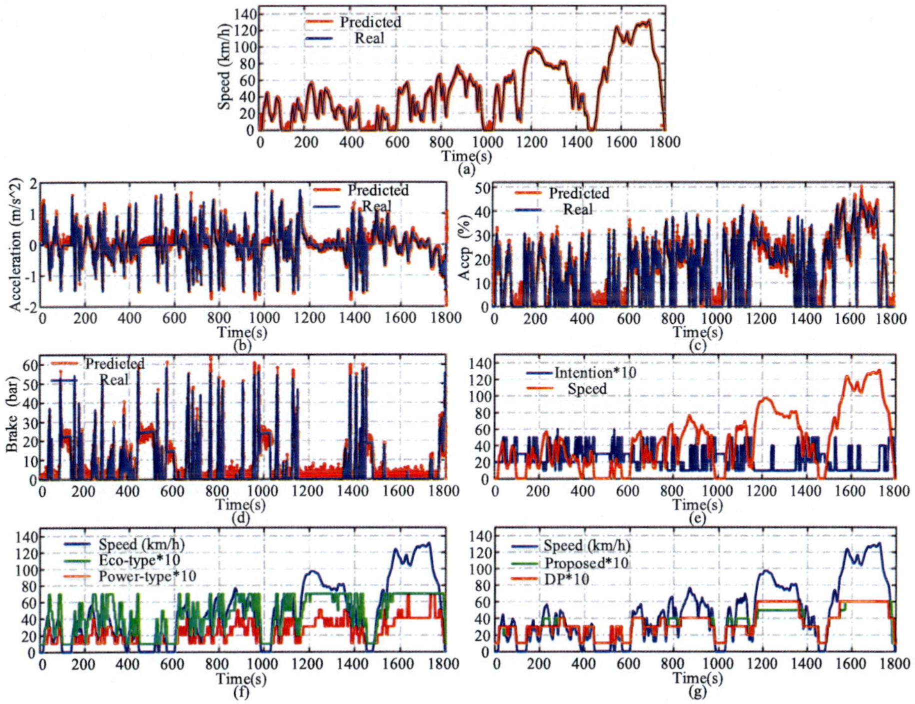

Fig. 6. Simulation results. (a) Speed prediction. (b) Acceleration prediction. (c) Accelerator opening prediction. (d) Brake oil pressure prediction. (e) Driving intention recognition. (f) The results of power-type and economic-type gear-shift decision. (g) The gear-shift results of proposed strategy and DP strategy.

6 Conclusion

Combined with the model predictive control theory, this paper proposes an online decision-making method of gear-shift considering the driving intention, which has good real-time performance. The simulation results show that the proposed short-term prediction method and driving intention identification method can accurately predict and recognize the vehicle states and driving intention, which provides driving behavior information for real-time gear-shift point decision. By comparing four different shift strategies, it shows that the proposed online gear-shift decision-making method can satisfy the adaptability of driving behavior as much as possible under the condition of low fuel consumption, improve the vehicle power performance, reduce the number of shifts, and significantly improve the driving quality.

References

1. Hui, J., Anlin, G.: A new vehicle intelligent shift system. Adv. Mech. Eng. **2013**(3), 1–9 (2013)

2. Liu, Y.G., Qin, D.T., Lei, Z.Z.: Intelligent correction of shift schedule for dual clutch transmissions based on different driving conditions. Appl. Mech. Mater. **121–126**, 3982–3987 (2011)
3. He, C.R., Qin, W.B., Ozay, N.: Optimal gear shift schedule design for automated vehicles: hybrid system based analytical approach. IEEE Trans. Control Syst. Technol. **26**(6), 2078–2090 (2017)
4. Zhu, B., Zhang, N., Walker, P.: Gear shift schedule design for multi-speed pure electric vehicles. Proc. Inst. Mech. Eng. Part D J. Autom. Eng. **229**(1), 70–82 (2015)
5. Chengsheng, M., Liu, H.: Three-parameter transmission gear-shifting schedule for improved fuel economy. Proc. Inst. Mech. Eng. Part D J. Autom. Eng. **232**(4), 521–533 (2018)
6. Fan, J., Hui, J.: Slope shift strategy for automatic transmission vehicle based on the road gradient. Int. J. Autom. Technol. **19**(3), 509–521 (2018)
7. Lei, Y., Liu, K., Zhang, Y.: Adaptive gearshift strategy based on generalized load recognition for automatic transmission vehicles. Math. Prob. Eng. **2015**(PT.16), 1–12 (2015)
8. Zhang, M., Cao, Q.: Fuzzy adaptive shift schedule of tractor subjected to random load. Math. Prob. Eng. Theory Methods Appl. 2017(SI)
9. Eckert, J.J., Santiciolli, F.M., Yamashita, R.Y.: Fuzzy gear shifting control optimization to improve vehicle performance, fuel consumption and engine emissions. IET Control Theory Appl. **13**(16), 2658–2669 (2019)
10. Li, G.H., Hu, J.J.: Modeling and analysis of shift schedule for automatic transmission vehicle based on fuzzy neural network. In: Proceedings of the 8th World Congress on Intelligent Control and Automation July 6–9, Jinan, China (2010)
11. Guo, L., Gao, B., Chen, H.: Online shift schedule optimization of 2-speed electric vehicle using moving horizon strategy. IEEE/ASME Trans. Mechatron. **21**(6), 1 (2016)
12. Guo, L., Gao, B., Liu, Q.: On-line optimal control of the gearshift command for multi-speed electric vehicles. IEEE/ASME Trans. Mechatron. **22**(4), 1519–1530 (2017)
13. Goodfellow, I., Bengio, Y., Courville, A.: Deep Learning. MIT Press, Cambridge (2016)

Analysis of the Influence of Powertrain Mount System on the Longitudinal Dynamic Features of DCT Vehicle Under Typical Working Conditions

Zheng Guo[1], Datong Qin[1](✉), Ju Wu[2], Changzhao Liu[1,4], Yonggang Liu[1], Xin Wang[3], and Xiaotao Zhang[3]

[1] State Key Laboratory of Mechanical Transmission, Chongqing University, Chongqing 400044, China
[2] Chongqing Chang'an New Energy Automobile Technology Co., Ltd., Chongqing 401120, China
[3] Chongqing Chang'an Automobile Co., Ltd., Chongqing 400000, China
[4] Department of Mechanical Engineering, University of Utah, Salt Lake City, UT 84112, USA

Abstract. A powertrain mount system exhibits an obvious influence on the starting and shifting performances of dual clutch transmission (DCT) vehicles. However, the law and mechanism of influence have not been clarified. In this study, a dynamic model of a DCT vehicle including a powertrain mount is built. In addition, the law and mechanism of influence of a powertrain mount on the longitudinal dynamics of a DCT vehicle under typical working conditions, such as starting and shifting are studied. The results show that the powertrain mount has an obvious additional harmonic impact on the longitudinal impact of the DCT vehicle. The stiffness of the mount mainly affects the frequency of the additional harmonics of the impact, and the damping of the mount mainly affects the amplitude of the additional harmonics. The influence of mounting on the longitudinal impact of the DCT vehicle is mainly related to the change of torque acting on the powertrain. The findings of this study lay the foundation for the design optimization of the stiffness and damping parameters of the powertrain mounting system and the improvement of longitudinal dynamic performance of DCT vehicles.

Keywords: Dual clutch transmission · Powertrain mount · Starting · Shifting

1 Introduction

The powertrain mount is a component connecting the powertrain and frame of a vehicle. It mainly supports the powertrain and isolates the vibration transmission between the powertrain, frame, and ground, and optimizes the noise, vibration, and harshness (NVH) performance of the vehicle [1]. Typical working conditions such as starting and shifting cause deformation and torsion movement of the powertrain mount, which affects the dynamic performance of the vehicle, causing the vehicle to shake, and may affect the comfort of the vehicle in severe cases. In the research and development of DCT vehicles,

N. V. Khang et al. (Eds.): ASIAN MMS 2021, MMS 113, pp. 179–191, 2022.
https://doi.org/10.1007/978-3-030-91892-7_17

it has been found that the mount characteristics have great influence on the shaking and shrugging of DCT vehicles in typical working conditions such as starting and shifting and cannot be ignored. However, the existing research has not theoretically clarified the law and mechanism of the influence of the mount; it is necessary as it could lay the foundation for parameter optimization of the powertrain mounting system of DCT vehicles and performance improvement of DCT vehicles under typical working conditions such as starting and shifting.

At present, in the dynamic modeling of vehicle starting and shifting performance, the interaction between the mount and vehicle is usually not considered, and the influence of the powertrain mount on vehicle starting and shifting dynamic performance is ignored [2–6]. Research on powertrain mounting systems mainly focuses on vibration isolation, and seldom considers the influence of the powertrain mount on the longitudinal dynamic performance of a vehicle. Jeong et al. [7] proposed a torque shaft decoupling method that was applied to the design and development of a mounting system to improve the vehicle vibration isolation performance. Park et al. [8] established an analysis model of the powertrain hydraulic mount system by using the transfer function method and the mechanical model method, and studied the influence of the frequency variation characteristics of the hydraulic mount on the mode of the powertrain mounting system. LV Zhaoping et al. [9] realized vibration isolation and motion control of the powertrain by using the force displacement nonlinear characteristics of the mount in three directions of the elastic principal axis. With the improvement of vehicle NVH performance requirements, many scholars have carried out research on vehicle NVH performance, such as the method of NVH transfer path analysis [10]. According to the dynamic characteristics of different mounts, the appropriate mount is selected reasonably to optimize the vehicle NVH [11].

In the existing modeling research of the powertrain mounting system, the mount models are reasonably simplified, mainly considering the mount as a 3-degree-of-freedom (DOF) model, 6-DOF model, and 16-DOF model [12–14]. Previous research [15–18] studied the influence of the powertrain mount on vehicle vibration under engine startup and stop conditions, and proposed a design and optimization method for a mounting system to reduce vehicle vibration under these conditions. Enrico et al. [19] compared simulation results of a DCT vehicle dynamic model with and without mounting under the tip-in condition. The results show that neglecting the powertrain mounts may lead to underestimation of the real vibration level. However, the influence of mounting on other typical working conditions was not considered in their study.

A systematic study has not yet been conducted on the influence of powertrain mounts on the longitudinal dynamic characteristics of DCT vehicles under the conditions of starting and shifting. However, under these typical working conditions, the transmitted torque of the power transmission system changes significantly, and the mount significantly affects the vehicle's longitudinal dynamic performance. In addition, the development practice of vehicle enterprises shows that the influence of the powertrain mount on the dynamic performance under typical working conditions, such as vehicle starting and shifting, cannot be ignored. However, there is a lack of relevant theoretical bases and understanding of the law of influence. Therefore, to clarify the law and mechanism of influence of the powertrain mount on a vehicle's longitudinal dynamic characteristics under typical working conditions, this study established a vehicle dynamics model by considering the powertrain mount and analyzing the interaction between the powertrain mount and vehicle. In addition, we studied the effect of the law and mechanism of the powertrain mount on a vehicle under typical working conditions, such as starting and shifting. Next, we analyzed the influence of the mount stiffness and damping on the longitudinal dynamic performance of the vehicle to provide a theoretical basis for the design optimization of the mounting system and performance improvement of DCT vehicles.

2 Mount-Based Vehicle Dynamics Model

2.1 Dynamic Model of a Powertrain Mounting System

To establish the coupled dynamic model of a DCT vehicle including a powertrain mount, a 3-DOF model of the powertrain mounting system was developed. Any coupling and interaction between the vehicle and transmission system was not considered.

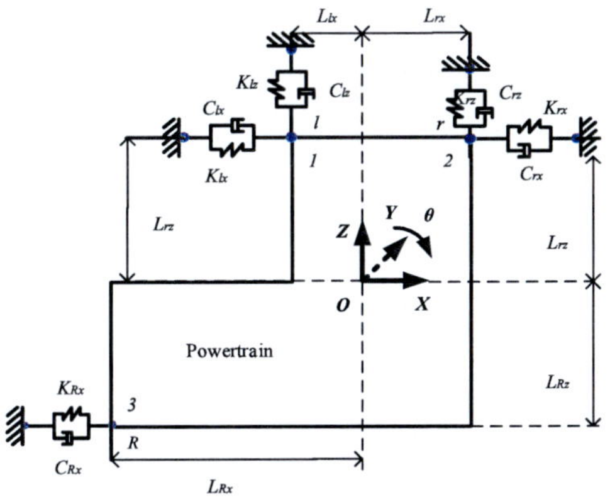

Fig. 1. Dynamic model of a powertrain mounting system

The powertrain mounting system is represented by three mounts with concentrated stiffness and damping measured using the lumped parameter method.

In Fig. 1, the left mount (subscript l) and right mount (subscript r) are represented by the numbers 1 and 2 and have displacements in the longitudinal and vertical Z directions, respectively. The rear mount (subscript R) is represented by 3, and only has displacement along the X direction. The motion of the powertrain mounting system is formulated as follows:

$$\mathbf{M}\ddot{\mathbf{q}} + \mathbf{C}\dot{\mathbf{q}} + \mathbf{K}\mathbf{q} = 0, \tag{1}$$

where $\mathbf{M}$ is the mass matrix, $\mathbf{C}$ is the damping matrix, $\mathbf{K}$ is the stiffness matrix, and $\mathbf{q}$ represents the generalized coordinates $\{X_p, Z_p, \theta_p\}^{\mathrm{T}}$, in which θ_p is the angular displacement of the powertrain.

2.2 Interaction Between Powertrain and Vehicle

The interaction between the powertrain and vehicle is reflected in terms of two aspects of the mount.

1) The powertrain and vehicle frame are connected by the mount, which allows relative movement between these two entities, and force is transmitted through the mount. Figure 2 shows the interaction between the powertrain and DCT vehicle through the mount.

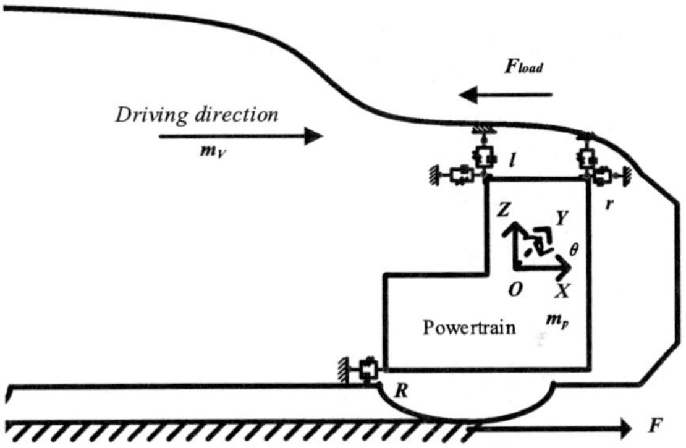

Fig. 2. Longitudinal interaction between the powertrain and vehicle

The longitudinal dynamic equation of the interaction between the powertrain and vehicle is expressed as follows:

$$\begin{cases} m_v a_{v,x} = F - F_{load} - F_{mou} \\ m_p a_{v,x} = F_{mou} \\ F_{mou} = F_{Rx} + F_{rx} + F_{lx} \end{cases} \tag{2}$$

where m_v is the mass of the vehicle without the powertrain; F is the driving force of the vehicle; F_{load} is the vehicle total drag force (including rolling resistance, wind resistance, slope resistance, and acceleration resistance); F_{mou} is the total force transmitted by the mount; $a_{v,x}$ is the longitudinal acceleration of the vehicle (excluding the powertrain); $a_{p,x}$ is the longitudinal acceleration of the powertrain; and F_{ix} ($i = r, l, R$) is the force transmitted by each mount in the longitudinal direction.

2) When the power transmission system is started, the reaction torque of the powertrain is strictly related to the engine output torque and DCT transfer torque and is represented as

$$T_p = T_e + \sum\nolimits_i T_{gear,i} + \sum\nolimits_j T_{B,i}, \tag{3}$$

where T_e is the reaction torque induced by the engine; $T_{gear,i}$ ($i = 1,2$) is the reaction torque induced by the torque transmitted through clutch 1 or 2; and $T_{B,j}$ is the friction torque of the bearing, which is usually several orders of magnitude lower than other torques and can be ignored.

2.3 Vehicle Dynamics Model with a Mount

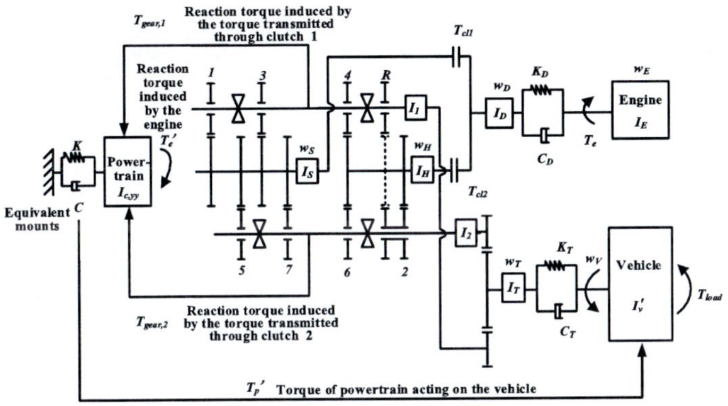

Fig. 3. Vehicle dynamics model considering the powertrain mounting system

According to the interaction between the powertrain mounting system and vehicle, the vehicle dynamics model with the mount is established using the lumped parameter method. As shown in Fig. 3, the powertrain mount is represented as a torsion spring with 1 DOF for the sake of simplicity and clarity. However, in the actual calculation, it is considered as a 3-DOF model, as shown in Fig. 1.

The motion of the system is described by **q** as follows:

$$q = \left[X_p\, Z_p\, \theta_p\, \theta_E\, \theta_D\, \theta_T\, \theta_V\right], \tag{4}$$

where X_p and Z_p are the displacements of the mount in the X and Z directions, respectively.

The coupled motion equation between the vehicle transmission system and powertrain mount is expressed as follows:

$$\begin{cases} m_p\ddot{X} + C_{11}(\dot{X} - r_w\dot{\theta}_V) + K_{11}(X - r_w\theta_V) + C_{13}\dot{\theta} + K_{13}\theta = 0 \\ m_p\ddot{Z} + C_{22}\dot{Z} + K_{22}Z + C_{23}\dot{\theta} + K_{23}\theta = 0 \\ I_{c,yy}\ddot{\theta}_p + C_{33}\dot{\theta}_p + K_{33}\theta_p + C_{13}(\dot{X} - r\dot{\theta}_V) + K_{13}(X - r\theta_V) + C_{23}\dot{Z} + K_{23}Z = -T_p \\ I_E\ddot{\theta}_E + K_D(\theta_E - \theta_D) + C_D(\dot{\theta}_E - \dot{\theta}_D) = T_e \\ I_D\ddot{\theta}_D - K_D(\theta_E - \theta_D) - C_D(\dot{\theta}_E - \dot{\theta}_D) = -T_{cl1} - T_{cl2} \\ I_{eq}\ddot{\theta}_T + K_T(\theta_T - \theta_V) + C_T(\dot{\theta}_T - \dot{\theta}_V) = T_{cl1}\, i_1 i_{a1} + T_{cl2} i_2 i_{a2} \\ I'_V\ddot{\theta}_V - K_T(\theta_T - \theta_V) - C_T(\dot{\theta}_T - \dot{\theta}_V) \\ -r_w\left[C_{11}(\dot{X} - r\dot{\theta}_V) + K_{11}(X - r\theta_V) + C_{13}\dot{\theta}_p + K_{13}\theta_p\right] = -T_{load} \end{cases} \tag{5}$$

where I_E is the moment of inertia of the engine crankshaft, I_D is the equivalent moment of inertia of the driving plate of the clutch, I_{eq} is the equivalent moment of inertia of the driven plate of the clutch i_{a1} is the ratio of final drive 1, i_{a2} is the ratio of final drive 2, K_D and C_D are the torsional stiffness and damping of the DMF respectively, K_T and C_T are the torsional stiffness and damping of the output shaft respectively, T_{load} is the load torque, τ is the transmission speed ratio, and r_w is the wheel radius.

2.4 Evaluating Indicator

Jerk refers to the change rate of vehicle acceleration, that is, the derivative of acceleration to time. If the jerk is too large, it will affect the starting smoothness and comfort of the vehicle; the jerk is too small, the clutch engagement time is too long, and the starting process is too slow. The mathematical expression of jerk j is:

$$j = \frac{da}{dt} = \frac{d^2v}{dt^2}$$

Where, j represents vehicle jerk, a represents vehicle acceleration, v represents vehicle speed and t represents time.

The impact degree not only truly reflects people's real feeling of comfort, but also excludes the impact of bounce and bumpy acceleration caused by road conditions and non shift operation, and truly reflects the load change of vehicle transmission system and vehicle motion state during starting and shifting.

Different national standards on vehicle impact give different impact limits. The jerk standard value in Germany is 10m/s^3. The jerk standard index of Germany, which is more stringent and more general in the world, can be selected to evaluate the comfort of the whole vehicle. Therefore, the upper limit $j_{\max}$ of the impact index with reference to the International German standard is 10 m/s^3.

3 Analysis of Typical Working Conditions

In the process of driving, a significant change is observed in the transmission torque of the whole vehicle under typical working conditions of starting and shifting. Mounts have a large deformation, which causes the impact and towering of the vehicle and affects its longitudinal dynamic performance and ride comfort. Therefore, the influence of the mount on the longitudinal dynamic characteristics of a DCT vehicle under typical working conditions must be studied.

3.1 Starting Condition

The engine adopts constant-speed control in the starting process. In this study, the DCT vehicle dynamics models with and without a mount were used to simulate the starting process. The simulation results are shown in Fig. 4. Figure 4 (a) shows the driving speed and driven ends of clutch 1 in the starting process and Fig. 4 (b) shows the jerking behavior of the vehicle.

Figure 4 (b) shows the starting process of the entire vehicle. Compared with the vehicle without the mount, the DCT vehicle with the mount produces an obvious additional harmonic impact. The maximum fluctuation amplitude is 4.5 $\text{m}\cdot\text{s}^{-3}$ and the maximum jerk exceeds 10 $\text{m}\cdot\text{s}^{-3}$, which does not meet the jerk criterion under the German standard, thus affecting the driving comfort of the vehicle in the starting process.

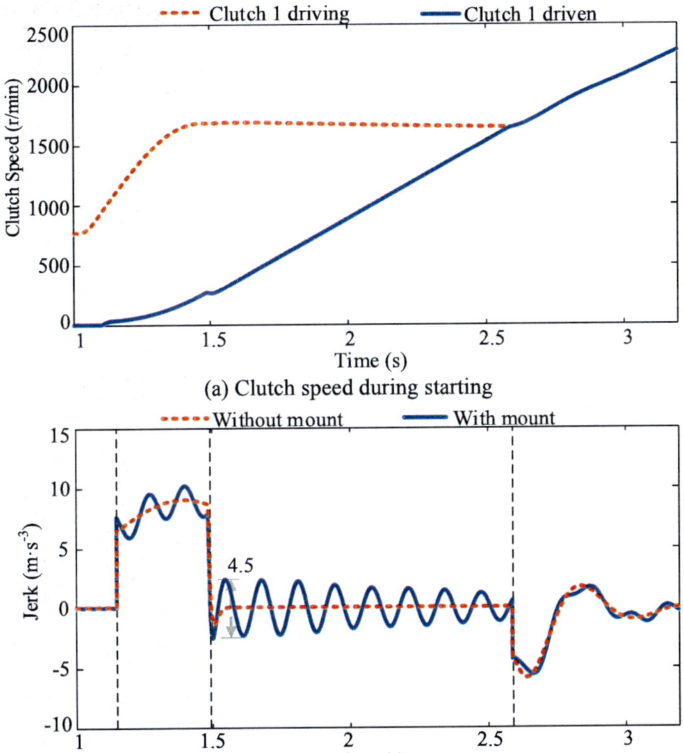

(a) Clutch speed during starting

(b) Vehicle jerk during starting

Fig. 4. Simulation results of the influence of mounting under the starting condition

3.2 Shifting Conditions

In the experiment, the throttle was set at 20% to simulate the shifting process for the dynamic models of the DCT vehicles with and without mounting, as shown Fig. 5. Figure 5 (a) shows the clutch speed during the upshift from 1st to 2nd gear and the downshift from 2nd to 1st gear. Figure 5 (b) shows the jerk of the vehicle during upshifting and downshifting.

As shown in Fig. 5 (b), in the process of shifting from 1st to 2nd gear, the jerking of the vehicle with the influence of the mount obviously produces additional harmonics compared with that of the vehicle without the mount, and the maximum fluctuation amplitude is 12.4 $m \cdot s^{-3}$, which is 18% higher than the vehicle without the mount. Furthermore, in the process of shifting from 2nd to 1st gear, the impact fluctuation of the vehicle with the mount increases and the maximum fluctuation amplitude is 26.5 $m \cdot s^{-3}$, which is 9.5% higher than that without a mount. Here, the maximum impact of the vehicle is greater than 10 $m \cdot s^{-3}$, which does not meet the impact criterion of the German standard and affects the driving comfort.

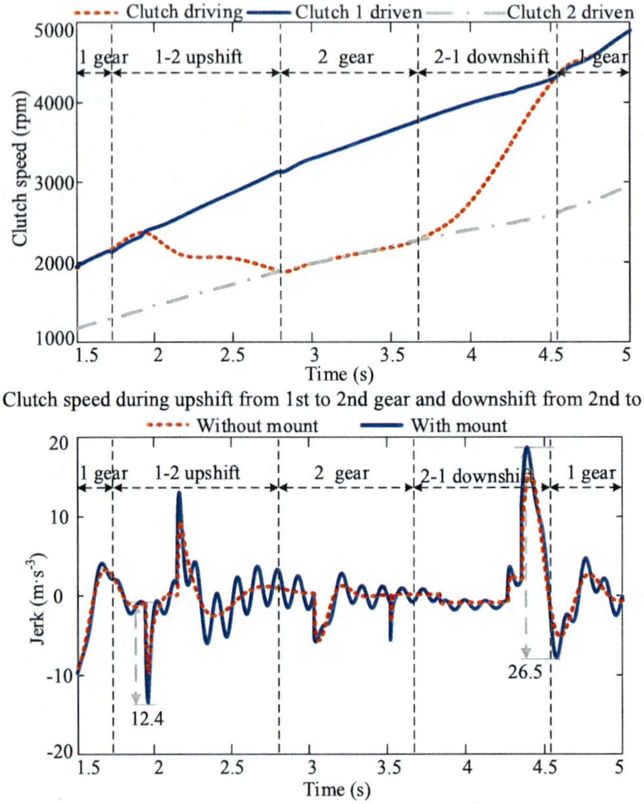

(a) Clutch speed during upshift from 1st to 2nd gear and downshift from 2nd to 1st gear

(b) Jerking of the vehicle during upshifting from 1st to 2nd gear and downshifting from 2nd to 1st gear

Fig. 5. Simulation results of the influence of mounting on the shifting conditions

4 Influence of Stiffness and Damping Parameters of the Mount on Vehicle Longitudinal Dynamic Characteristics

The vehicle models are used to determine the influence of the stiffness and damping parameters of the mount on the longitudinal dynamics of the DCT vehicle. Because of the limitation of length, this paper presents only the influence measurement during the starting process. The results of other working conditions show the same trend and are not discussed in detail. Figure 6 (a) shows the vehicle jerk in the starting process when the mount damping is constant and the stiffness is set at 0.5, 1, and 3 times, respectively. Figure 6 (b) shows the vehicle jerk when the mount stiffness is kept constant and the damping is set to 0.5, 1, and 3 times, respectively.

According to Figs. 6 (a), the fluctuation frequency of the jerk is 5.3 Hz, 7.5 Hz, and 13 Hz when the mount stiffness is 0.5, 1, and 3 times the initial mounting stiffness. The results show that the additional harmonic frequency of the vehicle jerk is related to the mount stiffness, and the additional harmonic frequency of impact increases with the increase in mounting stiffness. Figure 6 (b) shows that the harmonic amplitude of the jerk increases by 26.4% at 0.5 times the damping value compared with that for the damping of 1, and the amplitude of the additional harmonics at 3 times the damping is 58.9% less than that of the damping of 1. These results show that the additional harmonic amplitude of the vehicle impact degree is related to the mount damping, and the additional harmonic amplitude of the impact degree decreases with an increase in mount damping.

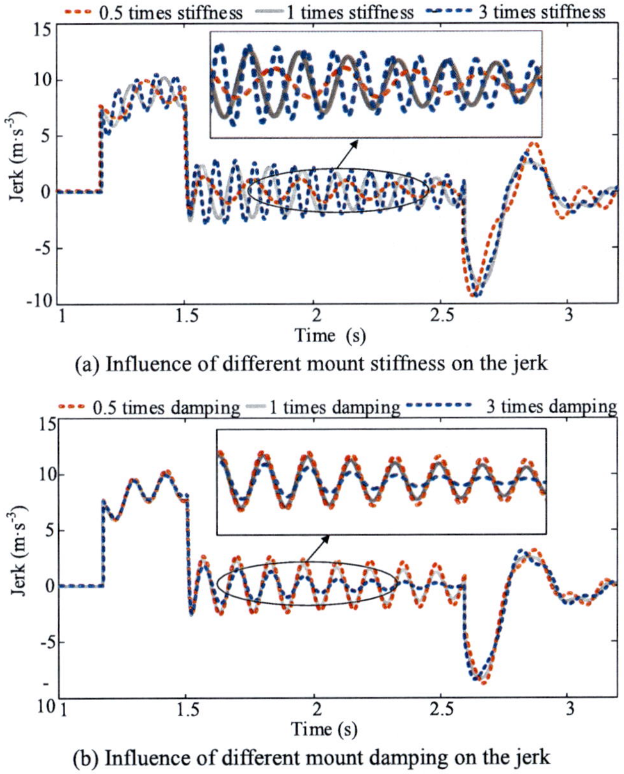

Fig. 6. Simulation results of different stiffness and damping of the mount on the starting process

5 Analysis of the Influence Mechanism of the Mount on Vehicle Longitudinal Dynamic Characteristics

To analyze the influence mechanism of the mount on vehicle longitudinal dynamic characteristics, the torque $T_p{'}$ of the powertrain acting on the whole vehicle by mounting and the whole torque $T_V{'}$ and T_V with and without mounting are compared and analyzed as they change with time.

The torque of the powertrain acting on the vehicle by mounting can be determined via Eq. (5):

$$T_p^{'} = r_w\left[C_{11}(\dot{X} - r\dot{\theta}_V) + K_{11}(X - r\theta_V) + C_{13}\dot{\theta}_p + K_{13}\theta_p\right] \tag{6}$$

The total torque equation of $T_V{}'$ considering the driving force, driving resistance, and the force of mounting on the vehicle is, according to the vehicle dynamics Eq. (5), as follows:

$$T_V' = K_T(\theta_T - \theta_V) + C_T(\dot{\theta}_T - \dot{\theta}_V) - T_{load} + T_p' \tag{7}$$

The total torque equation of T_V without the mount considering the driving force and driving resistance on the vehicle is as follows:

$$T_V = K_T(\theta_T - \theta_V) + C_T(\dot{\theta}_T - \dot{\theta}_V) - T_{load} \tag{8}$$

By considering the continuous starting process of a DCT vehicle (from 1st gear up to 2nd gear, and then back to 1st gear) as an example, the changes of $T_p{}'$, $T_V{}'$, and T_V with time are compared and analyzed. As shown in Fig. 7 (a), the torque $T_p{}'$ of the powertrain acting on the vehicle under the continuous process with the influence of the mount changes the total torque $T_V{}'$ acting on the vehicle with the mount compared with the total torque T_V acting on the vehicle without considering the mount. Figure 7 (b) shows the jerk change of the whole vehicle with and without the influence of the mount in the continuous process.

Taking the continuous process of DCT vehicle starting, 1st gear up to 2nd gear, and 2nd gear down to 1st gear as an example, the influence mechanism of a mount on vehicle longitudinal dynamic characteristics is analyzed. As shown in Fig. 7 (a), the torque $T_p{}'$ of the powertrain acting on the vehicle under the continuous process with the influence of the mount changes the total torque $T_V{}'$ acting on the vehicle with the mounting compared with the total torque T_V acting on the vehicle without considering the mount. Figure 7 (b) shows the jerk change of the whole vehicle with and without the influence of the mount in the continuous process.

It can be seen from Fig. 7 (a) that in this continuous starting process, due to the large driving torque of the whole vehicle, the torque $T_p{}'$ of the powertrain acting on the vehicle through the mount is large, resulting in large deformation and torsional movement of the mount. When the engine power is constant, the torque transmitted by the power system of the 1st gear is greater than that of the 2nd gear. Therefore, the $T_p{}'$ of the 1st gear is greater than that of the 2nd gear. In the process of upshift and downshift, $T_p{}'$ fluctuates obviously due to the great change of system torque. At the same time, according to Eqs. (7) and (8) and the torque $T_p{}'$ of the powertrain acting on the vehicle through the mounting, the magnitude and change rate of the total torque $T_V{}'$ and T_V acting on the vehicle vary greatly both with and without the mount in the above continuous process. When the mount is considered, the total torque $T_V{}'$ acting on the vehicle is less than the total torque $T_V{}'$ when the mount is not considered. At the same time, due to the deformation and torsional movement of the mount, the torque $T_p{}'$ of the powertrain acting on the whole vehicle produces impact and fluctuation, thus the total torque $T_V{}'$ acting on the vehicle has an obvious fluctuation.

It can be seen from Figs. 7 (a) and (b) that, when the mount is considered or not, the size and rate of total torque of $T_V{}'$ and T_V acting on the vehicle have changed, and the vehicle experiences fluctuating jerk. However, due to the influence of the deformation and torsion of the mount, the torque $T_p{}'$ of the powertrain on the vehicle has obvious additional harmonic fluctuation.

As a result, there are additional harmonic fluctuations in the total torque $T_V{}'$ acting on the vehicle, which results in significant additional harmonic fluctuations in the jerk of the vehicle. Especially when the torque of the whole vehicle transmission system changes greatly, the additional fluctuation amplitude of vehicle impact is larger.

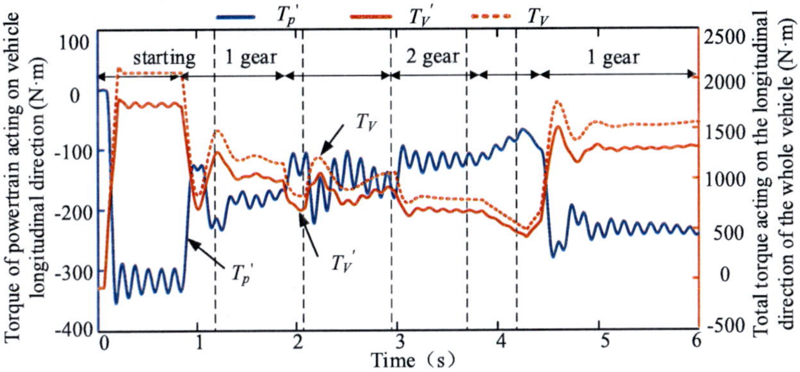

(a) Torque $T_p{}'$, $T_V{}'$, and T_V acting on the vehicle in the continuous process

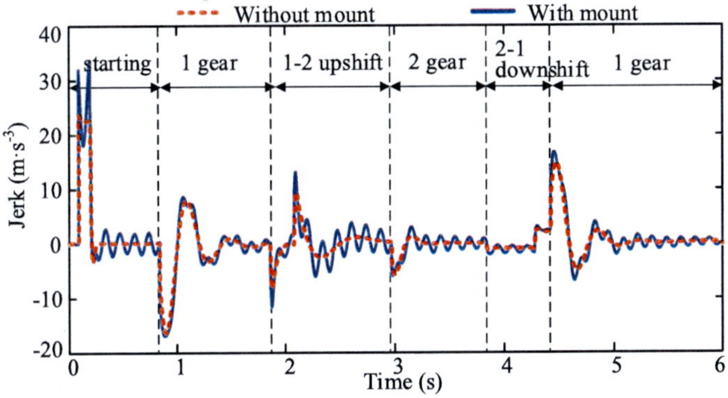

(b) Vehicle impact during starting and shifting

Fig. 7. Simulation results of the influence of mounting on the starting and shifting processes

6 Conclusion

In this study, a dynamic model of a DCT vehicle with a powertrain mount was established. In addition, we simulated and analyzed the influence of the powertrain mount on the longitudinal dynamic characteristics of the DCT vehicle under typical working conditions, such as starting and shifting.

(1) The powertrain mount has obvious influence on the longitudinal dynamic characteristics of a DCT vehicle under typical working conditions such as starting and shifting, which makes the jerk of the DCT vehicle produce obvious additional harmonics under starting and shifting conditions.

(2) The influence of the stiffness and damping of the powertrain mount on the longitudinal dynamic characteristics of the vehicle under typical working conditions are as follows. With increasing mounting stiffness, the frequency of the additional fluctuation of the vehicle jerk accelerates. In addition, with an increase in the mounting damping, the amplitude of the additional fluctuation of the jerk decreases. Moreover, parameter optimization of the stiffness and damping of the powertrain mount can improve the longitudinal dynamic performance of DCT vehicles.
(3) The influence of a mount on the longitudinal dynamics of a DCT vehicle is mainly related to the change of transmission torque of the whole vehicle. The greater the torque change of the transmission system, the more obvious the impact of the mount on the vehicle impact.

The findings of this study lay the foundation for the optimization design of the stiffness and damping parameters of the powertrain mounting system and the improvement of longitudinal dynamic performance of DCT vehicles.

Acknowledgments. This work has been supported by The State Key Laboratory of Mechanical Transmission, Chongqing University, China; the financial support of this work by the National Natural Science Foundation of China (grant no. U1764259), and Chongqing Technology Innovation and Application Development Project (grant no. cstc2019jscx-zdztzxX0047).

References

1. Pollack, M.L., Govindswamy, K., Wellmann, T.I., et al.: NVH refinement of diesel powered sedans with special emphasis on diesel clatter noise and powertrain harshness. SAE Tech. Paper Series, **15**(5) (2007)
2. Rondinelli, E., Velardocchia, M., Galvagno, E.: Electro-mechanical transmission modelling for series-hybrid tracked tanks. Int. J. Heavy Veh. Syst. **19**(3), 256–280 (2012)
3. Walker, P.D., Zhang, N.: Modelling of dual clutch transmission equipped powertrains for shift transient simulations. Mech. Mach. Theory **60**, 47–59 (2013)
4. Galvagno, E., Morina, D., Sorniotti, A., Velardocchia, M.: Drivability analysis of through-the-road-parallel hybrid vehicles. Meccanica **48**(3), 351–366 (2013)
5. Walker, P.D., Zhang, N.: Numerical investigations into shift transients of a dual clutch transmission equipped powertrains with multiple nonlinearities. J. Vib. Control **21**(8), 1473–1486 (2015)
6. Galvagno, E., Velardocchia, M., Vigliani, A.: Dynamic and kinematic model of a dual clutch transmission. Mech. Mach. Theory **46**(6), 794–805 (2011)
7. Jeong, T., Singh, R.: Analytical methods of decoupling the automotive engine torque roll axis. J. Sound Vib, **234**(1), 85–114 (2000)
8. Park, J.Y., Singh, R.: Role of spectrally varying mount properties in influencing coupling between powertrain motions under torque excitation. J. Sound Vib. **329**(14), 2895–2914 (2010)
9. Lv, Z.P., Yang, X.: The design method of the nonlinear stiffness for automotive powertrain mounting system. Automot. Eng. **33**(07), 581–585 (2011)
10. Muller, M., Eckel, H., Leibach, M., et al.: Reduction of noise and vibration in vehicles by an appropriate engine mount system and active absorbers. SAE Technical Paper Series (1996)

11. Janssens, K., Gajdatsy, P., Gielen, L., et al.: A new transfer path analysis method based on parametric load models. Mech. Syst. Signal Pr. **25**(4), 1321–1338 (2011)
12. Yu, Y.H., Naganathan Nagi Dukkipati, G., Rao, V.: A literature review of automotive vehicle engine mounting systems. Mech. Mach. Theory, **36**(1), 123–142 (2001)
13. El Hafidi, A., Martin, B., Loredo, A., et al.: Vibration reduction on city buses: Determination of optimal position of engine mounts. Mech. Syst. Signal Pr. **24**(7), 2198–2209 (2010)
14. Agarwal, K., Hazra, S., Kolage, V.: Virtual analysis of engine mount stiffness and stopper gap tuning for better NVH performance. SAE Tech. Paper Series (2017)
15. Rao, M., Moorthy, S., Raghavendran, P.: NVH analysis of powertrain start/stop transient phenomenon by using wavelet analysis and time domain transfer path analysis. SAE Tech. Paper Series (2015)
16. Sugimura, H., Takeda, M., Yamaoka, H., et al.: Development of HEV engine start shock prediction technique combining motor generator system control and multi-body dynamics models. SAE Technical Paper Series (2013)
17. Wang, D.: Study on the automotive PMS with semi-active hydraulic damping strut. South China University of Technology, Guangzhou, vol. 33, no. 7, pp. 581–585 (2017)
18. Wang, D.Y., Jiang, M., He, K.F., Li, X.J., Li, F.J.: Study on vibration suppression method of vehicle with engine start-stop and automatic start-stop. Mech Syst Signal Pr, 142 (2020)
19. Enrico, G., Pablo, G., Mauro, V., et al.: A theoretical investigation of the influence of powertrain mounts on transmission torsional dynamics. SAE Int. J. Veh. Dyn. Stabil. NVH (2) (2017)

Energy Management and Design Optimization for a Novel Hybrid Powertrain Based on Power-Reflux CVT

Guanlong Sun[1], Dongye Sun[1](✉), Datong Qin[1], Ke Ma[1], and Junlong Liu[2]

[1] Chongqing University, Chongqing 400044, China
dongye_sun@cqu.edu.cn
[2] Shandong University of Science and Technology, Qingdao 266590, China

Abstract. This paper presents a novel hybrid powertrain based on power-reflux continuously variable transmission (CVT) and explore its performance potential by optimizing energy management and design. Firstly, its structure and working principle were described, and the mode of power coupling and the working state of the clutches under different operation modes were explained. Secondly, the transmission ratio and transmission efficiency characteristics of the system were analyzed. Then, an iterative bi-loop multi-objective optimization method was proposed to conduct a simultaneous optimization on control strategy and component sizing to reduce the drivetrain cost and fuel consumption while improving the vehicle acceleration capacity, in which the outer loop is for selecting powertrain with different sizes while the inner loop performs the energy management strategy based on Pontryagin's minimum principle with the parameters selected. Finally, the Pareto optimal solution set is obtained by simulation under NEDC conditions. The results show that compared with original vehicle, the economical and dynamical performance of the vehicle corresponding to Pareto optimal solution set are all obviously improved. In addition, Pareto optimal solution set provide many options to choose in preliminary design according to the requirements of designers.

Keywords: Power-reflux · Continuously variable transmission · Energy management · Multi-objective optimization · Pareto optimal solution

1 Introduction

The application of the conventional metal belt CVT in plug-in hybrid electric vehicles (PHEVs) is limited due to its small speed ratio range and low torque transmission capability [1]. To expand the adaptability of conventional CVT, a novel power-reflux CVT characterized by a planetary gear train and a fixed-ratio (FR) gear installed on the conventional CVT was built. To determine the performance potential of a PHEV with power-reflux CVT (PRHEV), its optimal design and energy management need to be investigated.

To improve fuel economy, the DIRECT algorithm was used by He et al. for the optimization of engine power, motor power and battery capacity [2]. For similar optimization parameters, the PSO algorithm was adopted by YILDIZ et al. to obtain the

N. V. Khang et al. (Eds.): ASIAN MMS 2021, MMS 113, pp. 192–203, 2022.
https://doi.org/10.1007/978-3-030-91892-7_18

minimum fuel consumption [3]. Also using PSO algorithm, Lin et al. optimized the control rules to improve the vehicle performance [4]. However, these researchers analyzed independently either component sizing or control strategy in HEVs. These two aspects, in practice, are intimately coupled. To probe their interplay, Zeng et al. applied quantum genetic algorithm to optimize the power system parameters and control strategy parameters simultaneously, relying on the rule-based control strategy [5], however, a non-optimal energy management may lead to an inappropriate component sizing. To obtain a global optimal design, DP algorithm has been introduced to conduct a simultaneous optimization on them at the cost of tremendous computational complexity [6]. To improve computational efficiency, an integrated convex programming framework was built in [7], however, the approximate processing of the powertrain model with the convex function may cause analysis errors.

In this study, a combined energy management strategy and component sizing design optimization problem for a PRHEV is formulated and solved in an iterative bi-loop method, where Pontryagin's minimum principle (PMP) is applied for the energy management in the inner loop while the component sizing is optimized in the outer loop by evolutionary algorithm, simultaneously.

2 System Structure and Working Principle

2.1 Structure

The structure of the proposed PRHEV is shown in Fig. 1. It consists of engine, clutch L1, motor and power-reflux CVT. In the power-reflux CVT, the driving gear of the fixed-ratio (FR) gears are coaxial with the driving pulley of the metal belt CVT; the driven gear of the FR gears is connected to the planet carrier through the clutch L2; one end of the sun gear is fixedly connected to the driven pulley of the CVT, and the other end is connected to the final drive through the clutch L3; and the ring gear is directly connected to the final drive. Thus, there are two ways of power transmission of the PRHEV: One is flowed through ring gear and the other is output by sun gear. Its main parameters of vehicle were listed in Table 1.

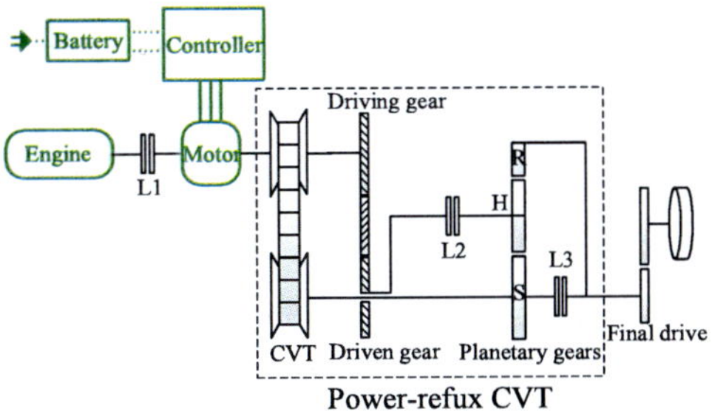

Fig. 1. Structure of PRHEV.

Table 1. Main parameters of vehicle and basic parameters of the PRHEV powertrain

Parameters	Value	Parameters	Value
Vehicle mass (m)/kg	1337	Rolling friction coefficient (f)	0.015
Aerodynamic drag coefficient (C_D)	0.25	Wheel radius (r) /m	0.287
frontal area (A)/m2	2.64	Final drive ratio (i0)	3.3

2.2 Working Principle

Through the disconnection and engagement of clutches L1, L2, and L3, PRHEV can realize multiple working modes. Among them, Clutch L1 was mainly employed for switching between the driving modes. If this clutch was disconnected, the vehicle was operated using only the motor. Once clutch L1 was engaged, the engine was also involved in the operation. The vehicle could be operated using only the engine similar to a traditional car or the dual power source according to the demand. Clutch L1 is used for the decision of the basic drive mode, and clutches L2 and L3 are adopted to switch the speed regulation modes.

If L2 is disconnected but L3 is engaged under the determined drive mode, all the power output by the power sources is directly transmitted to the final drive through the CVT and then to the wheels to meet the needs of the vehicle as shown in Fig. 2(a). Therefore, this speed regulation mode is called the pure-CVT speed regulation mode.

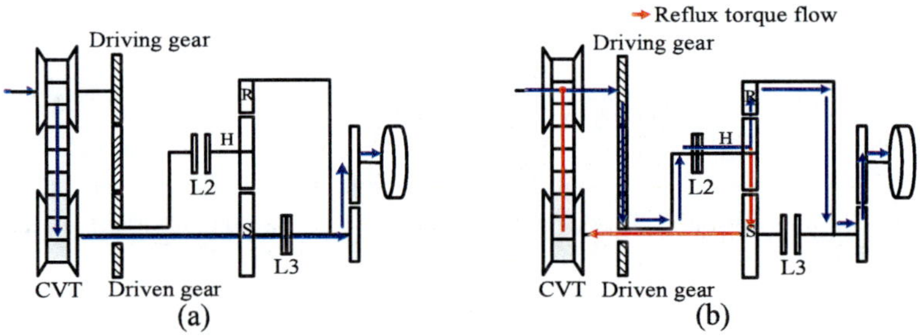

Fig. 2. Power flow of speed regulation modes

If system engages L2 and disconnects L3 under the determined drive mode, the power is transmitted to the planet carrier through the FR gears and L2. However, the power was returned partially to the input shaft via the planet gear, sun gear, and the CVT device, where it was combined with the output of the power source to drive the vehicle jointly as shown in Fig. 2(b). Therefore, this speed regulation mode is called the power-reflux speed regulation mode.

Considering various combinations of driving modes and speed regulation modes, the PRHEV can realize eight working modes. The working states of clutch and motor under different operation modes are listed in Table 2.

Table 2. Working modes and components state

Drive mode	Operation mode	Clutches states		
		L1	L2	L3
Pure electric driving	electric_pure-CVT	×	×	√
	electric_power-reflux	×	√	×
Pure engine driving	engine_pure-CVT	√	×	√
	engine_power-reflux	√	√	×
Hybrid driving	hybrid_pure-CVT	√	×	√
	hybrid_power-reflux	√	√	×
Charging while driving	charging_pure-CVT	√	×	√
	charging_power-reflux	√	√	×

* √ refers to clutches engagement, whereas × refers to clutches disengagement.

2.3 Transmission Characteristics

The transmission efficiency and ratio of the final drive can be regarded as constants, therefore, the transmission characteristics of the PRHEV depend on transmission efficiency and ratio changing regular of power-reflux CVT. Since there are two speed regulation modes in the system, transmission ratio and transmission efficiency characteristics of the power-reflux CVT need to be discussed separately in each mode.

Transmission Ratio Characteristic. In the power-reflux mode, the rotational speed characteristics of PRHEV can be obtained according to the rotational speed characteristics of planetary rows [8] and the transmission mechanism of the CVT as follows:

$$\begin{cases} n_s + kn_q - (1 + \mathrm{k})n_j = 0 \\ n_{in} = i_d n_j = i_{cvt} n_s \\ n_{out} = n_q \end{cases} \tag{1}$$

where n_j, n_q, n_s are the planet carrier, ring gear, and sun gear rotational speed, respectively; n_{in}, n_{out} are the input and output rotational speed of the PRCVT, respectively; k is the ring/sun gear ratio, and its value is 2.6; i_d is the ratio of the FR gears, and its value is 2,4704; i_{cvt} is the ratio of the CVT.

Therefore, the power-reflux CVT transmission ratio i_g in the power-reflux mode can be obtained as Eq. (2)

$$i_g = \frac{n_{in}}{n_{out}} = \frac{\mathrm{k}}{(1 + \mathrm{k})/\mathrm{i_d} - 1/\mathrm{i_{cvt}}} \tag{2}$$

In the pure CVT mode, the power is transferred through CVT, thus, $i_g(t) = i_{cvt}(t)$.

Efficiency Characteristic. The torque characteristic of planetary rows and the torque relations of the transmission system in the power-reflux mode are shown in the following:

$$\begin{cases} T_s : T_q : T_j = 1 : k : (1 + k) \\ T_{in} = \frac{(1+k)i_{cvt} - i_d \eta_{cvt}}{(1+k)i_{cvt} i_d} T_j \\ T_{out} = T_q \end{cases} \tag{3}$$

where T_j, T_q, T_s are the planet carrier, ring gear, and sun gear torque, respectively; T_{in}, T_{out} are the input and output torque of the power-reflux CVT; $\eta_{cvt}(t)$ is the instantaneous CVT efficiency.

Therefore, the power-reflux CVT efficiency η_g in the power-reflux mode can be obtained through output and input power according to Eq. (4)

$$\eta_g = \frac{P_{out}}{P_{in}} = \frac{T_{out}}{T_{in} i_g} = \frac{(1 + k)i_{cvt} - i_d}{(1 + k)i_{cvt} - \eta_{cvt} i_d} \tag{4}$$

In the pure CVT mode, η_g (t) equals $\eta_{cvt}(t)$. The two speed regulation modes should be switched at the transmission ratio synchronous point the as far as possible to avoid the significant impact caused by the sudden change of transmission ratio. Thus,

$$\frac{k i_{cvt}(t) i_d}{(1 + k)i_{cvt}(t) - i_d} = i_{cvt}(t) \tag{5}$$

The solution of the Eq. (5) is i_g $(t) = i_{cvt}$ $(t) = i_d$. From the perspective of efficiency, the efficiency of the power-reflux mode is not outstanding at high speed, however, the efficiency of pure CVT mode is much lower than that of power-reflux mode at low speed. Therefore, the power-reflux mode should be used when i_g (t) is larger than i_d, otherwise, the pure-CVT mode should be used.

Table 3. Transmission system parameters of the PRHEV

Components	Parameters	Value
FR gears	i_d	2.4704
Planetary row	k	2.6
CVT	i_{max}(Maximum speed ratio)	2.502
	i_{min}(Minimum speed ratio)	0.498

According to the parameters determined in Table 3 and its transmission characteristics, the power-reflux CVT model is established in Fig. 3.

With the same parameters, the conventional CVT transmission ratio variation range was observed to be only 0.498–2.502. Obviously, the power-reflux CVT can realize a wider speed ratio variation range, which is contribute to better matching the power sources operating points.

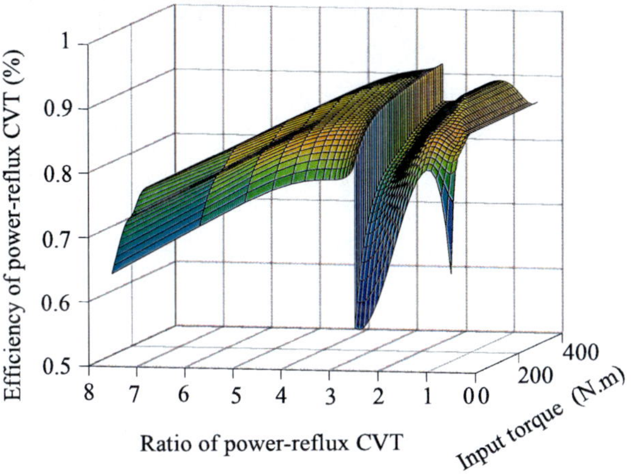

Fig. 3. PR-CVT model

3 Combined Optimization Scheme

In this paper, the optimal energy management strategy is embedded in the size optimization process to obtain the global optimal size parameters to form an iterative bi-loop optimization method. The inner optimization loop guarantees the optimality of the control strategy for the selected specific parameters, based on PMP algorithm. The performance of the system obtained from the inner loop will be converted into the fitness function and feed back to the outer loop to adjust the parameters. Therefore, this iterative bi-loop optimization method can conduct a simultaneous optimization on component sizing and energy management. In view of the computational complexity, evolutionary algorithm is used to solve the optimal parameters size.

3.1 Optimization Objective

The drivetrain cost, fuel consumption and acceleration capacity have been taken as the objectives to optimize the system parameters:

$$\begin{cases} \min\limits_{X \in \Omega} F(X) = [COST_{tatal}(X), T_{acc}(X), m(X)] \\ s.t.\ h_j(X) \geq 0, j = 1, 2, \cdots, n \end{cases} \tag{6}$$

where $COST_{purchase}$ (x) is the drivetrain cost; T_{acc} (x) is the vehicle acceleration time from 0 to 100 km/h; $m(x)$ is the fuel consumption per hundred kilometers.

Purchase Cost. The drivetrain cost only includes the component cost of the engine, motor, battery pack and their corresponding accessories:

$$COST_{purchase} = \frac{100}{Ly_s}\left(COST_{eng} + COST_{mot} + COST_{bat}\right) \tag{7}$$

where y_s is vehicle lifetime (15 years), L is yearly traveled distance (15000 km), $COST_{eng}$,$COST_{bat}$ and $COST_{mot}$ are the cost of the engine, motor and battery, calculated by Eq. (8)

$$COST_j = COST_{j\_0} + COST_{j\_1} \cdot PQ_j \tag{8}$$

where the values for $COST_{j\_0}$, $COST_{j\_1}$, and PQ_j are given in Table 4.

Table 4. Parameters of the cost model

j	$COST_{j\_0}$	$COST_{j\_1}$	PQ_j
Engine	\$424	\$12	P_{engmax}
Motor	\$425	\$21.775	P_{motmax}
Battery	\$680	\$651.2	Q_{bat}

Acceleration Time. Using the vehicle longitudinal dynamics, the vehicle acceleration can be computed:

$$a = \frac{dv}{dt} = \frac{F_t - \left(F_f + F_w\right)}{\delta m} \tag{9}$$

where F_t, F_f and F_w are the propelling force, rolling resistance force and aerodynamic drag force, respectively.

The acceleration time required to accelerate the vehicle from 0 to 100 km/h can be obtained, as follows:

$$\begin{aligned} T_{acc} &= \int_0^{100} \frac{1}{a} dv \\ &= \int_0^{100} \frac{\delta m}{(P_e + P_m)\eta_g \eta_0 / v - \left(mgf \cos\theta + C_D A v^2 / 21.15\right)} dv \end{aligned} \tag{10}$$

where P_e and P_m are the power of engine and motor.

3.2 Optimization Parameters

Considering the computational complexity of the optimization problem, only the system parameters closely related to the optimization objectives are selected for optimization. The engine maximum power, the motor maximum power, the battery capacity and the instantaneous battery output power are used as the system parameters to be optimized, that is optimization variables $X = [P_{engmax}, P_{motmax}, C_{bat}, P_b]$.

Subject to:

$$\begin{cases} P_{\text{engmax}} \in [36.5, 109.5] \\ P_{\text{motmax}} \in [30, 90] \\ C_{bat} \in [25,75] \\ P_b \in \left[-\eta_{\text{mot}} P_{\text{motmax}}, P_{\text{motmax}} / \eta_{\text{mot}}\right] \end{cases} \tag{11}$$

3.3 Optimization Constrains

It is necessary to ensure that PRHEV meets the requirements of dynamic performance, in the process of solving the multi-objective optimization problem. Reference to the china national standard"GBT19752-2005 the test method of hybrid electric vehicle's performance", as shown Table 5.

Table 5. Constraints on dynamic performance

Items	Description
Maximum velocity	≥100 km/h in pure electric driving mode
	≥160 km/h in engine driving mode
Acceleration ability	Time for 0–100 km/h ≤ 15 s in hybrid driving mode
	Time for 0–50 km/h ≤ 10 s in pure electric driving mode
Climbing capacity	The maximum climbing degree is 30% at 15 km/h
Endurance mileage	Electric driving mileage ≥ 50 km under ECE cycle

3.4 Optimization Algorithm

This study applies a multi-objective evolutionary algorithm based on the Pareto principle to solve the economic and dynamic multi-objective optimization problem, overcoming the shortcomings of the traditional weighted sum method. The algorithm introduces fast non-dominated sorting operators, individual crowding distance operators and elite strategy selection operators to create a better population in each iteration. The parameter optimization and optimal controls are computed in a nested way as illustrated in Fig. 4.

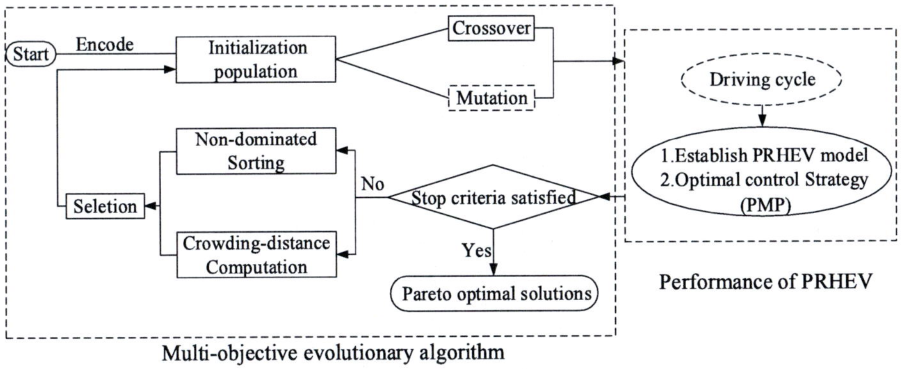

Fig. 4. Integrated multi-objective optimization algorithm scheme

4 Energy Management Strategy Based on PMP Algorithm

The fuel consumption can be computed as Eq. (12):

$$J(x, u, t) = \int_{t_0}^{t_{cycle}} \dot{m}(u, t)\, dt \tag{12}$$

where the control variable is $u = P_b$; the system state is $x = SOC(t)$.

According to the PMP algorithm, the Hamiltonian optimal objective function is constructed based on the fuel consumption function (12):

$$H(x, u, \beta, t) = \dot{m}(u, t) + \beta(t)\, \dot{SOC}(t) \tag{13}$$

The co-state equation of the system is:

$$\dot{\beta}(t) = -\frac{\partial H(SOC, u, \beta, t)}{\partial SOC} = \beta(t)\frac{\partial \dot{SOC}(t)}{\partial SOC} \tag{14}$$

It is assumed that the change rate of SOC is approximately 0, thus $\dot{\beta}(t) = 0$, then:

$$\beta(t) = \beta(t_0) = \beta_0 \tag{15}$$

The state equation of the system is:

$$\dot{x}(t) = -\frac{I}{Q_0} = -\frac{IV_{oc}}{Q_0 V_{oc}} = -\frac{P_b(t)}{Q_0 V_{oc}} = -\frac{u(t)}{Q_0 V_{oc}} \tag{16}$$

The boundary conditions are:

$$SOC(t_0) = SOC\left(t_f\right) = 0.55 \tag{17}$$

The condition for obtaining the minimum value of Hamilton function is:

$$H\left(x^*, u^*, \beta^*, t\right) = \min_{u \in \Omega} H\left(x^*, u, \beta^*, t\right) \tag{18}$$

The optimal control variable is

$$u^*(t) = \arg\min\{H(x(t), u(t), \beta(t), t)\} \tag{19}$$

Substituting Eqs. (15) and (16) into Eq. (13), the Hamilton function can be expressed as:

$$H(\mathrm{x}, \mathrm{u}, \lambda, \mathrm{t}) = \dot{m}_b(u(t), t) - \left(\frac{\beta_0}{Q_0 V_{oc}} u(t)\right) \tag{20}$$

It can be seen from Eq. (20) that the local optimal $u*$ can be applied to solve the optimal fuel economy of the system as long as the optimal $u*$ in the feasible region of the Hamiltonian function is found under the constraint conditions.

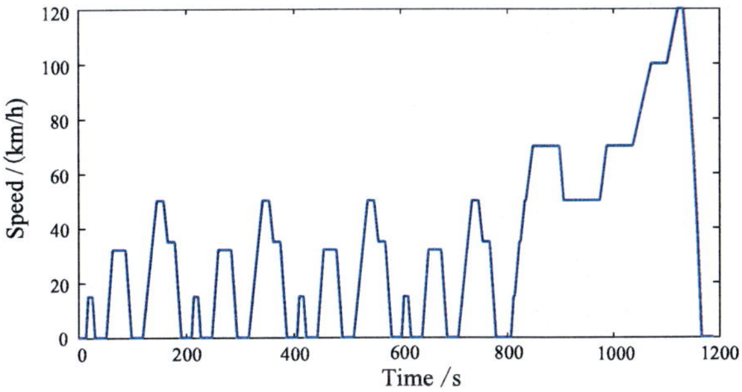

Fig. 5. NEDC driving cycle

5 Results and Analysis

In the optimization process, 10 NEDC driving cycles were chosen as test cycles, and the NEDC driving cycle is shown as Fig. 5. The main parameters of the adopted multi-objective evolutionary algorithm were set as follows, the population size was 100, the crossover probability was 0.9, the mutation probability was 0.1, and the maximum evolution generation was set to 150 to obtain Pareto optimal solutions.

Table 6 lists 10 representative Pareto optimal solutions, which represent the optimal matching system under different weights of three objectives.

Table 6. Pareto optimal solutions

Solutions	P_{engmax}	P_{motmax}	C_{bat}	*Fuel* (L/100 km)	*Cost* ($)	*Time* (s)
0	73	40	75	4.16	20565.1	12.083
1	81.09	69.52	32.23	3.905	11457.1	8.007
2	68.77	69.25	56.1	3.756	16799.3	8.179
3	81.51	75.51	39.61	3.874	13293.8	7.971
4	62	42.88	45.64	3.727	13732.9	11.471
5	61.91	41.95	55.02	3.693	15871.6	11.72
6	71.16	61.66	35.06	3.784	11816.8	8.427
7	61.911	61.58	68.79	3.767	19475.4	8.641
8	84.03	73.39	50.19	3.948	15713.5	7.949
9	66.59	42.91	44.43	3.726	13510.9	11.357
10	81.16	73.97	36.67	3.915	12580.2	10.963

Solution 0 shows the original parameter matching of the PRHEV drivetrain which is designed based on the constraints of vehicle dynamics, and its corresponding performance. Solutions 1—10 are Pareto optimal solutions. It can be observed from Table 6 that the economical and dynamical performance corresponding to the Pareto optimal solutions have been significantly improved after optimization. Optimization results achieve a 29.86% drivetrain cost average decrease, an 8.43% fuel consumption average decline and a 21.64% 0–100 km/h acceleration time average reduction from the original powertrain design.

Table 7. Optimization objectives improvement

Optimization objectives	Cost	Fuel	Acceleration time
Maximum improvement /%	44.29%	11.63%	34.21%
Average improvement /%	29.86%	8.43%	21.64%

Pareto optimal solutions can provide many parameters combinations for designers to choose according to their preferences. Solution 1 characterized by small size of battery capacity, motor maximum power, and engine maximum power, represents the minimum drivetrain cost design which is 44.29% lower than original value; Solution 5 has the largest reduction in fuel consumption by 11.63%, in which the battery capacity are large and the engine maximum power are low; Solution 8 with high motor maximum power and engine maximum power has the highest acceleration capability, so that the dynamical performance of the original vehicle is improved by 34.21%. The vehicle performance of the other solutions is balanced, and the specific improvement of Pareto optimal solutions are shown in Table 7.

6 Conclusions

A novel PRHEV powertrain has been proposed in this paper, which features a planetary gear train and a fixed-ratio gear installed on the conventional CVT for multiple speed regulation modes and further develops into 8 operation modes by combining with different driving modes. It greatly expanded the adaptability of the CVT in terms of broadening the speed range, and improving the transmission efficiency. To take full advantage of the distinctive PRHEV structure and advantages of CVT, an iterative bi-loop multi-objective optimization problem for the combined power management and design of the PRHEV is formulated and solved, with trade-off between economical and dynamical performance. Multi-objective evolutionary algorithm is applied to search the optimal component parameters in the outer loop, while PMP is applied to guarantee the optimality of the control strategy in the inner loop. Finally, 10 optimal Pareto solutions are obtained. Compared to the original vehicle, optimization results of the drivetrain cost, fuel consumption and 0–100 km/h acceleration time decreased by 17.38, 27.96 and 11.16%, respectively to the greatest extent. In addition, the Pareto optimal solutions also provide a wide range of choices in preliminary design.

References

1. Yang, Y., Zhao, X.F., Qin, D.T.: Design and operation mode coupling characteristics analysis on the new transmission system of hybrid electric vehicle. Automot. Eng. **34**(011), 968–975 (2012)

2. He, H.W., Huang, X.G., Zhang, Y.M.: Study on parameters optimization of hybrid electric vehicle based on DIRECT algorithm. Chin. J. Autom. Eng. **001**(001), 35–41 (2011)
3. Yildiz, E.T., Farooqi, Q., Anwar, S.: Nonlinear constrained component optimization for the powertrain configuration of a plug-in hybrid electric vehicle powertrain. J. Autom. Saf. Energy **3**(1), 64–70 (2012)
4. Lin, X.Y., Lin, H.B., Zhai, L.Q.: PSO-fuzzy multi-objective control strategy based on PHEV charge-sustaining mode. China J. Highw. Transp. **29**(10), 132–139 (2016)
5. Zeng, Y.: Parameter optimization of plug-in hybrid electric vehicle based on quantum genetic algorithm. Clust. Comput. **22**(6), 14835–14843 (2018). https://doi.org/10.1007/s10586-018-2424-4
6. Pourabdollah, M., Murgovski, N., Grauers, A.: Optimal sizing of a parallel PHEV powertrain. IEEE Trans. Veh. Technol. **62**(6), 2469–2480 (2013)
7. Zou, Y., Sun, F., Hu, X.: Combined optimal sizing and control for a hybrid tracked vehicle. Energies **5**(11), 4697–4710 (2012)
8. Zhang, Y., Lin, H.: Performance modeling and optimization of a novel multi-mode hybrid powertrain. J. Mech. Des. **128**(1), 79–89 (2006)

Influence of Centrodes Coefficient on the Characteristic of Gear Ratio Function of the Compound Non-circular Gear Train with Improved Cycloid Tooth Profile

Nguyen Hong Thai[1(✉)], Phung Van Thom[1], and Nguyen Thanh Trung[1,2]

[1] Department of Mechanical Design and Robotics, School of Mechanical Engineering, Hanoi University of Science and Technology, Hanoi, Vietnam
thai.nguyenhong@hust.edu.vn

[2] Vietnam National Research Institute of Mechanical Engineering, Hanoi, Vietnam

Abstract. In this paper, the authors evaluate the influence of the centrodes coefficients on the gear ratio function characteristics of the compound non-circular gear (CNCG) train formed by two pairs of external non-circular gears (NCGs). First, a mathematical model describing the mating centrodes for the CNCG train has been establishing. From there, evaluate the influence of centrode coefficients μ_1 and μ_2 of NCGs in the CNCG train on the gear ratio function characteristic for the CNCG train. A design of a CNCG train with an improved cycloid tooth profile by a rack cutter. The evaluation results show that when increasing μ_1, transmission size tends to decrease, but the speed and torque variation range rises to 54.31%. While reducing μ_2, the speed variation and torque range increase, but the mating centrodes of NCGs 4 is a concave curve that cannot gear shaping by a rack cutter or hob cutter must be by shaper cutter.

Keywords: Centrode coefficient · Gear ratio function · Non- circular gear · Compound non- circular gear train

1 Introduction

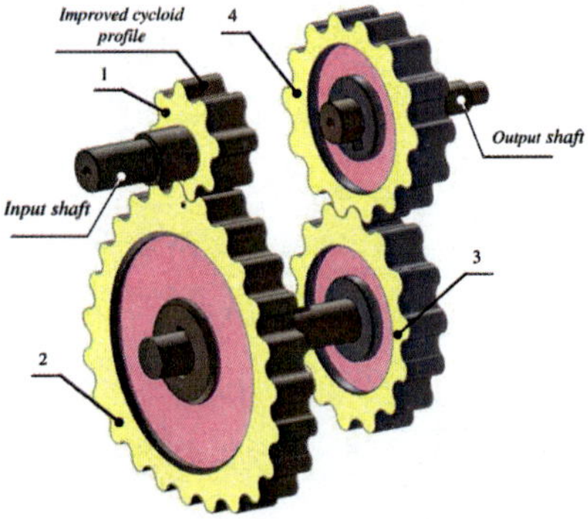

Fig. 1. The structure of the CNCG train.

Non-circular gear (NCG) trains with the advantage of creating speed converters according to a rule in practice with a simple mechanical structure. There has been much research applying NCG in creating machines and equipment such as Xu et al. [1] used the compound non-circular gear (CNCG) train in combination with cylindrical gears in designing the gear drive of the new coal seam gas (CSG) drainage machine in the CSG mining industry; Zhao et al. [2, 3] improved the CNCG train in the rice transplanter

N. V. Khang et al. (Eds.): ASIAN MMS 2021, MMS 113, pp. 204–214, 2022.
https://doi.org/10.1007/978-3-030-91892-7_19

of Guo and developed it to transplant other leafy vegetables; Zhou et al. [4] applied a simple NCGs train to the design and manufacture of rice potted seedling trans-planting mechanisms with a transplanting efficiency of up to 180 plants/min; Application simple oval gear drives for driving and as the rotor of Roots-type hydraulic machine [5–7]; Manufacturing a pair of the elliptical gears in the flow meter [8]; The design of an NCG pair for modifying the crank-slider kinematics of a nail machine [9]; The application of NCG pair with mating centrodes is convex and concave curves in the mechanical driving system of a pressing machine [10] etc. But most of the above research applies two types of curves as tooth profiles: (*i*) The involutes curve of a circle [2–4, 8–10] and (*ii*) The Novikov-type arc [11, 12]. In addition, other types of curves applied as tooth profiles for cylindrical gears with constant gear ratios such as cycloid curve [13, 14, 27], improved cycloid curve [15, 16] or involute ellipse curve [17] etc. Therefore, in this research, the authors apply the improved cycloid curve from the generated ellipse to the tooth profile design of the CNCG train, as shown in Fig. 1. Wherein gear 1 is the eccentric gear, compound gears 2–3 are NCGs and gear 4 is oval gears. Evaluate the influence of the centrodes coefficient of gear 1 and 4 on the gear ratio function characteristics of the CNCG train. In this research following issues are to be solved: (*1*) Model the mathematical model of the mating centrodes in the CNCG train has the structure as described in Fig. 1, assuming that the mating centrode of gears 1 and 4 have given; (*2*) Evaluate the influence of the centrodes coefficient $\mu_1 = e_1/R_1$ (R_1 is the radius and e_1 is the eccentricity of the eccentric circle) and the centrodes coefficient $\mu_2 = b_4/a_4$ (a_4 is the semi-major axis and b_4 is the semi-minor axis of the oval 4) to the gear ratio function characteristics of the CNCG train; (*3*) Numerical calculation and design of a CNCG train with tooth profile is improved cycloid by the rack cutter.

2 Mathematical Model of the Mating Centrodes of a Compound Non-circular Gear Train

To establish the mathematical model of the CNCG train as described in Fig. 1, we proceed to separate each pair NCGs as follows:

2.1 Mathematical Model of the Mating Centrodes for the Non-circular Gears Pair 1–2

As described in the introduction, gear 1 is an eccentric gear with an eccentric mating centrode denoted Σ_1. According to [18], the equation is written in polar coordinates:

$$\rho_{P_1}(\varphi_1) = (R_1^2 - e_1^2 \sin^2 \varphi_1)^{0.5} + e_1 \cos \varphi_1 \tag{1}$$

Where: R_1 is the eccentric circle radius; e_1 is the eccentricity, and $\varphi_1 \in [0 \div 2\pi]$ is the polar angle of Σ_1.

Thus, from Fig. 2, if Σ_2 is called the centrode of the NCG 2 meshing to the eccentric gear. Using the calculation method in [19], then the mathematical equation describing centrode Σ_2 is determined by:

$$\begin{cases} \rho_2(\varphi_2(\varphi_1)) = a_{12} - \rho_1(\varphi_1) \\ \varphi_2(\varphi_1) = \int\limits_0^{\varphi_1} (\rho_1(\varphi_1)/a_{12} - \rho_1(\varphi_1)) d\varphi_1 \end{cases} \tag{2}$$

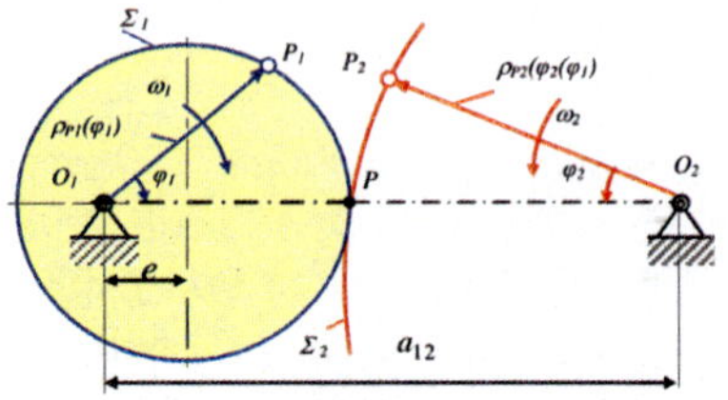

Fig. 2. The geometric relationship between two mating centrodes of the NCGs 1–2 pair.

Where: $\rho_2(\varphi_2(\varphi_1))$, $\varphi_2(\varphi_1)$ are the polar radius and polar angle of Σ_2 at point P_2, respectively.

The a_{12} shaft distance is given by:

$$f(n_1, e_1, R_1) = n_1 \int_0^{2\pi} (\rho_1(\varphi_1)/(a_{12} - \rho_1(\varphi_1)))d\varphi_1 - 2\pi = 0 \tag{3}$$

Where: n_1 is the number of revolutions of eccentric gear Σ_1 for one revolution of gear centrode Σ_2. Solve Eq. (3) by Litvin's method [20] the shaft distance $a_{12}(n_1, e_1, R_1)$ obtained as:

$$f(a_{12}) = a_{12}(n_1, e_1, R_1) = R_1(1 + n_1)(1 - ((n_1 - 12)e_1^2)/(4n_1R_1^2)) \tag{4}$$

2.2 Mathematical Model of the Mating Centrodes for the Non-circular Gears Pair 3–4

In this case, the NCGs 4 is assumed to have centrode Σ_4 as the oval as described in Fig. 3, and the mathematical equation describing Σ_4 is given as polar coordinates [17]:

$$\rho_4(\varphi_4) = 2a_4b_4((a_4 + b_4) - (a_4 - b_4)\cos(2\varphi_4))^{-1} \tag{5}$$

Wherein: a_4, b_4, $\varphi_4 \in [0 \div 2\pi]$ are the semi-major axis, semi-minor axis and pole angle of centrode oval Σ_4, respectively.

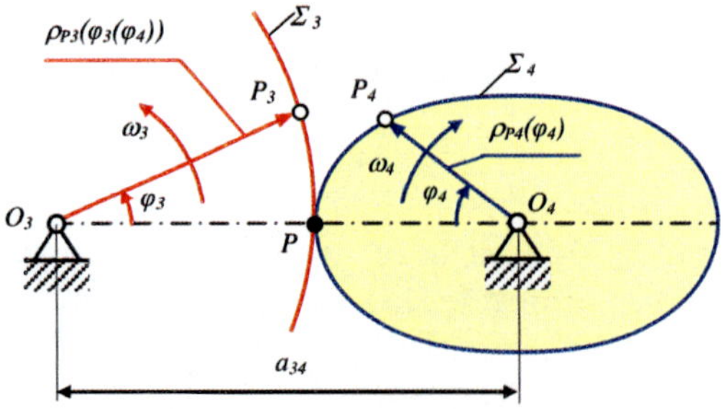

Fig. 3. The geometric relationship between two mating centrodes of the NCG 3–4 pair.

Same as for the pair NCGs 1–2 and from Fig. 3, mathematical equations for centrode Σ_3 are conjugation to Σ_4:

$$\begin{cases} \rho_3(\varphi_3(\varphi_4)) = a_{34} - \rho_4(\varphi_4) \\ \varphi_3(\varphi_4) = \int_0^{\varphi_4} \rho_4(\varphi_4)/(a_{34} - \rho_4(\varphi_4))d\varphi_4 \end{cases} \quad (6)$$

Where: $\rho_3(\varphi_3(\varphi_4))$, $\varphi_3(\varphi_4)$ are the pole radius and pole angle of the centrode Σ_3, respectively.

The a_{34} shaft distance is determined by:

$$f(n_4, a_4, b_4) = n_4 \int_0^{2\pi} \rho_4(\varphi_4)/(a_{34} - \rho_4(\varphi_4))d\varphi_4 - 2\pi = 0 \quad (7)$$

Wherein n_4 is the number of revolutions of oval centrode Σ_4 performed for one revolution of gear centrode Σ_3. Solving Eq. (7) by integral Dwight [19, 21] can determine the shaft distance $a_{34}(n_4, a_4, b_4)$ as follows:

$$f(a_{34}) = a_{34}(a_4, b_4, n_4) = 1/2((a_4 + b_4) + ((a_4 + b_4)^2 - 4a_4b_4(1 - n_4^2))^{0.5}) \quad (8)$$

2.3 Kinematics of the Compound Non-circular Gear Train

In this case ϕ the rotation angle of the gear shafts will be substituted for the pole angle φ in the mating centrodes section. Thus, similar to the cylindrical gear train, we also have the gear ratio function of the CNCG train following equation:

$$i_{14}(\phi_1) = \omega_1/\omega_4 = d\phi_1/d\phi_4 = i_{12}(\phi_1)i_{34}(\phi_4(\phi_3)) \quad (9)$$

In the formula (9):

$$i_{12}(\phi_1) = \omega_1/\omega_2 = d\phi_1/d\phi_2 = \rho_2(\phi_2(\phi_1))/\rho_1(\phi_1) = (a_{12} - \rho_1(\phi_1))/\rho_1(\phi_1)$$

$$i_{34}(\phi_4(\phi_1)) = \omega_3/\omega_4 = d\phi_3/d\phi_4 = \rho_4(\phi_4)/\rho_3(\phi_3(\phi_4)) = \rho_4(\phi_4)/(a_{34} - \rho_4(\phi_4))$$

After transforming Eq. (9) is rewritten as:

$$i_{14}(e_1, R_1, a_4, b_4) = (2a_4b_4(a_{12} - \rho_1(\phi_1)))/(\rho_1(\phi_1)(a_{34}((a_4 + b_4) - (a_4 - b_4)\cos(2\phi_4)) - 2a_4b_4)) \quad (10)$$

From that, the relationship of the angular velocity ω_2 of shaft NCGs 2 according to input shaft angular velocity ω_1:

$$\omega_2(\omega_1) = \left(\left((R_1^2 - e_1^2\sin^2\phi_1)^{0.5} + e_1\cos\phi_1\right)/a_{12} - \left((R_1^2 - e_1^2\sin^2\phi_1)^{0.5} + e_1\cos\phi_1\right)\right)\omega_1 \quad (11)$$

The angular velocity of oval gear 4 according to angular velocity ω_1 of eccentric gear 1:

$$\omega_4(e_1, R_1, a_4, b_4) = i_{14}(\phi_1)^{-1}\omega_1 \quad (12)$$

3 Influence of Centrodes Coefficient on the Characteristic of Gear Ratio Function

From Eqs. (10) and (12), it can be seen that the gear ratio function i_{14} and the output angular velocity ω_4 are functions that depend on the design parameters (e_1, R_1) and (a_4, b_4). Therefore, to evaluate the influence of the above parameters on the characteristics of the gear ratio function, set the centrodes coefficient $\mu_1 = e_1/R_1$, $\mu_2 = b_4/a_4$ and investigate the following cases:

3.1 Influence of Centrodes Coefficient μ_1 on the Characteristic of Gear Ratio Function

In this case $a_4 = 60$ mm, $b_4 = 40$ mm, $n_4 = 2$, $\omega_1 = 5$ rad/s, the centrode coefficient μ_1 is calculated so that the circumference Σ_1 constant with the original Σ_1 having parameters $R_1 = 50$ mm, $n_1 = 3$ and satisfy Eq. (4). After numerical calculation using Matlab, the calculated data are statistics in Table 1.

Table 1. Design parameters for the centrodes of the CNCG train according to the coefficient μ_1.

The designs	μ_1	e_1 (mm)	a_{12} (mm)	a_{34} (mm)
1	0.06	3	199.94	148.49
2	0.1	5	199.83	148.49
3	0.2	10	199.33	148.49

From Table 1, after calculating the numbers by Matlab, we have the graph of the gear ratio function according to μ_1 and ω_1 as shown in Fig. 4, and Fig. 5 is the centrodes design of the corresponding CNCG train in Table 1.

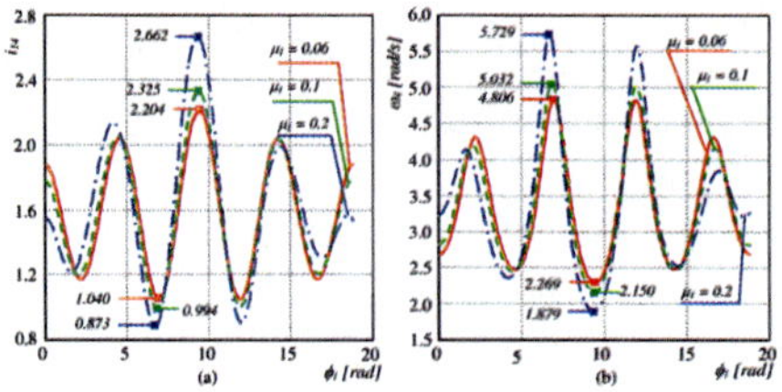

Fig. 4. Characteristic curve of the gear ratio function and ω_4 of the CNCG train according to μ_1 and ω_1.

To evaluate the coefficient μ_1 on the gear function characteristics of the CNCG train and the angular velocity ω_4 of the output shaft. Set: $\delta i_{14} = i_{14\,\max} - i_{14\,\min}$, $\delta\omega_4 = \omega_{4\,\max} - \omega_{4\,\min}$ are the amplitude variation of the gear ratio function i_{14} and the output angular velocity ω_4, respectively. With the above definition, Fig. 6 below is

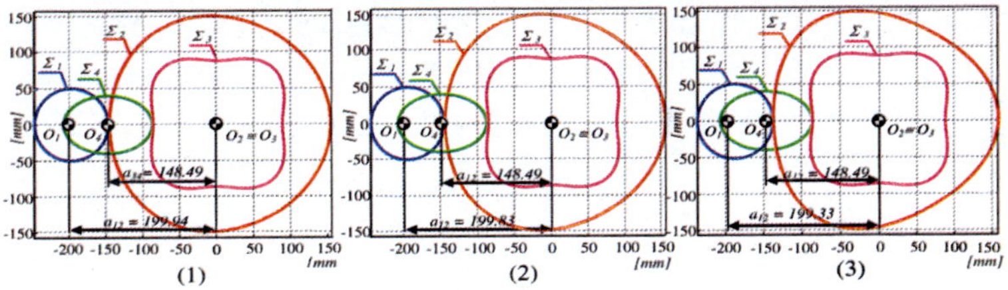

Fig. 5. Designs for the conjugation centrodes of the CNCG train according to the coefficient μ_1.

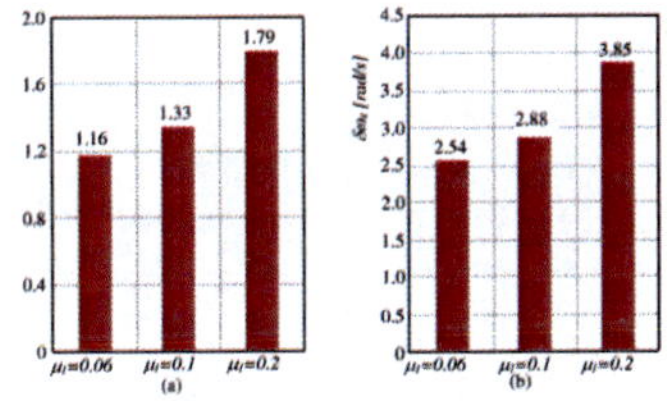

Fig. 6. Variable range of the gear ratio and output shaft angular velocity according to the coefficient μ_1.

a survey chart of the range of variation of gear ratio and output shaft angular velocity according to the coefficient μ_1.

From Table 1 and Fig. 4, 5 and 6, we have the following comments: With the centrode circumference Σ_1 constant, as μ_1 increases from 0.06 to 0.2, the amplitude range and output shaft angular velocity increases while the shaft distance a_{12} decreases. That means that, when increasing the eccentricity of gear 1, the size of the CNCG train decreases while increasing the torque and speed variation range. In specific cases, with the design option corresponding to $\mu_1 = 0.2$ compared to the design option $\mu_1 = 0.06$, the gear ratio δi_{14} increases is 54.31% and the amplitude of the speed angle $\delta\omega_4$ increased is 51.57%.

3.2 Influence of Centrodes Coefficient μ_2 on the Characteristic of Gear Ratio Function

Same as Sect. 3.1. In this case = *50* mm, $e_1 = 5$ mm, $n_1 = 3$, $n_4 = 2$, $\omega_1 = 5$ rad/s, the centrode coefficient μ_2 is calculated so that the circumference Σ_4 is constant with the original Σ_4 having parameters $a_4 = 60$ mm, $b_4 = 40$ mm and satisfying Eq. (8). After numerical calculation using Matlab, the calculated data are statistics in Table 2.

From Table 2, after calculating the numbers by Matlab, we have the graph of the gear ratio function according to μ_2 and ω_1 as shown in Fig. 7, and Fig. 8 is the conjugation centrodes design of the corresponding CNCG train in Table 2 (Fig. 9).

From Table 2, the graph of Fig. 7 and the conjugation centrodes design of Fig. 8. We can see that the centrode circumference Σ_4 constant, when μ_2 decreases from 0.85 to 0.3, the amplitude range of the gear ratio and speed output shaft angle increases. Specifically, the design corresponding to $\mu_2 = 0.3$ compared with the design plan $\mu_2 = 0.85$, the amplitude of the gear ratio δi_{14} increased is 375.95%, while the amplitude

Table 2. Design parameters for the centrodes of the CNCG train according to the coefficient μ1.

The designs	μ_2	a_4 (mm)	b_4 (mm)	a_{12} (mm)	a_{34} (mm)
1	0,85	55	46,60	199,83	152,14
2	0,49	65	31,64	199,83	140,54
3	0,3	70	20,93	199,83	125,85

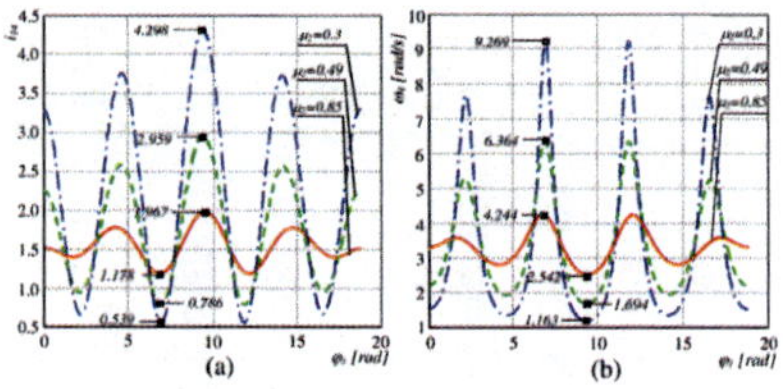

Fig. 7. Characteristic curve of the gear ratio function and ω_4 of the CNCG train according to μ_2 and ω_1.

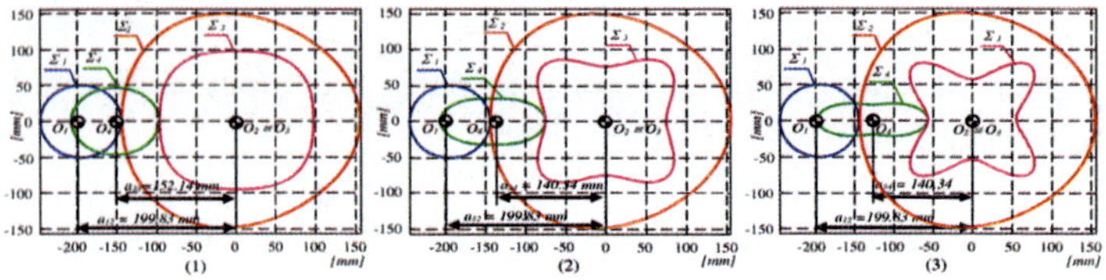

Fig. 8. Designs for the conjugation centrodes of the CNCG train according to the coefficient μ_2.

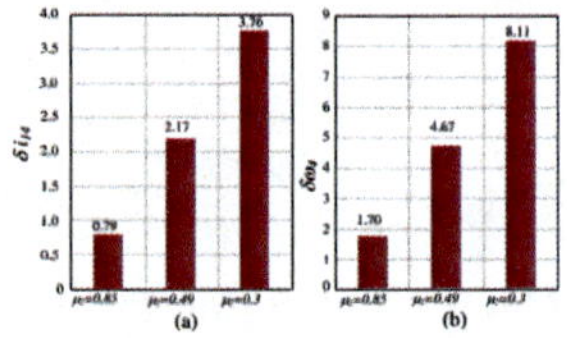

Fig. 9. Variable range of the gear ratio and output shaft angular velocity according to the coefficient μ_2.

of the angular velocity $\delta\omega_4$ increased is 377.06%. In addition, from Fig. 8, it can be seen that when μ_2 decreases, the gear ratio amplitude increases. But the centrode of the NCGs 4 is a concave curve, which leads to shape the tooth profile by the shaper cutter.

4 Examples of Computer-Based Design

From the discussion and evaluation in Sect. 3, to shape the tooth profile of the NGCs in the CNCG train by rack cutter, the design parameters of the mating centroids in the CNCG train must be convex curves. Thus, after the selection, is statistics in Table 3 below. For a pair of NCGs 1–2 select $\mu_1 = 0.2$, $e_1 = 10.0$ mm (see Table 1) and a pair of NCGs 3–4 select $\mu_2 = 0.85$, $a_4 = 55$ mm, $b_4 = 46.6$ mm (see Table 2).

Table 3. Design parameters for the mating centrodes of the CNCG train.

Parameters	Notation	Gears pair 1–2		Gears pair 3–4	
		Eccentric gear	NCGs2	NCGs3	Oval gear 4
The radius of the eccentric circle	R_1	50.0	–	–	–
The eccentricity	e_1	10.0	–	–	–
Circumference coefficient	n	3.0		2.0	
The semi-major axis	a_4	–	–	–	55.0
The semi-minor axis	b_4	–	–	–	46.6

From the design parameters in Table 3, after programming numerical calculations on Matlab, we have the mating centrodes of the CNCG train as shown in Fig. 10a, and Fig. 10b is the graph of the gear ratio function.

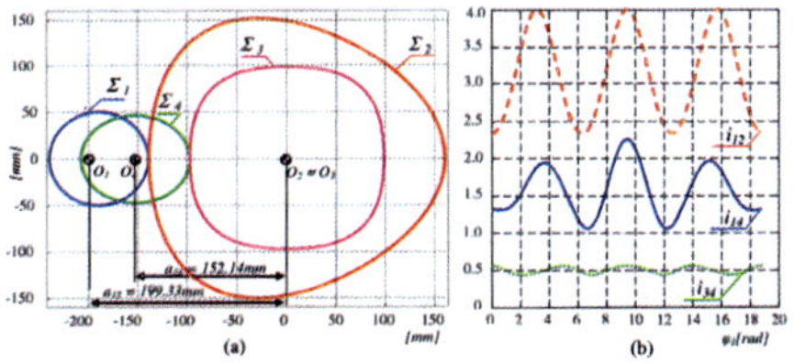

Fig. 10. Design of the mating centrodes for the CNCG train with (a) is the mating centrodes and (b) is the gear ratio function.

The mating centroids designed in Fig. 10b gear shaping the tooth profile for the NCGs in the CNCG train by the rack cutter has to profile is an improved cycloid curve as shown Fig. 11.

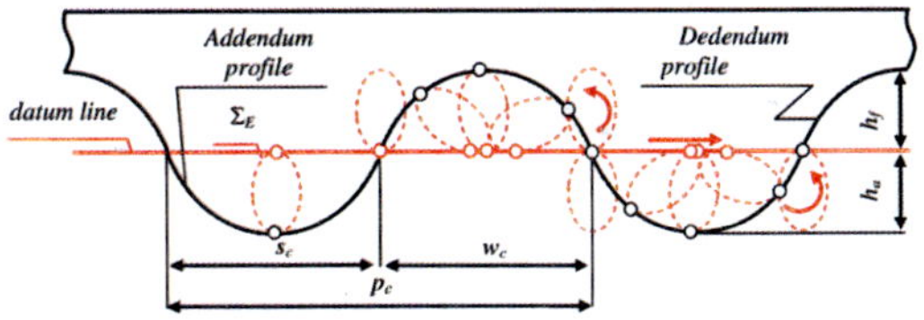

Fig. 11. The rack cutter with Σ_E the ellipse generating the tooth profile of the rack cutter [22].

According to [22–26] process of gear shaping the tooth profile of NCGs in the CNCG train is done as follows: (*i*) Given the datum line of the rack cutter roll without slipping outside centrodes of the NCG; (*ii*) The NCGs pairs after shaping to be able to meshing, each NCGs pairs must be shaped with a rack cutter. With the gear shaping principle as above, the rack cutter's design parameters are selecting so that the mating centrode circumference of the NCG divided by the tooth pitch on the datum line of the rack cutter must be a positive integer. Thus, the parameters of the rack cutter after numerical calculation are statistics in Table 4. In which rack cutter 1 is used to shape the pair NCG 1–2, rack cutter 2 is used to gear shaping the pair NCG 3–4.

Table 4. Design parameters of the rack cutter.

Parameter	Notation	The rack cutter 1	The rack cutter 2
The semi-major axis of Σ_E	a_{Ei}	1.50	1.27
The semi-minor axis of Σ_E	b_{Ei}	1.35	1.194
Module	m	5.55	5.09
Tooth thickness (mm)	s_c	8.73	8.00
Width of space (mm)	w_c	8.73	8.00
Tooth pitch (mm)	p_c	17.45	16.00
Tooth addendum (mm)	h_a	3.00	2.70
Tooth dedendum (mm)	h_f	3.00	2.70
Whole depth (mm)	h	6.00	5.40

From the mating centrodes design parameters in Table 3, Fig. 10b and the rack cutter parameters in Table 4, after numerical calculation by Matlab, we have the design parameters of the CNCG train as described in Table 5 below, and Fig. 12 is the design of the CNCG train.

Table 5. Design parameters of the CNCG train.

Parameter	Notation	Gear pair 1–2		Gear pair 3–4	
		Eccentric gear 1	NCG 2	NCG 3	Oval gear 4
Module	m	5.55	5.55	5.09	5.09
Circumference coefficient	n	3.00	–	–	2.00
Number of teeth	z	18.00	54.00	40.00	20.00
Tooth pitch (mm)	p	17.45	17.45	16.00	16.00
Tooth thickness (mm)	s	8.73	8.73	8.00	8.00
Width of space (mm)	w	8.73	8.73	8.00	8.00
Whole depth (mm)	h	6.00	6.00	5.40	5.40
Tooth addendum (mm)	h_a	3.00	3.00	2.70	2.70
Tooth dedendum (mm)	h_f	3.00	3.00	2.70	2.70

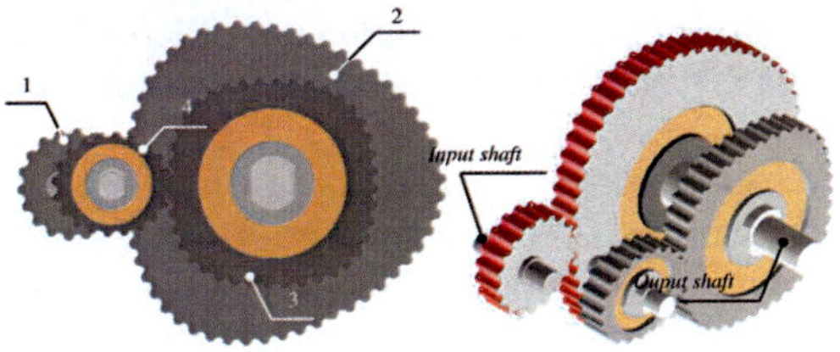

Fig. 12. Design of CNCG train with improved cycloid tooth profile.

5 Conclusions

From the above discussion and evaluation results, it shows that when designing CNCG train with the mating centrode 1 is an eccentric circle and the matting centrode 4 is an oval: (*i*) When the centrode coefficient μ_1 increases and the centrode coefficient μ_2 decreased, then the amplitude range of the train's speed and torque variation will increase from 1.54 to 4.76 times. But increasing μ_2, the mating centrode of NCGs 4 will become a concave curve that cannot gear shaping by the rack cutter. Therefore, if you want to gear shaping by the rack cutter while the design CNCG train, then need to select small μ_1 and large μ_2; (*ii*) When select small M2, the mating centrode of NCGs 4 has small curvature radius areas at the top (see Fig. 8 (2) and (3)), which leads to the inability to create a tooth profile. Therefore, when designing, it is necessary to compromise in the selection of design parameters.

Acknowledgements. This work was supported by Project of Ministry of Education and Training, Vietnam. Under grant Number: B2019 - BKA – 09.

References

1. Xu, G., Hua, D., Dai, W., Zhang, X.: Design and performance analysis of a coal bed gas drainage machine based on incomplete non-circular gears. Energies **10**, 2–19 (2017)
2. Zhao, X., Chu, M., Ma, X., Dai, L., Ye, B., Chen, J.: Research on design method of non-circular planetary gear train transplanting mechanism based on precise poses and trajectory optimization. Adv. Mech. Eng. **10**(12), 1–12 (2018)
3. Zhao, X., Ye, J., Chu, M., Dai, L., Chen, J.: Automatic scallion seedling feeding mechanism with an asymmetrical high-order transmission gear train. Chin. J. Mech. Eng. **33**(1), 1–14 (2020). https://doi.org/10.1186/s10033-020-0432-9
4. Zhou, M., Yang, Y., Wei, M., Yin, D.: Method for generating non-circular gear with addendum modification and its application in transplanting mechanism. Int. J. Agric. Biol. Eng. **13**(6), 68–75 (2020)
5. Tien, T.N., Thai, N.H., Long, N.D.: Effects of head gaps and rotor gap on flow rate and hydraulic leakage of a novel non-contact rotor blower. Vietnam J. Sci. Technol. **57**(4A), 125–140 (2019)
6. Tien, T.N., Thai, N.H.: A novel design of the roots blower. Vietnam J. Sci. Technol. **57**(2), 249–260 (2019)
7. Thai, N.H., Long, N.D.: A new design of the Lobe pump is based on the meshing principle of elliptical gear pairs. Sci. Technol. Dev. J. Eng. Technol. **4**(2), 861–871 (2021)

8. Hernández, C.G., Marín, R.M.G., Huertas-Talón, J.L., Efkolidis, N., Kyratsis, P.: WEDM manufacturing method for noncircular gears, using CAD/CAM software. Strojniški vestnik J. Mech. Eng. **62**, 137–144 (2016)
9. Niculescu, M, Andrei, L., Cristescu, A.: Generation of noncircular gears for variable motion of the crank-slider mechanism. In: 7th International Conference on Advanced Concepts in Mechanical Engineering, Materials Science and Engineering 147 (2016). https://doi.org/10.1088/1757-899X/147/1/012078
10. Mundo, D., Danieli, G.A.: Use of non- circular gear in pressing machine driving systems. IASME Trans. **I**(1), 7–11 (2004)
11. Bair, B.W.: Computer aided design of non-standard elliptical gear drives. Proc. Inst. Mech. Eng. C J. Mech. Eng. Sci. **216**(4), 473–482 (2001)
12. Bair, B.W., Sung, M.-H., Wang, J.-S., Chen, C.-F.: Tooth profile generation and analysis of oval gears with circular-arc teeth. Mech. Mach. Theory **4**, 1306–1317 (2009)
13. Thai, N.H., Trung, N.T.: Establishing formulas for design of Roots pump geometrical parameters with given specific flow rate. Vietnam J. Sci. Technol. **53**(4), 533–542 (2015)
14. Thai, N.H., Giang, T.C.: Influence of geometrical dimensions on the profile slippage in the hypogerotor pump. Vietnam J. Sci. Technol. **56**(4), 482–491 (2018)
15. Hsieh, C.-F.: A new curve for application to the rotor profile of rotary lobe pumps. Mech. Mach. Theory **87**, 70–81 (2015)
16. Thai, N.H., Tien, T.N., Dung, P.T., Huy, N.Q.: Influence of the designing parameters on flow fluctuation and pressure of the improved roots blower. In: International Conference of Fluid Machinery and Automation Systems – ICFMAS 2018, pp. 196–203 (2018)
17. Thai, N.H.: Shaping the tooth profile of elliptical gear with the involute ellipse curve. Sci. Technol. Dev. J. Eng. Technol. **24**(3), 1048–1056 (2021)
18. Thai, N.H., Trung, N.T., Viet, N.H.: Research and manufacture of external non-circular gear-pair with improved cycloid profile of the ellipse. Sci. Technol. Dev. J. Eng. Technol. **4**(2), 835–845 (2021)
19. Litvin, F.L., Fuentes-Azna, A., Gonzalez-Perez, I., Hayasaka, K.: Noncircular Gears Design and Generation. Cambridge University Press, Cambridge (2009)
20. Litvin, F.L., Fuentes, A.: Gear Geometry and Applied Theory. Cambridge University Press, Cambridge (2004)
21. Litvin, F.L., Gonzalez-Perez, I., Fuentes, A., Hayasaka, K.: Design and investigation of gear drives with non-circular gears applied for speed variation and generation of functions. Comput. Methods Appl. Mech. Eng. **197**, 3783–3802 (2008)
22. Thai, N.H., Thom, P.V., Lam, D.B.: Effects of pressure angle on uneven wear of a tooth profile of an elliptical gear generated by ellipse involute. Sci. Technol. Dev. J. Eng. Technol. **24**(3), 2031–2043 (2021)
23. Fetvaci, C., Imrak, E.: Computer modeling and simulation of spur involute gears by generating method. Key Eng. Mater. **450**, 103–106 (2011)
24. Fetvaci, C.: Computer simulation of involute tooth generation, chap. 22. Mechanical Engineering, Intech (2013)
25. Dooner, D.B.: Kinematic Geometry of Gear. Wiley, New York (2012)
26. Radzevich, S.P.: Gear Cutting Tool Fundamentals of Design and Computaton. CRC Press, Boca Raton (2010)
27. Thai, N.H., Trung, N.T., Nghia, L.X., Duong, N.T.: Profile sliding phenomenon in the external non-circular gear-train with cycloidal profile. Eng. Technol. Sustain. Dev. **31**(2), 053–057 (2021)

Machine and Robot Design

Inverse Kinematics Analysis of 7-DOF Collaborative Robot by Decoupling Position and Orientation

Nguyen Quang Hoang[1(✉)], Do Tran Thang[2], and Dinh Van Phong[1]

[1] Hanoi University of Science and Technology, Hanoi, Vietnam
hoang.nguyenquang@hust.edu.vn
[2] Institute of Mechanics, Vietnam Academy of Science and Technology, Hanoi, Vietnam

Abstract. Redundant manipulators have some advantages in comparison to standard ones, namely higher flexibility, obstacle, and joint limits avoidance capability, and much more solutions of inverse kinematics. This paper presents a combination of Jacobian-based and analytical method for finding solution of inverse kinematics of a 7-DOF manipulator. In this way, the first four joint variables are determined with Jacobian-based method, and the last three ones based on an analytical method. Some numerical simulations are carried out to verify the effectiveness of the proposed approach.

Keywords: Redundant manipulator · Inverse kinematics · Decoupling position and orientation

1 Introduction

In the past twenty years, collaborative robots, commonly known as cobots, are experiencing rapid market growth in the robotic industry. This type of robot is able to physically interact with humans in a shared workspace thanks to sensors, intelligent controls, and other design features such as lightweight materials and rounded edges.

To operate dexterously like a human arm, the cobot is designed with 7 degrees of freedom (DOF) with a structure of anthropomorphic arm. As a redundant manipulator, inverse kinematics of a 7-DOF arm has attracted many researchers. The solution of this arm can be found in a closed form or a numerical one. An analytical solution can be found in case of a non-offset arm [1, 2]. In this case, the shoulder and wrist are considered as spherical joints. Hence, the elbow joint can freely rotate around an axis passing through two spherical joints. Based on this feature, the analytical solution can be found, depending on a freely chosen parameter – free internal motion of an elbow joint around an axis through a shoulder and a wrist that is called swivel angle. In [2], authors chose the swivel angle of the robot arm so that robot configuration is as same as a human arm as possible. One drawback of this approach is the way to deal with singularity and joint limit avoidance.

Another approach for inverse kinematics of a redundant robot is based on Jacobian matrix, in which the inverse problem is solved at velocity, acceleration, or jerk level. A

N. V. Khang et al. (Eds.): ASIAN MMS 2021, MMS 113, pp. 217–227, 2022.
https://doi.org/10.1007/978-3-030-91892-7_20

set of linear equations is dealed by pseudo-inverse of Jacobian matrix. The joint variables are then obtained by integrating joint velocities, acceleration, or jerks. In this approach, the advantages of redundancy such as trajectory optimization, obstacle, joint mechanical limits, and singularity avoidance are easily exploited by using null space of Jacobian matrix [3, 5]. This method is simple, however, for 7-DOF anthropomorphic arm the Jacobian matrix is rather large in size.

In this paper, the method based on Jacobian matrix and the analytical method are combined to find the solution of inverse kinematics of a 7-DOF anthropomorphic manipulator without offsets. In this way, the first four joint variables are determined with a Jacobian-based method, and the last three ones based on an analytical method. Some numerical simulations are carried out to verify the effectiveness of the proposed approach.

2 Kinematic Analysis

2.1 Direct Kinematics

Let's consider a 7-DOF manipulator as shown in Fig. 1. The direct kinematics can be solved systematicly by using the Denavit-Hartenberg (DH) method. The link coordinate systems established with the DH convention and the corresponding DH parameters table are shown in Fig. 1 and Table 1, respectively. In which q_i, $i = 1, 2, ..., 7$ represents the joint variables.

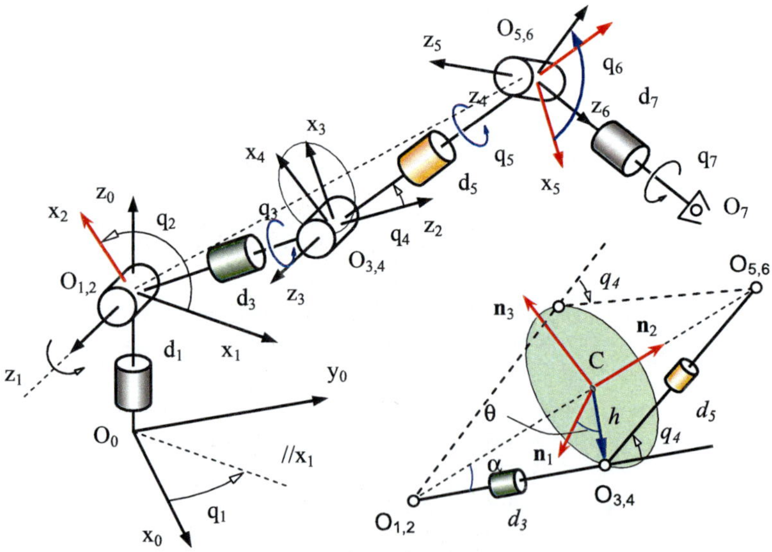

Fig. 1. 7-DOF manipulator and link-frames based on DH convention.

Table 1. DH parameters

Link i	θ_i(rad)	d_i(m)	a_i(m)	α_i(rad)
1	q_1	$d_1 = 0.28$	0	$\pi/2$
2	q_2	$d_2 = 0$	0	$\pi/2$
3	q_3	$d_3 = 0.33$	0	$-\pi/2$
4	q_4	$d_4 = 0$	0	$\pi/2$
5	q_5	$d_5 = 0.32$	0	$-\pi/2$
6	q_6	$d_6 = 0$	0	$\pi/2$
7	q_7	$d_7 = 0.10$	0	0

The relative homogeneous transformation matrices $\mathbf{A}_i^{i-1}(q_i)$ are calculated by substituting the DH parameters in Table 1 into the matrix equation for each joint:

$$\mathbf{A}_i^{i-1}(\theta_i) = \begin{bmatrix} \cos\theta_i & -\sin\theta_i\cos\alpha_i & \sin\theta_i\sin\alpha_i & a_i\cos\theta_i \\ \sin\theta_i & \cos\theta_i\cos\alpha_i & -\cos\theta_i\sin\alpha_i & a_i\sin\theta_i \\ 0 & \sin\alpha_i & \cos\alpha_i & d_i \\ 0 & 0 & 0 & 1 \end{bmatrix}. \quad (1)$$

The position and orientation of the kth link are given by:

$$\mathbf{T}_k^0(\mathbf{q}) = \mathbf{A}_1^0(q_1)\mathbf{A}_2^1(q_2)\ldots\mathbf{A}_k^{k-1}(q_k) = \begin{bmatrix} \mathbf{R}_k^0(\mathbf{q}) & \mathbf{r}_{Ok}^{(0)}(\mathbf{q}) \\ \mathbf{0} & 1 \end{bmatrix}, \quad k = 1, 2, \ldots, 7 \quad (2)$$

Some results of direct kinematic are given as follow:

$$\mathbf{r}_{O1}^{(0)}(\mathbf{q}) = \mathbf{r}_{O2}^{(0)}(\mathbf{q}) = \begin{bmatrix} 0 \\ 0 \\ d_1 \end{bmatrix}, \quad \mathbf{r}_{O3}^{(0)}(\mathbf{q}) = \mathbf{r}_{O4}^{(0)}(\mathbf{q}) = \begin{bmatrix} d_3\sin q_2\cos q_1 \\ d_3\sin q_2\sin q_1 \\ d_1 - d_3\cos q_2 \end{bmatrix} \quad (3)$$

$$\mathbf{r}_{O5}^{(0)}(\mathbf{q}) = \mathbf{r}_{O6}^{(0)}(\mathbf{q}) = \begin{bmatrix} d_3cq_1sq_2 + d_5[(cq_1cq_2cq_3 + sq_1sq_3)sq_4 + cq_1sq_2cq_4] \\ d_3sq_1sq_2 + d_5[(sq_1cq_2cq_3 - cq_1sq_3)sq_4 + sq_1sq_2cq_4] \\ d_1 - d_3cq_2 + d_5(sq_2cq_3sq_4 - cq_2cq_4) \end{bmatrix} \quad (4)$$

Remarks: We can see that the position of O_3 depends only on q_1 and q_2; the position of O_5 depends only on $q_1 \ldots q_4$ and not on q_5.

2.2 Singularity Analysis

For any robotic arm using rotational joints, there are always singular configurations in workspace. The kinematic singularities are independent of the coordinate frame. The singularity can be found based on calculating the determinant of Jacobian matrix $\det(\mathbf{J}(\mathbf{q}))$ $= 0$ or $\det(\mathbf{J}(\mathbf{q})\mathbf{J}^T(\mathbf{q})) = 0$ [5, 8]. However, with the structure of the manipulator as shown in Fig. 1, geometrically, we can see the singularities in some cases as shown in Fig. 2:

- When the origin $O_{5,6}$ is on the z_0 axis, the joint variable q_1 can have any value. This is called a shoulder singularity.
- When link 3 and 4 are stretched ($q_4 = 0$) out or folded ($q_4 = \pi$), axes z_2 and z_4 are collinear. The rotation of link 3 has no effect on the motion of the end-effector. This is called an elbow singularity.
- Similarly, when links 5 and 6 are stretched ($q_6 = 0$) out or folded ($q_6 = \pi$), axes z_4 and z_6 are collinear. The rotation of link 6 has no effect on the motion of the end-effector. This is called a wrist singularity.
- The other cases of singularities are the combination of the mentioned singularities.

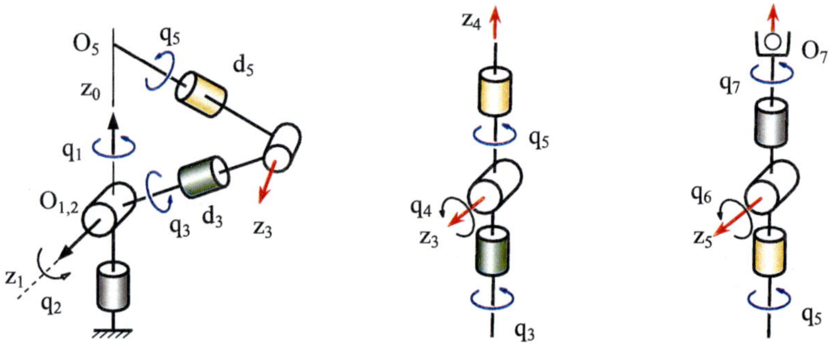

a) Shoulder singularity b) Elbow singularity c) Wrist singularity

Fig. 2. Singularities of 7-DOF anthropomorphic manipulator

2.3 Inverse Kinematics

Given a position and orientation of the end-effector $\mathbf{r}_7^{(0)} \& \mathbf{R}_7^0$, we need to find the joint variables q_k, $k = 1, ..., 7$.

Because three axes z_4, z_5, z_6 are concurrent at one point ($O_5 = O_6$), the Pieper's method (position and orientation decoupling) can be applied here for inverse kinematics [4]. The position of the wrist is determined from a given position and an orientation of the end-effector as:

$$\mathbf{r}_5^{(0)} = \mathbf{r}_6^{(0)} = \mathbf{r}_7^{(0)} - d_7\mathbf{R}_7^0\mathbf{k}_7^{(7)} = \left[x_5\ y_5\ z_6\right]^T, \quad \mathbf{k}_7^{(7)} = \left[0\ 0\ 1\right]^T \tag{5}$$

Analytical Solutions

Considering the triangle $O_2O_3O_5$, we can find joint angle q_4:

$$\begin{aligned} & d_3^2 - 2d_3d_5\cos\beta + d_5^2 = l_{2-5}^2 \\ & \Rightarrow \cos\beta = (d_3^2 + d_5^2 - l_{2-5}^2)/(2d_3d_5), \qquad \sin\beta = \sqrt{1-\cos^2\beta} \\ & \Rightarrow q_4 = \pm(\pi - \beta) = \pm(\pi - \operatorname{atan2}(\sin\beta, \cos\beta)) \end{aligned} \tag{6}$$

where $l_{2-5}^2 = \|\mathbf{r}_5^{(0)} - \mathbf{r}_2^{(0)}\| = (\mathbf{r}_5^{(0)} - \mathbf{r}_2^{(0)})^T(\mathbf{r}_5^{(0)} - \mathbf{r}_2^{(0)})$. In formula (6), we can take a positive sign or a negative sign depending on the upper or the lower configuration.

The angle $\alpha = \angle(O_5O_2O_3)$ is determined as:

$$\cos\alpha = \frac{l_{2-5}^2 + d_3^2 - d_5^2}{2d_3l_{2-5}}, \qquad \Rightarrow \alpha = \arccos\frac{d_3^2 + l_{2-5}^2 - d_5^2}{2d_3l_{2-5}}$$

From the structure of the manipulator, we can see that if the end-effector is fixed, point $O_{3,4}$ can move a along a circle with the radius of h and center C on the axis through O_2 and O_5 (see Fig. 1).

Let $\mathbf{R} = rot_{\vec{n}}(\theta)$ be a rotation matrix about axis $\vec{n}$ an angle θ:

$$\mathbf{R} = \mathbf{E} + \sin\theta\tilde{\mathbf{n}} + (1 - \cos\theta)\tilde{\mathbf{n}}\tilde{\mathbf{n}}.$$

Let $Cn_1n_2n_3$ be the coordinate system as shown in Fig. 1, $\vec{n}_2$ along the line O_2O_5, $\vec{n}_1 \perp \vec{z}_0$, $\vec{n}_1 = \vec{n}_2 \times \vec{z}_0/\|\vec{n}_2 \times \vec{z}_0\|$ and $\vec{n}_3 = \vec{n}_1 \times \vec{n}_2$.

If we know center C and radius h, then the position of point $O_{3,4}$ is determined as:

$$\mathbf{r}_{3,4}^{(0)} = \mathbf{r}_C^{(0)} + \mathbf{R}(\vec{n}, \theta)h\mathbf{n}_1. \tag{7}$$

Considering the right triangle O_2CO_3, we have:

$$O_2C = d_3\cos\alpha, \quad h = d_3\sin\alpha$$

Unit vector of O_2O_5 is calculated as:

$$\mathbf{n}_2 = (\mathbf{r}_5 - \mathbf{r}_2)/l_{2-5}$$

By freely choosing θ, we can find position of point O_3 as follows

$$\mathbf{r}_3 = \mathbf{r}_2 + O_2C\mathbf{n}_2 + \mathbf{R}(n_2, \theta)h\,\mathbf{n}_1 = \begin{bmatrix} x_3 \; y_3 \; z_3 \end{bmatrix}^T \tag{8}$$

Comparing (8) and (3) one yields

$$d_3 \sin q_2 \cos q_1 = x_3, \quad d_3 \sin q_2 \sin q_1 = y_3, \quad d_1 - d_3 \cos q_2 = z_3 \tag{9}$$

From (9), joint variable q_2 can be found as:

$$\begin{aligned} & d_1 - d_3 \cos q_2 = z_3 \Rightarrow \cos q_2 = (d_1 - z_3)/d_3, \qquad \sin q_2 = \pm\sqrt{1 - \cos^2 q_2} \\ & \Rightarrow q_2 = \operatorname{atan2}(\sin q_2, \cos q_2) \end{aligned} \tag{10}$$

Formula (10) gives out two solutions of q_2 depending on sign of $\sin(q_2)$. This sign will also decide joint variable q_1:

$$\begin{aligned} & d_3 \sin q_2 \cos q_1 = x_3 \quad \Rightarrow \quad \cos q_1 = x_3/(d_3 \sin q_2) \\ & d_3 \sin q_2 \sin q_1 = y_3 \quad \Rightarrow \quad \sin q_1 = y_3/(d_3 \sin q_2) \\ & q_1 = \operatorname{atan2}(y_3/\sin q_2, x_3/\sin q_2) \end{aligned} \tag{11}$$

We can see that if $\sin q_2 = 0$ (it means $q_2 = 0$ or $q_2 = \pm\pi$ and axis z_2 coincides with axis z_0), then we can not determine q_1, even though $q_1 + q_3$ is known.

If $\vec{n}_2 \| \vec{z}_0$ (it means $O_{5,6}$ lies on axis z_0), vector $\vec{n}_1$ is unidentified. Therefore, we can not find q_1 from (11). That is a singular configuration of this robot arm.

Joint variable q_3 is determined by comparing $\mathbf{r}_5^{(0)} - \mathbf{r}_4^{(0)} = \mathbf{r}_5^{(0)} - \mathbf{r}_3^{(0)}$. Knowing positions $\mathbf{r}_5^{(0)} \& \mathbf{r}_3^{(0)}$ of O_5 and O_3, we can find q_3 as follows:

$$\mathbf{r}_{O5}^{(0)}(\mathbf{q}) - \mathbf{r}_{O4}^{(0)}(\mathbf{q}) = \begin{bmatrix} d_5[(cq_1cq_2cq_3 + sq_1sq_3)sq_4 + cq_1sq_2cq_4] \\ d_5[(sq_1cq_2cq_3 - cq_1sq_3)sq_4 + sq_1sq_2cq_4] \\ d_5(sq_2cq_3sq_4 - cq_2cq_4) \end{bmatrix} = \begin{bmatrix} x_5 - x_3 \\ y_5 - y_3 \\ z_5 - z_3 \end{bmatrix} \tag{12}$$

Solving for $\sin q_3$ and $\cos q_3$, we get:

$$\begin{aligned} & \sin q_3 = \frac{-cq_1cq_2z_5 + cq_1cq_2z_3 - d_5cq_1cq_4 + x_5sq_2 - x_3sq_2}{d_5sq_1sq_2sq_4} \\ & \cos q_3 = \frac{d_5cq_2cq_4 + z_5 - z_3}{d_5sq_2sq_4} \\ & \Rightarrow q_3 = \operatorname{atan2}(sq_3, cq_3) \end{aligned} \tag{13}$$

So, the first four joint variables $q_1, \ldots, q_4$ are determined by (11), (10), (13) and (6), respectively. Now, we determine three last joint variables: q_5, q_6, q_7. The relative orientation of end-effector respect to link 4 is determined by:

$$\mathbf{R}_7^0 = \mathbf{R}_4^0(q_1, q_2, q_3, q_4)\mathbf{R}_7^4 \Rightarrow \mathbf{R}_7^4 = \mathbf{R}_4^{0T}(q_1, q_2, q_3, q_4)\mathbf{R}_7^0$$
$$\mathbf{R}_5^4(q_5)\mathbf{R}_6^5(q_6)\mathbf{R}_7^6(q_7) = \mathbf{R}_7^4$$

From direct kinematics we have:

$$\mathbf{R}_7^4 = \mathbf{R}_5^4(q_5)\mathbf{R}_6^5(q_6)\mathbf{R}_7^6(q_7)$$

$$= \begin{bmatrix} cq_5cq_6cq_7 - sq_5sq_7 & -cq_5cq_6sq_7 - sq_5cq_7 & cq_5sq_6 \\ sq_5cq_6cq_7 + cq_5sq_7 & -sq_5cq_6sq_7 + cq_5cq_7 & sq_5sq_6 \\ -sq_6cq_7 & sq_6sq_7 & cq_6 \end{bmatrix} = \begin{bmatrix} r_{11} & r_{12} & r_{13} \\ r_{21} & r_{22} & r_{23} \\ r_{31} & r_{32} & r_{33} \end{bmatrix}$$

Solving these equations, one gets q_5, q_6, q_7.

Remarks: free parameter θ can be chosen based on additional criteria such as singularity, obstacle, and joint limit avoidance.

Solution Based on Jacobian Matrix

Let $\boldsymbol{\xi}^T = [\mathbf{v}_E^{(0)}, \boldsymbol{\omega}_7^{(0)}]^T$ be the twist of the end-effector, we have:

$$\boldsymbol{\xi} = \mathbf{J}(\mathbf{q})\dot{\mathbf{q}} \quad \Rightarrow \quad \dot{\mathbf{q}} = \mathbf{J}^{\dagger}(\mathbf{q})\boldsymbol{\xi}, \quad \mathbf{J}^{\dagger}(\mathbf{q}) = \mathbf{J}^T(\mathbf{q})\Big[\mathbf{J}(\mathbf{q})\mathbf{J}^T(\mathbf{q})\Big]^{-1} \tag{14}$$

where $\mathbf{J}(\mathbf{q})$ and $\mathbf{J}^{\dagger}(\mathbf{q})$ are Jacobian matrix and its pseudo-inverse [6, 7].

In this paper, Eq. (14) is not used, since Jacobian of a 7-DOF anthropomorphic manipulator is quite bulky. Based on the arm structure, we can see that the redundancy is effective only on the position of O_5, when the end-effector is fixed. Therefore, we exploit the following relation instead of (14):

$$\mathbf{r}_5 = \mathbf{r}_5(\mathbf{q}_p) \quad \Rightarrow \quad \dot{\mathbf{r}}_5 = \mathbf{J}_5(\mathbf{q}_p)\dot{\mathbf{q}}_p \tag{15}$$

where $\mathbf{q}_p = [q_1, \ldots, q_4]^T$.

Applying pseudo-inverse, the solution of (15) is given as:

$$\dot{\mathbf{q}}_p = \mathbf{J}_5^{\dagger}\dot{\mathbf{r}}_5, \quad \mathbf{J}_5^{\dagger}(\mathbf{q}_p) = \mathbf{J}_5^T\Big[\mathbf{J}_5\mathbf{J}_5^T\Big]^{-1} \tag{16}$$

To avoid singularities, pseudo-inverse of $\mathbf{J}_5(\mathbf{q}_p)$ is modified by damped least square inverse as:

$$\mathbf{J}_5^{\dagger}(\mathbf{q}_p) = \mathbf{J}_5^T\Big[\mathbf{J}_5\mathbf{J}_5^T + k\mathbf{I}_s\Big]^{-1}, \quad k > 0 \tag{17}$$

$$\dot{\mathbf{q}}_p = \mathbf{J}_5^T\Big(\mathbf{J}_5\mathbf{J}_5^T + k\mathbf{I}\Big)^{-1}(\dot{\mathbf{r}}_5 + \mathbf{K}\mathbf{e}) + \Big(\mathbf{I} - \mathbf{J}_5^{\dagger}\mathbf{J}_5\Big)\mathbf{z}_0 \tag{18}$$

where $\mathbf{K} > 0$ is a gain matrix and $\mathbf{e} = \mathbf{r}_5(t) - \mathbf{r}_5^d(t)$, the error between actual and desired position of O_5, $\mathbf{z}_0 \in \mathbb{R}^4$ is an arbitrary vector. Vector $\mathbf{z}_0$ is normally chosen to exploit the redundancy such as singular, joint limits or obstacle avoidance.

Parameter k depending on $w(\mathbf{q}) = \sqrt{\det(\mathbf{J}(\mathbf{q}_p)\mathbf{J}^T(\mathbf{q}_p))}$, a manipulability measure, is chosen as following:

$$k = \begin{cases} \frac{1}{2}\varepsilon_0(1 + \cos(\pi w/w_0)), & w < w_0 \\ 0, & w \geq w_0 \end{cases} \tag{19}$$

where ε_0 and w_0 are small positive numbers.

The block diagram for inverse kinematics based on a combination of Jacobian matrix and analytical solution is shown in Fig. 3.

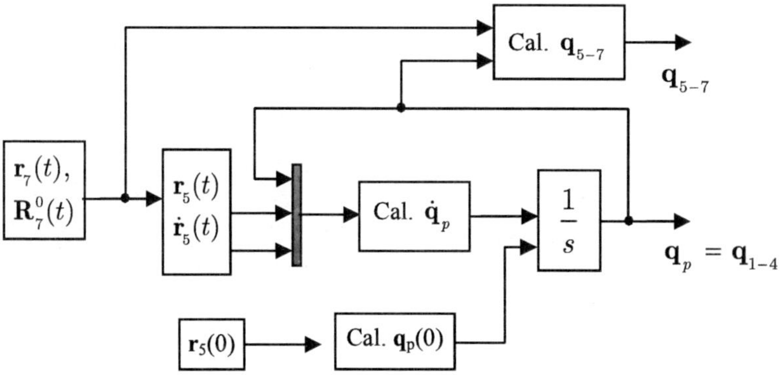

Fig. 3. Blockdiagram for inverse kinematics

3 Numerical Experiments

In order to confirm the validity of the algorithms proposed in this paper, some numerical simulations are carried. Two trajectories including rectilinear and curvilinear one from A to B are implemented by MATLAB:

$$\mathbf{r}_A = [0.5(d_3 + d_5 + d_7), \quad -0.4(d_3 + d_5), \quad d_1 + 0.0d_3]^T$$

$$\mathbf{r}_B = [0.5(d_3 + d_5 + d_7), \quad 0.4(d_3 + d_5), \quad d_1 + 0.6d_3]^T$$

The motion law along a trajectory is defined as following:

$$s(t) = s_i + \frac{s_f - s_i}{\pi}\left(\frac{\pi t}{t_f} - \frac{1}{2}\sin\frac{2\pi t}{t_f}\right), \quad 0 \leq t \leq t_f \tag{20}$$

Along a trajectory, the corresponding orientation is determined by the ZYZ Euler angles with:

$$\boldsymbol{\phi}_i = [\frac{\pi}{3}, \frac{\pi}{3}, \frac{\pi}{3}]^T, \quad \boldsymbol{\phi}_f = \boldsymbol{\phi}_i + \Delta\boldsymbol{\phi}, \quad \Delta\boldsymbol{\phi} = [\frac{5\pi}{18}, \frac{10\pi}{18}, \frac{20\pi}{18}]^T$$

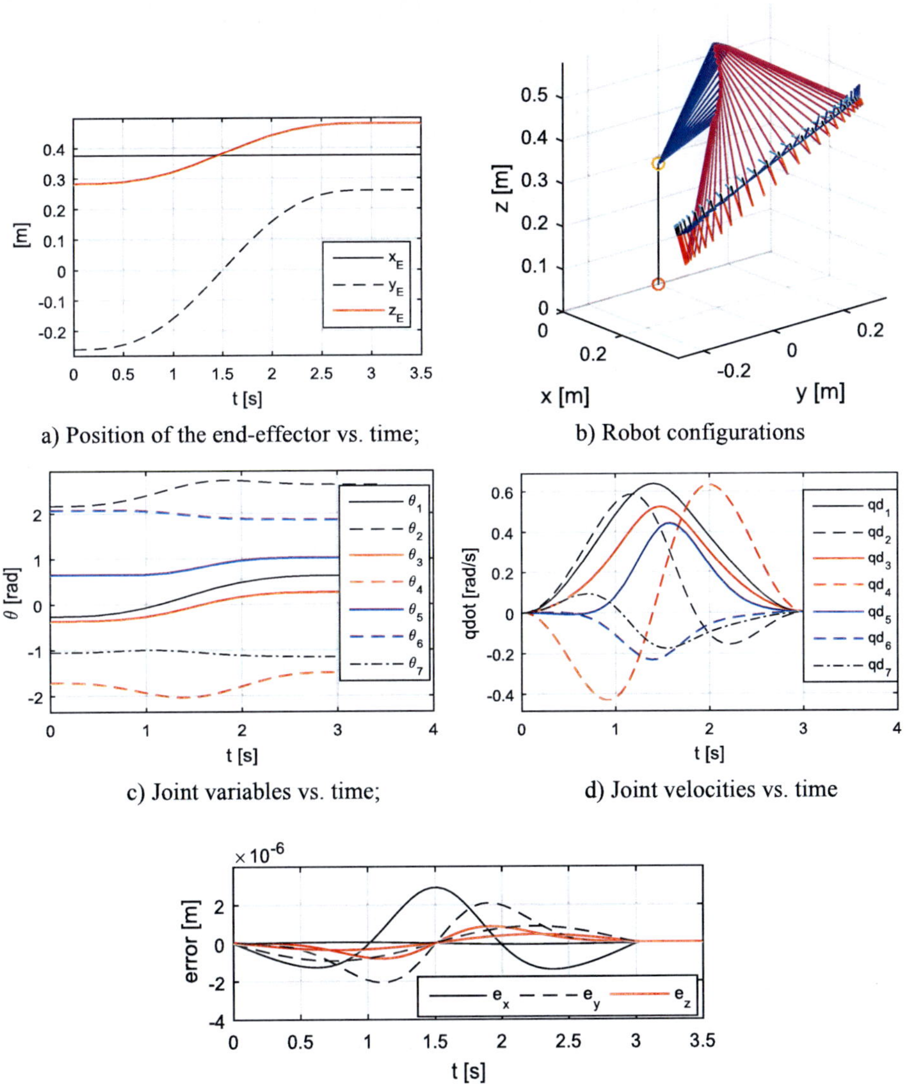

a) Position of the end-effector vs. time; b) Robot configurations

c) Joint variables vs. time; d) Joint velocities vs. time

e) Position error of end-effector vs. time

Fig. 4. Tracking of the end-effector along a rectilinear trajectory

The simulation results of the two cases are shown in Fig. 4 and Fig. 5, respectively. It can be seen from the simulation results that the pose of the manipulator changes uniformly, and the actual trajectory is consistent with the given trajectory, and the position errors of *x, y* and *z* are all within 2.5×10^{-6} m. All the joint variables change smoothly, and no sudden change appears. These results prove that the proposed inverse kinematics method can be applied to the control of continuous trajectory of the 7-DOF manipulator.

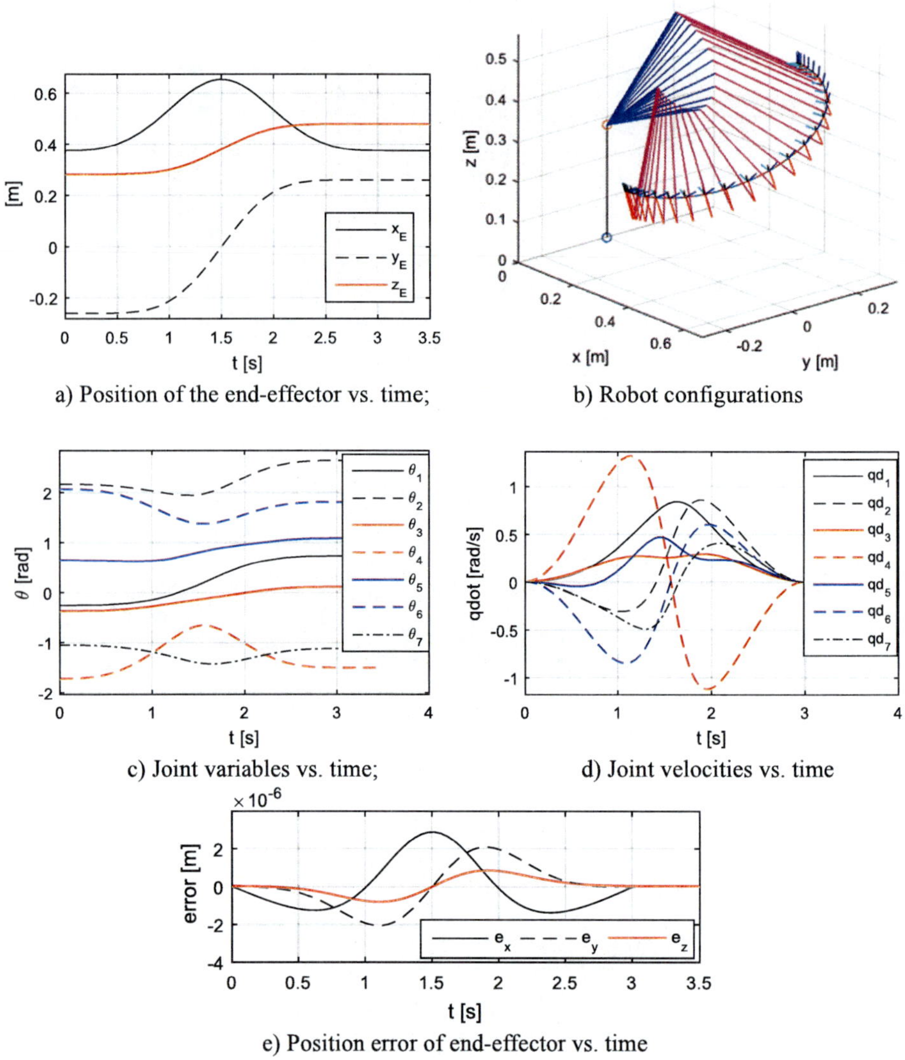

a) Position of the end-effector vs. time; b) Robot configurations

c) Joint variables vs. time; d) Joint velocities vs. time

e) Position error of end-effector vs. time

Fig. 5. Tracking of the end-effector along a curvilinear trajectory

To verify the ability to avoid singularities, a trajectory is chosen such that the origin O_5 pass through axis z_0. The simulation results in this case is shown in Fig. 6. It can be seen that $x_5 = y_5 = 0$ at time $t = 1.5$ s, but time histories of all joint variables are smooth, no sudden change of θ_1 occurs at this singularity.

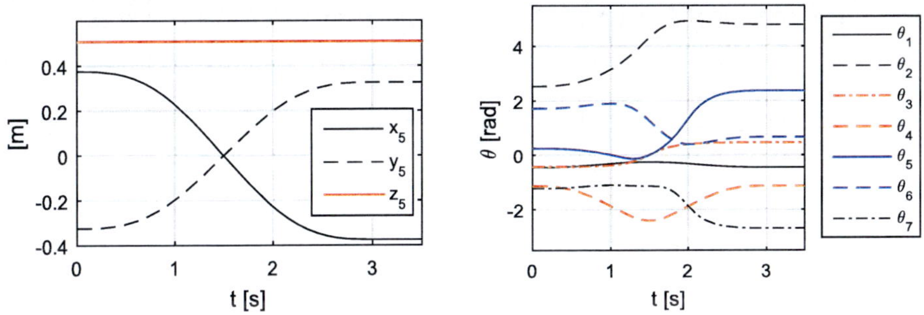

Fig. 6. Position of O_5 and joint variables vs. time

4 Conclusion

This paper combined successfully Jacobian-based and analytical method for an inverse kinematics of a redundant anthropomorphic manipulator. With the wrist equivalent to a spherical joint, we can decouple the position and orientation of an end-effector. The Jacobian-based method is applied for inverse kinematics of the position, that is a redundant case. In the proposed approach, the bulkiness of Jacobian matrix $\mathbf{J}_7$ is replaced by a simpler one $\mathbf{J}_5$. Therefore, the computational complexity is reduced. Additionally, the advantages of the Jacobian method such as singularity avoidance is retained. The numerical simulations verified the feasibility of the proposed approach.

Acknowledgment. This work was supported in part by the National Program: Support for research, development, and technology application of industry 4.0 (KC-4.0/19-25), under the grant for the project: Research, design and manufacture of Cobot applied in industry and some other fields with human-robot interaction (code: KC-4.0-35/19-35).

References

1. Tian, X., Xu, Q., Zhan, Q.: An analytical inverse kinematics solution with joint limits avoidance of 7-DOF anthropomorphic manipulators without offset. J. Franklin Inst. (2020). https://doi.org/10.1016/j.jfranklin.2020.11.020
2. Wang, Y., Artemiadis, P.: Closed-form inverse kinematic solution for anthropomorphic motion in redundant robot arms. Adv. Robot. Autom. **2**, 110 (2013). https://doi.org/10.4172/2168-9695.1000110
3. Wang, J., Li, Y., Zhao, X.: Inverse kinematics and control of a 7-DOF redundant manipulator based on the closed-loop algorithm. Int. J. Adv. Robot. Syst. **7**(4), 37 (2010)
4. Craig, J.J.: Introduction to Robotics: Mechanics and Control. Pearson/Prentice Hall, Hoboken (2005)
5. Siciliano, B., Sciavicco, L., Villani, L., Oriolo, G.: Robotics: Modelling Planning and Control. Springer, London (2010). https://doi.org/10.1007/978-1-84628-642-1
6. Nakamura, Y.: Advanced Robotics/Redundancy and Optimization. Addison-Wesley Publishing Company, Reading (1991)
7. Rao, C.R.: Generalized Inverse of Matrices and its Applications. Wiley, New York (1971)
8. Spong, M.W., Hutchinson, S., Vidyasagar, M.: Robot Modeling and Control. Wiley, New York (2006)

Inverse Kinematics and Dynamics of a 3$\underline{R}$RR Planar Parallel Manipulator in the Presence of Singularities

Nguyen Quang Hoang[1(✉)] and Vu Duc Vuong[2]

[1] Hanoi University of Science and Technology, Hanoi, Vietnam
hoang.nguyenquang@hust.edu.vn
[2] Thai Nguyen University of Technology, Thai Nguyen, Vietnam
vuducvuong@tnut.edu.vn

Abstract. The analysis of inverse kinematics and dynamics plays an important role in the design and control of parallel manipulators. At singular configurations, it is impossible to find an exact drive torque corresponding to the given motion of the moving platform. This paper presents a singular analysis of a 3$\underline{R}$RR planar parallel manipulator and a method for inverse kinematics and dynamics in the presence of singularities. The method of damped least square inverse is applied for inverse kinematics and the null space of Jacobian matrix is exploited for inverse dynamics. The numerical experiments are carried out to clarify the effectiveness of the proposed approach.

Keywords: Parallel manipulator · Inverse kinematics and dynamics · Singularity

1 Introduction

In the recent decades, the parallel robot manipulators are used widely in industrial applications due to its advantages such as higher accuracy and rigidity, higher loading rates on robot weight than those of serial robots. However, parallel robots also have some major disadvantages such as small workspaces and many singularities in the workspace due to their closed loop structure [1, 3, 5–7].

The problem of inverse kinematics plays an important role in the trajectory planning and the motion control of manipulators. This inverse problem should be solved as high accurate as possible. This inverse kinematic problem can be solved at three levels: position, velocity, and acceleration [4, 14]. Inverse trigonometric formulas are often applied to solve a problem at a position level. The robot configurations are dependent on these formulas. Based on the Jacobian matrix the problem is solved at velocity and acceleration. The advantage of this approach is that only the linear equations are considered. However, the solution of the problem is determined only if the Jacobian matrix is regular. The joint coordinates can be obtained by integrating its derivatives, which is the solution of linear equations. Simplicity is one of the advantages of this method.

N. V. Khang et al. (Eds.): ASIAN MMS 2021, MMS 113, pp. 228–237, 2022.
https://doi.org/10.1007/978-3-030-91892-7_21

Dynamic analysis of parallel manipulators have been attracted several researchers [2, 8–12]. In most of these works, authors have not considered inverse dynamics near the singular regions. In the neighbourhood of the drive singularities, Ider [11] used modified dynamic equations by using higher-order derivative information. Based on results in [11] Özdemir has shown the conditions, so that singularities of parallel manipulators can be removed to guarantee finite and continuous inverse dynamics solutions [13, 14]. In their approach, the L'Hopital's rule is applied to deal the case of zero by zero. To apply this approach, the trajectory should be planned such that the required derivatives exist. The third order derivative of constraint equations also made the problem more complicated. The method proposed by Ider [11] requires the singularity configurations must be known. Moreover, drive torques change significantly in the singular regions, but they are ignored in this method. That is another drawback of this approach.

This paper presents an inverse kinematics and dynamics of a 3RRR planar parallel manipulator in the presence of singularities. In the region closed to singularities, the optimal problem is introduced to overcome the singular configurations. Besides, the kinematic error feedback is used to stabilize the constraint equations. In this paper, the damped least square inverse is ultilized for inverse dynamics. This approach does not require the third or higher order derivative of constraint and does not take into account the conditions for trajectory as required in [13, 14].

This paper is organized as follows: Sect. 2 is kinematic analysis. Section 3 presents a singularity-free inverse dynamics based on null-space of Jacobian matrix. Some numerical simulations are shown in Sect. 4. Finally, the conclusion is given in Sect. 5.

2 Kinematic Model of a 3RRR Planar Parallel Manipulator

2.1 Constraint Equations and Jacobian Matrices

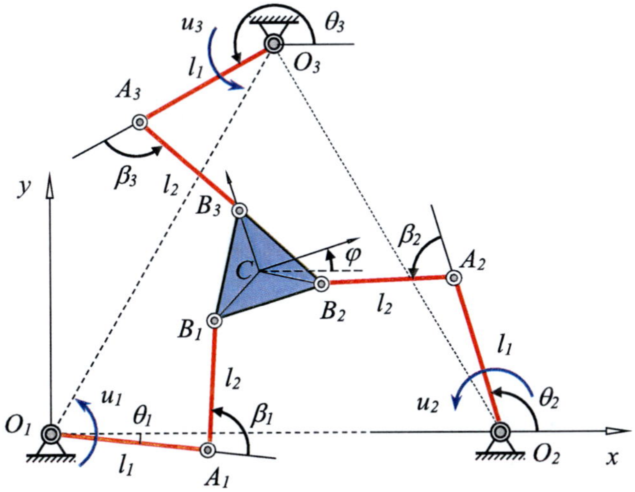

Fig. 1. The 3-DOF 3RRR planar parallel manipulator

Let consider a 3$\underline{R}$RR planar parallel robot with revolute joints as shown in Fig. 1. For this mechanism, the following the constructive characteristics are defined: the fixed platform, $O_1O_2O_3$ depicting an equilateral triangle with side L_0, the mobile platform $B_1B_2B_3$ forming another equilateral triangle with side a, the proximal links with $O_iA_i = l_1$, the distal links with $A_iB_i = l_2$.

Let $\boldsymbol{\theta} = [\theta_1, \theta_2, \theta_3]^T$ is a vector containing actuated joint variables, $\mathbf{x} = [x_C, y_C, \varphi]^T$ is a vector containing position and orientation variables of the mobile platform in the operational space, and $\boldsymbol{\beta} = [\beta_1, \beta_2, \beta_3]^T$ is a vector containing passive joint variables. The constraint equations are written as:

$$\mathbf{f}(\boldsymbol{\theta}, \boldsymbol{\beta}, \mathbf{x}) = 0, \ \mathbf{f} \in \mathbb{R}^6 \tag{1}$$

Based on Eq. (1) we have two kinematic problems: forward and inverse. The first problem, forward kinematics, is to determine the position and orientation of moving platform depending on the actuated joint variables, $\mathbf{x} = \mathbf{x}(\boldsymbol{\theta})$; and the second one, inverse kinematics, is to determine the actuated joint variables depending on the position and orientation $\mathbf{x}$ of moving platform, $\boldsymbol{\theta} = \boldsymbol{\theta}(\mathbf{x})$.

The Jacobian matrices are obtained by differentiating (1) with respect to time:

$$\mathbf{J}_x\dot{\mathbf{x}} + \mathbf{J}_\beta\dot{\boldsymbol{\beta}} + \mathbf{J}_\theta\dot{\boldsymbol{\theta}} = \mathbf{J}_x\dot{\mathbf{x}} + \mathbf{J}_y\dot{\mathbf{y}} = 0 \tag{2}$$

with $\mathbf{J}_x = \partial\mathbf{f}/\partial\mathbf{x}$, $\mathbf{J}_\beta = \partial\mathbf{f}/\partial\boldsymbol{\beta}$, $\mathbf{J}_\theta = \partial\mathbf{f}/\partial\boldsymbol{\theta}$, $\mathbf{y} = [\boldsymbol{\theta}^T, \boldsymbol{\beta}^T]^T$, $\mathbf{J}_y = [\mathbf{J}_\theta, \mathbf{J}_\beta]$.

2.2 Inverse Kinematics

There are some methods to solve this problem: analytical and numerical method, and method based on Jacobian matrices. The analytical method based on a geometry sketch can only be applied in some simple cases such as planar manipulators; and numerical methods are based on the iterative method of Newton-Raphson applying to nonlinear equations; the method based on the Jacobian matrix is used by many reseachers due to its simplicity.

Inverse Kinematics at Velocity Level – Jacobian Based Solution

The main ideas of this method is desribed as follows. Differentiating (1) with respect to time, one obtains:

$$\mathbf{J}_x\dot{\mathbf{x}} + \mathbf{J}_y\dot{\mathbf{y}} = 0 \tag{3}$$

Assuming that Jacobian $n_y \times n_y$ matrix $\mathbf{J}_y$ has a full rank. If $\mathbf{x}$ and $\mathbf{y}$ are known, Eq. (3) is a set of n_y linear algebraic equations with n_y unknown, which is a vector of joint velocity $\dot{\mathbf{y}}$. By solving Eq. (3) one gets:

$$\dot{\mathbf{y}} = -\mathbf{J}_y^{-1}\mathbf{J}_x\dot{\mathbf{x}} \tag{4}$$

By integrating and differentiating $\dot{\boldsymbol{\theta}}$ one gets the position and acceleration as follows:

$$\mathbf{y}(t) = \mathbf{y}(0) + \int_0^t \dot{\mathbf{y}}d\tau = \mathbf{y}(0) - \int_0^t \mathbf{J}_y^{-1}\mathbf{J}_x\dot{\mathbf{x}}(\tau)d\tau \tag{5}$$

$$\ddot{\mathbf{y}}(t) = -\tfrac{d}{dt}(\mathbf{J}_y^{-1}\mathbf{J}_x\dot{\mathbf{x}}) \tag{6}$$

Another way to determine the acceleration is to differentiate velocity Eq. (3) with respect to time:

$$\mathbf{J}_x\ddot{\mathbf{x}} + \dot{\mathbf{J}}_x\dot{\mathbf{x}} + \mathbf{J}_y\ddot{\mathbf{y}} + \dot{\mathbf{J}}_y\dot{\mathbf{y}} = 0$$

From this the acceleration of the actuated joints are obtained:

$$\ddot{\mathbf{y}} = -\mathbf{J}_y^{-1}\left(\mathbf{J}_x\ddot{\mathbf{x}} + \dot{\mathbf{J}}_x\dot{\mathbf{x}} + \dot{\mathbf{J}}_y\dot{\mathbf{y}}\right). \tag{7}$$

Stabilization with Position Error Feedback

Equations (4) and (7) show that the vector of generalized velocity $\dot{\mathbf{y}}(t)$ and acceleration $\ddot{\mathbf{y}}(t)$ can be determined. The following section presents a method to get joint variables. The generalized coordinates are determined by Eqs. (5). However, due to error accumulated the constraint equation may be violated, $\mathbf{f}(\mathbf{y}, \mathbf{x}) \neq 0$. In order to overcome this issue, the method of kinematic feedback is proposed. Instead of solving $\dot{\mathbf{y}}$ from $\dot{\mathbf{f}}(\mathbf{y}, \mathbf{x}) = 0$, the following kinematic equation is used:

$$\dot{\mathbf{f}} + \mathbf{Kf} = 0, \text{ with } \mathbf{K} > 0. \tag{8}$$

It is clear that the solution of (8) converge exponentially to zero, $\mathbf{f}(t) = \mathbf{f}(0)e^{-\mathbf{K}t} \to 0$. Equation (8) is rewritten as:

$$\dot{\mathbf{f}} + \mathbf{Kf} = \mathbf{J}_x\dot{\mathbf{x}} + \mathbf{J}_y\dot{\mathbf{y}} + \mathbf{Kf} = 0. \tag{9}$$

From this if $\det(\mathbf{J}_y) \neq 0$ one gets:

$$\dot{\mathbf{y}} = -\mathbf{J}_y^{-1}(\mathbf{J}_x\dot{\mathbf{x}} + \mathbf{Kf}) \tag{10}$$

2.3 Singular Configurations of a 3RRR Planar Parallel Manipulator

Singular configurations appear in the workspace or on its boundary. At the singularity, the degree of freedom of the mobile platform is reduced or the solution of the forward as well as inverse kinematics is undetermined. For the 3RRR planar parallel manipulator, singularities are all configurations of three connecting links that are parallel or concurrent (links intersect at a point); or when at least one of three legs is stretched (fully extended) or folded.

Algebraically, a configuration is singular when the rank of the Jacobian matrices is deficient. Based on Eq. (3), singularity of 3RRR planar parallel manipulators is classified into three types as follows:

The first type: (inverse kinematics singularity): $\det(\mathbf{J}_y) = 0$.
The second type: (forward kinematics singularity): $\det(\mathbf{J}_x) = 0$.
The third type: (mixed inverse and forward kinematics singularity): $\det(\mathbf{J}_y) = 0$ and $\det(\mathbf{J}_x) = 0$.

2.4 Singularity-Free Inverse Kinematics

In the previous section, the inverse kinematics is solved at a velocity level when $\det(\mathbf{J}_y) \neq 0$. This section deals with the case when the Jacobian matrix closed to singularity. In this case we can not find $\dot{\mathbf{y}}$ directly from (3) or (9), their solution is very sensitive to a small change of Jacobian matrix and it is undetermined at singularity. This section presents a method to find vector $\dot{\mathbf{y}}$ when $-d_0 \leq \det(\mathbf{J}_y) \leq d_0$, in which a vector $\dot{\theta}$ will be found so that the following quantainty is minimum:

$$L(\dot{\mathbf{y}}) = (\mathbf{J}_y\dot{\mathbf{y}} - \mathbf{b})^T(\mathbf{J}_y\dot{\mathbf{y}} - \mathbf{b}) + \kappa\dot{\mathbf{y}}^T\mathbf{E}\dot{\mathbf{y}} \rightarrow \min \;\text{ with } \mathbf{b} = -\mathbf{J}_x\dot{\mathbf{x}} \tag{11}$$

where $\kappa > 0$ is a parameter needed to choose.

Differentiating (11) with regard to $\dot{\mathbf{y}}$ and setting it to zero ones gets:

$$\dot{\mathbf{y}} = -(\mathbf{J}_y^T\mathbf{J}_y + \kappa\mathbf{E})^{-1}\mathbf{J}_y^T\mathbf{J}_x\dot{\mathbf{x}} = -\mathbf{J}_y^T(\mathbf{J}_y\mathbf{J}_y^T + \kappa\mathbf{E})^{-1}\mathbf{J}_x\dot{\mathbf{x}} \tag{12}$$

Combining with the kinematic error feedback, the above solution becomes:

$$\dot{\mathbf{y}} = -\mathbf{J}_y^T(\mathbf{J}_y\mathbf{J}_y^T + \kappa\mathbf{E})^{-1}(\mathbf{J}_x\dot{\mathbf{x}} + \mathbf{K}\mathbf{f}). \tag{13}$$

The parameter κ will be chosen as follows:

- $\kappa = \varepsilon$ at singularity, $\det(\mathbf{J}_y\mathbf{J}_y^T) = 0$;
- $\kappa = 0$ when the robot is far from singularity, $|\det(\mathbf{J}_y\mathbf{J}_y^T)| \geq d_0$;
- $0 \leq \kappa \leq \varepsilon$ when the robot is closed to singularity, $-d_0 \leq \det(\mathbf{J}_y\mathbf{J}_y^T) \leq d_0$.

In order to guarantee that the vector $\dot{\mathbf{y}}$ changes smoothly in the region closed to singularity, the parameter κ will be chosen as a function of $d = |\det(\mathbf{J}_y\mathbf{J}_y^T)|$, it means $\kappa = \kappa(d)$. In this paper, it is chosen as:

$$\kappa = \begin{cases} 0, & d \geq d_0 \\ \frac{1}{2}\varepsilon 1 + \cos(\pi d/d_0), & d < d_0 \end{cases} \tag{14}$$

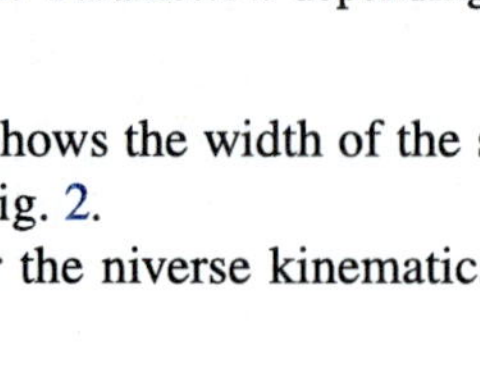

Fig. 2. Parameter κ depending on d

A small parameter $d_0 > 0$ shows the width of the singular boundary. The graph of κ depending on d is shown in Fig. 2.

The proposed algorithm for the niverse kinematics based on Jacobian matrices is illustrated in Fig. 3.

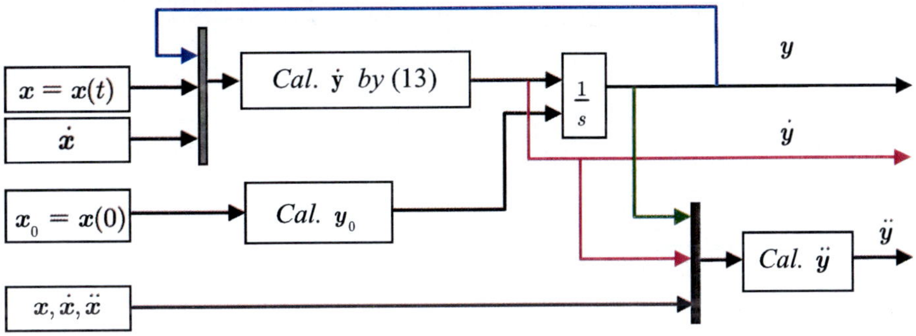

Fig. 3. Block diagram for inverse kinematic problem based on Jacobian matrices

3 Inverse Dynamics in the Presence of Singularities

Dynamic equations of a parallel manipulator can be established by many methods such as Lagrangian equations with multipliers, Newton-Euler equations, Kane equations, … With redundant coordinates $\mathbf{q}^T = [\boldsymbol{\theta}^T, \boldsymbol{\beta}^T, \mathbf{x}^T]$, the dynamic equations of a parallel manipulator are given as follows [2, 3, 8]:

$$\mathbf{M}(\mathbf{q})\ddot{\mathbf{q}} + \mathbf{C}(\mathbf{q}, \dot{\mathbf{q}})\dot{\mathbf{q}} + \mathbf{D}\dot{\mathbf{q}} + \mathbf{g}(\mathbf{q}) = \mathbf{B}\mathbf{u} + \mathbf{Q} + \boldsymbol{\Phi}^T(\mathbf{q})\boldsymbol{\lambda}, \tag{15}$$

$$\boldsymbol{\phi}(\mathbf{q}) = 0, \quad \boldsymbol{\Phi}(\mathbf{q}) = \frac{\partial \boldsymbol{\phi}(\mathbf{q})}{\partial \mathbf{q}}. \tag{16}$$

Based on these equations, direct and inverse dynamics can be solved. In this paper, only the latter issue is considered. With the given motion $\mathbf{q}(t)$, the drive torques $\mathbf{u}$ can be solved by rewritten (15) as following:

$$[\mathbf{B}\ \boldsymbol{\Phi}^T(\mathbf{q})]\begin{bmatrix}\mathbf{u}\\ \boldsymbol{\lambda}\end{bmatrix} = \mathbf{M}(\mathbf{q})\ddot{\mathbf{q}} + \mathbf{C}(\mathbf{q}, \dot{\mathbf{q}})\dot{\mathbf{q}} + \mathbf{D}\dot{\mathbf{q}} + \mathbf{g}(\mathbf{q}) - \mathbf{Q}. \tag{17}$$

The matrix $[\mathbf{B}\ \boldsymbol{\Phi}^T(\mathbf{q})]$ of size $m \times m$ is regular if the manipulator is not in singular configurations. Hence, from (17) one obtains $\mathbf{u}$ and $\boldsymbol{\lambda}$.

We can eliminate the multipliers $\boldsymbol{\lambda}$ from (17) by introducing the null space of Jacobian matrix defined by:

$$\mathbf{R} = \alpha\begin{bmatrix}\mathbf{E}\\ -[\boldsymbol{\Phi}_d^{-1}(\mathbf{q})\boldsymbol{\Phi}_i(\mathbf{q})]\end{bmatrix}, \quad \alpha \neq 0. \tag{18}$$

In which $\mathbf{q}_i \equiv \boldsymbol{\theta}$, $\mathbf{q}_d = [\boldsymbol{\beta}^T, \mathbf{x}^T]$ and $\boldsymbol{\Phi}_i = \partial\boldsymbol{\phi}/\partial\mathbf{q}_i$, $\boldsymbol{\Phi}_d = \partial\boldsymbol{\phi}/\partial\mathbf{q}_d$. It can be seen that $\boldsymbol{\Phi}(\mathbf{q})\mathbf{R} = \mathbf{0}$ and $\mathbf{R}^T\boldsymbol{\Phi}^T(\mathbf{q}) = \mathbf{0}$.

To eliminate $\boldsymbol{\lambda}$ from (15), we multiply $\mathbf{R}^T$ from left with (15), then rearrange, one obtains:

$$\mathbf{R}^T\mathbf{B}\mathbf{u} = \mathbf{R}^T(\mathbf{M}(\mathbf{q})\ddot{\mathbf{q}} + \mathbf{C}(\mathbf{q}, \dot{\mathbf{q}})\dot{\mathbf{q}} + \mathbf{D}\dot{\mathbf{q}} + \mathbf{g}(\mathbf{q}) - \mathbf{Q}) =: \mathbf{h}(\mathbf{q}, \dot{\mathbf{q}}, \ddot{\mathbf{q}}). \tag{19}$$

Solving (19), one gets:

$$\mathbf{u} = [\mathbf{R}^T\mathbf{B}]^{-1}\mathbf{R}^T(\mathbf{M}(\mathbf{q})\ddot{\mathbf{q}} + \mathbf{C}(\mathbf{q},\dot{\mathbf{q}})\dot{\mathbf{q}} + \mathbf{D}\dot{\mathbf{q}} + \mathbf{g}(\mathbf{q}) - \mathbf{Q}). \tag{20}$$

Note that, with a structure of $\mathbf{B} = [k\mathbf{E}_{n\times n}, \mathbf{0}_{n\times(m-n)}]^T$, then the product matrix $[\mathbf{R}^T\mathbf{B}]$ has a form of $\alpha k\mathbf{E}_{n\times n}$.

Singularity-Free Inverse Dynamics

In case the robot moves through or near the singular configurations, we can not use formula (18) because $\mathbf{\Phi}_d$ is singular. Hence, we need to find another matrix $\mathbf{R}$. Using the pseudo-inverse matrix, we can construct $\mathbf{R}$ as follows:

$$\mathbf{R} = \alpha[\mathbf{E}_m - \mathbf{\Phi}^+(\mathbf{q})\mathbf{\Phi}(\mathbf{q})] \text{ or } \mathbf{R} = \text{null}(\mathbf{\Phi}(\mathbf{q})), \tag{21}$$

with $\alpha \neq 0$ and $\mathbf{\Phi}^+$ is pseudo-inverse of $\mathbf{\Phi}$ satisfying $\Phi\Phi^+\mathbf{\Phi} = \mathbf{\Phi}$.

By using $\mathbf{R}$ defined by (21) in (19), the matrix $\mathbf{R}^T\mathbf{B}$ in (19) is not square anymore. Therefore, drive torques $\mathbf{u}$ is determined by:

$$\mathbf{R}^T\mathbf{B}\mathbf{u} = \mathbf{h}(\mathbf{q},\dot{\mathbf{q}},\ddot{\mathbf{q}}) \quad \Rightarrow \quad \mathbf{u} = \mathbf{A}^+\mathbf{h}(\mathbf{q},\dot{\mathbf{q}},\ddot{\mathbf{q}}), \tag{22}$$

with $\mathbf{A} = \mathbf{R}^T\mathbf{B},\ \mathbf{A}^+ = [\mathbf{A}^T\mathbf{A}]^{-1}\mathbf{A}^T$.

Note that, two formulae (21) and (22) are valid for both singular and regular configurations of the manipulator.

In case of singularity of direct kinematics, $\mathbf{A} = \mathbf{R}^T\mathbf{B}$ is a degenerate matrix, so we can not find $\mathbf{u}$ exactly satisfying (19). In this case, we can use damped least square inverse to find the best $\mathbf{u}$ as following:

$$\mathbf{u} = (\mathbf{A}^T\mathbf{A} + \kappa\mathbf{E})^{-1}\mathbf{A}^T\mathbf{h}(\mathbf{q},\dot{\mathbf{q}},\ddot{\mathbf{q}}). \tag{23}$$

4 Numerical Simulations

In this section, some simulations with a 3$\underline{\text{R}}$RR planar parallel manipulator which moves in the horizontal plane driven by 3 actuators are implemented. The parameters of the manipulator are chosen as:

$L_0 = 1.2, l_1 = 0.58, l_2 = 0.623, a = 0.2$ m,
$m_1 = 2.072, m_2 = 0.750, m_3 = 0.978$ kg,
$I_{C1} = 0.13, I_{C2} = 0.03, I_{C3} = 0.007$ kg.m^2.

In these simulations, the center of the mobile platform will be forced to move at the velocity of 0.8 m/s along a circular trajectory, while its orientation is constant, $\phi = 0$ rad. The trajectory has a center at $(x_C, y_C) = (1.039, 0.600)$ m and a radius of $r = 0.12$ m. The time history of platform motion is shown in Fig. 4.

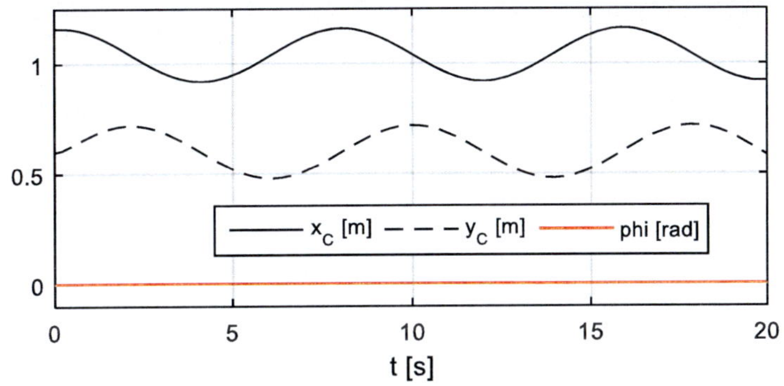

Fig. 4. Time history of position and orientation of the mobile platform

For the inverse kinematics, we choose the following parameters:

$$\varepsilon = 5.0\text{e}-3, \quad d_0 = 1.0\text{e}-4, \quad \mathbf{K} = 100\mathbf{E}_{6\times 6},$$

and the simulation results are shown in Fig. 5.

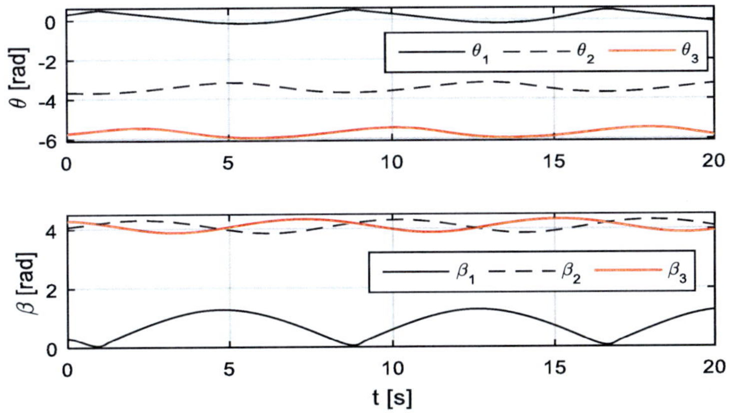

Fig. 5. Time history of active (θ) and passive (β) joint variables

We can see that angle β_1 sometime reaches very closed to zero. At this position, the first leg is fully extended, it means the robot is at a singular configuration. The diagram of $\det(\boldsymbol{\Phi}_d)$ shown in Fig. 6 reaches closed to zero when angle β_1 reaches closed to zero. Moreover, in the neighborhood of the singular position, the error of constraint functions increases, but it decreases fast to zero.

The actuator torques required to realize the desired motion are shown in Fig. 7. Near the singular position, only the torque of the first leg u_1 has a litle change, while u_2 and u_3 have almost no sudden changes.

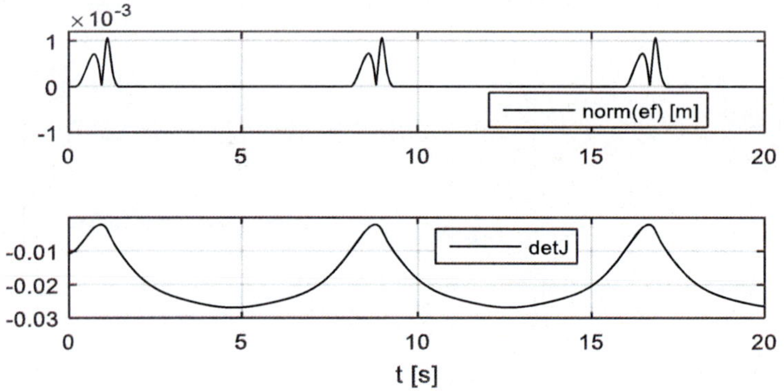

Fig. 6. Time history of error norm and det($\mathbf{\Phi}_d$)

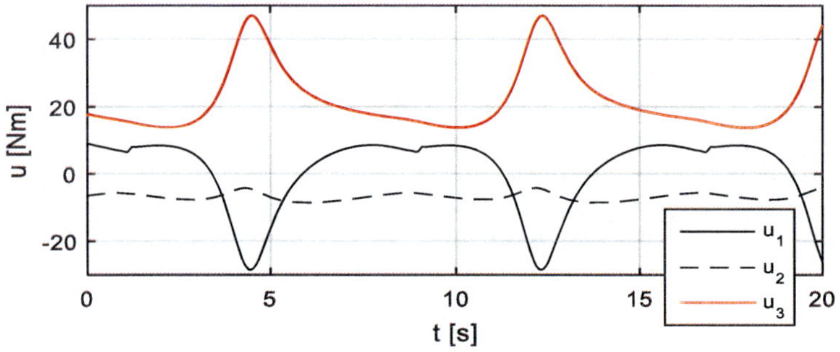

Fig. 7. Time history of drive torque

5 Conclusion

This paper presents algorithms to solve the inverse kinematics and dynamics of planar parallel manipulators, in which damped least square inverse is exploited to avoid singularities. The effectiveness of the proposed method is demonstrated by means of numerical experiments with the 3$\underline{R}$RR planar delta parallel manipulator. The simulation results show that there is no sudden change in drive torques. The problem of parallel robot control in the presence of singularities will be considered in the future study.

References

1. Ceccarelli, M.: Fundamentals of Mechanics of Robotic Manipulation. Springer, Netherlands (2004). https://doi.org/10.1007/978-1-4020-2110-7
2. Van Khang, N.: Dynamics of Multibody Systems. Science and Technology Publishing House, Hanoi (2007).(in Vietnamese)
3. Merlet, J.-P.: Parallel Robots, 2nd edn. Springer, Heidelberg (2006). https://doi.org/10.1007/1-4020-4133-0
4. Sciavicco, L., Siciliano, B.: Modelling and Control of Robot Manipulators, 2nd edn. Springer, London (2000). https://doi.org/10.1007/978-1-4471-0449-0

5. Liu, S., Qiu, Z., Zhang, X.: Singularity and path-planning with the working mode conversion of a 3-DOF 3-RRR planar parallel manipulator. Mech. Mach. Theory **107**, 166–182 (2017)
6. Gosselin, C.M., Wang, J.: Singularity loci of planar parallel manipulators with revolute actuators. Robot. Auton. Syst. **21**(4), 377–398 (1997)
7. Kotlarski, J., de Nijs, R., Abdellatif, H., Heimann, B.: New interval-based approach to determine the guaranteed singularity-free workspace of parallel robots. In: IEEE International Conference on Robotics and Automation, Kobe, Japan, pp. 1256–1261 (2009)
8. Staicu, S.: Dynamics of Parallel Robots. Springer, Heidelberg (2018). https://doi.org/10.1007/978-3-319-99522-9
9. Staicu, S.: Inverse dynamics of the 3-PRR planar parallel robot. Robot. Auton. Syst. **57**, 556–563 (2009)
10. Kemal Ider, S.: Inverse dynamics of parallel manipulators in the presence of drive singularities. Mech. Mach. Theory **40**, 33–44 (2005)
11. Kemal Ider, S.: Singularity robust inverse dynamics of planar 2-RPR parallel manipulators. Proc. Inst. Mech. Eng. Part C J. Mech. Eng. Sci. **218**, 721–730 (2004)
12. Tsai, L.-W.: Solving the inverse dynamics of a Stewart-Gough manipulator by the principle of virtual work. J. Mech. Des. **122**, 3–9 (2000)
13. Özdemir, M.: Removal of singularities in the inverse dynamics of parallel robots. Mech. Mach. Theory **107**, 71–86 (2017)
14. Özdemir, M.: High-order singularities of 5R planar parallel robots. Robotica **37**, 233–245 (2018)

Requirements and Design of a Hand for LARMbot Humanoid

Wenshuo Gao[1] and Marco Ceccarelli[2](✉)

[1] Graduate School, Chinese Academy of Agricultural Sciences, Beijing 100081, China
[2] Laboratory of Robotics Mechatronics, University of Rome "Tor Vergata", 00133 Rome, Italy
marco.ceccarelli@uniroma2.it

Abstract. The paper presents a design solution for a hand for LARMbot humanoid robot based on requirements that have been considered for a specific grasping application. Features are considered for integration and synergy in a humanoid arm by looking at the grasping aims in terms of motion and force. The proposed solution is presented based on three fingers interacting with a palm design provided of springs. Performance characteristics are evaluated through a simulation of a CAD design to check the feasibility of the proposed solution and to characterize the expected operation in the LARMbot humanoid.

Keywords: Humanoid robots · Robot hands · Grasping · Design · Simulation · Performance evaluation

1 Introduction

Humanoid hands are an important part of robots and studying multi-finger hand has become more and more popular in recent years. As manipulator end-effector, it can affect the working performance of robots [1]. In recent years, many scholars have conducted research on humanoid hands with different number of fingers. The more fingers, the more flexible and stable the hand is when grasping objects. However, multi-finger humanoid hands require complicated control system, which increases the requirements and economic costs for users [2–4]. Keeping a balance between the grasping performance and financial costs becomes important to develop a humanoid hand. A three-finger humanoid hand with linkage-driving mechanism can solve the problem and grasp the objects more stably and efficiently as suggested in [3].

Scholars at Laboratory of Robot Mechatronics (LARM) have made effort to improve the grasping stability, flexibility, mechanism portability, economy and energy-saving of the three-finger humanoid hand. To achieve suitable mechanical designs with high grasping efficiency, small-sized robust design, and light and low-cost devices [4]. For example, humanoid hands are developed by using linkage-driving mechanisms using a limited number of motors to obtain greater grasping performance of the hand. The grasping force generated by a linkage-driving mechanism is better than other kinds of driving mechanism, as pointed out for example in [1, 5, 6]. In addition, a one-finger underactuated humanoid hand with a movable palm was proposed to complete the closure

N. V. Khang et al. (Eds.): ASIAN MMS 2021, MMS 113, pp. 238–245, 2022.
https://doi.org/10.1007/978-3-030-91892-7_22

envelope grasping of the grasped objects by using one motor and strings with suitable grasping characteristics of stable, flexible and energy-saving operation, as reported in [7].

Based on previous achievements of humanoid hands at LARM2 in Rome, a three-finger humanoid hand with a movable palm is proposed by using a linkage-driving mechanism and based on design requirements for humanoid hands. A servomotor with a lead screw driving the linkage mechanism is applied to complete the grasping assignments. Force sensors are used to control grasping force, which can protect the grasped object on the basis of achieving stable grasping through force feedback system. Finally, performance characteristics are evaluated through simulation of a CAD design to check the feasibility of the proposed solution and to characterize the expected operation in the LARMbot humanoid.

2 Requirements for a Hand in LARMbot Humanoid

LARMbot humanoid is conceived as a low-cost humanoid robot with parallel structures. It is also designed to be manufactured through 3D printing, driven by commercial servomotors and linear actuators, and controlled by Arduino boards. A CAD model of the humanoid robot can be seen in Fig. 1a, while a prototype is shown Fig. 1b, [8, 9].

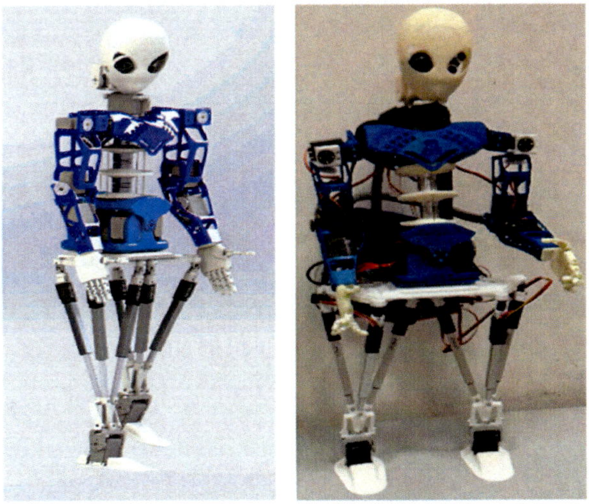

a) b)

Fig. 1. LARMbot design: a) CAD design; b) a built prototype

The locomotion module of the LARMbot prototype is less than 320 mm wide, 150 mm deep and 400 mm high. Its weight is 1.05 kg, considering both mechanical structure and electronics. The upper body (torso and manipulation module) is 320 mm wide, 150 mm deep and 450 mm high, and its weight is equal to 2.60 kg. Thus, the entire prototype is 850 mm tall and has a total weight of approximately 3.70 kg (approximately

2.00 kg for motors and 3D printed frame, and 1.70 kg for control boards, sensors and battery), making the entire system compact and lightweight. Its payload capability for manipulation is 0.85 kg, limited by the serial structure of the arm, while the parallel architecture of torso and legs allows for a payload up to 3.00 kg, [9, 10].

Humanoid hands are aimed at grasping objects for human-like tasks. Generally, these aspects should be taken into consideration when developing a humanoid hand, including grasping capability, grasping force and humanoid features, as summarized in Fig. 2.

As for humanoid hands for LARMbot humanoid, the size of fingers portion is expected to be about 30 mm. The workplace should be suitable for grasping a family of objects that are in the LARMbot humanoid working environment. In addition, the finger of humanoid hand can generate proper force with a grasped object by being checked by sensors. Meanwhile, human-like appearance helps the hands work flexibly. The limited number of actuators should drive the light-weight structure of hands to save costs. The integrated operation should be convenient and power consumption is expected to be low [9].

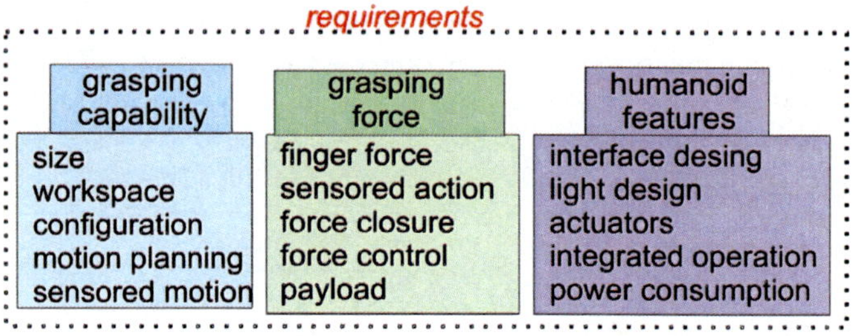

Fig. 2. Requirements for design and functionality of humanoid hands

3 A Design Solution

LARMbot humanoid can work in different kinds of tasks such as at serving for housework. A three-finger humanoid hand with a movable palm is proposed based on the former work at LARM2. A proposed linkage-driving mechanism is shown in Fig. 3. Based on this driving mechanism, the proposed humanoid hand is composed of a linkage-driving mechanism, three humanoid fingers, a movable palm, a lead screw, a servo motor, and platforms, as shown in Fig. 4a. The operation of the hand can be summarized as in the following. Firstly, the motor rotates forward to drive the lead screw. Then the movable platform moves upward as well as the rotating lead screw, which causes the connecting rod to move upward. After that, the linkage driving mechanism works. The humanoid fingers grasp the object. Finally, the humanoid hand completes the envelope grasping action and the object is grasped. The linkage driving mechanism can provide large grasping force. Combined with the movable palm in Fig. 4b, the humanoid hand is driven by linkage mechanism and can grasp more stably so that the grasped object can be protected from excessive force. The three-finger underactuated humanoid hand can grasp objects efficiently by using only one motor for the hand resulting low-cost. Consequently, the control system is fairly simple, which makes operation more convenient as user oriented. The whole hand will be made by using 3D manufacturing technology giving a light-weight solution.

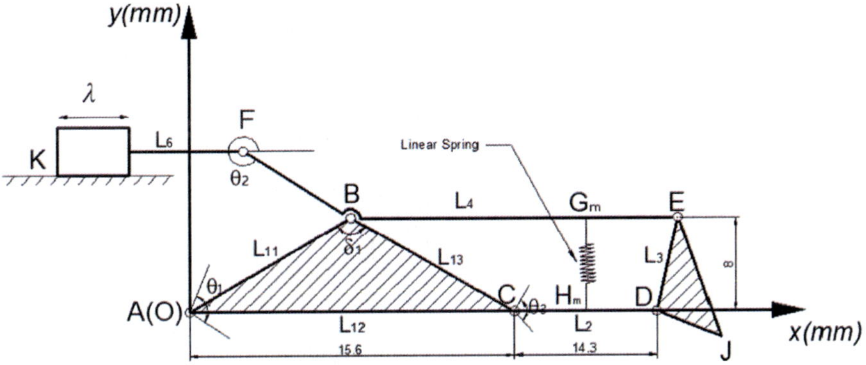

Fig. 3. The driving mechanism of the proposed three-finger humanoid hand with design parameters

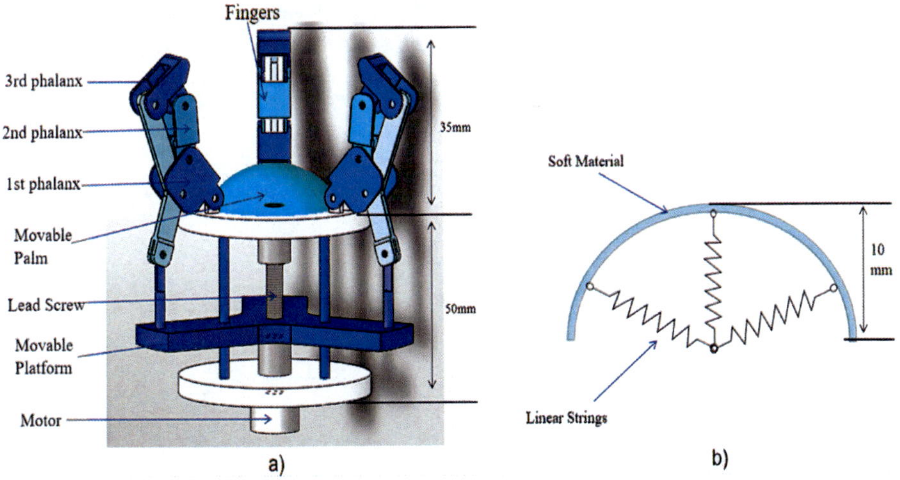

Fig. 4. Mechanical design of the proposed three-finger humanoid hand: a) a CAD design; b) structure of the movable palm

4 Performance Evaluation

Adams/View is used to conduct the grasping simulation to validate the grasping abilities of the proposed humanoid hand. The simulation model is shown in Fig. 5. The 3D model of the humanoid hand established by Solidworks is imported into Adams software. Some constraints are added to the model as well as to the motor. The function of the actuator is "time * 4", which is set in Adams software. Iteration times is 50. Then, a simulation taking 3 s is conducted to check whether the hand can move well without grasping objects. Finally, the speeds of three joints of the hand are obtained, as shown in Fig. 6.

Figure 6 shows results of the speeds of each joint when the hand is not grasping objects. The speeds of each joint are regular without sudden changes. Therefore, the proposed three-finger humanoid hand shows good motion performance and can meet design requirements for LARMbot humanoid as in Fig. 2.

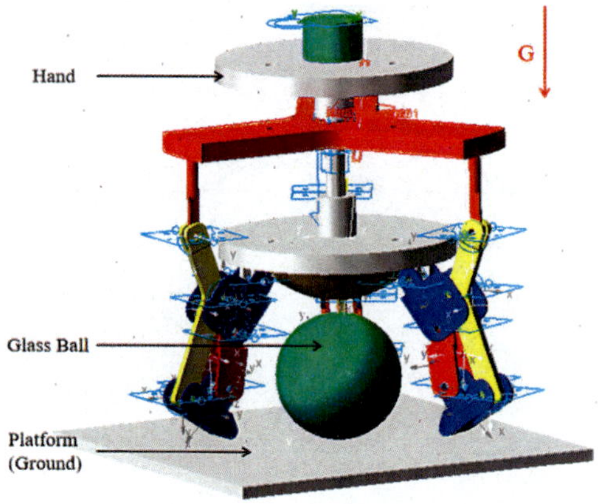

Fig. 5. The simulation model of the proposed humanoid hand in Fig. 4a

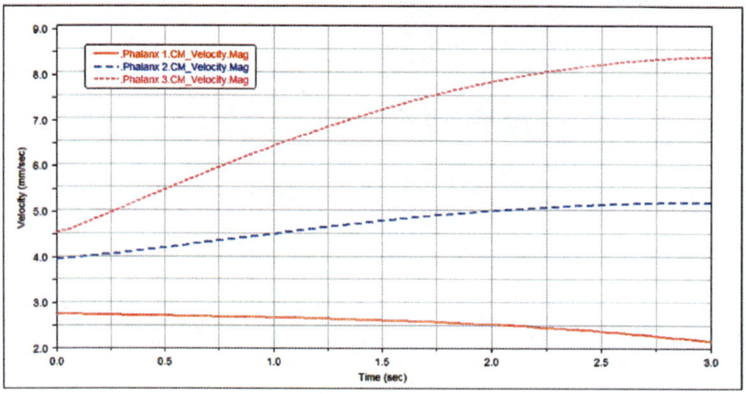

Fig. 6. The computed speeds of three phalanxes of the hand without grasping objects

The three-finger humanoid hand is designed for several grasping tasks when the LARMbot humanoid works in a specific circumstance. Therefore, a grasping simulation of the hand is conducted to test the grasping ability. The materials of the whole hand can be set in Adams software. Rubber is selected as the material for the soft palm as a first design solution, so figuring out that the most suitable material of the palm needs further work. Details are shown in Table 1. The mass of the designed hand is of about 0.25 kg calculated in Adams. The three-finger humanoid hand is simulated grasping a glass ball t on the ground, as an example of different objects with a maximum size of about 32 mm. Thus, in the example the diameter of the glass ball is set as 32 mm. In general, step function is applied in moving a motor in Adams. The function of actuator is "step(time, 0, 0, 2, 6) + step(time, 2, 6, 2.5, 6)". The simulation time is 2.5 s. Iteration times is 50 yet. The strings' coefficient is set as 5 N/mm. The simulated grasping process is as shown in Fig. 7. When the simulation time is 2.5 s, the grasping simulation is completed and the three-finger humanoid hand grasps glass ball stably.

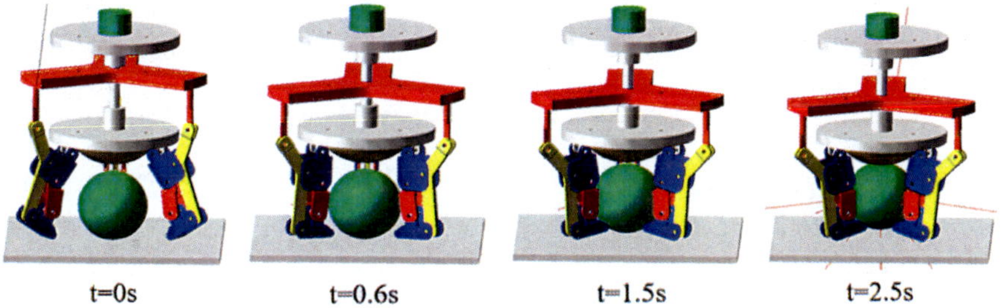

Fig. 7. Snapshot of the simulated grasping operation

Table 1. Details of the whole hand's materials set in Adams

Categories	Materials	Density (kg/mm^3)	Young's modulus (N/mm2)	Poisson's ratio
Fingers	Carbon_fiber_0_90	1.51 × 10–6	18.35 × 104	0.74
Palm	Rubber_belt	1.1 × 10–9	50.00	0.55
Ball	Glass	2.595 × 10–6	4.62 × 104	0.242

Based on assumed parameters set in Adams, the plots of the speed and grasping force of the hand are obtained as shown in Fig. 8 with changes of the speeds of each phalanx during the whole simulation. Because of the order of "step function", the actuator moves regularly as well as phalanxes work simultaneously. When the actuator stops, the fingers stop too. Th he speeds of each phalanx is obtained while the hand grasps objects, as shown in Fig. 8. The speeds of three phalanxes have similar trends with velocities from 0 to about 10 mm/s and then decreasing to 0 at the end of the grasping motion at the time of 2 s completing the grasping assignment.

The three-finger humanoid hand should grasp the objects with a suitable force not causing damage in the grasped objects. When the three-finger humanoid hand works, the grasping force between the grasped object and each phalanx of the fingers can change together with the movable palm. The changes of the grasping force between the hand and the grasped objects are shown in Fig. 9, as computed in the simulation.

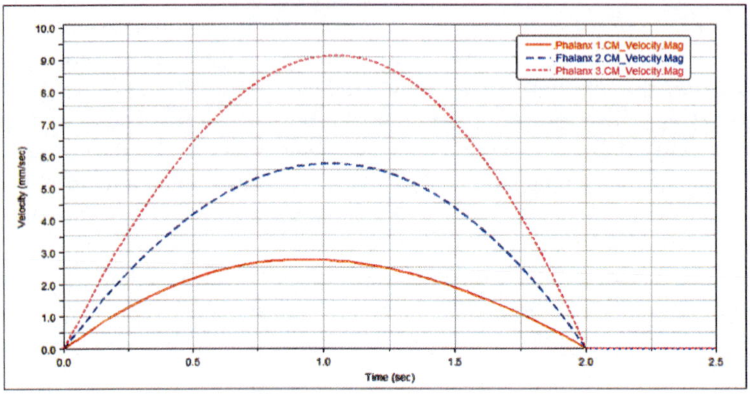

Fig. 8. The computed speeds of three phalanxes when grasping objects

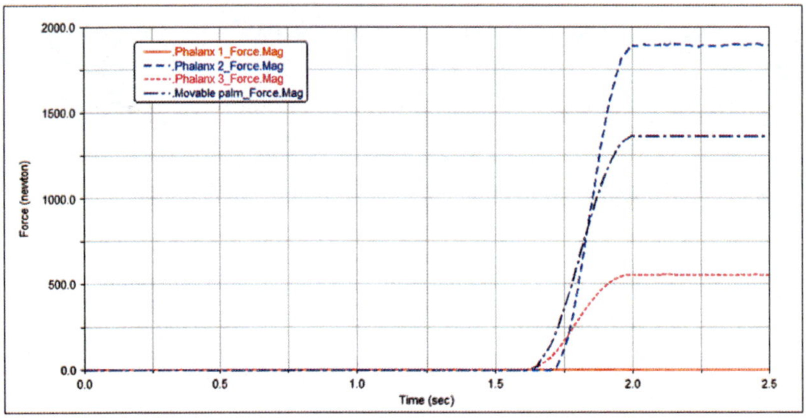

Fig. 9. The grasping force of three phalanxes and movable palm

During the whole grasping process, the 1st phalanxes of all 3 fingers do not contact with the grasped object and the force between it and the ball is zero. It is not until about 1.6 s that the phalanxes begin to contact the glass ball. According to the Fig. 8, the speed of the third phalanx can reach to 9 mm/s. Meanwhile, the sizes of the fingers is similar to the diameter of the glass ball. The 3rd phalanxes of all 3 fingers and movable palm contact the ball at the same time. After 0.1 s, the 2nd phalanxes of all 3 fingers start to contact the glass ball. When the time is about 1.95 s, the force of the third phalanx is stable to about 550 N. It is 2 s that the forces of the phalanx 3 and movable palm are in a stable situation. The results of grasping simulation illustrate that the developed three-finger humanoid hand has a good grasping performance.

5 Conclusion

In this paper, a new design of a three-finger humanoid hand is presented based on the LARM hand achievements. A CAD model of the hand is designed to simulate the grasping abilities of the proposed humanoid hand by using Adams software. According to the results of the simulation, the proposed humanoid hand can have a good grasping performance, cooperating with LARMbot arm. However, the structure and dynamics performance of the three-finger humanoid hand should be optimized in the future.

References

1. Ceccarelli, M., Tavolieri, C., Lu, Z.: Design considerations for underactuated grasp with a one D.O.F. anthropomorphic finger mechanism. In: International Conference on Intelligent Robotics and Systems, 9–15 October 2006, Beijing, China, pp. 1611–1616 (2006)
2. Xu, S., Xu, Y., Xu, X.: Structural design and kinematics analysis of SHU-hand II humanoid robotic hand. IOP Conf. Ser. Mater. Sci. Eng. **394**, 042068 (2018)
3. Zhang, X., Zhu, T., Yamayoshi, I., et al.: Dexterity, sensitivity and versatility: an under actuated robotic hand with mechanical intelligence and proprioceptive actuation. Int. J. Humanoid Robot. **17**, 2050006 (24 pages) (2020)
4. Ceccarelli, M.: Fundamentals of mechanics of robotic manipulation. XI, 312 p., Hardcover (2004)
5. Yao, S., Ceccarelli, M., Carbone, G., et al.: Grasp configuration planning for a low-cost and easy-operation underactuated three-fingered robot hand. Mech. Mach. Theory **129**, 51–69 (2018)
6. Ceccarelli, M., Zottola, M.: Design and simulation of an underactuated finger mechanism for LARM Hand. Robotica **35**, 483-497 (2015)
7. Espinosa-Garcia, F., Arias-Montiel, M.,Ceccarelli, M., et al.: Design and experimental characterization of a novel subactuated mechanism for robotic finger and movable palm. Int. J. Mech. Control **20**(2), 141–146 (2019)
8. Ceccarelli, M., Cafolla, D., Russo, M., et al.: LARMbot humanoid design towards a prototype. MOJ Appl. Bionics Biomech. **1**(2), 00008 (2017)
9. Ceccarelli, M., Russo, M., Morales-Cruz, C.: Parallel architectures for humanoid robots. Robot. Mechatron. **78**, 255–264 (2020)
10. Russo, M., Cafolla, D., Ceccarelli, M.: Design and experiments of a novel humanoid robot with parallel architectures. Robotics **7**(79), 1–14 (2018)

Design of a Compliant Bearing for Linear-Rotary Motors

Minh Tuan Nguyen, Van Tu Nguyen, and Minh Tuan Pham(✉)

Faculty of Mechanical Engineering, Ho Chi Minh City University of Technology, VNU-HCM, Ho Chi Minh City 700000, Vietnam
pminhtuan@hcmut.edu.vn

Abstract. This paper presents a novel design of the compliant bearing that can be used in a compliant linear-rotary motor for precision applications. With a couple of compliant bearings, a 2 degrees-of-freedom (DOF) motor with linear and rotary motions in a single axis can be achieved. The design requirements are defined based on the intended operating condition of the linear-rotary motor. Based on such requirements, the compliant bearing is designed based on analytical approach, and finite-element-analysis (FEA) is then employed to verify its actual performance. The good agreement between predicted data and simulation results demonstrates the correctness of the novel compliant bearing presented in this work and it can be further used to build a 2-DOF linear-rotary motor for industrial applications.

Keywords: Compliant mechanism · Compliant bearing · Linear-rotary motor

1 Introduction

Compliant mechanism is a particular structure that widely used in many precise mechanical and mechatronic systems due to numerous advantages, such as the frictionless and highly repeatable motions, lubrication-free and maintenance-free, etc. By employing elastic deformation of material, compliant mechanisms were created to eliminate the shortcomings of conventional mechanisms such as dry friction, backlash, wear and assembly error, etc. In addition, compliant mechanisms can be used in particularly harsh environments and be an ideal choice for devices with ultra-precise manipulations up to micro/nano scale. A number of compliant mechanisms with various degrees-of-freedom (DOF) have been presented in previous literatures, such as 1-DOF [1], 3-DOF planar-motion [2–4] and 3-DOF spatial-motion [5–7] mechanisms, etc. Fundamentals of compliant mechanism are used in this work to create a novel compliant bearing for the development of a compliant 2-DOF linear-rotary motor.

Nowadays, linear-rotary motor [8] is applied in many industrial fields with the functions of handling gantry, capping, pick and place, printing, etc., for the purpose of greater flexibility and higher productivity. Some common approaches of controlling linear and rotary motor are electromagnetic, pneumatic, hydraulic, etc. The driving principles are the use of linear voice-coil, linear motor, lead-screw, etc., for translational motion and the

N. V. Khang et al. (Eds.): ASIAN MMS 2021, MMS 113, pp. 246–257, 2022.
https://doi.org/10.1007/978-3-030-91892-7_23

use of servo-motor, stepper-motor, rotary voice-coil, etc., for rotational motion. However, these controlling and driving principles lead to a large size of motor the mismatch of output accuracy in ultra-precision and this paper focuses on the design and development of compliant bearing which is capable of the precise driving for linear-rotary motor up to micro/nano scale.

Similar to traditional bearings, compliant bearings also provide some main features, e.g., load bearing, spindle positioning, etc. Especially, it allows to deliver precise displacements in specific directions, this means when the inputs are force or torque, the corresponding outputs are the pure translational or angular displacements as desired with no unwanted motion. In order to achieve that, the compliant bearing must have a good stiffness property following its DOF, i.e., it must perform high flexibility in DOF directions to generate desired motions easily and high stiffness in other directions to resist against external disturbances. Compliant bearing and compliant mechanism are constructed by multiple flexures and rigid bodies, their stiffness can be represented by a compound spring system [9, 10]. Stiffness property of compliant mechanism can be analyzed based on 2 main approaches: finite element analysis (FEA) [10–13] and the traditional analytical method [13–19]. FEA approach is preferred due to the fast development of FEA software nowadays, but it is more suitable for verification than synthesis of compliant mechanisms. In contrast, the analytical method is ideal for synthesizing compliant mechanism in terms of stiffness characteristics [20]. As the positioning accuracy is also a key requirement of the linear-rotary motor, the output motions must be decoupled in order to generate the pure linear and rotary motions as required. Some design criteria for synthesizing compliant mechanisms with fully decoupled motions have been presented in [21] and they will be employed in this work, together with the analytical synthesis method presented in [20], to develop a novel compliant bearing for precise 2-DOF linear-rotary motors.

The remaining of the paper is organized as: Sect. 2 presents the synthesis of compliant bearing including problem definition, the design procedure of a compliant bearing and the stiffness calculation. Section 3 evaluates the results and determines the stroke for motor though FEM simulation. Section 4 concludes and orientates the motor research in the future.

2 Synthesis of the Compliant Bearing

2.1 Problem Definition

Figure 1a illustrates a general 2-DOF motor which produces a linear motion along the Z axis (ΔZ) and a rotary motion about the same axis (θZ). Note that the remaining motions, i.e., ΔX, ΔY, θX and θY, are constrained. As the shaft along the Z axis of the motor is supported by two compliant bearings at both ends as showed in Fig. 1b, the bearings must allow the motor to perform the requirements mentioned above by ensuring low stiffness in actuating directions (along and about the Z axis) and high stiffness in the non-actuating directions.

Figures 1b also illustrates the general construction of a compliant bearing. Each bearing includes 4 parallel legs which are symmetrical and 90° apart from the Z axis, each leg has one fixed end and one free end, the shaft of the motor locates along the Z

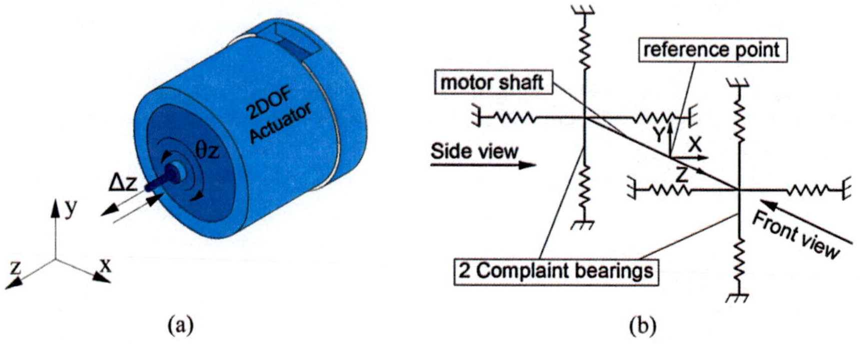

Fig. 1. (a) Principle and (b) general structure of a 2-DOF linear-rotary actuator.

axis and connects to the free ends of four compliant legs. To present the structure of the motor more visually, two projections (front view and side view) of Fig. 1b are included in Fig. 2 and Fig. 3 respectively.

Figure 2a and Fig. 2b illustrate 2 translational motions ΔX and ΔY that are produced by 2 external forces Fx and Fy along the X and Y axes respectively. Additionally, Fig. 2c shows the rotational motion θZ about the Z axis produced by the external torque Mz about the Z axis. The compliant bearing needs to perform θZ only while ΔX and ΔY must be constrained to achieve desired motions of the motor.

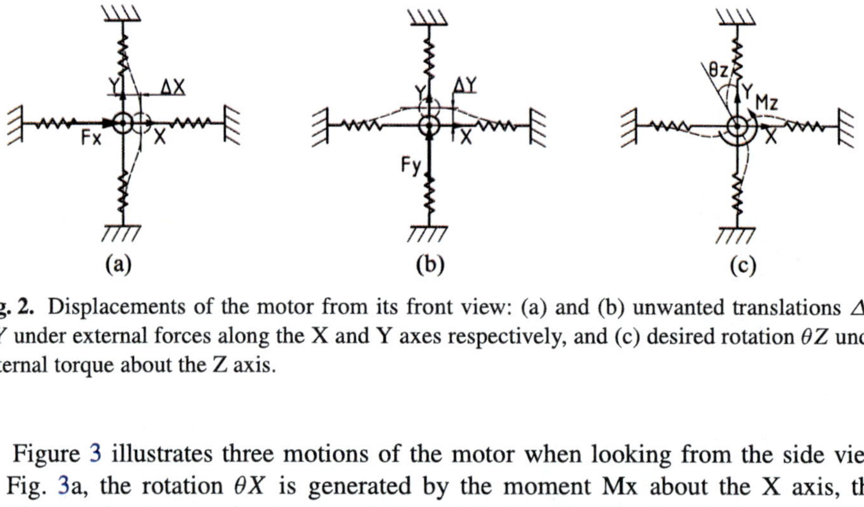

Fig. 2. Displacements of the motor from its front view: (a) and (b) unwanted translations ΔX, ΔY under external forces along the X and Y axes respectively, and (c) desired rotation θZ under external torque about the Z axis.

Figure 3 illustrates three motions of the motor when looking from the side view. In Fig. 3a, the rotation θX is generated by the moment Mx about the X axis, this rotation can be expressed to two translations ΔY along the Y axis at the two compliant bearings. Similarly, the rotation θY under the external moment My about the Y axis can be separated into two translations ΔX along the X axis at the two compliant bearings as shown in Fig. 3b. Lastly, Fig. 3c shows the desired translation along the Z axis when the motor acted by an external force Fz in the same direction. Because ΔX and ΔY are already constrained in each compliant bearing as mentioned above, two unwanted rotations θX and θY of the entire motor are always constrained independently on the rotations about the X and Y axes of the compliant bearings.

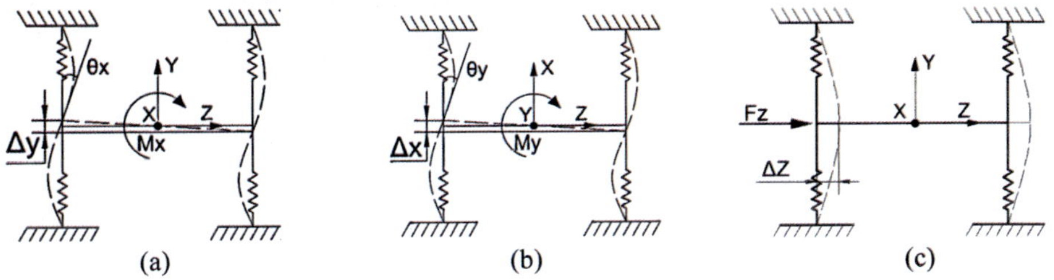

Fig. 3. Displacements of the motor from its side view: (a) and (b) unwanted rotations θX, θY under external moment about the X and Y axes respectively, and (c) desired translation ΔZ under external force along the Z axis.

Based on the analysis above, a compliant bearing can perform 4-DOF, i.e., a translation along the Z axis (ΔZ) and three rotations about the X, Y and Z axes (θX, θY and θZ respectively); while two translations along the X and Y axes (ΔX and ΔY respectively) must be constrained so that the 2-DOF linear-rotary motor can generate its desired motions.

2.2 Design of the Compliant Bearing

The design of the compliant bearing in this work is inspired by the existing designs of 1-DOF linear-motion compliant mechanism [22, 23] and 1-DOF rotary-motion compliant mechanism [1, 24]. The structures of the compliant bearings for linear and rotary motions are shown in Fig. 4a and 4b respectively.

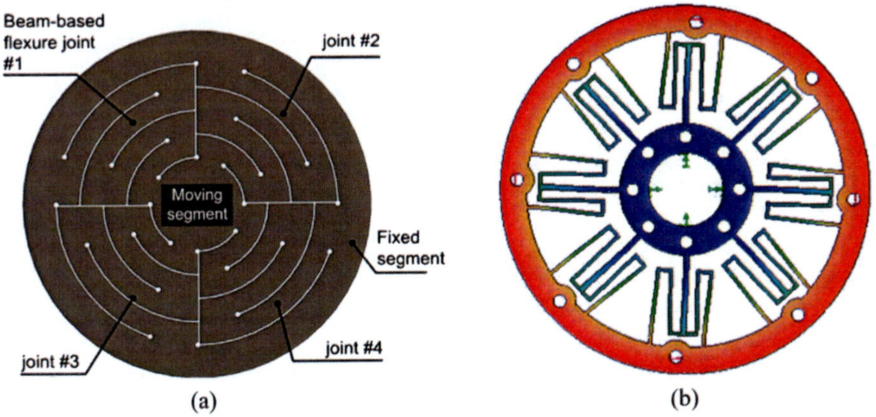

Fig. 4. Compliant bearing that allows (a) a linear motion [22] and (b) a rotary motion [1] along/about the normal axis.

The compliant bearing in Fig. 4a is constructed by a thin plate with some radial cuts to create four beam-based flexure joints, the bending deformation of four flexure joints generates the linear motion along the normal axis of the compliant bearing. The compliant bearing in Fig. 4b consists of eight symmetrical beam-based flexure joints, each has a thin rectangular cross-sectional area and their bending deformation under an external torque creates the rotary motion about the center axis. The combination of state-of-the-art mechanisms shown in Fig. 4 in a proper way creates a new compliant mechanism that is able to produce a translation and a rotation in the same axis simultaneously. In this work, the four-leg configuration is employed to construct the compliant bearing. In each leg, a linear flexure and a rotary flexure are connected serially so that the entire compliant bearing can generate two independent motions, i.e., translation ΔZ along and rotation θZ about the Z axis. The detailed design of the compliant bearing is shown in Fig. 5. With this design, it is seen that two independent motions, i.e., translation and rotation about the Z axis, can be obtained at the center of the compliant bearing that will be mounted to the motor shaft. Thus, a coupled of such compliant bearings can be used to support the moving shaft of a high accurate 2-DOF linear-rotary motor.

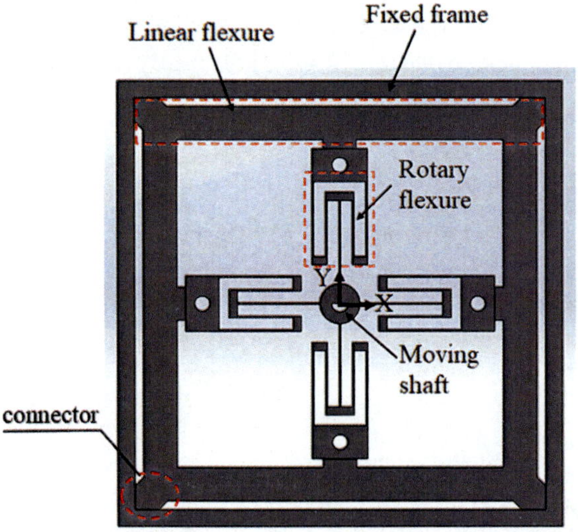

Fig. 5. Design of the compliant bearing for the linear-rotary motor containing a linear flexure and a rotary flexure in a leg to support the desired motion.

It is observed that the proposed compliant bearing can produce both the linear and rotary motions along the Z axis (ΔZ and θZ respectively) of the motor shaft as desired. It also allows the rotations about the X and Y axes (θX and θY respectively) as analyzed in Sect. 2.1, only two translations along the X and Y axes (ΔX and ΔY respectively) are constrained. In summary, each compliant bearing has 4-DOF (ΔZ, θX, θY and θZ); using two compliant bearings located along a specific distance along the Z axis as shown in Fig. 3 can create 2-DOF for the linear-rotary motor.

2.3 Stiffness Analysis Based on Analytical Method

The stiffness property of the compliant bearing shown in Fig. 5 is analyzed based on the analytical method. The basic principle of this method is to synthesize stiffness in all directions of the structure according to the connection of flexures, i.e., in serial or in parallel. For serial connection, the system stiffness is the sum of flexure's stiffness. For parallel connection, the system compliance (inverse of the system stiffness) is the sum of flexure's compliance. For complex flexure-based structures, the analytical method presented in [20] is a good solution to analyze the stiffness property of entire system. As the compliant bearing shown in Fig. 5 is constructed by beam-type flexures, its stiffness property can be derived based on the stiffness matrix of a standard cantilever beam. Figure 6 illustrates the general model of a cantilever beam-type flexure and its stiffness matrix is written in Eq. (1).

$$k_{beam} = \begin{bmatrix} \frac{AE}{L} & 0 & 0 & 0 & 0 & 0 \\ 0 & \frac{12EI_z}{L^3} & 0 & 0 & 0 & -\frac{6EI_z}{L^2} \\ 0 & 0 & \frac{12EI_y}{L^3} & 0 & \frac{6EI_y}{L^2} & 0 \\ 0 & 0 & 0 & \frac{GJ}{L} & 0 & 0 \\ 0 & 0 & \frac{6EI_y}{L^2} & 0 & \frac{4EI_y}{L} & 0 \\ 0 & -\frac{6EI_z}{L^2} & 0 & 0 & 0 & \frac{4EI_z}{L} \end{bmatrix} \tag{1}$$

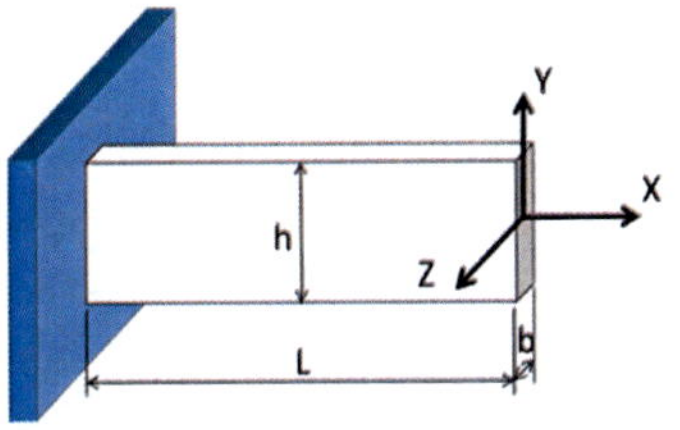

Fig. 6. An original cantilever beam

In this work, the aluminum alloy 6061-T6, a popular material has been used in many compliant mechanisms, is chosen as the building material. The mechanical properties of this material are given as: Young's modulus (E) is 69 GPa, Poisson ratio is 0,33 and yield stress is 276 MPa. The dimensions of each beam-type flexure are given in Table 1. In Eq. (2), $A = bh$ is the cross-sectional area of the beam, $I_y = \frac{b^3h}{12}$ and $I_Z = \frac{h^3b}{12}$ are the moments of inertia about the Y and Z axis respectively, $J = \frac{b^3h}{3}$ is the torsion constant for beams A, A' and B (with $h \gg b$) in the linear flexure while $J = \frac{h^3b}{3}$ is applied for beams a, a', b, b' and c (with $b \gg h$) in the rotary flexure.

The cross-sectional area of all beams is predefined as $10 \times 0.5\ \text{mm}^2$, the thickness of 0.5 mm chosen based on the fabrication ability. In addition, the overall dimensions of 132 mm × 132 mm × 10 mm of the compliant bearing are enough to create the required stroke and to ensure the compactness of the entire motor.

Table 1. The geometrical parameters of the beam elements.

Beam	Unit	A	A'	B	a	a'	b	b'	c
Length – L	mm	53,5 + 1,5		2,5	22		16,5		24,5
Height – h	mm	10		10	0,5		0,5		0,5
Width – b	mm	0,5		0,5	10		10		10

Figure 7a illustrates the modeling of the top leg of the compliant bearing (including leg 1 – the linear flexure and leg 1' – the rotary flexure connected together in series). In particular, this leg consists of 8 beam elements (A, A', B, a, a', b, b' and c) and several rigid links. As elastic deformation generates by beams while the rigid links are only used to connect beams together, the stiffness of rigid links are excluded. Moreover, the top leg is connected to the fixed frame by two connectors (Fig. 5) and these connectors are short and less deformed. To simplify the calculation process, the length of beams A, A' (shown in Table 1) are added by the equivalent length of connectors instead of calculating its stiffness directly. The connection scheme of the entire compliant bearing can be represented by Fig. 7b. Because the stiffness of rigid links in Fig. 7a is extremely high, their stiffness is eliminated from the stiffness modeling shown in Fig. 7b.

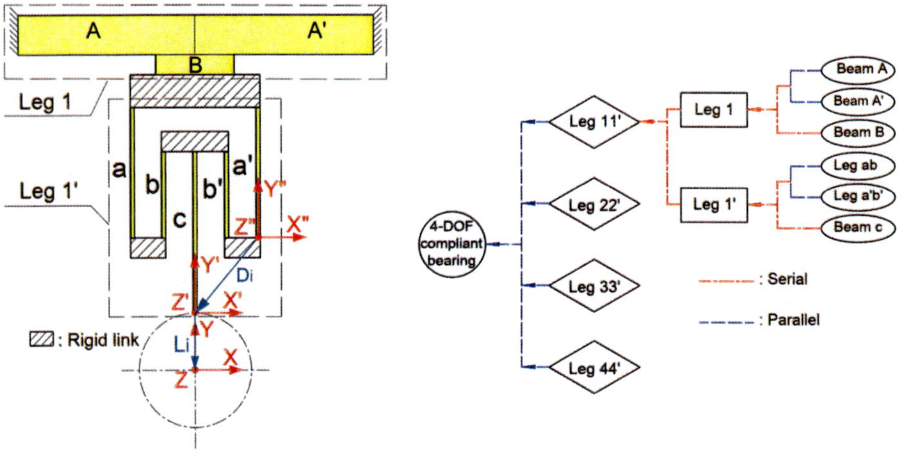

(a) (b)

Fig. 7. (a) Modeling of a leg synthesized by beam elements and (b) connection scheme of the compliant bearing.

The stiffness/compliance of each leg and the stiffness property of the entire structure can be analyzed by following steps:

Step 1: Beam stiffness. The stiffness matrix (k_e) and compliance matrix $\left(c_e = k_e^{-1}\right)$ of each beam element derived at the local frame $X"Y"Z"$ at the free end can be calculated based on Eq. (1) and parameters in Table 1.

Step 2: Leg stiffness. All the stiffness matrix of 8 beams (see Fig. 7a) is calculated referring to the local frame $X^{'}Y^{'}Z^{'}$ locates at the free end of this leg using following equations:

$$K_e = J_{Ke}\, R_e\, k_e\, (R_e)^{-1}\, (J_{Ke})^{-1} \tag{2}$$

$$C_e = J_{Ce}\, R_e\, c_e\, (R_e)^{-1}\, (J_{Ce})^{-1} \tag{3}$$

Here, K_e and C_e are the stiffness and compliance matrices of beams calculated at the frame $X^{'}Y^{'}Z^{'}$ respectively. J_{Ke} and J_{Ce} are the translation matrices for stiffness and compliance matrices of beams respectively, used to shift the reference point from $X"Y"Z"$ to $X^{'}Y^{'}Z^{'}$ used. R_e represents the orientation matrix of each beam about the reference point. The detailed expressions of these matrices are written as.

$$J_{Ke} = \begin{bmatrix} I & 0 \\ D_i & I \end{bmatrix} \text{ and } J_{Ce} = \begin{bmatrix} I & D_i \\ 0 & I \end{bmatrix}, \text{ where } D_e = \begin{bmatrix} 0 & r_{ze} & -r_{ye} \\ -r_{ze} & 0 & r_{xe} \\ r_{ye} & -r_{xe} & 0 \end{bmatrix} \tag{4}$$

$$R_e = \begin{bmatrix} R_{Xe}\, R_{Ye}\, R_{Ze} & 0 \\ 0 & R_{Xe}\, R_{Ye}\, R_{Ze} \end{bmatrix} \tag{5}$$

Where $R_{Xe} = \begin{bmatrix} 1 & 0 & 0 \\ 0 & \cos\gamma & -\sin\gamma \\ 0 & -\sin\gamma & \cos\gamma \end{bmatrix}$, $R_{Ye} = \begin{bmatrix} \cos\varphi & 0 & \sin\varphi \\ 0 & 1 & 0 \\ -\sin\varphi & 0 & \cos\varphi \end{bmatrix}$, $R_{Ze} = \begin{bmatrix} \cos\theta & -\sin\theta & 0 \\ \sin\theta & \cos\theta & 0 \\ 0 & 0 & 1 \end{bmatrix}$.

In Eq. (4), I represents an unit matrix; D_e is the vector presents the relative position of the frame $X''Y''Z''$ to the frame $X^{'}Y^{'}Z^{'}$ of the leg with the dimensions along the X, Y and Z axes are r_{xe}, r_{ye} and r_{ze} respectively. In Eq. (5), R_{Xe}, R_{Ye} and R_{Ze} are the rotational vector with the orientations of γ, φ and θ about the $X"$, $Y"$ and $Z"$ axes of the local frame respectively. When the calculation of stiffness and compliance matrices at the local frame $X^{'}Y^{'}Z^{'}$ of leg 11' are done, the stiffness of the whole leg is conducted to calculate based on the connection scheme (Fig. 7b) and is can be expressed as

Figure 8 illustrates the calculation process of the stiffness of leg 11' based on the connection scheme shown in Fig. 7b. Here, R_Y is the rotational matrix with $\varphi = 180°$ about the Y axis. C_1 and $C_{1'}$ are the compliance matrices of leg 1 and leg 1' (the linear and rotary flexures) in leg 11' respectively, C_{leg} and K_{leg} represent the compliance and stiffness matrices of the leg 11' respectively.

Step 3: The bearing stiffness. Referring to Fig. 7b, the stiffness matrix of the entire compliant bearing and is calculated as

$$K^m = \sum_{i=1}^{n} J_{Ki}^l R_i^l K^l \left(R_i^l\right)^{-1} \left(J_{Ki}^l\right)^T, \text{ where } C^m = \left(K^m\right)^{-1} \tag{6}$$

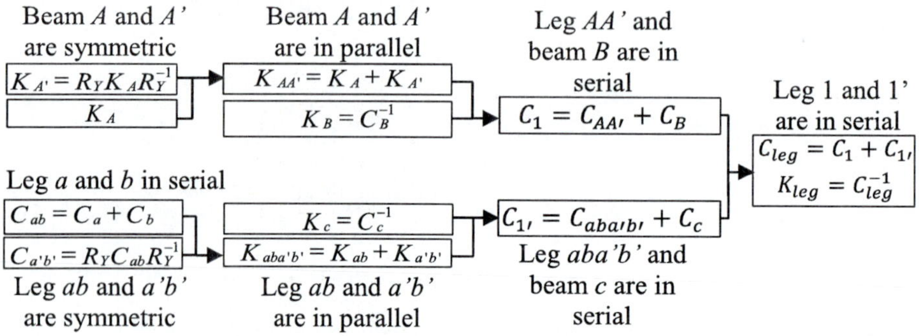

Fig. 8. The calculation diagram to calculate the stiffness of leg 11'.

Here, $n = 4$ indicates the number of legs. R_i^l represents the rotational matrix of the ith leg about the Z axis and θ_i correspond to the orientation of 0°, 90°, 180° and 270° of the ith leg ($i = 1, 2, 3, 4$) respectively. J_{Ki}^l represents the translation matrix of the ith leg from the local frame $X'Y'Z'$ to the global frame XYZ with three distances L_{xi}, L_{yi} and L_{zi} along the X, Y and Z axes respectively, written as

$$J_{Ki} = \begin{bmatrix} I & 0 \\ L_i & I \end{bmatrix}, \text{ where } L_i = \begin{bmatrix} 0 & L_{zi} & -L_{yi} \\ -L_{zi} & 0 & L_{xi} \\ L_{yi} & -L_{xi} & 0 \end{bmatrix} \quad (7)$$

By substituting actual dimensions of the compliant bearing into Eqs. (2)–(7), the resulting compliance and stiffness matrices of the entire compliant bearing is given as

$$C_m = diag[5,96.10^{-6}\ 5,96.10^{-6}\ 2,50.10^{-4}\ 14,77.10^{-2}\ 14,77.10^{-2}\ 16,04.10^{-2}] \quad (8)$$

$$K_m = diag[1,68.10^5\ 1,68.10^5\ 4,00.10^3\ 6,67\ 6,67\ 6,24] \quad (9)$$

Note that the units of components in Eq. (8) and Eq. (9) are in SI system. Equation (8) and Eq. (9) show that the 6 × 6 compliance/stiffness matrix of the compliant bearing has diagonal form with all non-diagonal components being zero. This diagonal stiffness matrix suggests that the compliant bearing is able to generate fully decoupled motions. The 4 DOF can be obtained by the four small values of K_{33}, K_{44}, K_{55} and K_{66} and two constraints correspond to two large values of K_{11} and K_{22}. In summary, the synthesized compliant bearing can achieve the desired DOF.

3 FEM Simulation Results

Figure 9 illustrates the FEM model of the synthesized compliant bearing. The outside frame of the compliant bearing is fixed, then unit loads ($F = 1N$) and unit moments ($M = 1N.m$) are applied to the end effector in three translational directions ($\Delta X, \Delta Y, \Delta Z$) and three rotational directions ($\theta X, \theta Y, \theta Z$) respectively. The

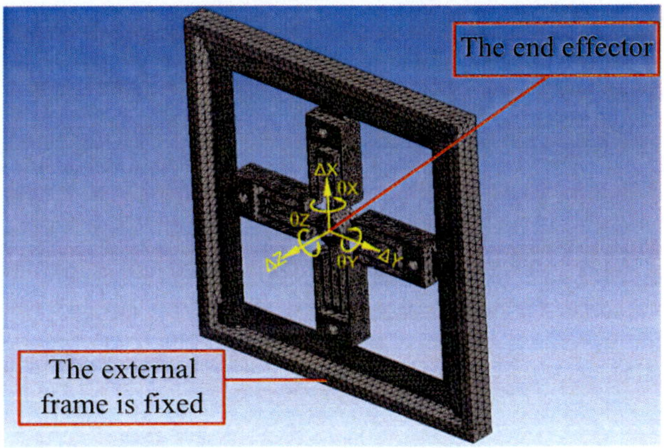

Fig. 9. FEM model of the compliant bearing.

displacements generated at global frame represent the compliance of the structure in corresponding directions.

Based on the FEM model shown in Fig. 9 and analysis processes as described above, the simulations via ANSYS Workbench have been carried out to investigate the performance of the synthesized compliant bearing in terms of stiffness and stroke, which are key specifications of a 2-DOF linear-rotary compliant motor.

3.1 Stiffness Property

The simulation results via ANSYS Workbench of the compliance of the bearing in six primary directions are recorded and compared to the predicted data calculated by the analytical method shown in Eq. (8). The detailed comparison is shown in Table 2.

Table 2. Comparison between predicted data and simulation results.

Method	$C_{\Delta x}$	$C_{\Delta y}$	$C_{\Delta z}$	$C_{\theta x}$	$C_{\theta y}$	$C_{\theta z}$
	Unit: m/N			Unit: rad/(N.m)		
Analytical	$5,956.10^{-6}$	$5,956.10^{-6}$	$2,500.10^{-4}$	0,1477	0,1477	0,1604
FEM	$5,455.10^{-6}$	$5,451.10^{-6}$	$2,645.10^{-4}$	0,1390	0,1389	0,1516
Error	8,41%	8,48%	5,48%	5,89%	5,96%	5,49%

Referring to Table 2, the highest error between the predicted data based on analytical method and the FEM simulation results is about 8,48%, and the errors of the 4 DOF of the compliant bearing are only less than 6%. Such small errors suggest that the synthesized compliant bearing has predictable stiffness property and is suitable to use in the linear-rotary motor. Additionally, the stiffness ratios between constraints and DOF in translational directions ($K_{\Delta X}/K_{\Delta Z}$ and $K_{\Delta Y}/K_{\Delta Z}$) are large with approximately 50 times in FEM and 42 times in analytical method, that represent the good stability of the compliant bearing under external mechanical disturbances.

3.2 Stroke of the Linear-Rotary Motor

With the yield stress of 276 MPa of Aluminum alloy 6061-T6, the simulation results show that the compliant bearing can perform the maximum displacements of 10.23 mm along the Z axis and 12.18° about the Z axis. This workspace is very large for current compliant mechanisms. The linear-rotary motor uses a couple of the proposed compliant bearing can reach more than 10 mm and 10° for linear and rotary motions respectively, fully decoupled motions and high repeatability. A motor with these characteristics is ideal for precise positioning systems.

4 Conclusion

In this paper, a compliant bearing with fully motion decoupling capability is synthesized by the analytical method. FEM simulation is used to verify the stiffness property and to define the workspace of the synthesized compliant bearing. The obtained results are good and reliable with the small deviations of less 8.5% between the prediction and simulation results, the high stiffness ratios demonstrate that the compliant bearing can resist external disturbances well.

By using a couple of the synthesized compliant bearings, a 2-DOF linear-rotary motor can be created. Such motor is able to produce large stroke of >10 mm for the linear motion and >10° for the rotary motion, fully decoupled and highly repeatable motions. Many precise positioning/alignment systems can be developed using the proposed linear-rotary motor.

The future work will focus on the design of entire 2-DOF compliant motor based on the compliant bearing presented in this paper, suitable actuators and sensors will be also allocated. A physical prototype of the motor will be developed, some experimental works will be carried out to evaluate the actual performance of the physical motor in terms of stiffness property, dynamic behavior, working range and resolution, etc.

References

1. Xu, Q.: Design of a large-range compliant rotary micropositioning stage with angle and torque sensing. IEEE Sens. J. **15**, 2419–2430 (2015)
2. Kim, H., Gweon, D.: Development of a compact and long range XYθz nano-positioning stage. Rev. Sci. Instrum. **83**, 085102 (2012)
3. Kim, H., Ahn, D., Gweon, D.: Development of a novel 3-degrees of freedom flexure-based positioning system. Rev. Sci. Instrum. **83**, 055114 (2012)
4. Bhagat, U., Shirinzadeh, B., Clark, L.: Design and analysis of a novel flexure-based 3-DOF mechanism. Mech. Mach. Theory **74**, 173–187 (2014)
5. Tanikawa, T., Arai, T., Koyachi, N.: Development of small-sized 3 DOF finger module in micro hand for micro manipulation. In: Proceedings of IEEE/RSJ International Conference on Intelligent Robots and Systems, vol. 2, pp. 876–881 (1999)

6. Li, Y., Xu, Q.: A totally decoupled piezodriven XYZ flexure parallel micropositioning stage for micro/nanomanipulation. IEEE Trans. Autom. Sci. Eng. **8**, 265–279 (2011)
7. Xiao, X., Yangmin, L.: Development and control of a compact 3-DOF micromanipulator for high-precise positioning. In: 2014 IEEE/ASME International Conference on Advanced Intelligent Mechatronics, pp. 1480–1485 (2014)
8. Design and Technology Portal Page. https://danielteodesigntechnology.wordpress.com/technologies/a-linear-rotary-electromagnetic-actuator/. Accessed 17 July 2021
9. Wikipedia Page. https://en.wikipedia.org/wiki/Hooke%27s_law. Accessed 17 July 2021
10. Wikipedia Page. https://en.wikipedia.org/wiki/Stiffness. Accessed 17 July 2021
11. Woronko, A., Huang, J., Altintas, Y.: Piezoelectric tool actuator for precision machining on conventional CNC turning centers. Prec. Eng. **27**, 335–345 (2003)
12. Zhang, S., Fasse, E.: A finite-element-based method to determine the spatial stiffness properties of a notch hinge. ASME J. Mech. Des. **123**, 141–147 (2001)
13. Xu, G., Qu, L.: Some analytical problems of high performance flexure hinge and micro-motion stage design. In: Proceedings of IEEE International Conference on Industrial Technology, China, pp. 771–776. IEEE (1996)
14. Yi, B.J., Chung, G., Na, H.: Design and experiment of a 3DOF parallel micro-mechanism utilizing flexure hinges. IEEE Trans. Robot. Autom. **19**, 604–612 (2003)
15. Lobontiu, N.: Compliant Mechanisms: Design of Flexure Hinges, 2nd edn. CRC Press, Boca Raton (2003)
16. Henein, S., Bottinelli, S., Clavel, R.: Parallel spring stages with flexures of micrometric cross-sections. In: Proceedings of the SPIE on Microrobotics and Micro-system Fabrication, vol. 3202, pp. 209–220 (1998)
17. Alici, G., Shirinzadeh, S.: Kinematics and stiffness analyses of a flexure-jointed planar micro-manipulation system for a decoupled compliant motion. In: Proceedings of IEEE International Conference on Intelligent Robots and Systems, USA, pp. 3282–3289. IEEE (2003)
18. Kim, W.K., Yi, B.J., Cho, W.: RCC characteristics of planar/spherical three degree-of-freedom parallel mechanisms with joint compliances. ASME J. Mech. Des. **122**, 10–16 (2000)
19. Simaan, N., Shoham, M.: Stiffness synthesis of a variable geometry six-degrees-of-freedom double planar parallel robot. Int. J. Robot. Res. **22**, 757–775 (2003)
20. Pham, H.H., Ming Chen, I.: Stiffness modeling of flexure parallel mechanism. Precis. Eng. **29**, 467–478 (2005)
21. Pham, M.T., Huat, Y.S., Joo, T.T., Pan, W., Ling, S.N.M.: Design and optimization of a three degrees-of-freedom spatial motion compliant parallel mechanism with fully decoupled motion characteristics. J. Mech. Robot. **11**, 051010 (2019)
22. Teo, T.J., Ming Chen, I., Yang, G., Lin, W.: A flexure-based electromagnetic linear actuator. Nanotechnology **19**, 315501 (2008)
23. Teo, T.J., Bui, V.P., Yang, G., Ming Chen, I.: Millimeters-stroke nanopositioning actuator with high positioning and thermal stability. IEEE/ASME Trans. Mechatron. **20**, 2813–2823 (2015)
24. Sung, E., Slocum, A.H., Ma, R., Jr., Bean, J.F., Culpepper, M.L.: Design of an ankle rehabilitation device using compliant mechanisms. J. Med. Devices **5**, 011001 (2011)

Design of Dynamically Isotropic Modified Gough- Stewart Platform Using a Geometry-Based Approach

Yogesh Pratap Singh[1], Nazeer Ahmad[2], and Ashitava Ghosal[1](✉)

[1] Indian Institute of Science, Bangalore, India
asitava@iisc.ac.in
[2] ISAC, Indian Space Research Organization, Bangalore, India

Abstract. A dynamically isotropic mechanism is useful for vibration isolation since such a mechanism can be used to attenuate the first six modes of vibration effectively and equally from a sensitive payload. A Gough-Stewart Platform (GSP) has been proposed, in literature, for six component vibration isolation. The conventional GSP, however, fails to give a dynamic isotropy; hence a Modified Gough-Stewart Platform (MGSP) is considered in this work. The force transformation matrix, together with a geometry-based approach, is used to obtain closed-form analytical solutions for dynamic isotropy in an MGSP at its neutral position. The more general case of variation of the centre of mass of the payload from the top platform is considered. The geometric approach presented in this work is simpler, faster, and more systematic than the previously existing methods. The configurations obtained with the above approach were successfully validated with the simulation results obtained using ANSYS®.

Keywords: Modified Gough-Stewart Platform (MGSP) · Dynamic isotropy · Natural frequency matrix · Force transformation matrix

1 Introduction

Several researchers have explored Stewart platform-based vibration isolator to attain desired micro-vibration isolation in spacecraft [1]. For effective vibration isolation, one of the primary design considerations is that the first six natural frequencies (translational and rotational) be nearly the same or ideally equal [1]. In such an ideal design, the first six modes (which contain a majority of the vibration energy) can be equally and effectively attenuated, thereby significantly reducing the vibration levels at the payload typically mounted on the top platform of a Stewart platform. In robotics literature, this condition is known as 'dynamic isotropy'. From a vibration isolation standpoint, if there is dynamic isotropy, all the peaks corresponding to the different degrees of freedom (DOFs) would be very close to each other in the amplitude versus frequency curve. As a consequence, the vibration isolation of one DOF does not interfere with that of another DOF and degrades the overall isolator's isolator performance.

N. V. Khang et al. (Eds.): ASIAN MMS 2021, MMS 113, pp. 258–268, 2022.
https://doi.org/10.1007/978-3-030-91892-7_24

Several researchers have worked on the geometric, stiffness, force, and velocity isotropies [1–7]. Unfortunately, the response of the control system cannot be evaluated by these performances [2]. A dynamically isotropic Stewart platform can not only simplify controls but also give information about stability. The lowest value of the natural frequency plays a crucial role in dynamic stability and ensuring all-natural frequencies to be nearly equal also implies that we maximize the lowest natural frequency, which is a favorable criterion for stability [3]. Additionally, the coupling among all the six DOFs of the Stewart platform complicates the controller design leading to a reduction in control accuracy [4]. Dynamic isotropy will also ensure that we can use decoupled controllers because a MIMO Stewart platform system can be ideally converted into several SISO systems (6 for a typical six DOF Stewart platform). These added advantages of dynamic isotropy over other types of isotropies are a result of its dependence on the payload's mass centre, the inertia properties along with the geometric and stiffness parameters.

A considerable amount of effort has been invested in studying various isotropies in a standard 6×6 GSP [1, 3, 6] leading to the conclusion that the dynamic isotropy for a standard GSP is not practically feasible due to the practical restriction in satisfying the inertia conditions $I_{ZZ} = 4I_{XX} = 4I_{YY}$ [2, 5]. Pertaining to the above constraint, a few researchers have studied the isotropic conditions for a Modified Gough–Stewart Platform (MGSP) [2, 5, 7]. In MGSP, the attachment points are on two radii on the top and bottom platform instead of a single radius in the conventional case (see Fig. 1(a)). Jiang et al. [5] described an MGSP using a pair of hyperboloids and investigated dynamic isotropy conditions. Yao et al. [6] explored spatial isotropy configuration for a MGSP based force sensor using a Jacobian matrix. Yi et al. [7] also introduced a two-parameter class of six-strut orthogonal GSP leading to isotropy. But most of the previous works assume that the motion reference point/centre of mass (COM) coincides with the top platform's geometric centre.

To the best of our knowledge, a general well-defined analytical solution in closed form for a dynamically isotropic MGSP is yet to be established. Though a few researchers have simplified the coupled dynamic equations for MGSP [2], their analytical expressions remain implicit and depend on multiple unknown variables that are themselves coupled. Hence, it is very tedious to calculate each of the unknowns' values and arrive at a design. To overcome these problems, we have developed a geometry-based method that will give a closed-form solution for MGSP in an explicit form and hence can determine the unknown parameters such as the radius and angle parameters easily. The above procedure was also applied to the general case, where the location of the motion reference point is not at the geometric centre of the top platform but is with some offset.

2 Formulation

A Gough-Stewart Platform (GSP) consists of a movable top platform, a fixed base, and six linearly actuated struts. In a MGSP, the anchoring points are placed at two radii on each platform as shown in Fig. 1(a). The three struts will be of equal length, and their anchoring points are uniformly spaced along the circumference forming a 120° angle between them. The same holds for other set of three struts. We use R_{bo}, R_{bi} to denote the outer and inner radii of the bottom or base platform while R_{to}, R_{ti} are the outer and inner radii of the top platform, respectively. The quantity $\{x_b, y_b, z_b\}$ represents coordinates of a point on the base frame $\{B\}$ and $\{x_p, y_p z_p\}$ denote the coordinates of a point on the top or moving frame $\{P\}$, respectively. Vector $\boldsymbol{oB}_1(R_{bo}$ line) is chosen along $\boldsymbol{x_b}$ direction, and $\alpha_{to}, \alpha_{bi}, \alpha_{ti}$ are angles made by vectors $\boldsymbol{c_o A}_1, \boldsymbol{oB}_4, \boldsymbol{c_o A}_4$ with $\boldsymbol{x_b}$, respectively. The variable H denotes the distance between the two platforms.

Assuming each platform as a rigid body, the stiffness matrix $[K_T]$ in the task space is given by

$$[K_T] = k[B][B]^T \tag{1}$$

where k is the elastic stiffness of each strut in the axial direction and the struts are assumed to have the same axial stiffness k for the simplicity of the problem. In Fig. 1(b) , let the vector ${}^P\boldsymbol{H_c}$ be the height of the motion reference point (COM) with respect to the top platform, ${}^B\boldsymbol{t}$ be a vector joining centers of two platforms and $\boldsymbol{S_j}(= l_j \boldsymbol{s_j})$ be a vector along the leg j with length l_j. Writing loop closure equation for this loop [OBAP], we get

$$\boldsymbol{x_j} + {}^B[R]_P({}^P\boldsymbol{p_j} - {}^P\boldsymbol{H_c}) - l_j \boldsymbol{s_j} - {}^B\boldsymbol{b_j} = 0 \tag{2}$$

From the above expression, the force transformation matrix, denoted by $[B]$, a 6×6 for the MGSP, can be derived as

$$[B] = \left[\begin{array}{c|c|c} \boldsymbol{s}_1 & \cdots & \boldsymbol{s}_6 \\ ({}^B[R]_P({}^P\boldsymbol{p}_1 - {}^P\boldsymbol{H_c}))\times \boldsymbol{s}_1 & & ({}^B[R]_P({}^P\boldsymbol{p}_6 - {}^P\boldsymbol{H_c}))\times \boldsymbol{s}_6 \end{array}\right] \tag{3}$$

where, for the legs $j = 1,...,6$, $\boldsymbol{s_j} = \frac{{}^B\boldsymbol{t} + {}^B[R]_P {}^P\boldsymbol{p_j} - {}^B\boldsymbol{b_j}}{l_j}$ are the unit vectors along leg j, ${}^P\boldsymbol{p_j}$ are location of the connection points on the top platform with respect to the frame $\{P\}$, and ${}^B\boldsymbol{b_j}$ are location of the connection points at the base with respect to the frame $\{B\}$. The force transformation matrix can be related to the Jacobian matrix ($[J]$) by $[B] = ([J]^{-1})^T$ [8], and the study of isotropy has been done by researchers using the Jacobian [2, 6] and the force transformation matrix [9]. In this work, we use the force transformation matrix to develop simple closed-form expressions for dynamic isotropy of an MGSP.

Let $[M]$ be the payload's mass matrix in the task space. The coordinate system can be chosen to coincide with the orientation of the principal axes of the payload to obtain a diagonal structure of $[M]$ matrix without any loss of generality. If m_p denotes the payloads' mass and I_{xx}, I_{yy}, I_{zz} denotes the moment of inertia along each direction with respect to its COM, then, $[M] = diag\left(\left[m_p\ m_p\ m_p\ I_{xx}\ I_{yy}\ I_{zz}\right]\right)$. Our formulations are based on the neutral pose of the platform where the top platform is parallel to the base (${}^B[R]_P = [I]$) with centers of the two platforms lying on same vertical line implying ${}^B\boldsymbol{t} = [0\ 0\ H]^T$. We also assume that the payload's center of mass lie on the same vertical line with ${}^P\boldsymbol{H_c} = [0\ 0\ Y]^T$. This is a fair assumption for operations requiring precise control such as vibration isolation and for camera pointing [1]. For the chosen neutral configuration, $[B]$ is a constant matrix.

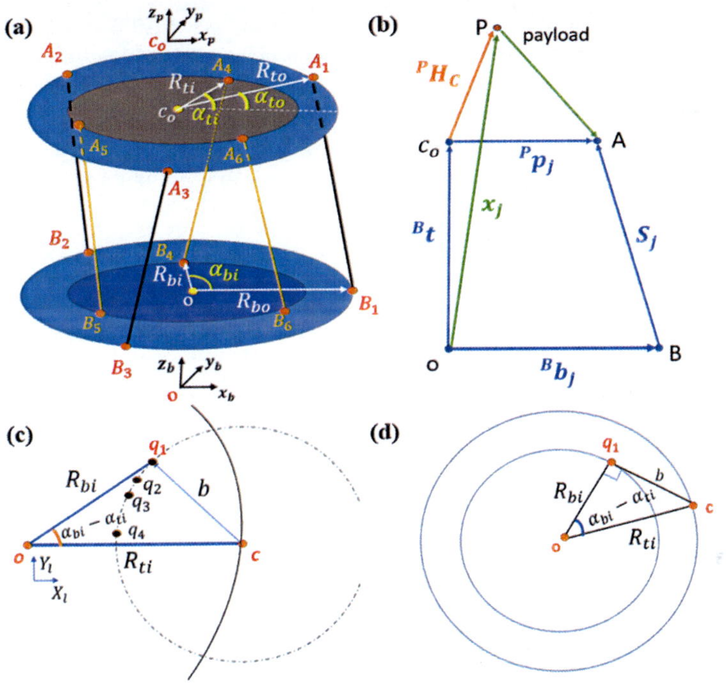

Fig. 1. a) Modified Gough-Stewart Platform b) Closed loop for MGSP c) Top view showing virtual circle and line R_{bi} intersections d) Tangency condition for virtual circle

For the dynamic isotropy, all six eigenvalues of the natural frequency matrix must be equal. Using Eqs. (1) and (3), we can write the natural frequency matrix $[G]$ in task space [3, 5] as:

$$[G] = [M]^{-1}[K_T] = [M]^{-1}k[B][B]^T = \begin{vmatrix} P_{3\times3} & R_{3\times3} \\ R^T_{3\times3} & Q_{3\times3} \end{vmatrix} \tag{4}$$

where the sub-matrices are given by

$$P_{3\times3} = \text{diag}(\lambda_1, \lambda_2, \lambda_3), Q_{3\times3} = \text{diag}(\lambda_4, \lambda_5, \lambda_6), R_{3\times3} = \begin{bmatrix} \mu_{11} & -\mu_{12} & 0 \\ \mu_{12} & \mu_{11} & 0 \\ 0 & 0 & \mu_{33} \end{bmatrix} \tag{5}$$

The relation between two set of legs, with a being the leg length ratio, is given by

$$l_2 = al_1 \tag{6}$$

where $l_1 = |\boldsymbol{S}_1| = |\boldsymbol{S}_2| = |\boldsymbol{S}_3| = \sqrt{R_{to}^2 + R_{bo}^2 - 2R_{to}R_{bo}\cos(\alpha_{to}) + H^2}$

$$l_2 = |\boldsymbol{S}_4| = |\boldsymbol{S}_5| = |\boldsymbol{S}_6| = \sqrt{R_{ti}^2 + R_{bi}^2 - 2R_{ti}R_{bi}\cos(\alpha_{bi} - \alpha_{ti}) + H^2}$$

If ω denotes the natural frequency, then for dynamic isotropy, $\lambda_1 = \lambda_2 = \lambda_3 = \lambda_4 = \lambda_5 = \lambda_6 = \omega^2$ and $\mu_{11} = \mu_{12} = \mu_{33} = 0$. The expressions for all λ's and μ's in Eqs. (4) and (5) can be referred from the appendix section at the end of this paper (see Eqs. (A1)). The dynamic isotropy condition are functions of variables R_{bo}, R_{to}, R_{bi}, R_{ti}, α_{to}, $(\alpha_{bi} - \alpha_{ti})$, a, Y, and H in the model of an MGSP and in the next section, we present our geometric approach to obtain closed-form expressions for these variables.

3 Design of MGSP

As mentioned, the unknowns in the model of an MGSP are R_{bo}, R_{to}, R_{bi}, R_{ti}, α_{to}, $(\alpha_{bi} - \alpha_{ti})$, a, Y, and H. The number of unknowns is more than the number of equations, and it is difficult to obtain a simple closed-form solutions to the above set of equations. To obtain the solution, we propose a *geometry-based method*. This is discussed below.

We start with the observation that for dynamic isotropy, $\lambda_1 = \lambda_3$ or $\lambda_2 = \lambda_3$ and using Eq. (6), we can get

$$b^2 = R_{ti}^2 + R_{bi}^2 - 2R_{ti}R_{bi}\cos(\alpha_{bi} - \alpha_{ti}), \text{ where } b^2 = \frac{H^2(3a^2+1)}{2} \tag{7}$$

The above equation follows the law of cosine in a triangle with sides R_{bi}, R_{ti} and b. On further extending our observation, if we project R_{ti} and α_{ti} in Fig. 1(a) to the bottom platform and visualize this from the top view, R_{bi} and R_{ti} can be seen making $(\alpha_{bi} - \alpha_{ti})$ angle between them as shown in Fig. 1(c). Hence, Eq. (7) and Δocq_1 in Fig. 1(c) gives a geometric relationship between the variables R_{bi}, R_{ti}, and the angle $(\alpha_{bi} - \alpha_{ti})$.

For a particular height H and ratio a, the variable b in Eq. (7) will be a constant. To fix this triangle, we need to fix an intersection point, for example $q_1, q_2, q_3...$ shown in Fig. 1(c). The intersection points are along a *line* R_{bi} having a slope $m = \tan(\alpha_{bi} - \alpha_{ti})$ with *a virtual circle* of radius b, and center c offset by $(R_{ti}, 0)$, see Fig. 1(c). $\{X_l, Y_l\}$ is a local coordinate system at o, and X_l is along $\boldsymbol{oc}(R_{ti})$ in the same figure. The intersection points $(q_1, q_2, q_3..)$ are obtained by solving for intersections at different slopes m. The solution exists when $m = \tan(\alpha_{bi} - \alpha_{ti}) \leq \sqrt{b^2/(R_{ti}^2 - b^2)}$. It can be further seen that the maximum value of m at equality corresponds to the *tangency condition*, which also implies that Δoq_1c is a right-angle triangle as shown in Fig. 1(d). We choose this right-angle triangle case and later generalize it for other intersections.

For the tangency condition, let x denote the ratio R_{ti}/H. Using Pythagoras theorem in Δoq_1c in Fig. 1(d), we get

$$R_{bi} = \sqrt{x^2H^2 - b^2} = \sqrt{\left(x^2 - \frac{(3a^2+1)}{2}\right)}H, \text{ and } \sin(\alpha_{bi} - \alpha_{ti}) = \frac{\sqrt{\frac{(3a^2+1)}{2}}}{x} \tag{8}$$

To take into account the variation in the payload's centre of mass (COM), denoted by variation Y, we consider two cases: *Case I)* $Y = 0$ (Payload COM is on the top platform) and *Case II)* $Y \neq 0$ (Payload COM is at some offset from the top platform). We first investigate the case of $Y = 0$ and then extend to the general case of $Y \neq 0$ and in both the cases, we aim to find the value of the intermediate variable x. It may be noted that for both these Cases, Eq. (8) are still valid.

Case I: Closed-form solution for $Y = 0$
Equating $\lambda_4 = \lambda_6$ or $\lambda_5 = \lambda_6$, $\mu_{11} = 0$ (see Eq. (A1) in Appendix), and using Eq. (6), on substitutions, we get

$$R_{to} = \sqrt{\left\{2K\left(\frac{a^2+1}{a^2}\right)\left(\frac{3a^2+1}{2}\right)\left(x^2 - \frac{(3a^2+1)}{2}\right) - x^2\right\}\frac{H}{a}} \quad (9)$$

where $K = I_{xx}/I_{zz}$ and denotes the inertia ratio of the payload. The payload properties are known with $I_{xx} = I_{yy}$ a necessary condition for satisfying $\lambda_4 = \lambda_5$. This is also a valid assumption in many practical applications, especially in a symmetrical payload.

For the requirements of $\lambda_1 = \lambda_3$ or $\lambda_2 = \lambda_3$, $\mu_{12} = 0$ (see Eq. (A1) in Appendix), and using Eq. (6), on substitutions, we get

$$R_{bo} = \sqrt{\left\{2K\left(\frac{a^2+1}{a^2}\right)\left(\frac{3a^2+1}{2}\right)\left(x^2 - \frac{(3a^2+1)}{2}\right) - x^2 + \left(\frac{7a^2}{2} + \frac{5}{2}\right)\right\}\frac{H}{a}} \quad (10)$$

Equating $\mu_{11} = 0$ or $\mu_{33} = 0$ (see Eq. (A1) in Appendix), and using Eq. (6), we get

$$\sin(\alpha_{to}) = R_{bi}R_{ti}\sin((\alpha_{bi} - \alpha_{ti}))/(a^2 R_{bo}R_{to}) \quad (11)$$

Using the identity $\sin^2(\alpha_{to}) + \cos^2(\alpha_{to}) = 1$ in $\mu_{11} = 0$ and $\mu_{12} = 0$, we get the value for x, and then equating $\lambda_3 = \lambda_4$, we can obtain H as

$$x = \sqrt{\frac{K(a^2+3)\left(\frac{3a^2+1}{2}\right)^2}{K(a^2+3)\left(\frac{3a^2+1}{2}\right) - 2a^2}} \;\&\; H = \sqrt{\frac{Q\left(K(a^2+3)\left(\frac{3a^2+1}{2}\right) - 2a^2\right)}{\left\{2K\left(\frac{3a^2+1}{2}\right)^2\right\}}} \quad (12)$$

where $Q = I_{xx}/m_p$ denotes the ratio of inertia to the payload mass The above solutions for H and x are in an explicit form and H can be easily chosen to satisfy our geometric constraint. The variables K and Q are known to us. From Eqs. (7) and (12), the value of the intermediate radius variable b^2 can be given as:

$$b^2 = \frac{Q\left(K(a^2+3)\left(\frac{3a^2+1}{2}\right) - 2a^2\right)}{\left\{2K\left(\frac{3a^2+1}{2}\right)\right\}} \quad (13)$$

In summary, for $Y = 0$, the *design procedure* involves selecting desired H from variable a and then finding x from Eq. (12) using the same a. Once H, a, and x are known, substitute them to find R_{ti}, R_{bi}, $(\alpha_{bi} - \alpha_{ti})$, R_{to}, R_{bo}, α_{to} in the respective order (Eqs. (8) - (12)). Also, for case $a = 1$ (when legs are equal), it can be shown that $R_{to} = R_{ti}$ as shown by other researchers after simplifications [6], and this is consistent

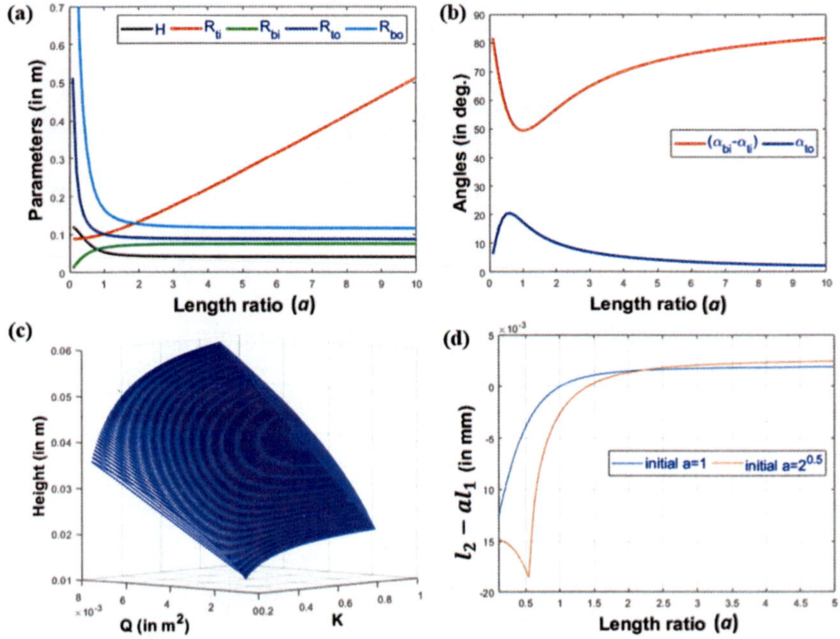

Fig. 2. a) Variation of geometric parameters for Case I, b) Variation of angles for Case I, c) Variation of height with payload properties K and Q for $a = 1$, d) $(l_2 - al_1)$ for Case I for initial a i.e., $a_{\,o} = 1,\sqrt{2}$.

with our formulation. Figure 2(a) and 2(b) shows plot for different parameters for a typical payload. As observed from the plots, for $a = 1, R_{to} = R_{ti}$, for high value of a, R_{ti} is greater, which makes l_2 larger and for small value of a, R_{bo} is larger making l_1 larger.

Interestingly, $a = 1$ is the point of transition after which the outer to outer radius connection (from bottom to top platform) is converted into outer radius to inner radius connection (cross leg type), as for $a > 1$, $R_{ti} > R_{to}$. At $a = 1$, it can be observed from the closed-form solution that the top radii depends only on payload property Q i.e. $R_{to} = R_{ti} = \sqrt{2Q}$. Figure 2(c) shows the plot of isotropic H with the payload property K and Q at this point.

We can find a general solution in this case after all the variables corresponding to the tangency condition ($a = a_o$) are known. The virtual circle (as shown in Fig. 1(c)) can be fixed keeping radius b and offset $(R_{ti}, 0)$ same as the tangency condition and all other parameters including a can change like before. The Intersection of line R_{bi} with the virtual circle for a general slope m revealed that the *condition of tangency is the only solution,* which is evident from Fig. 2(d) where the difference $(l_2 - al_1)$ is zero only at single a value for which the circle was fixed initially (*i.e.* a_o). Two cases for $a_o = 1$ and $\sqrt{2}$ nullify the difference $(l_2 - al_1)$ only at $a = 1$ and $\sqrt{2}$, respectively. Multiple solutions arise from the tangency condition for different circles (different a_o values).

Case II: Solution for $Y \neq 0$

The calculation for variables R_{ti}, R_{bi} and $(\alpha_{bi} - \alpha_{ti})$ remains the same as shown earlier. Taking consideration of Y now, and using the same sequence as in Case I, we can write:

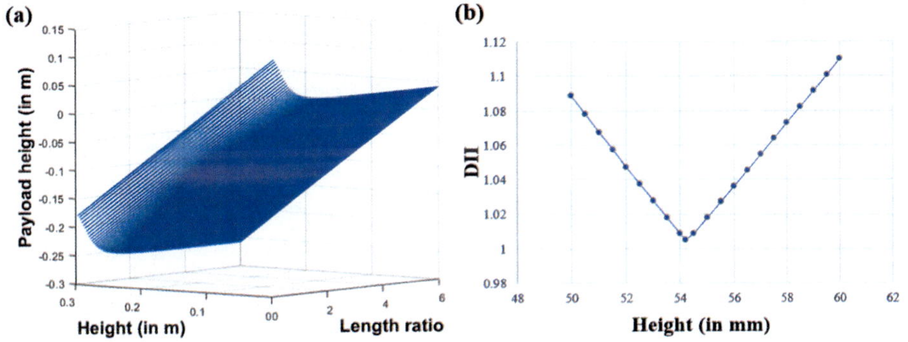

Fig. 3. a) Variation of H and Y for Case II, b) Sensitivity analysis for H with DII (for $a = 1$)

$$R_{to} = \sqrt{\left\{2K\left(\frac{a^2+1}{a^2}\right)\left(\frac{3a^2+1}{2}\right)\left(x^2 - \frac{(3a^2+1)}{2}\right) - x^2\right\}\frac{H^2}{a^2} + \frac{2Y^2(a^2+1)}{a^2}} \tag{14}$$

$$R_{bo} = \sqrt{R_{to}^2 + \frac{H^2}{a^2}\left(\frac{7a^2}{2} + \frac{5}{2}\right) + 4HY\left(\frac{a^2+1}{a^2}\right)} \tag{15}$$

Expression for calculating α_{to} remains the same as before and variable x and H can now be given as

$$x^2 = \frac{2Qa^2}{KH^2(3a^2+1)} + \frac{(3a^2+1)}{2} \tag{16}$$

$$Y^2 + 2YH + H^2 + \frac{4Qa^2}{K(3a^2+1)^2} - \frac{Q(a^2+3)}{(3a^2+1)} = 0 \tag{17}$$

In summary, the design procedure for $Y \neq 0$ is as follows:

a) Select a and H (within our constraints), from where Y and x can be obtained from Eqs. (16) and (17).

b) Substitute H, Y, a, and x to find R_{ti}, R_{bi}, $(\alpha_{bi} - \alpha_{ti})$, R_{to}, R_{bo}, α_{to} in the given order (refer to Eqs. (8), (11), (14), and (15)).

The variations of parameters in this case is given by Fig. 3(a) and the trends of other parameters are similar to that of Case I. The inclusion of non zero Y provides more flexibility to our solution procedure and even intersections beyond tangency conditions are possible contrary to the previous case. Another interesting conclusion that can be drawn from Eq. (17) is that H and Y can be interchanged due to the equation's symmetrical nature. This implies that interchanging H and Y will lead to a dynamically isotropic configuration corresponding to the tangency case.

In both the Cases I and II, natural frequency will depend only on k and m_p. For the dynamic isotropy, $\omega_1 = \omega_2 = \omega_3 = \omega_4 = \omega_5 = \omega_6 = \sqrt{2k/m_p}$. Hence the $[G]$ matrix will be a diagonal matrix for every solution with a diagonal value of $2k/m_p$.

Table 1. Showing DII comparison between results obtained using FEM (DII(s)) and closed-form solution (DII(t)). s denotes FEM, t denotes theoretical (closed-form solution discussed above).

Natural frequencies from ANSYS® (Configuration taken as in the closed form solution)										
	(mm)	Mode	1	2	3	4	5	6	DII(s)	DII(t)
1	Y=0, H=54.2		31.69	31.71	31.72	31.81	31.81	31.84	1.005	1.00
	$R_{bo}, R_{bi}, R_{to}, R_{ti}, \alpha_{to}, (\alpha_{bi} - \alpha_{ti}), a$					166.7, 65.6, 100, 100, 17.3959, 49.424. 1.0				
2	Y=33.2, H=40		31.60	31.69	31.72	31.81	31.81	31.84	1.007	1.00
	$R_{bo}, R_{bi}, R_{to}, R_{ti}, \alpha_{to}, (\alpha_{bi} - \alpha_{ti}), a$					265.7, 54.6, 184, 68.2, 7.2684, 36.7634, 0.6				
3	Y=40, H=33.2		31.53	31.66	31.69	31.80	31.82	31.83	1.009	1.00
	$R_{bo}, R_{bi}, R_{to}, R_{ti}, \alpha_{to}, (\alpha_{bi} - \alpha_{ti}), a$					265.7, 54.6, 197.6, 64.2, 5.611, 31.8027, 0.6				

4 Validation Through ANSYS®

Simulations were carried out in ANSYS® with top and bottom platforms treated as rigid bodies and struts as ideal springs to verify the geometry based closed form solutions. The closed-form solutions were obtained for a typical payload with $K = 3748/6343$, $Q = 5.089 * 10^{-3} m^2$, $m_p = 5$ Kg, and $k = 10^5$ N/m. Ideally, for dynamic isotropy, the ratio of the largest to the smallest natural frequency, denoted by DII, should be 1.0 and practically DII should be very close to 1. From the Case 2 and 3 in Table 1, it can be concluded that H and Y can be interchanged, and both will have at least one dynamic isotropic configuration with tangency condition. The DII obtained from closed-form expression and in the simulation in all the cases are very near to one (dynamic isotropic) and the natural frequency is around 31.83 Hz as obtained theoretically using $\sqrt{2k/m_p}$. For the Case for $Y = 0$ in Table 1 ($a = 1$), H was varied from the nominal value of 54.2 mm by approximately $\pm$ 10%, and Fig. 3(b) shows the variation of DII with H.. As expected, any deviation from the dynamic isotropic configuration (in this case height) tend to increase DII.

5 Conclusion

This paper deals with the design of a dynamically isotropic MGSP. Taking the variation in payloads' centre of mass (COM) and using the force transformation matrix, the natural frequency matrix was derived analytically. The closed-form solution in an explicit form was established using a *geometry-based approach* and can be used to design dynamically isotropic MGSP at its neutral position. For the case, when the payloads' COM is on the top platform, the tangency conditions for the virtual circle gives a complete set of solutions. When the COM is at some offset, the tangency condition is one of the solutions along with other intersection points of R_{bi} line with a virtual circle. The MGSP configurations obtained using the closed-form solution were validated via simulations using a finite element software. In future, we intend to incorporate damping and controls in our designs.

Conflict of Interest. The authors declare that they have no conflict of interest.

Appendix

The value of all λ's and μ's in Eq. (5) are given as:

$$\lambda_1 = \frac{3k(l_1^2\Psi_1 + l_2^2\Psi_2)}{2m_p l_1^2 l_2^2},\quad \lambda_2 = \frac{3k(l_1^2\Psi_1 + l_2^2\Psi_2)}{2m_p l_1^2 l_2^2},\quad \lambda_3 = \frac{3kH^2(l_1^2 + l_2^2)}{m_p l_1^2 l_2^2}$$

$$\lambda_4 = \frac{3kY^2(l_2^2\Psi_2 + l_1^2\Psi_1) + 6HkY(\Psi_3 l_1^2 + \Psi_4 l_2^2) + 3k\Psi_5 H^2}{2I_{xx} l_1^2 l_2^2},\quad \lambda_5 = \lambda_4 \frac{I_{xx}}{I_{yy}}$$

$$\lambda_6 = \frac{3k(\Psi_6^2 l_1^2 + \Psi_7^2 l_2^2)}{I_{zz} l_1^2 l_2^2},\quad \mu_{11} = \frac{-3kH(-\Psi_6 l_1^2 + \Psi_7 l_2^2)}{2l_1^2 l_2^2},\quad \mu_{33} = \frac{3kH(-\Psi_6 l_1^2 + \Psi_7 l_2^2)}{l_1^2 l_2^2}$$

$$\mu_{12} = \frac{3kH(\Psi_3 l_1^2 + \Psi_4 l_2^2) + 3kY(l_1^2\Psi_1 + l_2^2\Psi_2)}{2l_1^2 l_2^2} \tag{A1}$$

where,

$$\Psi_1 = R_{ti}^2 + R_{bi}^2 - 2R_{ti}R_{bi}\cos(\alpha_{bi} - \alpha_{ti}),\ \Psi_2 = R_{to}^2 + R_{bo}^2 - 2R_{to}R_{bo}\cos(\alpha_{to}),$$
$$\Psi_3 = R_{ti}^2 - R_{ti}R_{bi}\cos(\alpha_{bi} - \alpha_{ti}),\ \Psi_4 = R_{to}^2 - R_{to}R_{bo}\cos(\alpha_{to})$$
$$\Psi_5 = R_{ti}^2 l_1^2 + R_{to}^2 l_2^2,\ \Psi_6 = R_{ti}R_{bi}\sin(\alpha_{bi} - \alpha_{ti}),\ \Psi_7 = R_{to}R_{bo}\sin(\alpha_{to}) \tag{A2}$$

References

1. Shyam, R.B.A., Ahmad, N., Ranganath, R., Ghosal, A.: Design of a dynamically isotropic Stewart-Gough platform for passive micro vibration isolation in spacecraft using optimization. J. Spacecr. Technol. **30**(2), 1–8 (2019)
2. Tong, Z., He, J., Jiang, H., Duan, G.: Optimal design of a class of generalized symmetric Gough-Stewart parallel manipulators with dynamic isotropy and singularity-free workspace. Robotica **30**, 305–314 (2012)
3. Hong-zhou, J., Zhi-zhong, T., Jing-feng, H.: Dynamic isotropic design of a class of Gough-Stewart parallel manipulators lying on a circular hyperboloid of one sheet. Mech. Mach. Theor. **46**, 358–374 (2011)

4. Yun, H., Liu, L., Li, Q., Li, W., Tang, L.: Development of an isotropic Stewart platform for telescope secondary mirror. Mech. Syst. Signal Process. **127**, 328–344 (2019)
5. Jiang, H.Z., He, J.F., Tong, Z.Z., Wang, W.: Dynamic isotropic design for modified Gough-Stewart platforms lying on a pair of circular hyperboloids. Mech. Mach. Theor. **46**, 1301–1315 (2011)
6. Yao, J., Hou, Y., Wang, H., Zhou, T., Zhao, Y.: Spatially isotropic configuration of Stewart platform-based force sensor. Mech. Mach. Theor. **46**, 142–155 (2011)
7. Yi, Y., McInroy, J.E., Jafari, F.: Generating classes of locally orthogonal gough-stewart platforms. IEEE Trans. Rob. **21**(5), 812–820 (2005)
8. Ghosal, A.: Robotics Fundamental Concepts and Analysis. Oxford University Press, Oxford (2006)
9. Hanieh, A.A.: Active isolation and damping of vibrations via Stewart Platform. Ph.D. Thesis, ULB Active Structures Laboratory, 62–66 (2003)

Design and Control of a Double-Sarrus Mobile Robot

Phuong Thao Thai(✉) and Nguyen Ngoc Hai

Hanoi University of Science and Technology, Hanoi, Vietnam
thao.thaiphuong@hust.edu.vn

Abstract. Origami robot nowadays is used widely in both industrial and social life. This paper proposes a new origami robot model that is transformable and portable: Double-Sarrus Mobile robot. The Sarrus linkage is a mechanical linkage with 1 degree-of-freedom that can change a circular motion to a linear motion and vice versa. The new design takes advantage of Sarrus linkage to create a robot that can transform between a mobile box-shaped robot and a flat sleeping configuration with only one controlled motion. Therefore, only one motor is required for the transformation of the robot, making the robot lightweight and portable. The mechanism of robot is assembled by combining two Sarrus linkages. This paper analyzes the kinematic properties of double-connective-Sarrus linkages. Finally, robot's motion simulations are presented by MATLAB.

Keywords: Origami robot · Transformable robot · Sarrus linkage

1 Introduction

Origami robotics is the combination of origami art and robotics with the aim of creating robots that can transform into 3D model based on crease patterns [1]. With origami crease patterns, a robot is capable of altering different configurations easily with simple movement. Origami robotics is useful for packaging as it can be transformed from 2D to 3D model for active mode and returned to 2D model in sleep mode. This flexible feature is the advantage for robots when working in various working space. Up to now, there are several studies on robot transformation ability based on origami art, such as Miniature origami robot [2], Self-folding origami robot [3], Tribot [4], Origami-inspired robotic arm [5–7].

This paper proposes a new model that is transformable and portable: Double-Sarrus Mobile Robot. The Sarrus linkage is a mechanical linkage with 1 degree-of-freedom (DOF) that can change a circular motion to a linear motion and vice versa (Fig. 1). The new design takes advantage of Sarrus linkage to create a robot that can transform between a mobile box-shaped robot and a flat sleeping configuration with only one controlled motion. Therefore, only one motor is required for the transformation of the robot, making the robot lightweight and portable.

The following is a brief overview of this paper: Sect. 2 introduces the design of the transformable robot. The kinematics and dynamics analysis are discussed in Sect. 3. In Sect. 4, the control for the Sarrus mobile robot is proposed, and the simulations to illustrate our concept is analyzed in Sect. 5.

N. V. Khang et al. (Eds.): ASIAN MMS 2021, MMS 113, pp. 269–278, 2022.
https://doi.org/10.1007/978-3-030-91892-7_25

2 Design of Double-Sarrus Mobile Robot

2.1 Conceptual Design

With the idea of creating a robot that can transform from 2D to 3D, where the 3D configuration, which is also called "cubic mode" or "active mode", should have the shape of a box with wheels while in the 2D configuration ("flat mode" or "sleep mode"), the wheels are hidden, two Sarrus linkages are combined.

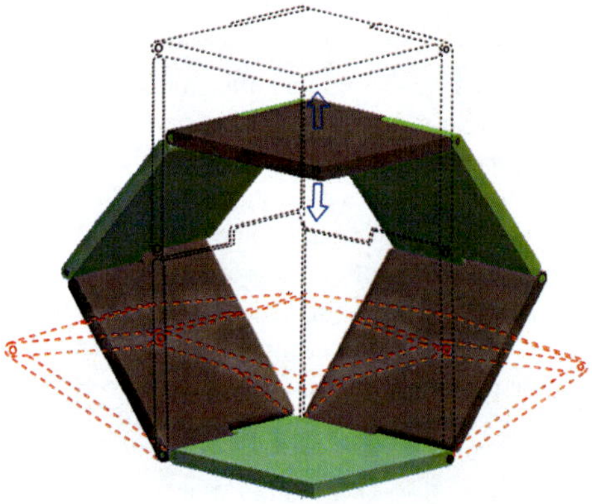

Fig. 1. Sarrus linkage

The schematic design of the first and second linkage are illustrated in Fig. 2a and Fig. 3a, the full model design are illustrated respectively in Fig. 2b and Fig. 3b.

The first linkage is a four-sided Sarrus linkage that forms the main body of the robot in the active mode (Fig. 2a and Fig. 2b). There is one based panel where the motor for transformation is mounted on. Four pairs of panels connect four sides of the based panel and those of the top panel. There are two type of the side panels, C1-side panel that is the below-sided panel and C2-side panel, which is the upper one. The sizes of the side panels are equal to satisfy the condition of a Sarrus linkage. All the joints are revolute joints, and only one of them is directly controlled by the motor while the others are passive joints.

The second linkage (Fig. 3a and Fig. 3b) is a variant of Sarrus linkage which let a underneath panel with wheels moves down during transformation into active mode. The two Sarrus linkages share the base panel as well as three C1-side panels adjacent to the base panel. Cuts are made on the base panel to make room for the three links that complete the second Sarrus linkage. The active joint is not a part of the second Sarrus linkage to save space for the motor for transformation. The sizes of the wheels should be small enough to be hidden in the flat mode.

In the active mode, three wheels are activated as they touch the ground: two active wheels controlled by two DC motors and one passive wheel for navigation. Table 1 is the basic parameters of the Sarrus mobile robot, Fig. 4 is the illustration of our design in 2 modes: Flat mode (Fig. 4a) and Cubic mode (Fig. 4b). The protoptype of the Double

Sarrus can be found in Fig. 5. The prototype is made by 3D printers, the material is PP plastic.

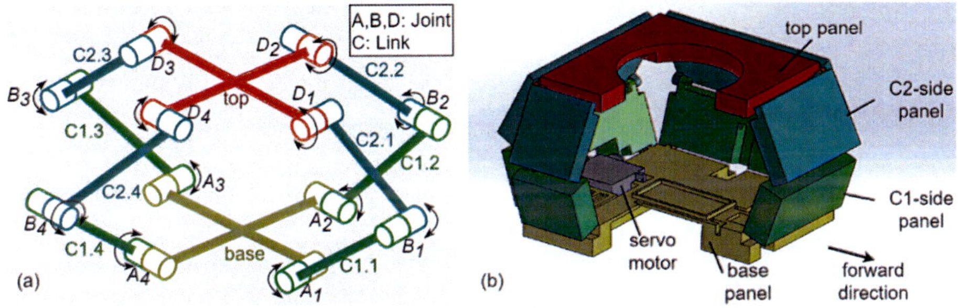

Fig. 2. a) The schematic of the first Sarrus linkage b) Design of linkage

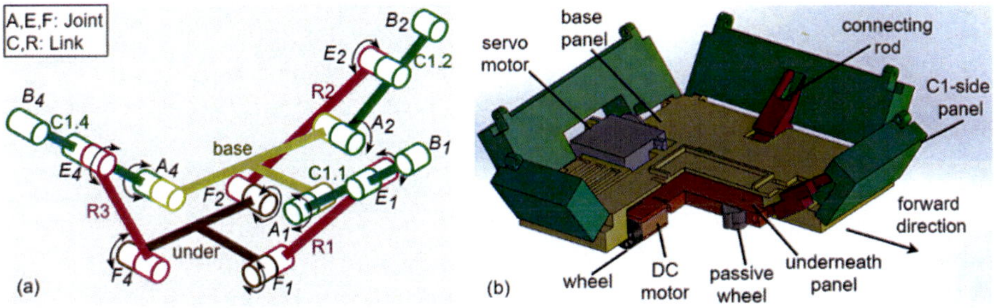

Fig. 3. a) The schematic of the second Sarrus linkage b) Design of linkage

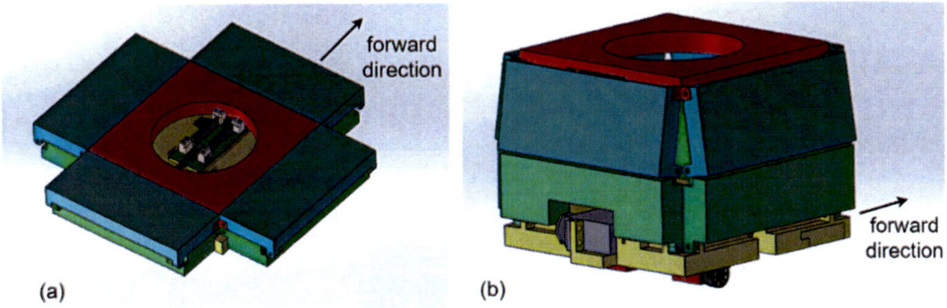

Fig. 4. a) Flat mode b) Cubic mode

2.2 Control System

The control system consists of a joystick on smartphone, a Bluetooth module HC05, an Arduino pro mini microcontroller, two DC motors, a DC motor driver L293D, and

Fig. 5. The protype of Double-Sarrus Mobile Robot

Table 1. Parameters of Double-Sarrus Mobile Robot

Robot	Dimensions in flat mode Dimensions in 3D mode Mass Material	256 × 256 × 36 mm 180 × 180 × 148 mm 1500 g Polylactic Acid
Actuator	Transformable mechanism Mobile mechanism	1 servo motor 2 DC servo motors
Control	Main controller Power	Arduino pro mini 2 Lipo batteries DC 3.7V

two Lipo power batteries. The joystick position is sent to the Arduino via the Bluetooth module. Then, Arduino program processes and sends PWM pulses to the DC motor driver. The PWM signal is converted into voltage applied to the two DC motors.

3 Analysis of Kinematics and Dynamics Modeling

In order to analyze the movement of transforming, a dynamic model has been created. The kinematic parameters, variables and coordinate system are given in Fig. 6. Dynamic parameters are given in Table 2.

3.1 Kinematics Analysis

When moving, the kinematics coordinates are connected by the link equations as:

$$h_1 - 2a \sin q_1 = 0 \tag{1}$$

$$d + k \cos q_1 - l \cos q_2 = 0 \tag{2}$$

$$k \sin q_1 - l \sin q_2 + h_2 = 0 \tag{3}$$

Table 2. Dynamic parameters of Double-Sarrus Mobile Robot

m_t	Mass of top panel
m_b	Mass of base panel
m_{c1}	Mass of C1-side panel
m_{c2}	Mass of C2-side panel
m_l	Mass of connecting rods
I_{c1}	Inertia moment of C1-side panel
I_{c2}	Inertia moment of C2-side panel
J_l	Inertia moment of connecting rod

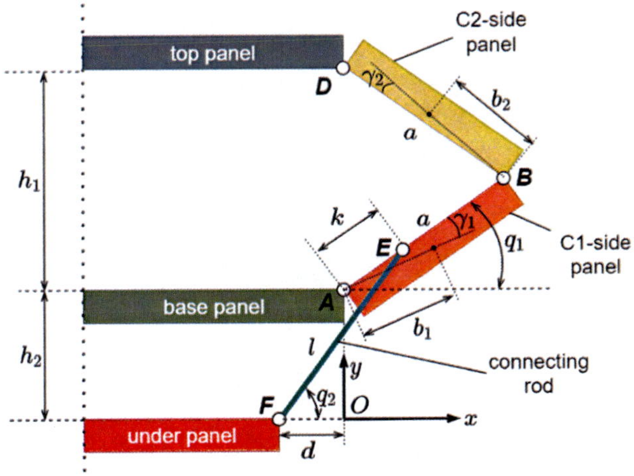

Fig. 6. Dynamic model

3.2 Dynamic Analysis

Kinetics and potential energy is calculated as follows:

$$\begin{aligned} T &= \frac{1}{2}m_b\dot{y}_b^2 + \frac{1}{2}m_t\dot{y}_t^2 + 2\left[I_{c1}\dot{q}_1^2 + m_{c1}\left(\dot{x}_{c1}^2 + \dot{y}_{c1}^2\right)\right] \\ &\quad + 2\left[I_{c2}\dot{q}_1^2 + m_{c2}\left(\dot{x}_{c2}^2 + \dot{y}_{c2}^2\right)\right] + \frac{3}{2}J_l\dot{q}_2^2 \\ &= 0.5m_\Sigma\dot{h}_2^2 + 2I_\Sigma\dot{q}_1^2 + K_1(q_1)\dot{h}_2\dot{q}_1 + K_2(q_1)\dot{q}_1^2 + 1.5J_l\dot{q}_2^2 \end{aligned} \tag{4}$$

$$\begin{aligned} \Pi &= m_bgy_b + m_tgy_t + 4g(m_{c1}y_{c1} + m_{c2}y_{c2}) + \frac{3}{2}m_lgl\ \sin q_2 \\ &= m_bgh_2 + m_tg(h_2 + 2a\ \sin q_1) + 4m_{c1}g\left[h_2 + b_1\ \sin(q_1 - \gamma_1)\right] \\ &\quad + 4m_{c2}g\left[h_2 + a\ \sin q_1 + b_2\sin(q_1 + \gamma_2)\right] + 1.5m_lgl\ \sin q_2 \end{aligned} \tag{5}$$

in which,
$m_\Sigma = m_b + m_t + 4(m_{c1} + m_{c2})$
$I_\Sigma = I_{c1} + I_{c2} + m_{c1}b_1^2 + m_{c2}(a^2 + b_2^2)$
$K_1(q_1) = (2m_t + 4m_{c1})a\cos q_1 + 4m_{c1}b_1\cos(q_1 - \gamma_1) + 4m_{c2}b_2\cos(q_1 + \gamma_2)$
$K_2(q_1) = 2m_t a^2\cos^2 q_1 + 4m_{c2}ab_2\cos(2q_1 + \gamma_2)$

The motion equations of robot is given by Lagrange multipliers, the generalized coordinates are q_1, q_1, h_2

$$\begin{aligned}&\left[4I_\Sigma + 2K_2(q_1)\right]\ddot{q}_1 + K_1(q_1)\ddot{h}_2 + K_2^{'}(q_1)\dot{q}_1^2 + K_1(q_1)g\\&-\lambda_1 k\ \sin\ q_1 + \lambda_2 k\ \cos\ q_1 = U\end{aligned} \tag{6}$$

$$3J_l\ddot{q}_2 + 1.5m_l gl\ \cos\ q_2 + \lambda_1 l\ \sin\ q_2 - \lambda_2 l\ \cos\ q_2 = 0 \tag{7}$$

$$m_\Sigma\ddot{h}_2 + K_1(q_1)\ddot{q} + K_1^{'}(q_1)\dot{q}_1^2 + m_\Sigma g + \lambda_2 = 0 \tag{8}$$

Simplifying λ_1, λ_2 in Eq. (7) (8), then replacing in Eq. (6). The Eq. (6) can be written as

$$\begin{aligned}&\left[4I_\Sigma + 2K_2(q_1) + h_q K_1(q_1)\right]\ddot{q}_1 + \left[K_1(q_1) + h_q m_\Sigma\right]\ddot{h}_2 + 3J_l f(q_1)\ddot{q}_2\\&+\left[h_q K_1^{'}(q_1) + K_2^{'}(q_1)\right]\dot{q}_1^2 + g\left[h_q m_\Sigma + K_1(q_1) + 1.5m_l\left(h_q + \cos\ q_1\right)\right] = U\end{aligned} \tag{9}$$

in which,

$$h_q = \frac{k(d\ \sin\ q_1 - h_2\ \cos\ q_1)}{h_2 + k\ \sin\ q_1}$$
$$f(q_1) = \frac{k\ \sin\ q_1}{h_2 + k\ \sin\ q_1}$$

The derivatives of the coordinates q_2, h_2 can be defined as the coordinate q_1 by Eqs. (2) (3). The general form of Eq. (9) is given as follows:

$$U = M(q)\ddot{q} + C(q, \dot{q})\dot{q} + G(q) \tag{10}$$

with $q = q_1$. The Eq. (10) is 1 DOF equation motion.

4 Control and Simulation

4.1 Transformation

The PD control torque is computed as follows:

$$U = M(q)\left(\ddot{q}_d + K_d\dot{e} + K_p e\right) + C(q, \dot{q})\dot{q} + G(q) \tag{11}$$

in which $e = q_d - q$, q_d is the desired angle. The motion of the desired angle is designed as in Table 3. The PD control parameters in simulation are given as follows: $K_d = 200$; $K_p = 10000$. The simulation has been done in two case, case 1 is calculated with cubic trajectory (Fig. 7 and Fig. 8) and case 2 is computed with step function (Fig. 9, Fig. 10 and Fig. 11). The requirement for controlling our robot the minimum control torque. Therefore, we choose the control coefficient K_p, K_d with large value. The value of control torque in case 1 is clearly less than case 2 (Fig. 8b and Fig. 11b). In addition, the control error in case 1 is less and the response time is faster than in case 2 (Fig. 8a and Fig. 11a).

Table 3. The motion of desired angle

Time	Desired angle (degree)
0 s–0.5 s	0
0.5 s–1.5 s	$q_d(t) = 270(t-0.5)^2 - 180(t-0.5)^3$ $\Delta t = 0.001$ s
1.5 s–2 s	90

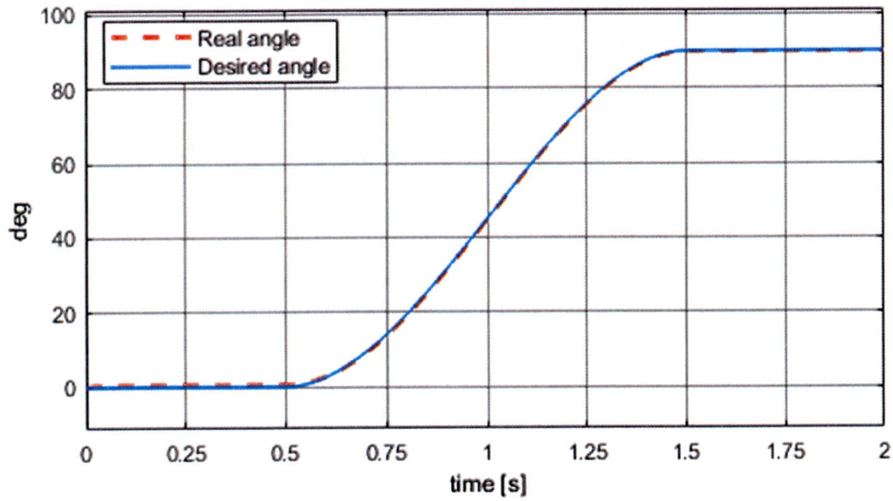

Fig. 7. Results of PD control with trajectory

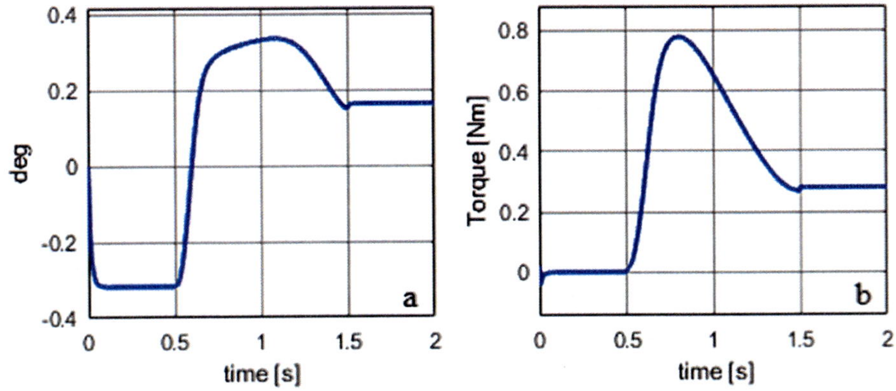

Fig. 8. Results of PD control with trajectory a) Error of control process, b) Control torque

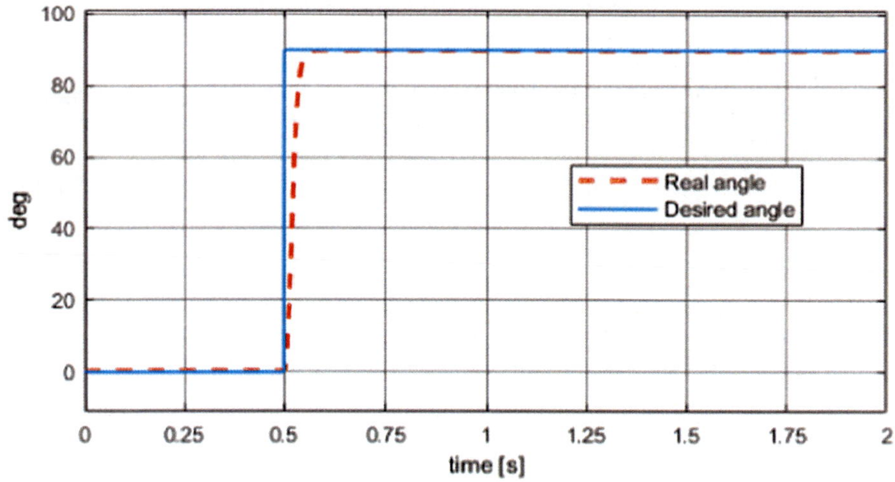

Fig. 9. Results of PD control without trajectory

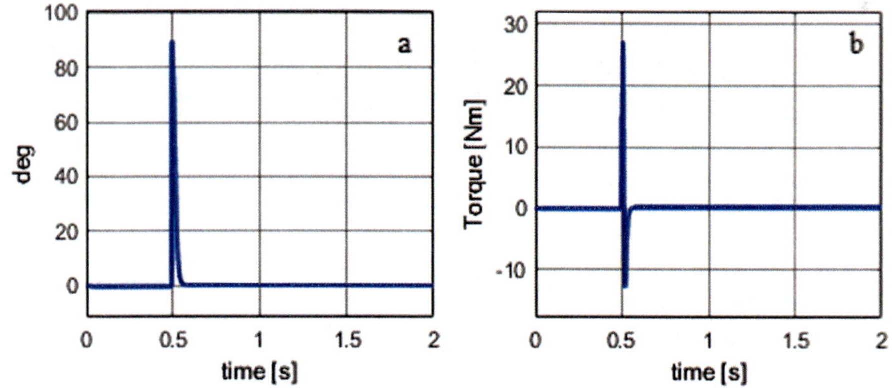

Fig. 10. Results of PD control without trajectory a) Error of control process, b) Control torque

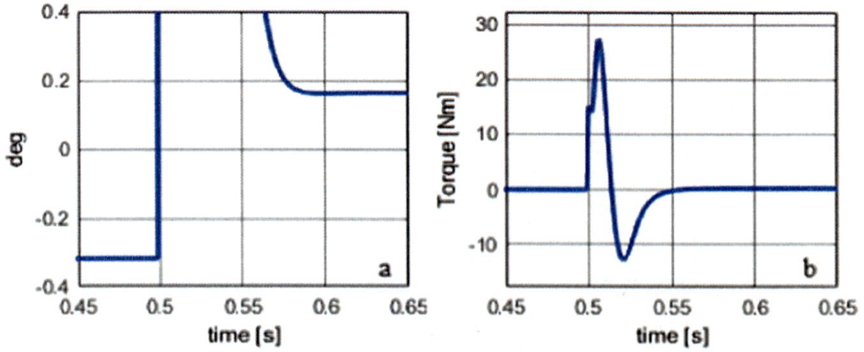

Fig. 11. Detailed results of PD control without trajectory a) Error of control process, b) Control torque

4.2 Mobility

When dragging the joystick on the phone, the program will send a signal of the joystick's position via Bluetooth. The joystick position is given by xy coordinates with a value from 0 to 255. The signal is received by the Arduino, processed by the control algorithm and a PWM pulse value from −255 to 255 is sent to the motors. Positive values are associated with forward motion and vice versa. Details can be found in Fig. 12 and Fig. 13.

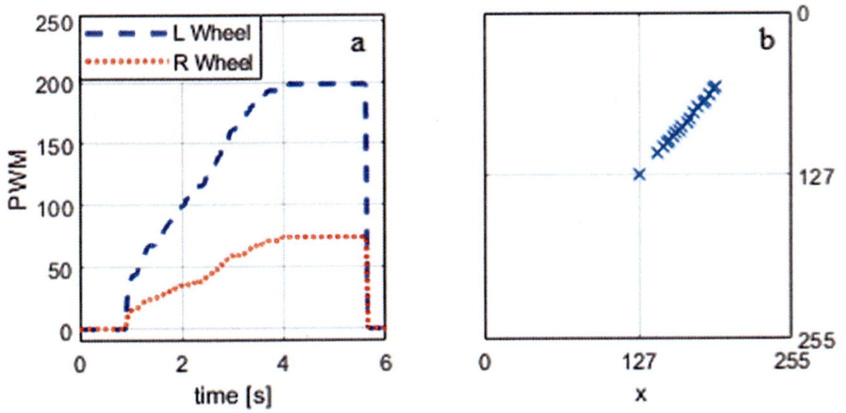

Fig. 12. a) PWM pulse for two wheels when moving forward to the right b) Joystick position

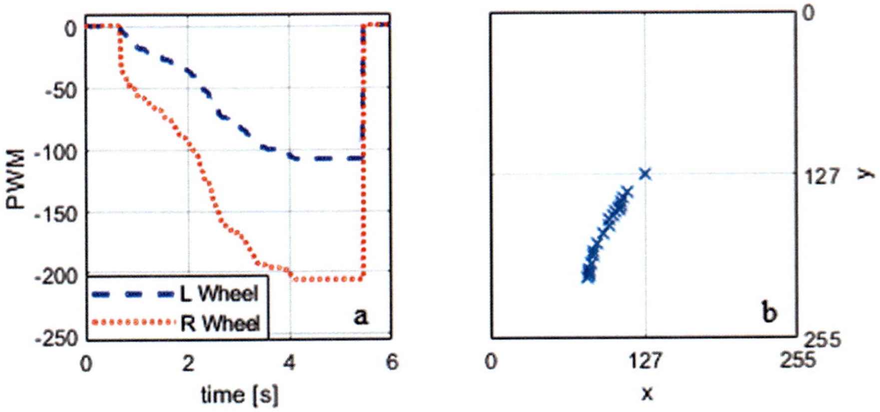

Fig. 13. a) PWM pulse for two wheels when moving backward to the right b) Joystick position

5 Conclusion

The paper proposes the design and control of a transformable Double-Sarrus Mobile Robot. The problem of movement and transformation has been investigated carefully.

This version is the first design and there are problems need to be solved later. For future work, it is necessary to consider attaching sensor system to make the robot transform automatically. The control concept should be improved to reduce the error more and make the motion faster and more accurate.

References

1. Rus, D., Sung, C.: Spotlight on Origami Robots. Sci. Robot. **3**(15), eaato938 (2018)
2. Miyashita, S., Guitron, S., Ludersdorfer, M., Sung, C.R., Rus, D.: An untethered miniature origami robot that self-folds, walks, swims, and degrades. In: 2015 IEEE International Conference on Robotics and Automation (ICRA), pp. 1490–1496. IEEE, Seatle, WA, USA (2015)
3. Felton, S., Tolley, M., Demaine, E., Rus, D., Wood, R.: A method for building self-folding machines. Science **345**(6197), 644–646 (2014)
4. Zhakypov, Z., Belke, C.H., Paik, J.: Tribot: a deployable, self-righting and multi-locomotive origami robot. In: 2017 IEEE/RSJ International Conference on Intelligent Robots and Systems (IROS), pp. 5580–5586. IEEE, Vancouver, BC, Canada (2017)
5. Kim, S., Lee, D., Jung, G., Cho, K.: An origami-inspired, self-locking robotic arm that can be folded flat. Sci. Robot. **3**(16), eaar2915 (2018)
6. Mintchev, S., Salerno, M., Cherpillod, A., Scaduto, S., Paik, J.: A portable three-degrees-of-freedom force feedback origami robot for human–robot interactions. Nat. Mach. Intell. **1**(12), 584–593 (2019)
7. Onal, C.D., Tolley, M.T., Wood, R.J., Rus, D.: Origami-inspired printed Robots. IEEE/ASME Trans. Mechatron. **20**(5), 1–8 (2018)

Effect of Static Load on Critical Speeds of a Shaft Supported by Two Symmetrically Arranged Bump-Foil Bearings in a Turbocharger

Minh-Quan Nguyen(✉), Minh-Hai Pham, and Van-Phong Dinh

Hanoi University of Science and Technology, Hanoi, Vietnam
quan.nguyenminh@hust.edu.vn

Abstract. Working as a self-acting bearing without lubrication, foil bearing has shown its potential applications into high-speed machines, including turbochargers. Numerous studies on various types of foil bearings have been conducted but bump-foil bearings have drawn the most attention. The role of bump-foil bearing in shaft performance has been analyzed through evaluating structural and technological parameters such as foil dimensions, nominal clearance, shaft weight... However, external parameters including static load have not been concerned in evaluating the shaft performance in a turbocharger. Therefore, in this theoretical study, a turbocharger shaft supported by two bump-foil bearings were investigated under the effect of an external static load at different positions on the shaft. Finite difference was employed to transform the governing equations, and then the system of ordinary differential equations was solved to obtain orbit of the shaft center, while Newton-Raphson algorithm was applied to investigate the shaft stability at equilibrium state. The results indicate significant influences of the load values as well as acting points on the critical speeds of the shaft. The research is believed to make further contribution to not only structural design for turbocharger shaft through load effect presented as bushes, shaft collars, ... but also shaft stability control through load control using piezoelectric material.

Keywords: Foil-air bearing · Reynold's equation · Shaft stability

1 Introduction

For the last few decades, oil-free technologies including foil bearings have been widely used in a variety of industries and engineering aspects. This type of bearing, which is a self-acting bearing and performs without oil supply, has demonstrated various advantages compared to traditional oil/air bearings [1]. Because the oil supply system is omitted, the maintenance work and machine weight can be significantly reduced. Working temperature of foil bearing depends only on materials of bearing foil and sleeve but does not suffer the restriction from oil quality. Moreover, with the compliancy of the foil structure, bigger shaft angular misalignment and centrifugal growth can be compensated than

N. V. Khang et al. (Eds.): ASIAN MMS 2021, MMS 113, pp. 279–291, 2022.
https://doi.org/10.1007/978-3-030-91892-7_26

traditional air bearing. Therefore, foil bearings have drawn many studies for application in high-speed machines, including turbochargers.

Turbocharger is a kind of turbine machine that utilizes the emission gas flow to generate more intake air into cylinders in a combustion engine as shown in Fig. 1. This generation is made through a shaft with two wheels assembled onto the two ends. Therefore, energy from the emission gas glow is transmitted to a compressor to provide more air into the chamber, improving the engine power.

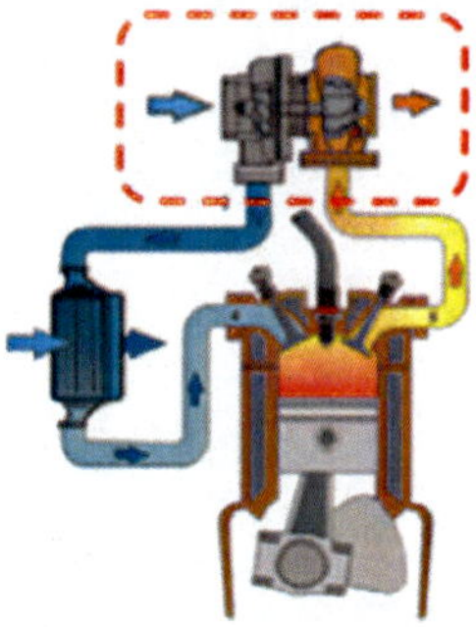

Fig. 1. Turbocharger structure [2]

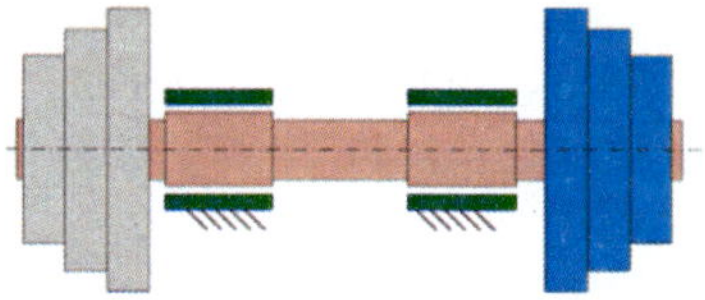

Fig. 2. Schematic diagram of a FB-shaft system

With the purpose of high-speed performance, the foil bearings have been developed to replace oil bearings in the transmission shaft as shown in Fig. 2; thus, dynamic response of the shaft-bearings system has been studied. J. F. Walton II et al. [3] designed a test rig to simulate a turbocharger rotor and investigated the performance of the rotor-bearings system under effect of hot air flow and shocking loads. The experimental results indicated that the rotor could return stability even when the speed reaches 150000 rpm without considerable temperature change inside the bearing. U. Borchert et al. even employed an exhaust gas turbocharger K26 from BorgWarner Turbo Systems [4] to evaluate the bearing operation. The bearing system was able to work well even with abrasion traces caused by acceleration and deceleration. However, some restrictions need to be made for high-efficient fluid energy machine. Effect of bearing clearance on rotor performance was investigated by K. Sim et al. for lobed gas foil bearing in turbocharger [5]. It was noted that static loads were also included in the study but through changes in the working clearance. On the theory, Hai Pham and P. Bonello developed a technique [6] of simultaneously solving a set of governing equations to analyze the stability and nonlinear dynamic of rotor in a turbocharger. Using finite element method with experiment [7, 8], M. Mahner et al. evaluated preload effects on the foil bearing performance. The effects were analyzed in two cases: rotating and non-rotating shaft. However, the load point was only assumed to be at the bearing, not at different positions on the shaft. Later, the authors conducted experiments to investigate the assembly preload on the hysteresis and the lift-off speed of three-pad air foil journal bearing. Using a new method of modelling elastic deformation of the foil structure [9], G. Zwyca et al. analyzed effect of assembly preload on the static and dynamic performance of the bump foil bearing. Effect of assembly preload was presented as different values of elastic deformation of the foils, which influences the stiffness of the foil as well as operating clearance of the bearing.

It can be seen from the literature that only assembly preload has been considered for the shaft stability. However, this type of load only exists at the bearings and the influence of this load can only be considered through changes in radial clearances of the bearing. It is also noted that the acting point of the static load on the shaft has not been investigated while this plays an important part because additional bending moment can have an influence on the rotor stability. Therefore, in this paper, an external static load was investigated to analyze the stability of the rotor through critical speeds. It can be noted that the static load was put at different positions on the shaft with various values. From that, numerical computation was conducted to evaluate speed ranges where the shaft remained stable. Effect of the static load and its acting point were analyzed then. This study is believed to make contribution to improving the shaft's performance in a turbocharger through parametric design and control.

2 Governing Equations

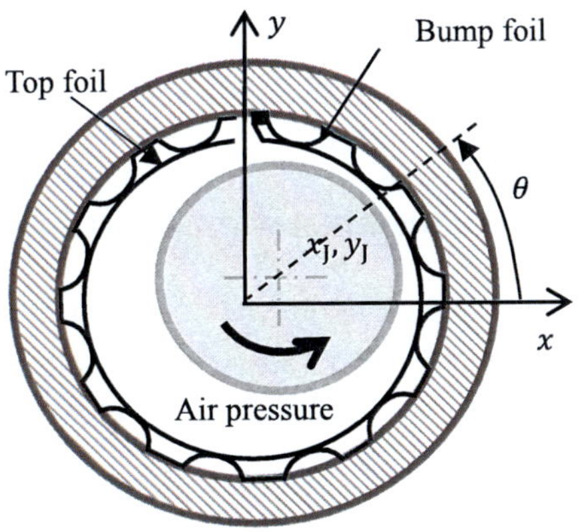

Fig. 3. Schematic diagram of a bump-foil bearing with a shaft

Structure of a bump-foil bearing is illustrated in Fig. 3 with a foil structure composed of a bump foil (corrugated sheet) and a top foil (bended sheet). Air pressure is generated during shaft's rotating time according to hydrodynamic lubrication theory. The pressure exerts on the shaft and makes eccentric displacements while the foil structure is supposed to be elastically deflected to change the air-film thickness between the shaft and the top foil. As can be seen, the performance of shaft-bearing assembly can be analyzed by modelling the interaction among three fields: air pressure variation, foil deflection and shaft dislocation. From that, governing equations can be established as follows.

2.1 Shaft Center Displacement

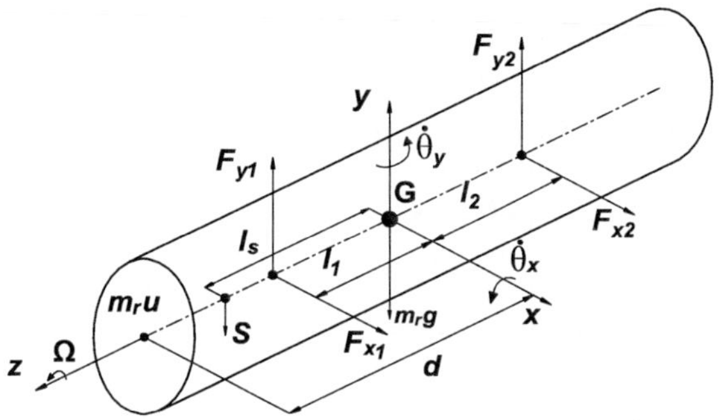

Fig. 4. Schematic diagram of a turbocharger shaft

A schematic diagram of a turbocharger shaft was illustrated in Fig. 4. In this structure, this shaft was considered as a rigid body. As can be seen in this diagram, the shaft is subjected to reaction forces F_x, F_y (generated from air pressure in two beatings) with distances between acting points l_1, l_2. Gravity force is also included with m_r denoting the shaft mass while influence of unbalance forces is introduced through unbalance u (normalized to the shaft mass) with the gap d from the center of gravity G. Moreover, a static load S with l_S from G is proposed in this diagram to investigate dynamic behavior of the rotating shaft. It is notable that the computation is made for different values of l_S corresponding to diverse acting points of the load on both sides of the shaft.

The differential equations governing shaft motion can be established accordingly:

$$\begin{aligned} m\ddot{x}_G &= F_{x1} + F_{x2} + m_r u_1 \Omega^2 cos(\Omega t + \varphi_1) + m_r u_2 \Omega^2 cos(\Omega t + \varphi_2) \\ m\ddot{y}_G &= F_{y1} + F_{y2} + m_r u_1 \Omega^2 sin(\Omega t + \varphi_1) + m_r u_2 \Omega^2 sin(\Omega t + \varphi_2) - m_r g - S \\ J_S \ddot{\theta}_x &= -l_1 F_{y1} + l_2 F_{y2} + l_S S - d_1 m_r u_1 \Omega^2 sin(\Omega t + \varphi_1) + d_2 m_r u_2 \Omega^2 sin(\Omega t + \varphi_2) - J_P \Omega \dot{\theta}_y \\ J_S \ddot{\theta}_y &= l_1 F_{x1} - l_2 F_{x2} + d_1 m_r u_1 \Omega^2 cos(\Omega t + \varphi_1) - d_2 m_r u_2 \Omega^2 cos(\Omega t + \varphi_2) + J_P \Omega \dot{\theta}_x \end{aligned} \quad (1)$$

Where J_P, J_S, Ω denote the transverse moment of inertia (about x and y axes), the polar moment of inertia (about z axis) and rotational speed of the shaft, respectively and φ_1, φ_2 are the phase difference between the two unbalances.

For more convenience in computation, the equation system (1) can be transformed into a concise form as follows (see the Appendix for more details):

$$\{\varepsilon''\} = \frac{4}{\Omega^2}[C]^{-1}([Q_F]\{F\} + [Q_S]\{S\}) + 4[C]^{-1}([Q_{uc}]cos2\tau + [Q_{us}]sin2\tau)\{U\} + 2[Q_G]\{\varepsilon'\} \quad (2)$$

In which $\left(^{\prime}\right) = \frac{d}{d\tau}(), \left(^{\prime\prime}\right) = \frac{d^2}{d\tau^2}(), \tau = \frac{\Omega}{2}t$

$$\{\varepsilon\} = \left\{\varepsilon_{x1}, \varepsilon_{y1}, \varepsilon_{x2}, \varepsilon_{y2}\right\}^T; \{F\} = \left\{F_{x1}, F_{y1}, F_{x2}, F_{y2}\right\}^T;$$

$$\{U\} = \{u_{1c}, u_{1s}, u_{2c}, u_{2s}\}^T; u_{ic} = u_i cos\varphi_i; u_{is} = u_i sin\varphi_i;$$

$$\{S\} = \begin{cases} \{0, -S - m_r g, |l_S|S, 0\}^T; l_S > 0 \\ \{0, -S - m_r g, 0, -|l_S|S\}^T; l_S < 0 \end{cases}$$

Where $\varepsilon_{xi} = \frac{x_i}{c_i}; \varepsilon_{yi} = \frac{y_i}{c_i}$ are non-dimensional eccentricities of the shaft. F_x, F_y are reaction forces acting onto the shaft from air film and can be determined by integrating the air pressure mesh over the bearing area including the inner circumference and the width L of the bearing.

2.2 Air Pressure Distribution

Air pressure distribution can be obtained by the Reynolds' equation for compressible air [10]:

$$\frac{\partial\psi}{\partial\tau} = \frac{1}{\Lambda}\left\{\begin{array}{c} \frac{\partial}{\partial\theta}\left[\psi\left(\tilde{h}\frac{\partial\psi}{\partial\theta} - \psi\frac{\partial\tilde{h}}{\partial\theta}\right)\right] \\ +\frac{\partial}{\partial\varsigma}\left[\psi\left(\tilde{h}\frac{\partial\psi}{\partial\varsigma} - \psi\frac{\partial\tilde{h}}{\partial\varsigma}\right)\right] - \frac{\partial\psi}{\partial\theta} \end{array}\right\} \tag{3}$$

where:

$$\psi = \tilde{p}.\tilde{h}; \tilde{h} = 1 - \varepsilon_x cos\theta - \varepsilon_y sin\theta + \tilde{w}; \tilde{p} = p/p_a$$

$$\tilde{w} = w/c; \Lambda = \frac{6\mu\Omega}{p_a}\left(\frac{R}{c}\right)^2$$

In which p_a is ambient pressure, c is radial clearance of the bearing, μ is air viscosity, R is the shaft radius and w is the radial deflection of the foil structure.

2.3 Foil Deflection

The foil deflection is governed by the equation [10]:

$$\frac{d\tilde{w}_{(\theta)}}{d\tau} = \frac{2}{\eta}\left(\frac{F_{(\theta)}}{K} - \tilde{w}_{(\theta)}\right) \tag{4}$$

where K is the effective stiffness of the foil structure, while η is the loss factor and $F_{(\theta)}$ is calculated by integrating pressure mesh along the axial direction.

3 Computation Method

The study on the stability of the shaft was conducted through solving the Eqs. (2–4). As done in the previous work [10], finite difference transformation was employed to convert the three sets of equations into a system of ordinary differential equations as follows:

$$\frac{ds}{d\tau} = f(\tau, s, \Omega, m_r, S, \mu, R, L, c, p_a, K, \theta_F) \tag{5}$$

In which $s = \{\psi^{(1)}, \psi^{(2)}, w^{(1)}, w^{(2)}, \{\varepsilon\}, \{\varepsilon'\}\}$ are state variables.

It is notable that the free edge of the foil structure was considered by setting the air pressure along the free edge equal to the ambiant pressure p_a. Moreover, the foil deflection was clearly set to be zero at the fixed edge while the deflection at the free edge $\theta = \theta_F$ was extraploated from a quadratic curve of three neighbouring points.

The system of ordinary differential equations was solved by using the *ode suite* integrated in Matlab® [11]. The solutions can be considered as the transient responses of the system at a defined rotating speed.

Moreover, as mentioned above, the stability problem of the system was also investigated to determine suitable speed ranges of the shaft under the effect of the static load with different arrangements. For that purpose, the Eqs. (5) was calculated with all time-derivative terms set to be zero:

$$f(\tau, s, \Omega, m_r, S, \mu, R, L, c, p_a, K, \theta_F) = 0 \tag{6}$$

These nonlinear equations were solved using Newton-Raphson algorithm. The system is assumed to be in unstable state when any of the eigenvalues of the Jacobian has a positive real part.

4 Results and Discussion

The analysis was conducted for a turbocharger shaft supported by two identical foil-air bearings with some parameters shown in Table 1.

Moreover, the effective stiffness K of the foil structure was employed from the previous study [12].

The shaft stability can be described in Fig. 5 through the variation of the real part of the leading eigenvalue of the Jacobian. As can be seen, the shaft remains stable at the speed up to 10500 revolution per minute (rpm) before transforming into unstable state at bigger speed (first Horf bifurcation). It was also presented in the previous work [10] that at unstable states, the shaft center moves in the orbit of a limit cycle through the transient responses. Based on that, the speed range of the stable state was evaluated through investigating the influence of the static load as well as bearing arrangement.

Table 1. Shaft and bearing parameters

Parameters	Value	Unit
R	19	mm
L	38	mm
μ	1.95 × 10–5	Ns/m^2
p_a	101325	Pa
c	32	Mm
η	0.25	
m_r	3	kg
θ_F	90	o
J_S	9.5×10^{-3}	kgm^2
J_P	2.3×10^{-4}	kgm^2
u	0.003	

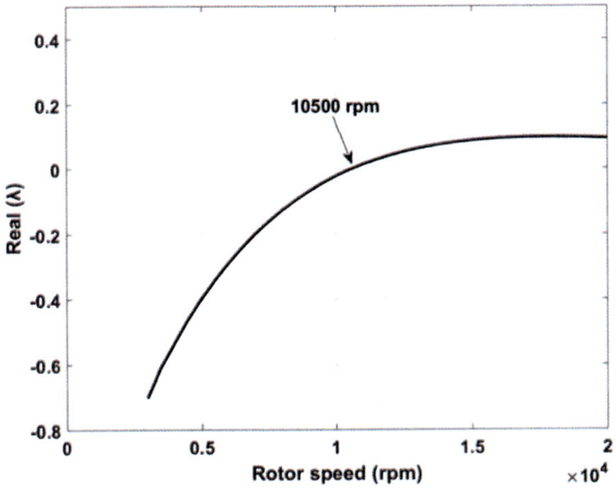

Fig. 5. Stability of the shaft

In this study, the static load was scrutinized for increasing values from 0 to 20 N at different positions on the turbocharger shaft. It is also noted that some dimensions need to be defined as follows: $d = 100$ mm and two foil-air bearings were arranged symmetrically with $l_1 = l_2 = 50$ mm.

Effect of load values on shaft stability is shown in Fig. 6. In this case, the acting point is in the middle (the center of gravity). It can be seen that under growing static loads from 0 to 20 N, the critical speed of the shaft expands from 10500 to 19000 rpm correspondingly, which means that it takes the shaft more time to work in stable state. It can be argued that bigger loads require higher air pressure (greater speed) to push

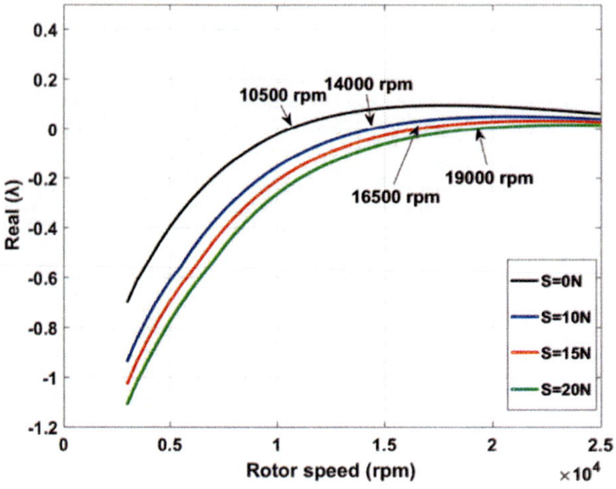

Fig. 6. Shaft stability under increasing static loads at $l_S = 0$

the shaft up, making a slower change in the gap between the shaft and foil (air film). Therefore, the stable state can be maintained longer at higher speed.

Moreover, as mentioned above, different acting points of the static load were investigated by altering the value of l_S as shown in Fig. 7, 8, 9 and 10.

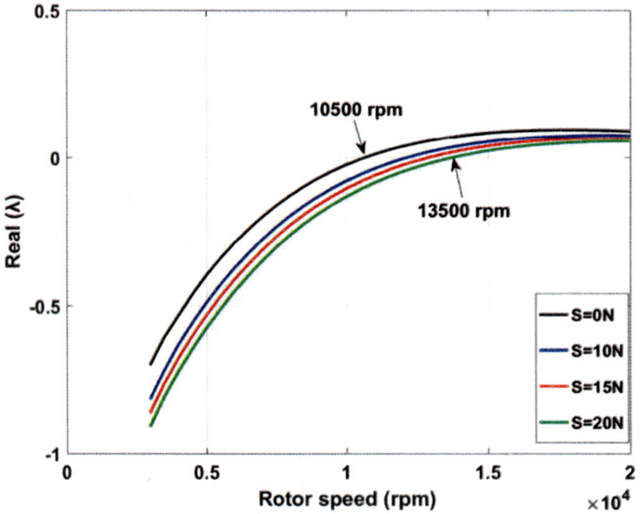

Fig. 7. Shaft stability under increasing static loads at $l_S = 30$

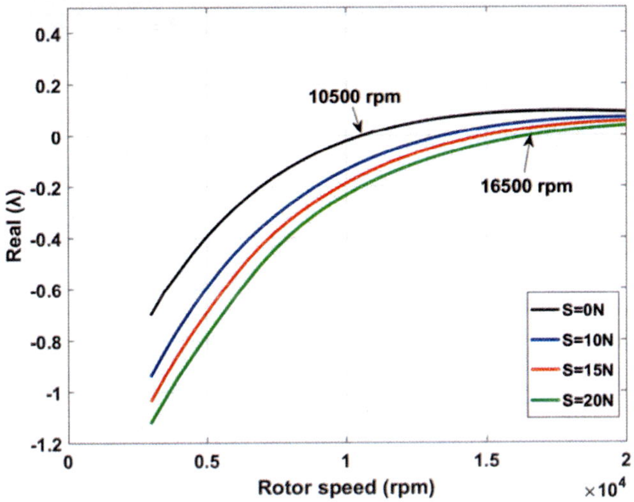

Fig. 8. Shaft stability under increasing static loads at $l_S = -30$

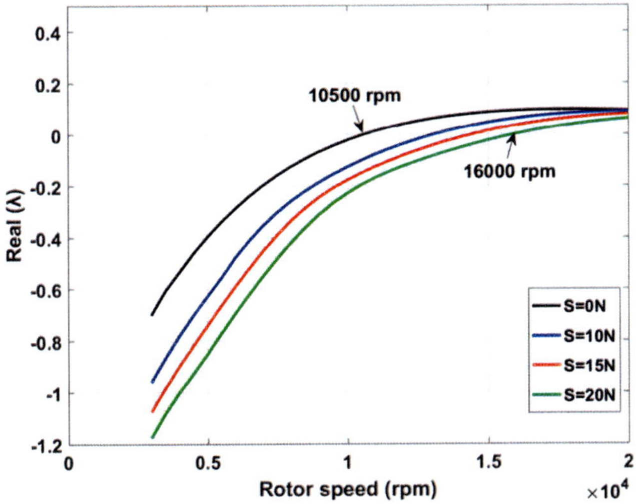

Fig. 9. Shaft stability under increasing static loads at $l_S = -60$

In Fig. 7, 8, 9 and 10, it can be seen that the same tendency is repeated when the critical speeds reach the highest at the load of 20 N. However, these speeds are a little smaller than the case $l_S = 0$. It can be reasoned by the appearance of the bending moment $|l_S|.S$ due to changes in the acting point. This bending moment can be assumed to cause the unstable state occuring earlier beacause the gap between the shaft and foil in one bearing is prone to change.

However, an opposite case is shown in Fig. 11–12 when it is easier for the shaft to turn into unstable state under bigger loads. With the load of 20 N, the critical speeds of

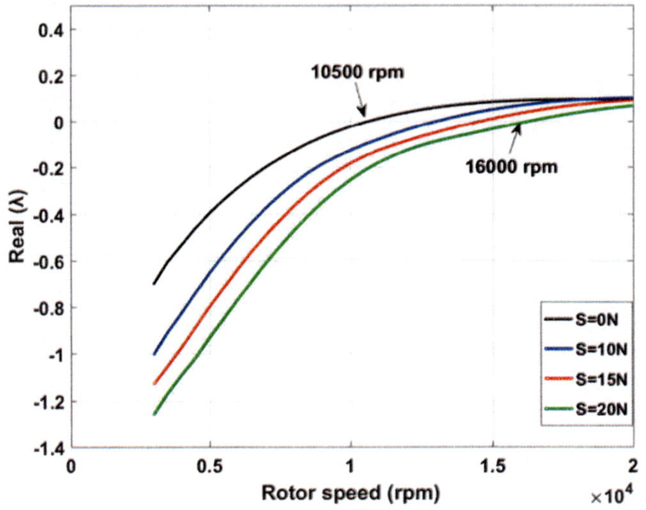

Fig. 10. Shaft stability under increasing static loads at $l_S = -90$

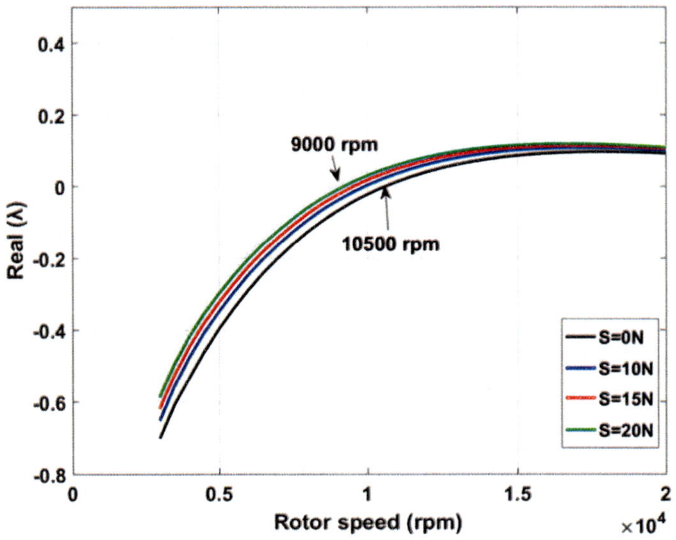

Fig. 11. Shaft stability under increasing static loads at $l_S = 60$

the shaft are 9000 and 5000 rpm in case $l_S = 60$, 90 mm respectively. This phenomenon can be explained by the unbalance radii u_1. As mentioned in Table 1, an unbalance force was set to appear at $d_1 = 50$ mm. Then, together with static load at acting points towards u_1 ($l_S > 0$), a moment of inertia was assumed to appear, causing the gap to change quickly and the shaft is possible to move to unstable state easily.

As can be seen from the figures above, speed range of the shaft at stable state bears a huge impact from the external static load. Based these results, parametric control with

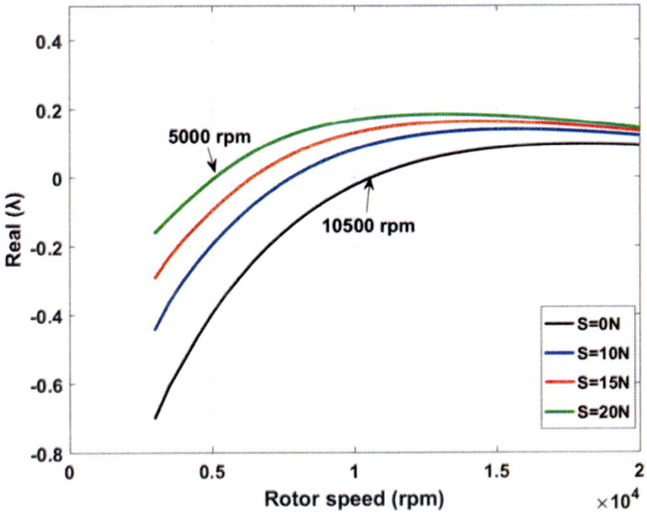

Fig. 12. Shaft stability under increasing static loads at $l_S = 90$

changing load values can be conducted with piezoelectric material to reach a required speed range of shaft at stable state. Moreover, this research together with previous parametric studies on geometry and technological parameters can initiate investigations on optimal design of a shaft-bearing system in a turbocharger.

5 Conclusion

In this paper, dynamic response of a turbocharger shaft supported by two symmetrically arranged foil-air bearings was evaluated under the effect of a static load. The study focused on the stability of the shaft demonstrated as critical speeds. Before that, differential governing equations were established to describe the interaction among the shaft, the foil structure and the air pressure. Finite difference was employed to obtain transient response of the shaft while Newton-Raphson algorithm was applied to solve the stability problem. From that, effect of the static load with distinct values and acting points was analyzed. The results indicate that bigger loads have significant influences on maintaining the shaft stability during rotation. This phenomenon can initiate potential methods of controlling the shaft performance through changing load values. Moreover, acting points of the loads were also investigated with different positions on the shaft. As can be seen, bending moment can be generated from these points and reduce the critical speed of the shaft with the same load values. The unbalance should also be noticed together with some neighboring acting points. In future research, the bearing arrangement of asymmetry and some experiments should be included for investigation.

Appendix

$$[L] = \begin{bmatrix} 1 & 0 & 0 & l_1 \\ 0 & 1 & -l_1 & 0 \\ 1 & 0 & 0 & -l_2 \\ 0 & 1 & l_2 & 0 \end{bmatrix}, [C] = \begin{bmatrix} c_1 & & & \\ & c_1 & & \\ & & c_2 & \\ & & & c_2 \end{bmatrix}, [L]\begin{Bmatrix} x_G \\ y_G \\ \theta_x \\ \theta_y \end{Bmatrix} = [C]\begin{Bmatrix} \varepsilon_{x1} \\ \varepsilon_{y1} \\ \varepsilon_{x2} \\ \varepsilon_{y2} \end{Bmatrix} = \begin{Bmatrix} x_1 \\ y_1 \\ x_2 \\ y_2 \end{Bmatrix}$$

$$[M] = \begin{bmatrix} m_r & 0 & 0 & 0 \\ 0 & m_r & 0 & 0 \\ 0 & 0 & J_S & 0 \\ 0 & 0 & 0 & J_S \end{bmatrix}, [L_F] = \begin{bmatrix} 1 & 0 & 1 & 0 \\ 0 & 1 & 0 & 1 \\ 0 & -l_1 & 0 & l_2 \\ l_1 & 0 & -l_2 & 0 \end{bmatrix}$$

$$[I_G] = \begin{bmatrix} 0 & 0 & 0 & 0 \\ 0 & 0 & 0 & 0 \\ 0 & 0 & 0 & -J_P \\ 0 & 0 & J_P & 0 \end{bmatrix}$$

$$[D_C] = \begin{bmatrix} 1 & 0 & 1 & 0 \\ 0 & 1 & 0 & 1 \\ 0 & -d_1 & 0 & d_2 \\ d_1 & 0 & -d_2 & 0 \end{bmatrix}, [D_S] = \begin{bmatrix} 0 & -1 & 0 & -1 \\ 1 & 0 & 1 & 0 \\ -d_1 & 0 & d_2 & 0 \\ 0 & -d_1 & 0 & d_2 \end{bmatrix}$$

$$[Q_F] = [L][M]^{-1}[L_F], [Q_S] = [L][M]^{-1}$$

$$[Q_{uc}] = m_r[C]^{-1}[L][M]^{-1}[D_C]$$

$$[Q_{us}] = m_r[C]^{-1}[L][M]^{-1}[D_S]$$

$$[Q_G] = [C]^{-1}[L][M]^{-1}[I_G][L]^{-1}[C]$$

References

1. DellaCorte, C., Bruckner, R.J.: Remaining technical challenges and future plans for oil-free turbomachinery. J. Eng. Gas Turbines Power, **133**(4) (2011)
2. Garrett by Honeywell, Turbocharger Guide, Volume 4, www.turbobygarrett.com
3. Walton, J.F., Heshmat, H., Tomaszewski, M.J.: Testing of a small turbocharger/turbojet sized simulator rotor supported on foil bearings. J. Eng. Gas Turbines Power, **130**(3) (2008)
4. Borchert, U., Delgado, A., Szymczyk, J.A.: Development and testing of air foil bearing system for an automotive exhaust gas turbocharger. Appl. Mech. Mater. **831**, 71–80 (2016)
5. Sim, K., Lee, Y.B., Kim, T.H.: Effects of mechanical preload and bearing clearance on rotordynamic performance of lobed gas foil bearings for oil-free turbochargers. Tribol. Trans. **56**(2), 224–235 (2013)
6. Bonello, P., Pham, H.: Nonlinear dynamic analysis of a turbocharger on foil-air bearings with focus on stability and self-excited vibration. Turbo Expo: Power Land, Sea, Air, **45776**, V07BT32A003. American Society of Mechanical Engineers (2014)

7. Mahner, M., Li, P., Lehn, A., Schweizer, B.: Numerical and experimental investigations on preload effects in air foil journal bearings. J. Eng. Gas Turbines Power, **140**(3) (2018)
8. Mahner, M., Bauer, M., Lehn, A., Schweizer, B.: An experimental investigation on the influence of an assembly preload on the hysteresis, the drag torque, the lift-off speed and the thermal behavior of three-pad air foil journal bearings. Tribol. Int. **137**, 113–126 (2019)
9. Żywica, G., Kiciński, J., Bogulicz, M.: Analysis of the rotor supported by gas foil bearings considering the assembly preload and hardening effect. In: Cavalca, K.L., Weber, H.I. (eds.) IFToMM 2018. MMS, vol. 60, pp. 208–222. Springer, Cham (2019). https://doi.org/10.1007/978-3-319-99262-4_15
10. Pham, M.-H., Nguyen, X.-H., Nguyen, M.-Q., Nguyen, D.-N.: A parametric study on the stability of a foil-air bearing-rotor system. In: The 2nd National Conference on Mechanical Engineering and Automation, pp. 232–238 (2016)
11. Shampine, L.F., Reichelt, M.W.: The Matlab ODE suite. SIAM J. Sci. Comput. **18**(1), 1–22 (1997)
12. Nguyen, M.Q., Hai, P.M., Van Phong, D.: Research on effects of geometry parameters on stability zone of a rotor supported by foil-air bearings using an improved foil model. J. Sci. Technol. **130**, 001–006 (2018)

Development of Data Gloves for Humanoid Robot Hand Simulation and Hand Posture Recognition

Do D. Khoa(✉), Ngo N. Vinh, and Nguyen M. Hieu

Hanoi University of Science and Technology, Hanoi 100000, Vietnam
khoa.dodang@hust.edu.vn

Abstract. Human hand movement data set is valuable in many practical applications. Scientists and engineers study human biometric movement data to understand how to interact with the surroundings to mimic the process in humanoid, to create prosthetic for the disables, or to visualize human's motions in entertaining products. Therefore, using a sensor integrated gloves is an effective method to collect human hand movement data. In this paper, the construction of data gloves is presented step by step from the hardware design, embedded data collecting to data processing to rebuild the human hand posture in a visualization tool. The gloves are applied a system of inertial measurement units (IMUs) to track the human hand movement to increase the accuracy, reliability, and the flexibility in comparison to those using other types of sensors. The glove built-in embedded system can collect and process the wrist and phalanges orientation data from the IMUs to model a simplified human hand of 22 degrees of freedom. Using the acquired human hand data, the Multilayer Perceptron algorithm (MLP) data is applied for hand posture recognition with fast and high accurate results.

Keywords: Data gloves · Inertial measurement unit · Hand posture recognition · Multilayer perceptron algorithm

1 Introduction

1.1 Development of Data Gloves

Compared to the current popular methods of using image processing with a camera or depth camera, the use of gloves with a sensor system seems less convenient. However, the conversion of image data to the angle of joints is often difficult to be precise because conventional cameras only collect 2-D data. Even if using multiple cameras or depth cameras is not effective due to many causes such as colored clothing of the user, unstable surrounding light sources. Thus, camera systems require a fixed mounting system and user position to be able to collect gesture and body data without being obstructed. Kangkan Wang, et al. [1] used the depth camera method to reconstruct the figure in their study in 2020, they also acknowledged that the use of the camera in general is greatly influenced by environmental noise. As a result, they were only able to reconstruct the

N. V. Khang et al. (Eds.): ASIAN MMS 2021, MMS 113, pp. 292–301, 2022.
https://doi.org/10.1007/978-3-030-91892-7_27

general figure without hand and finger movements in a limited working environment due to camera visibility. Srinath Sridhar [2] also used a depth camera that focused on collecting hand and finger shape data but only within the scan space of depth camera. Therefore, the use of a sensor integrated device such as gloves or clothing to collect human movements and especially hand and finger movements will bring accurate data needed for machine learning training.

The first glove prototypes to emerge included the Sayre glove, the Massachusetts Institute of Technology (MIT)-LED glove, and the Digital Entry Data Glove [3]. The Sayre Glove [4], which was developed in 1977, used flexible tubes with light source at one end and a photocell at the other, which are mounted along each other finger of the glove. Bending fingers result in decreasing the amount of light passed between the light source and the photodiode. The system thus detected the amount of finger bending using the voltage measured by the photodiode. From its developments in the early 1980s, MIT Data Glove has evolved dramatically offering different capabilities with different models. Currently developed under MIT spinoff company AnthroTronix, AcceleGlove [5], is a user programmable glove that records hand and finger movements in 3D. The other model includes 5DT's Data Glove for virtual reality that cost between $1000–$5000. The company initially developed Data Gloves for controlling robots. Their acceleGlove is also used in video games, sports training, or physical rehabilitation. As shown in Fig. 1a, an accelerometer rests just below each fingertip and on the back of the hand. The accelerometers can detect 3D orientation of the fingers and palm with respect to the gravity when a gesture or any movement is made. The measurement accuracy is within a few degrees that allow to detect slight changes in hand position. The glove openings for fingertips allow users to type or write while wearing the glove.

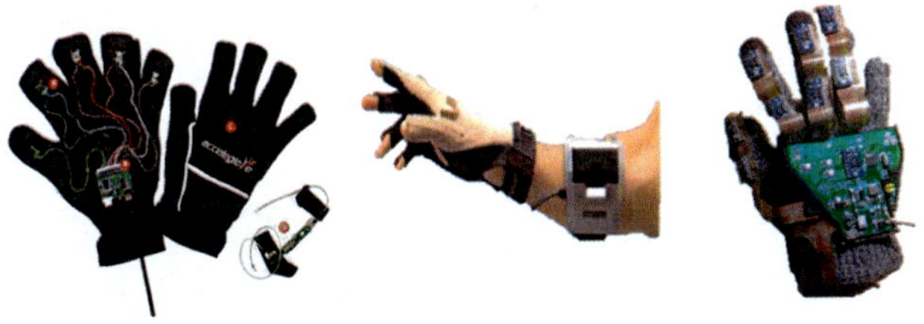

a) b) c)

Fig. 1. Data collecting Glove with multiple sensors

The CyberGlove [6] has been developed to deliver many data inputs due to different flexing of joints motion from other areas of the hand. The 18-sensor data glove features two bend sensors on each finger, four abduction sensors, and sensors measuring thumb crossover, palm arch, wrist flexion, and wrist abduction. A different version of this glove that contains 22 sensors has three flexion sensors per finger. abduction sensors, a palm-arch sensor, and sensors to measure wrist flexion and abduction. Each sensor is extremely thin and flexible making the sensors almost undetectable in the lightweight elastic glove. As shown in Fig. 1b, one version of the glove offers open fingertips that would allow a user to type, write, and grasp objects easily. The CyberGlove motion capture system has been used in many applications including digital prototype evaluation, virtual reality biomechanics, and animation. The motion of the hand can be measured by using optical fiber or some type of flexible resistive sensors. This measurement method utilizes changes in resistance of material due to the stretching and bending of fingers. However, this solution still has some disadvantages. The resistive sensors need a specific force to deliver displacement that makes a user hand movement displeasing. The most important factor in this method is the number of implemented sensors, the more flexible sensors, the more accurate measurement. Moreover, a flexible sensor can only measure the rate of change but cannot indicate the bending direction.

Another version of data gloves made by the Dynamics and control lab in HUST also used IMU sensors to collect the finger position and knuckle joint angles (Fig. 1c) [7]. However, the gloves still had several drawbacks such as flexible copper circuit boards to power and transfer data for sensors is very fragile and easily broken as well as expensive and hard to produce. This version of data glove measures the orientation of sensors and send the raw data to computer for further calculation. Based on the prototype, our team in the same lab delivered some improvements to the product.

1.2 Human Hand Simplified Model

The human hand is an amazing masterpiece that consists of 27 bones and many groups of muscles to control the hand movement. However, some connecting joints between fingers and carpal can hardly move or only move a small displacement and cannot be measured accurately. Therefore, tracking and simulating the precise motion of the human hand is a very difficult task. By simplifying the human hand structure into a multi-rigid-body system [8], such as 13 sensor-applicable parts in this paper, human hand movement can be captured and processed at the acceptable accuracy. Sensor modules are placed in the back of the hand, hand wrist, hand middle knuckles, and top knuckles as shown in the figure below.

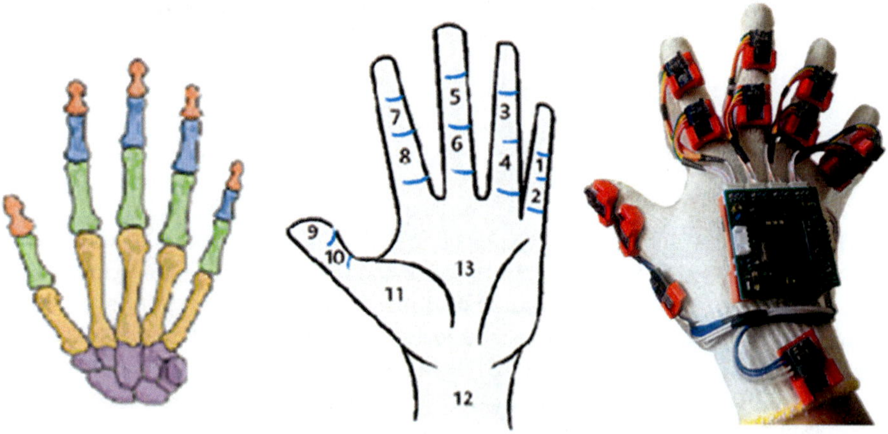

Fig. 2. Sensor position relative to human hand structure

These positions of the human hand are stable and large for sensor placement in the wool glove while delivering measurable relative orientation cooperating sensors in the nearby positions. The movement of the fingertips (red bones in Fig. 2) can be calculated from the relative orientation of middle and top knuckle joint angles (angle of blue and green bone in the Fig. 2.

2 Data-Collecting Glove Design

2.1 The Glove I/O Board Design Hardware Design and Assembly

The VR glove hardware board can be divided into several main blocks with specific functions such as sensor data collection, signal processing transmission to the host computer for further applications such as hand posture modeling, sign language detections or virtual reality environment interaction. Measured data can also be stored in the board micro-SD card. Every pair of sensors which have address 0x28 and 0x29 will be connected to one multiplexer channel via one header in the glove I/O board. For example, to measure the orientation of the index finger's phalanges, we use 2 sensors and place them on the proximal phalanx and middle phalanx. These 2 sensors will share the same SDA line and SCL line and will be wired to channel number 0 of the TCA9548 multiplexer. Like the other fingers, the back of the hand and the hand wrist position. To read the data from all sensors, the microcontroller will control the TCA9548 multiplexer to connect the data line of the controller to every channel. After collecting data from sensors connected to that channel, the next channel will be selected, and the collecting process is repeated [9] (Fig. 3).

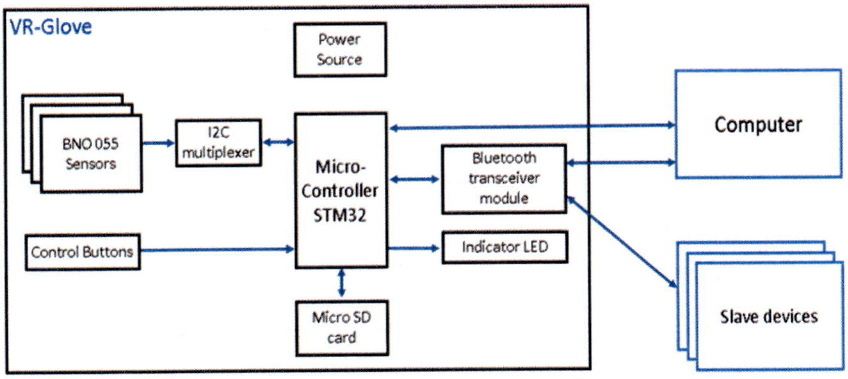

Fig. 3. Virtual reality glove block diagram

2.2 I/O Board Embedded Software

The MCU processes three main tasks: read and process the measured Quaternions the 13 sensors and send to other devices; read and calculate relative angles of sensors, process to send to other devices or store to the micro-SD card; monitor the working processes of sensors and display calibration level.

The first task of the glove I/O board is the process of collecting absolute quaternions of all the sensors, compressing it into a package, and sending it via serial communication port (Fig. 4).

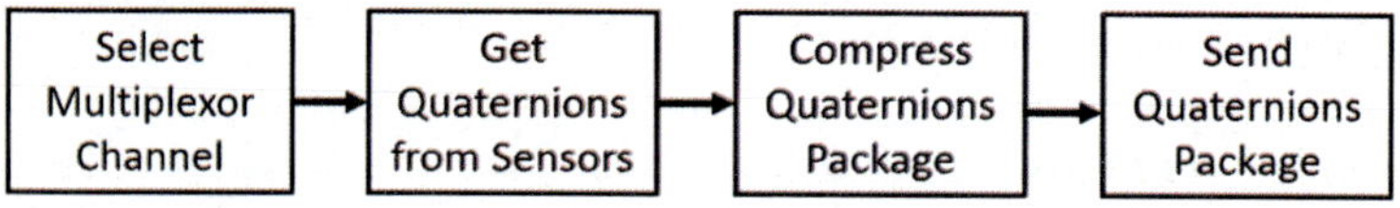

Fig. 4. Task 1: read and process quaternion flowchart

To calculate the relative angles, the microcontroller will do one similar process to the previous task but computing and compressing the relative angles package instead of quaternion package. In the computer, the relative angles are used to remodel the hand posture in the ROS Rviz visual simulation software [10] (Fig. 5).

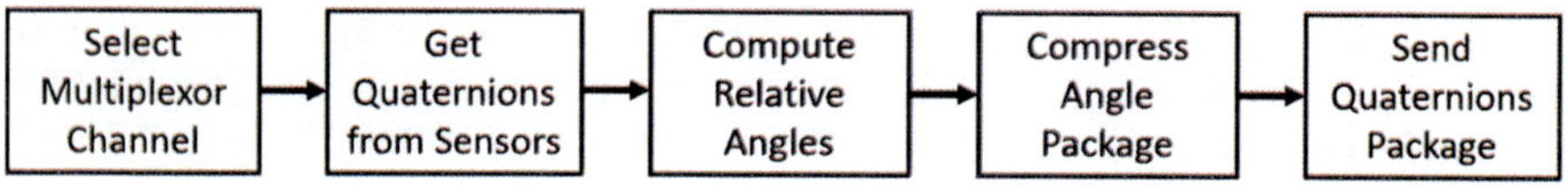

Fig. 5. Task 2: processing relative angles flowchart

3 Hand Posture Calculation

3.1 Composing Hand Posture

As being discussed in the previous section, the human hand model is simplified to study. For joint angles of fingers, each finger needs four relative angles. The first one is the relative angle between distal phalanx and middle phalanx; the second one is the relative angle between middle and proximal phalanx. However, these two angles are usually equal in real life because of our disability of control the distal phalanges separately from other phalanges.

Only two sensors are mounted to middle and proximal phalanges to calculate the relative angle between them, which represents the rotation angle about the X-axis. As shown in Fig. 2, the middle and proximal phalanx sensor of the index finger (sensor number 7 and 8) are used to compute the rotation about X-axis. In each base knuckle joint, there are two possible rotations. Considering the relative orientation of sensor number 8 and number 13 of human hand as denoted in Fig. 2 and Fig. 6, the two rotations are about the X and Z-axis of the sensor coordinate system (roll and yaw rotation). As a result, to represent one finger position, four relative angles are needed to calculate.

Human wrist joint motion can be described by two rotation angles about the X-axis and Z-axis. In this case, the sensor coordinate frame in the wrist position is considered

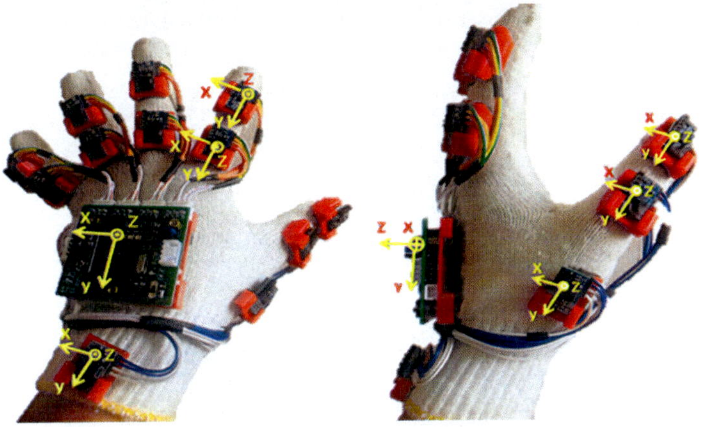

Fig. 6. Rotational orientation of fingers and hand wrist

to be the reference coordinate system. Because of the thumb motion is more flexible and more complex, we need to use one more sensor in the distal phalanx. We need to compute relative rotation angles in the Y-axis and X-axis from the backhand sensor to the thumb proximal phalanx. For the upper part, the X-axis relative angles are calculated. Totally, 4 angles needed to calculate for the thumb orientation. In conclusion, a total of 22 angles in each hand need to be measured and computed.

The method of reconstructing real angles between phalanges is presented by the computation of relative position between inertia measurement units (IMUs). To estimate the exact pose of each finger, two 9-axis IMUs are mounted to the hand's middle phalanx and proximal phalanx.

3.2 Calculating Relative Quaternion

Based on the data from the magnetometer, gyroscope, and accelerometer, the quaternion of each IMU is determined. Finally, the roll, pitch, yaw, or Euler angles showing the relationship of two IMUs are converted from the calculation of the relative position of two quaternions [11].

Assume that the measured quaternions of the 2 sensors are.

$$q_0 = q_{w0} + q_{x0}i + q_{y0}j + q_{z0}k \tag{1}$$

$$q_1 = q_{w1} + q_{x1}i + q_{y1}j + q_{z1}k \tag{2}$$

After obtaining the quaternion of each IMU, the relative quaternion could be calculated by using:

$$q_{01} = q_0^{-1}q_1 \tag{3}$$

The equation can be simplified using the multiplication table of quaternion basic elements, q_{01} can be denoted by:

$$q_{01} = q_{w01} + q_{x01}i + q_{y01}j + q_{z01}k \tag{4}$$

The following equations are used to convert the relative quaternion to Euler angles.

$$
\begin{aligned}
Roll(\varphi) &= \arctan\left(\frac{2(q_w q_x + q_y q_z)}{1 - 2\left(q_x^2 + q_y^2\right)}\right) \\
Pitch(\theta) &= \arcsin(2(q_w q_y + q_z q_x)) \\
Yaw(\psi) &= \arctan\left(\frac{2(q_w q_z + q_x q_y)}{1 - 2\left(q_y^2 + q_z^2\right)}\right)
\end{aligned} \tag{5}
$$

4 Sign Language Recognition

In the previous section, the use of the VR-Glove to collect and process hand posture data from sensors have been presented. The data could be useful for an application of Sign Language Recognition, which is helpful in teaching and communicating with disabled people or in controlling robots. The VR-Glove integrated recognition system will translate 22 angles calculated from any hand movement to corresponding words or characters. The main purpose of this application is to recognize more than 24 characters in the sign language alphabet table using one hand and to recognize some simple words by combination of two hands, then combine these words into a simple meaningful sentence. With one glove, the system can recognize more than 24 characters in the sign language alphabet table, but it is quite hard to detect a complex word which requires the gesture combined of two hands. To handle this problem, a second glove is designed and applied in this system. The combination of two gloves gives 44 relative angles (22 angles for each glove) and these angles could be considered as input data for the recognition system. The recognition task is supposed to be solved by Multilayer Perceptron algorithm which is an AI technique and the dataset collected from the relative angles recorded from hand gestures that need to detect.

4.1 Data Collection

Data is collected from 13 IMUs of the glove. Based on the data from magnetometer, gyroscope, and accelerometer, the quaternion of each IMU is determined and those quaternions are calculated and converted to 22 joint angles or Euler Angle. It means that each point of data could be represented by 22 features or a 22-dimensional vector and 44 features if two hands are used. In order to make our model more adaptive, a large amount of data including all possible orientations in which a user could perform to enact a sign language gesture were recorded. Each gesture made by different users might have minor offsets from the ideal finger positions. Hence, it is highly important to capture these slight deviations from each person for every gesture. After obtaining the filtered data, we collected 1000 samples for each alphabet gesture in the sign language. The dataset contains 34000 samples for all 24 alphabet signs and numbers from 0 to 9. Each sample is represented by a 22-dimensional vector.

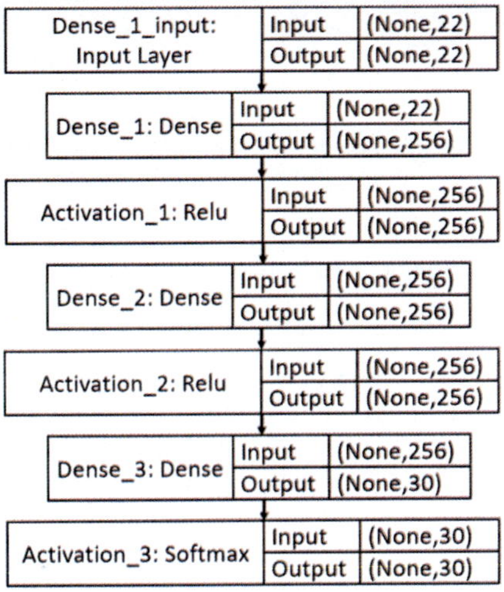

Fig. 7. Graphical description of the MLP classifier

4.2 Detection Using Multilayer Perceptron (MLP)

The Perceptron Learning algorithm is a case of a single-layer neural network with activation functions as sign functions. Meanwhile, Perceptron is a common name for Neural Network with only one input layer and one output layer, and no hidden layer. In addition to input layer and output layer, a multi-layer perceptron may have many hidden layers which are placed between input layer and output layer [12] (Fig. 7).

The input of recognition system will be an n dimensional vector which describe a hand gesture performing by the person wearing glove. "n" is dependent on number of gloves will be used. If one glove is used, n = 22 and if two gloves are used, n = 44. The number of features could be changed to 44 if two data gloves are applied. In the paper a one-dimensional feature vector that contains 22 features is fed into the MLP with three hidden layers. The first hidden layer has 256 output neurons and takes the input of 22 input neurons, which are represented by an input layer. The standard activation functions, Rectified Linear Unit (ReLu), is used to activate. ReLu is very much like a filter that allows positive inputs to pass through unchanged while clamping everything else to zero. The second and third hidden layer also have 256 output neurons and activates by means of ReLu. The number of 256 neurons are chosen since 128, 512 and 1024 neurons have lowered the performance metrics. This way, the model will be capable of learning the most important patterns, which helps generalizing to new data. Finally, the output layer, which has 30 output neurons (26 letters in alphabet and 4 hand signs) is activated by means of Softmax. The 30 neurons correspond to the 30 possible labels, classes, or categories. The idea of Softmax is very simple. It squashes the outputs into probabilities by normalizing the prediction. Here, each predicted output is a probability that the index is the correct label of the given input. The sum of all the probabilities for all outputs is 1.0.

The categorical cross entropy is used as the loss function and the model optimization is carried out by Stochastic gradient descent [13]. Next, the training data is fitted to the

```
8930/8930 [==================================] – 0s 30us/step – loss: 1.1912 –accuracy: 0.9268
Epoch 14/15
8930/8930 [==================================] – 0s 30us/step – loss: 1.0441 –accuracy: 0.9571
Epoch 15/15
8930/8930 [==================================] – 0s 30us/step – loss: 0.9147 –accuracy: 0.9832
Test loss: 0.8434
Test accuracy 0.9907
```

Fig. 8. The training process result

model and 15 epochs, or the number of iterations is selected before it stops training, a batch size of 128. The result shows training loss decreases rapidly as shown in Fig. 8. This is perfectly normal, as the model always learns most during the early stages of optimization.

4.3 Recognition Result

After getting the training model, the user could start predicting phase (testing phase). Data sent from 13 IMU sensors will be calculated and be used as input for predicting. In this paper, two options are offered for recognizing task:

- Single predict mode: Used for detecting a single character
- Multiple predict mode: Used for predicting words or even a sentence (Fig. 9).

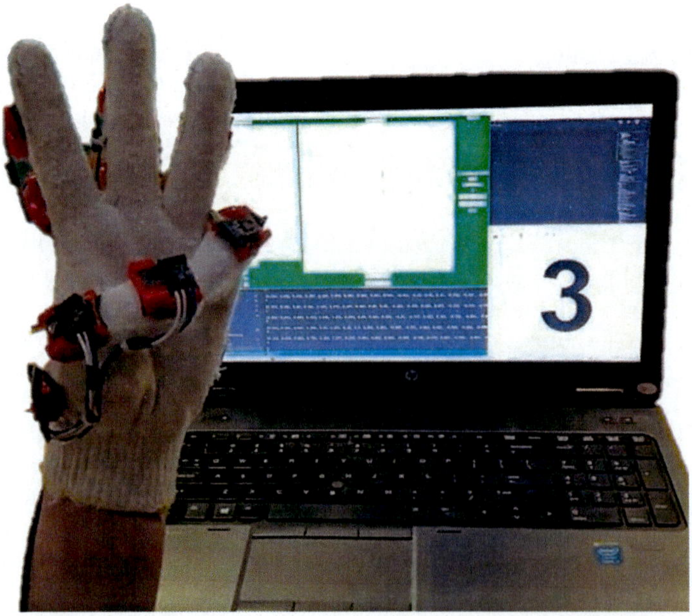

Fig. 9. Hand recognition results using the single predict mode

5 Conclusion

Applying the IMU to track the human hand movement is an excellent method to increase the accuracy and reliability. In the paper, the VR-Gloves design process is presented step by step, from hardware to software of the glove I/O board. Utilizing the data gloves, the sign language recognition application using data collected from the IMUs have been implemented.

In compared with the previous model of Sensorized-Glove [7], this version has more durability by replacing the flexible printed circuit board with wire connection and 3D printed sensor attachment while providing embedded data processing to calculate hand posture in the glove's mainboard. More than that, by utilizing Multilayer Perceptron Algorithm, sign language recognition can be carried out to both hands at the same time and improve the recognition performance.

Acknowledgement. This research is funded by Hanoi University of Science and Technology (HUST) under project number T2021-PC-038.

References

1. Wang, K., Zhang, G., Yang, J., Bao, H.: Dynamic human body reconstruction and motion tracking with low-cost depth cameras (2020)
2. Sridhar, S., Oulasvirta, A., Theobalt, C.: Fast tracking of hand and finger articulations using a single depth camera (2014)
3. Dipietro, L., Sabatini, A.M., Dario, P. : A survey of glove-based systems and their applications (2008)
4. Liu, H., Xi, X., Millar, M., Edmonds, M.: A glove-based system for studying hand-object manipulation. IEEE Comput. **14**(1), 30–39 (2016)
5. Asada, H.H., Mascaro, S.: Fingernail touch sensors: spatially distributed measurements and hemodynamic modeling, pp. 30–39 (1994)
6. Cyber Glove System (2020). https://www.cyberglovesystems.com/. Accessed June 2020
7. Quang, N.M., Anh, L.H., Long, N.D., Anh, N.D.T., Khoa, D.D.: Design and analysis of a sensorized glove with 9-axis IMU sensors for hand manipulation capture (2019)
8. Resell, J., Suarez, R., Garcia, C., Perez, J.A.: Motion planning for high DOF anthropomorphic hands (2018)
9. Kelvin, T.: Adafruit BNO055 absolute orientation sensor (2020)
10. Joseph, L.: Robot operating system (ROS) for absolute beginners (2018)
11. CH Robotics: Understanding quaternions (2012)
12. Aston, Z.: Dive to deep learning
13. Ruder, S.: An overview of gradient descent optimization algorithms (2017)

A New Class of Spring Four-Bar Mechanisms for Gravity Compensation

Vu Linh Nguyen[1](✉), Chin-Hsing Kuo[2], and Po Ting Lin[3,4]

[1] Department of Mechanical Engineering, National Chin-Yi University of Technology, Taichung 411030, Taiwan
vlnguyen@ncut.edu.tw

[2] School of Mechanical, Materials, Mechatronic and Biomedical Engineering, University of Wollongong, Wollongong, NSW 2522, Australia

[3] Department of Mechanical Engineering, National Taiwan University of Science and Technology, Taipei 10607, Taiwan

[4] Center for Cyber-Physical System Innovation, National Taiwan University of Science and Technology, Taipei 10607, Taiwan

Abstract. This paper presents a new class of spring four-bar mechanisms, with fifty-six different mechanism types, for gravity compensation. A synthesized mechanism is realized by linkage combination with a planar four-bar linkage and attachment of a linear spring. The main advantages of the proposed spring mechanisms are their simple architectures and high performances that possess great potential for practical applications. The parameters of the mechanisms for gravity compensation are derived from an optimization approach. Numerical examples are given to demonstrate the performances of the proposed spring mechanisms. The simulation results showed that the motor torques of the mechanisms could be reduced by up to ninety-nine percent with the spring attachment.

Keywords: Gravity compensation · Static balancing · Mechanism synthesis · Planar four-bar linkage · Zero-stiffness mechanism

1 Introduction

Gravity compensation of a mechanism or a mechanical system is a technique that allows reducing the gravity effect on its links and payload [1–4]. Complete gravity compensation can keep a mechanism immovable at any position without input forces and torques. It can also decrease energy consumption, the size of the actuators, the structural compliance of the mechanism, and improve safety and dynamic response [3].

Gravity compensation is usually realized by installing counterweights or/and springs onto a mechanism to eliminate the gravitational torques caused by the link masses and payload [2, 4]. First, counterweight methods intend to make the total center of mass of a mechanism stationary during operation [5, 6]. Second, spring methods exploit the elastic energy stored by springs to get rid of the variation of the potential energy [7–10]. As compared with counterweights, the use of springs is preferred as it has less mass and

N. V. Khang et al. (Eds.): ASIAN MMS 2021, MMS 113, pp. 302–312, 2022.
https://doi.org/10.1007/978-3-030-91892-7_28

inertia added to the mechanism [1, 3]. However, the biggest challenges of using springs are the resulting complex mechanical structure and large volume required for the spring implementation [1, 3].

Numerous spring methods have been applied to mechanisms for gravity compensation. For instance, springs could be implemented by using a noncircular pulley and wire mechanism [11], cam-follower mechanism [12], Cardan gear mechanism [13], noncircular-gear mechanism [14], inverted slider-crank mechanism [15], or geared five-bar mechanism [16]. However, each of the mechanisms described above undergoes its own limitations, such as the large volume required to implement the spring, many auxiliary components, low durability induced by the pulley and wire transmission, inadequate motion accuracy, or friction. Although intensive endeavors have been dedicated to developing gravity compensation mechanisms, the synthesis of such mechanisms has rarely been explored in the literature.

This paper synthesizes a new class of gravity compensation mechanisms from the linkage combination with planar four-bar linkages and spring attachment. The main advantages of the proposed mechanisms are their simple architectures and high performances of gravity compensation. Numerical examples are presented to demonstrate the performances of the proposed mechanisms.

2 Mechanism Construction

2.1 Linkage Combination

This section presents the combination of a single-degree-of-freedom (1-DoF) rotating link (with a mass to be gravity compensated) and a linkage (where a linear spring is attached). Since the rotating link has only one DoF motion, choosing a planar four-bar linkage is convenient. For example, a planar four-bar linkage with four revolute (R) joints is considered, as depicted in Fig. 1. Here, assume that the rotating link is fixed to link 4 as a crank whose motion is driven by the gravitational force. Links 1, 2, and 3 are named the fixed base, follower, and coupler link, respectively.

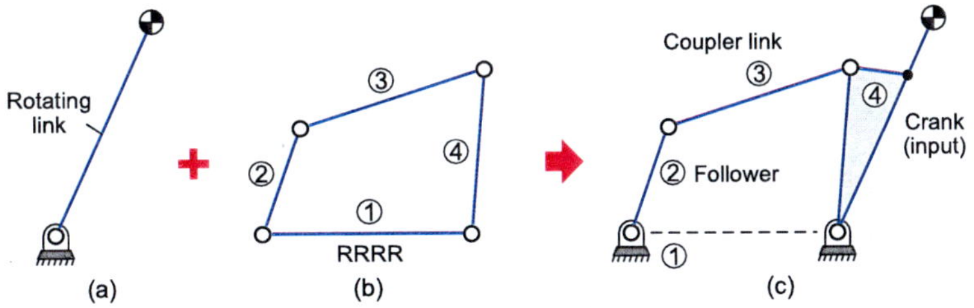

Fig. 1. Linkage combination: (a) a rotating link, (b) a planar four-bar (RRRR) linkage, and (c) a combined linkage with an attached weight to be gravity compensated.

Alternatively, other planar four-bar linkages can replace the four-bar (RRRR) linkage (Fig. 1(b)). One constraint for these linkages is that they must hold at least one revolute joint, which is supposed to be situated on link 1 and coincides with the pivot point

O of the rotating link. Six other planar four-bar linkages are satisfied with the above constraint, as shown in Fig. 2.

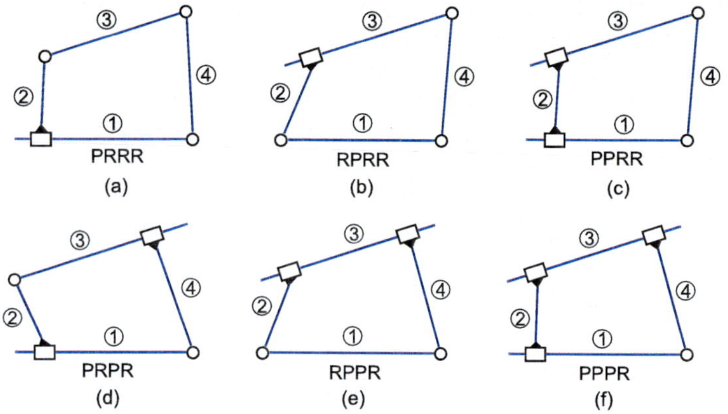

Fig. 2. Six other planar four-bar linkages for linkage combination.

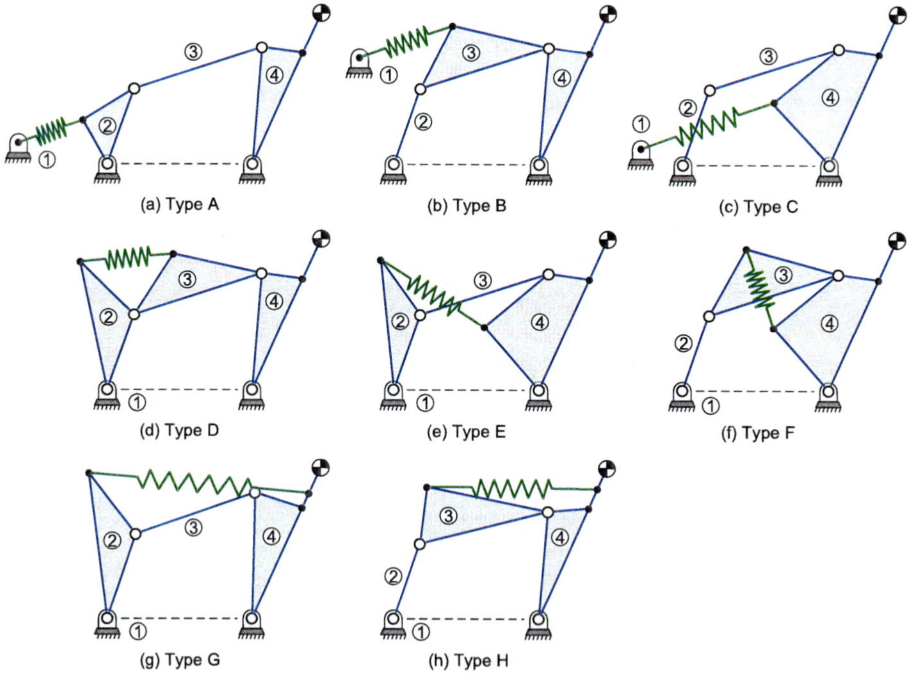

Fig. 3. Eight different types of spring four-bar mechanisms in Group I.

2.2 Spring Attachment

Let us consider the combined four-bar RRRR linkage (Fig. 1(c)). An extension spring is then attached to the mechanism to compensate for the gravitational force induced by the carried weight. As illustrated in Fig. 3, eight different types of spring mechanisms can be obtained. For ease of illustration, they can be grouped as Group I.

When another four-bar linkage (Fig. 2) is considered, the construction of its spring mechanisms can be realized similarly. This procedure can result in another group with eight other types of spring mechanisms. In sum, the above-described linkage combination and spring attachment can generate a class of spring four-bar mechanisms with seven groups and fifty-six different mechanism types.

3 Gravity Compensation Design

This section presents the gravity compensation design of the spring four-bar mechanisms. Here, types G and H in Group I (Figs. 3(g) and 3(h)) are studied, but for brevity, only type G is shown below.

3.1 Static Modeling

Figure 4 illustrates the free-body diagram of the spring four-bar mechanism (type G). Assume that the rotating link is limited within a range of joint angles $\theta = (\theta_{\text{int}}, \theta_{\text{end}})$.

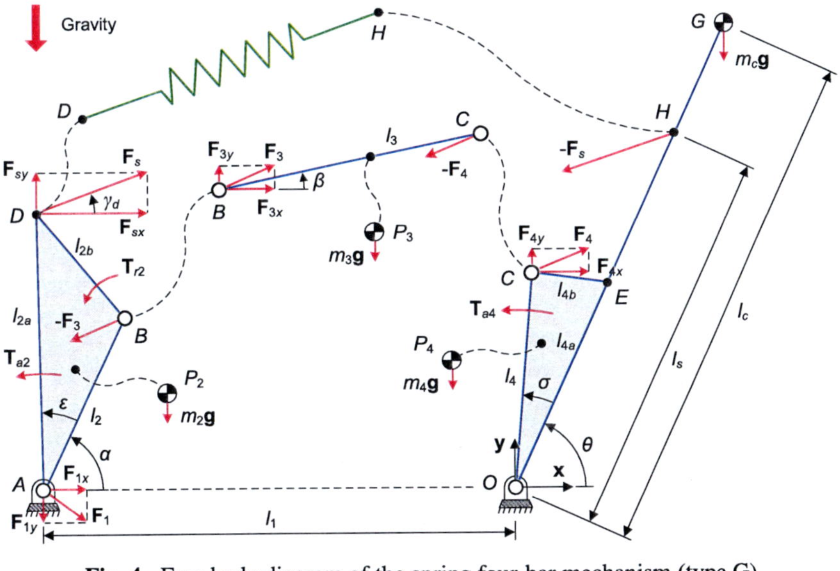

Fig. 4. Free-body diagram of the spring four-bar mechanism (type G).

Let F_s denote the spring force applied at point D on link 2. The same force F_s in inverse direction is also generated at point H on the rotating link. The generated spring force F_s can be calculated as:

$$F_s = k(s_1 + s_0) \quad (1)$$

where

$$s_1 = d_1 - d_0,\ d_0 = DH|_{\theta=\theta_{\text{int}}} = \|\mathbf{H} - \mathbf{D}\|_{\theta=\theta_{\text{int}}},\ d_1 = DH|_{\theta=(\theta_{\text{int}},\theta_{\text{end}})} = \|\mathbf{H} - \mathbf{D}\|_{\theta=(\theta_{\text{int}},\theta_{\text{end}})} \quad (2)$$

In Eqs. (1)–(2), k, s_0, and s_1 stand for the stiffness coefficient, the initial extension, and the instantaneous extension of the spring, respectively; d_0 and d_1 the lengths of the spring at the initial and instantaneous positions, respectively; l_i ($i = 1, 2, \ldots, 4$) the length of link i; l_{ia} and l_{ib} ($i = 2$ or 4) the lengths of non-straight link i; $l_s = OH$.

By observing Fig. 4, one can express the spring force vector $\mathbf{F}_s$ as:

$$\mathbf{F}_s = \begin{bmatrix} F_{sx} \\ F_{sy} \end{bmatrix} = \begin{bmatrix} F_s \cos \gamma_d \\ F_s \sin \gamma_d \end{bmatrix} \quad (3)$$

where

$$\gamma_d = \cos^{-1}\left(\frac{\mathbf{x} \cdot \boldsymbol{\mu}_s}{\|\mathbf{x}\| \|\boldsymbol{\mu}_s\|}\right),\ \boldsymbol{\mu}_s = \frac{\mathbf{H} - \mathbf{D}}{\|\mathbf{H} - \mathbf{D}\|},\ \mathbf{x} = \begin{bmatrix} 1 \\ 0 \end{bmatrix} \quad (4)$$

From the free-body diagram method [17], the equilibrium of forces for a body link i in the mechanism under static conditions can be expressed as:

$$\sum_j \mathbf{F}_{ij} + \sum_k \mathbf{F}_{ik} + \mathbf{F}_{ai} + m_i \mathbf{g} = 0 \quad (5)$$

where $\mathbf{F}_{ij}$, $\mathbf{F}_{ik}$, and $\mathbf{F}_{ai}$ represent the reaction force vector, the external force vector exerted on link i, and the input force vector from a linear actuator, respectively; m_i and $\mathbf{g}$ the mass of link i and the gravitational acceleration vector, respectively.

Next, the equilibrium of moments for link i can be written as:

$$\sum_j (\mathbf{r}_{ij} \times \mathbf{F}_{ij}) + \sum_k (\mathbf{r}_{ik} \times \mathbf{F}_{ik}) + m_i(\mathbf{r}_{ci} \times \mathbf{g}) + \mathbf{T}_{ai} + \mathbf{T}_{ri} = 0 \quad (6)$$

where $\mathbf{r}_{ij}$, $\mathbf{r}_{ik}$, and $\mathbf{r}_{ci}$ stand for the position vectors associated with the reaction force $\mathbf{F}_{ij}$, the external force $\mathbf{F}_{ik}$, and the mass center of link i, respectively; $\mathbf{T}_{ai}$ and $\mathbf{T}_{ri}$ the input torque vector and the reaction torque vector on link i, respectively.

Let $\boldsymbol{\Omega}_X = [F_{1x}, F_{1y}, F_{3x}, F_{3y}, F_{4x}, F_{4y}, T_{r2}]^T$ be a wrench vector representing the resulting forces and moments exerted on the mechanism. Applying the equilibrium equations of forces and moments (Eqs. (5) and (6)) to links (2, 3) and joints (A, B) can yield:

$$\boldsymbol{\Omega}_A \cdot \boldsymbol{\Omega}_X = \boldsymbol{\Omega}_B \quad (7)$$

where

$$\boldsymbol{\Omega}_A = \begin{bmatrix} 1 & 0 & 1 & 0 & 0 & 0 & 0 \\ 0 & 1 & 0 & 1 & 0 & 0 & 0 \\ 0 & 0 & 1 & 0 & -1 & 0 & 0 \\ 0 & 0 & 0 & 1 & 0 & -1 & 0 \\ 0 & 0 & y_B - y_A & x_A - x_B & 0 & 0 & 0 \\ y_A - y_B & x_B - x_A & 0 & 0 & 0 & 0 & 1 \\ 0 & 0 & 0 & 0 & y_C - y_B & x_B - x_C & -1 \end{bmatrix} \quad (8)$$

$$\boldsymbol{\Omega}_B = \begin{bmatrix} F_{sx} \\ F_{sy} - m_2 g \\ 0 \\ m_3 g \\ (y_D - y_A)F_{sx} - (x_D - x_A)F_{sy} + (x_{P2} - x_A)m_2 g \\ (y_D - y_B)F_{sx} - (x_D - x_B)F_{sy} + (x_{P2} - x_B)m_2 g \\ (x_{P3} - x_B)m_3 g \end{bmatrix} \quad (9)$$

When $\boldsymbol{\Omega}_A$ is nonsingular and invertible, $\boldsymbol{\Omega}_X$ can be derived from Eq. (7) as:

$$\boldsymbol{\Omega}_X = \boldsymbol{\Omega}_A^{-1} \cdot \boldsymbol{\Omega}_B \quad (10)$$

Once the wrench vector $\boldsymbol{\Omega}_X$ is found, the input torque T_{a4} at joint O can be determined by applying the equilibrium equation of moments (Eq. (6)) to link 4 as:

$$T_{a4} = -x_C F_{4y} + y_C F_{4x} + x_{P4} m_4 g + x_H F_{sy} - y_H F_{sx} + x_G m_c g \quad (11)$$

where m_c represents the mass of the rotating link.

3.2 Optimization Problem

Let T_m and T_{ms} denote the motor torques at joint O for actuating the rotating link only (uncompensated) and the entire spring four-bar mechanism (compensated), respectively. From Eq. (11), T_m and T_{ms} can be calculated as:

$$T_m = x_G m_c g \quad (12)$$

$$T_{ms} = T_{a4} = -x_C F_{4y} + y_C F_{4x} + x_{P4} m_4 g + x_H F_{sy} - y_H F_{sx} + x_G m_c g \quad (13)$$

From Eqs. (12) and (13), the torque reduction ratio (TRR) δ_t can be defined as:

$$\delta_t = \frac{\max(|T_{ms}|)}{\max(|T_m|)} \quad (14)$$

Equation (14) implies that the lower the TRR δ_t is, the better the torque reduction is received.

Then, the objective of the optimization problem is to minimize the TRR δ_t and is formulated as:

$$\begin{aligned}
&\text{Minimize } f(\mathbf{x}) = \delta_t|_{\theta_{\text{int}} \le \theta \le \theta_{\text{end}}} \\
&\text{with} \quad \mathbf{x} = [l_1,\ l_2,\ l_3,\ l_4,\ l_s,\ l_{2a},\ l_{2b},\ l_{4a},\ l_{4b},\ k,\ s_0]^T \in \Re \\
&\text{subject to } L_i \le x_i \le U_i\ (i = 1, 2, ..., 11)
\end{aligned} \tag{15}$$

where L_i and U_i represent the lower and upper bounds of the ith component in the design variable vector **x**, respectively.

4 Numerical Examples

This section gives numerical examples to illustrate the performance of the spring four-bar mechanisms (types G and H). The link masses of the mechanisms and the mass center of the rotating link are listed in Table 1. Assume that the rotating link is rotated within a range [0, π/2].

Table 1. Link masses and dimension of the spring four-bar mechanisms.

m_c (kg)	m_2 (kg)	m_3 (kg)	m_4 (kg)	l_c (m)
0.5	0.05	0.05	0.05	0.2

The bounds of the design variables of the two mechanisms are listed in Table 2. The optimization problem (Eq. (15)) was solved by using the Genetic Algorithm (GA) in MATLAB Toolbox. Ten independent runs for each type were performed, and the best objective (i.e., the TRR δ_t) among these ten runs is selected for representation, as presented in Table 2.

Table 2. Design variables and optimization results of the spring four-bar mechanisms.

Parameters	Unit	Bound		Optimal values	
		Lower	Upper	Type G	Type H
δ_t	–	–	–	0.015	0.02
l_1	m	0.02	0.06	0.0331	0.0593
l_2	m	0.02	0.06	0.0348	0.0495
l_3	m	0.02	0.06	0.0554	0.0570
l_4	m	0.02	0.06	0.0219	0.0463
l_s	m	0.06	0.18	0.1790	0.1736
l_{2a}	m	0	0.06	0.06	–

(continued)

Table 2. *(continued)*

Parameters	Unit	Bound		Optimal values	
		Lower	Upper	Type G	Type H
l_{2b}	m	0	0.06	0.0261	–
l_{3a}	m	0	0.06	–	0.0569
l_{3b}	m	0	0.06	–	0.0287
l_{4a}	m	0	0.06	0.0186	0.0198
l_{4b}	m	0	0.06	0.0033	0.0483
k	N/m	100	2000	740.7	817.5
s_0	m	0	0.06	0	0.0052

The convergence of the best objective and the motor torques of the two spring mechanisms are illustrated in Figs. 5(a) and 5(b), respectively. One can see that the two mechanisms can produce a similar best objective value, but type H converges at the iteration of 325, which is faster than type G with 590 iterations. These two mechanisms can provide compensated torques that are significantly lower than the uncompensated torque. The optimal configurations of the obtained mechanisms are plotted in Fig. 6.

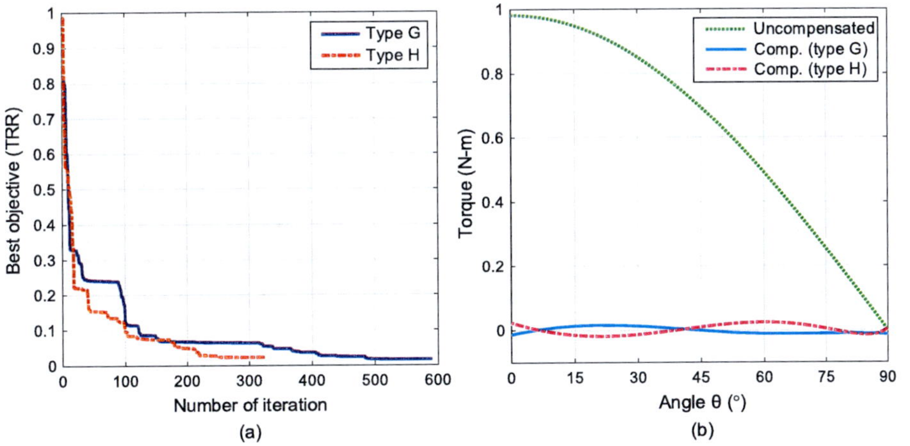

Fig. 5. Optimization results of the spring four-bar mechanisms: (a) convergence of the best objective and (b) uncompensated and compensated torques.

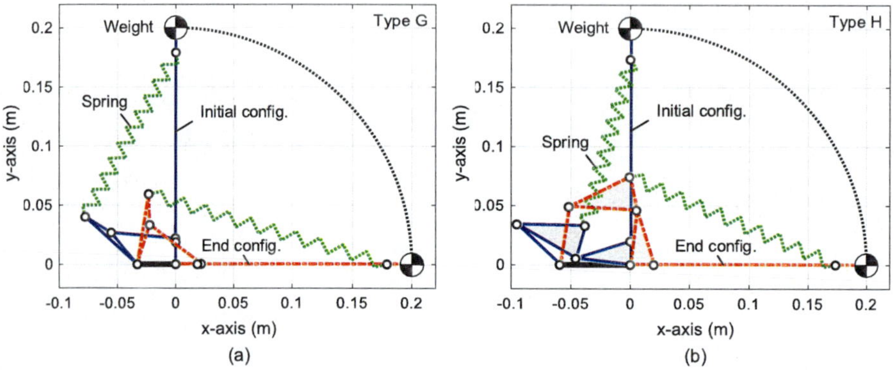

Fig. 6. Optimal configurations of the spring four-bar mechanisms: (a) type G and (b) type H.

MSC Adams software was used to validate the performance of the two above-obtained spring mechanisms. In this simulation, the mechanisms were set to move from the initial position $\theta = \pi/2$ to the end position $\theta = 0$ within 9 s, as shown in Fig. 7. The simulated motor torques of the mechanisms are exhibited in Fig. 8. As can be seen, the simulated torque curves of the two mechanisms look very similar to their calculated torque curves (Fig. 5(b)). The simulated TRR could also reach $\delta_t = 0.01$ by type G and $\delta_t = 0.02$ by type H.

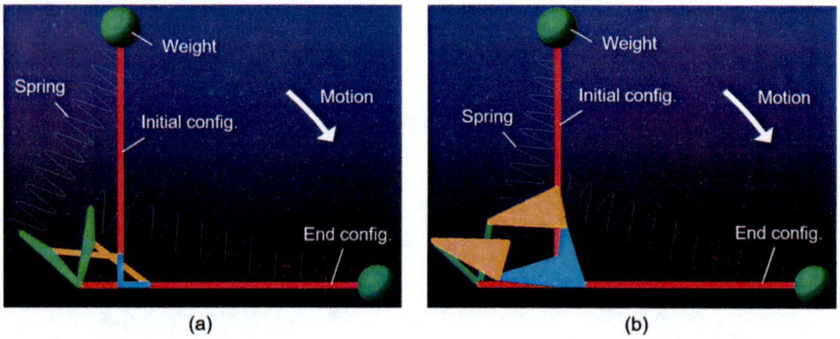

Fig. 7. Simulation models of the spring four-bar mechanisms: (a) type G and (b) type H.

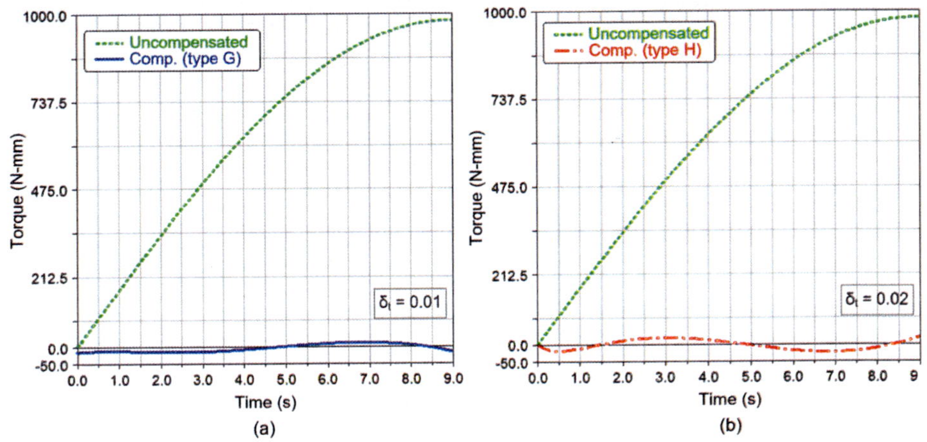

Fig. 8. Simulated torques of the spring four-bar mechanisms: (a) type G and (b) type H.

5 Conclusion

This paper presented a class of spring four-bar mechanisms for gravity compensation. Fifty-six different mechanism types were synthesized by linkage combination and spring attachment in sequence. An optimization approach was proposed to determine the parameters of the mechanisms for gravity compensation. Numerical examples were given to demonstrate the performances of the spring mechanisms. It was shown that the spring mechanisms require much less motor torque, and a torque reduction rate of up to ninety-nine percent was achieved by using type G.

Acknowledgment. This paper was supported by the Ministry of Science and Technology (MOST), Taiwan (grant number MOST 110-2222-E-167-004).

References

1. Arakelian, V.: Gravity compensation in robotics. Adv. Robot. **30**(2), 79–96 (2016)
2. Herder, J. L.: Energy-free systems: theory, conception and design of statically balanced spring mechanisms. Ph.D. thesis, Delft University of Technology, The Netherlands (2001)
3. Carricato, M., Gosselin, C.: A statically balanced Gough/Stewart-type platform: conception, design, and simulation. ASME J. Mech. Robot. **1**(3), 031005 (2009)
4. Arakelian, V., Briot, S.: Balancing of Linkages and Robot Manipulators: Advanced Methods with Illustrative Examples. Springer, Heidelberg (2015). https://doi.org/10.1007/978-3-319-12490-2
5. Van der Wijk, V.: Design and analysis of closed-chain principal vector linkages for dynamic balance with a new method for mass equivalent modeling. Mech. Mach. Theory **107**, 283–304 (2017)
6. Gosselin, C.M., Wang, J.: On the design of gravity-compensated six-degree-of-freedom parallel mechanisms. In: IEEE International Conference on Robotics and Automation (ICRA), Leuven, Belgium, 20 May 1998, pp. 2287–2294 (1998)

7. Essomba, T.: Design of a five-degrees of freedom statically balanced mechanism with multi-directional functionality. Robotics **10**(1), 11 (2021)
8. Nguyen, V.L., Lin, C.-Y., Kuo, C.-H.: Gravity compensation design of Delta parallel robots using gear-spring modules. Mech. Mach. Theory **154**, 104046 (2020)
9. Kuo, C.-H., Nguyen, V.L., Robertson, D., Chou, L.-T., Herder, J.L.: Statically balancing a reconfigurable mechanism by using one passive energy element only: a case study. ASME J. Mech. Robot. **13**(4), 040904 (2021)
10. Kuo, C.-H., Lai, S.-J.: Design of a novel statically balanced mechanism for laparoscope holders with decoupled positioning and orientating manipulation. ASME J. Mech. Robot. **8**(1), 015001 (2016)
11. Fedorov, D., Birglen, L.: Differential noncircular pulleys for cable robots and static balancing. ASME J. Mech. Robot. **10**(6), 061001 (2018)
12. Takesue, N., Ikematsu, T., Murayama, H., Fujimoto, H.: Design and prototype of variable gravity compensation mechanism (VGCM). J. Robot. Mechatron. **23**(2), 249–257 (2011)
13. Hung, Y.-C., Kuo, C.-H.: A novel one-DoF gravity balancer based on Cardan gear mechanism. In: Wenger, P., Flores, P. (eds.) New Trends in Mechanism and Machine Science. Mechanisms and Machine Science, vol. 43, pp. 261–268. Springer, Cham (2017). https://doi.org/10.1007/978-3-319-44156-6_27
14. Bijlsma, B.G., Radaelli, G., Herder, J.L.: Design of a compact gravity equilibrator with an unlimited range of motion. ASME J. Mech. Robot. **9**(6), 061003 (2017)
15. Arakelian, V., Zhang, Y.: An improved design of gravity compensators based on the inverted slider-crank mechanism. ASME J. Mech. Robot. **11**(3), 034501 (2019)
16. Nguyen, V.L., Lin, C.-Y., Kuo, C.-H.: Gravity compensation design of planar articulated robotic arms using the gear-spring modules. ASME J. Mech. Robot. **12**(3), 031014 (2020)
17. Budynas, R.G., Nisbett, J.K.: Shigley's Mechanical Engineering Design, 9th edn. McGraw-Hill, New York (2011)

Development and Identification of Working Parameters for Threshing Unit of Peanut Tuber Picking Machine

May Van Vo Tran[1], Lu Minh Le[2], Hoa Phan[1], Minh Vuong Le[3], Van-Hai Nguyen[4,5], and Tien-Thinh Le[4,5](✉)

[1] Hue University of Agriculture and Forestry, 102 Phung Hung, Thuan Thanh, Hue, Thua Thien Hue, Vietnam

[2] Faculty of Engineering, Vietnam National University of Agriculture, Gia Lam, Hanoi 100000, Vietnam

[3] Laboratoire Modélisation et Simulation Multi Echelle, Université Paris-Est, MSME UMR 8208 CNRS, 5 bd Descartes, 77454 Marne-la-Vallée, France

[4] Faculty of Mechanical Engineering and Mechatronics, PHENIKAA University, Yen Nghia, Ha Dong, Hanoi 12116, Vietnam
thinh.letien@phenikaa-uni.edu.vn

[5] PHENIKAA Research and Technology Institute (PRATI), A&A Green Phoenix Group JSC, No. 167 Hoang Ngan, Trung Hoa, Cau Giay, Hanoi 11313, Vietnam

Abstract. The threshing unit is the most important part in the peanut picking machine and determining its working principle is crucial. This research focuses on the development and design of a threshing unit with two counter-rotating tooth drums for harvesting fresh groundnuts. Such research of the threshing unit based on initial parameters involving the peanut plant are currently lacking in the literature. Improving the performance of the picking machine, especially the threshing unit can significantly improve the peanut yield. Based on experimental tests, we find that parameters such as rotational speed of the beating drum, the amount of supply and the gap between the two teeth of the beating drum all affect the productivity and quality of the harvesting process. The experimental tests also give suitable values of the drum speed, the gap between the two thresher teeth and the peanut supply as a basis for the multi-factor experiment to determine the optimal parameters and objective functions of groundnut picking machine.

Keywords: Picking systems · Picking machine · Single factor · Suitable ranges · Threshing unit · Peanuts

1 Introduction

The mechanization of peanut harvesting, especially the tuber picking, is an important factor to increase productivity and quality, lower product costs, and contributes to improving the competitiveness of peanut products in our country compared to other countries in the region and the world [1]. The technical requirements of the peanut threshing machine

N. V. Khang et al. (Eds.): ASIAN MMS 2021, MMS 113, pp. 313–323, 2022.
https://doi.org/10.1007/978-3-030-91892-7_29

during the harvest are high yield, guaranteed quality (residuality, cleanliness, especially limited tuber remaining, root peduncle stickiness after plucking and tuber breakage) and low energy costs [2, 3].

The threshing unit is the most important part in the peanut picking machine [4]. Determining the working principle of the threshing unit is important to the successful design and manufacture of a peanut picking machine. In the current practice of harvesting peanuts, many principles are used for the threshing unit. Zaied et al. [5] have developed a powered groundnut harvesting machine by optimizing the diggers by using ANSYS software. In another study, Azmoodeh-Mishamandani et al. [6] have determined effect of soil moisture content, forward speed and conveyor slope on pods loss in peanut harvesting. Similarly, they also compared the loss of peanut harvesting in manual and mechanical methods [7]. Wang et al. [8] have designed and tested a picking machine using spiral arc panel mechanism. Besides, Xiao Lian [9] have analyzed half-feed picking roller including structural parameters of picking blades. Finally, a comprehensive review of current situation and analysis on peanut picking technology and equipment can be referred to Huang et al. [10].

However, each principle has certain limitations in terms of residuals, product cleanliness, tuber breakage and high energy cost after extraction [6]. In the recent literature, the studies on improving the working performance of the threshing unit are quite lacking. An un-efficient threshing unit can affect the performance of the whole picking machine, leading to lower yield. Therefore, the study to propose the working principle of the threshing unit of the peanut peeling machine to improve productivity and ensure product quality after picking has both scientific and practical significance [11, 12].

The objective of this work is to determine the influence and optimal value range of crucial working parameters as a basis for designing and manufacturing of two-tooth peanut root picker that rotates in the opposite direction of supplying straight peanuts. For this purpose, experimental studies have been carried out to determine the working parameters of the machine involving the plant, tuber and picking forces. The performance of the threshing unit will then be studied based on these parameters.

2 Materials

There are many varieties of peanuts in Vietnam, but the most common variety is L14. The process of removing the groundnut tuber from the plant is done by the impact force (impact) between the smashing part and the tuber block when the peanut is inserted into the tuber. The groundnut tuber can be removed from the plant in three positions (Fig. 1): The contact position between the tuber and the root (whose force is F_{dc}), the contact position between the root and the stump (whose force is F_{dg}), and the contact position along the length of the root peduncle (whose force is F_{dr}).

3 Methods

3.1 Measurement of Critical Forces

The instruments used for measuring these forces were PASCO ME-8088 threshing force gauge; including PASCO CI-6746 force gauge, EXTECH 302213 converter, Science

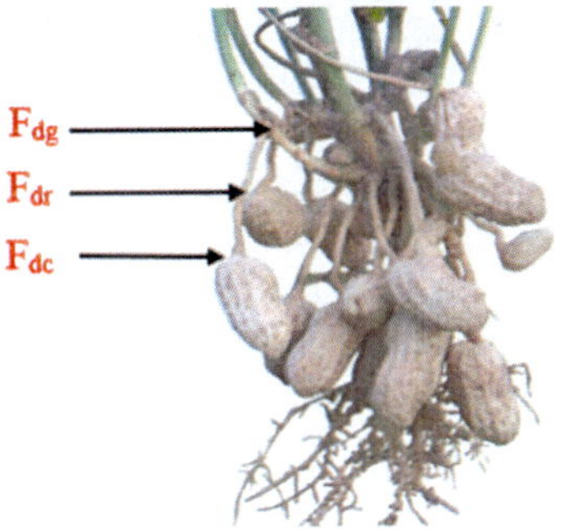

Fig. 1. Diagram for determination of different critical forces.

Workshop 750, microcomputer to run the software, mechanical force gauges NK-500, and electronic force gauges Sauter KF50 (50N/0.02N) (see also Fig. 2). The process of experimental arrangement and measurement of the rooting force were carried out at the laboratory of Department of Physics, Hue University of Education.

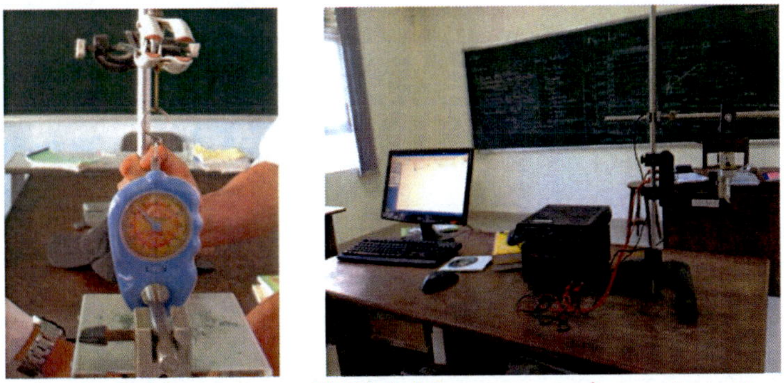

(a) Mechanical force gauge (b) Digital force gauge and computer system

Fig. 2. Instruments for determination of critical forces.

3.2 Selection of Working Principle and Working Parameters

The working principle of the threshing unit of the groundnut picking machine must ensure high productivity, the product must be clean after removing the tuber (avoiding root stalk on the tuber as well as the root and leaf stem); the tuber breakage rate must be low (reducing the number of broken tubers partial outer shell and broken tubers) and lower energy costs [4, 9]. To ensure the above requirements, the machine must have the following general principles:

- Only remove the tubers from the tree;
- Do not affect the trunk and leaves of peanuts and limit the impact on the root stalk;

- Minimize the amount of root peduncles plucked by tubers; limit roots, stems and leaves following the tuber block after being plucked;
- No damage to tubers.

In this way, the tubers will be clean, reducing energy costs when plucking, reducing energy costs after plucking and ensuring the quality of the rooting process.

The working principle of the threshing unit must ensure the removal of the tuber from the root so that the force of the tuber must act on the position between the tuber and the root peduncle (the shoulder of the tuber, along the length of the tuber). At that position, the force will be minimal and separate the tuber from the root, so it will limit the root stalk to follow the tuber as well as the root stalk and leaves, garbage following the tuber after plucking. Therefore, the working principle of the threshing unit of the peanut peeling machine has been selected consisting of two counter-rotating drums, round teeth, multi-cushioned tooth distribution, and the direction of supplying peanuts is vertical (as shown in Fig. 3s and 4). With this raw material, it will meet the requirements of peeling tubers and overcome the disadvantages and limitations of current peanut peeling machines.

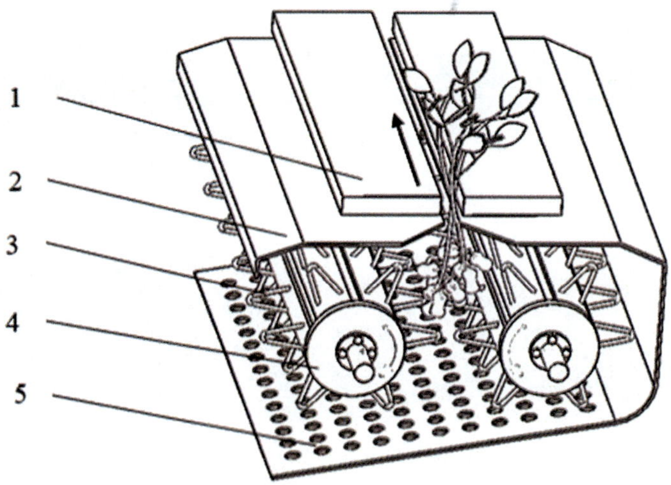

Fig. 3. Beater with 2 counter-rotating tooth drums: 1. Plant clamps; 2. Shields; 3. Teeth beat; 4. Drum beats; 5. Product collector

In order to have a basis for manufacturing threshing parts with this principle, it is necessary to define parameters such as:

- Diameter of drum;
- Drum length;
- Number of teeth;
- Shape and size of threshing teeth;
- The arrangement of teeth on the drum;
- Rotation speed of beating drum;

- Quantity of peanuts supplied to the tree;
- Tooth apex distance between two drums;

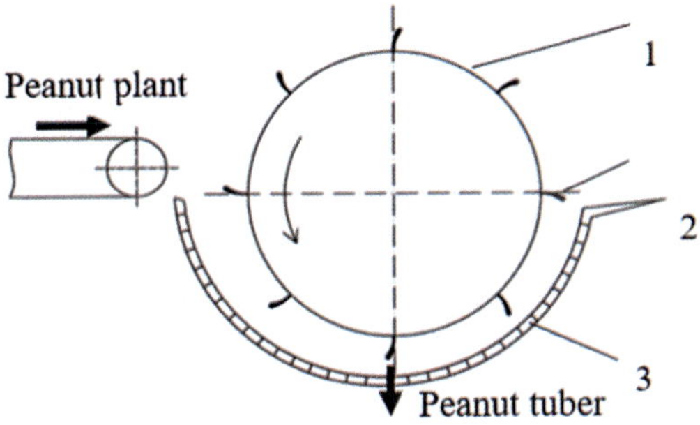

Fig. 4. Diagram of the principle of the drum-type shaft transverse beater: 1. Drum beats; 2. Teeth beat; 3. Drum trough

4 Results and Discussion

4.1 Results of Measuring the Force for Pulling the Tubers Out of the Tree

The results of the measurement of the rooting force are shown in Table 1.

Table 1. Values of picking force of rooting, root peduncle

Nr	Force	Notation	Value (N)
1	Picking force at the position between tuber and root peduncle	F_{dc}	16.76
2	Picking force at root peduncle	F_{dr}	32.45
3	Picking force at the position between the root peduncle and the main root	F_{dg}	18.85
4	Breaking force of tubers	F_{dv}	35.20

The experiments showed that the force of tuber separation at the position between the tuber and the root peduncle was the smallest among the three possible contact points to remove the tuber from the plant. Thus, to pluck peanuts out of the roots and plants, when using the threshing method, we find that: if we hit the right place of contact between the tuber and the root, the force is the lowest, only 16.76 N. This force also separates the tubers from the roots and plants right at the root and tuber positions, ensuring the cleanliness of the product after plucking and not sticking to the root of the product after separating the tubers. However, this force is very close to the force between the root peduncle and the main root (18.85 N), so it is necessary to impact at the contact position between the tuber and the root stem (the shoulder of the tuber) to pull the groundnut out of the root, not to turn it off. Therefore, when the peanut plant is supplied to the vertical tuber, the tuber will be arranged along the length of the stem and beaten by the tooth drum, with a reasonable arrangement of teeth plucked at the next position. This is the basis for determining the working principle of the rooting part and the method of supplying peanuts.

4.2 Working Parameters

To study the influence of working parameters and modes of the peanut threshing unit, we design and manufacture the experimental threshing unit with the main specifications determined on basis of analysis, selected from previous research results as well as results of determining physical and mechanical parameters of peanuts. In particular, from the width and height of the peanut tuber distribution on the tree, the diameter of the drum will be determined to ensure that with 2 drums, each drum will be able to break one side of the tuber block when it is put in. With single factor experiments to determine the influence and range of values of each parameter, it is the basis for determining the height, diameter of the beating teeth, drum speed, power and number of revolutions of the motor. At the same time, based on the peanut tuber diameter, the tuber mass ratio per plant and the yield of the threshing unit. We have determined the gap between the two drums (thereby determining the distance between the two tooth peaks of the two drums), the quantity of peanuts supplied to the threshing, etc. From that, the threshing is manufactured as shown in Fig. 3 and has the basic specifications as indicated in Table 2.

Table 2. Working parameters for threshing unit

Nr	Parameter	Unit	Value
1	Number of drums		2
2	Diameter of drum	mm	100
3	Length of drum	mm	500
4	Number of teeth		8
5	Height of teeth	mm	40
6	Width of tooth feet	mm	25
7	Diameter of steel wire for teeth	mm	4
8	Power of beating drum	kW	0.75
9	Rotation speed of beating drum	rpm	1450

4.3 Parametric Study to Determine Optimal Working Parameters

The parameters affecting the yield and quality of groundnut picking are mainly: Rotation speed of the threshing drum, the amount of supply and the gap between the teeth of the two threshing drums. According to preliminary exploratory research results, the values of these parameters are usually in the range:

- Rotation speed of the drum beater, n = 500–700 (rpm), equivalent to a long speed of 4.71–6.60 (m/s);
- Quantity supplied, q = 0.5–1.5 (kg/s);
- The clearance between the two ends of the drum teeth, δ = 5–25 (mm);
- Product quality is assessed by the following parameters:
- Product loss, including: survival rate (Y_S) is the number of groundnuts left on the tree after picking and tuber failure rate (Y_V) is the number of peanuts with broken outer shell or broken seeds after picking;
- Product cleanliness (Y_r) is the percentage of tubers still attached to the root stalk after plucking and the percentage of broken roots, stems and leaves with the product. However, in this trial, we focused on collecting data and determining the cleanliness according to the percentage of tubers that were still attached to the root after plucking. In case the groundnut tuber, after being plucked, still has a lot of roots attached to it, it will cost more energy and labor to separate the root stem from the tuber to prepare for the preliminary processing and subsequent processing of the groundnut.

To determine the degree of influence on the yield and quality of groundnut picking, we conducted a single factor test to determine the appropriate value range for each input parameter, namely the rotation speed of the drum threshing, the amount of supply and the gap between the two ends of the beating drum teeth. On that basis, a multifactorial experiment will be conducted to determine the optimal value of the input parameters of the peanut peeling machine.

4.3.1 Effect of Drum Beater Rotation Speed on Product Quality

Experiments to determine the effect of rotational speed of the tubers on yield and product quality were conducted with the following conditions: quantity supplied q = 1.0 kg/s; the gap between the two ends of the drum teeth, δ = 15 mm. The rotation speed of the drum beater is changed in 5 levels: 500, 550, 600, 650 and 700 rpm. The experimental results are shown in Fig. 5.

From Fig. 5, when the rotation speed of the drum is set at 500 rpm, the percentage of groundnut tubers still attached to the root is 20,05% and the percentage of missing leaves is also relatively large, 2.52%. At this speed, the tuber breakage rate was 0.40%. When increasing the speed of the drum beater to 550 rpm, the root peduncle adhesion rate decreased to 15.84%, the percentage of missed cuttings decreased by 2.33%, and the percentage of tuber breakage increased but not significantly. Continuing to increase the drum beater speed to 600 rpm, the percentage of tubers sticking to the root was reduced to 13.20%, the percentage of surviving bulbs also decreased to 1.95%, and the percentage of tubers breaking increased to 1.83%. With a drum speed of 650 rpm, the percentage of tubers still attached to the root and the percentage of tuber breakage increased but not much. When the speed of the drum was increased to 700 rpm, the percentage of tubers sticking to the rhizome increased to 15.05%, the percentage of missing increased slightly and the percentage of tubers broken increased but only 1.96%.

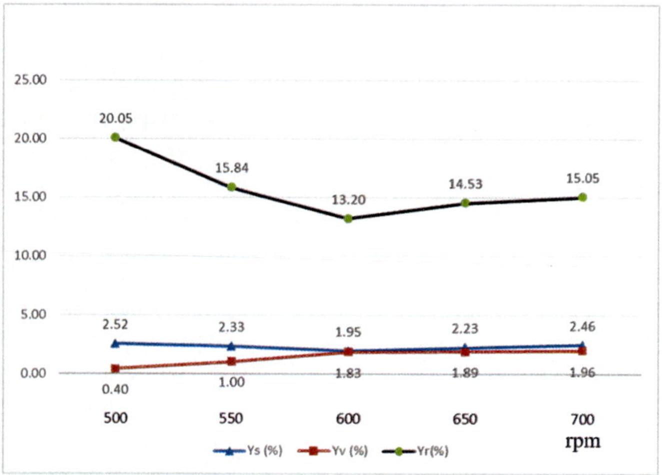

Fig. 5. Percentage of tubers remaining in function of rotation speed of the drum

Experimental results show that, with the speed of the drum beating in the range of 550–650 rpm, the rate of root stickiness and tuber survival is the lowest; tuber breakage rate remained below 2.0%. In conclusion, to ensure productivity and product quality, the appropriate value range of drum speed is n = 550–650 rpm.

4.3.2 Effect of Quantity Supplied (Yield) on Product Quality

Experiments to determine the effect of supply quantity on productivity and product quality are conducted with the following conditions: rotational speed of the threshing drum, n = 600 rpm; the gap between the two teeth of the beating drum, $\delta = 15$ mm. The amount of supply is changed in 5 levels: 0.5; 0.75; 1.0; 1.25 and 1.5 kg/s. The experimental results are shown in Fig. 6.

From Fig. 6, it can be seen that: when the amount supplied to the rooting part increased from 0.5 kg/s to 1.5 kg/s, the percentage of tubers still attached to the root stalk, the percentage of surviving tubers and broken tubers all increased. Specifically, the percentage of tubers still attached to the peduncle increased from 12.53% to 16.05%, the percentage of tubers left unrooted increased from 1.85% to 4.06%, the percentage of tubers broken increased from 0.93%. to 2.18%. But when the supply is in the range of 0.75 kg/s to 1.25 kg/s, the percentage of tubers still sticking to the root, the survival rate and the broken tuber is still within the allowable limit. Therefore, the range of values suitable for the quantity supplied is q = 0,75–1,25 kg/s.

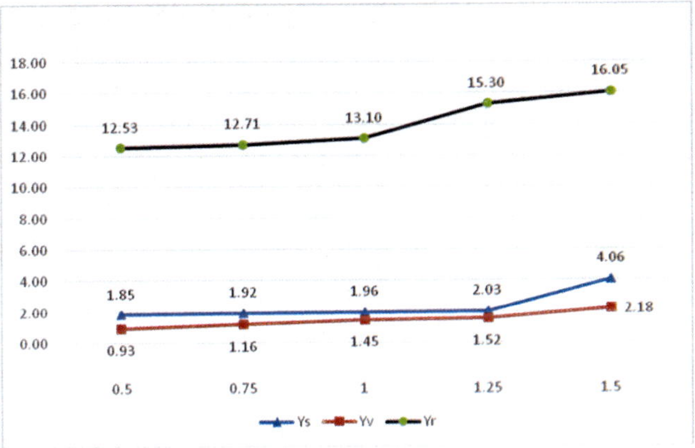

Fig. 6. Percentage of tubers remaining in function of the amount of supply

4.3.3 Effect of Clearance Between Two Ends of Drum Teeth on Product Quality

Experiments to determine the influence of the gap between the two ends of the drum teeth on the productivity and product quality are carried out with the following conditions: rotating speed of the beating drum, n = 600 rpm, quantity supplied, q = 1.0 kg/s. The gap between the two ends of the beating drum teeth is ranging between 5, 10, 15, 20 and 25 mm. The experimental results are shown in Fig. 7.

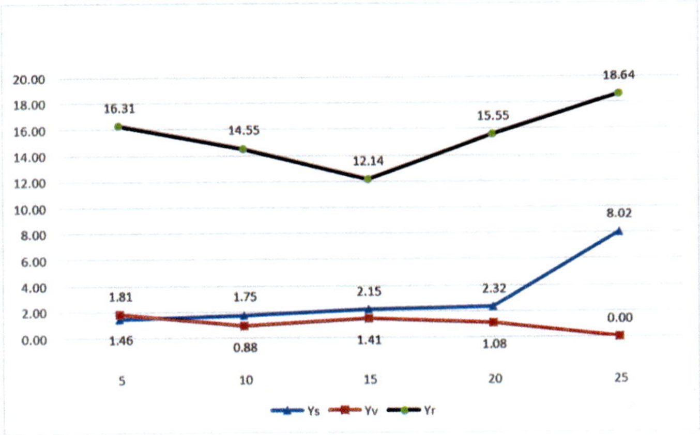

Fig. 7. Percentage of peanuts remaining in function of the gap between two ends of the drum teeth

From Fig. 7, when the gap between the two ends of the drum teeth is 5 mm, the percentage of peanuts that still stick to the root after plucking is 16.31% and the percentage of missing leaves is 1.46%. At this rate, the tuber breakage rate was 1.81%. When increasing the clearance to 10 mm, the percentage of root peduncle adhesion decreased to 14.55%, the rate of tuber breakage also decreased to 0.88%, and the percentage of missing shoots increased to 1.75%. When increasing the clearance to 15 mm, the percentage of root peduncles was greatly reduced, down to 12.14%, the rate of tuber breakage increased to 1.41%, and the percentage of missing shoots increased insignificantly (2.15%). When increasing the clearance to 20 mm, the percentage of root peduncle adhesion increased to 15.55%, the rate of tuber breakage decreased to 1.08% and the percentage of missing shoots increased to 2.32%. When increasing the clearance to 25 mm, the root peduncle adhesion rate skyrocketed to 18.64%, the percentage of missing shoots also increased to 8.02% and no more tuber breakage occurred. Therefore, to ensure productivity and product quality, the appropriate value range of the gap between the two ends of the beating drum teeth is $\delta = 10$–20 mm.

Thus, through the test results, we find that parameters such as rotational speed of the beating drum, the amount of supply and the gap between the two teeth of the beating drum all affect the productivity and quality of the process. In which, with the values of these parameters in the optimal region, the product will meet the quality requirements. Meanwhile, the percentage of tubers still attached to the root peduncle is less than 16.0%, the percentage of tubers remaining and breaking fluctuates between 2.0%, which is satisfactory for the process of rooting.

5 Conclusion

The experimental results of the peanut threshing unit with two counter-rotating tooth drums show that it is possible to give good results in terms of product quality and reduce specific energy costs. Through single-factor experiment, the reasonable values of the parameters affecting productivity and product quality have been determined: the drum beat speed is in the range of 550–650 rpm, the gap between the two teeth which is 10–20 mm and the supply is 0.75–1.25 kg/s. This is the basis for conducting multi-factor experimental planning to determine the optimal value of the working parameters of the peanut root picker.

However, these are only single factor experiments on the influence of parameters on the quality of groundnut picking. It is necessary to conduct further multifactorial experiments to have an accurate and comprehensive assessment of the parameters to the objective functions and the machine's ability to work. In addition, a computational model such as discrete element modeling can be used for the design of the peanut picking machine. It can significantly reduce the cost of experimental work and improve the quality of the machine. Moreover, the comparison between the drum velocity and the peanut size to be ripped out should be investigated in the further experimental works.

References

1. Aydin, C.: Some engineering properties of peanut and kernel. J. Food Eng. **79**, 810–816 (2007). https://doi.org/10.1016/j.jfoodeng.2006.02.045
2. Liu, C.-C., Tellez-Garay, A.M., Castell-Perez, M.E.: Physical and mechanical properties of peanut protein films. LWT Food Sci. Technol. **37**, 731–738 (2004). https://doi.org/10.1016/j.lwt.2004.02.012

3. Le, T.-T., Miclet, D., Heritier, P., Piron, E., Chateauneuf, A., Berducat, M.: Morphology characterization of irregular particles using image analysis. Application to solid inorganic fertilizers. Comput. Electron. Agric. **147**, 146–157 (2018). https://doi.org/10.1016/j.compag.2018.02.022
4. Alexandru, T., Glodeanu, M., Boruz, S., Popescu, S.: Experimental research on mechanized harvesting of peanuts in Romania. In: Engineering for Rural Development (Latvia). Latvia University of Agriculture (2011)
5. Zaied, M.B., Naim, A.M.E., Dahab, M.H., Mahgoub, A.S.: Development of powered groundnut harvester for small and medium holdings in North Kordofan State in Western Sudan. World J. Agric. Res. **2**, 119–123 (2014). https://doi.org/10.12691/wjar-2-3-6
6. Azmoodeh-Mishamandani, A., Abdollahpoor, S., Navid, H., Moghaddam, M.: Performance evaluation of a peanut harvesting machine in Guilan province. Iran. Int. J. Biosci. **5**, 94–101 (2014). https://doi.org/10.12692/ijb/5.10.94-101
7. Azmoodeh-Mishamandani, A., Abdollahpoor, S., Navid, H., Moghaddam Vahed, M.: Comparing of peanut harvesting loss in mechanical and manual methods. Int. J. Adv. Biol. Biomed. Res. **2**, 1475–1483 (2014)
8. Wang, D., Shang, S., Han, K.: Design and test of picking mechanism in 4HJL-2 peanut combines. Trans. Chin. Soc. Agric. Eng. **29**, 15–25 (2013)
9. Lü, X.L., Hu, Z.C., Peng, B.L.: Analysis and research on the picking roller of the half-feed peanut combine harvester. Appl. Mech. Mater. **597**, 502–506 (2014). https://doi.org/10.4028/www.scientific.net/AMM.597.502
10. Zhi-chao, H.: Present situation and analysis on peanut picking technology and equipment. Comput. Sci. (2012)
11. Kurt, C., Arioglu, H.: Physical and mechanical properties of some peanut varieties grown in mediterranean environment (2018). https://doi.org/10.2478/cerce-2018-0013
12. Le, T.-T.: Investigation of force transmission, critical breakage force and relationship between micro-macroscopic behaviors of agricultural granular material in a uniaxial compaction test using discrete element method. Part. Sci. Technol. **2021** (2021). https://doi.org/10.1080/02726351.2021.1983904

Mechanical Design of Drill Pipe Inspection Machine

Quoc Anh Tran[3], Van Tu Duong[1,2,3], Hoang Long Phan[1,2], Van Sy Le[4], and Tan Tien Nguyen[1,2,3](✉)

[1] Faculty of Mechanical Engineering, Ho Chi Minh City University of Technology (HCMUT), 268 Ly Thuong Kiet, District 10, Ho Chi Minh City, Vietnam
nttien@hcmut.edu.vn
[2] Vietnam National University Ho Chi Minh City, Linh Trung Ward, Thu Duc District, Ho Chi Minh City, Vietnam
[3] National Key Laboratory of Digital Control and System Engineering (DCSELab), HCMUT, 268 Ly Thuong Kiet, District 10, Ho Chi Minh City, Vietnam
[4] Petrovietnam Manpower Training College (PVMTC), Vũng Tàu, Vietnam

Abstract. This paper presents the mechanical designing process of a drill pipe inspection machine that applied the nondestructive inspection method using the electromagnetic fields. The distribution of this paper is to focus on analyzing the mechanical structures of the modules in the inspection machine which are designed to ensure the concentricity of the drill pipe, the defects inspection system, and the centering system during the operation process.

Keywords: Non-destructive inspection · Drill pipe · Inspection machine · Electromagnetic fields

1 Introduction

During the drilling process, drill pipes can be damaged by internal corrosion, vibration, and external pressure, which lead to cracking and failure of thread connections between the pipes in the drill string. The drill string is a series of drill pipes connected by tapered thread connections. A damaged drill pipe in a drill string may result in severe impacts to the human resources and the environment since that drill string cannot be repaired while operating at a depth of thousands of meters below sea level. Therefore, the drill pipes are required to detect the intrinsic cracks and repair them before usage. The commonly inspection methods consist of electromagnetic field inspection, visual inspection, X-rays inspection, magnetic particles, ultrasound, etc. For pipelines with large diameters, some existing studies presented the development of pipeline inspection robots [1–6]. However, for the pipelines with small diameters which cannot apply a pipeline robot, the well-known method to detect the corrosion of the pipeline is to utilize non-destructive inspection methods integrated with an intrusion system. The non-destructive inspection methods enable to detection of the intrinsic cracks of pipelines but not damaged during the inspection process of the pipes before installation and usage. This method is based on

N. V. Khang et al. (Eds.): ASIAN MMS 2021, MMS 113, pp. 324–334, 2022.
https://doi.org/10.1007/978-3-030-91892-7_30

the principle of electromagnetic inspection, specifically, magnetic flux leakage (MFL) [7, 8] method (Fig. 1). The inspection method comprises a permanent or electromagnet to magnetize the drill pipes, then the resulting magnetic field changes are recorded and analyzed to figure out the defects. A hall sensor is utilized to measure the magnitude of magnetic fields and it plays an important role in the machine since it detects the leakage of magnetic fields through the drill pipe's wall. In the electromagnetic inspection machine, the hall sensors have placed a part called sensor shoes (see Fig. 2).

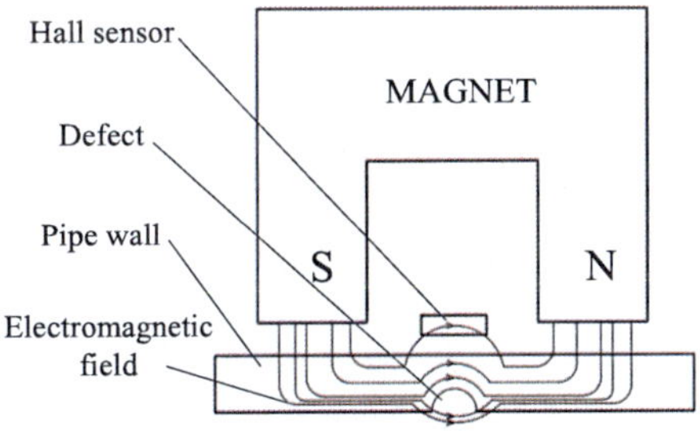

Fig. 1. Principle of Magnetic Flux Leakage (MFL)

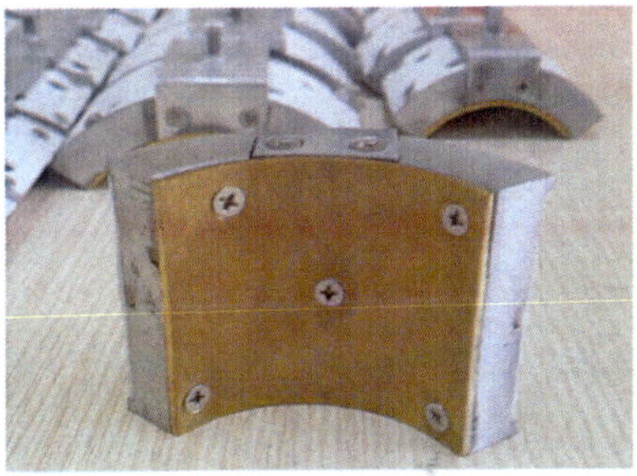

Fig. 2. Sensor shoe

The inspection machine measures the wall thickness of drill pipes and detects many types of defects like internal/external surface cracks in two directions (transverse and longitudinal), corrosion, leakage, etc. on drill pipes. It can be seen that the inspection machine requires to ensure the concentric between pipelines and the intrusion system.

This paper presents the mechanical design process of the intrusion system which is integrated with the MFL sensor systems to form the drill pipe inspection machine. The inspection machine is utilized to inspect drill pipes with different diameters varying from 27/8 in. (73,1 mm) up to 47/8 in. (114,3 mm). The intrusion system is required to stabilize the drill pipe and reduce the vibration which might cause several errors during the inspection process. Additionally, the sensor shoes are required to keep continuously a safety distance of about 2 to 3 mm away from the drill pipes' outer surface to prolong the service life of the sensor shoes.

2 Mechanical Design

The intrusion system comprises four main modules called the drill pipe centering module, the longitudinal defects inspection module, the transverse defects inspection module, and the lifting module.

2.1 Drill Pipe Centering Module

The kinematic diagram of the drill pipe centering module is illustrated in Fig. 3. There are three drill pipe centering modules equipped for the inspection machine that provides the same function of keeping the drill pipe stable during the inspection process. The drill pipe centering system consists of two rollers driven by hydraulic motors, seven revolute joints, and six links which form a parallelogram assembly in order to clamp the drill pipe when it is fed to the inspection machine. (CAD model shown in Fig. 4) (Fig. 5).

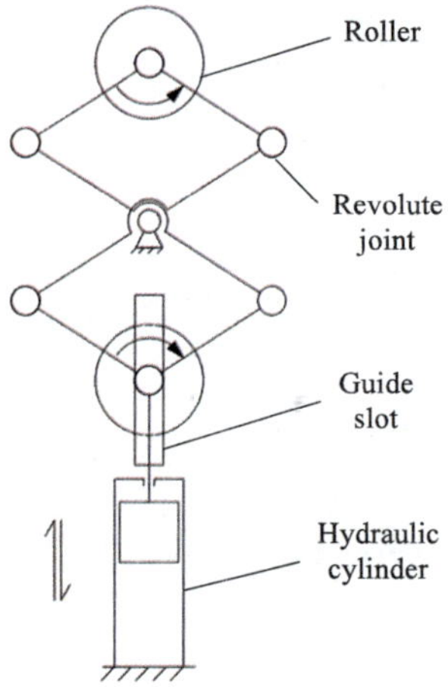

Fig. 3. Kinematic diagram of centering assembly

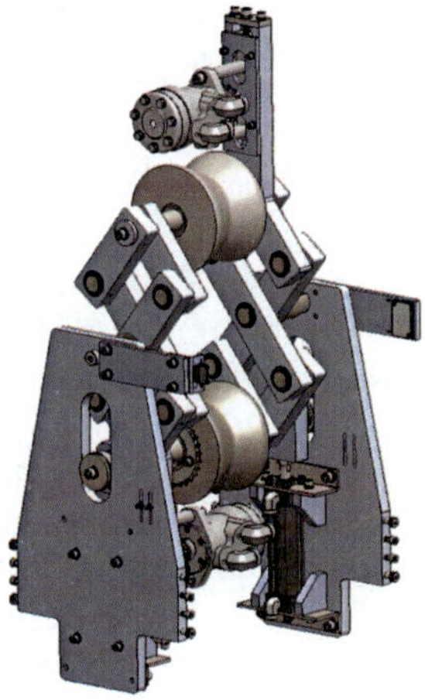

Fig. 4. CAD model of drill pipe centering module

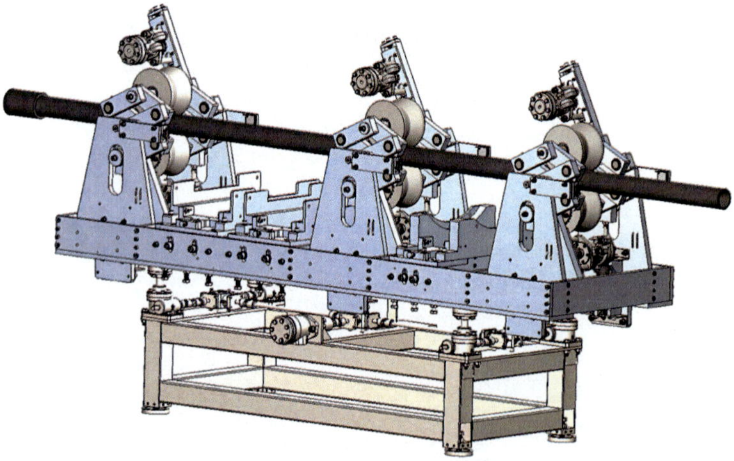

Fig. 5. Three drill pipe centering modules were assembled to the inspection machine

The driving rollers are required to match the following requirements: bearing the drill pipe payload, abrasive resistance, and providing the precise centering with various drill pipe diameters. The driving rollers are made of the C-45 steel material and are surface-hardened afterward to increase the hardness of the outer surface. The drill pipe centering module is divided into two sub-assemblies.

Firstly, the driving sub-assembly (Fig. 6) generates torque to feed the drill pipes along the inspection machine length during the inspection process. The driving sub-assembly includes the main components: The frame (1) is built from two rectangular aluminum plates and a trapezoidal-shaped aluminum plate to support the whole sub-assembly. The roller shaft (7) and (10) does not rotate thus there is no torque on these shafts, however, they are subjected to the bending moment due to the bearings acting on the rollers. The rollers (4) and (11) are driven by the hydraulic motor (5) and (12) through the chain drive (2) and (6), where the larger sprockets are installed on the roller (4) by a series of screws. The mechanisms of the upper and lower roller assemblies are quite similar with a slight difference in the lower one. The two ends of the lower roller shaft (10) are attached to the mountings of two hydraulic cylinders to perform the clamping operation.

Secondly, the parallelogram sub-assembly (Fig. 7) that clamps the drill pipes for the purpose of centering the drill pipes during the inspection process. While drill pipes are being clamped, their central axis will be perpendicular and coplanar with the central axis of the assembly. The parallelogram sub-assembly includes the main parts: Two support blocks (1) and (5) for hydraulic motors, six short hinge blocks (3), four long hinge blocks (6), two center-shafts (8) which are fixed on the frame of the assembly. The shafts (4) and (7) are the upper roller shaft and lower roller shaft, respectively. The shafts (2) are the hinge joints between the mentioned hinge blocks. When the hydraulic cylinder applies upward thrust to the lower roller shaft (7), the upper roller shaft (4) will move downward at the same speed as the shaft (7) with the appropriate clamping force to center the drill pipes and ensure that the central axis of the drill pipe is co-planar and perpendicular to the central axis of the center shaft (8) which is fixed on the inspection machine's frame.

2.2 Transverse Defect Inspection Module

There were earlier many proposed design structures of the transverse defect inspection module such as tension springs and revolute joints mechanism, and gearboxes mechanism. For the tension spring mechanism (Fig. 8), the sensor shoes are located at the end of the lever arms, and they are gathered together by a series of tension springs to ensure that they always stick to the outer surface of the pipe during the testing operation. However, the largest drawback of this mechanism is the friction between the sensor shoes and the outer surface of the pipes which reducing sensor service life. In terms of the gearbox's mechanism (Fig. 9), it includes the main drive motor which is connected to a worm gearbox and combined with a series of bevel gearboxes and universal joints to travel the torque through the ball screw shaft to move the sensor shoes towards the center of the drill pipe. The disadvantage of this mechanism is to require complex initial calibration and reparation, and it also needs to be recalibrated to ensure the allowed tolerances.

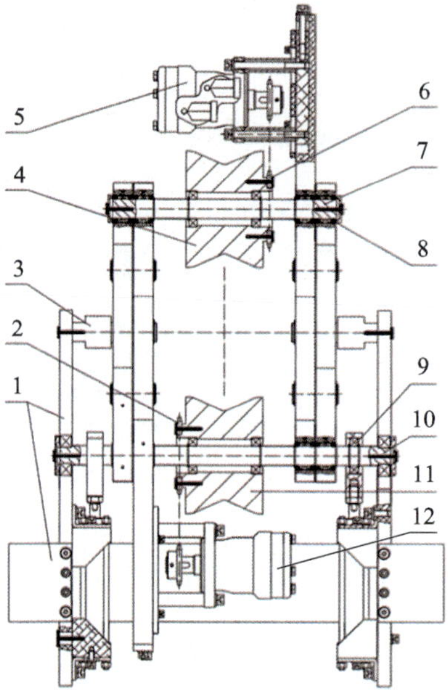

Fig. 6. Structure of the driving module

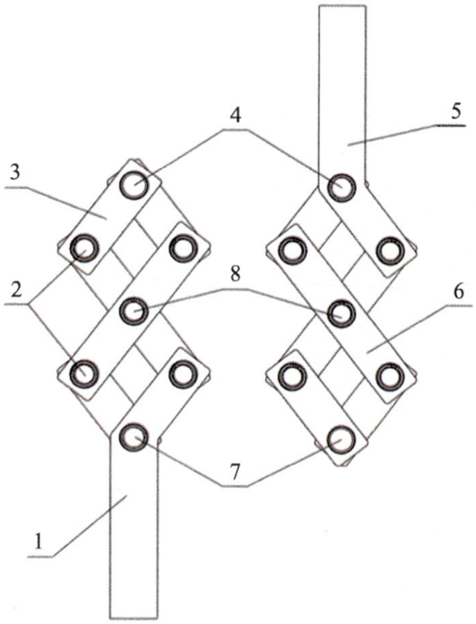

Fig. 7. Structure of the clamping assembly

Fig. 8. Tension spring mechanism

Fig. 9. Gearboxes mechanism

The drawbacks of the two mentioned mechanisms can be solved by the proposed cam mechanism of this paper. In our proposed design of the transverse defect inspection system, the Archimedean spiral is utilized to design the profile of the cam slot on the large driven pulley. This curve is used to ensure the uniform and continuous motion of the slider during the angular change of the driven pulley. The cam mechanism is mainly based on the motion transmission between the cam follower and the cam slot. The cam slots which are manufactured on the driven pulley are the motion paths of the cam followers. Linear guide rails are installed on a fixed part that is inserted into the large driven pulley. When the motor rotates and transmits the torque to the small driving pulley, through the timing belt drive, the large driven pulley also rotates in the same direction with the driving one but lower speed. Finally, the cam followers roll on the cam slots and move the sensor arms along the rails (towards the center of the drill pipes).

To calculate the length of the Archimedean spiral cam slot, the travel length of the slider concerning the angular change of the drive pulley needs to be figured out.

Firstly, according to Fig. 10, the diameters of the drill pipes needed to be inspected are from about 73,1 mm to 114,3 mm, hence their radiuses are about 36,6 mm (*OB*) and 57,2 mm (*OA*), respectively. It assumes that the safety distance from the sensor shoes to the outer wall of the largest drill pipe is 30 mm (*SA*).

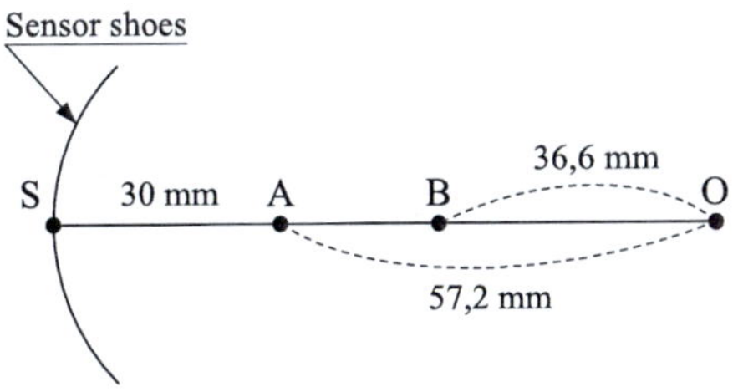

Fig. 10. Calculating sketch of the slider's travel length (*SB*)

It is easy to see that the total travel length of the slider (SB) is $SB = SA + (OA - OB) = 50{,}6$ (mm). In order to ensure the safety and prolong the service life of the cam mechanism, the total travel length is added by 15 mm (Assumed by the designers). Consequently, the final total travel length is about 65 mm. Secondly, the specifications of the Archimedean spiral for the pulley cam slot are assumed, with the steps: $\Delta l = 10$ mm and $\Delta\theta = 6°$, these features indicate that the slider moves along the rail at the distance of 10 mm every time the driven pulley rotates with the angle of 6°. Because the inner bore diameter of the driven pulley is 320 mm ($d_{pu} = 320$ mm), the start diameter of the Archimedean spiral must be equal or greater than it ($d_{start} \geq d_{pu}$), so the start diameter is chosen as $d_{start} = 400$ mm. The relationship between the slider's position along the rail l and the angular position θ is shown in Table 1 to determine the end diameter of the spiral d_{end}. (Steps: $\Delta d = 20$ mm and $\Delta\theta = 6°$).

Table 1. The relationship between l, θ, and d.

l (mm)	θ (°)	d (mm)	r (mm)	θ (°)	d (mm)
0	0	$d_{start} = 400$	50	30°	500
10	6°	420	60	36°	520
20	12°	440	65	39°	$d_{end} = 530$
30	18°	460	70	42°	540
40	24°	480	80	48°	560

Since the total travel length is about 65 mm, referring to Table 1, we figure out the end diameter of the Archimedean spiral $d_{end} = 530$ mm. With the calculated data of the spiral, we can draw an arc which is also the centerline of the spiral cam slot (Figs. 11, 12 and 13).

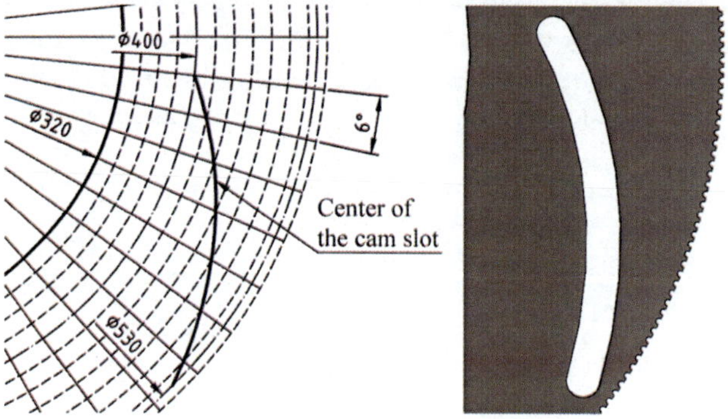

Fig. 11. Archimedean spiral cam slot design

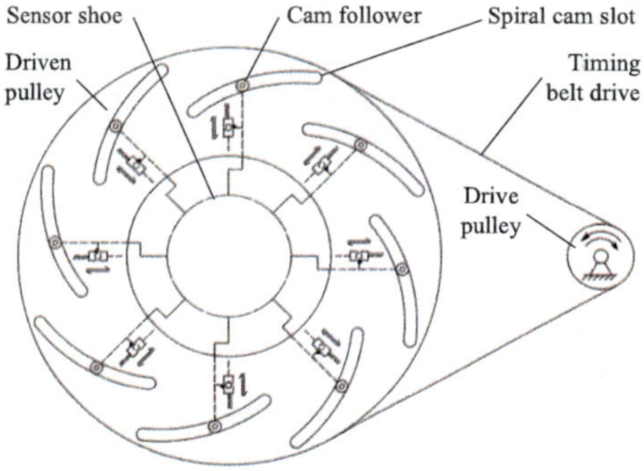

Fig. 12. Kinematic diagram of a transverse defect inspection module

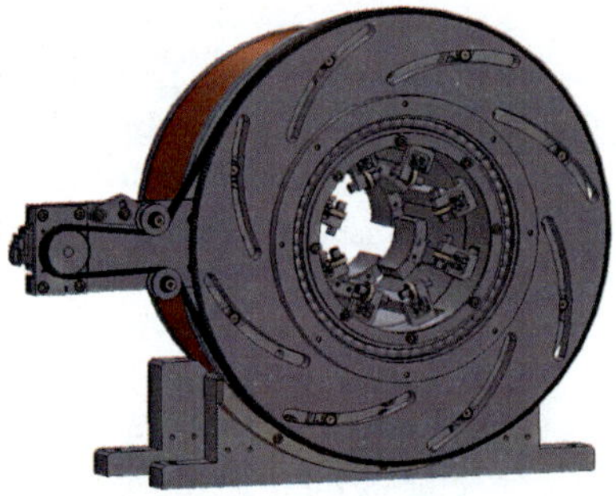

Fig. 13. The transverse defect inspection module

2.3 Longitudinal Defect Inspection Module

To ensure no area on the drill pipes' outer surface is missed during operation, the longitudinal defect inspection module is divided into two sub-assemblies which introduce the similar mechanism but they are arranged at an angle of 90° from each other. When the drill pipes are fed to the inspection machine, the pairs of sensor shoes and magnets (driven manually or by motors) are translated into the drill pipes' outer surface by a series of ball screws. The kinematic diagram of the longitudinal defect inspection module is shown in Fig. 14, whereas its CAD model is shown in Fig. 15.

Similarly, to the transverse defect inspection module, the sensor shoes of the longitudinal defect inspection module keep a safety distance ($2 \div 3$ mm) with the drill pipes' body, which leads to an increase in the service life of the sensor shoes. In order to match this requirement, ball bearings are inserted into the sensor arms where the sensor shoes and other linkages are attached (Fig. 16). The feature of the inspection process is quite

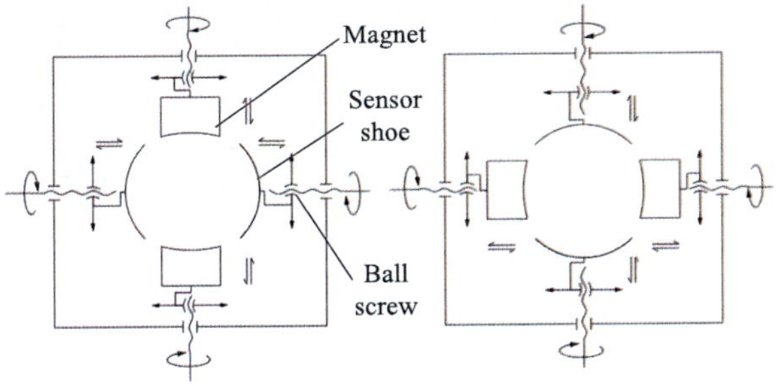

Fig. 14. Kinematic diagram of longitudinal defect inspection module

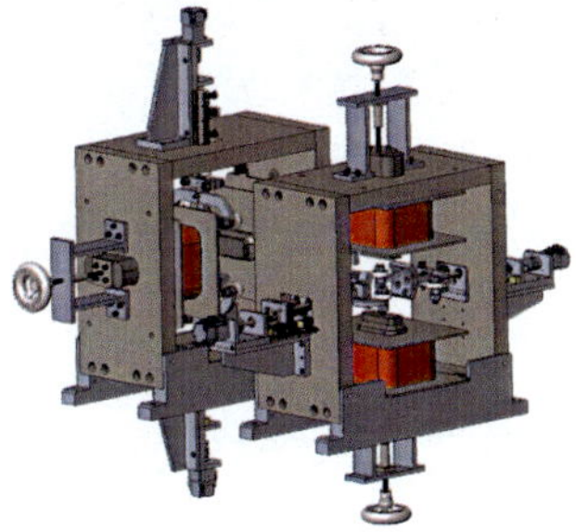

Fig. 15. Longitudinal defect inspection module

special, the measurement is based on the working principle of the electromagnetic fields, therefore, it is not appropriate to use steel ball bearings as usual since it may lead to the changes of electromagnetic fields result in severe measurement errors. Therefore, plastic ball bearings can be the alternative option since they are not affected by the magnetic field.

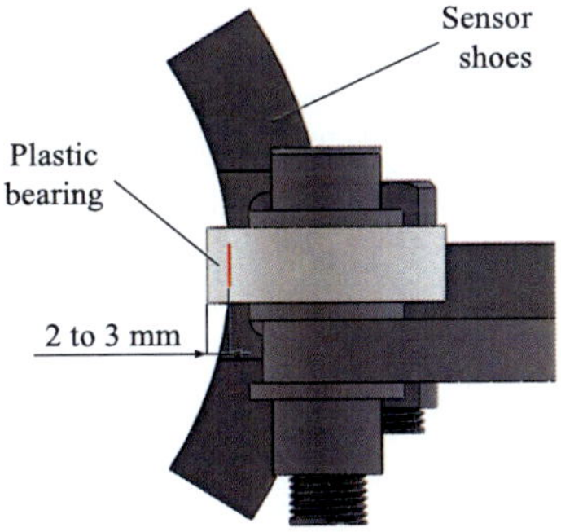

Fig. 16. The position of the bearing and the sensor shoes

Additionally, due to the vibration of the drill pipe when it travels along to the inspection machine, the drill pipe may act to the bearings which might lead to damage the bearings. Thus, a damping mechanism is utilized to reduce these impacts. The damping mechanism comprises two springs installed at the end of sensor arms' sliding shafts to absorb the shock exerted from the drill pipes. It is noted that the spring should be installed with the initial position where the spring is compressed about 5 mm compared to its natural length. One end of the sliding shaft is attached to the sensor shoes, and the other end is held by a split pin to prevent the sensor arm assembly from slipping off (see Fig. 17).

2.4 Lifting Module

The lifting module is designed to lift the entire upper assemblies including the two defect inspection modules, the drill pipe centering modules, and the supporting frame.

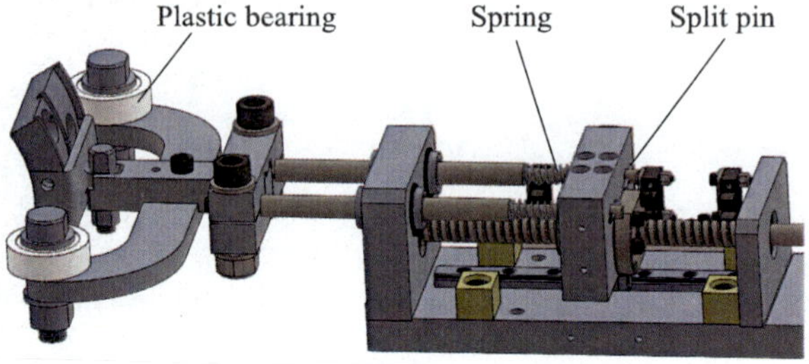

Fig. 17. Sensor arm assembly

The most important factors for designing the lifting module are the total weight and payloads of the upper assemblies, so the lifting module should be built lastly. The H-type screwjack lifting system includes four screwjack gearboxes in the corners, three bevel gearboxes that provide two bevel gears, and a series of universal joints linking the transmission shafts with gearboxes. Although the screwjack series features a limited lifting capacity compared to the hydraulic lifting system, its self-locking function is such a great advantage that designers should consider. This screwjack series also allows the torque from the motor applying on the worm gear drive is rejected because the lifting shaft will remain at the lifting position where it stops instead of moving downward, no matter how much load is on it. Noted that the maximum load of the lifting system can change corresponding to the input speed from the hydraulic motor and vice versa. The kinematic diagram of the lifting module is shown in Fig. 18 while the CAD model is shown in Fig. 19.

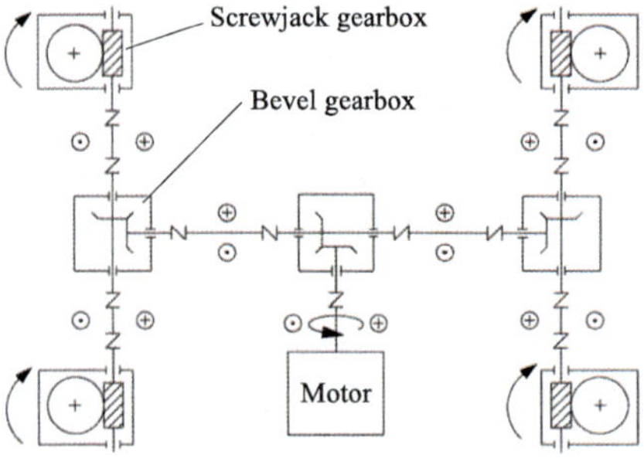

Fig. 18. Kinematic diagram of the H-type lifting module

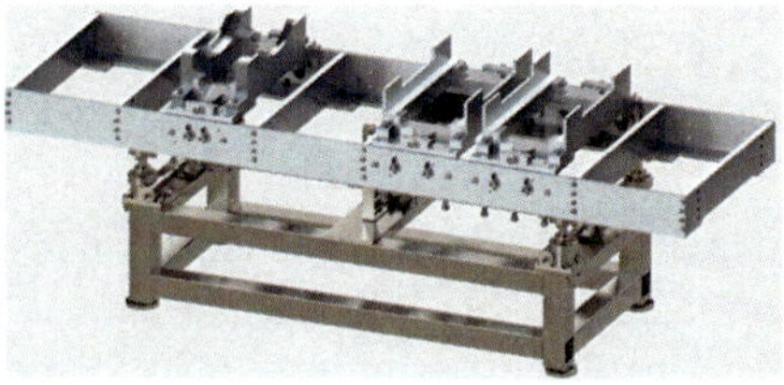

Fig. 19. H-type lifting module and machine's frame

The overall design of the drill pipe inspection machine is shown in 3D CAD model (in Fig. 20). To operate the inspection machine requires a hydraulic source to power for hydraulic cylinders used to lift the drill pipe centering modules and hydraulic motors used to drive the rollers. The level arms of the longitudinal defect inspection module are powered by AC servo motors to travel the sensor shoes towards the drill pipe. The cam mechanism of the transverse defect inspection module is also driven by a servo motor through the timing pulley and time belt mechanism. The level arms are driven along to the Archimedean spiral to travel the sensor shoes towards the drill pipe.

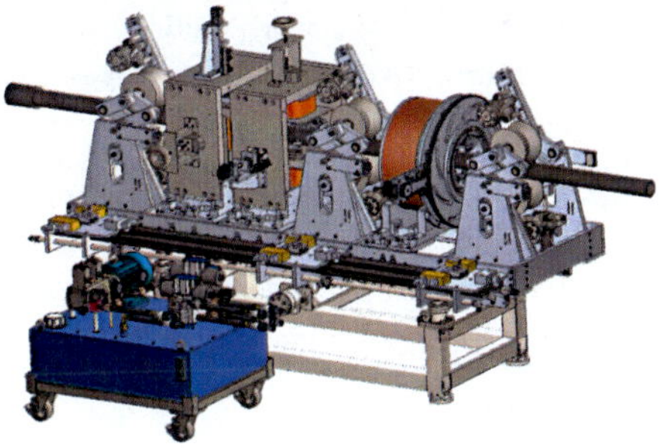

Fig. 20. Overall drill pipe inspection machine

3 Conclusion

This paper has been presented the mechanical design process of the drill pipe inspection machine consisting of the drill pipe centering module, the longitudinal defect inspection module, the transverse defect inspection module, and the lifting module. The design process is based on predefined parameters to match the requirements with various drill pipes' dimensions. The significant feature of the drill pipe centering module is that the roller is a possibility to clamp various drill pipes' diameter dimensions. Meanwhile, the transverse defect inspection module features the cam mechanism with Archimedean contour. Both the longitudinal defect inspection and the transverse defect inspection module proposed a damping mechanism to adsorb the vibration of the drill pipe to avoid damage to the sensor shoes. Finally, the lifting module enables that the inspection machine can serve various diameters of the drill pipes.

Acknowledgment. This research is supported by DCSELAB and funded by Vietnam National University Ho Chi Minh City (VNU-HCM) under grant number TX2021-20b-01. We acknowledge the support of time and facilities from Ho Chi Minh City University of Technology (HCMUT), VNU-HCM for this study.

References

1. Kwon, Y.S., Yi, B.J.: Design and motion planning of a two-module collaborative indoor pipeline inspection robot. IEEE Trans. Robot. **28**, 681–696 (2012). https://doi.org/10.1109/TRO.2012.2183049
2. Nguyen, T.T., Kim, S.B., Yoo, H.R., Rho, Y.W.: Modeling and simulation for PIG flow control in natural gas pipeline. KSME Int. J. **158**(15), 1165–1173 (2001). https://doi.org/10.1007/BF03185096
3. Nguyen, T.T., Kim, D.K., Rho, Y.W., Kim, S.B.: Dynamic modeling and its analysis for PIG flow through curved section in natural gas pipeline. In: Proceedings of IEEE International Symposium on Computational Intelligence in Robotics and Automation, CIRA, vol. 2001-January, pp. 492–497 (2001). https://doi.org/10.1109/CIRA.2001.1013250
4. Mohammed, M.N., et al.: Design and development of pipeline inspection robot for crack and corrosion detection. In: Proceedings - 2018 IEEE Conference on Systems, Process and Control, ICSPC 2018, pp. 29–32 (2018). https://doi.org/10.1109/SPC.2018.8704127

5. Kwon, Y.S., Lim, H., Jung, E.J., Yi, B.J.: Design and motion planning of a two-moduled indoor pipeline inspection robot. In: Proceedings - IEEE International Conference on Robotics and Automation, pp. 3998–4004 (2008). https://doi.org/10.1109/ROBOT.2008.4543825
6. Kwon, Y.S., Jung, E.J., Lim, H., Yi, B.J.: Design of a reconfigurable indoor pipeline inspection robot. In: ICCAS 2007 - International Conference on Control, Automation and Systems, pp. 712–716 (2007). https://doi.org/10.1109/ICCAS.2007.4406991
7. Pham, H.Q., et al.: Highly sensitive planar hall magnetoresistive sensor for magnetic flux leakage pipeline inspection. IEEE Trans. Magn. **54**, 1–5 (2018). https://doi.org/10.1109/TMAG.2018.2816075
8. Shi, Y., Zhang, C., Li, R., Cai, M., Jia, G.: Theory and application of magnetic flux leakage pipeline detection. Sens. (Basel) **15**, 31036 (2015). https://doi.org/10.3390/S151229845

Multi-mode Motion Analysis of Sideward Opening Aircraft Door Based on Position and Orientation Characteristic Theory

Borui Wu, Lubin Hang(✉), Jiyou Peng, Renhui Peng, and Xiaobo Huang

School of Mechanical and Automotive Engineering, Shanghai University of Engineering Science, Shanghai 201600, China

Abstract. As an important part of large passenger aircraft, the analysis of motion state and motion mode of passenger door is very important for the design of door function and motion mechanism. This paper studies the overall motion correlation of a sideward opening passenger door opening mechanism from the robot mechanisms. It is abstracted as a parallel mechanism composed of two SS branches and one ◇(R//R//R//R) ⊥ R//R branch for the first time. Based on the Position and Orientation Characteristic (POC) Theory, this paper analyzes the topological characteristics of the sideward opening passenger door mechanism, and obtains that the lifting and rotation process of the door mechanism has one degree of freedom and two independent motion modes. The independent motion modes of the door are different in different modes. The multi-mode of the sideward opening passenger door mechanism is realized under the scale-constrained that the projection size of the main hinge mechanism and the parallel link 2-SS mechanism in the horizontal plane is equal. According to the variable topological structure characteristics of the mechanism in the opening process of the sideward opening passenger door, the POC of different motion modes of the mechanism are analyzed, which provides a new idea for the design of the aircraft door.

Keywords: Position and orientation characteristic · Sideward opening aircraft door · Motion mode · Hinge arm mechanism · Parallel link mechanism

1 Introduction

As an important component of large passenger aircraft, the analysis of its function and motion mode is very important for the design of aircraft door mechanism [1]. There are many forms of aircraft passenger doors, including the door ladder integrated form, internal rotation opening form and sideward opening form.

N. V. Khang et al. (Eds.): ASIAN MMS 2021, MMS 113, pp. 335–341, 2022.
https://doi.org/10.1007/978-3-030-91892-7_31

Some early airliners in the world have adopted the door ladder integrated passenger door, including IL86/96 and some Boeing B747. However, the door ladder integrated passenger door occupies a lot of space, and this form of passenger door is no longer considered to be designed on airliners [1]. The early Boeing B737 passenger plane passenger door was opened by retraction and rotation [2]. When the passenger door was opened, it first moved inward for a certain angle and then turned outward. The upper and lower ends of the passenger door in the retraction and rotation opening mode were added with additional sealing door plates, which had poor sealing performance.

The latest Boeing airliners such as B777 are designed with sideward opening passenger door. The aircraft sideward opening passenger door mechanism has many functions, such as pressurization and pressure relief, safety slide release, gust lock and so on. Sideward opening passenger door has been more and more used because of its excellent opening structure, opening mode and saving internal space [3].

Further research is carried out on the basis of referring to the existing mature models in the world. For example, the dynamic analysis and parameter optimization design of the Internal rotation opening ARJ-21 passenger door are carried out [4]. In addition, In some literature, the mechanism of ARJ-21 emergency door is simulated and optimized [5]. COMAC has designed the passenger door of C919. At present, it has disclosed the design and simulation optimization of the scheme of the door guide slot mechanism, the design idea of the door latch mechanism, the parametric analysis of the door lifting principle, etc. However, they are only limited to the analysis of the mechanism design scheme with independent functions, and there is no motion analysis from the overall synergy of the passenger door mechanism.

In history, the aircraft passenger door has been misaligned or worn, resulting in the passenger door opening or closing not tight, which affects the normal take-off of flights [6]. The correlation between mechanisms will lead to component damage. For example, the wear of aviation bearing is often due to the offset or deformation of the connection between other components, resulting in the increase of force. Therefore, it is very important to analyze the form of passenger door movement from the overall correlation of passenger door mechanism.

Referring to the existing mature sideward opening passenger door scheme, based on the POC theory, this paper analyzes the motion input-output relationship and the motion modes of lifting process and rotary process from the integrity of the passenger opening door mechanism. The door movement mode is analyzed from the principle of multi-mode mechanism, which provides ideas for the design of driving mode and driving position of door, and lays a foundation for the design of new aircraft door.

2 Analysis of Sideward Opening Passenger Door Opening Process

The movement process of aircraft sideward opening passenger door is divided into lifting process and door rotation process. From the perspective of mechanism and the overall motion synergy of passenger door mechanism, the motion mode switching mode of passenger door lifting process and passenger door rotation process is analyzed.

The aircraft sideward opening passenger door opening mechanism is composed of hinge arm mechanism and parallel link mechanism, as shown in Fig. 1.

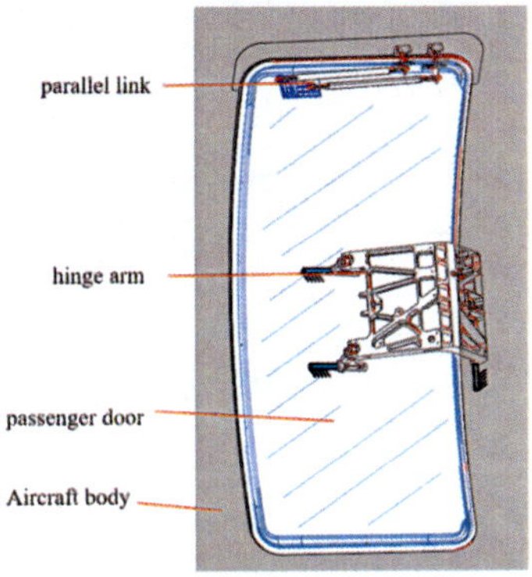

Fig. 1. Aircraft passenger door opening mechanism

The aircraft sideward opening passenger door is opened by the joint action of the hinge arm mechanism and the parallel link mechanism: the passenger door is connected to the hinge arm through the upper and lower rocker arms, and the parallel link is used together with the hinge arm to ensure that the door moves parallel to the fuselage when the passenger door is opened or closed. Its mechanism diagram is shown in Fig. 2.

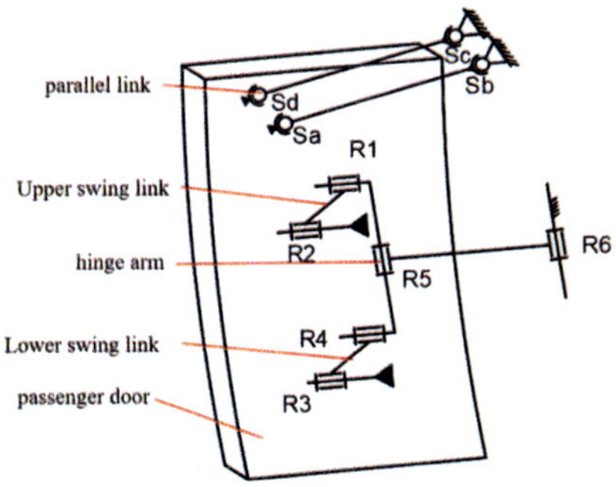

Fig. 2. Schematic diagram of aircraft passenger door opening mechanism

As shown in Fig. 3, the hinge arm mechanism is a very important structural load-bearing member in the passenger door. The hinge arm structure is composed of an upper rocker arm, a lower rocker arm, and upper and lower supports fixedly connected to the fuselage frame. One sideward of the hinge arm is connected with the machine body, and one sideward is connected with the door structure through the upper and lower lifting arms. After the passenger door is opened, the weight of the whole passenger door is supported by the hinge arm. During the opening and closing of the passenger door, the hinge arm needs to bear large additional bending moment and torque, and is transmitted to the fuselage structure through the upper and lower supports fixed on the door frame.

As shown in Fig. 4, the parallel link mechanism is composed of four ball hinges, which are respectively connected to the top of the passenger door and the fuselage to control the translational movement of the passenger opening door process and realize

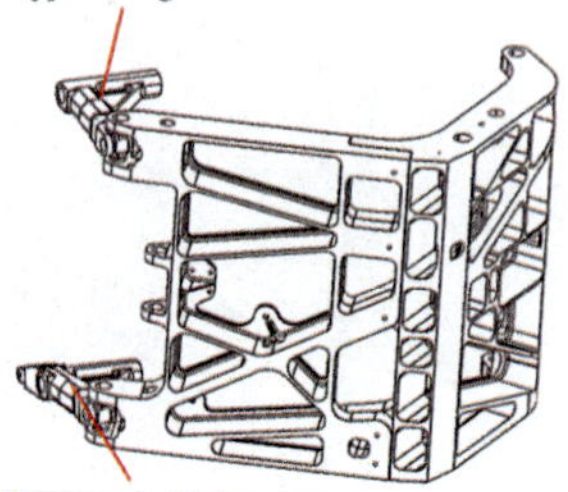

Fig. 3. Hinge arm mechanism

the movement of the passenger door parallel to the fuselage. The two links of the parallel link mechanism are equal in length and parallel to each other.

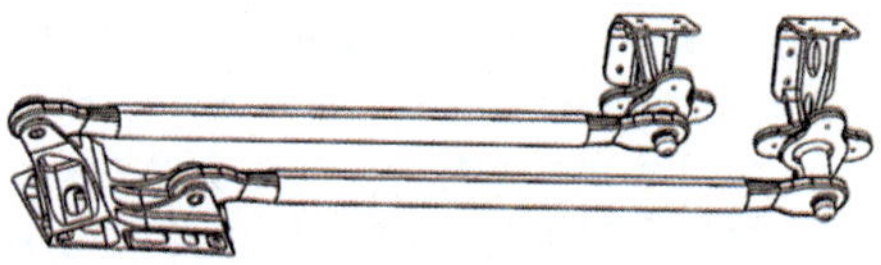

Fig. 4. Parallel link mechanism

3 Sideward Opening Passenger Door Branch Combination and Motion Degree of Freedom Analysis

3.1 Calculation of Degrees of Freedom of Parallel Mechanism for Passenger Door

Yang revealed the POC theory formed by the internal relationship among the mechanism scale constraint type [7], mechanism topology and independent displacement number, and creatively proposed a degree of freedom calculation formula reflecting the input and output characteristics of the mechanism:

$$\begin{cases} F = \sum_{i=1}^{m} f_i - \sum_{\mathrm{j}=1}^{v} \xi_{L_{\mathrm{j}}} \\ \xi_{L_{\mathrm{j}}} = \dim.\left\{ \bigcap_{j=1}^{j} M_{b_j} \bigcup M_{b_{(j+1)}} \right\} \end{cases} \tag{1}$$

Where:

F—Degree of freedom of mechanism;

f_i—Degrees of freedom of the ith kinematic pair;

$\xi_{L_{\mathrm{j}}}$—The number of independent displacement equations of the j-th independent loop;

M_{b_j}—POC of j branch end member.

Based on the POC and combined with the branch combination of the passenger door opening mechanism [8], the degrees of freedom and motion mode of the passenger opening door process are analyzed. The main steps are as follows:

The sub parallel mechanism DOF composed of three branches is:

$$F_{(2\text{-}3)} = \sum_{i=1}^{m} f_i - \sum_{j=1}^{1} \xi_{L_j} = 10 + 3 - 6 - 6 = 1 \tag{2}$$

Combined with the law of passenger door movement, there are two modes of passenger door movement: passenger door lifting and rotary.

The POC of parallel mechanism is calculated, and the following results can be obtained:

$$\begin{aligned} M_{Pa} &= \begin{bmatrix} t^3 \\ r^3 \end{bmatrix} \cap \begin{bmatrix} t^1(\perp R_6) \cup t^1(\perp R_1) \\ \left\{ r^1(\| R_1) \right\} \cup r^1(\| R_6) \end{bmatrix} \\ &= \begin{bmatrix} t^1(\perp R_6) \cup t^1(\perp R_1) \\ \left\{ r^1(\| R_1) \right\} \cup r^1(\| R_6) \end{bmatrix} \end{aligned} \tag{3}$$

From the calculation formula of degrees of freedom and POC M_{pa} of passenger opening door mechanism, it can be seen that there is a degree of freedom of passenger door, and the main motion modes are also different in different motion stages. The lifting process is rotation around R_1, and the translational process is rotation around R_6.

3.2 Scale Constraint Analysis of Motion Mode Switching During Door Opening

The process of rotating and opening the passenger door is mainly completed by the combined movement of hinge arm mechanism and parallel link mechanism. In order to ensure that the passenger door remains horizontal during the pushing process, prevent the collision between the roller and the fuselage guide groove during the rotation and opening of the passenger door, and ensure the stability of the opening track of the passenger door, it is necessary to stiffen the upper and lower rocker arms of the hinge arm during the rotation of the passenger door.

When the passenger door moves around the R_6 axis as an independent driving mode, the rotation radius of the passenger door movement is a_{56}. The motion trajectories of the two 2S branches shall meet the requirements for the rotation and opening of the passenger door to ensure that the passenger door is parallel to the fuselage during the opening process. When the passenger door is lifted to the highest position, in order to meet the design requirements for the opening of the passenger door, the projection length of the center distance between the two parallel link ball pairs and the dimension between the axes of the two rotation pairs of the passenger door hinge arm on the horizontal plane shall be the same and parallel, i.e. $a_{ab} = a_{cd} = a_{56}$ (as shown in Fig. 5), Otherwise, the door rotation opening process parallel to the fuselage cannot be met.

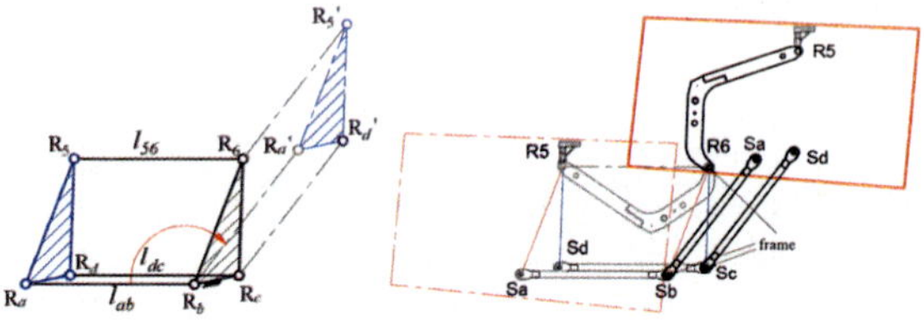

Fig. 5. Aircraft door mechanism and position of aircraft door rotation opening process (The door in the red box)

4 Sideward Opening Passenger Door Lifting Mechanism

By setting a stop at the end of the lifting process, the passenger door can accurately ensure the mode switching of the passenger door opening mechanism. The use of the stop to realize the mode switching of the passenger door can not only ensure the stability of the passenger door opening process, but also ensure small error and wear in the actual opening process of the passenger door. The passenger door stop and its position are shown in Fig. 6.

The door lifting process has an independent movement perpendicular to R_1. When the passenger door is at the highest horizontal position, the movement of the passenger door lifting chain is stopped by the stop limit control, the parallel link mechanism is at the upper end of the passenger door, and the two parallel links are parallel to the horizontal plane and the upper plane of the main hinge arm respectively.

The passenger door lifting is mainly controlled by the lifting chain of the opening handle. The lifting chain is composed of multiple motion circuits to drive the upper and lower rocker arms of the main hinge.

According to the calculation formula of degrees of freedom $F = 1$.

The lifting process of the passenger door is realized by the rotation of the lifting chain around the R_3 rotating shaft driven by the handle in the passenger door. During the lifting process, the passenger door has only one degree of freedom.

Open the handle, apply torque to the passenger door lifting mechanism, lift the passenger door and control the passenger door to move upward. The opening torque

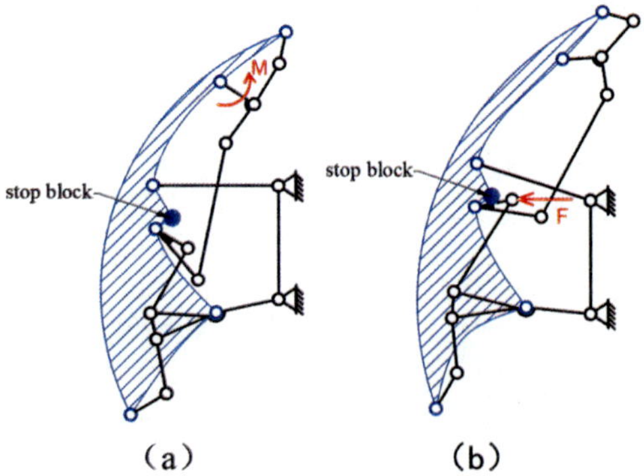

Fig. 6. Door lifting mechanism (a) Door closed position (b) door open mode switching position

after the passenger door is in contact with the stop block is converted into forward thrust to realize the rotary opening of the passenger door.

5 Conclusion

(1) In this paper, a topology analysis theory and method of parallel mechanism based on POC is proposed for the door motion mechanism of sideward open aircraft. The degree of freedom and multi-mode of the passenger door are analyzed in this paper.
(2) Based on the theory of POC, the motion mode of passenger door is analyzed in this paper. The motion state of passenger door is analyzed according to the transformation of independent motion modes and non-independent motion modes. In this paper, a new motion mode analysis method is proposed, which lays a foundation for the design of new aircraft passenger door.
(3) In this paper, the motion modes of each state of the passenger door are analyzed. The passenger door rotation process depends on the projection scale of parallel links and motion constraints.

References

1. Wu, S.: Analysis and application of opening and closing mechanism on door-stairs passenger door. Mech. Sci. Technol. Aerosp. Eng. 8–12 (2017)
2. Yan, Z.: Troubleshooting the abnormal operation of B737 passenger door. J. Civ. Aviat. Univ. China 26–30 (1988)
3. Ye, T.: Optimal design and analysis of a side opening passenger door used for airliner. Nanjing University of Aeronautics and Astronautics (2018)
4. Huang, Z.: Parameter optimization of hatch mechanism for airliner, pp. 2–16. Qinghua University (2011)
5. Xu, Q.: Research on simulation technique of aircraft emergency door motion mechanism. Nanjing University of Aeronautics and Astronautics (2012)
6. Chen, Y.: Troubleshooting the abnormal operation of B737 passenger door. Aviat. Maint. Eng. 60–62 (2011)
7. Yang, T., Liu, A., Jin, Q., Luo, Y., Shen, H., Hang, L.: Position and orientation characteristic equation for topological design of robot mechanisms. J. Mech. Des. (2009)
8. Yang, T., Liu, A., Shen, H., Luo, Y., Hang, L., Shi, Z.: On the correctness and strictness of the position and orientation characteristic equation for topological structure design of robot mechanisms. J. Mech. Robot. **5**, 021009 (2013)

Flexible Robots and Mechanisms

Modelling of Cable-Driven Continuum Robots with General Cable Routing: a Comparison

Soumya Kanti Mahapatra[1], K. P. Ashwin[2], and Ashitava Ghosal[1(✉)]

[1] Department of Mechanical Engineering, Indian Institute of Science, Bangalore 560012, Karnataka, India
{soumyam,asitava}@iisc.ac.in

[2] Department of Mechanical Engineering, BITS Pilani K.K. Birla Goa Campus, Goa 403726, India
ashwinp@goa.bits-pilani.ac.in

Abstract. Accurate modelling of continuum robots is required for predicting its motion and for model based control. In this work, we restrict our study to cable actuated continuum robots with a flexible backbone on which perforated disks are fixed and through which a cable is routed from the base to the tip. We compare two well known models, namely the optimization-based model and the more extensively used Cosserat rod model for kinematic analysis. Both the models can predict the shape and motion of a cable-driven continuum robot (CCR) with cable routed in a general manner. The paper focuses on comparing these two models based on their ease of modelling and simulation. It was found that the optimization-based method is comparatively faster than the Cosserat rod model for most of the real world cases. These two methods are also compared with experimental results obtained with a 3D printed prototype continuum robot. It is observed that both the methods can reasonably accurately predict the shape with the optimization-based method performing better by a small amount.

Keywords: Continuum robot · Cable-driven · Tendon-driven · Kinematics model · Generally routed cable

1 Introduction

With the need for better reachability, manoeuvrability and inherent compliance, continuum robots are becoming more common for applications involving congested spaces and handling of delicate objects. Continuum robots are characterized by an elastic backbone which is actuated in various ways – pneumatic, shape-memory alloys, pre-curved beams and cables are some the well-known ways. Cable-driven continuum robots (CCR) or tendon-driven continuum robots are widely studied for their simplicity in design and scalability and find use in fields such as medical devices, remote inspections, search and rescue, space

N. V. Khang et al. (Eds.): ASIAN MMS 2021, MMS 113, pp. 345–353, 2022.
https://doi.org/10.1007/978-3-030-91892-7_32

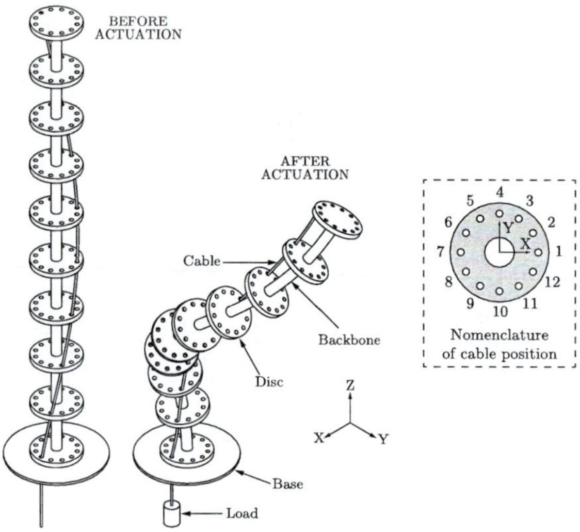

Fig. 1. Schematics of a CCR before and after actuation.

applications, biomimetics etc. [1]. The CCR usually consists of a slender rod like structure forming the *backbone* and circular disks, with holes arranged in a circle, are attached at equal intervals along the backbone (see Fig. 1). Cables are routed through these holes and attached to the top-most disk, and when these cables are actuated from the bottom, the whole CCR bends and takes various shapes depending on the routing of the cable.

Accurate modelling of CCR is essential in correctly predicting the shape and behaviour of a CCR. Detailed review of such models can be found in [2,3]. The simplest models employ geometry based methods that use the constant curvature approach [4]. More sophisticated models use finite element models [5,6], elasticity based methods considering cable friction [7], Euler-Bernoulli beam theory [8], Cosserat rod theory [9,10], pseudo-rigid body model [11], Euler-Lagrange formulation [12], principle of virtual power [13] and geometry-based optimization method [14] to name a few. Most of the models of CCRs focus on cables which are routed straight. Even though this makes the models simpler, with straight routing, a CCR cannot attain complicated shapes that a generally routed CCR can achieve. The optimization-based method and Cosserat rod model are capable of accurately predicting the shape of CCRs with generally routed cable. This paper focuses on these two approaches and presents the advantages and disadvantages in modelling and simulation of the kinematics of a generally routed CCR. We also compare the simulations results from these two models with experimental results obtained from a 3D printed prototype, again to judge their accuracy in predicting the shape of an actuated CCR.

The rest of the paper is organized as follows: In Sect. 2, a brief summary of the optimization-based and Cosserat rod modelling is presented. In Sect. 3, a comparison between the two methods are presented and then the results are com-

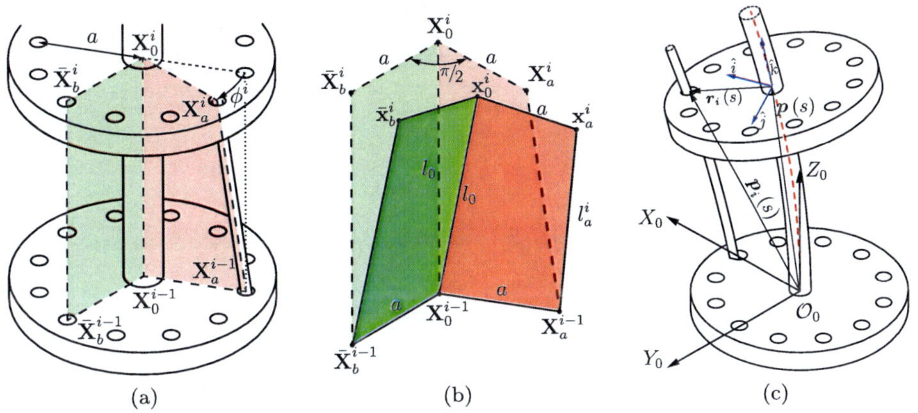

Fig. 2. (a) Nomenclature for optimization-based method, (b) Four bars before (in lighter shade) and after actuation (in darker shade), (c) Nomenclature for Cosserat rod model

pared with a 3D printed CCR prototype. Lastly in Sect. 4, concluding remarks along with scope for future work are presented.

2 Generally Routed CCR with a Single Cable

In this section, we present a brief overview of the optimization-based method and Cosserat rod model for the kinematics of a CCR with single generally routed cable. More details on the optimization and Cosserat rod modelling approaches can be found in references [9,14].

Optimization-based Model: In the optimization-based method, the CCR is discretized as four bar linkages, stacked one on top of the other with each four bar lying in between two consecutive disks. At any particular disk, the position of the center of a disk and location of the cable is denoted by $\mathbf{X}_0^{(\cdot)}$ and $\mathbf{X}_a^{(\cdot)}$, respectively with the superscript denoting the disk number starting with $i = 0$ for the base disk. Inside each segment an imaginary four-bar linkage is assumed consisting of the two disks, the backbone and the cable. Figure 2a shows the ith section with the four bar $\mathbf{X}_0^{i-1}\mathbf{X}_a^{i-1}\mathbf{X}_a^{i}\mathbf{X}_0^{i}$ shaded in red. A second imaginary four bar (seen in green) is assumed such that $\angle \mathbf{X}_a^i\mathbf{X}_0^i\bar{\mathbf{X}}_b^i = \pi/2$ and $\bar{\mathbf{X}}_b^{i-1}$ is directly below $\bar{\mathbf{X}}_b^i$. In the above mentioned four bars, the fixed linkages are $\mathbf{X}_0^{i-1}\mathbf{X}_a^{i-1}$ and $\mathbf{X}_0^{i-1}\bar{\mathbf{X}}_b^{i-1}$, and the couplers are $\mathbf{X}_0^i\mathbf{X}_a^i$ and $\mathbf{X}_0^i\bar{\mathbf{X}}_b^i$ with $\mathbf{X}_0^{i-1}\mathbf{X}_0^i$ being the first crank for both.

After actuation of the cable, the backbone deforms and the position vectors $\mathbf{X}_{(\cdot)}^i$ changes to $\mathbf{x}_{(\cdot)}^i$ (see Fig. 2b). The position after actuation can be found out by simultaneously minimizing both the coupler angles which can be mathematically stated as the following minimization problem:

$$\underset{\mathbf{x}_0^i,\mathbf{x}_a^i}{\arg\min}\left(\left[\arccos\left(\frac{\mathbf{A}}{\|\mathbf{A}\|}\cdot\frac{\mathbf{B}}{\|\mathbf{B}\|}\right)\right]^2+\left[\arccos\left(\frac{\mathbf{C}}{\|\mathbf{C}\|}\cdot\frac{\mathbf{D}}{\|\mathbf{D}\|}\right)\right]^2\right) \tag{1}$$

where, $\mathbf{A}=\mathbf{X}_0^{i-1}-\mathbf{X}_a^{i-1}$, $\mathbf{B}=\mathbf{x}_0^i-\mathbf{x}_a^i$, $\mathbf{C}=\mathbf{X}_0^{i-1}-\bar{\mathbf{X}}_b^{i-1}$, $\mathbf{D}=\mathbf{x}_0^i-\bar{\mathbf{x}}_b^i$

Subject to:

$$\begin{gathered}\|\mathbf{x}_0^i-\mathbf{X}_0^{i-1}\|=l_0,\quad \|\mathbf{x}_a^i-\mathbf{X}_a^{i-1}\|=l_a^i,\quad \|\mathbf{x}_0^i-\mathbf{x}_a^i\|=a,\\ \bar{\mathbf{X}}_b^i=a\frac{(\mathbf{X}_a^i-\mathbf{X}_0^i)\times(\mathbf{X}_0^i-\mathbf{X}_0^{i-1})}{\|(\mathbf{X}_a^i-\mathbf{X}_0^i)\times(\mathbf{X}_0^i-\mathbf{X}_0^{i-1})\|}\end{gathered} \tag{2}$$

Given data: $\mathbf{X}_0^i,\mathbf{X}_0^{i-1},\mathbf{X}_a^i,\mathbf{X}_a^{i-1},l_0,l_a^i,a$

Equation 1, along with the constraints, need to be solved starting from the base till the tip sequentially such that output of each step (the position of the coupler) becomes the base of the four bar for the next segment.

Cosserat Rod Model: In the Cosserat rod model [9], the backbone is assumed to be a parametric space curve with a parameter $s \in [0, L]$, where L is the total length of the CCR. The position of any point on the backbone can be represented as a vector $\boldsymbol{p}(s)$ and the orientation of the local coordinate system, similar to a Frenet–Serret frame, is represented as $\boldsymbol{R}(s)$ (see Fig. 2c). The location of the cable in the local coordinate system is denoted by $\boldsymbol{r}_i(s)$. It is assumed that $\boldsymbol{r}_i(s)$ does not change with actuation and the internal force is always tangent to the cable position, $\boldsymbol{p}_i(s)$. With $\boldsymbol{p}(s)$, $\boldsymbol{R}(s)$ and their respective rate of change with respect to s, $\boldsymbol{v}(s)$ and $\boldsymbol{u}(s)$, one can solve for a set of first order differential equations (ODEs) obtained from the constitutive relations to get the pose of the CCR after actuation. The first-order ODEs are given by

$$\begin{matrix}\dot{\boldsymbol{p}}=\boldsymbol{R}\boldsymbol{v},\\ \dot{\boldsymbol{R}}=\boldsymbol{R}\widehat{\boldsymbol{u}},\end{matrix}\quad \begin{bmatrix}\dot{\boldsymbol{v}}\\ \dot{\boldsymbol{u}}\end{bmatrix}=\begin{bmatrix}\boldsymbol{K}_{se}+\boldsymbol{A} & \boldsymbol{G}\\ \boldsymbol{G}^T & \boldsymbol{K}_{bt}+\boldsymbol{H}\end{bmatrix}^{-1}\begin{bmatrix}\boldsymbol{d}\\ \boldsymbol{c}\end{bmatrix} \tag{3}$$

where,

$$\boldsymbol{c}=-\widehat{\boldsymbol{u}}\boldsymbol{K}_{bt}\boldsymbol{u}-\widehat{\boldsymbol{v}}\boldsymbol{K}_{se}\left(\boldsymbol{v}-[0,0,1]^T\right)-\boldsymbol{b},\ \boldsymbol{b}=\sum_{i=1}^{n}\widehat{\boldsymbol{r}}_iA_i\left(\widehat{\boldsymbol{u}}\dot{\boldsymbol{p}}_i^b+\widehat{\boldsymbol{u}}\dot{\boldsymbol{r}}_i+\ddot{\boldsymbol{r}}_i\right)$$

$$\boldsymbol{d}=-\widehat{\boldsymbol{u}}\boldsymbol{K}_{se}\left(\boldsymbol{v}-[0,0,1]^T\right)-\boldsymbol{a},\ \boldsymbol{a}=\sum_{i=1}^{n}A_i\left(\widehat{\boldsymbol{u}}\dot{\boldsymbol{p}}_i^b+\widehat{\boldsymbol{u}}\dot{\boldsymbol{r}}_i+\ddot{\boldsymbol{r}}_i\right)$$

$$A_i=-\tau_i\left(\widehat{\dot{\boldsymbol{p}}_i^b}\right)^2/\left\|\dot{\boldsymbol{p}}_i^b\right\|^3,\ \boldsymbol{A}=\sum_{i=1}^{n}A_i,\ \boldsymbol{G}=-\sum_{i=1}^{n}A_i\widehat{\boldsymbol{r}}_i,\ \boldsymbol{H}=-\sum_{i=1}^{n}\widehat{\boldsymbol{r}}_iA_i\widehat{\boldsymbol{r}}_i$$

and $\widehat{(\cdot)}$ represents the skew-symmetric matrix made by the elements of $(\cdot)$, $\boldsymbol{K}_{se}$ and $\boldsymbol{K}_{bt}$ are the stiffness matrix for extension/shear and bending, respectively. τ_i is the load. Superscript 'b' denotes the vector in the local coordinate system.

The boundary conditions for $\boldsymbol{p}(s)$ and $\boldsymbol{R}(s)$ are known at the base. At the tip, the cable termination applies a force and moment on the backbone which produces the boundary conditions for $\boldsymbol{v}(s)$ and $\boldsymbol{u}(s)$. The ordinary differential equations and the boundary conditions form a boundary value problem.

3 Comparison Between Optimization and Cosserat Rod Models

We compare the optimization and the Cosserat rod model for a CCR of length, $L = 180$ mm with 9 equal segments (10 disks). Each disk has 12 holes, equally spaced along a circle of radius, $a = 8$ mm. Out of the several numerical simulations, the results for four of them are presented in Table 1. 'Cable positions' denote the location of the cable in a particular disk according to the nomenclature in the inset of Fig. 1, starting from the bottom to the tip of the CCR. For each case, the load applied was 400 g and the corresponding change in length of the cable inside the CCR (found out experimentally) denoted as '% Cable actuation' is presented in Table 1. All the simulations were performed using MATLAB in a PC with an Intel processor (3.1 GHz) and 16 GB RAM.

Table 1. Cable routings used for comparison

Case	Type	Cable position	% Cable actuation	RMS Error (mm)	
				Optimization	Cosserat
I	Straight	4-4-4-4-4-4-4-4-4-4	5.5	1.79	2.08
II	Helical	1-2-3-4-5-6-7-8-9-10	7.0	1.85	1.87
III	General	1-2-3-4-4-4-4-3-2-1	5.7	1.44	1.80
IV	General	10-10-10-10-10-10-11-12-1-2	5.3	1.37	1.90

Numerical Simulation Details: For the optimization-based method, Eq. 1 is solved using `fmincon` (available in MATLAB) with the in-built *interior-point* algorithm. For each section a total of 6 variables are solved. The boundary value problem for Cosserat rod model is solved using *shooting method* [15] for a total of 18 state variables. The cable positions, $\boldsymbol{r}_i(s)$ are provided as an 8th order Fourier approximations so that it is $\mathcal{C}^2$ continuous. For both cases, the initial guesses provided for the simulations are the values of the particular variables when the CCR was unactuated.

A sample simulation result for Case III is shown in Fig. 3a. For this simulation, the optimization based method took 2.20 s, where as it took 10.07 s for the Cosserat rod model.

Experimental Setup and Error Comparison: To compare both the models experimentally, a CCR prototype was 3D printed in ABS material ($E = 1.1$ GPa, $\nu = 0.3$) with the dimensions as specified in the beginning of Sect. 3. Thin nylon wires are used as cables for the CCR and 400 g weight is used to actuate these cables. The experimental setup for Case II before actuation is shown in Fig. 3b. The corresponding change in cable length inside the CCR is measured externally and is presented as '% Cable actuation' in Table 1. An image of the CCR is captured before and after actuation and then the simulation results are

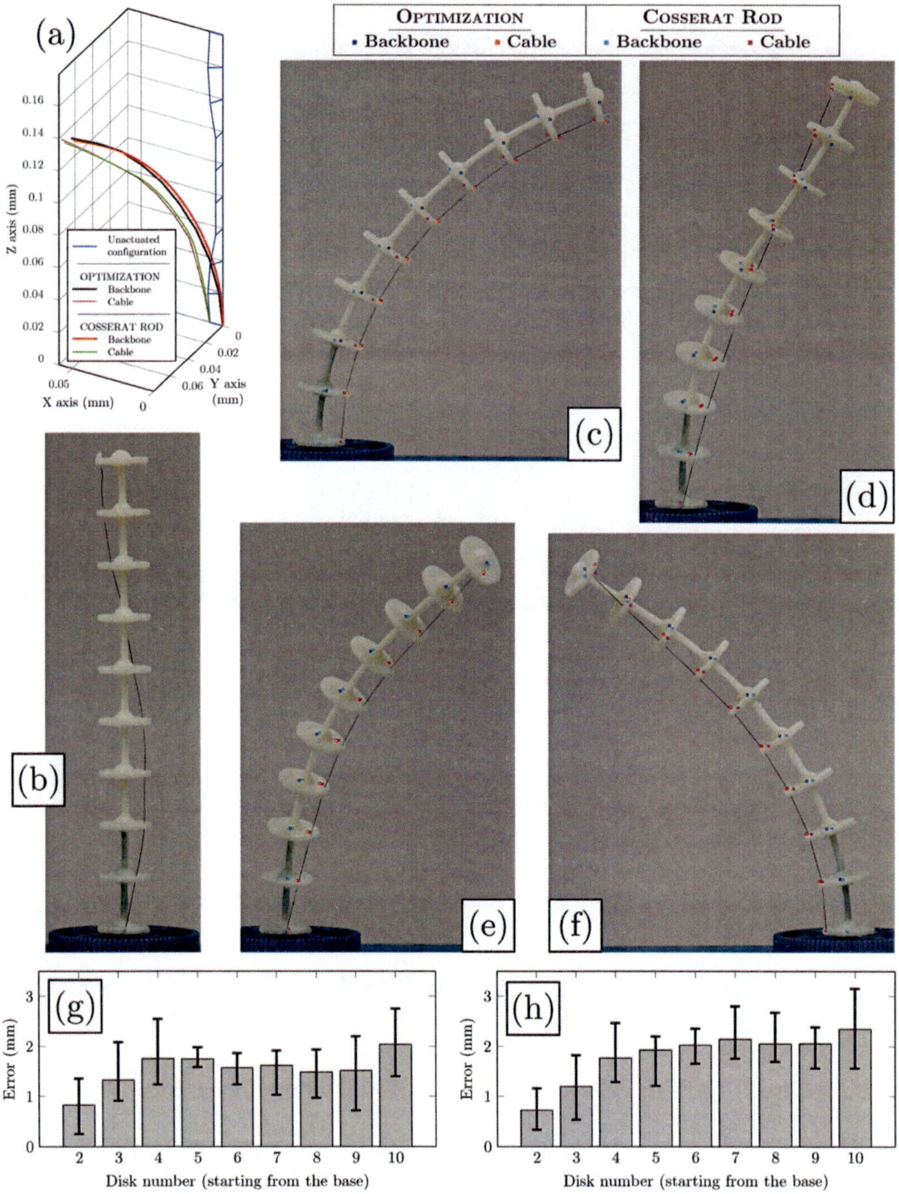

Fig. 3. (a) Simulation result for Case III. (b) Experimental setup. Comparison of experimental results for (c) Case I (d) Case II (e) Case III and (f) Case IV. Error at each disk as compared with the experiments for (g) Optimization-based and, (h) Cosserat rod model.

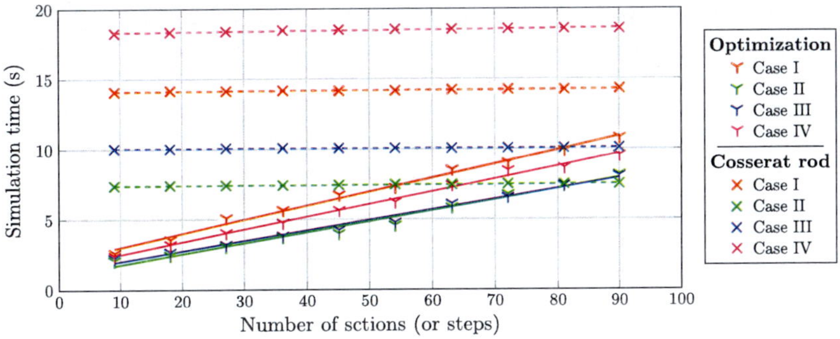

Fig. 4. Variation of simulation time vs number of sections (or step size)

superimposed on the captured image which are shown as blue, cyan, magenta and red dots in the Fig. 3(c)–(f).

The error in each case is found out by image processing. The mean error at each disk is shown in Fig. 3(g)–(h). The RMS error for each case is presented in Table 1. From both the error comparison, it can be seen that both the models can predict the shape of the robot accurately with maximum RMS error of 1.85 mm and 2.08 mm for optimization and Cosserat rod models, respectively. It was observed that in most cases the optimization-based method is more accurate in predicting the shape than the Cosserat rod model.

Comparison of Computation Time against Number of Sections: A comparison of computation time for simulation based on the number of sections (or number of steps for Cosserat rod model), for a fixed length of CCR, was performed for the routings presented in Table 1. Each model is simulated six times and gradually increasing the number of sections, keeping the length constant. The mean value of the results are presented in Fig. 4 as scatter plot and then a linear fit is applied to the data points. It can be seen that computation time increases linearly for the optimization-based model, but does not vary significantly for the Cosserat rod model. Hence, it is advantageous to use the Cosserat rod model when there is need for large number of sections. However, typically the number of sections is within 20, and the optimization-based method is faster is most of these cases. The simulation times for the optimization-based method can be further reduced if one provides the gradients as an input to `fmincon`.

4 Conclusion

In this paper we have presented a comparison of two modelling approaches for the shape-prediction of a cable-driven continuum robot (CCR). The optimization-based method is based on geometry, so it does not require the material properties as in the Cosserat rod model and it is significantly faster in most simulations. Even though Cosserat rod model requires elastic properties for kinematic analysis, the accuracy of the results are comparable with the optimization-based method. Based on this work, we can conclude that one should opt for the optimization-based method when the material properties are not known or hard to determine and when one needs faster simulation as in a possible model-based control scheme. However, for CCRs with large number of segments (more than 50), one may choose Cosserat rod model for calculations. Both the models predict the shape of the CCR reasonably accurately for the experiments performed in this work (with RMS error less than 2.1 mm). The optimization-based method is currently being extended to include other configurations of the CCR and to further reduce the simulation time for a model-based controller.

Acknowledgement. The 3D printing of the CCR was done at the Design Innovation Center (DIC), UTSAAH Laboratory both at the Centre for Product Design and Manufacturing, Indian Institute of Science, Bangalore. The authors thank the help provided at these labs.

Conflict of Interest. The authors declare that they have no conflict of interest.

References

1. Kolachalama, S., Lakshmanan, S.: Continuum robots for manipulation applications: a survey. J. Robot. **2020**, 1–19 (2020)
2. Webster, R.J., I., Jones, B.A.: Design and kinematic modeling of constant curvature continuum robots: a review. Int. J. Robot. Res. **29**(13), 1661–1683 (2010)
3. Rao, P., Peyron, Q., Lilge, S., Burgner-Kahrs, J.: How to model tendon-driven continuum robots and benchmark modelling performance. Front. Robot. AI **7** (2021)
4. Gravagne, I., Walker, I.: On the kinematics of remotely-actuated continuum robots. In: Proceedings 2000 ICRA. Millennium Conference. IEEE International Conference on Robotics and Automation. Symposia Proceedings, vol. 3, pp. 2544–2550 (2000)
5. Bieze, T.M., Kruszewski, A., Carrez, B., Duriez, C.: Design, implementation, and control of a deformable manipulator robot based on a compliant spine. Int. J. Robot. Res. **39**(14), 1604–1619 (2020)
6. Grazioso, S., Gironimo, G.D., Siciliano, B.: A geometrically exact model for soft continuum robots: the finite element deformation space formulation. Soft Robot. **6**(6), 790–811 (2019)
7. Yuan, H., Zhou, L., Xu, W.: A comprehensive static model of cable-driven multi-section continuum robots considering friction effect. Mech. Mach. Theory **135**, 130–149 (2019)
8. Oliver-Butler, K., Till, J., Rucker, C.: Continuum robot stiffness under external loads and prescribed tendon displacements. IEEE Trans. Robot. **35**(2), 403–419 (2019)
9. Rucker, D.C., Webster, R.J., III.: Statics and dynamics of continuum robots with general tendon routing and external loading. IEEE Trans. Robot. **27**(6), 1033–1044 (2011)
10. Dehghani, M., Moosavian, S.A.A.: Modeling of continuum robots with twisted tendon actuation systems. In: 2013 First RSI/ISM International Conference on Robotics and Mechatronics (ICRoM), pp. 14–19 (2013)

11. Mishra, A.K., Mondini, A., Del Dottore, E., Sadeghi, A., Tramacere, F., Mazzolai, B.: Modular continuum manipulator: analysis and characterization of its basic module. Biomimetics **3**(1), 3 (2018)
12. Dalvand, M.M., Nahavandi, S., Howe, R.D.: An analytical loading model for n -tendon continuum robots. IEEE Trans. Robot. **34**(5), 1215–1225 (2018)
13. Liu, Z., Zhang, X., Cai, Z., Peng, H., Wu, Z.: Real-time dynamics of cable-driven continuum robots considering the cable constraint and friction effect. IEEE Robot. Autom. Lett. **6**(4), 6235–6242 (2021)
14. Ashwin, K.P., Ghosal, A.: Profile estimation of a cable-driven continuum robot with general cable routing. In: IFToMM WC 2019. MMS, vol. 73, pp. 1879–1888. Springer, Cham (2019). https://doi.org/10.1007/978-3-030-20131-9_186
15. Till, J., Aloi, V., Rucker, C.: Real-time dynamics of soft and continuum robots based on cosserat rod models. Int. J. Robot. Res. **38**(6), 723–746 (2019)

Experimental Compliance Matrix Derivation for Enhancing Trajectory Tracking of a 2-DoF High-Accelerated Over-Constrained Mechanism

Erkan Paksoy(✉), Mehmet Ismet Can Dede, and Gökhan Kiper

Izmir Institute of Technology, Izmir, Turkey
{erkanpaksoy,candede,gokhankiper}@iyte.edu.tr

Abstract. If the positioning accuracy of the end-effector of a robot has high priority, compliance characteristics of the elements of its mechanism should be considered. Due to the external loading on the robot, the dimensions of the elements change and this leads to positioning errors for the end-effector. In this paper, an experimental test setup and an experimental procedure are described to derive the compliance characteristics of a planar 2-degree-of-freedom mechanism.

Keywords: Positioning accuracy · Parallel mechanism · Experimental compliance matrix derivation

1 Introduction

In order to express the positioning performance of a robotic manipulator, commonly the resolution, repeatability and accuracy of the robot are considered. The resolution is defined as the smallest incremental step that the robot's end-effector can move, and it mostly depends on the actuator and sensor capabilities. Repeatability is defined as the robot's ability to return to the same position and orientation. Accuracy is a measure of how accurately the robot can move to a desired location in the workspace [2].

Factors that affect the accuracy of a robotic manipulator named as inaccuracy factors and classified into two groups as geometrical errors and non-geometrical errors. Geometrical errors are due to three factors: manufacturing tolerances, assembly process and joint clearance. On the other hand, non-geometrical errors can be categorized into 5 subgroups which are compliance errors, measurement errors, environmental factors (temperature, humidity), control errors and the final one is the problems caused in the joint structure: friction, backlash and wear [6].

All 3 subgroups in the geometrical factors affect the accuracy of the robot and joint clearance errors have a dominant effect on the repeatability of the robot. By various calibration methods, the accuracy problems can be solved. To enhance the repeatability of a robot, an over-constrained kinematic structure can be used

N. V. Khang et al. (Eds.): ASIAN MMS 2021, MMS 113, pp. 354–362, 2022.
https://doi.org/10.1007/978-3-030-91892-7_33

so that the effect of the joint clearances is reduced. Over-constrained mechanisms have lower computed degrees of freedom (DoF) than practical degrees of freedom [3].

In this paper, we want to focus on how compliance information of the end-effector point can be obtained for a 2-DoF over-constrained planar parallel manipulator. Here, we assume that the other non-geometrical factors have small effect on the end-effector position compared to the effect of the compliance errors. The reason of assuming compliance errors have dominant effect on the end-effector location is that the mechanism includes links manufactured as a combination of aluminum parts and carbon fiber tubes connected to each other by glue and the end-effector accelerations are up to 5 g.

In order to determine the Cartesian compliance matrix or stiffness matrix of the manipulator, there are two methods classified as analytical and experimental stiffness modeling methods. In this paper, an experimental stiffness modeling method is described and the results of the experiments are presented for a 2-DoF over-constrained planar mechanism.

In experimental stiffness evaluation systems, generally the system consists of 2 elements; one of them is the displacement measurement sensor and the other one is the calibrated masses to create different set of force matrices. In an example experimental method, a formulation for numerical and experimental stiffness analysis and basic principles on how the stiffness matrix of a manipulator can be obtained experimentally are given. In a previous study, an experimental stiffness measuring system called Milli-CATRASYS system was produced to procure the stiffness characteristics of a parallel manipulator called CaPaMan. This system includes LVDT sensors on the steel wires in order to measure the end-effector displacements and calibrated masses on the end of each wire to create different set of force matrices [1].

In another experimental stiffness measurement method, a measurement system composed of cameras is used to measure the end-effector displacements of a haptic device [7].

In [4], for experimental validation of the analytical stiffness model of the R-CUBE mechanism, a test setup combined of a laser range sensor and a pulley-guide system is used.

In this paper, Faro Prime Measuring Arm 1.2 is used as the displacement measuring element and calibrated weights are used via a wire-pulley system to create different forces at the end-effector of a 2-DoF over-constrained mechanism designed for planar laser marking operation with high acceleration motion capability. The aim of this paper is to obtain compliant displacements of the center of gravity (CG) of the end-effector by applying a variety of forces at various locations of its workspace.

2 2-DoF Planar Over-Constrained Mechanism

The two mechanisms presented in Fig. 1a and 1b are kinematically equivalent when the positioning of point C is of concern given that the corresponding link

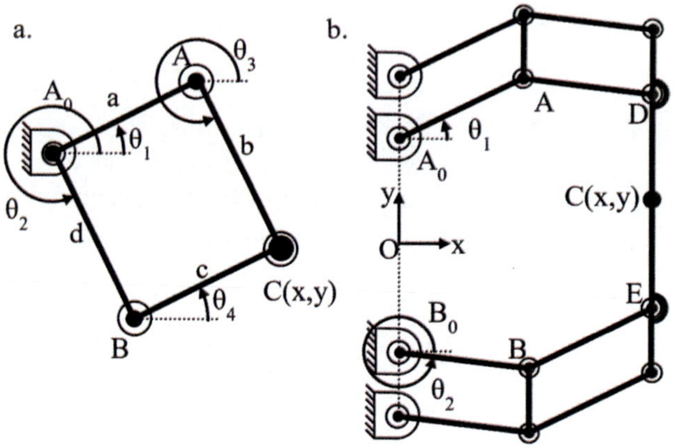

Fig. 1. a) 5R mechanism, b) 6R over-constrained mechanism [5]

lengths are equal to each other. An extensive description of this kinematic equivalence is presented in [5]. The advantage of having an over-constrained mechanism is to have better repeatability and enhanced stiffness performance. However, for the mechanism in Fig. 1b, there is no analytical inverse kinematics solution. For calibration and control purposes, the inverse kinematics of hidden robot model given in Fig. 1a can be used.

In Fig. 2, CAD model of the over-constrained mechanism is illustrated and important components of the mechanism are explained.

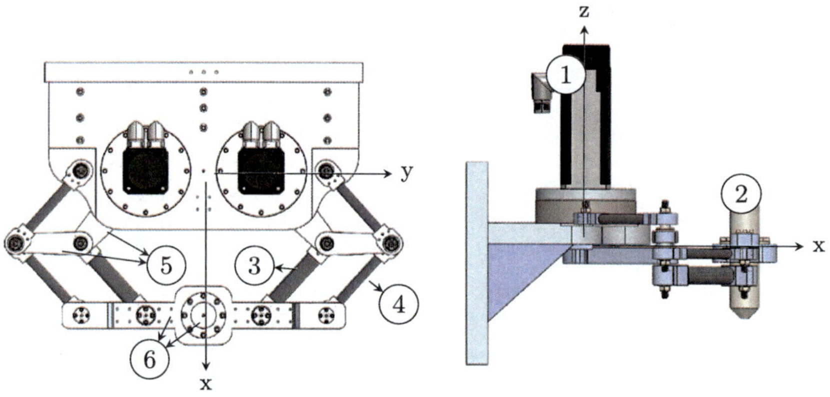

Fig. 2. a) Top-view of the 3D Model, b) Right-view of the 3D model; 1: Motor and reducers, 2: Replica of the laser-head end-effector, 3: Thick distal links, 4: Thin links, 5: Aluminum Links, 6: Platform including end-effector

In the following, position level forward kinematics of hidden robot model is given:

$$\left|\overrightarrow{A_0A}\right| = \left|\overrightarrow{A_0B}\right| = \left|\overrightarrow{AC}\right| = \left|\overrightarrow{BC}\right| = l = 150 \text{ mm} \tag{1}$$

$$\text{Because of the parallelogram loops; } \theta_1 = \theta_4, \theta_2 = \theta_3 \tag{2}$$

$$\vec{r_c} = \overrightarrow{A_0A} + \overrightarrow{AC} = \overrightarrow{A_0B} + \overrightarrow{BC} = l(e^{i\theta_1} + e^{i\theta_3}) = l(e^{i\theta_2} + e^{i\theta_4}) \tag{3}$$

$$\text{By using (2) in (3); } \vec{r_c} = l(e^{i\theta_1} + e^{i\theta_2}) \tag{4}$$

$$\text{By simplifying (4); } x = l[cos(\theta_1) + (cos(\theta_2)], y = l[sin(\theta_1) + (sin(\theta_2)] \tag{5}$$

3 Experimental Setup

In Fig. 3, the experimental test setup to measure the compliant displacement of the end-effector is given. To measure the compliant displacement of the

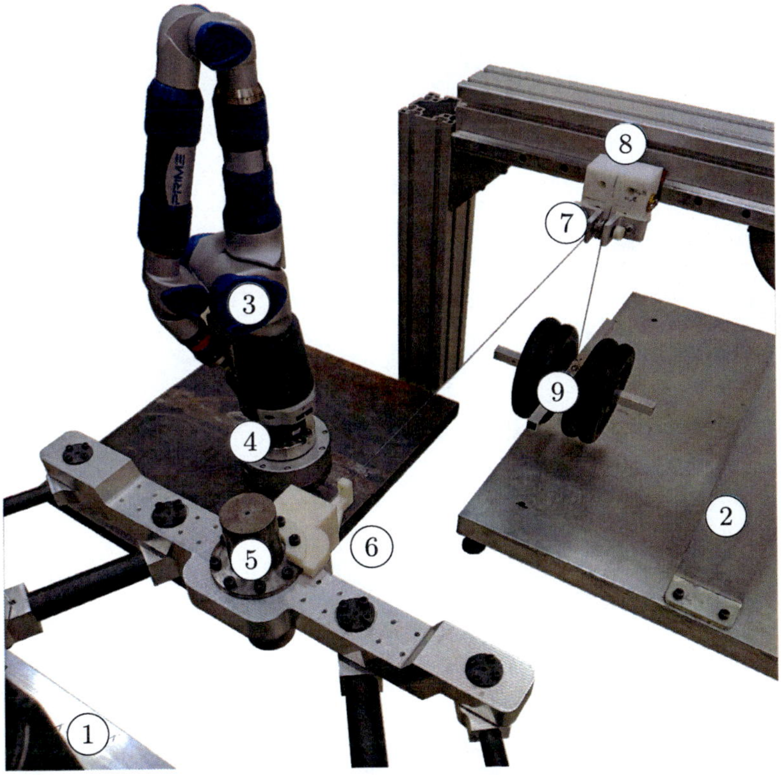

Fig. 3. Experimental test setup; 1: 2-DoF planar mechanism, 2: Translational mechanism to arrange the force direction, 3: Faro Prime Arm, 4: Magnetic base, 5: Replica of laser head, 6: Guide, 7: Pulley, 8: Linear rail, 9: Calibrated weight

end-effector point of the mechanism Faro Prime Arm (±23 µm measurement accuracy) is used and to exert a force at any point in the workspace of the mechanism, a system that includes a 3D translational mechanism, calibrated weight and steel wire is used. Faro Prime Arm is fixed to a 60 kg metal sheet by using a magnetic base ensures that the location of the Faro Prime Arm does not change while taking measurements. The center of gravity of the replica of the laser-head end-effector and the guide details are given in Fig. 4. For clarity, the CG of the end-effector is defined with respect to the whole moving platform including end-effector (see Fig. 2). Compliant displacement measurements at the CG can change with respect to mainly two factors. These factors are the a) measurement point on the mechanism, b) the force vector $\vec{F}$.

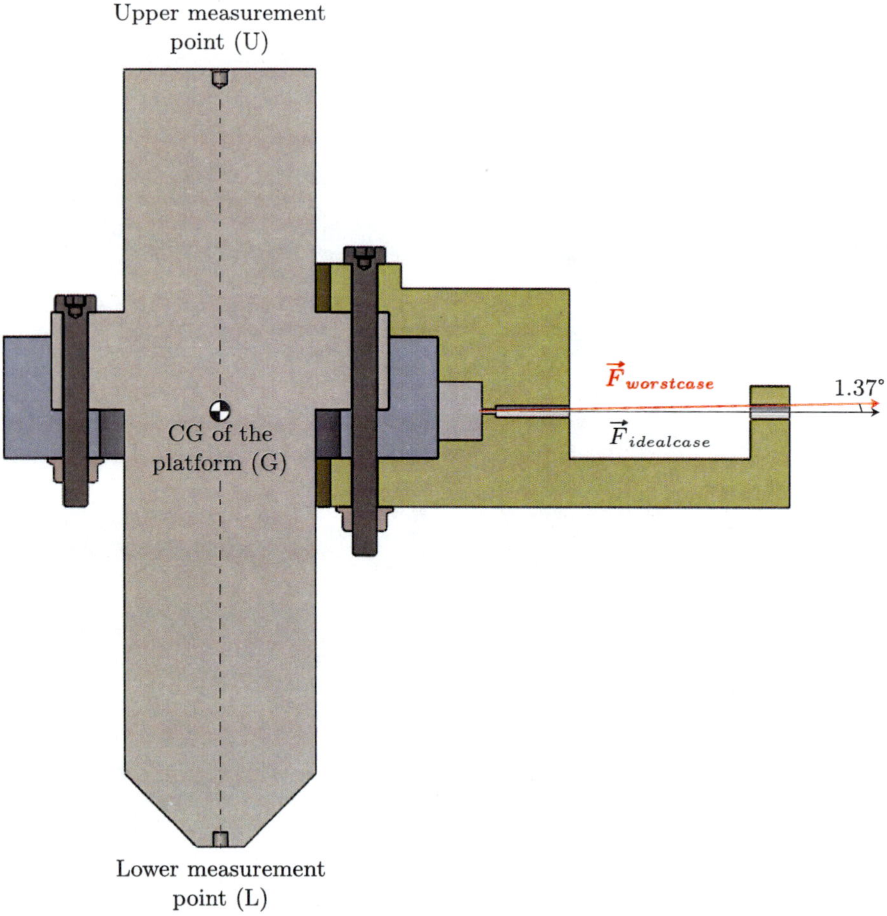

Fig. 4. Replica of laser head and guide details: $\left|\overrightarrow{UG}\right| = 69$ mm, $\left|\overrightarrow{UL}\right| = 160$ mm

To acquire the compliant displacements at the CG of the platform as a 6-DoF information (3 translational displacements and 3 angular displacements), replica of the laser-head that contains 2 measurement points is designed and located at the end-effector of the mechanism (Fig. 4). By using $\overrightarrow{UL}$ information presented in Fig. 4, compliance displacements at any point along the vertical-axis for the end-effector of the mechanism can be derived. In this way, we can calculate the positional deviation of the laser beam on the workpiece. Since the distance between the laser head and the workpiece can change with respect to the material, laser power and the thickness of the workpiece, in this study, we used the CG point for our calculations. To obtain repeatable tests, calibrated masses are used to generate the magnitude of the exerted force at the end-effector. The direction of the external force is regulated by the use of a guide that is presented in Fig. 4. The steel wire goes through the cylindrical hole ensures that the maximum deviation of force direction is 1.37°. That means, 99.97% of the loading will be in the desired direction if we neglect compliance of the mechanism because of the small forcing along other directions.

4 Experimental Procedure

The coordinate system of the robot is located between the two motors and the operational workspace of the robot is determined as 150 mm × 100 mm. In Fig. 5, the coordinate system of the robot and 15 measurement points on the workspace are given.

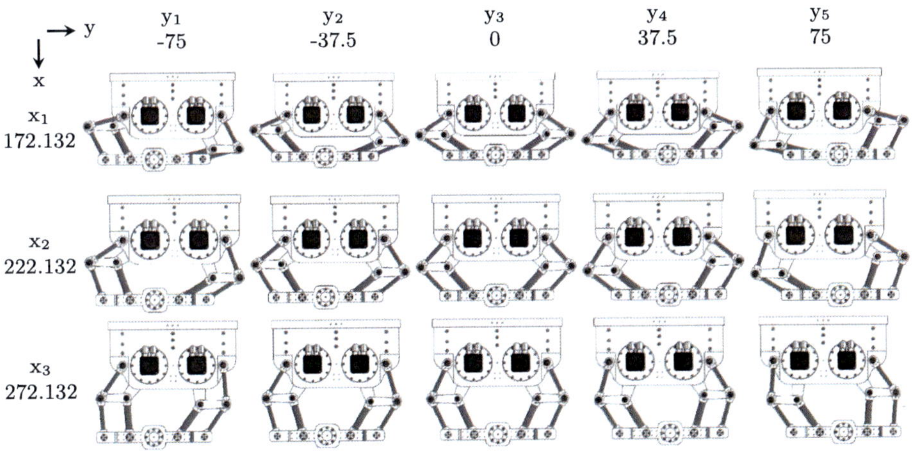

Fig. 5. Mechanism configurations and end-effector locations given in mm for the measurement points

At all measurement points of the workspace, the applied external forces are defined in steps in Table 1. In Step 0, there is a small amount of force because of the steel wire system. In each step, approximately 5 kg mass is added to system to exert a force and then, the coordinates of the upper measurement point (U) and Lower Measurement Point (L) are measured and recorded (Fig. 4).

Table 1. Applied external force at each test step

Step	Exerted force (kgf)	≈ in (N)
0	0.11	0
1	5	50
2	9.92	100
3	14.89	150
4	19.83	200
5	24.86	250

To determine the location of CG of the platform (G) Eqs. 6, 7 and 8 are used. If $\overrightarrow{UL}$ is known for each step, then both translational and angular compliant displacements of the platform can be found.

$$\vec{U} = U_x\vec{i} + U_y\vec{j} + U_z\vec{k}, \vec{L} = L_x\vec{i} + L_y\vec{j} + L_z\vec{k} \tag{6}$$

$$\overrightarrow{UL} = (L_x - U_x)\vec{i} + (L_y - U_y)\vec{j} + (L_z - U_z)\vec{k} \tag{7}$$

$$\vec{G} = \vec{U} + \frac{69}{160}\overrightarrow{UL} \tag{8}$$

5 Test Results

In Table 2, the translational compliant displacements under the forces (50–250 N) along $+X$ direction are given. As it was expected, because of the symmetrical structure of the mechanism, there are no ΔY displacement for the points that are located on $+X$-axis ($Y = 0$ mm) while the forces are increasing at $+X$-direction.

Table 2. Compliant displacements of CG under the forces in $+X$-direction (mm)

$\vec{F}$ in $\vec{X}$	ΔX	ΔY	ΔZ	ΔX	ΔY	ΔZ	ΔX	ΔY	ΔZ	ΔX	ΔY	ΔZ	ΔX	ΔY	ΔZ
	x_1y_1 (172.132, −75)			**x_1y_2 (172.132, −37.5)**			**x_1y_3 (172.132, 0)**			**x_1y_4 (172.132, 37.5)**			**x_1y_5 (172.132, 75)**		
50 N	0.105	0.061	-0.031	0.035	0.027	-0.009	0.027	0.002	-0.011	0.044	-0.024	-0.017	0.088	-0.058	-0.022
100 N	0.198	0.109	-0.039	0.085	0.047	-0.016	0.059	0.002	-0.018	0.091	-0.040	-0.028	0.182	-0.094	-0.028
150 N	0.289	0.164	-0.046	0.132	0.068	-0.030	0.094	0.000	-0.044	0.137	-0.059	-0.045	0.279	-0.151	-0.047
200 N	0.389	0.205	-0.060	0.180	0.080	-0.051	0.129	-0.006	-0.056	0.184	-0.077	-0.057	0.369	-0.190	-0.055
250 N	0.479	0.257	-0.080	0.226	0.096	-0.065	0.164	-0.009	-0.067	0.231	-0.099	-0.066	0.462	-0.232	-0.070
	x_2y_1 (222.132, −75)			**x_2y_2 (222.132, −37.5)**			**x_2y_3 (222.132, 0)**			**x_2y_4 (222.132, 37.5)**			**x_2y_5 (222.132, 75)**		
50 N	0.045	0.039	-0.020	0.027	0.019	0.008	0.024	-0.003	-0.023	0.027	-0.014	-0.012	0.044	-0.042	-0.015
100 N	0.085	0.083	-0.031	0.054	0.036	-0.021	0.054	-0.005	-0.041	0.058	-0.026	-0.030	0.089	-0.084	-0.041
150 N	0.131	0.118	-0.051	0.087	0.048	-0.026	0.083	-0.001	-0.059	0.091	-0.041	-0.031	0.133	-0.126	-0.054
200 N	0.176	0.151	-0.073	0.120	0.062	-0.047	0.111	0.007	-0.068	0.118	-0.056	-0.048	0.178	-0.153	-0.064
250 N	0.218	0.201	-0.081	0.147	0.077	-0.064	0.140	0.002	-0.079	0.147	-0.074	-0.070	0.220	-0.200	-0.091
	x_3y_1 (272.132, −75)			**x_3y_2 (272.132, −37.5)**			**x_3y_3 (272.132, 0)**			**x_3y_4 (272.132, 37.5)**			**x_3y_5 (272.132, 75)**		
50 N	0.039	0.088	-0.023	0.022	0.027	-0.023	0.012	0.002	-0.016	0.020	-0.029	-0.016	0.039	-0.085	-0.023
100 N	0.084	0.181	-0.052	0.050	0.070	-0.035	0.031	0.001	-0.030	0.043	-0.064	-0.037	0.084	-0.175	-0.041
150 N	0.124	0.275	-0.054	0.075	0.104	-0.070	0.048	0.000	-0.055	0.068	-0.101	-0.061	0.128	-0.270	-0.065
200 N	0.170	0.359	-0.077	0.098	0.138	-0.087	0.071	-0.006	-0.079	0.094	-0.140	-0.080	0.167	-0.348	-0.082
250 N	0.209	0.447	-0.104	0.117	0.175	-0.105	0.094	-0.003	-0.096	0.114	-0.164	-0.090	0.208	-0.430	-0.107

Also, absolute magnitudes of the translational displacement values are found to be symmetrical with respect to the $+X$-axis taking into account the resolution of the measurement system. These two observations are obtained from the measurement results to verify the test procedure's suitability.

Moreover, while forces are increasing linearly along $+X$-direction 50 N to 250 N, the displacements are also increasing linearly. For instance, the displacement values for the measurement point (172.123, −75) are increasing linearly as the external force is increased linearly (0.1 mm for 50 N, 0.2 mm for 100 N, etc.). These results suggest that for each point there is almost a linear relationship between the external force and compliant displacements along at the X- and Y-directions.

Table 3. Compliant displacements of CG under the forces in $-Y$-direction (mm)

$\vec{F}$ in $-\vec{Y}$	ΔX	ΔY	ΔZ	ΔX	ΔY	ΔZ	ΔX	ΔY	ΔZ	ΔX	ΔY	ΔZ	ΔX	ΔY	ΔZ
	x_1y_1 (172.132, −75)			**x_1y_2 (172.132, −37.5)**			**x_1y_3 (172.132, 0)**			**x_1y_4 (172.132, 37.5)**			**x_1y_5 (172.132, 75)**		
50 N	-0.069	-0.079	-0.006	-0.026	-0.091	-0.001	-0.003	-0.084	0.006	0.022	-0.078	-0.003	0.054	-0.086	0.007
100 N	-0.110	-0.190	-0.007	-0.047	-0.182	0.003	0.002	-0.172	0.008	0.051	-0.168	0.017	0.113	-0.182	0.016
150 N	-0.177	-0.283	-0.012	-0.074	-0.268	0.002	-0.013	-0.246	0.006	0.071	-0.256	0.005	0.160	-0.273	0.012
200 N	-0.243	-0.374	-0.019	-0.090	-0.357	0.001	-0.020	-0.326	0.010	0.088	-0.335	0.007	0.216	-0.359	0.013
250 N	-0.309	-0.473	-0.033	-0.115	-0.449	0.001	-0.019	-0.420	0.005	0.117	-0.434	0.019	0.266	-0.449	0.015
	x_2y_1 (222.132, −75)			**x_2y_2 (222.132, −37.5)**			**x_2y_3 (222.132, 0)**			**x_2y_4 (222.132, 37.5)**			**x_2y_5 (222.132, 75)**		
50 N	-0.033	-0.133	-0.005	-0.031	-0.115	-0.009	-0.004	-0.110	0.004	0.020	-0.120	0.000	0.055	-0.122	0.011
100 N	-0.078	-0.258	-0.009	-0.040	-0.243	-0.006	0.001	-0.232	0.006	0.050	-0.236	0.006	0.087	-0.251	0.013
150 N	-0.114	-0.384	-0.015	-0.063	-0.366	-0.006	0.004	-0.357	0.013	0.065	-0.358	0.008	0.129	-0.375	0.021
200 N	-0.161	-0.509	-0.008	-0.077	-0.488	-0.012	0.002	-0.475	0.014	0.083	-0.473	0.000	0.174	-0.499	0.014
250 N	-0.204	-0.638	-0.016	-0.099	-0.609	-0.006	0.009	-0.604	0.009	0.099	-0.593	0.005	0.205	-0.628	0.014
	x_3y_1 (272.132, −75)			**x_3y_2 (272.132, −37.5)**			**x_3y_3 (272.132, 0)**			**x_3y_4 (272.132, 37.5)**			**x_3y_5 (272.132, 75)**		
50 N	-0.061	-0.346	-0.004	-0.039	-0.265	-0.009	-0.002	-0.246	0.000	0.038	-0.255	0.001	0.098	-0.319	0.000
100 N	-0.148	-0.685	-0.009	-0.071	-0.532	-0.010	0.009	-0.494	-0.023	0.077	-0.523	-0.005	0.193	-0.658	-0.010
150 N	-0.261	-1.001	-0.026	-0.105	-0.789	-0.014	0.013	-0.755	-0.014	0.118	-0.792	-0.002	0.275	-0.998	-0.005
200 N	-0.335	-1.349	-0.023	-0.140	-1.052	-0.015	0.003	-0.990	-0.014	0.155	-1.042	0.004	0.358	-1.331	-0.012
250 N	-0.421	-1.690	-0.028	-0.182	-1.319	-0.016	0.011	-1.251	-0.008	0.195	-1.305	0.002	0.458	-1.671	-0.009

In Table 3, the translational compliant displacements under the forces (50–250 N) along $-Y$-direction are given. This time, due to the symmetry there are no ΔX displacement at the $+X$-axis while the external load is applied $-Y$-direction. The symmetrical compliant displacement magnitudes with respect to the $+X$-axis are obtained and almost a linear relationship can be seen again.

6 Conclusion

As a result of this work, the compliance behavior of a 2-DoF planar over-constrained mechanism is gathered by using a test setup including Faro Prime arm as coordinate-measurement machine and a combination of 3D translational mechanism and steel-wire equipment as external force application system. This compliance information will be used in future studies to improve the positioning accuracy of the 2-DoF planar over-constrained mechanism during high acceleration (up to 5 g) operation.

Acknowledgement. The study is supported in part by The Scientific and Technological Research Council of Turkey via grant number 116M272. We thank Dr. Emre Uzunoğlu and Mr. İbrahimcan Görgülü, who contributed to the design of the experimental test setup.

References

1. Ceccarelli, M., Carbone, G.: Numerical and experimental analysis of the stiffness performances of parallel manipulators. In: 2nd International Colloquium Collaborative Research Centre, vol. 562, pp. 21–35. Citeseer (2005)
2. Conrad, K.L., Shiakolas, P.S., Yih, T.: Robotic calibration issues: accuracy, repeatability and calibration. In: Proceedings of the 8th Mediterranean Conference on Control and Automation (MED2000), Rio, Patras, Greece, vol. 1719 (2000)
3. Gogu, G.: Mobility of mechanisms: a critical review. Mech. Mach. Theory **40**(9), 1068–1097 (2005)
4. Görgülü, İ, Dede, M.İC., Carbone, G.: An experimental test procedure for validation of stiffness model: a case study for R-CUBE parallel mechanism. In: Kuo, C.-H., Lin, P.-C., Essomba, T., Chen, G.-C. (eds.) ISRM 2019. MMS, vol. 78, pp. 391–402. Springer, Cham (2020). https://doi.org/10.1007/978-3-030-30036-4_35
5. Kiper, G., Dede, M.İ.C., Uzunoğlu, E., Mastar, E.: Use of hidden robot concept for calibration of an over-constrained mechanism (2015)
6. Klimchik, A.: Enhanced stiffness modeling of serial and parallel manipulators for robotic-based processing of high performance materials. Ph.D. thesis, Ecole Centrale de Nantes (ECN); Ecole des Mines de Nantes (2011)
7. Taner, B., Dede, M.İC.: Image processing based stiffness mapping of a haptic device. In: Corves, B., Lovasz, E.-C., Hüsing, M., Maniu, I., Gruescu, C. (eds.) New Advances in Mechanisms, Mechanical Transmissions and Robotics. MMS, vol. 46, pp. 447–454. Springer, Cham (2017). https://doi.org/10.1007/978-3-319-45450-4_45

Modelling to Analyze the Vertical Oscillation of RHex with Flexible Body

Minh Khong[1,2] and Van-Luc Ngo[1,2](✉)

[1] Faculty of Mechanical Engineering and Mechatronics, Phenikaa University, Hanoi 12116, Vietnam
luc.ngovan@phenikaa-uni.edu.vn
[2] Phenikaa Research and Technology Institute (PRATI), A&A Green Phoenix Group JSC, No. 167 Hoang Ngan, Trung Hoa, Cau Giay, Hanoi 11313, Vietnam

Abstract. RHex stands for Robot Hexapod, which is a type of robot that imitates insects with six legs. There have been several previous studies on this type of robot. However, previous researches have mainly developed RHex with rigid body, that mean the whole robot body is a rigid block. RHex with rigid bodies can operate flexibly on flat surfaces but has many limitations when operating on rough surfaces. RHex with rigid body is also not optimal in used energy because each step of legs has to lift the whole robot body. In fact, insects all have flexible body, which are divided into parts and connected to each other by flexible joints. This study will firstly review previous research and development on RHex, and then built a model to analyze vertical oscillations of RHex with flexible body. The objective of this study is to develop a RHex model which can operate flexibly in rough surfaces with minimal energy consumption.

Keywords: Robot Hexapod · RHex with flexible body · Vertical oscillation · Optimal energy · Modeling of RHex

1 Introduction

Robot Hexapod is a type of robot designed by imitated insects with six legs and abbreviated as RHex. This type of robot has been studied since 1998 and several models have been developed depending on the using purpose until now. RHex Platform (Fig. 1a) is a robot model with a rigid body that has a compact structure and is very agile [1]. RHex Platform is able to walk and run at speeds of up to 2.7 m/s on flat surface as well as the ability to climb slopes with inclines up to 40°. Shelly RHex (Fig. 1b) is an amphibian robot which was developed from the RHex Platform model for the purpose of increasing operability [1, 2]. Shelly RHex is designed with a closed shell cover that can protect the electronical equipments from water for the ability to travel on land and walking in water. Rugged RHex (Fig. 1c) is a robot model developed from Shelly RHex for commercial purposes [1, 2]. Rugged RHex has several advanced features such as the ability to carry more heavy objects with a load of up to 2 kg, a travel distance of 2 km, very good in off-road and 6 h of operation time. Rugged RHex is also an amphibian robot with ability

N. V. Khang et al. (Eds.): ASIAN MMS 2021, MMS 113, pp. 363–369, 2022.
https://doi.org/10.1007/978-3-030-91892-7_34

to operate both on land and in water. AQUA (Fig. 1d) is a model of RHex, which is developed base on Rugged RHex for the purpose of operating underwater [1–3]. The legs of AQUA are designed in the form of paddles to be able to swim and dive underwater. AQUA can dive to a depth of 10 m, the observation system is capable of creating three-dimensional maps of the coral reefs under the sea [3]. Another RHex with rigid body can be mentioned as the X-RHex model (Fig. 1e) which is equipped a mass balance system so that the Robot has the ability to operate more flexibly on different terrains [4]. X-RHex has the ability to climb stairs, jump, roll, climb slopes and the ability to recognize 3D models of the surrounding environment. X-RHex Lite or XRL (Fig. 1f) is a development of X-RHex to improve operating skills and reduced size. Compared with X-RHex, the XRL robot has many outstanding advantages such as greater load carrying capacity, easier to manufacture and significantly reduced weight [5]. Edu robot (Fig. 1g) is another model of RHex with rigid body developed to study the effect of leg stiffness on robot mobility on different terrains [6]. The XJus robot (Fig. 1h) is a model of RHex that is designed with an adjustable body spine stiffness to assess energy consumption [7]. Experimental results with the XJus robot shown that the stiffness of robot spine leads to reduced motor torque when riding on flat terrain, as well as improved ability to overcome obstacles.

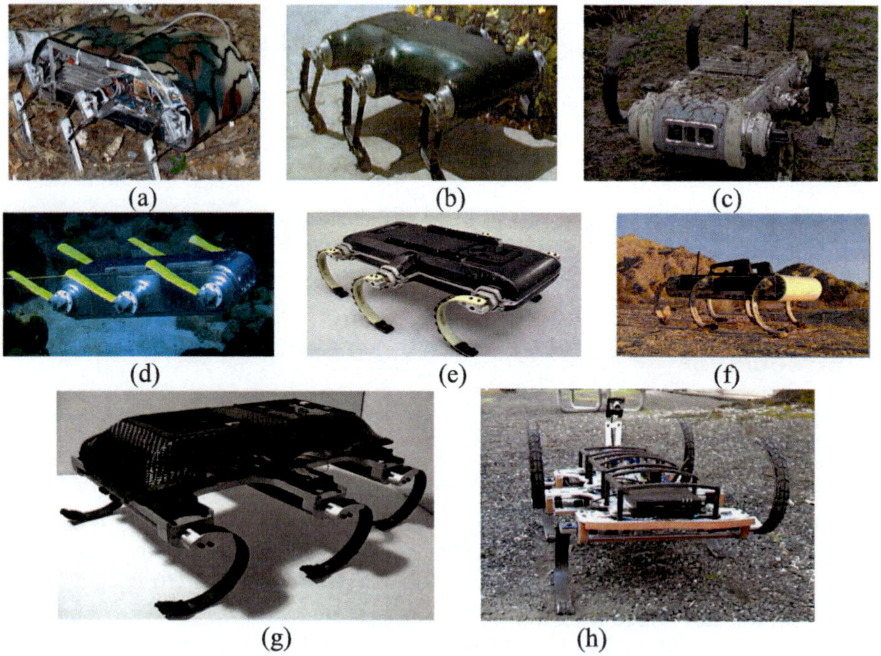

Fig. 1. Several models of RHex have researched and developed in previous studies

This research will create a diagram of rigid multiboby system dynamics for a model of RHex with flexible body to analyze the vertical oscillation during movement on rough surface. The diagram of rigid multiboby system dynamics is built in a general form that can be used for the research and manufacture of RHex with flexible body. The purpose of this research is to design and build a RHex with the ability to move flexibly on different terrains with a minimum of energy. For convenience of expression, the robot model in this study will be named RHex-FB (RHex with Flexible Body).

2 Modelling of RHex with Flexible Body (RHex-FB)

In this study, a simplified model of RHex-FB will be built to consider only vertical oscillations as shown in Fig. 3. The body of the RHex-FB is divided into 3 parts and connected by elastic bars, each body part connect with two legs. In order to calculating dynamics paramaters, whole body of RHex-FB will be considered as a diagram of rigid multiboby system dynamics. The elastic bars connecting the robot body parts are in the form of thin, short and wide plates so that the RHex-FB body is easily bent vertically but difficult to twist. Thus, the bars connecting the RHex-FB body parts can be modeled as shown Fig. 3a, in which the connecting bar is considered as a consonant beam and equivalent to a spring in diagram of RHex-FB. Similarly, the legs of RHex-FB is considered as a suspension in the calculating diagram. The connecting bars and legs RHex-FB are modeled equivalent to spring and suspension, thus, the diagram of the rigid multiboby system dynamics of RHex-FB is created as shown in Fig. 4.

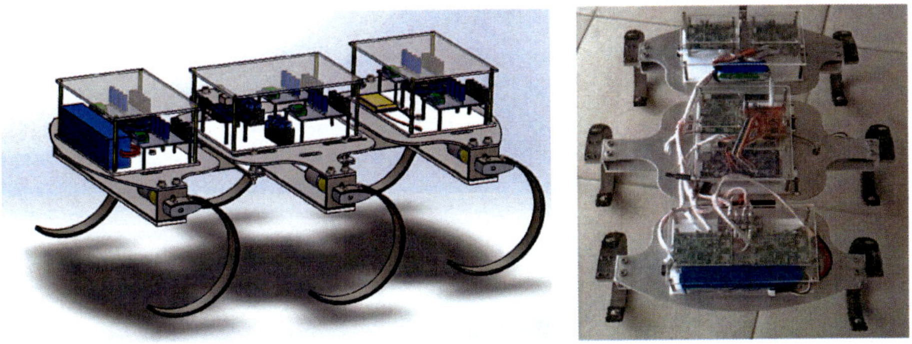

(a) (b)

Fig. 2. Model of robot design (a) and production (b) of RHex-FB

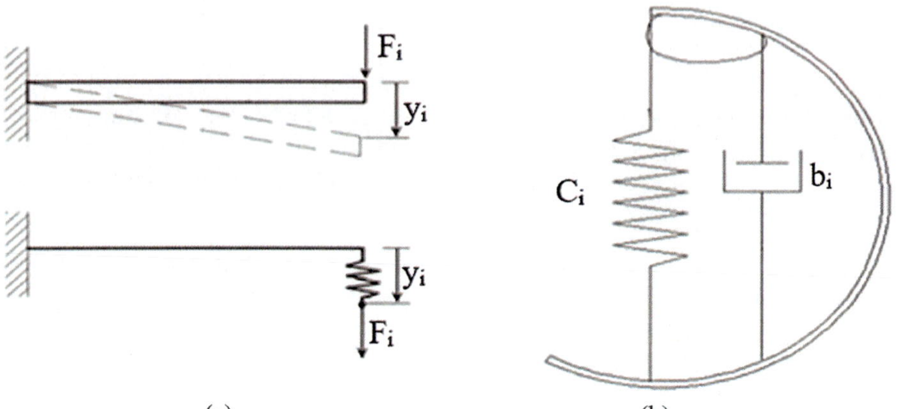

Fig. 3. Connecting bars and legs are modeled by spring and suspension

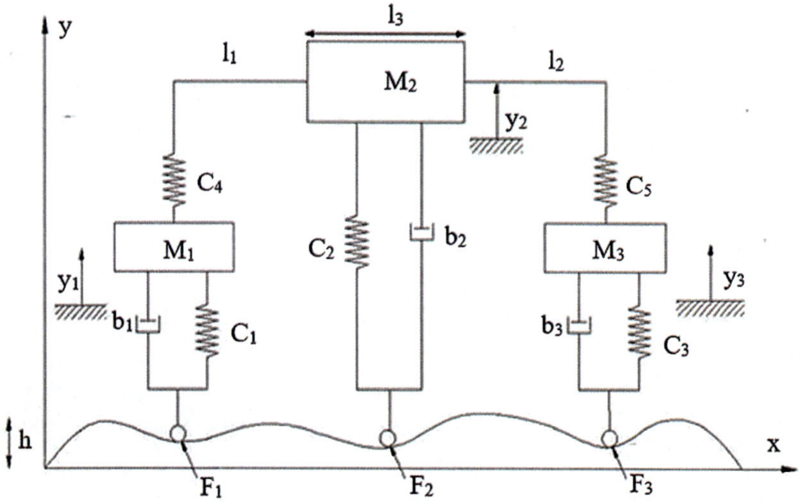

Fig. 4. RHex-FB are modeled by diagram of the rigid multiboby system dynamics

Figure 5 is the diagram to calculate the external forces acting on the RHex-FB at the contact positions between the legs and the road surface. The body of RHex-FB is solidified in diagram to calculate the external force. That meam the body of RHex-FB is considered as a rigid block. Then the formula to determine the external forces at the contact positions will be calculated as with RHex with rigid body [8]. The force components shown in Fig. 5 are included: F_{Si} is the elastic force determined proportional to the elongation and stiffness of legs; F_{Gi} is the friction force, $F_{Gi} = F_{Ai}$; F_{Ai} is the driving force calculated depend on the torque of the motor (τ_{ϕ}) and the distance between

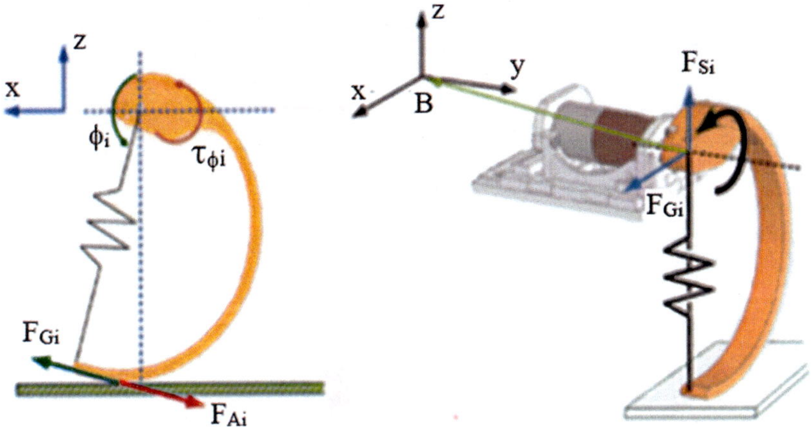

Fig. 5. The diagram to calculate the external forces acting on the RHex-FB

the contact point and the motor (d). The formula to calculate the external forces acting on the RHex-FB is expressed as following:

$$\begin{bmatrix} F_{xi} \\ F_{yi} \\ F_{zi} \end{bmatrix} = \begin{bmatrix} -\sin(\phi i) & -\cos(\phi i) \\ 0 & 0 \\ -\cos(\phi i) & \sin(\phi i) \end{bmatrix} * \begin{bmatrix} F_{Si} \\ F_{Gi} \end{bmatrix} \quad (1)$$

All components in diagram of the rigid multiboby system dynamics which modeling the RHex-Fb are determined depend on the each design. Thus, the diagram can be used to solve RHex-FB dynamics problems.

3 The Vertical Oscillation Equation of RHex with Flexible Body

Assuming that h and L are height and length of the undulations in road surface, the coordinates of the contact position (x*, y*) are determined as [9]:

$$y^* = \frac{h}{2}\left(1 - cos\frac{2\pi x^*}{L}\right) \quad (2)$$

With the notation that Mi is the mass of the body parts of RHex-FB; c_i and b_i are the stiffness of springs and viscous resistance of damper, respectively; l_1 is the distance between the first connecting bar and the the first leg; l_2 is the distance second connecting bar and the the third leg; l_3 is the distance between the two connecting bars. These parameters will be determined depending on the each design of the RHex-FB.

With t is the time variable; v is velocity of RHex-FB, set $\Omega = 2\pi\frac{v}{L}$ then the coordinates of the points of contact are determined as follows:

$$x_1^* = \text{vt}\,;\; y_1^* = \frac{h}{2}\left(1 - cos\frac{2\pi x_1^*}{L}\right) = \frac{h}{2}(1 - cos\Omega t) \quad (3)$$

$$x_2^* = \text{vt} + l_1 + \frac{l_3}{2};\ y_2^* = \frac{h}{2}\left(1 - cos\frac{2\pi x_2^*}{L}\right) = \frac{h}{2}\left(1 - \cos\left(\Omega t + \frac{2\pi\left(l_1 + \frac{l_3}{2}\right)}{L}\right)\right) \tag{4}$$

$$x_3^* = \text{vt} + l_1 + l_2 + l_3;\ y_3^* = \frac{h}{2}\left(1 - cos\frac{2\pi x_3^*}{L}\right) = \frac{h}{2}\left(1 - \cos\left(\Omega t + \frac{2\pi(l_1 + l_2 + l_3)}{L}\right)\right) \tag{5}$$

The diagram of the rigid multiboby system dynamics of RHex-FB will be separated each part corresponding to Mi to establish balanced equations for each part. Combining the balanced equations for each part, the system of balanced equations for whole diagram of RHex-FB can be determined as following:

$$\begin{cases} M_1\ddot{y}_1 + b_1\dot{y}_1 + c_4y_1 - c_4y_2 = F_{Z1} + c_1\frac{h}{2}(1 - \cos\Omega t) + b_1\frac{h}{2}\Omega\sin\Omega t \\ M_2\ddot{y}_2 + b_2\dot{y}_2 - c_4y_1 + (c_4 + c_5)y_2 - c_5y_3 = F_{z2} \\ \quad + c_2\frac{h}{2}\left(1 - \cos\left(\Omega t + \frac{2\pi\left(l_1 + \frac{l_3}{2}\right)}{L}\right)\right) + b_2\frac{h}{2}\Omega\sin\left(\Omega t + \frac{2\pi\left(l_1 + \frac{l_3}{2}\right)}{L}\right) \\ M_3\ddot{y}_3 + b_3\dot{y}_3 + c_5y_3 - c_5y_2 = F_{z3} \\ \quad + c_3\frac{h}{2}\left(1 - \cos\left(\Omega t + \frac{2\pi(l_1 + l_2 + l_3)}{L}\right)\right) + b_3\frac{h}{2}\Omega\sin\left(\Omega t + \frac{2\pi(l_1 + l_2 + l_3)}{L}\right) \end{cases} \tag{6}$$

This system of equations has the form:

$$M\ddot{y} + B\dot{y} + Cy = f(t) \tag{7}$$

Then the solution of the system of Eqs. (7) will have the following form:

$$\begin{cases} y_1 = A_0 + A_1\cos\Omega t + A_2\sin\Omega t + A_3\cos\omega t + A_4\sin\omega t \\ y_2 = B_0 + B_1\cos\Omega t + B_2\sin\Omega t + B_3\cos\omega t + B_4\sin\omega t \\ y_3 = C_0 + C_1\cos\Omega t + C_2\sin\Omega t + C_3\cos\omega t + C_4\sin\omega t \end{cases} \tag{8}$$

4 Results and Discussions

With the aim of creating a model to analyze vertical oscillations of RHex with flexible body, the product direction is a version of RHex-FB that acts like 6-legged insects. In this study, a model to design is proposed as described in Fig. 2, with this model the robot can bend in the vertical plane by an elastic plate joint. Based on this model, this study also built a diagram of the rigid multiboby system dynamics as described in Fig. 3. The general equation of vertical oscillation as show in formula (7) is created for above diagram of the rigid multiboby system dynamics, which has general solution as shown in formula (8). Particular solution can be found by analytic methods or using MathLab software with specifications of the model. In this study, specifications of the model such as dimensions, stiffness, mass, etc. have not been given, because these parameters need to be selected base results of analysis Eq. (7) to select the optimal parameters.

5 Conclusions and Recomemdations

This study has achieved the stated goal of building a model to analyze vertical oscillations of RHex with flexible body. The proposed model of robot can bend in the vertical plane by an elastic plate joint. The diagram of the rigid multiboby system dynamics and general equation of vertical oscillation were also established in this study. These results are basis for performing vertical oscillations analysis of RHex-FB to find the optimal design parameters. The next study will using the MathLab to analyze the vertical oscillations of RHex-FB base on the results in this study as well as using experimental models to find the most optimal design.

References

1. Altendorfer, R., et al.: RHex: a biologically inspired hexapod runner. Auton. Robot. **11**, 207–213 (2001)
2. Summary of the RHex robot platform. http://www.rhex.web.tr/
3. http://epitome.cim.mcgill.ca:8080/AQUA/index_html
4. Galloway, K.C., et al.: X-RHex: a highly mobile hexapedal robot for sensorimotor tasks. University of Pennsylvania. Technical reports (2010)
5. Galloway, K.C., et al.: X-RHex: a highly mobile hexapedal robot for sensorimotor tasks. http://kodlab.seas.upenn.edu/Kod/Xrhextech
6. Galloway, K.C., Clark, J.E., Koditschek, D.E.: Variable stiffness legs for robust, efficient, and stable dynamic running (edu robot). J. Mech. Robot. (2013)
7. Martirosyan, H., Hughes, G., Owlett, T., O'Leary, B., Payne, C.: xJus: a hexapedal robot with a passively flexible spine (2013). http://arks.princeton.edu/ark:/88435/dsp018k71nh20p
8. Panagou, D., Tanner, H.: Modeling of a hexapod robot; Kinematic equivalence to a unicycle. Department of Mechanical Engineering, University of Delaware Newark DE, 19716, p. 3 (2009)
9. Khang, N.V., Cau, T.M., Dien, N.P., Khiem, V.V., Le, N.N.: Engineering oscillation. Science and Technology Publisher, Vietnam (2004)

Design of Compliant Mechanism for Rectilinear Guiding with Non-conventional Optimization of Flexure Hinges

Dušan Stojiljković(✉), Nenad T. Pavlović, and Miloš Milošević

Department of Mechatronics and Control, Faculty of Mechanical Engineering, University of Niš, Aleksandra Medvedeva 14, Niš, Serbia
{dusan.stojiljkovic,nenad.t.pavlovic, milos.milosevic}@masfak.ni.ac.rs

Abstract. Flexure hinges are essential for the movement of compliant mechanisms. As compliant mechanisms get all their mobility from flexure hinges, and not from the rigid-body segments that connect these joints, much of the science of compliant mechanisms comes down to the analysis of these flexure hinges. The analysis of joints has so far been reduced to the use of different dimensions, materials and shapes of flexure hinges in order to obtain the best output parameters of the compliant mechanism that is designed and modelled. In this paper, by using undercut notch flexure hinges, a new analysis is given, which aims to show another factor that has an impact on the operation of compliant mechanisms. This factor is represented by the position of joints and its influence is shown through improving the accuracy of the coupler point rectilinear path of the Roberts-Chebyshev four-bar linkage. Hence, it will be described that the position of these flexure hinges and their geometry is a vital issue for performing an approximately rectilinear path. Therefore, several designs are investigated through the finite elements method (FEM) simulation. The most important part of the paper is the discussion of the results of structural analysis and position optimization as a tool for obtaining the best results of the initial designed compliant mechanism.

Keywords: Compliant mechanism · Flexure Hinges · Roberts-Chebyshev mechanism · Rectilinear path · FEA · Position optimization

1 Introduction

A mechanism is a system of interconnected (moving) bodies that transform the movement of one or more bodies into the forced movement of other bodies i.e., converts one type of movement into another. The basic function of the mechanism is the transmission of force and movement or guiding a point along a given path, that is, the body through given positions. Therefore, a functional division of the mechanisms can be made into transmission mechanisms and guidance mechanisms. Further division of mechanisms can be done based on the directions of the axes of rotation of the members of the mechanism and as such can be divided into straight (axes of rotation are parallel, members of

N. V. Khang et al. (Eds.): ASIAN MMS 2021, MMS 113, pp. 370–378, 2022.
https://doi.org/10.1007/978-3-030-91892-7_35

the mechanism move in parallel planes), spherical (all axes of rotation intersect at one point, and the movement of mechanism members is performed in concentric calottes) and spatial (axes of rotation pass by, and members of the mechanism realize spatial movement). A member of a mechanism is one or more elements that are firmly connected and move as one whole or one rigid body. It follows from the foregoing that a member of the mechanism is, as a rule, a rigid body [1]. However, sometimes compliant, flexible and elastic body elements can also be considered members. Such mechanisms are called compliant mechanisms. Compliant mechanism means movable materially coherent structures, which can transmit force or transform motion only due to the elasticity of the corresponding segments [2].

For the realization of high-precise motion (as required for example by precision engineering and micromechanics), compliant mechanisms are increasingly used because of their advantages over rigid-body mechanisms [3]. By implementing flexure hinges in the position of the classical joint, defects like friction, clearance, wearing and noise are eliminated. The shape of the flexure hinges determines how efficiently the compliant mechanism can perform a given task. As this is an important factor and the main disadvantage of flexure hinges, many papers deal with the study of the influence of different forms of flexure hinges on the operation of the compliant mechanism.

The basic nomenclature and classification of the flexure hinges have been established in the papers [3] and [4]. The paper [5] aims to analyze the guiding accuracy of the "coupler" point on the path segment for each of the three new-designed compliant mechanisms, as well as to analyze the influence of the input force acting point location on the guiding accuracy. In the papers [6–8] and [9] the influence of the flexure hinges geometry and rigid-body counterparts on the mobility of rectilinear guiding compliant mechanisms are described. The book [10] provides an introduction to the finite element method (FEM) and structural static analysis of the 3D CAM models. The paper [11] exhibit, with the help of the finite elements method (FEM), compliant straight-line mechanisms analysis and ways to improve the mechanical efficiency of these mechanisms. In kinematics, cognate linkages are linkages that ensure the same input-output relationship or coupler curve geometry, while being dimensionally dissimilar. A cognate compliant four-bar linkages for rectilinear guiding based on Roberts-Chebyshev four-bar linkage [5] is used in this paper. It aims to ameliorate rectilinear guiding of Roberts-Chebyshev cognate compliant mechanism by use of the specific undercut notch flexure hinge. In the paper, the discussion will be about how the position of these flexure hinges is a vital issue for performing an approximately rectilinear path. Hence, the influence of the position will be described in this paper through a finite element analysis (FEA) and optimization by which the best design for the most accurate mechanism guiding will be obtained.

2 Design of Undercut Flexure Hinges

The most commonly used joints as moving parts of compliant mechanisms are notch-shaped flexure hinges [2]. The shapes of these notches can be different, but with limiting flexure hinges to those with concentrated flexibility, with one degree of freedom (only one axis of elasticity) and those that are symmetrical, the division can be seen in papers

[3] and [4]. All these forms of flexure hinges can also be presented as asymmetric, where the notch is located only on one side of the joint (looking at the neutral axis). In this area of asymmetric flexure hinge with a notch, one interesting type of joints stands out. These are flexure hinges with an undercut notch. The only difference in geometry between these joints and the undercut joints is that it is impossible to represent these joints as symmetrical (see Fig. 1).

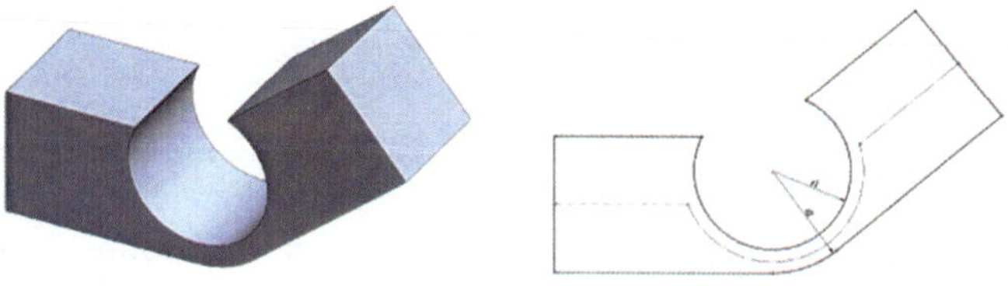

Fig. 1. Undercut notch flexure hinge

By positioning the thinnest segment of the flexure hinge at the joints of rigid-body mechanisms, which is implied for all types of flexure hinges, poor results are obtained of the compliant mechanism with this joint design [11]. This puts them in the background at the beginning, but with the additional analysis of these joints, one new favorable feature can be noticed, which stands out to all the previously mentioned flexure hinges. This feature excludes the geometry of flexure hinges as a factor in improving the precision of compliant mechanisms, which has been the idea in the synthesis of compliant mechanisms but defines the improvement of compliant mechanisms with a change in the position of flexure hinges relative to the initial set position of joints.

3 Roberts-Chebyshev Cognate Compliant Mechanism with Undercut Flexure Hinges

The basic Roberts-Chebyshev compliant mechanism can be seen in the papers [5, 6] and [11]. The same principle of model generation was used in this paper but in the case the first cognate Roberts-Chebyshev compliant mechanism from paper [5]. The design of this model is done so that all other influences are neutralized so that the movement of the undercut flexure hinges in the plane of the mechanism comes to the fore. This was obtained by placing basic symmetrical flexure hinges with a semi-circular notch in the places of the fixed supports (supports A_0 and B_0 in Fig. 2). The undercut flexure hinges were modelled at the joints A_1 and B_1. These joints are defined as a counterpart to the joints in supports A_0 and B_0, which means that they are parametrically dimensioned the same (with the same thickness of the thinnest part of the notch) in their geometric design. The reason that the undercut flexure hinges are not placed on the positions of supports A_0 and B_0 is that they are placed in places where there are possibilities for allowed movement, which is important in the later analysis. Also, if this type of flexure hinges were placed in these places, there would be a big deviation from the initial design of the mechanism.

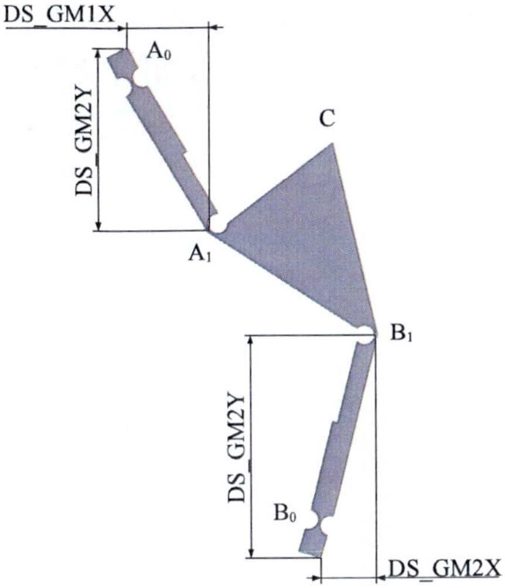

Fig. 2. Roberts-Chebyshev cognate compliant mechanism with parametrization design

Parametrization of the displacement of the undercut flexure hinges in X (represented as DS_GM1X and DS_GM2X in Fig. 2) and Y direction (presented as DS_GM1Y and DS_GM2Y in Fig. 2) was performed, which will be the main subject in positional optimization. These displacement dimensions were defined from the position of the fixed supports A_0 and B_0 retrospectively for both undercut flexure hinges to the middle of the circle notch of the undercut hinges. In this way, the mechanism with all its parameters is fully defined.

4 FEA and Position Optimization of Roberts-Chebyshev Cognate Compliant Four-Bar Linkage

The CAD program Solidworks 2016 was used to create the model of the Roberts-Chebyshev cognate compliant mechanism as shown in the previous chapter, while ANSYS 19.2 was used for the FEA process. For FEA and positional optimization to be performed, the geometric constraint of all flexure hinges and the compliant mechanism must be fully defined. In addition to the already mentioned geometric constraints, it is necessary to define the static and dynamic characteristics of the mechanism. In Fig. 3, these limitations can be seen in the form of defining the boundary conditions (fixed supports B and C) and the load (force A) acting on the mechanism. Using the "Edge Sizing" function, a refined mesh of finite elements is created at the places of flexure hinges (see Fig. 3). By stress convergence up to one percent of the repeatability of the results, the performed simulation is considered valid. The resulting mesh of FE model was composed of 7642 elements and 24335 nodes. The material that was used is Polylactide/polylactic acid (PLA) with the modulus of elasticity $E = 3450\ N/mm^2$ and bending strength of $\sigma bs = 54.1\ N/mm^2$. After successfully defining the analysis and simulation of the compliant mechanism model, the next step is to define the position optimization parameters.

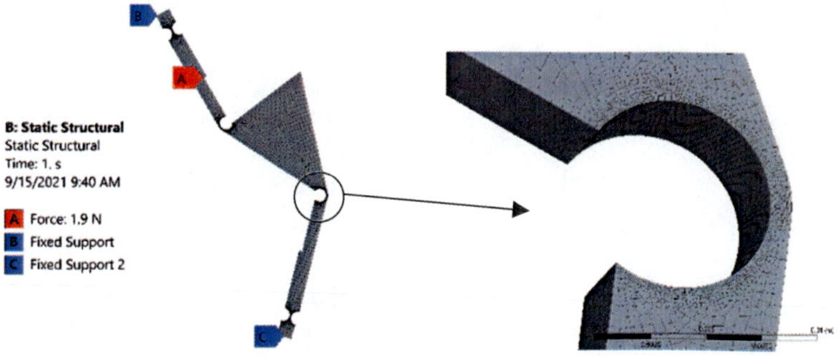

Fig. 3. The boundary conditions and load with finite element structure of flexure hinge

The goal of optimization is to determine the optimal position of the undercut flexure hinges and thus the dimensions of the mechanism as a whole. In the "Solutions" branch, the following requirements are set for determination:

a) Maximum equivalent elastic strain (P5)
b) Maximum displacement of point C in the X direction (P6)
c) Maximum displacement of point C in the Y direction (P7)
d) Maximum equivalent stress (P8)
e) Minimum safety factor (P9)

These values also represent the optimization output parameters. As two parameters are required for positioning of one joint (positioning in a vertical and horizontal direction), this model, therefore, has four input parameters:

a) Displacement of the undercut flexure hinge A1 in the Y direction (P1)
b) Displacement of the undercut flexure hinge A1 in the X direction (P2)
c) Displacement of the undercut flexure hinge B1 in the Y direction (P3)
d) Displacement of the undercut flexure hinge B1 in the X direction (P4)

The characters given in parentheses, in addition to the names of the input and output parameters, represent their identifications in ANSYS.

In order to select those input parameters whose change has the greatest impact on the output parameters, a sensitivity analysis was first performed (within the "Parameters Correlation" module in ANSYS). Hence, it was necessary to set the upper and lower limits within which the parameters can move (Table 1).

Based on the sensitivity analysis a quadratic parameter determination matrix is obtained. It presents the influence of the change of input parameters on the output parameters, values close to one indicate a close relationship between the parameters, and zeros indicate a weak relationship between the parameters. This influence can be seen more clearly in the Fig. 4 which shows the table segment that provides an overview related to important and less important parameters. Hence, the displacement of the undercut flexure hinge A1 in the Y direction and the X direction (P1 and P2) have the greatest influence on the output parameters as shown in Fig. 4. The next parameter in terms

Table 1. Intervals in which input parameters can move

Parameter	Units	Minimum	Maximum
P1 - DS_GM1Y	mm	72.9	89.1
P2 - DS_GM1X	mm	32.4	39.6
P3 - DS_GM2Y	mm	88.2	107.8
P4 - DS_GM2X	mm	21.6	26.4

of the relevance of the influence on the output parameters is the displacement of the undercut flexure hinge A2 in the Y direction (P3). Although the difference between the influences of input on output parameters is noticeable, the influence of P4 parameters is still not negligible, since its correlation is 0.41. Therefore, all four input parameters will be included in further optimization.

Major Input Parameters				
Input Parameter	Best Relationship With Filtering Output Parameter			
	Relevance	Output Parameter	R2 Contribution	Correlation Value
P1 - DS_GM1Y@Sketch2@V2.RCH I cognate.Part	1	P6 - Deformation Probe Maximum X Axis	0.85034	-0.92378
P2 - DS_GM1X@Sketch2@V2.RCH I cognate.Part	1	P5 - Equivalent Elastic Strain Maximum	0.37641	-0.61482
P3 - DS_GM2Y@Sketch2@V2.RCH I cognate.Part	0.87881	P5 - Equivalent Elastic Strain Maximum	0.25548	0.53248
Minor Input Parameters				
Input Parameter	Best Relationship With Filtering Output Parameter			
	Relevance	Output Parameter	R2 Contribution	Correlation Value
P4 - DS_GM2X@Sketch2@V2.RCH I cognate.Part	0.41131	P9 - Safety Factor Minimum	0.02634	-0.14587

Fig. 4. Quadratic determination matrix

To perform structural optimization within ANSYS, it is necessary to define the design of the experiment. "Central Composite Design" was used as the design type of the experiment (the "Face Centred" option was also selected here, while "Enhanced" was used as the template type). In the next step was to create response surface (type "Genetic Aggregation" was used here). The obtained response surface (Fig. 5) has the values of the coefficient of determination, which indicates that as a Meta model, this response surface represents a good approximation. It can be seen from Fig. 5 that for all defined values of the parameters that most affect the equivalent elastic strain, it is in the range of up to 2% so that it will not be taken as an additional condition for optimization.

The following objectives and constraints of optimization forces have been adopted:

a) Minimization of the parasitic displacement of point C in the Y direction
b) The displacement of point C in the X direction is ≤ -5 mm

This is done so that the optimal positions of the compliant mechanism are obtained and to limit the optimization. Within the "Response Surface Optimization" module, the MOGA optimization method is selected.

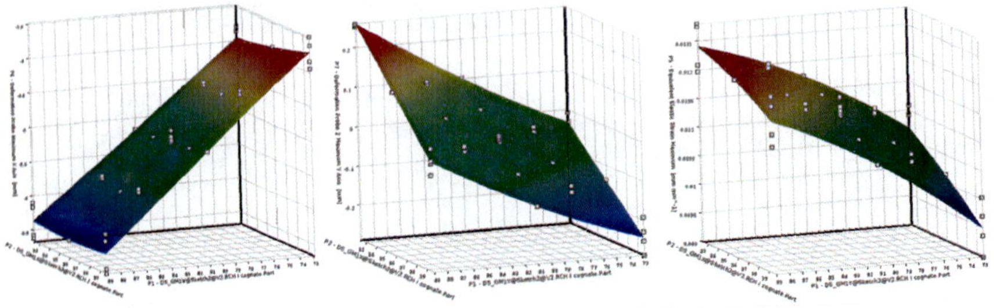

(a) (b) (c)

Fig. 5. Response surface of: (a) Displacement of point C in the X direction (P6), (b) displacement of point C in the Y direction (P7), (c) maximum equivalent stress (P8), in correlation with parameter P1 and P2 (only this example is shown, as the others are)

5 Results

After optimization, three combinations of dimensions (candidate points) are offered that best meet the mentioned goals and constraints (Fig. 6). By verification, it can be seen that points candidate two and three meet the given conditions (see Table 2). During the verification of the obtained candidates, the error occurred on the 6th decimal place for the displacement in the Y direction, which is acceptable, while the displacement in the direction X match.

Optimization Study			
Seek P7 = 1E-05 mm	Goal, Seek P7 = (Default Importance)		
P6 <= -5 mm	Strict Constraint, P6 values less than or equals to -5 mm (Default importance)		
Optimization Method			
MOGA	The MOGA method (Multi-Objective Genetic Algorithm) is a variant of the popular NSGA-II (Non-dominated Sorted Genetic Algorithm-II) based on controlled elitism concepts. It supports multiple objectives and constraints and aims at finding the global optimum.		
Configuration	Generate 4000 samples initially, 800 samples per iteration and find 3 candidates in a maximum of 20 iterations.		
Status	Converged after 7665 evaluations.		
Candidate Points			
	Candidate Point 1 DP 93	Candidate Point 2 DP 94	Candidate Point 3 DP 95
P1 - DS_GM1Y@Sketch2@V2.RCH I cognate.Part	88.418	86.555	83.357
P2 - DS_GM1X@Sketch2@V2.RCH I cognate.Part	39.135	38.383	36.357
P3 - DS_GM2Y@Sketch2@V2.RCH I cognate.Part	94.192	105.35	91.667
P4 - DS_GM2X@Sketch2@V2.RCH I cognate.Part	22.793	24.205	23.735
P6 - Deformation Probe Maximum X Axis (mm)	-6.1589	-6.1604	-5.3604
P7 - Deformation Probe 2 Maximum Y Axis (mm)	1.2462E-05	1.9067E-05	-1.3669E-05

Fig. 6. Candidate points determined on the basis of the MOGA method

The results clearly show an improvement in the accuracy of the operation of the basic mechanism (Roberts-Chebyshev four-bar linkages). The success of this approach to the design of compliant mechanisms can be seen with a comparison with other mechanisms that perform the same or similar movements. For example, in the papers [3, 5] and [11] retrospectively the best given guidance accuracy is 1.29 μm, 0.9 μm and 0.4725 μm for 5 mm of axial translation while in this paper the guiding accuracy is 91,9 nm and axial translation 6,16 mm.

Table 2. Obtained candidates with parameter correlation

Parameter	Units	Candidate 1	Candidate 2	Candidate 3
P1 - DS_GM1Y	mm	88.417575	86.554575	83.35710706
P2 - DS_GM1X	mm	39.13507969	38.38273594	36.35686828
P3 - DS_GM2Y	mm	94.19208573	105.3527582	91.66728264
P4 - DS_GM2X	mm	22.79252416	24.2047546	23.73452187
P5 - Max. Def. X Axis	mm	−6.16076088	−6.159910679	−5.360627174
P6 - Max. Def. Y Axis	mm	0.000127062	−0.0000919	0.0000966

6 Conclusions

Unlike the frequent synthesis for designing compliant mechanisms that were based on geometry and material changes of flexure hinges, improvement of compliant mechanisms designed with undercut flexure hinges comes from their position also. In this paper, by FEA and applied position optimization for undercut flexure hinges parameters on modelled Roberts-Chebyshev cognate compliant mechanism, the accuracy of the coupler point rectilinear path is improved in this way. The verification results of position optimization show that with the suitable position of undercut flexure hinges, compliant mechanisms guidance deviation can be measured in nanometers (parasitic displacement is decreased to $\Delta y_C = 91{,}9$ nm) with a possible additional increase in the path motion (from given $\Delta x_C = 5$ mm for the basic mechanism to $\Delta x_C = 6{,}16$ mm for complaint cognate mechanism). The results of this study can be used in the possible design of micro grippers. This can be done using two symmetrical mechanisms that resemble the mechanism in the paper. Further studies can be directed to the creation of such a compliant gripper.

References

1. Pavlović, N.D., Milošević, M.: Polužni mehanizmi. Mašinski fakultet u Nišu, Niš (2012). [in Srbian]
2. Pavlović, N.D., Pavlović, N.T.: Gipki mehanizmi. Mašinski fakultet, Univerziteta u Nišu, Niš (2013). [in Serbian]
3. Linß, S., Henning, S., Zentner, L.: Modeling and design of flexure hinge-based compliant mechanisms. In: Kinematics - Analysis and Applications. IntechOpen (2019)
4. Howell, L.L., Magleby, S.P., Olsen, B.M.: Handbook of Compliant Mechanisms. Wiley, Hoboken (2013)
5. Pavlović, N.T., Pavlović, N.D.: Design of compliant cognate mechanisms. In: Proceedings of the 4th International Conference Mechanical Engineering in XXI Century, Niš, pp. 253–258 (2018)
6. Pavlović, N.T., Pavlović, N.D.: Mobility of the compliant joints and compliant mechanisms. Theoret. Appl. Mech. **32**, 341–357 (2005)
7. Linß, S., Erbe, T., Theska, R., Zentner, L.: The influence of asymmetric flexure hinges on the axis of rotation. In: Proceedings of 56th International Scientific Colloquium, Ilmenau (2011)

8. Linß, S., Milojević, A.: Model-based design of flexure hinges for rectilinear guiding with compliant mechanisms in precision systems. Mechanismentechnik in Ilmenau, Budapest und Niš Ilmenau (2014)
9. Pavlović, N.T., Pavlović, N.D., Milošević, M.: Selection of the optimal rigid-body counterpart mechanism in the compliant mechanism synthesis procedure. Mech. Mach. Sci. **45**, 127–138 (2016)
10. Korunović, N., Trajanović, M.: Modeliranje za analizu metodom konačnih elemenata. Univerzitet u Nišu, Mašinski fakultet, Niš (2020). [in Serbian]
11. Stojiljković, D., Pavlović, T.N.: Influence of flexure hinges design on guiding accuracy of Roberts-Chebyshev compliant mechanism. In: Proceedings of the 5th International Conference Mechanical Engineering in XXI Century, University of Nis - Faculty of Mechanical Engineering, Niš, pp. 217–220 (2020)

Bi-directional Soft Robotic Finger Actuated Mechanisms

Sugandhana Shanmuganathan, V. Prasanna Venkatesh, Devarshi Pandey, and Rajeevlochana G. Chittawadigi(✉)

Department of Mechanical Engineering, Amrita School of Engineering, Bengaluru, Amrita Vishwa Vidyapeetham, Bengaluru, India
rg_chittawadigi@blr.amrita.edu

Abstract. Pressure actuated soft robotic fingers are commonly used in different structures, shapes and orientations for predominantly grasping applications. But soft robot application with two pressure chambers on either side of the restraining material that facilitates curling of the soft robotic finger in two directions when the chambers are actuated alternatively is only found in an attempt to mimic fish tail for its locomotion. This project outlines how the two-chamber finger with bi-directional motion capabilities can actuate mechanisms. It is an advantage to use pressure difference actuated pneumatic tubes over conventional heavy motor actuated rigid cranks in circumstances like tethered robots, ceiling robots, etc. An attempt is made to explain how the soft robot can act as a crank or input link for planar mechanisms such as four-bar mechanism and slider-crank mechanism, through analytical and numerical approach. Simulations of the soft robot has been used to obtain the input (pressure) vs output (curling of the robotic finger) characteristics. The tip of the finger is considered as input to the forward kinematic analysis of the mechanism, i.e., four-bar and slider-crank mechanism. Similarly, inverse kinematics formulations are obtained to obtain the pressure required as input to result in desired rotation of the follower link in a four-bar mechanism and the displacement of the slider in a slider-crank mechanism. Proof of concept prototype of the bi-directional soft robotic gripper has been developed and tested.

Keywords: Bi-directional soft actuator · Soft robotics · Pneumatic actuated mechanism

1 Introduction

Robots made of highly compliant materials can be classified as soft robots. Increased flexibility, adaptability, and protection are some of its characteristics. Soft robot actuators are made up of flexible materials that change their shape, size, orientation with response to stimuli of energy application like pressure difference, electricity, chemical, heat, etc. These have many applications including prosthesis [1] and grasping of uneven shaped objects [2]. Soft pneumatic actuator is a sought-after research area due to softness, lightweight, safe human machine interaction and low fabrication cost. Solidified elastic material with a hollow space can be injected with compressed air, like a balloon, and

N. V. Khang et al. (Eds.): ASIAN MMS 2021, MMS 113, pp. 379–388, 2022.
https://doi.org/10.1007/978-3-030-91892-7_36

the material tries to expand in all possible directions, i.e., in the longitudinal and the lateral direction. This can be explained by considering Fig. 1, where an elastic material in the form of a finger is subjected to compressed air. The expansion is uniform in all directions. If some form of restraint is placed so that the radial or transverse expansion is restricted, the volume is allowed to expand only longitudinal direction, as shown in Fig. 1(b). These restraints are known as fiber-reinforcements.

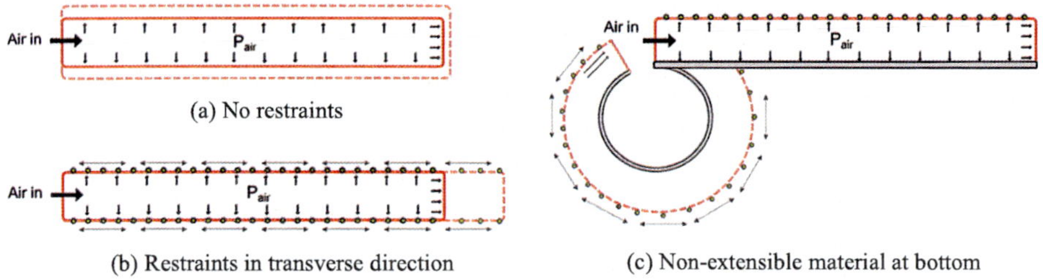

Fig. 1. Soft actuator with fiber-reinforcement [3]

In addition to these reinforcements, one of the face/sides of the finger is stuck with a non-extensible material. The material on the above part has a provision to expand whereas the bottom side is restricted. Hence, the finger curls down, as shown in Fig. 1(c). The research work in [4] showed better success rates in formation of actuators using PneuNets methodology in which the finger has several chambers and passage for compressed air through them. Such soft actuators are conveniently used to grasp objects of uneven shapes. One such demonstration of grasping vegetables and other agricultural products was attempted in [5]. In this paper, a novel application of bi-directional soft actuator is proposed.

Basic planar mechanisms are part of almost every mechanical machine. These mechanisms are actuated using rotary or linear actuators. Generally, these actuators are driven using electricity or hydraulics. Pneumatics is one form of energy which can also be used to drive the mechanisms. However, out-of-the-box application of pneumatics will not result in position control of the actuator. It can only have binary effect of either fully deployed or fully closed. Using a soft actuator, the authors propose a novel idea of achieving position control of the actuator link.

The approach is derived out of a soft robotic fish [6], where the tail-fin is made up of soft material with two volumes or chambers separated by a restraining material in between them. Alternate inlet of compressed air in the chambers result in the alternate curling motion of the tail-fin, which further propels the soft robotic fish. Similar setup of two chambers separated by a restraining medium between them can be used to curl a soft robotic finger or actuator in both the directions. If one can derive the relationship between the pressure of compressed air, and the curling radius and hence the tip of the finger, the soft finger can itself act as the input link, albeit a flexible one. With the knowledge of the position of the finger tip, the position and orientation of all other connected links of a mechanism can be determined using analytical or geometric methods. Here, this novel concept of a soft-actuated flexible link is used as input to drive planar mechanisms such as four-bar and slider-crank mechanisms. First, simulation is performed to determine the input to output relationship and then a physical prototype of the proposed bi-directional actuator is developed and tested to curl in both directions.

2 Relationship of Pressure and Curling

The proposed bi-directional bending soft actuator was first simulated in ABAQUS software. The CAD model of PneuNets based molds were created in SolidWorks software, using the dimensions considered in [5]. However, the finger mold used for simulation had hollow region that would act as chambers for compressed air, as shown in Fig. 2(a). A flat plate is created separately as shown in Fig. 2(b).

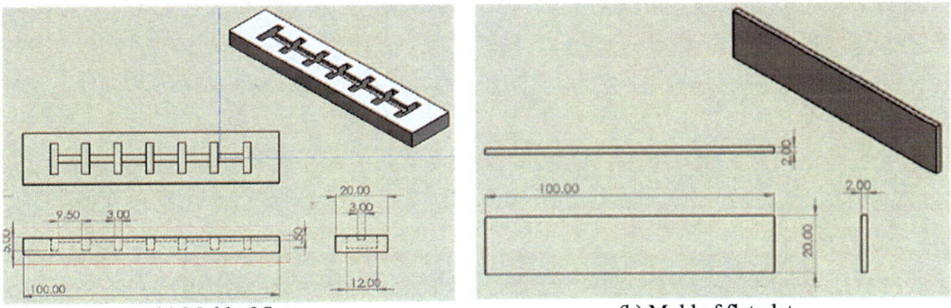

(a) Mold of finger (b) Mold of flat plate

Fig. 2. CAD models of finger and flat plate in SolidWorks (all dimensions in mm)

These two CAD files are then imported in ABAQUS software as STEP files and assembled as shown in Fig. 3(a). Note that there are two instances of finger mold and flat plate, each. They are placed such that the flat plats share a common face, which later will be used as restraining material. The finger molds are symmetric with respect to the common face, i.e., one of them is flipped.

The material property of finger and flat plate molds are set to that of Elastosil, which has the closest behavior of bending, required in a soft actuator. The common face is assigned property of a restraining material, such as paper, such that it has minimal expansion upon application of pressure. The inner walls of the chamber are all selected (Fig. 3(a)) and the pressure inside the volume is set as a load (Fig. 3(b)). The meshing is done in the software, as shown in Fig. 3(c). The pressure inside the volume is increased at certain incremental value. The software does the finite element analysis (FEA) at each pressure load and the deformation of the finger is displayed to the user. Since there are two fingers, pressure inside only one of them is applied at a time. An example for pressure in upper finger is shown in Fig. 3(d), which results in curling of the finger downward. However, if the bottom finger is injected with compressed air, the curling would happen upwards.

The relationship between the pressure in the finger's chamber and the curling of the finger has to be determined. This can be done by measuring the coordinates of the tip of the finger, along the center line. For the design under consideration, the coordinates of the tip were measured for a range of pressure input from 0 to 0.28 MPa. The values of pressure (P) and the coordinates of the tip (x_{tip}, y_{tip}) in mm and were fed into MATLAB's polyfit function, which returns the equations of polynomial equations as

$$x_{tip} = 1000\left(-2.3071\left(P^3\right) + 0.1740\left(P^2\right) - 0.0096(P) + 0.1\right) \tag{1}$$

$$y_{tip} = 10000\left(1.3058\left(P^4\right) - 0.5682\left(P^3\right) + 0.0232\left(P^2\right) - 0.0131(P)\right) \tag{2}$$

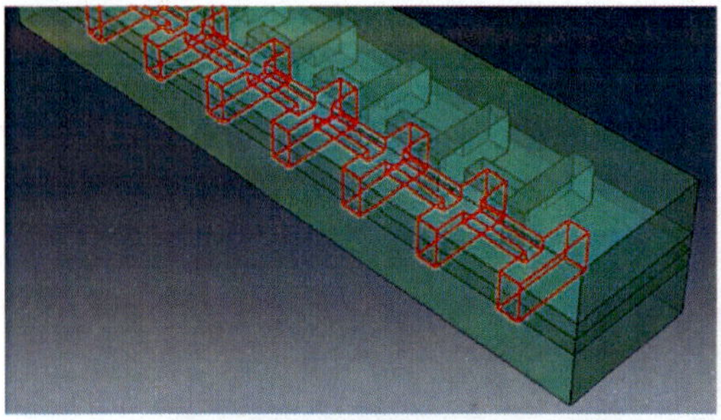

(a) Assembly of components and selection of inner surfaces

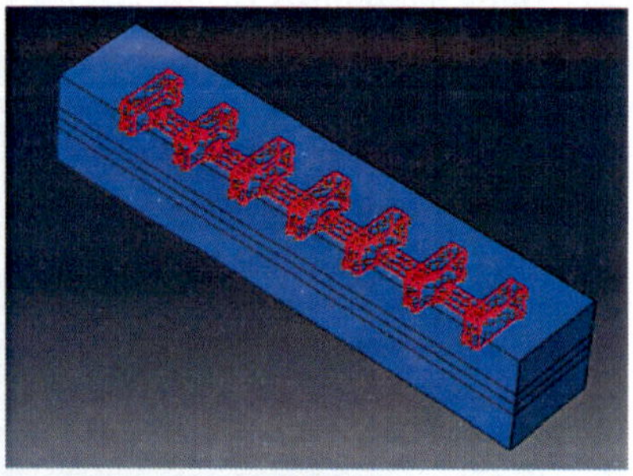

(b) Pressure as load inside chambers

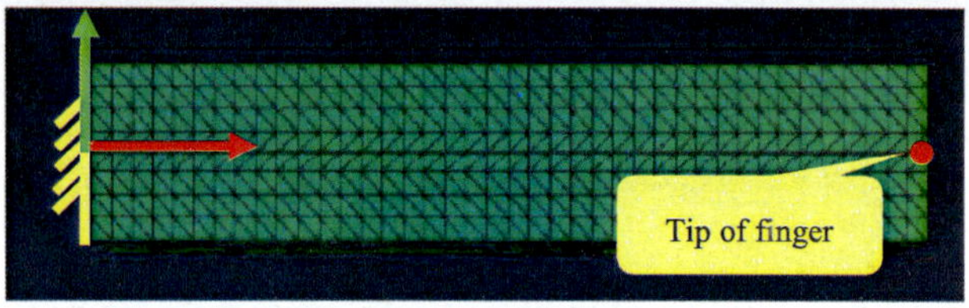

(c) Meshing of the elastic material in the software

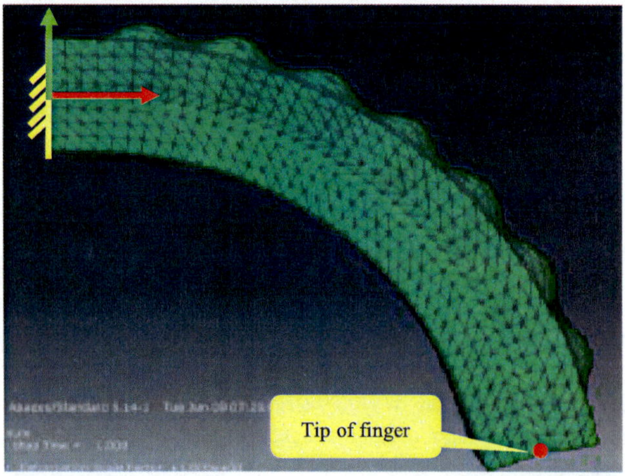

(d) Left side is rigidly fixed and the pressure in the chamber is increased to 0.28 MPa. The coordinates of the tip of the finger are measured and noted down for further steps

Fig. 3. The CAD files of finger and flat plate are imported in ABAQUS software for FEA analysis

The plots of x_{tip} v/s pressure obtained using Eq. (1) for various input pressure is shown in Fig. 4(a). Similarly, for y_{tip} using Eq. (2) is in Fig. 4(b). The points obtained from the polynomial equation was having a maximum error of 1% when compared to the input data used to fit the equations in the first place. Once these polynomials are obtained, for a given design and material, one can use that information for remaining tasks, such as in the kinematic analysis of few mechanisms, as explained next.

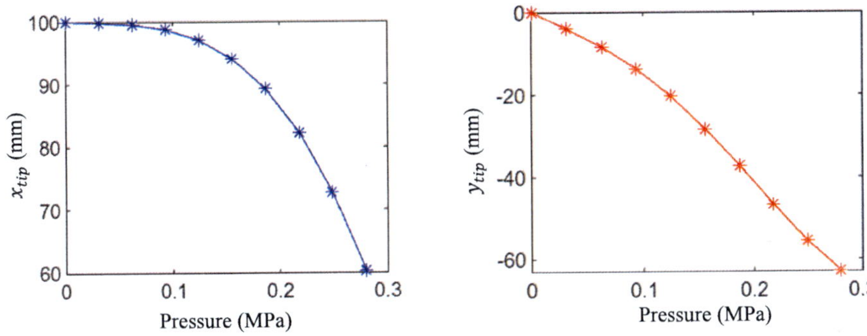

(a) x_{tip} v/s Pressure (b) y_{tip} v/s Pressure

Fig. 4. Relationship between pressure inside chamber and the displacement of the tip of the finger

3 Kinematic Analysis of Slider-Crank Mechanism

Slider-crank mechanism is one the most commonly used mechanisms in real-life applications. It consists of four rigid links, of which one is assumed to be fixed or grounded. The remaining three links have connections using three revolute joints and one prismatic or translatory joint, as shown in Fig. 5(a). The mechanism has a one degree-of-freedom (DOF) which means that one input is required to drive the mechanism. If the input is in the form of rotation of the crank (θ), the position and orientation of all other links can be determined. However, if the slider displacement (d) is considered as the input, the crank angle (θ) can also be determined using the standard kinematic equations, found in a standard text book on theory of machines. If the former is considered, a rotary driving unit or motor is required to actuate the mechanism. These motors are usually heavy and if the whole unit is to be mounted on a tethered system, the mass of the motor also has to be considered for the tethered system. Hence, alternative means of actuating the slider-crank mechanism is necessary to reduce the total weight of the tethered system.

One way to replace the motor is to use pneumatics. However, out-of-the box, it is difficult to obtain position control of a rotary or a linear actuator using pressurized air. Hence, a bi-directional soft robotic finger, explained in the previous section, can also act as an alternative actuator. If the tip of the bi-directional finger is considered as the starting point of the connecting-rod, shown in Fig. 5(b), and thereafter a slider is connected. The inlet pressure (P) in the bi-directional finger controls the coordinates of the tip of the finger, which in turn can be used to move the slider to have displacement.

Using Eqs. (1–2), the coordinates of the tip (A) were determined and then for the incremental values of pressure (0–0.280 MPa), position of the slider were determined. This can be obtained by drawing a circular arc centered at A with radius l (length of connecting rod) and its intersection with the sliding direction is the slider point B. The displacement (d) as a function of inlet pressure was again determined using polyfit function of MATLAB to be

$$d = 1000(-2.3960(P^3) + 0.0903(P^2) - 0.0031(P) + 0.399 \quad (3)$$

The inverse of the equation can also be represented as a polynomial equation where the pressure required to achieve an offset distance (d') from the maximum displacement can be determined using

$$P = \left(\left(-0.000212\left(d'^3\right)\right) - \left(0.004026\left(d'^2\right)\right) - 0.03275\left(d'\right) + 0.04875\right) \quad (4)$$

Equation (3) corresponds to the forward kinematics of the soft-finger actuated slider-crank mechanism and Eq. (4) is for the inverse kinematics. Both have been plotted in Fig. 6. By using pressure regulators or solenoids, one can control the inlet pressure and thus achieve the desired displacement of the slider.

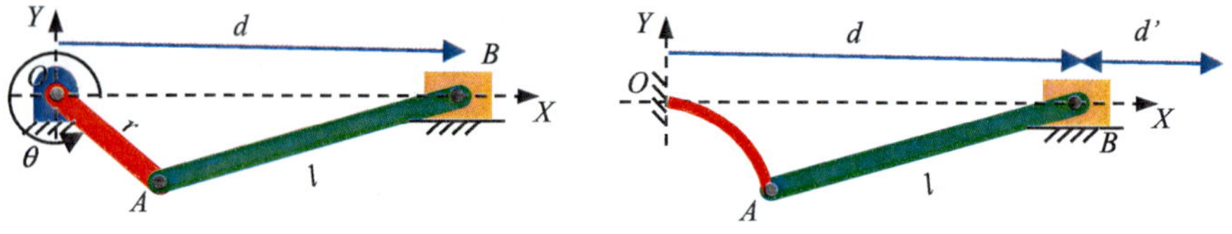

(a) Rotary actuator (b) Soft-finger actuator

Fig. 5. Conventional and proposed soft-finger actuated slider-crank mechanisms

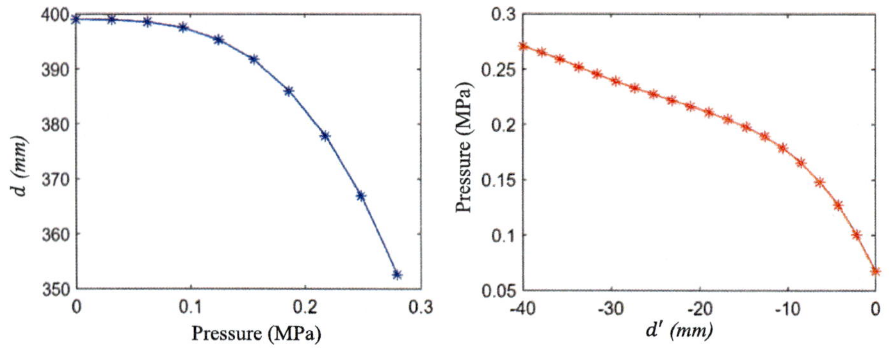

(a) Displacement (d) v/s Pressure (P) (b) Pressure (P) v/s Offset (d')

Fig. 6. Relationship between inlet pressure (P) and the displacement (d) of the soft-finger actuated slider-crank mechanism

4 Kinematic Analysis of Four-Bar Mechanism

Similar to a slider-crank mechanism, one can actuate a four-bar mechanism, another commonly found mechanism, using bi-directional soft finger. A conventional and the proposed four-bar mechanisms are shown in Fig. 7(a) and (b), respectively. Here, the rotation of the output link (θ'_4) can be determined for incremental values of inlet pressure (P) and the corresponding polynomial equation for the considered system is obtained as

$$\theta_4' = 1000\left(-1.6922\left(P^3\right) - 0.4165\left(P^2\right) - 0.0567(P) + 0.0877\right) \tag{5}$$

Similarly, the inverse function can also be obtained as a polynomial so that for a desired angle of the output link or follower, the corresponding inlet pressure (P) can be supplied to the bi-directional soft finger. For the given system, this equation is

$$P = ((-0.1663(10^{-7})(\theta_4'^{4})) + (0.2382(10^{-5})(\theta_4'^{3})) - (0.0001208(\theta_4'^{2})) + (0.00037(\theta_4')) + 0.2787145 \tag{6}$$

Equations (5) and (6) are plotted in Fig. 8(a) and (b), respectively.

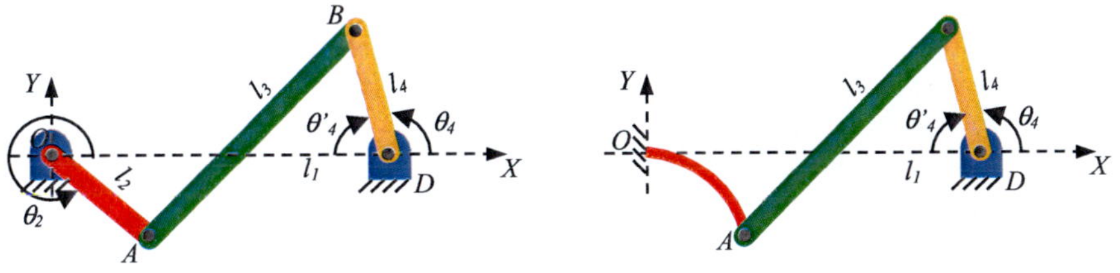

(a) Rotary actuator (b) Soft-finger actuator

Fig. 7. Conventional and proposed soft-finger actuated four-bar mechanisms

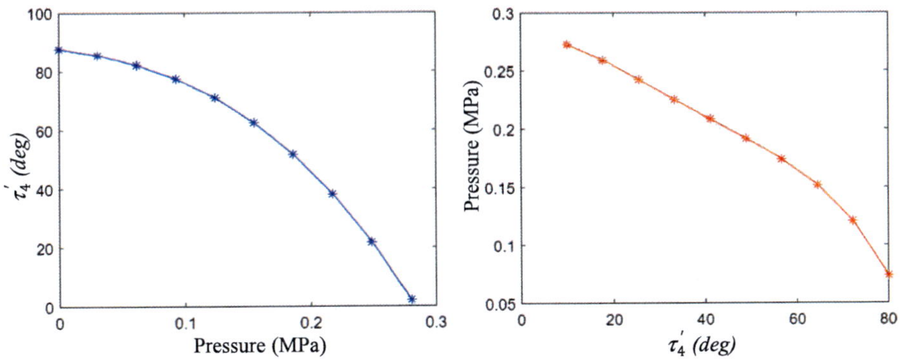

(a) Follower angle (τ_4') v/s Pressure (P) (b) Pressure (P) v/s Follower angle (τ_4')

Fig. 8. Relationship between inlet pressure (P) and the displacement (d) of the soft-finger actuated four-bar mechanism

5 Fabrication and Testing of Bi-directional Soft Finger

Physical prototyping of the soft robotic finger has been done as explained here. In, [5] the mold was designed using 3D CAD software Autodesk Inventor in PneuNets methodology and was 3D printed. Eco-Flex 00-30 was the material used to make the soft actuator. The steps followed in making the soft actuator are explained in Fig. 9. These processes are repeated twice to get two sets of soft actuators. Then a restraining material is placed in the middle and the entire body is covered up with an additional round of material and let to cure. The result was a soft actuator that has two entry points for two different PneuNet chambers respectively, as shown in Fig. 10.

The bi-directional actuator was tested to curl in both the directions for varying pressure as input. The tip of the actuator has to be made such that it can be connected to the next link in the Fig. 5(b) and 7(b) to obtain the physical prototypes of the mechanisms. These will be attempted in the future and the findings shall be reported once they are

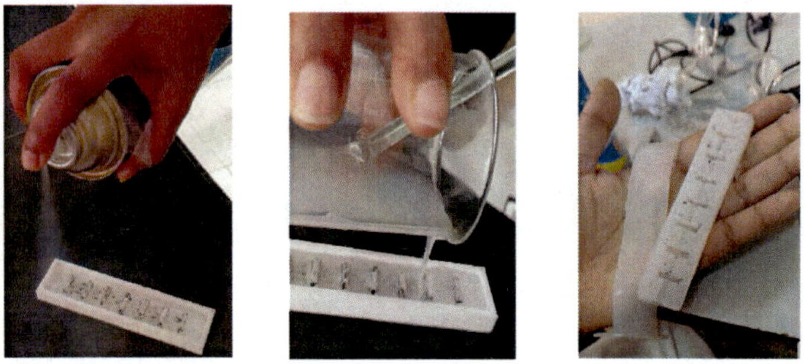

(a) (b) (c)

Fig. 9. (a) Spraying silicon mold spray on the mold (b) Pouring mixed solution in the mold (c) Peeling off after curing.

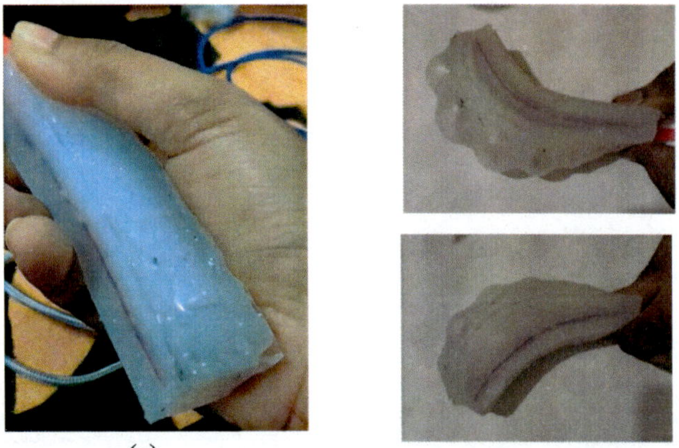

(a) (b) (c)

Fig. 10. (a) The two-way soft finger developed after curing; (b) & (c) Actuation of soft finger through both the entry points.

successfully achieved. In the next section, the simulation done for the proposed model using the polynomial equations obtained, is reported.

6 Simulation of Soft-Finger Actuated Mechanisms

The forward and inverse kinematics formulations for the soft-finger driven actuators, derived in Sects. 3 and 4 have been implemented as a Visual C# Windows form application. The slider-crank mechanism simulation takes pressure inlet as the input and the coordinates of the finger tip and the slider are determined for the given link length and polynomial equations in Sect. 3. The user can also move the pressure slider and accordingly the displacement of the slider is updated in the software. These are illustrated in Fig. 11(a).

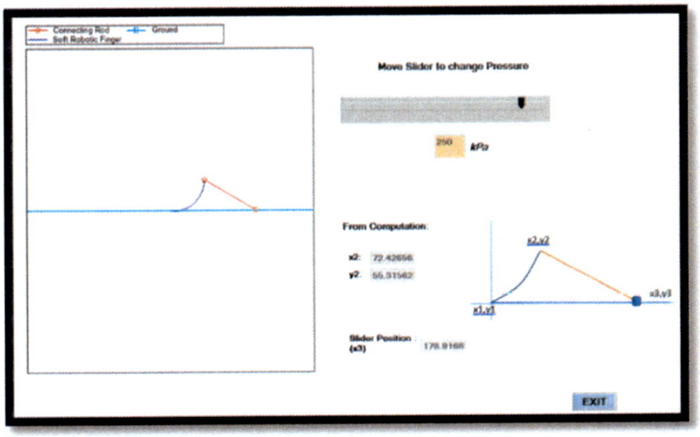

(a) Slider-crank mechanism

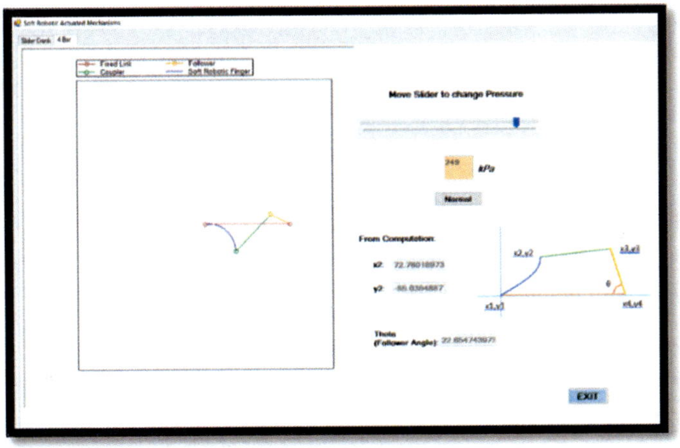

(b) Four-bar mechanism

Fig. 11. Simulation of bi-directional soft actuator driving planar mechanisms

Similarly, the equations of Sect. 4 are implemented for the considered four-bar mechanism, where the inlet pressure acts as input data and the inclination of the follower link is determined and updated in the 2D model of the mechanism, shown in Fig. 11(b).

Similar interfaces can be developed that can also determine the pressure required to achieve a desired slider position or follower inclination, in the above two mechanisms, respectively. The proposed methodology of actuating planar mechanisms with bi-directional soft robotic finger can be validated with a physical prototype, which will be carried out by the authors in the near future.

7 Conclusion

Mechanisms are generally actuated using electric or hydraulic actuators. In this paper, a possible arrangement using pressure regulated pneumatic supply is proposed which uses a bi-directional soft robotic finger. The CAD model of the finger is simulated in ABAQUS software for a range of input pressure inside the chambers of the finger, resulting in the curling of the finger. The curling data is noted down for the input pressure and a polynomial equation curve fitting is performed. The polynomial equations are further used in the forward and inverse kinematics to determine the relationship of the slider position, in case of a slider-crank mechanism and the angle of the follower link, in case of a crank-rocker four-bar mechanism. The forward and inverse kinematics of these two mechanisms has been implemented in a software application developed using Visual C# to visualize the motion of the mechanism, for varying input pressure value. An attempt to make the bi-directional soft robotic actuator was made and has been reported in the paper. Further work on connecting this link to the other links of the given mechanisms will be attempted in future and subsequently reported.

References

1. Devi, M.A., Udupa, G., Sreedharan, P.: A novel underactuated multi-fingered soft robotic hand for prosthetic application. Robot. Auton. Syst. **100**, 267–277 (2018)
2. Udupa, G., Sreedharan, P., Sai Dinesh, P., Kim, D.: Asymmetric bellow flexible pneumatic actuator for miniature robotic soft gripper. J. Robot. **2014** (2014)
3. Soft Robotics Toolkits. www.softroboticstoolkit.com
4. Sun, Y., Song, Y.S., Paik, J.: Characterization of silicone rubber based soft pneumatic actuators. In: IEEE/RSJ International Conference on Intelligent Robots and Systems, pp. 4446–4453 (2013)
5. Vidwath, S.M.G., Rohith, P., Dikshithaa, R., Nrusimha Suraj, N., Chittawadigi, R.G., Sambandham, M.: Soft robotic gripper for agricultural harvesting. In: Kumar, R., Chauhan, V.S., Talha, M., Pathak, H. (eds.) Machines, Mechanism and Robotics. LNME, pp. 1347–1353. Springer, Singapore (2022). https://doi.org/10.1007/978-981-16-0550-5_128
6. Katzschmann, R., de Maille, A., Dorhout, D., Rus, D.: Cyclic hydraulic actuation of soft robotic devices (2016)

Dynamics and Control of Machines and Robots

Research on the Control of the Mechanical System of Satellite Monitoring Antenna in Different Environmental Conditions

Duong Xuan Bien(✉), Pham Quoc Hoang, Nguyen Van Nam, Nguyen Tai Hoai Thanh, Pham Van Tuan, and Dinh Duc Manh

Le Quy Don Technical University, Hanoi, Vietnam
duongxuanbien@lqdtu.edu.vn

Abstract. This paper presents the results of position control and path control for the mechanical systems of satellite surveillance antenna systems under different environmental conditions. A traditional and highly reliable PID control system is considered and used for antenna system works in the conditions of wind and without wind effect. The control parameters are found based on the strong toolbox module of MATLAB software. The contents of this study are the next development of the results of the kinematics and dynamics analysis that have been published previously. The mathematical model and the differential equations of motion of the system are established based on the multi-bodies mechanics' theory and the Lagrange - Euler equations. The control analysis results show that there is a clear difference in the different operating conditions of the system. This is an important basis in the fabrication of the antenna system in reality.

Keywords: Monitoring antenna · Satellite · Dynamics · Control

1 Introduction

As mentioned in previous publications as [1] and [2], the designing, manufacturing, and controlling of satellite surveillance antennas is a specific scientific field and depends heavily on the development level of each country. Currently, only a few countries have mastered this space technology [3–9]. The issue of space technology transfer between countries is quite difficult due to national security and technology secrets. Therefore, developing countries like Vietnam need to be proactive in all of the researching problems related to the antenna and satellite system themselves.

A few works related to solving the dynamics problem of the geostationary satellite antennas are published such as in [10–14] and they have been reviewed and described in [2]. The PID control of the Antenna system is considered in [15] to ensure the delay compensation of the received and transmitted signals over a large area. This Antenna system connects to the NIGCOMSAT-1 satellite. The Bang-Bang control method is used in [16] to control the position of a geostationary satellite surveillance antenna system. The control system used is a microcontroller and the remote-control method. An antenna position control system with location data stored and sent via infrared signals

N. V. Khang et al. (Eds.): ASIAN MMS 2021, MMS 113, pp. 391–400, 2022.
https://doi.org/10.1007/978-3-030-91892-7_37

is described in [17]. Antenna position control system with satellite recognition capability is considered in [18]. Accordingly, the satellite is identified through its radiation signal, from which the position of the satellite will be input to the antenna control system. The main problem with this system is that the recognition ability will decrease if affected by weather or environmental conditions. Another antenna position control system is considered in [19]. The satellite position to be tracked is sent via the smartphone and decoded by the Raspberry Pi system. The control quality of the system is mainly affected by the signal delay. The PID control system incorporating LQR was developed in [20] to control the antenna position through the DC Servo motor. The result is a faster response time than a traditional PID system. The antenna position control system in [21] is built on the basis of the traditional PID system with the parameters found by the Ziegler-Nichols method and the 2nd order optimization system. The PID control system incorporating the H system was built to control the antenna and ensure the stability of the system is described in [22].

This paper presents the results of position and trajectory control for a geostationary satellite monitoring antenna system. A traditional and highly reliable PID control system is used with found control parameters based on the strong toolbox module of MATLAB software. The control problem is considered in case the system is affected by the wind and without the wind. Calculation results and numerical simulations show that when the antenna is affected by the wind, the control error is large and the actuators for the joints must maintain a state of continuous operation with high intensity to ensure the accuracy and stability of the antenna system.

2 Research Contents

2.1 Dynamics Modeling

The dynamic model of the satellite surveillance antenna in Fig. 1 and Fig. 2 has been clearly described in [2] with two main parts which are the rotating cluster (directional cluster) and the satellite pan cluster. The movement of the rotating cluster is done by the rotating joint q_1 which is driven by motor 1. The satellite pan cluster height is performed by rotating joint q_2. This joint is driven by motor 2 through the translational joint A and rotational joint B. However, these joints are only responsible for driving the joint q_2, so it is not considered in the kinematic problem. The center of the rotating cluster is G_1, the satellite pan assembly center is G_2.

Select a fixed coordinate system $(OXYZ)_0$ attached to the ground. The $(OXYZ)_i$ coordinate systems are respectively mounted at the positions shown in Fig. 2. In particular, taking point G_2 is the end-effector point representing the satellite pan cluster.

The position of G_2 point is determined according to the fixed coordinate system based on the setup of the DH parameter table and the homogeneous transformation matrix [2] as follows:

$$
\begin{aligned}
x_{G2} &= \cos q_1(a_2 + a_4 \cos q_2 - d_5 \sin q_2) \\
y_{G2} &= \sin q_1(a_2 + a_4 \cos q_2 - d_5 \sin q_2) \\
z_{G2} &= d_0 + d_1 + d_3 + a_4 \sin q_2 + d_5 \cos q_2
\end{aligned} \tag{1}
$$

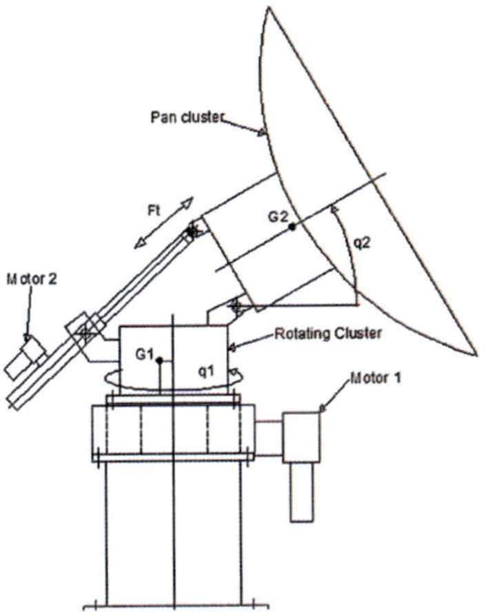

Fig. 1. Preliminary mechanical system

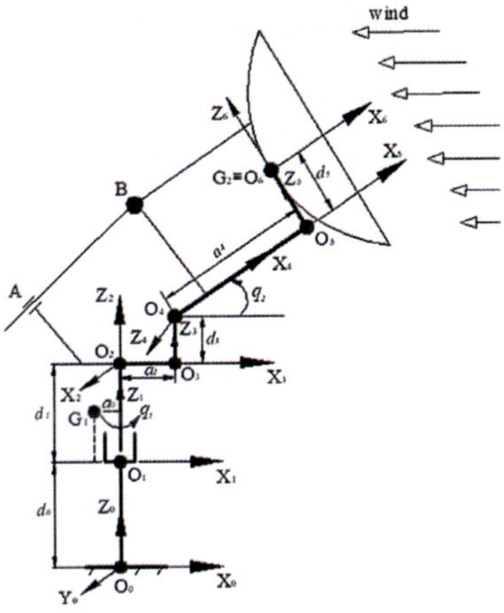

Fig. 2. Mathematical model

The position of the center mass G_1 of the rotating cluster ($\mathbf{R}_{C1}$) and the center mass G_2 of the pan cluster ($\mathbf{R}_{C2}$) in the fixed coordinate system are shown as follows

$$\mathbf{R}_{C1} = \begin{bmatrix} -a_0 \cos q_1 & -a_0 \sin q_1 & d_0 + \frac{d_1}{2} \end{bmatrix}^T, \ \mathbf{R}_{C2} = \begin{bmatrix} x_{G2} & y_{G2} & z_{G2} \end{bmatrix}^T \tag{2}$$

The dynamic equations of the antenna system are presented as

$$\mathbf{M}(\mathbf{q})\ddot{\mathbf{q}} + \mathbf{C}(\mathbf{q}, \dot{\mathbf{q}})\dot{\mathbf{q}} + \mathbf{g}(\mathbf{q}) = \boldsymbol{\tau} + \mathbf{f}^* \tag{3}$$

Where, $\mathbf{M}(\mathbf{q})$ is the mass matrix, $\mathbf{C}(\mathbf{q}, \dot{\mathbf{q}})$ is the Coriolis matrix which is determined in [14, 15], $\mathbf{g}(\mathbf{q})$ is the gravity potential energy vector, $\boldsymbol{\tau} = \begin{bmatrix} \tau_1 & \tau_2 \end{bmatrix}^T$ is the driven torque vector at the joints q_1, q_2 and $\mathbf{f}^* = \begin{bmatrix} Q_1^*, Q_2^* ..., Q_n^* \end{bmatrix}^T$ is the external force vector. In this study, the external force acting on the antenna system is the wind load. It is assumed that the influence of wind on the rotational cluster is ignored, only considering the effect of wind load on the pan cluster. The value of force caused by the wind on the pan cluster is calculated specifically in [23]. In this case, in order to reduce the calculating complexity, assumed that the wind force is always parallel and opposite to the axial direction OX. The distributed wind load is assumed to be concentrated force acting at the center of mass G_2. The maximum value of the wind load on OX axis is $F_{x\max} = 22174(N)$ and on OY axis is $F_{y\max} = 30670(\mathrm{N})$ in the case $q_2 = 0$ with velocity of the wind is $25(\mathrm{m/s})$ [23]. The driven torque vector can be found as follows

$$\boldsymbol{\tau}(t) = \mathbf{M}(\mathbf{q})\ddot{\mathbf{q}} + \mathbf{C}(\mathbf{q}, \dot{\mathbf{q}})\dot{\mathbf{q}} + \mathbf{g}(\mathbf{q}) - \mathbf{f}^* \tag{4}$$

2.2 PID Control System

The PID controller is a popular control system, which is used a lot in industry because of its simplicity, stability and ability to be controlled in many different working conditions.

The efficiency of PID system is limited in the nonlinear system. However, the PID controller is highly effective and stable for the systems which is explicitly mathematical. The quality of controlling depends on selecting the suitable control parameters including k_P, k_I, k_D. With the desired position and trajectory of the joint variable $\mathbf{q}_{desired}(t) = \begin{bmatrix} q_{1d} & q_{2d} \end{bmatrix}^T$, the PID control system needs to ensure that it finds a value of $\boldsymbol{\tau}(t)$ that meets the above requirements with the error $\mathbf{e}(t) = \mathbf{q}_{desired}(t) - \mathbf{q}_{actual}(t)$ moving towards zero with $\mathbf{q}_{actual}(t)$ being the actual value of the joint variable response. Thus, the control law is built as follows

$$\boldsymbol{\tau}(t) = \mathbf{K}_P\mathbf{e}(t) + \mathbf{K}_D\frac{d\mathbf{e}(t)}{dt} + \mathbf{K}_I\int_0^t \mathbf{e}ds \tag{5}$$

Where $\mathbf{K}_P = \begin{bmatrix} k_{p1} & k_{p2} \end{bmatrix}^T$, $\mathbf{K}_D = \begin{bmatrix} k_{d1} & k_{d2} \end{bmatrix}^T$, $\mathbf{K}_I = \begin{bmatrix} k_{i1} & k_{i2} \end{bmatrix}^T$ are the parameter vectors of the PID control system corresponding to each joint variable q_1, q_2. The control diagram of the antenna system is shown in Fig. 3.

2.3 Numerical Simulation Results

The basic parameters of the expected antenna system to be designed and manufactured are as follows

$a_0 = 0.33(m)$, $a_2 = 0.39(m)$, $d_0 = 3.62(m)$, $d_1 = 0.87(m)$, $d_3 = 0.2(m)$, $a_4 = 1.15(m)$, $d_5 = 0.21(m)$, $m_1 = 1050(kg)$, $m_2 = 1350(kg)$.

The given paths of joint variables are $q_1 = \frac{\pi}{4}\sin(\frac{t}{4})(rad)$ and $q_2 = \frac{\pi}{4}\sin(\frac{t}{6})(rad)$ (Fig. 5). The inverse dynamics problem solving diagram is described in Fig. 6.

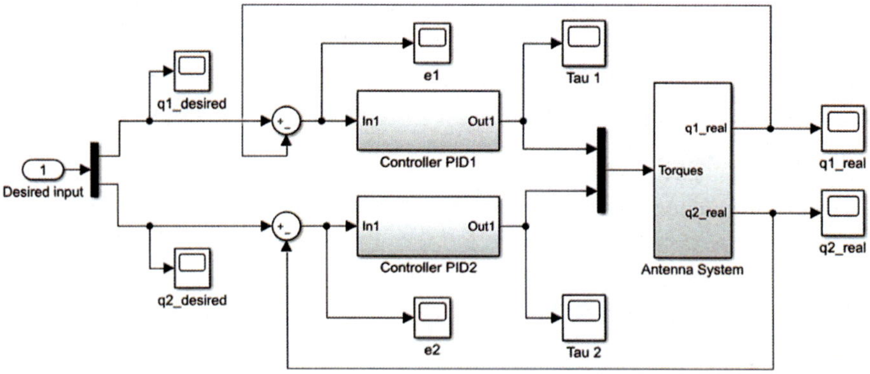

Fig. 3. The PID control diagram of the antenna system

Block "Antenna System" is established as in Fig. 4 with wind force vector $\mathbf{F}_{wind} = \begin{bmatrix} F_{ox} & F_{oy} \end{bmatrix}^T = \begin{bmatrix} 22174 & 30670 \end{bmatrix}^T (N)$.

The dynamics and control problems are divided into 2 cases corresponding to 2 different working environmental conditions.

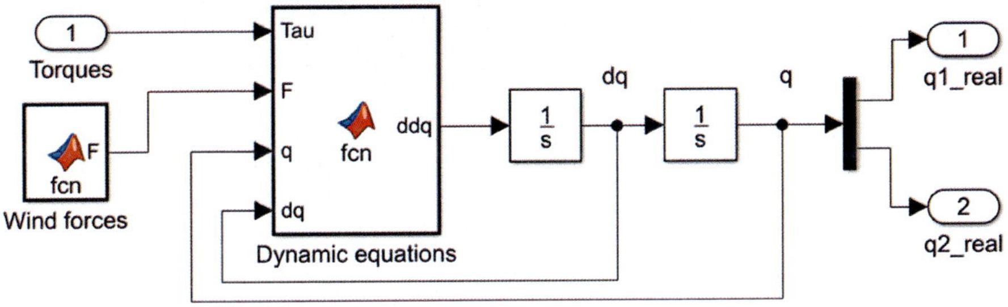

Fig. 4. The mathematical diagram in block "Antenna System"

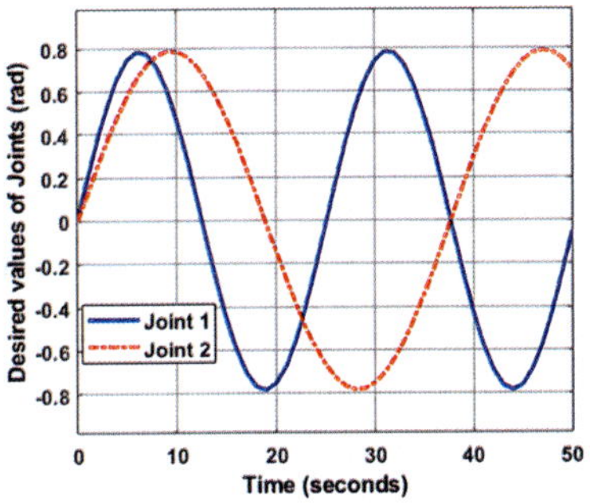

Fig. 5. The given joint variable paths

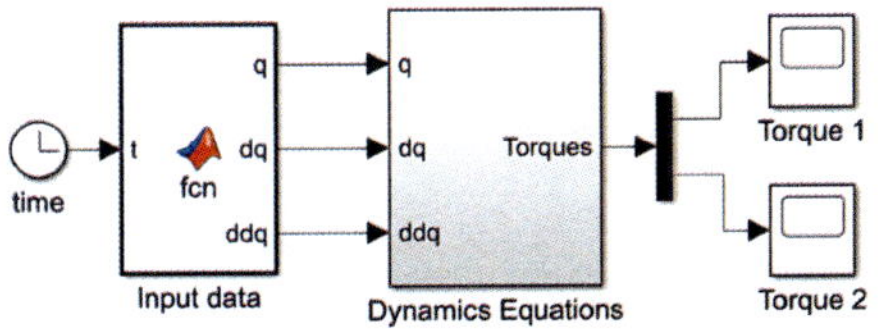

Fig. 6. The inverse dynamics solving diagram

Case 1: the antenna system works in the conditions of wind influence ($\mathbf{F}_{wind} = 0$)
Case 2: the antenna system works in the condition without the wind ($\mathbf{F}_{wind} \neq 0$)

The results of the inverse dynamics problem are shown in Fig. 7 and Fig. 8.

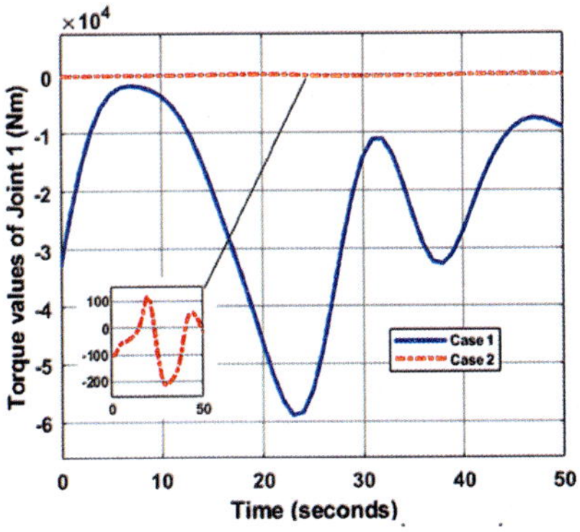

Fig. 7. The torque value of joint 1

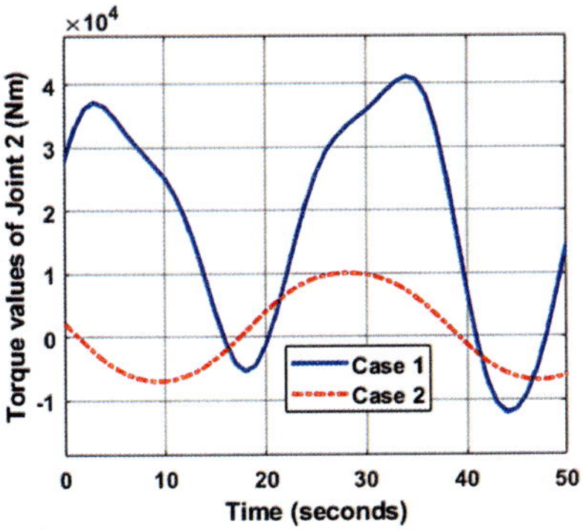

Fig. 8. The torque value of joint 2

Accordingly, the driving torque value in case 1 is higher than in case 2. The laws of torque values are also changed due to the influence of the wind with the largest value at joint 1 is $5.8 \times 10^4 (Nm)$ (Fig. 7), joint 2 is $4.2 \times 10^4 (Nm)$ (Fig. 8). Without the wind, the torque value in both joints is much lower.

The position and trajectory control parameters of the PID system in the two cases are shown in Table 1.

Table 1. PID parameters

PID control	Case 1	Case 2
Position control	$k_{p1} = 2.72 \times 10^5, k_{i1} = 2.64 \times 10^5,$ $k_{d1} = 0.62 \times 10^5, k_{p2} = 7.95 \times 10^5,$ $k_{i2} = 21.2 \times 10^5, k_{d2} = 0.58 \times 10^5$	$k_{p1} = 3.2 \times 10^4, k_{i1} = 1.3 \times 10^4,$ $k_{d1} = 1.9 \times 10^4, k_{p2} = 7.95 \times 10^5,$ $k_{i2} = 21.2 \times 10^5, k_{d2} = 0.58 \times 10^5$
Trajectory control	$k_{p1} = 2.95 \times 10^5, k_{i1} = 3.75 \times 10^5,$ $k_{d1} = 0.57 \times 10^5, k_{p2} = 2.02 \times 10^5,$ $k_{i2} = 4.8 \times 10^5, k_{d2} = 3.9 \times 10^5$	$k_{p1} = 2.72 \times 10^5, k_{i1} = 3.5 \times 10^5,$ $k_{d1} = 0.49 \times 10^5, k_{p2} = 5.6 \times 10^5,$ $k_{i2} = 12.8 \times 10^5, k_{d2} = 0.58 \times 10^5$

The results of the position control problem with input $q_1 = \frac{\pi}{5}(rad), q_2 = \frac{\pi}{7}(rad)$ are shown from Fig. 9, Fig. 10, Fig. 11 and Fig. 12.

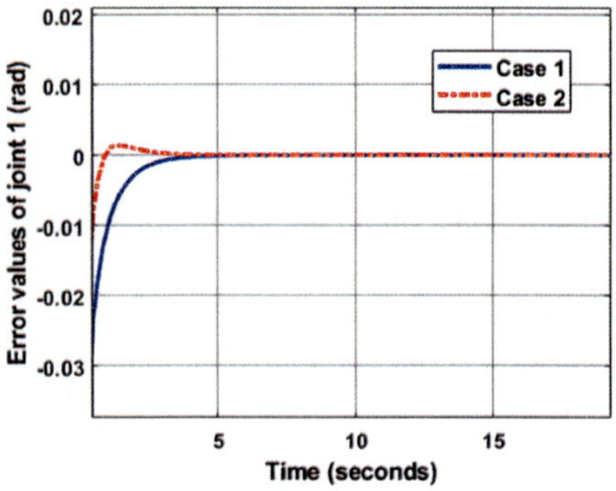

Fig. 9. The control error in joint 1

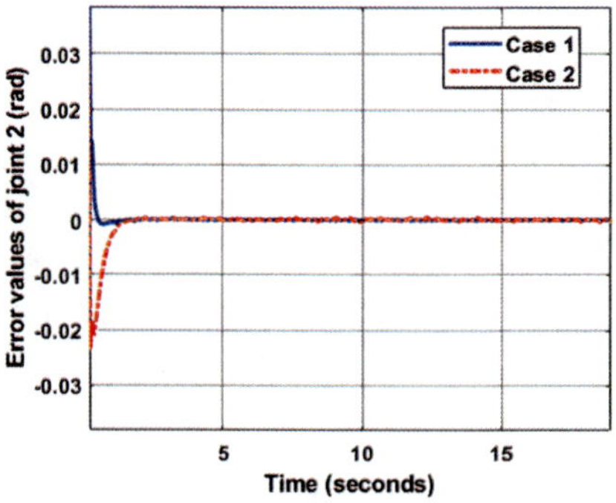

Fig. 10. The control error in joint 2

With joint 1 (Fig. 9), the control error value in case 1 at the initial time is larger and longer stable than in case 2. With joint 2 (Fig. 10), the control error value in case 1 is bigger and fast stable than in case 2.

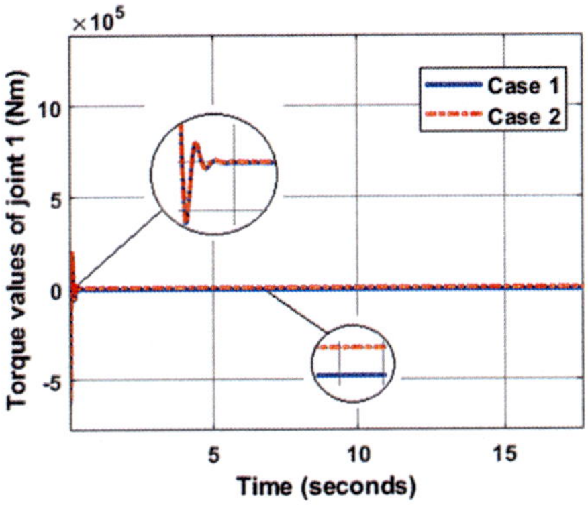

Fig. 11. The torque value of joint 1

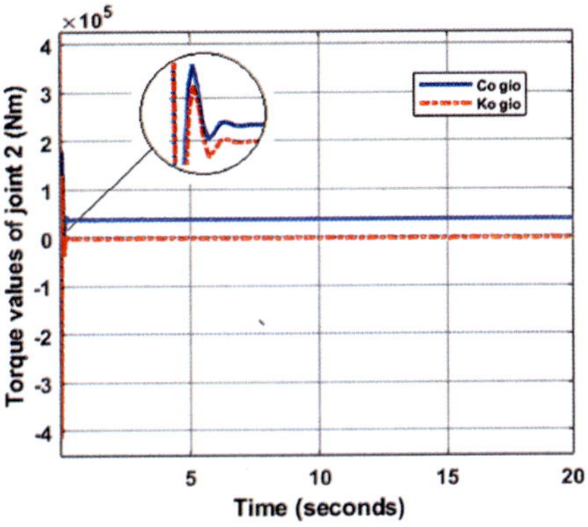

Fig. 12. The torque value of joint 2

Figure 11 and Fig. 12 show that the antenna is driven to the desired position with the driving torque value at the joints being stable. However, in case 1, the torque at the joints still needs to maintain a value to keep the system in the desired position due to the continuous influence of the wind. The driving torque in case 1 is greater than in case 2.

The joint trajectories $q_1 = \frac{\pi}{4}\sin(\frac{t}{4})(rad)$ and $q_2 = \frac{\pi}{4}\sin(\frac{t}{6})(rad)$ are input data to the trajectory control problem and the results are shown in Fig. 13, Fig. 14, Fig. 15 and Fig. 16.

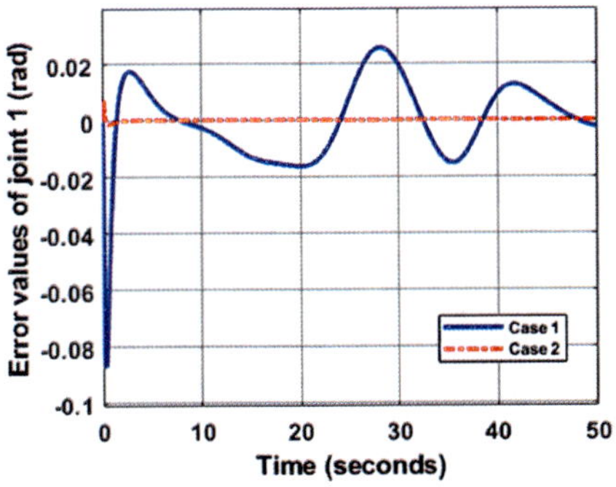

Fig. 13. The control error in joint 1

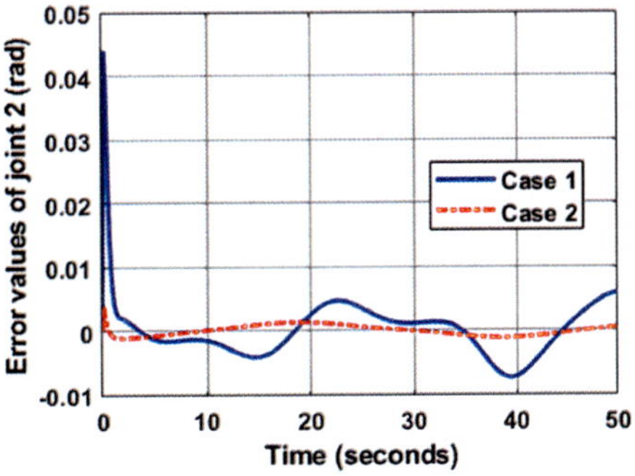

Fig. 14. The control error in joint 2

Figure 13 and Fig. 14 show that the value of control error in case 2 (without wind) is smaller than in case 1. The error value in case 2 has a clearer law than that in case 1. Similarly, the driving torque value at the corresponding joints in case 1 also varies with a large difference and is not as clear-cut as in case 2 (Fig. 15 and Fig. 16). The torque values at the joints are quite similar to the results in the inverse dynamics problem.

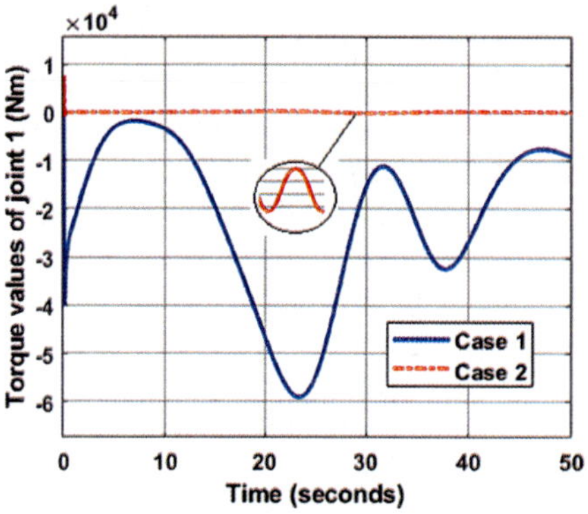

Fig. 15. The torque value of joint 1

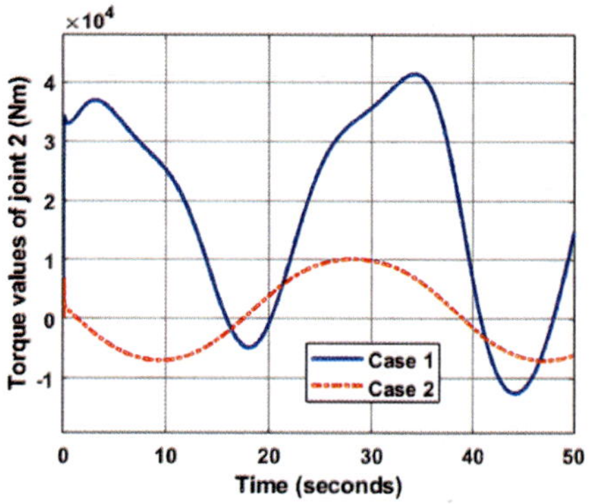

Fig. 16. The torque value of joint 2

3 Conclusion

In summary, this paper has presented some research results on the position and trajectory control of the satellite surveillance antenna with a traditional PID control system. The control system parameters are found through the efficient support of tools on MATLAB software. The control problem is built for antennas operating under two different environmental conditions. The numerical simulation results show that the control error of the antenna system under the influence of wind is higher than in the case without wind. The driving torque at the joints must also maintain a certain value to ensure compensation for deviations due to wind action. The results of this study have important value in calculating and selecting the transmission system with suitable joints, ensuring the antenna system can move in the right position and desired trajectory in case of continuous environmental impact. In fact, based on the research results on dynamics and control, the mechanical antenna system has been completely built and put into operation in reality (Fig. 17).

Fig. 17. A monitoring satellite antenna was fabricated and controlled

References

1. Hoang, P.Q., et al.: Kinematics modeling analysis of the geostationary satellite monitoring antenna system. Sci. Technol. Dev. J. Eng. Technol. **4**(1), 704–712 (2021)
2. Pham, Q.H., et al.: Kinematics and dynamics analysis of geostationary satellite antenna system. In: Long, B.T., Kim, YH., Ishizaki, K., Toan, N.D., Parinov, I.A., Vu, N.P. (eds.) Proceedings of the 2nd Annual International Conference on Material, Machines and Methods for Sustainable Development (MMMS2020). MMMS 2020, pp. 1009–1017. Lecture Notes in Mechanical Engineering. Springer, Cham (2021). https://doi.org/10.1007/978-3-030-69610-8_133
3. Maral, G.: Satellite Communications Systems - Systems, Techniques and Technology Fifth Edn. John Wiley and Sons Ltd Publication, Hoboken (2009)
4. Satcom and Antenna technologies division. http://www.cpii.com. Accessed 28 December 2020
5. Vincor™ Product Data and Specification Archives. http://www.catalog.vincor.com. Accessed 28 December 2020
6. Hubble Space Telescope. http://www.nasa.gov. Accessed 28 December 2020
7. Featured SATCOM Products. http://www.satcom-services.com. Accessed 28 December 2020
8. Antenna Systems. http://www.viasat.com/products/antenna-systems. Accessed 28 December 2020
9. Satellite Communications equipment. http://www.digisat.org. Accessed 28 December 2020
10. Bindi, Y., et al.: Dynamics analysis and control of a spacecraft mechanism with joint clearance and thermal effect. Precision Motion Systems: Modeling, Control, and Applications, Elsevier Inc, pp. 163–215 (2019)
11. Ogundele, D.A., Akoma, H.E.C.A., Adediran, Y.A.: Mathematical modelling of antenna look angles of geostationary communications satellite using two models of control stations. In: 3rd International Conference on Advanced Computer Theor and Engineering (ICACTE), pp. 236–240 (2010)
12. Ogundele, D.A., Aiyeola, S.Y., Adediran, Y.A., Oyedeji, E.O., Oseni, O.F.: Model validation and analysis of antenna look angles of a geostationary satellite. In: International Conference on Computer Science and Automation Engineering (CSAE), pp. 509–513. IEEE (2012)

13. Lida, T.: Satellite Communications Antenna Concepts and Engineering, Handbook of Satellite Applications. Springer Science+Business Media, New York (2015)
14. Shankar, S.G., Reddy, K.V.: Design and simulation of horn antenna in x-Ku band for satellite communications. Int. J. Res. Sci. Technol. (IJRST) **1**(10) (2014)
15. Ajiboye, A.T., Ajayi, A.R., Ayinla, S.L.: Effects of PID controller on the performance of dish antenna position control for distributed mobile telemedicine nodes. AZOJETE **15**(2), 304–313 (2019)
16. Waghmare, S., Pathak, P., Meshram, P., Pawar, V.: Satellite dish positioning system. Int. J. Innov. Res. Sci. Technol. (IJIRST) **4**, 50–54 (2017)
17. Shubham, P., Sankalp, P.T., Tanmay, W., Nandanwar, V.S.: Satellite dish positioning control by geared motor using RF module. Int. J. Adv. Res. Sci. Eng. Technol. **4**, 3388–3393 (2017)
18. Rahane, S.D., Mhaske, S.A., Shingate, S.R.: Design of advanced antenna positioning system. Int. J. Res. Advent Technol. **6**, 251–253 (2018)
19. Amritha, M.A.S., Divyasree, M.V., Jesna, P., Kavyasree, S.M., Keerthana, V.: Microcontroller based wireless 3d position control for antenna. Int. J. Sci. Res. (IJSR) **6**, 924–927 (2017)
20. Aloo, L.A., Kihato, P.K., Kamau, S.I.: DC servomotor-based antenna positioning control system design using hybrid PID-LQR controller. Eur. Int. J. Sci. Technol. **5**, 17–31 (2016)
21. Ahmad, M., Jiya, J.D., Anene, E.C., Haruna, Y.S.: Position control of parabolic dish antenna using feedback, zeigler-nichols and quadratic optimal regulator methods. Cont. J. Eng. Sci. **6**, 7–1 3 (2011)
22. Kuseyri, I.S.: MIMO H∞ control of three-axis ship-mounted mobile antenna systems. Int. J. Control (2017). https://doi.org/10.1080/00207179.2017.1279755
23. Hoang, P.Q, Tung, D.M, Toan, B.H.: Study to determine the wind load acting on the antenna serving the calculation and design of the mechanical system controlling the 7.6-diameter geostationary satellite monitoring antenna. In: National Scientific and Technology Conference on Mechanical Engineering-Dynamics, pp. 307–312 (2017)

Adaptive Control Using Barrier Lyapunov Functions for Omnidirectional Mobile Robot with Time-Varying State Constraints

Hoa Thi Truong and Xuan Bao Nguyen(✉)

The University of Danang - University of Technology and Education, 48 Cao Thang, Danang City, Vietnam

tthoa@ute.ute.vn

Abstract. An omnidirectional mobile robot's trajectory tracking control problem under the condition of full state constraints is studied. The Barrier Lyapunov function combined with the backstepping design method was developed to deal with the state constraints that exist during the tracking of the omnidirectional mobile robot and ensure that all the state variables will not exceed the range state constraints. Lyapunov theoretical analysis proves that the closed-loop system is stable. All signals can be guaranteed to be consistent and limited. Finally, the effectiveness and robustness of the method are verified by simulation and comparison with a controller that does not consider input saturation and state constraints and a classical proportional derivative controller.

Keywords: Adaptive control · Barrier Lyapunov · State constraints

1 Introduction

The omnidirectional mobile robot (OMR) is a special type of robot different from the traditional two-wheeled robot because of its flexibility. In recent years, with the rapid development of mobile robot technology, OMRs have received wide attention and are used in various fields such as industrial production, logistics, transport, military surveillance, and environmental detection. The problem of dynamic modeling has been studied extensively [1–3]. The controller needs to be designed to ensure that the robot moves in the desired trajectory. Many proposed controllers for omnidirectional robots include PID, fuzzy, and sliding controller [4–10]. However, these controllers did not consider the parameter uncertainty and external interference problems that exist widely in the real system, which would make the controller unable to achieve the desired performance in real applications. An adaptive control technique is the adaptive backstepping controller that has received much attention in the past decade. The controller's main advantages are smoothing, robustness, fast response, and good transient concerning system uncertainties and external disturbances. Besides, an adaptive tuner based on the Lyapunov synthesis method [11–13] was added in Controller to guarantee the controlled system's stability.

N. V. Khang et al. (Eds.): ASIAN MMS 2021, MMS 113, pp. 401–410, 2022.
https://doi.org/10.1007/978-3-030-91892-7_38

However, the state constraints have not been studied in the above documents in the actual system due to the limitation of robot motion space, safe speed. Ignoring these limitations will cause the designed controller to fail to achieve the desired tracking performance and even cause collision damage to the robot. Therefore, in practical application, it is necessary to study the method control method for tracking the trajectory of an omnidirectional mobile robot with state constraints.

Recently, for state-constrained systems, the Barrier Lyapunov function effectively deals with state constraints [14–17]. Bai et al. [15] proposed an adaptive controller to solve the full state constraint for DC motor. Meng et al. [17] designed an adaptive controller based on a radial basis function neural network, realizing the time-varying state constraint of a multi-input multi-output nonlinear system. Ding et al. [18] approximate the unknown wheeled robot model through the neural network and propose a state-limited fully adaptive neural network controller that can realize the effective tracking of the reference trajectory of a two-wheeled differential mobile robot. In summary, so far, there is no related research report on trajectory tracking and control of omnidirectional mobile robots in limited conditions input state requirements of the omnidirectional mobile robot during path tracking. State constraint conditions are introduced into the controller design process to ensure that the robot's position, posture, speed, and other motion states are always within the limits of the specified constraints. So that the robot can operate efficiently and safely; on the other hand, by designing an anti-saturation compensator satisfying the state constraints.

2 Kinematics and Dynamics of OMR

The following equation gives the kinematics of the robot is

$$\begin{bmatrix} \dot{\emptyset}_1 \\ \dot{\emptyset}_2 \\ \dot{\emptyset}_3 \end{bmatrix} = \begin{bmatrix} \omega_1 \\ \omega_2 \\ \omega_3 \end{bmatrix} = \frac{1}{r} H(\theta) \begin{bmatrix} \dot{x} \\ \dot{y} \\ \dot{\theta} \end{bmatrix} \tag{1}$$

with

$$H = H(\theta) = \begin{bmatrix} -S_\theta & C_\theta & l \\ -S_{\theta+\frac{2\pi}{3}} & C_{\theta+\frac{2\pi}{3}} & l \\ -S_{\theta-\frac{2\pi}{3}} & C_{\theta-\frac{2\pi}{3}} & l \end{bmatrix} \tag{2}$$

where r and ω_i, $i = 1, 2, 3$ are the radius and the angular velocities of the wheels. S_θ is $\sin\theta$, C_θ is $\cos\theta$, $S_{\theta+\frac{2\pi}{3}}$ is $\sin\left(\theta + \frac{2\pi}{3}\right)$, and $C_{\theta+\frac{2\pi}{3}}$ is $\cos\left(\theta + \frac{2\pi}{3}\right)$ (Fig. 1).

Matrix R is the non-singular matrix for each $\theta \in R$. The inverse matrix R^{-1} is expressed as following,

$$H^{-1} = \frac{2}{3} \begin{bmatrix} -S_\theta & -S_{\theta+\frac{2\pi}{3}} & -S_{\theta-\frac{2\pi}{3}} \\ C_\theta & C_{\theta+\frac{2\pi}{3}} & C_{\theta-\frac{2\pi}{3}} \\ \frac{1}{2l} & \frac{1}{2l} & \frac{1}{2l} \end{bmatrix}, \tag{3}$$

(a)

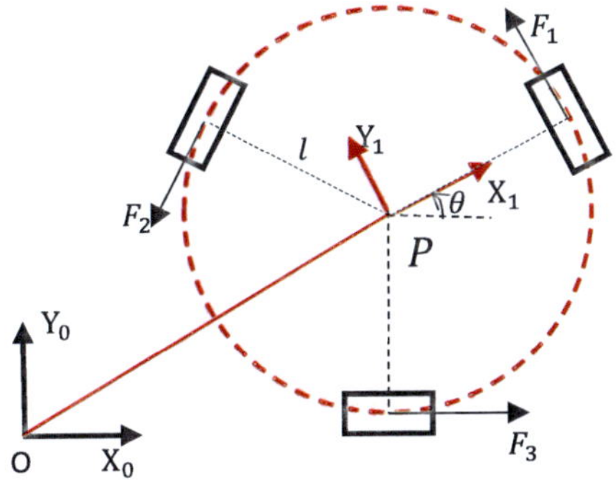

(b)

Fig. 1. Omnidirectional mobile robot: (a) the photo, (b) motion coordinates.

The translation can be obtained by Newton's second law of motion from which the linear and angular momentum balance equations are expressed as,

$$\begin{bmatrix} \ddot{x}_P \\ \ddot{y}_P \end{bmatrix} = \frac{1}{m}\sum_{i=1}^{3}\vec{F}_i \tag{4}$$

$$\ddot{\theta} = \frac{l}{J}\sum_{i=1}^{3}F_i \tag{5}$$

where F_i are the tractive force of i'th wheel, respectively. Define $P = \left[x_P y_P \theta\right]^T$ is the robot position in coordinate system, $F = [F_1 F_2 F_3]^T$ it yields

$$M\ddot{P} = H^T F \tag{6}$$

To drive the wheels, the torque τ_i and driving voltage u_i for the DC motor is expressed as,

$$\tau_i = \alpha_1 u_i - \beta_i \omega_i \tag{7}$$

and

$$F_i = \frac{\tau_i}{r}, F_i = \frac{\alpha_i u_i - \beta_i \omega_i}{r} \tag{8}$$

Note that $\omega = \frac{1}{r}R^T\dot{C}$, the Eq. (6) can be rewritten

$$\ddot{\mathrm{P}} = \frac{1}{r}H^T au - \frac{1}{r^2}H^T bH\dot{\mathrm{P}} \tag{9}$$

where

$$u = [u_1 u_2 u_3]^T,$$
$$a = \begin{bmatrix} \alpha_1/m & 0 & 0 \\ 0 & \alpha_2/m & 0 \\ 0 & 0 & \alpha_3/J \end{bmatrix}, \quad b = \begin{bmatrix} \beta_1/m & 0 & 0 \\ 0 & \beta_2/m & 0 \\ 0 & 0 & \beta_3/J \end{bmatrix}, \tag{10}$$

3 Controller Design

3.1 Proposed Control Algorithm

A robust adaptive controller is designed for an OMR with uncertain parameters.

Step 1: Define the state variables as $x_1 = \mathrm{P}$ and $x_2 = \dot{P}$ then we write the model of system (12) in state-space form as follows

$$\begin{cases} \dot{x}_1 = x_2 \\ \dot{x}_2 = \frac{1}{r}H^T au - \frac{1}{r^2}H^T bRx_2 \\ y \ = x_2 \end{cases} \tag{11}$$

The inequality is given as follows

$$\log\frac{k^2}{k^2 - x^2} \le \frac{x^2}{k^2 - x^2} \quad (|x| < k) \tag{12}$$

Assume that the desired trajectory satisfies $|\xi_d| < k_d$. The system tracking error is defined as,

$$e_1 = x_1 - \xi_d, \ \dot{e}_1 = x_2 - \dot{\xi}_d \tag{13a}$$

By introducing $e_2 = x_2 - \nu_1$, where $\nu_1\nu_1$ is the designed virtual control to be designed later on

$$e_2 = x_2 - \nu_1, \ \dot{e}_1 = e_2 + \nu_1 - \dot{\xi}_d, \ \dot{e}_2 = \dot{x}_2 - \dot{\nu}_1 \tag{13b}$$

The virtual control is designed as

$$\nu_1 = -k_1\left(k_{a1}^2 - e^2\right)e + \dot{\xi}_d \tag{14}$$

To ensure the stability, Barrier Lyapunov function candidate

$$V_1 = \frac{1}{2}\log\frac{k_{a1}^2}{k_{a1}^2 - e_1^2} \tag{15}$$

The time derivative

$$\dot{V}_1 = \frac{e_1\dot{e}_1}{k_{a1}^2 - e_1^2} = \frac{e_1(e_2 + \nu_1 - \xi_d)}{k_{a1}^2 - e_1^2}$$
$$= \frac{e_1\left(e_2 - k_1\left(k_{a1}^2 - e^2\right)e + \dot{\xi}_d - \dot{\xi}_d\right)}{k_{a1}^2 - e_1^2} = -ke_1^2 + \frac{e_1 e_2}{k_{a1}^2 - e_1^2} \tag{16}$$

where $\frac{e_1 e_2}{k_{a1}^2 - e_1^2}$ will be cancelled in the following step.

Step 2: In the second step, the actual control input will be described. The time derivative of the error variable $e_2 = x_2 - \nu_1$ is

$$\dot{e}_2 = \dot{x}_2 - \dot{\nu}_1 \tag{17}$$

From Eq. (14), we obtain

$$\dot{e}_2 = \frac{1}{r}H^T au - \frac{1}{r^2}H^T bHx_2 - \dot{\nu}_1 \tag{18}$$

Design a desired control input

$$u^* = r\left[H^T a\right]^{-1}\left(-k_2 e_2 + \frac{1}{r^2}H^T bRx_2 + \dot{\nu}_1\right) \tag{19}$$

Because the parameters aa and bb are not available, u*u* cannot be implemented in practice. To solve this problem, the estimated parameters are used to approximated of u*u*.

Define that $\tilde{a} = \hat{a} - a, \tilde{b} = \hat{b} - b$. The proposed adaptive control algorithm is expressed as,

$$u = r\left[H^T \hat{a}\right]^{-1} \times \left(-\frac{e_1(k_{a2}^2 - e_2^2) + k_2 e_2}{k_{a1}^2 - e_1^2} + \frac{1}{r^2}H^T \hat{b}Hx_2 + \dot{\nu}_1\right) \tag{20}$$

The proposed adaptive algorithms are describe as,

$$\hat{a} = \frac{1}{r}\frac{e_2}{k_{a2}^2 - e_2^2}H^T u \tag{21}$$

$$\hat{b} = \frac{e_2}{k_{a2}^2 - e_2^2}\frac{1}{r^2}H^T Rx_2 \tag{22}$$

3.2 Stability Analysis

Theorem: Consider the dynamic system (11) with the error tracking given by (13a, b) under the novel adaptive controller (20), the update control laws (21–22) such that all signals are bounded in the system, then the stability is achieved.

Proof: The Lyapunov function is defined as follows for entire system,

$$V_2 = V_1 + \frac{1}{2}\log\frac{k_{a2}^2}{k_{a2}^2 - e_2^2} + \frac{1}{2}\tilde{a} + \frac{1}{2}\tilde{b} \tag{23}$$

The time derivative of V_2 and

$$\begin{aligned}\dot{V}_2 &= \dot{V}_1 + \frac{e_2\dot{e}_2}{k_{a2}^2 - e_2^2} + \tilde{a}\dot{\hat{a}} + \tilde{b}\dot{\hat{b}} \\ &= -ke_1^2 + \frac{e_1 e_2}{k_{a1}^2 - e^2} + \frac{e_2\dot{e}_2}{k_{a2}^2 - e_2^2} + \tilde{a}\dot{\hat{a}} + \tilde{b}\dot{\hat{b}} \\ &= -ke_1^2 + \frac{e_1 e_2}{k_{a1}^2 - e^2} + \frac{e_2\left(\frac{1}{r}H^T(\hat{a} - \tilde{a})u - \frac{1}{r^2}H^T bRx_2 \ - \dot{v}_1\right)}{k_{a2}^2 - e_2^2} + \tilde{a}\dot{\hat{a}} + \tilde{b}\dot{\hat{b}}\end{aligned} \tag{24}$$

From proposed controller Eq. (20)

$$\begin{aligned}&= -ke_1^2 - k_2e_2^2 + \frac{e_2\left(\frac{1}{r}H^T\tilde{a}u + \frac{1}{r^2}H^T\hat{b}Hx_2 - \frac{1}{r^2}H^T bHx_2\right)}{k_{a2}^2 - e_2^2} + \tilde{a}\,\hat{a} + \tilde{b}\,\hat{b} \\ &= -ke_1^2 - k_2e_2^2 + \frac{e_2\frac{1}{r}H^T\tilde{a}u}{k_{a2}^2 - e_2^2} + \tilde{a}\,\hat{a} + \frac{e_2\frac{1}{r^2}H^T\tilde{b}Hx_2}{k_{a2}^2 - e_2^2} + \tilde{b}\,\hat{b}\end{aligned} \tag{25}$$

From the adaptive algorithm Eq. (22, 23), we have

$$V_2 = -ke_1^2 - k_2e_2^2 < 0 \tag{26}$$

Remark For the system (11) of omnidirectional mobile robots with full-state constraints, if the initial conditions satisfy $e_1(0) \in \overline{\Omega}_1 \triangleq \{|x_1| < k_{a1}, i = 1, 2, 3\}$ and $e_2(0) \in \overline{\Omega}_2 \triangleq \{|x_2| < k_{a2}, i = 1, 2, 3\}$, all signals will be uniformly bounded with the controller (20) and the adaptive (21, 22). The desired trajectory is bound $|\xi_d| < k_d$, $\left|\dot{\xi}_d\right| < k_{d2}$, so v_1 is also bounded by the Eq. (14). Introduce that $k_{b1} = k_{a1} + k_d$, $k_{b2} = k_{a2} + k_v$. From the Eq. (13a, b), we conclude that the state constraints will never be violated with suitable parameters, which means $|x_1| < k_{b1}$, $|x_2| < k_{b2}$. The tracking error will converge to a sufficiently small compact set to zero. The objectives of the controller are fulfill.

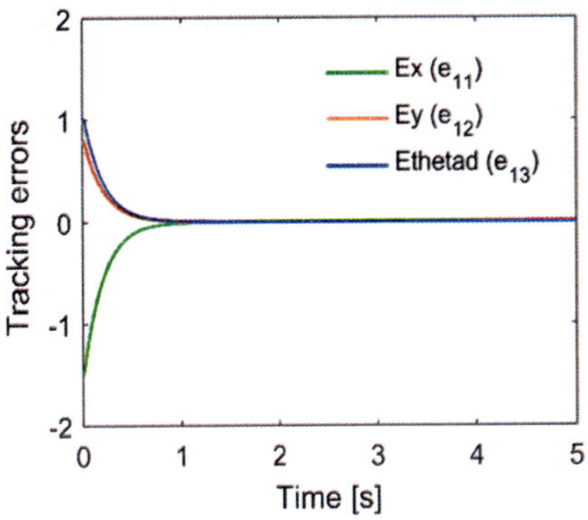

Fig. 2. The tracking error

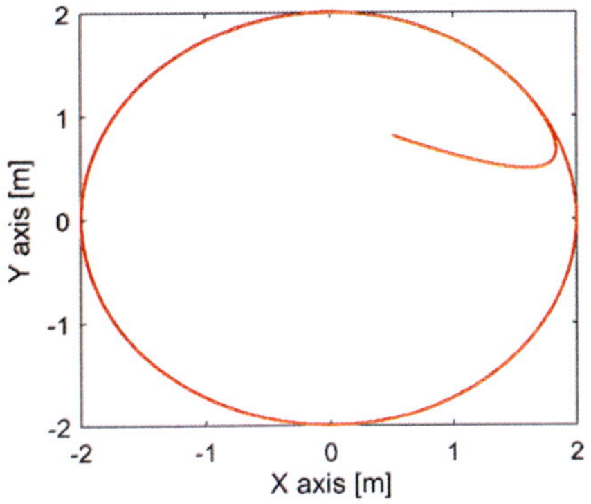

Fig. 3. The path tracking

4 Simulation Results

The actual and nominal values of the parameters in the omnidirectional mobile robot model can be used for the dynamics model. The parameters of the omnidirectional mobile robot model are $m = 4$ kg,$J = 10.0$ kgm^2, $r = 0.1$ m,$l = 0.5$ m,$d = 0.1$ m, $\delta = 0.1$, $a_1 = 1$, $a_2 = 1$, $a_3 = 1$, $b_1 = 0.1$, $b_2 = 0.1$, $b_3 = 0.1$, $\lambda_0 = 0.1$, $\Lambda = 2I$, $\phi = \pi/6$, $k_s = 1.0$. The initial state of the wheeled robot is chosen as $[X(0)\ Y(0)\ \theta(0)]^T = [0.5\ 0.8\ 0.1\ 0.0\ 0.0\ 0.0]^T$. The desired trajectory is a circle given by $C_d = [2\cos(1t), 2\sin(1t)]$.

Figure 2 shows the trajectory tracking errors of the robot. The figure indicate that the errors can converge to the small neighborhood of the origin, and the errors is always in the bounded interval. Figure 3 shows that the robot tracks along the desired trajectory at initial and full time. Figure 4 shows that the robot's angle has a quick response at the initial moment. The robot's angle is 1 rad at the starting point, and the desired angle is 0 rad. It only takes about 1 s for the robot to follow the desired angular trajectory. The proposed tracking controller is the effective and quick responsive. Figure 5 shows the voltages applied to the three motors of the three wheels. The voltage signal is smooth and reaches a maximum value of 80 V. As the robot track desired trajectory, the applied voltage values are limited to –2.5 V to 2, 5 V. It is proven that the proposed controller works efficiently. Figures 6 and 7 show the changes in the estimated values of the adaptive parameters $\hat{a}_i$ and $\hat{b}_i$. From the figures it can be seen that the parameters of the model update to the stable value after 0.05 s. To illustrate the effectiveness of the controller (20) in handling state constraints. The error convergence rate of e_2 is greatly reduced, as shown in Fig. 8. In Fig. 8, e_2 can converge to the vicinity of the origin in the range of 0–1.5 s. Therefore, from the controller (20) it is shown that the Barrier Lyapunov method avoids the occurrence of conditions that exceed the state constraints, and ensures the implementation of the desired trajectory tracking of the robot under state constraints.

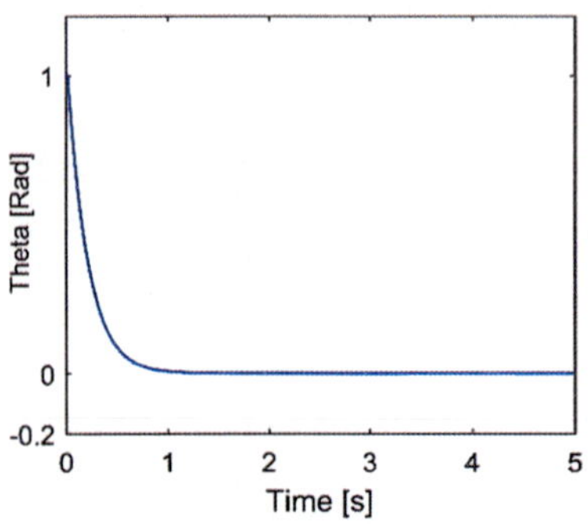

Fig. 4. Path tracking of the theta (θ).

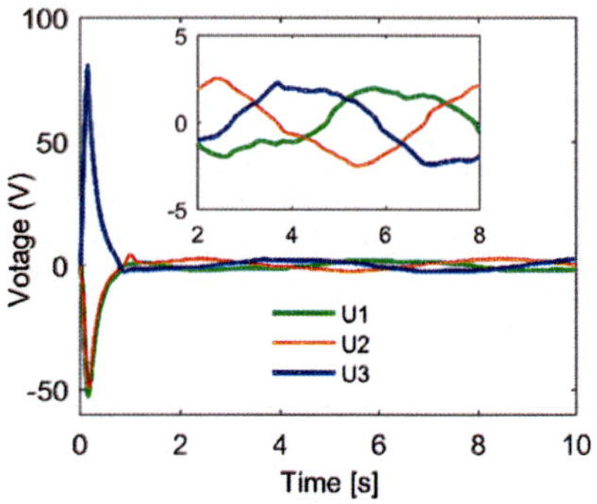

Fig. 5. The control voltage for motors

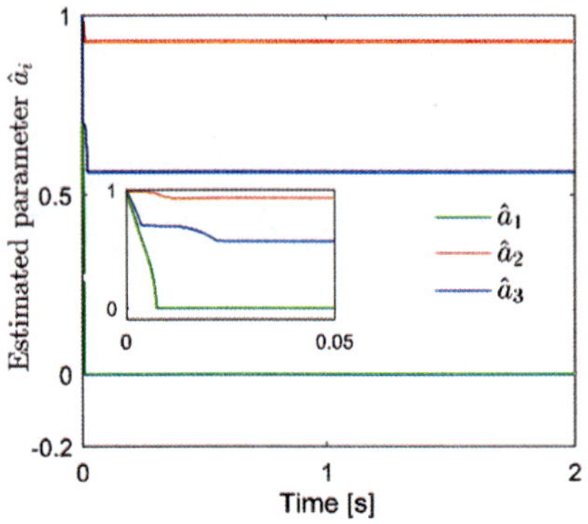

Fig. 6. The estimated parameter $\hat{a}_i$

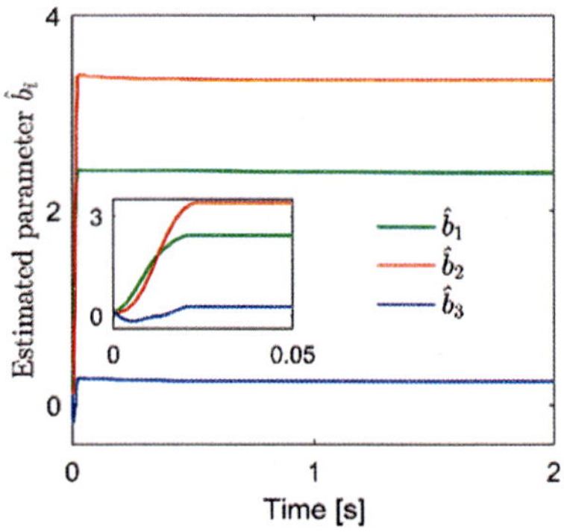

Fig. 7. The estimated parameter $\hat{b}_i$

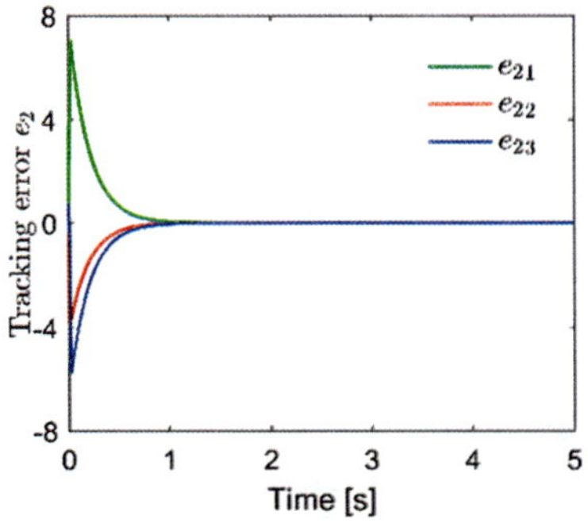

Fig. 8. The tracking error e_{2i}

5 Conclusions

To solve the problem of state constraint in the trajectory tracking process of an omnidirectional mobile robot, this paper proposes an adaptive tracking control method, which solves the problem of state constraint during the tracking trajectory of the robot with Barrier Lyapunov function. At the same time, an adaptive algorithm was proposed to estimate the unknown parameters of the system. Through simulation analysis, it is shown that the proposed method has better robustness and tracking trajectory. The next work is to design the tracking controller of an omnidirectional mobile robot when there are time-varying state constraints and input saturation, and combine it with perturbation observation nonlinearities to improved system dynamics.

References

1. Coelho, P., Nunes, U.: Path following control of mobile robots in presence of uncertainties. IEEE Trans. Robot. **21**(2), 252–261 (2005)
2. Kalmár-Nagy, T., D'Andrea, R., Ganguly, P.: Near-optimal dynamic trajectory generation and control of an omnidirectional vehicle. Robot. Auton. Syst. **46**(1), 47–64
3. Muir, P.F., Neuman, C.P.: Kinematic Modeling of Wheeled Mobile Robots (1986)
4. Nguyen, X.B., Truong, H.T.: Dynamics and robust adaptive controller of three wheel omnidirectional mobile robot. In: 2020 Applying New Technology in Green Buildings (ATiGB), pp. 110–114
5. Nguyen, X.B., Komatsuzaki, T., Iwata, Y., Asanuma, H.: Fuzzy semiactive vibration control of structures using magnetorheological elastomer. Shock and Vibration, Hindawi, vol. 2017, p. 15
6. Tong, S.C., Sheng, N., Li, Y.M.: Adaptive fuzzy control for nonlinear time-delay systems with dynamical uncertainties. Asian J. Control **14**, 1589–1598 (2012)
7. Nekoukar, V., Erfanian, A.: Adaptive fuzzy terminal sliding mode control for a class of MIMO uncertain nonlinear systems. Fuzzy Sets Syst. **179**, 34–49 (2011)
8. Wong, L.K., Leung, F.H.F., Tam, P.K.S.: A fuzzy sliding controller for nonlinear systems. IEEE Trans. Ind. Electron. **48**, 32–37 (2001)
9. Tao, C.W., Chan, M.L., Lee, T.T.: Adaptive fuzzy sliding mode controller for linear systems with mismatched time-varying uncertainties. IEEE Trans. Syst. Man Cybern. Part B **33**, 283–294 (2003)
10. Nguyen, X.B., Komatsuzaki, T., Iwata, Y., Asanuma, H.: Fuzzy semi-active control of multi-degree-of-freedom structure using magnetorheological elastomers ASME Proceedings. In: No: ASME 2017 Pressure Vessels and Piping Conference Volume 8: Seismic Engineering, 2017, pp. 1–10 (2017)
11. Slotine, J.J., Coetsee, J.A.: Adaptive sliding controller synthesis for nonlinear systems. Int. J. Control **43**(6), 1631–1651 (1986)
12. Nguyen, X.B., Komatsuzaki, T., Iwata, Y., Asanuma, H.: Modeling and semi-active fuzzy control of magnetorheological elastomer-based isolator for seismic response reduction. Mech. Syst. Signal Process. **101**, 449–466 (2018)
13. Nguyen, X.B., Komatsuzaki, T., Iwata, Y., Asanuma, H.: Robust adaptive controller for semi-active control of uncertain structures using a magnetorheological elastomer-based isolator. J. Sound Vib. **343**, 192–212 (2018). ScienceDirect
14. Tee, K.P., Ge, S.S., Tay, E.H.: Barrier Lyapunov functions for the control of output-constrained nonlinear systems. Automatica **45**(4), 918 (2009)

15. Bai, R.: Neural network control-based adaptive design for a class of DC motor systems with the full state constraints. Neurocomputing **168**, 65 (2015)
16. Nguyen, X.B., Komatsuzaki, T., Truong, H.T.: Novel semiactive suspension using a magnetorheological elastomer (MRE)-based absorber and adaptive neural network controller for systems with input constraints. Mech. Sci. **11** (2), 465–479
17. Meng, W.C., Yang, Q.M., Sun, Y.X.: Adaptive neural control of nonlinear MIMO systems with time-varying output constraints. IEEE Trans. Neural Netw. Learn. Syst. **26**(5), 1074 (2015)
18. Ding, L., Li, S., Liu, Y.J., et al.: Adaptive neural network-based tracking control for full-state constrained wheeled mobile robotic system. IEEE Trans. Syst. Man Cybern. Syst. **47**(8), 2410 (2017)

Trajectory Tracking Using Linear State Feedback Controller for a Mecanum Wheel Omnidirectional

Nguyen Hong Thai[1(✉)], Trinh Thi Khanh Ly[2], Nguyen Thanh Long[1], and Le Quoc Dzung[2]

[1] Department of Mechanical Design and Robotics, School of Mechanical Engineering, HanoiUniversity of Science and Technology, Hanoi, Vietnam
thai.nguyenhong@hust.edu.vn

[2] Faculty of Automation Technology, Electric Power University, Hanoi, Vietnam

Abstract. This paper presents a method to design a linear state feedback controller for the mecanum wheel omnidirectional to track a NURBS curve trajectory. Firstly, the robot's kinematic model and kinematic error were established to design a linear state feedback controller. Then, determine the parameters of the controller according to the varying angular velocity for the desired trajectory. An omnidirectional robot platform has been experimental designed and manufactured for experimental verification of the proposed controller. The simulation and experimental results show that the kinematic controller devel-oped by this study has high-performance trajectory tracking with minor tracking errors.

Keywords: Linear state feedback controller · Omnidirectional mobile robot · Trajectory tracking · NURBS curve · Kinematic error model

1 Introduction

In recent years, omnidirectional robots with mecanum wheels widely applied in many different fields, such as the unmanned rescue vehicle of Andrzej et al. [1]. The omnidirectional AGV model of L. Schulze et al. [2] uses mecanum wheels with the conveying or Michael Göller et al. [3, 4] with mecanum wheels for supermarket customers. In addition, several other research [5, 6] show different AGV robots applications in industrial fields.

Improving accuracy, including position and orientation, has always been a challenge, so it is a topic of interest by many researchers. Research in this area includes Li et al. [7] studied the omnidirectional mecanum wheeled robot motion trajectories generation expressed in polynomial functions passing through a given set of via points to minimize the energy consumption. Wang et al. [8] used model predictive control the trajectory tracking of an omnidirectional robot with mecanum wheels based on the kinematics model of the robot. Liu et al. [9] control the trajectory tracking of an omnidirectional mobile robot by linear control based on the kinematic and dynamic. Also, there are some

N. V. Khang et al. (Eds.): ASIAN MMS 2021, MMS 113, pp. 411–421, 2022.
https://doi.org/10.1007/978-3-030-91892-7_39

other similar studies [10, 11] on trajectory tracking of mobile robots with simple straight and curved trajectories.

In addition, one problem regarding the robot's trajectory is that it is necessary to smooth the bends to avoid sudden changes in the robot's speed. Research in this area currently has the following meth-ods: Dubins [12] using arcs to fillet the folded segments; use B-spline to path planning [13]; Using the NURBS curve to trajectory design control of robotic manipulator drilling blasting holes in tunnel construction [14]; Or use the NURBS curve to design trajectories for AGVs in production workshops [15, 16] etc. Nevertheless, when the trajectory is a NURBS curve, the control problem of tracking kinematic becomes complex and nonlinear according to the time-varying parameters. Invariant linear control theory is inefficient must be used as Intelligent controllers or complex nonlinear controls. This research presents a linear state feedback controller with a time-varying parameter with a simple control law for high convergence speed while ensuring accuracy to overcome the above drawback. First, A tracking linear controller with a pole position approach has been developed based on the linearization kinematic model. Then, the robot's kinematic model and kinematic error are establishing. On that basis, the linearized kinematic model is used to develop a tracking linear controller with a pole placement approach. Finally, verified the results under simulations and experiments of the omnidirectional robots with mecanum wheels platforms.

2 Mecanum Wheel Omnidirectional Robot Model

2.1 Kinematics of the Robot

The kinematic diagram of the mecanum wheeled robot is shown in Fig. 1. The robot parameters dimensions are defined as follows: r is the wheel's radius, ω_{i} is the angular velocity of No. i wheel with i represents the order of the mecanum wheel, V_{i} denotes the velocity of No. i wheel, $V_{L\mathrm{i}}$ is wheel velocity, V is the centroid velocity of robot, Ω is the angular velocity of robot, L is the distance in the $\mathrm{x_r}$-axis of two wheels, d is the distance in the $\mathrm{y_r}$-axis of two wheels, ξ is the desired travel trajectory.

From Fig. 1, Set $\vartheta_{\mathrm{f}}\{\mathrm{O_f x_f y_f}\}$ represents the global coordinate axis, $\vartheta_{\mathrm{R}}\{\mathrm{G x_R y_R}\}$ the coordinate system attached to the robot. Thus, $\mathbf{q}_d = \left[x_d\ y_d\ \varphi_d\right]^T$ is the robot positioning parameters in coordinate system ϑ_{f}; $\mathbf{q} = \left[x\ y\ \varphi\right]^T$ is the robot positioning parameters in coordinate system ϑ_{R}. Therefore, the motion equations of the omnidirectional mecanum wheeled robot are as follows:

$$\dot{\mathbf{q}} = \mathbf{Q}^{\mathbf{T}}(\varphi)\dot{\mathbf{q}}_d \tag{1}$$

Wherein: $\mathbf{Q}(\varphi) = \begin{bmatrix} \cos\varphi & \sin\varphi & 0 \\ -\sin\varphi & \cos\varphi & 0 \\ 0 & 0 & 1 \end{bmatrix}$ is defined to the ϑ_{f} into the ϑ_{R}.

According to [15], the relationship between the angular velocity (ω_i for $i = 1,\ldots, 4$) of the wheels and the motion of the omnidirectional robot is given by:

$$\boldsymbol{\omega} = \mathbf{J}\dot{\mathbf{q}} \tag{2}$$

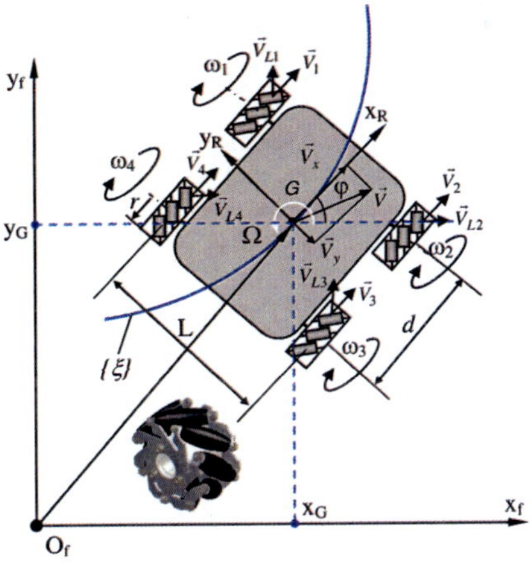

Fig. 1. Diagram of Omnidirectional mecanum wheeled robot.

Where: $\boldsymbol{\omega} = \left[\omega_1\ \omega_2\ \omega_3\ \omega_4\right]^T$, $\dot{\mathbf{q}} = \left[\dot{x}\ \dot{y}\ \dot{\varphi}\right]^T = \left[V_x\ V_y\ \Omega\right]^T$ and

$$\mathbf{J} = \begin{bmatrix} 1/r & 1/r & -(L+d)/2r \\ 1/r & -1/r & (L+d)/2r \\ 1/r & 1/r & (L+d)/2r \\ 1/r & -1/r & -(L+d)/2r \end{bmatrix}$$

2.2 Tracking Error Model of the Omnidirectional Robot

Defining the trajectory tracking error following equation:

$$\mathbf{e} = \left[e_x\ e_y\ e_\varphi\right]^T = \mathbf{q}_d - \mathbf{q} \tag{3}$$

The error equation in robot coordinate $\vartheta_{\mathrm{R}}\{\mathrm{Gx_R y_R}\}$ is given by:

$$\mathbf{e}_R = \mathbf{Q}^T(\varphi)\mathbf{e} \tag{4}$$

From Eq. (3), the differential equation of tracking error described by Eq. (5):

$$\dot{\mathbf{e}} = \dot{\mathbf{q}}_d - \dot{\mathbf{q}} \tag{5}$$

Where: $\dot{\mathbf{q}}_d = \left[\dot{x}_d\ \dot{y}_d\ \dot{\varphi}_d\right]^T = \left[V_{xd}\ V_{yd}\ \Omega_d\right]^T$, $\dot{\mathbf{q}} = \left[V_x\ V_y\ \Omega\right]^T$.
On the other hand, the derivative of Eq. (4) can be described as:

$$\dot{\mathbf{e}}_R = \mathbf{Q}^T(\varphi)\dot{\mathbf{e}} + \dot{\mathbf{Q}}^T(\varphi)\mathbf{e} \tag{6}$$

Combining Eqs. (6) and (3, 5) we have:

$$\dot{\mathbf{e}}_R = \begin{bmatrix} \dot{e}_{Rx} \\ \dot{e}_{Ry} \\ \dot{e}_{R\varphi} \end{bmatrix} = \begin{bmatrix} V_{xd}\cos e_\varphi - V_{yd}\sin e_\varphi - V_x + \Omega e_{Ry} \\ V_{xd}\sin e_\varphi + V_{yd}\cos e_\varphi - V_y - \Omega e_{Rx} \\ \Omega_d - \Omega \end{bmatrix} \tag{7}$$

Equation (7) is written in matrix form:

$$\dot{\mathbf{e}}_R = \begin{bmatrix} 0 & \Omega & 0 \\ -\Omega & 0 & 0 \\ 0 & 0 & 0 \end{bmatrix}\begin{bmatrix} e_{Rx} \\ e_{Ry} \\ e_{R\varphi} \end{bmatrix} + \begin{bmatrix} \cos e_\varphi & -\sin e_\varphi & 0 \\ \sin e_\varphi & \cos e_\varphi & 0 \\ 0 & 0 & 1 \end{bmatrix}\begin{bmatrix} V_{xd} \\ V_{yd} \\ \Omega_d \end{bmatrix} - \begin{bmatrix} V_x \\ V_y \\ \Omega \end{bmatrix} \tag{8}$$

Equation (8) is the nonlinear kinematic error model of the omnidirectional robot.

3 Controller Design

Equation (8) is the nonlinear error model. Linearizing Eq. (8) around the reference trajectory x_d, y_d, φ_d. We obtain the linear model of error in Eq. (9):

$$\dot{\mathbf{e}}_R = \begin{bmatrix} 0 & \Omega_d & -V_{yd} \\ -\Omega_d & 0 & V_{xd} \\ 0 & 0 & 0 \end{bmatrix}\mathbf{e}_R + \mathbf{I}_{3\times 3}\mathbf{u} \tag{9}$$

With $\mathbf{u} = \begin{bmatrix} u_1 & u_2 & u_3 \end{bmatrix}^T = \begin{bmatrix} e_{Vx} & e_{Vy} & e_\Omega \end{bmatrix}^T = \begin{bmatrix} V_{xd} - V_x & V_{yd} - V_y & \Omega_d - \Omega \end{bmatrix}^T$ is control input.

In this research, the controller design has three states, namely $x(t)$, $y(t)$, $\varphi(t)$ and three inputs, namely u_1, u_2, u_3. The general equation linear state space controller:

$$\mathbf{u} = \mathbf{K}\mathbf{e}_R \tag{10}$$

Therefore, the feedback control law becomes:

$$\begin{bmatrix} u_1 & u_2 & u_3 \end{bmatrix}^T = -diag(k_1, k_2, k_3)\begin{bmatrix} e_{Rx} & e_{Ry} & e_{R\varphi} \end{bmatrix}^T \tag{11}$$

And it is wished to solve the design problem of finding an error feedback gain matrix **K** such that the resulting closed-loop error system is stable with poles placed in some desired region of the complex plane.

In this model, the combination between feedforward control and linear feedback control is base on the pole placement approach, so generated a new control input as:

$$V_x(t) = V_{xd} - u_1,\ V_y(t) = V_{yd} - u_2,\ \Omega(t) = \Omega_d - u_3 \tag{12}$$

Where: $\Omega_d = V_d/\rho$, ρ is the curvature of the reference NURBS path (See Sect. 4). A desired closed-loop characteristic equation:

$$(s + 2\omega_n\zeta)(s^2 + 2\omega_n s\zeta + \omega_n^2) = 0 \tag{13}$$

Having constant eigenvalues (one negative real at $-2\zeta\omega n$ and a complex pair with natural angular frequency $\omega_n = (\Omega_d^2 + gV_d^2)^{0.5}$ and damping coefficient $0 < \zeta < 1$.

The gains in Eqs. (11) can be determined:

$$\begin{aligned} &k_1 = \zeta(\Omega_d^2 + gV_d^2)^{0.5} + (\zeta^2\Omega_d^2 + gV_d^2(\zeta^2 - 1))^{0.5}, \\ &k_2 = \zeta(\Omega_d^2 + gV_d^2)^{0.5} - (\zeta^2\Omega_d^2 + gV_d^2(\zeta^2 - 1))^{0.5},\ k_3 = 2\zeta(\Omega_d^2 + gV_d^2)^{0.5} \end{aligned} \tag{14}$$

Wherein: $g \le \zeta^2/((1 - \zeta^2)\rho_{\max}^2)$.

Figure 2 shows the schematic diagram of the closed-loop system composed of the state space controller, the reference trajectory and the omnidirectional robot.

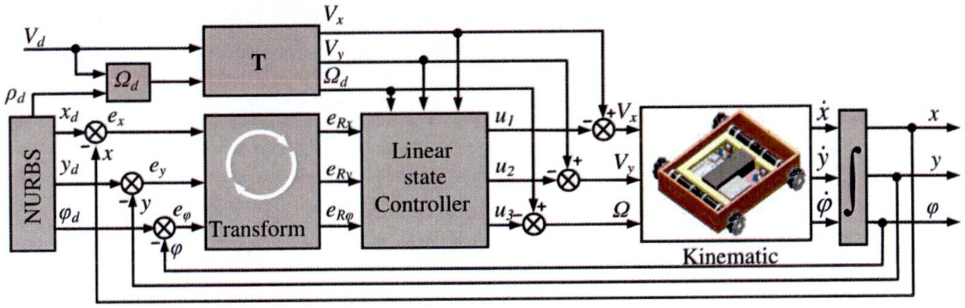

Fig. 2. Schematic diagram of the close-loop system.

4 NURBS Curve Trajectory Tracking Design

According to [14, 16, 17], the NURBS curve was designed by interpolating the control points according to the following equation:

$$P(\tau) = \sum_{i=1}^{n+1} B_i R_{i,k}(\tau) \tag{15}$$

Wherein: B_i is the vertex of the ith control polygon, k the order of the curve, τ is the parameter $(0 \leq \tau \leq 1)$ that are divided to determine the accuracy of the curve, $R_{i,k}(\tau)$ is the ith basis the B-Spline with k order and is given by:

$$R_{i,k}(\tau) = \frac{h_i N_{i,k}(\tau)}{\sum_{j=1}^{n+1} h_j N_{j,k}(\tau)} \tag{16}$$

$N_{i,k}(\tau)$ is the base irrational B-Spline function defined by:

$$N_{i,1}(\tau) = \begin{cases} 1 & x_i \leq \tau \leq x_{i+1} \\ 0 & \end{cases}, \quad N_{i,k}(\tau) = \frac{(\tau - x_i)N_{i,k-1}(\tau)}{x_{i+k-1} - x_i} + \frac{(x_{i+k-t} - \tau)N_{i+1,k-1}(\tau)}{x_{i+k} - x_{i+1}} \tag{17}$$

The node vector $[X]$ is defined by: $x_i = 0$ with $1 \leq i \leq k$; $x_i = i$ - k with $k + 1 \leq i \leq n + 1$

$x_i = n - k + 2$ with $n + 2 \leq i \leq nk + 1$.

From the above NURBS curve design method with control points given by Table 1, we have the NURBS curve trajectory shown in Fig. 3.

Table 1. NURBS curve trajectory control points.

B1	B_2	B_3	B_4	B_5	B_6	B_7	B_8	B_9	B_{10}
$\begin{bmatrix} 1 \\ 0 \end{bmatrix}$	$\begin{bmatrix} 2 \\ 0 \end{bmatrix}$	$\begin{bmatrix} 2 \\ 1 \end{bmatrix}$	$\begin{bmatrix} 3 \\ 1 \end{bmatrix}$	$\begin{bmatrix} 3 \\ 2 \end{bmatrix}$	$\begin{bmatrix} 1 \\ 2 \end{bmatrix}$	$\begin{bmatrix} 1 \\ 1 \end{bmatrix}$	$\begin{bmatrix} 0 \\ 1 \end{bmatrix}$	$\begin{bmatrix} 0 \\ 0 \end{bmatrix}$	$\begin{bmatrix} 1 \\ 0 \end{bmatrix}$

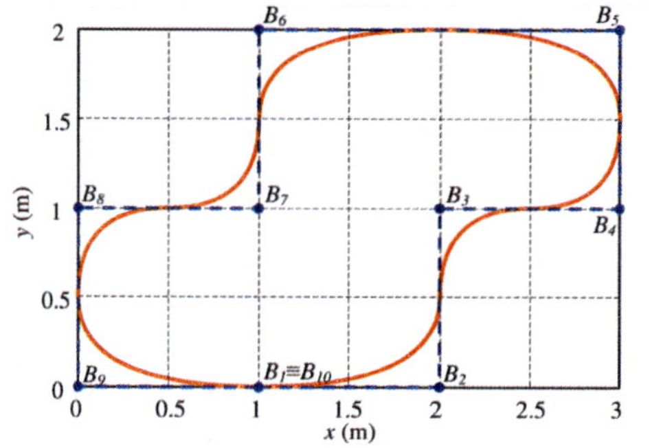

Fig. 3. NURBS curve trajectory of the omnidirectional robot.

5 Simulation and Experimental Results

Figure 4 is the robot platform manufactured by this study. Wherein: (1) Shell material: is 8 mm thick aluminium alloy A6061 to ensure lightness and rigidity; (2) Drive: Four drive wheels connected to integrated DC servo motor with integrated planetary gear reducer with gear ratio 1:14 and encoder 500CPR (2 channels A-B, 500 pulses); (3) Hardware: The central processor is an embedded computer Raspberry Pi 4 model b, control module Arduino ATmega 2560 R3; (4) Communication: connect between host and robot by module wifi WHR-G301N; (5) The INS and obstacle detection system includes a Waveshare 10 DOF IMU sensor and two LiDAR Delta 2A positioning scanners; (6) Battery source is Rechargeable Sealed lead acid battery 12 V 2.2Ah/20HR; (7) Robot control software developed on ROS operating system.

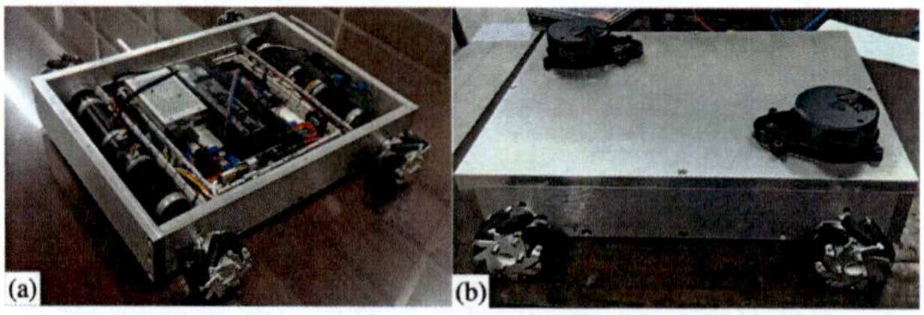

Fig. 4. The omnidirectional robot platform manufactured (a) The hardware structure and (b) The robot prototype.

With the parameters of the omnidirectional robot and the wheel are as follows: $L = 250$ (mm), $d = 300$ (mm), radius of wheel $r = 30$ (mm).

Setting parameters as follows:

Step 1: The motion trajectory of the omnidirectional robot is the NURBS curve given in Fig. 3 with the database in Table 1;

Step 2: Movement parameters: (i) The desired velocity $V_d = 0.3$ (m/s) and (ii) The angular velocity of the omnidirectional robot Ω is determined as follows: $\Omega_d = V_d/\rho$, with ρ is the radius of curvature of the desired NURBS trajectory and is given by:

$$\rho_i = \left| \frac{(\dot{x}_i^2 + \dot{y}_i^2)^{3/2}}{\dot{x}_i\ddot{y}_i - \dot{y}_i\ddot{x}_i} \right| \quad \text{with } i = 1, 2, \ldots, m\,, \dot{x}_i = \frac{\Delta x}{\Delta t} = \frac{x_{i+1} - x_i}{t_{i+1} - t_i}\,, \dot{y}_i = \frac{\Delta y}{\Delta t} = \frac{y_{i+1} - y_i}{t_{i+1} - t_i}$$

$\ddot{x}_i = \frac{\Delta \dot{x}}{\Delta t} = \frac{\dot{x}_{i+1} - \dot{x}_i}{t_{i+1} - t_i}$, $\ddot{y}_i = \frac{\Delta \dot{y}}{\Delta t} = \frac{\dot{y}_{i+1} - \dot{y}_i}{t_{i+1} - t_i}$, while the time at *P(i)* ($1 \le i \le m$, m is the number of elements of P) is determined by: $t_i = \sum_{j=1}^{i} \Delta t_j = \sum_{j=1}^{i} \frac{\Delta S_j}{V_d} = \sum_{j=1}^{i} \frac{P_j P_{j+1}}{V_d}$ $i = 1, 2, \ldots, m$.

Thus, putting $\begin{bmatrix} V_{xd} & V_{yd} & \Omega_d \end{bmatrix}^T = T \begin{bmatrix} V_d \\ \Omega_d \end{bmatrix}$ with $T = \begin{bmatrix} 0\,0\,1 \\ 1\,0\,0 \end{bmatrix}^T$ (see Fig. 2);

Step 3: Controller parameter: After evaluate the mean error, choose $\zeta = 0.9$, $g = 0.00266$ instead Eq. (14) the parameters K_1, K_2, K_3 of the controller can be determined. Figure 5 below is the motion trajectory of the controlled robot tracking the desired NURBS trajectory.

The omnidirectional robot controls the desired trajectory NURBS as described in Fig. 5 with the above parameters. Then Fig. 6 is the moving speed of the robot, including the velocity V and the angular velocity Ω of the robot when the radius ρ of curvature of the desired trajectory ξ varies from 0.358 m to 4 m (see Fig. 5b).

From Fig. 5 and 6, since the desired angular velocity Ω determined from V and ρ, the angular velocity error of the robot at each time $\Delta\Omega = \Omega - \Omega_d$ varies from 0 rad/s to 0.8 rad/s (see Fig. 6b). The reason at positions ①, ②, ③, ④, ⑤, ⑥ is that the robot positions change direction from clockwise to counterclockwise to follow the desired trajectory ξ as depicted in Fig. 6a. The tracking error between the movement trajectory and the desired trajectory is shown in Fig. 7.

From Fig. 5a and 7, it is easy to see that, when the robot moves: (1) From the starting point to ①, there will be a change of direction from clockwise to counterclockwise and, in turn, follow tracking from the inside to out of the desired trajectory with a position error is 1 mm and posture error of approximately 1°; (2) From position ① to position ② on the trajectory, the robot starts to track from the outside to the inside of the desired trajectory reversing the direction to the inward starting from a position ⑦) with the position error increasing from 1.2 mm to 2.8 mm and the error of direction is about 4.6°;

(3) From position ② to position ③, the robot reverses the tracking direction from the outside to the inside of the desired trajectory and vice versa at position ⑧, the error in the x-direction is 1.8 mm, at position (15) is the reverse direction with an angular error

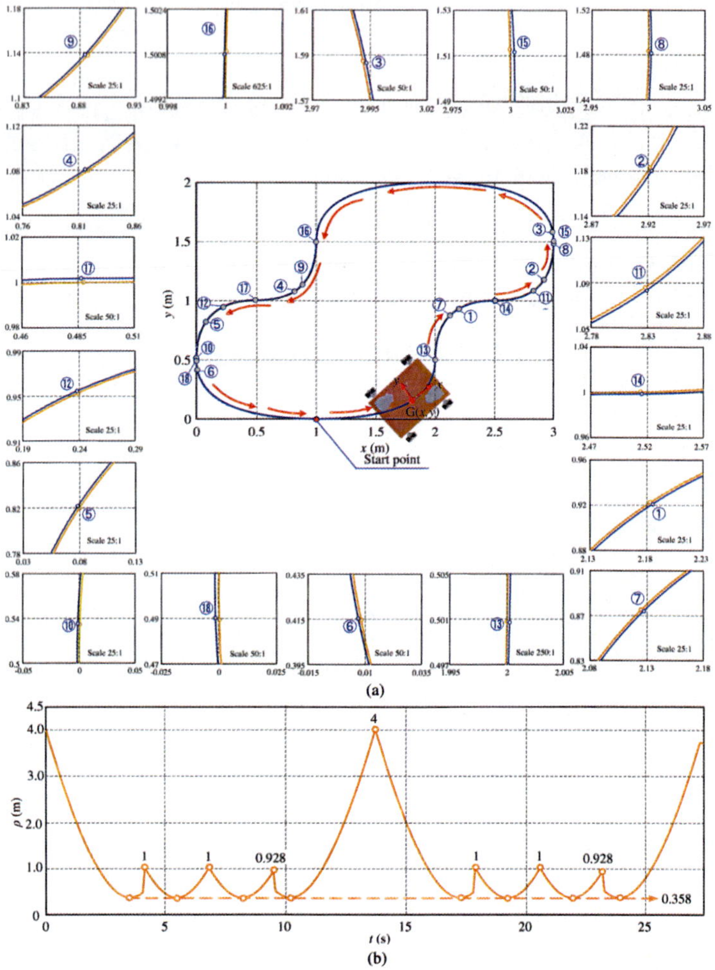

Fig. 5. The movement trajectory of the omnidirectional robot with (a) Actual and desired robot path trajectories and (b) Radius of curvature of desired path trajectory.

of 4.59° produces an error in the y-direction of 2.2mm; (4) From position ③ to position ④, the robot follows tracking from the outside to the inside of the desired trajectory with the maximum error of 1.5 mm in the x-direction, 0.8mm in the y-direction, and 8.6° in the reverse posture; (5) From position ④ to position ⑤, the robot follows tracking from the outside to the inside of the desired trajectory with the maximum error of 1.9 mm in the x-direction, 2.4 mm in the y-direction, and 1.1° in the reverse posture; (6) From position ⑤ to position ⑥, the robot follows tracking from the outside to the inside of the desired trajectory through points ⑩ and (18), Point ⑨ reverses direction with an angular error of 5.16°, making point ⑩ an error in the x-direction is 1.8 mm; (7) From ⑥ to starting point, the robot follows tracking inside to the outside of the desired trajectory.

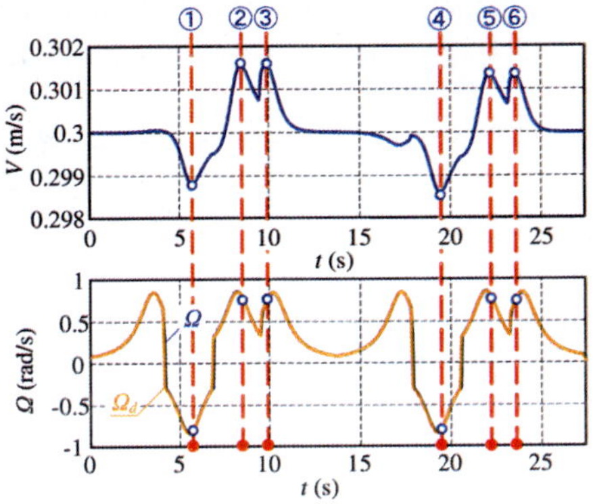

Fig. 6. The moving speed of the omnidirectional robot with (a) The moving speed of the robot and (b) The actual and desired angular velocity when the robot moves the trajectory tracking.

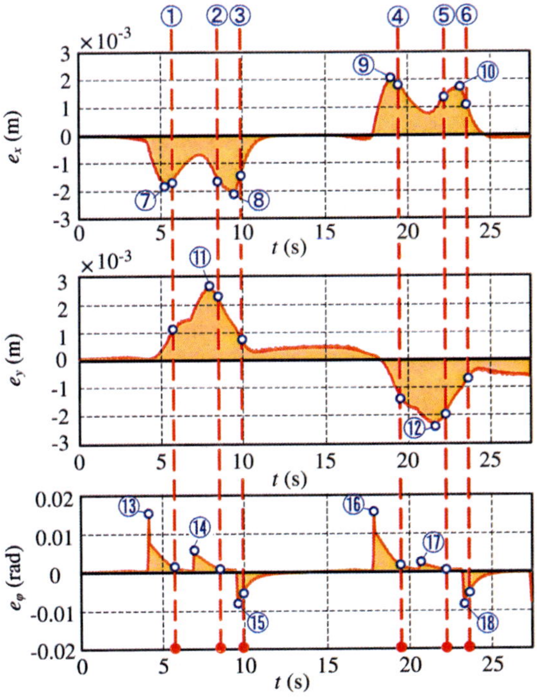

Fig. 7. Error in position and posture of the robot during movement the trajectory tracking.

Thus, the graph of Fig. 8 is the speed control parameter of the four wheels for the robot to tracking follow the desired NURBS trajectory.

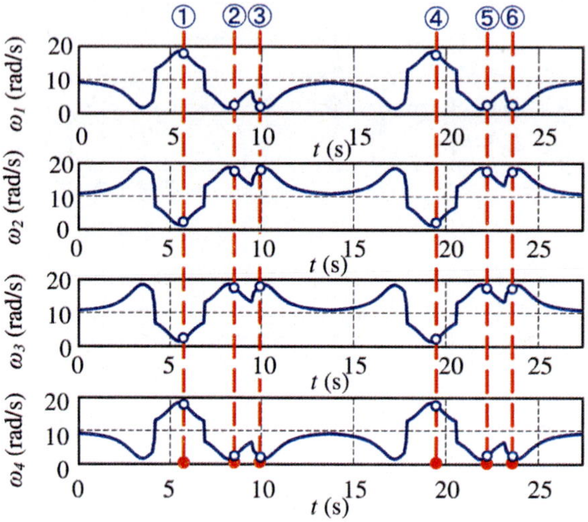

Fig. 8. Angular velocities of wheel 1, wheel 2, wheel 3 and wheel 4 as the robot moves along the trajectory.

6 Conclusions

This paper has solved controlling a mecanum wheeled omnidirectional mobile robot to track the desired NURBS trajectory. The proposed solution is to design a static state feedback controller with a poles placement approach based on a linear kinematic error model. A mecanum wheels omnidirectional mobile robot has been experimented, validated and manufactured with the following results:

1. A linear state feedback kinematics controller developed for the mecanum wheels omnidirectional mobile robot can produce high-performance tracking with minor tracking error.
2. Design the desired path of the robot by the NUBRS interpolation method to ensure that the path is a smooth curve to avoid sudden speed changes of the robot.
3. A robot platform to be manufacturing with defined hardware for experimental verification of the controller proposed by this study.

Thereby, the research results presented in this paper can be applied in many different fields: in hospitals, residential settings, industrial environments, etc.

Acknowledgements. This research was funded by the Ministry of Industry and Trade in a ministeriallevel scientific and technological research project, conducted in 2020, code: DTKHCN.076/20.

References

1. Typiak, A., Łopatka, M.J., Rykała, Ł., Kijek, M.: Dynamics of omnidirectional unmanned rescue vehicle with mecanum wheels. In: AIP Conference Proceedings, vol. 1922, p. 120005 (2018). 1–10

2. Schulze, L., Behling, S., Buhrs, S.: Development of a micro drive- under tractor research and application. In: Proceedings of the International Multi Conference of Engineers and Computer Scientists, vol. II, pp. 16–18, March 2011
3. Goller, M., et al.: Setup and control architecture for an interactive shopping cart in human all day environments. In: Proceeding of the International Conference on Advanced Robotics 2009, pp 1–6 (2009)
4. Göller, M., et al.: Haptic control for the interactive behavior operated shopping trolley InBOT. In: International Conference on Advanced Robotics (2009)
5. Puppim de Oliveira, D., Pereira Neves dos Reis, W., Morandin Junior, O.: A qualitative analysis of a USB camera for AGV control. Sensors **19**(19), 4111 (2019)
6. Gfrerrer, A.: Geometry and kinematics of the Mecanum wheel. Comput. Aided Geometr. Design **25**, 784–791 (2008)
7. Xie, L., Stol, K., Weiliang, X.: Energy-optimal motion trajectory of an omni-directional mecanum-wheeled robot via polynomial functions. Robotica **38**(8), 1400–1414 (2020)
8. Wang, C., Liu, X., Yang, X., Hu, F., Jiang, A., Yang, C.: Trajectory tracking of an omni-directional wheeled mobile robot using a model predictive control strategy. Appl. Sci. **8**, 1–16 (2018)
9. Liu, Y., Zhu, J. J., Williams II, R. L., Wu, J.: Omni-directional mobile robot controller based on trajectory linearization. Robot. Auton. Syst. **56**, 461–479 (2008)
10. Mohan Rayguru, M., et al.: A path tracking strategy for car-like robots with sensor unpredictability and measurement errors. Sensors (MDPI) **20**(11), 3077 (2020)
11. Sun, Z., Xie, H., Zheng, J., Man, Z., He, D.: Path-following control of Mecanum-wheels omnidirectional mobile robots using nonsingular terminal sliding mode. Mech. Syst. Signal Process. (147), 107128 (2021)
12. Cook, G.: Mobile Robots Navigation Control and Remote Sensing. Wiley, Hoboken (2011)
13. Sun, Y., Zhang, C., Liu, C.: Collision-free and dynamically feasible trajectory planning for omnidirectional mobile robots using a novel B-spline based rapidly exploring random tree. Int. J. Adv. Robot. Syst. **18**(3), 1–16 (2021)
14. Nguyen, T.H., Nguyen, T.Q.: A kinematic control algorithm for blasthole drilling robotic arm in tunneling. Sci. Technol. Dev. J. **20**(K5), 13–22 (2017)
15. Ly, T.T.K., Thai, N.H., Dzung, L.Q., Thanh, N.T.: Determination of Kinematic Control Parameters of Omnidirectional AGV Robot with Mecanum Wheels Track the Reference Trajectory and Velocity. In: Sattler, KU., Nguyen, D.C., Vu, N.P., Long, B.T., Puta, H. (eds.) Advances in Engineering Research and Application. ICERA 2020. Lecture Notes in Networks and Systems, vol. 178, pp. 319–328. Springer, Cham (2021). https://doi.org/10.1007/978-3-030-64719-3_36
16. Thai, N.H., Trinh, L.T.K., Dzung, L.Q.: Roadmap, routing and obstacle avoidance of AGV robot in the static environment of the flexible manufacturing system with matrix devices layout. Sci. Technol. Dev. J. Eng. Technol. **24**(3), 2091–2099 (2021)
17. Rogers, D.F.: An Introductron to NURBS. Morgan Kaufmann Publishers, Burlington (2001)

Design and Implementation of a Digital Twin to Control the Industrial Robot Mitsubishi RV-12SD

Minh Duc Vu[1], The Nguyen Nguyen[2], and Chu Anh My[2](✉)

[1] Department of Aerospace Engineering, Le Quy Don Technical University, Hanoi, Vietnam
[2] Institute of Simulation Technology, Le Quy Don Technical University, Hanoi, Vietnam
myca@lqdtu.edu.vn

Abstract. This paper presents a case study of using a digital twin model to control and monitor a real industrial robot. The digital twin and user interfaces are designed and implemented in the Unity 3D environment. The inverse kinematics of the robot is analytically calculated. Both the digital twin and the physical model (the robot Mitsubishi RV-12SD) is successfully integrated, and the data flow can be exchanged one another to control both the models to work together. The real system integration and the collaborating scenarios demonstrate the potential and advantage of the proposed method to develop a smarter programming, controlling and monitoring system for industrial robots.

Keywords: Digital twin · Welding robot · Inverse kinematics

1 Introduction

In recent years, the welding robot plays an important role in several manufacturing industries such as automotive industry, shipbuilding industry, etc. The use of the welding robot for welding complex mechanical parts is to increase the productivity of the production and reduce the use of manpower for manufacturing enterprises. Generally, in order to generate a program (G-code file) for a welding robot, the programmers usually use two main methods: (i) the teaching method and (ii) the CAD-based offline programming method. The teaching method is mostly used in practice to prepare programs for a welding robot rather than the CAD-based offline programming method. However, for welding complex parts with continuous seam welds, the use of the teaching method is not flexible and even impossible. For example, when teaching a robot to weld a seam curve of very high curvature, it is impossible to orient exactly the welding torch of the robot. In this situation, the CAD-based offline programming method is applicable to produce more precise command blocks for a welding robot.

With the advances in new-generation information technologies, digital twin, AI and bigdata analytics are considered as the key technologies to achieve more smart welding robotic systems. To make full use of these technologies, in making more smart decisions, such as self-optimization of real time welding process and online monitoring of real time

N. V. Khang et al. (Eds.): ASIAN MMS 2021, MMS 113, pp. 422–432, 2022.
https://doi.org/10.1007/978-3-030-91892-7_40

status for welding systems, which is to help enterprises to improve the flexibility and efficiency of the productions, and to increase the productivity and the quality of products, is challenging. The benefits of using digital twin for smart welding robot is enormous, however its adoption in real manufacturing is still in nascent stage. In the literature, most of the published papers focus on the framework design, implementation and pilot applications of a digital twin for a manufacturing process, e.g. [1–9]. An overview for incorporating digital twin technology into manufacturing models was presented in [1]. The potential use of a digital twin for diagnostics and prognostics of a corresponding physical twin was studied in [2]. The idea for using a sensor data fusion to construct a digital twin with a virtual machine tool was shown in [3]. The use of the machine learning approach and other data-driven optimization methods to improve the accuracy of a digital twin was investigated in [4–6]. The dynamic modelling of a digital twin based on the approach of discrete dynamic systems was studied in [7]. Some case studies about digital twin were presented in [8, 9].

In this paper, we present a novel technique to design a digital twin incorporated with an inverse kinematic modelling to control a real industrial robot. First, a digital twin connecting with the physical twin – the robot arm Mitsubishi RV-12SD - is designed with the support of CAD tools and Unity 3D software. Second, the inverse kinematics model of the robot is formulated and incorporated with the digital twin. Last, the connection between the digital twin and physical twin is implemented in a manner that the physical twin and the digital twin can be operated in unified scenarios, and the robot can be programed, controlled and monitored from the digital twin and its interfaces.

2 Design of a Digital Twin in Unity 3D

Unity Game Engine (Unity 3D) is a professional software for making games directly in real time. This software can be used to create user interfaces (UI) that helps users to easily manipulate and build applications (Fig. 1).

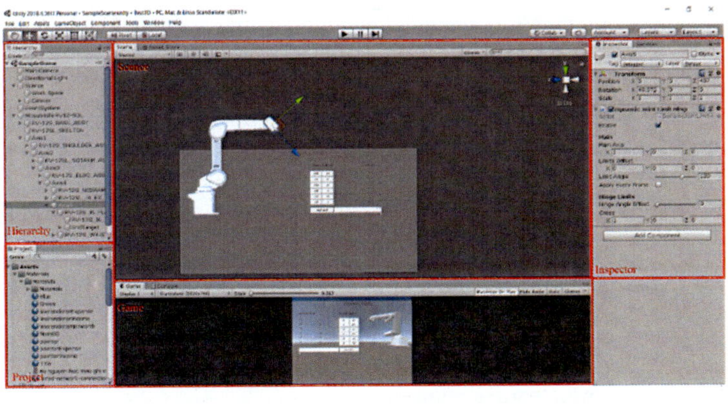

Fig. 1. Unity 3D interfaces (1. Scene 2. Game 3. Inspector 4. Hierarchy 5. Project)

A Unity's interface usually consists of scene, game, inspector, hierarchy and project. Scene window displays objects, in which the objects can be selected, dragged and dropped, zoomed in/out, and rotated. Game window which is viewed from a camera in a game, is a demo game screen. Inspector window is to view all components and their properties. Tab Hierarchy shows game objects and tab project displays assets in a game. One important thing is that the output data of a digital model and the simulation scenarios of the model can be used to communicate with a corresponding physical model. Generally, Unity 3D provides many convenient and effective utilities for 3D model building and simulation, the simulation scenarios of a digital model connected with a physical model can be implemented all together in real time. With these advantages, Unity 3D provides an ideal environment build a digital twin. Therefore, in this study, we use Unity 3D to design a digital twin for the industrial Robot Mitsubishi RV-12SD.

Building a digital twin in Unity 3D includes 2 steps:

Step 1. Building 3D model using CAD Software

In this study, we use SolidWork software to design all components of the 3D model of the robot Mitsubishi RV-12SD. However, the files in STL/STP format is not available in Unity 3D (.fbx,.obj formats). Therefore, to import 3D model from SolidWorks 2020 into Unity 3D requires a middle-ware (CAD Exchanger software) convert STL/STP file to.fbx (filmbox) file. Figure 2 shows the 3D model designed for the robot Mitsubishi RV-12SD.

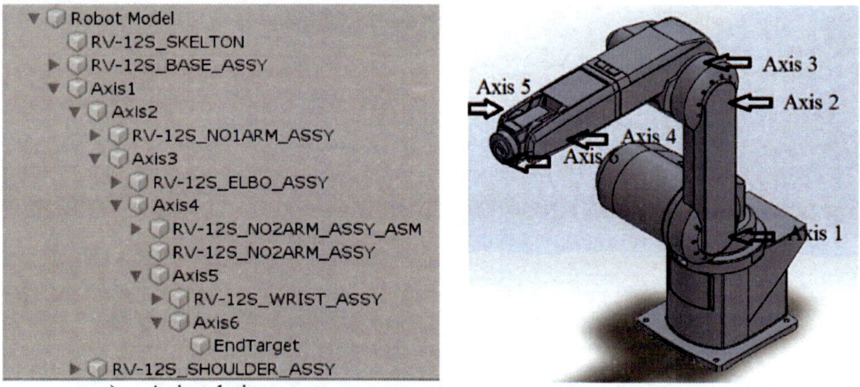

a) Axis relation b) Axis of 3D model

Fig. 2. 3D model in Unity3D

Step 2. Building a Digital Model

After converting and rendering the 3D model into a virtual environment, the model can be used to program in the virtual environment and interact with the physical robot. Unity 3D creates virtual environment to control and monitor the real industrial robot, it provides tools for designing user interfaces. In this manner, a real robot operation can be controlled and monitored from designed UIs. In this case study, a control panel has been designed (Fig. 3) to show parameters (joint angles) feedbacked from the real robot. The values of the joint angles (q1 to q6) are transferred to the virtual model to simulate the robot's motion in parallel with the real robot operation. To program the digital model, the C# programming language is used.

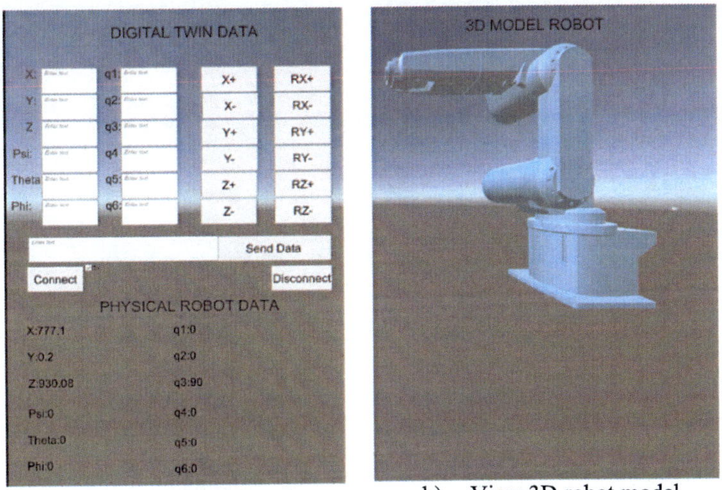

a) Control panel b) View 3D robot model

Fig. 3. Design of user interfaces

To control the robot, the inverse kinematic model of the robot is formulated, which uses the position (X, Y, Z) and the rotation (RX, RY, RX) of end-effector to calculate joint angles of the physical robot.

An important aspect when building a digital model is the connection between virtual and real environment. In this paper, the connection method based on the RS232 communication is selected which exchanges data between the physical robot and the virtual space. The information is exchanged in both-ways that include joint angles and actual robot status signals.

3 Inverse Kinematic Modelling of the Robot

The inverse kinematic modeling of a robot plays an important role in calculating the values of the joint variables (q_1, q_2, q_3, q_4, q_5, q_6) for control the physical robot in connection with a Digital Twin [10–14]. In this section, we present a analytical solution to the inverse kinematic problem of the robot Mitsubishi RV-12SD. Figure 4 show the feasible workspace of the robot, and Fig. 5 shows the kinematic diagram of the robot mechanism.

Note that the Denavit-Hatenberg method is used to define all the link frames as usual. With the kinematic parameters listed in Table 1, the transformation matrices and calculated as follows.

$$\mathbf{R}_1^0 = \begin{pmatrix} \cos(q_1) & 0 & -\sin(q_1) \\ \sin(q_1) & 0 & \cos(q_\cdot) \\ 0 & -1 & 0 \end{pmatrix}; \ \mathbf{R}_2^1 = \begin{pmatrix} \cos(q_2) & -\sin(q_2) & 0 \\ \sin(q_2) & \cos(q_2) & 0 \\ 0 & 0 & 1 \end{pmatrix} \tag{1}$$

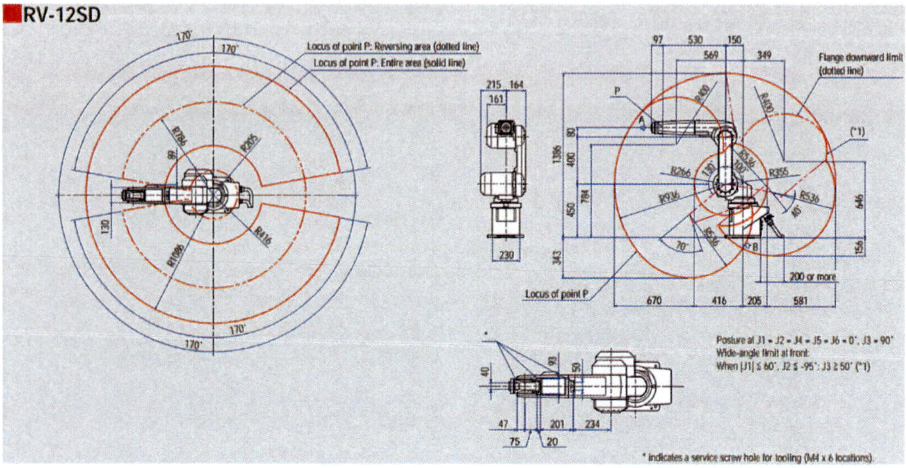

Fig. 4. The workspace of robot RV-12SD

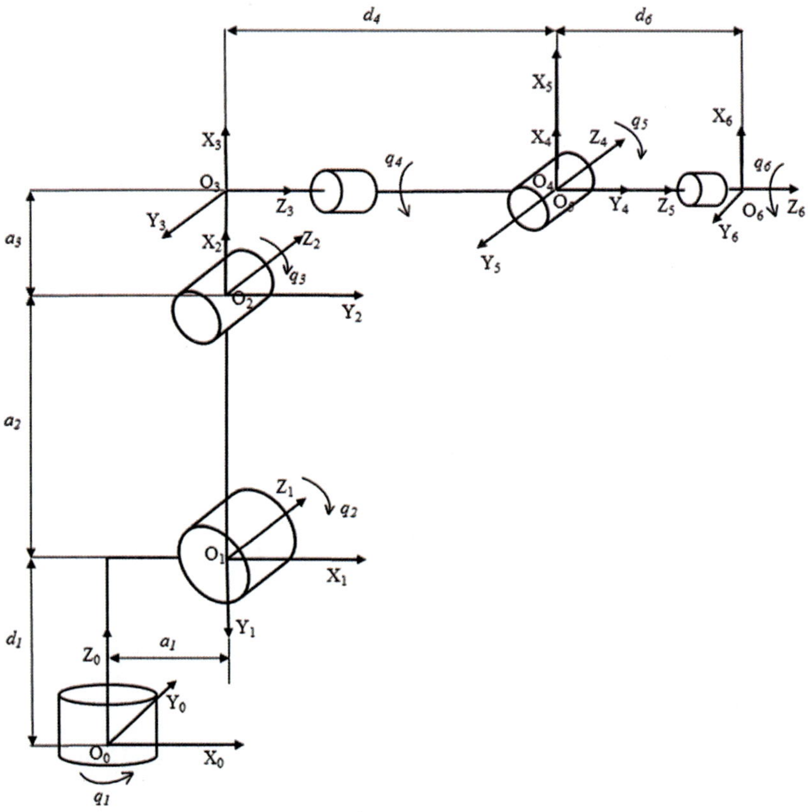

Fig. 5. Kinematic diagram of robot RV-12SD

Table 1. DH parameters

Link	θ_i	d_i	a_i	α_i
1	q_1	d_1	a_1	$-\frac{\Pi}{2}$
2	q_2	0	a_2	0
3	q_3	0	a_3	$-\frac{\Pi}{2}$
4	q_4	d_4	0	$\frac{\Pi}{2}$
5	q_5	0	0	$-\frac{\Pi}{2}$
6	q_6	d_6	0	0

$$\mathbf{R}_3^2 = \begin{pmatrix} \cos(q_3) & 0 & -\sin(q_3) \\ \sin(q_3) & 0 & \cos(q_3) \\ 0 & -1 & 0 \end{pmatrix}; \mathbf{R}_4^3 = \begin{pmatrix} \cos(q_4) & 0 & \sin(q_4) \\ \sin(q_4) & 0 & -\cos(q_4) \\ 0 & 1 & 0 \end{pmatrix} \tag{2}$$

$$\mathbf{R}_5^4 = \begin{pmatrix} \cos(q_5) & 0 & -\sin(q_5) \\ \sin(q_5) & 0 & \cos(q_5) \\ 0 & -1 & 0 \end{pmatrix}; \mathbf{R}_6^5 = \begin{pmatrix} \cos(q_6) & -\sin(q_6) & 0 \\ \sin(q_6) & \cos(q_6) & 0 \\ 0 & 0 & 1 \end{pmatrix} \tag{3}$$

$$\mathbf{R}_E = \mathbf{R}_6^0 = \begin{bmatrix} a_{11} & a_{12} & a_{13} \\ a_{21} & a_{22} & a_{23} \\ a_{31} & a_{32} & a_{33} \end{bmatrix} \tag{4}$$

In Eq. (4),

$$\begin{aligned} a_{11} &= \cos(\psi).\cos(\theta).\cos(\varphi) - \sin(\psi).\sin(\varphi) \\ a_{12} &= -\cos(\varphi).\sin(\psi) - \cos(\psi).\cos(\theta).\sin(\varphi) \\ a_{13} &= \cos(\psi).\sin(\theta) \\ a_{21} &= \cos(\theta).\cos(\varphi).\sin(\psi) + \cos(\psi).\sin(\varphi) \\ a_{22} &= \cos(\psi).\cos(\varphi) - \cos(\theta).sin(\psi).\sin(\varphi) \\ a_{23} &= \sin(\psi).\sin(\theta);\ a_{31} = -\cos(\varphi).\sin(\theta) \\ a_{32} &= \sin(\theta).\sin(\varphi);\ a_{33} = \cos(\theta) \end{aligned} \tag{5}$$

With a desired posture of Robot arm, the end effector position could be defined via x_E, y_E, z_E and orientation is ψ, θ, φ. From that, rotation matrix $\mathbf{R}_E = \mathbf{R}_6^0$ can be written to find q_1, q_2 and q_3.

It is clearly seen that the values of the three joint q_4, q_5, q_6 do not affect the center wrist (O_5) position. On one hand, O_5 position in coordinate O_6 X_6 Y_6 Z_6 is 0, 0, $-d_6$, so the position of the wrist center can be calculated from the position of end effector (x_E, y_E, z_E) as

$$\begin{bmatrix} x_w \\ y_w \\ z_w \end{bmatrix} = \begin{bmatrix} x_{O_5} \\ y_{O_5} \\ z_{O_5} \end{bmatrix} = \begin{bmatrix} x_E - d_6.R_{E13} \\ y_E - d_6.R_{E23} \\ z_E - d_6.R_{E33} \end{bmatrix} \tag{6}$$

The position of O_5 on X_0Y_0 plane is described as shown in Fig. 6.

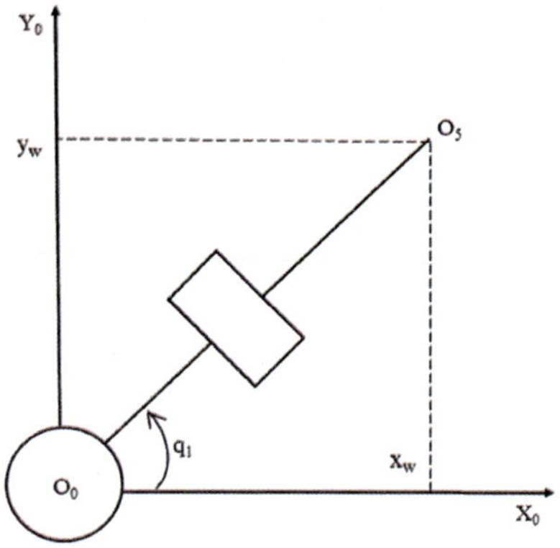

Fig. 6. Projection of O_5 on plane X_0Y_0

In this manner, q_1 can be calculated as

$$q_1 = \text{atan2}(x_w, y_w) \tag{7}$$

On the plane of Link1 and Link2, the joint q_1 and q_2 are involved as shown in Fig. 7.

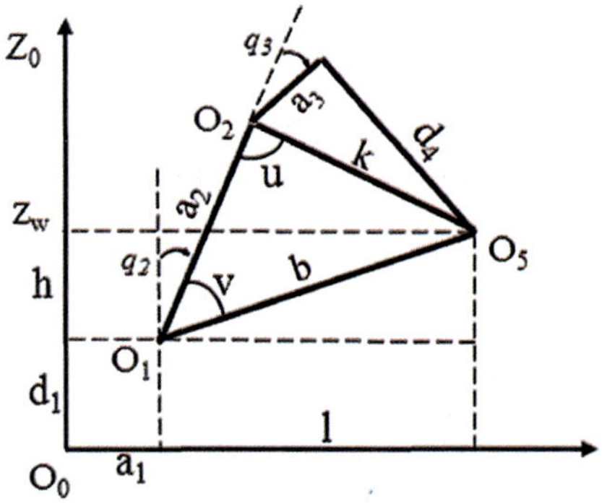

Fig. 7. Projection on the plane of Link2 and Link3

As seen in Fig. 7, q2 and q3 can be easily calculated. Note that $l = \sqrt{x_w^2 + y_w^2} - a_1$; $h = z_w - d_1$; $k = \sqrt{a_3^2 + d_4^2}$; $b = \sqrt{l^2 + h^2}$.

$$\cos(u) = \frac{a_2^2 + k^2 - b^2}{2.a_2.k} \tag{8}$$

$$u = a\tan 2(2.a_2.k, a_2^2 + k^2 - b^2) \tag{9}$$

$$q_3 = \Pi - a\tan\ 2(a_3, d_4) - u \tag{10}$$

$$q_2 = a\tan 2(h, l) - v \tag{11}$$

$$\cos(v) = \frac{a_2^2 + b^2 - k^2}{2.a_2.b} \tag{12}$$

$$q_2 = a\tan 2(h, l) - a\tan 2(2.b.a_2, a_2^2 + b^2 - k^2) \tag{13}$$

In order to calculate q_4, q_5 and q_6, the rotation matrix $\mathbf{R}_3^0$ and $\mathbf{R}_6^3$ must be calculated.

$$\mathbf{R}_3^0 = \mathbf{R}_1^0.\mathbf{R}_2^1.\mathbf{R}_3^2 = \mathbf{R}_P \tag{14}$$

$$\mathbf{R}_3^0 = \begin{pmatrix} \cos(q_1).\cos(q_2+q_3) & \sin(q_1) & -\cos(q_1).\sin(q_2+q_3) \\ \cos(q_2+q_3).\sin(q_1) & -\cos(q_1) & -\sin(q_1).\sin(q_2+q_3) \\ -\sin(q_2+q_3) & 0 & \cos(q_2+q_3) \end{pmatrix} \tag{15}$$

$$\mathbf{R}_6^3 = (\mathbf{R}_3^0)^{-1}.\mathbf{R}_6^0 = \mathbf{R}_P^{-1}.\mathbf{R}_E = \mathbf{R}_A = \begin{pmatrix} R_{A11} & R_{A12} & R_{A13} \\ R_{A21} & R_{A22} & R_{A23} \\ R_{A31} & R_{A32} & R_{A33} \end{pmatrix} \tag{16}$$

On the other hand,

$$\mathbf{R}_6^3 = \mathbf{R}_4^3.\mathbf{R}_5^4.\mathbf{R}_6^5 = \begin{bmatrix} b_{11} & b_{12} & b_{13} \\ b_{21} & b_{22} & b_{23} \\ b_{31} & b_{32} & b_{33} \end{bmatrix} \tag{17}$$

Hence

$$\begin{aligned} b_{11} &= \cos(q_6).\cos(q_4).\cos(q_5) - \sin(q_4).\sin(q_6); \\ b_{12} &= -\cos(q_4).\cos(q_5).\sin(q_6) - \sin(q_4).\cos(q_6); \\ b_{13} &= -\cos(q_4).\sin(q_5); \\ b_{21} &= \cos(q_6).\sin(q_4).\cos(q_5) + \cos(q_4).\sin(q6); \\ b_{22} &= -\sin(q_4).\cos(q_5).\sin(q_6) + \cos(q_4).\cos(q_6); \\ b_{23} &= -\sin(q_4).\sin(q_5);\ b_{31} = \cos(q_6).\sin(q_5); \\ b_{32} &= -\sin(q_5).\sin(q_6);\ b_{33} = \cos(q_5) \end{aligned} \tag{18}$$

Note that $R_6^3 = R_A$, hence we have

$$\tan(q_4) = \frac{R_{A23}}{R_{A13}};\ q_4 = a\tan 2(R_{A13}, R_{A23}) \tag{19}$$

$$\cos(q_5) = R_{A33};\ q_5 = a\tan 2(R_{A33} \pm \sqrt{1 - R_{A33}^2}) \tag{20}$$

$$\tan(q_6) = \frac{-R_{A32}}{R_{A31}};\ q_6 = a\tan 2(R_{A31}, -R_{A32}) \tag{21}$$

Finally, the values of all the joint variables q_1, q_2, q_3, q_4, q_5 and q_6 are analytically calculated with Eqs. (7, 10, 13, 19, 20, 21).

4 Connection of Both the Twins and Control the Robot

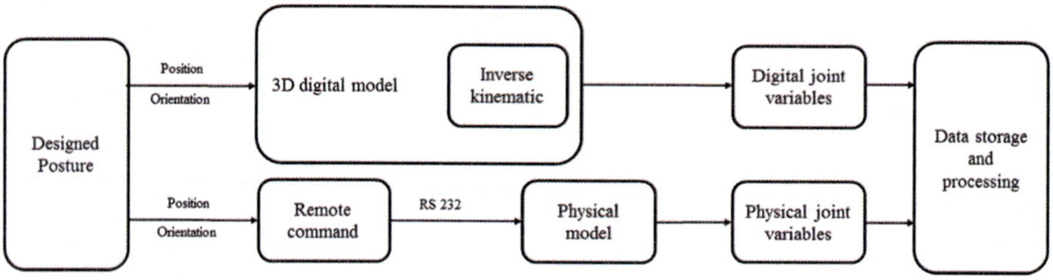

Fig. 8. Data flow for the communication between the robot and the digital twin model.

To transfer the kinematic data from a control computer to RV-12SD robot controller, command blocks containing a frame of characters are sent to the robot controller through a RS232 port. This protocol makes it easy for constructing a data flow throughout the software and hardware to enable the twin models (3D digital on computer and physical robot). In this digital-physical system, the kinematic data will be stored and used for other control and monitoring purposes when both the twins operate together in real time (Fig. 8).

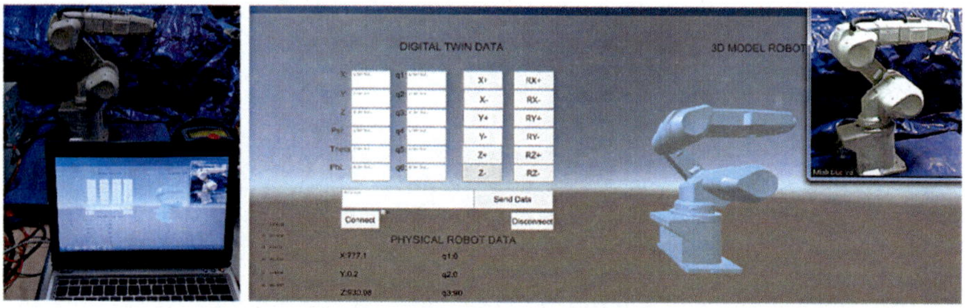

Fig. 9. Testing digital twin model

Figure 9 shows the integration of both the designed digital twin and the physical twin (the robot). After several testing scenarios, it has shown that both the models can collaborate well. By using the control panel with the digital twin model, it is able to control real robot from a virtual environment. Note that by using buttons (X+, X–, Y+, Z+, Z–, RX+, RX–, RY+, RZ+, RZ–) with the added window containing a camera signal of the real robot behavior, the real robot movement interacted with the digital twin simulation can be monitored effectively.

5 Conclusion

In this paper, a case study about design and implementation of a digital twin for control and monitoring a real industrial robot was presented. The 3D digital twin is designed in

Unity 3D software, which provides advantages for the collaboration and the real time data connection between the physical-digital models. The inverse kinematics of the real robot is studied and an analytical inverse kinematic solution is yielded, which makes it more effectively and precisely for the control of the robot in cooperation with the digital twin. The implementation and testing results shows that controlling and monitoring a real robot with the support of a collaborating digital twin model is feasible and applicable. This method makes it possible to improve the smartness of a robotic system by using the advancement of the recent development of ICT technology. Using this physical-digital twins for offline programming, testing-simulation, control and monitoring when a robot is required to weld a complex path of high curvature is the future work of this study.

Acknowledgement. This research is funded by Vietnam National Foundation for Science and Technology Development (NAFOSTED) under grant number 107.01-2020.15.

References

1. Madni, A.M., Madni, C.C., Lucero, S.D.: Leveraging digital twin technology in model-based systems engineering. Systems **7**(1), 7 (2019)
2. Booyse, W., Wilke, D.N., Heyns, S.: Deep digital twins for detection, diagnostics and prognostics. Mech. Syst. Signal Process. **140**, 106612 (2020)
3. Cai, Y., Starly, B., Cohen, P., Lee, Y.S.: Sensor data and information fusion to construct digital-twins virtual machine tools for cyber-physical manufacturing. Procedia Manuf. **10**, 1031–1042 (2017)
4. Cronrath, C., Aderiani, A.R. and Lennartson, B.: Enhancing digital twins through reinforcement learning. In: 2019 IEEE 15th International Conference on Automation Science and Engineering (CASE), pp. 293–298. IEEE (August 2019)
5. Dröder, K., Bobka, P., Germann, T., Gabriel, F., Dietrich, F.: A machine learning-enhanced digital twin approach for human-robot-collaboration. Procedia Cirp **76**, 187–192 (2018)
6. Ganguli, R., Adhikari, S.: The digital twin of discrete dynamic systems: initial approaches and future challenges. Appl. Math. Model. **77**, 1110–1128 (2020)
7. Guivarch, D., Mermoz, E., Marino, Y., Sartor, M.: Creation of helicopter dynamic systems digital twin using multibody simulations. CIRP Ann. **68**(1), 133–136 (2019)
8. He, R., Chen, G., Dong, C., Sun, S., Shen, X.: Data-driven digital twin technology for optimized control in process systems. ISA Trans. **95**, 221–234 (2019)
9. Liu, J., Zhou, H., Tian, G., Liu, X., Jing, X.: Digital twin-based process reuse and evaluation approach for smart process planning. Int. J. Adv. Manuf. Technol. **100**(5–8), 1619–1634 (2018). https://doi.org/10.1007/s00170-018-2748-5
10. My, C.A., et al.: Mechanical design and dynamics modelling of RoPC robot. In: Proceedings of International Symposium on Robotics and Mechatronics, Hanoi, Vietnam, pp. 92–96 (September 2009)
11. My, C.A.: Inverse dynamic of a N-links manipulator mounted on a wheeled mobile robot. In: 2013 International Conference on Control, Automation and Information Sciences (ICCAIS), pp. 164–170. IEEE (2013, November)
12. Chu, A.M., et al.: A novel mathematical approach for finite element formulation of flexible robot dynamics. Mech. Based Des. Struct. Mach. 1–21 (2020)

13. My, C.A., Makhanov, S.S., Van, N.A., Duc, V.M.: Modeling and computation of real-time applied torques and non-holonomic constraint forces/moment, and optimal design of wheels for an autonomous security robot tracking a moving target. Math. Comput. Simul. **170**, 300–315 (2020)
14. Toai, T.T., Chu, D.-H., My, C.A.: Development of a new 6 DOFs welding robotic system for a specialized application. In: Balas, V.E., Solanki, V.K., Kumar, R. (eds.) Further Advances in Internet of Things in Biomedical and Cyber Physical Systems. ISRL, vol. 193, pp. 135–150. Springer, Cham (2021). https://doi.org/10.1007/978-3-030-57835-0_11

General Approach for Parameterization of the Inverse Dynamic Equation for Industrial Robot and 5-axis CNC Machine

Chu Anh My[1](✉), Nguyen Cong Dinh[1], Truong Quoc Hung[1], Pham Hoang Hung[1], Nguyen The Nguyen[1], Dao Van Luu[2], Duong Xuan Bien[2], Vu Minh Duc[3], and Chi H. Le[4]

[1] Institute of Simulation Technology, Le Quy Don Technical University, Hanoi, Vietnam
[2] Center of Advanced Technology, Le Quy Don Technical University, Hanoi, Vietnam
[3] Department of Aerospace Engineering, Le Quy Don Technical University, Hanoi, Vietnam
myca@lqdtu.edu.vn
[4] Faculty of Engineering and Science, University of Greenwich, London, UK

Abstract. In practice, using the 5-axis CNC machines and industrial welding robots for fabricating more and more complex mechanical parts is an increasing demand. In this context, the tooltip of the machines/robots is usually required to move along a complex toolpath represented by a parametric equation such as Bezier, Spline or NURBS. Hence, computation of the inverse kinematics and dynamics e.g. velocities, accelerations, jerks and applied torques/forces in the joint space of a machine/robot when the tooltip tracks such as high curvature parametric curve is challenging. Therefore, this paper presents a general approach for the parameterization of the kinematic and dynamic equations of the machine/robot. By using the proposed method, the kinematics and dynamics model of a machine/robot can be formulated in the parametric domain, which is advantageous and applicable for the dynamic analysis of the 5-axis CNC machine/robot and other purposes.

Keywords: Robot dynamics · 5-axis CNC machine · Parameterization of dynamic equation · Dynamics of 5-axis CNC machine

1 Introduction

In recent decades, the dynamic modelling and analysis of 5-axis CNC machine and robot manipulator have been extensively studied and well documented [1–4], which play an important role in several technical issues such as the machine/robot design and analysis, the control design and optimization, the actuator selection for a machine/robot, the feed rate optimization, etc. For example, in order to optimize the feed rate of the cutter for a 5-axis CNC machine (or the welding speed of the welding torch for a welding robot), it is necessary to formulate all the relevant kinematic and dynamic constraints of the optimization model. In other words, the calculation of the maximized feed rate/speed must be subject to the limit of all the machine/robot drives such as the limit of torques, jerks, accelerations and velocities.

N. V. Khang et al. (Eds.): ASIAN MMS 2021, MMS 113, pp. 433–442, 2022.
https://doi.org/10.1007/978-3-030-91892-7_41

In recent years, the 5-axis CNC machine and industrial robot are demanded to carry out more and more complex tasks. The 5-axis CNC machine is usually required to machine more complex parts including freeform surfaces such as molds and dies. The welding robot is required to weld complex paths designed with 3D freeform parametric curves such as Spline, Bezier or NURBS. In this situation, in order to increase the effectiveness and the efficiency of the toolpath interpolation, the feed rate optimization, the control computation, etc. it is necessary to map the kinematic and dynamic equations of a robot/machine into the parametric domain of the designed toolpath.

Decades ago, there has been a large volume of published studies focusing on the kinematic and dynamic modelling of the 5-axis CNC machine/robot, e.g. [5–20]. The authors in [5–11] studied the kinematic modelling for individual 5-axis CNC machines. The inverse kinematic modelling of the machine DMU 50e/70e was presented in [5–7, 15], three families of commonly used 5-axis CNC machines were considered in [8]. The generic kinematics model of the 5-axis CNC machine was formulated in [11, 12]. It can be seen that the kinematics model of the 5-axis CNC machine in most of the previous works was formulated to develop the postprocessor and optimize the feed rate for the machine. A little attention has been paid to transformation of the kinematics and dynamics model into a parametric domain.

Regarding to the dynamic modelling of a robot manipulator, there has been a large amount of research published in recent decades, e.g. [16–20]. The fundamentals of the kinematic and dynamic modelling of robot were presented in [1–4].

It is clearly seen that though the kinematic and dynamic modelling of the 5-axis CNC machine/robot manipulator has been extensively studied and well documented, little attension has been paid to the parameterization of the kinematic and dynamic equation for the machine/robot. Hence, in this paper, we present a general approach for the parameterization of the dynamic equation of the machine/robot. Given the kinematic and dynamic of a machine/robot represented in the usual fashion, analytical transformations are presented to map the equations into the parametric domain. Example of the parameterization of the dynamic modelling for a real 5-axis CNC machine is illustrated to demonstrate the proposed method.

2 Parameterization of the Inverse Kinematic and Dynamic Equations for 5-axis CNC Machines and Industrial Robots

The family of 5-axis CNC machines includes several types of machine configurations. In general, a 5-axis CNC machine is constructed with two serial kinematic chains, one chain carries the cutter and one chain carries the part [12]. Each 5-axis CNC machine has three revolute joints and two prismatic joints. In other words, a 5-axis CNC machine looks like two collaborating robot manipulators. For this reason, the 5-axis CNC machine is so called the 5-axis milling robot. In terms of mechanism theory, a 5-axis CNC machine is a 5-DOFs mechanism consisting of five links with three revolute joints and two prismatic joints.

Figure 1 shows a general description of the 5-axis CNC mechanisms, where $q_1 \div q_n$ are joint variables on the part carrying chain, and $q_{n+1} \div q_5$ are the joint variables on the tool carrying chain. $O_0x_0y_0z_0$, $O_wx_wy_wz_w$ and $O_tx_ty_tz_t$ are the reference frame, the

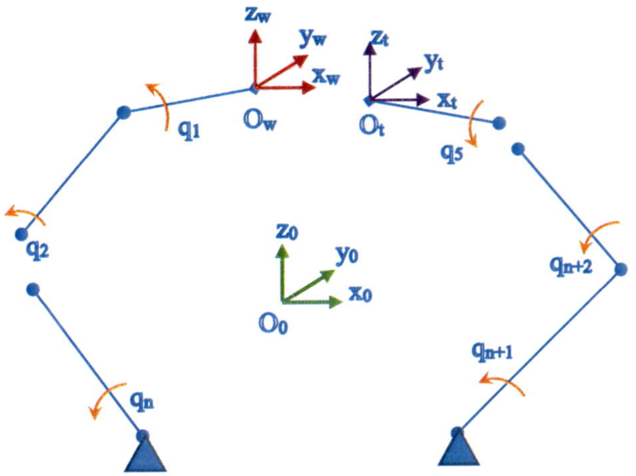

Fig. 1. General mechanism of the 5-axis CNC machines

part frame and the tool frame. By using Denanvit – Hartenberg approach, the kinematics model of a 5-axis machine can be formulated, that is similar to the kinematic modelling for a robot manipulator. In this manner, the inverse kinematic equation for both the 5-axis machine and the robot arm can be represented as follow.

$$\mathbf{q}(t) = \mathbf{f}^{-1}[\mathbf{p}(t)] \tag{1}$$

Where $\mathbf{p}$ denotes the position and orientation of the tool (the end effector) for the machine/robot. $\mathbf{q}$ is the vector of join variables.

The dynamic equation for the machine/robot can be written as follows [1–4].

$$\begin{cases} \boldsymbol{\tau}(t) = \mathbf{M}(\mathbf{q}(t))\ddot{\mathbf{q}}(t) + \mathbf{C}(\mathbf{q}(t), \dot{\mathbf{q}}(t))(t)\dot{\mathbf{q}}(t) + \mathbf{G}(\mathbf{q}(t)) - \mathbf{J}^T\mathbf{F} \\ \mathbf{q}(t) = \mathbf{f}^{-1}(\mathbf{p}(t)) \end{cases} \tag{2}$$

where $\boldsymbol{\tau}$ is a vector of the applied torques/forces, $\mathbf{J} = \left(\frac{\partial \mathbf{f}}{\partial \mathbf{q}}\right)$ is the Jacobian matrix, and $\mathbf{F}$ is the vector of the forces reacting to the end point of the cutter (the end effector).

Now suppose that a designed toolpath for the cutter (the end effector) of a machine/robot is given, which is represented by a freeform parametric curve $\mathbf{p}(u)$ where $u \in [0, 1]$ is the prameter, the following components $\mathbf{q}(u)$, $\dot{\mathbf{q}}(u)$, $\ddot{\mathbf{q}}(u)$, $\dddot{\mathbf{q}}(u)$, and $\tau(u)$ must be calculated. Note that the velocity of the cutter (the end effector) along the curve is also prescribed and represented by $f(t) = \dot{s}(t)$ where t is time and s is the arc length of $\mathbf{p}(s)$.

In a generalized case, suppose that $\mathbf{p} = \left[\mathbf{c}^T\ \mathbf{r}^T\right]$ is the posture of the tool in the task space, where $\mathbf{c} = \left[x\ y\ z\right]^T$ is the tooltip position, and $\mathbf{r} = \left[\phi\ \varphi\ \gamma\right]^T$ is the tool axis orientation. The geometric toolpath can be represented in NURBS form as follow.

$$\mathbf{c}(u) = \sum_{i=0}^{n} N_{i,k}(u)w_i\mathbf{c}_i \Big/ \sum_{i=0}^{n} N_{i,k}(u)w_i \tag{3}$$

The orientation of the tool axis is calculated as follow.

$$\mathbf{r}(u) = (\mathbf{o}(u) - \mathbf{c}(u))/|(\mathbf{o}(u) - \mathbf{c}(u))| \tag{4}$$

where $\mathbf{o}(u)$ is given by Eq. (5), which is the trajectory of an arbitrary point different from the tip point on the tool axis.

$$\mathbf{o}(u) = \sum_{i=0}^{n} N_{i,k}(u) w_i \mathbf{o}_i \bigg/ \sum_{i=0}^{n} N_{i,k}(u) w_i \tag{5}$$

In the parametric domain $u \in [0, 1]$, the inverse kinematic equation can be rewritten as:

$$\mathbf{q}(u) = \mathbf{f}^{-1}[\mathbf{p}(u)] \tag{6}$$

Taking time derivative of the following forward kinematic equation

$$\mathbf{p} = \mathbf{f}(u) \tag{7}$$

yields

$$\dot{\mathbf{p}} = \left(\frac{\partial \mathbf{f}}{\partial \mathbf{q}}\right)\dot{\mathbf{q}} \tag{8}$$

Hence

$$\begin{aligned} \dot{\mathbf{q}} &= \mathbf{J}^{-1}\dot{\mathbf{p}} \\ &= \mathbf{J}^{-1}\frac{d\mathbf{p}}{ds}\frac{ds}{dt} \\ &= \mathbf{J}^{-1}\mathbf{p}_s^{'}\dot{s} \end{aligned} \tag{9}$$

In terms of the differential geometry, the differential arc length of a curve, ds, can be approximated as

$$ds \simeq \left|\frac{d\mathbf{p}}{du}\right| du \tag{10}$$

Hence, the derivative of $\mathbf{p}(s)$ can be derived as follow.

$$\begin{aligned} \mathbf{p}_s^{'} &= \frac{d\mathbf{p}}{ds} \\ &= \frac{d\mathbf{p}/du}{ds/du} \\ &= \frac{d\mathbf{p}/du}{|d\mathbf{p}/du|} \\ &= \frac{\mathbf{p}_u^{'}}{|\mathbf{p}_u^{'}|} \end{aligned} \tag{11}$$

Recalculating Eq. (9) yields

$$\dot{\mathbf{q}}(u) = \mathbf{J}^{-1}\frac{\mathbf{p}_u^{'}}{\left|\mathbf{p}_u^{'}\right|}\dot{s} \tag{12}$$

With Eq. (12), we can calculate the joint velocities for the machine/robot in the parametric domain $u \in [0, 1]$, with respect to the feed rate of the cutter (the end effector) $\dot{s}$, the geometrical derivative of the parametric toolpath $\mathbf{p}_u^{'}$, and the analytical Jacobian $\mathbf{J}$. It is clearly seen that, given an identical feed rate profile $\dot{s}$, the joint velocities of a machine/robot only depend on the geometrical properties (derivatives, curvature, etc.) of the designed toolpath.

In order to calculate $\ddot{\mathbf{q}}(u)$, we continuously take time derivative of Eq. (8):

$$\ddot{\mathbf{p}} = \mathbf{J}\ddot{\mathbf{q}} + \dot{\mathbf{J}}\dot{\mathbf{q}} \tag{13}$$

Thus

$$\ddot{\mathbf{q}} = \mathbf{J}^{-1}(\ddot{\mathbf{p}} - \dot{\mathbf{J}}\dot{\mathbf{q}}) \tag{14}$$

Note that

$$\begin{aligned}\dot{\mathbf{p}} &= \frac{d\mathbf{p}}{dt}\\ &= \frac{d\mathbf{p}}{ds}\frac{ds}{dt}\\ &= \mathbf{p}_s^{'}\dot{s}\end{aligned} \tag{15}$$

Hence

$$\begin{aligned}\ddot{\mathbf{p}} &= \frac{d\mathbf{p}_s^{'}}{ds}\dot{s}^2 + \mathbf{p}_s^{'}\ddot{s}\\ &= \mathbf{p}_s^{''}\dot{s}^2 + \mathbf{p}_s^{'}\ddot{s}\end{aligned} \tag{16}$$

In Eq. (16), we need to derive $\mathbf{p}_s^{'}$ and $\mathbf{p}_s^{''}$ in the parametric domain. Rewriting Eq. (10) yields

$$s_u^{'} = \left|\mathbf{p}_u^{'}\right| \tag{17}$$

Hence,

$$\mathbf{p}_s^{'} = \frac{\mathbf{p}_u^{'}}{s_u^{'}} \tag{18}$$

Differentiating both sides of Eq. (18) yields

$$\mathbf{p}_s^{''} = \frac{(d\mathbf{p}_u^{'}/ds)s_u^{'} + \mathbf{p}_u^{'}(ds_u^{'}/ds)}{\left|\mathbf{p}_u^{'}\right|^2} \tag{19}$$

Note that

$$\frac{d\mathbf{p}_u^{'}}{ds} = \frac{\mathbf{p}_u^{''}}{s_u^{'}} = \frac{\mathbf{p}_u^{''}}{|\mathbf{p}_u^{'}|} \tag{20}$$

And,

$$\frac{ds_u^{'}}{ds} = \frac{s_u^{''}}{s_u^{'}} = \frac{(\mathbf{p}_u^{'})^T\mathbf{p}_u^{''}}{|\mathbf{p}_u^{'}|^2} \tag{21}$$

Hence, Eq. (19) becomes

$$\mathbf{p}_s^{''} = \frac{\mathbf{p}_u^{''}}{|\mathbf{p}_u^{'}|^2} + \frac{\mathbf{p}_u^{'}\left[(\mathbf{p}_u^{'})^T\mathbf{p}_u^{''}\right]}{|\mathbf{p}_u^{'}|^4} \tag{22}$$

From Eqs. (16, 18, 22), recalculating Eq. (14) yields

$$\ddot{\mathbf{q}}(u) = \mathbf{J}^{-1}\left(\left(\frac{\mathbf{p}_u^{''}}{|\mathbf{p}_u^{'}|^2} + \frac{\mathbf{p}_u^{'}((\mathbf{p}_u^{'})^T\mathbf{p}_u^{''})}{|\mathbf{p}_u^{'}|^4}\right)\dot{s}^2 + \frac{\mathbf{p}_u^{'}}{|\mathbf{p}_u^{'}|}\ddot{s} - \dot{\mathbf{J}}\dot{\mathbf{q}}\right) \tag{23}$$

In the similar way, the jerk $\dddot{\mathbf{q}}$ can be calculated in the parametric space as follows.

$$\dddot{\mathbf{q}}(u) = \mathbf{J}^{-1}\left\{\begin{array}{l} \frac{\mathbf{p}_u^{'''}}{|\mathbf{p}_u^{'}|^3}\dot{s}^3 + \frac{\mathbf{p}_u^{''}\left[(\mathbf{p}_u^{'})^T\mathbf{p}_u^{''}\right] + \mathbf{p}_u^{'}\left[(\mathbf{p}_u^{''})^T\mathbf{p}_u^{''} + (\mathbf{p}_u^{'})^T\mathbf{p}_u^{'''}\right]}{|\mathbf{p}_u^{'}|^5}\dot{s}^3 \\ + \left[\frac{\mathbf{p}_u^{''}}{|\mathbf{p}_u^{'}|^2} + \frac{\mathbf{p}_u^{'}((\mathbf{p}_u^{'})^T\mathbf{p}_u^{''})}{|\mathbf{p}_u^{'}|^4}\right]\dot{s}\ddot{s} - 2\dot{\mathbf{J}}\ddot{\mathbf{q}} - \ddot{\mathbf{J}}\dot{\mathbf{q}} \end{array}\right\} \tag{24}$$

By substituting all computed components $\mathbf{q}(u), \dot{\mathbf{q}}(u), \ddot{\mathbf{q}}(u)$ into Eq. (2), the forces/torques driving the joints are calculated in the parametric domain with the following equation.

$$\boldsymbol{\tau}(u) = \mathbf{M}(\mathbf{q}(u))\ddot{\mathbf{q}}(u) + \mathbf{C}(\mathbf{q}(u), \dot{\mathbf{q}}(u))\dot{\mathbf{q}}(u) + \mathbf{G}(\mathbf{q}(u)) - \mathbf{J}^T\mathbf{F} \tag{25}$$

Note that

$$\dot{\mathbf{J}} = \begin{bmatrix} \frac{\partial J_{11}}{\partial \mathbf{q}}\dot{\mathbf{q}} & \frac{\partial J_{12}}{\partial \mathbf{q}}\dot{\mathbf{q}} & \cdots & \frac{\partial J_{1n}}{\partial \mathbf{q}}\dot{\mathbf{q}} \\ \frac{\partial J_{21}}{\partial \mathbf{q}}\dot{\mathbf{q}} & \frac{\partial J_{22}}{\partial \mathbf{q}}\dot{\mathbf{q}} & \cdots & \frac{\partial J_{2n}}{\partial \mathbf{q}}\dot{\mathbf{q}} \\ \cdots & & & \\ \frac{\partial J_{n1}}{\partial \mathbf{q}}\dot{\mathbf{q}} & \frac{\partial J_{n2}}{\partial \mathbf{q}}\dot{\mathbf{q}} & \ldots & \frac{\partial J_{nn}}{\partial \mathbf{q}}\dot{\mathbf{q}} \end{bmatrix} = \left[\frac{\partial J_{ij}}{\partial \mathbf{q}}\right] \otimes \dot{\mathbf{q}} \tag{26}$$

and

$$\ddot{\mathbf{J}} = \left[\frac{\partial}{\partial \mathbf{q}}\left(\left[\frac{\partial J_{ij}}{\partial \mathbf{q}}\right] \otimes \dot{\mathbf{q}}\right)\right] \otimes \begin{bmatrix} \dot{\mathbf{q}} \\ \cdots \\ \dot{\mathbf{q}} \end{bmatrix}_{1\times n} + \left[\frac{\partial J_{ij}}{\partial \mathbf{q}}\right] \otimes \ddot{\mathbf{q}} \tag{27}$$

Example
Let's consider the 5-axis CNC machine DMU70e as shown in Fig. 2. The kinematic modelling of the machine and the kinematic parameters were presented in our previous works [5, 6]. Table 1 shows the dynamic parameters of the machine.

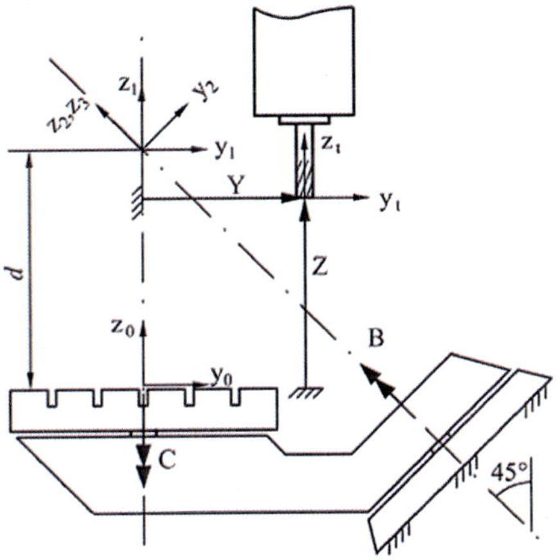

Fig. 2. The 5-axis CNC machine DMU70e

Table 1. Dynamics parameters of Spinner U5-620

Machine components	Center of mass $\mathbf{r}_{Ci}(m)$			Mass $m_i(kg)$	Moments of inertia $\mathbf{I}_i(kg.m^2)$		
	x_C	y_C	z_C		I_{xx}	I_{yy}	I_{zz}
X	0.0	0.075	0.0	530	9.3	10.15	10.75
Y	0.0	0.0	0.0	255	3.3	2.8	4.3
Z	0.0	0.0	0.25	205	3.6	2.8	4.35
B	0.0	–0.55	–0.3	490	8.2	10.15	10.75
C	0.0	0.0	0.55	320	2.4	3.1	4.3

The position and orientation of the tool are represented by $\mathbf{p} = \left[x\ y\ z\ i\ j\ k\right]^T$, where x, y, and z are the tool tip coordinates, symbols i, j and k are cosines of the tool axis vector. $\mathbf{q} = \left[X\ Y\ Z\ B\ C\right]^T$ is the vector of five joint variables. The resistance force vector and torque vector of axes are $\mathbf{F} = \left[F_t\ F_b\ F_n\ 0\ 0\right]^T$ and $\boldsymbol{\tau} = \left[F_x\ F_y\ F_z\ \tau_B\ \tau_C\right]^T$. Without loss the generality, let's consider a Bezier toolpath as

$$\mathbf{C}(u) = (1-u)^3\mathbf{P}_0 + 3(1-u)^2u\mathbf{P}_1 + 3(1-u)u^2\mathbf{P}_2 + \mathbf{P}_3u^3 \tag{28}$$

where $\mathbf{P}_0 = \left[0\ 0.05\ 0.05\right]^T$, $\mathbf{P}_1 = \left[0.1\ 0.05\ 0.15\right]^T$, $\mathbf{P}_2 = \left[0.2\ 0.05\ 0.05\right]^T$, and $\mathbf{P}_3 = \left[0.3\ 0.05\ 0.15\right]^T$. In this example, the resistance forces are given as $F_t = 150\,\mathrm{N}$,

$F_n = 70\,\text{N}$ and $F_b = 50\,\text{N}$. The constant feed rate is prescribed as $f = \dot{s} = 0.1$. The inverse kinematics and dynamics calculated in the parametric domain are shown in Figs. (3, 4, 5, 6, 7, 8).

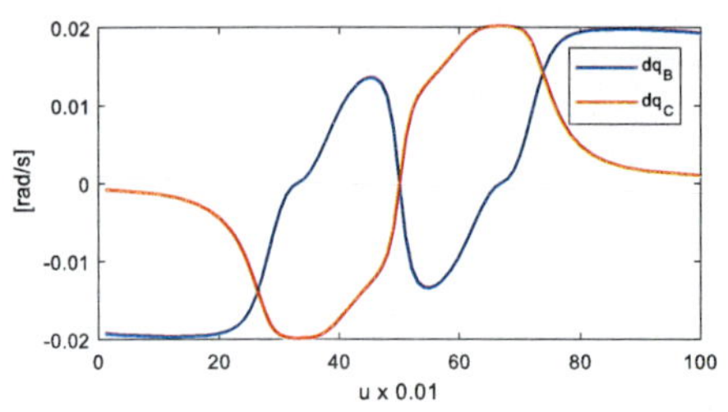

Fig. 3. The angular velocity of B and C axes

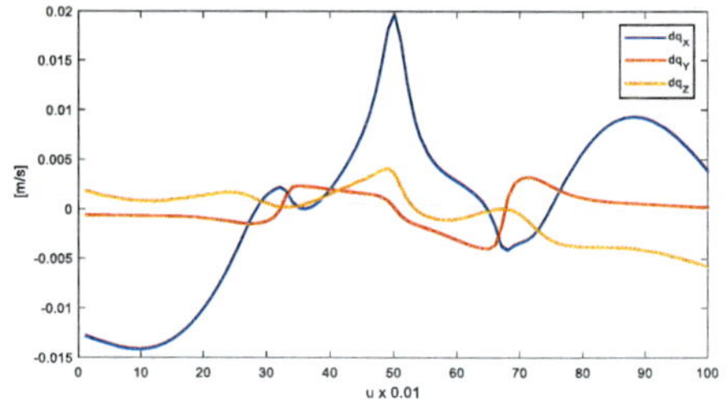

Fig. 4. The linear velocity of X, Y and Z axes

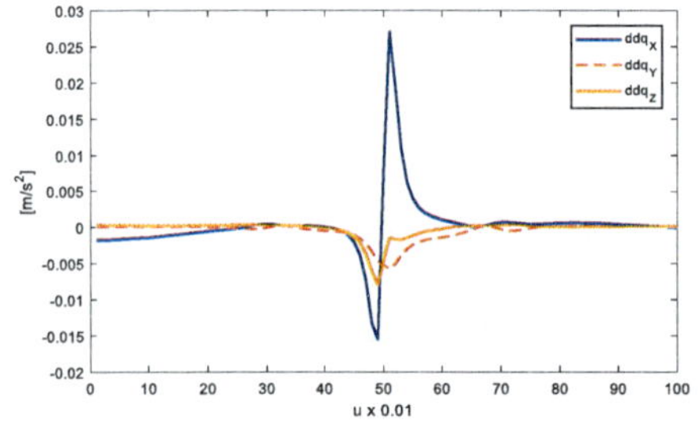

Fig. 5. The accelerations of X, Y and Z axes

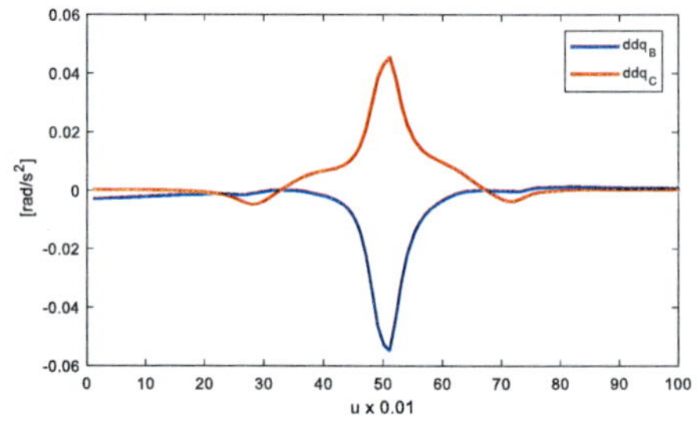

Fig. 6. The angular accelerations of B and C axes

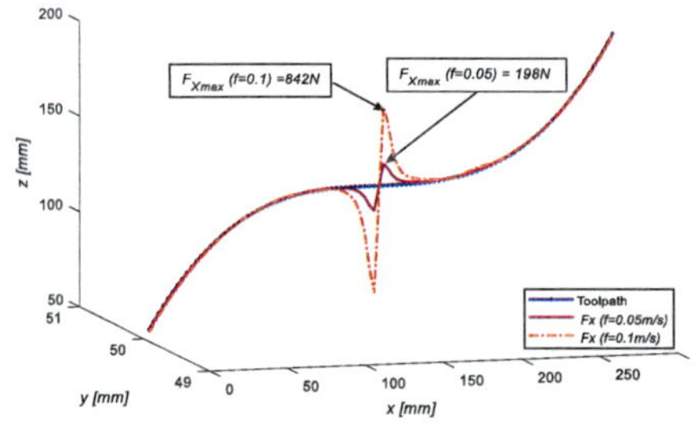

Fig. 7. The toolpath and the driving force Fx

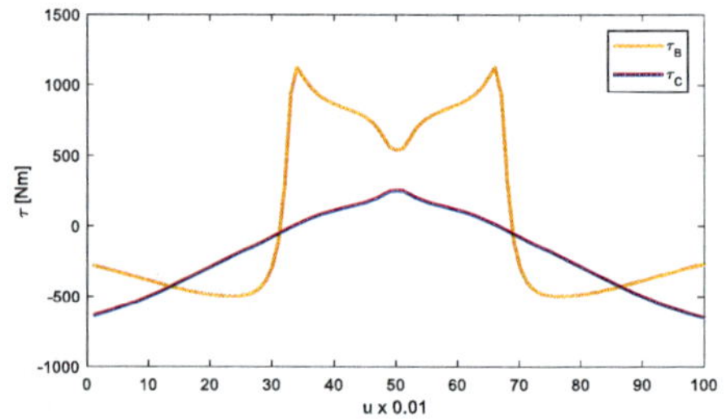

Fig. 8. The driving torques of B and C axes

3 Conclusion

In this paper, a general approach was presented for parametrization of the kinematic and dynamic equations for the robot manipulator/5-axis CNC machine. It has been shown that the inverse kinematic and dynamic equations are analytically transformed into the parametric domain of the parametric toolpath. This makes it possible to calculate the velocities, accelerations and applying torques/forces for all the active joints of a robot/machine, even in the case that a very complex trajectory represented in the parametric equation is given. By using the proposed method, several complex problems such as NURSB curve interpolation for welding robot, feed rate optimization for 5-axis CNC machine, etc. when the tooltip of a robot/machine travels a long a complex parametric curve could be solved effectively. Interpolating the feed rate and optimizing the welding parameters for a welding robot, in the case that a high curvature weld curve designed in NURBS representation is required to weld, will be the future work of this study.

Acknowledgement. This work was supported by the Research Foundation of Le Quy Don Technical University. This work was funded by Gia Lam Urban Development and Investment Company Limited, Vingroup and supported by Vingroup Innovation Foundation (VINIF) under project code VINIF.2019.DA08.

References

1. Craig, J.J.: Introduction to Robotics: Mechanics and Control, 3rd edn.. Pearson Education, Upper Saddle River, NJ, USA (2005)
2. Ceccarelli, M.: Fundamentals of the mechanics of robots. In: Fundamentals of Mechanics of Robotic Manipulation. International Series on Microprocessor-Based and Intelligent Systems Engineering, vol. 27, pp. 73–240. Springer, Dordrecht (2004). https://doi.org/10.1007/978-1-4020-2110-7_3
3. Merlet, J.P.: Parallel Robots. Kluwer Academic Publishers, London (2000)
4. Khang, N.V., My, C.A.: Fundamentals of Industrial Robots. Vietnam Education Publisher, Text Book (2011).(in Vietnamese)
5. My, C.A.: Integration of CAM systems into multi-axes computerized numerical control machines. In: 2010 Second International Conference on Knowledge and Systems Engineering, pp. 119–124. IEEE (2010)
6. My, C.A., Hai, V.X.: Generalized pseudo inverse kinematics at singularities for developing five-axes CNC machine tool postprocessor. Vietnam J. Mech. **35**, 147–155 (2013)
7. Sørby, K.: Inverse kinematics of five-axis machines near singular configurations. Int. J. Mach. Tools Manuf. **47**, 299–306 (2007)
8. Lee, R.S., She, C.H.: Developing a postprocessor for three types of five-axis machine tools. Int. J. Adv. Manuf. Technol. **13**, 658–665 (1997)
9. Xu, H.Y., Hu, L.A., Hon-Yuen, T., Shi, K., Xu, L.: A novel kinematic model for five-axis machine tools and its CNC applications. Int. J. Adv. Manuf. Technol. **67**, 1297–307 (2013)
10. My, C.A., Bohez, E.L.: Multi-criteria optimization approach for five-axis CNC tool path planning: modeling methodology. In: Proceedings of the 3rd Asian Conference on Industrial Automation and Robotics, pp. 5–11 (2003)
11. She, C.H., Chang, C.C.: Design of a generic five-axis postprocessor based on generalized kinematics model of machine tool. Int. J. Mach. Tools Manuf. **47**, 537–545 (2007)
12. My, C.A., Bohez, E.L.: A novel differential kinematics model to compare the kinematic performances of five-axis CNC centers, Int. J. Mech. Sci. **163**, 105117 (2019)
13. My, C.A., Bohez, E.L.J., Makhanov, S.S.: Critical point analysis of 3D vector field for five-axis tool path optimization. In: Proceedings of the 4th Asian Conference on Industrial Automation and Robotics, ACIAR, pp. 11–13 (2005)
14. My, C.A., et al.: Real time inverse kinematics of five-axis CNC centers. In: Long, B.T., Kim, Y.-H., Ishizaki, K., Toan, N.D., Parinov, I.A., Vu, N.P. (eds.) MMMS 2020. LNME, pp. 991–999. Springer, Cham (2021). https://doi.org/10.1007/978-3-030-69610-8_131
15. My, C.A., Cong, N.V., Hong, N.M., Bohez, E.L.: Transformation of CAM data for 5-axis CNC machine spinner U5–620. Int. J. Mech. Eng. Robot. Res. **9**(2) (2020)

16. My, C.A., et al.: Mechanical Design and Dynamics Modelling of RoPC Robot, pp. 92–96. In: Proceedings of International Symposium on Robotics and Mechatronics, Hanoi, Vietnam (2009)
17. My, C.A.: Inverse dynamic of a N-links manipulator mounted on a wheeled mobile robot. In: 2013 International Conference on Control, Automation and Information Sciences (ICCAIS), pp. 164–170. IEEE (2013)
18. My, C.A., et al.: A novel mathematical approach for finite element formulation of flexible robot dynamics. Mech. Based Des. Struct. Mach. 1–21 (2020)
19. My, C.A., Makhanov, S.S., Van, N.A., Duc, V.M.: Modeling and computation of real-time applied torques and non-holonomic constraint forces/moment, and optimal design of wheels for an autonomous security robot tracking a moving target. Math. Comput. Simul. **170**, 300–315 (2020)
20. Toai, T.T., Chu, D.-H., My, C.A.: Development of a new 6 DOFs welding robotic system for a specialized application. In: Balas, V.E., Solanki, V.K., Kumar, R. (eds.) Further Advances in Internet of Things in Biomedical and Cyber Physical Systems. ISRL, vol. 193, pp. 135–150. Springer, Cham (2021). https://doi.org/10.1007/978-3-030-57835-0_11

Finite-Time Formation Convergence of Vision-Based Nonholonomic Systems Without Explicit Communication

Jishnu Keshavan(✉)

Indian Institute of Science, Bangalore 560012, India
kjishnu@iisc.ac.in

Abstract. This study considers the problem of synthesizing a decentralized control policy for leader-follower formation tracking of a group of nonholonomic agents in the absence of explicit inter-agent communication. Each follower robot is equipped with a perspective camera, and by tracking a single point feature attached to the leader agent, the leader-follower formation kinematics is developed in the image space of the follower agent. An adaptive continuous sliding mode controller is then proposed for the follower agent, and it is shown that the resulting closed-loop observer-controller system achieves global finite-time convergence of the tracking error to a small uniform ultimate bound around the origin in the absence of either explicit measurement/estimation of the leader agent's position and velocity. Furthermore, it is shown that the size of this error bound can be arbitrarily reduced through a sufficiently large choice of the controller gain. Simulation results are used to validate and demonstrate robust performance of the proposed scheme in the presence of measurement noise.

Keywords: Perspective dynamical system · Robotics and automation · Leader-follower formation control

1 Introduction

The problem of formation control of nonholonomic mobile robots has attracted considerable interest in recent years. Most studies typically rely on direct global orrelative position measurements/estimates to regulate the global position of each robot [1–6]. Other studies accomplish formation control under the assumption that velocity information is available and can be transmitted to other agents without time delay [7–9], which is a difficult assumption to satisfy for most practical systems. In order to overcome this drawback, more recent approaches instead rely on the estimation of leader velocity for accomplishing leader-follower formation control [10,11]. Thus, a common approach of these earlier studies is to achieve exponential formation convergence to the desired pattern by relying on either measurement or estimation of the leader agent's relative position and velocity.

N. V. Khang et al. (Eds.): ASIAN MMS 2021, MMS 113, pp. 443–454, 2022.
https://doi.org/10.1007/978-3-030-91892-7_42

In contrast, this study considers the development of a novel formation control scheme that guarantees global finite-time convergence of leader-follower formation to the desired pattern in the absence of any knowledge of the leader agent's relative position and velocity. The leader-follower kinematics is formulated in the image space of the perspective camera (attached to each follower robot) that tracks a single point feature attached to the leader robot. For the case where feature height is assumed known, an adaptive continuous sliding mode-based controller is then proposed that yields exact global finite-time convergence of the tracking error to the origin. Furthermore, for the more challenging case of the feature height being unknown, the aforementioned adaptive control strategy is modified to ensure global finite-time convergence of the tracking error to a small uniform bound around the origin, where the size of the error bound can be reduced for a sufficiently large choice of the controller gain. Simulation results are used to validate the performance of proposed strategies in the presence of significant measurement noise. Thus, in contrast with most prior studies that only guarantee exponential formation convergence by relying on measurement/estimation of leader's position/velocity [7–9,11], an important contribution of this study is that formation convergence is achieved in finite-time without requiring explicit estimation of leader's position/velocity, and explicit communication between the leader-follower pair.

The rest of this paper is structured as follows. Section 2 formulates the formation control problem by modeling the leader-follower kinematics in the image space of the follower agent. Section 3 presents the adaptive sliding-mode control strategy that is used to drive visual measurements to the desired values in finite-time when feature height is assumed unknown. Section 4 presents numerical results to validate the performance of the proposed control scheme in the presence of measurement noise. Section 5 presents the conclusions of this study.

2 Problem Definition

Consider the motion of a group of labeled unicyle-type robots, where one of the robots (designated R_1) is the lead robot. This robot's motion is usually prescribed *apriori*, and the motion of individual followers in the formation is then defined in reference to the lead robot. The formation tracking objective is achieved when the relative position of each pair of leader-follower members are regulated to their desired values, which in turn are specified by the desired reference formation pattern for the group. Complex formation patterns can then be generated by specifying different reference relative positions for each leader-follower pair. Thus, it is sufficient to achieve the formation control objective for a pair of leader-follower robots in order to solve the problem of formation control for a group of multiple robots.

To this end, the leader-follower kinematic model can be obtained as [9],

$$\dot{\boldsymbol{r}}_{ij} = \left[v_i \cos\theta_{ij} - v_j + y_{ij}\omega_j,\ v_i \sin\theta_{ij} - x_{ij}\omega_j\right]^\top . \quad (1)$$

Now, consider a perspective camera mounted on the follower robot R_j. The reference frame for the camera is given by $\mathcal{C}_j = \{x_{cj}, y_{cj}, z_{cj}\}$, with the optical center of the camera located at the origin of the reference frame $\mathcal{B}_j$, the camera's positive vertical axis pointing straight up and perpendicular to the ground plane, and the camera's optical axis (z_{cj}-axis) aligned with the x_{bj} axis. As considered in [11], a single point feature rigidly located on the z_{bi} axis of the leader robot R_i is used to derive feedback information and accomplish formation control for the leader-follower pair. The Cartesian coordinates of the point feature P in the reference frame $\mathcal{B}_i$ are given by $\{0, 0, z_i\}^\top$ with z_i being a positive constant, and its coordinates in the camera's reference frame $\mathcal{C}_j$ can be derived as $\boldsymbol{r}_{Pj} := \{x_{Pj}, y_{Pj}, z_{Pj}\}^\top = \{y_{ij}, z_i, x_{ij}\}^\top$. For practical perspective systems, it is reasonable to assume that the point feature is located ahead of the camera's image plane, so that $x_{ij} > 0$.

The measurements of the image-plane pixel coordinates of the point feature under perspective projection, denoted as $\boldsymbol{\pi} := \{\pi_1, \pi_2\}^\top = \{u_{ij}, v_{ij}\}^\top$, are then given by,

$$\begin{bmatrix} \boldsymbol{\pi} \\ 1 \end{bmatrix} = \frac{1}{z_{Pj}} \boldsymbol{A} \boldsymbol{r}_{Pj}, \; \boldsymbol{A} = \begin{bmatrix} \alpha_u & \alpha_{uv} & u_0 \\ 0 & \alpha_v & v_0 \\ 0 & 0 & 1 \end{bmatrix}, \tag{2}$$

where $\boldsymbol{A}$ is the intrinsic camera calibration matrix with parameters α_u and α_v as the scaling factors along the x_{cj} and y_{cj} axes respectively, α_{uv} is the skew factor between these axes, and u_0 and v_0 denote the pixel coordinates of the principal focal point of the camera. Noting that the calibration matrix is invertible, the normalized coordinates of the point feature $\boldsymbol{m} := \{n_{ij}, s_{ij}\}^\top$ can be easily obtained as,

$$\begin{aligned} &\begin{bmatrix} \boldsymbol{m} \\ 1 \end{bmatrix} = \frac{1}{z_{Pj}} \boldsymbol{r}_{Pj} = \boldsymbol{A}^{-1} \begin{bmatrix} \boldsymbol{\pi} \\ 1 \end{bmatrix}, \text{ or,} \\ &n_{ij} = y_{ij}/x_{ij}, \; s_{ij} = z_i/x_{ij}. \end{aligned} \tag{3}$$

Observe that, with z_i, $x_{ij} > 0$, we have $s_{ij} > 0$.

Finally, differentiating (2), and substituting (1) and (3) in the resulting equation, the formation kinematics in the image space can be derived as,

$$\begin{aligned} \begin{bmatrix} \dot{u}_{ij} \\ \dot{v}_{ij} \end{bmatrix} &= \begin{bmatrix} \alpha_u \sin\theta_{ij} + (u_0 - u_{ij}) \cos\theta_{ij} \\ (v_0 - v_{ij}) \cos\theta_{ij} \end{bmatrix} \frac{s_{ij}}{z_i} v_i \\ &+ \begin{bmatrix} (u_{ij} - u_0) \frac{s_{ij}}{z_i} & (u_0 - u_{ij}) n_{ij} - \alpha_u \\ (v_{ij} - v_0) \frac{s_{ij}}{z_i} & (v_0 - v_{ij}) n_{ij} \end{bmatrix} \begin{bmatrix} v_j \\ \omega_j \end{bmatrix}, \\ \dot{\theta}_{ij} &= \omega_i - \omega_j. \end{aligned} \tag{4}$$

Motivated by practical considerations, the perspective system (4) needs to satisfy the following mild assumption.

Assumption 1: The pixel measurements, leader forward and angular velocity components, and follower forward and angular velocity components are smooth bounded functions of time, i.e., $\boldsymbol{\pi}(t)$, $\boldsymbol{m}(t)$, $\dot{\boldsymbol{\pi}}(t)$, $\dot{\boldsymbol{m}}(t), v_i(t)$, $v_j(t)$, $\omega_i(t)$, $\omega_j(t) \in \mathcal{L}_\infty$[1].

The formation control problem considered in this letter is described next. Assume that the relative position ($\boldsymbol{r}_{ij}$) and orientation (θ_{ij}) of the leader-follower pair of robots cannot be measured, and that the leader velocities v_i, ω_i are unknown. Then, given the desired formation pattern specified by the reference relative position $\boldsymbol{r}_{ij}^d = \{x_{ij}^d, y_{ij}^d\}^\top$ between the leader-follower pair, the objective is to design control inputs v_j, ω_j of the follower based on image feedback such that the relative position error $\tilde{\boldsymbol{r}}_{ij} = \boldsymbol{r}_{ij} - \boldsymbol{r}_{ij}^d$ is driven to zero in finite-time. From (3), observe that a one-to-one correspondence exists between the relative position of the leader-follower pair $\boldsymbol{r}_{ij}$ and the feature point coordinates $\boldsymbol{m}$ (or equivalently $\boldsymbol{\pi}$), and thus the desired formation pattern is characterized by the desired normalized coordinates of the point feature $\boldsymbol{m}^d = \{y_{ij}^d/x_{ij}^d, z_i/x_{ij}^d\}^\top$. The image feature tracking error $\boldsymbol{\pi} - \boldsymbol{\pi}^d$ is then regarded as the surrogate of the relative position error $\tilde{\boldsymbol{r}}_{ij}$, and subsequently referred to as the formation tracking error. Thus, the finite-time formation control objective is to find a dynamic control law in the form

$$[v_j, \omega_j]^\top = \varsigma_1(\boldsymbol{\pi}, \boldsymbol{\pi}^d, \boldsymbol{m}, \boldsymbol{\sigma}, \dot{\boldsymbol{\sigma}}),\ \dot{\boldsymbol{\sigma}} = \varsigma_2(\boldsymbol{\pi}, \boldsymbol{\pi}^d, \boldsymbol{m}, \boldsymbol{\sigma}), \tag{5}$$

where $\boldsymbol{\sigma}$ is an internal state to be designed, and $\varsigma_1(\cdot)$, $\varsigma_2(\cdot)$ are sufficiently smooth functions such that $\boldsymbol{\pi}(t) - \boldsymbol{\pi}^d(t)$, $\theta_{ij}(t)$ are bounded for all $t \geq 0$, and there exist positive constants v, T such that

$$||\boldsymbol{\pi}(t) - \boldsymbol{\pi}^d(t)|| \leq v, \qquad \forall t \geq T, \tag{6}$$

in the presence of uncertain variables $\theta_{ij}(t), v_i(t), \omega_i(t)$, and in the absence of explicit communication between the leader-follower pair of robots.

3 Controller Synthesis

In this section, two adaptive observer-controller systems are presented for accomplishing vision-based formation-convergence in finite-time. The first case assumes that the feature height is known apriori, while the second case considers the more difficult challenge of finite-time convergent control when the feature height is unknown.

3.1 Formation Convergence with Known Feature Height

The simpler case of controller synthesis when the feature height is assumed known is considered here. For this case, we begin by separating the known and unknown parts of (4), so that,

---

[1] For a function $f(t) \in \mathbb{R}^n\, \forall n \in [1, \infty)$, $f(t) \in \mathcal{L}_\infty$ when $\sup_t ||f(t)||_2 < \infty$, where $||.||_2$ denotes the 2-norm in $\mathbb{R}^n$.

$$
\begin{aligned}
\dot{\boldsymbol{\pi}} &= \boldsymbol{\Phi}\boldsymbol{u} + \boldsymbol{x},\ \boldsymbol{u} = \left[v_j, \omega_j\right]^\top, \\
\boldsymbol{\Phi} &= \begin{bmatrix} (u_{ij} - u_0)\frac{s_{ij}}{z_i} & (u_0 - u_{ij})n_{ij} - \alpha_u \\ (v_{ij} - v_0)\frac{s_{ij}}{z_i} & (v_0 - v_{ij})n_{ij} \end{bmatrix}, \\
\boldsymbol{x} &= \begin{bmatrix} \alpha_u \sin\theta_{ij} + (u_0 - u_{ij})\cos\theta_{ij} \\ (v_0 - v_{ij})\cos\theta_{ij} \end{bmatrix} \frac{s_{ij}}{z_i} v_i.
\end{aligned} \tag{7}
$$

Then, by differentiating (7) with respect to time, and invoking Assumption 1, it is straighforward to conclude that $\dot{x}(t) \in \mathcal{L}_\infty$, such that $\sup_t ||\dot{\boldsymbol{x}}|| \leq L$, with L as a positive constant. Furthermore, note that the determinant of Φ satisfies $\text{Det}\left[\Phi\right] = \alpha_u(v_{ij} - v_0)s_{ij}/z_i \neq 0$, ensuring that the matrix Φ is invertible.

To find a solution of the form (5), we introduce the internal states $\hat{\boldsymbol{x}}, \boldsymbol{y} \in \mathbb{R}^2$, and propose the following adaptive control law:

$$
\begin{aligned}
\boldsymbol{u} &= \boldsymbol{\Phi}^{-1}\left[\boldsymbol{y} - \hat{\boldsymbol{x}}\right],\ \dot{\hat{\boldsymbol{x}}} = h_2 \begin{bmatrix} \text{sign}(\pi_1 - \pi_1^d) \\ \text{sign}(\pi_2 - \pi_2^d) \end{bmatrix}, \\
\boldsymbol{y} &= -h_1 \begin{bmatrix} |\pi_1 - \pi_1^d|^{1/2}\text{sign}(\pi_1 - \pi_1^d) \\ |\pi_2 - \pi_2^d|^{1/2}\text{sign}(\pi_2 - \pi_2^d) \end{bmatrix},
\end{aligned} \tag{8}
$$

with controller gains $h_1, h_2 \in \mathbb{R}^+$.

Next, define the following error variables,

$$
\tilde{\boldsymbol{\pi}} := [\tilde{\pi}_1, \tilde{\pi}_2]^\top = \boldsymbol{\pi} - \boldsymbol{\pi}^d,\ \tilde{\boldsymbol{x}} := [\tilde{x}_1, \tilde{x}_2]^\top = \boldsymbol{x} - \hat{\boldsymbol{x}}. \tag{9}
$$

From (7) and (8), the error dynamics can then be obtained as,

$$
\begin{aligned}
\dot{\tilde{\pi}}_k &= -h_1|\tilde{\pi}_k|^{1/2}\text{sign}(\tilde{\pi}_k) + \tilde{x}_k\,, \\
\dot{\tilde{x}}_k &= -h_2\text{sign}(\tilde{\pi}_k) + \dot{x}_k,\ k = 1,\, 2.
\end{aligned} \tag{10}
$$

Notice that the dynamics of the error variables $\tilde{\pi}_1, \tilde{x}_1$ are completely decoupled from and similar to the dynamics of $\tilde{\pi}_2, \tilde{x}_2$. Consequently, we denote the variables in (10) as $\tilde{\pi}, \tilde{x}$, which can be then treated as scalar quantities, by discarding the subscript k with the understanding that $k \in \{1, 2\}$.

Next, in order to formally derive a closed-form expression for the controller gain parameters and convergence time, we propose the following theorem.

Theorem 1: Suppose that $v_i > 0$ and $|\theta_{ij}| < \pi$. Then, under Assumption 1, there exist control gains h_1, h_2 such that the control law (8) provides exact global finite-time convergence of the tracking error to zero, the internal dynamics is locally stable i.e. the relative orientation $\theta_{ij}(t)$ is bounded, and the formation control objective specified by (6) (with $v = 0$) is achieved.

Proof: Consider the change of variable

$$
\boldsymbol{\xi}^\top := [\xi_1,\ \xi_2] = \left[|\tilde{\pi}|^{1/2}\text{sign}(\tilde{\pi}), \tilde{x}\right]. \tag{11}
$$

The time-derivative of the vector $\boldsymbol{\xi}$ is then given by,

$$\dot{\boldsymbol{\xi}} = \frac{1}{|\tilde{\pi}|^{\frac{1}{2}}} \boldsymbol{B}\,\boldsymbol{\xi},$$
$$\boldsymbol{B} = \begin{bmatrix} -h_1/2 & 1/2 \\ -h_2 + \nu(t) & 0 \end{bmatrix}, \ \nu(t) = \dot{x}(t)\text{sign}(\tilde{\pi}). \tag{12}$$

The rest of the proof follows directly from the proof of Theorem 2 in [12] for the following choice of the matrix $\boldsymbol{P}$ and gain parameters h_1, h_2,

$$h_1 = \tau_2 \left[\frac{2\alpha L}{(1-\beta)\tau_1}\right]^{1/2}, \ h_2 = \frac{1+\beta}{1-\beta} L,$$
$$\tau_1 = \frac{\beta - 1/\alpha}{1-\beta^2}, \ \tau_2 = \frac{\beta^2 - 2\beta/\alpha + 1}{(1-\beta^2)},$$
$$p_{11} = 1, p_{22} = \frac{(1-\beta)\tau_1}{2L}, p_{12} = p_{21} = -(p_{22}/\alpha)^{1/2}. \tag{13}$$

where α, β are user-defined positive constants that satisfy $0 < \beta < 1$, $\alpha > 1/\beta$. Thus, global finite-time convergence of the tracking error to the origin is assured, with the convergence time smaller than $T_c = 2\big(V^{1/2}(\boldsymbol{\xi}(0))/\sigma$.

Finally, we need to prove that the orientation dynamics is locally stable. To this end, substituting (8) in the second part of (4), and using the fact that $\dot{\boldsymbol{\pi}} = \boldsymbol{y} + \tilde{\boldsymbol{x}}$, it is straightforward to show that,

$$\dot{\theta}_{ij} = -\frac{s_{ij} v_i}{z_i} \sin\theta_{ij} + \varphi(t),$$
$$\varphi(t) = \frac{1}{\alpha_u} \dot{u}_{ij} - \frac{u_{ij} - u_0}{\alpha_u (v_{ij} - v_0)} \dot{v}_{ij} + \omega_i. \tag{14}$$

From the boundedness of $\boldsymbol{\pi}(t)$, $\dot{\boldsymbol{\pi}}(t)$, and ω_i, we can show that $\varphi(t)$ is uniformly bounded. Then, with $s_{ij} > 0$, $v_i > 0$, $z_i > 0$ and $|\theta_{ij}(0)| < \pi$, using the theory of nonlinear systems with persistent disturbances and similar to the results in [8] and [11], it follows that there exists a positive constant θ_0 and some finite time instant $T > 0$ such that $|\theta_{ij}(t)| < \theta_0$, $\forall t \geq T$. Hence local stability of the internal dynamics is then guaranteed. This ends the proof. ■

Remark 2: The controller gain parameters in (13) require that a conservative estimate of the upper bound L is known, which is a reasonable assumption for most practical visual robotic systems. Although other approaches exists that adaptively tune the control gain parameters to estimate the upper bound L online [13], in the simulation results shown below, it was found that a conservative estimate of the upper bound was sufficient to realize robust controller performance in the presence of uncertainty, and thus the case of synthesizing adaptive gains was not considered in this study.

Remark 3: Note that the studies in [7–9] require inter-agent communication to realize stable control laws that only guarantee global exponential convergence of the formation tracking errors to zero. While the study in [11] is designed to achieve formation convergence in the absence of inter-agent communication, it suffers from the drawback of relying on the Perspective-n-Point approach or Homography decomposition, either of which usually require tracking more than one point feature for extracting the camera's relative pose θ_{ij}, to synthesize a stable adaptive law that only guarantees global exponential convergence of the formation tracking error to a uniform ultimate bound. Moreover, establishing formation control in [11] requires an estimation scheme for explicitly estimating leader's velocity v_i. In contrast, the proposed scheme helps achieve exact global finite-time formation convergence without relying on relative position and velocity measurements of the leader robot, and in the absence of explicit inter-agent communication. Furthermore, the proposed scheme is less unwieldy in that it does not rely on apriori knowledge of the relative orientation θ_{ij} and explicit estimation of the state v_i, which renders the current approach easier to implement compared to the control schemes in [7–9,11].

3.2 Formation Convergence with Unknown Feature Height

In the previous section, exact global finite-time formation convergence was demonstrated provided the feature height z_i was known apriori through an accurate offline identification process. In this section, we relax this assumption and consider the case of the feature height as being unknown. The objective then is to demonstrate finite-time formation convergence in the presence of uncertainty in the value of z_i, which affords more flexibility in formation control design.

To this end, we begin by separating the known and unknown parts of (4), so that,

$$\begin{aligned} \Theta &= \begin{bmatrix} (u_{ij} - u_0)s_{ij} & (u_0 - u_{ij})n_{ij} - \alpha_u \\ (v_{ij} - v_0)s_{ij} & (v_0 - v_{ij})n_{ij} \end{bmatrix}, \overline{\boldsymbol{u}} = \begin{bmatrix} \overline{v}_j \\ \omega_j \end{bmatrix}, \\ \dot{\boldsymbol{\pi}} &= \Theta\overline{\boldsymbol{u}} + \boldsymbol{x} + \Psi\overline{v}_j\frac{\tilde{z}_{ij}}{z_i}, \ \Psi := \begin{bmatrix} \Psi_1 \\ \Psi_2 \end{bmatrix} = \begin{bmatrix} u_{ij} - u_0 \\ v_{ij} - v_0 \end{bmatrix} s_{ij}. \end{aligned} \tag{15}$$

where $v_j = \hat{z}_{ij}\overline{v}_j$, $\tilde{z}_{ij} = \hat{z}_{ij} - z_i$, $\hat{z}_{ij}$ is an estimate of z_i, and $\boldsymbol{x}$ is defined in (7). Note that the determinant of Θ satisfies $\mathrm{Det}\big[\Theta\big] = \alpha_u(v_{ij} - v_0)s_{ij} \neq 0$, ensuring that the matrix Θ is invertible.

To find a solution of the form (5), we propose the following adaptive control law:

$$\overline{\boldsymbol{u}} = \Theta^{-1}\big[\boldsymbol{y} - \hat{\boldsymbol{x}}\big], \ \dot{\hat{z}}_{ij} = -h_3\overline{v}_j\Psi^\top \begin{bmatrix} \mathrm{sign}(\pi_1 - \pi_1^d) \\ \mathrm{sign}(\pi_2 - \pi_2^d) \end{bmatrix}. \tag{16}$$

where controller gain $h_3 \in \mathbb{R}^+$, and the internal states $\boldsymbol{y}(t), \hat{\boldsymbol{x}}(t)$ are updated as before in (8). Observe that the structure of (16) is very similar to (8) along with an update equation for the estimation of the unknown state z_i. We now have the following theorem.

Theorem 2: Suppose that $v_i > 0$ and $|\theta_{ij}| < \pi$. Then, under Assumption 1, there exist control gains h_1, h_2, h_3 such that the control law (16) provides global finite-time convergence of the tracking error to a uniform ultimate bound, the internal dynamics is locally stable i.e. the relative orientation $\theta_{ij}(t)$ is bounded, and the formation control objective specified by (6) is achieved, where the magnitude of the ultimate bound υ can be sufficiently reduced for a suitably large choice of control gain h_1.

Proof: In order to demonstrate finite-time convergence in the presence of uncertainty in z_i, we consider the full-state vector $\boldsymbol{\zeta} := \left[\boldsymbol{\zeta}_1^\top, \boldsymbol{\zeta}_2^\top\right]^\top$, where,

$$\boldsymbol{\zeta}_k^\top := \left[|\tilde{\pi}_k|^{1/2}\text{sign}(\tilde{\pi}_k), \tilde{x}_k\right],\ k = 1,\ 2, \tag{17}$$

with the error variables defined in (9). The time-derivative can then be obtained as,

$$\dot{\boldsymbol{\zeta}}_k = \frac{1}{|\tilde{\pi}_k|^{1/2}}\left[\boldsymbol{B}_k\boldsymbol{\zeta}_k + \frac{1}{2}\boldsymbol{\rho}_k\right],\ \boldsymbol{\rho}_k(t) = \left[\Psi_k\overline{v}_j\tilde{z}_{ij}/z_i,\ 0\right]^\top, \tag{18}$$

$k = 1,\ 2$, with the rest of the matrices as given in (12).

To facilitate subsequent analysis, we now consider the Lyapunov function candidate $W(\boldsymbol{\zeta}_z) = \boldsymbol{\zeta}_z^\top \tilde{\boldsymbol{P}}_z\boldsymbol{\zeta}_z$, with $\boldsymbol{\zeta}_z^\top := \left[\boldsymbol{\zeta}_1^\top, \boldsymbol{\zeta}_2^\top, \tilde{z}_{ij}\right]$, and the block diagonal matrix $\tilde{\boldsymbol{P}}_z = \text{diag}\{\boldsymbol{P}_z, \boldsymbol{P}_z, \frac{h_3^{-1}}{2z_i}\}$, where $\boldsymbol{P}_z = [p_{z_{lm}}]$ $(l,\ m \in \{1,2\})$ is a constant, symmetric, positive-definite matrix defined below. The Lyapunov function is an absolutely continuous function that is continuously differentiable everywhere except on the sets $\mathcal{M}_2 = \{\boldsymbol{\zeta}_z \in \mathbb{R}^5 | \tilde{\pi}_1 = 0\}$, $\mathcal{M}_3 = \{\boldsymbol{\zeta}_z \in \mathbb{R}^5 | \tilde{\pi}_2 = 0\}$.

The elements of the matrix $\boldsymbol{P}_z$ and the gain parameters $h_1,\ h_2$ are now obtained from (13) by replacing the user-defined parameter $\alpha = \alpha_0/\epsilon^2$, where $\alpha_0 > \epsilon^2/\beta$ and ϵ is a small positive constant such that $\epsilon << 1$. Then, we have,

$$\begin{aligned}
h_1 &= \frac{\tau_2}{\epsilon}\left[\frac{2\alpha_0 L}{(1-\beta)\tau_1}\right]^{1/2},\ h_2 = \frac{1+\beta}{1-\beta}L, \\
\tau_1 &= \frac{\beta - \epsilon^2/\alpha_0}{1-\beta^2},\ \tau_2 = \frac{\beta^2 - 2\epsilon^2\beta/\alpha_0 + 1}{(1-\beta^2)},\ p_{z_{11}} = 1, \\
p_{z_{22}} &= \frac{(1-\beta)\tau_1}{2L}, p_{z_{12}} = p_{z_{21}} = -\epsilon(p_{z_{22}}/\alpha_0)^{1/2}.
\end{aligned} \tag{19}$$

Then, for every point in $\mathbb{R}^5 \backslash \mathcal{M}_2 \cup \mathcal{M}_3$, and similar to the proof of Theorem 1 above, it can be shown that,

$$\begin{aligned}
\dot{W} \leq &-\sum_{k=1}^{2}\frac{1}{|\tilde{\pi}_k|^{\frac{1}{2}}}\boldsymbol{\zeta}_k^\top\boldsymbol{Q}_z\boldsymbol{\zeta}_k + h_3^{-1}\frac{\dot{\tilde{z}}_{ij}\tilde{z}_{ij}}{z_i} \\
&+\overline{v}_j\frac{\tilde{z}_{ij}}{z_i}\sum_{k=1}^{2}\frac{1}{|\tilde{\pi}_k|^{1/2}}\Psi_k\left[|\tilde{\pi}_k|^{1/2}\text{sign}(\tilde{\pi}_k) + p_{z_{12}}\tilde{x}_k\right],
\end{aligned} \tag{20}$$

where $\boldsymbol{Q}_z = [q_{z_{lm}}]$ $(l, m \in \{1, 2\})$ is a constant, symmetric and positive-definite matrix given by [12],

$$
\begin{aligned}
q_{z_{11}} &= -h_1 p_{z_{12}} - 2p_{z_{12}}^2 (h_2+L) - 2L(1-h_1 p_{z_{12}}) p_{z_{22}}, \\
q_{z_{12}} &= q_{z_{21}} = (h_1 p_{z_{12}} - 1)/2 + (h_2 + L) p_{z_{22}},\ q_{z_{22}} = 1.
\end{aligned} \tag{21}
$$

Then, from (16), we have $\dot{\hat{z}}_{ij} = \dot{\tilde{z}}_{ij} = -h_3 \overline{v}_j \big(\Psi_1 \mathrm{sign}(\tilde{\pi}_1) + \Psi_2 \mathrm{sign}(\tilde{\pi}_2)\big)$, so that,

$$
\begin{aligned}
\dot{W} &\leq -\sum_{k=1}^{2} \frac{1}{|\tilde{\pi}_k|^{\frac{1}{2}}} \left[\boldsymbol{\zeta}_k^\top \boldsymbol{Q}_z \boldsymbol{\zeta}_k - \overline{v}_j \frac{\tilde{z}_{ij} p_{z_{12}}}{z_i} \Psi_k \tilde{x}_k \right] \\
&\leq -\sum_{k=1}^{2} \frac{1}{|\tilde{\pi}_k|^{\frac{1}{2}}} \left[\boldsymbol{\zeta}_k^\top \boldsymbol{Q}_z \boldsymbol{\zeta}_k - \delta \eta \, ||\boldsymbol{\zeta}_k|| \right],\ \delta = \epsilon \delta_0, \\
\delta_0 &= \sqrt{\frac{p_{z22}}{\alpha_0}} \left[\sup_t |\overline{v}_j| \frac{|\tilde{z}_{ij}|}{z_i} \right],\ \eta = \sup_t ||\Psi(t)||.
\end{aligned} \tag{22}
$$

Then, from (22), it follows that,

$$
\begin{aligned}
\dot{W} &\leq -\Big[\lambda_{\min}\{\boldsymbol{Q}_z\} \big[\kappa ||\boldsymbol{\zeta}|| + (1-\kappa) ||\boldsymbol{\zeta}|| \big] - 2\delta\eta \Big] \\
&\leq -\kappa \lambda_{\min}\{\boldsymbol{Q}_z\} ||\boldsymbol{\zeta}||,\ \ \forall ||\boldsymbol{\zeta}|| \geq \mu,\ 0 < \kappa < 1,
\end{aligned} \tag{23}
$$

where $\mu = \epsilon \mu_0$, $\mu_0 = \frac{2\delta_0 \eta}{(1-\kappa)\lambda_{\min}\{\boldsymbol{Q}_z\}}$.

Then, as $||\boldsymbol{\zeta}|| \leq ||\boldsymbol{\zeta}_z|| \leq ||\boldsymbol{\zeta}|| + |\tilde{z}_{ij}|$, it follows that,

$$
\begin{aligned}
\dot{W} &\leq -\kappa \lambda_{\min}\{\boldsymbol{Q}_z\} \frac{||\boldsymbol{\zeta}||}{||\boldsymbol{\zeta}|| + |\tilde{z}_{ij}|} ||\boldsymbol{\zeta}_z||,\ \ \forall ||\boldsymbol{\zeta}||,\ ||\boldsymbol{\zeta}_z|| \geq \mu, \\
&\leq -\frac{\kappa}{\gamma} \lambda_{\min}\{\boldsymbol{Q}_z\} ||\boldsymbol{\zeta}_z||,\ \ \forall ||\boldsymbol{\zeta}||,\ ||\boldsymbol{\zeta}_z|| \geq \mu, \\
&\leq -\sigma_z W^{1/2}(\boldsymbol{\zeta}_z),\ \ \forall ||\boldsymbol{\zeta}_z|| \geq \mu,
\end{aligned} \tag{24}
$$

with $\gamma = \sup_t [1 + \frac{|\tilde{z}_{ij}|}{||\boldsymbol{\zeta}||}]$, and $\sigma_z = \frac{\kappa \lambda_{\min}\{\boldsymbol{Q}_z\}}{\gamma \lambda_{\max}^{1/2}\{\tilde{\boldsymbol{P}}_z\}}$.

Then, any trajectory enters the ball $\mathcal{W} = \left\{ \boldsymbol{s} \in \mathbb{R}^5 | W(\boldsymbol{s}) \leq \lambda_{\max}\{\tilde{\boldsymbol{P}}_z\} \epsilon^2 \mu_0^2 \right\}$ in finite-time and stays in $\mathcal{W}$ for all future time. The time taken to enter the ball $\mathcal{W}$ is not smaller than,

$$
T_{c_z}(\boldsymbol{\zeta}(0), \tilde{z}_{ij}(0)) = \frac{2}{\sigma_z} \Big[W^{1/2}(\boldsymbol{\zeta}_z(0)) - \epsilon \lambda_{\max}^{1/2}\{\tilde{\boldsymbol{P}}_z\} \mu_0 \Big]. \tag{25}
$$

Note that the size of the ball $\mathcal{W}$ can be made arbitrarily small for a sufficiently small choice of ϵ, so that $||\boldsymbol{\zeta}_z(t)|| \leq \frac{\lambda_{\max}^{1/2}\{\tilde{\boldsymbol{P}}_z\}}{\lambda_{\min}^{1/2}\{\tilde{\boldsymbol{P}}_z\}} \epsilon \mu_0$, $\forall t > T_{c_z}$. Thus, for a sufficiently small value of ϵ (thus a large value of the gain h_1), global finite-time convergence of the tracking error $||\tilde{\boldsymbol{\pi}}||$ to a small uniform ultimate bound is assured. Furthermore, the local stability of the orientation dynamics can then be guaranteed in a similar manner as given in the proof of Theorem 1. This ends the proof. ■

4 Numerical Results

In this section, the performance of the proposed vision-based formation tracking and control scheme is validated through numerical results in the absence of direct position and velocity measurements of the mobile robots. As in [11], we consider a group of five mobile robots $R_1 - R_5$ with leader-follower relationships as shown in Fig. 1. The initial states of the mobile robots are given by $\boldsymbol{r}_1(0) = \{3, 4\}^\top$ m and $\theta_1(0) = 0$ rad, $\boldsymbol{r}_2(0) = \{1, 6\}^\top$ m and $\theta_2(0) = -\pi/3$ rad, $\boldsymbol{r}_3(0) = \{1, 1\}^\top$ m and $\theta_3(0) = \pi/6$ rad, $\boldsymbol{r}_4(0) = \{-3, 8\}^\top$ m and $\theta_4(0) = -\pi/3$ rad, and $\boldsymbol{r}_5(0) = \{-3, -2\}^\top$ m and $\theta_5(0) = \pi/6$ rad. The feature heights of robots R_1, R_3 are fixed as $z_1 = z_3 = 0.3$ m. The camera intrinsic parameters are chosen as $u_0 = 320$ pixels, $v_0 = 240$ pixels, $\alpha_u = \alpha_v = 700$ pixels, and $\alpha_{uv} = 0$ pixels. The desired formation pattern is specified by placing the robots in the desired relative positions as $\boldsymbol{r}^d_{12} = \{2, -1\}^\top$ m, $\boldsymbol{r}^d_{13} = \{2, 1\}^\top$ m, $\boldsymbol{r}^d_{34} = \{1, -0.5\}^\top$ m, $\boldsymbol{r}^d_{35} = \{1, 0.5\}^\top$ m. The corresponding desired pixel coordinates for each leader-follower pair are then obtained from (2). The velocities of group leader R_1 are given by $v_1 = 2 + \sin 0.2t$ m/s and $\omega_1 = 0.5 + 0.2 \sin 0.2t$ rad/s. The image measurements are assumed to be corrupted by additive white Gaussian noise with an intensity of 5%.

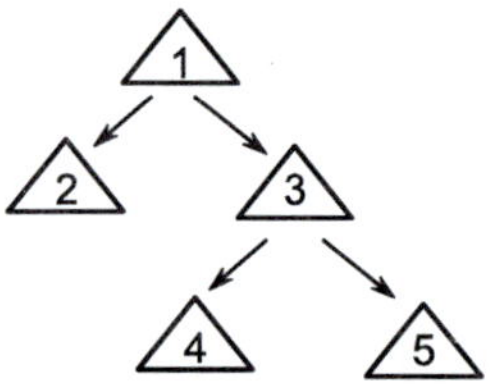

Fig. 1. Leader-follower formation graph for robots $R_1 - R_5$. R_1 is the group leader robot, R_2 and R_3 track R_1, and R_4 and R_5 track R_3.

The controller gain parameters for the control laws in (8), (16) are chosen as $\beta = 0.75$, $\alpha = 16$, and $h_3 = 0.0001$. The upper bound L is conservatively chosen as $L = 100$. The results are shown below for visual tracking of robot R_1 by R_2, and robot R_3 by R_5. The formation errors that result from visual tracking of robot R_1 by R_3 and robot R_3 by R_4 are very similar, and thus not shown. Moreover, the time-history trajectories of formation errors using the control law (8) are very similar to the time-history trajectories of formation errors obtained using (16) that assumes feature height as unknown, and thus not shown.

The initial feature height estimates are chosen as $\hat{z}_{12}(0) = \hat{z}_{35}(0) = 0$. The time-history plots of the formation errors both in configuration space $\tilde{\boldsymbol{r}}_{ij}$ and in image space $\tilde{\boldsymbol{\pi}}_{ij}$ are shown below in Fig. 2. It is apparent that the formation errors in configuration space and in image space are guaranteed to converge to a very small bound around the origin in finite-time, which confirms the conclusions of Theorem 2. Additionally, the time-history plots of the corresponding height

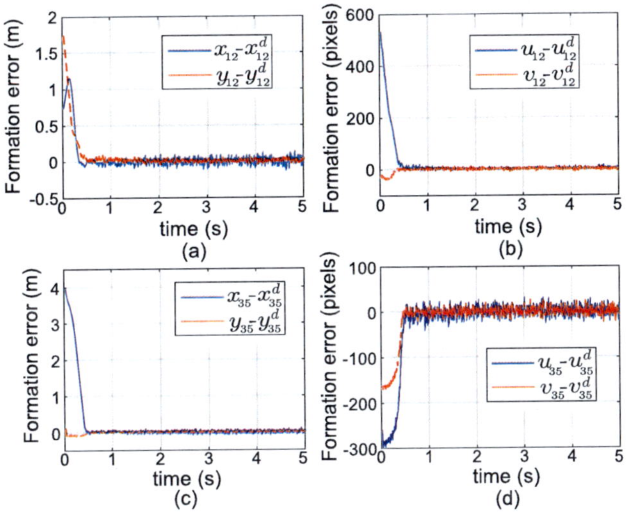

Fig. 2. Time-history trajectories of formation errors obtained using the control law (16) with sensor output corrupted by noise. (a) formation errors in configuration space $\tilde{r}_{12}$. (b) formation errors in image space $\tilde{\pi}_{12}$. (c) formation errors in configuration space $\tilde{r}_{35}$. (d) formation errors in image space $\tilde{\pi}_{35}$.

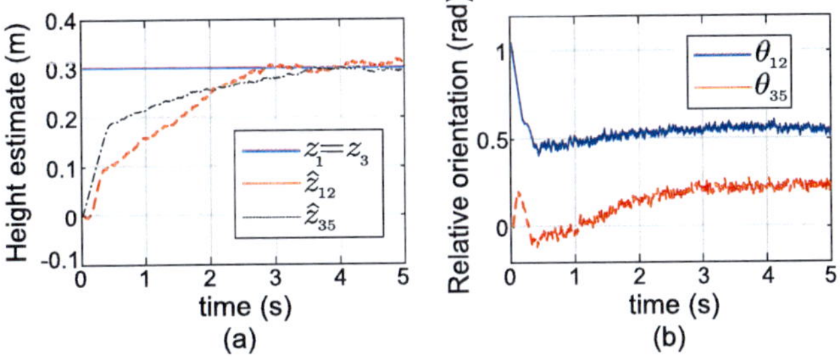

Fig. 3. Time-history trajectories of (a) height estimates $\hat{z}_{12}$ and $\hat{z}_{35}$, and (b) relative robot orientation θ_{12} and θ_{35} obtained using the control law in (16) with sensor output corrupted by noise.

estimates and relative robot orientation are shown in Fig. 3. Again, consistent with the proof of Theorem 2, it is apparent that the height estimates converge to a small bound around the ground truth in finite-time, and the relative orientation states θ_{12} and θ_{35} are bounded, thus ensuring local stability of the orientation dynamics of the robot formation.

5 Conclusion

In this study, a continuous sliding-mode controller is presented for finite-time convergence of leader-follower formation to the desired pattern without relying on measurements/estimation of leader agent's relative position and velocity. Each follower agent is equipped with a perspective camera, and visual measurements of a single tracked feature attached to the leader agent are used to formulate the leader-follower kinematics in the image space of the follower agent. Lyapunov analysis is then used to guarantee global stability of the closed-loop observer-controller system, and for the case where the feature height is assumed unknown, it is shown that the tracking error converges to a small bound around the origin, where the size of this error bound can be reduced for a sufficiently large choice of controller gain. Thus, finite-time formation convergence is achieved without explicit inter-agent communication. Simulation results are used to demonstrate the efficacy of the proposed scheme in the presence of measurement noise.

References

1. Mastellone, S., Stipanovic, D.M., Graunke, C.R., Intlekofer, K.A., Spong, M.W.: Formation control and collision avoidance for multi-agent nonholonomic systems: theory and experiments. Int. J. Robot. Res. **27**(1), 107–126 (2008)
2. Kwon, J.W., Chwa, D.: Hierarchical formation control based on a vector field method for wheeled mobile robots. IEEE Trans. Robot. **28**(6), 1335–1345 (2012)
3. Chen, J., Sun, D., Yang, J., Chen, H.: Leader-follower formation control of multiple non-holonomic mobile robots incorporating a receding- horizon scheme. Int. J. Robot. Res. **29**(6), 727–747 (2010)
4. Wang, P., Ding, B.: Distributed RHC for tracking and formation of nonholonomic multi-vehicle systems. IEEE Trans. Autom. Control **59**(6), 1439–1453 (2014)
5. Wang, W., Huang, J., Wen, C., Fan, H.: Distributed adaptive control for consensus tracking with application to formation control of nonholonomic mobile robots. Automatica **50**(4), 1254–1263 (2014)
6. Oh, K.K., Ahn, H.S.: Formation control of mobile agents based on distributed position estimation. IEEE Trans. Autom. Control **58**(3), 737–742 (2013)
7. Das, A.K., Fierro, R., Kumar, V., Ostrowski, J.P., Spletzer, J., Taylor, C.J.: A vision-based formation control framework. IEEE Trans. Robot. Autom. **18**(5), 813–825 (2002)
8. Mariottini, G.L., et al.: Vision-based localization for leader-follower formation control. IEEE Trans. Robot. **25**(6), 1431–1438 (2009)
9. Liang, X., Liu, Y.H., Wang, H., Chen, W., Xing, K., Liu, T.: Leader-following formation tracking control of mobile robots without direct position measurements. IEEE Trans. Autom. Control **61**(12), 4131–4137 (2016)
10. Panagou, D., Kumar, V.: Cooperative visibility maintenance for leader-follower formations in obstacle environments. IEEE Trans. Robot. **30**(4), 831–844 (2014)
11. Liang, X., Wang, H., Liu, Y.H., Chen, W., Liu, T.: Formation control of nonholonomic mobile robots without position and velocity measurements. IEEE Trans. Robot. **34**(2), 434–446 (2018)
12. Moreno, J.A., Osorio, M.: Strict Lyapunov functions for the super-twisting algorithm. IEEE Trans. Autom. Control **57**(4), 1035–1040 (2012)
13. Edwards, C., Shtessel, Y.: Adaptive dual-layer super-twisting control and observation. Int. J. Control **89**(9), 1759–1766 (2016)

Optimal Control of a 3-link Robot Manipulator Moving to a Prescribed Point

Do Dang Khoa[1](✉), Tran Si Kien[2], Phan Dang Phong[2], and Do Sanh[1]

[1] Hanoi University of Science and Technology, N.1 Dai Co Viet Road, Hanoi, Vietnam
khoa.dodang@hust.edu.vn

[2] National Research Institute of Mechanical Engineering, N.4 Pham Van Dong, Hanoi, Vietnam

Abstract. This paper presents a method to determine a point-to-point trajectory of a 3-dof manipulator while minimizing the control variables. This problem is often studied in many industries. The proposed solution includes the following steps. First, the method of transformation matrix is used to derive the ordinary differential equations (ODE) of motion of the mechanical system. Then, the Hamiltonian function is formed with the functional to minimize the control inputs and fix the end effector's target point. The necessary condition for optimality is derived by the Pontryagin's maximum principle. The obtained equations of two-point boundary value problem are solved to determine the optimal control using Maple software.

Keywords: Pontryagin's maximum principle · Method of transformation matrix · Point-to-point trajectory

1 Introduction

Point-to-point control problems of manipulators have been studied quite early in many industries such as welding, logistics, and mining, etc. In such applications, a manipulator is not only required to move between prescribed points but also needs to meet some optimal criteria such as minimizing time, control forces, power, or maximizing payload carrying [1–3]. One approach to such optimal problem is to discretize or linearize the equations of motion and applies the optimization methods such as linear quadratic regulator (LQR), model predictive control (MPC) [3–4]. Another approach is an open-loop optimal control based on building Hamiltonian function and Pontryaghin's maximum principle. The latter has some advantages in dealing with nonlinear dynamics systems and generating the necessary conditions for optimality analytically as a set of two-point boundary differential equations which is possibly solved numerically [5]. In the paper, a method is proposed to integrate the terminal-point constraint into the Hamilton function as in [6] then the Pontryagin's maximum principle is applied to solve for the minimum of control forces.

N. V. Khang et al. (Eds.): ASIAN MMS 2021, MMS 113, pp. 455–462, 2022.
https://doi.org/10.1007/978-3-030-91892-7_43

2 Problem Statement

Considering a model of a 3-link manipulator as shown in the Fig. 1. Assume the links are straight bar of length l_1, l_2, and l_3, of mass m_1, m_2, and m_3. The link centers of mass C_1, C_2, and C_3 are located at distances c_1, c_2, and c_3 ($c_1 = 0$ due to C_1 coincides with its rotation axis, $c_2 = AC_2$, $c_3 = BC_3$). The link moments of inertia with respect to the axis through its center of mass are J_1, J_2, and J_3, respectively. The generalized coordinates are q_1, q_2, q_3, respectively. The end effector is required to move a payload of mass m to a prescribed point.

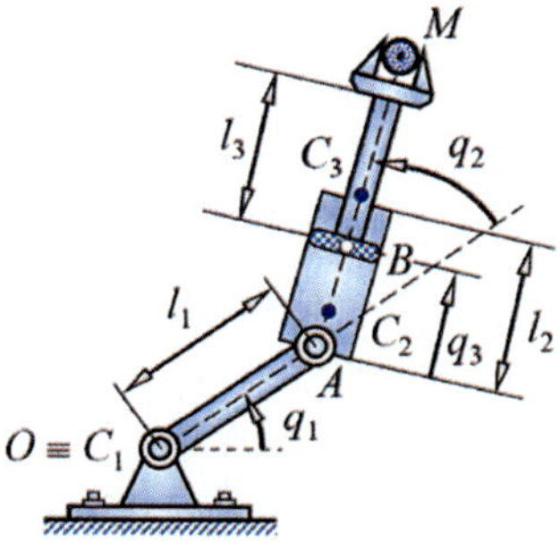

Fig. 1. Three-link robot manipulator

3 Dynamics Equations of the Manipulator

The problem is to determine the motor torques on each link (the control variables) for the end effector (point M) moving from the start point M_0, corresponding to the time t_0, to the end point M_f corresponding to the time t_f. From now on, matrices are written in bold letters and a vector is considered as a matrix (3 × 1), symbol T in the upper right corner of a character indicating matrix transposition. The motion equations of the manipulator have the following form [7, 8]

$$\mathbf{A}\ddot{\mathbf{q}} = \mathbf{Q} + \mathbf{Q}^{qt} \tag{1}$$

Where A is the inertial matrix of size (3 × 3), matrix **A** is symmetric and positive definite:

$$\mathbf{A} = \begin{bmatrix} a_{11} & a_{12} & a_{13} \\ a_{12} & a_{22} & a_{23} \\ a_{13} & a_{23} & a_{33} \end{bmatrix} \tag{2}$$

Q: the (3 × 1) matrix of generalized forces of the active forces.
$\mathbf{Q}^{qt}$: the (3 × 1) matrix of generalized forces of the inertial forces.
$\ddot{\mathbf{q}}$: the (3 × 1) matrix of generalized accelerations.

To calculate the quantities in Eq. (1), it is necessary to set up the following transformation matrices

$$t_1 = \begin{bmatrix} \cos(q_1) & -\sin(q_1) & 0 \\ \sin(q_1) & \cos(q_1) & 0 \\ 0 & 0 & 1 \end{bmatrix};\ t_2 = \begin{bmatrix} \cos(q_2) & -\sin(q_2) & l_1 \\ \sin(q_2) & \cos(q_2) & 0 \\ 0 & 0 & 1 \end{bmatrix};\ t_3 = \begin{bmatrix} 1 & 0 & q_3 \\ 0 & 1 & 0 \\ 0 & 0 & 1 \end{bmatrix};$$

$$t_{11} = \begin{bmatrix} -\sin(q_1) & -\cos(q_1) & 0 \\ \cos q_1) & -\sin(q_1) & 0 \\ 0 & 0 & 0 \end{bmatrix};\ t_{21} = \begin{bmatrix} -\sin(q_2) & -\cos(q_2) & 0 \\ \cos q_2) & -\sin(q_2) & 0 \\ 0 & 0 & 0 \end{bmatrix};\ t_{31} = \begin{bmatrix} 0 & 0 & 1 \\ 0 & 0 & 0 \\ 0 & 0 & 0 \end{bmatrix};$$

$$r_2 = \left[c_2\ 0\ 1\right]^T;\ r_3 = \left[c_3\ 0\ 1\right]^T;\ r = \left[l_3\ 0\ 1\right]^T$$
$$\dot{\mathbf{q}} = \left[\dot{q}_1\ \dot{q}_2\ \dot{q}_3\right]^T;\quad \ddot{\mathbf{q}} = \left[\ddot{q}_1\ \ddot{q}_2\ \ddot{q}_3\right]^T \tag{3}$$

The components of the inertial matrix **A** are calculated as follows:

$$\begin{aligned}
a_{11} &= J_1 + J_2 + J_3 + m_2 r_2^T t_2^T t_{11}^T t_{11} t_2 r_2 + m_3 r_3^T t_3^T t_2^T t_{11}^T t_{11} t_2 t_3 r_3 + m r t_3^T t_2^T t_{11}^T t_{11} t_2 t_3 r \\
&= m_2(l_1^2 + c_2^2 + 2c_2 l_1 \cos(q_2)) + m_3(l_1^2 + l_3^2 + 2l_3 q_3 + q_3^2 + 2q_3 l_1 \cos(q_2) \\
&+ 2l_1 q_3 \cos(q_2)) \\
&+ m(l_1^2 + l_3^2 + 2l_3 q_3 + q_3^2 - 2l_3 l_1 \cos(q_2) + 2l_1 q_3 \cos(q_2)) + J_1 + J_2 + J_3; \\
a_{12} &= J_1 + J_2 + m_2 r_2^T t_{21}^T t_1^T t_{11} t_2 r_2 + m_3 r_3^T t_3^T t_{21}^T t_1^T t_{11} t_2 t_3 r_3 + m r t_3^T t_{21}^T t_1^T t_{11} t_2 t_3 r \\
&= J_1 + J_2 + m_2(c_2^2 + l_1 c_2 \cos(q_2)) + m_3(c_3^2 + 2c3q3 + q_3^2 + l_1 c_3 \cos(q_2) \\
&+ l_1 q_3 \cos(q_2) \\
&+ m_3(l_3^2 + 2l3q3 + q_3^2 + l_1 l_3 \cos(q_2) + l_1 q_3 \cos(q_2)); \\
a_{13} &= m_3 r_3^T t_{31}^T t_2^T t_1^T t_{11} t_2 t_3 r_3 + m r t_{31}^T t_2^T t_1^T t_{11} t_2 t_3 r = -(m_3 + m) l_1 \sin(q_2); \\
a_{22} &= J_2 + J_3 + m_2 r_2^T t_{21}^T t_1^T t_1 t_{21} r_2 + m_3 r_3^T t_3^T t_{21}^T t_1^T t_1 t_{21} t_3 r_3 + m r t_3^T t_{21}^T t_1^T t_1 t_{21} t_3 r \\
&= J_2 + J_3 + m_2 c_2^2 + m_3(c_3^2 + 2c_3 q_3 + q_3^2) + m(l_3^2 + 2l_3 q_3 + q_3^2); \\
a_{23} &= 0;\quad a_{33} = m + m_3
\end{aligned} \tag{4}$$

The end effector's coordinates have the expression as:

$$\mathbf{r}_C = \mathbf{t}_1 \mathbf{t}_2 \mathbf{t}_3 \mathbf{r} \tag{5}$$

The generalized coordinates corresponding to the time t_0 and t_f of the end effector M are:

$$\begin{aligned}
x_M(t_0) &= (l_3 + q_{30}) \cos(q_{10} + q_{20}) + l_1 \cos(q_{10}) \\
y_M(t_0) &= (l_3 + q_{30}) \sin(q_{10} + q_{20}) + l_1 sin(q_{10}); \\
x_M(t_f) &= (l_3 + q_{3f}) \cos(q_{1f} + q_{2f}) + l_1 \cos(q_{1f}) \\
y_M(t_f) &= (l_3 + q_{30}) \sin(q_{1f} + q_{2f}) + l_1 sin(q_{1f});
\end{aligned} \tag{6}$$

where: $q_{i0} \equiv q_i(t_0) : q_{if} \equiv q_i(t_f); i = \overline{1,3}$ and $x_M(t_f) \equiv x_f; y_M(t_f) \equiv y_f$.

The problem is to determine the control variables u_1, u_2, and u_3 of motors attached in each link for the payload M moving from the start point $x_M(t_0)$, $y_M(t_0)$ to the prescribed end point M_f (x_f, y_f).

The generalized forces of active forces $\mathbf{Q}$ is represented by a matrix (3×1) including the controlling moments u_1, u_2, and u_3. The conservative forces are calculated from the potential function π (neglect the viscosity friction forces)

$$\pi = \left\{g[(m_2c_2 + m_3(c_3 + q_3) + m(l_3 + q_3))\sin(q_1 + q_2) + (m + m_2 + m_3)l_1\sin(q_1)]\right\} \tag{7}$$

The motor torques have the following form

$$M_1 = u_1 - k_1\dot{q}_1\,; \quad M_2 = u_2 - k_2\dot{q}_2\,; \quad F = u_3 - k_3\dot{q}_3 \tag{8}$$

The coefficients $k_i; i = \overline{1,3}$ are the curvature of the motor static curves. The generalized forces of active forces are:

$$\begin{aligned}
Q_1 &= u_1 - k_1\dot{q}_1 - \frac{\partial\pi}{\partial q_1} = u_1 - k_1\dot{q}_1 \\
&\quad + \{(m_2 + m_3 + m)l_1\cos(q_1) + [m_2c_2 + m_3(q_3 + c_3) + m(q_3 + l_3)]\cos(q_1 + q_2)\}g; \\
Q_2 &= u_2 - k_2\dot{q}_2 - \frac{\partial\pi}{\partial q_2} = u_2 - k_2\dot{q}_2 + \{[m_2c_2 + m_3(c_3 + q_3) + m(l_3 + q_3)]\cos(q_1 + q_2)\}g; \\
Q_3 &= u_3 - k_3\dot{q}_3 - \frac{\partial\pi}{\partial q_3} = u_3 - k_3\dot{q}_3 - g(m_3 + m)\sin(q_1 + q_2)
\end{aligned} \tag{9}$$

The calculation of generalized forces of inertial forces $\mathbf{Q}^{qt}$ is based on the following matrices:

$$\partial_i\mathbf{A} = \begin{bmatrix} \frac{\partial a_{11}}{\partial q_i} & \frac{\partial a_{12}}{\partial q_i} & \frac{\partial a_{13}}{\partial q_i} \\ \frac{\partial a_{12}}{\partial q_i} & \frac{\partial a_{22}}{\partial q_i} & \frac{\partial a_{23}}{\partial q_i} \\ \frac{\partial a_{13}}{\partial q_i} & \frac{\partial a_{23}}{\partial q_i} & \frac{\partial a_{33}}{\partial q_i} \end{bmatrix} \quad (i = \overline{1,3})\,; \quad \bar{\dot{\mathbf{q}}}_i = \begin{bmatrix} \dot{q}_1\dot{q}_i \\ \dot{q}_2\dot{q}_i \\ \dot{q}_3\dot{q}_i \end{bmatrix}; \dot{\mathbf{q}} = \begin{bmatrix} \dot{q}_1 \\ \dot{q}_2 \\ \dot{q}_3 \end{bmatrix} \tag{10}$$

The inertial force $\mathbf{Q}^{qt}$ is written in the form:

$$\begin{aligned}
\mathbf{Q}^{qt} &= \mathbf{Q}^{qt1} - \mathbf{Q}^{qt2} \\
\mathbf{Q}^{qt1} &= \left[Q_1^{qt1}\ Q_2^{qt1}\ Q_3^{qt1}\right]^T; \quad Q_i^{qt1} = 0.5\dot{\mathbf{q}}^T\,\partial_i\mathbf{A}\dot{\mathbf{q}} \quad (i = \overline{1,3}); \\
\mathbf{Q}^{qt2} &= \sum_{i=1}^{3}\partial_i\mathbf{A}\bar{\dot{\mathbf{q}}}_i
\end{aligned} \tag{11}$$

The dynamic equations of the manipulator are:

$$\begin{aligned}
eq1 &= a_{11}\ddot{q}_1 + a_{12}\ddot{q}_2 + a_{13}\ddot{q}_3 - u_1 + k_1\dot{q}_1 + \frac{\partial\pi}{\partial q_1} - Q_1^{qt1} + Q_1^{qt2}; \\
eq2 &= a_{12}\ddot{q}_1 + a_{22}\ddot{q}_2 + a_{23}\ddot{q}_3 - u_2 + k_2\dot{q}_2 + \frac{\partial\pi}{\partial q_1} - Q_2^{qt1} + Q_2^{qt2};
\end{aligned}$$

$$eq3 = a_{13}\ddot{q}_1 + a_{23}\ddot{q}_2 + a_{33}\ddot{q}_3 - u_3 + k_3\dot{q}_3 + \frac{\partial \pi}{\partial q_3} - Q_3^{qt1} + Q_3^{qt2}; \quad (12)$$

Equation (12) will be used to compute the accelerations $\ddot{q}_1, \ddot{q}_2$, and $\ddot{q}_3$ as a function of the quantities $(q_1, q_2, q_3, \dot{q}_1, \dot{q}_2, \dot{q}_3, u_1, u_2, u_3, \ldots)$ and thereby converting the Lagrange variable set $\{q_i, \dot{q}_i\}$; $i = \overline{1, n}$ to a conjugate variable set $\{q_j, p_j\} : j = \overline{1, 2n}$.

4 Optimal Control Problem

To find the control variables u_1, u_2, and u_3 for the motion of the manipulator end effector moving to a prescribed point (x_f, y_f), the following optimal condition is proposed as a target function

$$I = \int_0^{t_f} f_0 dt \to \min \quad (13)$$

The functional I is the optimization criteria, where f_0 is a function of control variables (u_1, u_2, u_3) and $\{q_j\}$; $j = \overline{1, 6}$. According to the Pontryagin's maximum principle [5] the motion of the system is described by the conjugate variables $\{q_1, q_2, q_3, q_4, q_5, q_6, p_1, p_2, p_3, p_4, p_5, p_6\}$. A part of the conjugate variables obtained from the Lagrange variables q_1, q_2, and q_3; the variables q_4, q_5, and q_6 are converted from generalized velocity variables ($\dot{q}_1 = q_4, \dot{q}_2 = q_5, \dot{q}_3 = q_6$), the variables $p_j (j = \overline{1, 6})$ are determined from the function Hamilton H:

$$H = -f_0 + p_1 f_1 + p_2 f_2 + p_3 f_3 + p_4 f_4 + p_5 f_5 + p_6 f_6 \quad (14)$$

Where:

$$f_1 = q_4;\ f_2 = q_5, f_3 = q_6;\ f_4 = \ddot{q}_1;\ f_5 = \ddot{q}_2;\ f_6 = \ddot{q}_3 \quad (15)$$

The function f_0 is chosen to minimize the control variables as well as to fix the terminal point as follows

$$f_0 = 0.5\left\{(u_1 - k_1 q_4)^2 + (u_2 - k_2 q_5)^2 + (u_3 - k_3 q_6)^2 + (x_M - x_f)^2 + (y_M - y_f)^2\right\}; \quad (16)$$

In general, it can be a function of conjugate variables $\{q_1, q_2, q_3, q_4, q_5, q_6\}$. According to the Pontryagin's maximum principle, for the functional (12) to reach minimum, the control variables u_1, u_2, and u_3 are selected from the maximum condition of the H function [9, 10]. Control variables are selected from the condition:

$$\frac{\partial H}{\partial u_1} = 0; \quad \frac{\partial H}{\partial u_2} = 0; \quad \frac{\partial H}{\partial u_3} = 0; \quad (17)$$

Control variables u_1, u_2, and u_3 solved from (15) are replaced in

$$\frac{dq_1}{dt} = f_1;\ \frac{dq_2}{dt} = f_2;\ \frac{dq_3}{dt} = f_3;\ \frac{dq_4}{dt} = f_4;\ \frac{dq_5}{dt} = f_5;\ \frac{dq_6}{dt} = f_6;\ \frac{dz}{dt} = f_0$$

$$\frac{dp_1}{dt} = -\frac{\partial H}{\partial q_1}; \frac{dp_2}{dt} = -\frac{\partial H}{\partial q_2}; \frac{dp_3}{dt} = -\frac{\partial H}{\partial q_3}; \frac{dp_4}{dt} = -\frac{\partial H}{\partial q_4}; \frac{dp_5}{dt} = -\frac{\partial H}{\partial q_5}; \frac{dp_6}{dt} = -\frac{\partial H}{\partial q_6} \quad (18)$$

with boundary conditions:

$$\left\{\begin{matrix} q_1(0) = 0, q_2(0) = 0.2, q_3(0) = 0, q_1(t_f) = 1, q_2(t_f) = 0.5, q_3(t_f) = 0.5, \\ p_1(t_f) = h_{1f}, p_2(t_f) = h_{2f}, p_3(t_f) = h_{3f}, p_4(t_f) = 0, p_5(t_f) = 0, p_6(t_f) = 0, z(0) = 0 \end{matrix}\right\} \quad (19)$$

will minimize the functional z(t) as

$$z = \int_0^t f_0 dt \quad (20)$$

and t_f is the movement time of the manipulator. The variables h_{1f}, h_{2f}, h_{3f} are:

$$h_{if} = \frac{\partial G}{\partial q_j}; \quad j = \overline{1,3}; \quad (21)$$

Where:

$$G \equiv 0.5[(x_M - x_f)^2 + (y_M - y_f)^2] \quad (22)$$

Simulation Results

The two-point boundary value Eqs. (18) is solved by Maple software with the following data:

$l_1 = 1\ (m),\ l_3 = 2l_1,\ c_1 = 0\ (m),\ c_2 = 0.75\ (m),\ c_3 = 0.5l_3,\ m = 10\ (kg),\ m_2 = 1\ (kg),$
$m_3 = 1\ (kg),$
$J_1 = 25\ \left(kgm^2\right),\ J_2 = 25\ \left(kgm^2\right),\ J_3 = 2.5\ \left(kgm^2\right),\ k_1 = 0.5,\ k_2 = 0.5,\ k_3 = 0.25,$
$t_f = 1.5(s)$

The end effector terminal point: $M_f\{x_f = 0.71714531;\ y_f = 3.33520845\}\ (m)$. The numerical simulation of the problem is shown in the following figures (Figs. 2, 3, 5, 6):

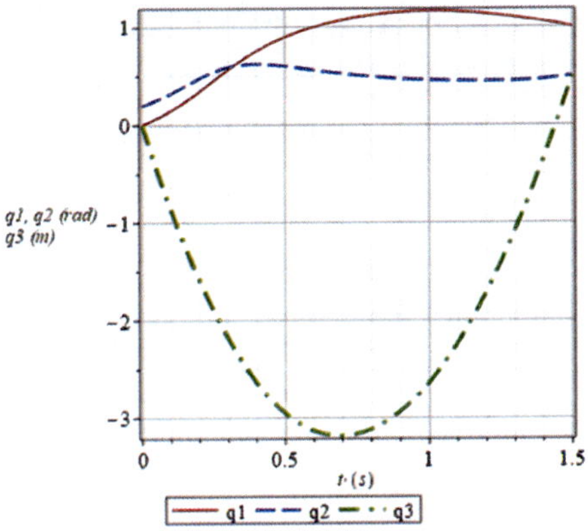

Fig. 2. Generalized coordinates

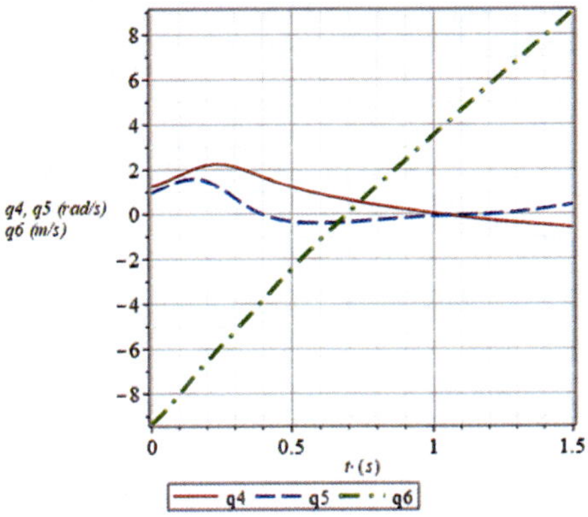

Fig. 3. Generalized velocities

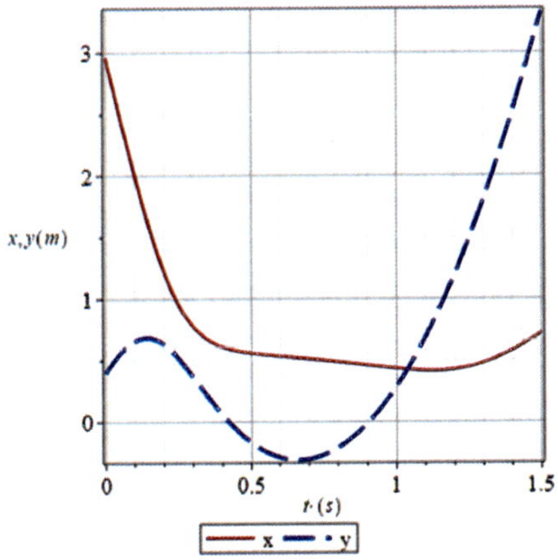

Fig. 4. The position of end effector

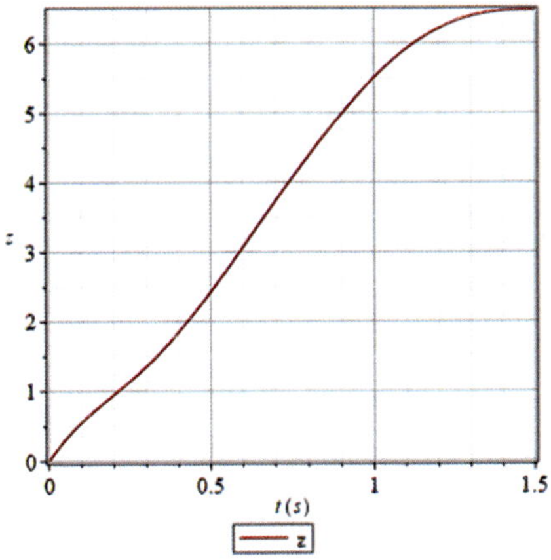

Fig. 5. The optimal criteria

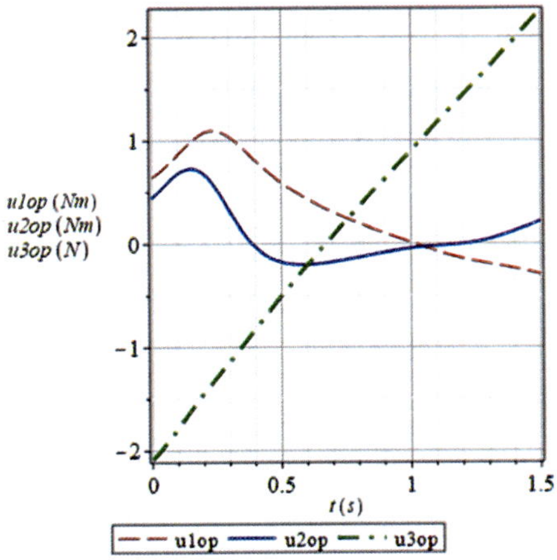

Fig. 6. The optimal control u_1, u_2, u_3

As shown in the Fig. 4, the end effector moves from the start point M_0 (2.95, 0.45) to the terminal point M_f (0.717, 3.335) in 1.5 s as planned. The cost function is also kept at low value. The simulation results prove the credibility of the proposed method.

5 Conclusion

In the paper, the optimal control problem of a manipulator whose end effector moves to a target point is studied. The challenging issue is to find the minimum control forces to drive the end effector to a fixed terminal point for a specific time. Another challenging problem is to determine the boundary and initial conditions of conjugate variables of the Hamiltonian function. To overcome those difficulties, the paper has proposed a modified cost function combined the minimum effort and terminal point constraints and the boundary conditions in (21) and (22). The simulation results prove the proposed method suitable to solve the optimal control problem for nonlinear systems.

References

1. Korayem, M.H., Nikoobin, A.: Maximum payload path planning for redundant manipulator using indirect solution of optimal control problem. Int. J. Adv. Manuf. Technol. **44**, 725 (2009)
2. Korayem, M.H., Nohooji, H.R., Nikoobin, A.: Optimal motion generating of VV nonholonomic manipulators with elastic revolute joints in VVVVV generalized point-to-point task. Int. J. Adv. Des. Manuf. Technol. 3 (2010)
3. Saraf, P., Ponnalagu, R.N.: Modeling and Simulation of a Point to Point Spherical Articulated Manipulator using Optimal Control (2020)
4. Garimella, G., Kobilarov, M.: Towards model-predictive control for aerial pick-and-place. In: 2015 IEEE International Conference on Robotics and Automation (ICRA). IEEE (2015)
5. Pontryagin, L.S., et al.: The Mathematica Theory of Optimal Processes, trans. by Trirogoff, K., Interscience Publishers, John Wiley and Sons. Inc., NewYork (1962)
6. Shiller, Z.: Time-energy optimal control of articulated systems with geometric path constraints. In: Proceedings of the 1994 IEEE International Conference on Robotics and Automation, vol. 4, pp. 2680–2685 (1994)
7. Sanh, D., Do, D.K.: Method of transmission matrix applying for investigation of motion of planar mechanism. Mach. Dyn. Res. **34**(4), 5–22. Varsaw (2010)
8. Do Sanh, D.V.P., Do Dang Khoa, T.D.: A method for solving the motion of constrained systems. In: Proceedings of the 16th Asian Pacific Vibration Conference, pp. 803–811, Hanoi, Vietnam (2015)
9. Sanh, D., Do, D.K.: Optimal Control of Dynamics Systems-Program Motion and Optimal Control. Bach Khoa Publisher, Vietnam (2010)
10. Sanh, D., Do, D.K., Phong, P.D.: Control of Machine Dynamics and Robotics, Bach Khoa Publisher, Vietnam (2020)

Optimal Control of a 3-DOF Robot Manipulator Subject to a Prescribed Trajectory

Do Dang Khoa[1(✉)], Tran Si Kien[2], Phan Dang Phong[2], and Do Sanh[1]

[1] Hanoi University of Science and Technology, N.1 Dai Co Viet Road, Hanoi, Vietnam
khoa.dodang@hust.edu.vn

[2] National Research Institute of Mechanical Engineering, N.4 Pham Van Dong, Hanoi, Vietnam

Abstract. This paper presents a method to solve an optimal control problem of a 3-DOF manipulator whose end effector is required to move along a straight line while minimizing the control variables. This issue belongs to an optimal control problem of constrained mechanical systems. The proposed solution includes the following steps. First, the principle of compatibility and the method of transformation matrix is used to derive the equations of motion of the constrained mechanical system in the form of ordinary differential equations (ODEs). Then, the Pontryagin's maximum principle is applied to determine the optimal control to minimize the control forces. The obtained equations establish a two-point boundary value problem which is solved by Maple software.

Keywords: Pontryagin's maximum principle · Principle of compatibility · Method of transformation matrix

1 Introduction

Optimal control problems in manipulator motion have been studied quite early in many industries such as palletizing, transportation, mining, etc. In such problems, a manipulator is not only required to follow a prescribed path but also needs to meet some optimal criteria, for example, energy or time criteria [1–3]. Optimal control problems of mechanical systems are usually approached by direct and indirect methods. The indirect methods based on Pontryagin's maximum principle have some advantages in establishing necessary conditions for optimality analytically as a set of two-point boundary differential equations which is possibly solved numerically [4]. When dealing optimization with constraints, a general method is usually to integrate the constraints into the Hamilton function as in [2] or to build the constrained equations of motion with Lagrange multipliers that often leads to the differential algebraic equations (DAEs) [3]. In this paper, a method to derive ODEs of motion of constrained dynamics system using the principle of compatibility and transformation matrix method is utilized [5–8]. Then the Pontryagin's maximum principle is applied to solve the energy optimal control issue [4].

N. V. Khang et al. (Eds.): ASIAN MMS 2021, MMS 113, pp. 463–471, 2022.
https://doi.org/10.1007/978-3-030-91892-7_44

2 Problem Statement

Considering a model of 3-link manipulator as shown in Fig. 1. Assume the links are straight bar of length l_1, l_2, and l_3, of mass m_1, m_2, and m_3. The link centers of mass C_1, C_2, and C_3 are located at distances c_1, c_2, and c_3 ($c_1 = 0$ due to C_1 coincides with its rotation axis, $c_2 = AC_2$, $c_3 = BC_3$). The link moments of inertia about its centers of mass are J_1, J_2, and J_3, respectively. The generalized coordinates are q_1, q_2, and q_3, respectively. The elasticity of two revolute joints and one prismatic joint are modeled as 2 torsion springs of stiffness coefficients k_1, and k_2 and a linear spring of stiffness coefficient k_3.

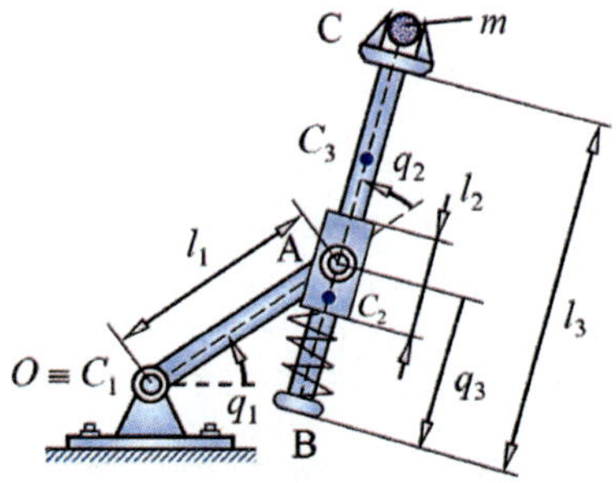

Fig. 1. Three-link robot manipulator

The end effector C of the last link is required to move a payload of mass m along a straight line described by an equation

$$G \equiv y_C + ax_C - b = 0 \tag{1}$$

The problem is to determine the forces acting on the links for Eq. (1) to be realized while satisfying an optimal condition. This problem belongs to the program motion type [10]. The 3-DOF manipulator configuration is described by three generalized coordinates q_1, q_2, and q_3. However, with the geometric constraint (1), the system is reduced to only two degrees of freedom and therefore only two independent control variables are chosen. The control variables u_1 and u_2 which are the motor torques on link 1 and link 2 are selected. The forces exerting on the translational link need to be preselected with the value **F** (propulsive force). The optimal condition is defined as

$$I = \int_0^{t_f} f_0 dt \rightarrow \min \tag{2}$$

The functional I is the optimization criteria, where f_0 is a function of the control variables u_1 and u_2. Thus, the problem is to determine the control variables $u_1(t)$, $u_2(t)$ for the end effector to fulfill the conditions (1) and (2). Two steps will be proceeded as follows: - first study the program motion of the manipulator and then to solve the optimal control problem [10, 11].

3 Equations of Motion of the Constrained System

In this section, the principle of compatibility and method of transformation matrix is applied to build the ordinary differential motion equations of a constrained mechanical system. From now on, matrices are written in bold letters and a vector is considered as a matrix (3 × 1), symbol T in the upper right corner of a character indicating matrix transposition. The motion equation of the manipulator has the following form [10, 11]

$$\mathbf{DA}\ddot{\mathbf{q}} = \mathbf{D}(\mathbf{Q} + \mathbf{Q}^{qt}) \tag{3}$$

Where **A** is the inertial matrix of size (3 × 3), matrix **A** is symmetric and positive definite:

$$\mathbf{A} = \begin{bmatrix} a_{11} & a_{12} & a_{13} \\ a_{12} & a_{22} & a_{23} \\ a_{13} & a_{23} & a_{33} \end{bmatrix} \tag{4}$$

Q, and $\mathbf{Q}^{qt}$ - the matrix of generalized forces of the active and inertial forces.

$\ddot{\mathbf{q}}$ - the matrix of generalized accelerations.

D - the matrix obtained when representing all the generalized velocities in terms of independent generalized velocities.

To calculate the quantities in Eq. (4), it is necessary to set up the following transformation matrices [8]:

$$t_1 = \begin{bmatrix} \cos(q_1) & -\sin(q_1) & 0 \\ \sin(q_1) & \cos(q_1) & 0 \\ 0 & 0 & 1 \end{bmatrix};\ t_2 = \begin{bmatrix} \cos(q_2) & -\sin(q_2) & l_1 \\ \sin(q_2) & \cos(q_2) & 0 \\ 0 & 0 & 1 \end{bmatrix};\ t_3 = \begin{bmatrix} 1 & 0 & -q_3 \\ 0 & 1 & 0 \\ 0 & 0 & 1 \end{bmatrix};$$

$$t_{11} = \begin{bmatrix} -\sin(q_1) & -\cos(q_1) & 0 \\ \cos q_1) & -\sin(q_1) & 0 \\ 0 & 0 & 0 \end{bmatrix};\ t_{21} = \begin{bmatrix} -\sin(q_2) & -\cos(q_2) & 0 \\ \cos q_2) & -\sin(q_2) & 0 \\ 0 & 0 & 0 \end{bmatrix};\ t_{31} = \begin{bmatrix} 0 & 0 & -1 \\ 0 & 0 & 0 \\ 0 & 0 & 0 \end{bmatrix};$$

$$r_2 = \begin{bmatrix} c_2 & 0 & 1 \end{bmatrix}^T;\ r_3 = \begin{bmatrix} c_3 & 0 & 1 \end{bmatrix}^T;\ r = \begin{bmatrix} l_3 & 0 & 1 \end{bmatrix}^T;$$

$$\dot{\mathbf{q}} = \begin{bmatrix} \dot{q}_1 & \dot{q}_2 & \dot{q}_3 \end{bmatrix}^T;\quad \ddot{\mathbf{q}} = \begin{bmatrix} \ddot{q}_1 & \ddot{q}_2 & \ddot{q}_3 \end{bmatrix}^T \tag{5}$$

The components of the inertial matrix **A** are calculated as follows:

$$\begin{aligned} a_{11} &= J_1 + J_2 + J_3 + m_2 r_2^T t_2^T t_{11}^T t_{11} t_2 r_2 + m_3 r_3^T t_3^T t_2^T t_{11}^T t_{11} t_2 t_3 r_3 + m r t_3^T t_2^T t_{11}^T t_{11} t_2 t_3 r \\ &= m_2(l_1^2 + c_2^2 + 2c_2 l_1 \cos(q_2)) + m_3(l_1^2 + l_3^2 - 2l_3 q_3 + q_3^2 - 2q_3 l_1 \cos(q_2) \\ &- 2l_1 q_3 \cos(q_2)) \\ &+ m(l_1^2 + l_3^2 - 2l_3 q_3 + q_3^2 - 2l_3 l_1 \cos(q_2) - 2l_1 q_3 \cos(q_2)) + J_1 + J_2 + J_3; \\ a_{12} &= J_2 + m_2 r_2^T t_{21}^T t_1^T t_{11} t_2 r_2 + m_3 r_3^T t_3^T t_{21}^T t_1^T t_{11} t_2 t_3 r_3 + m r t_3^T t_{21}^T t_1^T t_{11} t_2 t_3 r \\ &= J_2 + m_2(c_2^2 + l_1 c_2 \cos(q_2)) + m_3(c_3^2 - 2c_3 q_3 + q_3^2 + l_1 c_3 \cos(q_2) \\ &- l_1 q_3 \cos(q_2) \\ &+ m_3(l_3^2 - 2l_3 q_3 + q_3^2 + l_1 l_3 \cos(q_2) - l_1 q_3 \cos(q_2)); \end{aligned}$$

$$a_{13} = m_3 r_3^T t_{31}^T t_2^T t_1^T t_{11} t_2 t_3 r_3 + m r t_{31}^T t_2^T t_1^T t_{11} t_2 t_3 r = -(m_3 + m) l_1 \sin(q_2);$$
$$a_{22} = J_2 + J_3 + m_2 r_2^T t_{21}^T t_1^T t_1 t_{21} r_2 + m_3 r_3^T t_3^T t_{21}^T t_1^T t_1 t_{21} t_3 r_3 + m r t_3^T t_{21}^T t_1^T t_1 t_{21} t_3 r$$
$$= J_2 + J_3 + m_2 c_2^2 + m_3 (c_3^2 - 2 c_3 q_3 + q_3^2) + m(l_3^2 - 2 c_3 q_3 + q_3^2);$$
$$a_{23} = 0; \; a_{33} = m + m_3 \tag{6}$$

To calculate the matrix **D**, let's write the constraint Eq. (1) in the form of generalized coordinates. The end effector's coordinates have the expression as

$$\mathbf{r}_C = \mathbf{t}_1 \mathbf{t}_2 \mathbf{t}_3 \mathbf{r} \tag{7}$$

Rewrite Eq. (7) in the Cartesian coordinates, we have:

$$x_C = (l_3 - q_3)\cos(q_1 + q_2) + l_1 \cos(q_1)$$
$$y_C = (l_3 - q_3)\sin(q_1 + q_2) + l_1 sin(q_1) \tag{8}$$

The constraint Eq. (1) is written in generalized coordinates:

$$G \equiv a(l_3 - q_3)\cos(q_1 + q_2) + (l_3 - q_3)\sin(q_1 + q_2) + l_1(a\cos(q_1) + \sin(q_1)) - b; \tag{9}$$

Let's calculate the following quantities:

$$h_1 = \frac{\partial G}{\partial q_1} = a(l_3 - q_3)\sin(q_1 + q_2) + (l_3 - q_3)\cos(q_1 + q_2)$$
$$+ l_1(\cos(q_1) - a\sin(q_1));$$
$$h_2 = \frac{\partial G}{\partial q_2} = (l_3 - q_3)(\cos(q_1 + q_2) - a\sin(q_1 + q_2));$$
$$h_3 = \frac{\partial G}{\partial q_3} = -(a\cos(q_1 + q_2) + \sin(q_1 + q_2)); \tag{10}$$

The matrix (2 × 3) **D** will take the form:

$$\mathbf{D} = \begin{bmatrix} 1 \; 0 - \frac{h_1}{h_3} \\ 0 \; 1 - \frac{h_2}{h_3} \end{bmatrix} \tag{11}$$

The generalized forces of active forces **Q** is represented by a matrix (3 × 1), include the control torques u_1 and u_2, propulsive force F and the conservative forces calculated from the potential function π

$$\pi = \begin{Bmatrix} g[(m_2 c_2 + m_3(c_3 - q_3) + m(l_3 - q_3))\sin(q_1 + q_2) + (m + m_2 + m_3) l_1 \sin(q_1)] \\ +0.5(k_1 q_1^2 + k_2 q_2^2 + k_3(q_3 - l_0)^2) \end{Bmatrix}$$

$$Q_1 = u_1 - \frac{\partial \pi}{\partial q_1} = u_1 - k_1 q_1$$
$$+ \{(m_2 + m_3 + m) l_1 \cos(q_1) + [m_2 c_2 + m_3(q_3 - c_3) + m(q_3 - l_3)]\cos(q_1 + q_2)\} g;$$

$$Q_2 = u_2 - \frac{\partial \pi}{\partial q_2} = u_2 - k_2 q_2 + \{[m_2 c_2 + m_3(c_3 - q_3) + m(l_3 - q_3)]\cos(q_1 + q_2)\}g;$$

$$Q_3 = F - \frac{\partial \pi}{\partial q_3} = F - k_3(q_3 - l_0) - g(m_3 + m)\sin(q_1 + q_2) \tag{12}$$

The calculation of generalized forces of inertial forces $\mathbf{Q}^{qt}$ is based on the following matrices:

$$\partial_i \mathbf{A} = \begin{bmatrix} \frac{\partial a_{11}}{\partial q_i} & \frac{\partial a_{12}}{\partial q_i} & \frac{\partial a_{13}}{\partial q_i} \\ \frac{\partial a_{12}}{\partial q_i} & \frac{\partial a_{22}}{\partial q_i} & \frac{\partial a_{23}}{\partial q_i} \\ \frac{\partial a_{13}}{\partial q_i} & \frac{\partial a_{23}}{\partial q_i} & \frac{\partial a_{33}}{\partial q_i} \end{bmatrix} \quad (i = \overline{1,3}); \quad \dot{\bar{\mathbf{q}}}_i = \begin{bmatrix} \dot{q}_1 \dot{q}_i \\ \dot{q}_2 \dot{q}_i \\ \dot{q}_3 \dot{q}_i \end{bmatrix}; \tag{13}$$

The inertial force $\mathbf{Q}^{qt}$ is written in the form:

$$\mathbf{Q}^{qt} = \mathbf{Q}^{qt1} - \mathbf{Q}^{qt2}$$

$$\mathbf{Q}^{qt1} = \left[Q_1^{qt1} \; Q_2^{qt1} \; Q_3^{qt1} \right]^T; \quad Q_i^{qt1} = 0.5\dot{\mathbf{q}}^T \partial_i \mathbf{A} \dot{\mathbf{q}} \quad (i = \overline{1,3});$$

$$\mathbf{Q}^{qt2} = \sum_{i=1}^{3} \partial_i \mathbf{A} \dot{\bar{\mathbf{q}}}_i \tag{14}$$

Using the following expression:

$$eq1 = a_{11}\ddot{q}_1 + a_{12}\ddot{q}_2 + a_{13}\ddot{q}_3 - u_1 + \frac{\partial \pi}{\partial q_1} - Q_1^{qt1} + Q_1^{qt2};$$

$$eq2 = a_{12}\ddot{q}_1 + a_{22}\ddot{q}_2 + a_{23}\ddot{q}_3 - u_2 + \frac{\partial \pi}{\partial q_1} - Q_2^{qt1} + Q_2^{qt2};$$

$$eq3 = a_{13}\ddot{q}_1 + a_{23}\ddot{q}_2 + a_{33}\ddot{q}_3 - F + \frac{\partial \pi}{\partial q_3} - Q_3^{qt1} + Q_3^{qt2}; \tag{15}$$

Equation (3) is written as follows:

$$EQ1 \equiv h_3 * eq1 - h_1 * eq3 = 0;$$
$$EQ2 \equiv h_3 * eq1 - h_2 * eq3 = 0; \tag{16}$$

Take the second derivative of Eq. (9) we have:

$$\begin{aligned} EQ3 \equiv\ & h_1\ddot{q}_1 + h_2\ddot{q}_2 + h_3\ddot{q}_3 + \\ & + \left(\frac{\partial h1}{\partial q_1}\dot{q}_1 + \frac{\partial h_1}{\partial q_2}\dot{q}_2 + \frac{\partial h_1}{\partial q_3}\dot{q}_3\right)\dot{q}_1 + \left(\frac{\partial h_2}{\partial q_1}\dot{q}_1 + \frac{\partial h_2}{\partial q_2}\dot{q}_2 + \frac{\partial h_2}{\partial q_3}\dot{q}_3\right)\dot{q}_2 \\ & + \left(\frac{\partial h_3}{\partial q_1}\dot{q}_1 + \frac{\partial h_3}{\partial q_2}\dot{q}_2 + \frac{\partial h_3}{\partial q_3}\dot{q}_3\right)\dot{q}_3; \end{aligned} \tag{17}$$

Equation (16) and (17) will be used to compute the accelerations $\ddot{q}_1, \ddot{q}_2, \ddot{q}_3$ as a function of the quantities $(q_1, q_2, q_3, \dot{q}_1, \dot{q}_2, \dot{q}_3, u_1, u_2, F, \ldots)$ and thereby converting the Lagrange variable set to the conjugate variable set.

4 Optimal Control Problem

To find the control variables u_1 and u_2 such that the motion of the manipulator satisfies Eq. (1) and (2), we can use the Pontryagin's maximum principle [4]. According to this principle, the motion of the system is described by the conjugate variable set $\{q_1, q_2, q_3, q_4, q_5, q_6, p_1, p_2, p_3, p_4, p_5, p_6\}$. A part of the conjugate variable set obtained from the Lagrange variable set is q_1, q_2, and q_3 the variables q_4, q_5, and q_6 is converted from generalized velocity variables ($\dot{q}_1 = q_4, \dot{q}_2 = q_5, \dot{q}_3 = q_6$), the variables $p_j (j = \overline{1, 6})$ are determined from the Hamilton function H:

$$H = -f_0 + p_1 f_1 + p_2 f_2 + p_3 f_3 + p_4 f_4 + p_5 f_5 + p_6 f_6 \tag{18}$$

Where:

$$f_0 = 0.5\left\{(u_1 - k_1 q_1)^2 + (u_2 - k_2 q_2)^2\right\};$$

$$f_1 = q_4; f_2 = q_5, f_3 = q_6; f_4 = \ddot{q}_1; f_5 = \ddot{q}_2; f_6 = \ddot{q}_3$$

and $\ddot{q}_1, \ddot{q}_2$, and $\ddot{q}_3$ are represented through the conjugate variables by applying motion equations of the manipulator in Sect. 3. Hence the H function is a function of the conjugate variables. According to the Pontryagin's maximum principle, for the function (2) to reach minimum, the control variables u_1 and u_2 are selected from the maximum condition of the H function. Control variables are selected from the condition:

$$\frac{\partial H}{\partial u_1} = 0 : \frac{\partial H}{\partial u_2} = 0 \tag{19}$$

Control variables u_1 and u_2 solved from (19) are plug into the following equations:

$$\begin{aligned}&\frac{dq_1}{dt} = f_1; \frac{dq_2}{dt} = f_2; \frac{dq_3}{dt} = f_3; \frac{dq_4}{dt} = f_4; \frac{dq_5}{dt} = f_5; \frac{dq_6}{dt} = f_6; \frac{dz}{dt} = f_0\\ &\frac{dp_1}{dt} = -\frac{\partial H}{\partial q_1}; \frac{dp_2}{dt} = -\frac{\partial H}{\partial q_2}; \frac{dp_3}{dt} = -\frac{\partial H}{\partial q_3}; \frac{dp_4}{dt} = -\frac{\partial H}{\partial q_4}; \frac{dp_5}{dt}\\ &= -\frac{\partial H}{\partial q_5}; \frac{dp_6}{dt} = -\frac{\partial H}{\partial q_6}\end{aligned} \tag{20}$$

With the boundary conditions:

$$\left\{\begin{matrix} q_1(0) = 0, q_2(0) = \pi/6, q_3(0) = 0.05 l_3, q_4(0) = 0, q_5 0) = 0, q_6(0) = 0, \\ p_1(t_f) = h_{1f}, p_2(t_f) = h_{2f}, p_3(t_f) = h_{3f}, p_4(t_f) = 0, p_5(t_f) = 0, p_6(t_f) = 0, z(0) = 0 \end{matrix}\right\} \tag{21}$$

will minimize system efforts by the cost function z(t)

$$z = \int_0^{t_f} 0.5\left\{(u_1 - k_1 q_1)^2 + (u_2 - k_2 q_2)^2\right\} dt \tag{22}$$

and t_f is the movement time of the manipulator.

$$h_{1f} = h_1(q_1 = q_1(t_f), q_2 = q_2(t_f), q_3 = q_3(t_f))$$

$$h_{2f} = h_2(q_1 = q_1(t_f), q_2 = q_2(t_f), q_3 = q_3(t_f))$$
$$h_{3f} = h_3(q_1 = q_1(t_f), q_2 = q_2(t_f), q_3 = q_3(t_f))$$

Simulation Result: The system of differential equations is solved by Maple software with the following data:

$$l_1 = 1.5(m), l_2 = 0.2(m), l_3 = 2l_1\sqrt{3}(m), c_1 = 0, c_2 = 0.1(m), c_3 = 0.05l_3,$$
$$m = 25(kg), m_2 = 15(kg), m_3 = 10(kg), J_1 = 25(kgm^2), J_2 = 15(kgm^2),$$
$$J_3 = 10(kgm^2),$$
$$k_1 = 60(Nm/rad), k_2 = 60(Nm/rad), k_3 = 100(N/m), l_0 = 0.02l_3$$
$$F = 300(N), a = \sqrt{3}(m), b = 3l_1\sqrt{3}(m), g = 10(m/s^2), t_f = 2.5(s)$$

The numerical simulation is represented in the following figures (Figs. 2, 3):

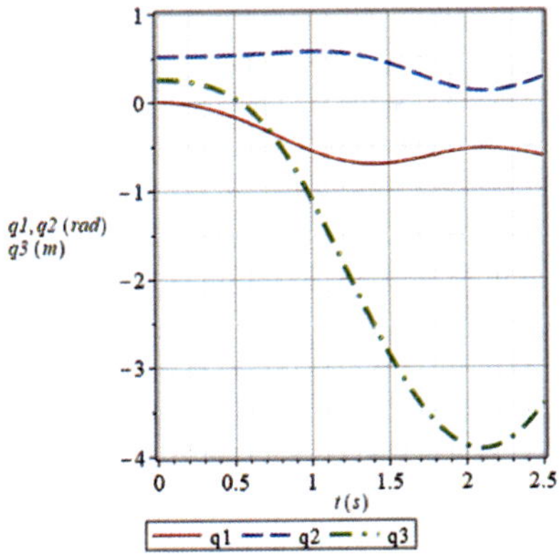

Fig. 2. Generalized coordinates

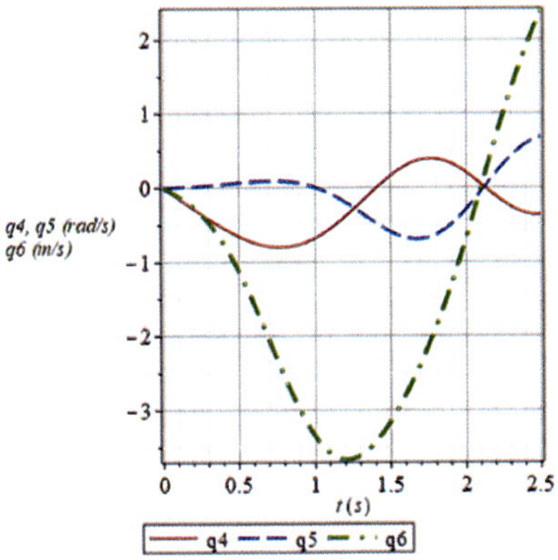

Fig. 3. Generalized velocities

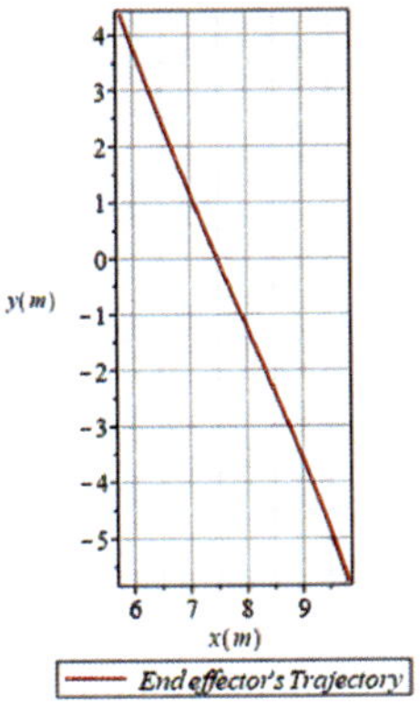

Fig. 4. Required trajectory

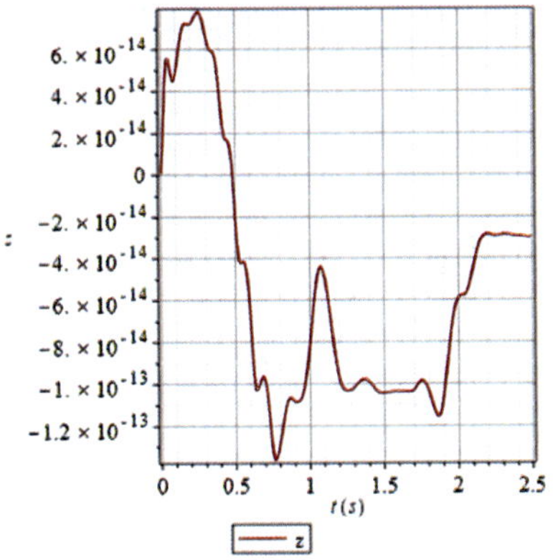

Fig. 5. The optimal criteria

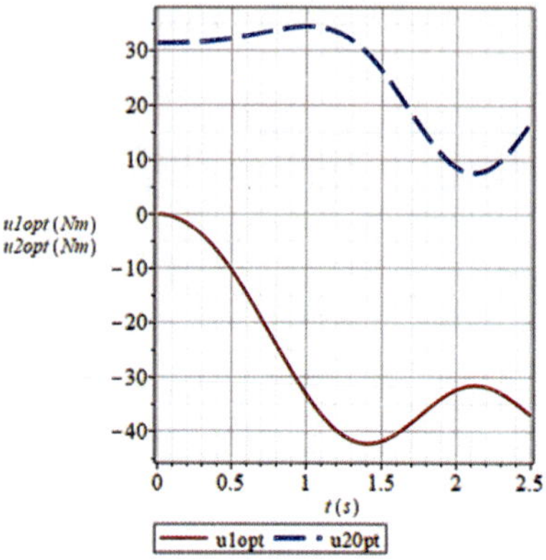

Fig. 6. The optimal control u_1, u_2

It is obviously seen that the end effector moves exactly along the prescribed trajectory (1) as shown in Fig. 4 while keeping the cost function minimum as shown in Fig. 5. The optimal control variables u_1, u_2 are also kept bounded during the simulation time in Fig. 6.

5 Conclusion

The optimal control problem of a 3 DOF manipulator following a required trajectory belongs to one of the most challenging issues. The first challenge is to study the optimal problem of constrained mechanical system. The equations of motion are built in Lagrangian variables, while the optimal approach uses the conjugate variables. Thus, the transformation from the Lagrangian variables into the conjugate variables is a difficulty. The second challenge is to determine the boundary and initial conditions of conjugate variables. In addition, the amount of computation is quite large. To overcome those difficulties, the paper has proposed a solution to use the principle of compatibility and the method of transformation matrix to build the motion equations of a constrained mechanical system in matrix form [7, 8]. The simulation results prove the end effector follow the prescribed trajectory while minimizing the control forces as shown in Fig. 5.

References

1. Gourdeau, R., Schwartz, H.M.: Optimal control of a robot manipulator using a weighted time-energy cost function. In: Proceedings of the 28th IEEE Conference on Decision and Control, vol. 2, pp. 1628–1631 (1989)
2. Shiller, Z.: Time-energy optimal control of articulated systems with geometric path constraints. In: Proceedings of the 1994 IEEE International Conference on Robotics and Automation, vol. 4, pp. 2680–2685 (1994)
3. Bobrow, J.E., Dubowsky, S., Gibson, J.S.: Time-optimal control of robotic manipulators along specified paths. Int. J. Robotics Res. **4**(3), 3–17 (1985)
4. Pontryagin, L.S., et al.: The Mathematical Theory of Optimal Processes, trans. by Triorgoff, K., Interscience Publishers, John Wiley and Sons. Inc., New York, US (1962)

5. Sanh, D.: On the principle of compatibility and equations of motion of constrained. In: Mechanical Systems ZAMM 60, pp. 210–212, Berlin (1980)
6. Sanh, D.: On the motion of controlled mechanical systems. Adv. Mech. **2**(7), 3–24 (1984). Varsawa
7. Sanh, D.: On the Motion of Constrained Mechanisms, Thesis of Doctorate in Science, Hanoi University of Science and Technology, Vietnam (1984)
8. Do, S., Do, D.K.: Method of transmission matrix applying for investigation of motion of planar mechanism. Mach. Dyn. Res. **34**(4), 5–22 (2010). Varsaw
9. Do Sanh, D.V.P., Do Dang Khoa, T.D.: A method for solving the motion of constrained systems. In: Proceedings of the 16th Asian Pacific Vibration Conference, pp. 803–811. Hanoi, Vietnam (2015)
10. Sanh, D., Do, D.K.: Optimal Control of Dynamics Systems-Program Motion and Optimal Control. Bach Khoa Publisher, Vietnam (2010)
11. Sanh, D., Do, D.K., Phong, P.D.: Control of Machine Dynamics and Robotics. Bach Khoa Publisher, Vietnam (2020)

A Novel Control Strategy of Gripper and Thrust System of TBM Based on Model Analysis

Yuxi Chen, Guofang Gong(✉), and Xinghai Zhou

Zhejiang University, Hangzhou 310027, China
{kiwicyx,gfgong}@zju.edu.cn

Abstract. The control system based on restricted thrust force and constant support force in gripper and thrust system (TFRSFCGT, short for thrust-force-restricted support-force-constant gripper-thrust), and the control system based on constant thrust force and constant support force in gripper and thrust system (TFCSFCGT, short for thrust-force-constant support-force-constant gripper-thrust) were proposed, in order to solve the problems that the gripper shoes equipped in TBM failed to support the surrounding rock under a strong thrust load impact or crushed the surrounding rock in a soft rock formation. The coupling relationship among the output force of support cylinder, the output force of thrust cylinder and the type of surrounding rock at the place of supporting was derived after establishing the mechanical model of the gripper and thrust mechanism. The electro-hydraulic systems of these two control strategies were designed, then their hydraulic system models were set up in AMESim software and their control system models were set up in Matlab software, respectively. Finally, the performance of these two control strategies was compared and analyzed by using co-simulation method. Results showed that the TFR (short for TFRSFCGT) system and the TFC (short for TFCSFCGT) system can maintain the support force acting on the surrounding rock under the premise of providing required propulsion, and the maximum overshoot were 0.928% and 0.378%, respectively. The TFC system had a faster tunneling speed, while the TFR system had better adaptability to load impact under the same load.

Keywords: Tunnel Boring Machine (TBM) · Pressure limitation · Gripper and thrust system · Propulsion force · Support force · Co-simulation

1 Introduction

Full face hard rock tunnel boring machine (TBM) is a kind of large-scale tunnel excavation equipment which is widely used in hard rock tunnels construction due to the advantages of safety, environment protection and high efficiency [1, 2]. Single-pair horizontally supported open type TBMs are suitable for tunnels with good rock integrity and are most widely used because they are fast, inexpensive, and easy for maintenance. The gripper and thrust system is one of the critical subsystems of TBM, which provides the propulsion force for the cutterhead to move forward, its gripper shoes connected with the thrust cylinders and the gripper cylinder push against the surrounding rock with

N. V. Khang et al. (Eds.): ASIAN MMS 2021, MMS 113, pp. 472–485, 2022.
https://doi.org/10.1007/978-3-030-91892-7_45

certain pressure to provide reactive propulsion force for the thrust system and reactive torque for the cutterhead system during tunneling [3, 4]. In the current construction site, the support cylinder pressure of the gripper system is set manually by means of dirvers' experience, the speed control mode is used in the thrust system, without establishing the coupling relationship between the output force of the support cylinder of the gripper system and the thrust cylinder of the thrust system. In the speed control mode of the thrust system, the propulsion force is related to the rock characteristics of the stratum, and when TBM suddenly encounters a hard rock layer during the tunneling process, causing the propulsion force exceeds the propulsion reaction force provided by the gripper, the actual support force provided by the surrounding rock is lower than the theoretical support force required for tunneling, and the gripper will not be able to hold up the surrounding rock, thus causing skidding accidents; when TBM gripper is in a soft stratigraphic section, the safety grounding ratio of the surrounding rock is low, and the actual pressure of the gripper acting on the surrounding rock is large enough to collapse the surrounding rock; under unstable surrounding rock conditions, large fluctuations in the support force will also cause the tunnel sidewalls to collapse [5–8]. Therefore, it is necessary to solve the coupling relationship between support force and propulsion force for safe and efficient construction.

Lots of scholars have conducted a series of studies on TBM support and thrust system. Huo et al. [9] analyzed the stability of surrounding rock under different support forms of TBM, Rao et al. [10] proposed the theory of real-time adjustment of support cylinder output force to keep the support force constant, Sun et al. [11] studied the stability of surrounding rock under different support shoe contact area and the adaptability of horizontal support under different geological conditions; Wu et al. [12] established the stiffness model of the thrust hydraulic cylinder and analyzed the variation law of the effect of frequency domain on the velocity stiffness through simulation, Yao et al. [13] proposed the intelligent control theory based on neuroendocrine for the propulsion velocity control and verified its effectiveness through simulation. Most of the current studies focus on the mapping relationship between the support mechanism and the surrounding rock properties in the gripper system, the modeling of the thrust mechanism and the analysis of the dynamics. However, there is no research on the impact of thrust load changes on the gripper system and few studies on the coordination relationship between the TBM gripper system and the thrust system.

In this study, we built the mechanical model of the support and thrust mechanism, solved the coupling relationship between support force and propulsion force in a specific surrounding rock environment, designed the hydraulic system and control system in AMESim and Matlab software respectively, designed the simulation parameters by combining the indexes of Robbins' products. A detailed computing method is also developed to get the output force of gripper cylinder and the thrust limitation force. Then compared and analyzed control effects of the TFR system, TFC system and the conventional propulsion speed control based support and thrust system (TSCGT, short for thrust-support-constant gripper-thrust) by means of co-simulation method.

2 Mechanical Model of Gripper and Thrust Mechanism

The open type TBM consists of support and thrust system, cutterhead drive system and tail support system, etc. Its structure is shown in Fig. 1. When working, the gripper shoe supports the tunnel sidewall with a certain pressure, the tail support is lifted, and the hydraulic cylinders of thrust system push the main beam to make the cutterhead squeeze the rock in front, while the motor drives the cutterhead to rotate and cut the rock, thus realizing the one-time formation of the tunnel.

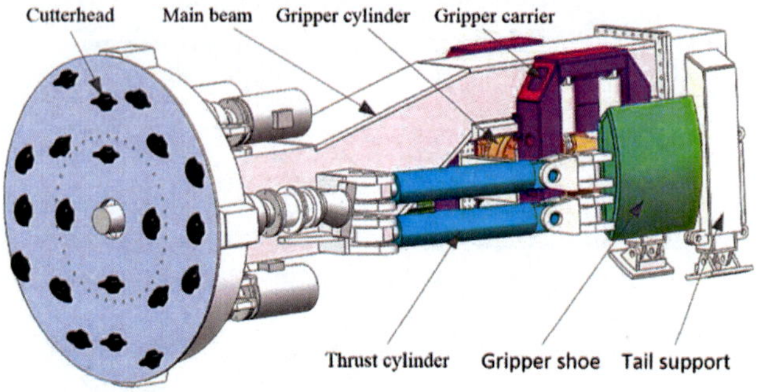

Fig. 1. Simplified structure of an open-type TBM

2.1 Modeling of Support and Thrust Mechanism

The mechanical analysis of gripper and thrust mechanical structure is shown in Fig. 2. The thrust structures and gripper structures on both sides of the main beam are symmetrical, but when the cutterhead rotates in an counterclockwise direction, the direction of the reactive torque on the right gripper shoe is the same as that of the gravity direction, which causes the mechanical force larger in the z axis direction. Therefore, considering the safety and effectiveness, the force analysis of the right gripper shoe was adopted, and the force analysis is simplified as shown in Fig. 2.

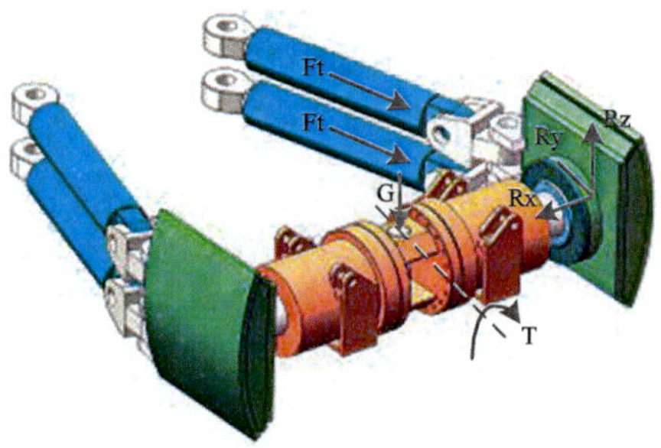

Fig. 2. Mechanical analysis of gripper and thrust mechanism

The force equilibrium equations of the right gripper shoe can be expressed as

$$\begin{cases} F_{Rx} = 2F_t \sin\theta + F_c \\ F_{Ry} = 2F_t \cos\theta \\ F_{Rz} = \frac{T}{l} + \frac{G}{2} \end{cases} \quad (1)$$

Where F_{Rx} is the supporting force of gripper shoe from the surrounding rock, F_t is propulsion force, F_c is the output force of gripper cylinder, F_{Ry} and F_{Rz} are the friction force in y and z direction respectively from the surrounding rocks, T is the reactive torque from cutterhead, G is the gravity acting on the gripper cylinders, l is the distance between left gripper shoe and right gripper shoe, θ is the angle between the axis of thrust cylinder and thrust direction.

The real ground pressure p_g^r of gripper shoe can be written as

$$p_g^r = \frac{F_{Rx}}{S} = \frac{2F_t \sin\theta + F_c}{S} \quad (2)$$

Where S is the vertical projection area along the axial direction of gripper cylinder.

The open type TBM consists of two sets of left and right thrust cylinders acting on the main beam to provide propulsion for the whole machine, each set consists of two hydraulic cylinders connected together in parallel. The diagram of the main beam is shown in Fig. 3.

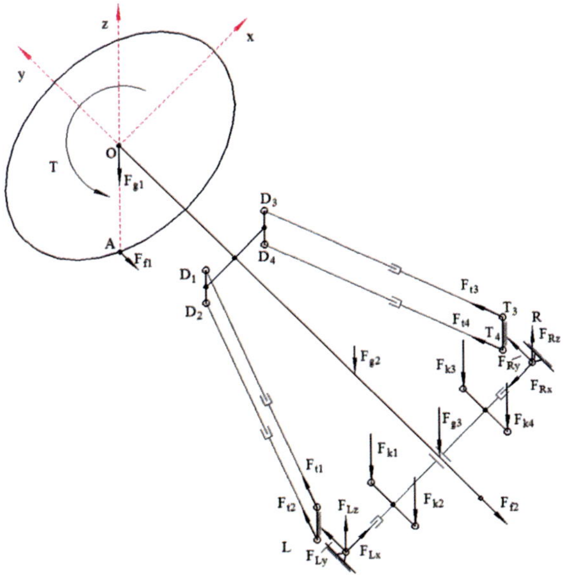

Fig. 3. Mechanical analysis of the right gripper mechanism

In the horizontal plane, the TBM main beam is subjected to the propulsion force $4F_t$ generated by the left and right sets of thrust cylinders and the load force F_l during the digging process. Where $F_l = F_{l0}+F_f+F_1+B_v v$, F_{l0} represents resistance generated by the surrounding rock during excavation, F_f is the total friction force, $B_v v$ is the damping force for the advance of the whole machine.

The equilibrium equation in the direction of digging can be established as

$$4F_t \cos\theta = F_{l0} + F_f + F_1 + B_v v \tag{3}$$

Where is the advance speed in the direction of TBM digging.

Assuming that s_t is the distance between hinged joint near the rodless chamber and hinged joint near the rod chamber of thrust cylinder, s_y is the projection of s_t in the direction of the digging, then $s_t^2 = s_y^2 + a^2$ was established. Taking the derivative for both sides of the equation gives $s_t' = \frac{s_y}{s_t} s_y'$, where $s_t' = v_t = v_y \cos\theta$.

Thus, the propulsion force of single cylinder of thrust system is

$$F_t = \frac{1}{4\cos\theta}\left(F_{10} + F_f + F_1 + B_v \frac{v_t}{\cos\theta}\right) \tag{4}$$

For thrust system, the propulsion load is equal to the propulsion force. Assuming that the load on the single cylinder is $F_{0y} = (F_{l0} + F_f + F_1)/4$, the damping factor is $B_{vy} = B_v/4$, then $F_{tl} = \frac{F_{0y}}{\cos\theta} + \frac{B_{vy}}{\cos^2\theta} v_t$.

2.2 Force Limitation and Analysis

When TBM is in the complex rock formations composed of IV and V types rocks, p_g^r should be less than the safe ground pressure of the complex rock, avoiding crushing the surrounding rock. At the same time, F_{Rx} should be large enough to provide reactive propulsion force for thrust cylinder and provide reactive torque for cutterhead. Therefore, some relations can be established as

$$\begin{cases} p_g^r \le \frac{p_g^{max}}{s_g} \\ \mu F_{Rx} \ge \sqrt{F_{Ry}^2 + F_{Rz}^2} \end{cases} \tag{5}$$

Where s_g is security coefficient, μ is friction coefficient between surrounding rock and gripper shoe.

According to formula (1), (2) and (3), it is easy to get

$$\begin{cases} 2F_t \sin\theta + F_c \le \frac{p_g^{max}}{s_g} \\ 2F_t \sin\theta + F_c \ge \frac{1}{\mu}\sqrt{(2F_t\cos\theta)^2 + (\frac{T}{l} + \frac{G}{2})^2} \end{cases} \tag{6}$$

Suppose the required supporting force is F_{Rx}^t, then

$$F_{Rx}^t = \frac{1}{\mu}\sqrt{(2F_t\cos\theta)^2 + (\frac{T}{l} + \frac{G}{2})^2} \tag{7}$$

Suppose the load force changes very little, as the piston rods of thrust cylinders are extended, θ and F_{Rx} decreases gradually, but F_{Rx}^t increases. Therefore, on the beginning of thrusting, the force acting on surrounding rock is maximal, and when the piston rods reach maximum stroke, the required supporting force is maximal.

2.3 Calculation Method

In order to avoid crushing the surrounding rock and meet the demand for reactive propulsion force and reactive torque simultaneously from the start of stroke to the end, a method to get the output force of gripper cylinder and the maximum propulsion force is proposed.

The supporting action for each thrust stroke is performed after the end of the previous thrust stroke, thus the surrounding rock conditions at the place where the gripper shoe acts is available, namely $p_g^{\max}$ is known before the supporting action.

Suppose the safe supporting force is F_{Rx}^s, then

$$F_{Rx}^s = \frac{S}{s_g} \cdot p_g^{\max} \tag{8}$$

The required supporting force is maximal at the end of thrusting, so let

$$\sqrt{(2F_{t\max}^{'} \cos\theta_{\min})^2 + (\frac{T}{l} + \frac{G}{2})^2} = \mu F_{Rx}^s \tag{9}$$

then

$$F_{t\max}^{'} = \frac{1}{2\cos\theta_{\min}} \sqrt{(\mu F_{Rx}^s)^2 - (\frac{T}{l} + \frac{G}{2})^2} \tag{10}$$

The output force of gripper cylinder can be computed as

$$F_c = F_{Rx}^s - 2F_{t\max}^{'} \sin\theta_{\max} \tag{11}$$

So, the working pressure of gripper system p_c can be got

$$p_c = \frac{F_c}{s_c} = \frac{4F_c}{\pi D_c^2} \tag{12}$$

Where s_c is the sectional area of gripper cylinder, D_c is the piston diameter of gripper cylinder.

Besides, the force acting on surrounding rock is minimal, so $F_{t\max}$ can be calculated from the following formula

$$F_c + 2F_{t\max}\sin\theta_{\min} - \frac{1}{\mu}\sqrt{(2F_{t\max}\cos\theta_{\min})^2 + (\frac{T}{l} + \frac{G}{2})^2} = 0 \tag{13}$$

Thus, the limited pressure of the thrust system $p_{t\max}$ can be computed as

$$p_{t\max} = \frac{F_{t\max}}{s_t} = \frac{4F_{t\max}}{\pi D_t^2} \tag{14}$$

Where s_t is the sectional area of thrust cylinder, D_t is the piston diameter of thrust cylinder.

3 Novel Control Strategy Design for Gripper and Thrust System

3.1 Design of TFR Control Strategy

In order to solve the problem of gripper slippage caused by excessive propulsion load during TBM propulsion, an electric proportional relief valve can be introduced in the thrust system to limit the propulsion pressure; the support force of the gripper acting on the surrounding rock is composed of the division of the propulsion on the support axis and the force of the support cylinder, so adjusting the output force of the support cylinder according to the actual propulsion can not only avoid the collapse of the surrounding rock, but also reduce the fluctuation of the support force on the surrounding rock. The type of surrounding rock to be excavated can be identified according to the excavation parameters, and the purpose of this study is to analyze the coupling characteristics between the thrust system and the gripper system, so the type of surrounding rock at the support is assumed to be known, and the simplified control strategy of the TFR system is shown in Fig. 4.

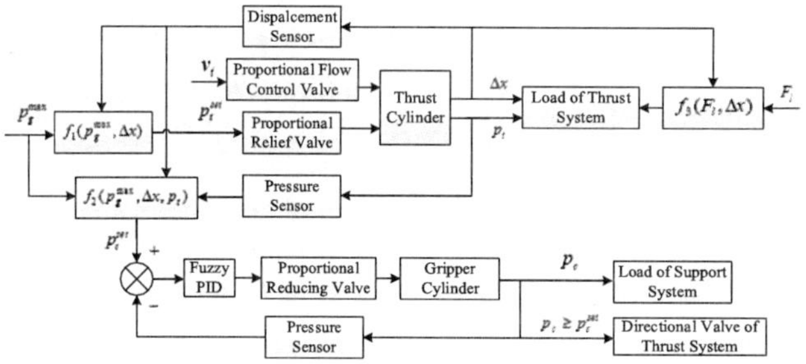

Fig. 4. Simplified diagram of TFR system

The expression for the safety support force of the surrounding rock be described as

$$F_{\mathrm{su}}^{\mathrm{s}} = \frac{S}{s_{\mathrm{g}}} p_{\mathrm{g}}^{\max} \tag{15}$$

The function f_1 in the diagram is

$$p_{\mathrm{t}}^{\mathrm{set}} = \frac{4F_{\mathrm{t}}}{\pi D_{\mathrm{t}}^2} = \frac{4}{\pi D_{\mathrm{t}}^2} \cdot \frac{1}{2\cos\theta} \left[(\mu F_{su}^{t})^2 - (\frac{T}{l} + \frac{G}{2})^2 \right]^{1/2} \tag{16}$$

Where s_{g1} is the safety coefficient, D_t is the diameter of the rodless cavity section, and $F_{su}^{t} = F_{su}^{s}/s_{g1}$.

The function f_2 in the diagram is

$$p_{\mathrm{c}}^{\mathrm{set}} = \frac{4F_{\mathrm{c}}}{\pi D_{\mathrm{c}}^2} = \frac{4}{\pi D_{\mathrm{c}}^2} \left(F_{\mathrm{su}}^{\mathrm{r}} - 2F_{\mathrm{t}} \sin\theta \right) \tag{17}$$

Where s_{g2} is the safety coefficient, D_c is the diameter of the rodless cavity section, and $F_{su}^{r} = F_{su}^{s}/s_{g2}$.

3.2 Design of TFR Hydraulic System

According to the control strategy of TFR system, design the support and thrust hydraulic system based on constant support force and propulsion force limitation, and its simplified hydraulic system schematic diagram is shown in Fig. 5.

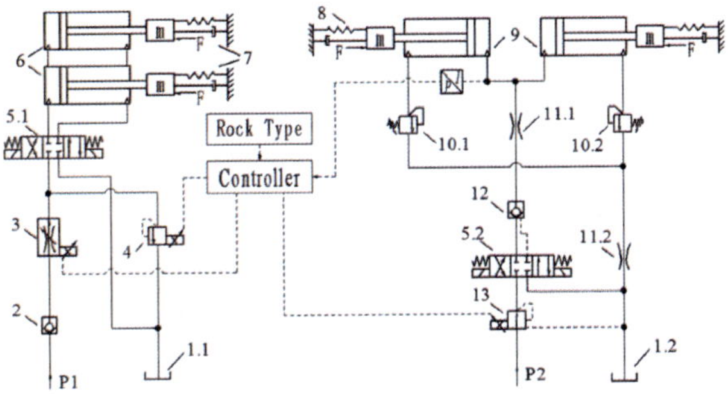

Fig. 5. Schematic diagram of TFR hydraulic system

1-Oil tank; 2-Check valve; 3-Proportional flow control valve; 4-Proportional relief valve; 5-Proportional reversing valve; 6-Thrust system cylinders; 7-Propulsion load; 8-Support load; 9-Gripper system cylinders; 10-Proportional relief valve; 11-Orifice; 12-Hydraulic one-way valve; 13-Proportional pressure reducing valve.

The controller in the figure mainly completes four parts of work: (1) According to the known surrounding rock type and the parameter calculation method in the control strategy, calculate the limited pressure value of the rodless chamber of the thrust cylinder, and transmit the signal to the proportional relief valve 4; (2) Collect the pressure signal and the displacement signal of the thrust cylinder, then calculate the set pressure value of the rodless chamber of the support cylinder, and transmit the signal to the proportional pressure reducing valve 13; (3) Collect the rod chamber and rodless chamber pressure signals of the support cylinder to achieve closed-loop control and improve the control accuracy of the output force; (4) Adjust the propulsion load force according to the collected displacement signal to simulate the actual load model of the thrust cylinder.

3.3 Design of TFC Control Strategy

In the whole digging stroke of the TBM, limiting the change of digging force along the digging direction can reduce the impact on the cutterhead. The thrust system of the designed TFC system uses a three-way proportional pressure reducing valve to control the propulsion pressure to ensure that the propulsion pressure is controllable in real time, and combined with the displacement signal, the propulsion force in the digging direction can be controlled in real time, and the output force of the support cylinder can be adjusted in real time according to the change of the propulsion force to make the actual support force of the surrounding rock remain constant. The simplified control strategy of the TFCSFCGT system is shown in Fig. 6.

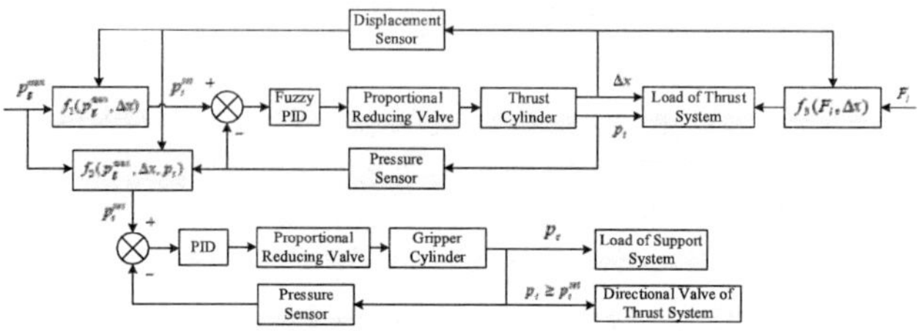

Fig. 6. Simplified diagram of TFC system

3.4 Design of TFC Hydraulic System

According to the control strategy of TFC system, the hydraulic system of support propulsion is designed with constant propulsion force and constant support force in the digging direction, and the schematic diagram of the hydraulic system is shown in Fig. 7. In the diagram, the controller collects the displacement signal of the digging cylinder, calculates the pressure setting value of the propulsion cylinder in real time combined with the surrounding rock category, and adjusts the propulsion load force in real time; it collects the pressure signal of the large and small chambers of the support cylinder, and realizes closed-loop control of the pressure of the large chamber of the support cylinder.

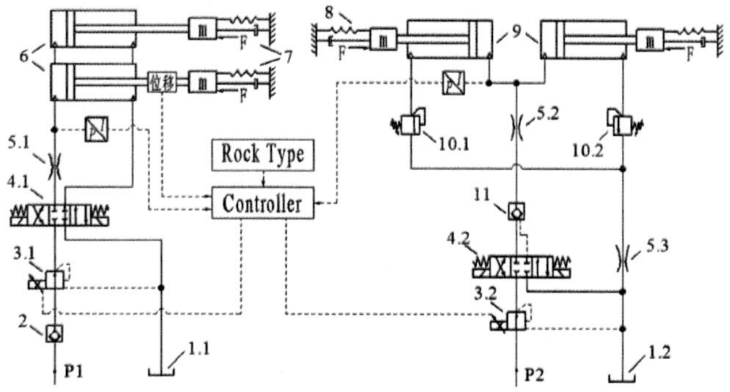

Fig. 7. Schematic diagram of TFR hydraulic system

1-Oil tank; 2-Check valve; 3-Proportional reducing valve; 4-Proportional relief valve; 5-Proportional reversing valve; 6-Thrust system cylinders; 7-Propulsion load; 8-Support load; 9-Gripper system cylinders; 10-Proportional relief valve; 11- Hydraulic one-way valve.

4 CO-Simulation Analysis

To validate the effectiveness of the gripper and thrust electrohydraulic system designed in the last chapter, their simulation models are constructed in AMEsim and Matlab software respectively. The set value of the proportional reducing valve in the gripper system and the set value of the proportional relief valve in the thrust system should be calculated and initialized before the start of the thrust stroke.

4.1 Simulation Parameters

According to the parameters of the ϕ 8.03 m Robbins TBM 264-311, some parameters are designed as shown in Table 1.

Table 1. Simulation parameters

Parameters	Value
Thrust Speed ($v/(mm/s)$)	(0,2.5)
Cutter Torque ($T/(kN \cdot m)$)	(0,150)
Maximum Propulsion force ($F/(kN)$)	2000
Thrust Cylinder Displacement ($\Delta s/\text{mm}$)	1070
Size of Thrust Cylinders (mm)	$\phi 180/125$
Size of Gripper Cylinder (mm)	$\phi 380/250$

Compared with the change of propulsion force, the change of cutterhead torque has little effect on the required support force. Therefore, it is wise to assume that the cutterhead torque is maximal during calculating. In the gripper system simulation model, the elastic coefficient is set to $2.0 \times 10^9 \text{N/m}$.

In the thrust system simulation model, the thrust speed increases gradually from original 1.46 mm/s at 100 s, and reaches the 1.90 mm/s at 250 s, and then the speed is gradually reduced to 1.46 mm/s at 400 s. The damping coefficient is set to $3.2 \times 10^8 \text{N}/(\text{m/s})$. The constant load force increases linearly from 320 kN at 400 s, and reaches the 450 kN at 475 s, and then reduces to 300 kN at 575 s. At 625 s, the constant load force increases suddenly to 450 kN and reduces gradually until the end of thrust stroke.

4.2 Simulation Analysis

As shown in Fig. 8, for the TFR system, in the periods of 42.6–57.4 s, 152–242 s, 300–400 s and 502 –601 s, the flow rate provided by the system is higher than the required flow rate for propulsion, and the excess flow rate is discharged through the proportional relief valve, where the trend of flow rate change is almost the same as that of the TFC system, the difference is that the TFC system produces obvious flow impact under the disturbance; when the propulsion force of TFR system does not reach the overflow limit, the propulsion flow of TFR system is consistent with the setting value of the proportional relief valve, while the propulsion flow of TFC system is obviously higher than that of TFR system. However, it can be seen from 100 and 242 s that the propulsion flow rate of the TFC system will change with the change of propulsion load. Combined with the analysis of the control strategies of the two systems, it can be seen that the thrust system pressure setting value of the TFC system is equal to the pressure limit value of the TFR system; under the same load characteristics, the TFC system can complete the thrust

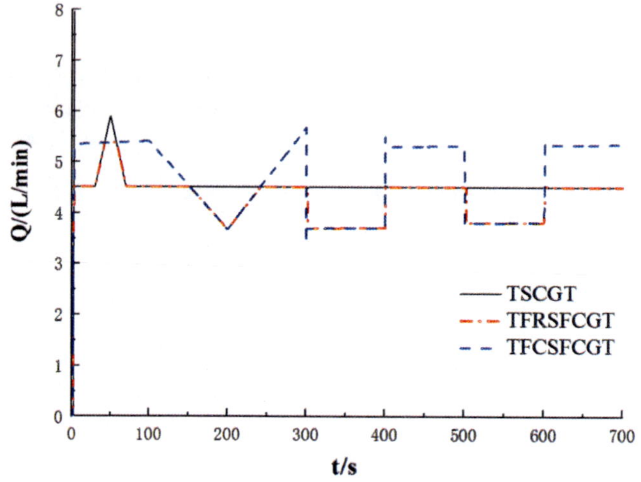

Fig. 8. Flow response of different systems

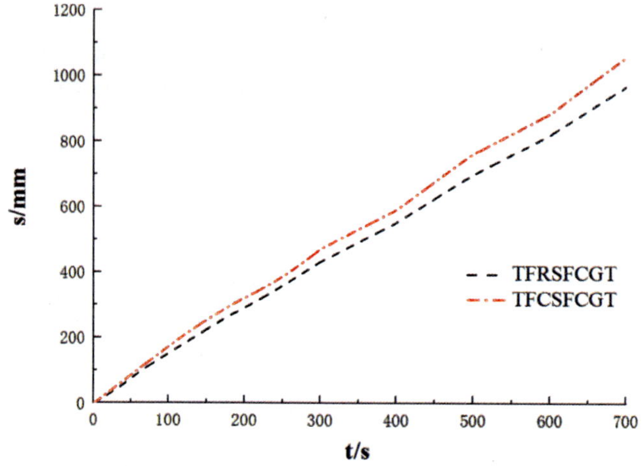

Fig. 9. Displacement response of different systems

stroke faster than the TFR system, and the thrust displacement comparison curves of the two systems are shown in Fig. 9.

The output force change curve of the conventional thrust speed control of the gripper and thrust system (TSCGT) is shown in Fig. 10. As can be seen from the figure, the thrust force change is consistent with the load change, and the theoretical support force changes significantly with the propulsion change, and the theoretical support force obviously exceeds the actual support force provided by the surrounding rock in the periods of 46–55 s, 161–240 s, 301 –403 s and 502– 602 s. During the whole digging stroke, the output force of the support cylinder remains constant, but the digging force changes with the load, and the angle between the thrust cylinder and the digging direction decreases continuously during the digging process, so the actual support force of the gripper on the surrounding rock is reduced overall. At the beginning of the digging stroke, the actual support force is closest to the safe support force, and the change of the thrust force will cause the actual support force to be larger than the safe support force, thus crushing the surrounding rock, as shown in Fig. 10 from 47 to 53 s.

The output force change curve of TFR is shown in Fig. 11. As can be seen from the figure, the theoretical support force is constantly lower than the actual support force, and the actual support force is constantly lower than the safe support force during the whole digging stroke, and the actual support force remains constant with the maximum overshoot of 0.928%. In order to avoid excessive thrust force, TFR system uses electric proportional relief valve to limit the propulsion force. By comparing the propulsion limiting force from 42.6 to 57.4 s, 502 s and 601 s, it can be seen that the propulsion limiting force becomes smaller during the whole digging stroke. The output force of the support cylinder changes with the change of the thrust force on the support axis, thus ensuring the actual support force remains constant.

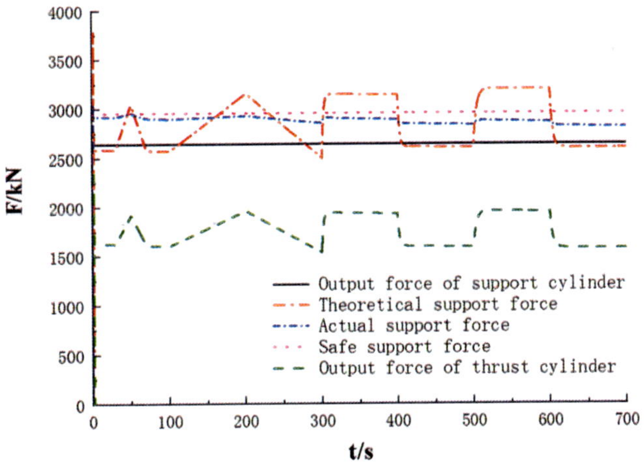

Fig. 10. Output force of support and thrust cylinder of TSCGT

The output force change curve of TFC is shown in Fig. 12. As can be seen from the figure, the actual support force remains constant and lower than the safe support force during the whole digging stroke, and the maximum overshoot is 0.378%; the theoretical support force remains constant and lower than the actual support force, and the maximum overshoot is 0.216%. The angle between the thrust cylinder and the digging direction keeps decreasing, so the output force of the thrust cylinder keeps decreasing to ensure the constant theoretical support force, and the output force of the support cylinder increases accordingly to ensure the constant actual support force.

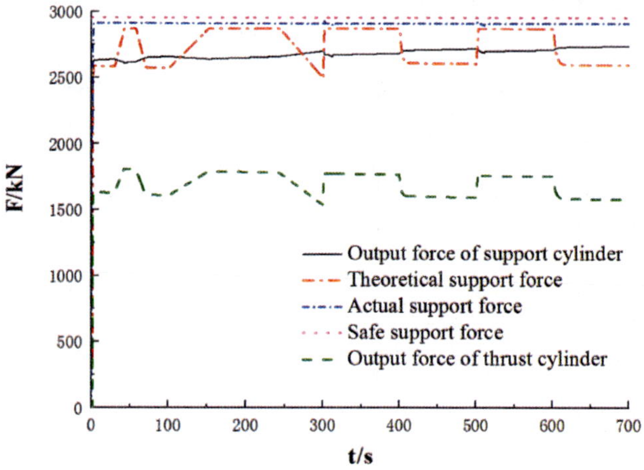

Fig. 11. Output force of support and thrust cylinder of TFR

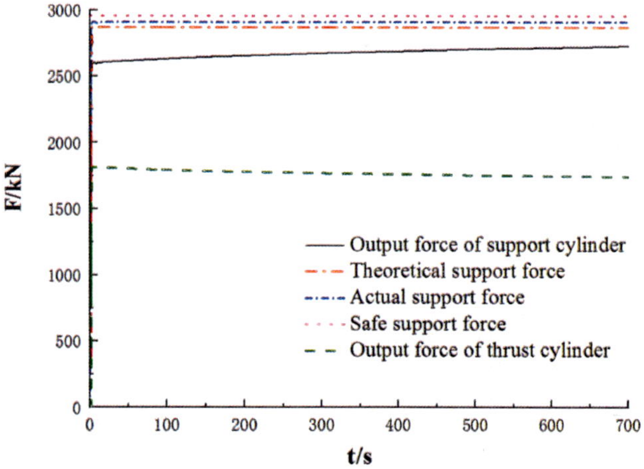

Fig. 12. Output force of support and thrust cylinder of TFC

5 Conclusion

(1) Compared with the traditional TSCGT system, both the TFR system and the TFC system designed in this article effectively avoid the occurrence of two situations in which the theoretical support force of the surrounding rock is higher than the actual support force of the surrounding rock and the actual support force of the surrounding rock is higher than the safe support force of the surrounding rock during the tunneling process, and effectively reduce the fluctuation of the support force of the surrounding rock, and the overshoot can be controlled within 1%.

(2) The TFC system takes relatively short time to complete the digging stroke, but it generates flow shock when encountering sudden change of propulsion load, and the compliance to sudden change of load is not as good as TFR system.

(3) Both novel gripper and thrust systems are easy to realize engineering applications and can effectively solve engineering problems such as slippage of the support shoe and crushing of the surrounding rock caused by unreasonable support and propulsion force settings, which can improve the reliability and safety of TBM construction.

Acknowledgments. This work is supported by Hunan Province Major Science and Technology Projects (Grant No. 2019GK1013). Their support is gratefully appreciated.

References

1. Meng-shu, W.: Construction technique of open TBM for long railway tunnels in very hard or soft rock strata. Chin. Civil Eng. J. **38**(5), 54–58 (2005)
2. Yan-liang, D.U., Li-jie, D.U.: Full Face Hard Rock Tunnel Boring Machine System Principles and Integrated Design, pp. 1–16. Huazhong University of Science and Technology Press, Wuhan (2011)
3. Huang, T., et al.: Force analysis of an open TBM gripping–thrusting–regripping mechanism. Mech. Mach. Theory **98**, 101–113 (2016)
4. Huayong, Y., et al.: Electro-hydraulic proportional control of thrust system for shield tunneling machine. Autom. Constr. **18**(7), 950–956 (2009)
5. Farrokh, E., Rostami, J.: Effect of adverse geological condition on TBM operation in Ghomroud tunnel conveyance project. Tunn. Undergr. Space Technol. **24**(4), 436–446 (2009)
6. Paltrinieri, E., Sandrone, F., Zhao, J.: Analysis and estimation of gripper TBM performances in highly fractured and faulted rocks. Tunn. Undergr. Space Technol. **52**, 44–61 (2016)
7. Liu, J., et al.: Effects of discontinuities on penetration of TBM cutters. J. Cent. South Univ. **22**(9), 3624–3632 (2015)
8. Shi, H., et al.: Determination of thrust force for shield tunneling machine. J. Zhejiang Univ. (Eng. Sci.) **45**(1), 126–131 (2011)
9. Huo, J., et al.: Stability analysis of surrounding rock under different TBM supporting forms. J. Northeastern Univ. (Nat. Sci.) **35**(11), 1602–1606 (2014)
10. Rao, Y., et al.: Research on geologic adaptive control of ground pressure for single gripping TBM. In: International Conference on Control, Automation and Information Sciences, pp. 7–11. IEEE (2015)
11. Wei, S., et al.: Stability analysis of surrounding rock of TBM gripper with different contact areas. J. Harbin Eng. Univ. **50**(21), 92–98 (2014)
12. Wu, Y., et al.: Stiffnessmodeling of thrustcylinder in hardrock tunnel boring machine. In: International Conference on Fluid Power and Mechatronics, pp. 157–162. IEEE (2015)
13. Yao, J., Yimin, X., Yongliang, C., Huan, Z.: Endocrine intelligent control of thrust hydraulic system for TBM. In: Lee, J., Lee, M.C., Liu, H., Ryu, J.H. (eds.) Intelligent Robotics and Applications. ICIRA 2013. LNCS, vol. 8103, pp. 720–726. Springer, Berlin, Heidelberg (2013). https://doi.org/10.1007/978-3-642-40849-6_71

Dynamics and Control of Multibody Systems

Stability Control of Dynamical Systems Described by Linear Differential Equations with Time-Periodic Coefficients

Dinh Cong Dat[1,2](✉), Nguyen Van Khang[1], Nguyen Quang Hoang[1], and Nguyen Van Quyen[1]

[1] Hanoi University of Science and Technology, Hanoi, Vietnam
[2] Hanoi University of Mining and Geology, Hanoi, Vietnam

Abstract. The analysis of dynamical stability is an important problem in the design and control of vibrating structures which are described by a linear differential equation system with time-periodic coefficients. For this kind of system, stable criteria according to the Floquet multipliers is given. In case of an unstable system, a PD controller is added, and its optimal parameters are determined by the Taguchi method.

Keywords: Linearization · Flexible manipulator · Floquet theory · Taguchi method · Stability

1 Introduction

Mathematically, the motion of a multibody system with f degrees of freedom can be described by the following nonlinear differential equation [1–3]

$$\mathbf{M}(\mathbf{q}, t)\ddot{\mathbf{q}} + \mathbf{k}(\dot{\mathbf{q}}, \mathbf{q}, t) = \mathbf{h}(\dot{\mathbf{q}}, \mathbf{q}, t) \tag{1}$$

where $\mathbf{M}(\mathbf{q}, t)$ is the symmetric $f \times f$ inertia matrix, $\mathbf{k}(\dot{\mathbf{q}}, \mathbf{q}, t)$ is the $f \times 1$ vector of the generalized gyroscopic forces, $\mathbf{h}(\dot{\mathbf{q}}, \mathbf{q}, t)$ is the $f \times 1$ vector of the generalized applied forces, and $\mathbf{q}$, $\dot{\mathbf{q}}$, $\ddot{\mathbf{q}}$ are the vectors of generalized position, velocity, acceleration variables, respectively [1]. It is very difficult or impossible to find the analytical solution of Eq. (1). Hence, the numerical methods are the efficient way to solve the problem [1, 2]. The solution of Eq. (1) can be used to simulate the dynamic behavior of multibody systems that undergo large movements.

It is well-known that technical systems work mostly in a neighbourhood of its desired motion which is called the fundamental motion. For instance, the fundamental motion of a driver system is the motion of working components, so that the driver output rotates uniformly, and all components are assumed to be rigid. The fundamental motion of a flexible robotic systems usually described through state variables determined by prescribed motions of the end-effector.

Equation (1) is usually linearized about the fundamental motion to use the linear analysis tools [3–7] for analysing the behavior of the multibody system in the vicinity

N. V. Khang et al. (Eds.): ASIAN MMS 2021, MMS 113, pp. 489–500, 2022.
https://doi.org/10.1007/978-3-030-91892-7_46

of the fundamental motion. The result of this linearization process leads to a set of linear differential equations with time-varying coefficients. For the case of a flexible manipulator in steady-state motions [5–7], these equations can be written in the matrix form

$$\mathbf{M}_L(t)\ddot{\mathbf{y}} + \mathbf{C}_L(t)\dot{\mathbf{y}} + \mathbf{K}_L(t)\mathbf{y} = \mathbf{h}_L(t), \tag{2}$$

where $\mathbf{M}_L(t),\ \mathbf{C}_L(t),\ \mathbf{K}_L(t)$ and $\mathbf{h}_L(t)$ are time-periodic with period T.

Equation (2) can then be expressed in the compact form as

$$\dot{\mathbf{x}} = \mathbf{P}(t)\mathbf{x} + \mathbf{f}(t) \tag{3}$$

where the state variable $\mathbf{x}$, the matrix $\mathbf{P}(t)$ and vector $\mathbf{f}(t)$ are defined by:

$$\mathbf{x} = \begin{bmatrix} \mathbf{y} \\ \dot{\mathbf{y}} \end{bmatrix}, \quad \dot{\mathbf{x}} = \begin{bmatrix} \dot{\mathbf{y}} \\ \ddot{\mathbf{y}} \end{bmatrix} \tag{4}$$

$$\mathbf{P}(t) = \begin{bmatrix} 0 & \mathbf{E} \\ -\mathbf{M}_L^{-1}\mathbf{K}_L & -\mathbf{M}_L^{-1}\mathbf{C}_L \end{bmatrix}, \quad \mathbf{f}(t) = \begin{bmatrix} 0 \\ \mathbf{M}_L^{-1}\mathbf{h}_L \end{bmatrix}. \tag{5}$$

In this study, the optimal design of control parameters for linear differential systems with time-periodic coefficients is addressed. Firstly, an overview of the numerical algorithm for calculating stable conditions of linear differential systems with time-periodic coefficients is presented in Sect. 2. In the next sections, a procedure based on Taguchi method for optimal design of the stable parameters of a system described by Eq. (2) is proposed with some concluding remarks. The proposed approach is then applied to a single-link flexible manipulator that perform a simple harmonic motion.

2 Numerical Calculation of Stable Conditions of Linear Differential Systems with Time-Periodic Coefficients: A Review

Consider a system of homogeneous differential equations as

$$\dot{\mathbf{x}} = \mathbf{P}(t)\mathbf{x} \tag{6}$$

where $\mathbf{P}(t)$ is a continuous T-periodic matrix with $n \times n$. According to Floquet theory [12–16], the characteristic equation of Eq. (6) is independent from the fundamental solutions. Therefore, the characteristic equation can be formulated by the following way.

Firstly, we specify a set of n initial conditions $\mathbf{x}_i(0)$ for $i = 1, ..., n$ with the following elements

$$x_s^{(i)} = \begin{cases} 1, & s = i \\ 0, & s \neq i \end{cases} \tag{7}$$

and $[\mathbf{x}_1(0), \mathbf{x}_2(0), \ldots, \mathbf{x}_n(0)] = \mathbf{I}$, where $\mathbf{I}$ denote $n \times n$ identity matrix. Taking numerical integration of Eq. (6) within interval $[0, T]$ for n given initial conditions respectively, we obtain n vectors $\mathbf{x}_i(T)$, $i = 1, \ldots, n$. Matrix $\boldsymbol{\Phi}(t)$ defined by

$$\boldsymbol{\Phi}(T) = [\mathbf{x}_1(T), \mathbf{x}_2(T), \ldots, \mathbf{x}_n(T)] \tag{8}$$

is called the monodromy matrix [15] of Eq. (6). The characteristic equation of Eq. (6) can then be written in the form

$$|\mathbf{\Phi}(T) - \rho\mathbf{I}| = \begin{vmatrix} x_1^{(1)}(T) - \rho & x_2^{(1)}(T) & \dots & x_n^{(1)}(T) \\ x_1^{(2)}(T) & x_2^{(2)}(T) - \rho & \dots & x_n^{(2)}(T) \\ \ddots & \ddots & \dots & \ddots \\ x_1^{(n)}(T) & x_2^{(n)}(T) & \dots & x_n^{(n)}(T) - \rho \end{vmatrix} = 0 \tag{9}$$

The Eq. (9) yields a n order polynomial equation

$$\rho^n + a_1\rho^{n-1} + a_2\rho^{n-2} + \ldots + a_{n-1}\rho + a_n = 0 \tag{10}$$

Roots $\rho_i, i = 1, \ldots, n$ of Eq. (10) are called Floquet multipliers of (6). Based on these *Floquet multipliers* stability criteria of (6) are given as following:

- If the moduli of all the *Floquet multipliers* of the characteristic Eq. (10) are less than the unity, then the periodic system (6) is asymptotically stable at the origin.
- If even one of *Floquet multipliers* of the characteristic equation has a modulus larger than unity then the periodic system (6) is asymptotically unstable at the origin.
- If there is no *Floquet multiplier* of the characteristic Eq. (10) with a modulus greater than unity, but there is a *Floquet multiplier* with a modulus equal to the unity, then the solution of the system of differential Eqs. (6) may be stable, and may also be unstable, depends on the nonlinear terms.

The problem of the stability control of linear differential equation systems with time-periodic coefficients is as follows. In the general case, the solution to the characteristic Eq. (10) is a function of m parameters $u_1, u_2, ..., u_m$. Based on the stability criteria according to the Floquet multipliers [12, 13], we derive the definition of the parameter vector as follows.

Definition: The parameter vector

$$\mathbf{u} = [u_1\ u_2 \ldots u_m]^T \tag{11}$$

of the differential equation system (6) is called asymptotic stable parameter vector if all the Floquet multipliers $\rho_k(u_1, u_2, \ldots, u_m)$ of the characteristic Eq. (10) have modulus less than unity. Conversely, if at least one Floquet multiplier of the characteristic Eq. (10).

In case the parameter vector $\mathbf{u}$ is not stable, we add a simple PD controller to the system to force it stable. The m freely selectable parameters $u_1, u_2, \ldots, u_m$ of the coefficients of the linear differential equation system (6) determine an m-dimensional solution space. It is limited in engineering so that these parameters can only be changed in a certain specified domain. Thus, there are obtained the following constraints:

$$u_{i\min} \le u_i \le u_{i\max}\ (i = 1, \ldots, m) \tag{12}$$

3 Dynamic Model of a Single-Link Flexible Manipulator

Considering a single-link flexible manipulator as shown in Fig. 1, link OE of length l with a payload at the free end rotates about vertical axis O. The tip mass m_E is attached at E. The link is considered as a homogeneous beam with parameters are shown in Table 1.

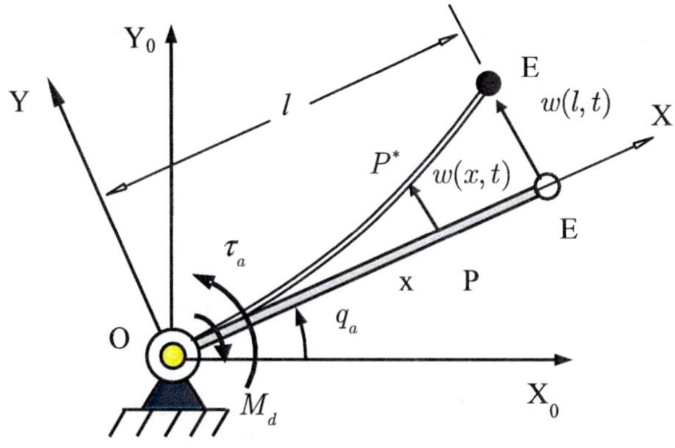

Fig. 1. Single-link flexible manipulator

Table 1. Parameters of the manipulator

Parameters of the model	Variable and unit	Value
Length of link	l (m)	0.9
Sectional area of beam	A (m^2)	4×10^{-4}
Density of beam	ρ (kg/m^3)	2700
Area moment of inertia	I (m^4) = $bh^3/12$	1.3333×10^{-8}
Modulus	E (N/m^2)	7.11×10^{10}
Mass moment of inertia of link 1 (including the hub)	J_1 (kg.m^2)	5.86×10^{-5}
Tip mass	m_E (kg)	0.1
Damping coefficient	α $(N.m.s/rad)$	0.01

The fundamental motion of the manipulator corresponding to applied torque $\boldsymbol{\tau}^R(t)$ is described by $\mathbf{q}^R(t)$, in which the beam is considered as a rigid link. The generalized coordinate of a manipulator is

$$\mathbf{q}^R(t) = [\,q_a^R(t)\; q_e^R(t)\,]^T = [\,q_a^R(t)\; 0\,]^T. \tag{13}$$

and the torque $\boldsymbol{\tau}^R(t)$ is

$$\boldsymbol{\tau}^R(t) = [\,\tau_a^R\; \tau_e^R\,]^T = [\,\tau_a^R\; 0\,]^T \tag{14}$$

In Eqs. (13) and (14), $q_e^R(t)$ denotes the elastic generalized coordinate and $\tau_e^R(t)$ the elastic torque of the virtual rigid link. The differential equations of a single-link flexible manipulator can be expressed in the following matrix form [16]

$$\mathbf{M}(\mathbf{q})\ddot{\mathbf{q}} + \mathbf{C}(\mathbf{q}, \dot{\mathbf{q}})\dot{\mathbf{q}} + \mathbf{g}(\mathbf{q}) = \boldsymbol{\tau}(t) \tag{15}$$

where $\mathbf{q}$, $\dot{\mathbf{q}}$ and $\ddot{\mathbf{q}}$ are vectors of generalized coordinate, velocity and acceleration, respectively

$$\mathbf{q} = [q_a, q_e]^T, \ \boldsymbol{\tau}(t) = [\tau_a(t), \tau_e(t)]^T = [\tau_a(t), 0]^T \tag{16}$$

Let Δq_a and Δq_e be the difference between the real motion $\mathbf{q}(t)$ and the fundamental motion $\mathbf{q}^R(t)$, it follows that

$$q_a(t) = q_a^R(t) + \Delta q_a(t) = q_a^R(t) + y_1(t) \tag{17}$$

$$q_e(t) = q_e^R(t) + \Delta q_e(t) = y_2(t) \tag{18}$$

where y_1 and y_2 are called the perturbed motions. Similarly, we have

$$\boldsymbol{\tau}(t) = [\tau_a(t), \tau_e(t)]^T = [\tau_a(t), 0]^T \tag{19}$$

Substituting Eqs. (17), (18) into Eq. (15) and using Taylor series expansion around the fundamental motion, then neglecting nonlinear terms, we obtain a system of linear differential equations with time-varying coefficients for the manipulator as follows [16]

$$\mathbf{M}_L(t)\ddot{\mathbf{y}} + \mathbf{C}_L(t)\dot{\mathbf{y}} + \mathbf{K}_L(t)\mathbf{y} = \mathbf{h}_L(t). \tag{20}$$

Matrices $\mathbf{M}_L(t)$, $\mathbf{C}_L(t)$, $\mathbf{K}_L(t)$ and vector $\mathbf{h}_L(t)$ in Eq. (20) have the following form [16]

$$\mathbf{M}_L(t) = \begin{bmatrix} J_1 + m_E l^2 + \frac{1}{3} m_{OE} l^2 & \rho A D_1 + m_E l X_1(l) \\ m_E l X_1 + \rho A D_1 & m_E X_1^2(l) + \rho A m_{11} \end{bmatrix} \tag{21}$$

$$\mathbf{C}_L(t) = \begin{bmatrix} \alpha & 0 \\ 0 & 0 \end{bmatrix} \tag{22}$$

$$\mathbf{K}_L(t) = \begin{bmatrix} k_{11} & k_{12} \\ k_{21} & k_{22} \end{bmatrix} \tag{23}$$

where

$$\begin{aligned} k_{11} &= -l \sin q_a^R(t) m_E g - \frac{m_{OE} g l \sin q_a^R(t)}{2}, \\ k_{12} = k_{21} &= -m_E g X_1(l) \sin q_a^R(t) - \mu g \sin q_a^R(t) C_1, \\ k_{22} &= -m_E [\dot{q}_a^R(t)]^2 X_1^2(l) - \rho A [\dot{q}_a^R(t)]^2 m_{11} + EIk_{11}^*. \end{aligned}$$

and

$$\mathbf{h}_L(t) = \begin{bmatrix} 0 \\ -m_E g X_1(l)\cos q_a^R(t) - \mu g \cos q_a^R(t) C_1 - m_E l X_1 \ddot{q}_a^R(t) - \rho A D_1 \ddot{q}_a^R(t) \end{bmatrix} \quad (24)$$

where fundamental motion is choosen as $q_a^R(t) = 0.5\pi + 0.5\pi\,\sin(\Omega t)$ and C_1, D_1, X_1, m_{11}, k_{11}^* are constants. It should be noted that matrices $\mathbf{M}_L(t)$, $\mathbf{C}_L(t)$, $\mathbf{K}_L(t)$ and vector $\mathbf{h}_L(t)$ in this example are time-periodic with period T.

In order to investigate the dynamic stability of an elastic single-link robot, we consider the homogeneous differential equation corresponding to Eq. (20)

$$\mathbf{M}_L(t)\ddot{\mathbf{y}} + \mathbf{C}_L(t)\dot{\mathbf{y}} + \mathbf{K}_L(t)\mathbf{y} = 0. \quad (25)$$

For numerical simulation, the calculating parameters of the considered manipulator are listed in Table 1.

It follows from the parameters in Table 1 that

$$C_1 = -0.704632, \quad D_1 = -0.460710,$$
$$m_{11} = 0.899850, \quad k_{11}^* = 16.955151, \quad X_1 = -1,9987$$

Some calculation results of the maximum value of the Floquet multiplier are listed in Table 2.

Table 2. Modulus of Floquet multiplier for four cases

Case 1: $\Omega = 2\pi$	$\lvert\rho_1\rvert = 13.7797,\ \lvert\rho_2\rvert = 0.0706,\ \lvert\rho_3\rvert = 0.5651,\ \lvert\rho_4\rvert = 0.5651.$
Case 2: $\Omega = 4\pi$	$\lvert\rho_1\rvert = 3.7506,\ \lvert\rho_2\rvert = 0.2628,\ \lvert\rho_3\rvert = 0.7517,\ \lvert\rho_4\rvert = 0.7517.$
Case 3: $\Omega = 6\pi$	$\lvert\rho_1\rvert = 2.4175,\ \lvert\rho_2\rvert = 0.4097,\ \lvert\rho_3\rvert = 0.8268,\ \lvert\rho_4\rvert = 0.8268.$
Case 4: $\Omega = 8\pi$	$\lvert\rho_1\rvert = 1.9396,\ \lvert\rho_2\rvert = 0.5119,\ \lvert\rho_3\rvert = 0.8674,\ \lvert\rho_4\rvert = 0.8674.$

With the initial condition

$$t = 0 : \mathbf{x}(0) = [0\ 0\ 0.25\pi\ \ 0]^T \quad (26)$$

we calculate transient vibration of the flexible manipulator with the parameters given in Table 2. Some calculation results of the transient vibration are shown in Figs. 2, 3, 4 and 5.

From Table 2 and Figs. 2, 3, 4 and 5 we can see that in the investigated cases the maximum values of the Floquet multiplier are greater than 1 and the transition oscillations tend to increase gradually. Therefore, the study of dynamic stability conditions is necessary in controlling the flexible manipulator.

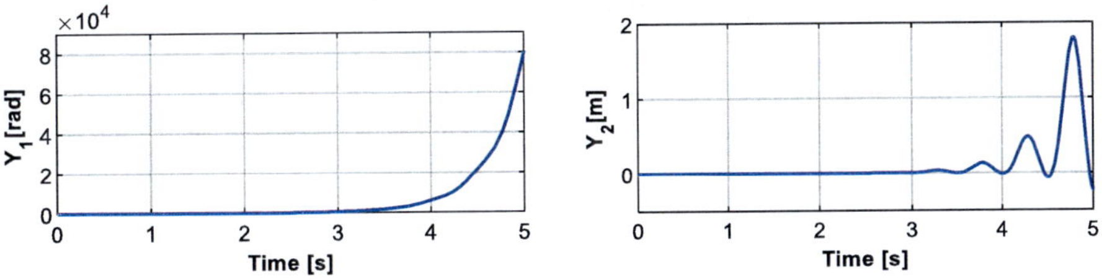

Fig. 2. Transient vibration of the flexible manipulator with $\Omega = 2\pi$

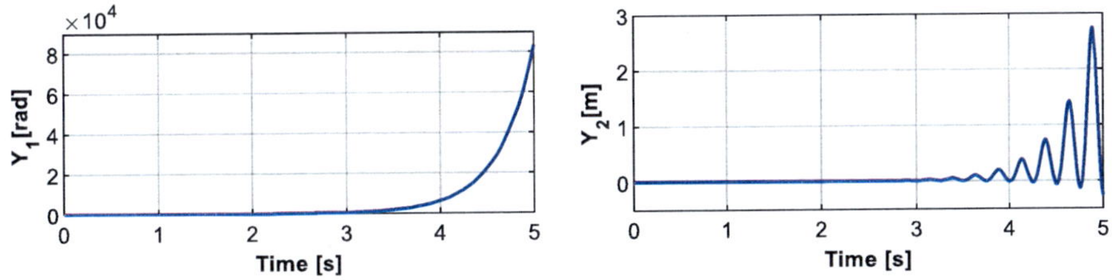

Fig. 3. Transient vibration of the flexible manipulator with $\Omega = 4\pi$

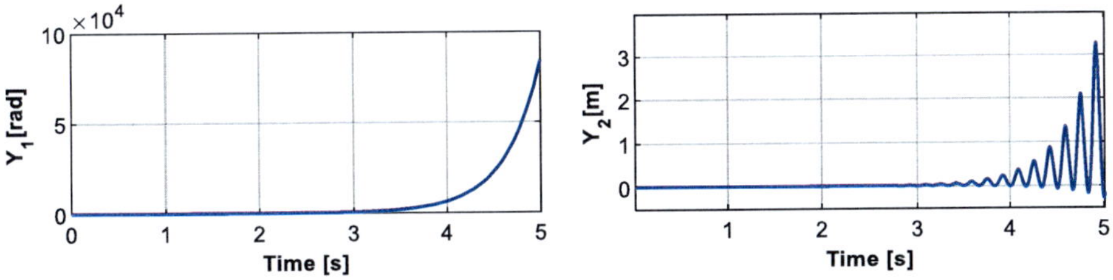

Fig. 4. Transient vibration of the flexible manipulator with $\Omega = 6\pi$

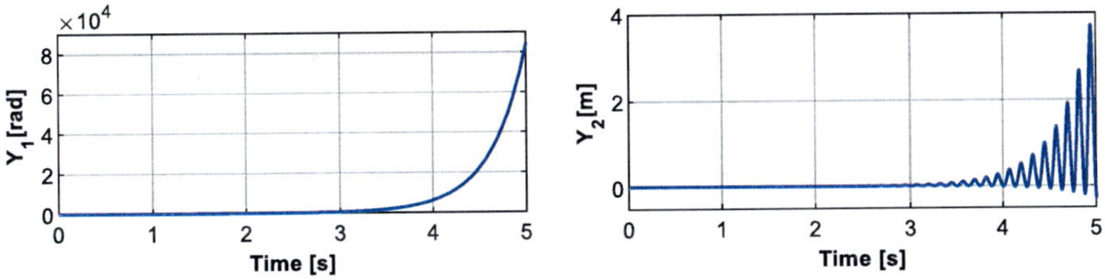

Fig. 5. Transient vibration of the flexible manipulator with $\Omega = 8\pi$

4 Dynamic Stability Control of a Flexible Manipulator Using the Taguchi Method

In the harmonic fundamental motion of a flexible manipulator, the matrices and vector of the linear differential Eq. (20) are time-periodic with period T. In this section, we propose an algorithm to control the dynamic stability of a flexible manipulator.

4.1 The PD Controller

It should be noted that for the stability control of a flexible manipulator, we can design a PD controller as follows

$$\Delta\tau_a = -k_{d1}(\dot{q}_a - \dot{q}_a^R) - k_{p1}(q_a - q_a^R) = -k_{d1}\dot{y}_1 - k_{p1}y_1 \tag{27}$$

The linearized equation according to Eq. (20) now takes the expression

$$\mathbf{M}_L(t)\ddot{\mathbf{y}} + \mathbf{C}_L(t)\dot{\mathbf{y}} + \mathbf{K}_L(t)\mathbf{y} = \mathbf{h}_L(t) - \mathbf{K}_D\dot{\mathbf{y}} - \mathbf{K}_P\mathbf{y}, \tag{28}$$

where $\mathbf{K}_D$ and $\mathbf{K}_P$ are diagonal matrices with positive elements as

$$\mathbf{K}_D = \begin{bmatrix} k_{d_1} & 0 \\ 0 & 0 \end{bmatrix}; \ \mathbf{K}_P = \begin{bmatrix} k_{p_1} & 0 \\ 0 & 0 \end{bmatrix}. \tag{29}$$

It follows from Eqs. (28) that

$$\mathbf{M}_L(t)\ddot{\mathbf{y}} + [\mathbf{C}_L(t) + \mathbf{K}_D]\dot{\mathbf{y}} + [\mathbf{K}_L(t) + \mathbf{K}_P]\mathbf{y} = \mathbf{h}_L(t) \tag{30}$$

Equation (30) can then be written in the form

$$\mathbf{M}_L^{(1)}(t)\ddot{\mathbf{y}} + \mathbf{C}_L^{(1)}(t)\dot{\mathbf{y}} + \mathbf{K}_L^{(1)}(t)\mathbf{y} = \mathbf{h}_L^{(1)}(t) \tag{31}$$

where

$$\mathbf{M}_L^{(1)}(t) = \mathbf{M}_L(t), \mathbf{K}_L^{(1)}(t) = \mathbf{K}_L(t) + \mathbf{K}_P, \mathbf{C}_L^{(1)}(t) = \mathbf{C}_L(t) + \mathbf{K}_D, \mathbf{h}_L^{(1)}(t) = \mathbf{h}_L(t) \tag{32}$$

It should be noted that, the Eq. (31) can then be expressed in the compact form as Eq. (2). To study the dynamic stability conditions of the manipulators, the homogeneous linear differential system corresponding to Eq. (31) can be written in the following form

$$\dot{\mathbf{x}} = \mathbf{P}(t)\mathbf{x}, \tag{33}$$

where $\mathbf{P}(t)$ is a matrix of periodic elements with period T.

4.2 Determination of Gain Values According to Floquet Multipliers by the Taguchi Method

Using the Taguchi method [8, 9], Khang et al. proposed an algorithm to determine the optimal parameters of the TMDs to reduce the vibrations of the mechanical systems described by the system of linear differential equations with constant coefficients [10, 11]. The main problem in the papers [10, 11] is to calculate the eigenvalues of the constant matrix. In this paper, we use the Taguchi method to determine the gain values of the PD controller for the system the system of linear differential equations with periodic coefficients (31). The main task of the problem of determining the control parameters of the periodic system of linear differential equations is to determine the Floquet multipliers of the periodic matrix. The problem of determining the control parameters of the periodic system of linear differential equations is a new problem. The problem of determining the Floquet multipliers of a periodic matrix is much more difficult than the problem of determining the eigenvalues of a constant matrix.

The target function needs to minimized is defined by:

$$f(\mathbf{u}) = \max_i |\rho_i(\mathbf{u})| - \rho_d \to \min, \text{ with } \mathbf{u} = [k_{p1}, k_{d1}]^T. \tag{34}$$

In which $\max_i |\rho_i(\mathbf{u})|$ is the biggest modulus of Floquet multipliers in the ith experiment, and ρ_d is the target Floquet multiplier. The desired value of the target Floquet multiplier is usually chosen empirically. Some calculation results of the maximum value of the Floquet multipliers are presented in Table 3.

Table 3. Control parameters and Floquet multipliers

Ω	ρ_d	k_{p1}	k_{d1}	$\lvert\rho\rvert$
2π	0.3	37.1617	29.2410	$\lvert\rho_1\rvert = 0.3,\ \lvert\rho_2\rvert = 0,\ \lvert\rho_3\rvert = 0,\ \lvert\rho_4\rvert = 0$
4π	0.3	28.7617	11.7501	$\lvert\rho_1\rvert = 0.3,\ \lvert\rho_2\rvert = 0,\ \lvert\rho_3\rvert = 0,\ \lvert\rho_4\rvert = 0$
6π	0.3	22.2666	6.7208	$\lvert\rho_1\rvert = 0.3,\ \lvert\rho_2\rvert = 0.0042,\ \lvert\rho_3\rvert = 0,\ \lvert\rho_4\rvert = 0$
8π	0.4	23.0147	6.8628	$\lvert\rho_1\rvert = 0.4,\ \lvert\rho_2\rvert = 0.0148,\ \lvert\rho_3\rvert = 0,\ \lvert\rho_4\rvert = 0$

4.3 Simulation Results

Using the initial condition

$$t = 0 : \mathbf{x}(0) = [0\ 0\ 0.25\pi\ \ 0]^T \tag{35}$$

we can calculate transient vibration of the flexible manipulator with the parameters given in Table 3. Some calculation results of the transient vibration are shown in Figs. 6, 7, 8 and 9.

From Figs. 6, 7, 8 and 9, we can see that with the selected control parameter, the transient vibration of the flexible manipulator decreases rapidly to zero. In other words, the dynamic stability of the flexible manipulator is guaranteed by a simple PD controller.

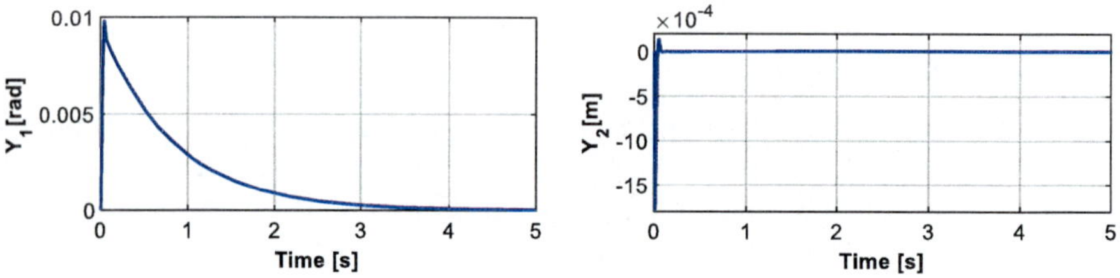

Fig. 6. Transient vibration of the flexible manipulator with control torque case $\Omega = 2\pi$

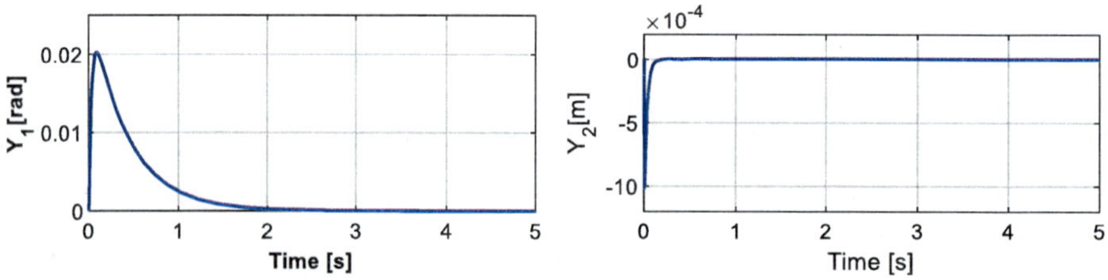

Fig. 7. Transient vibration of the flexible manipulator with control torque case $\Omega = 4\pi$

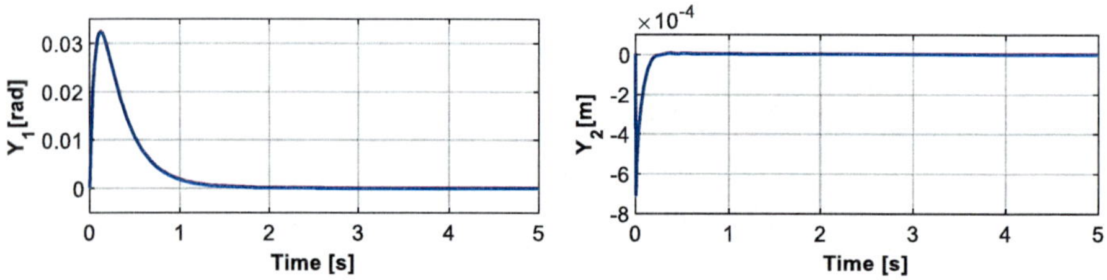

Fig. 8. Transient vibration of the flexible manipulator with control torque case $\Omega = 6\pi$

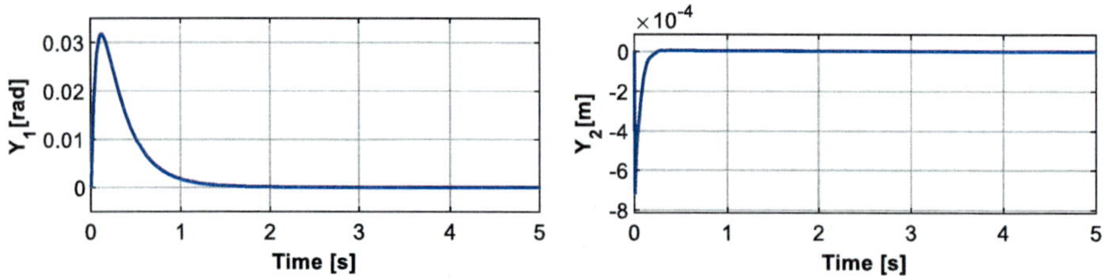

Fig. 9. Transient vibration of the flexible manipulator with control torque case $\Omega = 8\pi$

5 Conclusions

The selection of the stable parameters for the linear differential equation system is an important part of the dynamic problem for flexible manipulators. This paper presented a procedure for the optimal design of control parameters of the homogeneous linear differential equations with time-periodic coefficients. The new findings made in this study are summarized as follows:

1) Using the Taguchi method, a procedure to optimally design the stability control parameters of a system of homogeneous linear differential equations with periodic coefficients over time has been proposed.
2) Numerical calculation of the dynamic stability properties of a single-link flexible manipulator according to the Taguchi method has been implemented.
3) The method proposed in this paper can be used to calculate control parameters for multi-link flexible robots.

Acknowledgement. This paper was completed with the financial support of the Vietnam National Foundation for Science and Technology Development (NAFOSTED) under grant number 107.04-2020.28.

References

1. Schiehlen, W., Eberhard, P.: Applied Dynamics. Springer, Switzerland (2014). https://doi.org/10.1007/978-3-319-07335-4
2. Ceccarelli, M.: Fundamentals of Mechanics of Robotics Manipulation. Springer, Dodrecht (2004). https://doi.org/10.1007/978-1-4020-2110-7
3. Van Khang, N.: Dynamics of Multibody Systems. 2 edn. Science and Technic House, Hanoi (2017). (in Vietnamese)
4. Van Khang, N.: Dynamische Stabilität und periodische Schwingungen in Mechanismen. Diss.B, TH Karl-Marx-Stadt (1986)
5. Van Khang, N., Dien, N.P., Cuong, H.M.: Linearization and parametric vibration analysis of some applied problems in multibody systems. Multibody Syst. Dyn. **22**, 163–180 (2009)
6. Van Khang, N., Dien, N.P.: Parametric vibration analysis of transmission mechanisms using numerical methods. In: Advances in Vibration Engineering and Structural Dynamics, Edited by F.B. Carbajal, Intech, Croatia, pp.301–331 (2012)
7. Van Khang, N., Sy Nam, N., Van Quyen, N.: Symbolic linearization and vibration analysis of constrained multibody systems. Arch. Appl. Mech. **88**(8), 1369–1384 (2018). https://doi.org/10.1007/s00419-018-1376-8
8. Roy, R.K.: A Primer on the Taguchi Method. Society of Manufacturing Engineers, USA (2010)
9. Zambanini, R.A.: The application of Taguchi's method of parameter design to the design of mechanical systems. Master thesis, Lehigh University (1992)
10. Van Khang, N., Phuc, V.D., Duong, D.T., Van Huong, N.T.: A procedure for optimal design of a dynamic vibration absorber installed in the damped primary system based on Taguchi's method. Vietnam J. Sci. Technol. VAST **55**(5), 649–661 (2018)

11. Van Khang, N., Duong, D.T., Van Huong, N.T., Dinh, N.D.T.T., Phuc, V.D.: Optimal control of vibration by multiple tuned liquid dampers using Taguchi method. J. Mech. Sci. Technol. **33**(4), 1563–1572 (2019)
12. Hale, J.K.: Oscillations in Nonlinear Systems, 3rd edn. Dover Publications Inc., New York (2015)
13. Demidovich, B.P.: Lectures on the Mathematical Stability Theory. Nauka, Moscow (1967). (in Rusian)
14. Mitropolskii, Y., Van Dao, N.: Applied Asymptotic Methods in Nonlinear Oscillations. Klawer Academic Publisher, Dordrecht (1997)
15. Dao, N.V.: Stability of Dynamic Systems. Vietnam National University Publishing House, Hanoi (1998)
16. Van Khang, N., Dat, D.C., Nam, N.S.: Stability control and inverse dynamics of a single-link flexible manipulator. In: Proceedings of the National Conference on Dynamics and Control, Danang 2019, pp. 167–176. Natural Sciences and Technology Publishing House, Hanoi (2019). (in Vietnamese)

Design of a Cubic Drone, a Foldable Quadcopter that Can Rotate Its Arm Vertically

Chu Tan Cuong[1] and Phuong Thao Thai[2](✉)

[1] Hanoi University of Science and Technology, 1 Dai Co Viet, Hanoi, Vietnam
[2] Department of Applied Mechanics, Hanoi University of Science and Technology, 1 Dai Co Viet, Hanoi, Vietnam
thao.thaiphuong@hust.edu.vn

Abstract. Several studies on the flexible designs of unmanned aerial vehicles that can transform their shapes during flights to adapt to the environment have been conducted recently. In this paper, we propose a morphing design for a quadcopter that can fold itself into a cube when it is disabled and can deploy its arms holding the propellers in the active mode. When all the arms are stretched to the horizontal position, the cubic drone works like a conventional quadcopter. The arms morph to adapt to the environment or serve special purposes, such as flying through a narrow gap, relieving the effect of wind or creating a flyable assembled formation of drones. Considering the stretching angles, the dynamics of the cubic drone is different from that of a conventional drone. The equations of motion of the cubic drone are carried out to study its dynamics. The equations are applied to simulate an illustrative example via MATLAB. The results show that it is a step forward to improve drones with better versatility and more applications.

Keywords: Cubic drone · Quadcopter · Adaptive morphology

1 Introduction

Unmanned aerial vehicles (UAVs), also known as drones, are continuously being developed and applied to various fields, such as surveillance, inspections, rescue, military, entertainment, etc. Thanks to its flying ability, the drone can easily maneuver in large space overhead. But the higher it flies, the more it is affected by the wind, so the drone needs to ensure its balance and rigidity to avoid somersaults. Additionally, the bulkier the drone is, the more energy will be consumed which results in less flight time. To solve those drawbacks, many studies on the ability to transform and assemble drones in the air have been carried out, which can open new potentials for drones.

In terms of transformation ability, current commercial drones, such as the DJI's drone series [1], are designed to be compact and foldable. However, this foldability is used to save space when the drone is disabled, not during flights. Likewise, Dufour had proposed a drone with origami-inspired folding wings [2] but it can only be implemented for flat-wing aircrafts. In terms of assembly, several studies on combining drones and controlling them in formation have shown the potential for drone application in complex tasks.

N. V. Khang et al. (Eds.): ASIAN MMS 2021, MMS 113, pp. 501–511, 2022.
https://doi.org/10.1007/978-3-030-91892-7_47

Zhao et al. had introduced an aerial robot dragon [3, 4], which is an assembly of two-dimensional multi-links and can morph during flights. ModQuad in [5], a combination of multiple drones linked by magnets, can self-assemble in the air by repositioning the frames in a horizontal plane with respect to the drone.

We took inspiration to design and create a cubic drone, a quadcopter that not only can fold itself into a cube but also combine with other cubic drones to create different shapes as building blocks. While the foldable drone investigated by Falanga et al. [6] can transform itself into different morphologies by adding new four horizontal angles of the arms, our cubic drone adopts four vertical angles of the arms instead (see Fig. 1). We believe that the combination of foldability and assembly promises to not only bring a breakthrough in the versatility of UAVs but also allow them to perform difficult tasks.

This article proposes our cubic drone design and build its dynamics controller. The structure of this article is organized as follows. In Sect. 2, the design of our cubic drone will be demonstrated. The differential equations of motion are demonstrated next in Sect. 3. Section 4 presents the simulations of illustrative examples.

2 The Cubic Drone Design

The mechanical design of our cubic drone consists of three main sections (see Fig. 1): (1) a fuselage, (2) four arm frames and (3) four four-bar-linkage mechanisms.

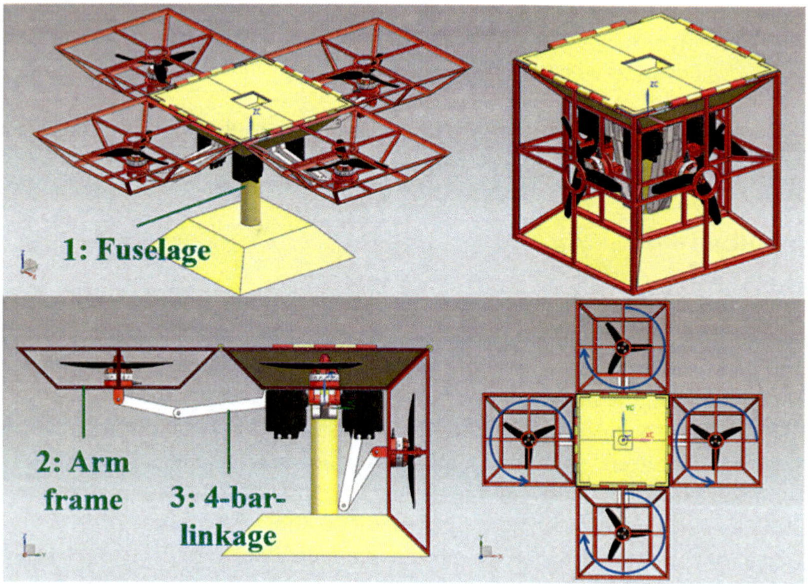

Fig. 1. The Computer Aided Design (CAD) model of our cubic drone

The fuselage is designed in a shape of a hand dumbbell, with the upper and lower parts containing electrical components. Each arm frame links to the upper part of the fuselage by a swivel joint, which allows the arm to swing up (or down to fold the drone into a cube). The planar four-bar linkage mechanism is used to control the arm angle. With the same bar lengths and input angle, there are two possible positions of pivot C (Fig. 2). Since we want to control the input bar AB to rotate in the same direction as the output bar DC, the ABCD configuration must form a convex polygon. This mechanism has only 1 degree of freedom; hence, for a given input angle α_1 about the anchor pivot A, the output angle α_3 about the anchor pivot D is calculated as follows

$$\alpha_3 = 180^o - \widehat{ADB} - \widehat{BDC} = 180^o - \sin^{-1}\left(\frac{AB}{DB}\sin\alpha_1\right) - \cos^{-1}\left(\frac{DB^2 + DC^2 - BC^2}{2 \cdot DB \cdot DC}\right) \quad (1)$$

where $DB^2 = AD^2 + AB^2 - 2 \cdot AB \cdot AD \cdot \cos\alpha_1$.

Assuming the bar AB rotates with the input angular velocity ω_A, the output angular velocity ω_D of the bar DC is defined by derivation the Eq. (1) with respect to time

$$\omega_D = \omega_A \cdot \frac{AB}{DC} \cdot \frac{\sin(\alpha_1 - \alpha_2)}{\sin(\alpha_3 - \alpha_2)} \quad (2)$$

where $\alpha_2 = \tan^{-1}\left(\frac{DC \cdot \sin\alpha_3 - AB \cdot \sin\alpha_1}{AD + DC \cdot \cos\alpha_3 - AB \cdot \cos\alpha_1}\right)$.

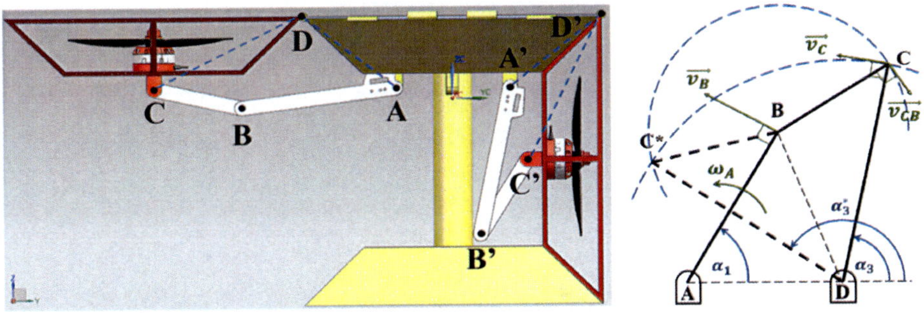

Fig. 2. The four-bar linkage mechanism (left) and its kinematics model (right)

In short, with the input angle α_1 and input angular velocity ω_A controlled by the servo, we can measure the output angle α_3 and output angular velocity ω_D.

3 Differential Equations of Motion of the Cubic Drone

The simplified model of the cubic drone (Fig. 3) includes the inertial frame $Ox_0y_0z_0$ and the body frame $Bx_By_Bz_B$. Origin B is the geometric center of the top surface and its position, translational speeds and acceleration in the inertial frame sequentially are

$$^{\mathbf{0}}\mathbf{r_B} = \begin{bmatrix} {}^0x_B \\ {}^0y_B \\ {}^0z_B \end{bmatrix}; {}^{\mathbf{0}}\mathbf{v_B} = {}^{\mathbf{0}}\dot{\mathbf{r}}_\mathbf{B} = \begin{bmatrix} {}^0\dot{x}_B \\ {}^0\dot{y}_B \\ {}^0\dot{z}_B \end{bmatrix}; {}^{\mathbf{0}}\mathbf{a_B} = {}^{\mathbf{0}}\ddot{\mathbf{r}}_\mathbf{B} = \begin{bmatrix} {}^0\ddot{x}_B \\ {}^0\ddot{y}_B \\ {}^0\ddot{z}_B \end{bmatrix} \quad (3)$$

The rotational movement of drone along x, y, z axes is described by Roll (φ), Pitch (θ) and Yaw (ψ) angles. Accordingly, the rotation matrix ${}^{0}\mathbf{R_B}$ from the body frame to the inertial frame is determined by rotating sequentially Roll, Pitch and Yaw angles as [7]

$$ {}^{0}\mathbf{R_B} = {}^{0}\mathbf{R_z}(\psi){}^{0}\mathbf{R_y}(\theta){}^{0}\mathbf{R_x}(\varphi) = \begin{bmatrix} c_\psi c_\theta & -s_\psi c_\varphi + c_\psi s_\theta s_\varphi & s_\psi s_\varphi + c_\psi s_\theta c_\varphi \\ s_\psi c_\theta & c_\psi c_\varphi + s_\psi s_\theta s_\varphi & -c_\psi s_\varphi + s_\psi s_\theta c_\varphi \\ -s_\theta & c_\theta s_\varphi & c_\theta c_\varphi \end{bmatrix} \tag{4} $$

where c_ψ, s_ψ stand for cosψ, sinψ and the other angles are applied similarly.

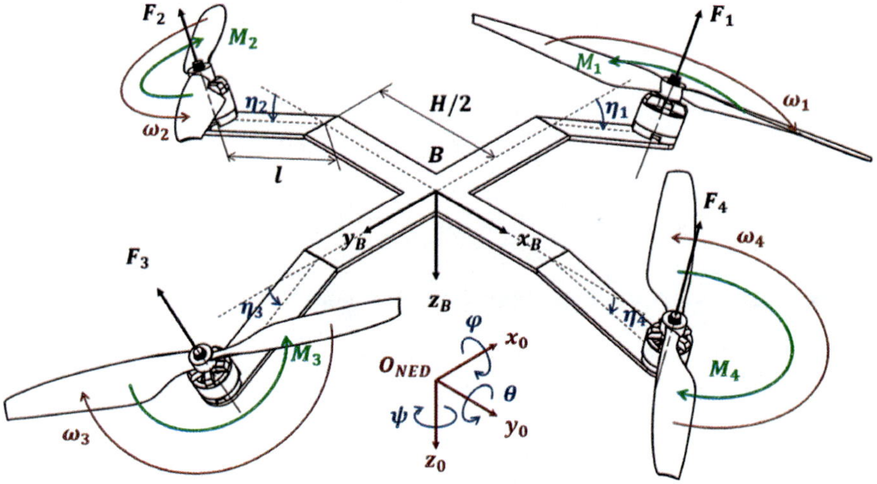

Fig. 3. The simplified model of the proposed cubic drone

The angular velocities in the body frame ${}^{\mathbf{B}}\boldsymbol{\omega}$ can be calculated as in [7] as

$$ {}^{\mathbf{B}}\boldsymbol{\omega} = \begin{bmatrix} {}^{B}\omega_x \\ {}^{B}\omega_y \\ {}^{B}\omega_z \end{bmatrix} = \begin{bmatrix} 1 & 0 & -s_\theta \\ 0 & c_\varphi & c_\theta s_\varphi \\ 0 & -s_\varphi & c_\theta c_\varphi \end{bmatrix} \begin{bmatrix} \dot{\varphi} \\ \dot{\theta} \\ \dot{\psi} \end{bmatrix} \tag{5} $$

The morphology of our cubic drone has a direct impact on (i) the location of the center of mass C_{drone}, (ii) the inertial tensor ${}^{\mathbf{B}}\mathbf{I}$ at the center of mass (CoM), and (iii) the effect each rotor thrust has on the drone's dynamics. To simplify, we consider the cubic drone as a multibody system consisting of a fuselage and four arm frames, ignoring the small weight of the four-bar mechanisms. Assuming the arms have rotated with the angles $\boldsymbol{\eta} = \begin{bmatrix} \eta_1 & \eta_2 & \eta_3 & \eta_4 \end{bmatrix}^T$, the location of C_{drone} in the body frame is defined as

$$
{}^{\mathbf{B}}\mathbf{r_C} = \frac{m_{body} \cdot {}^{\mathbf{B}}\mathbf{r}_{C_{body}} + m_{arm} \cdot \sum_1^4 {}^{\mathbf{B}}\mathbf{r}_{C_{armi}}}{m_{body} + 4 \cdot m_{arm}} = \frac{\begin{bmatrix} m_{arm}\left({}^{B}z_{G_{arm}}(s_{\eta 2} - s_{\eta 4}) - l(c_{\eta 2} - c_{\eta 4})\right) \\ m_{arm}\left({}^{B}z_{G_{arm}}(s_{\eta 1} - s_{\eta 3}) - l(c_{\eta 1} - c_{\eta 3})\right) \\ m_{body} \cdot {}^{B}z_{G_{body}} + m_{arm}\left({}^{B}z_{G_{arm}} \cdot \sum_1^4 c_{\eta_i} + l \cdot \sum_1^4 s_{\eta_i}\right) \end{bmatrix}}{m} \quad (6)
$$

Subsequently, the acceleration of C_{drone} in the inertial frame is calculated as in [7] as

$$
{}^{0}\ddot{\mathbf{r}}_{\mathbf{C}} = {}^{0}\ddot{\mathbf{r}}_{\mathbf{B}} - {}^{\mathbf{B}}\widetilde{\mathbf{r}}_{\mathbf{C}} \cdot {}^{\mathbf{B}}\dot{\boldsymbol{\omega}} + \left(\widetilde{{}^{\mathbf{B}}\widetilde{\mathbf{r}}_{\mathbf{C}} \cdot {}^{\mathbf{B}}\boldsymbol{\omega}}\right) \cdot {}^{\mathbf{B}}\boldsymbol{\omega} \quad (7)
$$

in which ${}^{\mathbf{B}}\widetilde{\mathbf{r}}_{\mathbf{C}}$ is the skew-symmetric matrix of vector ${}^{\mathbf{B}}\mathbf{r}_{\mathbf{C}}$.

With the above assumption, the drone's inertial tensor can be calculated by using the parallel axis theorem. Thus, it is vital to convert the principal inertia axes of each part to be parallel to the body frame. We figured out that the moment of inertia (MoI) about the origin B of the body frame depends on the rotation angle of each arm as follows

$$
{}^{\mathbf{B}}\mathbf{I}(\boldsymbol{\eta}) = {}^{\mathbf{B}}\mathbf{I}_{\mathbf{arm\,1}} + {}^{\mathbf{B}}\mathbf{I}_{\mathbf{arm\,2}} + {}^{\mathbf{B}}\mathbf{I}_{\mathbf{arm\,3}} + {}^{\mathbf{B}}\mathbf{I}_{\mathbf{arm\,4}} + {}^{\mathbf{B}}\mathbf{I}_{\mathbf{body}} \quad (8)
$$

where

$$
{}^{\mathbf{B}}\mathbf{I}_{\mathbf{arm\,1}} = {}^{\mathbf{B}}\mathbf{R}_{\mathbf{x}}(-\eta_1) \cdot \mathbf{I}_{\mathbf{arm}} \cdot {}^{\mathbf{B}}\mathbf{R}_{\mathbf{x}}^{\mathbf{T}}(-\eta_1) - m_{arm} \cdot {}^{\mathbf{B}}\widetilde{\mathbf{r}}^{2}_{C_{\mathbf{arm\,1}}}
$$

$$
{}^{\mathbf{B}}\mathbf{I}_{\mathbf{arm\,2}} = {}^{\mathbf{B}}\mathbf{R}_{\mathbf{y}}(-\eta_2) \cdot \mathbf{I}_{\mathbf{arm}} \cdot {}^{\mathbf{B}}\mathbf{R}_{\mathbf{y}}^{\mathbf{T}}(-\eta_2) - m_{arm} \cdot {}^{\mathbf{B}}\widetilde{\mathbf{r}}^{2}_{C_{\mathbf{arm\,2}}}
$$

$$
{}^{\mathbf{B}}\mathbf{I}_{\mathbf{arm\,3}} = {}^{\mathbf{B}}\mathbf{R}_{\mathbf{x}}(\eta_3) \cdot \mathbf{I}_{\mathbf{arm}} \cdot {}^{\mathbf{B}}\mathbf{R}_{\mathbf{x}}^{\mathbf{T}}(\eta_3) - m_{arm} \cdot {}^{\mathbf{B}}\widetilde{\mathbf{r}}^{2}_{C_{\mathbf{arm\,3}}}
$$

$$
{}^{\mathbf{B}}\mathbf{I}_{\mathbf{arm\,4}} = {}^{\mathbf{B}}\mathbf{R}_{\mathbf{y}}(\eta_4) \cdot \mathbf{I}_{\mathbf{arm}} \cdot {}^{\mathbf{B}}\mathbf{R}_{\mathbf{y}}^{\mathbf{T}}(\eta_4) - m_{arm} \cdot {}^{\mathbf{B}}\widetilde{\mathbf{r}}^{2}_{C_{\mathbf{arm\,4}}}
$$

$$
{}^{\mathbf{B}}\mathbf{I}_{\mathbf{body}} = \mathbf{I}_{\mathbf{body}} - m_{body} \cdot {}^{\mathbf{B}}\widetilde{\mathbf{r}}^{2}_{C_{\mathbf{body}}}
$$

Let ω_i be the angular velocity of the ith rotor. Each rotor generates a thrust $F_i = C_F\omega_i^2$ with C_F being a thrust coefficient [8], and an aerodynamics torque M_i around the rotor axis defined as $M_i = C_M\omega_i^2$ with C_M being a torque coefficient [8]. The direction of vectors $\overrightarrow{M_i}$ can be determined by applying the right-hand principle

$$
\boldsymbol{\omega}_{\mathbf{rotor}} = \begin{bmatrix} \omega_1 \\ \omega_2 \\ \omega_3 \\ \omega_4 \end{bmatrix}; \begin{bmatrix} F_1 \\ F_2 \\ F_3 \\ F_4 \end{bmatrix} = C_F \begin{bmatrix} \omega_1^2 \\ \omega_2^2 \\ \omega_3^2 \\ \omega_4^2 \end{bmatrix}; \begin{bmatrix} M_1 \\ M_2 \\ M_3 \\ M_4 \end{bmatrix} = C_M \begin{bmatrix} \omega_1^2 \\ \omega_2^2 \\ \omega_3^2 \\ \omega_4^2 \end{bmatrix} \quad (9)
$$

The combined thrusts of rotors create a total force $^{\mathbf{B}}\mathbf{F}$ in the body frame are

$$^{\mathbf{B}}\mathbf{F} = \begin{bmatrix} ^BF_x \\ ^BF_y \\ ^BF_z \end{bmatrix} = \begin{bmatrix} \sum_1^4 F_{ix} \\ \sum_1^4 F_{iy} \\ \sum_1^4 F_{iz} \end{bmatrix} = C_F \begin{bmatrix} -\omega_2^2 s_{\eta 2} + \omega_4^2 s_{\eta 4} \\ -\omega_1^2 s_{\eta 1} + \omega_3^2 s_{\eta 3} \\ -\omega_1^2 c_{\eta 1} - \omega_2^2 c_{\eta 2} - \omega_3^2 c_{\eta 3} - \omega_4^2 c_{\eta 4} \end{bmatrix} \tag{10}$$

The total torques $^{\mathbf{B}}\mathbf{M}$ with respect to B about the main axes of the body frame are

$$^{\mathbf{B}}\mathbf{M} = \begin{bmatrix} ^BM_x \\ ^BM_y \\ ^BM_z \end{bmatrix} = \begin{bmatrix} \omega_1^2 C_F\left(\frac{H}{2} c_{\eta 1} + l\right) + \omega_2^2 C_M s_{\eta 2} - \omega_3^2 C_F\left(\frac{H}{2} c_{\eta 3} + l\right) - \omega_4^2 C_M s_{\eta 4} \\ -\omega_1^2 C_M s_{\eta 1} - \omega_2^2 C_F\left(\frac{H}{2} c_{\eta 2} + l\right) + \omega_3^2 C_M s_{\eta 3} + \omega_4^2 C_F\left(\frac{H}{2} c_{\eta 4} + l\right) \\ -\omega_1^2 C_M c_{\eta 1} + \omega_2^2 C_M c_{\eta 2} - \omega_3^2 C_M c_{\eta 3} + \omega_4^2 C_M c_{\eta 4} \end{bmatrix} \tag{11}$$

Ignoring air resistance or wind force from the environment, the differential equations for the translational motion of the cubic drone in the inertial frame are

$$m^{\mathbf{0}}\ddot{\mathbf{r}}_{\mathbf{C}} = {}^{\mathbf{0}}\mathbf{P}_{\mathbf{drone}} + {}^{\mathbf{0}}\mathbf{F} = \left[0\ 0\ mg\right]^T + {}^{\mathbf{0}}\mathbf{R}_{\mathbf{B}}\left[^BF_x\ ^BF_y\ ^BF_z\right]^T \tag{12}$$

The Euler's equations for the cubic drone with respect to B are defined as in [7] as

$$^{\mathbf{B}}\mathbf{I}(\boldsymbol{\eta}) \cdot {}^{\mathbf{B}}\dot{\boldsymbol{\omega}} + {}^{\mathbf{B}}\tilde{\boldsymbol{\omega}} \cdot {}^{\mathbf{B}}\mathbf{I}(\boldsymbol{\eta}) \cdot {}^{\mathbf{B}}\boldsymbol{\omega} + m \cdot {}^{\mathbf{B}}\tilde{\mathbf{r}}_{\mathbf{C}} \cdot {}^{\mathbf{0}}\ddot{\mathbf{r}}_{\mathbf{B}} = {}^{\mathbf{B}}\mathbf{M} \tag{13}$$

According to Eq. (12) and (13), we have

$$\begin{bmatrix} ^{\mathbf{0}}\ddot{\mathbf{r}}_{\mathbf{B}} \\ ^{\mathbf{B}}\dot{\boldsymbol{\omega}} \end{bmatrix} = \begin{bmatrix} m \cdot {}^{\mathbf{B}}\tilde{\mathbf{r}}_{\mathbf{C}} & ^{\mathbf{B}}\mathbf{I}_{\mathbf{C}}(\boldsymbol{\eta}) \\ m \cdot \mathbf{I}_3 & -m \cdot {}^{\mathbf{B}}\tilde{\mathbf{r}}_{\mathbf{C}} \end{bmatrix}^{-1} \begin{bmatrix} ^{\mathbf{B}}\mathbf{M} - {}^{\mathbf{B}}\tilde{\boldsymbol{\omega}} \cdot \underbrace{{}^{\mathbf{B}}\mathbf{I}(\boldsymbol{\eta}) \cdot {}^{\mathbf{B}}\boldsymbol{\omega}} \\ ^{\mathbf{0}}\mathbf{P}_{\mathbf{drone}} + {}^{\mathbf{0}}\mathbf{F} - m \cdot \left({}^{\mathbf{B}}\tilde{\mathbf{r}}_{\mathbf{C}} \cdot {}^{\mathbf{B}}\boldsymbol{\omega}\right) \cdot {}^{\mathbf{B}}\boldsymbol{\omega} \end{bmatrix} \tag{14}$$

To sum up, the motion in space of the cubic drone is described by a system of 12 first-order differential Eqs. (6 dynamics and 6 kinematics ones) as follows

$$\begin{cases} ^{\mathbf{0}}\dot{\mathbf{r}}_{\mathbf{B}} = \frac{\mathbf{d}}{\mathbf{dt}} {}^{\mathbf{0}}\mathbf{r}_{\mathbf{B}} \\ \dot{\mathbf{q}}_{\mathbf{RPY}} = Q(\mathbf{q}_{\mathbf{RPY}}) \cdot {}^{\mathbf{B}}\boldsymbol{\omega} \\ \begin{bmatrix} ^{\mathbf{0}}\ddot{\mathbf{r}}_{\mathbf{B}} \\ ^{\mathbf{B}}\dot{\boldsymbol{\omega}} \end{bmatrix} = \begin{bmatrix} m \cdot {}^{\mathbf{B}}\tilde{\mathbf{r}}_{\mathbf{C}} & ^{\mathbf{B}}\mathbf{I}_{\mathbf{C}}(\boldsymbol{\eta}) \\ m \cdot \mathbf{I}_3 & -m \cdot {}^{\mathbf{B}}\tilde{\mathbf{r}}_{\mathbf{C}} \end{bmatrix}^{-1} \begin{bmatrix} ^{\mathbf{B}}\mathbf{M} - {}^{\mathbf{B}}\tilde{\boldsymbol{\omega}} \cdot \underbrace{{}^{\mathbf{B}}\mathbf{I}(\boldsymbol{\eta}) \cdot {}^{\mathbf{B}}\boldsymbol{\omega}} \\ ^{\mathbf{0}}\mathbf{P}_{\mathbf{drone}} + {}^{\mathbf{0}}\mathbf{F} - m \cdot \left({}^{\mathbf{B}}\tilde{\mathbf{r}}_{\mathbf{C}} \cdot {}^{\mathbf{B}}\boldsymbol{\omega}\right) \cdot {}^{\mathbf{B}}\boldsymbol{\omega} \end{bmatrix} \end{cases} \tag{15}$$

4 Simulation

The Eqs. (15) describe the motion of a rigid body. Thus, when considering the arm angles to be constant, we use these 12 state variables to design our controller

$$\mathbf{X} = \left[{}^0x_B\ {}^0y_B\ {}^0z_B\ {}^0\dot{x}_B\ {}^0\dot{y}_B\ {}^0\dot{z}_B\ \varphi\ \theta\ \psi\ {}^B\omega_x\ {}^B\omega_y\ {}^B\omega_z\right]^T \tag{16}$$

A quadcopter is an underactuated mechanism with six degrees of freedom, but its rotational and translational motion are produced and controlled by only four rotors. Let $\gamma_i = \omega_i^2$ and we chose them as the control variables for our controller

$$\mathbf{u}(t) = \left[\gamma_1\ \gamma_2\ \gamma_3\ \gamma_4\right]^T = \left[\omega_1^2\ \omega_2^2\ \omega_3^2\ \omega_4^2\right]^T \tag{17}$$

Our cubic drone will be equipped with a gyroscope and an altitude sensor; as a result, we obtain φ, θ, ψ from the gyroscope and ${}^0z_B$ from the altitude sensor. From this, we now design a PID controller to examine the complicated dynamics of the cubic drone:

$$\begin{bmatrix}\ddot{\varphi}\\ \ddot{\theta}\\ \ddot{\psi}\\ {}^0\ddot{z}_B\end{bmatrix} = \begin{bmatrix} K_{P\varphi}.e_\varphi + K_{I\varphi}\int e_\varphi dt + K_{D\varphi}.\dot{e}_\varphi \\ K_{P\theta}.e_\theta + K_{I\theta}\int e_\theta dt + K_{D\theta}.\dot{e}_\theta \\ K_{P\psi}.e_\psi + K_{I\psi}\int e_\psi dt + K_{D\psi}.\dot{e}_\psi \\ K_{P,z_B}.e_{z_B} + K_{I,z_B}\int e_{z_B} dt - K_{D,z_B}.\dot{e}_{z_B}\end{bmatrix} \tag{18}$$

where $e_\varphi = \varphi_{cmd} - \varphi$, $e_\theta = \theta_{cmd} - \theta$, $e_\psi = \psi_{cmd} - \psi$, $e_{z_B} = {}^0z_{Bcmd} - {}^0z_B$.

Taking derivative of Eq. (5) with respect to time, we have:

$$\begin{bmatrix}\ddot{\varphi}\\ \ddot{\theta}\\ \ddot{\psi}\end{bmatrix} = \begin{bmatrix} 0 & \dot{\varphi}t_\theta c_\varphi + \dot{\theta}\frac{s_\varphi}{c_\theta^2} & -\dot{\varphi}t_\theta s_\varphi + \dot{\theta}\frac{c_\varphi}{c_\theta^2} \\ 0 & -\dot{\varphi}s_\varphi & -\dot{\varphi}c_\varphi \\ 0 & \dot{\varphi}\frac{c_\varphi}{c_\theta} + \dot{\theta}\frac{s_\theta s_\varphi}{c_\theta^2} & -\dot{\varphi}\frac{s_\varphi}{c_\theta} + \dot{\theta}\frac{s_\theta c_\varphi}{c_\theta^2}\end{bmatrix} {}^{\mathbf{B}}\boldsymbol{\omega} + \begin{bmatrix} 1 & t_\theta s_\varphi & t_\theta c_\varphi \\ 0 & c_\varphi & -s_\varphi \\ 0 & \frac{s_\varphi}{c_\theta} & \frac{c_\varphi}{c_\theta}\end{bmatrix} {}^{\mathbf{B}}\dot{\boldsymbol{\omega}} \tag{19}$$

Let $\mathbf{W}_{(4\times 6)}$ be the last 4 rows of $\begin{bmatrix} m\cdot{}^{\mathbf{B}}\widetilde{\mathbf{r}}_{\mathbf{C}} & {}^{\mathbf{B}}\mathbf{I}_{\mathbf{C}}(\boldsymbol{\eta}) \\ m\cdot\mathbf{I}_3 & -m\cdot{}^{\mathbf{B}}\widetilde{\mathbf{r}}_{\mathbf{C}}\end{bmatrix}^{-1}$ and $\mathbf{W}^+_{(6\times 4)} = \left(\mathbf{W}^T\mathbf{W}\right)^{-1}\mathbf{W}^T$ be the pseudo inverse matrix of $\mathbf{W}$. Equation (14) can be converted to

$$\begin{bmatrix}{}^{\mathbf{B}}\mathbf{M}\\ {}^{\mathbf{0}}\mathbf{F}\end{bmatrix} = \begin{bmatrix} {}^{\mathbf{B}}\widetilde{\boldsymbol{\omega}}\,.{}^{\mathbf{B}}\mathbf{I}(\boldsymbol{\eta})\cdot{}^{\mathbf{B}}\boldsymbol{\omega} \\ m\cdot\left({}^{\mathbf{B}}\widetilde{\mathbf{r}}_{\mathbf{C}}\cdot{}^{\mathbf{B}}\boldsymbol{\omega}\right)\cdot{}^{\mathbf{B}}\boldsymbol{\omega} - {}^{\mathbf{0}}\mathbf{P}_{\mathbf{drone}}\end{bmatrix} + \mathbf{W}^+\cdot\left[{}^{\mathbf{0}}\ddot{z}_B\ {}^B\dot{\omega}_x\ {}^B\dot{\omega}_y\ {}^B\dot{\omega}_z\right]^T \tag{20}$$

On the other hand, the system of Eqs. (11) and (12) can be demonstrated as:

$$\begin{bmatrix}{}^{\mathbf{B}}\mathbf{M}\\ {}^{\mathbf{0}}\mathbf{F}\end{bmatrix} = C_F\begin{bmatrix} \frac{H}{2}c_{\eta 1} + l & \frac{C_M}{C_F}s_{\eta 2} & -\left(\frac{H}{2}c_{\eta 3} + l\right) & -\frac{C_M}{C_F}s_{\eta 4} \\ -\frac{C_M}{C_F}s_{\eta 1} & -\left(\frac{H}{2}c_{\eta 2} + l\right) & \frac{C_M}{C_F}s_{\eta 3} & \frac{H}{2}c_{\eta 4} + l \\ -\frac{C_M}{C_F}c_{\eta 1} & \frac{C_M}{C_F}c_{\eta 2} & -\frac{C_M}{C_F}c_{\eta 3} & \frac{C_M}{C_F}c_{\eta 4} \\ R^{12,S1}_{13,C1} & R^{11,S2}_{13,C2} & R^{-12,S3}_{13,C3} & R^{-11,S4}_{13,C4} \\ R^{22,S1}_{23,C1} & R^{21,S2}_{23,C2} & R^{-22,S3}_{23,C3} & R^{-21,S4}_{23,C4} \\ R^{32,S1}_{33,C1} & R^{31,S2}_{33,C2} & R^{-32,S3}_{33,C3} & R^{-31,S4}_{33,C4}\end{bmatrix}\cdot\mathbf{u}(t) \tag{21}$$

in which $R^{mn,Si}_{ml,Ci}$ stands for $\left(-R_{mn}s_{\eta_i} - R_{ml}c_{\eta_i}\right)$, with R_{mn} being an element in row m and column n of the matrix ${}^{\mathbf{0}}\mathbf{R}_{\mathbf{B}}$.

To simulate the drone's motion via Matlab, we adopt parameters displayed in Table 1.

In all examples below, the initial conditions are set as

$$\mathbf{X} = \left[0\ 0\ -10(m)\ 0\ 0\ 0\ 0\ 0\ 0\ 0\ 0\ 0\right]^T \tag{22}$$

The angular speed of each rotor cannot exceed a certain threshold. For our case, the control variables are set as $0 \le \gamma_i \le 2600^2\left(\text{rad}^2\text{s}^{-2}\right)$. For the drone to fly, the lift must

Table 1. The cubic drone parameters

Symbol	Description	Value
H	Side length of the cube	200 mm
l	Distance from a swivel joint to the relevant rotor's axis	100 mm
m_{body}	Mass of the fuselage	0.6 kg
m_{arm}	Mass of an arm	0.1 kg
${}^{B}z_{G_{body}}$	z-coordinate of the fuselage's CoM	100 mm
${}^{B}z_{G_{arm}}$	z-coordinate of an arm's CoM when $\eta_i = 0$	20 mm
C_F	Thrust coefficient	$1.277 \cdot 10^{-6}\,\mathrm{Ns^2rad^{-2}}$
C_M	Torque coefficient	$1.678 \cdot 10^{-8}\,\mathrm{Ns^2rad^{-2}}$
$\mathbf{I_{body}}$	Principal MoI of the fuselage	$\mathrm{diag}(7465, 7465, 4135)$ kg.mm^2
$\mathbf{I_{arm}}$	Principal MoI of an arm	$\mathrm{diag}(180, 180, 230)$ kg.mm^2

exceed the load, $\left|{}^{0}F_z\right| \geq mg$. It can be seen from Eq. (10) that the lift decreases when the arm angles increase. Hence, we limit the arm angles to

$$4 \cdot C_F \cdot \gamma_{max} \cdot c_{\eta_i} \geq mg \Rightarrow 0 \leq \eta_i \leq 73.5^o, i = \overline{1, 4} \tag{23}$$

Two cases are examined: when $\boldsymbol{\eta} = \mathbf{0}$ and when some arms are rotated. The two first examples are the basic rotation with a roll angle (Fig. 4 and 5), and a yaw angle (Fig. 6 and 7).

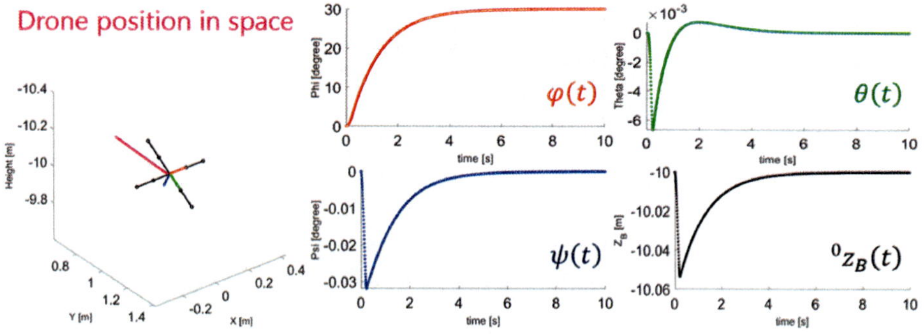

Fig. 4. The stabilization of φ when $\left[\varphi, \theta, \psi, {}^{0}z_B, \eta_1, \eta_2, \eta_3, \eta_4\right] = [30, 0, 0, -10, 0, 0, 0, 0]$

The spatial displacement of B is illustrated by the pink dots while the parameters $\varphi, \theta, \psi, {}^{0}z_B$ are sequentially displayed in red, green, blue, black in the plots. If the drone rotates with φ_{cmd} (θ_{cmd}), it will drift a distance along the y-axis (x-axis). If the drone rotates with ψ_{cmd}, it will fly up some distance due to the increment in velocities of each

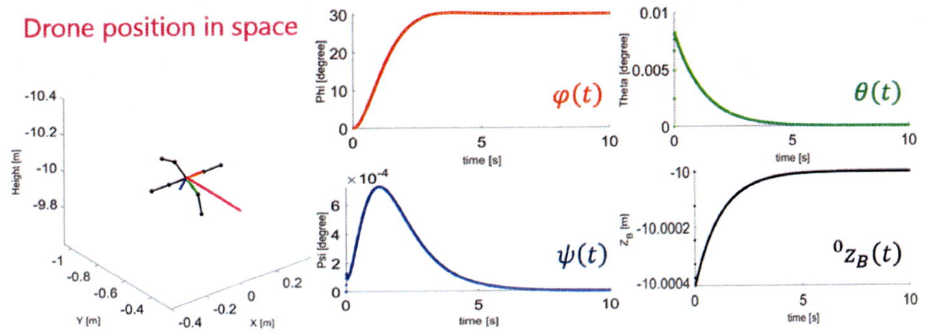

Fig. 5. The stabilization of φ when $\left[\varphi, \theta, \psi, {}^0z_B, \eta_1, \eta_2, \eta_3, \eta_4\right] = [30, 0, 0, -10, 45, 0, 45, 0]$

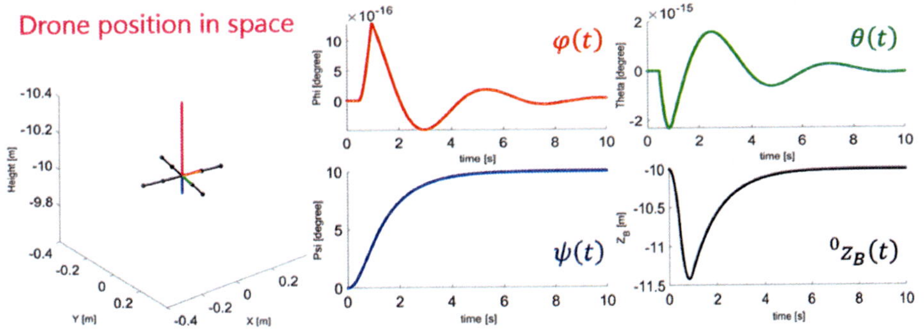

Fig. 6. The stabilization of ψ when $\left[\varphi, \theta, \psi, {}^0z_B, \eta_1, \eta_2, \eta_3, \eta_4\right] = [0, 0, 10, -10, 0, 0, 0, 0]$

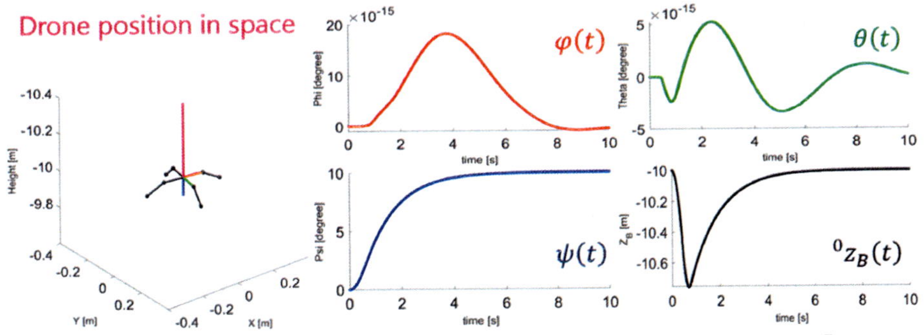

Fig. 7. The stabilization of ψ when $\left[\varphi, \theta, \psi, {}^0z_B, \eta_1, \eta_2, \eta_3, \eta_4\right] = [0, 0, 10, -10, 45, 30, 45, 30]$

rotor. Since the rotation and height of the cubic drone can stabilize expeditiously, we will control them all at once as in Fig. 8 and 9.

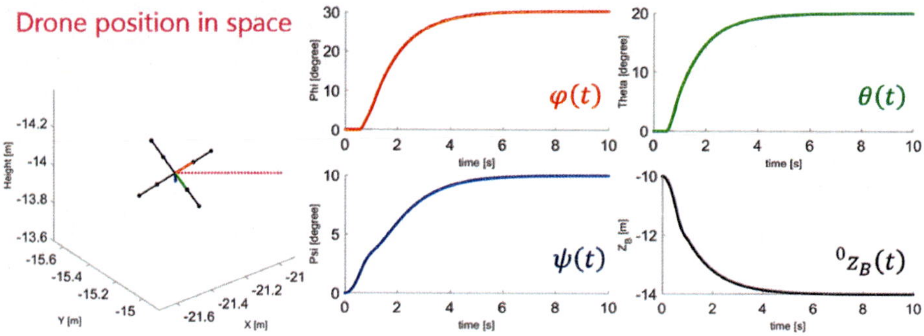

Fig. 8. Drone stabilization when $\left[\varphi, \theta, \psi, {}^0z_B, \eta_1, \eta_2, \eta_3, \eta_4\right] = [30, 20, 10, -14, 0, 0, 0, 0]$

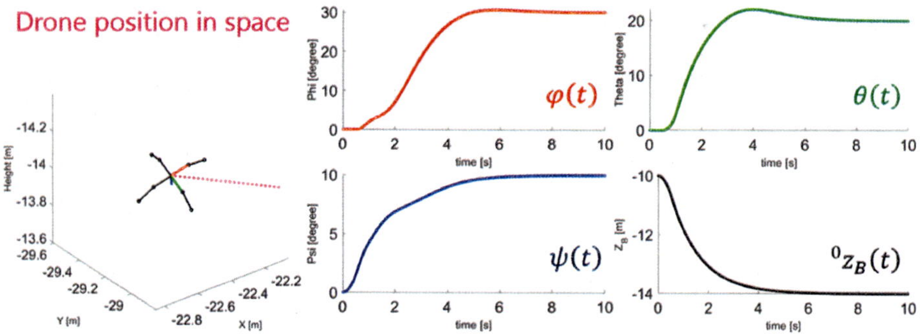

Fig. 9. Drone stabilization when $\left[\varphi, \theta, \psi, {}^0z_B, \eta_1, \eta_2, \eta_3, \eta_4\right] = [30, 20, 10, -14, 45, 30, 45, 30]$

The PD parameters $K_P = 50$ and $K_D = 60$ are applied to all simulations above. While $K_{I\varphi} = 0.6$ and $K_{I\theta} = 0.6$ are set when $\boldsymbol{\eta} = \mathbf{0}$, we set $K_{I\varphi} = 0.09$ and $K_{I\theta} = 0.27$ when $\boldsymbol{\eta} = [45, 30, 45, 30]$. Through the above simulations, even though the PID parameters are figured out heuristically, we have successfully controlled the movement of the cubic drone with the errors of φ_{cmd}, θ_{cmd}, ψ_{cmd} and ${}^0z_{B\,cmd}$ swiftly converge to 0.

5 Conclusion and Discussion

In conclusion, we have proposed an adaptive morphology drone whose arms are able to rotate vertically so that it can fold itself into a cube. The transformability directly affects the kinematics and dynamics of the cubic drone, in which the complicated equations of motion have been shown in this article. We have designed and implemented a PID controller to examine the maneuvers of the cubic drone. The simulations have shown that the cubic drone's movements are controllable. Overall, if only some fundamental rotations are performed, folding the arm frames down will reduce the drift of the cubic drone while maintaining the commanded state. However, in some cases, the cubic drone will maneuver in the horizontal plane faster than a regular quadcopter if we fold the arms down. These simulation results motivate us to continue to develop the cubic drone with the aim of improving the versatility and applications of UAVs.

References

1. Shenzhen DJI Sciences and Technologies Ltd Homepage (2021). https://www.dji.com
2. Dufour, L., Owen, K., Mintchev, S., Floreano, D.: A drone with insect-inspired folding wings. In: 2016 IEEE/RSJ International Conference on Intelligent Robots and Systems (IROS), pp. 1576–1581 (2016)
3. Zhao, M., Kawasaki, K., Chen, X., Noda, S., Okada, K., Inaba, M.: Whole-body aerial manipulation by transformable multirotor with two-dimensional multilinks. In: IEEE International Conference on Robotics and Automation (ICRA) (2017)
4. Zhao, M., Anzai, T., Shi, F., Chen, X., Okada, K., Inaba, M.: Design, modeling, and control of an aerial robot dragon: a dual-rotor-embedded multilink robot with the ability of multi-degree-of-freedom aerial transformation. IEEE Robot. Autom. Lett. **3**, 1176–1183 (2018)
5. Saldaña, D., Gabrich, B., Li, G., Yim, M., Kumar, V.: ModQuad: the flying modular structure that self-assembles in midair. In: 2018 IEEE International Conference on Robotics and Automation (ICRA), pp. 691–698 (2018)
6. Falanga, D., Kleber, K., Mintchev, S., Floreano, D., Scaramuzza, D.: The foldable drone: a morphing quadrotor that can squeeze and fly. IEEE Robot. Autom. Lett. **4**(2), 209–216 (2019)
7. Nguyễn, K.: Động lực học hệ nhiều vật – dynamics of multibody systems. NXB Khoa học và kỹ thuật (2007)
8. Hoffman, G., Huang, H., Waslander, S.L., Tomlin, C.J.: Quadrotor helicopter flight dynamics and control: theory and experiment. In: The Conference of the American Institute of Aeronautics and Astronautics. Hilton Head, South Carolina (2010)

A Study on Localization of Floating Aquaculture Sludge Collecting Robot

Cong Nguyen Nguyen[3], Thi Ngoc Lan Le[3], Van Dong Nguyen[3], Thien Phuc Tran[1,2], and Tan Tien Nguyen[1,2,3](✉)

[1] Faculty of Mechanical Engineering, Ho Chi Minh City University of Technology (HCMUT), 268 Ly Thuong Kiet, District 10, Ho Chi Minh City, Vietnam
nttien@hcmut.edu.vn

[2] Vietnam National University Ho Chi Minh City, Linh Trung Ward, Thu Duc District, Ho Chi Minh City, Vietnam

[3] National Key Laboratory of Digital Control and System Engineering (DCSELab), HCMUT, 268 Ly Thuong Kiet, District 10, Ho Chi Minh City, Vietnam

Abstract. The recirculating aquacultural system discharges the wastewater in the form known as aquaculture sludge which composes of suspended solids, nitrogen, and phosphorus. Aquaculture sludge is recognized as the significant factor that causes the pollutant water, resulting in a decrease in productivity. Thus, aquaculture sludge needs to be removed from the aquaculture pond frequently due to its harmful effect. This paper studies the localization method of a floating aquaculture sludge collecting robot. The localization method is conducted by fusing the signals of a global positioning system (GPS), a Real-Time Kinematic (RTK), and a custom universal joint attached potentiometer to localize the real-time position of the floating aquaculture sludge collecting robot. The experimental result of the proposed positioning method is carried out to validate the effectiveness.

Keywords: Aquaculture sludge · Robot localization · Real-time kinematic · Universal joint

1 Introduction

In recent years, the aquaculture has been emerging as one of the key industries of Vietnam, bringing a huge amount of benefit to the farmers. Pangasius farming in aquaculture pond is a popular recirculating farming model which enables to be easy fish care with simple required techniques, especially gain high economic efficiency. However, the most significant issue that the farmers have to be encountered during aquafarming is to treat aquaculture sludge. Aquaculture sludge is a solid type produced by wasted nutrients, phytoplankton, and animals waste that composes of nitrogenous compound, phosphorus and organic carbon that could endanger the cultured animals if the concentration present is higher than usual [1, 2].

Traditionally, the aquaculture sludge of pangasius ponds is manually collected by divers who has to dive into the pond bottom and carry a suction tool powered by a

N. V. Khang et al. (Eds.): ASIAN MMS 2021, MMS 113, pp. 512–518, 2022.
https://doi.org/10.1007/978-3-030-91892-7_48

diesel engine dredger (see Fig. 1). According to a survey was carried out at the standard pangasius pond of Dai Thanh Seafood Company Limited (Bau Mua Hamlet, Thanh Hung Commune, Kien Tuong Town, Long An province, Vietnam), the operation time for each section depends on the amount of food provided to the fish, the rate per m^2 is 12 to 15 kg, the average thickness of aquaculture sludge is from 10 cm to 15 cm. Two divers are employed to collect aquaculture sludge with a cycle of 12 to 15 days for one suction time, each time with a fee of \$0.032/m^2, equivalent to \$379/10,000 m^2 with a duration of 4 days and 8 h per day, not to mention the cost of the manual dredger as well as its maintenance. Aquaculture sludge with manual process, in a long term, will consume a lot of manpower and a great amount of money. Therefore, the collecting of aquaculture sludge method by using a floating autonomous robot will be enable to reduce the cost, manpower, and resulting in an increase of production.

Fig. 1. Manual diesel dredger

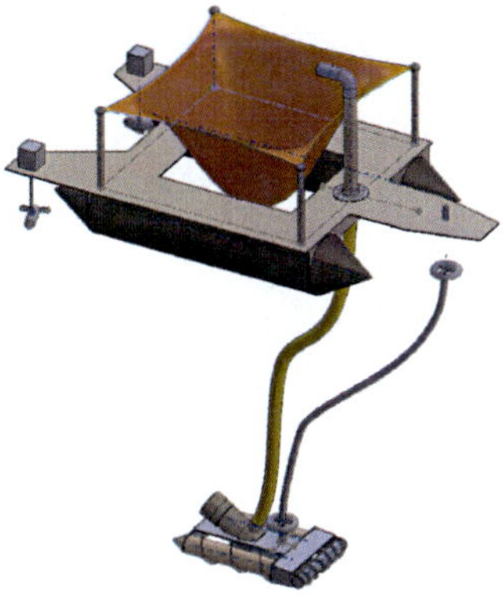

Fig. 2. 3D model concept of aquaculture sludge collecting robot

There were some existing studies in earlier that presented in both remote manpower and engine power of sewage collecting robots [3–16]. These robots proposed a mechanical structure that enables the robot working on the water surface like a boat, while the robot is remote controlled through Bluetooth signal [15], Internet of Thing [13], and radio frequency [10]. The characteristic of sewage collecting robot is operated by human who have to search the sewage on rivers. Therefore, a remote control is well-suited for these application robots. However, the aquaculture sludge collecting robot is required to move and collect the sludge over the aquaculture pond. Thus, an autonomous robot will be efficiency equipment to perform this task frequently.

This paper introduces a concept of aquaculture sludge collecting robot (Fig. 2) consisting of two clusters known as a floating sludge container and a sludge suction robot like remotely operated vehicle (ROV). The sludge suction robot walks on the pond and collect the aquaculture sludge then traveling to the floating sludge container. Both of the floating sludge container and the sludge suction robot are required to operate autonomously to achieve the high efficiency and optimize the operating costs. The encountered problem is to determine the location of both clusters. There, this paper proposes a GPS integrated

RTK localization method for the floating sludge container. As well, a customized universal joint attached a potentiometer is designed to measure the relative pitch and roll angle of the sludge suction robots' pose in comparison to the floating sludge container.

2 Material and Method

Our research proposes a customized universal joint attached two potentiometers to measure the roll and pitch angles. The floating sludge container and the sludge suction robot is connected by a unstretched rope denoted as L (see Fig. 3). The working of the universal joint is described as follows:

When the sludge suction robot does not move, the unstretched rope L is shortest distance. This implies that the position of the floating sludge container concentrates the position of the sludge suction robot ($\theta = 0$). Conversely, when the floating sludge container moves, the unstretched rope form with the z axis by an angle θ. The value of the angle θ is measured by the position of the first potentiometer. From there, it is calculated and analyzed the angle by the unstretched rope on the two planes Oyz and Oxz (α, β), combined with the floating sludge container movement direction relative to the direction (δ). Then, the difference between the measured value of angle and the initial value of that can be used to estimate the current position of the sludge suction robot.

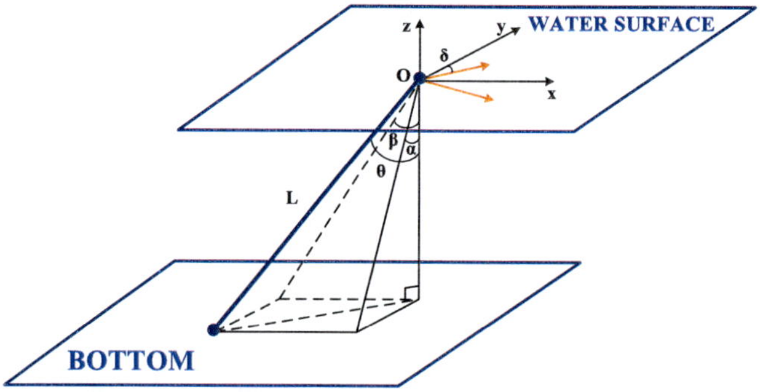

Fig. 3. Describe the principle of underwater positioning with universal joint

3 Experimental Setup

This proposed localization method adopts a customized universal joint attached two potentiometers to measure the roll and pitch angle of the unstretched rope. This method features a low-cost solution but it can achieve acceptable measurement and high accurate. However, it cannot directly localize the position of the sludge suction robot. Therefore, this study has come up with a solution that is to indirectly determine the location of the sludge suction robot through the location of the floating sludge container on the water surface with the utilization of as Real-time kinematic positioning (RTK) sensor named as GNSS C099 – F9P and the universal joint.

This solution comprises an RTK-enabled GPS receiver, called the 'rover', that is installed on the floating sludge container, and an GPS-RTK network is also composed of a stationary GPS reference receiver, called a base station, and a radio link between the base station and the floating sludge container. The base station transmits a radio signal to the user's receiver which contains correction data, called carrier phase information, and is the basis for high accuracy positioning.

The C099-F9P application board allows efficient evaluation of ZED-F9P, the U-Blox high precision positioning module. The ZED-F9P module provides multi-band GNSS positioning and comes with built-in RTK technology providing centimeter level accuracy to users. A test was performed using GNSS C099-F9P integrated with ZED-F9P chip combined with RTK to survey the accuracy of this GPS device by calculating the distance of a point from a predetermined path.

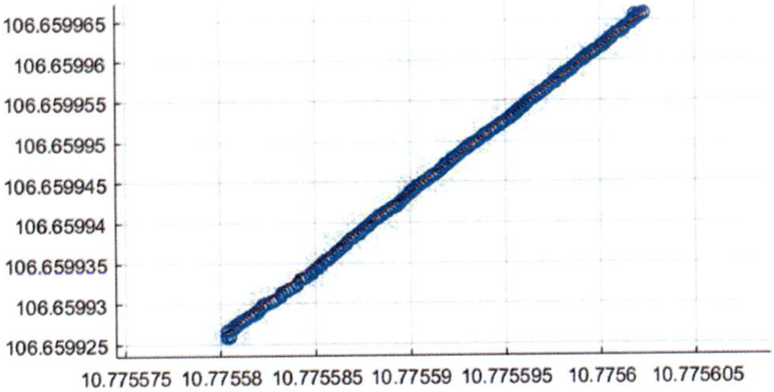

Fig. 4. GNSS C099 - F9P pre-test with a straight line

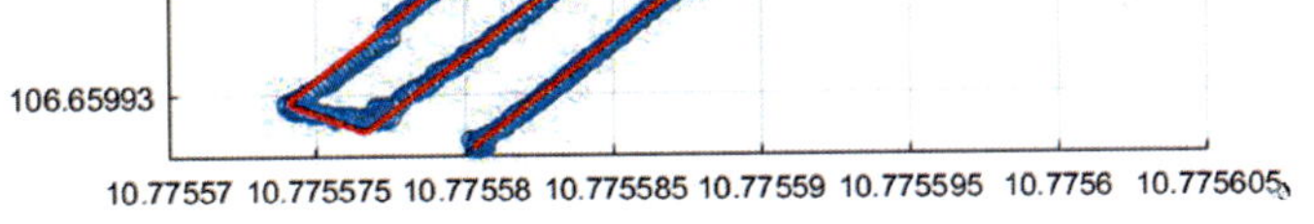

Fig. 5. GNSS C099 - F9P pre-test with a zig-zag path

To validate the RTK-enabled GPS, a first pre-test is carried out by moving the sensor module in a straight line 5 m on the ground at a speed of 10 cm/s at 24.8 m in attitude. The result of the first pre-test is shown in Fig. 4. The measured value is compared to the marked line on the ground to evaluate the measurement error. The experimental pre-test results of this sensor module shows that the maximum error is about 5 cm throughout the straight line.

The second pre-test is conducted by moving the sensor module along a zig-zag path as shown in Fig. 5 with the same parameters described as the first pre-test. The second pre-test shows the largest error of 8.2 cm, in which there is a 7 cm segment of signal interruption.

From the pre-test results, GNSS C099-F9P fully meets the requirements for quality and cost of collecting the position of the sludge suction robot. But the disadvantage of this device is that it is not directly integrated with the sludge suction robot because it is not capable of receiving signals when in the water environment. Therefore, it needs to be added an intermediate device attached to the sludge suction robot to be able to determine the exact position of the sludge suction robot. The design of the floating sludge container with integrated sensors and microcontrollers is carried out for research purposes.

To validate the effectiveness of the sensor module in the large range, a revolved buoy is used to inserted the sensor module because the design of revolved buoy enables to maintain the center of gravity and to prevent itself from tipping over when forces are applied. The universal joint is placed at the center of the buoy's body to keep the stable input signal. The universal joint, Arduino, and GNSS antenna are put in a 3D printed box, designed to have clamps which have the same profile of the inner tube to tighten the inner tube from the inside. Since the box contains electronic microchips such as universal joint and Arduino, it is necessary to waterproof by using a rubber film with silicone glue around the open part of the box. An antenna is mounted on the lid of the box to conveniently receive GNSS signal (Fig. 6).

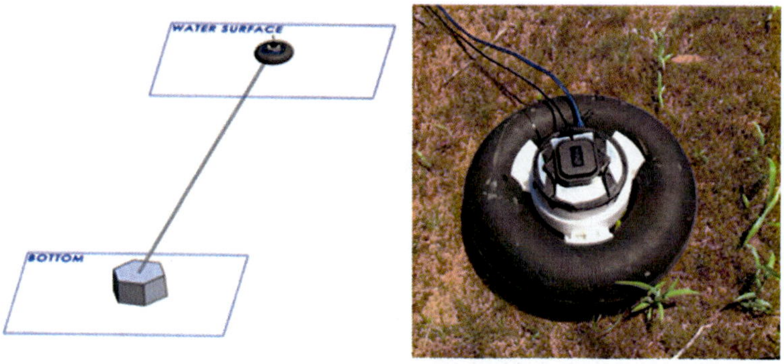

Fig. 6. 3D and realistic model

4 Experiment Results

An experiment was made on a lake surface similar to that of pangasius ponds. The altitude is -1.1 m above sea level, the lake depth is 2.3 m. The experiment was carried out under sparsely cloudy weather conditions which makes well received satellite signal.

Two antennas are set up with a distance of 20 m. The buoy was moved in a straight line back and forth to the origin in a radius of about 1 m. The results obtained from the above experiment are shown in Fig. 7 and Fig. 8. The experiment results show that the maximum position error of the sludge suction robot is 0.137 m.

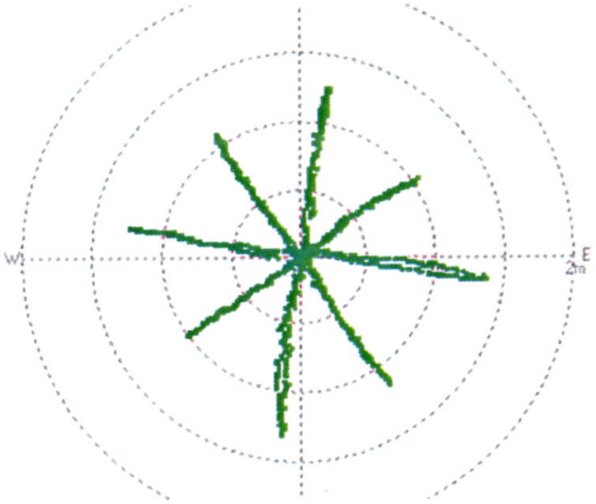

Fig. 7. The received position of the sludge suction robot on the support software of GNSS C099-F9P

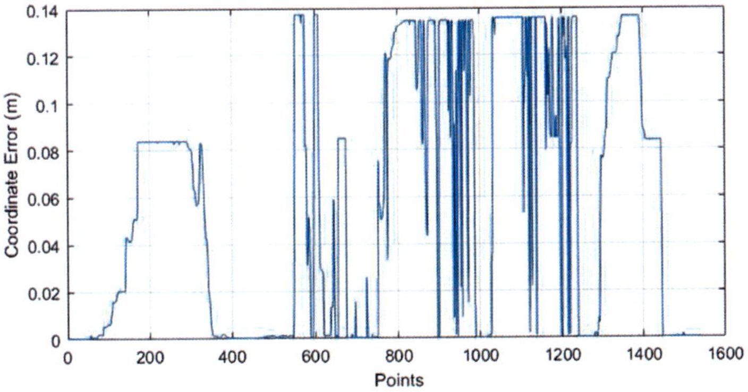

Fig. 8. Location error statistics

5 Conclusion

In terms of performance, the experiment fully satisfied our requirements. The experiment conducted on the combination of GNSS application board C099-F9P and the universal joint has demonstrated that accurate positioning can be achieved for this aquaculture sludge collecting robot. Therefore, with this proposed localization method, it is able to upgrade the positioning of the sludge suction robot with low-cost and high accuracy.

Acknowledgment. This research is supported by DCSELAB and funded by Vietnam National University Ho Chi Minh City (VNU-HCM) under grant number TX2021-20b-01. We acknowledge the support of time and facilities from Ho Chi Minh City University of Technology (HCMUT), VNU-HCM for this study.

References

1. Junge-Berberovic, R.: Possibilities and limits of wastewater-fed aquaculture. In: Ecological Sanitation-Symposium, pp. 112–122 (2000)
2. Jasmin, M.Y., Syukri, F., Kamarudin, M.S., Karim, M.: Potential of bioremediation in treating aquaculture sludge: review article. Aquaculture **519**, 734905 (2020). https://doi.org/10.1016/j.aquaculture.2019.734905
3. Nair, S.S., Sudheer, A.P., Joy, M.L.: Design and fabrication of river cleaning robot. In: ACM International Conference on Proceeding Series, pp. 472–479 (2019). https://doi.org/10.1145/3352593.3352663
4. Kandare, D.N., Kalel, A.N., Jamdade, A.S., Jawale, G.P.: Design & construction of river cleaning mechanism. Int. J. Innov. Sci. Res. Technol. **3**, 428–432 (2018)
5. Ranjan, C., Akhtar, S.J., Shaw, D., Nikhil, S., Babu, B.S.: Design and demonstration of manual operated pedalo water boat for garbage collection from lake. In: AIP Conference on Proceedings 2200 (2019). https://doi.org/10.1063/1.5141267
6. Ravindrabhai, M., Vaibhav, D.: Design and fabrication of river waste collector. Int. J. Adv. Eng. Res. Dev. **5**, 736–740 (2018)
7. Raikar, P.E., Rahul, P., Shridhar, H.C., Tharun, S.T., Rajashekhar, B.: Lake health monitoring and waste collecting aquabot. Int. J. Adv. Res. Sci. Eng. **8**, 54–64 (2019)
8. Sane, S., Sawant, M.: Environment sustainability through smart waste collector. J. Adv. Res. Comput. Graph. Multimed. Technol. **2**, 20–22 (2020)
9. Magdum, P.S., et al.: Solar based river cleaning machine. Int. J. Anal. Exp. modal Anal. **12**, 1029–1035 (2020)
10. Idhris, M.M., Parthi, M.E., Kumar, C.M., Vathy, N.N., Waran, K.S., Kumar, S.A.: Design and fabrication of remote controlled sewage cleaning machine. Int. J. Eng. Trends Technol. **45**, 63–65 (2017). https://doi.org/10.14445/22315381/ijett-v45p214
11. Sharma, T., Sharma, S., Sharma, P., Jain, S., Joshi, A.: Design of river cleaning machine. SSRN Electron. J. 1–11 (2020). https://doi.org/10.2139/ssrn.3621962
12. S, N.S., V, P.S.: River cleaning machine. Int. Res. J. Eng. Technol. **6**, 228–229 (2019)
13. Saran Raj, B., Murali, L., Vijayaparamesh, B., Sharan Kumar, J., Pragadeesh, P.: IoT based water surface cleaning and quality checking boat. J. Phys. Conf. Ser. **1937**, 012023 (2021). https://doi.org/10.1088/1742-6596/1937/1/012023
14. Khekare, G.S., Dhanre, U.T., Dhanre, G.T., Yede, S.S.: Design of optimized and innovative remotely operated machine for water surface garbage assortment. Int. J. Comput. Sci. Eng. **7**, 113–117 (2019). https://doi.org/10.26438/ijcse/v7i1.113117
15. Kumar, N.K., Vivek, G., Reddy, A.P.: Bluetooth based garbage collection robot using aurduino microcontroller. J. Emerg. Technol. Innov. Res. **5**, 656–661 (2018)
16. Nair, S.S., Sudheer, A.P., Joy, M.L.: Design and fabrication of river cleaning robot. In: ACM International Conference on Proceeding Series, vol. 6, pp. 29–46 (2019). https://doi.org/10.1145/3352593.3352663

Simulation of Engineering Systems

On the Formation of Protrusion and Parameters in the Tube Hydroforming

Vu Duc Quang[1,2], Dinh Van Duy[2(✉)], and Nguyen Dac Trung[2]

[1] Economics - Technology for Industries, Hanoi, Vietnam
[2] Hanoi University of Science and Technology, Hanoi, Vietnam
duy.dinhvan@hust.edu.vn

Abstract. Tube hydrostatic forming is one of the attractive forming processes for manufacturing hollow tubes using high fluid pressure. The number of tube hydrostatic forming applications is known in the automotive, aerospace, and shipbuilding industries. More intricate geometries are formed from tubes with fewer operations, lighter weight, and enhanced mechanical properties. They are characterized by the use of tubes, thus allowing expansion to a variety of shapes. The design and production of tubular components require knowledge about tube material behavior and the effects of technological parameters during hydrostatic forming. The hydrostatic forming operation itself should be controlled. The tube hydrostatic forming process is a relatively complex manufacturing process; the performance depends on various factors and requires proper part design, material selection, and boundary conditions. Due to the complex nature of the process, the behavior of the process will be done by simulation. The main object of the proposed study is to examine the influence of technological parameters like internal pressure, axial force, counterforce, axial displacement, and friction on protrusion height by using DEFORM-3D software. The research results help technologists optimize process parameters in the hydrostatic forming process of products with protrusion from tube billet.

Keywords: Tube hydroforming · Process parameters · Protrusion height

1 Introduction

Tube hydrostatic forming is a unique metal forming process whereby complicated protrusion shapes are created by using a high-pressure fluid source instead of traditional mechanical force. This process has outstanding advantages over conventional manufacturing processes. These include reduced weight, uniform thickness of the wall, reduced tool cost, increased structural strength and stiffness, reduced secondary operations, improved dimensional accuracy, less scrap rate [1–4]. These advantages primarily stem from the ability of the working fluid to exert pressure evenly over the entire surface of a material and for the equipment to vary this fluid pressure during the forming cycle based upon an optimized load path. So, tube hydroforming establishes it in a wide range of specific applications in the automotive, bicycle, aircraft, chemical, gas, oil, power plant construction, home appliance industries, and more [1–3].

N. V. Khang et al. (Eds.): ASIAN MMS 2021, MMS 113, pp. 521–530, 2022.
https://doi.org/10.1007/978-3-030-91892-7_49

In recent years, the development trend of metal forming technology has been to use advanced methods, such as numerical simulation, to solve industrial problems. Numerical simulation of tube hydrostatic forming using 3D finite element models is truly state of the art. Simulation study uses finite element method-based virtual prototyping techniques to characterize material flows, gain insight into strain mechanics, and predict shaped part's mechanical properties. The application of finite element analysis (FEA) and another computer-aided engineering (CAE) techniques to product design at an early stage of the design cycle has significantly reduced the time to take a vehicle from concept phase to production [5].

Figure 1 shows a tube billet is inserted in between lower and upper die halves. The shape of the die cavity corresponds to the external shape of the produced part. These two die halves are separated in the longitudinal direction, dies are closed by hydraulic load, and axial punches seal the two ends of the tube. The minimum loads required to sealing the tube ends are equal to the force resulting from the cross-sectional area of the tube and internal pressure. But if the forming process needs, the axial load may be increased to the highest possible value. During the process, the pressure may be increased till the tube comes in contact with the inner surface of the die cavity. Compared with the other theoretical methods, the FEM has more advantages in solving general problems with complex shapes of the formed parts. Because of the criticality involved in Y-shaped bends, it is considered for this study to analyze the forming behavior by using the explicit solver of DEFORM-3D.

Technological parameters considered to analyze their influence on the quality of the product of hydroformed Y-shape tube included: internal pressure, axial loading, counter loading, and contact friction coefficient. Material flow, distribution of effective stress, distribution of effective strain, damage factor, etc., may be obtained in the product are also analyzed. The forming laws of tube billets may be predicted.

2 Materials and Methods

As a highly efficient and low-cost analytical method, the finite element simulation method is used in this study to analyze the hydrostatic forming of Y-shaped tubes. Simulations were conducted to hydroformed Y-shape tube from SUS304 tube billet with 160 mm length, 22 mm outer diameter, and 1.3 mm wall thickness. The diameter of protrusion was equal to that of the main tube. The geometry of the die and the tube were modeled to be consistent with the experimental. Only the billet's deformation behavior was studied and meshed with 3D tetrahedral elements in the present analysis. The models were built in six parts: flexible tube billet, rigid lower die half, rigid upper die half, rigid left punch, rigid right punch, and rigid counterpunch (Fig. 1). The FE simulations were accurate and convergent; the billet was set as plastic deformation type and meshed by tetrahedral elements with absolute mesh type. The shear friction model is used to describe the friction between the workpiece and the die, and the friction coefficient for the lubricated cold forming process is used according to the recommended Deform-3D software. The mechanical properties of the material and the specifications of the Y-tube part are shown in Table 1, Table 2.

The simulations were conducted with different boundary conditions, constraints, loading. Initial boundary conditions were calculated using theatrical equations to study

Table 1. Room temperature mechanical properties of tube billet [6].

Alloy	Temperature (°C)	Yield strength (MPa)	Ultimate Tensile Strength MPa)	Elongation (%)	Hardness (HB)	Density (kg/m^3)
SUS304	24	230	480	40	183 max	7930

Table 2. Y-shape tube parameters.

	Parameters	Symbol/unit	Value
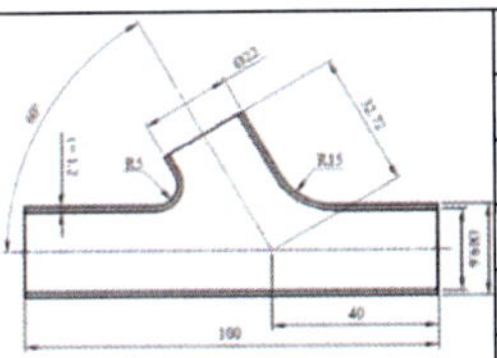	Outside tube diameter	D_0 (mm)	22
	Inside tube diameter	d_I (mm)	19.6
	Protrusion angle	δ (^{0})	60
	Initial tube thickness	t_0 (mm)	1.2
	Height protrusion	H_p (mm)	to be designed
	Protrusion diameter	D_p (mm)	22
	Component length	L_c (mm)	100

and analyze process parameters such as internal pressure, axial loading, counter loading, and contact friction on the process behavior.

The fluid pressure in the tube is one of the essential technological parameters in the tube hydrostatic forming process, and other parameters are also affected by this pressure parameter. The yield strain pressure $(P_i)y$ is the minimum pressure required to initiate plastic deformation in hydrostatic forming. The degree of this pressure varies depending on the material and shape of the tube part. The yield strain pressure is calculated by Eq. 1 [7]. Although the computed yield pressure is accurate only for simple tube expansion with fixed ends, it is also a good initial guess for hydrostatic shaping of more complex parts. (i.e., Y shape) with an axial feed applied.

Internal pressure at yielding is calculated by Eq. 1 [7]:

$$(P_i)_y = \sigma_y * \frac{2t_0}{(D - t_0)} \tag{1}$$

Maximum internal pressure may be calculated by Eq. 2 [7]:

$$(P_i)_b = \sigma_u * \frac{2t_0}{(D - t_0)} \tag{2}$$

Where: σ_y = Yield strength of the tube material, N/mm^2; σ_u = Ultimate tensile strength of the tube material, N/mm^2; t_0 = Initial tube thickness, mm; D = Mean tube diameter, mm.

The numerical simulation research method gives the ability to determine some key technological parameters with intricate details with higher accuracy due to the setting of boundary conditions close to experimental and actual production. Technological parameters determined by theoretical research methods will then be tried and adjusted through repeated FEA simulations until satisfactory results. When the initial technological parameters are reasonably estimated, the number of repeated simulations to achieve good results will be reduced. The study of the influence of some technological parameters of hydrostatic forming of tube parts with Y-shaped protrusion from tube billet will be discussed here.

The shape of the tube billet and the setting of the tool components when simulated by Deform - 3D are as shown in Fig. 1.

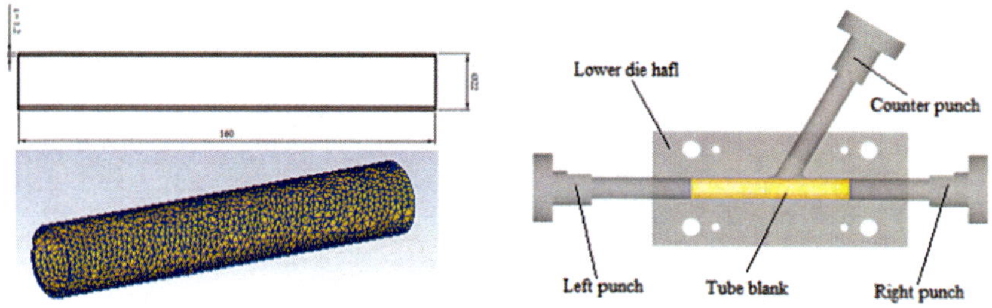

Fig. 1. Workpiece and tools used in FEM simulation.

3 Results and Discussion

Simulations were conducted with different conditions (different process parameters) to study the effect of various parameters on protrusion height and thickness at the maximum protrusion point.

3.1 Effect of Internal Pressure and Counter Punch Force

The simulations were conducted with internal pressure p_i ranging from 40MPa to 130 MPa, friction coefficient $\mu = 0.03$ and speed of axial displacement $v_{left} = 2$ mm/s; $v_{right} = 1$ mm/s were kept constant. The final protrusion height development and the change in counterforce at different internal pressures were studied for the part expansion and wall thickness variation at the top of the protrusion.

Figure 2 showed the effect of internal pressure and counterforce on protrusion height, filling quality of protrusion. From Fig. 2, it was observed that when the liquid pressure p_i increased, the result was a small change in the axial punch force (left axial punch force $\approx$17,4 $\div$ 17,9 Ton; right axial punch force $\approx$14,7 $\div$ 15,4 Ton), counterforce increased strongly (1.5 $\div$ 3.5 Ton) and protrusion achieved better height, better material thickness distribution. The counterforce also increased sharply with the internal pressure p_i while improving the quality of protrusion forming.

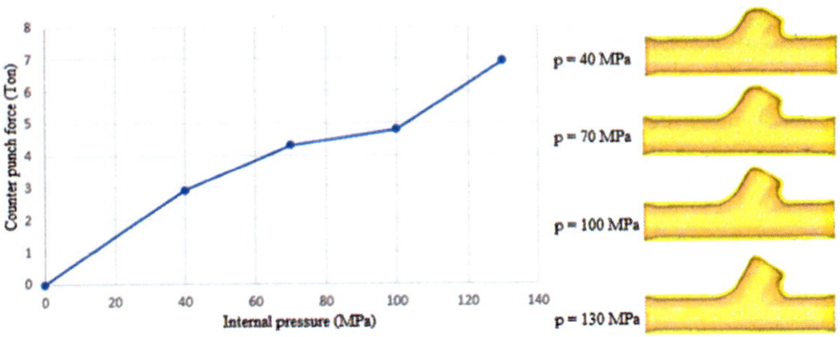

Fig. 2. Effect of internal pressure and counter punch force on protrusion height.

3.2 Effect of Axial Punch Displacement Velocity

Simulations were conducted with the speed change of 2 axial punches ($v_{left} = 2^*v_{right}$; $v_{left} = 2 \div 10$ mm/s; $v_{right} = 1 \div 5$ mm/s), friction coefficient $\mu = 0, 03$ and internal pressure $p_i = 130$ MPa were kept constant. The final protrusion height development and the speed of 2 axial punches were studied for the part expansion and wall thickness variation at the top of the protrusion.

Table 3. Effect of axial displacement velocity on protrusion height.

Right punch velocity (mm/s)	Right punch force (Ton)	Left punch velocity (mm/s)	Left punch force (Ton)	Counter punch force (Ton)	Tube- after deformation
1	15.4	2	17.9	3.47	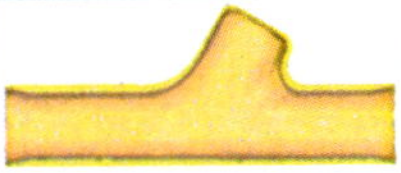
3	15.8	6	18.8	4.52	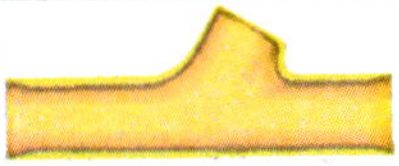
5	15.9	10	18.9	4.45	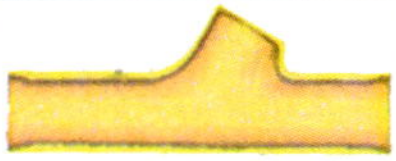

Table 3 shows the influence of axial punch travel velocity on protrusion height and protrusion filling quality. It may be seen that the protrusion height shaping results are relatively similar when changing the travel speed of the axial punches, but the material thickness distribution is different. When increasing the travel speed of the axial punches, the material will be sufficiently fed into the plastic deformation area to form the shape of the protrusion and give a better thickness of the protrusion. Thus, from the results, it may be confirmed that the travel speed of the axial punch has little effected on the formation of the protrusion height of the part, but only a tiny change on the axial force and counterforce when forming.

3.3 Effect of Friction and Axial Force

The interaction of the tube surface and the lubricant is a critical parameter in the characterization of hydrostatic forming, and the goal is to reduce friction for better material flow. Friction helps control material flow by reducing thinning in critical areas.

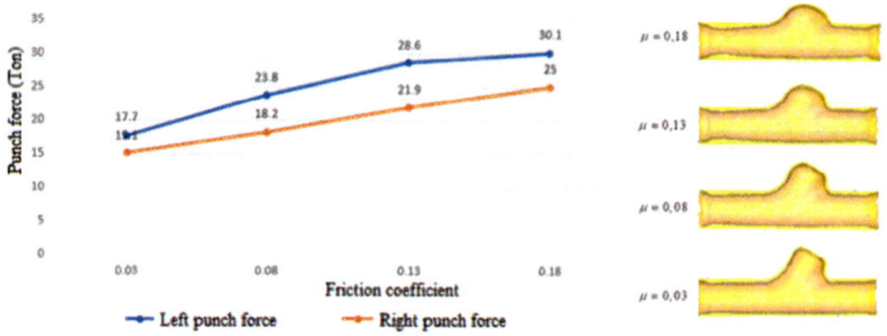

Fig. 3. Effect of friction and axial force on protrusion height.

The friction coefficient is significant; it affects product quality, surrounding physical parameters. The simulations were conducted with the friction coefficients between the upper and lower dies and the outer surface of the tube in order of low to high friction coefficient: $\mu = 0{,}18$; 0.13 (oil films); 0.08 (emulsifiable pastes); 0.03 (dry films) [8], internal pressure pi = 130 MPa was kept constant. Figure 3 illustrates the influence of friction coefficient and axial force on the protrusion forming quality. From Fig. 3, it is observed that the protrusion height decreases with increasing friction coefficient. The plastic deformation zone can be swelling up to a certain limit when reducing the friction coefficient without forming cracks. These limits are further increased by the material being fed as the punches move in the axial direction. When reducing friction coefficient, the axial punch force decreases (left axial punch force $\approx$30,1 $\div$ 17,7 Ton; proper axial punch force $\approx$25,0 $\div$ 15,1 Ton), the counterforce increases sharply (0.0 $\div$ 2.4 Ton) and the protrusion achieve better height and material thickness distribution.

3.4 Optimization of Technological Parameters in the Hydrostatic Forming Process

Based on the analysis of research results that simulate the influence of some of the above technological parameters, the research team obtained the most optimal parameters to proceed to experiment with the hydrostatic forming process to create the protrusion of the Y-shape tube part (Fig. 4) as shown in Table 4.

From Fig. 4 ($t_0 = 1.2$ mm), it is observed that the thickness variation of the hydrostatic formed product is irregular: the guided zone was most strongly thickened and reached 2.1 mm thickness, protrusion apex area had a minimum thickness of 0.69 mm. For more uniform workpiece thickness, we may use the counterpunch to a lower position at the beginning to increase the thickness of the protrusion.

Figure 5 shows the results of the evaluation of the damage factor on the SUS304 hydrostatic stamping product. The damage factor reached the maximum value of 1.2, but

Table 4. Optimal parameter set.

No	Parameters	Symbol/unit	Value
1	Internal pressure	P_i (MPa)	130
2	Right punch force	F_R (Ton)	15.4
3	Left punch force	F_L (Ton)	17.9
4	Counterpunch force	F_C (Ton)	3.47
5	Left punch velocity	v_{left} (mm/s)	2
6	Right punch velocity	v_{right} (mm/s)	1
7	Friction coefficient	μ	0.03

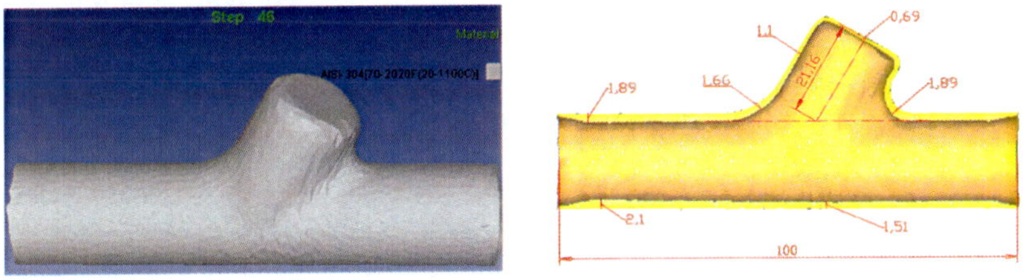

Fig. 4. Dimensional survey of product with optimal parameters.

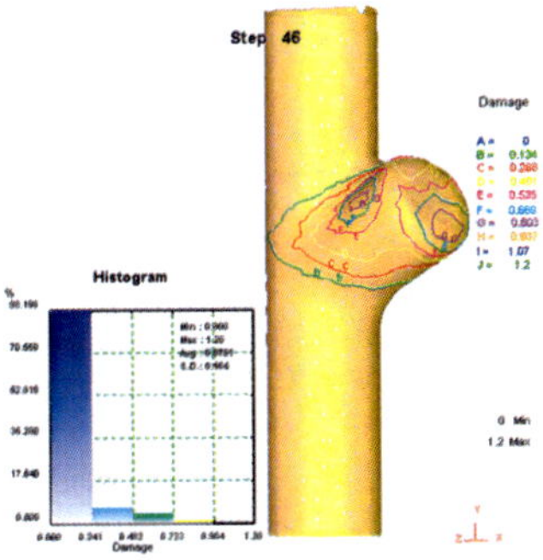

Fig. 5. Damage factor

the average value was minimal at 0.0781. Figure 5 also shows that the breaking capacity is only around the protrusion and the top of the protrusion; the top of the protrusion is the most thinned area, but the value of the failure coefficient is still at a safe level by design. The results may be concluded that the hydrostatic formed Y-tube part without any risk of wrinkling or fracture after the plastic deformation forming process.

The metal flow in the plastic deformation process is shown in Fig. 6, and the results showed that during the pressing process of the two axial punches, the material at the two ends of the tube was compressed into the middle region and flowed to the die cavity and forms a protrusion of detail. The left tube billet head moves 40 mm with speed v_{left}

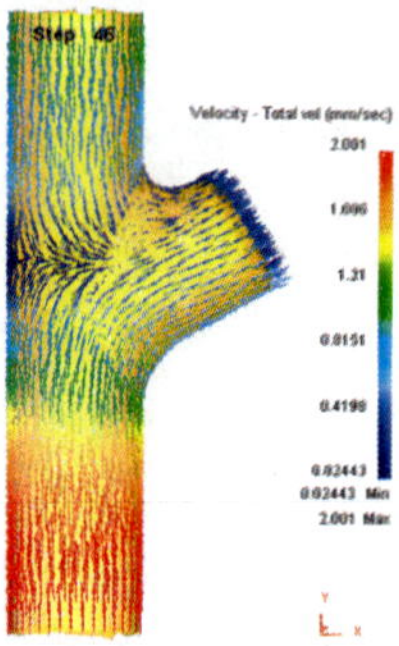

Fig. 6. Metal flowing in deformation process

= 2 mm/s with the left axial punch, so the material is pulled more than the right tube billet head moves 20 mm with the speed v_{right} = 1 mm/s with the right axial punch. The material flows to the center of the tube billet at different speeds (shown in different colors), so it tends to gather and move slowly.

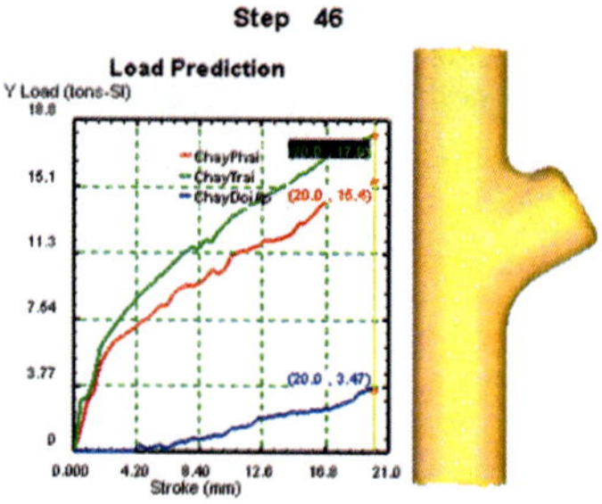

Fig. 7. The impact force of 3 punches during hydrostatic forming.

The force graph in Fig. 7 shows that the impact force of 3 punches on the pipe billet during hydrostatic stamping increases gradually over time. There is no sudden decrease so that the system will work stably and safely. The different load paths of the three punches will require more complex control.

Some failure forms of protrusions after hydrostatic forming should be noted, as shown in Table 5.

Table 5. Some failure forms of protrusions.

Parameter	Protrusion shape of hydrostatic forming product
High forming pressure	
Low forming pressure	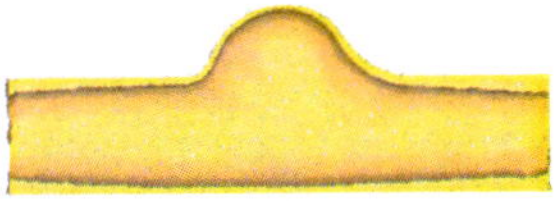
High counter pressure	

4 Conclusion

Coordination of internal pressure, axial feed, and lubrication were crucial in achieving the maximum height of the protrusion without any defects and suitable thickness distributions. These simulations show that an increase in the pressure path affects axial punch force and counter punch force increase. Due to the accumulation of internal pressure acting on the front of the punches and the tube billet's inner surface, the punches' stroke causes material thickening and hardening. The main thinning was at the top of the protrusion, whereas the thickening was at both ends. The top protrusion is susceptible to excessive tension in the absence of support from the counter. The influence of some critical technological parameters on the protrusion formability of Y-tube detail in the hydrostatic forming process has been considered. The optimal set of technological parameters has been found in this process.

References

1. Pham, V.N.: Hydrostatic Forming Technology, 1st edn. Bach Khoa Publishing House, Hanoi (2007)
2. Koç, M.: Hydroforming for Advanced Manufacturing, 1st edn. Woodhead Publishing Limited, Cambridge (2008)
3. Singh, H.: Fundamentals of Hydroforming, 1st edn. The Society of Manufacturing Engineers (2003)
4. Bell, C., Corney, J., Zuelli, N., Savings, D.: A state of the art review of hydroforming technology. Int. J. Mater. Form. **13**(5), 789–828 (2019). https://doi.org/10.1007/s12289-019-01507-1
5. Singh, H.: Tubular hydroforming process and tool design optimization using computer simulation. In: Tube/Pipe Fabricating Conference. Tube and Pipe Association, International, Rockford (1999)

6. Gale, W.F., Totemeier, T.C.: Smithells Metals Reference Book-Butterworth-Heinemann, 8th edn. Elsevier and The Materials Information Society (2004)
7. Bogoyavlensky, K.N., Vagin, V.A., Kobyshev, A.N., et al.: Hydro-plastic Processing of Metals. Mashinostroenie, Moscow; Teknika, Sophia, U.S.S.R (1988)
8. Brownback, P.: How Lubricant Selection Affects the Cost of Hydroforming, no. 1. D.A Stuart Company, Warrendale (2001)

Dynamics of Closed-Loop Planar Mechanisms with Coulomb Friction in Prismatic Joints

Minh-Tuan Nguyen-Thai(✉)

Hanoi University of Science and Technology, Hanoi, Vietnam
tuan.nguyenthaiminh@hust.edu.vn

Abstract. Simulation of systems with stick-slip phenomenon is crucial in various applications. While Coulomb friction law looks simple, it is a challenging task to carry out integration of equations utilizing this dry friction law. This report focuses on closed-loop planar mechanisms with all prismatic and revolute joints, which means a differential-algebraic equations system should be solved, making the problem more complicated. The switching conditions are examined carefully to determine whether stiction occurs or slip phase continues when relative velocity passes through zero. It turns out that, with the appearance of dry friction only at prismatic joints, the equations are piecewise-linear in terms of generalized accelerations and reaction forces in both stick and slip phases. The numerical instabilities are suppressed by means of Baumgarte stabilization. Examples using an eccentric slider-crank mechanism are performed. It shows that the simulation results are not reliable without stabilization. More appropriate results are obtained with Baumgarte stabilization and a fixed-step explicit integrator.

Keywords: Coulomb friction · Piecewise system · Baumgarte stabilization

1 Introduction

The consideration of dry friction is important in applications where both stick and slip may occur at a connection such as impact oscillators with drift [1], bipedal robots [2], grippers [3], and servo mechanisms [4]. Due to its simple mathematic formulations as well as the low number of parameters, Coulomb friction should be the go-to model when dealing with dry friction in joints, especially when there is a possibility of stiction. However, numerical simulation of dynamical systems employing Coulomb friction may show unexpected results where stick is not perfectly described but a type of velocity oscillation is seen [4]. Moreover, for closed-loop mechanisms, a system of differential-algebraic equations (DAE) should be solved, which makes the problems more complicated. Such difficulties in calculation are usually overcome by choosing alternative friction models that are differentiable [5] or at least continuous [6]. The problem of alternative friction models is that they require more parameters than Coulomb friction law. Those parameters are either artificial (which means they do not have physical meaning) or difficult to experimentally determine. Another approach of determining switching times between slip phase and stick phase is also used, but the difference between static and kinetic

N. V. Khang et al. (Eds.): ASIAN MMS 2021, MMS 113, pp. 531–540, 2022.
https://doi.org/10.1007/978-3-030-91892-7_50

coefficients of friction is not considered [7]. Hence, a practical way to solve the forward dynamics problem of closed-loop mechanisms with Coulomb friction is the topic of the current study.

This report focuses on closed-loop planar mechanisms with Coulomb friction in prismatic joints. The non-smooth characteristics not only lie in the friction model but also in the change of the senses of the contact normal forces. The unexpected numerical behaviors caused by the non-smooth friction model is stabilized simply by finding switching times. The conditions of switching are carefully checked to determine whether stiction occurs, and if stiction occurs, the equations are changed accordingly. The difficulty posed by DAE systems is suppressed using Baumgarte stabilization [8, 9]. Examples are carried out with a slider-crank mechanism and the results with and without stabilizations are compared.

2 Establishment of Stabilized Equations

Consider a closed-loop planar mechanism of rigid bodies connected by revolute joints or prismatic joints. Dry friction appears in some prismatic joints. The other joints are either frictionless or subjected to linear viscous damping. It is assumed that there is no significant clearance at the joints.

A prismatic joint restricts two independent planar motions so that there should be two ideal reaction forces. Thus, there are two sets of normal and friction forces (Fig. 1a). It should be noted that each normal force can change its sense of direction. This change also leads to the change in the point of action. This report is restricted to the case where the prismatic joint is associated with a slider that can be considered a point mass. In this case, the reaction forces are only a set of normal and friction forces acting at the center of the slider (Fig. 1b).

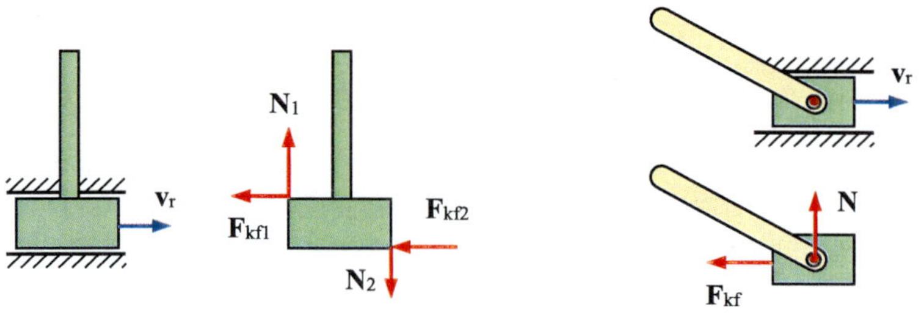

a) Nonideal reaction forces at a prismatic joint

b) Reaction forces at a point-mass slider

Fig. 1. Slider-crank mechanism

A dry friction force depends on the associated normal force. Thus, to establish equations of motion, one should separate all joints that have dry friction. In this way, the normal forces and friction forces can be determined through the equations of motion and equations of Coulomb friction. Additionally, the subsystems after separation should be serial systems; hence, more joints should be separated if necessary.

The motion of the system is governed by a set of DAE [10]

$$\begin{array}{c}\mathbf{M}(\mathbf{q})\ddot{\mathbf{q}} + \boldsymbol{\Phi}_q^T(\mathbf{q})\boldsymbol{\lambda} = \boldsymbol{\tau}(\mathbf{q}, \dot{\mathbf{q}}, \boldsymbol{\lambda}, t) \\ \boldsymbol{\Phi}(\mathbf{q}) = \mathbf{0}\end{array}, \tag{1}$$

where $\mathbf{q}$ is the vector of n generalized coordinates, $\mathbf{M(q)}$ is the symmetric mass matrix of the size $n \times n$, $\boldsymbol{\lambda}$ is the vector of $n - f$ Lagrange multipliers with f is the degree of freedom (DOF) of the system, $\boldsymbol{\tau}$ is an $n \times 1$ vector, $\boldsymbol{\Phi}(\mathbf{q})$ is $n - f$ constraint equations, and $\boldsymbol{\Phi}_{\mathrm{q}}$ satisfies

$$\boldsymbol{\Phi}_q = \frac{\partial \boldsymbol{\Phi}}{\partial \mathbf{q}}. \tag{2}$$

All the normal forces should be included in $\boldsymbol{\lambda}$ and since the dry friction forces depend on the normal forces, $\boldsymbol{\tau}$ depends on $\boldsymbol{\lambda}$. To solve (1), the constraint equations are usually differentiated twice.

$$\dot{\Phi} = \boldsymbol{\Phi}_q \dot{\mathbf{q}} = \mathbf{0}. \tag{3}$$

$$\ddot{\Phi} = \boldsymbol{\Phi}_q \ddot{\mathbf{q}} - \mathbf{h}(\mathbf{q}, \dot{\mathbf{q}}) = \mathbf{0}. \tag{4}$$

Replacing the constraint equations with theirs second derivatives, one obtains the equations of motion of the whole system

$$\begin{bmatrix} \mathbf{M} & \boldsymbol{\Phi}_q^T \\ \boldsymbol{\Phi}_q & \mathbf{0} \end{bmatrix} \begin{bmatrix} \ddot{\mathbf{q}} \\ \boldsymbol{\lambda} \end{bmatrix} = \begin{bmatrix} \boldsymbol{\tau}(\mathbf{q}, \dot{\mathbf{q}}, \boldsymbol{\lambda}, t) \\ \mathbf{h}(\mathbf{q}, \dot{\mathbf{q}}) \end{bmatrix}. \tag{5}$$

Equation (5) is not differentiable to obtain a system of ordinary differential equations (ODE) because $\boldsymbol{\tau}$ is non-smooth due to dry friction.

With initial conditions satisfying constraint equations, an exact solution of Eq. (5) always satisfies the constraint equations since the second derivatives of constraint equations are kept zero. However, a numerical solution is not an exact solution, errors in the second derivatives are usually inevitable, leading to errors in constraint equations. Such errors increase over time because constraint equations do not explicitly appear in Eq. (5); therefore, Baumgarte stabilization should be used [8, 9]. The stabilized constraint equations read

$$\ddot{\Phi} + 2\alpha\dot{\Phi} + \beta^2\boldsymbol{\Phi} = \boldsymbol{\Phi}_q\ddot{\mathbf{q}} - \mathbf{h}(\mathbf{q}, \dot{\mathbf{q}}) + 2\alpha\boldsymbol{\Phi}_q\dot{\mathbf{q}} + \beta^2\boldsymbol{\Phi} = \mathbf{0}, \tag{6}$$

where α and β are chosen positive constants. The equations of motion of interest are therefore

$$\begin{bmatrix} \mathbf{M} & \boldsymbol{\Phi}_q^T \\ \boldsymbol{\Phi}_q & \mathbf{0} \end{bmatrix} \begin{bmatrix} \ddot{\mathbf{q}} \\ \boldsymbol{\lambda} \end{bmatrix} = \begin{bmatrix} \boldsymbol{\tau}(\mathbf{q}, \dot{\mathbf{q}}, \boldsymbol{\lambda}, t) \\ \mathbf{h}(\mathbf{q}, \dot{\mathbf{q}}) - 2\alpha\boldsymbol{\Phi}_q\dot{\mathbf{q}} - \beta^2\boldsymbol{\Phi} \end{bmatrix}. \tag{7}$$

3 Coulomb Friction, Piecewise Equations and Switching Conditions

3.1 Slip Phase

For the ease of explanation, the case with only one dry friction force is considered first. Without focusing on the switching conditions between stick and slip, the kinetic friction force in Fig. 1b can be written handily using sign function

$$F_{kf} = \text{sign}(N)\text{sign}(v_r)\mu_k N. \tag{8}$$

It is only acceptable in simulation when stick does not occur and numerical results of normal force N and relative velocity v_r do not oscillate. In the slip phase, it is helpful in writing the equations of motion. Note that N is included in the Lagrange multiplier so it can be determined as

$$N = \mathbf{n}^T \boldsymbol{\lambda} \tag{9}$$

with a suitable constant vector $\mathbf{n}$.

Assume that the relative velocity v_r is linearly dependent on generalized velocities

$$v_r = \mathbf{J}_{f,\mathbf{q}}\dot{\mathbf{q}}, \tag{10}$$

with

$$\mathbf{J}_{f,\mathbf{q}} = \frac{\partial f}{\partial \mathbf{q}}, \tag{11}$$

where

$$f(\mathbf{q}) = const \tag{12}$$

is the additional constraint when stiction occurs and it will be mentioned later.

The virtual work done by friction force is

$$-F_{kf}\delta f = \delta \mathbf{q}^T(-\mathbf{J}_{f,\mathbf{q}}^T F_{kf}). \tag{13}$$

Using Eqs. (8), (9), (10) and (13), one can rewrite Eq. (7) as

$$\begin{bmatrix} \mathbf{M} & \boldsymbol{\Phi}_q^T \\ \boldsymbol{\Phi}_q & \mathbf{0} \end{bmatrix} \begin{bmatrix} \ddot{\mathbf{q}} \\ \boldsymbol{\lambda} \end{bmatrix} = \begin{bmatrix} \boldsymbol{\tau}(\mathbf{q}, \dot{\mathbf{q}}, t) - \text{sign}(\mathbf{n}^T\boldsymbol{\lambda})\text{sign}(\mathbf{f}_q\dot{\mathbf{q}})\mathbf{J}_{f,\mathbf{q}}^T \mu_k \mathbf{n}^T\boldsymbol{\lambda} \\ \mathbf{h}(\mathbf{q}, \dot{\mathbf{q}}) - 2\alpha\boldsymbol{\Phi}_q\dot{\mathbf{q}} - \beta^2\boldsymbol{\Phi} \end{bmatrix}. \tag{14}$$

With a simple alteration, the right-hand side is free from $\boldsymbol{\lambda}$

$$\begin{bmatrix} \mathbf{M} & \boldsymbol{\Phi}_q^T + \text{sign}(\mathbf{n}^T\boldsymbol{\lambda})\text{sign}(\mathbf{J}_{f,\mathbf{q}}\dot{\mathbf{q}})\mathbf{J}_{f,\mathbf{q}}^T \mu_k \mathbf{n}^T \\ \boldsymbol{\Phi}_q & \mathbf{0} \end{bmatrix} \begin{bmatrix} \ddot{\mathbf{q}} \\ \boldsymbol{\lambda} \end{bmatrix} = \begin{bmatrix} \boldsymbol{\tau}(\mathbf{q}, \dot{\mathbf{q}}, t) \\ \mathbf{h}(\mathbf{q}, \dot{\mathbf{q}}) - 2\alpha\boldsymbol{\Phi}_q\dot{\mathbf{q}} - \beta^2\boldsymbol{\Phi} \end{bmatrix}. \tag{15}$$

Equation (15) is a piecewise-linear function of the generalized accelerations and $\boldsymbol{\lambda}$ so that if the values of sign functions are known and there is no singularity, the generalized accelerations and $\boldsymbol{\lambda}$ can be easily solved from known values of generalized coordinates and velocities.

If there are two or more dry friction forces, the equations when slipping occurs at all joints are

$$\begin{bmatrix} \mathbf{M} & \boldsymbol{\Phi}_q^T + \text{sign}(\mathbf{N}^T\boldsymbol{\lambda})\text{sign}(\mathbf{J}_{\mathbf{f},\mathbf{q}}\dot{\mathbf{q}})\mathbf{J}_{\mathbf{f},\mathbf{q}}^T\mathbf{Y}_k\mathbf{N}^T \\ \boldsymbol{\Phi}_q & \mathbf{0} \end{bmatrix} \begin{bmatrix} \ddot{\mathbf{q}} \\ \boldsymbol{\lambda} \end{bmatrix} = \begin{bmatrix} \boldsymbol{\tau}(\mathbf{q}, \dot{\mathbf{q}}, t) \\ \mathbf{h}(\mathbf{q}, \dot{\mathbf{q}}) - 2\alpha\boldsymbol{\Phi}_q\dot{\mathbf{q}} - \beta^2\boldsymbol{\Phi} \end{bmatrix}, \tag{16}$$

where $\mathbf{N}$ is an identity matrix if $\boldsymbol{\lambda}$ contains nothing other than the normal forces of prismatic joints with dry friction, $\mathbf{Y}_k$ is a diagonal matrix of coefficients of kinetic friction

$$\mathbf{Y}_k = \text{diag}(\mu_{k_1}, \mu_{k_2}, \ldots), \tag{17}$$

and the diagonal sign matrix of a vector is defined by

$$\text{sign}([a_1, a_2, \ldots]^T) = \text{diag}(\text{sign}(a_1), \text{sign}(a_2), \ldots). \tag{18}$$

3.2 Stick Phase

In stick phase, the dry friction force becomes static friction force F_{sf} which is an ideal reaction force, and constraint Eq. (12) is valid. Equation (12) is differentiated twice and stabilized using Baumgarte stabilization

$$\mathbf{J}_{f,\mathbf{q}}\ddot{\mathbf{q}} + \dot{\mathbf{J}}_{f,\mathbf{q}}\dot{\mathbf{q}} + 2\alpha\mathbf{J}_{f,\mathbf{q}}\dot{\mathbf{q}} + \beta^2(f(\mathbf{q}) - C) = 0, \tag{19}$$

in which C is a constant determined by the initial conditions of the stick phase.

The equations in the stick phase are modified from Eq. (15)

$$\begin{bmatrix} \mathbf{M} & \boldsymbol{\Phi}_q^T & \mathbf{J}_{f,\mathbf{q}}^T \\ \boldsymbol{\Phi}_q & \mathbf{0} & \mathbf{0} \\ \mathbf{J}_{f,\mathbf{q}} & \mathbf{0} & 0 \end{bmatrix} \begin{bmatrix} \ddot{\mathbf{q}} \\ \boldsymbol{\lambda} \\ -F_{sf} \end{bmatrix} = \begin{bmatrix} \boldsymbol{\tau}(\mathbf{q}, \dot{\mathbf{q}}, t) \\ \mathbf{h}(\mathbf{q}, \dot{\mathbf{q}}) - 2\alpha\boldsymbol{\Phi}_q\dot{\mathbf{q}} - \beta^2\boldsymbol{\Phi} \\ -\dot{\mathbf{J}}_{f,\mathbf{q}}\dot{\mathbf{q}} - 2\alpha\mathbf{J}_{f,\mathbf{q}}\dot{\mathbf{q}} - \beta^2(f(\mathbf{q}) - C) \end{bmatrix}. \tag{20}$$

If there are two or more prismatic joints with dry friction as in Eq. (16), and one, some or all of them are in stick phase with static friction forces $\mathbf{F}_{\text{sf}}$, the equations of motion read

$$\begin{bmatrix} \mathbf{M} & \boldsymbol{\Phi}_q^T & \mathbf{J}_{\mathbf{f},\mathbf{q}}^T \\ \boldsymbol{\Phi}_q & \mathbf{0} & \mathbf{0} \\ \mathbf{J}_{\mathbf{f},\mathbf{q}} & \mathbf{0} & \mathbf{0} \end{bmatrix} \begin{bmatrix} \ddot{\mathbf{q}} \\ \boldsymbol{\lambda} \\ -\mathbf{F}_{sf} \end{bmatrix} = \begin{bmatrix} \bar{\boldsymbol{\tau}}(\mathbf{q}, \dot{\mathbf{q}}, \boldsymbol{\lambda}, t) \\ \mathbf{h}(\mathbf{q}, \dot{\mathbf{q}}) - 2\alpha\boldsymbol{\Phi}_q\dot{\mathbf{q}} - \beta^2\boldsymbol{\Phi} \\ -\dot{\mathbf{J}}_{f,\mathbf{q}}\dot{\mathbf{q}} - 2\alpha\mathbf{J}_{f,\mathbf{q}}\dot{\mathbf{q}} - \beta^2(f(\mathbf{q}) - C) \end{bmatrix}. \tag{21}$$

3.3 Piecewise Equations and Switching Conditions

Consider the case with only one dry friction force. To treat the discontinuities, three logic functions are used: the sign or the potential sign of the relative velocity γ_v (which now replaces the sign of the relative velocity in the equations), the sign of the normal force γ_n, and γ_f which indicates whether the static friction force reach the threshold value and, if yes, it is the sign of the friction force

$$\begin{cases} \gamma_f = 0 & \text{if } -\mu_s|N| < F_{sf} < \mu_s|N| \\ \gamma_f = 1 & \text{if } F_{sf} \geq -\mu_s|N| \\ \gamma_f = -1 & \text{if } F_{sf} \leq -\mu_s|N| \end{cases} . \tag{22}$$

In additional, the current phase γ_p is always kept track of to help the solver choose which equations to integrate.

Provided that the relative velocity is continuous, the necessary condition for the transition from slip to stick is the change in the sign of the relative velocity, which means there is a time point t_v that the relative velocity is zero. There is a possibility that the relative velocity reaches zero but immediately takes a nonzero value so that stiction does not occur due to zero dwell time and static coefficient of friction should not be used. Such scenarios happen when, for instance, precalculated control forces are applied to get a desired sinusoidal motion [7]. The condition for the slip phase to continue is that either of the following cases happen at t_v: (i) choose the potential sign of relative velocity $\gamma_\mathrm{v} = 1$ and the calculated relative acceleration has the same sign $a_\mathrm{r} > 0$, (ii) $\gamma_\mathrm{v} = -1$ and $a_\mathrm{r} < 0$. The use of logic function γ_v instead of the sign of the relative velocity also helps avoid unrealistic zero friction. Thus, the sufficient condition for stiction to begin is that $\gamma_\mathrm{v} = 0$ because it cannot take any other value, and t_v coincides with t_p where γ_p changes from "slip" to "stick".

During stick phase, γ_v plays no role. However, when there is a change in γ_f at time point t_f, the slip phase starts and γ_v should be chosen to have the same value as $\gamma_\mathrm{f}(t_\mathrm{f})$ to continue the simulation, and t_f coincides with t_p where γ_p changes from "stick" to "slip".

The sign of the normal force γ_n is necessary for simulation during slip phase including t_f. During stick phase, γ_n is not needed. At t_f, γ_n takes the same value as γ_f. Numerically, γ_n at any time point in slip phase is assumed to have the value of the previous time point so that the normal force can be solved from a system of linear algebraic equations. Then, γ_n is rechecked to see if the calculated normal force has the assumed sign. If not, that means there is a time point t_n at which the normal force changes its sign and γ_n should take the opposite value beyond that time point.

In simulation, the switching times are usually not exactly found which may cause numerical instability. For instance, when the system enters stick phase at γ_v but v_r is not exactly zero, the joint that should be sticked is not sticked in the simulation result. Such instability can also be suppressed by Baumgarte stabilization.

In short, the following piecewise equations should be solved

$$\begin{cases} \begin{bmatrix} \mathbf{M} & \mathbf{\Phi}_q^T + \gamma_n \gamma_v \mathbf{J}_{f,\mathbf{q}}^T \mu_k \mathbf{n}^T \\ \mathbf{\Phi}_q & \mathbf{0} \end{bmatrix} \begin{bmatrix} \ddot{\mathbf{q}} \\ \boldsymbol{\lambda} \end{bmatrix} = \begin{bmatrix} \overline{\boldsymbol{\tau}}(\mathbf{q}, \dot{\mathbf{q}}, t) \\ \mathbf{h}(\mathbf{q}, \dot{\mathbf{q}}) - 2\alpha \mathbf{\Phi}_q \dot{\mathbf{q}} - \beta^2 \mathbf{\Phi} \end{bmatrix} \text{if } \gamma_p = \text{"slip"} \\ \begin{bmatrix} \mathbf{M} & \mathbf{\Phi}_q^T & \mathbf{J}_{f,\mathbf{q}}^T \\ \mathbf{\Phi}_q & \mathbf{0} & \mathbf{0} \\ \mathbf{J}_{f,\mathbf{q}} & \mathbf{0} & 0 \end{bmatrix} \begin{bmatrix} \ddot{\mathbf{q}} \\ \boldsymbol{\lambda} \\ -F_{sf} \end{bmatrix} = \begin{bmatrix} \overline{\boldsymbol{\tau}}(\mathbf{q}, \dot{\mathbf{q}}, t) \\ \mathbf{h}(\mathbf{q}, \dot{\mathbf{q}}) - 2\alpha \mathbf{\Phi}_q \dot{\mathbf{q}} - \beta^2 \mathbf{\Phi} \\ -\dot{\mathbf{J}}_{f,\mathbf{q}} \dot{\mathbf{q}} - 2\alpha \mathbf{J}_{f,\mathbf{q}} \dot{\mathbf{q}} - \beta^2 (f(\mathbf{q}) - C) \end{bmatrix} \text{if } \gamma_p = \text{"stick"} \end{cases} \tag{23}$$

with the consideration of the logic functions γ_{v}, γ_{n}, γ_{f}, γ_{p}, and switching times t_{v}, t_{n}, t_{f}, t_{p} as explained above.

4 Examples and Discussion

Consider the slider-crank mechanism with the parameters in Fig. 2a. The slider is considered a point mass subjected to dry friction between it and the base.

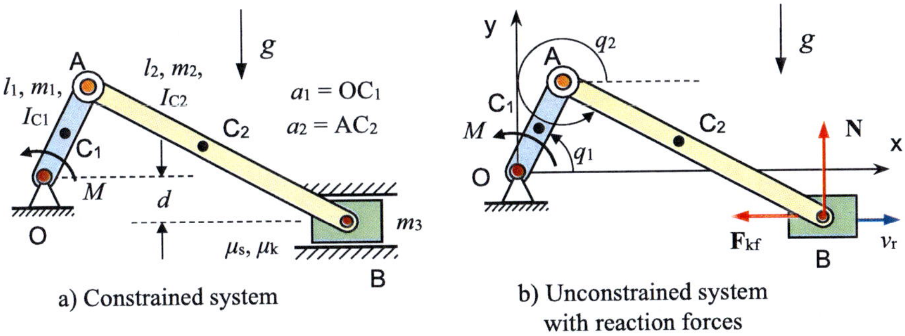

Fig. 2. Slider-crank mechanism

The connection between the slider and the base is replaced by reaction forces (Fig. 2b), and the equations of motion of the system is established accordingly.

The equations of motion of the form (23) are define by

$$\mathbf{M} = \begin{bmatrix} I_{C1} + m_1 a_1^2 + (m_2 + m_3) l_1^2 & (m_2 a_2 + m_3 l_2) l_1 \cos(q_2 - q_1) \\ (m_2 a_2 + m_3 l_2) l_1 \cos(q_2 - q_1) & I_{C2} + m_2 a_2^2 + m_3 l_2^2 \end{bmatrix}, \tag{24}$$

$$\overline{\boldsymbol{\tau}}(\mathbf{q}, \dot{\mathbf{q}}, t) = \begin{bmatrix} M + (m_2 a_2 + m_3 l_2) l_1 \sin(q_2 - q_1) \dot{q}_2^2 - (m_1 a_1 + m_2 l_1 + m_3 l_1) g \cos q_1 \\ -(m_2 a_2 + m_3 l_2) l_1 \sin(q_2 - q_1) \dot{q}_1^2 - (m_2 a_2 + m_3 l_2) g \cos q_2 \end{bmatrix}, \tag{25}$$

$$\mathbf{\Phi} = [l_1 \sin q_1 + l_2 \sin q_2 + d],\ \mathbf{\Phi}_q = [\, l_1 \cos q_1 \ l_2 \cos q_2 \,],$$
$$\mathbf{h} = [l_1 \dot{q}_1^2 \sin q_1 + l_2 \dot{q}_2^2 \sin q_2], \tag{26}$$

$$\mathbf{n} = 1, \boldsymbol{\lambda} = N, \tag{27}$$

$$f = l_1 \cos q_1 + l_2 \cos q_2, \tag{28}$$

$$\mathbf{J}_{f,\mathbf{q}} = \begin{bmatrix} -l_1 \sin q_1 & -l_2 \sin q_2 \end{bmatrix}, \dot{\mathbf{J}}_{f,\mathbf{q}} = \begin{bmatrix} -l_1 \dot{q}_1 \cos q_1 & -l_2 \dot{q}_2 \cos q_2 \end{bmatrix}. \tag{29}$$

$$v_r = \mathbf{J}_{f,\mathbf{q}} \dot{\mathbf{q}} = -l_1 \dot{q}_1 \sin q_1 - l_2 \dot{q}_2 \sin q_2. \tag{30}$$

The values of the parameters are given in Table 1.

Table 1. Values of the parameters

Parameters	m_1 kg	l_1 m	a_1 m	I_{C1} kgm^2	m_2 kg	l_2 m	a_2 m	I_{C2} kgm^2	m_3 kg	d m	μ_s	μ_k
Values	1	0.1	0.5	$1/12 \times 10^{-2}$	2	0.4	0.2	$4/3 \times 10^{-2}$	2	−0.2	0.55	0.35

Table 2. Applied moment

Case	1	2
$M(t)$ Nm	0.03	$\begin{cases} 0 & t \in [0, 0.9) \cup [1, 1.9) \cup [2, 2.9) \ldots \text{ s} \\ 3 & \text{otherwise} \end{cases}$

The initial conditions are

$$\mathbf{q}(0) = \begin{bmatrix} 0 \\ \pi/6 \end{bmatrix}, \dot{\mathbf{q}}(0) = \begin{bmatrix} 0 \\ 0 \end{bmatrix}. \tag{31}$$

Two examples are simulated with two different moment rules applied to the crank (Table 2). Chosen integrator is the classic Runge-Kutta method with a step-size of 0.001 s. The stabilization parameters are $\alpha = \beta = 0$ (no stabilization) and $\alpha = \beta = 100$.

In the first case, the applied moment is so small that the system eventually rests at an equilibrium state. Without stabilization, q_1 drifts away from that state, resulting in a qualitatively incorrect result, which is avoided by Baumgarte stabilization (Fig. 3).

In the second case, the applied moment is periodically switch on and off. The results with and without stabilization are again qualitatively different (Fig. 4 and Fig. 5). The error of the constraint equation without stabilization diverges while with stabilization, it is lower and exhibits a periodically behavior which agrees with the motion as well as excitation. The low and bounded errors show that the proposed method is reliable in the illustrative examples. As opposed to the methods presented in the literature [5, 6], the proposed method does not require artificial parameters in the friction law so that it can be applied handily if the friction coefficients are known.

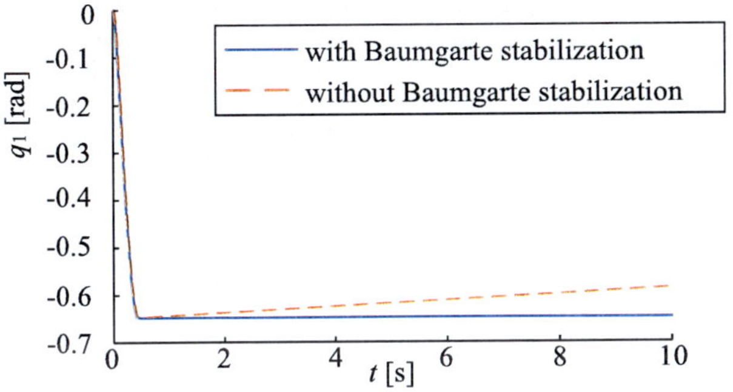

Fig. 3. Simulation results of case 1: small constant moment.

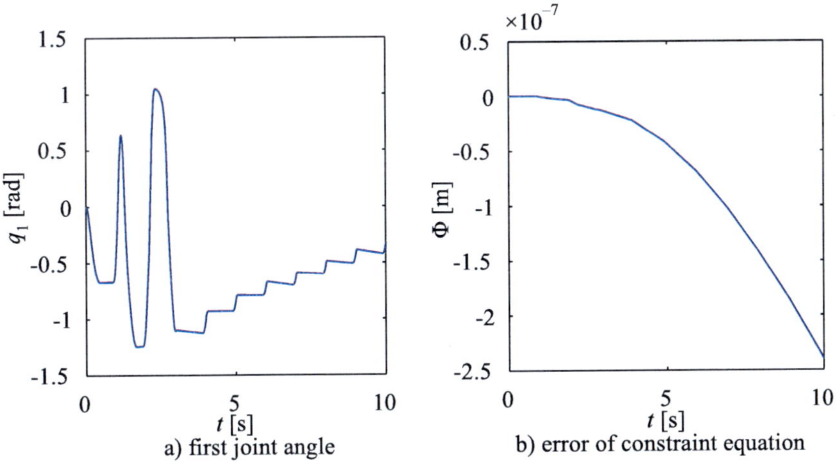

Fig. 4. Simulation results of case 2: on-off moment without stabilization.

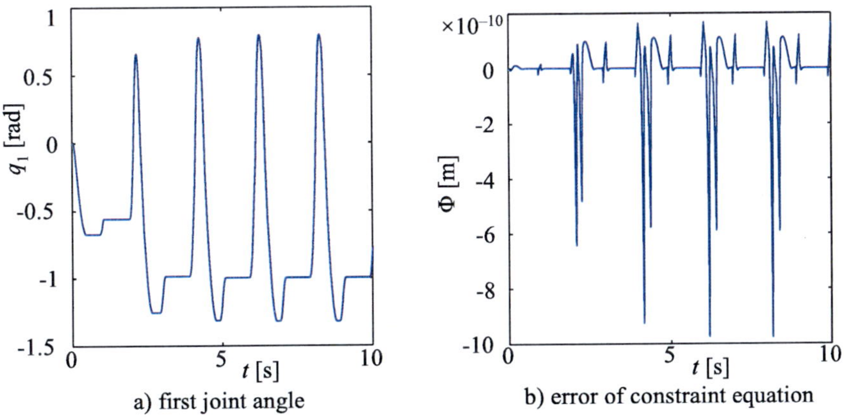

Fig. 5. Simulation results of case 2: on-off moment with stabilization.

5 Conclusions and Outlook

Piecewise equations as well as switching conditions of closed-loop planar mechanisms with Coulomb friction at prismatic joints are established. The equations are piecewise-linear in terms of generalized accelerations and reaction forces in both stick and slip phases.

Examples with a slider-crank mechanism show that Baumgarte stabilization can be used to avoid qualitatively incorrect results.

Future work may study the combination of Baumgarte stabilization or maybe other stabilization methods and root-finding strategies to improve the performance of the integrator. More complicated systems will also be considered in the future.

Acknowledgements. This research is funded by Hanoi University of Science and Technology (HUST) under project number T2021-PC-037.

References

1. Pavlovskaia, E., Wiercigroch, M.: Analytical drift reconstruction for visco-elastic impact oscillators operating in periodic and chaotic regimes. Chaos Solit. Fractals **19**(1), 151–161 (2004)
2. Boone, G.N., Hodgins, J.K.: Slipping and tripping reflexes for bipedal robots. Auton. Robot. **4**(3), 259–271 (1997)
3. Veiga, F., Peters, J., Hermans, T.: Grip stabilization of novel objects using slip prediction. IEEE Trans. Hapt. **11**(4), 531–542 (2018)
4. Borello, L., Dalla Vedova, M.D.L.: Dry friction discontinuous computational algorithms. Int. J. Eng. Innov. Technol. (IJEIT) **3**(8), 1–8 (2014)
5. Haug, E.J.: Simulation of spatial multibody systems with friction. Mech. Based Design Struct. Mach. **46**(3), 347–375 (2018)
6. Eich-Soellner, E., Führer, C.: Numerical Methods in Multibody Dynamics, vol. 45. Teubner, Stuttgart (1998)
7. Farhat, N., Mata, V., Page, A., Díaz-Rodríguez, M.: Dynamic simulation of a parallel robot: coulomb friction and stick-slip in robot joints. Robotica **28**(1), 35–45 (2010)
8. Baumgarte, J.: Stabilization of constraints and integrals of motion in dynamical systems. Comput. Methods Appl. Mech. Eng. **1**(1), 1–16 (1972)
9. Flores, P., Machado, M., Seabra, E., da Silva, M.T.: A parametric study on the Baumgarte stabilization method for forward dynamics of constrained multibody systems. J. Comput. Nonlinear Dyn. **6**(1), 011019, 9 p. (2011). https://doi.org/10.1115/1.4002338
10. Schiehlen, W., Eberhard, P.: Applied Dynamics, vol. 57. Springer, Berlin (2014)

Developing Geometric Error Compensation Software for Five-Axis CNC Machine Tool on NC Program Based on Artificial Neural Network

Van-Hai Nguyen[1,2] and Tien-Thinh Le[1,2](✉)

[1] Faculty of Mechanical Engineering and Mechatronics, PHENIKAA University, Yen Nghia, Ha Dong, Hanoi 12116, Vietnam
thinh.letien@phenikaa-uni.edu.vn

[2] PHENIKAA Research and Technology Institute (PRATI), A&A Green Phoenix Group JSC, No. 167 Hoang Ngan, Trung Hoa, Cau Giay, Hanoi 11313, Vietnam

Abstract. This paper aims to develop a geometric error compensation software for five-axis CNC machine tool basing on NC program and machine learning artificial neural network. These modules are used in this software such as Human Machine Interface (HMI), data preprocessing, NC rebuild, and checking NC program. To determine error compensation, the inverse error algorithm from ideal points (from inputted NC program) and machine learning based artificial neural network model are calculated. While, artificial neural network model is optimized based on mean squared error cost function. The performance of the prediction model is confronted with experimental data points and various regression analyses. The geometric compensation software is then constructed and appended to this manuscript for a practical application purpose.

Keywords: Geometric error · Error compensation software · Five-axis machine tool · Inverse error algorithm

1 Introduction

In the modern industry, the accuracy of product is the most important performance specification on numerical control (NC) machine tool. Normally, the accuracy of machine tools is affected by geometric errors; thermal effects; static/dynamic load effects and, as well as workpiece and tool-related error while geometric errors is the largest source caused machine tool inaccuracy [1–4]. Therefore, for improving machining accuracy of CNC tools, many researchers have been proposed different compensation geometric error on the various type of machine tool focusing on software compensation and hardware compensation. The hardware compensation method has appeared many limitations [5], while software compensation method has presented more advantage such as lower cost, easier to realize. This is motivative for researchers launched some solutions to improve machining accuracy such as some correct error methods for NC program [6–9].

Supplementary Information The online version contains supplementary material available at https://doi.org/10.1007/978-3-030-91892-7_51.

N. V. Khang et al. (Eds.): ASIAN MMS 2021, MMS 113, pp. 541–548, 2022.
https://doi.org/10.1007/978-3-030-91892-7_51

The results of this literature show that, there are many different error compensation methods for many types of machine tool. However, calculating and predicting errors over the entire position in the machine tool workspace were relying on calculus methods. In this work, we propose to develop a machine learning based artificial neural network model for the prediction of geometric errors for five-axis CNC machining tool. The result of this model is used on developing a software for geometric error compensation of NC program. This also is the aim of this study.

2 Materials and Methodology

2.1 Prediction Model: Machine Learning - Artificial Neural Network

In this work, a machine learning model based on Artificial neural network was developed for the prediction of geometric error of five-axis CNC machining tool. The model was early concepted in 1950s [10] based on the information processing method of biological neuron systems. It is made up of a large number of elements (called neurons) connected together through links (called link weights) that work as a whole to solve a particular problem. In essence, it is the process of adjusting the linkage weight between neurons so that the error function value is minimal. The basic structure of an ANN usually consists of neurons grouped into input data layers, output data and one or more hidden layers. The process of adjusting weights so that the network knows the relationship between the input and the desired outcome is called learning or training. Currently, the mathematical algorithm used to adjust the performance of the artificial neural network is now widely used as the backpropagation algorithm. The backpropagation algorithm uses a set of input and output values to find the desired neural network. A set of inputs is put into a certain preset system to calculate the output value, then this output value is compared with the actual value measured. If there is no difference, then there is no need to perform a test, otherwise the weights will be changed during back propagation in the neural network to reduce the difference. The backpropagation network usually has one or more hidden layers with sigmoid-like neurons and the output layer is neurons with linear transfer function.

In this work, the database for the development of artificial neural network model was collected from the available literature [11], in which the volumetric errors for three axes of X, Y and Z were measured by an API Radian-20 laser tracker for a five-axis CNC machining tool.

2.2 Framework

In this work, the geometric error compensation software for NC program is develop with MATLAB. In the previous research, the acquiring error data from measured error experiment, kinematic error modelling and geometric error model have been performed. Current work inherits the result from previous research, an ideal NC program is generated by CAM software, this is understood that the set of ideal points movement of machine tool. The actual NC program is the set of adjusted points which is corrected according to the ideal points trajectory and predicted model by the inverse error compensation algorithm, as shown in Fig. 1.

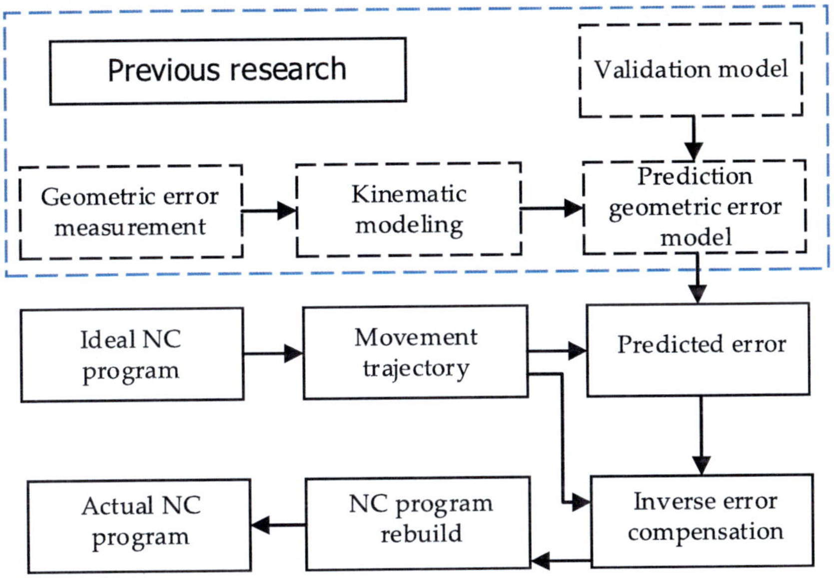

Fig. 1. Flowchart of error compensation according to NC program rebuild.

2.3 Setting Equation

Supposing, one position of tool tip can be expressed as $P_{err}(E_x, E_y, E_z)$ and the corresponding the position that defined in five axis machine tool tip or NC code is $G(x, y, z, \alpha, \gamma)$. From the perspective of the black box, system models can always be demonstrated as a function $P_{err} = f(G)$, in which G is the vector of several independent inputs is determined by initial points (X-axis, Y-axis, Z-axis, C-rotational and A-rotational); and P_{err} are dependent outputs which is chosen to be analysed.

Supposing $P_i(x_i, y_i, z_i)$ are the ideal points from ideal NC program, and $P_{err,i} = [E_{xi}\, E_{yi}\, E_{zi}]^T$ are volumetric error of the set of n ideal points; $E_{xi}\, E_{yi}$ and E_{zi} is predicted error at i point. From that, the real points from real cutting tool tip is $P_{ri}(x_{ri}, y_{ri}, z_{ri})$ [5]:

$$P_{ri} = P_i + P_{err,i} \tag{1}$$

Equation (1) is detailed below:

$$x_{ri} = x_i + E_{xi}(P_i) \tag{2}$$

$$y_{ri} = y_i + E_{yi}(P_i) \tag{3}$$

$$z_{ri} = y_i + E_{zi}(P_i) \tag{4}$$

From Eqs. (3), (4), and (5), the uncompensated trajectory points can be obtained by mapping the ideal trajectory points into real trajectory point is $P_{uci}(x_{uci}, y_{uci}, z_{uci})$

$$P_{uci} = P_i - P_{err,i} \tag{5}$$

Equation (5) is detailed below:

$$x_{uci} = x_i - E_{xi}(P_i) \quad (6)$$

$$y_{uci} = y_i - E_{yi}(P_i) \quad (7)$$

$$z_{uci} = z_i - E_{zi}(P_i) \quad (8)$$

The real trajectory points after compensation is $P_{ci}(x_{ci}, y_{ci}, z_{ci})$ is described

$$P_{ci} = P_{uci} + E_{ci} = P_i - P_{err,i} + E_{ci} \quad (9)$$

where $E_{ci}\left(E_{cxi}, E_{cyi}, E_{czi}\right)$ is compensation error for ideal point i^{th}. Equation (9) is detailed below:

$$x_{ci} = x_{uci} + E_{xi}(P_{uci}) = x_i - E_{xi}(P_i) + E_{xi}(P_{uci}) \quad (10)$$

$$y_{ci} = y_{uci} + E_{yi}(P_{uci}) = y_i - E_{yi}(P_i) + E_{yi}(P_{uci}) \quad (11)$$

$$z_{ci} = z_{uci} + E_{zi}(P_{uci}) = z_i - E_{zi}(P_i) + E_{zi}(P_{uci}) \quad (12)$$

We must determine the E_{ci}. The Eq. (9) can be rewritten under general equation form:

$$P_{ci}^{k} = F(P_i, Q_i) \quad (13)$$

where $F(P_i, Q_i)$ is inverse equation this have 2 variables P_i and predicted equation Q_i, which is the artificial neural network model presented in Sect. 2.1.

3 Results and Discussion

3.1 Prediction Performance of Artificial Neural Network Model

In this work, a neural network model exhibiting one hidden layer with ten neurons was applied [12–14]. A visualization of the model architecture is shown in Fig. 2. In such an architecture, the input layer includes five variables, which are coordinates at X, Y, and Z-axes and angles A and C, respectively. The hidden layer includes ten neurons, using an activation of sigmoid. The output layer exhibits 3 neurons, using a linear activation function. Finally, the outputs of the model are geometric errors along X, Y, and Z-axes, respectively.

Based on the database collected from the available literature [11], the artificial neural network model was optimized in weights and bias. Quantity of sample for training was set as 70% of all data points, and the rest (30%) was used for validation and testing. The sample for training was randomly selected in the input space based on a uniform distribution. The procedure of optimization is illustrated in Fig. 3a, showing the value of mean squared error cost function as a function of iterations. It is seen that after 22

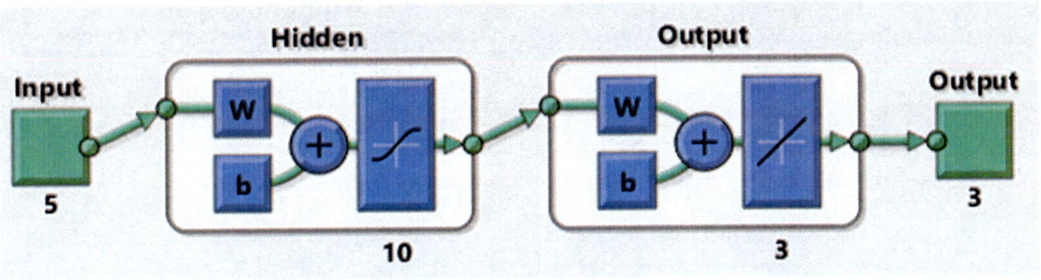

Fig. 2. Architecture of artificial neural network model.

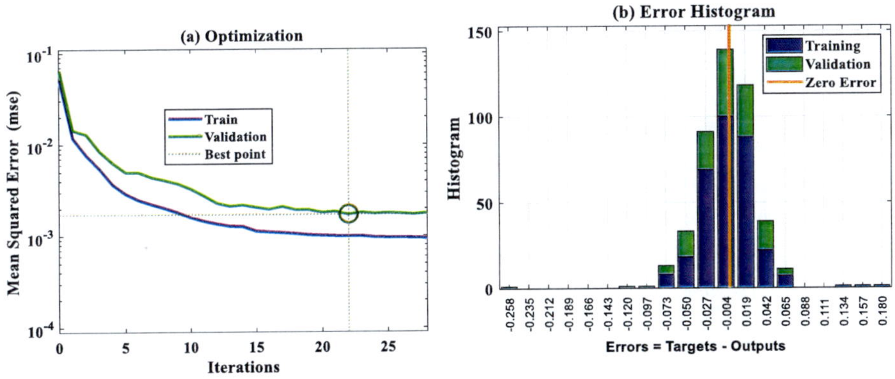

Fig. 3. (a) Optimization of artificial neural network model. (b) Error distributions

iterations, the weights and bias of the model are optimized, showing the smallest value of mean squared error for the validation data points.

The performance of artificial neural network model is shown in Fig. 3b for distributions of error and Fig. 4 as regression analysis. It is seen in Fig. 3b that the error distribution for both training and validation data points is centered at zero, confirming the relevance of the model. The same remark can be deduced by observing Fig. 4a and 4b, which show the scatter points between experimental and predicted data points, for training and validation parts, respectively. It is obtained that the coefficient of correlation $R = 0.9633$ and $R = 0.9518$ for training and validation parts, respectively. Moreover, the linear fit is also close to the diagonal line, showing the good performance of the model. Finally, all data points are distributed almost equally around the linear fit, and there is no concentration zone. This means that the model is not over or under fitted.

Based on these analyses, the artificial neural network model is employed as prediction function in the next sections.

3.2 Designed Function of Error Compensation Software

Figure 5 show the error compensation system basing on rebuild NC program, this include four modules: Human machine interface (HMI), data preprocessing, NC rebuild, and checking NC program. Where, HMI module is interactive environment, this is convenient for user interaction with tool function and visual representation of the system's

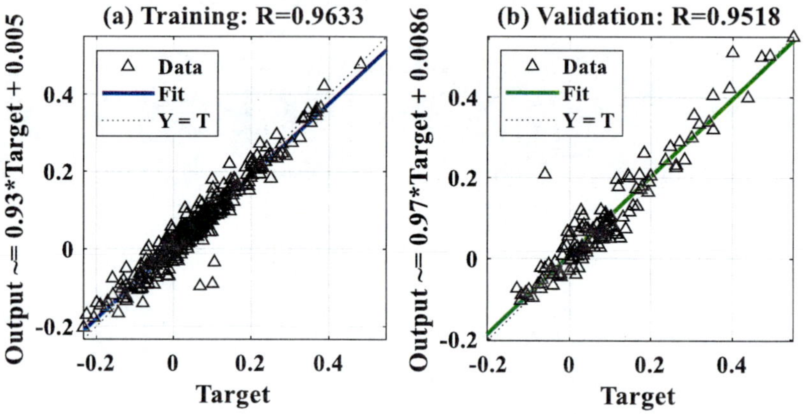

Fig. 4. Regression analysis using artificial neural network model.

processing results. Data preprocessing is used for input data, reading data, and determining data command that contain geometric positions such as G01, G02, G03. NC program module rebuild contain the inverse calculation the compensation error, and corrected the NC program. Finally, Checking NC program include the checking the corrected NC program algorithm, this notify the error compensation values by windows frame for each highlight NC command.

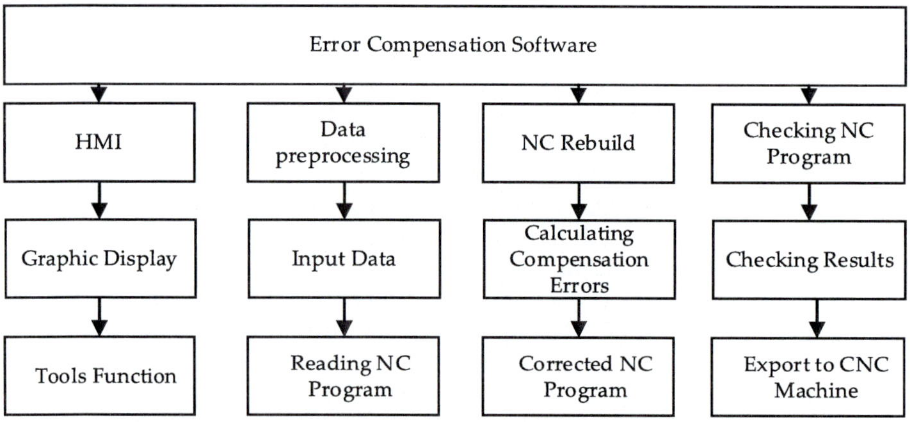

Fig. 5. Function tree of geometric error compensation software.

3.3 NC Rebuild Program

The NC rebuild program was constructed based on Matlab 2018a, using AppDesigner tool, for a practical application purpose. The artificial neural network prediction model is also included in this software for direct prediction of geometric error. Figure 6 presents a visualization for the Graphical User Interface of the software, version 0.1. Such a Graphical User Interface is appended to this manuscript as a supplementary material. Interested users can download and use. However, this software is now only validated for five-axis CNC machining tool, including the ranges of values of variables used in this study.

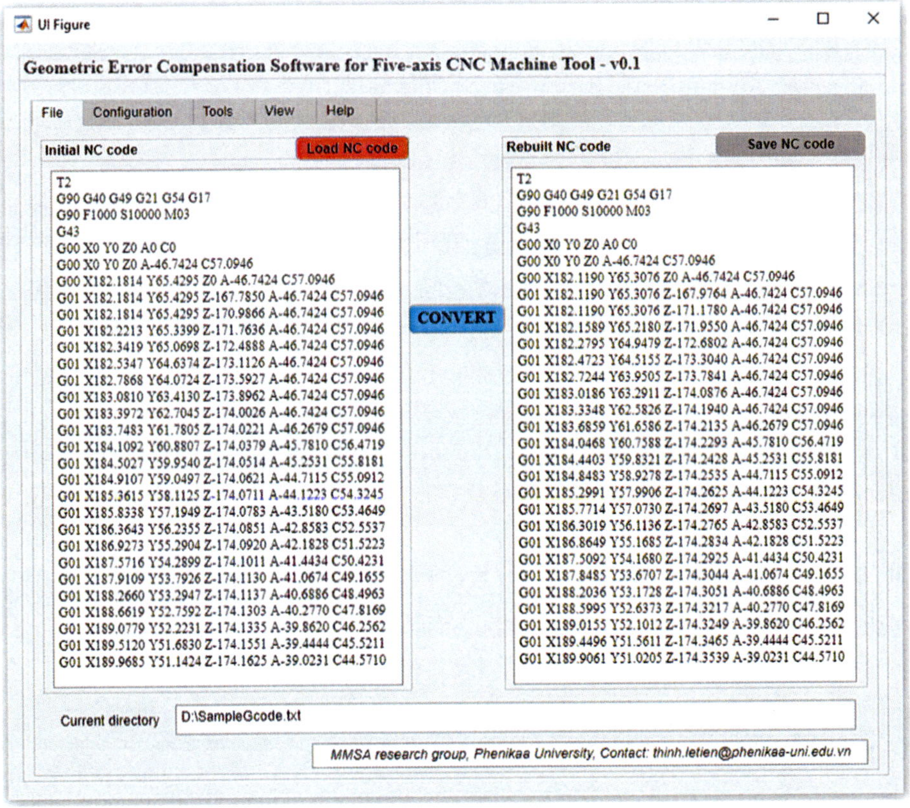

Fig. 6. Graphical user interface for geometric error compensation software for five-axis CNC machining tool developed in this work.

4 Conclusion

In this work, we propose to develop a machine learning based artificial neural network model for the prediction of geometric error for five-axis CNC machining tool. Parameters of the model was optimized using mean squared error cost function and available database in the literature. The performance of the prediction model was analyzed through various quality assessment indices including relative error, coefficient of correlation and linear fitting. For practical application, to overcome the "black-box" meaning of machine learning model, a Graphical User Interface including the prediction model was developed and appended to this paper for interested users. The software can assist researchers/engineers/students on research and development for error compensation of CNC machining tools.

In further researches, more works should be done, including exploring a larger database for training machine learning models, together with different types of CNC machining tools. Besides, different machine learning techniques should be tested in order to increase the performance of prediction. Moreover, the current Graphical User Interface should be enhanced for a more attraction purpose. Finally, experiments should be carried out to improve the performance of the model.

References

1. Creamer, J., Sammons, P.M., Bristow, D.A., Landers, R.G., Freeman, P.L., Easley, S.J.: Table-based volumetric error compensation of large five-axis machine tools. J. Manuf. Sci. Eng. **139**, 021011 (2017)
2. Shen, H., Fu, J., He, Y., Yao, X.: On-line asynchronous compensation methods for static/quasi-static error implemented on CNC machine tools. Int. J. Mach. Tools Manuf. **60**, 14–26 (2012)
3. Treib, T., Matthias, E.: Error budgeting—applied to the calculation and optimization of the volumetric error field of multiaxis systems. CIRP Ann. **36**, 365–368 (1987)
4. Nguyen, V.-H., et al.: Applying Bayesian optimization for machine learning models in predicting the surface roughness in single-point diamond turning polycarbonate. Math. Probl. Eng. **2021**, e6815802 (2021). https://doi.org/10.1155/2021/6815802
5. Cui, G., Lu, Y., Li, J., Gao, D., Yao, Y.: Geometric error compensation software system for CNC machine tools based on NC program reconstructing. Int. J. Adv. Manuf. Technol. **63**, 169–180 (2012)
6. Rahman, M., Heikkala, J., Lappalainen, K.: Modeling, measurement and error compensation of multi-axis machine tools. Part I: theory. Int. J. Mach. Tools Manuf. **40**, 1535–1546 (2000)
7. Li, Y., Shen, X., Wang, A.: Study on the method of the correcting NC command for the error compensation of NC machine. J. Shaanxi Univ. Sci. Technol. Nat. Sci. Ed. **6** (2009)
8. Jing, H.J., Yao, Y.X., Chen, S.D., Wang, X.P.: Machining accuracy enhancement by modifying NC program. In: Key Engineering Materials, pp. 71–75. Trans Tech Publ (2006)
9. Vu, D.T., Tran, X.-L., Cao, M.-T., Tran, T.C., Hoang, N.-D.: Machine learning based soil erosion susceptibility prediction using social spider algorithm optimized multivariate adaptive regression spline. Measurement **164**, 108066 (2020). https://doi.org/10.1016/j.measurement.2020.108066
10. Shanmuganathan, S., Samarasinghe, S. (eds.): Artificial Neural Network Modelling. Springer, New York (2016). https://doi.org/10.1007/978-3-319-28495-8
11. Li, Q., Wang, W., Zhang, J., Shen, R., Li, H., Jiang, Z.: Measurement method for volumetric error of five-axis machine tool considering measurement point distribution and adaptive identification process. Int. J. Mach. Tools Manuf. **147**, 103465 (2019)
12. Le, T.-T., Asteris, P.G., Lemonis, M.E.: Prediction of axial load capacity of rectangular concrete-filled steel tube columns using machine learning techniques. Eng. Comput. (2021). https://doi.org/10.1007/s00366-021-01461-0
13. Thanh Duong, H., Chi Phan, H., Le, T.-T., Duc Bui, N.: Optimization design of rectangular concrete-filled steel tube short columns with balancing composite motion optimization and data-driven model. Structures **28**, 757–765 (2020). https://doi.org/10.1016/j.istruc.2020.09.013
14. Ho, N.X., Le, T.-T., Le, M.V.: Development of artificial intelligence based model for the prediction of Young's modulus of polymer/carbon-nanotubes composites. Mech. Adv. Mater. Struct. 1–14 (2021). https://doi.org/10.1080/15376494.2021.1969709

Artificial Intelligence and Robots

A Gradient-Based Learning Algorithm for Mobile Robot Path Planning in Environment Exploration

Zhiliang Wu(✉) and Yiqiang Wang

School of Mechanical Engineering, Tianjin University, Tianjin 300354, People's Republic of China
{zhlwu,wyq2019}@tju.edu.cn

Abstract. Path planning for mobile robots is of paramount importance in environment exploration and monitoring. The core issue is to decide where to go for the next observation. In this paper, we use an intuitive gradient-based algorithm to select a preferred action. The algorithm is developed in the reinforcement learning paradigm, where the reward as a feedback from the environment is modeled by the gradient of the features and the value of actions are estimated using a weighted combination of historical rewards. The robot then greedily selects the next move. Numerical simulation has been conducted on exploration of Gaussian distributed features over a region of interest. The performance of the presented algorithm has been addressed in four metrics to evaluate error in prediction and accuracy in anomaly identification. Comparison with three different algorithms shows that the presented algorithm performs well for simple feature distribution. Further investigation is needed to extend the utilization in more complex environment.

Keywords: Path planning · Environment exploration · Reinforcement learning

1 Introduction

Mobile robots have been increasingly applied in environment exploration and monitoring due to their invulnerability to hostile or extreme environments. Examples of such applications include space exploration [1], ocean exploration [2], and pollution monitoring [3, 4]. However, the utility of mobile robots is largely confined by the limited power on board. One approach to alleviating the energy problem is to elaborately design or plan a path, following which the robot would be able to efficiently complete its task.

Path planning for mobile robots focuses on generating an optimal moving path, usually given the initial and the goal states. The generated path may be an obstacle-free path or one that meets certain optimization objectives, such as minimum distance, minimum time, and minimum energy consumption. But in the cases of environment exploration and monitoring, the goal state of the mobile robot is not necessarily preset. The mobile robot moves in the region of interest to collect data, with which the features in the environment could be eventually reconstructed.

N. V. Khang et al. (Eds.): ASIAN MMS 2021, MMS 113, pp. 551–558, 2022.
https://doi.org/10.1007/978-3-030-91892-7_52

In environment exploration and monitoring, the quality of the collected data directly affects the accuracy of feature reconstruction. Where to go for data collection is of paramount importance in robotic path planning. Since the features in the environment are usually fully or partially unknown, it would be desirable that the mobile robot could wisely decide online its next move while taking measurement. The fundamental concept for robotic path planning under such circumstances is to maximize information gain during the course of data collection [5]. One popular strategy is using Gaussian processes based exploration techniques [6]. Marchant et al. [7] devised a layered Bayesian optimization approach that uses two Gaussian Processes to optimize sampling over continuous paths. Morere et al. [5] formulated the path planning problem within a Partially Observable Markov Decision Process, including uncertainty information when reasoning. Hitz et al. [8] used an evolutionary strategy for path optimization and a re-planning scheme according to the measurement. Hollinger et al. [9] introduced uncertainty modeling for an autonomous underwater vehicle path planning in ship hull inspection. Li et al. [10] presented a long-term adaptive algorithm for scalar field monitoring using cross-entropy optimization. The success of an environment exploration task may sometimes be judged by accurate reconstruction of the anomalous features presented in the environment [11]. Neumann et al. [12] proposed an adaptive strategy based on artificial potential field approach to locate gas source. Russell [13] developed a hex-path algorithm for locating underground chemical source.

In this paper, we present an intuitive gradient-based path planning algorithm for unknown environment exploration. Decision on where to go for the next move is made based on the gained knowledge from previous observations. Numerical simulation is conducted for algorithm verification. Comparisons are presented with different planning strategies. The rest of this paper is organized as follows. Section 2 describes the underlying learning mechanism for mobile robot path planning. Section 3 presents the numerical simulation of unknown environment exploration. Simulation results and discussion are presented in Sect. 4. Conclusions are summarized in Sect. 5.

2 Gradient-Based Exploration Strategy

In environment exploration applications where obstacle avoidance is not the primary concern, the essential task of exploration usually aims at building an accurate map of the features in the region of interest. This task could be decomposed into anomaly identification and sample coverage. Assuming that no prior knowledge on the environment is available, decision on where to go is based on the interaction between the robot and the environment under reinforcement learning paradigm [14].

2.1 Exploration Formulation

Suppose a 2D region is assigned for feature exploration, as indicated in Fig. 1. Define an undirected graph $G = \langle V, E\rangle$, where V denotes the waypoint set and $E \subseteq V \times V$ denotes the path set that the robot can move along. The exploration problem can be formulated as follows. The robot starts to explore by an initial random walk. It may have several potential actions, indicating different moving direction. Its interaction with the environment is characterized by observation and the immediate reward of robot action is defined by the gradient of the feature. The next favorite action will then be selected according to the estimate of action values.

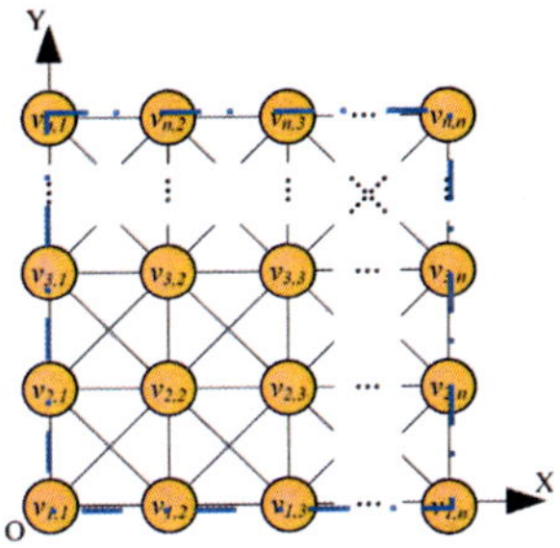

Fig. 1. Schematic of the region of interest.

2.2 Action Selection

During the course of exploration, historical observations are the fundamental information that the robot can learn to have an understanding of the unknown environment. In the reinforcement learning framework, such understanding is gained through a rewarding mechanism to evaluate the available robot actions. The robot selects the most preferred action as the next move.

We use the gradient of the features to define the immediate reward of taking an action, as indicated in Eq. 1. Assuming that the features are continuous, the value of action is estimated by summarizing weighted historical rewards, as indicated in Eq. 2, where the weight is assigned according to the distance between the current and historical positions of the robot.

$$R_{t-1}(s_{t-1}, a_{t-1}) = z_t - z_{t-1} \tag{1}$$

where R denotes the immediate reward, s and a represent robot's state and action, z denotes observation, and t is the time step.

$$Q(a_t) = \frac{1}{N}\sum_{i=1}^{t-1} \omega_i R_i(s_i, a_t) \tag{2}$$

where $Q(a_t)$ denotes the value of action a_t, ω is the weight of historical reward, N is the total number of taking action a_t until time step t. $R_i(s_i, a_t)$ is formulated to filter out the rewards when action a_t is not taken at time step i as:

$$R_i(s_i, a_t) = \begin{cases} R_i(s_i, a_i), & a_i = a_t \\ 0, & else \end{cases} \tag{3}$$

The action selection is myopic. Myopic algorithms are preferred for online path planning mostly because they have lower computational complexity compared to non-myopic approaches. As the robot gets to a new state, we calculate the value of action for all potential actions. The preferred next action is then select by a greedy action selection strategy. The robot will take the action that has the highest value of action.

3 Numerical Simulation

In this section, a mobile robot is assigned to explore an unknown region with anomalous features. The region is a 40×40 square with features ranging from 0 to 1.0. Two cases are presented with different environmental features, which are modeled by Gaussian function and Morokoff & Caflisch function [15] respectively, as shown in Eq. 4 and Eq. 5. The contour plots of the two feature functions are illustrated in Fig. 2. In the numerical simulation, the mobile robot starts the exploration from the lower left corner of the region, whose coordinate is indicated as (0, 0) in Fig. 2. While the robot is traveling across the region, it is supposed to have eight possible actions. At each state, the robot can move toward north, south, east, west, northeast, northwest, southeast, and southwest.

$$f(x,y) = (2\pi\sigma_1\sigma_2)^{-1} exp\left\{ -\frac{1}{2} \cdot \left[\frac{(x-\mu_1)^2}{\sigma_1{}^2} + \frac{(y-\mu_2)^2}{\sigma_2{}^2} \right] \right\} \tag{4}$$

where $\mu_1 = \mu_2 = 28$ and $\sigma_1 = \sigma_2 = 7$, and (x, y) denotes positions in the exploring region.

$$f(x,y) = \frac{9}{4}\sqrt{xy} \tag{5}$$

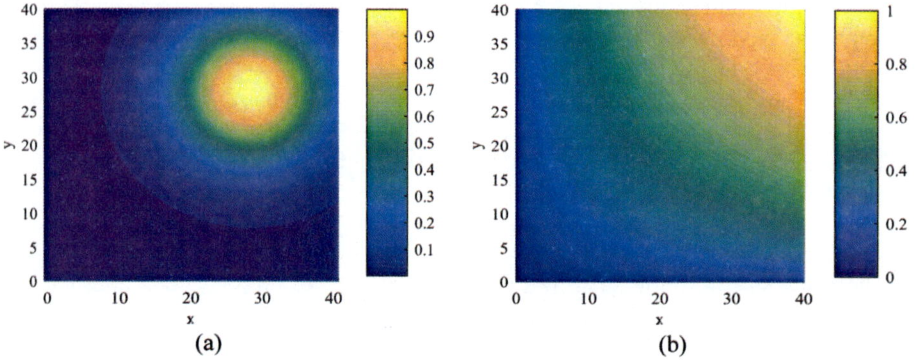

Fig. 2. Contour plots of the features in the environment: (a) Gaussian function; (b) Morokoff & Caflisch function.

To assess the performance of the algorithm, the following metrics are adopted:

i) Root mean square error (RMSE):

$$rmse = \sqrt{\frac{1}{M}\sum_{i=1}^{M}\left[f(x_i,y_i) - g(x_i,y_i)\right]^2} \tag{6}$$

where $g(x_i, y_i)$ denotes the predicted feature function and M is the number of locations used to evaluate the error.

ii) Weighted root mean square error (WRMSE):

$$wrmse = \sqrt{\frac{1}{M}\sum_{i=1}^{M}\lambda \cdot \left[f(x_i, y_i) - g(x_i, y_i)\right]^2} \tag{7}$$

where $\lambda = \frac{g(x_i, y_i) - min[g(x_i, y_i)]}{max[g(x_i, y_i)] - min[g(x_i, y_i)]}$.

iii) Distance to the extreme feature:

$$l = \|(x_f, y_f) - (x_g, y_g)\| \tag{8}$$

where (x_f, y_f) and (x_g, y_g) denote the locations of the extreme feature.

iv) Difference in the extreme value:

$$\Delta = |f^*(x, y) - g^*(x, y)| \tag{9}$$

where f^* and g^* denote the extreme values.

4 Results and Discussion

Figure 3 and Fig. 4 present the simulation results of the two exploring cases via the four performance metrics addressed in Sect. 3. The results are compared with three different algorithms: uncertainty sampling [16], complete coverage by zigzag trajectories [17], and random sampling. Uncertainty sampling is a Gaussian process-based planning technique. It is focused on find a most informative observation. In the complete coverage algorithm, the path is planned as a zigzag trajectory that usually taken by lawnmowers. Random sampling strategy could be literally understood as selecting actions randomly.

Figure 3 shows that our gradient-based algorithm performs far better than complete coverage and random selection algorithms and its performance is comparative to uncertainty sampling strategy when the environmental features indicate a Gaussian distribution. It is shown in Fig. 3(a) and Fig. 3(b) that our algorithm and the uncertainty sampling outperform complete coverage and random selection in RMSE and WRMSE, indicating that these two algorithms are fairly more computationally efficient. With regard to anomaly identification, we look into the deviation in location and the error in predicted values. As indicated in Fig. 3(c), our algorithm, uncertainty sampling, and random sampling deliver much better performance in localization of the extreme feature than the complete coverage algorithm. In Fig. 3(d) where the error of the predicted extreme feature is presented, it is apparent that our algorithm has a more accurate result than the other three algorithms. The performance of the complete coverage algorithm would be greatly affected by the robot's initial state. Fast and accurate anomaly identification could be achieved if the extreme features are close to the robot's initial position. But very poor performance would be resulted otherwise.

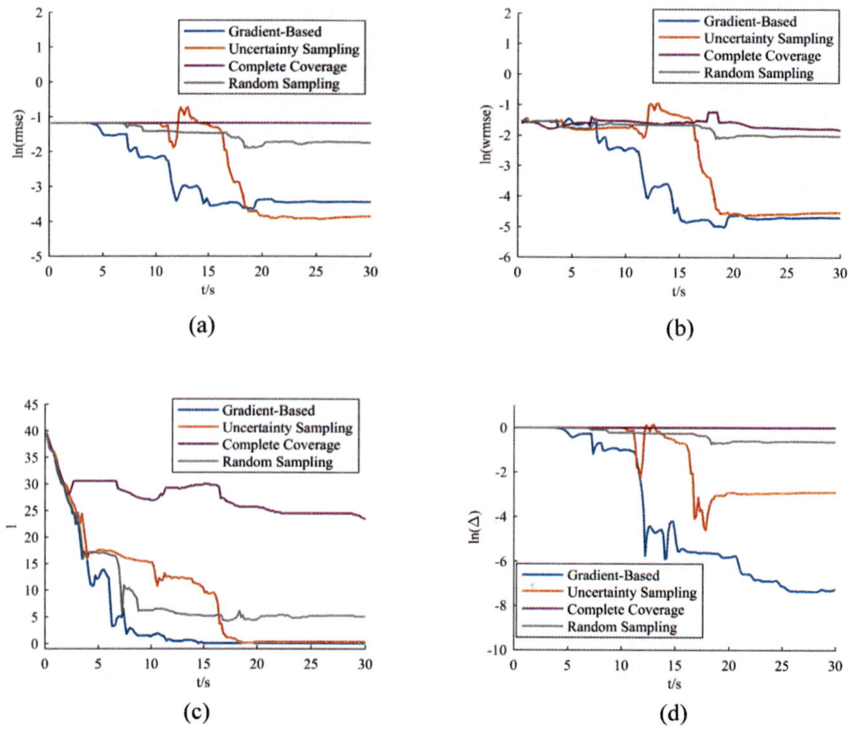

Fig. 3. Comparison of algorithm performance in a Gaussian environment: (a) Root mean square error; (b) Weighted root mean square error; (c) Distance to the extreme feature; (d) Difference in the extreme value.

In our second case, the environmental feature is described by Morokoff & Caflish function. A good agreement with the Gaussian case on algorithm performance could be observed through comparison of the results presented in Fig. 4. It is shown in Fig. 4(a) and Fig. 4(b) that our algorithm and uncertainty sampling outperform complete coverage and random selection in RMSE and WRMSE. Figure 4(c) shows that our algorithm and uncertainty sampling are much more efficient in locating the extreme feature than random sampling and complete coverage. The comparison in Fig. 4(d) illustrates that a favorable prediction on the extreme feature could be obtained using our algorithm.

The simulation results indicate that our intuitive gradient-based algorithm works well in environments with Gaussian and Morokoff & Caflisch feature distributions. Such environmental features are relatively simple with merely one extreme. Further investigation will be conducted on more complicated features.

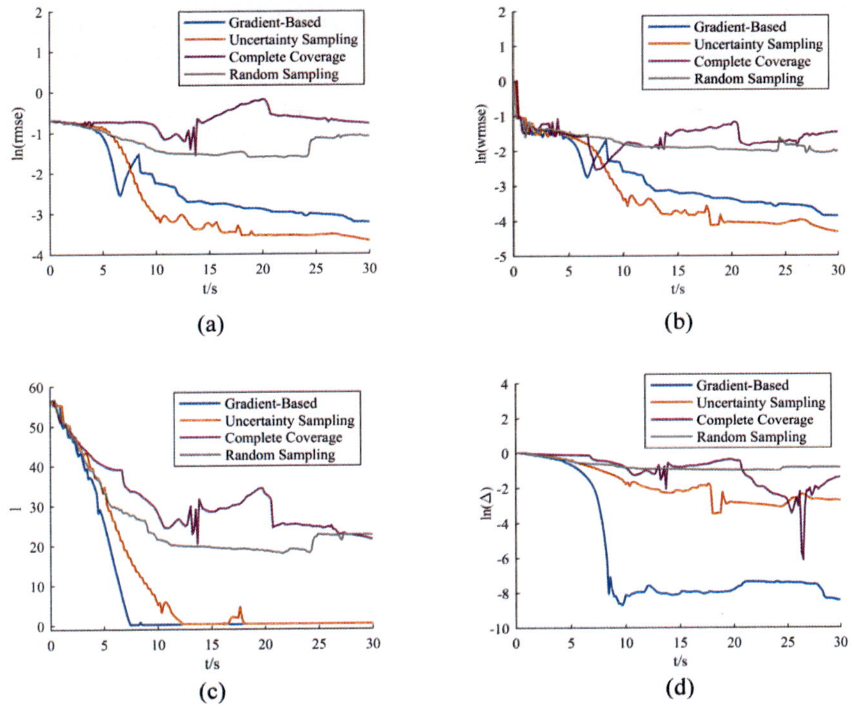

Fig. 4. Comparison of algorithm performance in Morokoff & Caflisch function: (a) Root mean square error; (b) Weighted root mean square error; (c) Distance to the extreme feature; (d) Difference in the extreme value.

5 Conclusion

In this paper, we tried an intuitive gradient-based learning algorithm for mobile robot path planning to explore unknown environment. The algorithm is developed under the framework of reinforcement learning. The next move of the robot is optimized by greedy selection according to evaluation of the value of actions. Simulation results show that the presented algorithm performs well for simple feature distribution. Accuracies on feature prediction and anomaly identification are comparable to or better than current approaches.

References

1. Lehner, P., et al.: Mobile manipulation for planetary exploration. In: 2018 IEEE Aerospace Conference, pp. 1–11. IEEE (2018)
2. Girdhar, Y., Dudek, G.: Modeling curiosity in a mobile robot for long-term autonomous exploration and monitoring. Auton. Robot. **40**(7), 1267–1278 (2015). https://doi.org/10.1007/s10514-015-9500-x
3. Ishida, H., Wada, Y., Matsukura, H.: Chemical sensing in robotic applications: a review. IEEE Sens. J. **12**, 3163–3173 (2012)

4. He, X., et al.: Autonomous chemical-sensing aerial robot for urban/suburban environmental monitoring. IEEE Syst. J. **13**(3), 3524–3535 (2019)
5. Morere, P., Marchant, R., Ramos, F.: Sequential Bayesian optimization as a POMDP for environment monitoring with UAVs. In: 2017 IEEE International Conference on Robotics and Automation, pp. 6381–6388. IEEE (2017)
6. Ouyang, R., Low, K.H., Chen, J., Jaillet, P.: Multi-robot active sensing of non-stationary Gaussian process-based environmental phenomena. In: Proceedings of the 2014 International Conference on Autonomous Agents and Multi-agent Systems, pp. 573–580. IFAAMAS (2014)
7. Marchant, R., Ramos, F.: Bayesian optimisation for informative continuous path planning. In: 2014 IEEE International Conference on Robotics and Automation, pp. 6136–6143. IEEE (2014)
8. Hitz, G., Galceran, E., Garneau, M., Pomerleau, F., Siegwart, R.: Adaptive continuous-space informative path planning for online environmental monitoring. J. Field Robot. **34**(8), 1427–1449 (2017)
9. Hollinger, G.A., Englot, B., Hover, F., Mitra, U., Sukhatme, G.S.: Uncertainty-driven view planning for underwater inspection. In: 2012 IEEE International Conference on Robotics and Automation, pp. 4884–4891. IEEE (2012)
10. Li, Y., Cui, R., Yan, W., Xu, D.: Long-term adaptive informative path planning for scalar field monitoring using cross-entropy optimization. Sci. China Inf. Sci. **62**, 50208 (2019)
11. Blanchard, A., Sapsis, T.: Informative path planning for anomaly detection in environment exploration and monitoring. arXiv:2005.10040 (2020)
12. Neumann, P.P., Asadi, S., Lilienthal, A.J., Bartholmai, M., Schiller, J.H.: Autonomous gas-sensitive microdrone: wind vector estimation and gas distribution mapping. IEEE Robot. Autom. Mag. **19**(1), 50–61 (2012)
13. Russell, R.A.: Chemical source location and the RoboMole project. In: Proceedings of the 2003 Australasian Conference on Robotics & Automation, pp. 1–6. ARAA, Brisbane (2003)
14. Sutton, R.S., Barto, A.G.: Reinforcement Learning: An Introduction, 2nd edn. The MIT Press, Cambridge (2020)
15. Virtual Library of Simulation Experiments: Test Functions and Datasets, http://www.sfu.ca/~ssurjano. Accessed 28 Sept 2021
16. MacKay, D.J.C.: Information-based objective functions for active data selection. Neural Comput. **4**(4), 590–604 (1992)
17. Hsu, P.M., Lin, C.L.: Optimal planner for lawn mowers. In: 2010 IEEE 9th International Conference on Cybernetic Intelligent Systems, pp. 1–7. IEEE (2010)

Application of MobileNet-SSD Deep Neural Network for Real-Time Object Detection and Lane Tracking on an Autonomous Vehicle

Van-Son Vu[1] and Duc-Nam Nguyen[2,3](✉)

[1] University of Transport and Communications, Hanoi 12116, Vietnam
[2] Phenikaa University, Hanoi 12116, Vietnam
nam.nguyenduc@phenikaa-uni.edu.vn
[3] Phenikaa Research and Technology Institute (PRATI), Hanoi 11313, Vietnam

Abstract. Recent development in perception of autonomous vehicle is mainly originated by deep learning. To apply in real-time application, the object detector of learning model is required to run on a high-end GPU platform. Therefore, there is always a great challenge to process deep learning algorithm on an embedded system. In this study, we present an application of recent lightweight network, namely MobileNet – SSD, to detect and classify objects of an autonomous vehicle. The vehicle is self-piloted along a lane using an image processing algorithm. We employ the neural network on small power consumption NVIDIA Jetson Nano Development Kit platform. The accuracy of detection is up to 96% at the 42 fps, while the model can archive 26.7 ms timing latency and 74% mAP (mean average precision) via input dataset of 800 Full-HD resolution pictures. The prototype vehicle is capable of upgrading to large-scale autonomous vehicles which are required to work in special conditions such as quarantined or lockdown areas, poisonous and dangerous area, warehouse, factory, airport, etc.

Keywords: MobileNet · Deep learning · Embedded system · Object detection · Object classification · Lane tracking

1 Introduction

Mobile robots, or Autonomous Guided Vehicles (AGV) have been studied and applied in industry for years with notable improvement of intelligent functions [1]. One of the most stability pilot solution for AGVs is using the magnetic line integrating with LIDARS to 3-dimentional mapping the working environment. Characteristics of working environment of AGV is small amount of moving objects. Hence, object detection and classification model are not necessary. The ultimate level of autonomous vehicle is the driverless car which can operate in the real-time regardless of various type of impact conditions such as weather, road, moving speed, pedestrians, etc. The driverless cars developed by Google, Uber or Tesla are usually equipped with lots of sensors such as camera, LIDARS, Radar, thermal image sensor, etc. Recent models of driverless cars have been integrated with high-ends sensor fusion algorithms to realize and predict the perceptional situations. In

N. V. Khang et al. (Eds.): ASIAN MMS 2021, MMS 113, pp. 559–565, 2022.
https://doi.org/10.1007/978-3-030-91892-7_53

these applications, deep learning-based convolution neural network (CNN) is the core of the object detection or object classification model. However, implementation such complex object detection network model into an embedded platform while obtaining a considerable accuracy and stability is still a great challenge. Considering a gap between working areas of AGVs and driverless car, including large scale warehouse, semi-auto factory, quarantined or lockdown areas because of COVID-19 pandemic, or nuclear effected area, there is a must for an autonomous vehicle which can pilot along a fixed lane while still can make decision based on the traffic light or traffic sign. This type of autonomous vehicle is required to process a lightweight object detection model in an embedded system with limited computational resources.

Object detection algorithms have been studied and developed for decades along with the fast development of machine learning techniques [2]. The object detection algorithms can be categorized as two-stage and one-stage algorithms. Several of the most used two-stage object detection algorithms are Faster R-CNN, R-FCN, FPN and Casecade R-CNN with the mean average precision (mAP) can be up to 95.32% [3]. However, one of the drawbacks of two-stage algorithms is computational cost which is not suitable for an embedded platform, real-time application and low-power consumption system. The one-stage algorithm is limited by lower mAP (up to 75.9%); however, it is possible to implement on a low-cost embedded computer. Recently, Single Shot Detector (SSD) and YOLO [4] are two of the most popular one-stage detectors. Compared to YOLO, the SSD provides less accuracy but much lower computational cost while still remains an impressive mAP value (74.3 mpA, 59 fps).

In this research, we employed the recent neural network model for mobile applications namely MobileNet-SSD on a Jetson Nanoboard platform. The MobileNet was first reported by a Google team [5] for mobile and embedded vision applications which is capable of implantation on a low-cost embedded system. Therefore, in this study, we do not develop neither a new control algorithm nor neural network for a real-time embedded system; instead, we integrate the current state-of-art in computer vision such as object detection, object classification and line tracking for an autonomous vehicle to improve the potential application in dedicated conditions. The target of this research is focusing on two major tasks: evaluating the Real-time Object Detection and piloting a small-size autonomous vehicle using the NVIDIA Jetson Nano platform.

2 Hardware and Implementation of the Project

2.1 Hardware

The hardware of the autonomous vehicle consists of a RC car frame with a size ratio of 1:18, two LiPo 5400 mAh batteries, a servo motor to steering the front wheels, camera Raspberry Pi V2 8 MP for capturing the object, and a 5MP webcam to detect and track the lane. The NVIDIA Jetson Nano Development Kit with 128 – core Maxwell GPU, quad-core ARM Cortex A57 CPU, 4 GB 64bit LPDDR4 RAM was selected to implement the MobileNet-SSD Deep Neural Network. The vehicle is tested in the 3×1 meter simulated map as shown in the Fig. 1. The traffic light, STOP and CAUTION sign are located along of the lane.

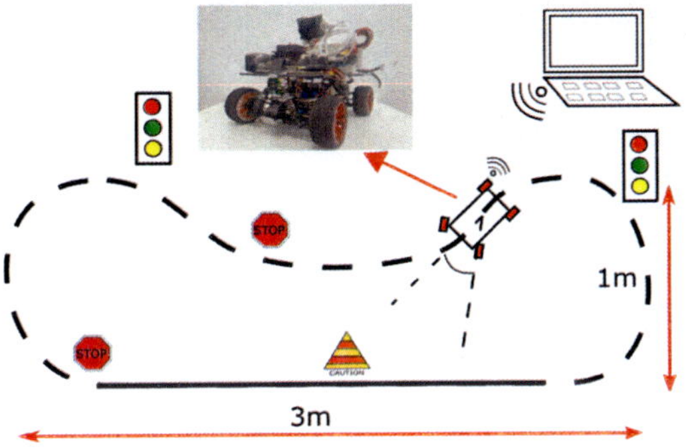

Fig. 1. The configuration of simulated road to validate the performance

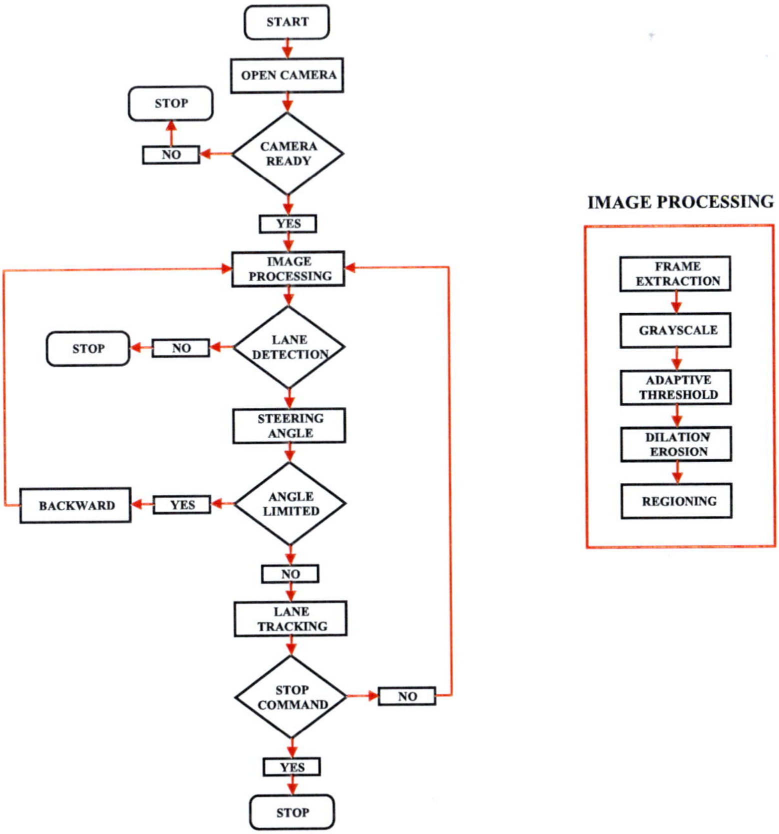

Fig. 2. Flowchart of the lane tracking algorithm

2.2 Control System and Lane Tracking Algorithm

In this study, we employed classic PID control algorithm to enhance the smoothness of motion along the curve and at the transition between the curve and the straight path. The lane tracking algorithm is shown in Fig. 2.

To increase the processing speed, the frame is divided into smaller boxes and only the box-frame containing the lane was selected. Because of the limitation of the steering angle, whenever the calculated steering angle is over-control, the vehicle will move backward and literately adjust to the proper angle. We also add more functions to the program to monitor and take control the program on a laptop via Wi-Fi connection. This wireless control process will not only compensate the possibility of missing lane tracking but also open the potential for tele-operation missions.

2.3 Implementation of MobileNet SSD Deep Neural Network on Jetson Nano Development Board

MobileNet is a class of CNN that was open-sourced by Google [5] and was the TensorFlow's first mobile computer vision model. MobileNet uses depthwise separable convolutions to significantly reduces the number of parameters. As the results, MobileNet is considered as a lightweight deep neural networks. The core of MobileNet is based on depthwise separable convolutions which allow a filter's depth and spatial dimension can be separated. The depthwise separable convolution is a depthwise convolution followed by a pointwise convolution as sketched in Fig. 3. The operation cost is:

$$D_K \bullet D_K \bullet M \bullet D_F \bullet D_F + M \bullet N \bullet D_F \bullet D_F \tag{1}$$

where M: number of input channels, N: number of output channels, D_K: Kernel size, D_F: feature map size.

The operation cost of standard convolution is calculated as:

$$D_K \bullet D_K \bullet M \bullet N \bullet D_F \bullet D_F \tag{2}$$

Comparing to operation cost of standard convolution, that of MobileNet is 8 to 9 times less while the accuracy is reduced slightly.

$$\frac{D_K \bullet D_K \bullet M \bullet D_F \bullet D_F + M \bullet N \bullet D_F \bullet D_F}{D_K \bullet D_K \bullet M \bullet N \bullet D_F \bullet D_F} = \frac{1}{N} + \frac{1}{D_K^2} \tag{3}$$

The detail of network can be found in details in [5].

To implement the MobileNet-SSD Deep Neural Network on Jetson Nano Development Kit, the flowchart algorithm is processed as Fig. 4. By optimizing the layer and input parameter, we obtain the process latency of 26.7 ms at 42 fps.

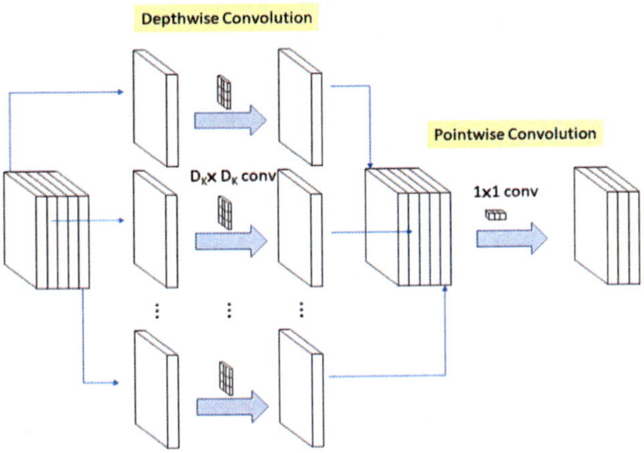

Fig. 3. Depthwise separable convolution and pointwise convolution – reprinted from [6]

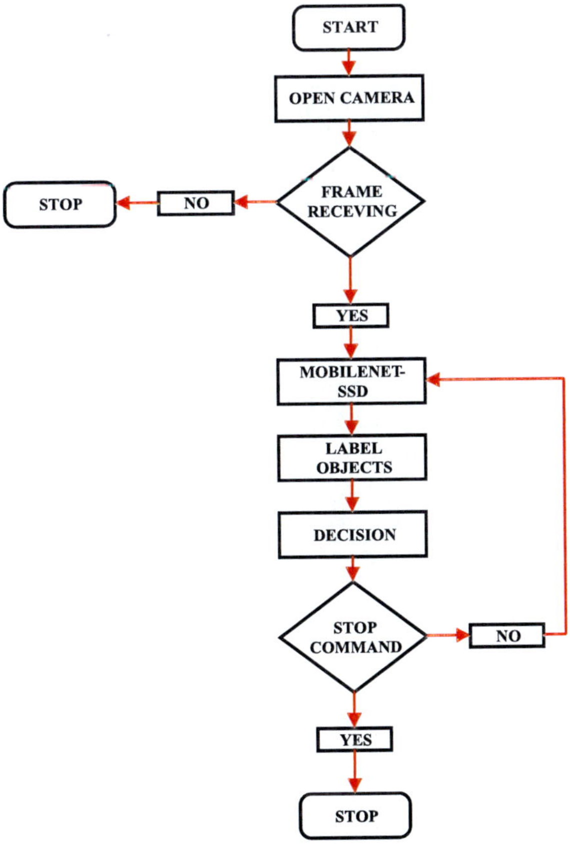

Fig. 4. Flowchart of the object detection and classification

3 Results and Discussion

The autonomous vehicle is capable of detecting and recognizing several fundamental traffic lights (Red – Green – Yellow), STOP sign, NOTICE sign while keeps tracking a fixed lane (Fig. 5). By employing the MobileNet-SSD Deep Neural Network, we achieve an average of 74% mAP, 42 fps with the maximum moving speed of 1.26 km/h. The possibility of maneuvering via WiFi opens a wide application for tele-operated mission. However, there are still several limited regarding to the optimization of the power source to increase the operation time. More recent advanced networks such as MobileNet – V2 will be tried in the future for better detection accuracy and less computational cost. This prototype is the first step for development of large-scale autonomous car working in special conditions such as quarantined or lockdown areas, poisonous and dangerous area, warehouse, factory, airport, etc.

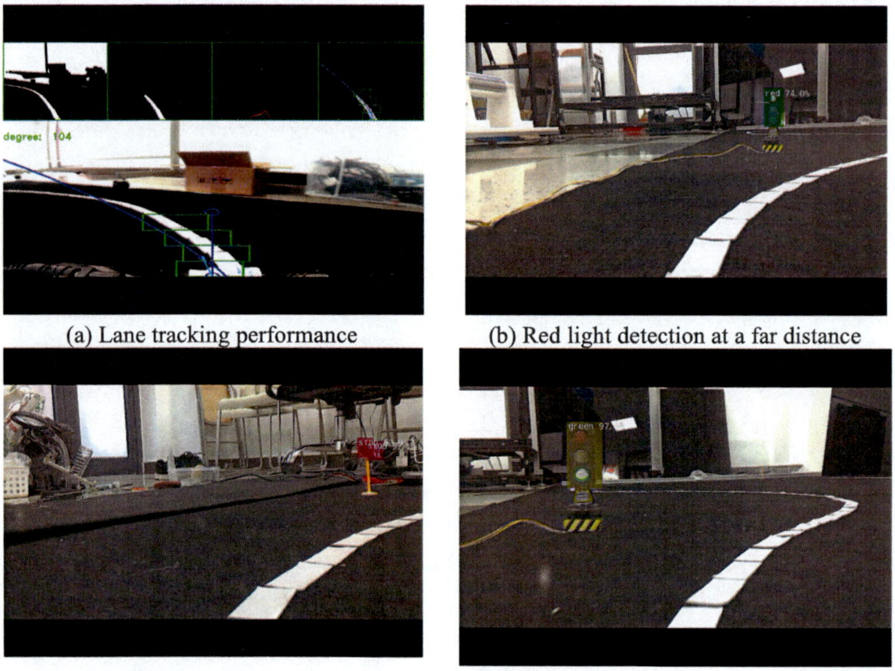

(a) Lane tracking performance (b) Red light detection at a far distance

(c) STOP sign detection (d) Green light detection at a close distance

Fig. 5. Capture images from recorded camera to validate the algorithm

The demo video can be viewed in URL [7, 8].

References

1. Feng, D., et al.: Deep multi-modal object detection and sematic segmentation for autonomous driving: datasets, methods, and challenges. arXiv preprint arXiv:1902.07830 (2020)

2. Chuixin, C., Hanxiang, C.: AVG robot based on computer vision and deep learning. In: 3rd IEEE International Conference on Robotics and Automation Sciences, pp. 28–34 (2019)
3. Du, L., Zhang, R., Wang, X.: Overview of two-stage object detection algorithms. J. Phys.: Conf. Ser. **1544**, 012033 (2020)
4. Liu, W., Anguelov, D., Erhan, D., Szegedy, C., Reed, S.: SSD: single shot multibox detector. arXiv preprint arXiv:1512.02325 (2016)
5. Howard, A.G., et al.: MobileNets: efficient convolutional neural networks for mobile vision applications. arXiv preprint arXiv:1704.04861 (2017)
6. Image Classification with MobileNet. https://medium.com/analytics-vidhya/image-classification-with-mobilenet-cc6fbb2cd470
7. Lane Tracking demo video. https://youtu.be/uoZA2yvDMFQ
8. Object Detection and Recognition demo video. https://youtu.be/-RtPcnTm8CQ

Actuators and Sensors Involving Mechanics

Calibration of a Dual-Telecentric Fringe Projection System Using a Planar Calibration Target

Zhuoran Wang, Xianmin Zhang, Hai Li(✉), and Shuiquan Pang

Guangdong Key Laboratory of Precision Equipment and Manufacturing Technology, South China University of Technology, Guangzhou, Guangdong, China
lihai@scut.edu.cn

Abstract. The development of modern manufacturing techniques increases the precision of tiny components. It is of great significance to accurately obtain the three-dimensional (3D) shape information of objects. In this paper, a telecentric fringe projection system for 3D measurement at micron-millimeter scales is constructed, and a calibration method for the system based on a planar calibration target is proposed. The calibration of the system is divided into two parts: the calibration of the telecentric camera and the calibration of the telecentric projector. Specifically, the telecentric camera parameters are obtained by using the characteristics of the rotation matrix and the repair marks of the positioning platform. The projector is calibrated by using the reversibility principle of the optical path. Experimental results show that the reprojection error of the calibration method is 0.2pixel, which proves its feasibility and high measurement accuracy. Application experiments show that the dense point cloud of small objects can be generated by using the proposed system, which shows that it has good potential application prospects in 3D measurements of small objects.

Keywords: Telecentric imaging · System calibration · 3D measurement · Point cloud · 3D reconstruction

1 Introduction

With the development of modern industrial manufacturing technologies, the high-precision acquisition of 3D information has become increasingly important in many fields, such as reverse engineering, computer-aided design (CAD) and 3D measurement [1]. To realize the precision measurement of cross-scale objects, which size on the order of micrometer to the millimeter, the traditional method is point-by-point measurement using the coordinate measuring machine (CMM). The contact way limits the speed of the gauge and the hardness of objects. The advantage of optical microscopic imaging techniques is that non-contact measurement can be easily achieved. However, the cross-scale measurement is hard to be realized due to the limit of magnification of the optical lens. Additionally, the large parallax problem of ordinary lenses poses a great challenge to

N. V. Khang et al. (Eds.): ASIAN MMS 2021, MMS 113, pp. 569–579, 2022.
https://doi.org/10.1007/978-3-030-91892-7_54

high-precision micro-measurement. Compared with traditional optical lenses, the telecentric lenses have a constant amplification rate, high resolution, and low distortion [2], which has more advantages in precision optical measurement systems.

The device equipped with traditional optical lenses is generally regarded as the pinhole imaging model, and many calibration methods have been proposed, such as Tsai's [3] two-step method, Zhang's [4] calibration method, etc. Due to the different geometric models of the traditional measuring devices and the proposed measurement system, the calibration methods of traditional devices cannot be directly applied to the calibration of the proposed measurement system. It is necessary to reconstruct the telecentric imaging model and find a way to calibrate the system composed of the camera and projector with telecentric lenses. Yin et al. [5] used the general imaging model method to calibrate the telecentric profilometry. The general imaging model has good versatility in the calibration of the optical imaging system, but it needs to collect a lot of calibration information and the calibration results may not be accurate. Chen et al. [6] were inspired by Zhang's calibration method, and obtained the closed-form solution of the telecentric camera by using the characteristics of the rotation matrix. This calibration method can obtain high-precision parameters, which is very enlightening for the calibration of the telecentric fringe projection system. Li et al. [7] directly calibrated the structure light telecentric system with known intrinsic parameters. This method obtained high precision parameters, but the application was limited. Haskamp et al. [8] used the nonlinear optimization method to estimate camera parameters, but the selection of initial value may lead to the result falling into a local minimum. Li et al. [9] present a novel method to calibrate a microscopic structured light system using a camera with a telecentric lens. The projector using a long working distance lens, which complicated the calibration of the system. Liu et al. [10] analyzed the problem that the parameters were not unique when the projector and camera replaced the telecentric lens at the same time, but there was no detailed calibration process. Rao et al. [11] proposed an improved two-step method using the traditional two-dimensional planar for calibration. However, the sign of extrinsic parameters obtained by the traditional planar calibration method may lead to a problem, when the parameter values close to zero may get a wrong sign.

In this research, a complete calibration method for the telecentric fringe projection system is proposed. The matrix characteristics of telecentric imaging have some similarities with pinhole imaging. So we take a similar idea with Zhang by taking advantage of the characteristics of the rotation matrix, obtaining the closed-form solution of the telecentric imaging process. Compared with other calibration methods, the closed-form solution method can obtain the camera parameters with higher precision. Meanwhile, the problem of parameter γ symbol is solved through control Z-axis coordinate transformation in the calibration process, which further improves the accuracy of calibration. The feasibility of this method can be proved by the calibration reprojection error and point cloud reconstruction results. This structured light imaging method based on telecentric calibration has a good application prospect in small and micro object measurement and cross-scale measurement.

2 Description of the Proposed System and Calibration Method

2.1 The Proposed System and Telecentric Imaging Model

The telecentric measurement system consists of a camera and a projector both equipped with a telecentric lens, as shown in Fig. 1. The telecentric camera consists of a GS3-U3-41C6M-C PointGrey camera (resolution 2048 × 2048, sensor format 1″, pixel size 5.5 μm) and an OPT-5M03-110 telecentric lens (working distance 110 mm, angle of view 37.55 mm × 55 mm, depth of field is about 6 mm). The telecentric projector consists of a DLP4710 projector (resolution 1920 × 1080, micromirror size 5.4 μm) and an OPT-5M03-110 telecentric lens. Halcon7 × 7 dot ceramic plate is used in calibration (dot diameter 1.25 mm, center spacing 2.5 mm, accuracy ± 0.001 mm).

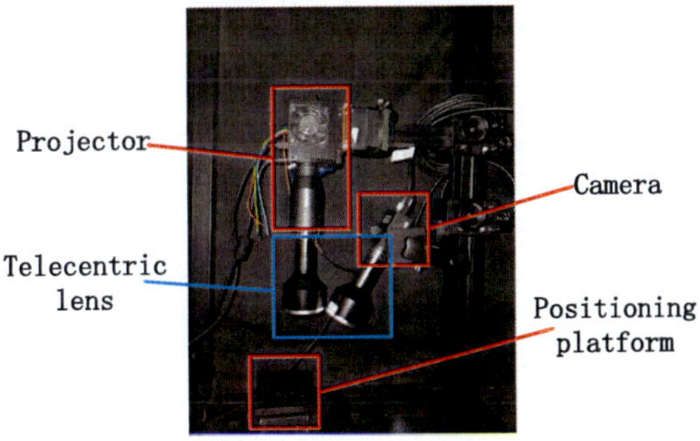

Fig. 1. The proposed telecentric fringe projection system.

Compare with a traditional optical lens, the aperture stop is set at the focal point of a telecentric lens, which only allows parallel light to pass through, ensures the stability of the image magnification in the range of depth of field (DOF). The telecentric lens produces an orthographic projection of the object, and the imagery object will maintain constant size projection without being susceptible to distance.

Figure 2 shows the projection process of telecentric imaging [2]. In optical imaging, the projection from the world coordinate system to the camera coordinate system is a rigid body transformation process, which can be denoted by a rotation matrix R and a translation vector T. In imaging path, the telecentric lens loses depth information. Then the transformation from image coordinate system to pixel coordinate system can be divided into two parts: 2D rigid body transformation and pixelate [12]. Combine those transformations, the transfer matrices matrix between the world coordinate and the image coordinate can be described as follows:

$$\begin{pmatrix} u \\ v \\ 1 \end{pmatrix} = \begin{pmatrix} \alpha & \gamma & 0 & u_0 \\ 0 & \beta & 0 & v_0 \\ 0 & 0 & 0 & 1 \end{pmatrix} \begin{pmatrix} r_{11} & r_{12} & r_{13} & t_x \\ r_{21} & r_{22} & r_{23} & t_y \\ r_{31} & r_{32} & r_{33} & t_z \\ 0 & 0 & 0 & 1 \end{pmatrix} \begin{pmatrix} X_w \\ Y_w \\ Z_w \\ 1 \end{pmatrix} = AE \begin{pmatrix} X_w \\ Y_w \\ Z_w \\ 1 \end{pmatrix} = M \begin{pmatrix} X_w \\ Y_w \\ Z_w \\ 1 \end{pmatrix}, \quad (1)$$

where α, β and γ are the intrinsic parameters. The parameters (u_0, v_0) represent the principal point of the camera. For the telecentric imaging model, the principal point

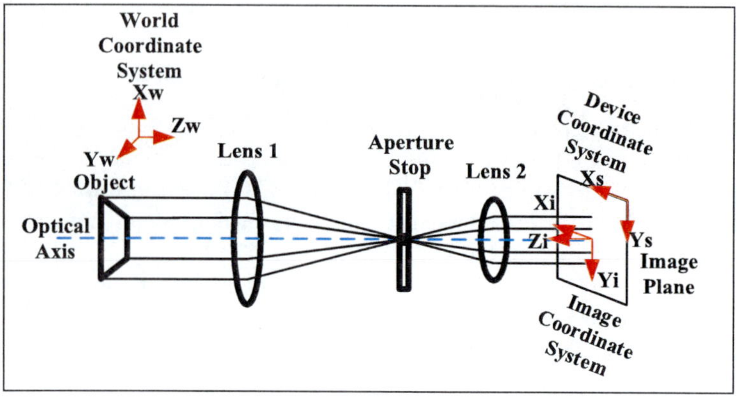

Fig. 2. Telecentric imaging model.

(u_0, v_0) do not have practical meaning. Therefore, it can be set as $(0, 0)$. E including a rotation matrix R and a translation vector T. Only the first two rows of the whole RT matrix have practical significance cause the third column of A is 0. The homography matrix M between the image coordinate system and the world coordinate system can be obtained by combining intrinsic and extrinsic parameters.

2.2 Camera Calibration

Without loss of generality, the world coordinate system of the plane where the calibration plate is located can be taken as $Z_w = 0$ in Eq. 1. The first step is to simplify the relational matrix. By establishing the homography matrix from the world coordinates to the pixel coordinates of the camera after eliminating the influence of Z_w, the relationship between the camera coordinate (u_c, v_c) and the world coordinate (X_w, Y_w) can be described as follows:

$$\begin{pmatrix} u_c \\ v_c \\ 1 \end{pmatrix} = \begin{pmatrix} \alpha & \gamma & u_0 \\ 0 & \beta & v_0 \\ 0 & 0 & 1 \end{pmatrix} \begin{pmatrix} r_{11} & r_{12} & t_x \\ r_{21} & r_{22} & t_y \\ 0 & 0 & 1 \end{pmatrix} \begin{pmatrix} X_w \\ Y_w \\ 1 \end{pmatrix} = A_s E_s \begin{pmatrix} X_w \\ Y_w \\ 1 \end{pmatrix} = H \begin{pmatrix} X_w \\ Y_w \\ 1 \end{pmatrix}, E_s = \begin{pmatrix} R_s & T_s \\ 0 & 1 \end{pmatrix}, \quad (2)$$

where A_s is the simplified intrinsic matrix, E_s include R_s and T_s is the simplified extrinsic matrix. The homography matrix H can be obtained by image calibration. From camera coordinate system to world coordinate system transform, the rigid-body matrix R of the complete extrinsic is a unit orthogonal matrix. The partitioning of the matrix can be obtained. When the part of H associated with the rotation matrix R is separated, the RT matrix can be solved as:

$$\left(A_s^{-1}H\right)^{\mathrm{T}} A_s^{-1}H = H^{\mathrm{T}} A_s^{-T} A_s^{-1}H = \begin{bmatrix} R_s^T R_s & R_s^T T_s \\ T_s^T R_s & T_s^T T_s + 1 \end{bmatrix}, H = \begin{pmatrix} h_{11} & h_{12} & h_{13} \\ h_{21} & h_{22} & h_{23} \\ 0 & 0 & 1 \end{pmatrix}. \quad (3)$$

According to the unit orthogonality of the rotation matrix R, we can obtain:

$$R^T R = \begin{bmatrix} R_s^T & Q^T \\ P & d \end{bmatrix} \begin{bmatrix} R_s & P^T \\ Q & d \end{bmatrix} = \begin{bmatrix} R_s^T R_s + Q^T Q & R_s^T P^T + Q^T d \\ PR_s + dQ & PP^T + d^2 \end{bmatrix} = E_{3\times3}, \quad (4)$$

where Rs is the second-order principal and child form of R, then can get the norm:

$$\det(R_s^T R_s - E_{2\times2}) = -\det(Q^T Q) = -\left| \begin{bmatrix} r_{31} & r_{32} \end{bmatrix}^T \begin{bmatrix} r_{31} & r_{32} \end{bmatrix} \right| = 0. \quad (5)$$

Take orthogonal matrix $B = A^{-T} A^{-1} = \begin{pmatrix} B_{11} & B_{12} & B_{13} \\ B_{21} & B_{22} & B_{23} \\ B_{31} & B_{32} & B_{33} \end{pmatrix}$,

where $B_{12} = B_{21}$. The values of B_{11}, B_{12} and B_{22} are only related to the camera's intrinsic parameters α, β and γ, r_{13} and r_{23} are lose sign, and the values of the main points u_0 and v_0 do not affect the calibration of the telecentric camera. Transform the equation to isolate the specific coefficients as follows:

$$g_1 l_1 + g_2 l_2 + g_3 l_3 - l_4 = k, \quad (6)$$

where $g_1 = (h_{11}^2 + h_{12}^2)$, $g_2 = (h_{21}^2 + h_{22}^2)$, $g_3 = 2h_{11}h_{21} + h_{12}h_{22}$, $k = (h_{12}h_{21} - h_{11}h_{22})^2$,
and $l_1 = \frac{B_{11}}{(B_{11}B_{22} - B_{12}^2)}$, $l_2 = \frac{B_{22}}{(B_{11}B_{22} - B_{12}^2)}$, $l_3 = \frac{B_{12}}{(B_{11}B_{22} - B_{12}^2)}$, $l_4 = \frac{1}{(B_{11}B_{22} - B_{12}^2)}$.

Solve each of the three parameters of matrix B, the intrinsic parameters can be recovered:

$$\alpha = \sqrt{\frac{l_4}{l_1}}, \beta = \sqrt{\frac{l_1 l_4}{l_1 l_2 - l_3^2}}, \gamma = -\frac{l_3}{l_1}\sqrt{\frac{l_1 l_4}{l_1 l_2 - l_3^2}}. \quad (7)$$

The extrinsic parameter matrix can be obtained by solving the H. Making the calibration plate be changed by the positioning platform with a ΔZ. Then the sign of r_{13} and r_{23} solution can be get by the corresponding relational equation. Through higher precision closed-form solution and solving the sign ambiguity, the complete parameter matrix can be obtained:

$$\begin{pmatrix} u + \Delta u \\ v + \Delta v \\ 1 \end{pmatrix} = \begin{pmatrix} m_{11} & m_{12} & m_{13} & m_{14} \\ m_{21} & m_{22} & m_{23} & m_{24} \\ 0 & 0 & 0 & 1 \end{pmatrix} \begin{pmatrix} X_w \\ Y_w \\ Z_w + \Delta Z \\ 1 \end{pmatrix}. \quad (8)$$

To minimize the distances between the predicted points and actual points, the lens distortion parameter can be used to improve the model accuracy. After constructing the minimization equation, the intrinsic and extrinsic parameters of the camera can be obtained by using the iterative Levenberg-Marquardt optimization method.

2.3 Projector Calibration

Under the same world coordinate system, the projector can be regarded as a reversible camera in the optical path. The calibration plate images of the projective fringes are captured by the calibrated camera and get the homography matrix between the projector and characteristic points. Using Zhang's calibration method from paper [13], good camera calibration results can ensure the calibration quality of the projector.

First, it is necessary to anti-tangent the three different pixel period stripes to obtain the corresponding wrapping phase. Then the unwrapped phase can be obtained by two steps heterodyning. To minimize the noise effect, the step-by-step backward phase unwrapping could be using the longer fringe period to unwrap an equivalent short period phase to obtain the unwrapped phase of it. We extract the sub-pixel position of each circle center (u_c, v_c) from the camera plane image. The absolute phases of the point in the two directions φ_v, φ_h can be obtained by the phase-shifting algorithm and linear interpolation. (φ_v, φ_h) is the absolute phase of the projector, which establishes the correspondence between the camera coordinates (u_c, v_c) and the projector coordinates (u_p, v_p). Then the homography matrix from the projector image plane to the world coordinate can be obtained by establishing a one-to-one correspondence between the corner points of the calibration board and the pixel coordinates of the phase diagram as shown in Fig. 3.

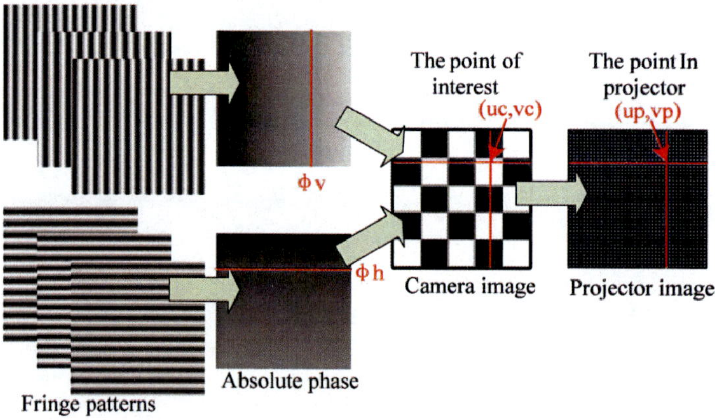

Fig. 3. Generation of projector calibration image.

During the telecentric projector calibration, a single calibration place $Z_w = 0$ will result in loss of depth information as the same problem we meet in camera calibration. The complete homography matrix can be obtained through the displacement of ΔZ on the Z-axis with the help of the positioning platform as Eq. 8. The telecentric projector calibration can be combined with telecentric camera calibration. The homography matrix from the projector pixel coordinates to the world coordinates can be established by collecting structured light images with the help of the camera. In this process, the calibration accuracy of the telecentric camera determines the accuracy of the projector homography matrix.

3 Experiments and Analysis

3.1 Calibration Results

The first experiment completes the calibration of the telecentric camera. By collecting 2 + 20 images of the calibration board Fig. 4, which all images to get the most parameter. And the two specific pictures show that the calibration plate only has a fixed displacement $\triangle Z$ on the Z-axis by controlling the positioning platform.

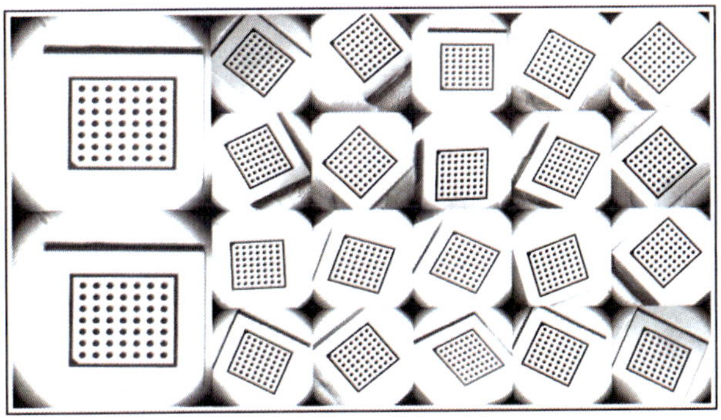

Fig. 4. Camera calibration images $N + 2$.

Figure 5 shows the re-projection errors of the proposed calibration method. As can be seen that the average reprojection errors in the X-axis and Y-axis are 0.2246 pixel and 0.2082 pixel, respectively. It proves that the proposed method has high accuracy.

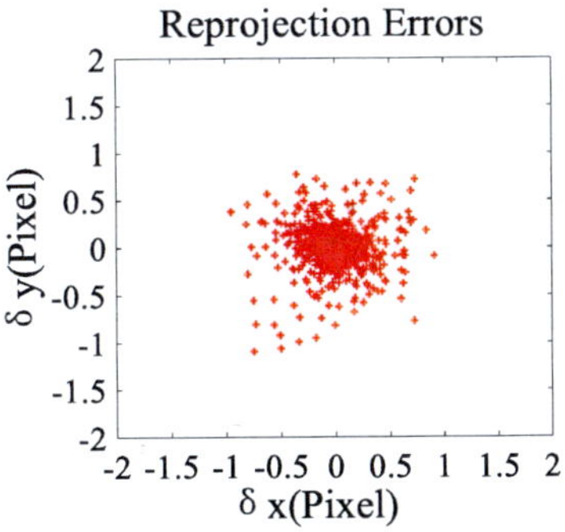

Fig. 5. Re-projection errors.

By using the proposed calibration method, the intrinsic and extrinsic parameter matrix of the telecentric camera can be obtained. Then the corresponding homography matrix is obtained. The actual lens distortion is very small, which conforms to the physical characteristics of the telecentric lens. The calibration results of the camera are shown in Table 1.

Table 1. Calibration results of the camera.

Item	Parameters
Camera intrinsic matrix	$\begin{pmatrix} 55.056 & -0.0095 & 0 \\ 0 & 55.056 & 0 \\ 0 & 0 & 1 \end{pmatrix}$
Camera extrinsic matrix	$\begin{pmatrix} 0.0125 & 0.9999 & -0.0042 & 20.687 \\ -0.8778 & 0.0130 & 0.4788 & 20.768 \\ 0 & 0 & 0 & 1 \end{pmatrix}$
Camera homography matrix	$\begin{pmatrix} 0.6964 & 55.051 & -0.2246 & 1138.8 \\ -48.352 & 0.7141 & 26.376 & 1144.0 \\ 0 & 0 & 0 & 1 \end{pmatrix}$
Distortion coefficients	$k1 = -6.7407 \times 10^{-11}$, $k2 = 2.0475 \times 10^{-17}$

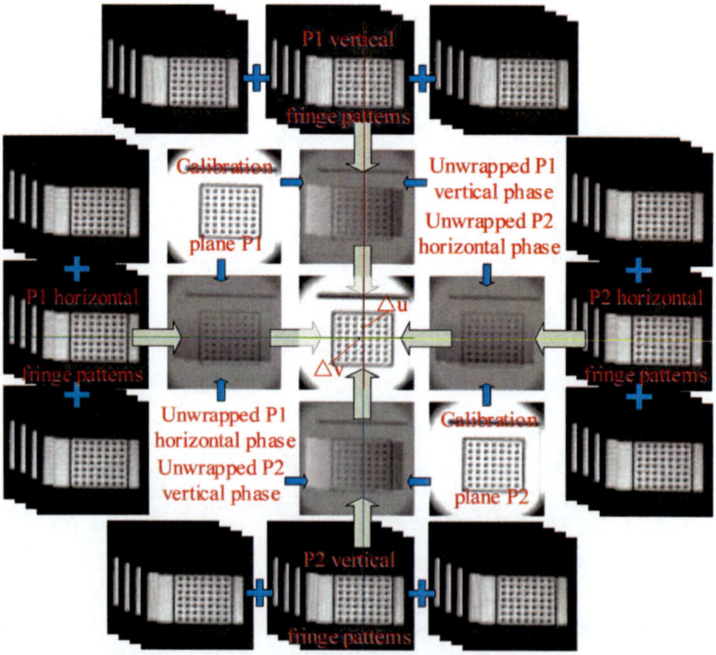

Fig. 6. Projector calibration.

The second experiment completed the calibration of telecentric projection. In the first pose calibrated by the camera and the special spatial pose achieved by the micro motion positioning platform, the vertical and horizontal three frequency four-step phase-shifting fringes are projected on the calibration plate respectively, and the images are collected. Using the telecentric projector calibration method, the complete telecentric projector homography matrix is obtained in Fig. 6 and Table 2.

Table 2. Calibration results of the projector.

Item	Parameters
Projector homography matrix	$\begin{pmatrix} -11.683 & -44.161 & -3.9764 & 1423.6 \\ -54.812 & 0.7188 & -99.422 & 1235.5 \\ 0 & 0 & 0 & 1 \end{pmatrix}$

3.2 3D Clouds Reconstruction Results

After obtaining the complete telecentric system parameter matrix, the three-dimensional measurement of the object is realized by using the spatial position relationship between the projector and the camera. We get the point cloud of real objects (coins, small chips) in small size as Fig. 7.

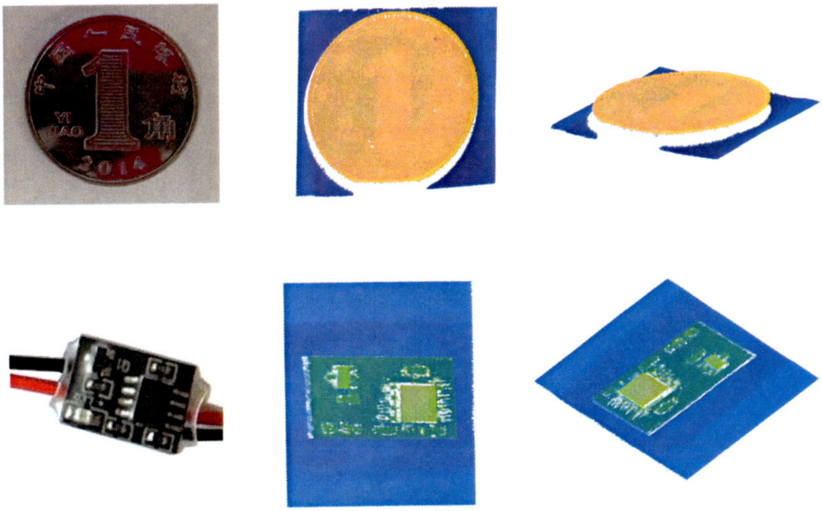

Fig. 7. Point clouds of small objects.

From the above experimental results, it can be concluded that the structured light system based on the principle of telecentric imaging can realize the 3D morphology measurement of tiny objects. The main advantages of the method include: 1. Combined the structured light method and the telecentric imaging technology to achieve high precision detection for small objects. 2. Obtain the exact system parameters by using a high-precision positioning platform and finish the complete system calibration. 3. Telecentric imaging system has good ductility. Based on the telecentric imaging model, it can combine macro and micro vision, multi-ocular vision and other methods. The current disadvantages are: 1. The system needs a positioning platform which increases the cost. 2. The calibration process of the whole system is complex, and the calibration speed will be limited compared with the traditional camera.

4 Conclusion

A telecentric fringe projection system and a complete calibration method of the system are proposed in this paper. The high-precision intrinsic and extrinsic parameters of the camera are obtained by using the properties of the closed-form solution, and then the parameters are completed by obtaining the depth information by the positioning platform. The problem of parameter γ symbol is solved in the calibration process, which further improves the accuracy of calibration. The projector light path is regarded as a reverse camera light path, and the complete homography matrix of the projector is obtained by the multifrequency phase-shifting algorithm. The feasibility of this method can be proved by the calibration reprojection error and point cloud reconstruction results. This structured light imaging method based on telecentric calibration has a good application prospect in small and micro object measurement and cross-scale measurement.

Acknowledgement. This work has been funded by National Natural Science of China (No. 51820105007, 51905176), the Guangdong Basic and Applied Basic Research Foundation (2021A1515012418), Guangdong HUST Industrial Technology Research Institute, Guangdong Provincial Key Laboratory of Manufacturing Equipment Digitization (2020B1212060014).

References

1. Manske, E., et al.: Recent developments and challenges of nanopositioning and nanomeasuring technology. Meas. Sci. Technol. **23**(7), 074001 (2012)
2. Opto Engineering: Telecentric lenses tutorial: basic information and working principles. http://www.opto-engineering.com/resources/telecentric-lenses-tutorial.
3. Tsai, R.: A versatile camera calibration technique for high-accuracy 3D machine vision metrology using off-the-shelf TV cameras and lenses. IEEE J. Robot. Autom. **3**(4), 323–344 (1987)
4. Zhang, Z.: A flexible new technique for camera calibration. IEEE Trans. Pattern Anal. Mach. Intell. **22**(11), 1330–1334 (2000)
5. Yin, Y., et al.: Fringe projection 3D microscopy with the general imaging model. Opt. Express **23**(5), 6846–6857 (2015)
6. Chen, Z., Liao, H., Zhang, X.: Telecentric stereo micro-vision system: calibration method and experiments. Opt. Lasers Eng. **57**, 82–92 (2014)
7. Li, D., Liu, C., Tian, J.: Telecentric 3D profilometry based on phase-shifting fringe projection. Opt. Express **22**(26), 31826–31835 (2014)

8. Haskamp, K., Kästner, M., Reithmeier, E.: Accurate calibration of a fringe projection system by considering telecentricity. In: Optical Measurement Systems for Industrial Inspection VII, vol. 8082. International Society for Optics and Photonics (2011)
9. Li, B., Zhang, S.: Flexible calibration method for microscopic structured light system using telecentric lens. Opt. Express **23**(20), 25795–25803 (2015)
10. Liu, H., Lin, H., Yao, L.: Calibration method for projector-camera-based telecentric fringe projection profilometry system. Opt. Express **25**(25), 31492–31508 (2017)
11. Rao, L., et al.: Flexible calibration method for telecentric fringe projection profilometry systems. Opt. Express **24**(2), 1222–1237 (2016)
12. Andrew, A.M.: Multiple view geometry in computer vision. Kybernetes **30**(9), 1331–1341 (2001)
13. Zhang, S.: High-speed 3D Imaging with Digital Fringe Projection Techniques. CRC Press, Boca Raton (2018)

Model-Based Fault Detection and Isolation of Speed Sensors in Dual Clutch Transmission

Jinchao Mo, Datong Qin(✉), and Yonggang Liu

State Key Laboratory of Mechanical Transmission,
Chongqing University, Chongqing 400044, China
dtqin@cqu.edu.cn.com

Abstract. Speed sensors in a dual clutch transmission play an important role in the launch and gear-shift processes. Speed sensor faults seriously affect vehicle performance, resulting in riding discomfort, apart from being a threat to the safety of the system. To diagnose speed sensor faults in a dual clutch transmission vehicle, this paper proposes a fault detection and isolation method for the speed sensors based on the dynamic of the dual clutch transmission powertrain. First, a control-oriented dynamic model of a dual clutch transmission was established. Considering the model uncertainties and external disturbance, an unknown input observer was constructed to realize robust sensor fault detection. Second, a bank of observers was designed for sensor fault isolation. Finally, simulation results show that the proposed method can realize robust fault detection and isolation for dual clutch transmission speed sensors, which may further facilitate active fault-tolerant control of a dual clutch transmission powertrain.

Keywords: Fault detection and isolation · Speed sensor · Dual clutch transmission

1 Introduction

Dual clutch transmission (DCT) has the advantages of high efficiency and good gear-shifting performance, and it has been extensively used in automatic transmission vehicles. The key technology of DCT is the control strategy, and many methods have been proposed in previous studies to achieve driving comfort [1, 2]. To achieve smooth launch and gear shifting, speed sensors are extremely important for closed-loop control of the system. Speed sensor faults may lead to noise and impact in the gearbox. Moreover, the launch or shifting process may fail in the presence of a speed sensor fault. Speed sensors are also vital for synchronizer engagement control. In the pre-shift control process, the speed sensor signal is used to determine whether the speed is synchronized [3]. Because of the poor working environment and the increase in working hours, speed sensors are prone to faults, such as deviation, constant output value, and drifting [4]. To ensure the reliability and stability of a DCT system, it is imperative to diagnose speed sensor faults in a DCT powertrain.

N. V. Khang et al. (Eds.): ASIAN MMS 2021, MMS 113, pp. 580–589, 2022.
https://doi.org/10.1007/978-3-030-91892-7_55

There are three methods to diagnose sensor faults, which are typically classified as model-based, signal-based, and knowledge-based methods [5]. Samara et al. [6] used the covariance of the sensing signals as the feature, and presented a statistical method for detecting abrupt sensor faults in aircraft control systems. However, the main limitation of signal-based fault diagnosis is the need for prior knowledge of the symptoms of health systems. Furthermore, knowledge-based fault diagnosis methods require a large amount of historical fault data, and the underlying fault feature can be extracted by many machine learning algorithms. The characteristics of knowledge-based fault diagnosis lend itself to some large and complex systems. Elnokity et al. [7] proposed an artificial neural network-based sensor fault detection, isolation, and estimation for a nuclear process. In model-based fault diagnosis, the system model of is required, which can be obtained using physical principles and system identification techniques. Model-based fault diagnosis has been widely researched for various engineering problems [8]. Disturbance and model error is inevitable in practice, fault detection and isolation (FDI) methods have to be robust to those uncertainties. Unknown input observer (UIO) shows a good state estimation performance in case of unknown input disturbance, and it has been widely used for fault detection and isolation. Li et al. [9] proposed an unknown input observer to detect an angular displacement sensor fault on the actuator of an automated manual transmission (AMT).

The traditional fault diagnosis method for automatic transmission speed sensors is signal range checking. The output of the speed sensor varies between the lower and upper bounds before a fault occurs. When a fault occurs, the signal deviates from the normal range. However, the speed of a DCT varies over a large range, and it is difficult to diagnose sensor faults by range checking, especially for incipient speed sensor faults. Another method for automatic transmission speed sensor fault diagnosis is based on analytical redundancy. When the clutch is locked, the engine speed is equal to the input shaft speed, and the input shaft speed and output shaft speed are related to the current gear. Analyzing the redundancy relationship can effectively diagnose the speed sensor faults. Qin et al. [10] proposed an analysis redundancy-based fault diagnosis for an AMT powertrain speed sensor. As a result, the engine speed, input shaft speed, and output shaft speed sensor faults could be detected and isolated using a diagnostic logic table. However, when the vehicle is launching or shifting, the clutch slips. The analytical redundancy method failed to diagnose the speed sensor fault. Recently, model-based fault diagnosis methods have been used for automatic transmission speed sensors. Wang et al. [11, 12], presented a parity space based sensor faults detection for DCT vehicles, and a data-driven design of parity space-based fault detection and isolation for AMT vehicles was proposed.

To diagnose speed sensor faults in a DCT powertrain, this study investigates the speed sensor fault detection and isolation method based on an unknown input observer. The main contributions of this study are as follows: Based on the dynamic model of a DCT powertrain, an unknown input observer was designed to realize the robust fault detection of the speed sensor in a DCT powertrain. A bank of observers was established to realize fault isolation of the speed sensor.

The paper is organized as follows. The simplified control-oriented dynamic model of a DCT is established in Sect. 2. In Sect. 3, considering the model error and external disturbance, unknown input observer-based sensor fault detection and isolation has been proposed. Simulation result is shown in Sect. 4. Conclusions are provided in Sect. 5.

2 Dynamic Model of a DCT

The schematic structure of the DCT is shown in Fig. 1. The gearbox is composed of two wet clutches, connected to two primary shafts and two secondary shafts that carry four synchronizers and eight gears (1st–7th and reverse). Combining different gears pairs positioned on the two primary shafts K1 and K2, secondary shafts K3 and K4 and the output shaft together with the four synchronizers S1, S2, S3, and S4, different gear ratios can be achieved. Gear shifting is realized by simultaneously engaging the oncoming clutch and releasing the off-going clutch. The dynamic characteristics of DCT have been widely investigated, as shown in various studies [13, 14]. A control-oriented simplified model is shown in Fig. 2. The reason for the simplification is that the shocks and vibrations that may occur when the teeth of the two gears come into contact need not be observed. Therefore, the stiffness of the gear and shafts are infinite. A simple mean-value engine model is provided, in which the engine torque is relative to the engine speed and engine throttle angle. The torque capacity of the wet multi-plate clutch is determined by the piston axial force which is related to the oil pressure in the clutch chamber. The oil pressure is control by an electromagnetic valve. However, the control of clutch pressure is excluded from this study. As a result, the torque transmitted by the clutch is referred to as the input of the DCT. From the engine to the end of the driveline, the powertrain consists of a dual clutch, gearbox, output shaft, and wheels. The dynamics of the powertrain can be described using the following equations:

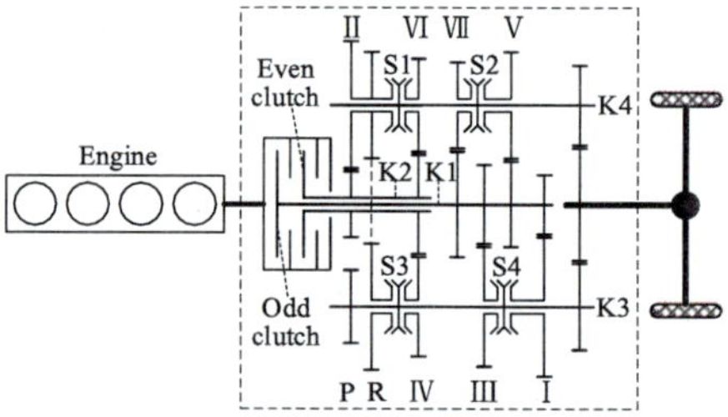

Fig. 1. Structure diagram of a DCT

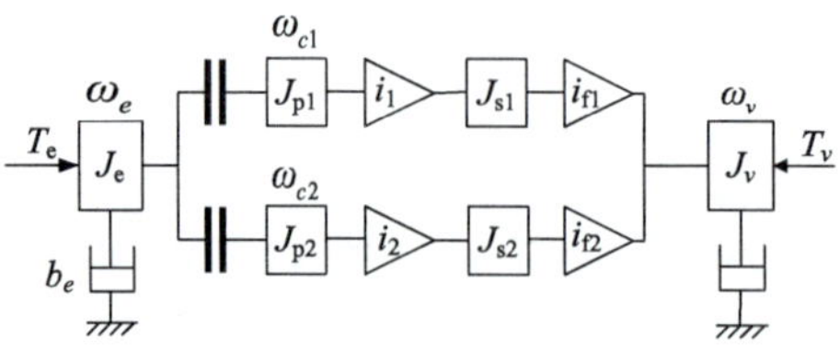

Fig. 2. Simplified dynamic of a DCT

$$\begin{cases} J_e \dot{\omega}_e = T_e - T_{c1} - T_{c2} - b_e \cdot \omega_e \\ \omega_{c1} = i_1 \cdot i_{f1} \cdot \omega_v \\ \omega_{c2} = i_2 \cdot i_{f2} \cdot \omega_v \\ J_{v\_eq} \dot{\omega}_v = i_1 T_{c1} + i_2 T_{c2} - T_v - b_f \cdot \omega_v \\ T_e = f(\alpha, \omega_e) \\ T_v = (mg \sin\varphi + \frac{\rho C_d A}{2} v^2 + \mu mg + m\dot{v}) r_w \\ J_{v\_eq} = J_v + (i_1 \cdot i_{f1})^2 J_{p1} + i_{f1}^2 J_{s1} + (i_2 \cdot i_{f2})^2 J_{p2} + i_{f1}^2 J_{s2} \end{cases} \quad (1)$$

where ω_e, ω_{c1}, ω_{c2}, ω_v, are the engine crankshaft angular speed, odd clutch angular speed, even clutch angular speed, and output angular speed, respectively. i_1 and i_2 are the gear ratio from the two primary shafts to the two secondary shafts, i_{f1} and i_{f2} are the two final drive ratios from the respective secondary shaft to the driven shaft. J_e includes all the inertia before the clutch. J_{p1}, J_{p2}, J_{s1}, J_{s2} are the inertia of the primary and secondary shafts respectively, referred by their own shaft. J_v is the inertia of the output shaft, and J_{v_eq} is the equivalent vehicle inertia referred to the output shaft. T_e, T_{c1}, T_{c2}, T_v are the engine torque, odd clutch torque, even clutch torque, and vehicle resistive torque, respectively. b_e is the viscosity of the input shaft and b_f is the wheel slip. The other parameters are as follows: engine throttle openingα, vehicle mass m, vehicle speed v, wheel radius r_w, road grade φ, air density ρ, air drag coefficient C_d, vehicle cross-sectional area A, rolling resistance coefficient μ, and gravitational acceleration g.

For speed sensor fault diagnosis using an unknown input observer, the launching process is used as a case study. The model of a DCT vehicle during launch is established as follows:

$$\begin{cases} \dot{\omega}_e = \frac{1}{J_e} T_e - \frac{1}{J_e} T_{c1} - \frac{1}{J_e} b_e \omega_e \\ \dot{\omega}_{c1} = \frac{(i_1 \cdot i_{f1})^2}{J_{v\_eq}} T_{c1} - \frac{(i_1 \cdot i_{f1})}{J_{v\_eq}} T_v - \frac{(i_1 \cdot i_{f1})}{J_{v\_eq}} b_f \omega_v \\ \dot{\omega}_v = \frac{1}{J_{v\_eq}} i_1 i_{f1} T_{c1} - \frac{1}{J_{v\_eq}} T_v - \frac{1}{J_{v\_eq}} b_f \omega_v \end{cases} \quad (2)$$

The state variables are chosen as the engine angular speed ω_e, odd clutch angular speed ω_{c1}, and output angular speed ω_v. The control inputs are the engine torque T_e, odd clutch torque T_{c1}, and vehicle resistive torque T_v. The output of the entire system is the three angular speeds, which are measured by three speed sensors in a product vehicle. The state-space representation of model (2) is as follows:

$$\begin{cases} \dot{x} = Ax + Bu \\ y = Cx \end{cases} \quad (3)$$

where $x = [\omega_e, \omega_{c1}, \omega_v]^{\mathrm{T}}$, $u = [T_e, T_{c1}, T_v]^{\mathrm{T}}$, $y = x = [\omega_e, \omega_{c1}, \omega_v]^{\mathrm{T}}$, and

$$A = \begin{bmatrix} -\frac{b_e}{J_e} & 0 & 0 \\ 0 & 0 & -\frac{(i_1 \cdot i_{f1}) b_f}{J_{v\_eq}} \\ 0 & 0 & -\frac{b_f}{J_{v\_eq}} \end{bmatrix} B = \begin{bmatrix} \frac{1}{J_e} & -\frac{1}{J_e} & 0 \\ 0 & \frac{(i_1 \cdot i_{f1})^2}{J_{v\_eq}} & -\frac{(i_1 \cdot i_{f1})}{J_{v\_eq}} \\ 0 & \frac{(i_1 \cdot i_{f1})}{J_{v\_eq}} & -\frac{1}{J_{v\_eq}} \end{bmatrix} C = \begin{bmatrix} 1 & 0 & 0 \\ 0 & 1 & 0 \\ 0 & 0 & 1 \end{bmatrix}$$

3 UIO-Based Speed Sensor FDI for a DCT

3.1 UIO-Based Robust Sensor Fault Detection

In model (2), the stiffness and damping of the gearbox are neglected; the effect of which can be referred as internal disturbance in the system. Additionally, the resistive torque T_v is subjected to disturbance, such as aerodynamic load force and road roughness. Moreover, T_e is the mean value of the actual engine torque where the harmonic torque is neglected, the harmonic torque can be modeled as disturbance of the input. In addition, the clutch torque T_{c1} has a relation with the axis piston force and the friction coefficient among the multi-plate discs. The axis force may suffer from disturbance due to the hydraulic clutch system, and the friction coefficient among the multi-plate discs is time vary variable as the wear of the clutch discs with the working hours increasing. All modeling errors and internal and external disturbances can be equivalent to an unknown input d in the system. Then, system (3) can be modified as follows:

$$\begin{cases} \dot{x} = Ax + Bu + Ed \\ y = Cx \end{cases} \tag{4}$$

where $x \in R^n$ is the state vector, $y \in R^m$ is the output vector and $u \in R^r$ is the input vector, $A \in R^{n\times n}$, $B \in R^{n\times r}$, $C \in R^{m\times n}$ are known matrices as shown in (3), $E \in R^{n\times q}$ is the distribution matrix of the external disturbance and $d \in R^q$ is the equivalent disturbance. For the DCT powertrain system, E is a 3rd order unit diagonal matrix, and $d = [d_1, d_2, d_3]^{\mathrm{T}}$ where d_1, d_2, d_3 is the function of the equivalent disturbance of the engine torque, the clutch torque and the vehicle resistance torque respectively. Moreover, the disturbance d is unknown. The disturbance plays a role of unknown input in the system. For robust sensor fault diagnosis in the DCT powertrain system, a proper observer can be designed to alleviate the unknown disturbance.

The following unknown input observer [15] is proposed for fault detection.

$$\begin{cases} \dot{z} = Mz + NBu + Ky \\ \hat{x} = z + Hy \end{cases} \tag{5}$$

where z and $\hat{x}$ denote the state of the observer system and the estimation of x, respectively. Matrices M, N, K, and H are of appropriate dimensions to be determined.

Denote the estimated error as $e = x - \hat{x}$, the dynamic of the estimated error is

$$\begin{aligned} \dot{e} = \dot{x} - \dot{\hat{x}} &= (A - HCA - K_1C)e + [(A - HCA - K_1C) - M]z \\ &+ [(A - HCA - K_1C)H - K_2]y + [(I - HC) - N]Bu + (I - HC)Ed \end{aligned} \tag{6}$$

where $K = K_1 + K_2$.

If the following equations hold

$$\begin{cases} (HC - I)E = 0 \\ N = I - HC \\ M = A - HCA - K_1C \\ K_2 = MH \end{cases} \tag{7}$$

the dynamic error can be described as $\dot{e} = Me$. The dynamic error e asymptotically converges to zero if matrix M is a Hurwitz matrix. The matrices in the observer can be obtained by solving Eqs. (7).

There exists an unknown input observer (5) for system (4) if the following condition holds [15]:

(i) rank $(C \cdot E) = \text{rank}\ (E)$;
(ii) (C, A_1) is a detectable pair, where $A_1 = (I - H \cdot C)A$

Considering sensor fault, the system (4) can be described as follows:

$$\begin{cases} \dot{x} = Ax + Bu + Ed \\ y = Cx + f \end{cases} \tag{8}$$

where $f \in \mathrm{R}^{s \times 1}$, and s is the number of faulty sensors in the system.

The following residual is generated for sensor fault detection.

$$r = y - \hat{y} = (I - HC)y - Cz \tag{9}$$

For the system (8), the dynamic estimation error and the residual are as follows:

$$\begin{cases} \dot{e} = Me + K_1 f - H\dot{f} \\ r = Ce + f \end{cases} \tag{10}$$

It is shown in (10) that the dynamic estimation error and the residual have no nothing to do with the unknown input d. To detect sensor faults, a threshold should be set appropriately, and the detection logic is given by the following:

$$\begin{cases} \|r\| < J, & \text{fault - free} \\ \|r\| \geq J, & \text{fault} \end{cases} \tag{11}$$

where $\|r\|$ is the Euclidean norm of vector r, and J is the predetermined threshold.

3.2 UIO-Based Robust Sensor Fault Isolation

When a speed sensor in a DCT powertrain is faulty, the residual may exceed the predefined threshold. To determine the location of the faulty sensor, a bank of observers was designed for sensor fault isolation. It is intuitive to build an observer for each speed sensor [16].

Grouping the output of the system (8) as

$$\begin{cases} \dot{x} = Ax + Bu + Ed \\ y^j = C^j x + f^j \\ y_j = C_j + f_j \end{cases} \tag{12}$$

where y^j is the output of all the sensors except the j-th sensor; y_j is the output of the j-th sensor; C^j is the matrix C without the j-th row; C_j is the j-th row of the matrix C; f^j is the fault vector f without the j-th element; f_j is the j-th element of the fault vector f.

The observer and residual are as follows:

$$\begin{cases} \dot{z}^j = M^j z^j + N^j Bu + K^j y^j \\ r^j = (I - C^j H^j) y^j - C^j z^j \end{cases} \tag{13}$$

The observer can be obtained by solving the following equations:

$$\begin{cases} H^j C^j E = E \\ N^j = I - H^j C^j \\ M^j = N^j A - K_1^j C^j \\ K_2^j = M^j H^j \\ K^j = K_1^j + K_2^j \end{cases} \tag{14}$$

where M^j is Hurwitz.

The fault isolation logic for the j-th speed sensor fault is as follows:

$$\begin{cases} \left\| r^j \right\| < J^j \\ \left\| r^k \right\| \geq J^k,\ k = 1, \cdots j-1, j+1, \cdots m \end{cases} \tag{15}$$

When the j-th speed sensor is faulty, the residual $\|r^j\|$ lies below its threshold along, whereas the others do not.

4 Simulation Results

The launch process of a DCT vehicle is simulated under 30% engine throttle opening. Figure 3 shows the engine speed and clutch speed during the vehicle launch process without a sensor fault. Before 0.4 s, the odd clutch speed is zero with a low friction torque, and the output shaft torque is too low to drive the vehicle. The output shaft torque increases with an increase in the odd clutch torque. At 0.4 s, the output shaft torque is greater than the vehicle resistive torque, and the vehicle starts to move. Subsequently, the speed of the odd clutch increases gradually. About 1.5 s, the engine speed is synchronized with the odd clutch speed (also known as the input shaft speed), and the launch process is completed. Thereafter, the vehicle ran in the first gear.

Figure 4 shows the UIO-based residuals when speed sensor faults in a DCT occur. $\|r_1\|$, $\|r_2\|$, and $\|r_3\|$ denote the residuals when a fault occurs at the engine speed sensor, input speed sensor and output speed sensor, respectively. Before the faults occur, the residual remains zero, even in the presence of an unknown input d. At 0.8 s, when a fault occurs in one of the speed sensors in a DCT, the residual will significantly deviate from zero. Speed sensor faults can be detected without a time delay.

Bias fault is a typical fault type of speed sensor in a DCT vehicle as the increase of working hours. A bias fault with a magnitude of 100 revolutions per minute (rpm), 80 rpm, and 10 rpm is set for speed sensor, input speed sensor and output speed sensor, respectively. Figure 5 shows the UIO-based residuals for the sensor fault isolation. When the engine speed sensor occurs bias fault at 0.8 s, the residual $\|r^1\|$ remains zero, while $\|r^2\|$ and $\|r^3\|$ significantly deviate from zero. When the input speed sensor and output speed sensor become faulty, the corresponding residual $\|r^2\|$ or $\|r^3\|$ remains zero, respectively, while the other two residuals rise significantly. The fault isolation of the speed sensors in a DCT can be achieved by the three UIOs based residuals.

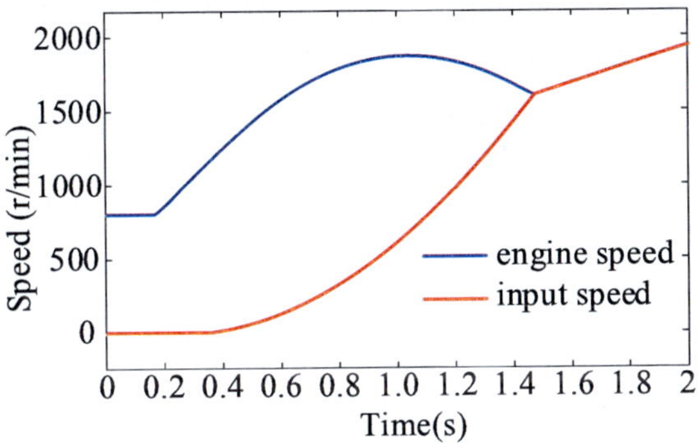

Fig. 3. Engine speed and input speed during the launch process of a DCT vehicle

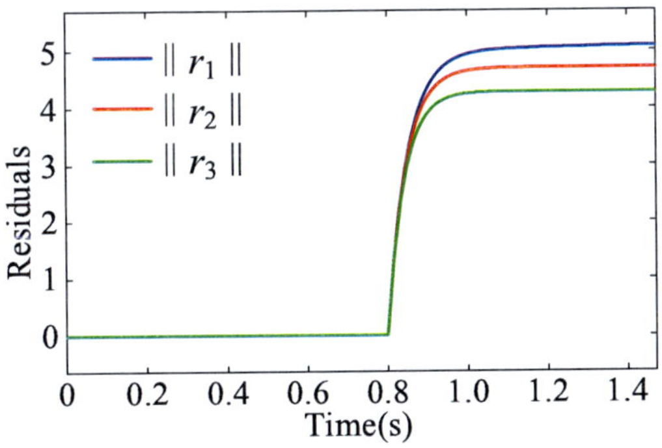

Fig. 4. Residuals of speed sensor fault

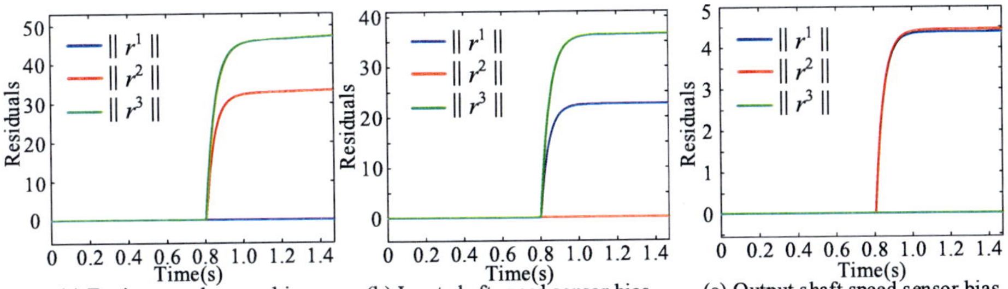

Fig. 5. Residuals of speed sensor bias fault

5 Conclusion

In this study, an unknown input observer-based FDI for speed sensors of a DCT powertrain is proposed. First, a simplified control-oriented dynamic model of a DCT is established. Based on the model, a robust fault detection method based on an unknown input observer was proposed. By establishing a bank of unknown input observers, the fault isolation for DCT speed sensors was realized. Simulation results show that the proposed method can detect and isolate speed sensor fault in a DCT powertrain. The structure of the AMT transmission powertrain is similar to that of the DCT, and the proposed method may also be suitable for the sensor fault diagnosis of AMT. Future work will study the UIO-based fault diagnosis method for DCT systems with different driving conditions, verify the fault diagnosis method through hardware-in-the-loop simulation and a real DCT vehicle, and design an active fault-tolerant control strategy for speed sensor faults.

Acknowledgement. The authors are very grateful to the China government by the support of this work through the National Natural Science Fund of the Peoples Republic of China (Grant No. U1764259).

References

1. Kulkarni, M., Shim, T., Zhang, Y.: Shift dynamics and control of dual-clutch transmissions. Mech. Mach. Theory **42**(2), 168–182 (2007)
2. Li, G., Görges, D.: Optimal control of the gear shifting process for shift smoothness in dual-clutch transmissions. Mech. Syst. Signal Process. **103**, 23–38 (2018)
3. Walker, P.D., Zhang, N.: Engagement and control of synchroniser mechanisms in dual clutch transmissions. Mech. Syst. Signal Process. **26**, 320–332 (2012)
4. Blanke, M., Kinnaert, M., Lunze, J., Staroswiecki, M.: Diagnosis and Fault-Tolerant Control, 2nd edn. Springer, Heidelberg (2006). https://doi.org/10.1007/978-3-540-35653-0
5. Gao, Z., Cecati, C., Ding, S.X.: A survey of fault diagnosis and fault-tolerant techniques—part I: fault diagnosis with model-based and signal-based approaches. IEEE Trans. Ind. Electron. **62**(6), 3757–3767 (2015)
6. Samara, P.A., Fouskitakis, G.N., Sakellariou, J.S., et al.: A statistical method for the detection of sensor abrupt faults in aircraft control systems. IEEE Trans. Control Syst. Technol. **16**(4), 789–798 (2008)
7. Elnokity, O., Mahmoud, I.I., Refai, M.K., et al.: ANN based sensor faults detection, isolation, and reading estimates – SFDIRE: applied in a nuclear process. Ann. Nucl. Energy **49**, 131–142 (2012)
8. Isermann, R.: Model-based fault-detection and diagnosis – status and applications. Annu. Rev. Control **29**(1), 71–85 (2005)
9. Li, L., He, K., Wang, X., et al.: Sensor fault-tolerant control for gear-shifting engaging process of automated manual transmission. Mech. Syst. Signal Process. **99**, 790–804 (2018)
10. Qin, G., Ge, A., Li, H.: On-board fault diagnosis of automated manual transmission control system. IEEE Trans. Control Syst. Technol. **12**(4), 564–568 (2004)
11. Wang, Y., Liu, Q., Li, K., et al.: Resilient fault and attack detection of DCT vehicles using parity space approach. In: 2019 Chinese Automation Congress (CAC), pp. 431–436 (2019)

12. Wang, Y., Gao, B., Chen, H.: Data-driven design of parity space-based FDI system for AMT vehicles. IEEE/ASME Trans. Mechatron. **20**(1), 405–415 (2015)
13. Walker, P.D., Zhang, N.: Modelling of dual clutch transmission equipped powertrains for shift transient simulations. Mech. Mach. Theory **60**, 47–59 (2013)
14. Galvagno, E., Velardocchia, M., Vigliani, A.: Dynamic and kinematic model of a dual clutch transmission. Mech. Mach. Theory **46**(6), 794–805 (2011)
15. Chen, J., Patton, R.J., Zhang, H.: Design of unknown input observers and robust fault detection filters. Int. J. Control **63**(1), 85–105 (1996)
16. Patton, R.J., Chen, J.: Robust Model-Based Fault Diagnosis for Dynamic Systems. Springer, New York (1999). https://doi.org/10.1007/978-1-4615-5149-2

Hysteresis Modeling of Twisted-Coiled Polymer Actuators Using Long Short Term Memory Networks

Tuan Luong, Sungwon Seo, Kihyeon Kim, Jeongmin Jeon, Ja Choon Koo, Hyouk Ryeol Choi, and Hyungpil Moon(✉)

Sungkyunkwan University, Suwon, South Korea
{luongtuan,hyungpil}@g.skku.edu

Abstract. Twisted-coiled polymer actuators (TCAs) are a new promising type of artificial muscles that is light-weight, low-cost and provides high stroke. This paper presents a dynamic model based on long short term memory (LSTM) networks to predict the nonlinear behavior of an antagonistic joint driven by the hybrid TCA bundle made from Spandex and nylon fibers. Different from our previous work, the current model considers pre-strains of TCAs as its inputs, therefore, any change in pre-strains of both TCA actuators will not require new data and training. It was verified from the experimental results that when there are pre-strain changes, the present model has better performance in capturing the system's behavior compared with that of the previous one. The proposed LSTM based model can estimate the joint angle with the mean error of 0.06° compared with that of 1.57° using the previous one within the working range of 30% of the TCA.

Keywords: Twisted-coiled actuators · Antagonistic mechanism · Long short term memory · Hysteresis modeling

1 Introduction

Twisted-Coiled Polymer Actuators (TCAs) are built by a simple manufacturing process that starts from a polymeric wire (precursor). The precursor is twisted using a motor and the fiber is highly over-twisted, then coils are formed spontaneously to minimize strain energy [1], which provides remarkable tensile actuation, therefore linear actuators can be obtained. Thermal treatment is then employed to stabilize the final twisted and coiled shape. When applying heat, there is expansion or contraction of a twisted polymeric fiber in longitudinal or radial directions, therefore, TCA can be activated by using thermal actuation by using air or water flows at controlled temperature, or by Joule heating thanks to the conductive surface of the fiber or heating wires. To date, precursors made

This work was supported by the National Research Foundation of Korea (NRF) grant funded by the Korea government (MSIT) (No. 2020R1A4A1018227).

N. V. Khang et al. (Eds.): ASIAN MMS 2021, MMS 113, pp. 590–599, 2022.
https://doi.org/10.1007/978-3-030-91892-7_56

of many different materials such as nylon 6, nylon 6.6, polyethylene, Kevlar, spandex have been tested. A hybrid type of TCAs made from Spandex yarn (Creora®, Hyosung Inc.) and nylon conductive sewing thread (ShieldexTM No. 200121235343B) has been chosen in our work because of its remarkable advantages such as low-cost, lightweight, high power-to-weight ratio and high actuation stroke over the same type made from nylon.

The properties of the Spandex-Nylon TCAs were discovered in [2]. Experimental results have shown that the dependence of the produced force over temperature appears very linear. It was also discovered that variable stiffness can also be obtained with varying temperatures, which was utilized to realize simultaneous position and stiffness control through controlling the actuator's temperature [3]. However, as other types of actuators, such as Shape Memory Alloys (SMAs) [4] or McKibben Pneumatic Actuators [5], this type of actuator also shows hysteresis behavior between elongation and force relationships, which is a critical issue affecting control performance.

Several models, both linear and nonlinear, have been developed for the TCA actuator. Phenomenal linear models have been developed for TCAs by Yip and Niemeyer [6], Arakawa [7] and Sutton [8]. Although those models provide simple mathematical expressions for control of the actuator, they could not exhibit the nonlinear properties of TCAs. Using a physics-based approach, a nonlinear dynamic model has been developed to capture the relationship between temperature, force, and displacement of the actuator [9], however a large number of parameters such as material properties were still needed. To catch the hysteresis behaviors of the actuator, nonlinear models were developed for TCAs such as a static nonlinear model by Luong [10, 11], differential hysteresis model by Luong [12], augmented Preisach model, augmented generalized Prandtl-Ishlinskii model by Zhang [13]. However, the parameter identification processes are complex and require a significant amount of time for exhaustive experiments. The development of an efficient modeling method for TCAs that can predict the actuator's nonlinear behavior precisely and has a simple identification procedure is still an open research problem.

In this work, we address the problems by using the Long Short Term Memory (LSTM) neural networks. In addition to high prediction performance, using the proposed model, the number of tests needed to obtain the model parameters are reduced significantly compared with the above mentioned ones. Compared with our previous work [2], we considered pre-strains of the actuator into the model, so that the model is more robust to the change in the system's configuration. An experimental setup was developed to collect training and testing data and the model was verified with different levels of pre-strains.

The remaining of this paper is organized as follows. Section 2 presents the related works involving in modeling problems of TCAs. LSTM model of the antagonistic joint and modeling results are presented in Sect. 3. We conclude the paper in Sect. 4.

2 Dynamics of the TCA and Related Works

The dynamics of a TCA-driven antagonistic joint can be separated into a thermo-electrical and a thermo-mechanical model. The thermo-electrical model represents the relationship between the applied power to each actuator and its temperature. The thermo-mechanical model was used to represent the temperature-force-strain relationship of the TCA.

2.1 Thermo-Electrical Model

The thermo-electrical process of the TCA actuator can be modeled using the energy approach [14] as follows. Heat transfer from the body to the ambient at a given time is

$$\Delta q_1 = hA_s(T - T_{amb}) \tag{1}$$

where h is the heat transfer coefficient, and A_s is the surface area. The heat provided by heating is

$$\Delta q_2 = \frac{V^2}{R} \tag{2}$$

where V and R are the voltage applied to the actuator and the resistance of the actuator. The change in temperature of the body is given by

$$\Delta q_3 = \rho c_p Vol \frac{dT}{dt} \tag{3}$$

where ρ is the density and c_p is the specific heat of the actuator. Vol is the body volume of the actuator. From Eq. (1)–(3)

$$\rho c_p Vol \frac{dT}{dt} = \frac{V^2}{R} - hA_s(T - T_{amb}) \tag{4}$$

It can be rewritten as

$$\frac{dT}{dt} + \frac{1}{\tau}T = \frac{V^2}{R\rho c_p Vol} + \frac{1}{\tau}T_{amb} \tag{5}$$

where the time constant is

$$\tau = \frac{C_{th}}{\lambda} \tag{6}$$

where $C_{th} = \rho c_p Vol$, and $\lambda = hA_s$

From Eq. 6, it is seen that the time constant of the TCA can be adjusted by changing the specific heat of the actuator, which is normally related to the modification of the actuator's material, or by adjusting the heating coefficient through the cooling method. In our work, the time constant is changed by varying the heating coefficient.

From Eq. 5, the thermo-electric model for a TCA actuator can be rewritten as:

$$C_{th} \frac{dT(t)}{dt} = P(t) - \lambda(T - T_{amb}) \tag{7}$$

where C_{th} is the thermal mass of the actuator, $P(t) = \frac{V^2}{R}$ is the Joule heating that is applied to the actuator, and λ is the absolute thermal conductivity of the actuator in its ambient environment. Another model considering higher orders of the temperature was also developed in [15].

2.2 Thermo-Mechanical Model

A linear model of the Spandex-Nylon TCA was developed in [16] as follows

$$F = b\dot{x} + kx + c(T - T_{amb}) \tag{8}$$

where b, k and c are the damping, spring stiffness of the actuator, and temperature effect on the actuator force, respectively. However, since TCAs are highly nonlinear, a linear model can not exhibit the actuator's behavior. An enhanced nonlinear model was developed in [17]. The model could capture the nonlinear behavior of the actuator and took variable stiffness property of the actuator into account. To consider hysteresis behaviors of TCAs, some other complex models have been developed such as augmented Preisach model, augmented generalized Prandtl-Ishlinskii model in [13], differential hysteresis model [12].

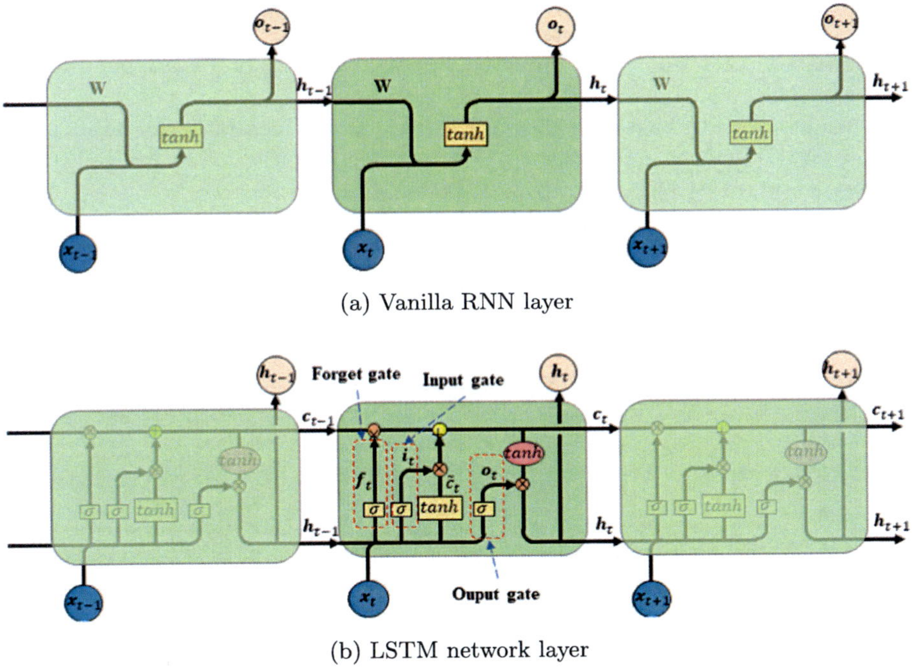

(a) Vanilla RNN layer

(b) LSTM network layer

Fig. 1. Layer structure of an LSTM network and a Vanilla RNN

3 Dynamic Modeling of TCAs Using LSTM Networks

One of the disadvantages of the above mentioned models is the complex identification process. For examples, to identify the parameters C_{th} and λ in Eq. 7 and b, k and c in Eq. 8, multiple experiments are needed such as heating, cooling, isothermal, iso-metric, dynamic tests. The identification processes, therefore, are normally tedious and time-consuming. The problem can be improved by utilizing deep learning.

Considering the hysteretic dynamics of the TCA, we propose a dynamic model based on LSTMs to predict the behavior of the TCAs driven antagonistic system. First, the capability of LSTM networks to model hysteresis behaviors will be briefly explained. Then, an LSTM network based model of the TCAs-driven antagonistic joint will be presented.

3.1 LSTM Neural Networks

Long short term memory (LSTM), a kind of RNN, is well-known for its capability of learning long-term dependencies. The difference between a typical Vanilla RNN and an LSTM network can be seen on the Fig. 1.

The hidden-to-hidden recurrence structure of a Vanilla RNN as shown in Fig. 1a provides it with a better capability of learning memory-related behaviors over feed-forward neural networks. However, the sharing of the hidden-to-hidden weight $\boldsymbol{W}$ induces numerical stability [18], which are gradient exploding and gradient vanishing, due to repeatedly multiplying by the same matrix W during back-propagation. When gradients are exploded, bad updates (or even divergence) will happen. When gradients vanishes, deep layers are not learned, which limits the network's long term dependency learning capability. While the gradient exploding problem can be fixed by gradient clipping [19], gradient vanishing is a more difficult problem to solve.

LSTM networks provide a notable solution by adding internal mechanisms called gates that can help regulate the flow of information. An LSTM layer consisting of multiple connected LSTM units is shown in Fig. 1b. It produces cell vector $\boldsymbol{c}_t \in \mathbb{R}^n$ given the inputs $\boldsymbol{x}_t, \boldsymbol{h}_{t-1}$ and $\boldsymbol{c}_{t-1}$ by using the following operations:

$$\boldsymbol{i}_t = \sigma(\boldsymbol{W}^i \boldsymbol{x}_t + \boldsymbol{U}^i \boldsymbol{h}_{t-1} + \boldsymbol{b}^i) \tag{9}$$

$$\boldsymbol{f}_t = \sigma(\boldsymbol{W}^f \boldsymbol{x}_t + \boldsymbol{U}^f \boldsymbol{h}_{t-1} + \boldsymbol{b}^f) \tag{10}$$

$$\boldsymbol{o}_t = \sigma(\boldsymbol{W}^o \boldsymbol{x}_t + \boldsymbol{U}^o \boldsymbol{h}_{t-1} + \boldsymbol{b}^o) \tag{11}$$

$$\tilde{\boldsymbol{c}}_t = tanh(\boldsymbol{W}^g \boldsymbol{x}_t + \boldsymbol{U}^g \boldsymbol{h}_{t-1} + \boldsymbol{b}^g) \tag{12}$$

$$\boldsymbol{c}_t = \boldsymbol{f}_t \odot \boldsymbol{c}_{t-1} + \boldsymbol{i}_t \odot \tilde{\boldsymbol{c}}_t \tag{13}$$

$$\boldsymbol{h}_t = \boldsymbol{o}_t \odot tanh(\boldsymbol{c}_t) \tag{14}$$

where $\sigma(\cdot)$ and $tanh(\cdot)$ are the element-wise sigmoid and hyperbolic tangent activation functions. $\odot$ represent element-wise product. $\boldsymbol{i}_t \in \mathbb{R}^n, \boldsymbol{f}_t \in \mathbb{R}^n$ and

$\boldsymbol{o}_t \in \mathbb{R}^n$ denote input gates, forget gates and output gates, respectively. $\boldsymbol{W}^k, \boldsymbol{U}^k$ and $\boldsymbol{b}^k$ where $k \in \{i, f, o\}$ are parameters for each of the gates.

From Eq. 13, it can be seen that by controlling the forget gate, we can control how much the past cell state $\boldsymbol{c}_{t-1}$ will be copied to the next cell state, and by controlling the input gate, we can control how much of the new state candidate $\tilde{\boldsymbol{c}}_t$ will be added to the next cell state. When $\boldsymbol{f}_t = \mathbf{1}$, the past memory cell will be saved over time and passed to the next step, which helps alleviate the gradient vanishing problem and better capture long-range dependencies.

3.2 LSTM Based Model for Antagonistically Driven TCAs

In our previous work [2], the LSTM model was trained and tested with a fixed TCA configuration. For the best prediction performance, any change in the configuration such as pre-strains of both TCA actuators will require new data and training. To overcome the limitation, in this work we consider TCA configuration parameters as inputs to the LSTM network. The proposed LSTM modeling structure is shown in Fig. 2. Inputs 1 and 3, consisting of the current power to be applied, corresponding fan signal and previous state, are passed through one LSTM layer followed by a fully connected layer with ReLU to predict the temperatures of actuators 1 and 2. The remaining input includes the previous states of the joint angles and the pre-strain information l_0. After passing through an LSTM layer and a fully connected layer with ReLU, the output is combined with the predicted temperature values and they are passed through another fully connected layer with ReLU to estimate the joint angle.

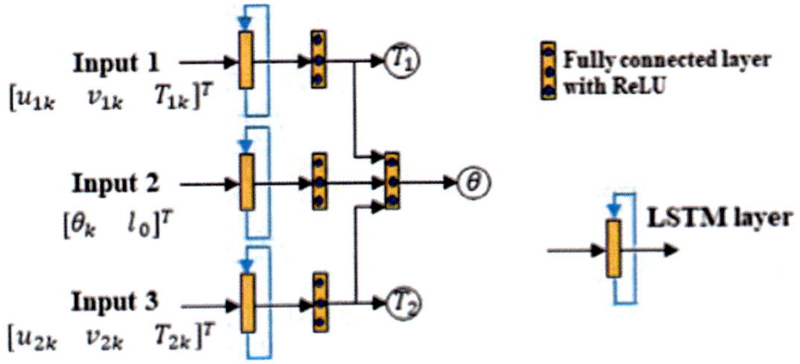

Fig. 2. LSTM network structure for modeling TCAs considering pre-strains

The mean squared error (MSE) function is used as a metrics to evaluate the loss between the predicted and the measured joint angle, temperature 1 and 2. The cost function for training is the summation of the three losses. To prevent the model from overfitting, 33% of the training data was chosen as a validation set and a test dropout rate of 50% is applied to each LSTM layer. The hyperparameters used for LSTM networks are the same as our previous work [2].

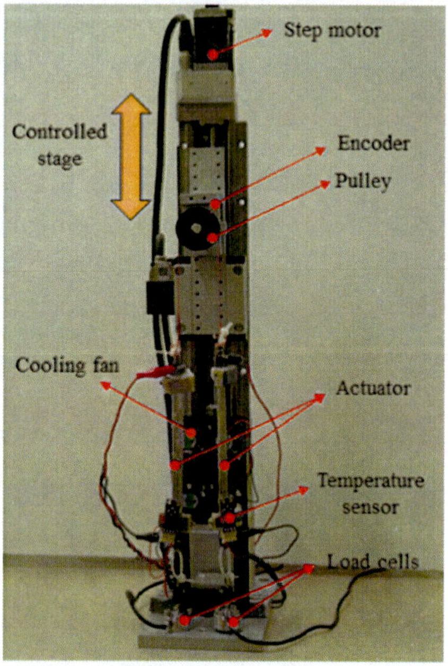

Fig. 3. Experimental setup for LSTM based modeling considering pre-strains

To collect data for training and testing the network, an new experimental setup was developed as in Fig. 3. In which, a motorized stage with a resolution of $1\mu m$ was used to change the initial strain of the antagonistic actuators. The joint encoder is measured using (ATM203-512). The data were collected 15 Hz. At each pre-strains of 10%, 14%, 19%, 24% and 29%, random inputs have been applied to the antagonistic joint. In the case of using LSTM without considering the pre-strain of the actuator, the LSTM network was trained with data collected at the pre-strain value of 19%, and was tested with data collected at the pre-strain value of 24%. In the case of using LSTMs considering pre-strains, the LSTM network was trained with data collected at the pre-strain values of 10%, 14%, 19%, 29%, and was tested with data collected at the pre-strain value of 24%.

3.3 Modeling Results

The testing results using LSTMs without and with considering pre-strains were shown in Fig. 4a and Fig. 4b, respectively, and quantitative results were listed in Table 1. It was observed that while the temperature modeling was not much affected, the modeling error for position prediction was much larger when pre-strains were not considered. Using the proposed LSTM based model, the joint angle can be estimated with the mean error of 0.06° compared with that of 1.57° using the previous one.

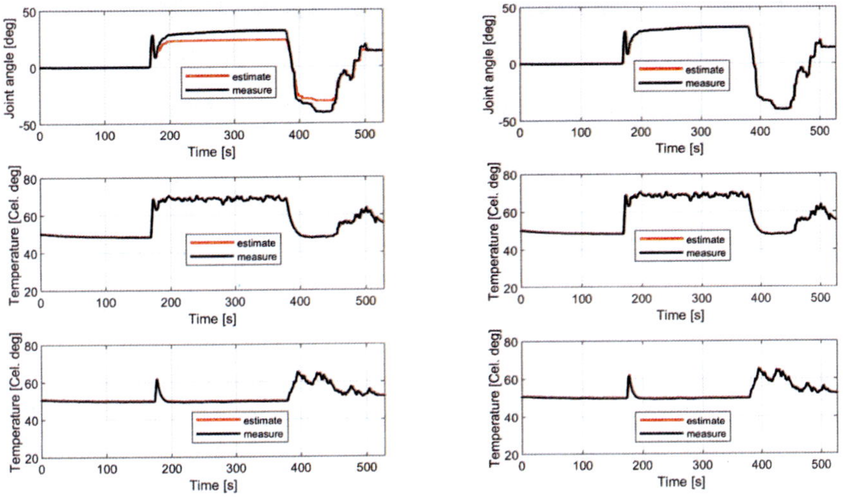

(a) Modeling results without considering pre-strains

(b) Modeling results considering pre-strains

Fig. 4. Modeling results using LSTMs with and without considering pre-strains

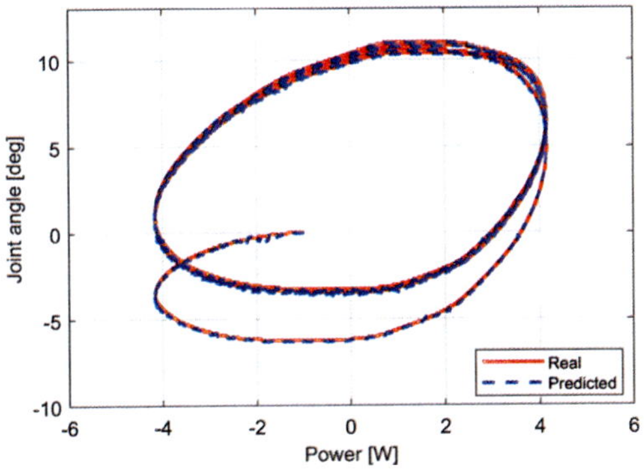

Fig. 5. Hysteresis behavior prediction performance with sinusoidal waveform power inputs at the frequency of 0.01 hz and the duty cycle amplitude of 0.2 (negative power values represent $-u_1$ and positive values represent u_2

Moreover, to demonstrate the capability of the LSTM model to capture the complex hysteretic dynamics of the TCA actuator-driven joint in its working range, a sinusoidal waveform of input powers has been applied with a frequency of 0.01 Hz and a duty cycle amplitude of 0.2. The prediction performance is shown in Fig. 5. It is seen that the model can predict well the value of the joint angle, therefore showing the hysteresis modeling capability of the proposed LSTM model.

Table 1. LSTM model prediction errors considering pre-strains

	Max	Mean	Std. Dev.
Angle error (pres-train) (deg)	1.4887	0.0615	0.0340
Angle error (no pre-strain) (deg)	5.7460	1.5690	1.8368
Temp. 1 error (pres-train) (Cel. deg.)	0.769	0.062	0.002
Temp. 1 error (no pre-strain) (Cel. deg.)	0.7417	0.067	0.01
Temp. 2 error (pres-train) (Cel. deg.)	1.097	0.086	0.005
Temp. 2 error (no pre-strain) (Cel. deg.)	1.021	0.092	0.009

4 Conclusions

An LSTM network based model was presented to model the hysteresis behavior of antagonistic joint driven by Spandex-Nylon TCAs. The model has shown its capability to predict the actuator's temperature and the join angle precisely. Moreover, by considering pre-strains, the approach took into account the configuration-dependent behavior of the antagonistic joint, providing better modeling capability without re-training when there are configuration variations. The proposed method also shows the possibility to apply deep learning methods to the development of faster and more precise tools in modeling complex systems like TCAs.

References

1. Ghatak, A., Mahadevan, L.: Solenoids and plectonemes in stretched and twisted elastomeric filaments. Phys. Rev. Lett. **95**(5), 057801 (2005)
2. Luong, T., et al.: Long short term memory model based position-stiffness control of antagonistically driven twisted-coiled polymer actuators using model predictive control. IEEE Robot. Autom. Lett. **6**(2), 4141–4148 (2021)
3. Luong, T., et al.: Simultaneous position-stiffness control of antagonistically driven twisted-coiled polymer actuators using model predictive control. In: 2020 IEEE/RSJ International Conference on Intelligent Robots and Systems (IROS), pp. 8610–8616. IEEE (2020)
4. Pruski, A., Kihl, H.: Shape memory alloy hysteresis. Sens. Actuators, A **36**(1), 29–35 (1993)
5. Chou, C.-P., Hannaford, B.: Measurement and modeling of McKibben pneumatic artificial muscles. IEEE Trans. Robot. Autom. **12**(1), 90–102 (1996)
6. Yip, M.C., Niemeyer, G.: High-performance robotic muscles from conductive nylon sewing thread. In: 2015 IEEE International Conference on Robotics and Automation (ICRA), pp. 2313–2318. IEEE (2015)
7. Arakawa, T., Takagi, K., Tahara, K., Asaka, K.: Position control of fishing line artificial muscles (coiled polymer actuators) from nylon thread. In: SPIE Smart Structures and Materials+ Nondestructive Evaluation and Health Monitoring, pp. 97982W–97982W. International Society for Optics and Photonics (2016)

8. Sutton, L., Moein, H., Rafiee, A., Madden, J.D., Menon, C.: Design of an assistive wrist orthosis using conductive nylon actuators. In: 2016 6th IEEE International Conference on Biomedical Robotics and Biomechatronics (BioRob), pp. 1074–1079. IEEE (2016)
9. Abbas, A., Zhao, J.: A physics based model for twisted and coiled actuator. In: 2017 IEEE International Conference on Robotics and Automation (ICRA), pp. 6121–6126. IEEE (2017)
10. Luong, T.A., Cho, K.H., Song, M.G., Koo, J.C., Choi, H.R., Moon, H.: Nonlinear tracking control of a conductive supercoiled polymer actuator. Soft Robot. **5**, 190–203 (2017)
11. Luong, T., et al.: Impedance control of a high performance twisted-coiled polymer actuator. In: 2018 IEEE/RSJ International Conference on Intelligent Robots and Systems (IROS), pp. 8701–8706. IEEE (2018)
12. Luong, T.A., Seo, S., Koo, J.C., Choi, H.R., Moon, H.: Differential hysteresis modeling with adaptive parameter estimation of a super-coiled polymer actuator. In: 2017 14th International Conference on Ubiquitous Robots and Ambient Intelligence (URAI), pp. 607–612. IEEE (2017)
13. Zhang, J., Iyer, K., Simeonov, A., Yip, M.C.: Modeling and inverse compensation of hysteresis in supercoiled polymer artificial muscles. IEEE Robot. Autom. Lett. **2**(2), 773–780 (2017)
14. Luong, T., et al.: Modeling and position control of a high performance twisted-coiled polymer actuator. In: 2018 15th International Conference on Ubiquitous Robots (UR), pp. 73–79. IEEE (2018)
15. Masuya, K., Ono, S., Takagi, K., Tahara, K.: Nonlinear dynamics of twisted and coiled polymer actuator made of conductive nylon based on the energy balance. In: 2017 IEEE International Conference on Advanced Intelligent Mechatronics (AIM), pp. 779–784. IEEE (2017)
16. Luong, T.A., Moon, H., et al.: Impedance control of a high performance twisted-coiled polymer actuator. In: International Conference on Intelligent Robots and Systems (IROS) (2018)
17. Luong, T., et al.: Realization of a simultaneous position-stiffness controllable antagonistic joint driven by twisted-coiled polymer actuators using model predictive control. IEEE Access **9**, 26071–26082 (2021)
18. Zhang, A., Lipton, Z.C., Li, M., Smola, A.J.: Dive into deep learning. arXiv preprint arXiv:2106.11342 (2021)
19. Goodfellow, I., Bengio, Y., Courville, A.: Deep Learning. MIT Press, Cambridge (2016)

A Differential Mechanism to Enhance the Scalability of a SMA-Wire-Bundle Linear Actuator

Andres Osorio Salazar(✉), Yusuke Sugahara, and Yukio Takeda

Tokyo Institute of Technology, 2-12-1, Ookayama, Meguro-ku, Tokyo 152-8552, Japan
osorio.a.aa@m.titech.ac.jp

Abstract. Scalability of actuators is the ability to change their output characteristics on demand. In a previous research, we proposed a Wet-activated Shape Memory Alloy (SMA) wire-bundle linear actuator that can scale its maximum output force in an easy manner. However, it was determined that the bundling technique we used produces a nonuniform stress distribution, related to the length difference between bundles. For this reason, we propose in this paper the use of a differential wire mechanism that joins the wire bundles in pairs and compensates for this length difference. This mechanism was tested for different prestresses and number of wire bundles in parallel, to determine if its scaling factor or maximum output force suffer any variation. We observed that the scaling factor and the maximum output force increase when using the differential mechanism, as compared to simply assembling the wire bundles directly, an effect more prominent with lower prestress. It was also observed that if the differential mechanism is used, the scaling factor does not significantly vary between the prestresses that we tested.

Keywords: Shape Memory Alloy · Linear actuator · Scalability · Wet-activated · Wire mechanism

1 Introduction

Scalability of actuators is the capability of modifying their input vs. output energy ratio, altering simultaneously the relationship of the controller signal to the output of the actuator, therefore, changing the resolution of its response [1]. If an actuator is able to do this modification on demand, it has a new level of usability. There are two types of scalability for actuators:

1. **Scalability with a *constant* scaling factor.** Effective for power improvement. In general, once the scaling factor is decided, it is difficult to modify.
2. **Scalability with a *variable* scaling factor.** Effective for usability enhancement. This is the focus of this research. The dynamic selection of a scaling factor depending on the use case brings further usability.

N. V. Khang et al. (Eds.): ASIAN MMS 2021, MMS 113, pp. 600–609, 2022.
https://doi.org/10.1007/978-3-030-91892-7_57

Previous studies have presented approaches for scalable actuators. Of those able to obtain a "variable scaling factor", the majority use multiple actuators in series or in parallel. For example, the actuator proposed by Britz et al. [2], where the maximum output angle of a Shape Memory Alloy (SMA) rotational actuator is determined by the number of modules arranged in series, with the objective of being "adapted to various application specifications by customizing the rotational angle and the output torque." On the other hand, actuators that change their energy conversion elements for achieving scalability have been proposed as well. For example, in the concept proposed by Wang et al. [3], their flexible pneumatic actuators can change dimensions according to its input pressure. Their maximum size is determined by the dimensions of some external rigid components. By changing their dimensions, different relations of input vs. output energy could be obtained.

Both examples show the conventional idea of scalability as a design variable that decides the (semi-)permanent behavior of the actuator, but can not react to different use cases instantaneously. Considering the previous situation, the long-term purpose of this research is to develop an actuator that can simultaneously scale force and displacement, to take advantage of the increased usability offered by scalable actuators. The used scaling technique should be easy enough to be able to achieve a variable, situation-dependent scale factor.

In the next chapter, we will explain the design concept that we proposed in a previous work [1,4], which basically consists of SMA wire bundles arranged mechanically in parallel, and activated simultaneously using a thermal conducting liquid (wet activation). We will focus on a specific problem that arises when using multiple bundles of wires subjected to tensile force, which is the nonuniform stress distribution that occurs due to the length variation between the elements of the group. We propose a differential mechanism that equally distributes the tensile force to the individual SMA bundles in a group, regardless of their length, making the input and output stresses uniform across bundles of SMA wires. Finally, we describe the maximum output force evaluation experiments that we used to test the implemented mechanism, along with its findings and conclusions.

2 Design Concept

In a previous work, we presented a design concept of a scalable actuator based on SMA wires. It consists of a variable number of SMA wires assembled mechanically in parallel and immersed in a liquid used to transmit heat for activation from a separate heat source, without the use of Joule heating, based on the Wet SMA activation technique proposed by Park [5]. In our concept, more wires increase the maximum force, and longer wires increase the maximum displacement; both quantities can be independently modified to scale the output of the actuator. We tested the force scalability of different actuators with multiple configurations. To determine their scalability, we compared the obtained maximum output force with the number of wires in parallel. The slope of this relationship is called in our work the "scaling factor". It was found that their maximum output force can be scaled by changing the number of SMA wires placed mechanically in parallel. The change of scaling factor is simple; an automatic mean of changing it could be implemented to obtain an actuator with a variable scaling factor.

The type of SMA we used, unidirectional SMA, contracts when a heat is applied, from a deformed shape (elongated by a prestress) to a base length. For this reason, a prestress is required to set the initial deformed state of the SMA's cycle. Nevertheless, with the crimp-based SMA wire bundles that we used, which can be seen in Fig. 1, a length dispersion is common. As an example, in our case, the length of the bundles that we used can be seen in Table 1, with an average total length of 105.96 mm and a standard deviation of 0.92 mm. In comparison, in ideal conditions, i.e., if the lengths of the bundles were equal and assuming a uniform wire diameter, the theoretical total deformation caused by a standard prestress of 70 MPa would be 0.264 mm, considering a martensite phase modulus of elasticity of 28 GPa [6]. Evidently, a prestress of such magnitude would produce a total deformation smaller than the standard deviation of length between the bundles we used.

In practice, in a blocked-force isometric experiment, the stress would reach the bundles in steps, starting from the shortest one, deforming it until it reaches the length of the second shortest, and so on, in a process repeated for all remaining bundles. As such, the stress required to make all wires the same length (after which point the stress would be equally distributed) can be represented with the sequence described in the following Eq. 1, where i and j are respectively the indexing variables for the length and stress of the individual bundles, starting from the shortest bundle, n is the total number of bundles and E is the SMA Young's modulus:

$$\sigma_j = E \sum_{i=j}^{n-1} \frac{l_{i+1} - l_i}{l_i} \tag{1}$$

This equation assumes that the stress is not larger than the austenite start stress σ_{As}, in which case the Young's modulus E would be dependent on the martensite fraction ξ. For shorter wires (those with a lower index i), the number of addends is larger, thus, the magnitude of the stress is larger than longer wires (those with a higher index i).

Thus, the purpose of this paper is to experimentally identify the effect of the length differences between SMA wire bundles on the maximum output force and scaling factor of the actuator, and to propose a mechanism that can reduce it. For this purpose, we present a wire-based differential mechanism that we tested with various configurations of number of bundles in parallel and with different values of prestress for evaluating its basic performance.

The proposed mechanism is similar to the "movable pulley", a fundamental example of a differential mechanism explained by Hirose, a device in which "dynamic inputs from three ports act in balance". The basic principle is shown in Fig. 2 (b), where an arrangement with two SMA bundles is exemplified, but a similar concept was used for three and four bundles. In this proposed mechanism, the three ports are the pulley and both extremities of the wire of the mechanism. The tensile forces have a $a_1 : a_2 : a_3 = 1 : 0.5 : 0.5$ relationship between them, where a_i is the variable corresponding to the reduction ratio of its ports [7].

Fig. 1. Crimping method for SMA wire bundles

Table 1. Total length specifications of the used SMA wire bundles.

ID	B1.1	B1.3	B1.4	B1.5
Length [mm]	105.8	106.5	104.55	107

When using the differential mechanism, the length of the two connected bundles or group of bundles is averaged, since the wire connecting them can slide across the pulley. By doing so, and by using a differential wire mechanism per each pair of bundles in a cascading arrangement (the total number of parallel wire mechanisms is $n-1$, where n is the total number of bundles), the numerator in Eq. 1 ($l_{i+1} - l_i$) is equal to 0, and so is the stress required to make the wires the same length. In other words, the arrangement skips the staggered stage of the prestress application.

To test the proposed differential wire mechanism, we constructed an actuator, called "constant flow SMA linear actuator, version 2" (CFSMALA-V2). As can be seen in Fig. 2 (a), the actuator consists of two 3D-printed resin coated plates with water inlet ports, a geometry for mounting pulleys, and a sealing method that uses a compression seal with an High-Density Polyethylene (HDPE) compliant tube.

3 Experiment

In our previous work [1], we presented an apparatus to input the thermal fluid to the actuator, which can be seen in Fig. 3. However, it was determined that the transient behavior of the flow rate of a pump directly subjected to an electric step impulse, like the arrangement that we used, superimposes a dead time and a large time constant to the measured rise and fall time. For this reason, we modified the experiment arrangement, as can be seen in Fig. 4, to include solenoid valves to provide the temperature step input to the actuator. With this arrangement, both the time constant and the dead time of the arrangement were minimized. While the measurement of the time response is not relevant for this paper, in a future work we will show how the variation of the flow-rate can change the performance of a wet-activated SMA linear actuator, both in terms of maximum force and activation/deactivation time.

In the arrangements illustrated in Figs. 3 and 4, the pumps are indicated by P, temperature sensors T, flow meters FM, force transducer FT, and solenoid valves SV, etc. The solenoid valves SVH1 and SVC1 are used to independently

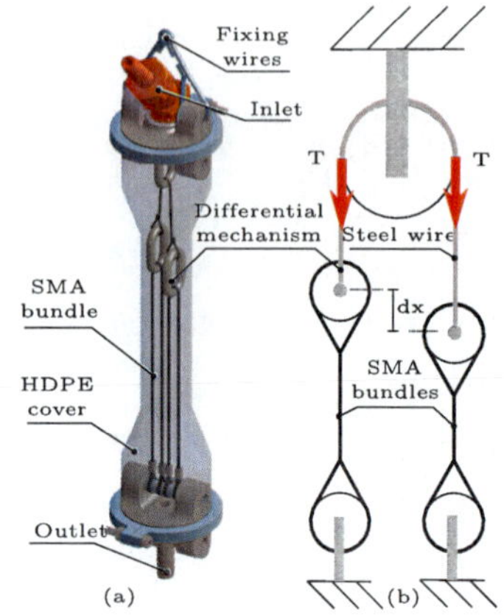

Fig. 2. Construction details of the CFSMALA-V2: (a) Actuator assembly (b) Differential wire mechanism

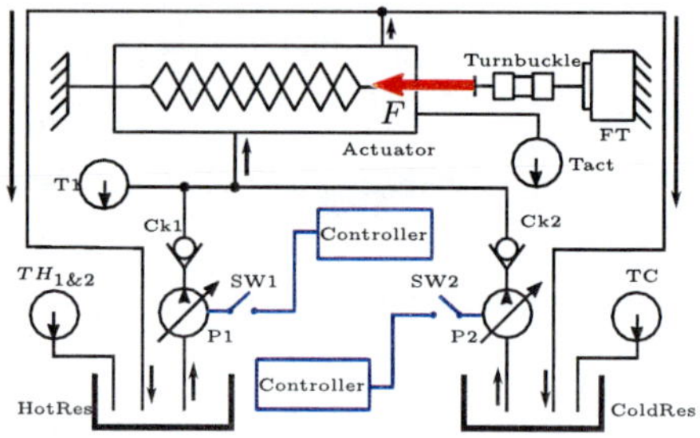

Fig. 3. Constant-flow version 1 (CFSMALA-V1) hydraulic circuit, proposed in [1]

stop the flow from the "hot reservoir" or "cold reservoir" respectively. This way, the pump can reach or continue to maintain a set flow rate without liquid going inside the actuator chamber, instead going through SVH2 and SHC2, which are open in this stage, to return to the "hot reservoir" or "cold reservoir". Once the flow rate is sufficiently stabilized, SVH1 or SVC1 can be activated and SVH2 or SHC2 deactivated to suddenly introduce to the actuator chamber a hot or cold liquid. This preparation step eliminates any effect of the transients of the pump on the measurements.

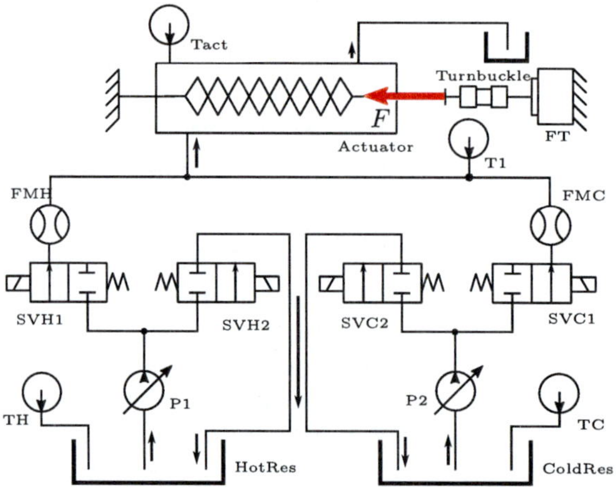

Fig. 4. Constant-flow version 2 (CFSMALA-V2) hydraulic circuit

We fixed one side of the actuator to a 6-axis force/torque sensor, which was rigidly attached to an aluminum frame. On the other side of the actuator, we prepared two turnbuckles to be able to set a prestress, necessary since this SMA actuator uses uni-directional SMA. The turnbuckles were fixed to the frame using a 3 mm stainless steel wire.

Additionally, we used a FLIR C2 thermography camera to observe the transient temperature behavior of the wall of the actuator, with the purpose of doing a future study on shape optimization for maximum heat transmission using a CFD (Computer Fluid Dynamics) model, and to observe any heat concentrations and hot-spots that need to be avoided to insure isothermal conditions inside the SMA chamber.

The experiments we did were isometric blocked-force experiments, and their details can be found in the following subsection.

3.1 Maximum Force Evaluation: Method

The purpose of this evaluation can be divided in three parts:

- To determine the scalability characteristics of the actuator
- To measure the effect of the prestress on its scaling factor
- To find out the effect of the length differences between bundles on the maximum output force.

For these purposes, we used the experiment setup explained previously with the variations included in Table 2, and detailed as follows:

1. Using a turnbuckle, apply prestress with a slope of 10 N/min, stop when the measured tension required for the prestress remains stable for five minutes
2. Open SVH2 and SVC2, start P1 and P2, wait until the speed of P1 and P2 >600 rad/s
3. Open SVC1, close SVC2, so the temperature in the SMA chamber (T_{SMA}) is equal to the temperature in the cold reservoir (TC). Wait 5 s
4. Close again SVC1 and open SVC2. Wait 3 s. This step allows the actuator to depressurize and reduce its water volume
5. Open SVH1 and close SVH2, to input the hot liquid step impulse. Wait until the output stress is greater than 270 MPa or $T_{SMA} \approx TH$
6. Close SVH1 and open SVH2; open SVC1 and close SVC2, to input the cold liquid step impulse. Wait until $T_{SMA} \approx TC$.

This procedure was executed automatically with a software developed in Labview© 2019. During the whole process, temperature, force, and flow rate measurements were acquired with a sample frequency of $f_s = 10$ Hz, with the components detailed in Table 3. The temperatures of "Hot reservoir" and "Cold reservoir" were 97 and 20 °C, respectively.

Similarly than our previous research, the data from this experiment can be used to analytically determine the maximum force per configuration, obtained as the intersection of the heating and cooling cycle hysteresis curves. From this

Table 2. Experiment variations, repeated three times for each iteration

Parameter	Tested values	Observed variables
No. of parallel SMA wires	10, 20, 30, 40	Scalability, scaling factor
Prestress [MPa]	35, 70[a]	Effect to scaling factor
Differential Mechanism	With, without	Effect to scaling factor

[a]The values for the tested prestresses were decided using as reference the recommendations of the manufacturer (70 MPa, recommended by Dynalloy [6] as the basic "cooling deformation force")

Table 3. CFSMALA-V2 experiment components and specifications

Component	Parameters		Component	Parameters	
Pump	Model	M510S-V	SMA	Diameter [μm]	375
	Flow rate range [l/min]	3.5–8.4		Activation temperature [°C]	70
	Type	Brushless	Temperature sensor	Type	K-type thermocouple
Solenoid	Model	CKD AB41		Diameter [mm]	0.2

information the scaling factor for all experiments was determined as the slope of the relationship of the total area of the bundles (the number of wires in parallel) vs. the maximum output force.

3.2 Maximum Force Evaluation: Results and Discussion

The graphs included in Figs. 5 and 6 show that the linear correlation of number of wires and the maximum output force remains similar for all tested variations. In this figure, the standard deviation (σ) error bars for the maximum output (vertical bars) for each configuration are shown. As it can be seen in the fitting equations in Figs. 5 and 6, the differential mechanism was effective, increasing the actuator's scaling factor by 42%, with 35 MPa of prestress, from 151.54 to 215.25 N per bundle (10 wires) in parallel, but only 14.18% with a 70 MPa of prestress, from 200.55 to 229 N per bundle in parallel. The increase is most likely because of the non-uniform stress distribution along the SMA wires caused by the variation in length of the bundles, which was decreased with the addition of the differential wire mechanism. More than the mechanism is more effective for 35 MPa than with 70 MPa of prestress, the results show that the actuator without the mechanism performs worse with 35 MPa. This indicates that a lower prestress is not sufficient to produce the required initial elongation uniformly across the bundles if they have a length variation. An insufficient deformation results in a lower maximum output force.

On the other hand, considering the experiments done with the differential mechanism, we can observe that the scaling factor in both tested prestress magnitudes remained approximately the same. From this preliminary results we can hypothesize that, since a larger prestress does not contribute to the improvement of the scaling factor, the measured value can be regarded as the upper boundary of the scaling factor of this particular actuator, with the tested SMA wire bundles. The differential mechanism reduces the magnitude of the prestress required for the activation cycle of the actuator and normalizes its effect to the scaling factor. In other words, the prestress can simply be regarded as a vertical shift of the scalability curve when using the differential mechanism. Future experiments with more values of prestress are required to properly characterize this behavior, both with larger and lower magnitudes of prestress. An optimal prestress magnitude could potentially be derived as a function of maximum output force vs. the maximum contraction of the actuator (both dependent on the prestress), for which an isotonic experiment (constant loading condition throughout the experiment) would be required.

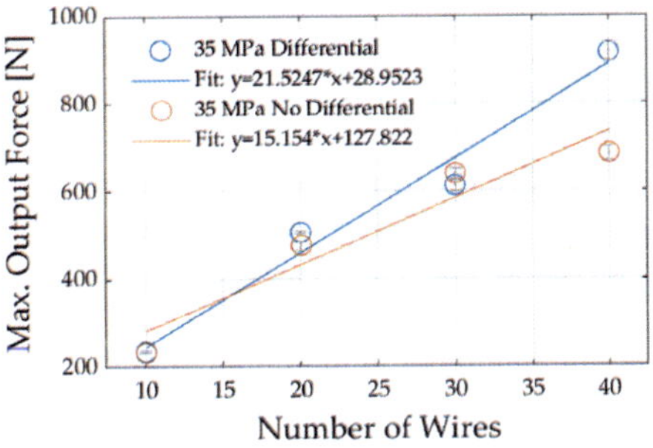

Fig. 5. Maximum output force vs. number of wires, with/without differential mechanism, 35 MPa of prestress.

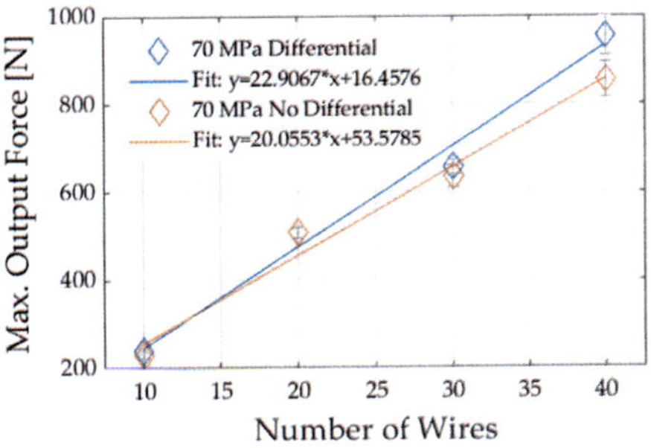

Fig. 6. Maximum output force vs. number of wires, with/without differential mechanism, 70 MPa of prestress.

Additionally, Fig. 7 shows the typical heating and cooling cycle of the CFSMALA-V2 with a configuration of 40 wires captured with a thermography camera, when the output force is at 90% and 10% of the output range, for the cooling and heating cycle. This figure highlights the hot spots, information that will be useful for improving the actuator's actuation frequency in a future study. These hot spots are more pronounced in the inlet and outlet areas of the actuator, where there are brass barb fittings to connect to the hydraulic circuit, which will be replaced with a material with lower thermal conductivity. The footage from the thermography camera will also be used in a future work to verify and tune a CFD model employed to optimize the shape of the actuator to maximize the heat transfer coefficient to the SMA bundles.

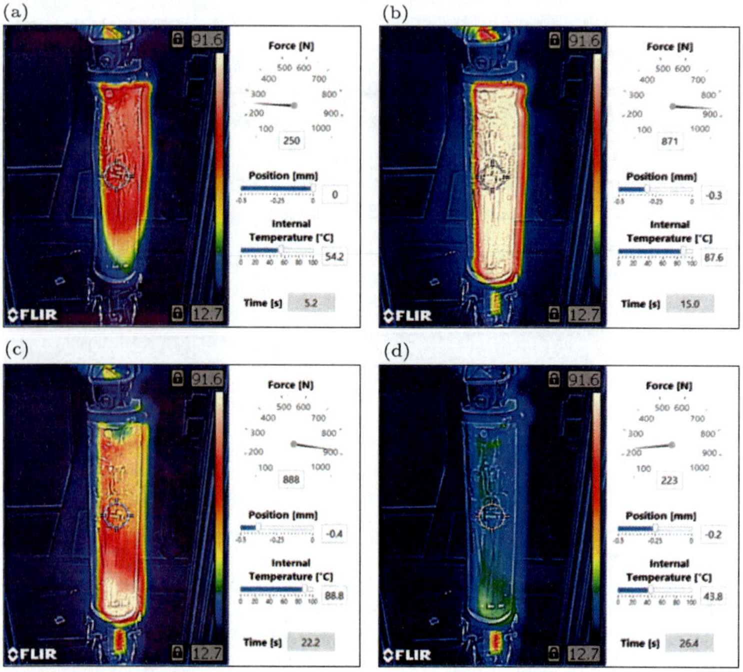

Fig. 7. Thermography images of the VFSMALA-V2 actuator

4 Conclusions and Future Work

In this work, we proposed a differential mechanism that aims to adapt to real-world conditions to wire bundle-based SMA actuators, like manufacturing tolerances, loading conditions, etc. We constructed an actuator, CFSMALA-V2, to test the proposed mechanism, to which we did a maximum force evaluation experiment with a blocked-force isometric arrangement, to determine its basic scalability performance under various prestress magnitudes and configurations.

The differential wire mechanism that we tested reveals that the small differences in length of the SMA wire bundles have an impact on the maximum output force of the actuator, reducing its magnitude. This reduction is more pronounced for a prestress of 35 MPa than for 70 MPa, because the lower force associated with a lower prestress is not enough to produce sufficient deformation to some individual bundles. A length difference like the one present in the tested bundles is most likely introduced during its manufacture and can not be entirely avoided. We demonstrated that the presented differential mechanism reduces the effect of this length difference by equalizing the tensile force between parallel bundles.

We also showed that the scalability factor of the actuator remains unchanged for two different magnitudes of prestress when using the differential mechanism, a result not observed if the differential was not used. This means that a lower magnitude of prestress would be required for the activation cycle of the actuator if the differential is used, and inconsistent values of prestress do not affect the scaling factor.

The capabilities of the CFSMALA-V2 presented in this work will be expanded in the future in the following ways:

1. A more precise model of the staggered prestress distribution in bundles with length differences with simulations to clarify the effect on the output stress.
2. Higher and lower values of prestress to determine a relationship of scaling factor and prestress.
3. A study on its displacement and energy scalability capabilities.
4. A study on the relationship between the input flow rate and the maximum output force of the actuator.
5. A variable scaling factor by automatically changing the number of wires arranged in parallel or the length of the active sections.

A scalable actuator such as the one presented in this work brings flexibility to robotic systems because they can be optimized for each specific use case and circumstance, considering design constrains such as step resolution, weight reduction, response frequency, and maximum required output.

Acknowledgment. This work was partially supported by JKA and its promotional founds from KEIRIN RACE.

References

1. Osorio Salazar, A., Sugahara, Y., Matsuura, D., Takeda, Y.: Scalable output linear actuators, a novel design concept using shape memory alloy wires driven by fluid temperature. Machines **9**(1), 14 (2021)
2. Britz, R., Motzki, P., Seelecke, S.: Scalable bi-directional SMA-based rotational actuator. Actuators **8**(3), 60 (2019)
3. Wang, N., Chen, B., Ge, X., Zhang, X., Chen, W.: Design, kinematics, and application of axially and radially expandable modular soft pneumatic actuators. J. Mech. Robot. **13**(2) (2021)
4. Osorio Salazar, A., Sugahara, Y., Matsuura, D., Takeda, Y.: A novel, scalable shape memory alloy actuator controlled by fluid temperature. In: Niola, V., Gasparetto, A. (eds.) IFToMM ITALY 2020. MMS, vol. 91, pp. 617–625. Springer, Cham (2021). https://doi.org/10.1007/978-3-030-55807-9_69
5. Park, C.H., Choi, K.J., Son, Y.S.: Shape memory alloy-based spring bundle actuator controlled by water temperature. IEEE/ASME Trans. Mechatron. **24**(4), 1798–1807 (2019)
6. Blair, D.: Technical characteristics of flexinol actuator wires. Technical report TCF1140RevJ, Dynalloy Inc., 1562 Reynolds Avenue, Irvine, California 92614 USA (2018). Accessed 15 June 2021
7. Hirose, S.: Connected differential mechanism and its application. In: Proceedings of the 1985 International Conference on Advanced Robotics, pp. 319–326 (1985)

Reduce Phase-Lead Effect in an Active Velocity Feedback by Frequency Range Selector

La Duc Viet[1] (✉), Nguyen Van Hai[1], and Nguyen Tuan Ngoc[2]

[1] Institute of Mechanics, Vietnam Academy of Science and Technology, 264 Doi Can, Hanoi, Vietnam
ldviet@imech.vast.vn

[2] Freyssenet Vietnam, 11 Tran Hung Dao, Hanoi, Vietnam

Abstract. The high pass filters are essential to avoid the integrator's drift in control loop. However, the phase-lead of high pass filter in an active velocity feedback vibration isolation limits the controller gain. To overcome the limitation, this paper considers a frequency range selector combining the isolated mass's velocity and displacement. This approach suppresses low frequency control without additional phase-lead. The controller's effectiveness under harmonic disturbance is illustrated through numerical simulations.

Keywords: Velocity feedback · Active isolation · Phase-lead · Practical integrator

1 Introduction

Vibration isolation reduces the displacement transmitted from vibratory foundation to sensitive equipment. A passive isolator has some limitations, for e.g. unchangeable transmissibility of the invariant point [1, 2]. Active isolator can improve the performance but the instability can occur due to energy injected. The use of complex system of sensor, actuator and microprocessor requires simple mechanism structures and efficient algorithms. Some strategies such as PI-controller, velocity feedback, state feedback, static output feedback, H∞ controller, force feedback and acceleration feedback were discussed in [2, 3]. However, only the absolute velocity feedback control has been widely used in commercial active vibration isolation [3] because of its simplicity and robustness [4]. It is shown in [5] that the exact velocity feedback controller is unconditionally stable. However, some high-pass filters are often required to avoid drift and actuator saturation. These filters introduce the phase-lead, which is a source of instabilities at low frequencies [6].

To decrease the controller gain at low frequencies, some types of filter should be considered. However, the conventional filters can not be used because they add more phase-lead to the loop, which causes more instability. Some recent switching type filters [7, 8] have used to reduce the phase-lag but not the phase-lead. This paper proposes a filter in the form of frequency range selector. The proposed selector use the velocity and displacement data obtained from integrated acceleration. The sign of the selector is

N. V. Khang et al. (Eds.): ASIAN MMS 2021, MMS 113, pp. 610–616, 2022.
https://doi.org/10.1007/978-3-030-91892-7_58

frequency-dependent while the phase-lead is not developed. To gain insight into how the selector frequency governs the instabilities, active control of a single-degree-of-freedom system is studied. To support the theoretical analysis, some numerical simulations are presented.

2 Instability Due to Phase-Lead

This paper studies in Fig. 1 a base excited system, where the mass is isolated by a velocity feedback isolation.

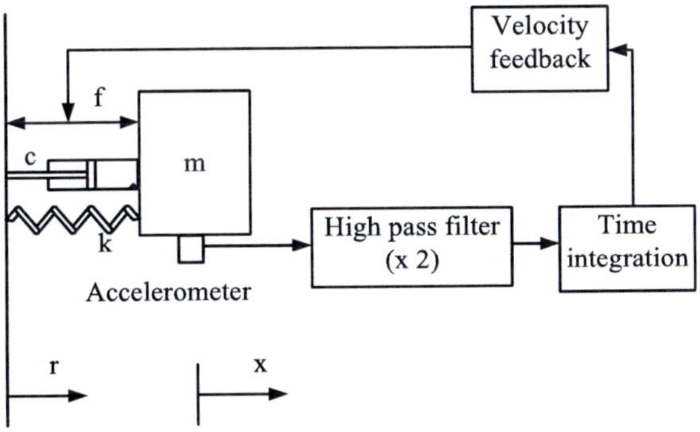

Fig. 1. Active vibration isolator with high pass filter

The isolation's parameters in Fig. 1 include: stiffness k, damping c, mass m, control force f, foundation motion r and absolute displacement x. Because accelerometers are often used in practice, velocity and displacement are obtained by time integrations. The high-pass filters are often used in series with ideal integrator to reduce the drift. Unfortunately, these filters cause the phase-lead, which can destabilize the system. In Fig. 1, two high-pass filters are considered to obtain the velocity and displacement.

The equation of motion is written as [2]:

$$m\ddot{x} + c\dot{x} + kx = c\dot{r} + kr + f \tag{1}$$

The frequency response of the transmissibility is given by:

$$T = \frac{X}{R} = \frac{j\omega c + k}{\left(k - m\omega^2\right) + j\omega c + jg\omega H(j\omega)} \tag{2}$$

In which $x = Xe^{j\omega t}$, $r = Re^{j\omega t}$ and $H(j\omega)$ is the transfer function of the high pass filter. Let us consider the high pass filter in the form:

$$H(j\omega) = \left(\frac{j\frac{\omega}{\omega_h}}{1 + j\frac{\omega}{\omega_h}}\right)^2 \tag{3}$$

where ω_h is the cut-off frequency of the high pass filter. Using the standard notations: $\omega_n = \sqrt{k/m}$(natural frequency), $\zeta = c/\left(2\sqrt{km}\right)$ (damping ratio), $\Omega = \omega/\omega_n$ (normalized forcing frequency), $\alpha = \omega_n/\omega_h$ (normalized cut off parameter), the transmissibility reduces to:

$$T = \frac{1 + j2\zeta\Omega}{1 - \Omega^2 + j2\zeta\Omega + \frac{g}{m\omega_n} j\Omega\left(\frac{j\alpha\Omega}{1+j\alpha\Omega}\right)^2} \tag{4}$$

The limits on stability are obtained by setting the denominator of (4) to zero. Setting the imaginary and real parts the denominator of (4) to zeros gives

$$\begin{cases} 1 - \Omega^2 - \dfrac{g}{m\omega_n}\dfrac{2\alpha^3\Omega^4}{\left(1+\alpha^2\Omega^2\right)^2} = 0 \\ 2\zeta\Omega - \dfrac{g}{m\omega_n}\dfrac{\alpha^2\Omega^3\left(1-\alpha^2\Omega^2\right)}{\left(1+\alpha^2\Omega^2\right)^2} = 0 \end{cases} \tag{5}$$

Solving (5) yields the critical frequency Ω_c and the critical gain g_c as:

$$\left(1 - \Omega_c^2\right)\left(1 - \alpha^2\Omega_c^2\right) = 4\alpha\zeta\Omega_c^2 \tag{6}$$

$$\frac{g_c}{c_{crit}} = \frac{\zeta\left(1-\alpha^2\Omega_c^2\right) + \left(1-\Omega_c^2\right)\alpha}{\alpha^2\Omega_c^2} \tag{7}$$

where $c_{crit} = 2m\omega_n$ is the critical damping. When the gain reaches the critical value g_c, at the critical frequency Ω_c, the transmissibility tends to infinity. The calculations in [6] showed that, if α is large, the critical frequency and critical gain are approximated as $1/\alpha$ and α, respectively. The instability occurs at low frequency and due to the phase-lead.

3 Frequency Range Selector

To reduce the unstable behavior without sacrificing the high gain, it is desired to produce a frequency-dependent gain. The low-gain is expected at low frequency. Intuitively, this can be done by adding more high-pass filters to the loop. However, the phase-lead due to high-pass filter make the unstable even worse. Base on the idea of [9, 10], this paper considers the following controller:

$$f_m(t) = \begin{bmatrix} -g\dot{x}(t) & \dot{x}^2(t) > \omega_s^2 x^2(t) \\ 0 & \text{otherwise} \end{bmatrix} \tag{8}$$

in which ω_s is a selector frequency. If the forcing frequency ω is larger than ω_s, over a period, the first condition of (8) is held more frequently, and conversely. This behavior is illustrated in Figs. 2 and 3. While the desired frequency-dependent behavior is obtained, there is no phase difference with the original active control force. Indeed, when $\dot{x}(t)$ reaches its extreme values, as seen from (8), the first condition is held, the modified control force $f_m(t)$ is in phase with the original one $f(t)$ $(= -g\dot{x}(t))$ for any excitation frequency.

Some properties of the selector can be drawn as:

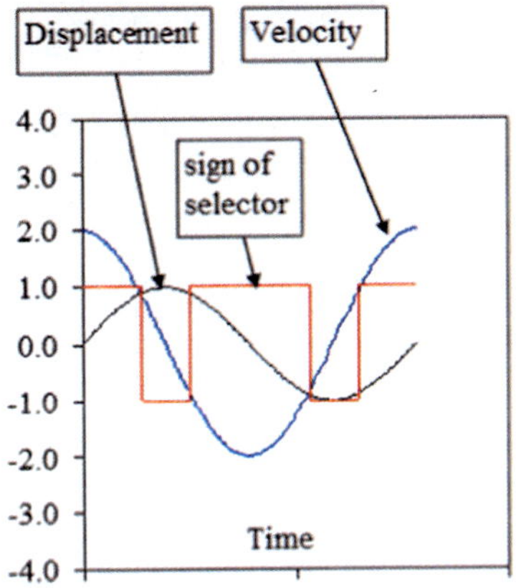

Fig. 2. Sign of selector over time when $\omega > \omega_s$

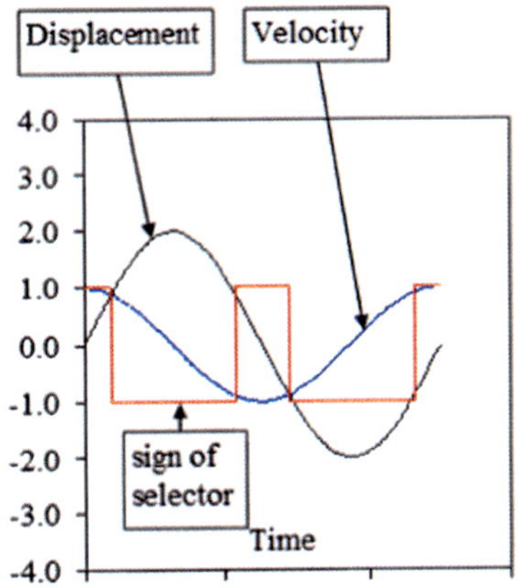

Fig. 3. Sign of selector over time when $\omega < \omega_s$

- The lower frequency keeps the second condition of (8) more frequently, which reduces the effective gain and then reduces the unstable effect due to the phase-lead.
- Two limit cases of ω_s, equal to 0 or ∞, correspond to the full active and passive ones, respectively.
- The frequency-dependent characteristic (8) gives the same effect as a high-pass filter but adds no more phase-lead to the control loop.

4 Numerical Verification

The differential Eq. (1) with the control force (8) is solved numerically. The frequency response is obtained by the procedure presented in [9]:

- The sinusoidal disturbance r with a predefined frequency and with unit amplitude is applied to the equation.

- For each predefined frequency, the simulation time is taken large enough (50 natural periods) to eliminate the transient responses.
- The discrete Fourier Transform is used to compute the amplitude of the output at the corresponding frequency.

Various cases of normalized cut off parameter α is considered in the simulations. The normalized gains are taken from (7) to reveal the unstable behavior. Figures 4, 5, 6 show the frequency responses of the transmissibility for varying α and ω_S.

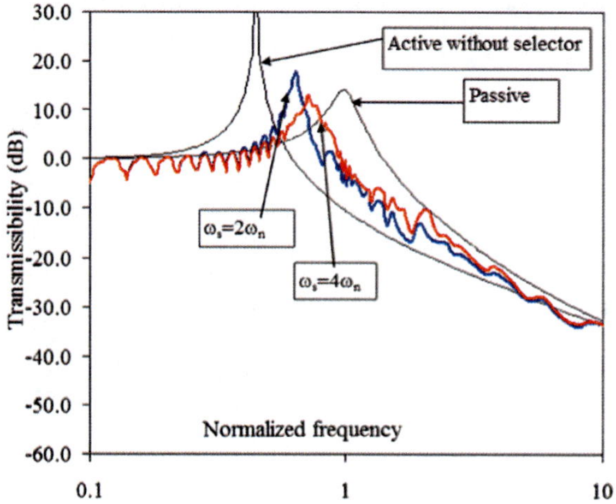

Fig. 4. Frequency response of transmissibility, $\alpha = 2$

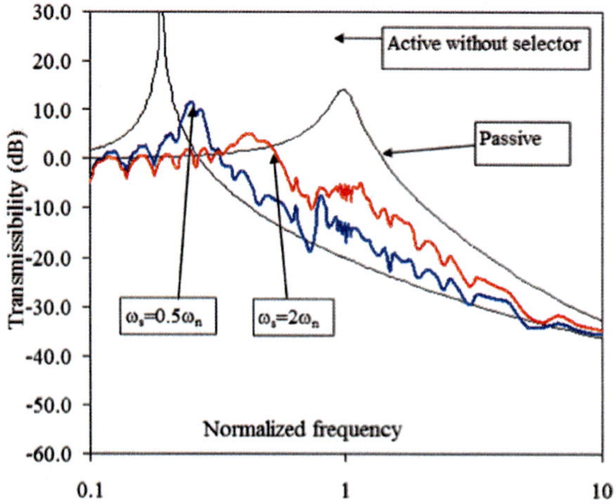

Fig. 5. Frequency response of transmissibility, $\alpha = 5$

It is seen that the larger frequency selector ω_s move the curve closer to that of the passive case, and also reduce the unstable behaviors in the low frequency range.

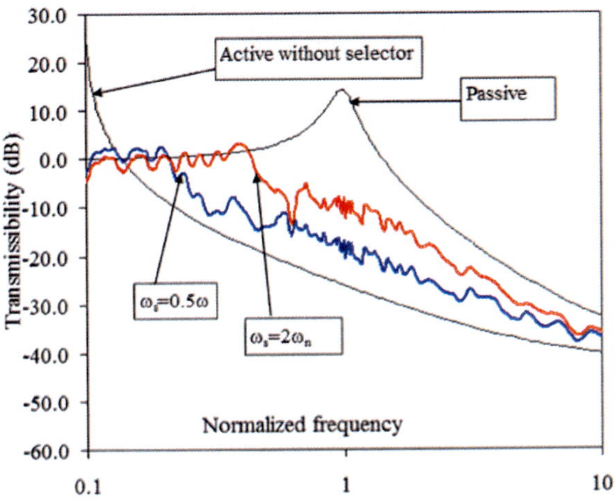

Fig. 6. Frequency response of transmissibility, $\alpha = 10$

This effect was discussed above. By choosing the appropriate selector frequency, the effectiveness of active high gain can be achieved at high frequency while the instability due to phase-lead at low frequency can be avoided.

5 Conclusion

This paper proposes a simple frequency-range-selector to modify controller gain of an active velocity feedback isolation. The proposed selector works like a filter to reduce the unstable behavior due to low-frequency phase-lead of the inherent high-pass filters. The filter shape is controlled by the selector frequency. The advantage of the approach is the in-phase behavior of the modified control force with the original one. Numerical results are presented to validate the effectiveness of the proposal.

Acknowledgement. This paper is funded by Vietnam Academy of Science and Technology under grant number "VAST01.04/21-22" and "NVCC03.08/21-21".

References

1. Rao, S.S.: Mechanical Vibrations. Prentice Hall, Hoboken (2010)
2. Preumont, A.: Vibration Control of Active Structures. Springer, Heidelberg (2011). https://doi.org/10.1007/978-94-007-2033-6
3. Kerber, F., Hurlebaus, S., Beadle, B.M., Stobener, U.: Control concepts for an active vibration isolation system. Mech. Syst. Signal Process. **21**, 3042–3059 (2007)
4. Balas, M.J.: Direct velocity feedback control of large space structures. J. Guid. Control **2**(3), 252–253 (1979)
5. Preumont, A.: An introduction to active vibration control. In: Preumont, A. (ed.) Responsive Systems for Active Vibration Control, pp. 1–42. Kluwer Academic Publishers, Dordrecht (2002)

6. Brennan, M.J., Ananthaganeshan, K.A., Elliott, S.J.: Instabilities due to instrumentation phase-lead and phase-lag in the feedback control of a simple vibrating system. J. Sound Vib. **304**, 466–478 (2007)
7. Saikumar, N., Sinha, R.K., Hossein Nia Kani, H.: Constant in gain lead in phase element - application in precision motion control. IEEE/ASME Trans. Mechatron. **24**(3), 1176–1185 (2019)
8. van den Eijnden, S.J.A.M., Knops, Y., Heertjes, M.F.: A hybrid integrator-gain based low-pass filter for nonlinear motion control. In: 2018 IEEE Conference on Control Technology and Applications, Copenhagen, Denmark, 21–24 August 2018 (2018)
9. Savaresi, S.M., Poussot-Vassal, C., Spelta, C., Sename, O., Dugard, L.: Semi-Active Suspension Control Design for Vehicles. Butterworth-Heinemann, Oxford (2010)
10. Savaresi, S., Spelta, C.: A single sensor control strategy for semi-active suspension. IEEE Trans. Control Syst.Technol. **17**(1), 143–152 (2009)

Biomechanics and Medical/Healthcare Devices

Performance Analysis of a Cable-Driven Ankle Assisting Device

Marco Ceccarelli[1], Matteo Russo[2](✉), and Margarita Lapteva[1]

[1] LARM2: Laboratory of Robot Mechatronics, Dept of Industrial Engineering, University of Rome Tor Vergata, 00133 Roma, Italy
marco.ceccarelli@uniroma2.eu, margarita.lapteva@students.uniroma2.eu

[2] Faculty of Engineering, University of Nottingham, Nottingham NG8 1BB, UK
matteo.russo@nottingham.ac.uk

Abstract. The paper presents a cable-driven parallel manipulator as a device for the motion assistance of the human ankle. The proposed design solution is discussed with design and operation requirements as from consideration of biomechanics and human-machine interaction. A performance evaluation is carried out to check the feasibility of the proposed design and to characterize its operation efficiency. Simulation results are discussed to give numerical estimation for a future prototype construction.

Keywords: Cable parallel manipulators · Motion assistance · Design · Ankle biomechanics · Simulation

1 Introduction

The ankle joint of the human body is susceptible to musculoskeletal and neurological injures. For muscular rehabilitation, patients perform training exercises with the supervision of a therapist, which are a time-consuming and repetitive process that requires constant attention from both patient and therapist. Moreover, training sessions are short, while a continuous muscular activity of the injured joint would shorten rehabilitation time and improve joint motion recovery. Robotic system such as exoskeletons can make the process less intensive for the therapist and increase the duration and frequency of training sessions, so that rehabilitation can be provided in a better and faster way.

According to this idea, many concepts of parallel rehabilitation robots have been presented in literature [1]. In [2], a parallel-platform-based robot is presented. This robot provides 3 rotational degrees of freedom (DOF) for the ankle with a good interaction between the robot, patient, and therapist. In [3] a wearable robot is proposed for post-stroke lower limb rehabilitation, but its usefulness is limited by large volume, heavy mass, and complicated structure. In [4], another wearable rehabilitation mechanism with 6 DOF is presented and evaluated, but it is heavy and difficult to move. Conversely, the portable robot in [5], based on spherical mechanism, provides 3 DOF and it can be used at home and fitness for ankle muscle relaxation.

N. V. Khang et al. (Eds.): ASIAN MMS 2021, MMS 113, pp. 619–627, 2022.
https://doi.org/10.1007/978-3-030-91892-7_59

A device which is based on a cable-driven manipulator, already applied to fields such as space, biomechanics, transporting, safety systems, and rescue operations [6], could solve the problems of safety [7], non-portability, complicated actuation, device encumbrance and weight, as reported in [8]. Thus, a cable-driven parallel manipulator can apply the necessary load to the foot by applying a constant tension to the cables, in order to rotate the ankle joint enough to activate muscles, but in a controlled motion to avoid further pain or injury.

In this paper, the novel device first proposed in [8] is analyzed and characterized with numerical simulation, in order to evaluate both its kinematic and static behavior.

2 Requirements for Ankle Motion Assistance

The ankle anatomy is shown in Fig. 1a, whereas its characteristic motion parameters are indicated in Fig. 1b.

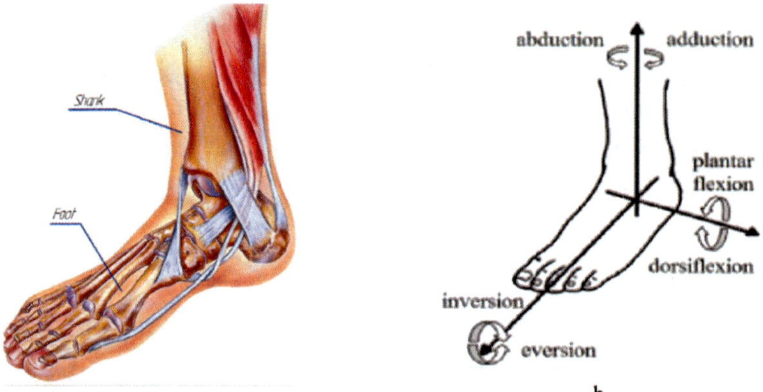

a. b.

Fig. 1. The foot-ankle system: a. Anatomy; b. A scheme of ankle motion.

As shown in Fig. 1b, the ankle joint allows different types of independent motions such as inversion-eversion, abduction-adduction, and plantarflexion-dorsiflexion around three perpendicular axes. For rehabilitation, operating these independent rotations is usually enough, without considering the limited (negligible) translations allowed by the ankle joint. An ankle rehabilitation device should provide angles within motion range limits, in order to avoid pain or injury for the patient and train effectively after being weakened by injury or inactivity. The limitations for human motion are assumed as from –8° to 8° for inversion-eversion and from –32° to 40° for plantarflexion-dorsiflexion [9]. Additional requirements for an ankle rehabilitation device also include parameters such as portability, weight, comfort, size, and user-friendliness, as explained in [10]. Finally, the user interface should provide the option to change the maximum loading condition of the ankle.

3 The Proposed Design

The proposed device, introduced in [8], can be represented by two platforms which are connected by four cables, as shown in Fig. 2. The upper platform, representing the shank, is fixed, and the lower platform, representing the foot, moves within the motion range by variating the lengths of four actuation cables. The kinematic behavior of the ankle is considered as a spherical joint, as proposed in [8].

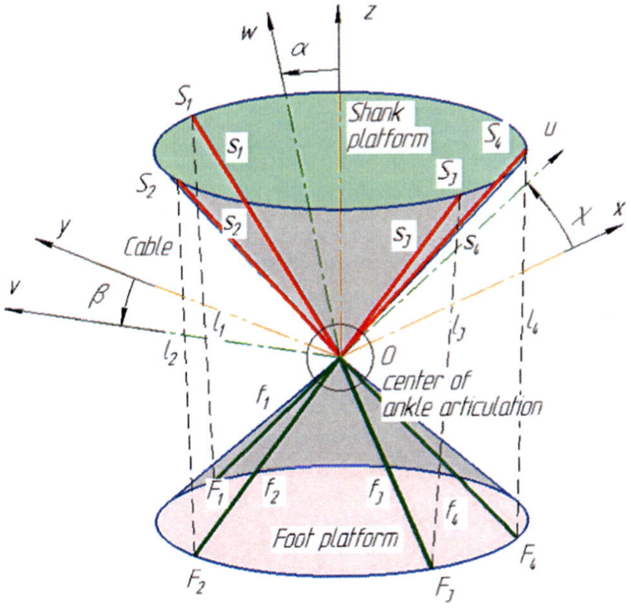

Fig. 2. A kinematic scheme of ankle with design parameters

The following hypotheses are assumed:

- All the cables are always kept in tension during a controlled motion,
- The attachment point of each cable is assumed as a spherical joint,
- The varying length of the cable can be modeled as an actuated cylindrical joint with a negligible axial deformation.

A CAD solution of the mechanical device is presented on the Fig. 3a. Servomotors actuate the lengths of cables l_1, l_2, l_3 and l_4 to provide a proper rotation of the foot platform.

In Fig. 2, the reference frame of the shank platform is O_{xyz} and the reference frame of foot is O_{uyw}. The ith cable's upper extremity is attached to point S_i of the shank platform and its lower point is attached by point F_i of the foot platform. A general relative motion of the ankle joint can be described by three consecutive rotations around three axes as in Fig. 1b. Thus, the orientation of the foot platform can be expressed as $\boldsymbol{R}^{SF} = \boldsymbol{R}_z(\alpha)\boldsymbol{R}_y(\beta)\boldsymbol{R}_x(\gamma)$.

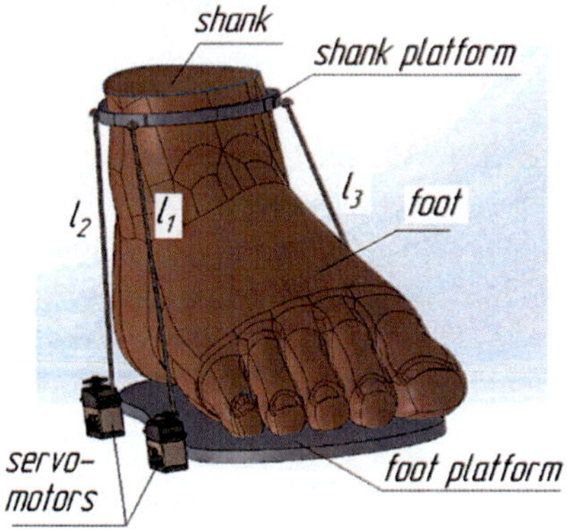

Fig. 3. A CAD solution of the mechanical design

The position of the ith cable attachment is $\boldsymbol{s}_i^s = \left(s_{ix}s_{iy}s_{iz}\right)^{\mathrm{T}}$ for the shank platform and $\boldsymbol{f}_i^f = \left(f_{xi}f_{yi}f_{zi}\right)^{\mathrm{T}}$ for the foot platform. The ankle pose can be described as $\boldsymbol{x} = (\alpha\ \beta\ \gamma)^{\mathrm{T}}$, where α, β and γ are the rotations around z-, y-, x- axes, as it was previously defined. The actuation can be expressed by $\boldsymbol{q} = (l_1\ l_2\ l_3\ l_4)^{\mathrm{T}}$, where l_i is the distance between points S_i and F_i of each cable, computed as magnitude of vectors l_i in Fig. 2. Vectoral loop-closure equations for each cable can be expressed as

$$\boldsymbol{OS}_i + \boldsymbol{S}_i\boldsymbol{F}_i + \boldsymbol{F}_i\boldsymbol{O} = 0 \tag{1}$$

The cable vector is computed from (1) as:

$$^{S}\boldsymbol{l}_i = {}^{SF}\boldsymbol{R}\,{}^{F}\boldsymbol{f}_i - {}^{S}\boldsymbol{s}_i \tag{2}$$

Then, by computing the scalar product of each side by itself, the length of each cable can be expressed as

$$l_i = \sqrt{{}^{S}\boldsymbol{s}_i^T\,{}^{S}\boldsymbol{s}_i + {}^{F}\boldsymbol{f}_i^T\,{}^{F}\boldsymbol{f}_i - 2\,{}^{S}\boldsymbol{s}_i^T\,{}^{SF}\boldsymbol{R}\,{}^{F}\boldsymbol{f}_i} \tag{3}$$

The device with a controlled smooth motion should provide no stress, pain, or damage to the patient. Because of the negligible moving mass and the slow speed, dynamics and inertial effects can be neglected so that performance analysis can be modelled by a static model where the weight of the foot and the platform are applied to the ankle joint itself. When external forces are considered, the actuation should balance them to maintain equilibrium. To this aim, reaction force F_A and reaction moment M_A of a reference point which is considered as a center of ankle actuation can be used to maintain to keep the whole system in static equilibrium. F_W and M_W are the external force and torque, in this case representing the foot's own weight. A force diagram of the moving platform is shown in Fig. 4.

Force equilibrium can be expressed as

$$\sum_{i=1}^{4} \boldsymbol{T}_i + \boldsymbol{F}_a + \boldsymbol{F}_w = 0 \tag{4}$$

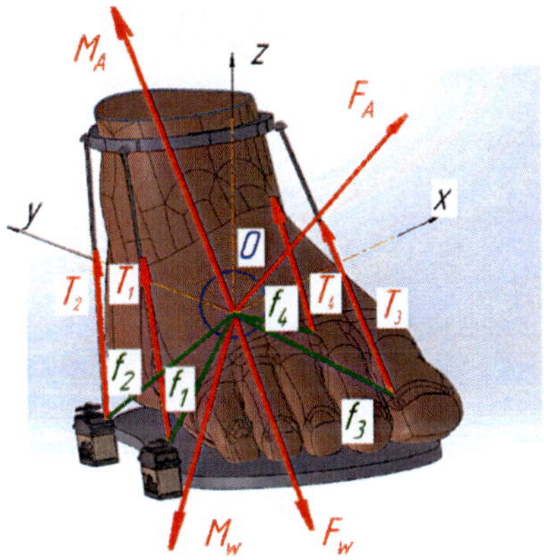

Fig. 4. A model for static equilibrium of the proposed solution.

Moment equilibrium can be expressed as

$$\sum_{i=1}^{4} \boldsymbol{f}_i \times \boldsymbol{T}_i + \boldsymbol{M}_a + \boldsymbol{M}_w = 0 \tag{5}$$

By using cable unit vector $\mathbf{u}_i$ the tension of cable can be expressed as $\boldsymbol{T}_i = -\boldsymbol{T}_i \boldsymbol{u}_i$. The actuation vector $\mathbf{T}_i$ with components (T_1, T_2, T_3, T_4) can be used to rewrite Eqs. (4) and (5) as

$$\boldsymbol{U}^{\mathrm{T}}\boldsymbol{T} - \boldsymbol{F}_A = \boldsymbol{F}_W \tag{6}$$

$$\boldsymbol{A}^{\mathrm{T}}\boldsymbol{T} - \boldsymbol{M}_A = \boldsymbol{M}_W \tag{7}$$

where $\boldsymbol{U}^{\mathrm{T}} = \left[\boldsymbol{u}_1\ \boldsymbol{u}_2\ \boldsymbol{u}_3\ \boldsymbol{u}_4\right]$ and $\boldsymbol{A}^{\mathrm{T}} = \left[\boldsymbol{f}_1 \times \boldsymbol{u}_1 \cdots \boldsymbol{f}_4 \times \boldsymbol{u}_4\right]$.

Reaction moment $\boldsymbol{M}_A$ is a null vector when the ankle joint is modelled as a spherical joint, and the full equilibrium can be formulated as

$$\begin{bmatrix} \boldsymbol{U}^{\mathrm{T}} & -\mathbf{I}_3 \\ \boldsymbol{A}^{\mathrm{T}} & 0_3 \end{bmatrix} \begin{pmatrix} \boldsymbol{T} \\ \boldsymbol{F}_A \end{pmatrix} = \begin{pmatrix} \boldsymbol{F}_W \\ \boldsymbol{M}_W \end{pmatrix} \tag{8}$$

to determine the actuation vector with its components when an input motion is given.

4 Results for Performance Analysis

In order to evaluate the performance of the proposed mechanism, a motion analysis has been performed with a multibody simulation tool on Solidworks. This motion analysis studies typical rehabilitation motion, which are characterized by the angular motion laws in Fig. 5. The results for reaction force of the reference point of the mechanism (on the ankle joint) are reported in Fig. 6, while Fig. 7 indicates the change in cable lengths. The simulation results have been computed by assuming an appropriate load onto cable

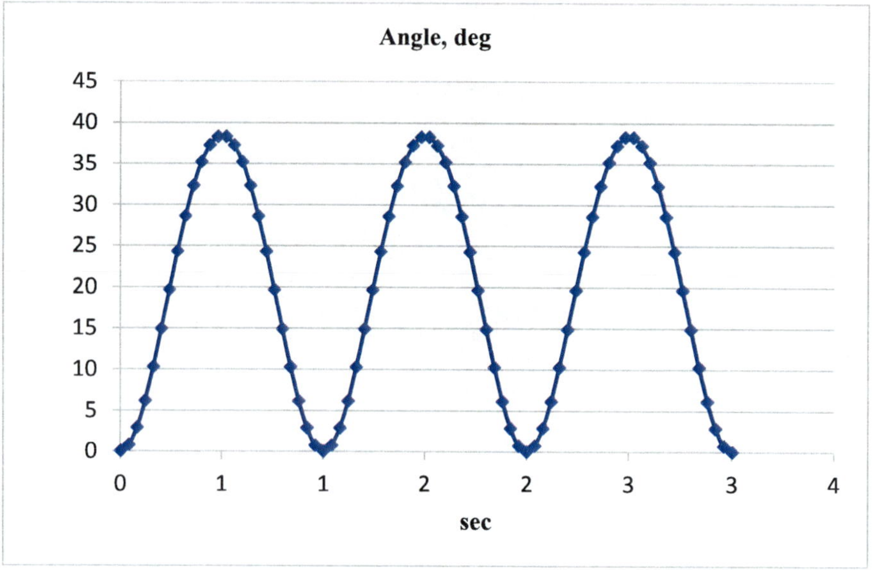

a.

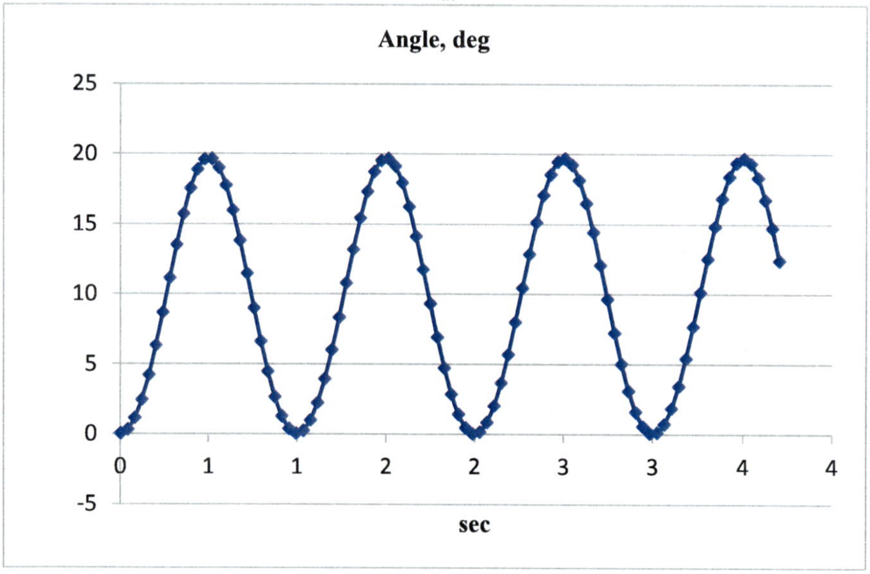

b.

Fig. 5. Evaluated change of angle for an harmonical load: a. Dorsiflexion/plantarflexion; b. Inversion/eversion.

as tension to obtain the desired motion. The foot mass has been estimated as 0,860 kg for an exercise of the ankle in the air. The criteria for defining the tension force onto cables required foot reaction force to be smaller than the maximum load sustainable by an injured foot, which is about 50 N [11].

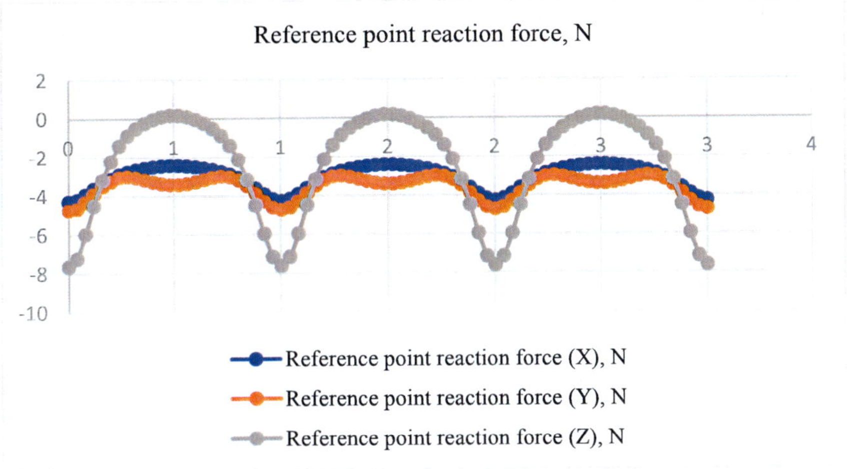

a.

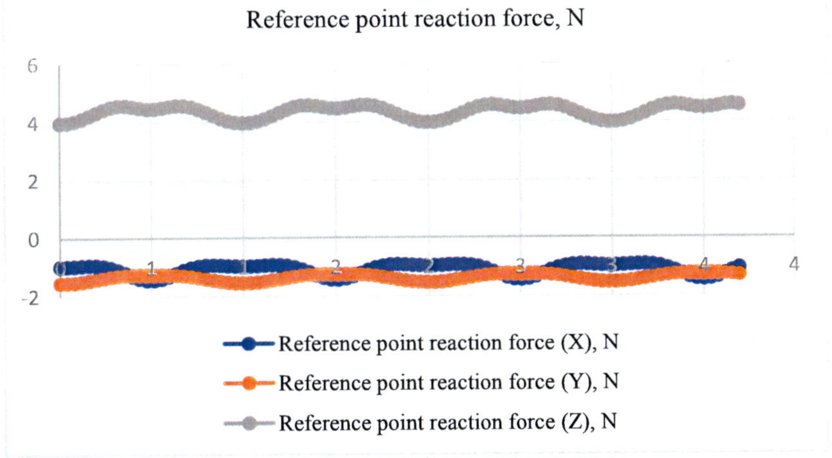

b.

Fig. 6. Reaction force of the reference point for harmonical load: a. Dorsiflexion/plantarflexion; b. Inversion/eversion.

The performance has shown that, with these input parameters, the reaction force of the reference joint (ankle) for the abduction-adduction movement quickly reaches 50 N, which is the upper limit for the load on an injured human ankle. Thus, for performing at the full exercising angle, an increase of cable tension would be required, but it cannot be provided as it would result into a load higher than the maximum one that injured human ankle could bear [11]. Within a safe limited abduction-adduction motion, the maximum tension observed in the actuation cables is 16 N.

Conversely, for the reported dorsiflexion-plantarflexion exercise, the maximum value of the reaction force is 10 N, which is acceptable for an injured ankle. The maximum applied input force on the cable is 18 N. In the reported inversion-eversion movement

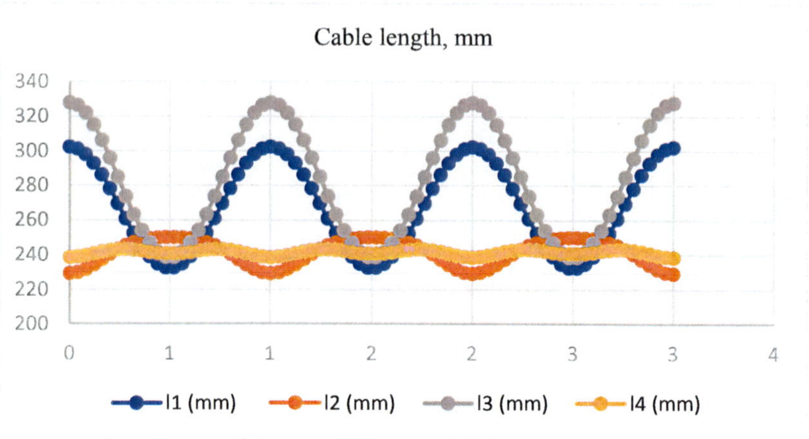

a.

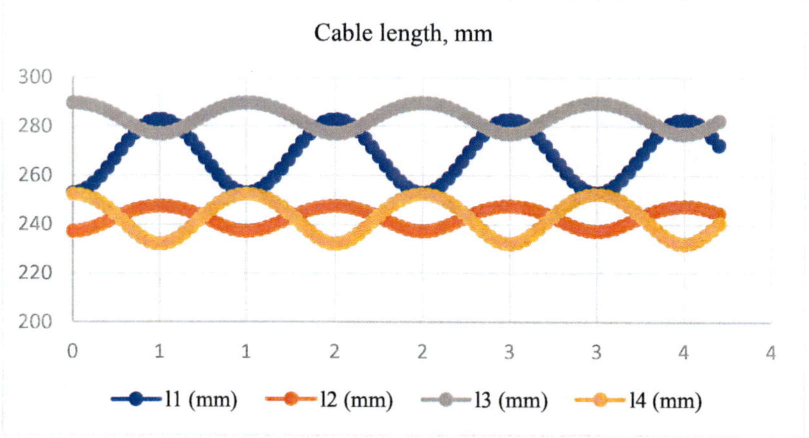

b.

Fig. 7. Cable lengths: a. Dorsiflexion/plantarflexion; b. Inversion/eversion/

the maximum value of the ankle joint reaction force is 5 N, and the applied input force on the cables to provide the rehabilitation motion is 5 N.

As shown in the results in Fig. 6, both these motion exercises result in a smooth distribution of the ankle reaction force in time, without sudden changes or discontinuities. This uniform rate is desirable in a rehabilitation system, as the lack of impulsive loads significantly reduces the chance of injuring the patient during exercising. Furthermore, this feature also reflects on control performance, as reported in Fig. 7, since the cable velocity and acceleration are limited throughout the entire operation and does not demand power peaks or high performance from the actuators. This enables the use of conventional servomotors to drive the system, reducing cost and improving the accessibility of the device to general patients.

Overall, the reported result prove the feasibility of the proposed cable-driven assistive device, by demonstrating its capability to guide a patient through most of the critical exercises for ankle rehabilitation. Furthermore, the numerical characterization can be used to define design requirement for a future prototype.

5 Conclusions

In this paper, a cable-driven parallel manipulator is proposed as a device for motion assistance of human ankle. The design solution is discussed with design and motion requirements that are derived by considering biomechanics and human-machine interaction. A performance evaluation is carried out to check the feasibility of the proposed design and to characterize its operation efficiency. Simulation results were obtained as a numerical estimation of device performance for a future prototype construction.

References

1. Dong, M., et al.: State of the art in parallel ankle rehabilitation robot: a systematic review. J. Neuroeng. Rehabil. **18**(1), 1–15 (2021). https://doi.org/10.1186/s12984-021-00845-z
2. Zhang, L., et al.: Design and workspace analysis of a parallel ankle rehabilitation robot (PARR). J. Healthcare Eng. (2019).https://doi.org/10.1155/2019/4164790
3. Zhang, X., Yue, Z., Wang, J.: Robotics in lower-limb rehabilitation after stroke. Behav. Neurol. (2017). https://doi.org/10.1155/2017/3731802
4. Zuo, S., Li, J., Dong, M., Zhou, X., Fan, W., Kong, Y.: Design and performance evaluation of a novel wearable parallel mechanism for ankle rehabilitation. Front. Neurorobot. **14**, 9 (2020)
5. Du, Y., Li, R., Li, D., Bai, S.: An ankle rehabilitation robot based on 3-RRS spherical parallel mechanism. Adv. Mech. Eng. **9**(8), 1687814017718112 (2017). https://doi.org/10.1177/1687814017718112
6. Qian, S., Zi, B., Shang, W.-W., Xu, Q.-S.: A review on cable-driven parallel robots. Chin. J. Mech. Eng. **31**(1), 1–11 (2018). https://doi.org/10.1186/s10033-018-0267-9
7. Cafolla, D., Russo, M., Carbone, G.: Design and validation of an inherently-safe cable-driven assisting device. Int. J. Mech. Control **19**(01), 23–32 (2018)
8. Russo, M., Ceccarelli, M.: Analysis of a wearable robotic system for ankle rehabilitation. Machines **8**(3), 48 (2020). https://doi.org/10.3390/machines8030048
9. Andrade, R.J., et al.: The potential role of sciatic nerve stiffness in the limitation of maximal ankle range of motion. Sci. Rep. **8**(1), 1 (2018). https://doi.org/10.1038/s41598-018-32873-6
10. Miao, Q, Zhang, M., Wang, C., Li, H.: Towards optimal platform-based robot design for ankle rehabilitation: the state of the art and future prospects. J. Healthcare Eng. (2018).https://doi.org/10.1155/2018/1534247
11. Huston, R.L.: Fundamentals of Biomechanics, p. 512. CRC Press, Boca Raton, FL (2013)

Fabrication of Airy, Lightweight Polymer Below-Elbow Cast by a Combination of 3D Scanning and 3D Printing

Chi-Vinh Ngo(✉), Quang Anh Nguyen, Nhan Le, Nguyen Lam Linh Le, and Quoc Hung Nguyen(✉)

Faculty of Engineering, Vietnamese-German University, Thu Dau Mot, Binh Duong, Viet Nam
{vinh.nc,hung.nq}@vgu.edu.vn

Abstract. In this research, an airy and lightweight polymer below-elbow cast is fabricated using 3D scanning and 3D printing. First, a 3D model of an arm and a 3D model of a below-elbow cast are created. Then, the effect of design patterns on the below-elbow casts' performance is investigated. Small samples with hole structures and with mesh-like structures are 3D printed to compare several properties such as airy performance, lightweight factor, printing time, and ultimate strength. The below-elbow cast with mesh-like structures showed better performance and then was fabricated.

Keywords: Airy and lightweight arm-cast · 3D scanning · 3D printing

1 Introduction

Injured forearms happen popularly, especially in children [1]. Conventional plaster casts, splints, and synthetic material casts have been utilized in rehabilitation to immobilize injured arms. A typical treatment period consists of implementing a cast and several follow-up visits for 4 to 6 weeks [2]. However, the conventional casts described above have still existed poor ventilation, improper fit, discomfort with itchy and smelly phenomena. Moreover, several unexpected cast complications may happen during rehabilitation, such as cutaneous diseases, bone injuries, and malunion [3]. Therefore, the current traditional casts still have limitations.

Recently, additive manufacturing or 3D printing has been significantly developed due to its unique characteristics such as weight saving, freedom of design and complexity, waste reduction, and product personalization. The 3D printing technology has provided enormous applications in manufacturing, medical, healthcare, and mechanical fields. In healthcare applications, the fabrications of 3D-printed orthopedic casts have been introduced [4–6]. Hui Lin et al. developed a rapid designing technique using several algorithms for creating the orthopedic cast [7]. Additionally, Yanjun Chen et al. employed finite element analysis and comparative clinical assessment to evaluate the 3D-printed orthopedic cast [8]. Recently, Pengcheng Lu et al. fabricated a 3D-printed cast for ankle fracture using MRI images for 3D modeling supported by finite element analysis and 3D

N. V. Khang et al. (Eds.): ASIAN MMS 2021, MMS 113, pp. 628–637, 2022.
https://doi.org/10.1007/978-3-030-91892-7_60

printing [9]. While some researchers have only developed algorithms and 3D modeling parts, others have fabricated the prototypes. The most common pattern design of these prototypes is the hole pattern. However, the effects of pattern designs on the ventilation and comfort of patients have not been studied yet.

In this research, a combination between 3D scanning and 3D printing is introduced to produce a below-elbow cast or an arm cast. The effects of two design patterns that compose a hole structure or a mesh-like structure on the product's performance are also investigated. Several factors such as airy performance, lightweight factor, printing time, and ultimate strength are evaluated. Then, the proper design of the below-elbow cast is suggested. A prototype with the proper design is finally produced with more advanced properties than conventional arm casts, such as good ventilation, lightweight, comfort and easy recycling. The obtained results could provide experts informative references when choosing a patient's pattern design of a below-elbow cast.

2 Experimental

2.1 Material

Commercial polylactic acid (PLA) filaments with a diameter of 1.75 mm were used to produce below-elbow casts because they are eco-friendly, plant-based, non-toxic, biocompatible, and biodegradable thermoplastic materials that can be applied in a profusion of healthcare applications [10]. Moreover, PLA filaments have also been widely utilized in the fused deposition modeling process, with several advantages such as higher strength and stiffness than acrylonitrile butadiene styrene (ABS) and nylon, low melting temperature, and minimal warping.

2.2 Fabrication Method

The fabrication method is shown in Fig. 1. First, an arm of a person who was assumed as a patient having a broken below-elbow part was scanned by an industrial 3D scanner (Go!Scan 50, Creaform), as shown in Fig. 1b. The scanner included three cameras. While the upper camera and the lower right camera were utilized to capture the deformation of the objects for making a point cloud, the lower left-right camera provided object information. Between the two bottom cameras, a light-structured emitter formed a set of four LEDs together with the three cameras. The scanner was connected to VXelements software through a computer to observe, modify, and export the scanning 3D file of the arm. Some white circle stickers that play a role as reference points were attached to the human arm to enhance the precision of the scanning process. In the Vxelements software, after setting up the resolution at 2.0, a "scan" button was clicked to start a scanning process. A colorful range from red to blue indicated distances between the 3D scanner and the object to alarm users for adjustment. During the scanner's control, the object should be in the middle of the color range, a green range to supply precise information to the software. After completing the scan of the whole object, a "stop scanning" button was clicked, and the removal of background function was chosen to delete unexpected information.

After completing the scanning process, the 3D file with ending type as STL was exported. Then, another software called Blender was used to create the CAD model of the below-elbow cast based on the 3D file of the arm exported in the VXelements software. After importing the STL file, a cover surface, which had the same shape as the 3D scanned file, was created using a shrinkwrap modifier and then smoothened using smooth and grab brushes. The surface was then extruded to form a thickness of the cast, and the 3D scanned model could be removed. Next, a pattern design (hole structure of mesh-like structure) on this surface was created. The 3D CAD models of the below-elbow cast were finally built. A Siemens NX software was then employed to retouch the CAD models of the below-elbow cast and add required components.

Next, the modified CAD model files were converted into STL printing files integrated with printing parameters, as shown in Table 1. Finally, STL files were inserted into a 3D printer (Dual-Feed, Builder) employing a fused deposition modeling technology to produce the below-elbow casts. The fused deposition modeling technology included several steps: loading of PLA filaments, PLA liquification, application of pressure to push melted PLA to a nozzle, extrusion, movement of an extruder to print layer by layer of the 3D object, bonding of the PLA to itself or the next PLA layer to form a 3D solid structure. Before the final below-elbow prototyped cast was produced, small samples with two different patterns (hole and mesh-like structures) were fabricated as the parameters in Table 1 to compare airy performances, lightweight, printing time, and ultimate strength properties.

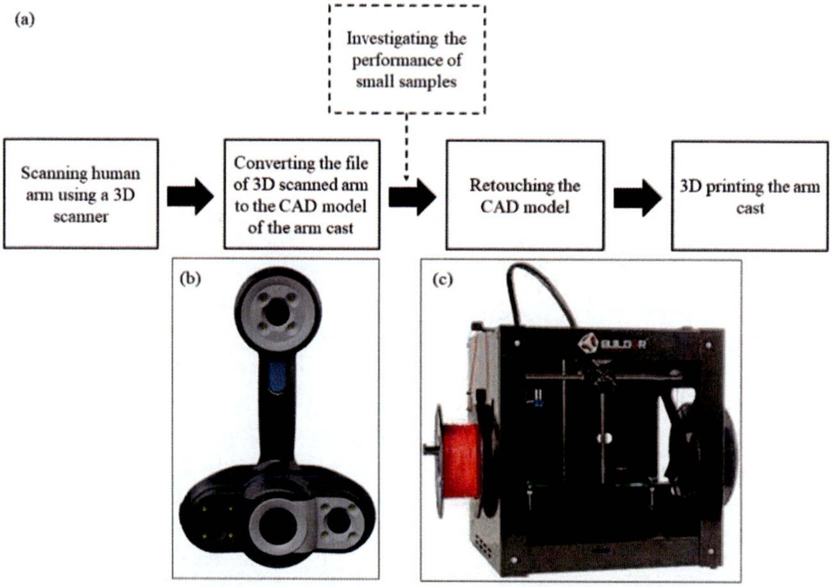

Fig. 1. Schematic diagram of the fabrication process combining 3D scanning and 3D printing.

Table 1. 3D printing parameters of the below-elbow cast.

Density	100%
Layer height	0.2 mm
Printing speed	80 mm/s
Operating temperature	210 °C
Shell thickness	0.8 mm
Bottom thickness	0.6 mm

2.3 Characterization

The performances of the printed samples with two different patterns were investigated to suggest a proper design for a below-elbow cast. Several criteria such as airy performance, weight, printing time, and ultimate strength were compared and then given scores on a scale from 1 to 5 varying from "not at all satisfied" to "extremely satisfied". The airy performance was evaluated based on survey questions, as shown in Table 2. Four people were asked to wear the small samples for 4 h and evaluated their feelings by answering the questionnaire.

Table 2. Human satisfaction questionnaire.

Questions	Rating scores				
	1	2	3	4	5
Q.1: Is the cast comfortable?	Highly uncomfortable	Uncomfortable	Occasionally irritating	Comfortable	Highly comfortable
Q.2: Is it enough airiness?	Extremely airless (itchy and smelly)	Stifling	Slightly stifling	Fairly airy	Highly airy

An electronic balance measured the weight of each patterned sample to compare the lightness. The printing time of each sample was shown in the 3D printer. The ultimate strengths of samples were also investigated by a tensile testing machine (AGX-Plus 50 KN, Shimadzu) with an applied force of 50 kN and a testing speed of 2.5 mm/min. The output data from the measurement includes the engineering stress and the engineering strain of the 3D printed sample, which were theoretically calculated by Eqs. (1) and (2), respectively.

$$s = \frac{F}{A_0} \tag{1}$$

$$e = \frac{L - L_0}{L_0} \tag{2}$$

where s = enginerring stress (MPa), F = applied force in the test (N), A_0 = original area of the sample (mm^2), e = engineering strain, L = length at any point during the elongation (mm), and L_0 = original gage length (mm).

3 Result and Discussion

3.1 Creation of Individual 3D Printable Arm-Cast Model

The results of creating arm-cast models are presented in Fig. 2. The 3D scanning supported to make a 3D below-elbow arm. This scanning approach can be applied to individual arms because each person has a different arm's size. Sizes of arm casts affect the protection of broken arms as well as patients' comfort during the recovery.

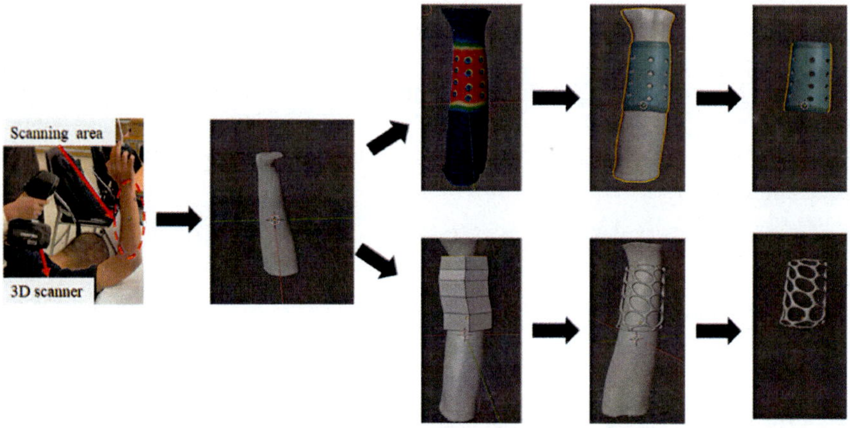

Fig. 2. Creation of 3D below-elbow cast model.

Different from typical scanning techniques such as the reputed Terrestrial laser scanners (TLS) [11] and the desktop scanners employing the triangulation principle [12], the 3D scanning in this research applied a positioning method to create the 3D below-elbow arm using highly accurate structured-light scanners. The method was demonstrated for achieving accurate results [13, 14]. The mechanism for this successful scanning approach is using good algorithms to analyze the obtained disordered scanning points. These disordered points were formed when scanning the human arm with different corners and edges, resulting in many images creating overlapping areas. Moreover, the movement from one position to another position during the scanning, which included a rotation and a translation, caused distortions in the lateral areas of the scanning human arm's images. Therefore, an iterative closet point (ICP) algorithm was used to optimize geometric transformations minimizing errors like the distortions [15, 16].

In addition, when the measurement noise is hard to investigate, this also can affect the precision of measurement. Therefore, a constrained minimization approach using Lagrange multipliers was used to evaluate each measurement point's maximum and minimum deviations [17], resulting in more precise information. The application of both algorithms explained the high measurement accuracy of structured-light scanning using the positioning method. On the other hand, the white circle stickers attached to the human arm helped better locate the expected scanning area positions.

After getting the 3D below-elbow arm, the two patterns of the cast (hole and mesh-like structures) were then designed to the 3D below-elbow arm model, following described steps in the fabrication part. The CAD models were finally created.

3.2 Effect of the Designed Patterns on the Performances of Below-Elbow Casts

Airy Performance: Four people wore the two patterns as shown in Fig. 3 and then provided their feedback as shown in Table 3. The obtained responses showed that the below-elbow cast with the mesh-like structure had higher airy performance than the hole structure. Most testers had the same opinion that the below-elbow cast with the mesh-like structure is comfortable to highly comfortable to users with scores of 4 or 5. On the other hand, most testers felt that the below-elbow casts with the hole structures, a typical pattern design in other researches, were extremely airless that caused itchy and smelly over time. The evaluation from the testers is essential for the prevention of itchy and smelly feelings during the treatment time of the patient with the broken arm.

Weight and Printing Time: The weight of the sample with the mesh-like structure was 88 g, which was lighter than the one with the hole structure (194 g). Moreover, the 3D printed samples with the two structures are much lighter than a conventional plaster cast (roughly 680 to 900 g). Therefore, the 3D printed product can satisfy the requirement for lightweight factors. The evaluated scores for the samples with the hole structure and the mesh-like structure can be 2 and 5, respectively. The sample with the mesh-like structure had a printing time of 8 h, while the one with the hole structure took around 11 h to complete the print. The evaluated scores can be 3 and 4 for the samples with the hole structure and the mesh-like structure, respectively.

Ultimate Strength: The ultimate strengths of the samples with hole structure and mesh-like structure are shown in Fig. 4. While the sample with a hole structure showed an ultimate strength at 19 MPa, the sample with a mesh-like structure had only the ultimate strength value at 9 MPa. The difference in ultimate strength may come from the sizes of the holes and the meshes. When the size of patterns becomes smaller, the samples' matrix area becomes larger, resulting in a stronger bonding between the PLA filaments. The mesh-like structures or the hole patterns in this research might be similarly considered as an infill density factor of a 3D-printed solid bulk. This assumption has a similar agreement that when the infill density reduced or the area lacking PLA matrix material increased, the ultimate strength of materials declined [18]. In this research, the mesh-like structures' dimensions were double those of the holes, resulting in a reduction of roughly 53%. Although other criteria showed that the mesh-like structure has more advantages than the hole structure, the sample with the hole structure had higher strength. Nevertheless, even the ultimate strength of the mesh-like structure was reduced, the ultimate strength values of both pattern designs were much greater than the one of plaster of Paris bandage, ranging from 2 to 6 MPA [19]. Therefore, the mesh-like structure can be effectively utilized in below-elbow cast applications because the plaster of Paris bandage is a common material to make conventional plaster casts. The evaluated scores based on comparing the ultimate strength of the plaster of Paris bandage can be considered as 5 and 3 for the samples with the hole structure and the mesh-like structure, respectively.

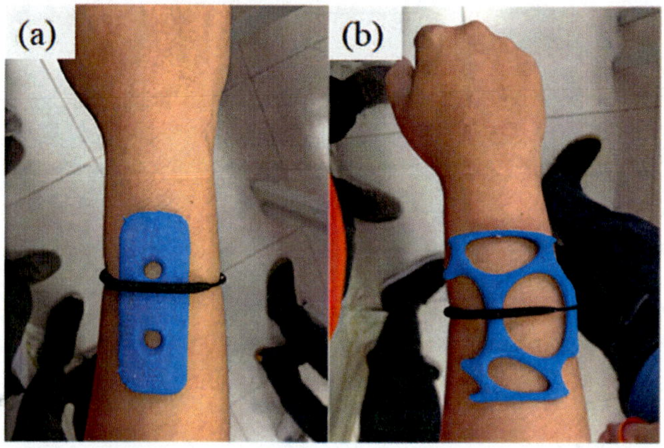

Fig. 3. Evaluation of airy performance for: (a) hole structure, and (b) mesh-like structure.

Table 3. Feedback from the testers for airy performance.

Tester	Hole structure			Mesh-like structure		
	Q.1	Q.2	Total scores	Q.1	Q.2	Total scores
A	4	1	2.5	3	5	4
B	1	1	1	3	3	3
C	2	1	1.5	4	4	4
D	3	3	3	5	5	5
Average scores			2			4

After investigating some criteria such as airy performance, weight, printing time, and ultimate strength of the 3D printed samples with the two different structures, the evaluated scores can be given in Table 4. With the obtained result, the mesh-like structure should be chosen to produce the below-elbow cast because it is airy, lightweight, and short fabrication time. Furthermore, although the strength of this structure is not high, it is sufficiently strong to support the broken arm of the patient, comparing to conventional plaster casts.

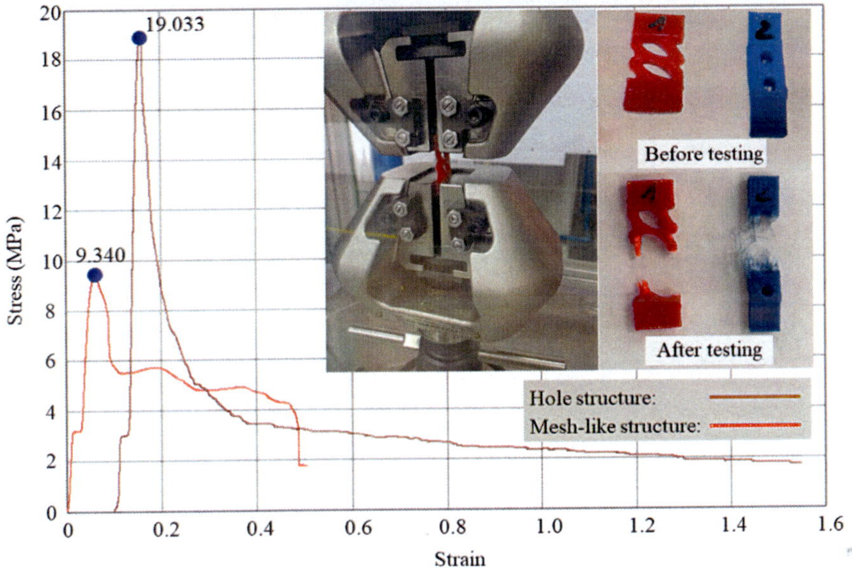

Fig. 4. Tensile test of the samples with hole structure and mesh-like structure.

Table 4. Performance evaluation of the two designed structures.

Criteria	Hole structure	Mesh-like structure
Airy performance	2	4
Weight	2	5
Printing time	3	4
Strength	5	3
Average scores	3	4

3.3 Creation of Individual 3D Printing Arm-Cast

To make the below-elbow cast facile to be 3D printed, the 3D below-elbow cast model with the mesh-like structure was slightly modified, as shown in Fig. 2. First, the model was divided into two parts (upper cast and lower cast), which can be fixed by the added locking parts and the support bars. The detailed modification was shown in Fig. 5a and b. The printing parts were then applied to the broken arm, as shown in Fig. 5c. After the broken arm is recovered, the polymer below-elbow cast can be easily recycled to fabricate other casts, showing feasibility for sustainable development.

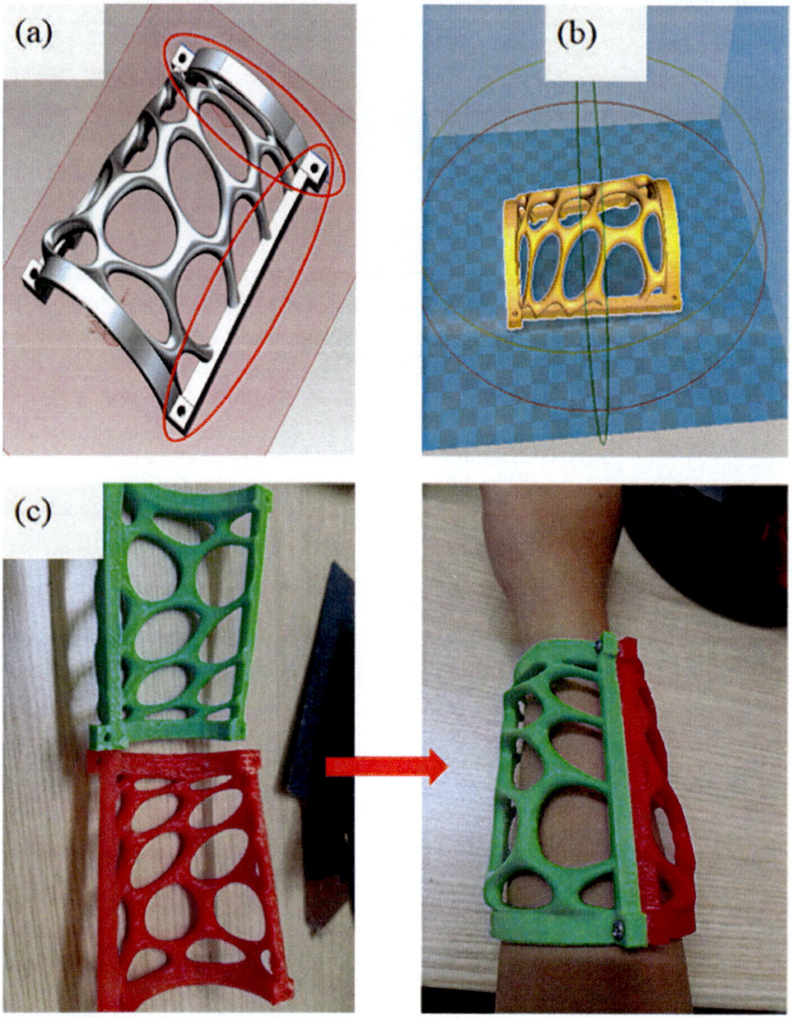

Fig. 5. (a) Modification of 3D arm-cast model, (b) Conversion of the 3D model into STL printing file, and (c) Implementation of the 3D printed arm cast.

4 Conclusion

The research has combined highly accurate structured-light 3D scanning using positioning method and fused deposition modeling 3D printing to design an individual below-elbow cast with more advanced properties than conventional plaster casts. Moreover, the effects of pattern designs on the product's performance, such as airy performance, lightweight factor, printing time, and tensile strength, were also investigated. The below-elbow cast with the mesh-like structure was chosen after the investigation. The obtained result brings feasibility to implement advanced techniques such as 3D scanning and 3D printing and choose patients' proper pattern designs in healthcare applications.

References

1. Chess, D.G., Hyndman, J.C., Leahey, J.L., Brown, D.C.S., Sinclair, A.M.: Short arm plaster cast for distal pediatric forearm fractures. J. Pediatr. Orthop. **14**(2), 211–213 (1994)
2. Egol, K.A., Walsh, M., Romo-Cardoso, S., Dorsky, S., Paksima, N.: Distal radial fractures in the elderly: operative compared with nonoperative treatment. J. Bone Joint Surg. **92**(9), 1851–1857 (2010)
3. Delasobera, B.E., Place, R., Howell, J., Davis, J.E.: Serious infectious complications related to extremity cast/splint placement in children. J. Emerg. Med. **41**(1), 47–50 (2011)
4. Fitzpatrick, A.P., Mohanned, M.I., Collins, P.K., Gibson, I.: Design of a patient specific, 3D printed arm cast. KnE Eng. **2**(1), 135–142 (2017)
5. Chen, Y.J., Lin, H., Zhang, X., Huang, W., Shi, L., Wang, D.: Application of 3D–printed and patient-specific cast for the treatment of distal radius fractures: initial experience. 3D Print. Med. **3**(1), 1–9 (2017)
6. Blaya, F., San Pedro, P., Silva, J.L., D'Amato, R., Heras, E.S., Juanes, J.A.: Design of an orthopedic product by using additive manufacturing technology: the arm splint. J. Med. Syst. **42**(3), 1–15 (2018)
7. Lin, H., Shi, L., Wang, D.: A rapid and intelligent designing technique for patient-specific and 3D-printed orthopedic cast. 3D Print. Med. **2**(1), 1–10 (2016)
8. Chen, Y., et al.: Application of 3D-printed orthopedic cast for the treatment of forearm fractures: finite element analysis and comparative clinical assessment. BioMed Res. Int. **2020** (2020). https://doi.org/10.1155/2020/9569530
9. Lu, P., et al.: Customized three-dimensional-printed orthopedic close contact casts for the treatment of stable ankle fractures: finite element analysis and a pilot study. ACS Omega **6**(4), 3418–3426 (2021)
10. DeStefano, V., Khan, S., Tabada, A.: Applications of PLA in modern medicine. Eng. Regen. **1**, 76–87 (2020)
11. Armesto-González, J., Riveiro-Rodríguez, B., González-Aguilera, D., Rivas-Brea, M.T.: Terrestrial laser scanning intensity data applied to damage detection for historical buildings. J. Archaeol. Sci. **37**(12), 3037–3047 (2010)
12. Perrone, R.V., Jr., Williams, J.L.: Dimensional accuracy and repeatability of the NextEngine laser scanner for use in osteology and forensic anthropology. J. Archaeol. Sci. Rep. **25**, 308–319 (2019)
13. Allard, P.H., Lavoie, J.A.: Differentiation of 3D scanners and their positioning method when applied to pipeline integrity. EITEP Institute (2014)
14. Liang, Y.B., Zhan, Q.M., Che, E.Z., Chen, M.W., Zhang, D.L.: Automatic registration of terrestrial laser scanning data using precisely located artificial planar targets. IEEE Geosci. Remote Sens. Lett. **11**(1), 69–73 (2013)
15. ISO, I., OIML, B.: Guide to the expression of uncertainty in measurement, Geneva, Switzerland, vol. 122, pp. 16–17 (1995)
16. Greenspan, M., Godin, G.: A nearest neighbor method for efficient ICP. In: Proceedings Third International Conference on 3-D Digital Imaging and Modeling, pp. 161–168. IEEE (2001)
17. Goulette, F.: Modélisation 3D automatique: outils de géométrie différentielle. Presses des MINES (1999)
18. Rismalia, M., Hidajat, S.C., Permana, I.G.R., Hadisujoto, B., Muslimin, M., Triawan, F.: Infill pattern and density effects on the tensile properties of 3D printed PLA material. J. Phys.: Conf. Ser. **1402**(4), 044041 (2019)
19. Adekola, F.A., Olosho, A.I., Adeleke, A.A., Eletta, O.A.A., Agaja, S.B.: Physico-mechanical assessment of plaster of Paris bandage produced from locally sourced materials. Bull. Mater. Sci. **42**(2), 1–6 (2019). https://doi.org/10.1007/s12034-019-1744-1

Research on the Characteristics of the Length, Breadth, and Diagonal Hand Dimensions of Male Students by Indirect Measurement Method

La Thi Ngoc Anh, Nguyen Thi Kim Cuc(✉), Pham Thanh Hien, Tran Thi Nga, Ta Van Doanh, and Tran Minh Hieu

Hanoi University of Science and Technology, No. 1 Dai Co Viet Street, Hanoi, Vietnam
cuc.nguyenthikim@hust.edu.vn

Abstract. Hands are essential parts of the human body. They support various functions such as grasping objects or doing complex activities in daily living and production. The direct measurement method has an impact force due to hand deformity, which causes errors and is also time-consuming. In contrast, the indirect measurement method does not cause hand deformity and saves time. This paper presents applying an indirect method to measure the hand's length, breadth and diagonal dimensions. The advantage of the indirect measuring system is that it can use the boundary determination algorithm to extract 29-point marks. Therefore, 16 dimensions were determined. The system was calibrated using a 25 mm ceramic gauge block at different positions in the measuring area. The average time was reduced from 144 s to 30 s. The hand length and handbreadth were proved to be critical dimensions. The results showed that hand dimensions at Hanoi University of Science and Technology (HUST) students were larger than that of Arab students.

Keywords: Hand dimensions system · Critical dimensions · Image analysis

1 Introduction

In Korea, Se Jin Park and his colleagues [1] studied Korean teenagers' hand dimensions by using indirect methods and direct methods. The direct method uses calipers and measuring tape, the indirect method uses a 3D scanner. Research results showed that there were no significant differences between the 3D scanning method and the traditional measurement method. However, the traditional measurement method is often time-consuming. The indirect method saves more time in measuring and be able to measure many dimensions on the hand.

Research by A. Yu and his colleagues [2] determined hand dimensions by using a 3D scanner model. The plaster hand was put on a rotatable disc and rotated 360° after each image capture. After ten rotations, the scanner obtained a 3D image of the hand, and the dimensions of the hand were extracted. There are also studies [3, 4] that also applied the 3D scanning method to determine hand size.

N. V. Khang et al. (Eds.): ASIAN MMS 2021, MMS 113, pp. 638–648, 2022.
https://doi.org/10.1007/978-3-030-91892-7_61

Although the indirect measurement by the 3D scanning method has complete hand images and greater accuracy, the installation of equipment and software set up are complicated and expensive, therefore, some other studies used the 2D image scanning method to extract hand size. This method has a much lower cost than the 3D scanning method, high repeatability, and relatively short measuring time. Moreover, the 2D image scanning method also ensures to minimize unnecessary movements of the hand during the capturing process [5, 6].

In Vietnam, there are not many studies regarding the hand size system. The results are not enough to provide hand dimensions to produce protective products. The researchers [7–10] use direct measurement to determine hand dimensions. Measuring tools used in these studies are compasses, calipers, cloth tape measurement. The curvature of the wrists, fingers, and joints can affect the results of hand measurements, especially when measurements are obtained by different people due to significant error. From the results of the overview research in the topic field, the research team aims to build an indirect measurement system by 2D image scanning method. Follow this, the system is applied to determine the hand dimensions of male students at HUST. The research was carried out in the following steps: determining the point marks, the length, and breadth dimensions; setting up an indirect measurement system for the required dimensions; conducting a survey measuring the hand size, and finally evaluating the results.

2 Methodology of Measurement

2.1 Determine Point Marks and Measuring Dimensions

"Cross-sectional study" method and direct measuring method were used to study male students' hand size characteristics. The point marks are defined by the corresponding skeletal and muscle anatomical landmarks [8]. In this study, 29-point marks were defined. There are three types of measurement: length, breadth, and diagonal dimension according to TCVN 5781:2009 [8] and Atlas of Vietnamese anthropometric in working age [7].

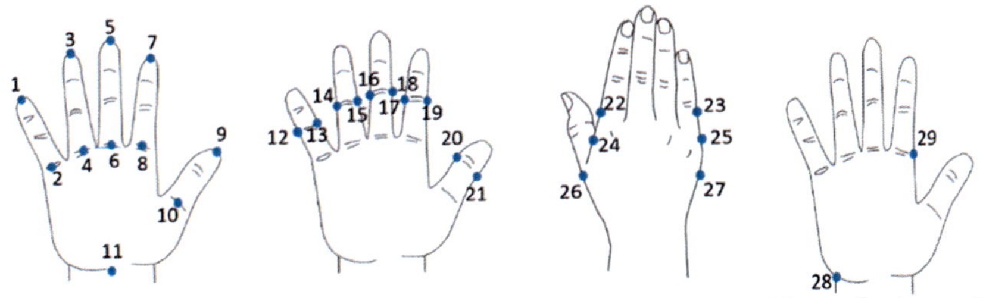

Fig. 1. Point marks of hand measurements

Figure 1 and Table 1 shows positions and a list of hand dimensions of this research.

Table 1. Description of hand measurements

Code	Dimension	Description image
D1	Little finger length	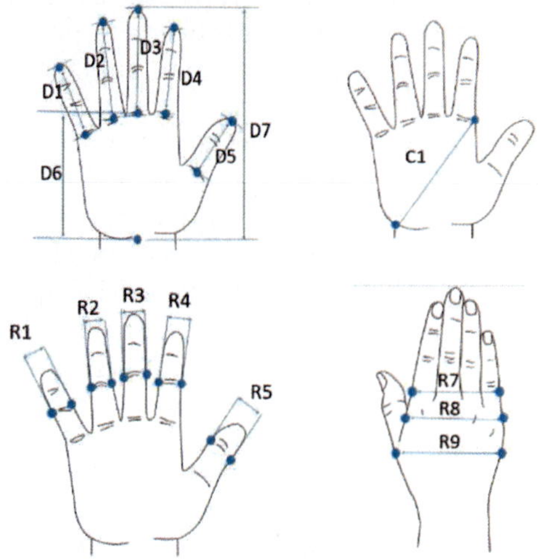
D2	Ring finger length	
D3	Middle finger length	
D4	Index finger length	
D5	Thumb length	
D6	Palm length	
D7	Hand length	
C1	Palm diagonal	
R1	Little finger breadth	
R2	Ring finger breadth	
R3	Middle finger breadth	
R4	Index finger breadth	
R5	Thumb breadth	
R7	Handbreadth at metacarpals	
R8	Palm breadth	
R9	Hand breadth	

2.2 Establish a System Measuring of Length, Breadth and Diagonal Dimensions

The working principle of the measuring system is presented in Fig. 2. The hands are put on the measuring area. The two dimensions (2D) images were captured by the camera and imported to a computer. Then, the computer analyzed data by established software. The boundary of hand images was detected, the landmarks and dimensions were extracted according to the description.

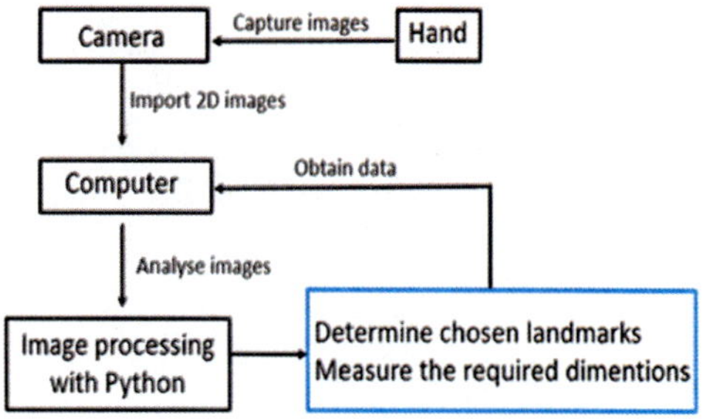

Fig. 2. The working principle of measuring system

The system should ensure safety requirements for human health; no effect of skin color during capturing processing; the shortest measuring time to minimize adverse hand movements during the measurement and the measurement space should be suitable with the hand shape.

To measure the hand dimensions, the limit of the horizontal and vertical measurement area ($w \times h$) is chosen to be larger than the maximum size of the hand according to the parameters measured by the direct measurement method. The depth (d) of space is calculated by the formula:

$$d = \frac{2.D.c.L^2}{f} \quad (1)$$

According to formula (1), for a defined measuring area ($w \times h$), the depth of the measuring area is determined by three factors: focal length f, aperture D, and distance of the camera to subject L. In optical systems, the scattering ring size c is arbitrary. In this study, the diameter of the scattering ring is chosen to be equal to pixel size. The software algorithm to extract the measure and hand size and the process of the measurement results was described as the algorithm diagram in Fig. 4. The computer analyzed the 2D images by the Threshold algorithm to detect the purple boundary of the hand (Fig. 3).

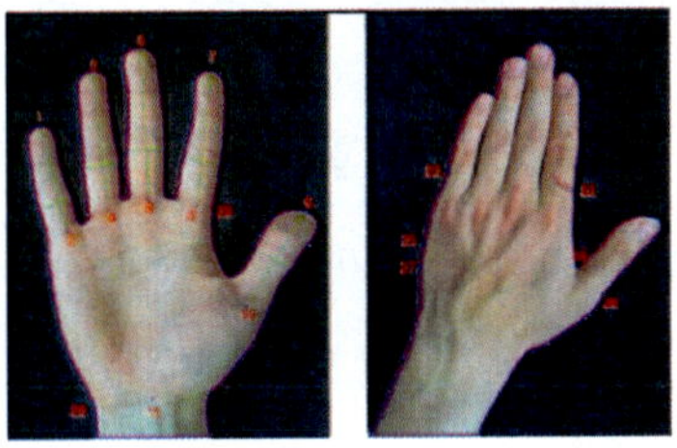

Fig. 3. Point marks and dimensions on 2D images by using Python

Based on the captured images, the land marks 1, 3, 5, 7 are determined to be the highest points on the finger arcs determined by the MIN/MAX algorithm; the landmarks 2, 4, 6, 8, 10, 11 are determined to be the midpoint of the line connecting the lowest points of the finger and hand arcs (minimum point); the land marks from 12 to 21 were at the positions with the largest breadth; the land marks from 22 to 29 were on the edges of the folds of the fingers or hands. The determined landmarks will be saved as data in the specified order with coordinates shown from $M_1(x_1, y_1)$ to $M_{29}(x_{29}, y_{29})$, respectively. The length and breadth dimensions will be calculated according to the Pythagorean theorem (2) through the coordinates of point marks in the coordinate system of the image sensor $M_i(x_i, y_i)$. After that, the obtained data is extracted to an excel sheet to provide the calculation of statistical features.

$$D_{1i} = \sqrt{(x_{1i} - x_{2i})^2 + (y_{1i} - y_{2i})^2} \quad (2)$$

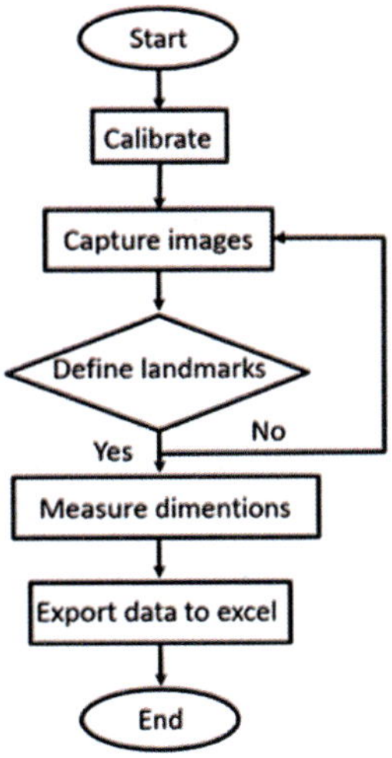

Fig. 4. Algorithm diagram

The process of extracting results is shown in Fig. 4. Before calibrating the measuring system, it is necessary to calibrate the intrinsic parameters of the camera including focal length, principal point, and pixel size. A is the camera intrinsic parameter matrix and was presented by:

$$A = \begin{bmatrix} f_u & \gamma & u_0 \\ 0 & f_v & v_0 \\ 0 & 0 & 1 \end{bmatrix} \tag{3}$$

Where: (u_0, v_0) is the principal point in the coordinate; defined as the image position where the optical axis intersects the image plane; f_u, f_v is the focal length of the camera along the u and v axes of the image plane in pixels; and γ is the parameter representing the tilt of the two image axes.

3 Experiment Results

3.1 Establishment of an Experimental System

The measuring system consists of the Microsoft LifeCam studio with a resolution of 1920 × 1080 at 30 fps and 75° angle of view, which was mounted with a custom bracket system. This system is connected with a high configuration computer (8 GB Ram, Intel Core i5-4460, CPU up to 3.20 GHz, a separate graphics card). The image of the measuring system is shown in Fig. 5.

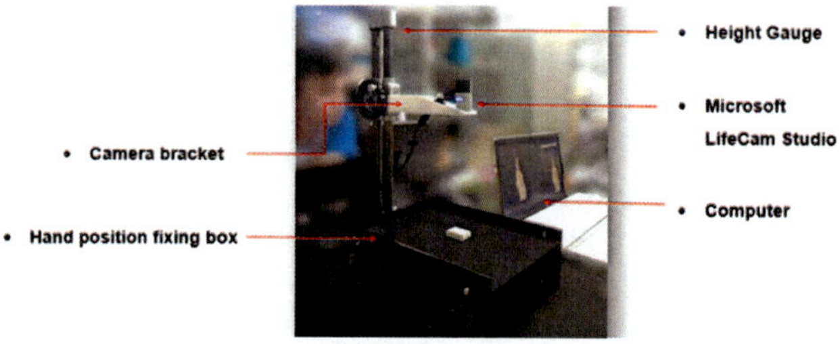

Fig. 5. Hand measuring system

The intrinsic parameter of the camera is calibrated with a checkerboard square. The checker size is 10 mm, the corner of the chessboard square is determined as the intersection of two black squares.

There are 10 × 7 corners of the corresponding size in the horizontal and vertical directions. For camera distortion measurements in the working range of $(h \times w \times d)$ (mm), each image was taken at the checkerboard position in the measurement area. Repeat these steps with 10 different chess square positions.

After capturing 10 groups of calibration images, the camera internal parameters were determined by the Eq. (3) as:

$$A = \begin{bmatrix} 1217.256 & 0 & 9483.6 \\ 0 & 1218.147 & 5351 \\ 0 & 0 & 1 \end{bmatrix}$$

The tilt value of γ is zero. The ratio of length and width $\frac{f_u}{f_v}$ is approximately equal to 1. Therefore, the pixels are square. The horizontal and vertical measurement area is 300 × 220 mm. The depth of space is calculated by the formula (1) $d = 105$ mm. In the coordinate system of the image sensor, the dimensions obtained by the software are in a pixel unit, therefore it is necessary to calculate the conversion ratio of the image size from pixel into mm. The Mitutoyo 25 mm ceramic gauge block is placed under the camera. The obtained value was 77 pixels. Thus, the conversion ratio from pixel to mm calculated is 25/77.

3.2 Evaluation of the Hand Measurement System by Indirect Measurement Method

The research team conducted a comparison between the direct measurement method and indirect measurement method to evaluate the newly established hand measurement system. The direct measurement method (*DM*) used a Mitutoyo electronic caliper with resolution: 0.01 mm and measuring range: 300 mm. The indirect measurement method (*IM*) used a hand size measurement system as described in Sect. 2.2.

Table 2. Comparison of hand dimensions measured by direct method and indirect method

Code	Dimension	M		SD		Cv	
		ID (mm)	DM (mm)	ID	DM	ID	DM
D1	Little finger length	6.29	6.37	0.10	0.19	1.60	3.06
D2	Ring finger length	7.95	7.77	0.14	0.25	1.78	3.17
D3	Middle finger length	8.54	8.35	0.19	0.27	2.20	3.27
D4	Index finger length	7.71	7.62	0.28	0.28	3.59	3.68
D5	Thumb length	6.04	6.29	0.31	0.33	5.15	5.10
D6	Palm length	11.44	11.29	0.39	0.34	3.45	2.97
D7	Hand length	19.67	19.63	0.47	0.43	2.41	2.20
R1	Little finger breadth	1.62	1.56	0.10	0.08	6.08	5.18
R2	Ring finger breadth	1.82	1.78	0.09	0.05	4.69	2.66
R3	Middle finger breadth	1.93	1.93	0.08	0.06	4.00	2.95
R4	Index finger breadth	1.92	1.92	0.10	0.09	5.21	4.93
R5	Thumb breadth	2.13	2.06	0.14	0.10	6.37	4.99
R7	Handbreadth at metacarpals	7.54	7.26	0.17	0.23	2.20	3.20
R8	Palm breadth	9.03	8.82	0.34	0.23	3.79	2.66
R9	Handbreadth	11.03	10.79	0.44	0.16	4.03	1.49
C1	Palm diagonal	12.71	12.52	0.56	0.38	4.41	3.04

This study measured 16 length, breadth, diagonal dimensions for 10 students with a height from 1.68 ÷ 1.79 m and weight from 50 ÷ 75 kg according to TCVN:2009 by both methods. Table 2 shows the measurement results of two methods of measuring direct and non-contact. The results show that the values measured by the direct measurement method are smaller than those of the indirect method. This proved that in the measuring process, a caliper's jaws have exerted a force on the measuring points on the hand, leading to change in position and reduce the measured values. On the other hand, the Standard Deviation (*SD*) and coefficient variation (*Cv*) values of the two methods were almost equal. To demonstrate the superiority of the indirect measuring method, the study performed a time-recording of the measurement. The results are obtained after 30 s. Each dimension measured by the direct measurement method takes 9 s. Total time to measure 16 dimensions is $16 \times 9 = 144$ s. Thus, regarding measuring time, the indirect measurement method saves up to 80% of the time compared to the direct measurement method.

The indirect measuring system after calibration by the 25 mm ceramic gauge block single and being compared with the direct measurement method, it was found that the values of the two measurements differ by no more than 3 mm, the standard deviation (*SD*) and the variation coefficient (*Cv*) are not significantly different. On the other hand, the measurement time of the indirect hand measurement system is significantly reduced from 144 s to 30 s. Thus, this system was fully qualified to be applied to the study of the hand length, hand breadth and palm diagonal dimensions characteristics.

3.3 The Study of Hand Length, Breadth, and Palm Diagonal of Male Students

Characteristics of Hand Dimensions: The sample size [11] is calculated based on the formula:

$$m = \frac{t \cdot SD}{\sqrt{n}} \Rightarrow n = \frac{t^2 \cdot SD^2}{m^2} \tag{4}$$

Where: n is the sample size; with probability $p = 95\%$; SD is the standard deviation. We selected the standard deviation $SD = 5$ cm; the error $m = 1\%$. It follows that $n = 96$. In fact, 96 male HUST students with height 1.68 ÷ 1.79 m, bodyweight from 50 ÷ 75 kg according to TCVN:2009 were randomly selected to conduct the survey. They have anthropometric characteristics of the hand representatively and suit to use in the research process. Statistical features of male student hand dimensions were calculated by Excel and SPSS 22.0 software. Table 3 presents the descriptive statistics of 16 hand dimensions for male HUST students.

Compared to the Hand Size Characteristics of Male Arab Students. The results in this paper were compared to the hand size characteristics of male Arab students [12] shown in Table 4.

The data of Table 4 shows that the length and breadth dimensions of the hands and fingers of Vietnamese students are larger than that of Arab students. There is no significant difference between the standard deviation of HUST students' hand size and that of Arab students' hand size. The coefficient variation of the data set obtained in this

Table 3. Statistical characteristics of 16 hand dimensions

Code	Dimension	Value (cm)						
		M	Me	Mo	Min	Max	SD	Cv
D1	Little finger length	5.86	5.86	5.50	5.06	6.95	0.42	7.16
D2	Ring finger length	7.61	7.60	7.55	6.82	8.52	0.36	4.83
D3	Middle finger length	8.22	8.23	7.78	7.39	9.30	0.40	4.86
D4	Index finger length	7.34	7.34	6.67	6.58	8.25	0.42	5.78
D5	Thumb length	5.89	5.86	6.37	5.22	6.58	0.41	6.89
D6	Palm length	11.37	11.37	10.39	10.39	12.43	0.48	4.25
D7	Hand length	19.30	19.31	19.74	17.78	20.61	0.71	3.66
R1	Little finger breadth	1.66	1.66	1.46	1.43	1.89	0.11	6.38
R2	Ring finger breadth	1.84	1.84	1.82	1.61	2.05	0.09	4.86
R3	Middle finger breadth	1.95	1.94	1.91	1.74	2.16	0.09	4.52
R4	Index finger breadth	1.93	1.94	1.91	1.70	2.24	0.10	4.98
R5	Thumb breadth	2.19	2.19	2.09	1.91	2.48	0.12	5.71
R7	Handbreadth at metacarpals	7.56	7.53	7.65	6.86	8.41	0.32	4.18
R8	Palm breadth	9.10	9.09	8.94	7.94	9.98	0.45	4.99
R9	Handbreadth	10.97	10.97	10.70	9.69	11.94	0.51	4.72
C1	Palm diagonal	12.80	12.85	12.04	10.59	14.06	0.75	5.89

study is lower than the coefficient of variation of the study [12]. This shows that the obtained data in the study of the authors have less variation than the mean.

Determine the Critical *Dimensions.* The critical dimension is the most significant dimension in the range of anthropometric dimensions and closely correlates with other dimensions in the same plane. There are many different methods to prove an anthropometric dimension is a critical dimension. In this study, the critical dimension is proven as follows [11]: the mean value (M), the median (Me), and the mode (Mo) are nearly equal; The skewness (SK) is in the range of (−1.1); The statistical distribution is bell-shaped; The significance level (Sig.) in the Kolmogorov-Smirnov test is greater than 0.05; The Normal Q-Q plot is a linear relationship.

Figures 6, 7, 8 and 9 indicated that the frequency graphs of the hand length and hand breadth were bell-shaped, asymptotic to the horizontal axis, and close to the graph of the standard normal distribution. The results in Table 5 showed that the empirical distribution of the hand length and hand breadth dimensions is suitable with the normal distribution. This allows concluding that hand length and hand breadth are the critical dimensions.

Table 4. Comparison of hand size characteristic of Vietnamese and Arab male students

Code	Dimension	M (cm)		SD (cm)		Cv	
		Vietnamese	Arab	Vietnamese	Arab	Vietnamese	Arab
D1	Little finger length	5.86	5.44	0.42	0.40	7.16	7.35
D2	Ring finger length	7.61	7.32	0.36	0.47	4.83	6.42
D3	Middle finger length	8.22	7.78	0.40	0.48	4.86	6.17
D4	Index finger length	7.34	6.82	0.42	0.46	5.78	6.74
D5	Thumb length	5.89	5.35	0.41	0.46	6.89	8.60
D6	Palm length	11.37	9.80	0.48	0.60	4.25	6.12
D7	Hand length	19.30	17.87	0.71	0.78	3.66	4.36
R1	Little finger breadth	1.66	1.45	0.11	0.09	6.38	6.21
R2	Ring finger breadth	1.84	1.68	0.09	0.09	4.86	5.36
R3	Middle finger breadth	1.95	1.83	0.09	0.09	4.52	4.92
R4	Index finger breadth	1.93	1.81	0.10	0.09	4.98	4.97
R5	Thumb breadth	2.19	1.89	0.12	0.12	5.71	6.35
R7	Handbreadth at metacarpals	7.56	7.67	0.32	0.35	4.18	4.56
R8	Palm breadth	9.10	–	0.45	–	4.99	–
R9	Handbreadth	10.97	–	0.51	–	4.72	–
C1	Palm diagonal	12.80	–	0.75	–	5.89	–

Table 5. Statistical parameters demonstrating the primary dimensions

Code	Dimension	M	Me	Mo	.Sig	SK
D7	Hand length	19.30	19.31	19.74	0.20	−0.34
R9	Hand breadth	7.56	7.53	7.65	0.20	−0.46

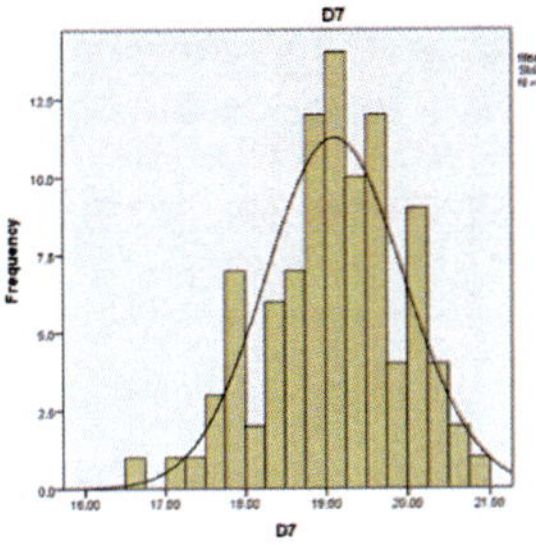

Fig. 6. The statistical distribution of D7 was bell shaped

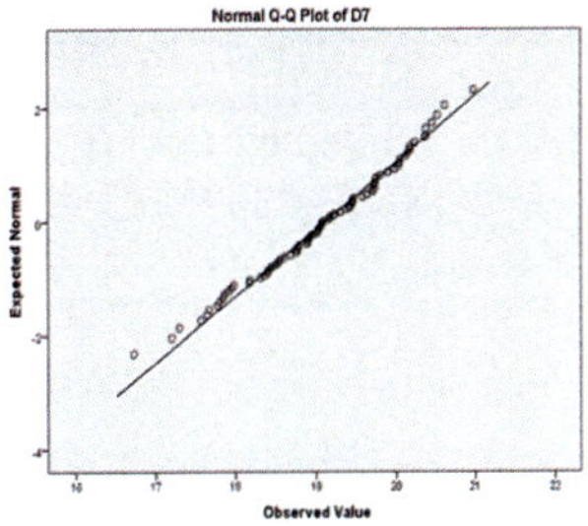

Fig. 7. The normal Q-Q plot of D7 was linear relationship

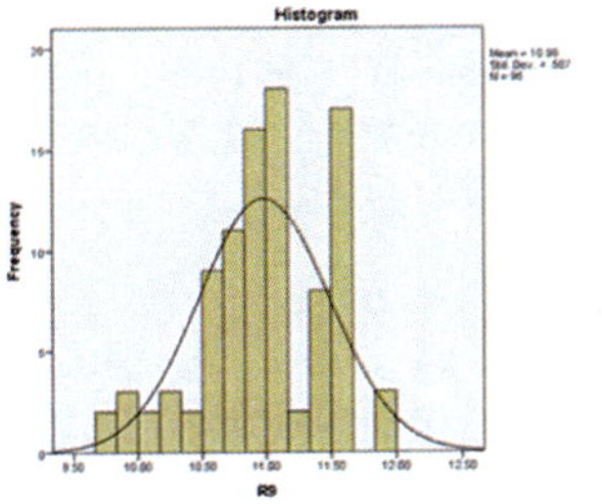

Fig. 8. The statistical distribution of R9 was bell shaped

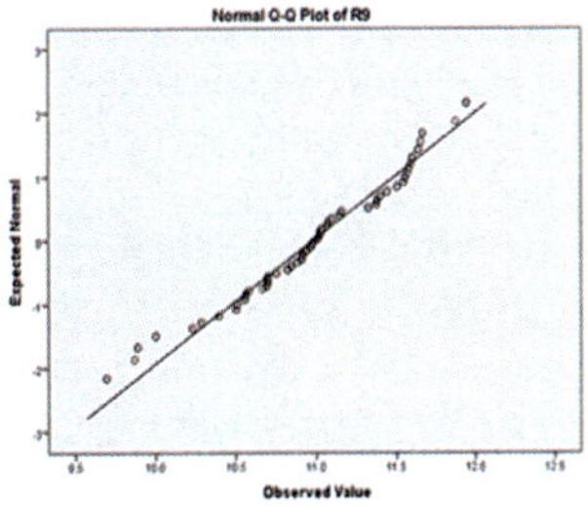

Fig. 9. The normal Q-Q plot of R9 was linear relationship

4 Conclusions

The study aimed to research the hand sizes of male students at Hanoi University of Science and Technology (HUST). By using an indirect measurement method, it has determined 29-mark points to evaluate 16 dimensions including length, breadth and diagonal of a hand. The research has established a reliable indirect measurement system to estimate the dimensions of a hand. The time needed for measuring a hand has decreased from 144 s to 30 s. It is initially shown that this method is more convenient than the direct measuring one, especially when estimating a hand having hard ascertained dimensions. Dimensions of hand length and hand breadth are the primary criteria to build the hand-sizing system and glove design. The average hand dimensions of male students at HUST are larger than that of Arabic students.

Acknowledgements. This research is funded by Hanoi University of Science and Technology (HUST) under project number T2021-PC-043.

References

1. Park, S.J., Min, S.N., Lee, H., Subramaniyam, M., Ahn, S.J.: 3D Hand anthropometry of Korean Teenager's and comparison with manual method. In: Stephanidis, C. (Ed.): HCII 2014 Posters, Part II, CCIS, vol. 435, pp. 491–495 (2014). https://doi.org/10.1007/978-3-319-07854-0_85
2. Yu, A., Yick, K.L., Ng, S.P., Yip, J.: 2D and 3D anatomical analyses of hand dimensions for custom-made gloves. Appl. Ergon. **44**, 381–392 (2013)
3. Yu, F., Zeng, L., Pan, D., Sui, X., Tang, J.: Evaluating the accuracy of hand models obtained from two 3D scanning techniques. Sci. Rep. **10**, 1–10 (2020)
4. González, A.G., Salgado, D.R., García Moruno, L., Sánchez Ríos, A.: An ergonomic customized-tool handle design for precision tools using additive manufacturing: a case study
5. Anuar, F.S., Soni, G.: A study of anthropometric measurement of hand length and their correlation with stature in university students. Malays. J. Forensic Sci. **8**, 32–38 (2018)
6. Han, H.S., Park, C.K.: Automatic Hand measurement system from 2D hand image for customized glove production. Fash. Text. Res. J. **18**(4), 468–476 (2016)
7. NILP: Atlas of anthropology for Vietnamese in working age - static anthropometric data (1986)
8. TCVN 5781 – 2009: Method of human body measuring
9. Ngan, P.T.B., Huyen, N.T., Hong, N.D., et al.: Basic of construction method of anthropometric Atlas of static and dynamic ergonomics for Vietnamese in the working age within the period 2017–2019. J. Saf.- Health Work. Environ. (4,5,6/2018), 21–30 (2019)
10. Anh, L.T.N.: Study on hand shape characteristics of male construction workers aged 20 to 30 years. J. Saf. – Health Work. Environ. (4,5,6/2020)
11. Khoa, N.Đ.: Statistic method applying in biology. General University (1975)
12. Mansour, M.A.A.: Hand anthropometric data for Saudi Arabia engineering students of aged 20–26 years at King Khalid University. Am. J. Eng. Appl. Sci. **9**(4), 877–888 (2016)

Mechanical Evaluation of the Large Cranial Implant Using Finite Elements Method

Nguyen Thi Kim Cuc(✉), Phan Dinh Hung, Bui Minh Duc, and Nguyen Hoang Anh

Hanoi University of Science and Technology, No. 1 Dai Co Viet Street, Hanoi, Vietnam
cuc.nguyenthikim@hust.edu.vn

Abstract. Skull defect reconstruction, used in Computer-Aided Design (CAD) and 3D printing software, helps evaluate durability, increases surgical accuracy, and reduces surgical costs and risks. The large cranial implants located on the two halves of the skull are asymmetrical and consist of fixation parts. The research aims to propose to investigate the numerical simulation of large cranial implants in 3D for Polyether-ether-ketone (PEEK). Using the finite element method, the asymmetric defect skull model was simulated, under a static load of 50 N, and intracranial pressure. The results of two investigated structures (perforated structure (PS) and non-perforated structure (NPS) are shown through deformation, Von-Mises stress, and Von-Mises strain. The designed of the PS cranial implant reduces the amount of fabrication material and meets the required mechanical characterizations.

Keywords: Finite elements method · Cranial implant · 3D printing · PEEK

1 Introduction

Skull defects, which come from many causes such as congenital, traumatic, brain tumors are a common problem in surgery [1]. Cranial reconstruction surgery is a big challenge because it involves the brain, eyes, and nerves [2]. There are many methods to reconstruct the skull, and the best method is using an autogenous bone. Bone grafts are associated with low cost and minimal risk. However, their use is limited by the limited availability of donor graft material, donor-specific disease, difficulty in determining graft formation, increased operative time and postoperative risks, as well as risks of infection, fragmentation, and resorption. Autologous bone grafts can work well if the defect is small and the contours are simple; however, it is challenging in the presence of larger and more complex shape defects. Therefore, it is necessary to have alternative materials in large defects surgery. Specifically, the design of the cranial implant must ensure two factors: the contour of the cranial implant is suitable for the defect of each patient and the material is biocompatible [3].

The bone tissue implanting method with alternative materials such as titanium alloys, Polymethyl methacrylate (PMMA), Polyether-ether-ketone (PEEK), [4, 5], etc., have been researched and the most suitable material to be used remains controversial. In material research [6], each material has its own properties and advantages. Previously,

N. V. Khang et al. (Eds.): ASIAN MMS 2021, MMS 113, pp. 649–658, 2022.
https://doi.org/10.1007/978-3-030-91892-7_62

titanium alloy was a popular material in cranial surgery by helping bones grow on implants, a low infection rate, and high strength. Titan alloy has a larger Young's modulus than natural bone, thus reducing stress shielding. However, the use of titanium alloy is associated with implant exposure, infection, magnetic resonance imaging, and the potential of overlying soft tissue coverage for poor long-term adaptability [7]. Several bone replacement studies using metal implants in other parts of the human body have also been studied [8–10]. PMMA is a thermoplastic approved for medical use. It has high biocompatibility. However, pure PMMA contains irritants [11]. Recently, due to its combination of strength and rigidity, PEEK has been known as an alternative to existing implantable materials. PEEK has high biocompatibility, minimizes stress shielding, and can be fabricated by 3D printing.

Advances in segmentation software have made it increasingly easy to extract the surface of structures of interest automatically or semiautomatically from 3D medical imaging data [12, 13]. After obtaining the 3D model, it is possible to use the de-signed defect compensation for the fabrication of implants that conform to the physical and mechanical requirements. The craniofacial region presents special problems for tissue engineering. The stresses and strains that the engineered tissues will experience have been mostly poorly studied. For tissue transplantation to be useful in ameliorating craniofacial malformations, a clear understanding of the growth activity coordination of transplanted and native tissues is required. Bone growth in response to loading conditions has also been studied. After the design, the patch pattern should be tested for the load. Therefore, the finite element method is a useful tool for evaluating the mechanical properties of a design [3].

The purpose of this study is to propose a design option with large asymmetrical cranial implants using 3D design tools. In addition, the study investigates the mechanical properties of two different structures of the cranial implants by numerical simulation (finite element method) with boundary conditions of intracranial pressure and intracranial pressure and 50 N external force.

2 Materials and Methods

2.1 Methods

In research [14], the authors are trying to enhance the modernity by considering more complex skull finite element model which is composed of more intracranial features. It is modeled using complex non-linear constitutive rules based on the original outer geometry, but segmenting it with sutures, and cortical bone, having variable thickness across different head sections. In this paper to simplify the model, continuous skull thickness and homogeneous skull material are assumed. The model is meshed directly in ANSYS software. The meshing process is divided into 3 separate parts according to the characteristics of the regions: body sizing, edge sizing, face sizing. Objects such as skulls and cranial implants have complex contours, so the elements are divided as triangular elements. The fixation parts are divided by quadrangular elements.

Model

Usually, on skull patches, doctors often punch holes in the surface to drain fluid and soft muscles can grow to feed the skin covering the grafted skull. Most current simulations are based on unpunched patch simulations. Therefore, in this study, we develop the perforated skull patch structure model from the designed skull patch model to compare with the original skull patch model. The 3D model we use is the cranium model and the artificial cranial part of a female patient with an accidental cranial defect. Here the model does not take the whole skull but will be extracted a part to reduce the computational volume. The 3D model of the skull with the left defect was imported into ANSYS 19.2 software (ANSYS Workbench; ANSYS, Inc., Canonsburg, Pennsylvania, United States) to simulate the stress distribution, deformation, and strain parameters exerted on the cranial artificial part as shown in Fig. 1.

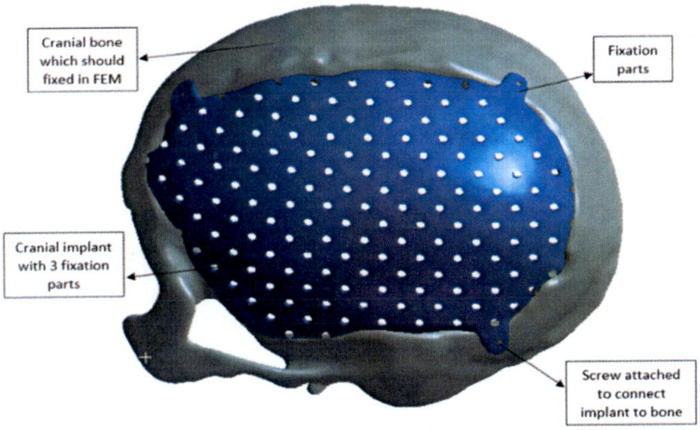

Fig. 1. Model of skull imported into ANSYS 19.2

For the left part of the patient's skull, the patch of the skull consists of 3 fixing parts attached to the skull by 3 cylindrical screws 2 mm in diameter, 4 mm long. Meshing data for two Perforated structure and non-perforated structure (solid) cases is shown in Table 1.

Table 1. Mesh data for two cases

	Nodes	Elements
Non-perforated structure (NPS)	1078590	559103
Perforated structure (PS)	1551699	840727

With the meshed model, the mean Skewness value is 0.24992 and the standard deviation is 0.15622. This index is within the acceptable range for mesh quality (0–0.5).

Material Properties
Table 2 shows the material parameters assigned to the finite element model, assigned to bones, screws and skull patches, respectively.

Table 2. Parameter of materials

	E (GPa)	ρ (Kg/m^3)	ν
Bone [15]	18	1810	0.3
Ti6Al4V	107	4405	0.323
PEEK	3.85	1310	0.4

E: Elastic modulus (GPa); ρ: Density (Kg/m^3); ν: Poisson's ratio

2.2 Boundary Conditions

Internal pressure: the normal internal pressure of a normal person $\leq$15 mmHg. In this simulation, a pressure corresponding to the maximum value of 15 mmHg $\approx$2000 Pa is placed over the entire face in the patch and bone [16]. An external force of 50 N, similar to the patient's head mass when laying down [17], is applied to the cranial implant and the skull. The skull is a fixed part, with zero displacement. According to research [18], the coefficient of friction between bone and Ti6Al4V under SBF body fluid simulation conditions is from 0.46 to 0.56. Therefore, the coefficient of friction between bone and Ti6Al4V in this study was set to 0.5. The coefficient of friction between the screw Ti-6Al4V and the cranial implant PEEK has been set to 0.3.

3 Simulation Results and Discussion

Simulation results comparing two structures: perforated structure and non-perforated structure, are presented through three parameters: deformation, stress, strain in two cases: (i) intracranial pressure 15 mmHg and (ii) intracranial pressure 15 mmHg and external force 50 N.

3.1 Cases (i): Intracranial Pressure

Simulation results of the first case with intracranial pressure of perforated and non-perforated structures are shown in Fig. 2 and Table 3.

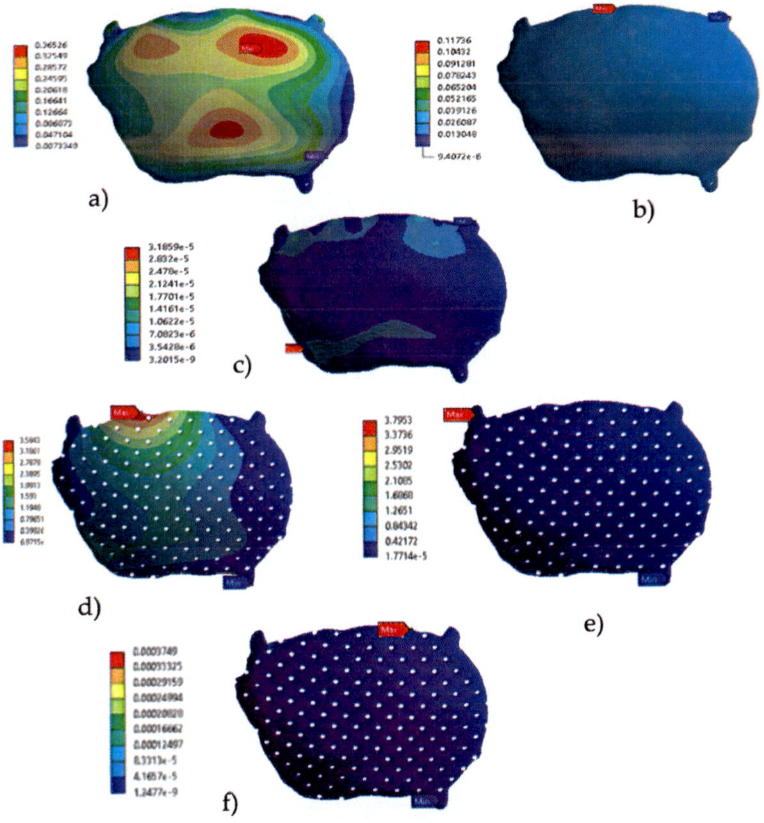

Fig. 2. Case (i) PS: (a) Deformation distribution; (b) von-Mises stress distribution; (c) von-Mises strain distribution; NPS: (d) Deformation distribution; (e) von-Mises stress distribution; (f) von-Mises strain distribution

Table 3. Simulation results case (i)

	Max. deformation (m)	Von Mises stress (MPa)	Von Mises strain
NPS	0.36526E–6	0.11736	3.1859E–5
PS	3.5843E–6	3.7953	3.749E–4

It can be observed that the deformation, stress, and strain values of PS are much larger than those of NPS. The deformation value of PS is about 10 times larger than the displacement value of NPS. The deformation values of the two structures NPS and PS are 0.36526E–6 and 3.5843E–6, respectively. The deformation distribution of NPS is concentrated in 3 main regions near fixation parts. Meanwhile, the deformation distribution of PS is concentrated at the edge of the cranial implant. The maximum stress value of PS is 3.7953 Mpa and is located at screw hole 1. The maximum stress value of NPS is 0.11736 Mpa and is located at the edge of the implant. The strain values of the NPS and PS structures are very small at 3.1859E–5 and 3.749E–4, respectively. The strain distribution of NPS is concentrated in three regions around the fixation parts. The deformation distribution of PS is uniform over the entire cranial implant and the maximum value is concentrated at the edge of the cranial implant.

In the study [19], the results of the NPS structure were almost like the case of PEEK and the intracranial pressure of 15 mmHg. Deformation and strain distributions are mainly around the positions of fixed screw holes. The results of the studies have small differences due to the implant position of the cranial implant, the number of screws, boundary conditions, meshing method, etc.

3.2 Cases (ii): Intracranial Pressure and External Force 50 N

An external force of 50 N is equivalent to the weight of a person's head which will be applied to the position with the maximum Deformation and von Mises strain value, direction of force from outside to inside as shown in the simulation results in case (i) in Fig. 3.

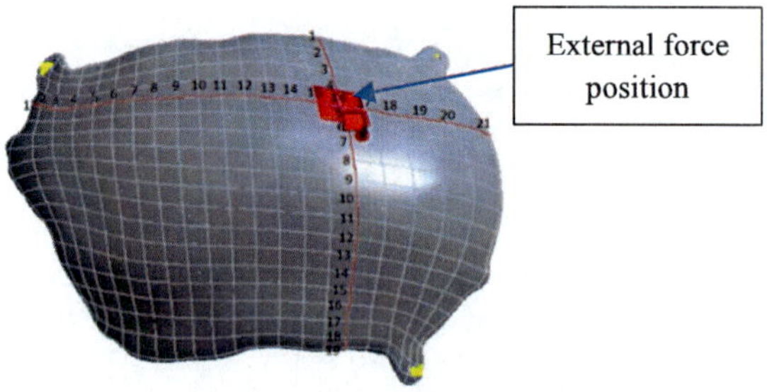

Fig. 3. Numbered points on the horizontal and vertical axes are investigated

Simulation results of the second case with intracranial pressure of perforated and non-perforated structures are shown in Fig. 4 and Table 4.

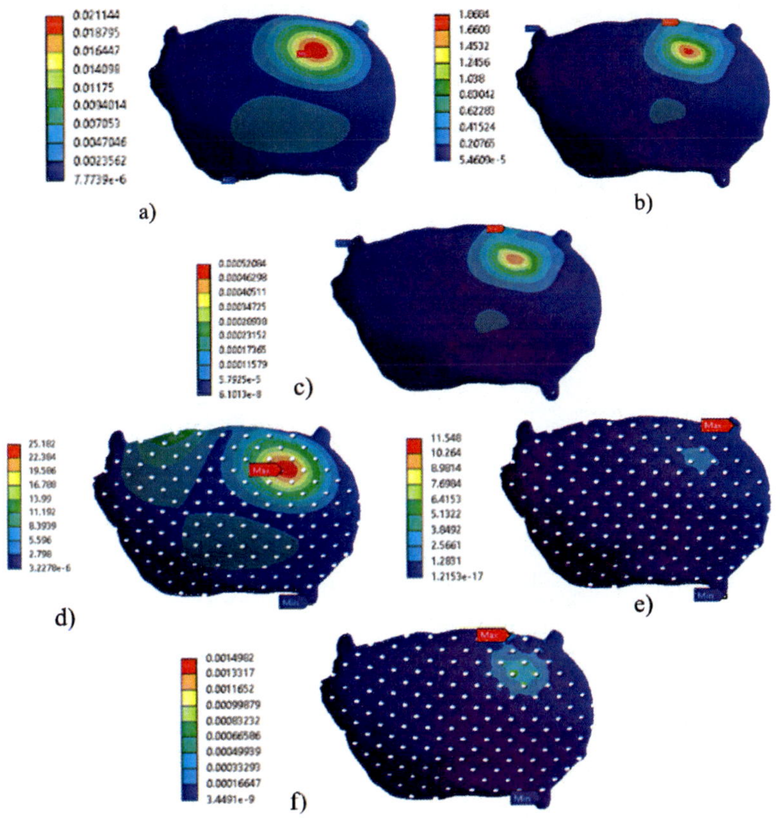

Fig. 4. Case (ii) PS: (a) Deformation distribution; (b) von-Mises stress distribution; (c) von-Mises strain distribution; NPS: (d) Deformation distribution; (e) von-Mises stress distribution; (f) von-Mises strain distribution

Table 4. Simulation results case (ii)

	Max. deformation (m)	Von Mises stress (MPa)	Von Mises strain
NPS	2.1144E–5	1.8684	5.2084E–4
PS	2.5182E–5	11.548	1.4982E–3

In general, the distribution of deformation, stress and strain of the two structures NPS and PS are concentrated at the site of external force. The deformation, stress and strain values of case (ii) are far larger than those of case (i). The maximum deformation

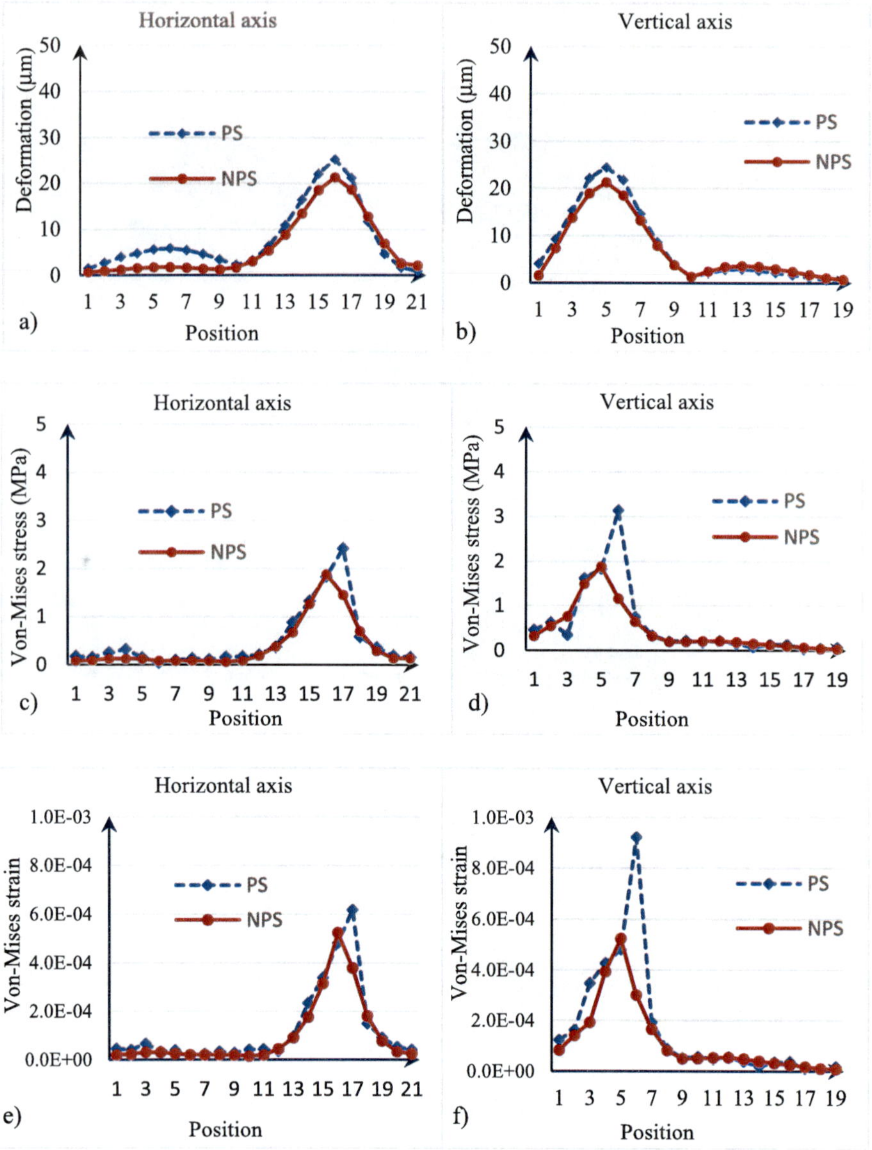

Fig. 5. Histogram of values along the horizontal and vertical axis: The deformation value on the horizontal axis (a) and vertical axis (b); The Von-Mises stress value on the horizontal axis (c) and vertical axis (d); The Von-Mises strain value on the horizontal axis (e) and vertical axis (f).

value of the two cases NPS and PS is about 20 μm both located at the position of external force. The maximum stress values of the two cases NPS and PS are 1.8684 MPa and 11,548 MPa, respectively. These values are much smaller than the Yield strength of PEEK (95 MPa) [20]. The maximum strain values of the two structures NPS and PS are very low, respectively, 5.2084E–4 and 1.4982E–3.

In addition, on two horizontal and vertical lines passing through the center of the external force (21 points on the horizontal axis and 19 points on the longitudinal axis numbered as Fig. 3), the magnitude of the displacement, stress, and strain are investigated. The center point where the external force is placed has coordinates (16; 5) (Fig. 5).

The results of this study are generally like those of existing studies. The results of the studies may differ due to the model, the implant location, the structure of the implant, the boundary conditions. Research is still limited as many perforated structures have not been investigated with different hole densities and different external force locations. The design method also opens the option of combining with the collection of sample skull characteristics and image processing technologies to improve design capabilities. In future studies, the proposed method will find a skull prototype with similar characteristics to a patient's skull from a data bank to make it easier and more accurate to design a cranial implant.

4 Conclusions

This study has compared the mechanical properties of two solid and perforated skull structures by numerical simulation (finite element method) with boundary conditions of 50 N external force and intracranial pressure. The perforated structure in this study meets the requirements of the material's durability, under the patient's basic conditions (intracranial pressure and external force 50 N). The perforated structure not only provides durability but also helps bone cells grow on the structure. In addition, the new perforated structure helps to save materials, thereby reducing surgical costs for the patient. Finally, the implant design ensures aesthetics, helping the patient improve the quality of life.

Acknowledgment. This research is funded by Hanoi University of Science and Technology (HUST) under project number T2020-TT-201.

References

1. Goiato, M.C., Anchieta, R.B., Pita, M.S., Dos Santos, D.M.: Reconstruction of skull defects: currently available materials. J. Craniofac. Surg. **20**(5), 1512–1518 (2009). https://doi.org/10.1097/SCS.0b013e3181b09b9a
2. Moiduddin, K., Darwish, S., Al-Ahmari, A., ElWatidy, S., Mohammad, A., Ameen, W.: Structural and mechanical characterization of custom design cranial implant created using additive manufacturing. Electron. J. Biotechnol. **29**, 22–31 (2017). https://doi.org/10.1016/j.ejbt.2017.06.005
3. El Halabi, F., Rodriguez, J.F., Rebolledo, L., Hurtós, E., Doblaré, M.: Mechanical characterization and numerical simulation of polyether-ether-ketone (PEEK) cranial implants. J. Mech. Behav. Biomed. Mater. **4**(8), 1819–1832 (2011). https://doi.org/10.1016/j.jmbbm.2011.05.039

4. Zanotti, B., Zingaretti, N., Verlicchi, A., Robiony, M., Alfieri, A., Parodi, P.C.: Cranioplasty: review of materials. J. Craniofac. Surg. **27**(8), 2061–2072 (2016). https://doi.org/10.1097/SCS.0000000000003025
5. Khader, B.A., Towler, M.R.: Materials and techniques used in cranioplasty fixation: a review. Mater. Sci. Eng. C **66**, 315–322 (2016). https://doi.org/10.1016/j.msec.2016.04.101
6. Kwarcinski, J., Boughton, P., Ruys, A., Doolan, A., van Gelder, J.: Cranioplasty and craniofacial reconstruction: a review of implant material, manufacturing method and infection risk. Appl. Sci. **7**(3), 1–17 (2017). https://doi.org/10.3390/app7030276
7. Al-Tamimi, Y.Z., et al.: Comparison of acrylic and titanium cranioplasty. Br. J. Neurosurg. **26**(4), 510–513 (2012). https://doi.org/10.3109/02688697.2011.633640
8. Aguilar-Pérez, L.A., Sánchez-Cruz, J.I., Flores-Campos, J.A., Torres-SanMiguel, C.R.: Numerical and experimental assessment of a novel anchored for intramedullary telescopic nails used in osteogenesis imperfecta fractures. Appl. Sci. **11**(12) (2021). https://www.mdpi.com/2076-3417/11/12/5422
9. Leal-Naranjo, J.A., et al.: Multi-objective optimization of a parallel manipulator for the design of a prosthetic arm using genetic algorithms. Lat. Am. J. Solids Struct. **15**(3) (2018). https://doi.org/10.1590/1679-78254044
10. Torres-Sanmiguel, C.R., Hernández-Gómez, J.J., Urriolagoitia-Sosa, G., Romero-ángeles, B., Martínez-Sáez, L.: Design and manufacture of a customised temporomandibular prosthesis. Revista Internacional de Metodos Numericos para Calculo y Diseno en Ingenieria **35**(1) (2019). https://doi.org/10.23967/j.rimni.2019.02.001
11. Becker, L.C., et al.: Final report of the cosmetic ingredient review expert panel safety assessment of polymethyl methacrylate (PMMA), methyl methacrylate crosspolymer, and methyl methacrylate/glycol dimethacrylate crosspolymer. Int. J. Toxicol. **30**(3) (2011). https://doi.org/10.1177/1091581811407352
12. Mohammed, M.I., Fitzpatrick, A.P., Malyala, S.K., Gibson, I.: Customised design and development of patient specific 3D printed whole mandible implant. In: Solid Freedom Fabrication 2016 Proceedings 27th Annual International Solid Freedom Fabrication Symposium - An Additive Manufacturing Conference SFF 2016, no. January 2018, pp. 1708–1717 (2016)
13. Bücking, T.M., Hill, E.R., Robertson, J.L., Maneas, E., Plumb, A.A., Nikitichev, D.I.: From medical imaging data to 3D printed anatomical models. PLoS ONE **12**(5), 1–10 (2017). https://doi.org/10.1371/journal.pone.0178540
14. Barbosa, A., Fernandes, F.A.O., de Sousa, R.J.A., Ptak, M., Wilhelm, J.: Computational modeling of skull bone structures and simulation of skull fractures using the YEAHM head model. Biology (Basel) **9**(9), 1–18 (2020). https://doi.org/10.3390/biology9090267
15. Pal, S.: Design of Artificial Human Joints & Organs. Springer, Heidelberg (2014)
16. Nagasao, T., Miyamoto, J., Jiang, H., Kaneko, T., Tamaki, T.: Biomechanical analysis of the effect of intracranial pressure on the orbital distances in trigonocephaly. Cleft Palate-Craniofac. J. **48**(2), 190–196 (2011). https://doi.org/10.1597/09-027
17. Yoganandan, N., Pintar, F.A., Zhang, J., Baisden, J.L.: Physical properties of the human head: mass, center of gravity and moment of inertia. J. Biomech. **42**(9), 1177–1192 (2009). https://doi.org/10.1016/j.jbiomech.2009.03.029
18. Wang, C., et al.: Tribological behavior of Ti-6Al-4V against cortical bone in different biolubricants. J. Mech. Behav. Biomed. Mater. **90**, 460–471 (2019). https://doi.org/10.1016/j.jmbbm.2018.10.031
19. Bogu, V.P., Ravi Kumar, Y., Khanara, A.K.: Modelling and structural analysis of skull/cranial implant: beyond mid-line deformities. Acta Bioeng. Biomech. **19**(1), 125–131 (2017). https://doi.org/10.5277/ABB-00547-2016-04
20. Schwitalla, A.D., Abou-Emara, M., Spintig, T., Lackmann, J., Müller, W.D.: Finite element analysis of the biomechanical effects of PEEK dental implants on the peri-implant bone. J. Biomech. **48**(1), 1–7 (2015). https://doi.org/10.1016/j.jbiomech.2014.11.017

Design and Performance of a Motion-Assisting Device for Ankle

Zhetenbayev Nursultan[1](✉), Alexander Titov[2,2], Marco Ceccarelli[2,2], and Gani Balbayev[3]

[1] Satbayev University, Almaty, Kazakhstan
[2] LARM2 Laboratory of Robot Mechatronics, University of Rome Tor Vergata, Rome, Italy
aleksandr.titov@students.uniroma2.eu,
marco.ceccarelli@uniroma2.eu
[3] Academy of Logistics and Transport, Almaty, Kazakhstan

Abstract. This paper presents a new design for a motion-assisting device as an exoskeleton for ankle articulation. The proposed design is characterized by a light weighted structure with adaptable geometry to different users with low-cost and easily wearing features. The exoskeleton assistance is obtained with three linear electric actuators to provide the three motions of an ankle. A CAD model is elaborated for design details and for simulation whose results give data for feasibility of the proposed design and its characterization in basic operation performance.

Keywords: Medical device · Biomechanics · Ankle exoskeleton · Design · Simulation

1 Introduction

The population with limited mobility increases every year. This factor affects the quality of life of people and their dependence on others. Physical therapy and motion exercise is required to treat these cases. With the help of robotic devices, the training of people's movements can be improved with controlled exercises. The shortcomings of existing robotic rehabilitation exercise solutions have revealed the need to develop low-cost devices that allow the rehabilitation exercise of patients with damaged limbs. An exoskeleton with artificial muscles can help people to restore the function of individual joints of the lower and upper extremities [1].

The ankle joint is crucial for locomotion, supporting the full-body load while simultaneously applying key forces during push-off, leg swing, and center-of-mass movement during human walk. Unfortunately, it is also one of the most frequently injured joints of the human body, often prone to sprains or fractures that hinder the patient's full ability to walk. To restore full mobility, rehabilitation and assistance in movement are necessary. Rehabilitation training is usually conducted by a physical therapist one - on-one with the patient, as it requires the full attention of the physical therapist [1]. Robotic rehabilitation has been proposed as an effective alternative, as it would allow a single physiotherapist

N. V. Khang et al. (Eds.): ASIAN MMS 2021, MMS 113, pp. 659–668, 2022.
https://doi.org/10.1007/978-3-030-91892-7_63

to provide care to multiple patients at the same time, potentially remotely controlling the robots. The feasibility of a robotic solution to assist patients in restoring gait functions has been proven by several research groups [2]. However, most ankle rehabilitation robots are based on static platform structures, as shown in [3–7]. In [8], rehabilitation is performed with a patient standing on two platforms with variable orientation. This solution requires the patient to be able to stand up, and it is characterized by a bulky non-portable device. Another reasonable design is proposed in [9] for a restorative exercise of one leg, like the prototype in [10] with a mechanism driven by pneumatic muscles. The evolution of the same construction is described in [11]. Smaller, but still grounded devices are presented in [12] and [13] as based on a medical service parallel architecture and a compatible mechanism, respectively.

All these robots are characterized by static structures, and most of them, such as the prototypes in [10–12], are fixed to the ground or difficult to move due to the large area and weight. To overcome this limit, few research groups have proposed wearable structures, such as ankle exoskeletons and orthoses [3]. These new rehabilitation systems are usually characterized by rigid or malleable connections and bulky enclosures, so they still require additional support for their considerable weight.

The here in proposed design offers a new solution focused on the implementation of a linear electric drive at the level of the ankle joint using an inexpensive one-legged mechanism. The appropriate profile of the linear electric drive is designed for proper movement during physical therapy.

This paper presents a new solution of the lower limb exoskeleton. The mechanism is based on an electric linear to assist the user in performing the movements.

The article consists of four sections, which provide information on other examples of exoskeletons for the ankle joint, kinematic, and modeling of the exoskeleton, and a numerical characterization as for check of design feasibility.

2 Ankle Anatomy and Motion Requirements

The ankle is a hinge-type synovial joint and is involved in lower limb stability [13]. The ankle and foot contain 26 bones connected by 33 joints, and more than 100 muscles, tendons, and ligaments. The ankle joint is located between the lower ends of the tibia and the fibula and the upper part of the talus. This joint is supported by ligaments such as the medial or deltoid and the lateral ligaments (anterior talofibular, posterior talofibular and calcaneofibular).

The ankle comprises basically two joints: the talocrural joint and the talocalcaneal joint. However, in biomechanical modelling it is usually treated as a single joint. There are two movements that occur around a transverse axis between malleoli, Fig. 1.

- Dorsal flexion (or dorsiflexion): this movement brings the foot dorsally to the anterior surface of the leg. The range of motion in this movement is up to 20°. The muscles involved in dorsiflexion are the tibialis anterior, extensor hallucis longus, extensor digitorum longus and fibularis tertius.
- Plantar flexion: this is the opposite movement to dorsiflexion. It occurs, for instance, when the toes are in contact with the ground and the heel is raised off the ground. The

range of motion in plantar flexion is from 40 to 50°. The muscles involved in plantar flexion are the gastrocnemius, soleus and plantaris.

The human ankle joint plays an important role in maintaining a person's balance while walking. This is the most difficult movement of the lower limb joint due to its 3 degrees of freedom (DOF) in the X, Y, and Z directions. Thus, it can move rotationally in all three anatomical planes, as shown in Fig. 1. Based on the morphological movement of the ankle joints Fig. 1, ankle sprain is the most common ankle injury. There are three types of ankle injuries: ankle sprain, ankle bone fracture, muscle, and tendon sprain [13]. Untreated ankle sprain can lead to chronic ankle instability. Severe ankle sprain injuries can include a serious bone fracture, which, if left untreated, can lead to alarming complications. Thus, the rehabilitation of a dislocated ankle should begin immediately with proper procedures.

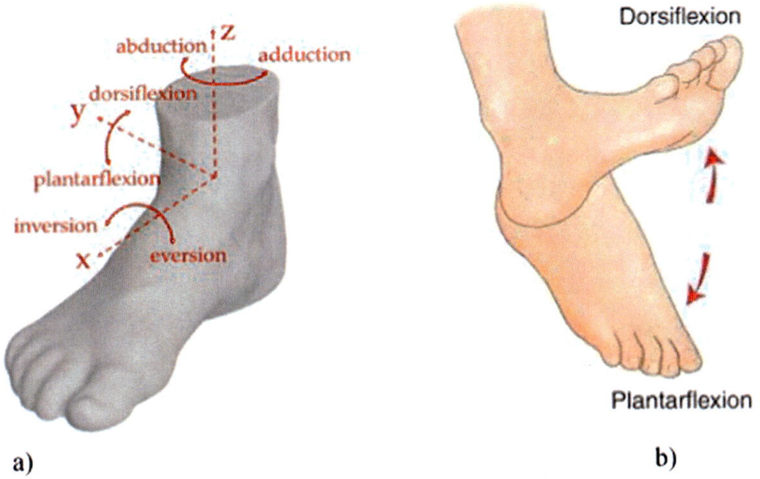

Fig. 1. Characteristic motions of a human ankle joint [2]; (a) reference frames; (b) an example

The ranges of motion of these modes are characterized by a significant variability between individuals, due to geographical/cultural differences, anatomical structures, and distinct data-acquisition methodologies (Table 1).

Table 1. Ranges of motion of the human ankle joint, Fig. 1(a) [2]

Motion	Dorsiflexion (deg)	Plantarflexion (deg)	Abduction/Adduction (deg)	Inversion/Eversion (deg)
Range limits	20	50	±10	±12

3 The Proposed Device

The main purpose of this work is to present a device that can meet the minimum in requirements for physical therapy exercises. The design offers a new solution aimed at implementing a linear electric drive at the level of the ankle of the foot using an inexpensive one-legged mechanism. The appropriate profile of the linear electric drive is designed for proper movement during physical therapy.

The advantage of this device design is the expected weight and volume of the device body. The device should be light, compact, comfortable to wear and active movements. The most important thing, the device helps the user to make movements. The device must provide sufficient torque and act quickly to maintain at least the minimum range of required movements.

The design of the mechanism in Fig. 2 works with four linear electric drives, which are installed in parallel with spherical hinges between the lower leg and the ankle joint, respectively. The translational movement of linear electric drives ensures the rotation of the ankle and ankle joints.

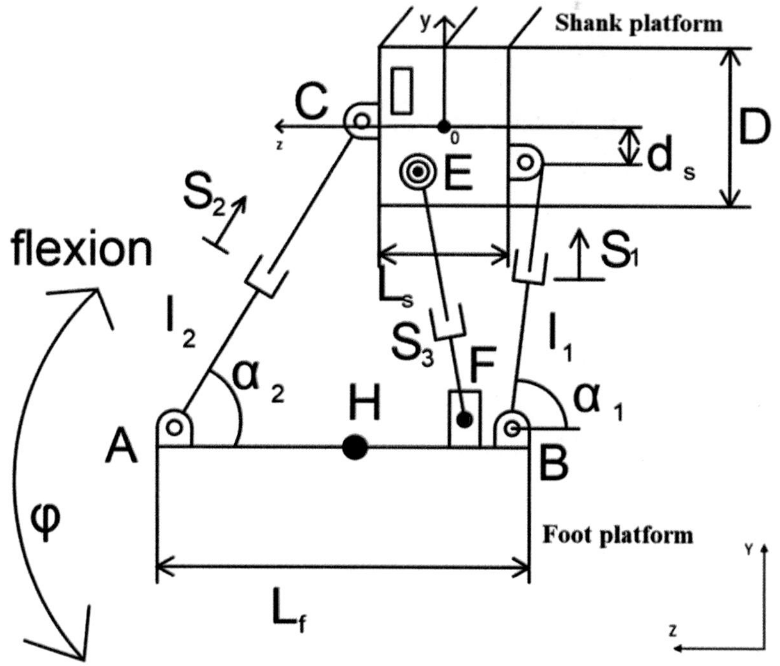

Fig. 2. The kinematic design of ankle exoskeleton

From the kinematics of the coupling solutions in the exoskeleton design in Fig. 2, the angles of articulation of the joints can be expressed as follows (Table 2):

A passive exoskeleton of the ankle joint, consisting of fasteners between the foot (A) and the lower leg (B) using a ball joint element (L4), the fastening of the human

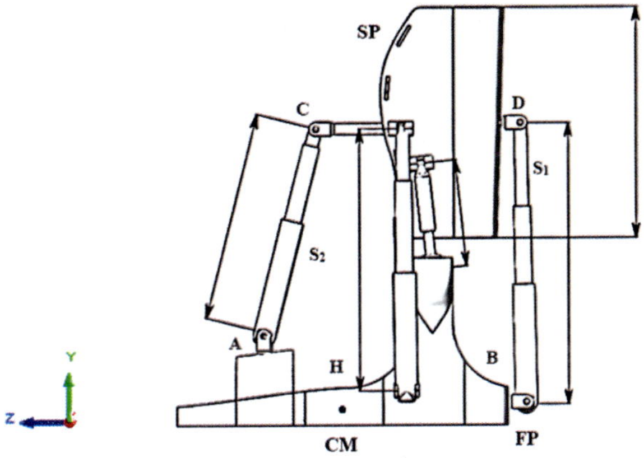

Fig. 3. A CAD design of the ankle motion - assisting exoskeleton in Fig. 2

Table 2. Design and operation parameters of a CAD solution in Fig. 3

Size	SP (mm)	FP (mm)	L_1 (mm)	L_2 (mm)	L_3 (mm)	L_4 (mm)
	200	265	243	226	181	93

foot consists of two parts connected to each other by means of loops and attached to the upper part from the front. (L1) (L3) electric linear actuator, fixed at the back and front of the ankle side.

Connections E and F of the support link the ball joint is the guide link between the shank and the foot platform.

The electric linear actuator consists of a system that synchronizes the movement of two interconnected skeletons and moves the ankle together with the exoskeleton.

4 Performance Analysis

A 3D modeling and simulation calculations were performed in a virtual environment using the Solidworks Simulation software, and the Motion simulation supplement.

Using Solidworks Simulation an electric linear actuator input is given, which generates movement of the ankle joint. Figure 4 shows the movement of the dorsal flexion plantar flexion ankle in a normal position. Figure 5 shows the dorsal flexion/plantarflexion obtained using Solidworks Simulation. Dorsal flexion the range of motion in this movement will be up to 20°, while the Simulation will bend to 15°. The range of motion in plantar flexion is from 40 to 50°, and in Simulation, bending is up to 20°.

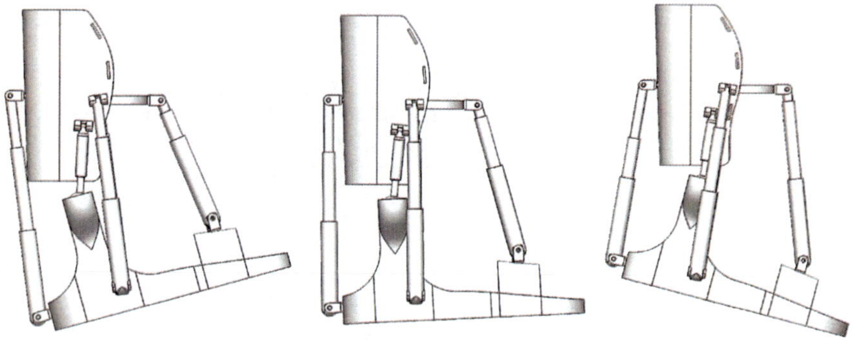

Fig. 4. A snapshot of simulated dorsiflexion – plantarflexion assisted motion

Figure 5 shows components of linear displacement of the platform. From the plot, the motions from all directions are with peaks of maximum values 100 deg/s^2. represented for X – component and less than 85 deg/s^2 for another components.

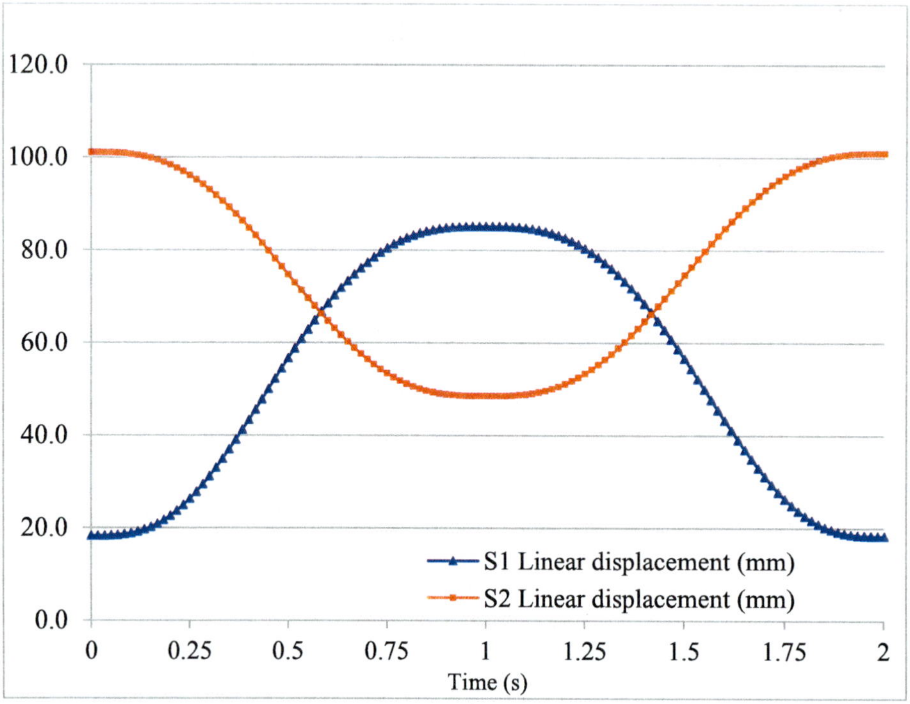

Fig. 5. Input data for the simulated motion in Fig. 4 in terms of displacement of linear actuators

The angular acceleration regarding the angle is shown in Fig. 6. The largest value of the acceleration, which equals to 240 deg/s^2, is near the top position, and another peak, which equals 150 deg/s^2, is near the bottom position.

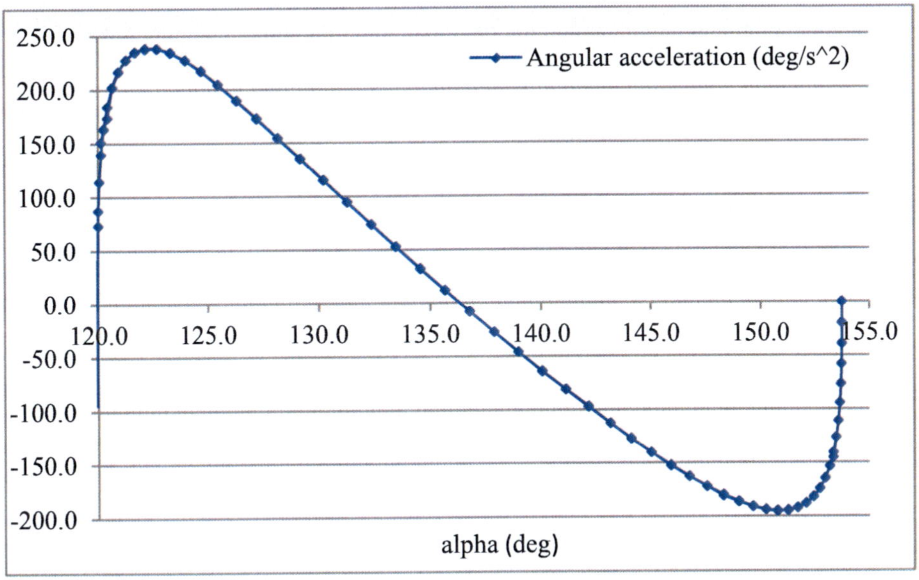

Fig. 6. Computed results of the simulated motion in Fig. 4 in terms of α angle of the foot platform

Figure 7 shows trajectory of point H the central point on the platform. The movements of this point along Z axis are equal to 25.5 mm, movements along Y axis reach 74.6 mm (Fig. 9).

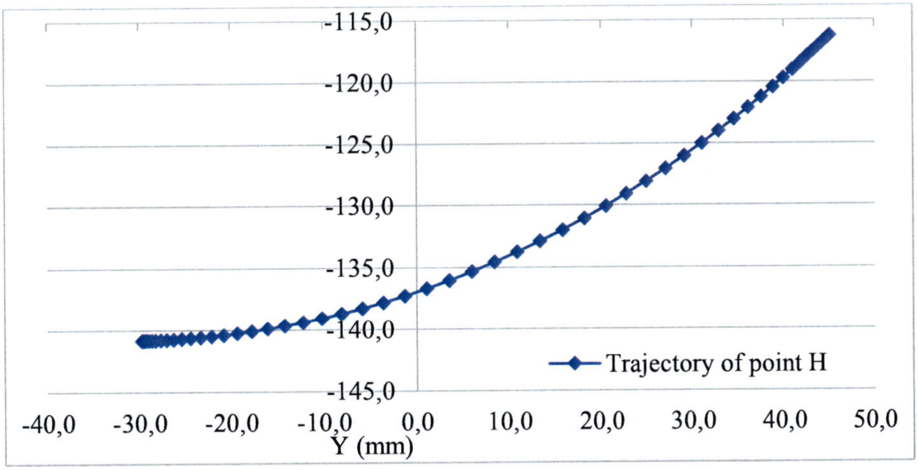

Fig. 7. Computed results of the simulated motion in Fig. 4 in terms of the trajectory of point H of the foot platform

Figure 8 shows driving force by the linear actuators: F1 is force of the linear actuator in front side of the leg, F2 is the force of the linear actuator in back side of the leg, respectively. The computing results of the forces F1 and F2 reach 0.8 N.

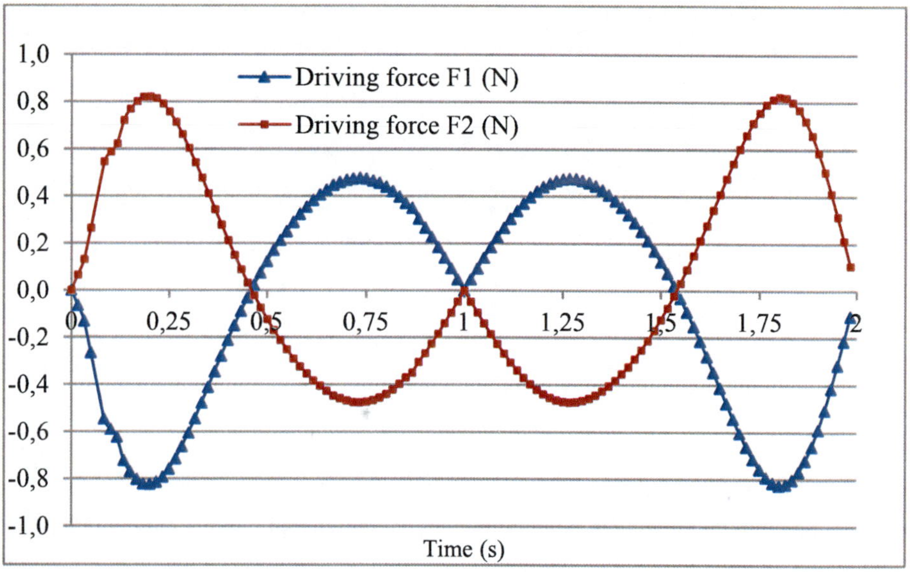

Fig. 8. Computed results of the simulated motion in Fig. 4 in terms of driving force by the linear actuators

5 Prototype Exoskeleton of the Ankle Joint

In accordance with the model in Fig. 3, a prototype was created and assembled, shown in Fig. 10, for future tests. The material of the parts is PLA. The model is driven by electric linear actuators with a maximum force of 50 N. As shown in Fig. 10, there are two electric linear actuators that operate when bending the dorsal flexion and plantar flexion. The prototype is not fully assembled, there will still be two electric drives and ball joints will be installed. This device will be tested in the future.

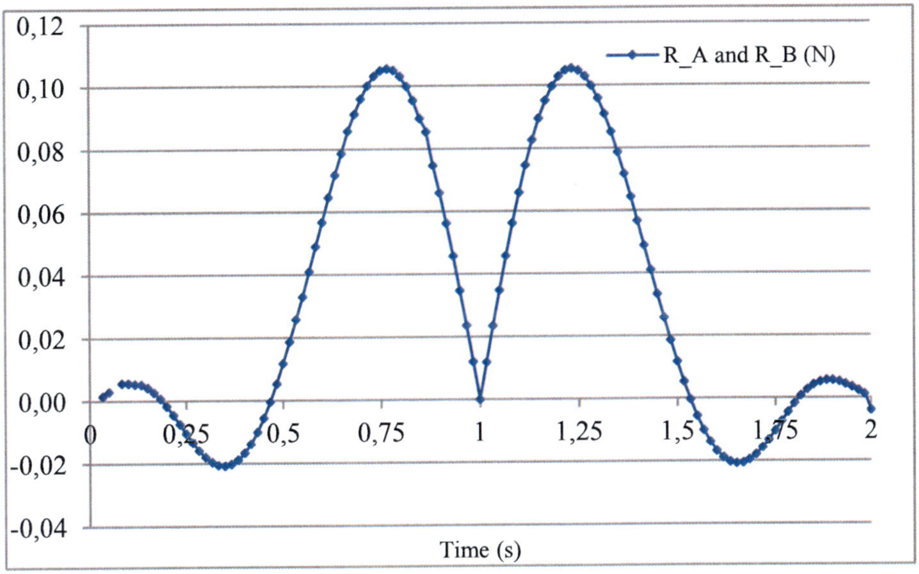

Fig. 9. Computed results of the simulated motion in Fig. 4 in terms of reaction force at the joint points A and B of a linear actuator

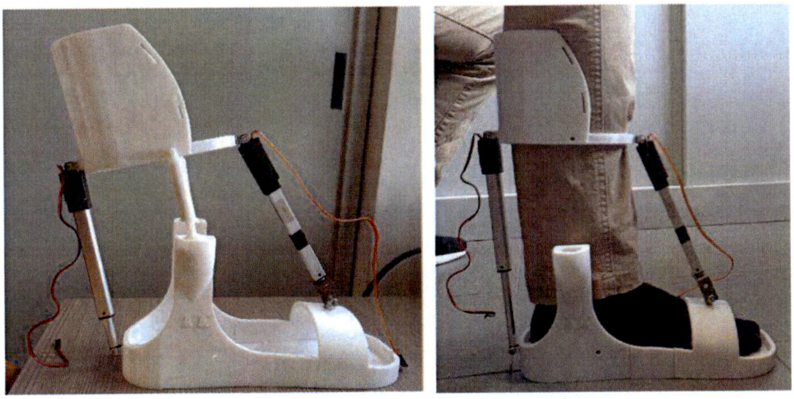

a) b)

Fig. 10. A first prototype at LARM2 in Rome: (a) a mockup; (b) worn on a volunteer

6 Conclusion

The presented ankle exoskeleton is designed using low-cost lightweight features with ankle mechanism for motion assistance. The model is developed and simulated in Solidworks, the results of which are discussed for performance characteristics. The prototype will be assembled to test the new design, and future work will give an experimental characterization.

References

1. Iancu, C.A., Ceccarelli, M., Lovasz, E.-C.: Design and lab tests of a scaled leg exoskeleton with electric actuators. In: Ferraresi, C., Quaglia, G. (eds.), Advances in Service and Industrial Robotics, pp. 719–726. Springer, Cham (2018). https://doi.org/10.1007/978-3-319-61276-8_76
2. Russo, M., Ceccarelli, M.: Analysis of a wearable robotic system for ankle rehabilitation. Machines **8**, 48 (2020). https://doi.org/10.3390/machines8030048. www.mdpi.com/journal/machines
3. Zhang, M., Davies, T.C., Xie, S.: Effectiveness of robot-assisted therapy on ankle rehabilitation–a systematic review. J. Neuroeng. Rehabil. **10**, 30 (2013)
4. Díaz, I., Gil, J.J., Sánchez, E.: Lower-limb robotic rehabilitation: literature review and challenges. J. Robot. **2011**, 759764 (2011)
5. Alvarez-Perez, M.G., Garcia-Murillo, M.A., Cervantes-Sánchez, J.J.: Robot-assisted ankle rehabilitation: a review. Disabl. Rehabil. Assist. Technol. **15**, 394–408 (2019)
6. Shi, B., et al.: Wearable ankle robots in post-stroke rehabilitation of gait: a systematic review. Front. Neurorobotics **13**, 63 (2019)
7. Yoon, J., Ryu, J., Lim, K.B.: Reconfigurable ankle rehabilitation robot for various exercises. J. Robot. Syst. **22**, S15–S33 (2006)
8. Roy, A., et al.: Robot-aided neurorehabilitation: a novel robot for ankle rehabilitation. IEEE Trans. Robot. **25**, 569–582 (2009)
9. Sung, E., Slocum, A.H., Ma, R., Bean, J.F., Culpepper, M.L.: Design of an ankle rehabilitation device using compliant mechanisms. J. Med. Devices **5**, 011001 (2011)
10. Lin, C.C., Ju, M.S., Chen, S.M., Pan, B.W.: A specialized robot for ankle rehabilitation and evaluation. J. Med. Biol. Eng. **28**, 79–86 (2008)
11. Zhang, M., McDaid, A., Veale, A.J., Peng, Y., Xie, S.Q.: Adaptive trajectory tracking control of a parallel ankle rehabilitation robot with joint-space force distribution. IEEE Access **7**, 85812–85820 (2019)
12. Chang, T.C., Zhang, X.D.: Kinematics and reliable analysis of decoupled parallel mechanism for ankle rehabilitation. Microelectron. Reliab. **99**, 203–212 (2019)
13. Zhang, M., Davies, T.C., Xie, S.: Effectiveness of robot – assisted therapy an ankle rehabilitation a systematic review. J. NeuroEng. Rehabil. **10**, 1–6 (2013)

Improved Exoskeleton Human Motion Capture System

Trung Nguyen[1](✉), Linh Tao[1], Tinh Nguyen[1], and Tam Bui[1,2]

[1] Hanoi University of Science and Technology, Hanoi, Vietnam
trung.nguyenthanh@hust.edu.vn
[2] Shibaura Institute of Technology, Tokyo, Japan

Abstract. Motion capture systems are becoming more and more important tools serving in the process of calculation, design process as well as evaluating the performance of the rehabilitation equipments. In addition, it is also a common device used in the performance evaluation of rehabilitation process during exercise stages. In this article, a cheap motion capture system measured 7 Degree of Freedom (DOF) of upper limb will be presented. It is an improvement from an earlier portable measuring device made of plastic material with 3D printing technology. The new, static device, constructed of aluminum, has proven to be more accurate than the previous one. After that, the study performed measurement experiments for some basic movements in daily living. In each movement, the study obtained 3 results including: the angles of the rotation angles of the measuring device, the trajectories of the end effector of the wrist joint, and the angle of the human arm joints.

Keywords: Human motion capture system · Exoskeleton · Degree of Freedom · Upper limb · Activity of daily living

1 Introduction

Demand for rehabilitation treatment tends to increase rapidly in the past years. According to annual statistics in the US [1] 795000 Stroke patients need to participate the rehabilitation training every year. Particularly for developing countries like Vietnam, Malaysia, Thailand, etc., the number of new patients is increasing very fast and tend to be younger. According to statistics, in 2017, for every 100,0000 people, more than 2,500 Vietnamese people died or become disabled due to stroke [2]. The number of stroke patients under 45-year-old are increasing rapidly. Consequences of improper treatment and rehabilitation influence greatly to the integration back to life as well as the work of those patients [3].

In stage of design and manufacture for the rehabilitation system, one of the key inputs for design process is anthropometry of the patients who would later use that device. Those parameters include factors such as the Range of Motion (ROM), speed, angular acceleration of the joints in the activity of Daily Ling (ADL). It can ask for more strict requirements with high-speed operation, such as sports exercise. Currently, the

N. V. Khang et al. (Eds.): ASIAN MMS 2021, MMS 113, pp. 669–679, 2022.
https://doi.org/10.1007/978-3-030-91892-7_64

parameters most are referred to the documents of the other authors. These are parameters that may not match with anthropometric parameters of those countries or regions in developing countries such as Vietnam, Thailand, Malaysia, etc.

Due to the above reasons, the design and manufacture of a Human Motion Capture System (HMCS) in the developing countries ensure the cost requirements as well as convenient usage are very necessary. In fact, there have been many studies on this HMCS system. It can be divided into 2 main groups based on the type of sensor used. They are HMCS system based on image sensor (V-HMCS) and HMCS system based on position sensor (P-HMCS) [4]. The V-HMCS uses several cameras and reflective markers that are attached to several necessary positions on the body. By modeling the human body as multi rigid-body system as the skeleton, these cameras can follow the motion of each segment at the same time. From the movements of each of these sections, the system extracts the kinetic parameters necessary for the body parts such as the head, torso, limbs or their smaller segments, such as movement of the joints. There are number of typical V-HMCS systems such as VICON, Qualisys motion capture system [5, 6]. The P-HMCS system uses several types of sensors such as inertial measurement units (IMU) and Encoders. These sensors allow accurate measurements of the rotation of segments on the human body. In addition, when performing measurements, the authors used filters such as the Kalman filter extension (EKF) [7] and complementary filter [8] to ensure the accuracy of the measurement. When using IMU sensors, the use of the above EKF or CF algorithms is essential to estimate the position of the joints. The velocity of the joints will be calculated based on angular information over time as well as acceleration information of the segments.

When comparing two types of V-HMCS and P-HMCS, the V-HMCS system results in more accurate measurement and higher resolution. Therefore, in order to evaluate the effectiveness and authenticity of the measurement methods, we often use this system to evaluate. V-HMCS is also known as the "gold standard" for evaluating the quality of HMCS systems [9]. However, the main disadvantage of this system is that it is quite expensive. Access to these systems in developing countries is still a significant obstacle. Besides, the process of installing and calculating thanks to image processing is also quite time-consuming and not very friendly. P-HMCS systems are often cheaper, more compact, and more portable than V-HMCS systems. Signal processing in P-HMCS systems is also easier than V-HMCS systems.

Document [9] reviews the P-HMCS types using to measure the (ROM) of the human upper limb. The sensors are attached to the segments by methods such as strap, tape, velcro. Few studies have used orthosis attachment method [4]. By attaching the sensor directly to the segments of the human arm, where there is a high degree of laxness due to the tendon structure, the skin, the flesh of the segments, etc. These methods have to use complex algorithms to filter. It also takes longer to set up and calibrate for each measurement and when changing objects. To make it easier to wear the device on the body, the author of the article [4] attached the sensors to an orthosis system. However, this orthosis structure is quite inflexible. It has no structure to change when the length varies from person to person. Besides, very few P-HMCS systems fully measure the movement of 7 DOF of the human arm.

In this paper, an improved version of the measuring device previously developed by the author [11] will be presented. The system is designed to have the same structure as the exoskeleton as in the previous version. However, in order to improve the accuracy of the measuring device, the new device is made of Aluminum material and is stationary. In the study, measurement experiments were also conducted to evaluate the quality of the device and some movements in human daily activities. The rest of the paper is organized as follows: the improved system design is shown in Sect. 2. Section 3, the experiment setup and results are performed. Finally, the conclusion is outline in Sect. 4.

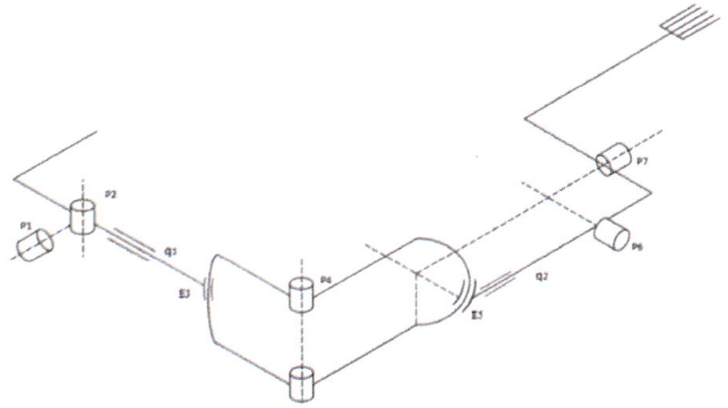

Fig. 1. Structure principal diagram of E-HMC [11]

2 New Improvement Design of E-HMCS

With the purpose of designing a measuring device for Asian's ROM as well as can be used to generate motion trajectories in later exercises. The system can then also be used as the master side in a master-slave system. In this system, a physiotherapist may wear this master side to generate reference motions for slave side according to the patient's current condition assessment. The mechanical system of the measuring device was proposed as exoskeleton type [11]. This exoskeleton structure is moved by healthy person when measuring the angular range. The Asian anthropological parameters [10] are also the input data to the process of designing the measuring equipment shown in Ref. [11].

The principal structure of the device is shown as Fig. 1. The corresponding P_1; P_2; E_3; P_4; E_5; P_6; P_7 Potentiometers and Encoders will collect data at the wrist, forearm, and shoulder joints. The length of the bar q1, q2 depends on the length of the person wearing the outer skeleton that varies accordingly. The author has designed, manufactured, and basically tested the E-HMCS device [10]. The Vr 1.0 version had advantages such as: low production cost, light weight, portable design. However, it contains some major limitations as follows:

- The process of don and doff the device still takes a lot of work.
- Due to the need to fasten the device to the upper body of human, the device causes discomfort to the user.
- Made from plastic materials by 3D printing technology, E-HMCS Vr. 1 has low accuracy. The reason is that the manufacturing accuracy, the assembly accuracy of the parts is low, especially the mechanical properties of the parts made from plastic

materials are weak and the elastic deformation is big, leading to the low-end position accuracy when measuring. In order to still use this device, the author in the study [11, 12] did not focus on assessing the accuracy of the actual endpoint of the measuring device. The author only took the sets of angles of the joints and then puts them into the forward kinematics problem to create the end point trajectory of the system. From there, this trajectory was fed back to the proposed algorithm to solve the inverse kinematics problem. In fact, when measuring and comparing, the author found that the actual endpoint deviation from the endpoint obtained from the measured angles could be over 100 (mm).

- Because of the above disadvantages, studies on this device have not been able to accurately determine the user's joint angles.

To overcome the above disadvantages, in this study the author has made some improvements as follows:

- Change the device from mobile to stationary form to easily facilitate the don and doff process as well as reduce the impact on the wearer's body.
- The device has been made from aluminum to increase the accuracy and rigidity of the system.
- After the system was switched to using metal materials, there were some structural changes such as: changing the structure to ensure the ability to process by metal cutting method; reduce the size in some details to reduce weight.

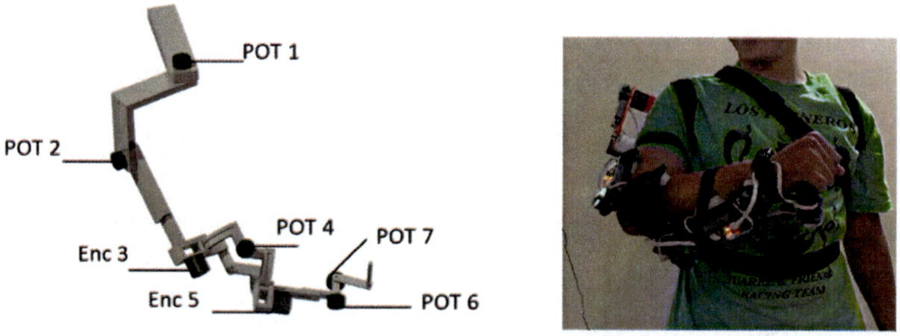

Fig. 2. Manufactured results of the previous E-HMCS model 11

For signal acquisition and processing, the system is using 5 Potentiometers of CPP-35 (Midori Precision Co., Ltd.) and 2 Encoders of E6B2-CWZ6C (Omron Corp.). The location of the sensors is shown as Fig. 2. In addition, the study used the Audruino Uno circuits to connect to the sensors as well as the Matlab software on the computer. For the purpose of checking the quality of the end-point position, the study only needed to focus on measuring and calculating the angles of the shoulder and elbow joints. These two joints are combined by 5 rotating joints: Shld. Flx./Ext; Shld. Abd./Add.; Shld. Int./Ext. Rot.; Elb. Flx./Ext.; Elb. Pro./Sup. Model of human arm combined with measuring device (Fig. 3).

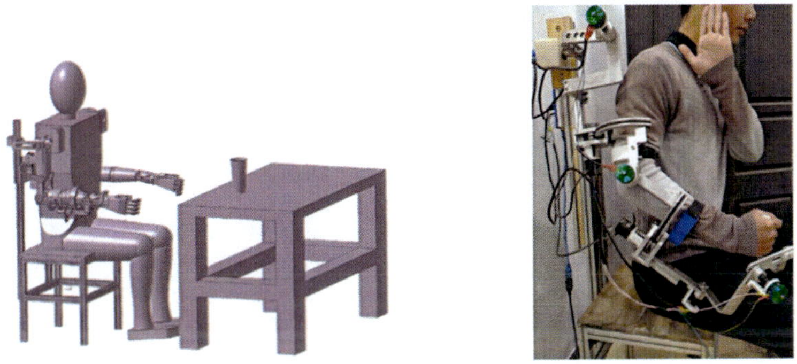

a) Virtual model in simscape environment b) Produced model

Fig. 3. Manufactured results of the improving E-HMCS

2.1 Human Arm Model Design

The purpose of designing a human arm model is to convert of rigid segments connected to each other by three anatomical joints (shoulder joint, elbow joint and wrist joint) while ignoring the movement of scapular and clavicle motions [13]. In this study, 7 Degree of Freedoms (DoFs) were analyzed including shoulder flexion/extension, shoulder abduction/adduction, shoulder lateral-medial rotation, elbow flexion/extension, elbow pronation/supination, wrist flexion/extension and wrist pronation/supination. In order to facilitate the research and manufacturing process of upper limb rehabilitation robot, we also proposed a 3D model that simulates the human arm to ensure the basic anthropometric parameters of Asians in general and especially pay attention to Vietnamese [14]. The human arm anthropometric parameters and ROM are shown detail in Ref. [15]. The segments' lengths of model were chosen based on the segments' length of the person who wore the E-HMCS. The human arm model wearing the E-HMCS measuring system is shown as Fig. 4.

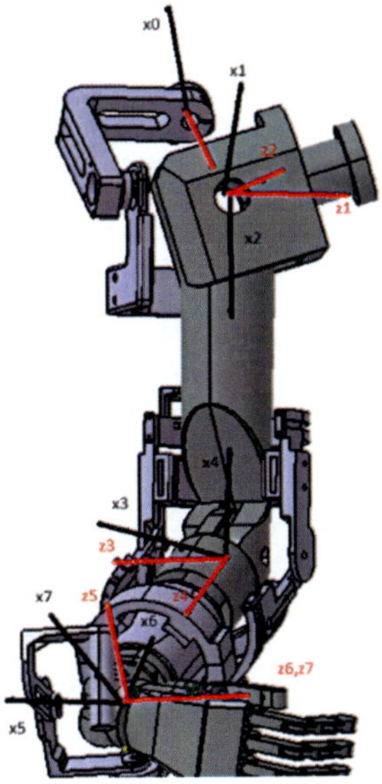

Fig. 4. Human arm model wearing the E-HMCS

2.2 Converting Measured Angle to Human Arm Angle

Measuring model E-HMCS and human arm model have different order of the first two joints as well as different origin coordinate system Fig. 4. Therefore, they will have different D-H parameters and the matching angles of 1st, 2nd, and 3rd joints of the 2 models are different. From angular of joint 4th onwards, the angle of the measured model will coincide with the angle of the arm model. To convert angles from gauge to human arm, it is needed to find $\{q'_1, q'_2, q'_3\}$ angle of human arm so that the direction of link 2 (elbow point) of 2 models are the same. The converting calculation was shown detail in research [15].

3 Experiments and Results

3.1 Experiments Setup

To evaluate the quality of the improved E-HMCS measuring device, ideally we should use higher accuracy devices such as VICON, Qualisys motion capture system [5, 6] to measure and evaluate price. However, under limited conditions, the study uses statistical methods to measure and evaluate. The way to set up the evaluation experiment is done as follows:

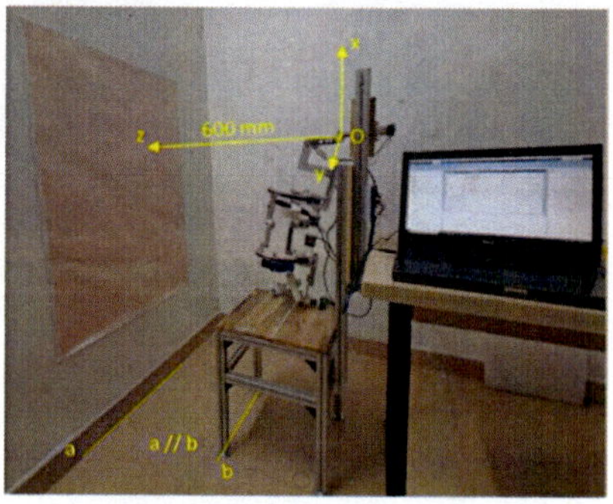

Fig. 5. Experimental setup

- Evaluate device accuracy through endpoint accuracy. It was done by comparing the position in real space with the position in the virtual model of the Simscape environment with the input angle values being the angle values measured from the improved E-HMCS device.
- Model of instrumentation was set up as shown in Fig. 5.
- A man of 22 years old, 1.72 m of height, 71 kg of weight wore this device to perform the experiments.
- Conducted two types of research:

  o Experiment to move to fixed points: The wearer of the E-HMCS device moved the end point to points on the wall as shown in the Fig. 5. For each point on the wall, the person repeated 10 times. The obtained results of the average deviation were calculated, on this basis, the device's accuracy was evaluated.

  o The ADL experiment: The wearer of the device has performed movement according to brushing tooth activity.

3.2 Results

In case the wearer of the E-HMCS device moves the end point to the fixed points on the plane, the experiment was performed to move to 3 points: A (−300; 0; 600); B (−250; −150; 600) and C (−350; 50; 600).

The results after the experiment are shown in Fig. 6. From the graphs, it can be seen that the amount of variation of values in the Z-axis was the smallest, followed by the Y and the largest variation was in the X-axis. Specifically, the values in the Z-direction had a small amount of variation. It ranged from only 599 to 602 for the A point; from 591 to 601 for point B and 595 to 597 for point C. The amount of variations of coordinate value along the Y axis at measuring points A; B; C were ($-13 \div 7$); ($-140 \div -175$);

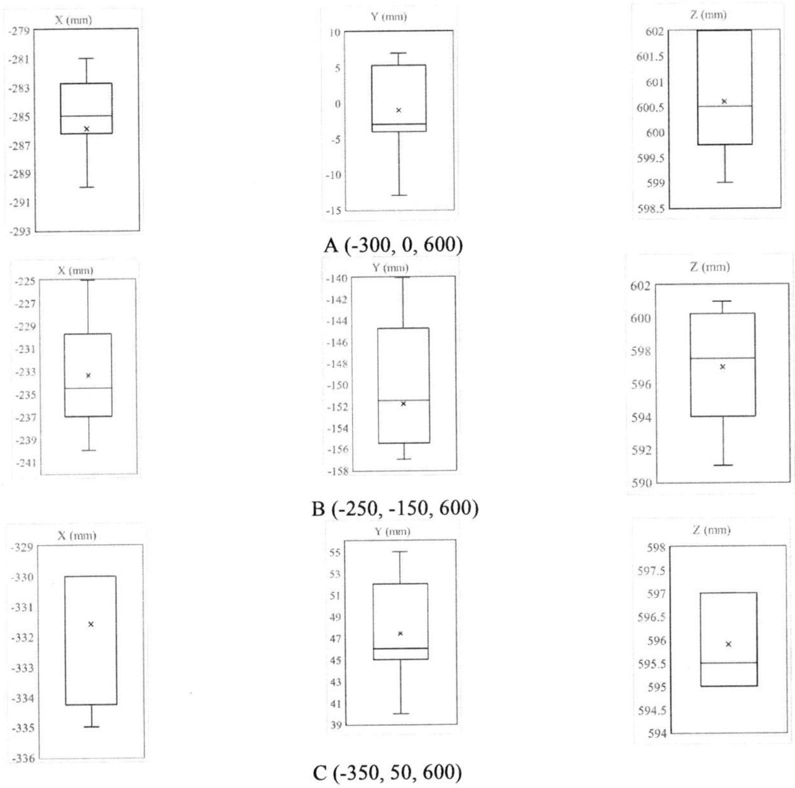

Fig. 6. Experimental results measuring independent points

(40 ÷ 55), respectively. The variations in the X coordinate value at the measuring points A; B; C were (−281 ÷ −300); (−230 ÷ −240); (−330 ÷ −355), respectively. This distribution represented the concentration or scatter of the data. From that, it is noted that the values in the Z direction give the most concentrated data; next are the coordinate values in the Y direction and the scattered data and largest fluctuation in the X direction. In terms of the average deviation in the X, Y, and Z directions, respectively, they were 16.4 mm; 3.2 mm and 2.5 mm. The large deviation in the X-direction can be explained by the elastic deformation of the system since the X-direction is in the same direction as the gravitational acceleration. Meanwhile, the Z and Y directions are perpendicular to the g direction, so the effect is less.

Next, the study performed an experiment to measure joint angles in a movement of ADL. Then, these device angle values will be converted to human arm joint variable angles. Research to measure, collect and calculate for the following drinking movement:

Drinking water activity includes two phases:

- First phase: Reaching phase.
- Second phase: Lifting phase.

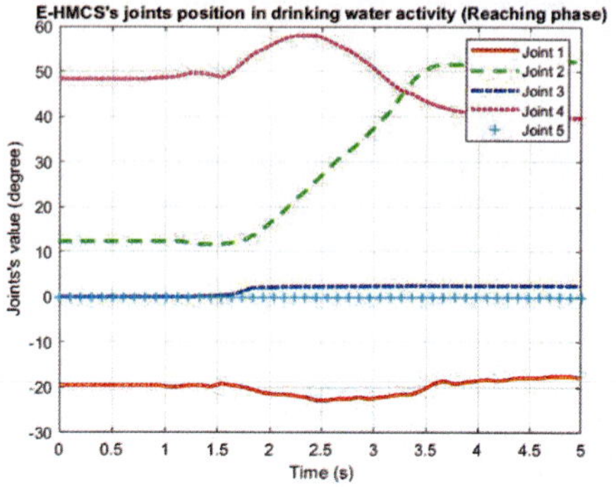

a) Joints' value of E-HMCS

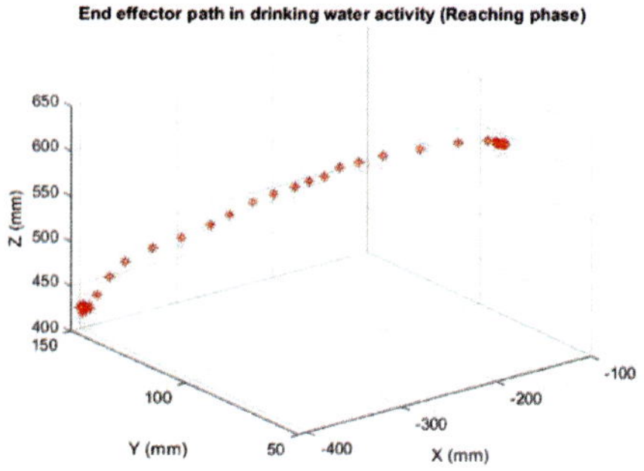

b) End effector trajectory

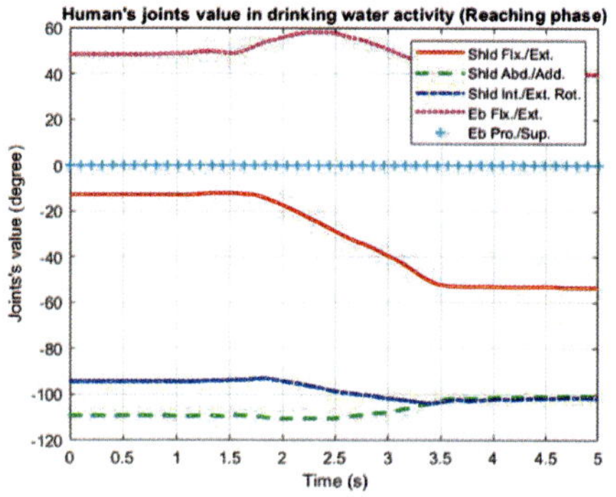

c) Human joints' value

Fig. 7. Reaching stage of drinking water activity

The measurement and calculation results for these movements were shown as Figs. 7 and 8. For each movement, the study had 3 results: the value of the angle of the measuring device; end-point trajectory and corresponding human arm's joint angles. Specifically, the results for the first phase of the drinking water activity were shown in Fig. 7. During this phase, the healthy person put on the measuring device and moved his arm from an initial position to one position of a cup on the table. The angle values of the E-HMCS

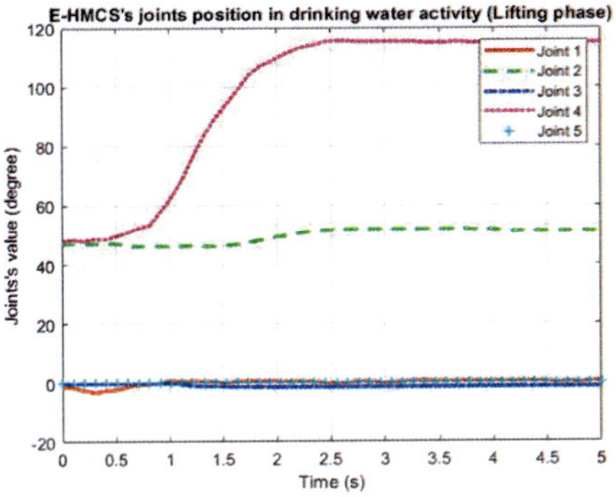

a) Joints' value of E-HMCS

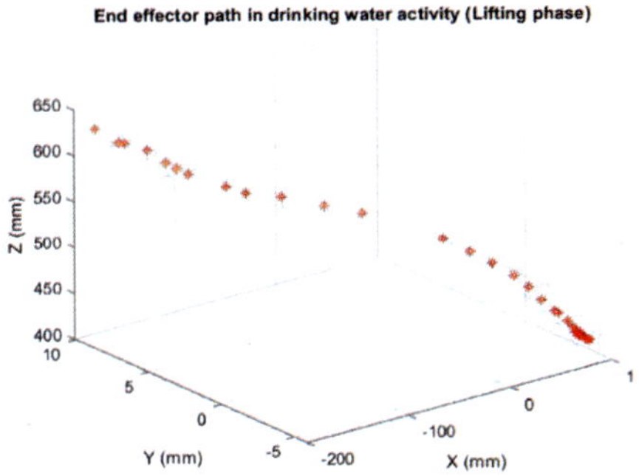

b) Endeffector trajectory

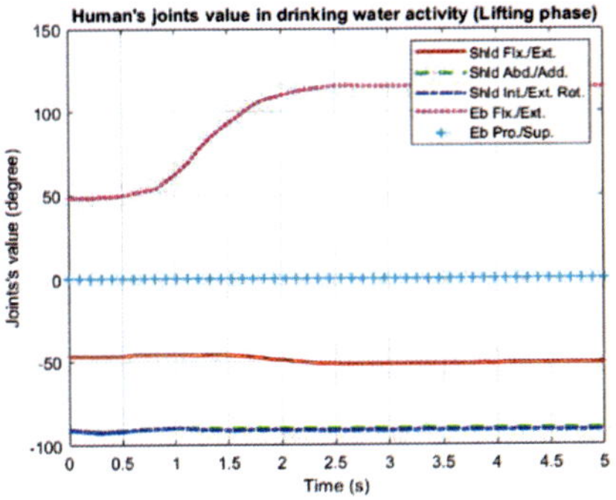

c) Human joints' value

Fig. 8. Lifting stage of drinking water activity

device were collected, processed, and displayed as shown in Fig. 7.a; These values were included in the forward kinematics values. From there, we got the endpoint trajectory as Fig. 7.b. In addition, when transferring the values of the measured angles into the formulas to calculate the values of the human arm joints, we got the results as Fig. 7.c. The total value of joints was calculated including shoulder flexion/extension, shoulder abduction/adduction, shoulder lateral-medial rotation, elbow flexion/extension, elbow pronation/supination.

During the second phase of drinking water activity (Lifting phase), the wearer of the E-HMCS device lifted the cup from the table position to his mouth. The results on the values of the measuring device angles; the results of the end-point movement trajectory (wrist joint) and the value of the human arm joint angles are shown as Fig. 8.a; b; c, respectively.

4 Conclusion

In this paper, the research has upgraded the previous exoskeleton type motion measuring device for human upper limb. Mechanical structure, data collection and processing were presented. To evaluate the quality of the measuring device, the study performed an experiment to measure and evaluate the accuracy of the endpoint locations. The results showed that the vertical position accuracy has the lowest accuracy, while the horizontal accuracy and the direction away from the person have the highest accuracy. The study then performed three experiments measuring movements during daily activity as well as athletic performance. Each operation gave us 3 results about the value of the measuring device joint angles, the endpoint coordinates, and the value of the human arm angles. The study showed the feasibility of the proposed instrumentation with low cost and minimize setup time in the process of measuring the kinetic data. These data are very valuable in the future calculation, design, and control of the rehabilitation robot. In addition, the device can also be used as a master device in a master slave system to control robot.

References

1. https://www.cdc.gov/stroke/facts.htm
2. http://www.healthdata.org/vietnam
3. Chaparro-Rico, B.D.M., Cafolla, D., Ceccarelli, M., Castillo-Castaneda, E.: NURSE-2 DoF device for arm motion guidance: kinematic, dynamic, and FEM analysis. Appl. Sci. (Switzerland) **10**(6), Art. no. 2139 (2020). https://doi.org/10.3390/app10062139
4. Kan, K., Wang, Y., Zhang, W., Lauren, W., Masayoshi, T.: A motion capture system based on human inertial sensing and a complementary filter. In: Proceedings of the 2013 ASME Dynamic Systems and Control Conference DSCC2013, 21 to 23 October 2013, Palo Alto, California, USA (2013)
5. VICON: (2013). http://www.qualisys.com/
6. Qualisys: (2013). http://www.vicon.com/
7. Sabatini, A.M.: Quaternion-based extended Kalman filter for inertial and magnetic quy´êt orientation by sensing. IEEE Trans. Biomed. Eng. **53**(7), pp. 1346–1356 (2006)
8. Fourati, H., Manamanni, N., Afilal, L., Handrich, Y.: A nonlinear filtering approach for the estimation attitude and dynamic body acceleration sensors based on inertial and magnetic: bio-logging application. IEEE Sens. J. **11**(1), pp. 233–244 (2011)
9. Corrin Walmsley, S.Y.P., Williams, S.A., Grisbrook, T., Elliott, C., Imms, C., Campbell, A.: Measurement of upper limb range of motion using wearable sensors: a systematic review. Sports Med. J. Springer Open **8** (2018). https://doi.org/10.1186/s40798-018-0167-7
10. David, A.W.: Motor and Control of Human Biomechanics Movement. University of Waterloo, Waterloo (2009). https://doi.org/10.1002/9780470549148

11. Nguyen, T.-T., Bui, N.-T., Hiroshi, H.: Design and manufacture a cheap equipment to measure human arm motion in developing countries. In: The 4th International Conference on Mechatronics Systems and Control Engineering (ICMSCE 2021), Chengdu, China, 5–7 March 2021 (2021)
12. Nguyen, T., Nguyen, H., Dang, K., Nguyen, P., Pham, H., Bui, A.: Simulation and experiment in solving inverse kinematic for human upper limb by using optimization algorithm. In: Nguyen, N.T., Chittayasothorn, S., Niyato, D., Trawiński, B. (eds.) Intelligent Information and Database Systems. ACIIDS 2021, pp. 556–568. Springer, Cham (2021).https://doi.org/10.1007/978-3-030-73280-6_44
13. Korein, J.U.: A Geometric Investigation of Reach. MIT Press, Cambridge (1986). ISBN:978-0-262-11104-1
14. Trung, N., Hiep, D., Thien, D., Tam, B., Dai, W.: Design a human arm model supporting the design process of upper limb rehabilitation robot. In: 14th South Asian Technical University Consortium Symposium 2020 (SEATUC 2020), 27–28 February 2020, Bangkok, Thailand, pp. 2186–7631 (2020)
15. Nguyen, T., Bui, T., Pham, H.: Using proposed optimization algorithm for solving inverse kinematics of human upper limb applying in rehabilitation robotic. Artif. Intell. Rev. (2021). https://doi.org/10.1007/s10462-021-10041-z

A LSTM-Based Fall Prediction Method Using IMU

Jing Peng, Xianmin Zhang(✉), and Hai Li

South China University of Technology, Guangzhou 510640, China
zhangxm@scut.edu.cn

Abstract. Injuries caused by falls greatly affect quality of the elder's life. Predicting and preventing a fall accident plays an important role in health care. Existing fall prediction methods can hardly be generalized into real-world applications since falling is a highly nonlinear and time-dependent process. The long-short term memory (LSTM) is known for its ability to learn from nonlinear sequential data. In this paper, a fall prediction method is proposed based on LSTM algorithm using a 9-axis Bluetooth inertial measurement unit (IMU). Tri-axial accelerations, angular velocities at the hip are collected by the IMU and used as inputs to train the fall prediction algorithm. By analyzing the sensor data of each motion sequences, fall risk curves are defined and serve as the output of the algorithm. By setting a proper threshold on the fall risk, a fall can be predicted. To validate the proposed algorithm, five activities of daily life (ADLs) and two common falls are performed by 10 participants for 5 times. 4/5 of the collected data are used for training the model while others are used for validation. Result shows that the accuracy, sensitivity and specificity are 97.1%, 100.0%, 96.0%, respectively. Meanwhile, it is found that a fall prediction can be made about 360 ms ahead of the collision, which verifies the applicability of the proposed algorithm.

Keywords: Fall prediction · LSTM · IMU · Sequence

1 Introduction

As the population ages, health care catches more and more attention. According to data published by World Health Organization in 2021, Falls are the second leading cause of accidental or unintentional injury deaths worldwide [1]. Adults older than 60 years of age suffer the greatest number of fatal falls. Besides fracturing bones, a fall might result in a long lie on the floor which can cause many other complication [2].

To reduce the damage caused by falls, many fall detection methods were proposed. According to the used sensor, the existed fall detection methods can be widely divided into three types, namely: the wearable sensor-based methods [3–8], wave reception

This research was supported by the National key technologies R&D program of China (Grant No. 2018YFB1307800).
Guangdong HUST Industrial Technology Research Institute, Guangdong Provincial Key Laboratory of Manufacturing Equipment Digitization (2020B1212060014).

N. V. Khang et al. (Eds.): ASIAN MMS 2021, MMS 113, pp. 680–690, 2022.
https://doi.org/10.1007/978-3-030-91892-7_65

sensor-based methods [9–11] and vision based methods [12–15, 24]. Due to the advantages of abundant information, multi-sensor-based fall detection has attracted much attention [16–18]. The algorithms used to detect falls mainly including threshold-based and machine learning algorithms, such as Support Vector Machine (SVM) [16], K-Nearest Neighbors (KNN) [9], Multi-Layers Perception (MLP) [15], Naive Bayes [19], Hidden Markov Model (HMM) [7].

Compared to wave-based and vision-based ways, use of wearable sensors is free from space constraints and its working cost is low. Portability and aesthetics improvements will facilitate wearable sensors to be widely used in fall detection.

As a typical sensing setup that can simultaneously sensing accelerations and posture information of a moving target, Inertial Measurement Unit (IMU) is an attractive sensor for fall detection [21]. A fall usually happens with quick posture changes and huge impact. During a fall, the tri-axial accelerations of hip would change sharply. Since accelerations change during most of the ADLs are small, threshold-based strategy and SVM are widely used to detect falls. Both algorithms are with low complexity, the false alarm rate inevitably becomes the biggest weakness. The number of accelerometers put in use makes little difference to detection accuracy, but two or more accelerometers can help the algorithm to identify specific activities, not just fall or ADL [19]. Gyroscope always cooperates with accelerometer, which appends tri-axial angular velocity information to fall detection model. Combining posture information and acceleration to judge activities can improve the algorithm performance, it can distinguish some fall-like motions from falls [3]. Considering mobile phone has equipped IMU, some research designed fall detection algorithm based on mobile phones, as well as the function model to alarm and call for help [20].

However, fall detection algorithms only give feedback after a fall when the injury had caused. Fall prediction strategies provide possibilities for avoiding harms. Tong et al. [7] proposed HMM-based method to predict falls using tri-axial accelerometer, the prediction responses 200–400 ms earlier than collision. The result shows 100% sensitivity and 88.75% specificity and the rate of mistaking ADLs for falls leaves room for improvement. Wrong alarms happen frequently could lead to unnecessary troubles which would lower usability and restrict promotion.

Human behavior is continuous in time and space, which provides the most important basis for the predictability of human motion. The value of sensor data at any moment may occur in several activities. That is why threshold-based method could bring high false alarm rate. To correctly predict falls, time properties of the data flow must be considered in algorithm. Hidden Markov model (HMM) is essentially a statistical model, which has poor generalization performance without massive training sets. In addition, the inputs of HMM are sequences with fixed length. So that sliding window and feature extraction should are used to the raw sequences, which sacrifices much time related information. Considering that a fixed length sliding window can't adapt to all activity and all people, a better way is proposed. Long short-term memory (LSTM) is an artificial recurrent neural network (RNN) architecture. It shows excellent performance in processing sequence data with different lengths and is suitable for making predictions based on time series.

Compared to previous researches, which are mainly for fall detection, the LSTM-based prediction model can capture trends at the beginning rather than the result at the ending of falls, and supplies the most important basis for fall prevention. Also, both as time-serials prediction algorithm, LSTM performs better in generalization than HMM.

This paper sought for a practical fall prediction method based on LSTM using IMU-collected information as input. Combining wireless IMU with belt promises portability. LSTM model can get useful information from previous data to make prediction. That means it can learn rules before falls, which increases reliability of predictions. The rest of paper is organized as follows. In Sect. 2, experiments details are introduced. Section 3 describes the basic theory of LSTM and gives the training and evaluating steps. Finally, conclusion is presented.

2 Experiment Design

2.1 Sensor Configuration

It has been reported that for fall prediction, hip is the optimal location for single IMU [19, 22]. As hip joint is close to the center of mass and the acceleration information at this location is more distinguishable for different activities.

In this paper, a wireless IMU (LPMS-B2) was used to collect accelerations and angular velocities. It is a 9-axis Bluetooth IMU, with the size of 39 × 39 × 8 mm, and the weight is 12 g. Rate of data transmission is up to 400 Hz. The working parameters are listed. Orientation range: roll: ±180°; pitch: ±90°; yaw: ±180°. Accuracy is less than 0.5°(static) and 2°(dynamic). In this experiment, we choose ±8 g for accelerometer, ±125°/s for gyroscope (16 bit of ADC) and set the sample frequency to 100 Hz.

2.2 Data Acquisition

Every subject puts on the belt with LPMS which is directly located in front of the waist. The initial state of all test action sequences is upright, cartesian coordinate system is defined as shown in Fig. 1. Data collected includes information of tri-axial accelerometer, gyroscope, and magnetometers, only accelerations and angular velocities are used to predict falls.

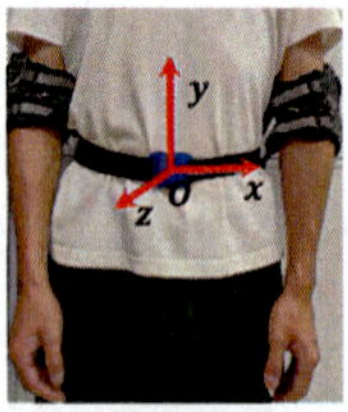

Fig. 1. Cartesian coordinate system

Fig. 2. Slippery floor simulation

Different from simulating falls, slippery floor is simulated to collect databases of real falls (see Fig. 2). Smooth multiple foam boards are placed in front of the sponge

Table 1. Activities selected for this experiment.

Fall (F)/ADL (A)	Activity	Subjects	Trials
F1	Fall forward	P1-P10	R1-R5
F2	Fall side	P1-P10	R1-R5
A1	Lie down & stand up	P1-P10	R1-R5
A2	Jogging	P1-P10	R1-R5
A3	Picking things up & put it down	P1-P10	R1-R5
A4	Sit down & stand up	P1-P10	R1-R5
A5	Upstairs & downstairs	P1-P10	R1-R5

P1-P10: participants number; R1-R5: repeat round number.

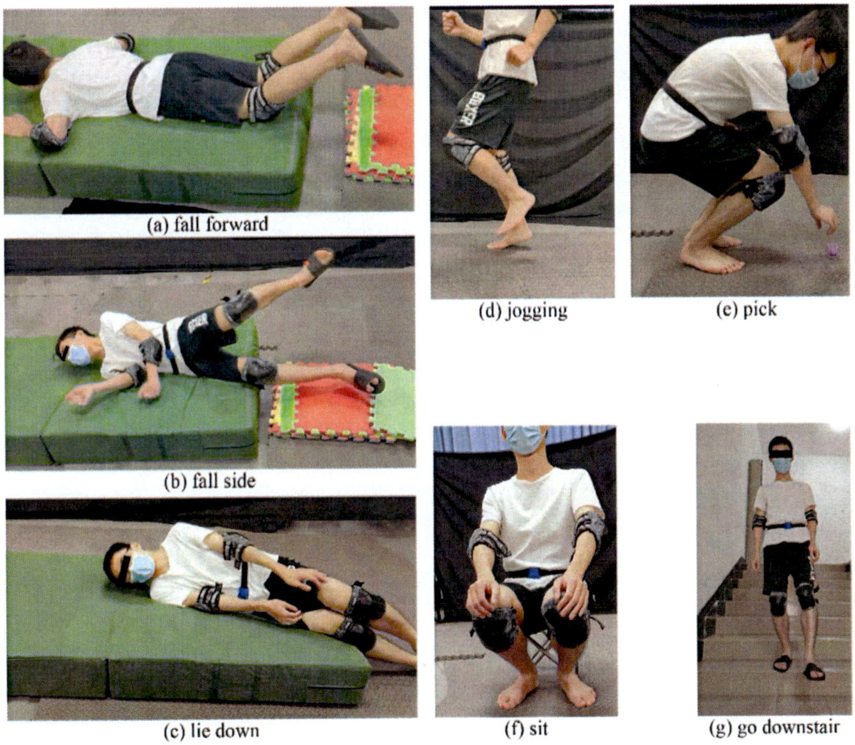

Fig. 3. Activities selected in this experiment

pad, participants will fall due to lack of friction when stepping on the foam board. Real fall data plays an important role in the process of training fall prediction model.

Before the experiment, the informed consent of all subjects was obtained. Sponge pads and protective gear sets for cushioning falls are provided. Ten young participants (aged 23–24 years) perform 7 activities including 2 types of falls and 5 types of ADL (see Fig. 3). Each activity is repeated for 5 times.

Fall forward and fall side are chosen since they are two most common fall styles [23]. ADLs shown in Table 1 are selected based on common activities, motions with similar acceleration trends with falls or high frequency and high amplitude exercise. The purpose of choosing these activities is to enhance the specificity of the model and reduce the false alarm rate. Before every activity except jogging, subjects were asked to walk for a short distance.

3 Fall Prediction Algorithm

3.1 Basic Theory of LSTM

Recurrent neural network (RNN) is a deep learning model and designed to model sequence data. Different from feed forward neural networks, current hidden unit is influenced by all the other hidden units before and influence decreases with the distance away from this node. RNN is capable to capture temporal dependence of a short sequence because of vanishing gradient problem.

An improved RNN model named LSTM overcomes the defect. LSTM unit replaces the original hidden unit. From Fig. 4, we can know that a LSTM unit consists of one memory cell and three gates including forget gate, input gate and output gate. Forget gate filters useless information left and discard it or lower its weight according to current input. The gate just reserves valid information and makes it possible for long-term memory. Input gate optionally adds new information got from current node to cell state. Output gate saves reorganized information to the hidden layer.

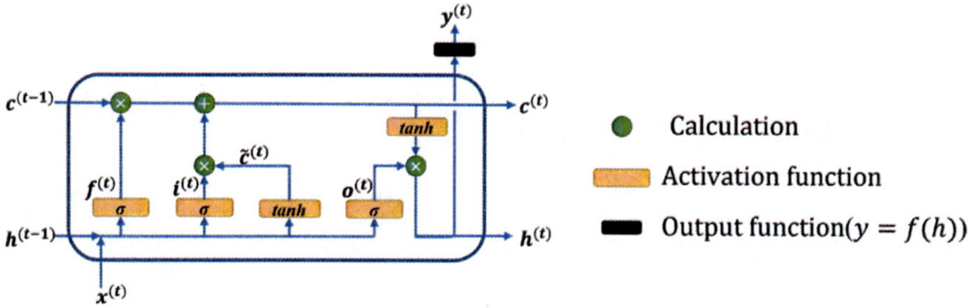

Fig. 4. A LSTM unit. t is the node rank in sequence, c means memory cell. f, *i*, o is the first letter of forget, input and output respectively.

3.2 Fall Prediction Based on LSTM

The process of training the prediction model includes data preprocessing and algorithm structure establish (Fig. 5). Model evaluation is discussed in next subsection.

The premise of accurately predicting a fall is to find the law of the data before the collision. The signal frequency of the falling process differs from person to person. LSTM is good to deal with different time-step sequential signals and has no limit to length of stack sequences. The input data of LSTM is a 6 × n matrix, each row representing the acceleration and angular velocity of the *x*, *y*, *z* axis, denoted as *ax*, *ay*, *az*, *wx*, *wy*, *wz*. The output of the prediction algorithm is a $1 \times (n - 1)$ vector representing predictive probability of falls. It is a regression model. Root mean square error (RMSE) of the fall risk curve is used to reflect performance of the prediction model.

Data Preprocessing. Every participates performs 5 rounds for each activity. Database is divided into two parts in 4:1 ratio, the large part is for training (10 * 7 * 4 trails) and another for testing (10 * 7 * 1 trails). Data normalization does make sense in algorithm optimization, it weak the two differences mentioned above. The RMSE decreased from 0.039 to 0.024 in testing database. Normalization progress is shown in Fig. 6.

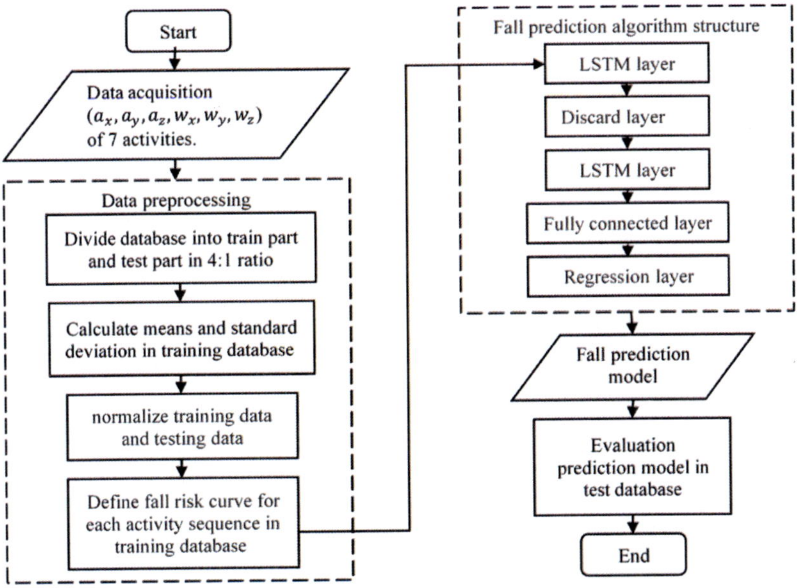

Fig. 5. Flowchart of obtaining fall prediction model

Equation 1 describes the normalization strategy, where *I_N* is normalized input sequences, *I* is raw data matrixes.

$$I\_N = (I - u)./\sigma \tag{1}$$

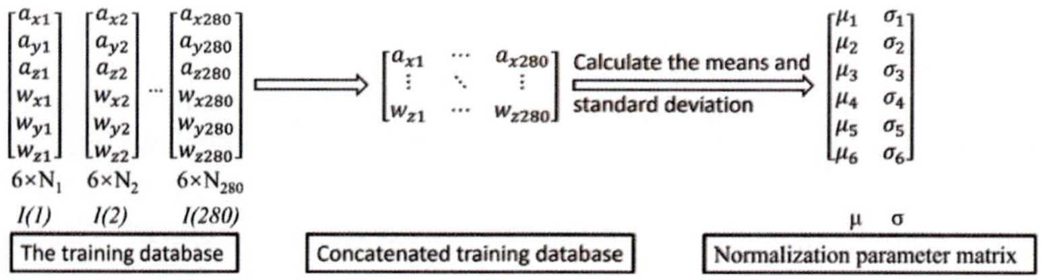

Fig. 6. The procedures to extract normalization parameters

Except normalization, the fall risk curve is defined for each motion sequence, presented in Fig. 7. Three turning time points (t_s, t_r, t_e) and two risk levels (l_s, l_h) are the keys for falls. Where t_s denotes the start of unbalance, t_r denotes the time point when the risk coefficient become the largest; t_e denotes the moment when the body collides with the ground. Before t_s, balance is detected and the risk keeps being l_s; When the unbalance becomes self-irreversible, tilt angle of the overall body is close to 45°, the risk increases rapidly to a high level—l_h and reduce to l_s when collision occurs.

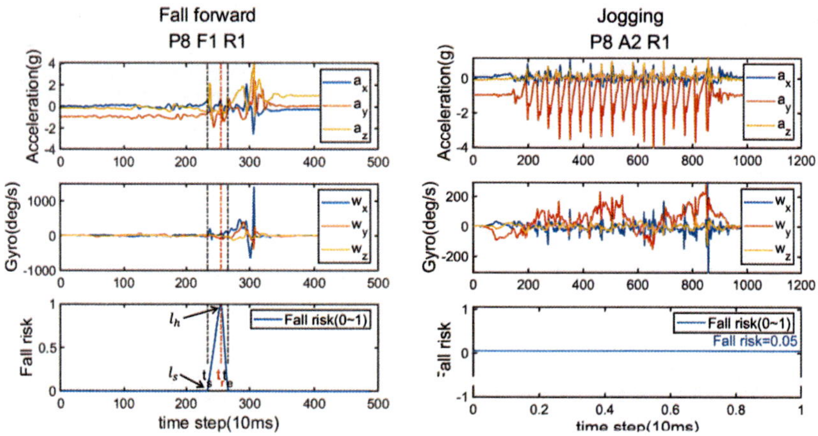

Fig. 7. Define the fall risk curves for the P8_F1_R1 and P8_A2_R1

Algorithm Structure. As Fig. 5 shows, the fall prediction model consists of five layers: two LSTM layers and a discard layer is added between them to avoid overfitting. A fully-connected layer condenses the outputs of the last LSTM layer to one—fall risk coefficient range 0–1. The last layer is a regression layer, which is to modify the model, so that the model outputs in training data is as close as possible to the training output.

3.3 Evaluation of Fall Prediction Model

Accuracy (ACC), sensitivity (SE) and specificity (SP) are selected to evaluate the performance of the proposed prediction model and can be obtained from Table 2.

Table 2. Evaluation parameters

Actual situation	Model prediction		Evaluation parameters	
	Fall	ADL		
Fall	TP	FN	$SE = TP/(TP + FN)$	$Acc = \frac{TP + TN}{TP + FN + TN + FP}$
ADL	FP	TN	$SP = TN/(FP + TN)$	

This fall prediction model works well in test database including 70 sequences collected from 7 activities of 10 subjects, prediction results are presented in Fig. 8. Risk prediction values of most ADLs are lower than 0.1, several jogging, pick up and go upstairs testing samples get higher risk value ranging from 0.26 to 0.38. The peaks of three lie down & get up samples reach up to 0.80, 0.74 and 0.47. As young adults act the lie down activity in completely relax state, which causes the impact. An intentional fall gets similar risk curve as a real fall. However, the elder could lie down more gently than the youth. All the risk curve peaks of fall sequences exceed 0.5. Thus, a threshold T1 = 0.5 was set to diminish false alarm. Prediction results are shown in Table 3. If falls are predicted and ADLs are detected as un-falls, they are correct predictions. Otherwise, wrong predictions. To display result more clearly and intuitively, resultant acceleration subtracted gravity substitute tri-axial acceleration and angular velocity to describe motion state and the corresponding fall risk prediction result is drawn in red continues line. The red dotted line shows the moment the fall prediction is captured.

Table 3. Evaluation results in test database

True situation	Model prediction		
	Correct prediction	Wrong prediction	Evaluation
Fall forward & fall side	20	0	$SE = 100.0\%$
5 ADLs (A2-A5)	48	2 (A1: Lie down)	$SP = 96.0\%$
All tests	68	2	$Acc = 97.1\%$

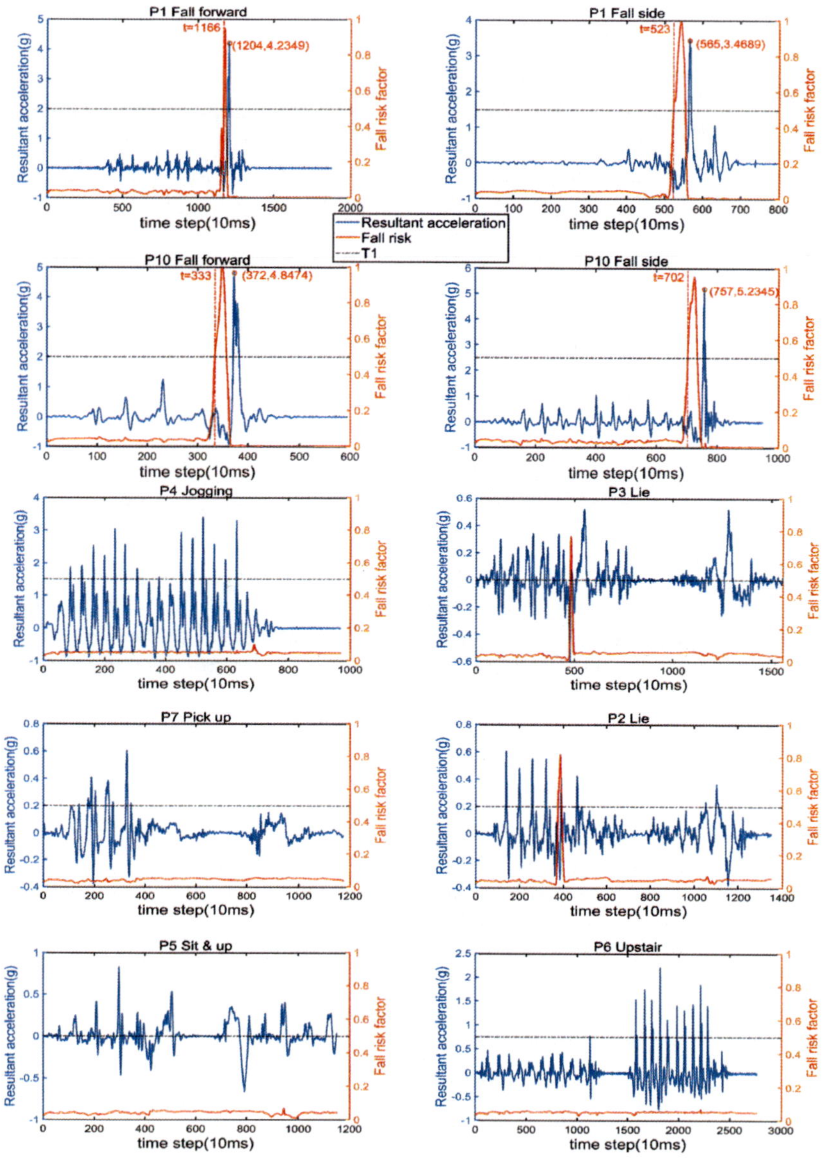

Fig. 8. Fall risk predictions on testing database

4 Conclusion

This paper proposes a fall prediction method based on LSTM using a wearable IMU. Database are collected from 2 falls and 5 ADLs acted by 10 subjects. Tri-axial acceleration and angular velocity are extracted as features of motion. Normalization applied to raw data sequences enhances the generalization ability of the model. The normalized data sequences are input to the LSTM layer to learn the law of the fall process.

Fully connected and regression layers convert outputs of LSTM to one-dimensional risk sequence.

The experiment results validate the excellent performance of LSTM-based method in fall prediction. There are 20 falls and 50 ADLs in testing dataset, falls are successfully predicted at least 160 ms before the collision of body, 360 ms for average, and considering delay in the neurological system is 100–200 ms [25], the prediction signal leaves enough time to trigger fall prevention equipment before elders try to balance their own. Only two lies are misjudged. Sensitivity and specificity, characterizing proportion of correctly detecting falls and correctly detecting ADL, achieved 100% and 96% separately.

It must be pointed out that characteristics of each movement between young people and older people are different. In the future, we will try to collect activity information of the elder and adjust the fall prediction method to application in the real world.

References

1. WHO Newsroom: https://www.who.int/news-room/fact-sheets/detail/falls. Accessed 18 June 2021
2. Rubenstein, L.Z., Josephson, K.R.: The epidemiology of falls and syncope. Clin. Geriatr. Med. **18**(2), 141–158 (2002)
3. Li, Q., Stankovic, J.A., Hanson, M.A., et al.: Accurate, fast fall detection using gyroscopes and accelerometer-derived posture information. In: Sixth International Workshop on Wearable and Implantable Body Sensor Networks 2009, pp. 138–143 (2009)
4. Kerdegari, H., Samsudin, K., Ramli, A.R., et al.: Evaluation of fall detection classification approaches. In: 4th International Conference on Intelligent and Advanced Systems (ICIAS2012), pp. 131–136 (2012)
5. Bourke, A.K., O'brien, J.V., Lyons, G.M.: Evaluation of a threshold-based tri-axial accelerometer fall detection algorithm. Gait Posture **26**(2), 194–199 (2007)
6. Lee, J.K., Robinovitch, S.N., Park, E.J.: Inertial sensing-based pre-impact detection of falls involving near-fall scenarios. IEEE Trans. Neural Syst. Rehabil. Eng. **23**(2), 258–266 (2014)
7. Tong, L., Song, Q., Ge, Y., et al.: HMM-based human fall detection and prediction method using tri-axial accelerometer. IEEE Sens. J. **13**(5), 1849–1856 (2013)
8. Sucerquia, A., López, J.D., Vargas-Bonilla, J.F.: SisFall: a fall and movement dataset. Sensors **17**(1), 198 (2017)
9. Popescu, M., Li, Y., Skubic, M., et al.: An acoustic fall detector system that uses sound height information to reduce the false alarm rate. In: 30th Annual International Conference of the IEEE Engineering in Medicine and Biology Society 2008, pp. 4628–4631 (2008)
10. Wang, H., Zhang, D., Wang, Y., et al.: RT-Fall: a real-time and contactless fall detection system with commodity WiFi devices. IEEE Trans. Mob. Comput. **16**(2), 511–526 (2016)
11. Su, B.Y., Ho, K.C., Rantz, M.J., et al.: Doppler radar fall activity detection using the wavelet transform. IEEE Trans. Biomed. Eng. **62**(3), 865–875 (2015)
12. Mastorakis, G., Makris, D.: Fall detection system using Kinect's infrared sensor. J. Real-Time Image Process. **9**(4), 635–646 (2014)
13. Rougier, C., Meunier, J., St-Arnaud, A., et al.: Robust video surveillance for fall detection based on human shape deformation. IEEE Trans. Circ. Syst. Video Technol. **21**(5), 611–622 (2011)
14. Deng, Z.F., Ming, W.D., Zou, S.: A fall detection method based on CNN and human ellipse contour motion features. J. Graph. **39**(06), 1042–1047 (2018). (In Chinese)

15. Foroughi, H., Aski, B.S., Pourreza, H.: Intelligent video surveillance for monitoring fall detection of elderly in home environments. In: 11th International Conference on Computer and Information Technology 2008, pp. 219–224 (2008)
16. Kwolek, B., Kepski, M.: Human fall detection on embedded platform using depth maps and wireless accelerometer. Comput. Methods Programs Biomed. **117**(3), 489–501 (2014)
17. Martínez-Villaseñor, L., Ponce, H., Brieva, J., et al.: UP-fall detection dataset: a multimodal approach. Sensors **19**(9), 1988 (2019)
18. Kwolek, B., Kepski, M.: Improving fall detection by the use of depth sensor and accelerometer. Neurocomputing **168**, 637–645 (2015)
19. Cleland, I., Kikhia, B., Nugent, C., et al.: Optimal placement of accelerometers for the detection of everyday activities. Sensors (Basel, Switzerland) **13**(7), 9183–9200 (2013)
20. Kau, L., Chen, C.: A smart phone-based pocket fall accident detection, positioning, and rescue system. IEEE J. Biomed. Health Inform. **19**(1), 44–56 (2015)
21. Cafolla, D., Chen, I.-M., Ceccarelli, M.: An experimental characterization of human torso motion. Front. Mech. Eng. **10**(4), 311–325 (2015)
22. Kangas, M., Konttila, A., Lindgren, P., et al.: Comparison of low-complexity fall detection algorithms for body attached accelerometers. Gait Posture **28**(2), 285–291 (2008)
23. O'neill, T.W., Varlow, J., Silman, A.J., et al.: Age and sex influences on fall characteristics. Ann. Rheum. Dis. **53**(11), 773–775 (1994)
24. Chaparro-Rico, B.D., Cafolla, D.: Test-retest, inter-rater and intra-rater reliability for spatiotemporal gait parameters using SANE (an eaSy gAit aNalysis systEm) as measuring instrument. Appl. Sci. **10**(17), 5781 (2020)
25. Di, P., et al.: Fall detection and prevention control using walking-aid cane robot. IEEE/ASME Trans. Mechatron. **21**(2), 625–637 (2016)

Micro and Nano Mechanisms and Systems

Single Mask and Low Voltage of Micro Gripper Driven by Electrothermal V – Shaped Actuator

Phuc Hong Pham(✉) and Dien Van Bui

Hanoi University of Science and Technology, No.1, Daicoviet, Hai Ba Trung, Hanoi, Vietnam
phuc.phamhong@hust.edu.vn

Abstract. This paper presents a design, calculation, simulation and fabrication of a micro gripper driven by the electrothermal V-shaped actuator. The finite difference method is applied to calculate more exactly the temperature distribution and displacement of V-beams. The advantages of this device are large displacement amplification factor, low driving voltage and simple manufacturing using only single mask. Simulation of the micro gripper in ANSYS multiphysics shows an average voltage-deviation of 4.45% in comparison with calculation at the same displacement. A maximum stroke of each jaw can be up to 40 μm at the measured and calculated voltages of 10 V and 14.42 V, respectively. The micro gripper has been fabricated successfully by utilizing SOI-MEMS technology and surface sputtering technique.

Keywords: Micro gripper · Electrothermal V-shaped actuator · SOI-MEMS technology · Sputtering technique

1 Introduction

Currently, MEMS (micro-electro-mechanical-system) is increasingly being developed in the fields of micro-assembly, micro-robot, material research, or integration structures. With the characteristics and performances at tiny scale, MEMS devices exhibit advantages such as light weight, low power consumption, stable operation, and easy integration. One of the typical systems that can be considered as a feature of MEMS devices is a micro gripper, with capabilities to grip, hold and move objects at the micro/nano scale only.

The types of micro gripper are classified mainly according to the physical effects used for the actuation, or driving actuators. They are categorized as electrothermal, electrostatic, piezoelectric, electromagnetic, and shape memory alloys - SMA [1]. In case of electrothermal actuators (ETAs), they can generate large displacement and force based on thermal expansion of thin beams while applying voltage [2]. The ETA is mostly smaller in size than electrostatic and electromagnetic actuators, and also simple fabrication/control. In addition, there is not significant delay like the SMA actuator when ETA is working. Therefore, the electrothermal-based micro grippers are suitable for manipulating in various applications such as micro assembling, micro robotics or in micro testing system. There are three types of beam structures in the electrothermal actuators

N. V. Khang et al. (Eds.): ASIAN MMS 2021, MMS 113, pp. 693–704, 2022.
https://doi.org/10.1007/978-3-030-91892-7_66

as V-shaped, U-shaped, and Z-shaped. The U-shaped actuator uses the asymmetrical thermal expansion of the hot and cold arms to create deflection and motion [3]. The V-shaped and Z-shaped use the total thermal expansion along the beams with one fixed-end to generate a force and motion [4]. At the same voltage, the V-shaped produces a larger displacement than the U-shaped [5]. With the similar size, the V-shaped produces larger displacement and force than the Z-shaped [6]. Besides, the working stability of the V-shaped actuator is also the best in comparison with others. Therefore, V-shaped actuator is usually preferred for driving MEMS devices such as micro motors or micro grippers.

In this paper, we propose a new design of micro gripper driven by the electrothermal V-shaped actuator (EVA), which is developed from our previous design in [7]. The micro gripper has outstanding advantages like low driving voltage, large jaw's displacement, simple fabrication, etc. The device is manufactured perfectly using the SOI-MEMS technology and surface sputtering technique.

2 Configuration, Operation and Heat Transfer Model

2.1 Configuration and Working Principle

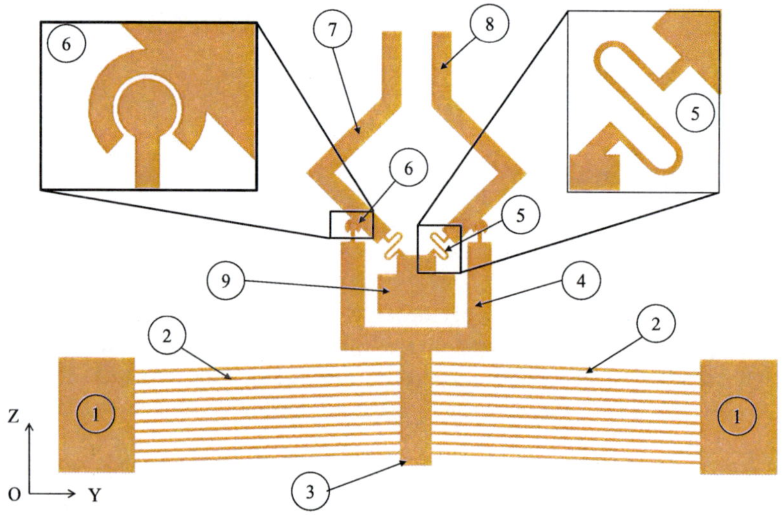

Fig. 1. Configuration of the micro gripper

Figure 1 shows the design and configuration of micro gripper. The micro gripper works on the principle of thermal expansion of V-beam. When voltage is being applied on two fixed electrodes ①, the long V-beams ② will be heated and expanded thermally. Here one end is connected to the electrodes ①, the other will push the shuttle ③ moving in Z-direction. The shuttle ③ is connected to U-shaped claw ④ and the claw can drive gripping arms ⑦ through two revolute joints ⑥. The end of the gripping arms connects to the fixed pad ⑨ by two elastic springs ⑤. When the shuttle ③ moving forward, the U-shaped claw pushes the both jaws ⑧ to move closely. If the driving voltage goes to zero, the temperature decreases and the V-beams will shrink. As a result, the shuttle moves to initial position and helps to open two jaws.

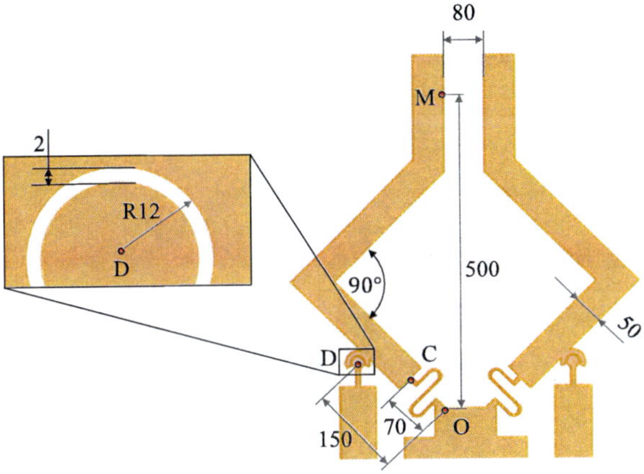

Fig. 2. Dimensions of the gripping arm

In our design, the micro gripper has advantages as: Reducing concentrated stress at the elastic elements by using revolute joints ⑥; increasing grip capacity by using elastic springs ⑤ with smaller stiffness; amplifying the displacement of the jaws ⑧ by locates the revolute joint on the gripping arm ⑦ near the spring (with amplification factor $k = \frac{OM}{OD} = \frac{500}{150} = 3.33$). The design dimensions of revolute joint, gripping arm and jaw are shown as in Fig. 2.

The geometrical dimensions of V-shaped actuator are shown in Fig. 3. Where n is the number of beam pairs; L, h, w are the length, thickness and width of a single beam, respectively; θ is the inclined angle of V-beam in Y-direction. This V-shaped structure is similar to the design in [7] and integrated to drive micro gripper as displayed in Fig. 1.

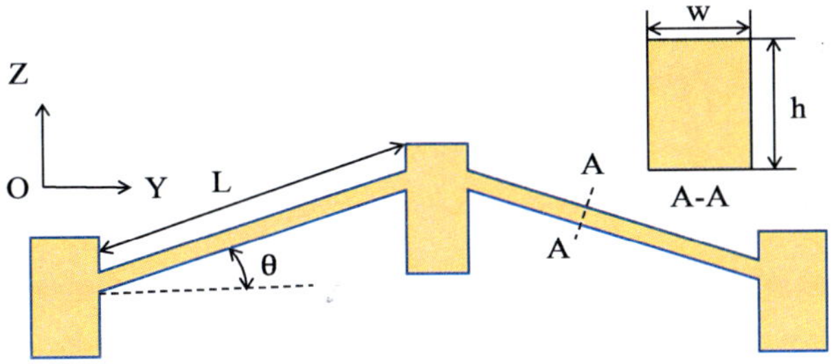

Fig. 3. Configuration of the electrothermal V-shaped actuator

2.2 Heat Transfer Model of the Electrothermal V-Shaped Actuator

The heat transfer in the V-beam system is a complex process. Here, we assume that the heat transfer in the beam is one-way and along the beam length of $2L$; neglect convection, radiation, and heat loss of the shuttle. According to [7], the energy balancing equation in Y-direction of a beam element i^{th} with a tiny section of Δx is as follow:

$$q_{i+1} + q_{i-1} + q_s + q_e = q_{st} \tag{1}$$

where q_{i+1} and q_{i-1} are heat energy conducting from elements $(i + 1)^{th}$ and $(i - 1)^{th}$ to i^{th}; q_s is heat energy radiating from the beam to substrate through the air gap g_a; q_e and q_{st} are Jul heating and heat energy stored in element i^{th} of the beam, respectively.

The equation for computing temperature within element i^{th} of the beam at the period of $(j + 1)$ is expressed as:

$$T_i^{j+1} = \frac{k_s}{C_p.D}\cdot\frac{\Delta t}{\Delta x^2}.T_{i-1}^j + \left(1 - 2.\frac{k_s}{C_p.D}\cdot\frac{\Delta t}{\Delta x^2} - \frac{k_a.S}{h.g_a}\cdot\frac{\Delta t}{C_p.D}\right).T_i^j + \frac{k_s}{C_p.D}\cdot\frac{\Delta t}{\Delta x^2}.T_{i+1}^j + \frac{k_a.S}{h.g_a}\cdot\frac{\Delta t}{C_p.D}.T_0 + \frac{U^2}{4\rho.L^2}\cdot\frac{\Delta t}{C_p.D} \tag{2}$$

Where, k_s and k_a are thermal conductivity of single crystal-silicon and air; T_i is the temperature of beam element i^{th}; $T_0 = 20\,^{\circ}\mathrm{C}$ is an environment temperature; Δt is a time increment; Δx is a length of beam-element; S is the shape factor of beam area section: $S = 0.6265\left(\frac{h}{w}\right) + 1.1188$ (determined by simulation); C_p and D are the specific heat and density of silicon; $\rho = \rho_0[1 + \lambda(T - T_0)]$ is the resistivity of silicon; h is a thickness of device layer; U is the driving voltage [8, 9].

2.3 Thermal Expansion Force

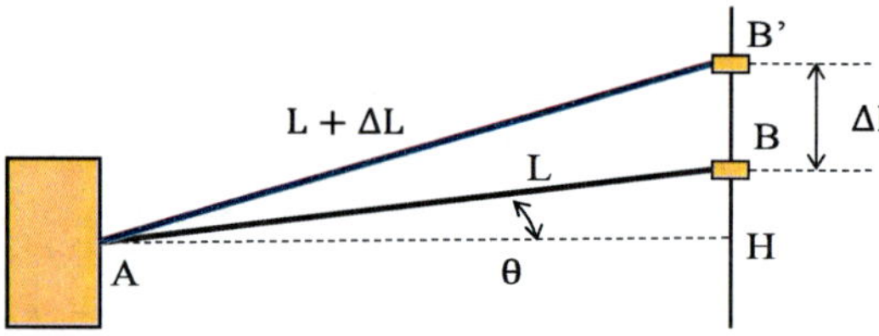

Fig. 4. The displacement of the shuttle

The total thermal expansion of the single beam can be calculated as [7, 9]:

$$\Delta L = \frac{1}{2}\sum_{i=1}^{m}\alpha.(\mathrm{T}_i - \mathrm{T}_0).\Delta x \tag{3}$$

Where α is a thermal expansion factor of silicon; m is total number of beam-element.

When applying the voltage, the single beam will expand and push top B moves to new position B'. The displacement of shuttle $\Delta D \approx BB'$ in Z-direction (Fig. 4) can be calculated as:

$$\Delta D = \sqrt{(L + \Delta L)^2 - (L\cos\theta)^2} - L\sin\theta \tag{4}$$

The thermal expansion force generates along the single beam is computed by equation:

$$F_b = A \cdot E \cdot \frac{\Delta L}{L} \tag{5}$$

Where, $A = h \cdot w$ is cross-section area of the beam; E is Young's modulus of silicon.

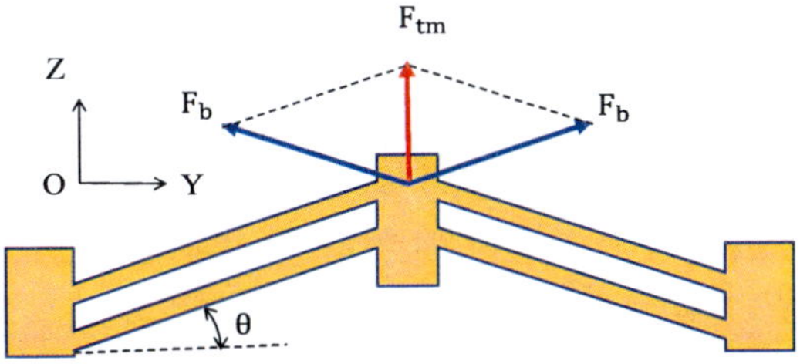

Fig. 5. Thermal expansion force acting on the shuttle

The total expansion force F_{tm} of V-beam system (including n beam pairs) acting on the shuttle in Z-direction (Fig. 5) is calculated as [9]:

$$F_{tm} = 2n.\, F_b.\, \sin\theta \tag{6}$$

3 Calculation and Simulation of Micro Gripper

3.1 Calculation of jaw's Stroke

The movement model of the jaw and gripping arm is shown in Fig. 6. Finally, we need to find the relationship between the jaw's stroke (ΔM) and driving voltage U of the EVA. Bases on dimensions illustrated in Figs. 2 and 6, the stroke ΔM is calculated by following equation:

$$\Delta M = k \cdot \Delta K = k \cdot \Delta K_z \cdot \cos\varphi = k \cdot (\Delta D - g_K) \cdot \cos\varphi \tag{7}$$

Where, ΔK is the displacement of the point D when the gripper arm rotates around the elastic neck of spring (i.e. the point O in Fig. 6); ΔK_z is the displacement of the point D on the gripping arm in Z-direction; g_K is fabrication gap of the revolute joint (in this design $g_K = 2$ μm); φ is an inclined angle between gripping arm and Y-direction.

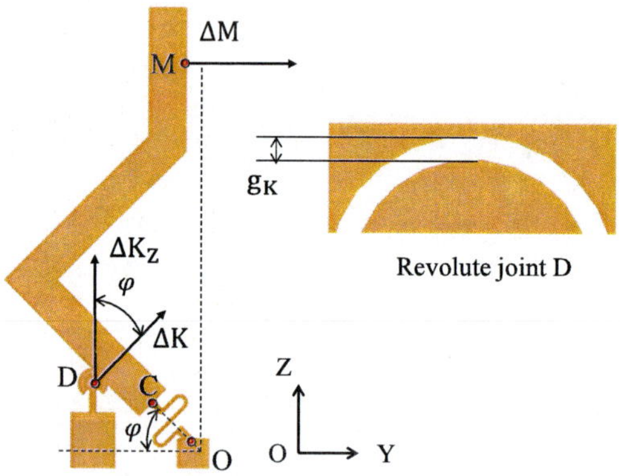

Fig. 6. Displacement of gripping arm and revolute joint

From Eqs. (4) and (7), the transverse displacement of the jaw is calculated according to the thermal expansion of a single beam as following:

$$\Delta M = k \cdot \left(\sqrt{(L + \Delta L)^2 - (L.\cos\theta)^2} - L.\sin\theta - g_K \right) \cdot \cos\varphi \tag{8}$$

From Eqs. (2), (3), and (8), the relationship between driving voltage U and jaw's stroke ΔM is defined. It is clear that ΔM to be amplified larger than the displacement ΔD of the shuttle when micro gripper is working.

Geometric dimensions and material properties of V-shaped actuator are given as: $n = 10$; $L = 700\ \mu\text{m}$; $h = 30\ \mu\text{m}$; $w = 6\ \mu\text{m}$; $\theta = 2°$; $\varphi = 45°$; $g_a = 4\ \mu\text{m}$.

Table 1. Material properties of silicon

$D\left(kg/m^3\right)$	$E(GPa)$	$k_a(W/mK)$	$C_p(J/kgK)$	$\rho_0(\Omega\mu m)$	$\lambda\left(K^{-1}\right)$
2330	169	0.0257	712	148	1.25×10^{-3}

Table 2. Material parameters depend on temperature

$T(K)$	300	400	500	600	700	800	900	1000	1100	1200	1300
$k_s(W/mK)$	156	105	80	64	52	43	36	31	28	26	25
$\alpha\left(10^{-6}\,K^{-1}\right)$	2.62	3.25	3.61	3.84	4.02	4.15	4.18	4.26	4.32	4.38	4.44

Based on material properties of silicon listed in Tables 1 and 2, the relation between the driving voltage U and the values of stroke ΔM are determined and shown in Fig. 9.

3.2 Calculate the Gripping Force

This section establishes the relationship between gripping force F_g generated on each jaw and driving voltage U. In order to move the gripping arms, the total thermal expansion force F_{tm} loaded on the shuttle has to be greater than the total hindering forces F_{hf}. These forces include the elastic forces F_{ef} of the V-beams and F_s of the springs (here, the friction force is ignored).

$$F_{tm} > \sum F_{hf} = F_{ef} + 2F_s \tag{9}$$

The simulation is used to determine the stiffness k_s (i.e. elastic force F_s) of the spring ⑤. The stiffness of V-beam system is determined by principle of equivalent displacement in Z-direction. Then, the total stiffness in Z-direction of n beam pairs is calculated as [7, 10]:

$$k_{td} = \frac{2n \cdot E \cdot \left(12I \cdot \cos^2\theta + A \cdot L^2 \cdot \sin^2\theta\right)}{L^3} \tag{10}$$

Where I is an inertia moment of beam's cross-section area.

When displacement of shutter $\Delta D_{min} = g_K = 2\ \mu\text{m}$, the U-shaped claw will overcome the gap g_K and begin to push on gripping arm. Therefore, the minimum driving voltage needs to cross this gap is:

$$U_{min} \approx 5.19\,\text{V} \tag{11}$$

In order to guarantee the micro gripper can grip the object, we need to calculate the force F_g generated on each jaw when applying various voltages. Figure 7 shows the diagram of the forces acting on gripping arm. Here F_d is the driving force from the claw acting on one arm; F_n is the element force acting on joint D of the arm in the direction perpendicular to the rotational radius OD; F_g is the gripping force at point M of each jaw while applying voltage U; F_s is the elastic force of spring located at point C.

These forces can be calculated as:

$$F_g = \frac{F_n \cdot OD - F_s \cdot OC}{OM} = \frac{F_n \cdot 150 - F_s \cdot 70}{500} \tag{12}$$

$$F_n = F_d \cdot \cos\varphi = \left(\frac{F_{tm} - F_{ef}}{2}\right) \cdot \cos 45^\circ \tag{13}$$

Where: $F_{ef} = k_{td} \cdot \Delta D$; k_s and F_s will be determined by simulation as in Sect. 3.3.

By solving the Eqs. (12); (13), the gripping force F_g corresponding to each driving voltage U will be found.

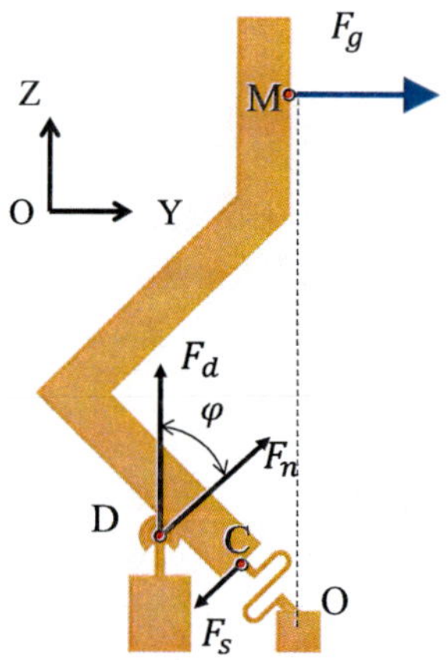

Fig. 7. Force acting on the gripping arm

3.3 Simulation

In order to confirm the calculation results and test the strength of the elastic elements, we use ANSYS software to simulate and predict displacement, temperature distribution of the micro gripper and maximum stress of the elastic components. Simulation results at $U = 12.74$ V (voltage for the jaws touching each other) are shown in Fig. 8: The maximum temperature on the V-beam is $T_m = 492.1$ °C (Fig. 8a); the displacement of shuttle is $\Delta D = 18.97$ μm (Fig. 8b); and the maximum stress of the elastic spring is 97.49 MPa (Fig. 8c). These values are safety and acceptable while V-shaped actuator is working.

Figure 9 compares the calculation and simulation results of jaw's stroke ΔM according to the driving voltage U. At the same displacement value, average deviation of voltage is about 4.45%, and maximum voltage-deviation is 11.65% at the maximal stroke $\Delta M = 40$ μm. The reason of error can be explained as: in the heat transfer problem mentioned at the Sect. 2.2, the convection, radiation, and heat loss of the shuttle have been neglected. Moreover, non-linear displacement of V-beam has been ignored while calculating displacement ΔD in Eq. (4).

4 Fabrication and Result

4.1 Fabrication Process

The micro gripper has been fabricated by the SOI-MEMS technology using SOI (silicon on insulator) wafer with three layers as: device layer of 30 μm; SiO_2—buffer layer of 4 μm and silicon substrate of 450 μm. The fabrication process (see Fig. 10) consists of six main steps as following: (1) cleaning SOI wafer; (2) photolithography (uses single mask) and developing; (3) deep reactive ion etching (DRIE); (4) cutting and removing photoresist; (5) vapor HF etching and (6) platinum sputtering step.

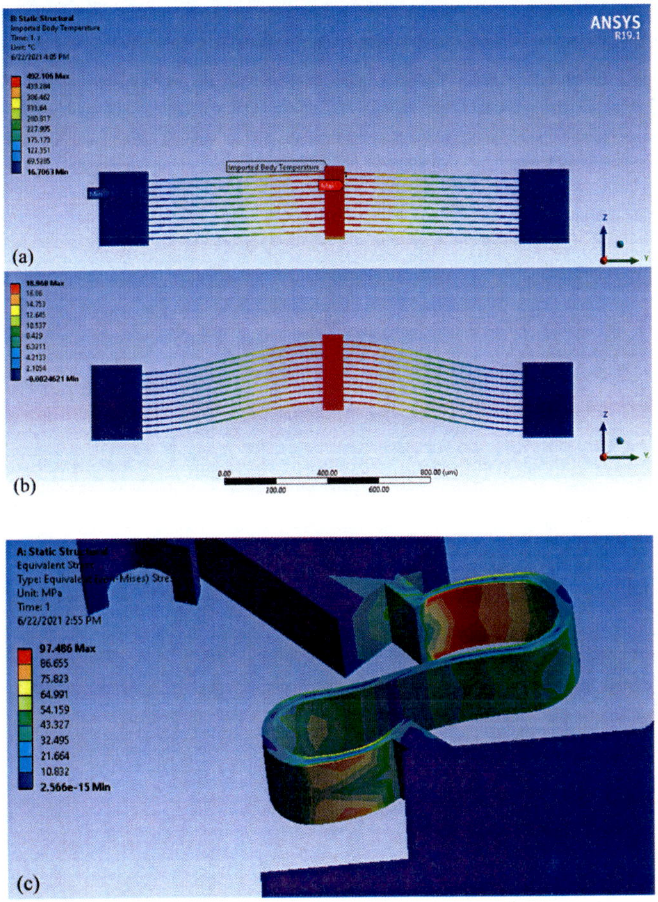

Fig. 8. Simulation results at $U = 12.74$ V: temperature (a); displacement (b) of the V-beam; and maximum stress of the spring (c)

In order to reduce the resistance of V-shaped beams, the structure after vapor HF etching is sputtered by a platinum thin layer (from 50 to 100 nm-thick) on the surface for the better electrical conductivity, because of lower resistivity of platinum in comparison with silicon as well as other semiconductors [10]. Here, the sputtering step helps to reduce applying voltage (i.e. reduce power consumption of the device) at the same displacement when comparing with non-sputtering silicon counterpart.

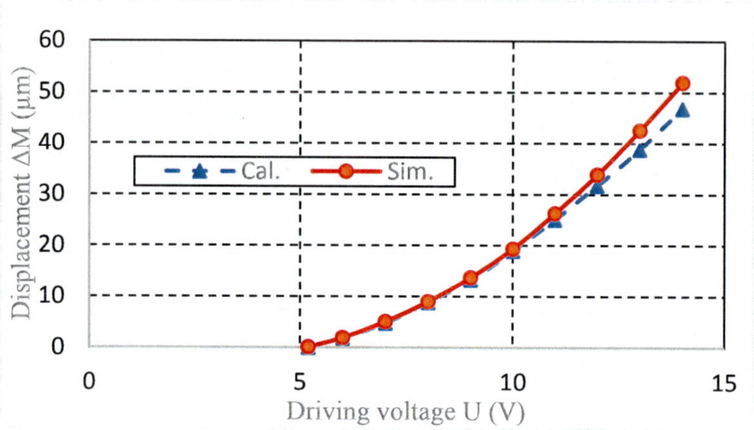

Fig. 9. Calculation and simulation of the jaw's displacement

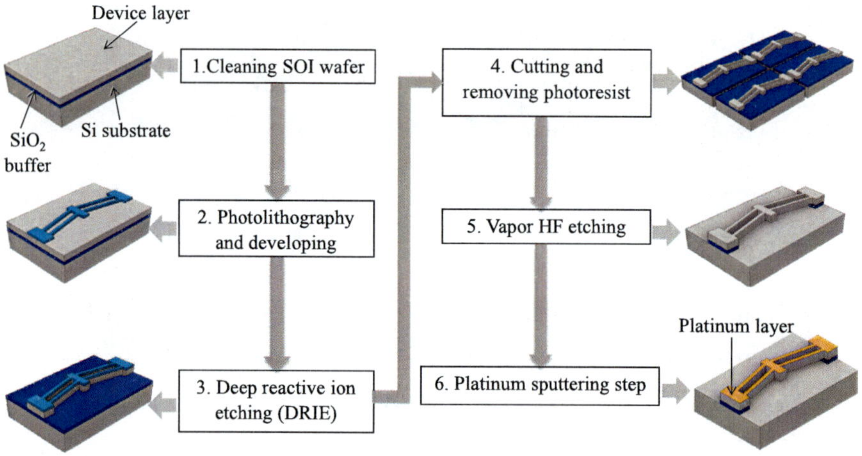

Fig. 10. SOI-MEMS fabrication process of the micro gripper

4.2 Evaluation and Result

The micro gripper after sputtering has been tested for actual operation and jaw's displacement on the specialized measuring system **4200-SCS** supported by Cascade Microtechnology Corporation (see Fig. 11). The preliminary result has explained that the device can work well at the low voltages ranging only from 2 V to 10 V.

At the driving voltage $U = 2$ V, the U-shaped claw overcomes 2 μm-gap of revolute joint and pushes two jaws starting to move, the initial distance between two jaws is 80 μm as shown in Fig. 12a. When increasing the voltage $U = 6$ V, both jaws move closely (see Fig. 12b) and touch each other if $U = 10$ V (Fig. 12c), i.e. each jaw can move a stroke of 40 μm gained by large amplification factor ($k = 3.33$). It confirmed the operation of the micro gripper matches to design as well as reduces driving voltage (i.e.

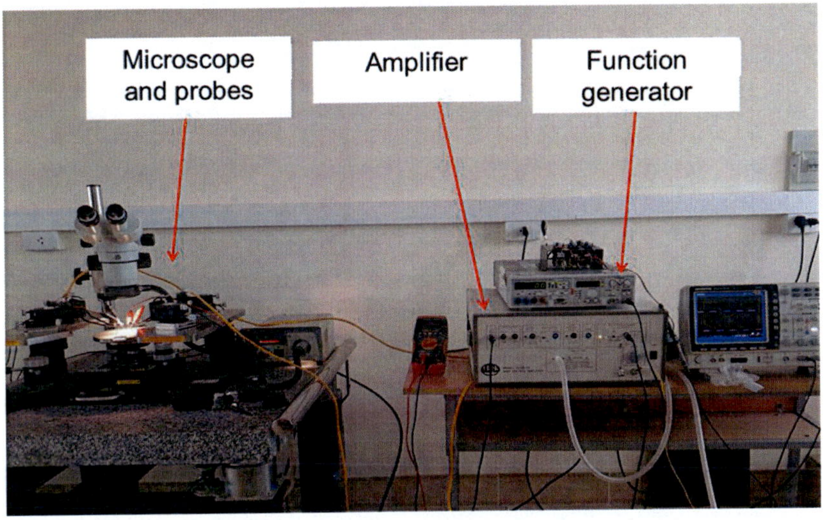

Fig. 11. Measuring system 4200-SCS

reduces power consumption) of the device after applying platinum sputtering step. The detailed measurement and evaluation at different driving voltages showed that the voltage can be reduced ultimately 30.6% corresponding to calculation. It may be explained that tolerance of side-etching effect in micromachining makes to fabricated-beam dimensions are smaller than design specimens and leads to reduce the stiffness k_{td} of V-beam system. Besides, by adding platinum sputtering step at the end of SOI-MEMS fabrication process, we can save energy consumption if comparing to non-sputtering device.

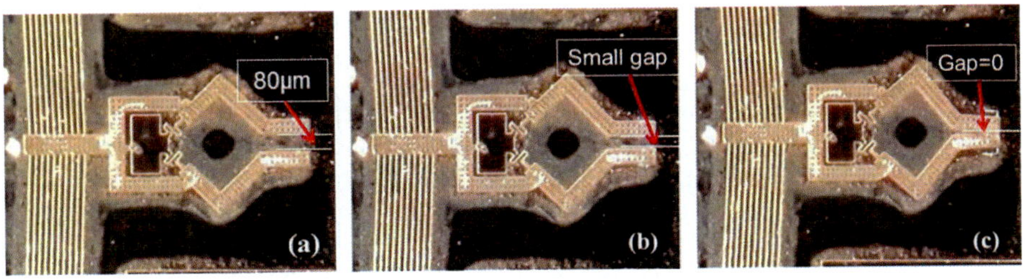

Fig. 12. Operation of the gripper at U = 2 V (a); U = 6 V (b); and U = 10 V (c)

5 Conclusion

The paper presented the design, calculation, simulation and trial fabrication of a novel micro gripper with the advantages like single photo mask, large amplification coefficient and low driving voltage. The temperature distribution, length expansion of V-beam and stress of elastic components were examined by finite difference method and demonstrated by ANSYS simulation. The simulation results are closely matching with theoretical calculations.

By using platinum sputtering process on the surface of devices, the 40 μm-measured stroke of each jaw can be obtained when applying voltage $U = 10$ V. This experiment shows that we can save applied voltage about 30.6% in comparison with calculating voltage (i.e. $U = 14.42$ V) at the same stroke.

In future, the improved micro gripper can be integrated in micro robot or micro assembling systems for flexible handling, lifting and moving micro samples from this place to another. Furthermore, lower driving voltage is strong advantage and supports this gripper may be used or participated easily into biomedical applications.

References

1. Yang, S., Xu, Q.: A review on actuation and sensing techniques for MEMS-based microgrippers. J. Micro-Bio Robot. **13**, 1–14 (2017)
2. Potekhina, A., Wang, C.: Review of electrothermal actuators and applications. Actuators **8**(4), (2019). https://doi.org/10.3390/act8040069
3. Varona, J., Tecpoyotl-Torres, M., Hamoui, A.A.: Modeling of MEMS thermal actuation with external heat source. In: Electronics, Robotics and Automotive Mechanics Conference – IEEE (2007)
4. Zhang, Z., Yu, Y., Liu, X., Zhang, X.: Dynamic modelling and analysis of V- and Z-shaped electrothermal microactuators. Microsyst. Technol. **23**(8), 3775–3789 (2017)
5. Luo, J.K., Flewitt, A.J., Spearing, S.M., Fleck, N.A., Milne, W.I.: Comparison of microtweezers based on three lateral thermal actuator configurations. J. Micromech. Microeng. **15**, 1294–1302 (2005)
6. Zhang, Z., Yu, Y., Liu, X., Zhang, X.: A comparison model of V- and Z-shaped electrothermal microactuators. In: International Conference on Mechatronics and Automation - IEEE, pp. 1025–1030 (2015)
7. Hoang, T.K., Nguyen, T.D., Pham, H.P.: Impact of design parameters on working stability of the electrothermal V-shaped actuator. Microsys. Technol. **26**(5), 1479–1487 (2020)
8. Shan, T., Qi, X., Cui, L., Zhou, X.: Thermal behavior modeling and characteristics analysis of electrothermal microactuators. Microsyst. Technol. **23**(7), 2629–2640 (2017)
9. Nguyen, T.D., Hoang, T.K., Pham, H.P.: Heat transfer model and critical driving frequency of electrothermal V-shaped actuator. In: International Conference on Engineering Research and Applications - ICERA2019, pp. 394–405 (2020)
10. Nguyen, T.D., Hoang, T.K., Pham, H.P.: Larger displacement of silicon electrothermal V-shaped actuator using surface sputtering process. Microsyst. Technol. **27**(5), 1985–1991 (2021)

Size Effects on Mechanical Properties of Single Layer Molybdenum Disulfide Nanoribbon

Danh-Truong Nguyen(✉)

Department of Mechanics of Materials and Structures, School of Mechanical Engineering, Hanoi University of Science and Technology, Hanoi 10000, Vietnam
truong.nguyendanh@hust.edu.vn

Abstract. This work investigates size effects on mechanical properties of single layer molybdenum disulfide ($SLMoS_2$) nanoribbon under uniaxial tension using molecular dynamics finite element method with Stillinger-Weber potential. For the square shaped nanoribbon of increasing size, Young's modulus of the armchair nanoribbon increases while that of the zigzag one decreases. Both of them tend to be size-independent and isotropic for large sizes. Fracture stress of square shaped armchair nanoribbon is nearly independent of size but that of zigzag one reduces as size increases. Poisson's ratio of them goes up as size increases but the larger nanoribbons have a slower increase. For the rectangle shaped nanoribbons, the aspect ratio has no influence on $SLMoS_2$ nanoribbons with fixed width but significantly affected on those with fixed length. Young's modulus, Poisson's ratio, fracture stress and strain of $SLMoS_2$ zigzag nanoribbons with fixed length decrease as width increases. With $SLMoS_2$ armchair nanoribbon of fixed length, narrower one has lower Young's modulus and higher Poisson's ratio but fracture stress and strain are almost unchanged.

Keywords: Molybdenum disulfide nanoribbons · Size effect · Mechanical properties · Atomistic finite element method

1 Introduction

Two dimensional (2D) materials become a hot topic in the past several decades because of its excellent mechanical, electrical and optical properties. Currently, graphene is the most widely studied 2D material. Besides, single layer molybdenum disulfide ($SLMoS_2$), one of the most potential 2D materials, exhibits a large intrinsic bandgap as well as excellent mechanical properties and has shown great potential for applications in semiconducting devices instead of graphene which has a band gap of zero [1–3]. Therefore, studying of mechanical properties of the SLMoS2 is necessary to its applications.

In general, it is too difficult to study mechanical properties of 2D materials by the conventional experimental methods. Using atomic force microscopy nanoindentation, the in-plane Young's modulus of $SLMoS_2$ have been determined to be 120 ± 30 N/m by Cooper et al. [4] which was slightly lower than 180 ± 60 N/m by Bertolazzi [5]. Besides experimental works, Jiang et al. used molecular dynamics (MD) simulations

N. V. Khang et al. (Eds.): ASIAN MMS 2021, MMS 113, pp. 705–715, 2022.
https://doi.org/10.1007/978-3-030-91892-7_67

to find mechanical properties of $SLMoS_2$ of dimension 100×100 Å with the in-plane Young's modulus being 96 N/m in the armchair direction, 97 N/m in the zigzag one and the Poisson's ratio being 0.27 in both of the directions [6]. Those values were also found to be 123 N/m and 0.25 respectively by first-principles calculations [7]. In addition, size-dependent elastic modulus of $SLMoS_2$ nanosheets was reported in Ref. [8], but size effects on Poisson's ratio, fracture stress and strain are not clearly explored. Hence, a more complete understanding of size effect on the mechanical behaviors of SLMoS2 nanosheets is necessary.

Although first-principles calculations and MD simulations are the most accurate theoretical approaches, they require tremendous computation effort. The size-dependent on the mechanical properties of $SLMoS_2$ has not been fully evaluated. Thus, this work will investigate size effect on the mechanical properties of $SLMoS_2$ nanoribbons by using molecular dynamics finite element method (MDFEM). Effects of size as well as length-width ratio on Young's modulus, Poisson's ratio, fracture stress and strain of the $SLMoS_2$ nanoribbons are estimated and discussed. Stillinger-Weber potential is adopted. Uniaxial tensile tests in the armchair and zigzag directions were simulated. Results will be compared with available data from the literature.

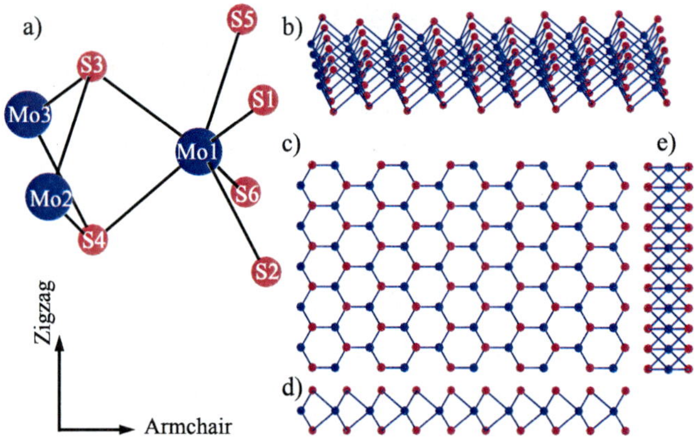

Fig. 1. Schematic illustration of atomic structure of $SLMoS_2$: a) zoom in of the nine nearest atoms; b) 3D view; c) top view; d) front view; e) side view of the sheet. Blue and magenta balls represent the Mo and S atoms, respectively.

2 Numerical Procedure

2.1 Model and Interatomic Potentials

Figure 1 shows the puckered configuration of $SLMoS_2$. Each Mo atom is surrounded by six S atoms, which are categorized into the top and bottom groups. Atoms S 1, 3, and 5 are from the top group, while atoms S 2, 4, and 6 are from the bottom group. The bond length between neighboring Mo and S atoms is 2.382 Å [6], and the angles (S-Mo-S)

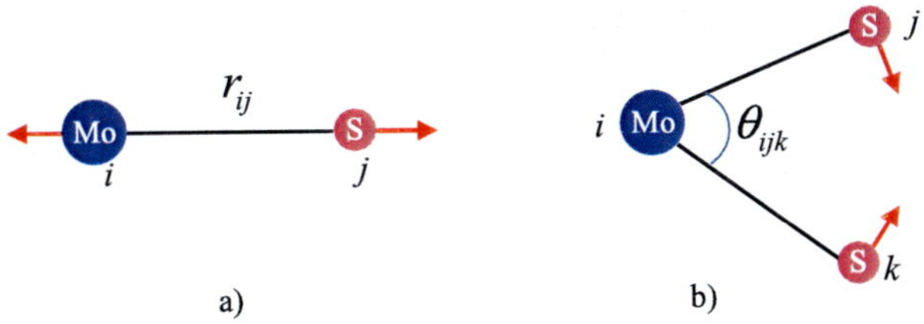

Fig. 2. Two types of elements in Stillinger-Weber potential: a) Two-body (bond-stretching) element and b) Three-body (angle-bending) element.

and (Mo-S-Mo) exhibit the same angle of 81.788° [6]. These geometric parameters are here adopted to establish the initial $SLMoS_2$ structure.

Table 1. Two-body (bond-stretching) Stillinger-Weber potential parameters.

	A (nN.Å)	ρ (Å)	B (Å^4)	r_{maxik} (Å)
Mo-S	11.058	1.252	17.771	3.16

Table 2. Three-body (angle-bending) Stillinger-Weber potential parameters. S-Mo–S indicates the bending energy for the angle with Mo as the apex.

	K (nN.Å)	θ_0 (degree)	ρ_{ij} (Å)	ρ_{ik} (Å)	r_{maxij} (Å)	r_{maxik} (Å)	r_{maxjk} (Å)
S-Mo-S	108.511	81.788	1.252	1.252	3.16	3.16	3.78
Mo-S-Mo	99.825	81.788	1.252	1.252	3.16	3.16	4.27

In this work, Stillinger-Weber potential is used to model the interatomic interactions [6]. The potential energy E of the atomic structure is the total of the bond stretching energy E_r and bond angle bending energy E_θ:

$$E = E_r + E_\theta, \tag{1}$$

$$E_r = \sum_{e=1}^{M} V_2, \quad E_\theta = \sum_{e=1}^{N} V_3, \tag{2}$$

$$V_2 = Ae^{[\rho/(r_{ij}-r_{\max ij})]}\left(B/r_{ij}^4 - 1\right), \tag{3}$$

$$V_3 = Ke^{[\rho_{ij}/(r_{ij}-r_{\max ij})+\rho_{ik}/(r_{ik}-r_{\max ik})]}\left(\cos\theta_{ijk} - \cos\theta_0\right)^2, \tag{4}$$

where V_2 corresponds to the bond-stretching and V_3 associates with the angle-bending. M and N denote the total numbers of bond-stretching and angle-bending element, respectively (Fig. 2). The cutoffs $r_{\max ij}$, $r_{\max ik}$ are geometrically determined by the material 's structure. A, K are energy parameters. ρ, B, ρ_{ij}, ρ_{ik} and θ_0 are five geometrical parameters. r_{ij}, r_{ik} are length of bond *ij* and *ik*, respectively. θ_{ijk} is angle between two bonds *ij* and *ik*. They are defined as follows:

$$r_{ij} = \sqrt{\left(x_j - x_i\right)^2 + \left(y_j - y_i\right)^2 + \left(z_j - z_i\right)^2}, \tag{5}$$

$$\cos\theta_{ijk} = \frac{\left(x_j - x_i\right)(x_k - x_i) + \left(y_j - y_i\right)(y_k - y_i) + \left(z_j - z_i\right)(z_k - z_i)}{r_{ij}.r_{ik}}, \tag{6}$$

here (x_i, y_i, z_i), $\left(x_j, y_j, z_j\right)$ and (x_k, y_k, z_k) are coordinates of atom *i*, *j* and *k*, respectively.

Stillinger-Weber potential parameters are taken from Ref. [6] for Mo-S interaction in molybdenum disulfide and tabulated in Table 1 and 2.

2.2 Molecular Dynamic Finite Element Method

Advanced computational techniques such as DFT calculations and MD simulations are time-consuming and costly. Molecular dynamic finite element method, sometime known as atomic-scale finite element methods, have been developed to analyze nanostructured materials in a computationally efficient way [9–13].

The total energy of the atomic structure, E_T, reads:

$$E_T = E - \sum_{i=1}^{N} \mathbf{f}_i \mathbf{u}_i, \tag{7}$$

where the potential energy E of the atomic structure is a function of atomic coordinates $E = E(\mathbf{x}_1, \mathbf{x}_2, \ldots, \mathbf{x}_N)$ with $\mathbf{x}_i$ is the position of atom *i*, and N is the number of atoms. The last term in Eq. (7) is the work of the external forces $\mathbf{f}_i$ where $\mathbf{u}_i$ is the displacement of atom *i*.

Equilibrium of the structure leads to the minimization of the total energy E_T, so the first derivative of the total energy E_T for the atomic displacements must be zero:

$$\frac{\partial E_T}{\partial \mathbf{u}_i} = 0, i = 1 \div N. \tag{8}$$

Atomic displacements can be iteratively estimated by solving the system of nonlinear Eqs. (8). Using the well-known Newton-Raphson method and following the previous works [9–13] we have:

$$\boldsymbol{K}^{(k)}.\boldsymbol{u}^{(k)} = \boldsymbol{F}^{(k)}, \tag{9}$$

where

$$\boldsymbol{K}_{ij}^{(k)} = \frac{\partial^2 E^{(k)}}{\partial \boldsymbol{u}_i \partial \boldsymbol{u}_j}, \boldsymbol{F}_i^{(k)} = -\frac{\partial E_T^{(k)}}{\partial \boldsymbol{u}_i} = \boldsymbol{f}_i - \frac{\partial E^{(k)}}{\partial \boldsymbol{u}_i} \tag{10}$$

At every step k, regarding Eqs. (9) and (10), the standard formalism of conventional finite element method can be adopted to assemble the global stiffness matrix $\mathbf{K}^{(k)}$ nodal displacement vector $\mathbf{u}^{(k)}$ and non-equilibrium force vector $\mathbf{F}^{(k)}$.

Atoms and atomic displacements are here considered as nodes and nodal displacements, respectively. 2-node and 3-node elements, see Fig. 2, are introduced to model atomic interactions according to Stillinger-Weber potentials in sheets. The number of degrees of freedom is 3 N. Relations between atomic (nodal) displacements and forces with respect to boundary conditions are derived by solving the system of linear equations, Eq. (9). $\mathbf{U}^{(0)}$ is taken as a guess at the initial step k = 0. The atomic displacements are then updated at every step as below:

$$\boldsymbol{x}^{(k+1)} = \boldsymbol{x}^{(k)} + \boldsymbol{u}^{(k)} \tag{11}$$

until $\left|\boldsymbol{F}^{(k)}\right| \leq \boldsymbol{\delta}$ where $\boldsymbol{\delta}$ is a given tolerance.

The present MDFEM is implemented by my own codes to compute nodal displacements under specified boundary conditions. Uniaxial tensions of the zigzag and armchair nanoribbons are simulated (Fig. 3). σ and ε denote the nominal axial stress (engineering stress) and nominal axial strain (engineering strain), respectively. Young's modulus Y and Poisson's ratio are determined from the stress-strain curve in the linear region, $\varepsilon \in [0, 0.01]$. Here Yt and σt denote 2D Young's modulus (or in- plane stiffness) and 2D stress (or in-plane stress), respectively (t is the sheet's thickness). However, in the remainder of this work Young's modulus Yt and axial stress σt will be used for short.

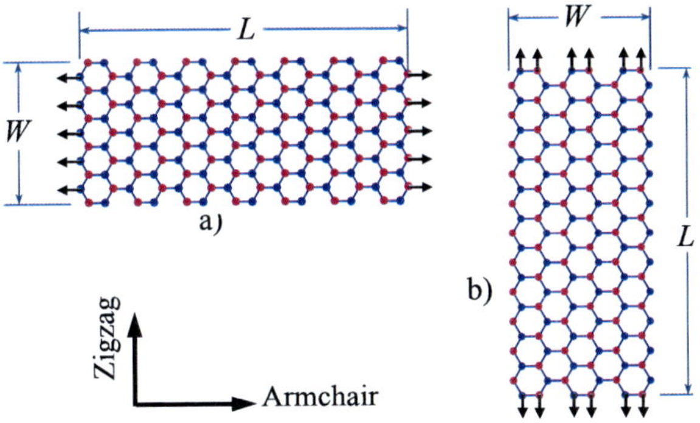

Fig. 3. Schematic illustration of uniaxial tension of $SLMoS_2$ a) armchair nanoribbon and b) zigzag nanoribbon.

3 Results and Discussion

3.1 Size Effects of Square Shaped $SLMoS_2$ Nanoribbon

In this subsection, the square $SLMoS_2$ nanoribbons with size between ~ 20 × 20 Å and ~ 200 × 200 Å are established. Figure 4 shows the stress – strain curves of some square shaped $SLMoS_2$ nanoribbons under tension. It can be seen that stress-strain curves of the armchair nanoribbons are almost identical up to fracture points, while those of the zigzag ones are slightly separate in the final stage. Fracture stress of the armchair nanoribbons always higher than that of the zigzag ones, which can be clearly observed in Fig. 5c.

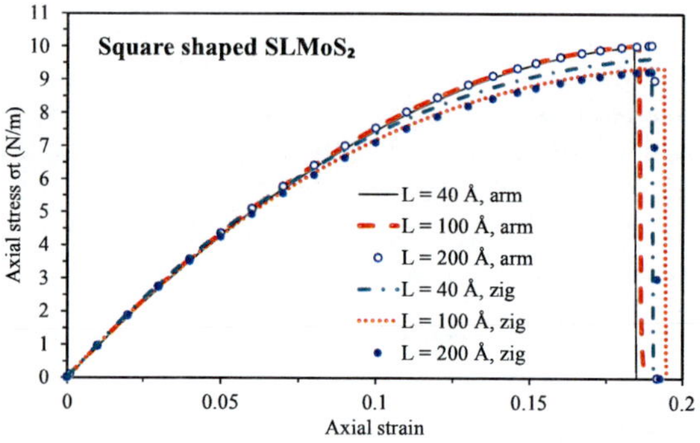

Fig. 4. Stress – strain curves of some square shaped $SLMoS_2$ nanoribbons under tension.

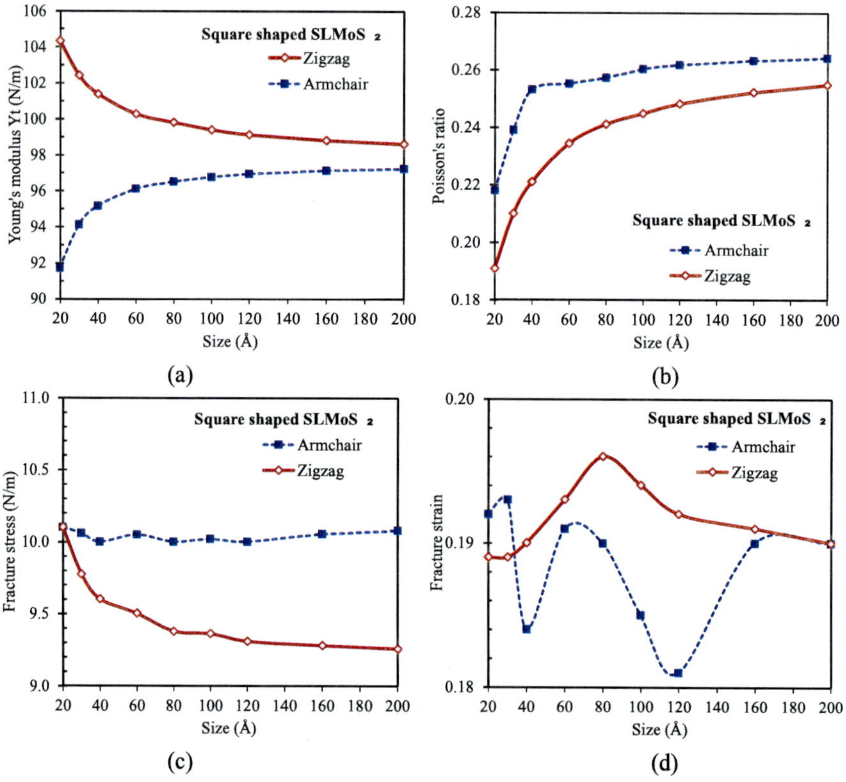

Fig. 5. Size effects on (a) Young's modulus; (b) Poisson's ratio; (c) fracture stress and (d) fracture strain of square shaped $SLMoS_2$ nanoribbons.

It can be seen in Fig. 5a that the increase of size gives rise to the increase of Young's modulus in the armchair nanoribbon (from 91.7 to 97.2 N/m) and the decrease of those in the zigzag one (from 104.3 to 98.6 N/m). These values are good agreement with results by MD simulations [6, 14] and experimental measurements [4]. As we can also see, Young's moduli in the armchair and zigzag nanoribbons have slighter change and tend to the same value in larger sheets. Similar phenomenon has been reported by Bao et al. in Ref. [8] using MD simulation. These findings show that the Young's modulus of the large square shaped $SLMoS_2$ sheet is isotropic.

For Poisson's ratio, Fig. 5b shows that the Poisson's ratios go up as size increases, but they have a gradually rise when the size is larger than 100 Å. At the square shaped nanoribbons of ~ 200 Å, the Poisson's ratios are 0.25 and 0.26 in the zigzag and armchair one, which are close to those by MD simulations [6] and DFT calculations [7].

Size effects on fracture stress and strain are given in Fig. 5c, d. While fracture stress in the armchair nanoribbons is nearly unchanged with the value being 10.1 N/m, that in the zigzag one goes down from 10.1 N/m to 9.3 N/m as size increases from 20 to 200 Å. Fracture strains fluctuate both in the armchair and zigzag nanoribbons with the average value being about 0.19.

Figure 6 shows the snapshot of the fracture nanoribbons under tension, where a bond is not drawn if its length excesses the maximum length (3.16 Å). Arrow dotted lines in Fig. 6 predicted the direction of fracture which can be explained by structure of $SLMoS_2$ and tensile directions. For $SLMoS_2$ zigzag nanoribbon, bonds make initially an angel of 30° with the zigzag direction, stretched much during tension, which leads to fracture taking place first at those bonds. Similar, fracture took place first at the most stretched bonds which are parallel to the armchair direction. Similar phenomenon is reported for graphene [15].

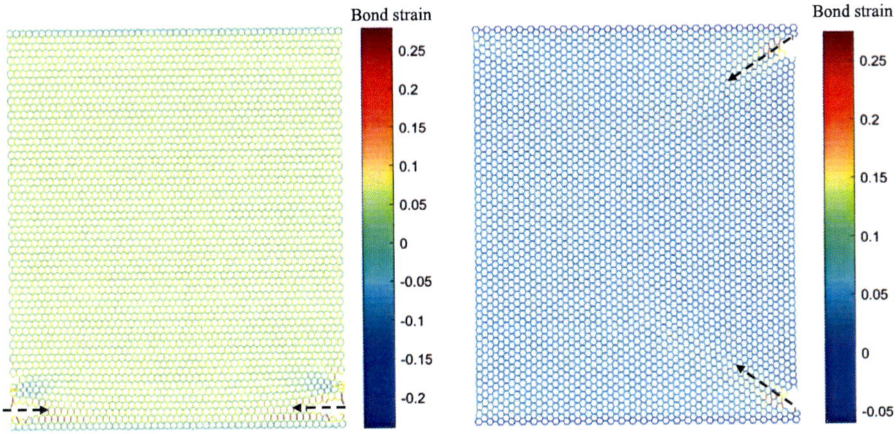

Fig. 6. Fracture of square shaped $SLMoS_2$ armchair (left) and zigzag (right) nanoribbon.

3.2 Aspect Ratio Effects on $SLMoS_2$ Nanoribbons of Fixed Length

In subsection, while $SLMoS_2$ armchair and zigzag nanoribbons have constant length of 198.1 Å and 199.6 Å, respectively, the length-width ratios vary from 1.0 to 22.2 to study aspect ratio effects on mechanical behaviors of $SLMoS_2$ nanoribbons. Stress – strain curves are shown in Fig. 7. Again, we can observe that the curves of the armchair nanoribbons are nearly identical. Fracture stress of the zigzag nanoribbons increases from 9.3 to 11.1 N/m by 19.4% as the width reduces, but that of the armchair ones is nearly unchanged with the value being about 10 N/m (see Table 3 and 4). While fracture strain of the zigzag nanoribbons goes up from 0.189 to 0.211 by 11.6% as the width reduces, that of the armchair one slightly fluctuates between 0.188 and 0.193.

Table 3 and 4 also showed the size effects on Young's modulus and Poisson's ratio. Young's modulus of the armchair nanoribbons gradually increases by 3.1% as the width goes up. By contrast, that of zigzag ones declines by 19%. Poisson's ratio in both the armchair and zigzag nanoribbon decreases with the former being from 0.35 to 0.26 and the latter being from 0.28 to 0.25.

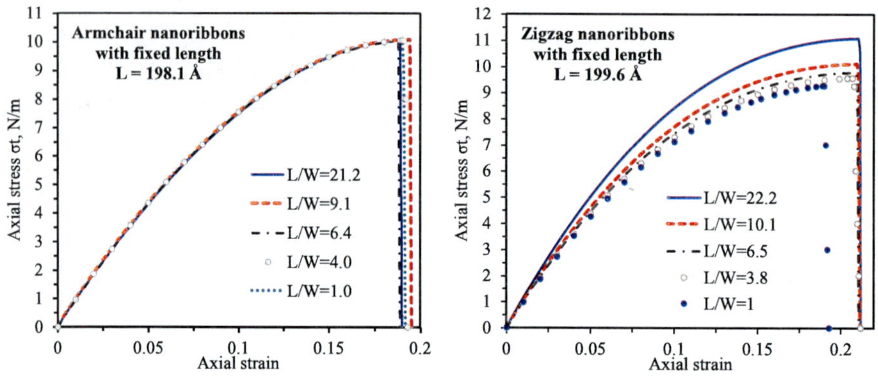

Fig. 7. Stress – strain curves of $SLMoS_2$ armchair (left) and zigzag (right) nanoribbons with fixed length under tension.

Table 3. Mechanical properties of armchair $SLMoS_2$ nanoribbons of fixed length of 198.1 Å.

W (Å)	L/W	Yt (N/m)	σ_f (N/m)	ε_f	Poisson's ratio
9.4	21.2	94.3	10.1	0.189	0.35
21.8	9.1	96.1	10.0	0.193	0.30
31.2	6.4	96.4	10.0	0.188	0.29
49.9	4.0	96.8	10.1	0.190	0.28
198.1	1.0	97.2	10.1	0.190	0.26

Table 4. Mechanical properties of zigzag $SLMoS_2$ nanoribbons of fixed length of 199.6 Å.

W (Å)	L/W	Yt (N/m)	σ_f (N/m)	ε_f	Poisson's ratio
9.0	22.2	115.9	11.1	0.211	0.28
19.8	10.1	106.0	10.1	0.210	0.25
30.6	6.5	103.1	9.8	0.209	0.25
52.2	3.8	100.9	9.6	0.207	0.25
199.6	1.0	97.3	9.3	0.189	0.25

3.3 Aspect Ratio Effects for $SLMoS_2$ Nanoribbons of Fixed Width

In the final part, to investigate aspect ratio effects on mechanical properties, the widths of $SLMoS_2$ armchair and zigzag nanoribbons remain unchanged at 21.8 Å and 19.8 Å, respectively. Four length-width ratio nanoribbons are chosen as listed in Table 5 and Table 6.

Table 5. Mechanical properties of armchair $SLMoS_2$ nanoribbons of fixed width of 21.8 Å.

L (Å)	L/W	Yt (N/m)	σ_f (N/m)	ε_f	Poisson's ratio
219.7	10.1	96.1	10.1	0.198	0.30
327.7	15.0	96.2	10.0	0.197	0.30
441.2	20.2	96.2	10.0	0.195	0.30
549.2	25.2	96.3	10.0	0.189	0.30

Table 6. Mechanical properties of zigzag $SLMoS_2$ nanoribbons of fixed width of 19.8 Å.

L (Å)	L/W	Yt (N/m)	σ_f (N/m)	ε_f	Poisson's ratio
199.6	10.1	106.0	10.1	0.211	0.25
299.4	15.1	106.0	10.1	0.211	0.26
399.2	20.2	106.0	10.1	0.211	0.26
499.0	25.2	106.0	10.1	0.211	0.26

As can be seen, Young's modulus, fracture stress, fracture strain of the armchair nanoribbons have a small change to be about 96.1–96.3 N/m; 10.0–10.1 N/m; and 0.189–0.198, respectively. While those of the zigzag nanoribbons remain unchanged the same values as length varies. Besides, length has also no clear effect on Poisson's ratio. These results show that tensile properties of $SLMoS_2$ nanoribbons are nearly independent of length-width ratios when their widths are fixed. Similar phenomenon is observed in BN [16] and graphene nanoribbons [15].

4 Conclusions

This work has used MDFEM with Stillinger-Weber potential to study the size effects and aspect ratio effects on the mechanical properties of $SLMoS_2$ nanoribbons under uniaxial tension. For the square shaped nanoribbons, fracture stress of the armchair nanoribbon is nearly independent of size but that of zigzag one reduces as size increases. Young's modulus goes up in the armchair nanoribbons and goes down in the zigzag one as size increases. Both of them tend to size-independent and isotropic for the large nanoribbons. Poisson's ratio of them increases as size increases but larger size has a slower increase. Aspect ratio has negligible effect on $SLMoS_2$ nanoribbons with fixed width. For case of fixed length, narrower $SLMoS_2$ armchair nanoribbons have lower Young's modulus and higher Poisson's ratio but fracture stress and strain are almost unchanged. The change of width has significant influence on Young's modulus, Poisson's ratio, fracture stress and strain of $SLMoS_2$ zigzag nanoribbons with fixed length. All of them decrease as width increases.

References

1. Radisavljevic, B., et al.: Single-layer MoS2 transistors. Nat. Nanotechnol. **6**(3), 147–150 (2011)
2. Lopez-Sanchez, O., et al.: Ultrasensitive photodetectors based on monolayer MoS2. Nat. Nanotechnol. **8**(7), 497–501 (2013)
3. Mak, K.F., et al.: Atomically thin MoS2: a new direct-gap semiconductor. Phys. Rev. Lett. **105**(13), 136–805 (2010)
4. Cooper, R.C., et al.: Nonlinear elastic behavior of two-dimensional molybdenum disulfide. Phys. Rev. B **87**(3), 035423 (2013). https://doi.org/10.1103/PhysRevB.87.035423
5. Bertolazzi, S., Brivio, J., Kis, A.: Stretching and breaking of ultrathin MoS2. ACS Nano **5**(12), 9703–9709 (2011)
6. Jiang, J.-W., Zhou, Y.-P.: Parameterization of Stillinger-Weber Potential for Two- Dimensional Atomic Crystals. In: Jiang, J.-W., Zhou, Y.-P. (eds.) Handbook of Stillinger-Weber Potential Parameters for Two-Dimensional Atomic Crystals. InTech, Chennai (2017). https://doi.org/10.5772/intechopen.71929
7. Yue, Q., et al.: Mechanical and electronic properties of monolayer MoS 2 under elastic strain. Phys. Lett. A **376**(12), 1166–1170 (2012)
8. Bao, H., et al.: Size-dependent elastic modulus of single-layer MoS2 nano-sheets. J. Mater. Sci. **51**(14), 6850–6859 (2016)
9. Liu, B., et al.: The atomic-scale finite element method. Comput. Methods Appl. Mech. Eng. **193**(17), 1849–1864 (2004)
10. Nasdala, L., Ernst, G.: Development of a 4-node finite element for the computation of nano-structured materials. Comput. Mater. Sci. **33**(4), 443–458 (2005)
11. Wang, Y., et al.: Atomistic finite elements applicable to solid polymers. Comput. Mater. Sci. **36**(3), 292–302 (2006)
12. Wackerfuß, J.: Molecular mechanics in the context of the finite element method. Int. J. Numer. Meth. Eng. **77**(7), 969–997 (2009)
13. Nasdala, L., Kempe, A., Rolfes, R.: The molecular dynamic finite element method (MDFEM). Comput. Mater. Continua. **19**(1), 57 (2010)
14. Jiang, J.-W.: Parametrization of Stillinger-Weber potential based on valence force field model: application to single-layer MoS2 and black phosphorus. Nanotechnology **26**(31), 315–706 (2015)

15. Chu, Y., Ragab, T., Basaran, C.: The size effect in mechanical properties of finite-sized graphene nanoribbon. Comput. Mater. Sci. **81**, 269–274 (2014)
16. Le, M.-Q.: Size effects in mechanical properties of boron nitride nanoribbons. J. Mech. Sci. Technol. **28**(10), 4173–4178 (2014). https://doi.org/10.1007/s12206-014-0930-8

FEM Simulation of the Thermo-Mechanical Behaviors of the Micro V-shaped Beams

Vu Van The(✉) and Hoang Trung Kien

Le Quy Don Technical University, Hanoi, Vietnam
vuvanthe@mta.edu.vn

Abstract. Calculation and simulation for the thermo-mechanical behavior of micro V-shaped beams using ANSYS Workbench software are introduced in this paper. The heat transfer model of the V-beam is built to determine the displacement of the mechanical system with different temperatures supplied, taking account of the dependence of material properties on temperature changes. The theoretical calculations for the deformation and the actuation force are carried out thanks to the formulas of the expansion of V-beams when the temperature is supplied to the whole system. Thermal simulation results were also performed using ANSYS Workbench software to compare with the theoretical calculation for the mechanical response when the temperature formed from the supplying the voltage between electrodes. These results are the basis for the design of micro V-shaped structures to be applied in later micro-motor systems.

Keywords: Micro-motor · V-shaped beam · Thermo-mechanical behavior

1 Introduction

The micro-motors are used commonly in MEMS (Micro-Electro-Mechanical System) technology to provide the activated force or initial displacement in microsystems [1, 2]. To drive micro-motors, it is possible to use actuators operating on physical effects such as piezoelectric, electrostatic, shape memory alloys, or electro-thermal effects. The electro-thermal effect is realized due to the expansion of the material while applying the voltages to the electrodes of the structure. The micro electro-thermal actuators have some advantages, such as large output force, low input energy, simple configuration, and low cost in fabrication [3]. They can be classified based on the shape of the elastic beams in the actuator structure: U-shaped [4, 5], Z-shaped [6], and V-shaped [7, 8]. The V-shaped beams are used more regularly because of creating a larger deformation and actuation force [7].

The operation of the V-shaped actuator is based on the thermal expansion of the V-shaped beams in the actuator structure with a lot of advantages such as simple structure, larger displacement and actuation force, lower driving voltage [7]. The displacement and active force in the V-shaped structure are determined by analytical formulas when the thermal expansion coefficient of the material is considered as constant. It can affect

N. V. Khang et al. (Eds.): ASIAN MMS 2021, MMS 113, pp. 716–723, 2022.
https://doi.org/10.1007/978-3-030-91892-7_68

partly the accuracy of the result describing the displacement and output force of the structure and create errors in controlling microsystems.

This paper introduces the electro-thermal structure with two couple of Silicon V-shaped beams. The displacement and actuation force is calculated by taking into account the dependence of the thermal expansion coefficient of Silicon on the temperature in both analytical and simulated methods. The thermos-mechanical behaviors of this structure are compared in both methods and are basic for the optimization of the configuration of V-shaped structure applying on the later micro-motors. In the research scope of the paper, the process of heat formation on the structure by thermo-electrical effects will not be considered.

2 Theoretical Basis

The configuration of the structure includes a slide bar - shuttle (1), two couple of V-beams (2), and the anchor (3). This configuration is showed in Fig. 1. Each V-beam couple consists of two beams inclined the α angle to the horizontal direction, one end connects to the anchor (3) and the other connects to the shuttle. Each beam is defined with L length, across-section $w \times t$ (t is the thickness of the structure). The anchor (3) is fixed with the substrate while shuttle (1) and V-beams (2) separates with the substrate by g gap. It allows the shuttle to slide along in the y direction when the beams expand to increasing temperature.

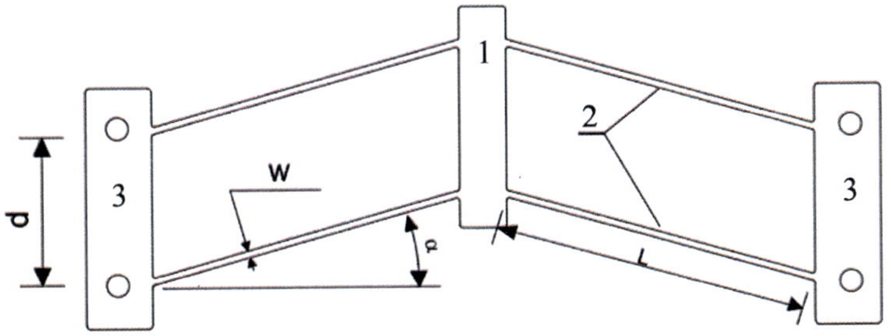

Fig. 1. The V-beams structure model

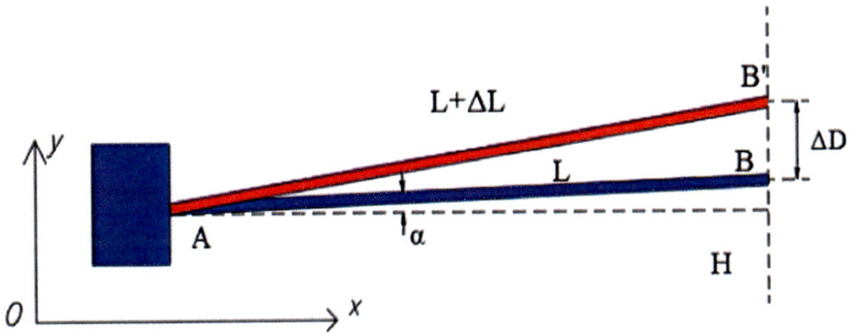

Fig. 2. Displacement of a single V-beam

The assumption in this paper is neglecting the thermal expansion of the shuttle and only taking into account the thermal expansion of the V-beams. The thermal expansion of one beam in a couple of V-beams is shown in Fig. 2 and is calculated by the compact heat transfer equation:

$$k\frac{d^2T}{dx^2} + J^2\rho = 0 \quad (1)$$

where J is current density, ρ is the resistivity of the beam, k is the coefficient of thermal conductivity.

Solving the Eq. (1), the heat distribution equation inside the V-beam is described:

$$T(x) = T_S + \frac{B}{A^2} + C_1e^{Ax} + C_2e^{-Ax} \quad (2)$$

where: $B = \frac{U^2}{l^2\rho_0 k}$, $A^2 = B\lambda$ và $l = 2L$;$C_1 = -\frac{1}{\lambda}\frac{(-e^{-2AL}+1)}{(e^{2AL}-e^{-2AL})}$; $C_2 = -\frac{1}{\lambda}\frac{(e^{2AL}-1)}{(e^{2AL}-e^{-2AL})}$; T_S = 25 °C; ρ_0 is the resistivity of the beam at temperature T_S (ambient temperature), and λ is the linear temperature coefficient.

From expression (2), the elongation of the single beam is defined as the following expression:

$$\Delta L = \int_0^L \alpha_T[T(x) - T_S]dx = \alpha_T\left[\frac{B}{A^2}L + \frac{C_1}{A}\left(e^{Al} - 1\right) - \frac{C_2}{A}\left(e^{-Al} - 1\right)\right] \quad (3)$$

However, within the scope of this paper's research, only the thermo-mechanical part is considered, the electrostatic part applied to the plates was shown in the related study [7]. Therefore, the reduced formula for the thermal elongation of a single beam (homogeneous, isotropic, uniform cross-section) has the form:

$$\Delta l = L.\alpha_T.\Delta T \quad (4)$$

with, α_T - coefficient of thermal expansion of Silicon materials shown in Fig. 3 [5], ΔT - the temperature difference between the initial and later state of the beam. Here, it is assumed that the temperature is evenly distributed throughout the beam.

Displacement of the end beam in the y direction is:

$$\Delta D = \sqrt{(L + \Delta l)^2 - (L\cos\alpha)^2} - L\sin\alpha \quad (5)$$

The actuation force N appeared by the thermal expansion of one V-beam is determined by the relationship between thermal elongation of the beam and the force through the expression:

$$\Delta l = \int_l \Delta dz = \int_l \frac{N}{E.A}dz \quad (6)$$

In the isotropic, uniform cross-section beam, the thermal elongation of the beam is also determined as the expression:

$$\Delta l = \frac{N.l}{E.A} \quad (7)$$

Therefore, the actuation force can be calculated as follow:

$$N = \frac{E.A.\Delta l}{l} \tag{8}$$

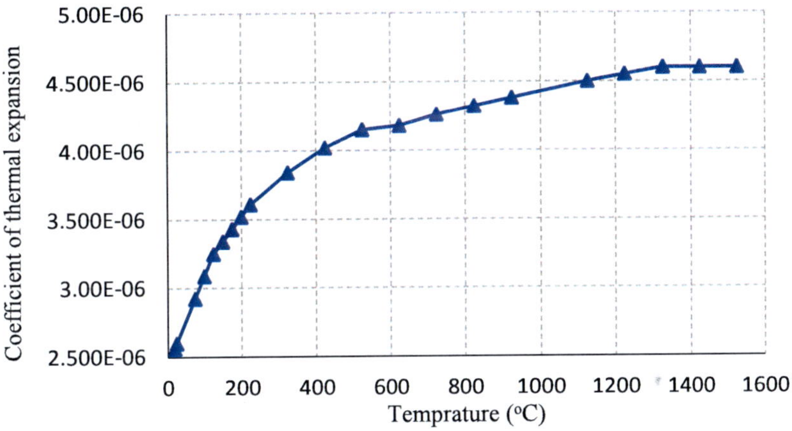

Fig. 3. Coefficient of thermal expansion of Silicon

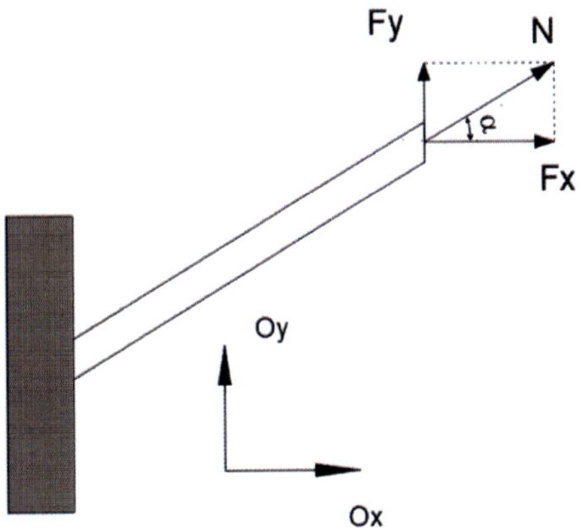

Fig. 4. Diagram of calculating actuation force

In the structure above, it consists of two couple of V-beams arranged symmetrically on two sides of the shuttle. Thence, the force components in the *Ox* axis are self-balancing while the force components in the *Oy* axis are compounded into the desired actuation force (see Fig. 4):

$$F = \sum F_y = 4AE\frac{\Delta L}{L}\sin\alpha \tag{9}$$

where A is the cross-section of the beam (μm^2); E is the elastic modulus of the beam material (Silicon).

3 Simulation of the Thermos-Mechanical Behaviors

To determine the thermo-mechanical behaviors of V-beams, the thermal simulation is carried out by ANSYS Workbench. The geometric parameters of the V-beams are chosen as: inclined angle of V-beam $\alpha = 2\ °$, the beam length $L = 300$ μm, the beam width $w =$ 5 μm, the thickness of the structure $t = 30$ μm, number of couple V-beams $n = 2$, initial temperature (room temperature) $T_0 = 25$ °C. The temperatures set on the structure in order as: 75 °C, 100 °C, 125 °C 150 °C, 175 °C, 200 °C, 225 °C, 325 °C, 425 °C, 525 °C. The material characteristics of Silicon as elastic modulus $E = 169.10^9$ Pa, density $D = 2330$ kg/m^3, α_T coefficient of thermal expansion of Silicon material is shown in Fig. 3. It can be assuming that the nonlinearity of the structure and material is without taken into account in this issue.

3.1 Displacement of the V-shaped Beams

Figure 5 describes the simulation result about the displacement of the shuttle at 75 °C.

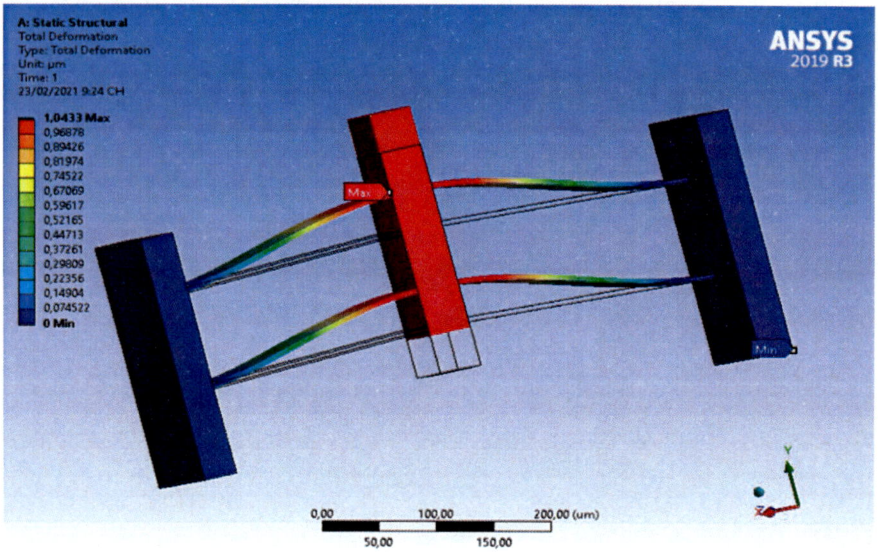

Fig. 5. Displacement of the structure at 75 °C

The displacement and actuation force of the V-shaped structure is shown in Fig. 6 and listed in Table 1. The result shows that the simulation displacement in the y direction is nearly linear to the applied temperature with an R-square factor of 0.9987. Besides that, the analytic result differs from the simulation as 27.5% maximum at 525 °C. In the temperature range 75 – 325 °C, the difference between the two methods is less than 12%. The higher temperature the larger difference because of the expansion of the Silicon to temperature.

Table 1. Displacement of the shuttle to supplied temperature.

Temperature (°C)	75	100	125	150	175	200	225	325	425	525
Analytics (μm)	1.19	1.83	2.50	3.12	3.75	4.38	5.01	7.34	9.51	11.5
Simulation (μm)	1.04	1.67	2.37	3.03	3.72	4.44	5.19	8.21	11.4	14.7
Error (%)	12.3	8.7	5	3	0.8	1.4	3.5	12	20	27.5

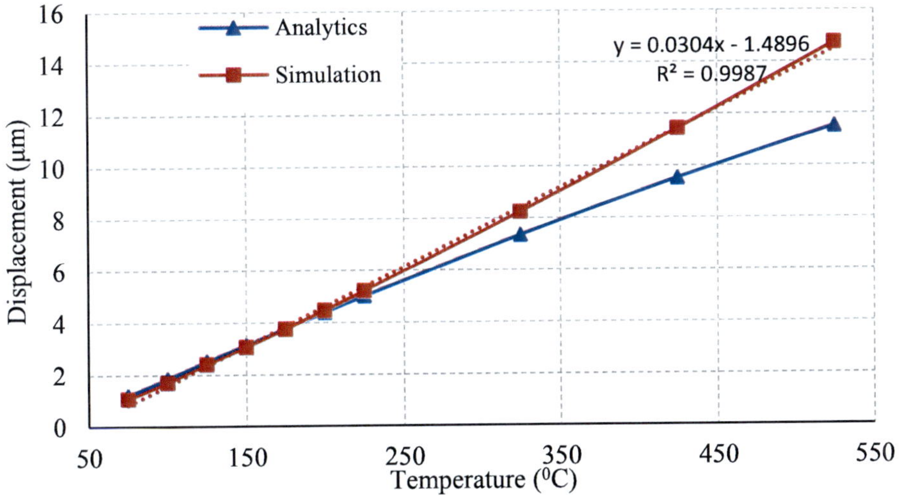

Fig. 6. Difference between simulation and analytic displacement

3.2 Actuation Force

The actuation force of the V-shaped structure is determined by both analytics and simulation. The results are listed in Table 2 and shown in Fig. 7. The increasing temperature causes the deformation of the V-beams, thence the actuation force appears on one end of the shuttle. The deviation in the value of force in the two methods is determined as about 10%–13%.

Table 2. The actuation force on the shuttle to temperature.

T (°C)	75	100	125	150	175	200	225	325	425	525
Analytics (N)	519	821.8	1153.4	1481.7	1826	2186	2562.3	4088.4	5706.8	7364.2
Simulation (N)	582.6	933	1322.7	1692.6	2078	2479.5	2896.5	4585.5	6370.5	8193.3
Error	11%	12%	13%	12%	12%	12%	12%	11%	10%	10%

In the low temperature (less than 250 °C), the value of the actuation force and the displacement of the structure in both two methods are fit together. In this temperature range, the results have demonstrated the correctness of the theoretical calculations and simulations of the thermo-mechanical behavior of the beam system model. However, at the higher temperature, the deviation between analytics and simulation increases. The major reason is the assumption of neglecting the thermal expansion of the anchors and the shuttle. Besides that, the higher temperature not only causes errors in calculating the displacement and actuation force but also impacts the mechanical properties as well as reduces the durability of the structure during working. Therefore, during design, it is necessary to pay attention to determining the maximum temperature (or critical voltage) to take heat to the structure in order to achieve the desired displacement while ensuring the durability of the structure.

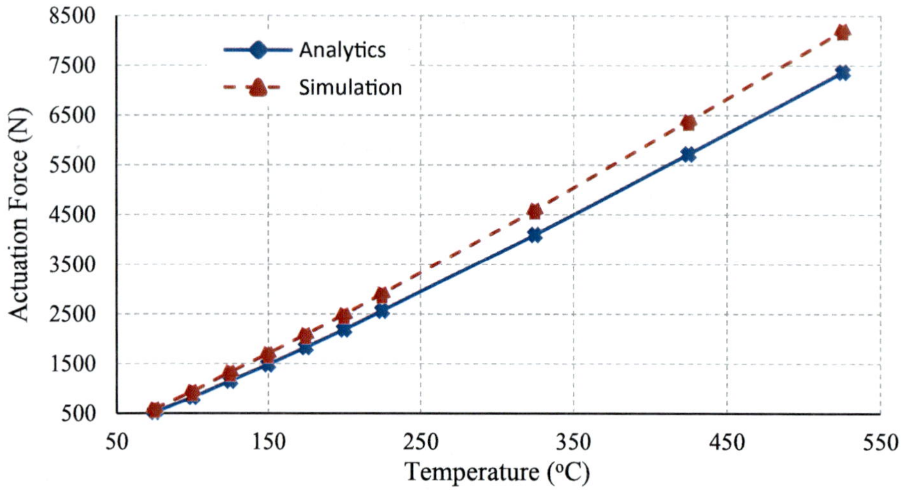

Fig. 7. Relation between actuation force and supplied temperature

To optimize the structure in increasing displacement and actuation force, besides temperature (or applied voltage), the other structural parameters (such as V-beam size, beam tilt angle, or the number of beam pairs) are also factors that need to be studied in the process of calculating and designing the V-shaped beam structure. These researches will be mentioned in the following studies.

4 Conclusions

The driving structure in the micro-size operating by the electric-thermal effect was introduced in this paper. The temperature supplied to the structure is generated by the voltage applied to the electrodes. The advantage of this structure is simplicity, effective operation. The calculations show that the displacement and actuation force of the structure increases linearly for supplied temperature with the R-square coefficient of 0.9987. The error in calculating and simulating the actuation force of the structure is about 10%–13%. The displacement of the shuttle in both theoretical and simulation methods in the range of temperature 25 °C–250 °C is agreeable together with a maximum error of 10%. With higher temperatures, larger errors appear when calculating displacement and actuation force. The simulation method can be used to rapidly predict the maximum temperature, actuation force, and displacement of V-shaped actuator with an acceptable accuracy. The results of this study are the basis for optimizing the design parameters of the structure to achieve the desired displacement and actuation force. Besides, these results show that it is necessary to determine the critical voltage so that the temperature supplied to the structure is within the appropriate operating range, both to ensure the calculation results, and to ensure thermal stability and mechanical durability of the structure.

References

1. Sarajlic E. et al: Three-phase electrostatic rotary stepper micromotor with a flexural pivot bearing. J. MicroElectroMech. Syst. **19**(2), 338–349 (2010). (Author, F.: Article title. Journal 2(5), 99–110 (2016))
2. Pham, P.H., Dao, D.V.: Micro transportation systems: a review. Mod. Mech. Eng. **01**(02), 31–37 (2011). https://doi.org/10.4236/mme.2011.12005
3. Dang, B.L., Vu, N.H., Pham, H.P.: Micro mechanisms in the micro robot systems: case studies of the electrostatic micro mechanisms. National Conference of Mechanic IX (2012)
4. Elsen R, Bharadwaj K, Ramesh T: A parametric study on electro thermally actuated compliant microgripper. In: SAE Technical Paper 2019–28–0032. (2019). https://doi.org/10.4271/2019-28-0032
5. Mankame, N.D., Ananthasuresh, G.K.: Comprehensive thermal modelling and characterization of an electro-thermo-compliant microactuator. J. Micromech. Microeng. **11**(5), 452–462 (2001)
6. Guan, C., Zhu, Y.: An electrrothermal microactuator with Z-shaped beams. J. Micromech. Microeng. **20**(8), 085014 (2010)
7. Hoang, K.T., Nguyen, D.T., Pham, P.H.: Impact of design parameters on working stability of the electrothermal V-shaped actuator. Microsyst. Technol. **26**, 1–9 (2019). https://doi.org/10.1007/s00542-019-04682-y
8. Chiorean, R.S.: Electro-thermor-mechanical modeling of V-beam actuator. In: 8th International Conference Interdisciplinarity in Engineering, pp. 56–61 (2014)

Determination of Fabrication Parameters for Fabrication of FOTURAN® II Glass Applied in Micro-channel Cooling System

Duc-Nam Nguyen[1,2,4](✉), Jeong Hyun Lee[3], and Wonkyu Moon[4]

[1] Phenikaa University, Hanoi 12116, Vietnam
nam.nguyenduc@phenikaa-uni.edu.vn
[2] Phenikaa Research and Technology Institute (PRATI), Hanoi 11313, Vietnam
[3] Plato Tech PTE. LTD., Serangoon 358043, Singapore
[4] Pohang University of Science and Technology (POSTECH), Pohang 37673, South Korea

Abstract. Photosensitive FOTURAN® II glasses have not been widely used in MEMS fabrication because of high cost and restricted fabrication conditions. Most of the studies related to FOTURAN® II glass utilized high power light source and complex systems such as femtosecond laser, nano-ultraviolet laser, and focused ion beam. In this study, a conventional and low-cost lithography process was demonstrated and targeted to determine etching rates of FOTURAN® II glass. A correlation between etching rates and fabrication parameters is presented which exhibits potentials to fabricate microchannel and its application such as microchannel cooling system.

Keywords: Photosensitive glass · FOTURAN® II glass · Micro-channels

1 Introduction

Recent trends in microprocessor design for the next generation are focusing on increasing the ability of heat dissipation and integrating of ICs with efficient cooling capability. Therefore, light-based photonic processor which is fabricated on glass could lead to ultra-fast data transfers capability and while remain low heat generation. A microchannel cooling system (MCS) based on Silicon-glass materials applying the heat pipe theory has been fabricated and exhibited excellent heat dissipation function. The next generation prototype of an MCS is expected to fabricate on glass material which is compatible with glass-glass based IC chips.

An MCS is composed of numbers of micro-channels with the optimized design having different channel depths ranged from tens to 240 μm while the smallest width is 20 μm. The working principle of an MCS was based on a circulation of heating dissipation flow (Fig. 1). The flow carries heat from the heated area (T_1) to heat dissipation area (T_2). The temperature difference between two positions is up to 30 °C. Therefore, a fabrication method for glass wafer which is capable of fabricating a high selectivity, high aspect ratio (10:1), deep etching (240 μm), and multiple etching depth capabilities

N. V. Khang et al. (Eds.): ASIAN MMS 2021, MMS 113, pp. 724–730, 2022.
https://doi.org/10.1007/978-3-030-91892-7_69

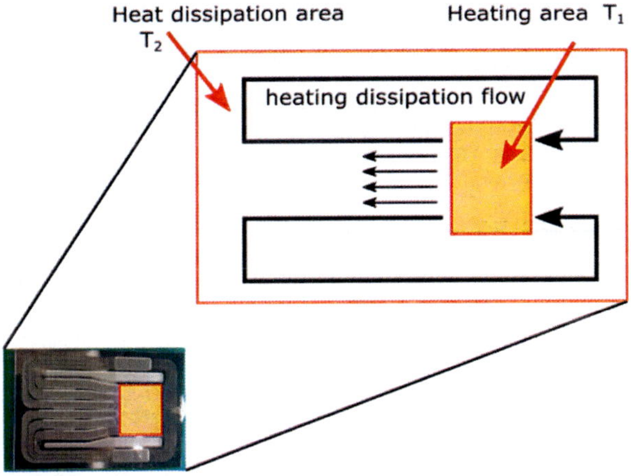

Fig. 1. Working principle of an MCS. Temperature at the dissipation area T_2 is measured 20 – 30 °C smaller than that at the heating area T_1

is needed. One of a potential solution is to use photosensitive glass such as FOTURAN® II glass [1].

The fabrication of FOTURAN® II consists of three major steps: (i) photosensitization: a UV-radiation step by photolithography technique without the use of photoresist layer to predefine the selective area in the glass, (ii) heat treatment: thermal annealing steps which crystallizes nuclei structures in the exposed area, (iii) etching: high selectivity etching step by standard wet etching solution such as HF 10% in DI water. To effectively machine microstructures in FOTURAN® II glass, alternative methods with advanced technologies have been proposed, including laser femtosecond radiation [2], UV excimer laser [3], and focus ion beam [4]. Because of the capability of precisely localized radiation, and highly absorbed wavelength with high energy density, laser-based fabrication methods are applied widely to modify the local crystalline phase and create designed structures.

The laser-based fabrication methods, however, have resulted in a large range of etch rates from 10 to 28 μm per min because of different fabrication parameters. High-cost equipment, limited in batch production, and complex system are drawbacks for the use of these methods. Since the first publication by Dietrich et al. in 2003 [1], most of the research has used FOTURAN® I for the experiment. In this paper, a conventional and low-cost lithography were utilized to investigate the fabrication parameters of FOTURAN® II glass. The initial effort of machining microchannel applied for MCS was also proceeded.

2 Experimental Procedure

All experiments were conducted on 6 in FOTURAN® II wafers with a thickness of 700 μm supplied by SCHOTT AG, Germany. The detail fabrication procedure is presented in Fig. 2.

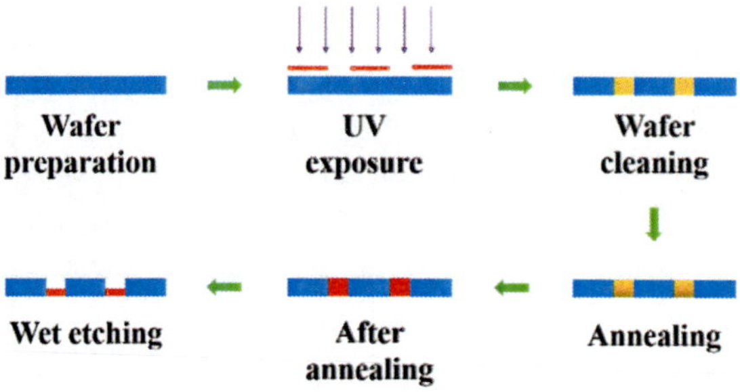

Fig. 2. Fabrication procedure of FOTURAN® II glass wafer in this study

2.1 Ultraviolet (UV) Exposure

The FOTURAN® II glass has the highest UV sensitivity at the wavelength from 290 to 320 nm which is different from an i-line wavelength of 365 nm generated by a standard lithography equipment commonly used for radiating photoresist materials such as SUSS MicroTec Mask Aligner MA6/BA6. Therefore, a Phillips UVB PL-S 9W/01/2P bulb was used to generate UV light with the wavelength band between 305 nm and 315 nm. The wavelength peak is 311 nm being compatible with the fabrication of the FOTURAN® II glass. For demonstration purpose, an Aluminum hard mask was prepared. A customized experimental setup was established based on an electric lamp with the time-controlled procedure (Fig. 3). The FOTURAN® II glasses were exposed in different time duration to investigate the contribution of UV dose to the fabrication parameters. Five new FOTURAN® II wafers were used to conduct the UV exposure experiment with different time durations, including 1 h, 2 h, 3 h, 4 h, and 8h respectively at a fixed distance of 5 cm. Five samples were marked as 1 h, 2 h, 3 h, 4 h and 8 h wafers for ease of classification.

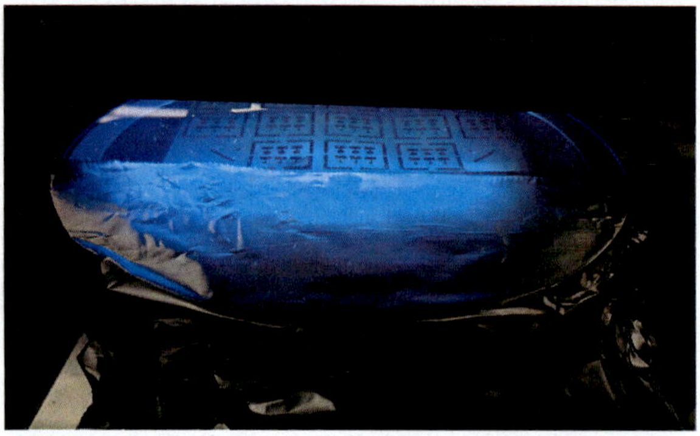

Fig. 3. Customized UV exposure experiments utilizing a UVB lamp at 311 nm wavelength and an Aluminum hard mask.

2.2 Thermal Treatment and Etching

The FOTURAN® II crystal size depends on the UV-exposure time, wavelength, the thermal annealing time and temperature [6]. Therefore, it is critical to apply a proper procedure to temper the exposed glass to archive as finest crystal size as possible after etching step. The exposed FOTURAN® II glass was mounted horizontally onto a chamber and annealed by an RTA system (Eurotherm, Schneider Electric) (Fig. 4). The chamber temperature was set-up and monitored by a customized program (iTools v.9.67, Eurotherm). Followed by results of J. Kim et al. [2] and J.M.F. Pradas et al. [3], the sample was first heated slowly from room temperature to 500 °C in 2.5 h (5 °C/min), then hold isothermally at 500 °C for 1 h to provoke nucleation. The temperature was then slowly increased to 600 °C (1 °C/min) and remained at that temperature level for 1 h to induce crystallization and grain growth [5]. Finally, the sample was slowly cooled to room temperature in 3 h to prevent thermal residue stress side effect. After thermal treatment, the glass wafer color had changed to red color because of the crystallization process (Fig. 5). It is also noticeable that after radiated by UV light, the color of FOTURAN® II wafer had also changed to orange color instead of the white color of an unexposed wafer.

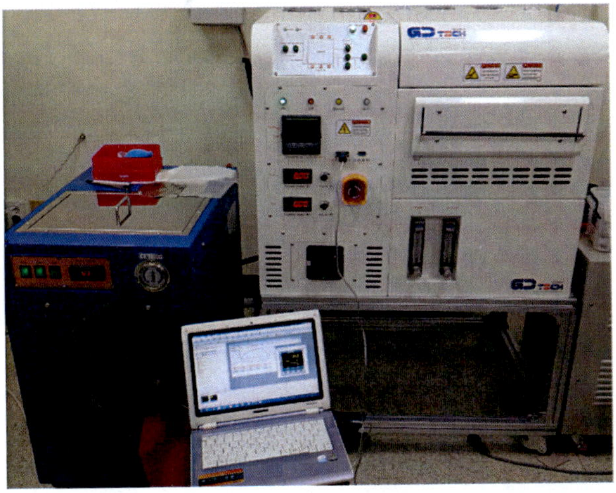

Fig. 4. Photograph of the RTA annealing system

The tempered wafers were cleaned and rinsed by Acetone and DI water respectively to remove residue contaminated particles. Each wafer was diced into small pieces before carrying out the wet etching experiment. The standard procedure of wet etching was to use 10% hydrofluoric acid (HF) in DI water to etch the exposed glass. The etching duration was timed in 5 min, 10 min, and 20 min. The etching depth was measured by Alpha-Step Surface Profilometer (Tencor).

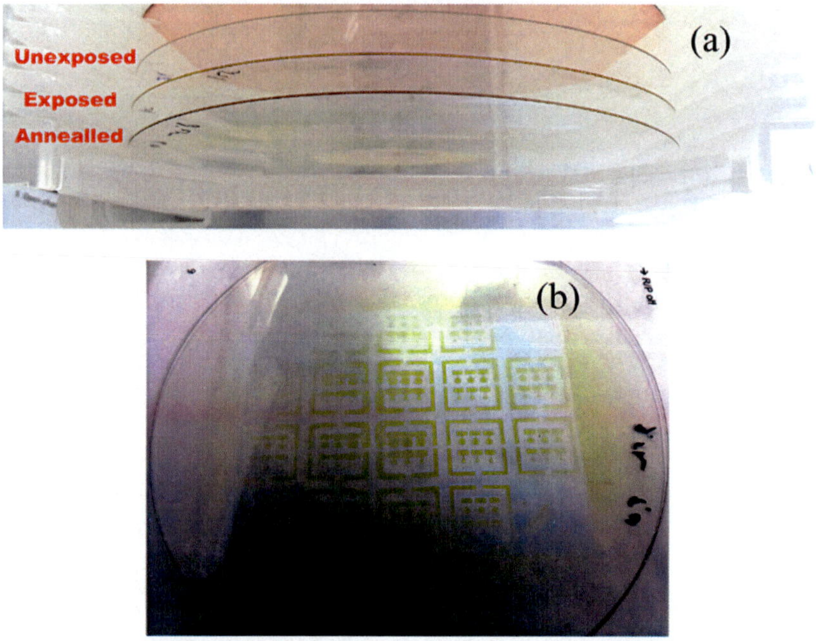

Fig. 5. (a) Color alternation of an unexposed (white), an exposed (red), and an annealed (red) FOTURAN® II glass wafer and (b) the pattern after thermal treatment (Color figure online)

3 Results and Discussion

3.1 Results

The data of Profilometer revealed an etch rates of 12 μm per min (Fig. 6) which is slower than the values reported by R. Tolke *et al.*(22.7 μm) [2], and J. Stillman *et al.*(18.7 μm) [3]. In the initial effort, ultrasonic etching bath was not used, becoming one of the reasons for low etch rates in acid HF solution. On the other hand, limitation of the light source and annealing condition have affected significantly to the etching step. We also observed the warping effect during the annealing process in air.

3.2 Discussion

Effect of UV source

The first process step in the fabrication of photosensitive glass is to illuminate the wafer by a sufficient energy density light source having a wavelength matching with the optics of material. The FOTURAN® II used in this research requires a light source having the minimum an energy density of 7 J/cm^2 or 7 W/cm^2 per second radiating with the wavelength between 290 and 320 nm. We had used a light bulb which is widely used for phototherapy purpose; therefore, the low electric power of 9 W can only produce a very small amount of energy density (approximately 72 μW/cm^2 at a distance of 5 cm).

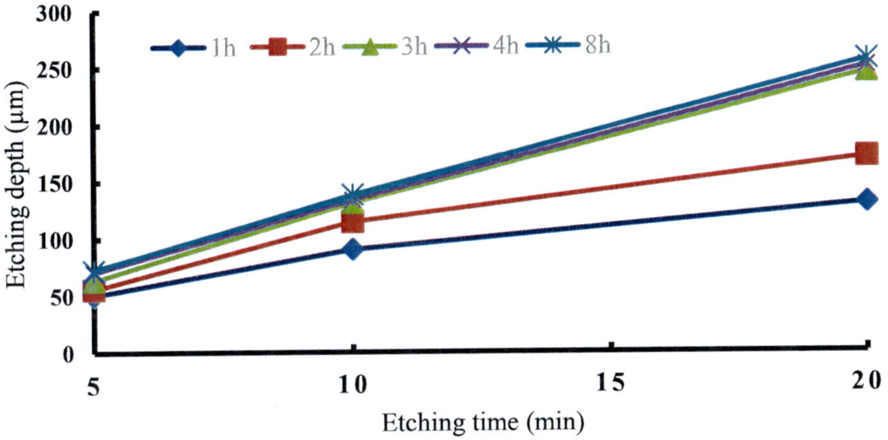

Fig. 6. Measured etching depths of 5 samples (1 h, 2 h, 3 h, 4 h and 8 h exposed wafer) in 5, 10 and 20 min respectively. It is obvious that the longer exposure time is, the better etching rate is.

In order to obtain enough exposure dose, exposure time must be more than 1 h in order to get the effect of irradiation. The optimal exposure time duration for our customized setup is at least 3 h.

Effect of Annealing Process

None of the crystallized patterns was observed when the top temperature was setup at 500 °C at first. Because of the configuration of the convection heating system, the real temperature on the surface of FOTURAN® II wafer was much lower than the needed value for grain growth. The temperature profile was afterward determined experimentally to obtain full crystallized UV-exposed patterns.

Trials of Radiating Through Soda-lime Glass Mask

A film mask printed on a soda-lime glass substrate was used for testing the possibility of processing lithography to fabricate MCS patterns. After 8h exposure, the result was poor as expected. Because the soda-lime glass absorbs remarkably wavelength below 330 nm; hence, only a small portion of 315 nm light can transmit through 2 mm thick soda-lime film mask (Fig. 7 (a), (b)). As the result, in order to use this customization setup, much longer exposure time and/or higher radiation power source should be employed.

Remaining Issues

Warping side-effect or mechanical residue stress after annealing process appeared and limited the tested wafers being unable to bond in wafer-wafer level. Hence, after etching steps, glass-glass microchannel had not been implemented up-to-date. More experiments are also on the way to complete the optimizations condition of fabrication FOTURAN® II glass including elucidation the processing parameters to aspect ratio and side-wall roughness.

(a) (b)

Fig. 7. (a) Film mask of an MCS patterns printed on a 2 mm thickness soda-lime glass and (b) patterns appeared on the glass after thermal treatment

References

1. Dietrich, T.R., Ehfeld, W., Lacher, M., Kramer, M., Speit, B.: Fabrication technologies for microsystems utilizing photoetchable glass. Microelectron. Eng. **30**, 497–504 (1996)
2. Kim, J.: Fabrication of microstructures in photoetchable glass ceramics using excimer and femtosecond lasers. J. Micro/Nanolithogr. MEMS MOEMS **3**(3), 478 (2004). https://doi.org/10.1117/1.1759330
3. Fernández-Pradas, J.M., Serrano, D., Morenza, J.L., Serra, P.: Microchannel formation through Foturan® with infrared femtosecond and ultraviolet nanosecond lasers. J. Micromech. Microeng. **21**(2), 025005 (2011). https://doi.org/10.1088/0960-1317/21/2/025005
4. Anthony, C.J., Docker, P.T., Jiang, K.C.: Microfabrication in FOTURAN photosensitive glass using focused ion beam. In: Proceedings of the World Congress on Engineering, Vol. II (2007)
5. Tolke, R., Hutter, A.B., Evans, A., Rupp, J.M., Gauckler, L.J.: Processing of Foturan glass ceramic substrates for micro-solid oxide fuel cells. J. Eur. Ceram. Soc. **32**, 3229–3238 (2012)
6. Evans, A., Rupp, J.L.M., Gauckler, L.J.: Crystallization of Foturan glass-ceramics. J. Eur. Ceram. Soc. **32**, 203–210 (2012)

Engineering Vibrations and Nonlinear Dynamics

A New Approach for Dynamic Modeling of Magneto–Rheological Dampers Based on Quasi–static Model and Hysteresis Multiplication Factor

Quoc–Duy Bui[1,2] and Quoc Hung Nguyen[2(✉)]

[1] Faculty of Civil Engineering, HCMC University of Technology and Education, Ho Chi Minh City, Vietnam
duybq.ncs@hcmute.edu.vn

[2] Faculty of Mechanical Engineering, Industrial University of Ho Chi Minh City, Ho Chi Minh City, Vietnam
{buiquocduy,nguyenquochung}@iuh.edu.vn

Abstract. It is well–known that magneto–rheological dampers (MRDs) are usually designed based on their quasi–static models. However, in dynamic response, the dampers exhibit nonlinear hysteresis phenomena. For connection between the design phase and dynamic modeling of the MRDs, in this research, a new dynamic model named QSHM model based on quasi–static (QS) model and a hysteresis multiplication (HM) factor is developed. After the new approach in experiment–based dynamic modeling of the MRDs is proposed, several hysteresis multiplication factors are introduced. The proposed approach is then implemented for a prototype MRD and the results are compared with a typical hysteresis model previously developed for MRDs. From the comparison results, advantages of the proposed approach are clarified.

Keywords: Dynamic modeling · Quasi–static model · Hysteresis · Magneto–rheological damper

1 Introduction

In recent years, MRDs have been increasingly attracted attention of scholars in the field of vibration control such as seismic protectors [1, 2], suspension systems [3–6]. Due to the ability of adjusting damping force rapidly, effectively and flexibly corresponding to applied magnetic field, MRDs have shown potential applicability to semi–active control systems. However, MRDs exhibit strong nonlinear hysteresis characteristics in dynamic response, for that an accurate model should be developed to exploit full advantages of the dampers.

It is noted that MRDs are usually designed based on quasi–static models [7]; nevertheless, this model category cannot adapt hysteresis behaviors at low velocities. Another category of dynamic models has been subsequently introduced in efforts to capture

N. V. Khang et al. (Eds.): ASIAN MMS 2021, MMS 113, pp. 733–743, 2022.
https://doi.org/10.1007/978-3-030-91892-7_70

the hysteresis phenomena, which can be classified into non–parametric and parametric models. Although the former models, such as polynomial model [8], neutral network model [9], fuzzy model [10], and division–based model [11], well fit practical hysteresis curves of MRDs, they have no physical meanings and cost much computation. Inversely, parametric dynamic ones are more favorable since physical meanings are maintained in their mechanical factors such as springs, dashpots and friction. However, initial suppositions and constraints should be established suitably for convergent solutions. Several models included in this group are Bingham model [12], biviscous model [13], viscoelastic–plastic model [14], and Bouc–Wen model [15, 16]. To enhance performance of hysteresis prediction in the origin vicinity, Spencer et al. [17] developed a phenomenological model based on the Bouc–Wen model, which has been generally mentioned in recent studies. Nevertheless, the model contains two differential equations with ten basic parameters resulting in complexity of parameters identification and control design. Many later attempts have also been made to further improve effectiveness of hysteresis models [18, 19].

Although the above–mentioned models can describe the hysteresis characteristics of MRDs, they are basically formulated from combinations of mechanical (dashpots, springs) and hysteresis factors to well adapt the practical shape of hysteresis curves. Therefore, the design phase (usually based on QS model) cannot be connected meaningfully to dynamic modeling of MRDs. In other words, the previously developed models are inadequate to reflect the physical nature of MRDs, which restricts their applicability and continuous improvability. Consequently, this research focuses on a new approach for dynamic modeling of MRDs. In the proposed approach based–QSHM model, the mechanical factors are inherited from the QS model such as viscosity, yield stress and friction while several sigmoid functions are considered for the HM factor. This structure helps MRDs retain physical nature and association between the design phase and dynamic modeling, also the model parameters identification is facilitated. To verify efficiency, the new approach is applied for hysteresis modeling of a prototype MRD. The results are then compared with the Spencer's phenomenological model and advantages of the proposed approach are enlightened.

2 Concept of the Proposed Approach Based–QSHM Model

Figure 1 shows schematic of the QSHM model. The proposed model is formulated considering the supplementation of a HM factor in the frame of QS model. The QS mechanical parameters represent the physical nature while the HM ones characterize the nonlinear dynamic responses of MRDs. Governing equation of the model is given by

$$F = F_{QS} * HM \tag{1}$$

where F is the damping force, F_{QS} is the damping force based on QS model and HM is the hysteresis multiplication factor.

For MRDs, the QS damping force can be decomposed into three components, as presented in Fig. 2

$$F_{QS} = F_{\eta} + F_{\tau} + F_f \tag{2}$$

in which F_η is the viscous damping force, F_τ is the damping force due to yield stress and F_f is the friction force.

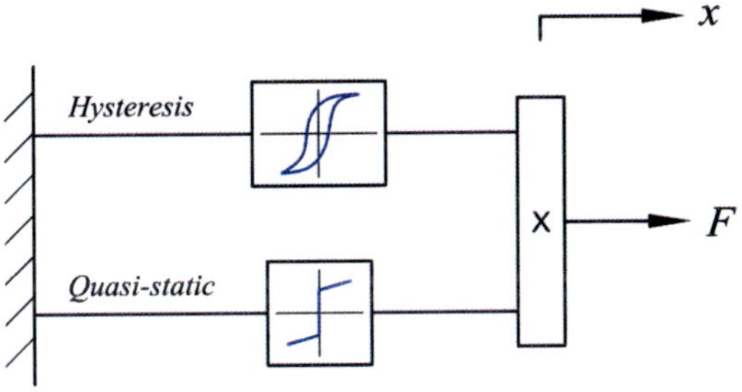

Fig. 1. Schematic of the proposed QSHM model.

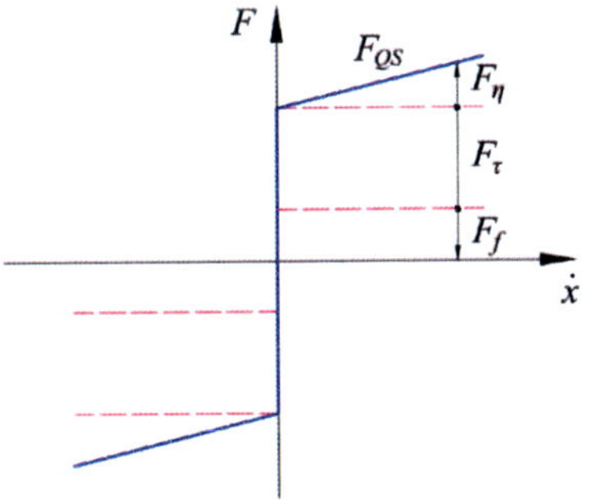

Fig. 2. Illustration of the QS damping force decomposition.

The HM factor expresses the "S" shape of practical hysteresis curves of MRDs at low velocities and is thus generally modeled by sigmoid functions. Figure 3 shows comparisons between some typical sigmoid functions. They are normalized in such a way that their principal value range is [–1, 1] and their slope at origin is 1 as follows.

$$\text{Error function: } f(x) = \operatorname{erf}\left(\frac{\sqrt{\pi}}{2}x\right) \tag{3}$$

$$\text{Hyperbolic tangent function: } f(x) = \tanh(x) \tag{4}$$

$$\text{Algebraic function: } f(x) = \frac{x}{\sqrt{1+x^2}} \tag{5}$$

$$\text{Arctangent function: } f(x) = \frac{2}{\pi}\arctan\left(\frac{\pi}{2}x\right) \tag{6}$$

From the figure, it is observed that the functions differ in their asymptotic speeds. The more parameters are added, the better the curve shape can be controlled, but at the same time the more complicated the model structure becomes. Therefore, depending on the practical hysteresis curve shape, an appropriately chosen sigmoid function will guarantee simplicity, precision and effectiveness of the model.

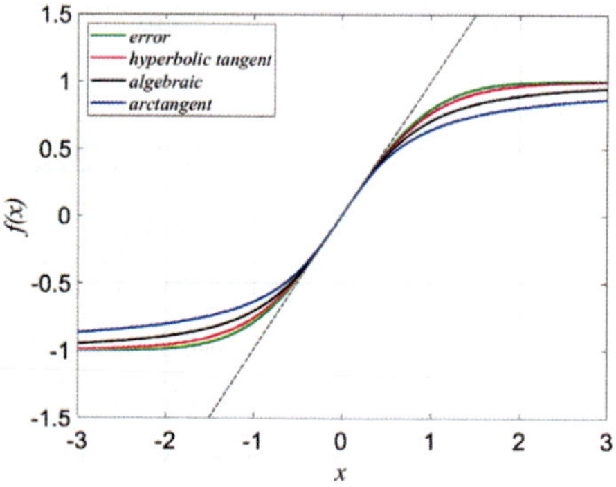

Fig. 3. Comparison between some normalized sigmoid functions.

3 Implementation of the QSHM Model

In this section, the QSHM model is applied for dynamic modeling of a shear–mode prototype MRD developed in Bui et al. [20]. Configuration of the damper is shown in Fig. 4. When the damper coils are electrically excited, magnetic flux is generated across the MR fluid (MRF) gap. As a result, the MRF solidifies and produces damping force via friction against the shearing shaft. Figure 5 shows experimental responses of the damper under 2 Hz frequency, 20 mm amplitude and different currents. It can be seen that the damping force increases with the applied magnetic field intensity. The experimental results are fluctuant and a little smaller than those from the QS model. The damper behavior also experiences strong hysteresis in the origin vicinity.

According to Fig. 4, the three components of the QS damping force can be established as follows. The friction force represents the Coulomb friction between the shaft and O–rings, which is an experimentally obtained constant. The damping force due to yield stress is written by

$$F_\tau = A_l \tau_y \tag{7}$$

where τ_y is the MR fluid yield stress coefficient and A_l is the lateral area of the cylindrical shaft part in the MRF region. By assuming a linear velocity profile of the MRF in the gap, the viscous damping force can be expressed as a function of velocity

$$F_\eta(\dot{x}) = \frac{A_l}{t_g}\eta\dot{x} \tag{8}$$

in which $\dot{x}$ is the velocity, η is the MRF viscosity coefficient and t_g is the MRF gap thickness. The parameters A_l and t_g can be obtained from the geometric dimensions of the damper.

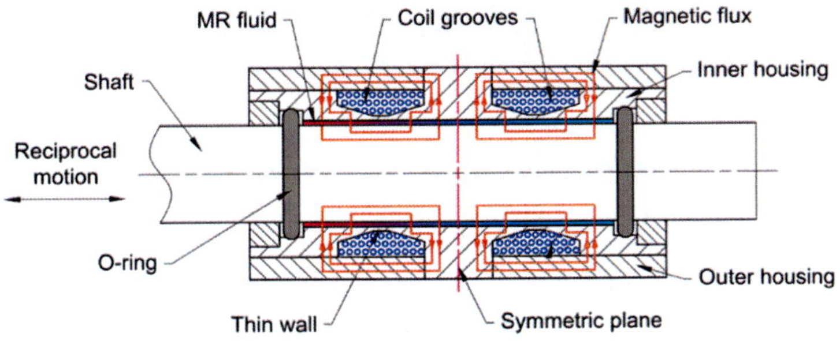

Fig. 4. Configuration of the shear–mode MRD [20].

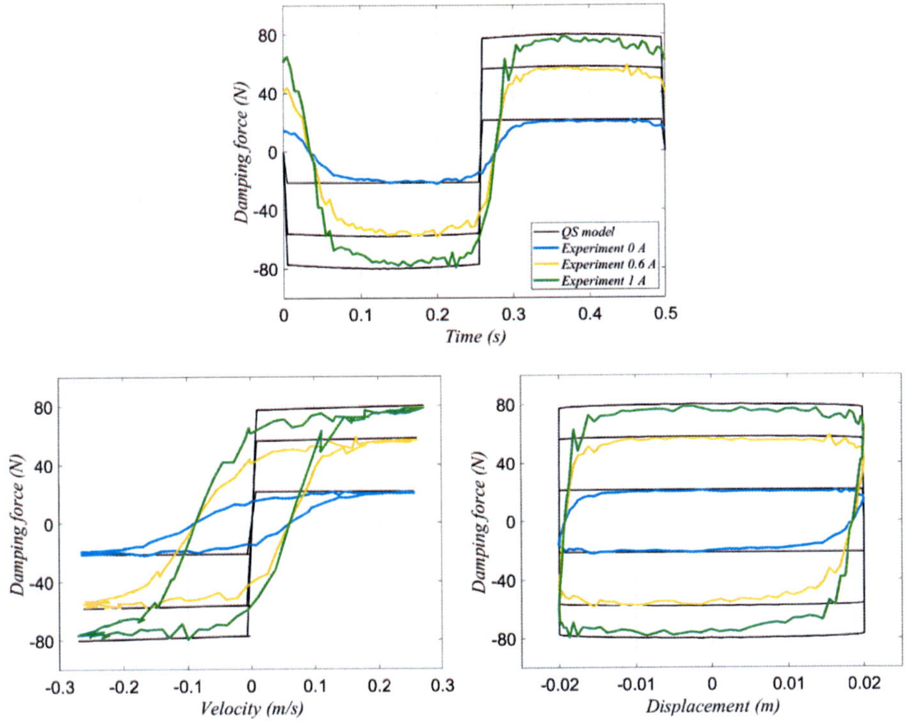

Fig. 5. Experimental responses of the MRD under 2 Hz frequency, 20 mm amplitude and different currents.

In this study, we consider the hyperbolic tangent and algebraic functions for the HM factor. The QSHM models featuring the hyperbolic tangent function (QSHM–tanh) and algebraic function (QSHM–algebraic) are formulated as follows.

$$\text{QSHM} - \tan\text{h}: F = F_{QS} * \tanh\{A[\dot{x} + B\text{sgn}(x)]\} \tag{9}$$

$$\text{QSHM} - \text{algebraic}: F = F_{QS} * \frac{A\left[\dot{x} + B\text{sgn}(x)\right]}{\sqrt{1 + \left[\dot{x} + B\text{sgn}(x)\right]^2}} \tag{10}$$

It can be realized from the equations that F_{QS} presents the peak value of damping force at magnetic saturation. For control of the curve slope at origin, the stiffness coefficient A is added to the HM factor. The product of the shift coefficient B and signum function of displacement *sgn(x)* is left to characterize the half width of the hysteresis curve in progression near zero velocity. Influence of the coefficients A and B on the appearance of hysteresis curve are illustrated in Fig. 6.

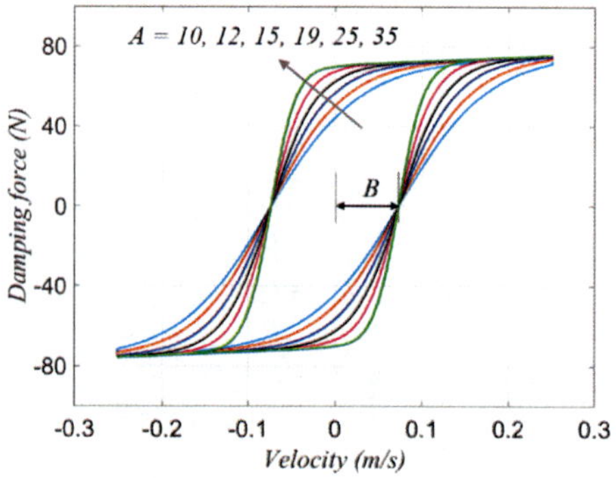

Fig. 6. Influence of the coefficients A and B on the appearance of hysteresis curve.

The coefficients of the QSHM model η, τ_y, A and B are determined using curve fitting combined with least square method in MATLAB. The objective function is defined by

$$OBJ = \sum_{i=1}^{n} \left(F_{m.i} - F_{exp.i}\right)^2 \tag{11}$$

where n is the number of data points, $F_{m.j}$ and $F_{exp.j}$ are respectively the ith estimated and experimental damping forces. It is noted that η and τ_y characterize the rheological properties of the MRF and hence take initial searching values from the QS model used in the design phase. With respect to the geometric coefficients A and B, the searches start at the points extracted from the experiment data of the hysteresis curve at the intersection with velocity axis.

The parallel combination between the QS and HM factors in the new approach of hysteresis model plays an important role in development of MRDs. First, it connects the design phase and dynamic modeling, for that the proposed QSHM model is valuable not only to dynamic responses prediction and system control but also to design and preliminary estimations of MRDs (by simplifying the HM factor). Second, the presence of the QS mechanical parameters helps maintain the physical nature of MRDs, which have not been clarified in most previous literature. Relying on this characteristic, the model specification is favorable since the QS coefficients can also be determined independently via QS tests. Without any differential equation, the QSHM model has another advantage of low computational cost that facilitates the inverse model identification and control design.

4 Results and Discussions

Figure 7 shows comparisons between the damping forces predicted by the two QSHM models (QSHM–tanh and –algebraic) and measured experimentally under 2 Hz frequency, 20 mm amplitude and different currents. It can be observed that the proposed models well track the time history of experimental responses as well as energy dissipation of the MRD.

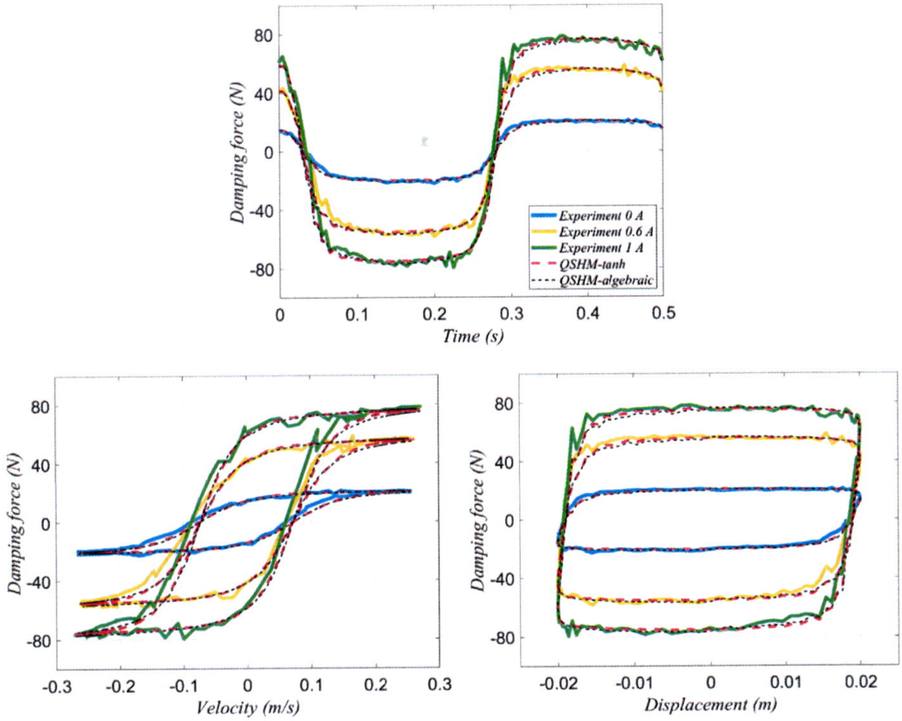

Fig. 7. Comparisons between the QSHM models and experimental data under 2 Hz frequency, 20 mm amplitude and different currents.

To evaluate effectiveness the proposed approach, performances of the two QSHM models are compared with that of the Spencer's phenomenological model, as shown in Fig. 8. It can be seen that all three models well describe the hysteresis responses of the MRD. However, in the force–velocity relationship, the practical hysteresis curve exhibits an asymmetry between the lower (positive acceleration) and upper (negative acceleration) branches, which results in differences between the prediction and measurement at low velocities. Adding more coefficients can improve the accuracy but at the same time increases the complexity of the models, as mentioned in Sect. 2. This trade–off is decided depending on requirements of the control system using MRDs.

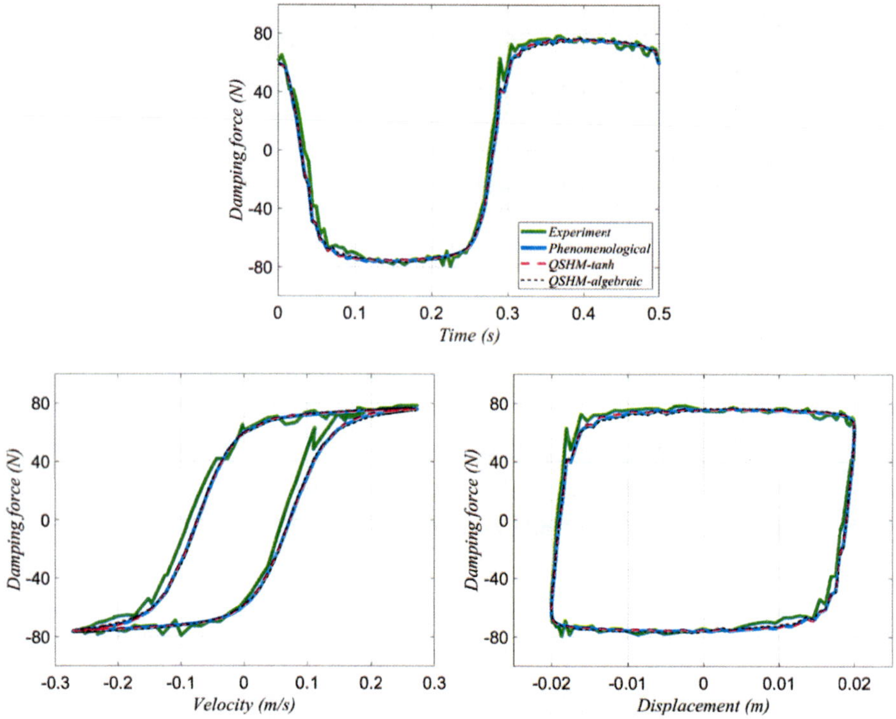

Fig. 8. Comparisons between the three models and experimental data under 2 Hz frequency, 20 mm amplitude and 1 A current.

In this work, normalized errors in relations to time, displacement and velocity between the simulated and experimental damping forces are used to assess precision of the models, which are respectively given by

$$E_t = \sqrt{\frac{\int_0^T (F_m - F_{\exp})^2 dt}{\int_0^T (\mu_{\exp} - F_{\exp})^2 dt}};\ E_x = \sqrt{\frac{\int_0^T (F_m - F_{\exp})^2 \left|\frac{dx}{dt}\right| dt}{\int_0^T (\mu_{\exp} - F_{\exp})^2 \left|\frac{dx}{dt}\right| dt}};\ E_v = \sqrt{\frac{\int_0^T (F_m - F_{\exp})^2 \left|\frac{d\dot{x}}{dt}\right| dt}{\int_0^T (\mu_{\exp} - F_{\exp})^2 \left|\frac{d\dot{x}}{dt}\right| dt}} \quad (12)$$

where F_m and F_{exp} are the simulated and experimental damping forces, respectively, and μ_{exp} presents the average value of the experimentally obtained forces over the cycle T. Results under the certain excitations (2 Hz frequency, 20 mm amplitude and different currents) are presented in Table 1. It can be seen that the precision of the phenomenological model is a little more than those of the two QSHM models. This is obvious because the phenomenological model employs ten coefficients against four ones of the proposed models. On an inspection, it is found that the force–velocity curve predicted by the QSHM–tanh model approximates closer to the experimental curve than that by the QSHM–algebraic one at high velocities, which correlates to the asymptotic characteristic of each sigmoid function analyzed in Sect. 2. For this reason, the QSHM–tanh model is more accurate than the QSHM–algebraic one with respect to the prototype MRD.

Table 1. Normalized errors between the three models under the certain excitations. (Phe: phenomenological, Q–tanh: QSHM–tanh, Q–alge: QSHM–algebraic)

Error	0 A			0.6 A			1 A		
	Phe	Q–tanh	Q–alge	Phe	Q–tanh	Q–alge	Phe	Q–tanh	Q–alge
E_t	0.0871	0.0872	0.0878	0.0939	0.0945	0.0947	0.085	0.0851	0.0867
Ex	0.0695	0.0695	0.0707	0.0740	0.0745	0.0748	0.0643	0.0647	0.066
Ev	0.0855	0.0858	0.0854	0.0893	0.0895	0.0901	0.0831	0.0834	0.0842

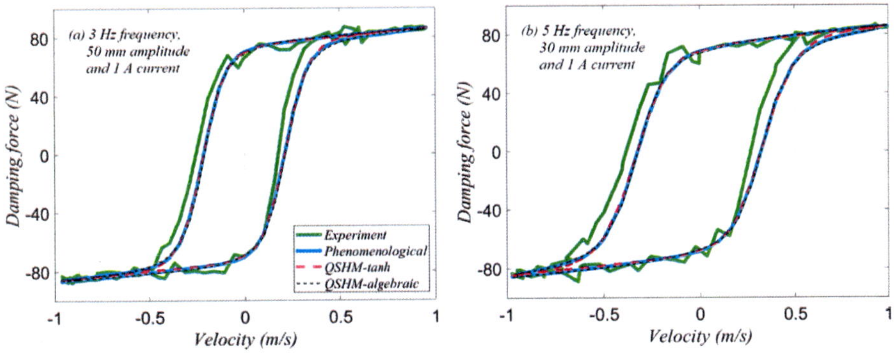

Fig. 9. Comparisons between the three models and experimental data under higher excitations.

The same remarks can be found out when the MRD is tested under higher excitation conditions, as shown in Fig. 9. It is also observed that the hysteresis region of the curve changes according to the excitations. This will be further investigated in the next research stages.

5 Conclusions

This research contributed a new approach for dynamic modeling of MRDs. In the proposed approach–based QSHM model, the mechanical factor inherited from the QS model reflects the physical nature while the intrinsic hysteresis behavior of MRDs is expressed by a HM factor. With this structure, the design phase is connected significantly to the dynamic modeling of MRDs and hence the model is beneficial to both initial design and dynamic analysis. In addition, the QS mechanical parameters can be determined separately via QS experiments or chosen as initial searching values of approximation methods. Two sigmoid functions, including hyperbolic tangent and algebraic functions, were considered for the HM factor. The results showed that the QSHM–tanh and QSHM–algebraic models well predicted the hysteresis responses of the prototype MRD. Compared with the typical Spencer's phenomenological model, the proposed models almost have equivalent precision but contain no differential equations and much less coefficients, which facilitate the model establishment. It was also observed from the results that the QSHM–tanh model is a little better than the QSHM–algebraic one. Because practical hysteresis curves of MRDs mainly differ in their "S" shape, a reasonably chosen sigmoid function will improve the model precision. Consequently, the QSHM model is effective and applicable for control systems based on MRDs such as seismic isolation structures of buildings and bridges, suspension systems of vehicles and automatic machines, force feedback system of haptic joysticks. In the next stage, dependence of the model on excitation conditions will be clarified.

Acknowledgements. This work was supported by research fund from Ministry of Education and Training (MOET) of Vietnam.

References

1. Dyke, S.J., Spencer, B.F., Sain, M.K., Carlson, J.D.: An experimental study of MR dampers for seismic protection. Smart Mater. Struct. **7**(5), 693–703 (1998)
2. Weber, F.: Semi–active vibration absorber based on real–time controlled MR damper. Mech. Syst. Signal Process. **46**(2), 272–288 (2014)
3. Nguyen, Q.H., Choi, S.B.: Optimal design of MR shock absorber and application to vehicle suspension. Smart Mater. Struct. **18**(3), 0350–12 (2009)
4. Bui, Q.D., Nguyen, Q.H., Nguyen, T.T., Mai, D.D.: Development of a magnetorheological damper with self–powered ability for washing machines. Appl. Sci. **10**(12), 4099 (2020)
5. Bui, Q.D., Nguyen, Q.H., Hoang, L.V., Mai, D.D.: A new self–adaptive magneto–rheological damper for washing machines. Smart. Mater. Struct. **30**(3), 037001 (2021)
6. Bui, Q.-D., Hoang, L.-V., Mai, D.-D., Nguyen, Q.H.: Design and testing of a new shear-mode magneto–rheological damper with self-power component for front–loaded washing machines. In: Long, B.T., Kim, Y.-H., Ishizaki, K., Toan, N.D., Parinov, I.A., Vu, N.P. (eds.) MMMS 2020. LNME, pp. 860–866. Springer, Cham (2020). https://doi.org/10.1007/978-3-030-69610-8_114
7. Phillips, R.W.: Engineering applications of fluids with a variable yield stress. Ph.D. thesis, University of California Berkeley, California (1969)
8. Choi, S.B., Lee, S.K., Park, Y.P.: A hysteresis model for the field–dependent damping force of a magnetorheological damper. J. Sound Vib. **245**(2), 375–383 (2001)
9. Wang, D.H., Liao, W.H.: Modeling and control of magnetorheological fluid dampers using neural networks. Smart Mater. Struct. **14**(1), 111–126 (2004)
10. Kim, H.S., Roschke, P.N.: Fuzzy control of base–isolation system using multi–objective genetic algorithm. Comput. Aided. Civil. Infrastruct. Eng. **21**(6), 436–449 (2006)
11. Yu, J.Q., Dong, X.M., Zhang, Z.L.: A novel model of magnetorheological damper with hysteresis division. Smart. Mater. Struct. **26**(10), 105042 (2017)
12. Stanway, R., Sproston, J.L., Stevens, N.G.: Non–linear modeling of an electrorheological vibration damper. J. Electrostat. **20**(2), 167–184 (1987)
13. Wereley, N.M., Pang, L., Kamath, G.M.: Idealized hysteresis modeling of electrorheological and magnetorheological dampers. J. Intell. Mater. Syst. Struct. **9**(8), 642–649 (1998)

14. Gamota, D.R., Filisko, F.E.: Dynamic mechanical studies of electrorheological materials: moderate frequencies. J. Rheol. **35**(3), 399–425 (1991)
15. Bouc, R.: Modele mathematique d'hysteresis. Acustica **24**, 16–25 (1971)
16. Wen, Y.K.: Method of random vibration of hysteretic systems. J. Eng. Mech. Div. **102**(2), 249–263 (1976)
17. Spencer, B.F., Dyke, S.J., Sain, M.K., Carlson, J.D.: Phenomenological model of a magnetorheological damper. J. Eng. Mech. **123**(3), 230–238 (1997)
18. Bai, X.X., Cai, F.L., Chen, P.: Resistor–capacitor (RC) operator–based hysteresis model for magnetorheological (MR) dampers. Mech. Syst. Signal. Process. **117C**, 157–169 (2019)
19. Bui, Q.D., Nguyen, Q.H., Bai, X.X., Mai, D.D.: A new hysteresis model for magneto–rheological dampers based on magic formula. Proc. Inst. Mech. Eng. C-J. Mec. Eng. Sci. **235**(13), 2437–2451 (2021)
20. Bui, D.Q., Hoang, V.L., Le, H.D., Nguyen, H.Q.: Design and Evaluation of a Shear-Mode MR Damper for Suspension System of Front-Loading Washing Machines. In: Hung, N.-X., Phuc, P.-V., Rabczuk, T. (eds.) ACOME 2017. LNME, pp. 1061–1072. Springer, Singapore (2017). https://doi.org/10.1007/978-981-10-7149-2_74

Development of a Novel Self–adaptive Shear–Mode Magneto–Rheological Shock Absorber for Motorcycles

Quoc–Duy Bui[1](✉) and Quoc Hung Nguyen[2](✉)

[1] Faculty of Mechanical Engineering, Industrial University of Ho Chi Minh City, Ho Chi Minh City, Vietnam
buiquocduy@iuh.edu.vn

[2] Faculty of Engineering, Vietnamese–German University, Thu Dau Mot City, Binh Duong Province, Vietnam
hung.nq@vgu.edu.vn

Abstract. This paper investigates a new magneto–rheological (MR) shock absorber in shear–mode for motorcycle suspensions which is self–adaptive to varying road disturbance sources. By disposing permanent magnets at the shaft ends, the damping force of the proposed shock absorber can be activated and increases with traveling stroke amplitude. This displacement–based damping characteristic is very adequate to motorcycles operation for simultaneous satisfaction of ride comfort and steering stability criteria. Compact structure and extremely low cost are also considerable advantages of the proposed shock absorber. To achieve optimum vibration isolation performance, the multi–magnet configuration of the MR shock absorber is studied. Subsequently, modeling of the motorcycle suspension featuring the MR shock absorber is implemented. Design optimization of the proposed shock absorber is then conducted and advances in damping characteristics are figured out via both simulation and experiment.

Keywords: Self–adaptive · Shear–mode · Magneto–rheological shock absorber · Motorcycle · Suspension · Vibration control

1 Introduction

It is well–known that suspension system is an essential component of motorcycles to mitigate unwanted vibrations from rough road profiles. A good suspension system helps improve the ride comfort, road holding, steering stability as well as fatigue life of motorcycle elements. Passive suspensions have simple design and cost–effective but have reached practical limits on the optimal trade–off between the two conflicting requirements of ride comfort and steering stability; therefore, restricted performances are inevitable in rather high frequency excitations. This shortcoming has been compensated by subsequent studies of active and semi–active vibration control systems. Compared with the active suspensions, the semi–active ones are more desirable as they are more

N. V. Khang et al. (Eds.): ASIAN MMS 2021, MMS 113, pp. 744–754, 2022.
https://doi.org/10.1007/978-3-030-91892-7_71

economical, simpler and safer while providing better performance against the passive ones.

A technology that has been increasingly attracted attention of scholars in the field of vibration control is to use MR shock absorbers. By adjusting the magnetic field applied to the MR fluid (MRF), the damping force of MR shock absorbers can be controlled effectively and versatilely between off– and active–states. Thus, MR shock absorbers have shown potential applicability to semi–active vibration isolation systems such as seismic protectors [1, 2], suspensions of washing machines [3, 4] and vehicles [5, 6]. For motorcycles, several studies of semi–active suspension systems using MR shock absorbers have been developed [7–10]. However, flow–mode configuration of these shock absorbers demands a large amount of MRF and hence increases production cost, also the off–state forces are relatively high, which cannot achieve optimum vibration control performance at high frequencies. On the other hand, the above conventional MR shock absorbers encounter issues of structural complexity, occupied installation space, high weight and cost resulting from associated sensors, power supply and controller that should be further speculated.

Recently, area of energy–harvesting has been exploited to combine with MR shock absorbers into self–powered ones [11–13]. For the first time of motorcycle suspensions, a self–powered MR shock absorber has been proposed by Chen et al. [14]. The above-mentioned shock absorbers have integrated structure, low cost and are self–adaptable to external excitations. Nevertheless, complicated structure coming from the energy–harvesting part, coils winding and connection still impedes manufacturing and maintenance. In addition, velocity–based damping characteristic of self–powered MR shock absorbers may not be appropriate for operating conditions of motorcycles. It is well–known that displacement and velocity of the motorcycle chassis are both high at resonance, so high damping level is sensible. At higher frequencies, however, the velocity is still high but the displacement may not, which results in unwanted active–state and increased force transmission of self–powered MR shock absorbers–based control systems. Therefore, the damping of MR shock absorbers for motorcycle suspensions should be tuned to vibratory excitation amplitude rather than excitation velocity.

It can be realized from the above analyses that previous semi–active suspensions featuring MR shock absorbers have disadvantages of complicated structure and high cost due to controller requirement while the self–powered ones meet with difficulties of winding coils and inappropriate velocity–based damping characteristic. Consequently, this study work develops a novel self–adaptive MR shock absorber for motorcycle suspensions. The shear–mode used in the proposed shock absorber provides a simple, compact design with low off–state force and low cost. In addition, the shock absorber possesses stroke–by–activated ability, which entirely conform to the operating principle of motorcycles. The design without winding coils and control system is also more economical for commercialization criterion. To improve damping efficiency and at the same time satisfy assembly space, size and production cost, optimization procedure is implemented for main geometrical dimensions of the proposed shock absorber. Then the advances in performance of the proposed shock absorber are figured out via both simulation and experiment with discussions.

2 Modeling of the Self–adaptive Shear–Mode MR Shock Absorber– Based Motorcycle Suspension

Figure 1 shows the half–motorcycle rigid tire suspension model featuring the MR shock absorber. In this model, the motorcycle tire is considered to connect in series with the suspension based on the fact that the tire stiffness is much greater than the suspension spring one so that it can be neglected. From the figure, governing equation of motion can be written as follows

$$m\ddot{x}_y(t) + c\dot{x}_y(t) + kx_y(t) = -m\ddot{y}(t) \quad (1)$$

In the above equation, m is the sprung mass, c and k are the suspension damping coefficient and stiffness, respectively, $x_y(t)$ is the relative deflection of the sprung mass forced by the road surface input $y(t)$. The damping ratio ξ, natural free–vibration frequency ω_n and damped frequency ω_d are calculated by

$$\xi = \frac{c}{2\sqrt{km}};\ \omega_n = \sqrt{\frac{k}{m}};\ \omega_d = \omega_n\sqrt{1-\xi^2} \quad (2)$$

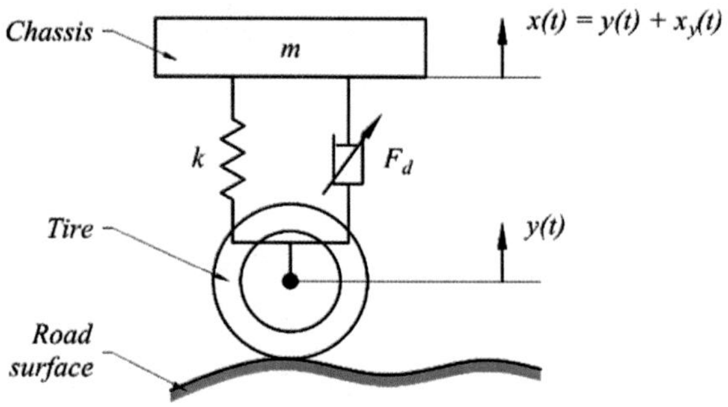

Fig. 1. Half–motorcycle rigid tire suspension model featuring the MR shock absorber.

Assuming that the road surface response subjects to a harmonic displacement with an excitation frequency of ω and an amplitude of Y

$$y(t) = Y\sin(\omega t) \quad (3)$$

then the acceleration of the road surface input is

$$\ddot{y}(t) = -\omega^2 Y\sin(\omega t) \quad (4)$$

From Eqs. (1) and (4), the steady–state relative displacement response of the sprung mass forced by the road surface input can be derived by

$$x_y(t) = \frac{m\omega^2 Y}{k} D\sin(\omega t - \theta) \quad (5)$$

where θ is the phase angle and D is the dynamic magnification factor

$$\theta = \arctan\left(\frac{2\xi r}{1-r^2}\right);\ D = \frac{1}{\sqrt{\left(1-r^2\right)^2 + (2\xi r)^2}} \tag{6}$$

in which r is the ratio of the excitation frequency to the natural frequency, $r = \omega/\omega_n$. By vectorially adding this relative motion to the road surface input $y(t)$ in Eq. (3), we can obtain the total steady–state response of the sprung mass m as follows

$$x(t) = y(t) + x_y(t) = YD\sqrt{1 + (2\xi r)^2}\sin(\omega t - \theta) \tag{7}$$

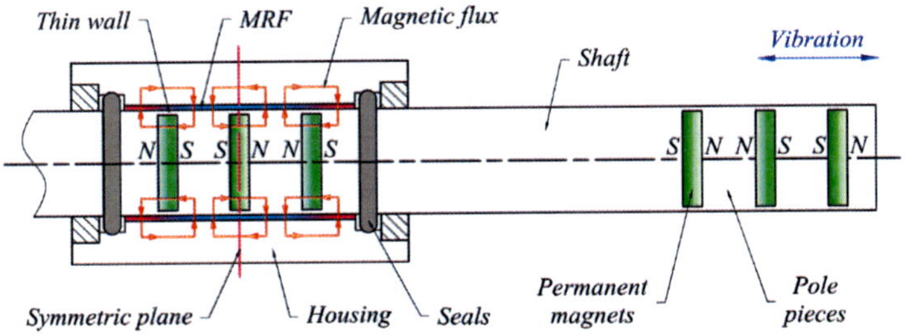

Fig. 2. Configuration and operating principle of the self–adaptive shear–mode MR shock absorber.

Figure 2 sketches configuration and operating principle of the self–adaptive shear–mode MR shock absorber. As shown in the figure, a layer of MRF is positioned between the shaft and inner surface of housing. The shear–mode comes from the fact that the shaft moves reciprocally due to vibration of washing machines and thereby developing damping force via direct shearing of the MRF. Among three operational modes of MR damper (shear–mode, flow–mode and their combination, mixed–mode), this shear–mode is proposed for our study due to its simple design, small off–state friction force and low cost. Inside each shaft end, permanent magnets and pole pieces are installed alternatively. A plain shaft part is left in the middle for the off–state of the shock absorber. From the figure, it is seen that the MR shock absorber off–state is maintained during small–amplitude vibration strokes since no magnet contacts the MRF so as to close the magnetic circuit. Under greater oscillation, the magnets start to move into the MRF region, which applies magnetic fields to the MRF. As a result, the higher the vibration stroke is, the more active–state MRF is and the higher the damping is generated to suppress the vibration. At resonant frequency of motorcycle rigid mode, the suspension amplitude is great and a large damping force is produced. At high frequencies, the damping force is however reduced due to small vibration strokes, which prevents force transmissibility to occupants. In order words, this configuration enables the self–adaptable ability of the shock absorber to external disturbances without any control. In order to increase the effective length of the MRF gap, the configurations with two or more magnets are considered. It is noted that in the optimal design of multi–magnet shock absorbers, the magnets and pole pieces are symmetrically distributed about the middle symmetric plane and two consecutive magnets are laid in cross–pole situation for expected magnetic flux directions.

3 Optimal Design of the Self–adaptive Shear–Mode MR Shock Absorber

Main geometry of the self–adaptive shear–mode MR shock absorber is shown in Fig. 3. Materials are assigned to the components as follows: NdFeB grade N35 for the axial disc–shaped permanent magnets, 70–durometer NBR rubber for the O–rings sealing the MRF gap, and commercial C45 steel for the shaft, housing and pole pieces. By assuming a linear velocity profile of the MRF in the gap, the damping forces in active–state and off–state (also referred as zero–field friction force) are determined as follows

$$F_d = A\left(\tau_y + \frac{\eta}{t_g}\dot{x}\right) + A_0\left(\tau_{y0} + \frac{\eta_0}{t_g}\dot{x}\right) + 2F_{or} \tag{8}$$

$$F_0 = A_0\left(\tau_{y0} + \frac{\eta_0}{t_g}\dot{x}\right) + 2F_{or} \tag{9}$$

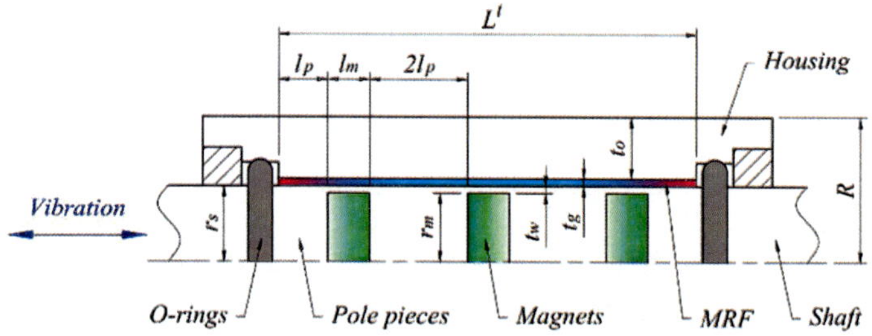

Fig. 3. Main geometry of the self–adaptive shear–mode MR shock absorber.

where the subscript *0* denotes the off–state, F_d is the damping force, η and τ_y are the post yield viscosity and yield stress of the MRF, respectively, F_{or} is the Coulomb friction force between each O–ring and the shock absorber shaft, t_g is the MRF gap size, and A is the lateral area of the cylindrical shaft, which is obtained by

$$A = 2\pi r_s L;\ A_0 = 2\pi r_s L_0 \tag{10}$$

in which r_s is the shaft radius, L and L_0 are the active and inactive MRF lengths, respectively. In this design

$$L = n_p l_p;\ L_0 = n_m l_m \tag{11}$$

where n_m, n_p are the numbers of magnets and pole pieces, and l_m, l_p are the lengths of each magnet and pole piece, respectively. The sum of these products presents the total operating length of the MR shock absorber L^t. The MRF used for the MR shock absorber is the 140–CG type produced by Lord Corporation. Its rheological characteristics (η and τ_y) depend on the applied magnetic density [15].

In this work, design optimization of the proposed self-adaptive shear-mode MR shock absorber is carried out based on the dynamic response of the motorcycle suspension and quasi-static model of the MR shock absorber presented above. From Fig. 1 and Eq. (7), let $X = YD\sqrt{1 + (2\xi r)^2}$ be the total response amplitude of the sprung mass m, the force transmissibility of the system is defined by

$$T_f = \frac{X}{Y} = D\sqrt{1 + (2\xi r)^2} \tag{12}$$

Supposing the sprung mass m is 150 kg and the spring stiffness k is 5 kN/m for the half-motorcycle suspension model, the damping ratio $\xi = 0.7$ is sufficient for an attenuation of force transmissibility at resonance [3, 4]. With the road surface input amplitude Y is about 0.5 m, the sprung mass total response amplitude X is found to be 0.0635 m. Then the equivalent expected damping force at the resonant frequency ratio $r = \sqrt{1 - \xi^2}$, which can be calculated by [3], is around 249.3 N. Thus, the expected maximum damping force for the proposed MR shock absorber is set by 250 N.

In this study, the optimization objective is to find optimal essential geometry of the proposed shock absorber which can generates the above expected maximum damping force while its off-state one is controlled small possibly. The housing thickness t_o, pole length l_p, magnet length l_m and radius r_m are considered as design variables. The thicknesses of the thin wall and MRF gap are empirically take the value 0.8 mm based on the manufacturing aspect. To satisfy the working principle presented in Fig. 2, the total operating length of the shock absorber L^t is established to equal to the sprung mass total response amplitude X, which is 63.5 mm. Coming from fitness to commercial suspension springs, a limit of 20 mm is also assigned to the shock absorber radius R. With this constraint, the self-adaptive shear-mode MR shock absorber is expected to be significantly more compact than conventional and self-powered MR ones. In summary, the design optimization problem of the proposed MR shock absorber can be addressed: Find optimal main geometrical dimensions of the shock absorber so that the off-state force F_0 is minimized while the maximum active force F_d should be greater than 250 N and the total radius R is limited not to exceed 20 mm.

4 Results and Discussions

In this section, the first-order technique combined with steepest gradient algorithm of commercial ANSYS software and finite element model based on 2D symmetric couple element (PLANE 13) are utilized to obtain optimal solution of the self-adaptive shear-mode MR shock absorber. Finite element analysis (FEA)-based procedure to achieve optimal design of MR devices can be found in details from previous literature [5].

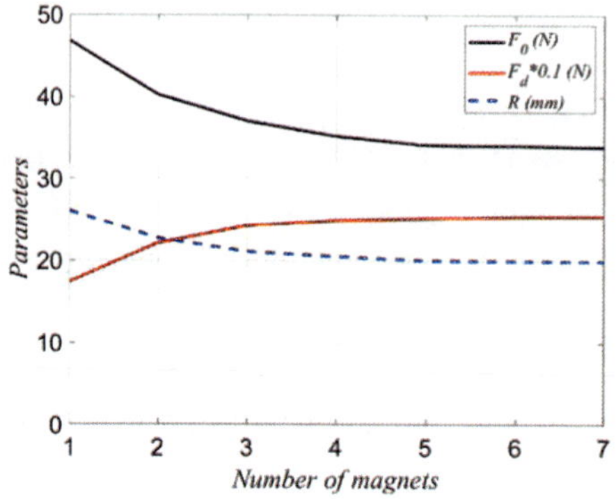

Fig. 4. Influence of the number of magnets on the optimal parameters.

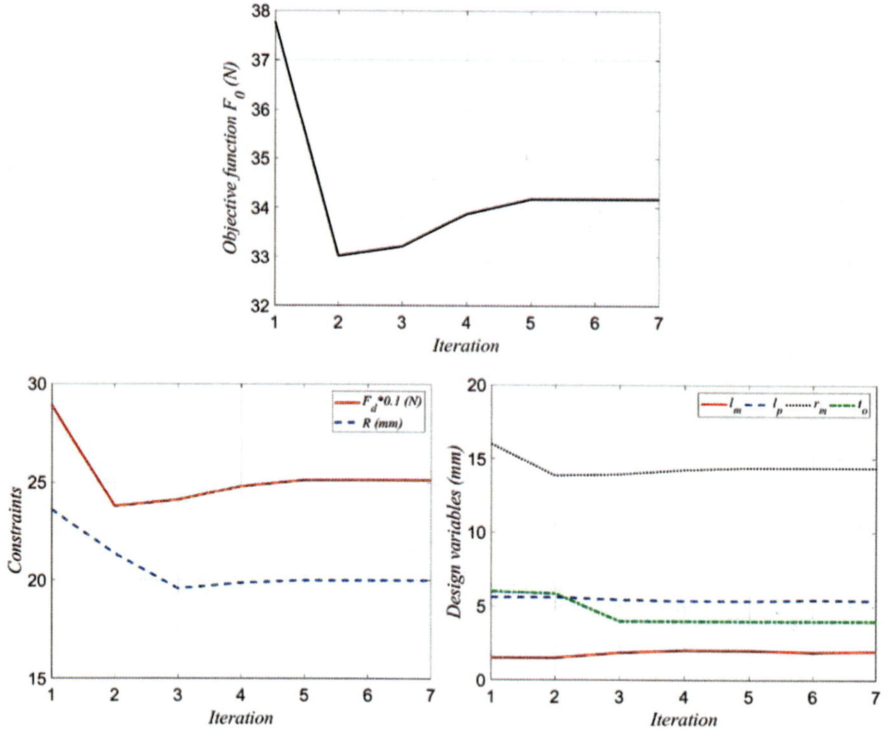

Fig. 5. Optimization process of the self–adaptive five–magnet MR absorber.

As mentioned in Sect. 2, increasing the number of permanent magnets results in damping performance improvement of the MR shock absorber. However, this increases the structure complication at the same time and is also restricted by the available total operating length of the shock absorber. For a most effective configuration, the number of magnets should be studied. Figure 4 shows influence of the number of magnets on the optimal results. It can be seen from the figure that the constraints of F_d and R start to be satisfied at the five–magnet configuration. In addition, the objective function F_0 almost has no significant improvement when the number of magnets is increased from 5 to 7. Therefore, the five–magnet configuration is optimum for this research.

Table 1. Optimal solution of the self–adaptive five–magnet MR shock absorber.

Design variables (mm)	Performance characteristics (N)
Magnets: length $l_m = 2$, radius $r_m = 14.4$, pole $l_p = 5.3$	Max. damping force $F_d = 251$
Shaft: thin wall $t_w = 0.8$, radius $r_s = 15.2$	Off–state force $F_0 = 34.2$
MRF gap: length $L^t = 63.5$, thickness $t_g = 0.8$	
Housing: thickness $t_o = 4$, outer radius $R = 20$	

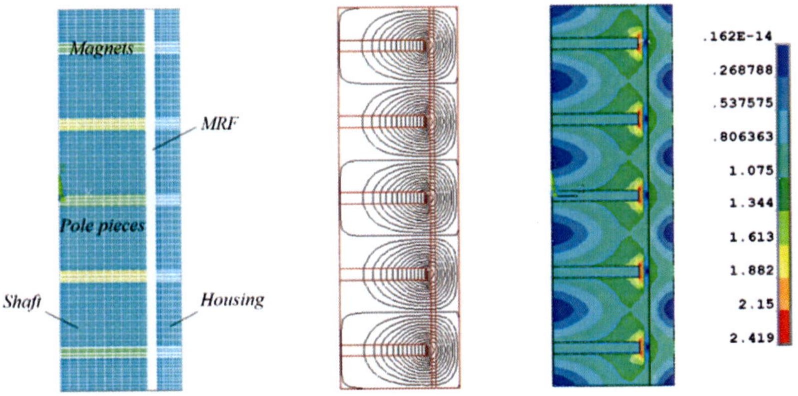

(a) finite element model (b) magnetic flux lines (c) magnetic flux density

Fig. 6. FEA results of the optimized self–adaptive five–magnet MR shock absorber.

Optimization process of the five–magnet MR shock absorber is shown in Fig. 5. It is observed that the solution converges after 7 iterations, at which the minimum objective function F_0 is 34.2 N. The optimal solution of the five–magnet MR shock absorber is summarized in Table 1. As shown in the table, the maximum damping force can reach up to 251 N as constrained for the resonance region while the off–state force is kept at 34.2 N which is small enough to prevent force transmissibility to occupants at high frequencies. It is also noted that the housing thickness t_o is not decreased smaller than 4 mm for an O–ring grooves manufacturing without warping. Remarkably, the outer radius of the proposed shock absorber (20 mm) is much smaller than those of the conventional MR shock absorbers [7, 9] (25.4 and 40 mm, respectively), which is a highly considered business criterion. FEA solution of magnetic flux of the optimized self-adaptive five–magnet shear–mode MR shock absorber is illustrated in Fig. 6. The figures show that the magnetic flux passing through the thin wall almost reaches to saturation. By optimizing design, the magnetic flux across the MRF gap can be distributed throughout the gap length. This increases the solidification rate of the MRF in the gap and hence improves the damping force considerably.

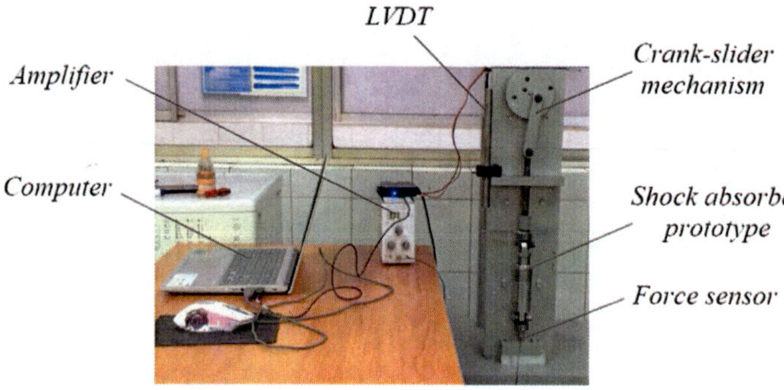

Fig. 7. Test rig to evaluate performance of the self–adaptive shock absorber prototype.

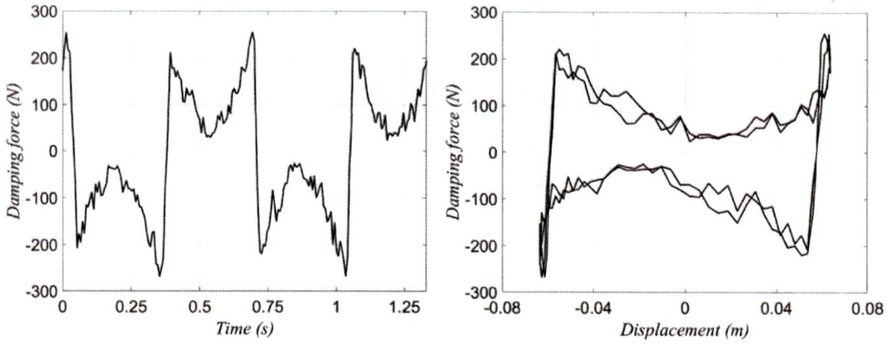

(a) force vs. time (b) force vs. displacement

Fig. 8. Responses of the self–adaptive shock absorber prototype under 1.5 Hz frequency.

Based on the optimal solution, a self–adaptive shear–mode MR shock absorber prototype is designed and manufactured. Figure 7 shows a test rig to evaluate output damping performance of the shock absorber. The experimental step responses under the resonant frequency of 1.5 Hz are obtained in Fig. 8. The average maximum damping force is 234.2 N, which is about 94% of the expected one. This mainly comes from the flux leakage between the magnetic parts. The average off–state force is 39.6 N, which is 116% of the simulated one. The difference is believed to rise from the partial magnetic impact of the magnets on the MRF even though they do not enter the region yet. In general, the experimental performance of the shock absorber prototype well correlates to the theoretical analyses. From the figure, it is also observed that the damping force increases with the displacement, which is consistent to the previous design principle. The same results are obtained in higher–frequency tests of the shock absorber, as shown in Fig. 9, except the damping force increases slightly due to inertia effect.

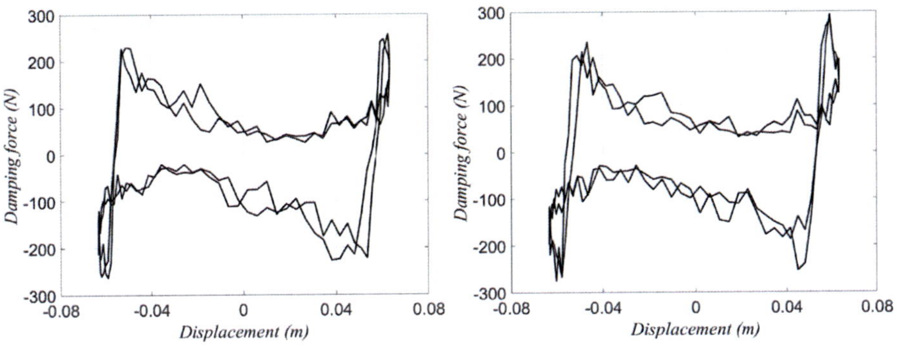

(a) 5 Hz (b) 15 Hz

Fig. 9. Responses of the self–adaptive shock absorber prototype under higher frequencies.

5 Conclusions

This paper developed a new MR shock absorber in shear–mode which can substitute effectively for commercial passive ones of motorcycle suspensions in the field of vibration isolation. The proposed shock absorber can tune damping level by itself to varying road disturbances. Compared with conventional and self–powered MR shock absorbers, the design without winding coils and control system of the proposed one facilitates manufacturing, maintenance and extremely reduces cost. Furthermore, its stroke–by–activated damping characteristic is highly compatible with working conditions of motorcycles. In this study, multi–magnet configurations were considered and an optimization process for the proposed shock absorber design was implemented based on the analyses of the motorcycle suspension dynamic model. The simulated results showed that the performance characteristics are most ideal with the five–magnet configuration, also the optimal geometric dimensions of the self–adaptive shock absorber were obtained. Based on the optimal solution, a five–magnet MR shock absorber prototype was designed, manufactured and tested. There was a good correlation between the experimental damping forces and theoretical analyses. To continue this work in the next phase, the proposed shock absorber will be evaluated on a motorcycle prototype.

References

1. Dyke, S.J., Spencer, B.F., Sain, M.K., Carlson, J.D.: An experimental study of MR dampers for seismic protection. Smart Mater. Struct. **7**(5), 693–703 (1998)
2. Weber, F.: Semi–active vibration absorber based on real–time controlled MR damper. Mech. Syst. Signal Process. **46**(2), 272–288 (2014)
3. Nguyen, Q.H., Choi, S.B., Woo, J.K.: Optimal design of magnetorheological fluid–based dampers for front–loaded washing machines. Proc. Inst. Mech. Eng. C-J. Mec. Eng. Sci. **228**(2), 294–306 (2014)
4. Bui, D.Q., Hoang, V.L., Le, H.D., Nguyen, H.Q.: Design and Evaluation of a Shear-Mode MR Damper for Suspension System of Front-Loading Washing Machines. In: Nguyen-Xuan, Hung, Phung-Van, Phuc, Rabczuk, Timon (eds.) ACOME 2017. LNME, pp. 1061–1072. Springer, Singapore (2017). https://doi.org/10.1007/978-981-10-7149-2_74

5. Nguyen, Q.H., Choi, S.B.: Optimal design of MR shock absorber and application to vehicle suspension. Smart Mater. Struct. **18**(3), 035012 (2009)
6. Bai, X.X., Hu, W., Wereley, N.M.: Magnetorheological damper utilizing an inner bypass for ground vehicle suspensions. IEEE Trans. Magn. **49**(7), 3422–3425 (2013)
7. Ericksen, E.O., Gordaninejad, F.: A magneto–rheological fluid shock absorber for an off–road motorcycle. Int. J. Veh. Des. **33**(1–3), 139–152 (2003)
8. Li, W.H., Du, H., Guo, N.Q.: Design and testing of an MR steering damper for motorcycles. Int. J. Adv. Manuf. Technol. **22**(3), 288–294 (2003)
9. Wang, D.H., Wang, T., Bai, X.X., Yuan, G.: A self–sensing magnetorheological shock absorber for motorcycles. In: 19th International Conference on Adaptive Structures and Technologies, pp. 639–649. ETH Zurich and Empa, Switzerland (2008)
10. Ahmadian, M., Sandu, C.: An experimental evaluation of magneto–rheological front fork suspensions for motorcycle applications. Int. J. Veh. Syst. Model. Test. **3**(4), 296–311 (2008)
11. Choi, Y.T., Wereley, N.M.: Self–powered magnetorheological dampers. J. Vib. Acous. **131**(4), 044501 (2009)
12. Bui, Q.D., Nguyen, Q.H., Nguyen, T.T., Mai, D.D.: Development of a magnetorheological damper with self–powered ability for washing machines. Appl. Sci. **10**(12), 4099 (2020)
13. Bui, Q.D., Nguyen, Q.H., Nguyen, T.T., Mai, D.D.: Design and testing of a new shear-mode magneto–rheological damper with self-power component for front–loaded washing machines. In: Long, B.T., Kim, Y.-H., Ishizaki, K., Toan, N.D., Parinov, I.A., Vu, N.P. (eds.) MMMS 2020. LNME, pp. 860–866. Springer, Cham (2020). https://doi.org/10.1007/978-3-030-69610-8_114
14. Chen, C., Chan, Y.S., Zou, L., Liao, W.H.: Self–powered magnetorheological dampers for motorcycle suspensions. Proc. Inst. Mech. Eng. D-J. Aut. Eng. **232**(7), 921–935 (2018)
15. Zubieta, M., Eceolaza, S., Elejabarrieta, M.J., Ali, M.M.B.: Magnetorheological fluids: characterization and modeling of magnetization. Smart. Mater. Struct. **18**(9), 095019 (2009)

Nonstationary Resonant Oscillations of a Gyroscopic Rigid Rotor with Nonlinear Damping and Non-ideal Energy Source

Zharilkassin Iskakov(✉), Nutulla Jamalov, and Azizbek Abduraimov

Institute of Mechanics and Engineering, Al-Farabi Kazakh National University, Almaty, Kazakhstan

Abstract. The article is concerned with the effect of nonlinear cubic damping of an elastic support on unsteady resonant vibrations of a gyroscopic rigid rotor when interacting with a non-ideal energy source. It is confirmed that nonlinear cubic damping can suppress not only the maximum amplitude, but also the amplitude of unsteady oscillations behind the rotation speed corresponding to the amplitude peak. It shifts the control parameter corresponding to the maximum amplitude, downward with a rigid nonlinear elastic characteristic of the support material, and upward with a soft nonlinear elastic characteristic of the support material. An increase in the nonlinear cubic damping coefficient can significantly weaken the Sommerfeld effect with a nonlinear jump in unsteady oscillations, up to its complete elimination. The difference in the values of the maximum amplitude and in the corresponding values of the control parameter in the resonance curves with an increasing and decreasing control parameter is explained by the difference in the values of the same parameters relating to the jumping effects during the acceleration and runout of the rotary machine.

Keywords: Gyroscopic rotor · Non-ideal source · Nonlinear damping · Nonstationary oscillation

1 Introduction

Vibration is commonly found in rotating equipment during start-up, operation, and shutdown.

It is very important to use properties and characteristics of the material of the supports for attenuation and damping of vibration in order to stabilize movement of an unbalanced rotor and vibration systems.

A convenient way to introduce attenuation to support bearings in a rotor system on viscoelastic flexible rubber supports [1]. In parallel with the development of viscoelastic material modeling, which helps to describe the complexity of material properties, the use of viscoelastic components in the dynamics of the rotor and vibration systems also increased as a whole, in particular with non-linear elastic characteristic and damping. So, for example, in works [2–5], the research results show that linear and nonlinear cubic damping can significantly suppress the maximum amplitudes, eliminate the jump-like

N. V. Khang et al. (Eds.): ASIAN MMS 2021, MMS 113, pp. 755–763, 2022.
https://doi.org/10.1007/978-3-030-91892-7_72

phenomena of a nonlinear system. In nonresonant regions, where the vibration frequency is higher than its resonant value, nonlinear cubic damping, unlike linear damping, can reduce the amplitude of the rotor vibration. Therefore, in all regions of oscillation frequency (rotation speed), only non-linear cubic damping can support the performance characteristics of the vibration isolator. [2] provides an excellent overview of research on passive vibration isolation systems.

In work [6], a nonlinear damping suspension can affect the stability of a flexible rotor in short support bearings, a numerical method is used to solve the equations of motion, and bifurcation diagrams, orbits, Poincaré maps, and amplitude spectra are used to display motions. The results of works [2–5] are confirmed.

In this paper, a gyroscopic nonlinear rotor system is considered on the assumption that it receives an action from an energy source with a limited power. Movement of the oscillatory system under the influence of such sources is accompanied by mutual influence on each other: the energy source and the oscillatory system. Interaction of an oscillatory system with an energy source should also manifest itself in non-stationary modes of motion, especially in such practically important cases as the case of passage of oscillatory systems through resonance.

The process of passage of an oscillatory system through resonance, taking into account its interaction with an energy source, was considered in [7, 8] and in other works. A complete review of articles on research concerning basic properties of vibrational non-ideal systems is given in work [9]. In particular, it has been found that under conditions when the power reserve of the energy source is small, the course of the process depends very much on the characteristics of the energy source. Changes in the frequency of vibrations are closely related to changes in the amplitude of vibrations. In the region of large amplitudes, the rate of passage of the system through the resonance sharply slows down; as the amplitude decreases, the rate of passage increases, if the excess power of the energy source is sufficient for this (a typical Sommerfeld effect). Such resonance capture can cause failure of the rotor shaft, bearings and other structural parts. Therefore, vibration isolation is very important for movement stabilization, and specifically with the help of complex linear and non-linear cubic damping of the elastic support, taking into account change in voltage supplied to the motor during acceleration and runout.

2 Equations of Motion

The following system is considered: gyroscopic rotor – elastic support – DC motor, the structural diagram of which is shown in Fig. 1. The rotor consists of a shaft with a length L, mounted vertically by means of a lower hinge and an upper elastic support spaced from it at a distance l_0 and a disk fixed at the free end of the shaft, having a mass m, a polar moment of inertia I_p and a transverse moment of inertia I_T, the same for any direction.With such an arrangement of the shaft with the disk relative to the supports and with a sufficiently high rotation speed of the shaft $\dot{\varphi}$, the rotor can be considered as a gyroscope. The elastic support has linear stiffness c_1, non-linear stiffness c_3, linear damping μ_{d1} and non-linear cubic damping μ_{d3}. The position of the geometric center of the disk S is determined by coordinates x, y in a fixed coordinate system $Oxyz$, and the position of the shaft and the rotor as a whole in space by the Euler angles α, β and the angle of rotation φ. The angles α, β are small, consequently, the movement of the rotor in the direction of the coordinate axis z is neglected. Next, denote the coordinates of the center of mass m of the disk through x_m and y_m. Assume also that the linear eccentricity e lies in the direction of the N axis of the *ONKZ* coordinate system. Restrict to small deviations of the rotor axis. Then $\sin\alpha \approx \alpha$, $\sin\beta \approx \beta$, $\cos\alpha \approx 1$, $\cos\beta \approx 1$.

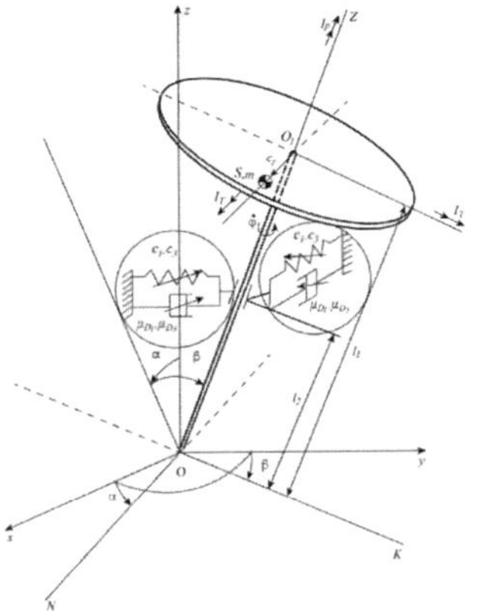

Fig. 1. Rotor geometry.

Express the projections of the angular velocity of the rotor in the coordinate axes of the *ONKZ* system, the coordinates of the center of mass of the disk and the coordinates of the upper support through the angular coordinates α, β and φ, and find expressions for determination of the kinetic energy, the potential energy of the rotor, the Rayleigh function and the projections of the moments of forces acting on the system and the motor torque with straight characteristic. Substituting them into the Lagrange equations of the second kind, and using the following dimensionless parameters

$$\begin{gathered} l = l_0/L;\ \bar{t} = t\omega_0;\ \bar{I}_P = I_P/(mL^2);\ \bar{I}_T = I_T/(mL^2); \\ \overline{C}_1 = c_1/\left(m\omega_0^2\right);\ e_r = e/\left[L\left(1+\bar{I}_T\right)\right];\ I_{P1} = \bar{I}_P/\left(1+\bar{I}_T\right); \\ \overline{G} = g/\left(L\omega_0^2\right);\ C_3 = c_3 l_0^4/\left[mL^2\omega_0^2\left(1+\bar{I}_T\right)\right]; \\ \mu_1 = \mu_{d1}/\left[mL^2\omega_0\left(1+\bar{I}_T\right)\right];\ \mu_3 = \mu_{d3}\omega_0/\left[mL^2\left(1+\bar{I}_T\right)\right]; \\ u_1 = C_M\Phi\omega_0 U/\left[RmL^2\left(1+\bar{I}_T\right)\right];\ u_2 = \left(C_M C_E\Phi^2/R + q_m\right)/\left[mL^2\omega_0\left(1+\bar{I}_T\right)\right], \end{gathered} \tag{1}$$

where $\omega_0 = \sqrt{\left(k_1 l_0^2 - mgL\right)/\left[mL^2 - \left(I_p - I_T\right)\right]}$ the natural frequency of the damped rotor system, $C_M = C_E$ is the mechanical (or electrical) constant, Φ is the magnetic flux of one pole, U is the voltage drop across the entire power circuit of the motor with resistance R, q_m is the coefficient of resistance to the rotation of the motor rotor, obtain the equations of the rotor motion in the form

$$\begin{aligned} \alpha'' + \omega_n^2\alpha &= e_r\varphi'^2\cos\varphi - I_{P1}\varphi'\beta' - \mu_1\alpha' - \mu_3\alpha'^3 - C_3\alpha^3, \\ \beta'' + \omega_n^2\beta &= e_r\varphi'^2\sin\varphi + I_{P1}\varphi'\alpha' - \mu_1\beta' - \mu_3\beta'^3 - C_3\beta^3, \\ \varphi'' &= [e_r(\alpha''\sin\varphi - \beta''\cos\varphi) - I_{P1}(\alpha''\beta + \alpha'\beta') + u_1 - u_2\varphi']/I_{p1}, \end{aligned} \tag{2}$$

where $\omega_n = \sqrt{(\overline{K}_1 l^2 - \overline{G})/(1 + \overline{I}_T)}$ the dimensionless natural frequency of the linear rotor system (2) at $\overline{I}_T \gg \overline{I}_P$, u_1 is the control parameter (controlling parameter), depending on the voltage on the motor; u_2- a parameter that depends on the type of energy source.

On the right-hand part of the system of Eqs. (2) perturbations containing φ'', were discarded, since in the region close to the resonance velocity $\varphi'' \ll \varphi'^2$, and perturbations having a parameter $\overline{I}_P$ (in what follows, assuming that $\overline{I}_P \ll \overline{I}_T$) and values of the second and higher orders of smallness with respect to α, β, their derivatives, and their combinations. The indicated disturbances are small in comparison with disturbances, the amplitudes of which are proportional to the angular velocity squared.

Equations (2) are a system of second order nonlinear ordinary differential equations with respect to α, β and φ.

3 Solutions of Motion Equations

Consider a rotor system close to a linear system. To use the Bogolyubov method, the following restrictions are taken to solve Eqs. (2). The projections of the moments of the damping forces $\mu_1\alpha'$, $\mu_1\beta'$ and $\mu_3\alpha'^3$, $\mu_3\beta'^3$, as well as the moment of the cubic component of the restoring force $C_3\alpha^3$, $C_3\beta^3$, the moments of the centrifugal force of the imbalance $e_r\varphi'^2\cos\varphi$, $e_r\varphi'^2\sin\varphi$ of mass and are considered small in comparison with the projections of the moments of the vibration inertia force and the linear restoring force acting in the system. Assuming that $\overline{I}_P \ll \overline{I}_T$ the projections of the moment of the passive gyroscopic force can also be considered small $I_{P1}\varphi'\alpha'$, $I_{P1}\varphi'\beta'$, limit also to considering a spinning rotor: $\varphi'^2 \gg \overline{G}$ and motion in the resonance range, where the frequency of free oscillations ω_n is close to the frequency of forced oscillations Ψ. Therefore, search for solutions (2) in the form:

$$\alpha = A\cos(\varphi + \chi), \beta = A\sin(\varphi + \chi). \tag{3}$$

Here variables A, χ, Ψ will be slowly changing functions of times $\overline{t}$. They represent the most essential parameters of motion: A- vibration amplitude, χ- phase shift angle between the angular coordinate α or β and the disturbing moment.

Following Bogolyubov's method, obtain a system of equations in relation to A, χ, Ψ, the approximate solutions of which can be represented in the form

$$\Psi = \Omega + \varepsilon E_1(\overline{t}, \Omega, a, \xi), A = a + \varepsilon E_2(\overline{t}, \Omega, a, \xi), \chi = \xi + \varepsilon E_3(\overline{t}, \Omega, a, \xi), \tag{4}$$

where $\varepsilon E_1(\overline{t}, \Omega, a, \xi)$, $\varepsilon E_2(\overline{t}, \Omega, a, \xi)$, $\varepsilon E_3(\overline{t}, \Omega, a, \xi)$ are small periodic functions of time $\overline{t}$, $\varepsilon \ll 1$ is the small parameter.

After averaging of the obtained equations of motion, which are equivalent to the system (2), obtain the system of equations for the transient process of the rotor in the following form

$$
\begin{gathered}
d\Omega/d\bar{t} = (u_1 - u_2\Omega + e_r a\omega_n\Omega\sin\xi)/I_{P1},\\
da/d\bar{t} = -\left(e_r\Omega^2\sin\xi/\omega_n + \mu_1 a + 3\mu_3\omega_n^2 a^3/4\right)/2,\\
d\xi/d\bar{t} = \omega_n - \Omega - e_r\Omega^2\cos\xi/(2\omega_n a) + 3C_3a^2/(8\omega_n),\\
d\varphi/d\bar{t} = \Omega.
\end{gathered}
\tag{5}
$$

4 Unsteady Oscillations

The system of equations for the unsteady process (5) was modeled in the Mathlab-Simulink package. The control parameter u_1 "slowly" increased uniformly ($\nu > 0$) and "slowly" decreased uniformly over ($\nu < 0$) time in accordance with the regularity $u_1 = u_{10} + \nu\bar{t}$ [10].

The general parameters of the system have been selected in accordance with various design parameters of the centrifuge used in [5] for experimental studies, and have the following dimensionless values: $e_r = 0.0346$, $\omega_n \approx 1$, $I_{P1} = 0.021$ ($\bar{I}_P = 0.026$, $\bar{I}_T = 0.213$). The values of the parameters C_3, μ_1, μ_3, u_1, u_2 and ν were chosen in the course of numerical experiments, and the values of μ_1 and μ_3, taking into account those values at which the jumping effects disappear, in accordance with the rest of the known design parameters necessary for creation of the effective vibration isolation for the centrifuge based on gyroscopic rotor. The values of the parameters for the initial conditions are borrowed from the frequency characteristics of the stationary oscillation at $\Omega_0 < \omega_n$ and $\Omega_0 > \omega_n$. Thus, for the parameters $u_2 = 1.245$, $\mu_1 = 0.01$ and $\mu_3 = 0.01, 0.02$, we take the following initial conditions: at $C_3 = 0.1$ for the case$\nu = 0.00025$: $\bar{t} = 0$, $\Omega_0 = 0.88$, $u_{10} = 1.096$, $a_0 = 0.11$, $\xi_0 = $ —.0420, for the case $\nu = -0.00025$: $\bar{t} = 0$, $\Omega_0 = 1.2$, $u_{10} = 1.50$, $a_0 = 0.12$, $\xi_0 = 0.0253$; for $C_3 = $ —0.1 for the case $\nu = 0.00025$: $\bar{t} = 0$, $\Omega_0 = 0.88$, $u_{10} = 1.096$, $a_0 = 0.112$, $\xi_0 = $ —0.0424, for the case$\nu = -0.00025$: $\bar{t} = 0$, $\Omega_0 = 1.2$, $u_{10} = 1.494$, $a_0 = 0.124$, $\xi_0 = 0.0253$.

The abscissa axis has two scales: the parameter scale u_1 and the corresponding time scale $\bar{t}$.

The graphs of dependence $a = a(u_1)$ non-stationary oscillations of the rotor, built according to the results of modeling Eq. (5), are shown in Figs. 2 and 3.

From these figures, first of all, the damping effect of the parameter μ_3 on the maximum value a_m and on the quasiperiodic variation of the oscillation amplitude behind the amplitude peak and the influence μ_3 on the value of the control parameter u_m corresponding to the maximum amplitude are obvious. Non-linear cubic damping shifts the control parameter u_m (shaft speed Ω_m) corresponding to the maximum amplitude at $C_3 \succ 0$ (Fig. 2a) downwards, and at $C_3 \prec 0$ (Fig. 3b) upwards, i.e. in both cases, the characteristics of the nonlinear stiffness of the support Ω_m approaches ω_n.

The values of the maximum amplitude and the corresponding control parameter in the resonance curves during acceleration (Figs. 2a and 3a) and runout (Figs. 2b and 3b) of the rotary machine, approximately determine the positions of the jumping effects. As the nonlinear cubic damping coefficient increases, the distance between these positions decreases, and its further increase can completely eliminate the jumping phenomena.

Thus, increasing the value of nonlinear cubic damping can significantly weaken the Sommerfeld effect with a nonlinear jump, up to its complete elimination.

The change in the characteristics of the nonlinear stiffness of the elastic support and the nature of the change in the controlling parameter significantly affects the description of the dependency graphs $a = a(u_1)$.

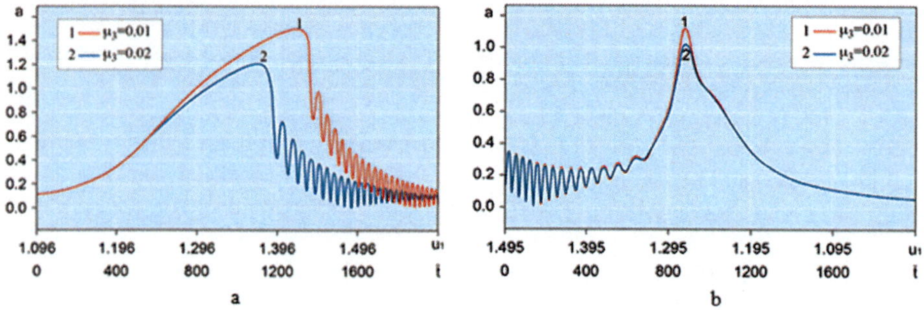

Fig. 2. Transition through resonance in the case of $C_3 = 0.1$ with different values of μ_3 and angular acceleration: a - $\nu = 0.00025$, b - $\nu = -0.00025$.

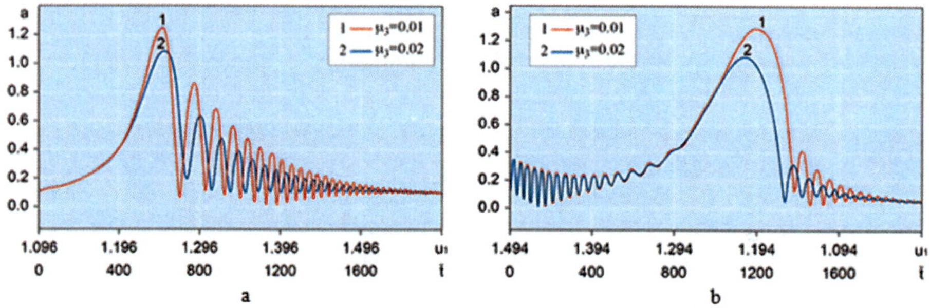

Fig. 3. Transition through resonance in the case of $C_3 = -1$ with different values of μ_3 and angular acceleration: a - $\nu = 0.00025$, b - $\nu = -0.00025$.

The difference in the values of the maximum amplitude, the values of the corresponding control parameter (shaft rotation speed), with increasing (Figs. 2a and/or 3a) and decreasing (Figs. 2b and/or 3b) control parameter is explained by jumping transitions with different values of these parameters during the take-off and runout of the rotary machine. In the case with $C_3 > 0$ and with $\nu > 0$ the jump is carried out from a higher amplitude to a lower one (Fig. 2a), with a $\nu < 0$ from a lower amplitude to a higher one (Fig. 2b), and with $C_3 < 0$ vice versa (Figs. 3b and 3a, respectively). This is usually observed in experimental studies during acceleration and deceleration of the machine [5]. With a rigid nonlinear characteristic of elasticity, the support $C_3 > 0$ jumps will be located in the area of the shaft rotation speed, where $\Omega > \omega_n$ (Fig. 2), with the soft nonlinear elasticity characteristic of the support $C_3 < 0$ - in the range of shaft rotation speed, where $\Omega < \omega_n$ (Fig. 3).

After the transition of the amplitude peak with a slow change in the control parameter, the response amplitude of the system, having made damped oscillations, tends to certain values, regardless of the values of the nonlinear cubic damping coefficient.

To confirm the analytical study, the system of Eq. (2) was solved directly numerically. Figure 4 shows the numerical results for passage through resonance with a rigid nonlinear elastic characteristic of the support of the support material and a “slowly” increasing value of the control parameter u_1. In this figure, the effects of damping by the parameter μ_3 of resonant oscillations are also observed after resonant damped beats. Damped beats occur due to the superposition of forced non-stationary oscillations and damped natural oscillations with frequencies closely matching in the vicinity of the resonance [11]. The results in Fig. 4 are consistent with previous analytical results shown in Figs. 2a and 3a. The differences are in the value of the maximum vibration amplitude and the offset of the corresponding value of the control parameter. Despite this, the basic behavior of the transient process persists.

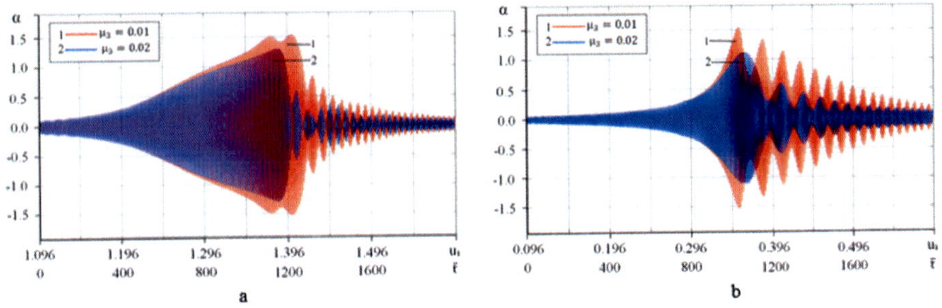

Fig. 4. Passage through resonance c $\nu = 0.00025$ according to the results of the numerical solution of Eqs. (2) for a - $C_3 = 0.1$, b - $C_3 = -0.1$.

To ensure the reliability of the process of transition through resonance, from Eq. (2), setting equal to zero the expression $e\Omega^2$ due to the moment of inertia of the mass unbalance and the damping coefficients μ_1 and μ_3, the equation of the reference line of the resonance curve will be obtained:

$$\Omega = \omega_n + \frac{3C_3a^2}{8\omega_n}, \tag{6}$$

Assuming that, $\nu \ll \Omega^2$ the maximum amplitudes and the corresponding rotational speeds of the resonance curves approximately satisfy Eq. (6). So, for example, at the $C_3 = 0.1, \nu = 0.00025, \mu_1 = 0.01, \mu_3 = 0.01, 0.02$ maximum amplitude $= 1.507, 1.20715$, the rotation speed corresponds $\Omega = 1.085, 1.040$, respectively (Fig. 2a), at $C_3 = -0.1, \nu = 0.00025, \mu_1 = 0.01, \ \mu_3 = 0.01, 0.02$ maximum amplitude $a = 1.316, 1.10792$ - the rotation speed $\Omega = 0.935, 0.954$, respectively (Fig. 3b).

5 Conclusions

Differential equations of motion of a gyroscopic rigid unbalanced rotor with nonlinear cubic damping, a non-ideal energy source are constructed and solved by the Bogolyubov method. Differential equations of unsteady oscillations of the rotor are obtained, which were solved numerically for the transient process through the resonance region.

It is shown that nonlinear cubic damping significantly suppresses the maximum amplitude, and the amplitude of unsteady oscillations after the value of the control parameter corresponding to the amplitude peak. It shifts this control parameter value downwards with the rigid non-linear elastic characteristic ($C_3 > 0$) of the support material and upwards with the soft non-linear elastic characteristic ($C_3 < 0$) of the support material.

Nonlinear cubic damping can significantly weaken the Sommerfeld jumping effect, up to its complete elimination.

The values of the maximum amplitude and the corresponding control parameter in the resonance curves with an increasing and decreasing control parameter approximately correspond to the values of the oscillation amplitude and the control parameter related to jumping effects during acceleration and runout of a rotary machine.

There is an agreement between the results of analytical solutions and numerical solutions of the equations of rotor motion.

The research results can be used in the manufacture of a vibration isolator, which significantly suppresses the peak amplitude, and the amplitude of oscillations behind the resonant rotation speed, for a vibrating system, incl. rotary one.

Acknowledgments. This research is funded by the the Ministry of Education and Science of the Republic of Kazakhstan (Grant No. AP08856763).

References

1. Zakaria, A.A., Rustighi, E., Ferguson, NS: A numerical investigation into the effect of the supports on the vibration of rotating shafts. In: Proceedings of the 11-th International Conference on Engineering Vibration, Ljubljana, Slovenia, pp. 539–552 (2015)
2. Peng, Z.K., Meng Lang, Z.Q., Zhang, W.M., Chu, F.L.: Study of the effects of cubic nonlinear damping on vibration isolations using harmonic balance method. Int. J. Nonlin. Mech. **47**(10), 1065–1166 (2012)
3. Ho, C., Lang, Z., Billings, S.A.: The benefits of non-linear cubic viscous damping on the force transmissibility of a Duffing-type vibration isolator. In: Proceedings of UKACC International Conference on Control, UK, Cardiff, UK, pp. 479–484 (2012)
4. Zhenlong, X., Xingjian, J., Li, C.: The transmissibility of vibration isolators with cubic nonlinear damping under both force and base excitations. J. Sound. Vib. **332**(5), 1335–1354 (2013)
5. Iskakov, Z., Bissembayev, K.: The nonlinear vibrations of a vertical hard gyroscopic rotor with nonlinear characteristics. Mech. Sci. **10**, 529–544 (2019)
6. Yan, S., Dowell, E.H., Lin, B.: Effects of nonlinear damping suspension on nonperiodic motions of a flexible rotor in journal bearings. Nonlinear. Dyn. **78**(2), 1435–1450 (2014)

7. Dimentberg, M.F., McGovern, L., Norton, R.L., Chapdelaine, J., Harrison, R.: Dynamics of an unbalanced shaft interacting with a limited power supply. Nonlinear Dyn. **13**(2), 171–187 (1997)
8. Samantaray, A.K., Dasgupta, S.S., Bhattacharyya, R.: Sommerfeld effect in rotationally symmetric planar dynamical systems. Int. J. Eng. Sci. **48**(1), 21–36 (2010)
9. Cvetićanin, L.: Dynamics of the non-ideal mechanical systems, a review. J. Serbian Soc. Comput. Mech. **4**(2), 75–86 (2010)
10. Warminski, J.: Reguler and chaotic vibrations of Van der Pol-Mathieu oscillator with non-ideal energy source. J. Theor. Appl. Mech. **2**(40), 415–433 (2002)
11. Grobov, V.A.: Asymptotic Methods for Calculating Bending Vibrations of Turbomachine Shafts. Publishing house of the Academy of Science, USSR, Moscow (1961)

Dynamic Impact Factor Analysis of the Prestressed Reinforced Concrete Girder Bridges Subjected to Random Vehicle Load

Toan Nguyen-Xuan(✉), Thuat Dang-Cong, Loan Nguyen-Thi-Kim, and Thao Nguyen-Duy

The University of Danang, University of Science and Technology, Danang, Vietnam
toan_nguyenxuan@dut.udn.vn

Abstract. This paper presents some analysis results of dynamic impact factor (DIF) in prestressed reinforced concrete girder bridge according to vehicle - bridge interaction model. The finite element method and the Monte-Carlo simulation method are applied to the behavior analysis of the bridge structure subjected to vehicle load. Based on the probability distribution of the vehicle load that have been reported at the Dau Giay weighing station, the random vehicle load is generated by the Pseudo - Random Process Generator method. The research results show that the DIF of prestressed girder bridge subjected to random vehicle loads is significantly increased compared to the value in the bridge design process of Vietnam and the United States. The cumulative probability distribution of DIF and the safety of the bridge are also evaluated on the basis of considering the threshold safety of the value 1 + IM. These research results help engineers have more information to analyze the design and check the safety during the operation of the bridge.

Keywords: Dynamic impact factor (DIF) · Finite element method · Monte - Carlo simulation method · Random vehicle load · Prestressed reinforced concrete girder bridge

1 Introduction

The reinforced concrete girder bridge is one of the most popular bridge types around the world. Many researchers have studied the dynamics vehicle – bridge interaction of girder bridges since the 1950s [1]. A significant amount of effort has been made to analyze the dynamic impact factor (DIF) of simply girder bridges due to moving vehicle ([2–5]). Based on analysis of existing continuous bridges, Fafard et al. [6] indicated that AASHTO standard specifications [7] tend to underestimate DIF for long-span continuous bridges. A review of the different DIF for bending moment and shear adopted by some bridge design codes can be found in Deng et al. [8]. Many researchers have shown that the DIF of different bridge responses is different, and some have found that the DIF obtained from different bridge responses should be treated differently ([6, 9–12]).

In addition, there are many publications for the analysis of structures subjected to deterministic moving forces. The most commonly model applied to beams or plates,

N. V. Khang et al. (Eds.): ASIAN MMS 2021, MMS 113, pp. 764–774, 2022.
https://doi.org/10.1007/978-3-030-91892-7_73

traversed by either concentrated ([13, 14]) or distributed loads [15]. The response of beams due to concentrated forces with random amplitudes and velocities was studied in [13], providing expressions for expected value and variance of the displacement. [16] analyzed a simple beam subjected to a concentrated moving load of the same magnitude, moving at a constant speed and at random intervals time delay using computer generated random data. A simply supported beam was loaded by a sequence of concentrated forces moving in the same direction, with random instants of arrival, continuous random speeds, and constant random amplitudes [17]. The results were obtained for the mean value, the spectral density function and the standard deviation of the displacement caused by random moving loads on the bridges. Most of the published results only consider the predetermined load, studying the actual random load has not been taken into account in the field of design and calculation bridge.

In the present study, the numerical simulations were performed to analyze the DIF of Suoi Tuong bridge in Phan Thiet - Dau Giay highway construction project, Vietnam. Actual data on vehicle loads crossing the bridge is collected through Dau Giay weighing station. On that basis, the authors use statistics and enrich random load values of three-axle vehicles to analyze of the DIF of the bridge.

2 The Model of Vehicle – Bridge Interaction and the Differential Equations of Motion

The reinforced concrete girder bridge with one span of I - beam subjected to a three-axle vehicle as in the Fig. 1.

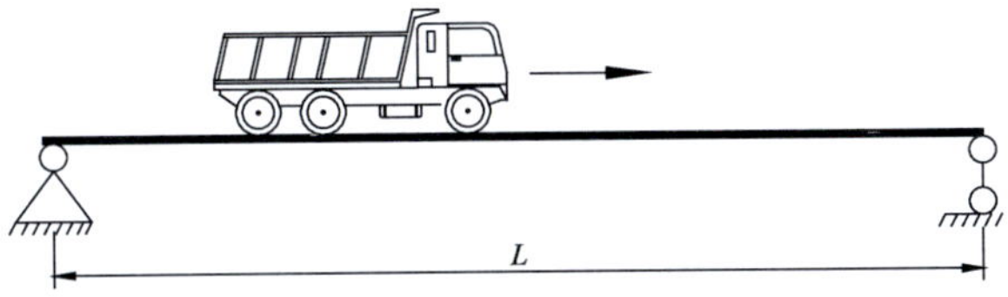

Fig. 1. Schematic of vehicle moving on girder bridge

The dynamic interaction model between a three-axle vehicle and a girder element is described as in the Fig. 2. Where $G_i.\sin\psi_i$ is the engine excitation force at i^{th} axle, it is assumed as a harmonic function; m_{1i} is the mass of the vehicle including the goods conveyed to the ith axle, is considered as a random quantity; m_{2i} is the mass of ith axles, respectively; k_{1i}, d_{1i} are the spring and daspot of suspension at i^{th} axle, respectively; k_{2i}, d_{2i} are the spring and damping of tire at i^{th} axle, respectively; L is the length of the girder elements; $w_i(x_i,t)$ is the vertical displacement of girder element at i^{th} axle of vehicle; z_{1i} is the vertical displacement of chassis at i^{th} axle of vehicle; z_{2i} is the vertical displacement of ith axle of vehicle; y_{1i} is the relative displacement between the chassis and i^{th} axle; y_{2i} is the relative displacement between i^{th} axle and girder element; x_i is the coordinate of the ith axle of the vehicle at time t (i = 1, 2, 3).

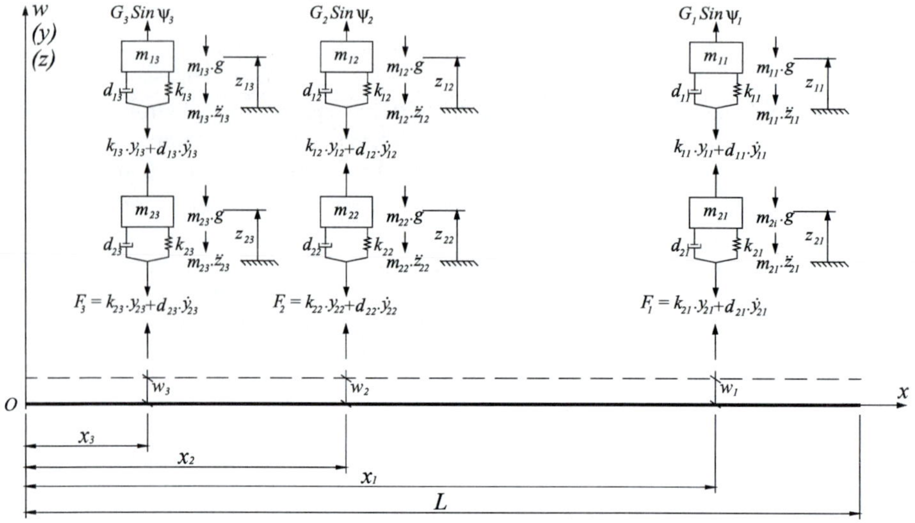

Fig. 2. Schematic of vehicle-bridge interaction

2.1 The Differential Equations of Girder Element Due to Moving Load

According to [18] and [19], the differential equations of girder element due to a distributed load p(x, z, t) taking into account the influence of internal and external friction as follows:

$$EJ_d.\left(\frac{\partial^4 w}{\partial x^4}+\theta.\frac{\partial^5 w}{\partial x^4.\partial t}\right)+\rho F_d.\frac{\partial^2 w}{\partial t^2}+\beta.\frac{\partial w}{\partial t}=p(x,z,t)$$

$$p(x,z,t)=\sum_{i=1}^{N}\xi_i(t).[G_i\sin\Psi_i-(m_{1i}+m_{2i}).g-m_{1i}.\ddot{z}_{1i}-m_{2i}.\ddot{z}_{2i}].\delta(x-a_i)$$

$$m_{1i}.\ddot{z}_{1i}+d_{1i}.\dot{z}_{1i}+k_{1i}.z_{1i}-d_{1i}.\dot{z}_{2i}-k_{1i}.z_{2i}=G_i.\sin\Psi_i-m_{1i}.g$$

$$m_{2i}.\ddot{z}_{2i}+(d_{1i}+d_{2i}).\dot{z}_{2i}+(k_{1i}+k_{2i}).z_{2i}-d_{1i}.\dot{z}_{1i}-k_{1i}.z_{1i}=-m_{2i}.g+d_{2i}\dot{w}_i+k_{2i}w_i \quad (1)$$

Where EJ_d is the bending stiffness of the girder element; ρF_d is the mass per unit length of girder; θ and β are the coefficients of internal friction and external friction of the girder element, respectively; *p(x, z, t)* is the uniform loading over the girder element horizontally; $\delta(x-a_i)$ is the Dirac delta function, i = 1 to N (N is number of axles, N = 3).

$$\xi_i(t)=\begin{cases}1 & khi\ \ t_i\le t\le t_i+T_i\\ 0 & khi\ \ t<t_i\ \ and\ \ t>t_i+T_i\end{cases};\quad T_i=\frac{L}{v_i} \quad (2)$$

The Galerkin method and Green theory are applied to Eqs. (1) to transform these equations into matrix forms, and the differential equations of girder element and vehicle can be written as follows:

$$M_e.\ddot{q} + C_e.\dot{q} + K_e.q = f_e \quad (3)$$

where M_e, C_e, K_e are the mass matrix, damping matrix, and stiffness matrix of the system including girder and vehicle, respectively.

$$M_e = \begin{bmatrix} M_{ww} & M_{wz1} & M_{wz2} \\ 0 & M_{z1z1} & 0 \\ 0 & 0 & M_{z2z2} \end{bmatrix}; \; C_e = \begin{bmatrix} C_{ww} & 0 & 0 \\ 0 & C_{z1z1} & C_{z1z2} \\ C_{z2w} & C_{z2z1} & C_{z2z2} \end{bmatrix}; \; K_e = \begin{bmatrix} K_{ww} & 0 & 0 \\ 0 & K_{z1z1} & K_{z1z2} \\ K_{z2w} & K_{z2z1} & K_{z2z2} \end{bmatrix} \quad (4)$$

$\ddot{q}, \dot{q}, q, f_e$ are the complex acceleration vector, complex velocity vector, complex displacement vector, complex forces vector of the system including girder and vehicle, respectively.

$$\{\ddot{q}\} = \begin{Bmatrix} \ddot{W} \\ \ddot{Z}_1 \\ \ddot{Z}_2 \end{Bmatrix}; \; \{\dot{q}\} = \begin{Bmatrix} \dot{W} \\ \dot{Z}_1 \\ \dot{Z}_2 \end{Bmatrix}; \; \{q\} = \begin{Bmatrix} W \\ Z_1 \\ Z_2 \end{Bmatrix}; \; \{f_e\} = \begin{Bmatrix} F_w \\ F_{z1} \\ F_{z2} \end{Bmatrix} \quad (5)$$

$W = \left[u_1 \; \varphi_1 \; u_2 \; \varphi_2\right]^T$ is the node displacement vector of the beam element in the local coordinate system. u_1, φ_1, u_2, φ_2 are the vertical displacement, rotation displacement on the left and right node of the beam element, respectively.

M_{ww}, C_{ww} and K_{ww} are mass, damping and stiffness matrices of the girder elements, respectively.

$$K_{ww} = \frac{EJ_d}{L^3}\begin{bmatrix} 12 & 6L & -12 & 6L \\ 6L & 4L^2 & -6L & 2L^2 \\ -12 & -6L & 12 & -6L \\ 6L & 2L^2 & -6L & 4L^2 \end{bmatrix} \quad M_{ww} = \frac{\rho F_d L}{420}\begin{bmatrix} 156 & 22L & 54 & -13L \\ 22L & 4L^2 & 13L & -3L^2 \\ 54 & 13L & 156 & -22L \\ -13L & -3L^2 & -22L & 4L^2 \end{bmatrix}$$

$$C_{ww} = \beta.\frac{M_{ww}}{\rho F_d} + \theta.K_{ww} \quad (6)$$

$$M_{z1z1} = \lceil m_{11} \quad m_{12} \dots m_{1i} \dots \quad m_{1N} \rfloor; \; M_{z2z2} = \lceil m_{21} \quad m_{22} \dots m_{2i} \dots \quad m_{2N} \rfloor \quad (7)$$

$$M_{wz1} = P.M_{z1z1}; \quad M_{wz2} = P.M_{z2z2} \quad (8)$$

$$P = \begin{bmatrix} P_{11} & P_{12} & \dots & P_{1i} & \dots & P_{1n} \\ P_{21} & P_{22} & \dots & P_{2i} & \dots & P_{2n} \\ P_{31} & P_{32} & \dots & P_{3i} & \dots & P_{3n} \\ P_{41} & P_{42} & \dots & P_{4i} & \dots & P_{4n} \end{bmatrix}_{(4\times n)} \quad P_i = \begin{Bmatrix} p_{1i} \\ p_{2i} \\ p_{3i} \\ p_{4i} \end{Bmatrix} = \frac{\xi_i(t)}{L^3}.\begin{Bmatrix} (L+2a_i)(L-a_i)^2 \\ L.a_i(L-a_i)^2 \\ a_i^2(3L-2a_i) \\ -L.a_i^2(L-a_i) \end{Bmatrix} \quad (9)$$

with P_i is the ith column of the matrix

$$C_{z1z1} = \lceil d_{11} \quad d_{12} \dots d_{1i} \dots \quad d_{1N} \rfloor; \; C_{z2} = \lceil d_{21} \quad d_{22} \dots d_{2i} \dots \quad d_{2N} \rfloor \quad (10)$$

$$\mathrm{C_{z1z2}} = \mathrm{C_{z2z1}} = -\mathrm{C_{z1z1}};\ \mathrm{C_{z2z2}} = \mathrm{C_{z1z1}} + \mathrm{C_{z2}};\ \mathrm{C_{z2w}} = (\mathrm{N_a.C_{z2}})^{\mathrm{T}}; \quad (11)$$

$$N_a = \begin{bmatrix} N_{11}\ N_{12}\ ...\ N_{1i}\ ...\ N_{1N} \\ N_{21}\ N_{22}\ ...\ N_{2i}\ ...\ N_{2N} \\ N_{31}\ N_{32}\ ...\ N_{3i}\ ...\ N_{3N} \\ N_{41}\ N_{42}\ ...\ N_{4i}\ ...\ N_{4N} \end{bmatrix}_{(4xN)} \text{ where } \left.\begin{aligned} N_{1i} &= \frac{1}{L^3}.(L^3 - 3.L.a_i^2 + 2.a_i^3) \\ N_{2i} &= \frac{1}{L^2}.(L^2.a_i - 2.L.a_i^2 + a_i^3) \\ N_{3i} &= \frac{1}{L^3}.(3.L.a_i^2 - 2.a_i^3) \\ N_{4i} &= \frac{1}{L^2}.(a_i^3 - L.a_i^2) \end{aligned}\right\} \quad (12)$$

$$K_{z1z1} = \lceil k_{11} \quad k_{12}\ ...\ k_{1i}\ ... \quad k_{1N} \rfloor;\ K_{z2} = \lceil k_{21} \quad k_{22}\ ...\ k_{2i}\ ... \quad k_{2N} \rfloor \quad (13)$$

$$\begin{aligned} &K_{z1z2} = K_{z2z1} = -K_{z1z1};\ K_{z2z2} = K_{z1z1} + K_{z2}; \\ &K_{z2w} = (N_a.K_{z2})^T + (\dot{N}_a.C_{z2})^T \end{aligned} \quad (14)$$

$$F_w = \left[F_1\ ...\ F_i\ ...\ F_6\right]^T;\quad F_i = \sum_{i=1}^{N} [G_i \sin \Psi_i - (m_{1i} + m_{2i}).g].P_i; \quad (15)$$

$$F_{z1} = \begin{Bmatrix} G_1 \sin \Psi_1 - m_{11}g \\ \vdots \\ G_i \sin \Psi_i - m_{1i}g \\ \vdots \\ G_N \sin \Psi_N - m_{1N}g \end{Bmatrix};\ F_{z2} = \begin{Bmatrix} -m_{21}.g \\ \vdots \\ -m_{2i}.g \\ \vdots \\ -m_{2N}.g \end{Bmatrix}; \quad (16)$$

The Runge-Kutta method is applied to solve the Eq. (3) by the direct step by step integration to obtain the responses of the girder elements in the real time.

3 Random Vehicle Load

In fact, the vibration of the bridge is influenced by many random factors such as the road unevenness, vehicle weight, vehicle speed, vehicle braking force,… Therefore, the bridge response is also a random quantity. The scope of this study only considers the case that the axle load $m_{axle\ i}$ is a random quantity taken from the vehicle weighing stations. The quantity $\mathrm{m}_{1i} = m_{axle\ i} - m_{2i}$ in the Eqs. (1), (7), (15), (16) is random, and other factors are considered as deterministic.

Based on vehicle weighing data at Dau Giay weighing station, N. Lan [20] estimated the probability rules for a three-axle vehicles and the load of the axles. From this PDF function, Pseudo - Random Process Generator method [21] is used to generate a dataset of 20,000 values for the load of the three-axle vehicle and the load of the axles. Figure 3 is a random representation of the third axle load.

From N. Lan [20], the graph of load distribution of vehicle load is shown in the Fig. 4. The results show a high fit of the generated data set for both cases, which is the load of the third axle and the total load of the three-axle vehicle trucks.

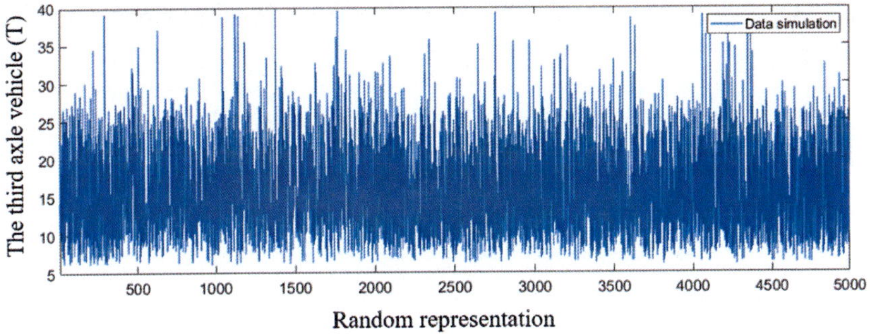

Fig. 3. Random representation of the third axle vehicle.

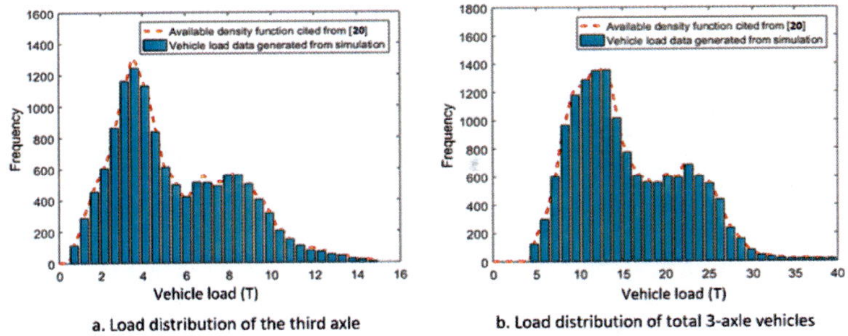

Fig. 4. Simulations of random vehicle load data

4 Application to Analysis of Suoi Tuong Bridge

4.1 Properties of Structure Bridge and Vehicle

Suoi Tuong Bridge belongs to the Phan Thiet - Dau Giay highway construction project, which is a single span bridge with a span length of 24 m. The cross-section of the bridge consists of 11 prestressed reinforced concrete I-beams as shown in the Fig. 5.

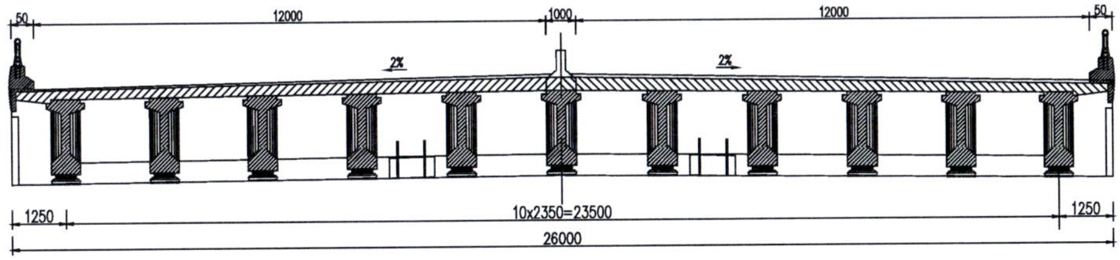

Fig. 5. Schematic of Suoi Tuong Bridge

The properties of girder beam Young's modulus of elasticity $E = 3230769.23$T/m2 ($E = 32307.6923$ Mpa), beam system $J_d = 0.30921\text{m}^4$, the mass of beams, slabs, wearing surface and pedestrian lane is 2.8T/m^2, friction coefficient $\theta = 0.027$ and $\beta = 0.01$.

The moving load with the values m_{11}, m_{12} and m_{13} are random quantities taken from a dataset based on vehicle weighing data at Dau Giay weighing station. The remaining data are taken from the type of vehicle popular in the area. The three – axle vehicle used in the numerical simulation is Foton dumper truck with the following parameters: $m_{21} = 0.26$T, $m_{22} = m_{23} = 0.87$T, $k_{11} = 120$T/m, $k_{12} = k_{13} = 260$T/m, $k_{21} = 240$T/m, $k_{22} = k_{23} = 380$T/m, $d_{11} = 0.7344$Ts/m, $d_{12} = d_{13} = 0.3672$Ts/m, $d_{21} = 0.4$Ts/m, $d_{22} = d_{23} = 0.8$Ts/m.

4.2 Application to DIF Analysis of Suoi Tuong Bridge

The dynamic impact factor (DIF) or (1 + IM) in AASHTO [7] and TCVN11823–13:2017 [22] is taken as the ratio of dynamic and static displacements responses as follows equation:

$$(1 + IM) = DIF = \frac{S_{D\max}}{S_{S\max}} \tag{17}$$

where S_{Dmax}, S_{Smax} - the absolute maximum displacements of dynamic and the static at the same point, respectively.

With each the value of vehicle weight, the Eq. (3) is solved by means of the finite element method and the Runge-Kutta method to obtain static and dynamic responses of girder elements. And then, DIF of bridge can be determined as the Eq. (17). After analyzing with the input random data set with N = 20000, the results obtained the DIF of vertical and rotation displacement at positions of 1/4 span (Fig. 6), 1/2 span (Fig. 7) and 3/4 span (Fig. 8) corresponding to node 2, 3 and 4.

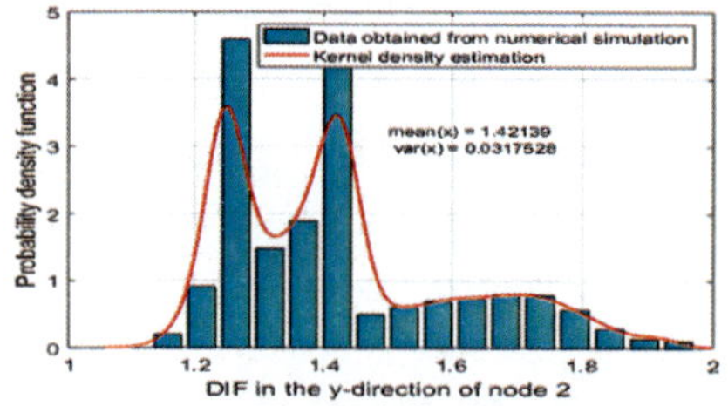

a. Vertical displacement

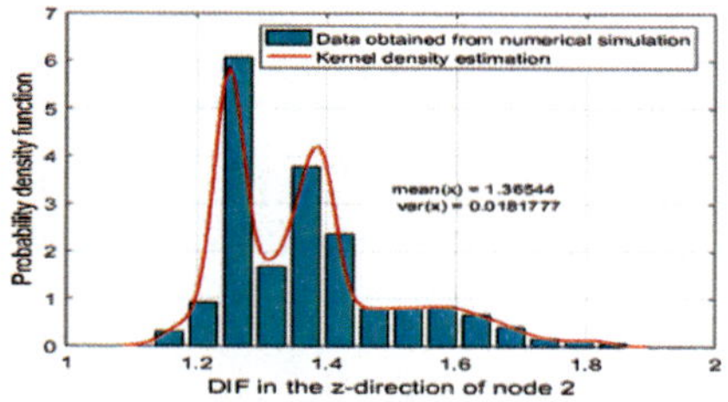

b. Rotation displacement

Fig. 6. DIF of node 2

Figure 9 describe the convergence of Monte Carlo simulations when determining the DIF at node 2. The results show that only 10,000 simulations are needed for the mean of DIF to converge well. The coefficient of variation c_v in the y and z directions, is 3.09% and 1.77%, respectively, both small and <5%. Similar analysis for nodes 3 and 4, the results also converge well with this number of simulations.

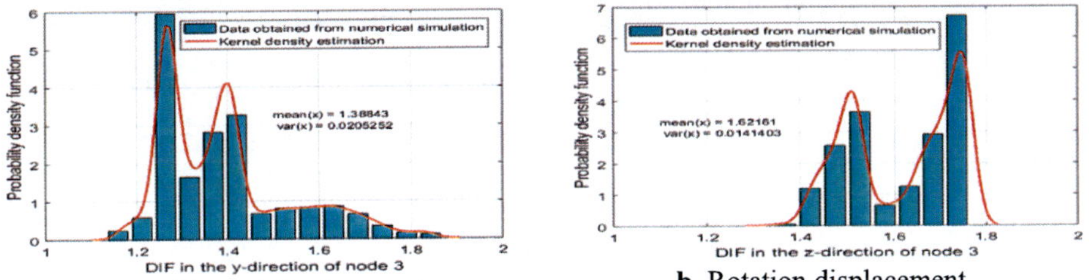

a. Vertical displacement **b.** Rotation displacement

Fig. 7. DIF of node 3

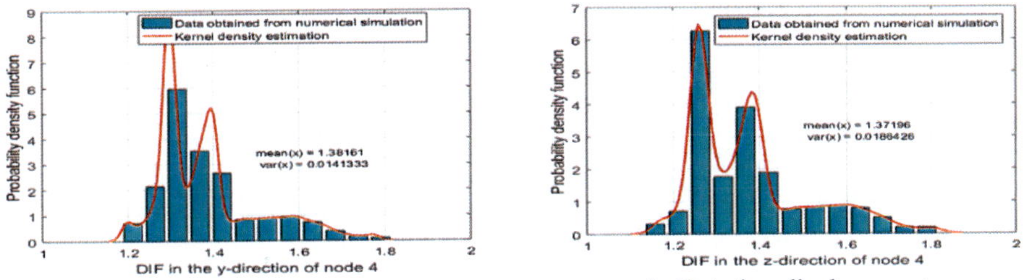

a. Vertical displacement **b.** Rotation displacement

Fig. 8. DIF of node 4

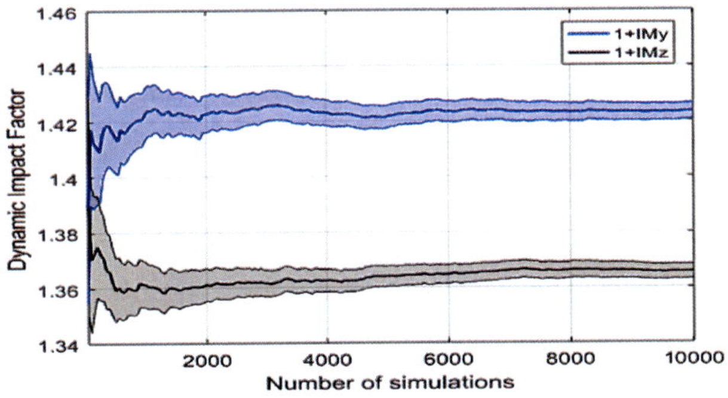

Fig. 9. Estimation of the convergence quantity of the Monte Carlo simulation at node 2

The research results also show the probability distribution of DIF random variables is also very complicated. It has two peaks and is not compatible with any normal probability rules, for example: normal, log-normal… only has a single peak. In addition, the rules of the input random variable (vehicle load) and the output random variable (DIF) do not have the same probability rule. This proves that the mapping relationship between them is not monotonous.

The mean and variance of the DIF of the vertical and rotational displacement at nodes 2, 3, and 4 are shown in Table 1.

Table 1. The mean and variance of the DIF of the displacement at nodes

Statistical	DIF of the vertical displacement			DIF of the rotational displacement		
	Node 2	Node 3	Node 4	Node 2	Node 3	Node 4
Mean	1.42139	1.38843	1.38161	1.36544	1.62161	1.37196
Var	0.0317528	0.0205252	0.0141333	0.0187777	0.0141403	0.0186426

The results shows that the increase of the DIF in the bridge structure subjected to random vehicle loads is very large and greater than those recommended by current bridge design codes as AASHTO [7] and Vietnamese Design of Road Bridges [22] is 1.33.

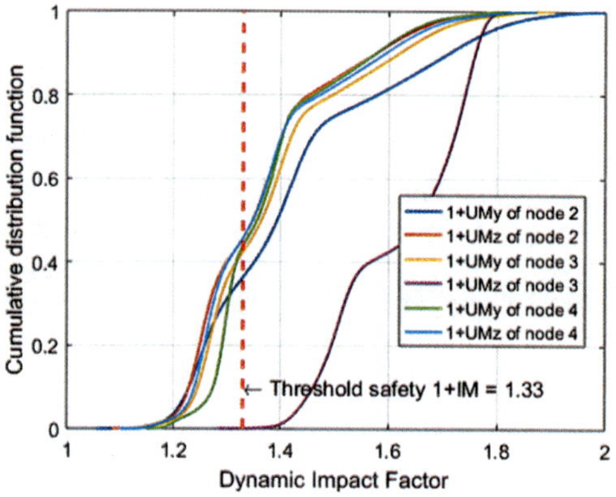

Fig. 10. The cumulative probability of DIF at nodes in the y and z directions

Using the DIF value to evaluate the reliability of the bridge structure is shown in the Fig. 10. The cumulative probability Prob(1 + IM $\leq$ 1.33) graph shows that, with a threshold value of 1.33 recommending by [7] and [22], the probability of exceeding the threshold value p_f = 1 - Prob(1 + IM $\leq$ 1.33) of the cases are relatively large (>50%). Detailed results are shown in the Table 2. Therefore, it is necessary to consider the influence of random vehicle loads on dynamic response analysis of bridge structure.

Table 2. Probability of exceeding the threshold value of the bridge in terms of DIF

DIF	Node 2		Node 3		Node 4	
	1 + IMy	1 + IMz	1 + IMy	1 + IMz	1 + IMy	1 + IMz
p_f (%) = 1 - Prob(1 + IM $\leq$ 1.33)	63.67	54.28	57.70	100	56.04	54.24

5 Conclusion

This paper analyzes the dynamic interaction between random vehicle loads and prestressed reinforced concrete girder bridge structures by finite element method and monte-carlo simulation method. Actual data on vehicle loads crossing the bridge was collected at DAU giay weighing station and applied to the analysis for Suoi tuong bridge. The cumulative probability distribution of dif and the safety level of the bridge are also evaluated on the basis of considering the safety threshold of the value 1 + IM. The research results show that the DIF of the prestressed girder bridge subjected to random vehicle load increases significantly compared to the value in the bridge design process of Vietnam and the United States. Therefore, the reliability of the bridge structure is significantly reduced, which is the one of reasons cause unsafety during the operation of the bridge.

References

1. Huang, D., Wang, T.L., Shahawy, M.: Impact analysis of continuous multigirder bridges due to moving vehicles. J. Struct. Eng. **118**(12), 3427–3443 (1992)
2. Chang, D., Lee, H.: Impact factors for simple-span highway girder bridges. J. Struct. Eng. **120**(3), 704–715 (1994)
3. Nguyen, X.T., Tran, V.D.: A finite element model of vehicle—cable stayed bridge interaction considering braking and acceleration. World. Congr. Adv. Civil. Environ. Mater. Res. **109**, 1–22 (2014)
4. Nguyen, X.T.: Dynamic interaction between the two-axle vehicle and continuous girder bridge with considering vehicle braking force. Vietnam J. Mech. **36**(1), 49–60 (2014). https://doi.org/10.15625/0866-7136/36/1/2949
5. Deng, L., Cai, C.S.: Development of dynamic impact factor for performance evaluation of existing multi-girder concrete bridges. Eng. Struct. **32**(1), 21–31 (2010)
6. Fafard, M., Laflamme, M., Savard, M., Bennur, M.: Dynamic analysis of existing continuous bridge. J. Bridge Eng. **3**(1), 28–37 (1998). https://doi.org/10.1061/(ASCE)1084-0702(1998)
7. AASHTO.: LRFD bridge design specifications, Washington, DC (2012)
8. Deng, L., He, W., Shao, Y.: Dynamic impact factors for shear and bending moment of simply supported and continuous concrete girder bridges. J. Bridge Eng. **20**(11), 04015005 (2015). https://doi.org/10.1061/(ASCE)BE.1943-5592.0000744
9. Huang, M.S.D., Wang, T.: Vibration of thin-walled box-girder bridges excited by vehicles. J. Struct. Eng. **121**, 1330–1337 (1995)
10. Wang, T., Huang, D., Shahawy, M.: Dynamic behavior of slant-legged rigid-frame highway bridge. J. Struct. Eng. **120**(3), 885–902 (1994)
11. Nguyen, X.-T., Tran, V.-D., Hoang, N.-D.: A study on the dynamic interaction between three-axle vehicle and continuous girder bridge with consideration of braking effects. J. Constr. Eng. **2017**, 1–12 (2017). https://doi.org/10.1155/2017/9293239
12. Toan, N.X., Van Duc, T.: Determination of dynamic impact factor for continuous girder bridge due to vehicle braking force by finite element method and experimental. Vietnam J. Mech. **39**(2), 149–164 (2017). https://doi.org/10.15625/0866-7136/8745
13. Sniady, P.: Vibration of a beam due to a random stream of moving forces with random velocity. J. Sound Vib. **97**(1), 23–33 (1984)
14. Turner, J.D., Pretlove, A.J.: A study of the spectrum of traffic-induced bridge vibration. J. Sound Vib. **122**(1), 31–42 (1988)

15. Sorrentino, S., Catania, G.: Dynamic analysis of rectangular plates crossed by distributed moving loads. Math. Mech. Solids **23**(9), 1291–1302 (2018)
16. Abu-Hilal, M.: Vibration of beams with general boundary conditions due to a moving random load. Arch. Appl. Mech. **72**(9), 637–650 (2003)
17. Sorrentino, S.: Power spectral density response of bridge-like structures loaded by stochastic moving forces. Shock Vibr. **2019**, 1–10 (2019). https://doi.org/10.1155/2019/1790480
18. Nguyen-Xuan, T., Kuriyama, Y., Nguyen-Duy, T.: Analysis of dynamic impact factors of bridge due to moving vehicles using finite element method. In: Hung, N.-X., Phuc, P.-V., Rabczuk, T. (eds.) ACOME 2017. LNME, pp. 1105–1119. Springer, Singapore (2017). https://doi.org/10.1007/978-981-10-7149-2_77
19. Clough, R.W.: Dynamics of structrures. McGraw-Hill, Inc., Singapore (1993)
20. Lan, N.: Research, evaluate and determine the allowable load to cross the bridge on the basis of the bridge test results, thesis of Doctor (2015)
21. Gentle, J.E.: Random Number Generation and Monte Carlo Methods, vol. 381. Springer, New York (2003)
22. Vietnamese National Standards: TCVN11823–13:2017 Design of Road Bridges (2017)

Calculating Transverse Vibration of a Continuous Beam on Nonlinear Viscoelastic Supports Under the Action of Moving Loads

Pham Thanh Chung and Nguyen Minh Phuong(✉)

Hanoi University of Science and Technology, Hanoi, Vietnam
phuong.nguyenminh@hust.edu.vn

Abstract. Models of continuous beam on viscoelastic supports are often applied in engineering, especially in the field of civil engineering. Linear models are usually used partly because of the availability of the solving methods. However, the supports should be nonlinear. Hence, this paper focuses on calculating the transverse vibration of a continuous beam on nonlinear viscoelastic supports (stiffness and damping coefficients are nonlinear functions) under the action of moving loads. To obtain the dynamic response of such model, the proposed approach consisting of two main steps. In the first step, the substructure method is used to establish the vibration differential equations of the system. Therefore, the motion of the beam is governed by equations of a free-free beam subjected to external forces. In the second step, the mode shapes of the free-free beam are used to construct the solution through a numerical process. Examples are given to demonstrate the applicability of the methods to real-world problems.

Keywords: Continuous beam · Nonlinear viscoelastic support · Moving load

1 Introduction

To reduce damages of earthquake loads on bridge structures, seismic isolators are often used. They are made of antivibration materials with nonlinear mechanical properties and can be modeled as an extension in the nonlinear domain for the non-viscous linear Kelvin-Voigt model [2]. They are designed with the purpose of earthquake resistance for bridges. However, in exploitation process, bridges are subjected to moving vehicles. Thus, the dynamic response of the bridges under the influence of such loadings should be examined.

In the literature, several different models of elastically supported beams are studied. They include: a fairly simple model of a beam with elastic supports at both ends subjected to moving forces [3–5]; a further developed model where the beam is subjected to moving vehicles [8]; a relatively complex and quite general model, consisting of a continuous beam on rigid and elastic supports under the action of moving bodies [6] but the supports at both ends are required to be rigid. This obstacle is overcome by the method presented in [7], but the proposed model here is limited to beam on linear elastic supports.

N. V. Khang et al. (Eds.): ASIAN MMS 2021, MMS 113, pp. 775–782, 2022.
https://doi.org/10.1007/978-3-030-91892-7_74

As a general rule, to analyze the vibration of the beam, eigenfunctions that satisfies the boundary conditions of the beam at all elastic supports are usually found and used in the calculation process. Such eigenfunctions can be found by considering the free vibration problem and finding the mode shapes. This approach is only convenient if all the supports are linear but not easily applied to models with nonlinear supports since the boundary conditions at the nonlinear supports are much more complicated.

Since the models of the supports should be nonlinear to be more realistic, this study focuses on the dynamics of a continuous beam bridge on nonlinear viscoelastic supports under the action of moving loads. Thus, a new approach is proposed to deal with the added nonlinearities. Firstly, the method of substructures is used to replace the supports with external forces, so that the model becomes a free-free beam. Then, the mode shapes of the free–free beam are used to construct the dynamic response of the model. Applied examples are presented to validate the proposed approach.

2 Vibration Model and Equations of Motion

Consider a continuous Euler-Bernoulli beam with length *l*, mass of length unit $\mu = const$, bending rigidity $EI = const$. It rests on *J* nonlinear viscoelastic supports with stiffness coefficients $c_j(z)$ and damping coefficients $b_j(z)$ that are nonlinear functions depending on the deformation of the supports, and subjected to *N* moving loads. The ith load ($i = \overline{1, N}$)[6, 7, 10] consists of a single mass-spring-damper system with mass m_i, linear stiffness k_i and linear damping coefficient d_i, moves along the beam with the velocity v_i. Assume that when the beam vibrates, the connections between the beam and the loads as well as the viscoelastic supports are not separated (Fig. 1).

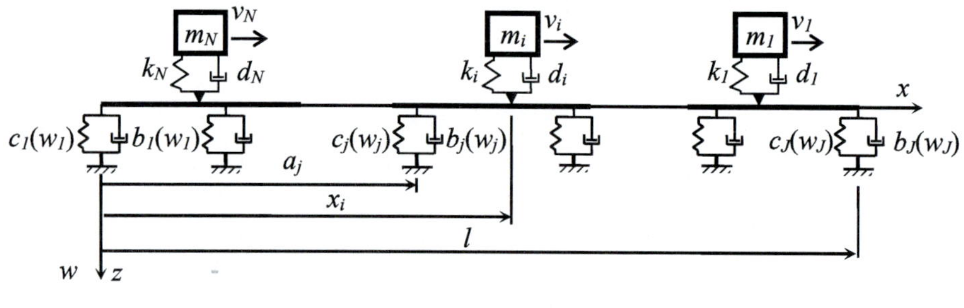

Fig. 1. Vibration model of a continuous beam under moving loads

Using the method of substructures to derive vibration equations of the beam and the bodies, we divide the system into $N + 1$ substructures: the beam and *N* mass-spring-damper systems (Fig. 2) [6, 7]. Here the viscoelastic supports are replaced by the reaction forces F_j^{dh}, the connections between the beam and the moving loads are replaced by the interaction forces F_i.

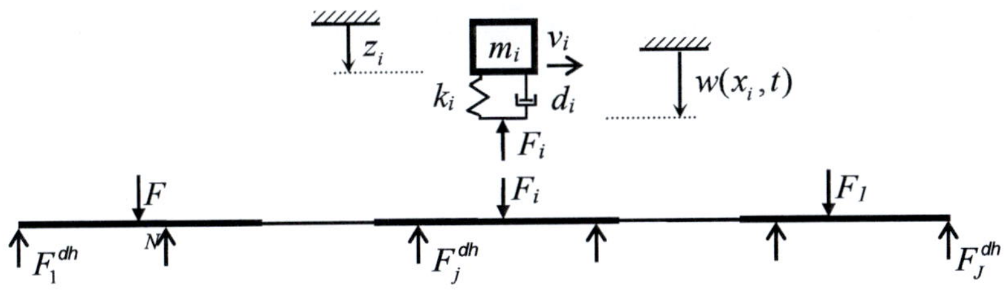

Fig. 2. Substructures

Vibration differential equations of substructures can be obtained by applying the basic principles of dynamics. The equation describing the vibration of the ith load has the following form [6]

$$m_i\,\ddot{z}_i + d_i\dot{z}_i + k_i z_i = m_i g + d_i\dot{w}(x_i, t) + k_i w(x_i, t) \tag{1}$$

where z_i is the absolute coordinate of the ith load in the vertical direction; $w(x_i,t)$ is the deflection of the beam at the position of the ith load.

The equation describing transverse vibration of beam with the consideration of internal friction can be written in the form [1, 6, 7]

$$\begin{aligned} EI\left(\frac{\partial^4 w}{\partial x^4} + \alpha\frac{\partial^5 w}{\partial x^4 \partial t}\right) + \mu\left(\frac{\partial^2 w}{\partial t^2} + \beta\frac{\partial w}{\partial t}\right) &= \sum_{i=1}^{N} L(x_i)F_i\,\delta(x - x_i) - \\ - \sum_{j=1}^{J} c_j(w_j)\,w(x,t)\,\delta(x - a_j) &- \sum_{j=1}^{J} b_j(w_j)\,\dot{w}(x,t)\,\delta(x - a_j) + \mu g \end{aligned} \tag{2}$$

in which α and β are damping constants; $\delta()$ is the Dirac-function; $L(x_i)$ is the logic signal – function, which is determined by the following relations

$$L(x_i) = \begin{cases} 1 & \text{when } 0 \le x_i \le l \\ 0 & \text{when } x_i < 0 \text{ or } x_i > l \end{cases}$$

3 Solution Using the Mode Shapes of the Free-Free Beam

After the system is divided into substructures, the beam becomes a free body subjected to external forces. Using the modal analysis approach, the solution of Eq. (2) is assumed to be a linear combination of normal modes of beam as

$$w(x, t) = \sum_{r=1}^{n} q_r(t)\, X_r(x) \tag{3}$$

where: n is the number of used mode shapes, $X_r(x)$ is the rth mode shape of the free–free beam [7, 9]

$$X_r(x) = \cos\frac{\lambda_r x}{l} + \cosh\frac{\lambda_r x}{l} + \left(\frac{\cos\lambda_r - \cosh\lambda_r}{\sinh\lambda_r - \sin\lambda_r}\right)\left(\sin\frac{\lambda_r x}{l} + \sinh\frac{\lambda_r x}{l}\right) \tag{4}$$

with: λ_r is the rth root of the frequency equation: $f(\lambda) = \lambda(1 - \cos\lambda\,\cosh\lambda) = 0$.

Because this equation has a double root $\lambda = 0$, the first two mode shapes of the beam corresponding to these values of λ are $X_1(x) = 1$ and $X_2(x) = x - \frac{l}{2}$.

By substituting relation (3) into Eq. (2) and transforming similarly to [6, 7], we receive a system of $n + N$ ordinary differential equations with unknowns q_r, z_i $\left(r = \overline{1,n};\ i = \overline{1,N}\right)$ as follows:

$$\ddot{q}_s = -\sum_{r=1}^{n}\left\{\delta_r^s\left[\frac{EI\alpha}{\mu}\left(\frac{\lambda_s}{l}\right)^4 + \beta\right] + \frac{1}{\mu h(s)}\left(\sum_{i=1}^{N} L(x_i)d_iX_r(x_i)X_s(x_i) + \sum_{j=1}^{J} b_j(w_j)X_r(a_j)X_s(a_j)\right)\right\}\dot{q}_r$$
$$-\sum_{r=1}^{n}\left\{\delta_r^s\frac{EI}{\mu}\left(\frac{\lambda_s}{l}\right)^4 + \frac{1}{\mu h(s)}\left(\sum_{i=1}^{N} L(x_i)\left[k_iX_r(x_i) + d_iv_iX_r'(x_i)\right]X_s(x_i) + \sum_{j=1}^{J} c_j(w_j)X_r(a_j)X_s(a_j)\right)\right\}q_r$$
$$+\frac{1}{\mu h(s)}\sum_{i=1}^{N} L(x_i)d_iX_s(x_i)\dot{z}_i + \frac{1}{\mu h(s)}\sum_{i=1}^{N} L(x_i)\,k_i\,X_s(x_i)\,z_i + \frac{g}{h(s)}\int_0^l X_s(x)dx \quad \left(s = \overline{1,n}\right) \tag{5}$$

$$\ddot{z}_i = L(x_i)\sum_{r=1}^{n}\left(\frac{d_i}{m_i}X_r(x_i)\right)\dot{q}_r - \frac{d_i}{m_i}\dot{z}_i + L(x_i)\sum_{r=1}^{n}\left(\frac{k_i}{m_i}X_r(x_i) + \frac{d_iv_i}{m_i}X_r'(x_i)\right)q_r - \frac{k_e}{m_e}z_i + g \quad \left(i = \overline{1,N}\right) \tag{6}$$

in which $\delta_r^s = \begin{cases} 1 & \text{when } r = s \\ 0 & \text{when } r \neq s \end{cases}$; $h(s) = \int_0^l X_s^2(x)dx$; $x_i = x_{0i} + v_it$;

(x_{0i} – Initial coordinate of the ith load).

After solving this system of equations by numerical methods, the deflection and stress of the beam can be calculated by the formulae:

$$w(x,t) = \sum_{r=1}^{n} q_r(t)X_r(x); \qquad \sigma(x,t) = \frac{-EI}{M_{ku}}\sum_{r=1}^{n}\left[q_r(t) + \alpha\,\dot{q}_r(t)\right]X_r''(x) \tag{7}$$

with: $M_{ku} = I/z_{max}$ is the elastic section modulus, z_{max} is the distance from the neutral axis to the most extreme fiber.

It should be noted that the free-free beam mode shapes can also be used to find the natural frequencies and mode shapes of the beam in undamped linear case. To do so, one should remove the nonlinear factors, damping coefficients and external forces from equation to obtain

$$\ddot{\boldsymbol{q}} + \boldsymbol{Cq} = \boldsymbol{0} \tag{8}$$

with: $\boldsymbol{q} = \left[q_1\ q_2 \cdots q_n\right]^T$

$$\boldsymbol{C} = [C_{sr}] : C_{sr} = \delta_r^s\frac{EI}{\mu}\left(\frac{\lambda_s}{l}\right)^4 + \frac{1}{\mu h(s)}\sum_{j=1}^{J} c_jX_r(a_j)X_s(a_j) \quad \left(s = \overline{1,n};\ r = \overline{1,n}\right)$$

Equation (8) is solved to find the natural frequencies ω_k and eigenvectors q_k $\left(k = \overline{1,n}\right)$. Then these eigenvectors are substituted into relation to get the normal modes of the beam on the linear elastic supports.

4 Applied Example

4.1 Parameters of the Model

Consider a continuous beam resting on three nonlinear viscoelastic supports and subjected to two moving loads (Fig. 3).

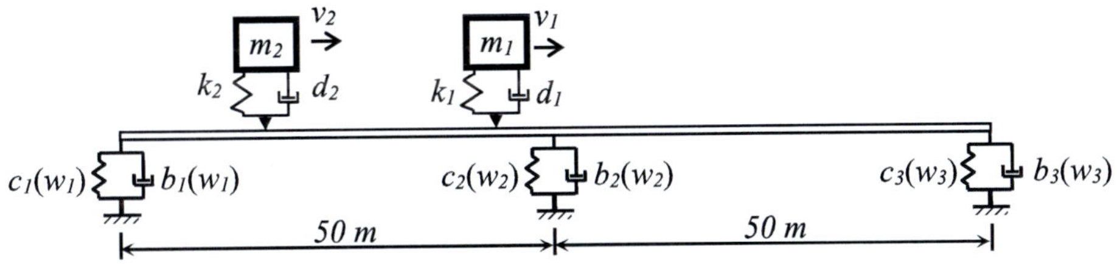

Fig. 3. Beam under the action of moving loads.

- The parameters of the beam and loads [7]: $EI = 29.074 \times 10^{10}$ Nm^2, $\mu = 27168$ kg/m, $l = 100$ m, $M_{ku} = 4.873$ m^3, $\alpha = 0.027$ s, $\beta = 0.001$ s^{-1}, $g = 9.81$ m/s^2, $a_1 = 0$ m, $a_2 = 50$ m, $a_3 = 100$ m, $n = 10$, $m_1 = m_2 = 17600$ kg, $k_1 = k_2 = 3500000$ N/m, $d_1 = d_2 = 42000$ Ns/m, $v_1 = v_2 = 19.444$ m/s, $x_{01} = 0$ m, $x_{02} = -30$ m.
- The parameters of the viscoelastic supports: stiffness coefficient $c(z)$ and damping coefficient $b(z)$ are determined by the formulae [2]:

$$c(z) = \frac{S_n}{h} E(z) \quad [\text{N/m}]; \qquad b(z) = 2m\omega_0 D(z) \quad [\text{Ns/m}] \tag{9}$$

where S_n is the cross-sectional area of the viscoelastic support. h is the support's height, m is the mass converted from the force of the beam acting on the support, ω_0 is the fundamental frequency of the system (which can be found from the linear problem). *E(z)* is the modulus function and *D(z)* is the damping function of the support material which are found experimentally [2]:

$$\begin{cases} E(z) = \left(25 + 75e^{-101.626z}\right)10^6 & [\text{N/m}^2] \\ D(z) = 15.023 - 14.524e^{-129.66z} & [\%] \end{cases} \tag{10}$$

Using the given data for beam and applying formulae (9), the stiffness coefficient $c(z)$ and damping coefficient $b(z)$ for the viscoelastic supports can be obtained as follows:

$$c_1(z) = c_2(z) = c_3(z) = 19174804.687\left(25 + 75e^{-101.626z}\right) \text{ N/m}$$

$$b_1(z) = b_3(z) = 130164.944\left(15.023 - 14.524e^{-129.66z}\right)\text{Ns/m}$$

$$b_2(z) = 433883.148\left(15.023 - 14.524e^{-129.66z}\right)$$

4.2 Linear Mode Shapes

In this example, the nonlinear factor of the supports is ignored, the stiffness coefficients of the resulted linear supports are: $c_1 = c_2 = c_3 = 1917475985.0$ N/m. All the damping coefficients are also ignored.

Because the stiffness coefficient of the linear elastic supports is quite high, we will compare the mode shapes of the beam in this example with the mode shapes of a continuous beam on rigid supports [6] with the same parameters. The proposed method using the mode shapes of the free-free beam is applied to demonstrate its usefulness even in linear problems (Fig. 4).

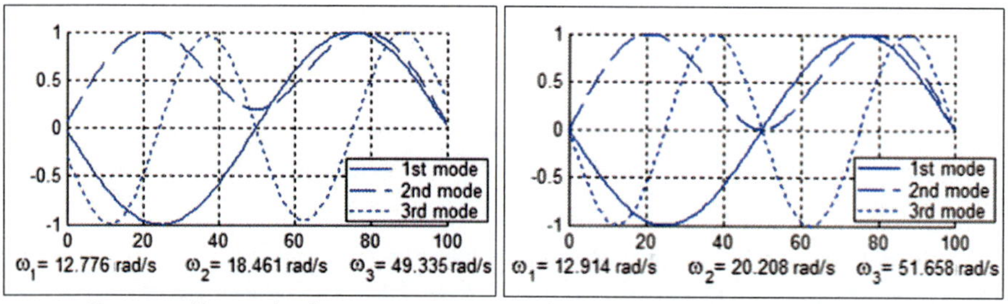

Beam on linear elastic supports   Beam on rigid supports

Fig. 4. Natural frequencies and mode shapes of the beams.

Comparing the graphs of the first three mode shapes and the associated natural frequencies shows that the mode shapes of the beam on the linear elastic supports look almost like the mode shapes of the beam on rigid supports, but the displacements of the beam at the elastic support positions do not equal zero (except for the antisymmetric mode, the displacement at the mid-beam equals zero). The natural frequencies of the beam on linear elastic supports are lower than those on rigid supports because the system is “softer”. The fundamental frequencies of the two beams are not much different from each other (about 1.07%), but higher differences are seen for higher order natural frequencies.

4.3 Forced Vibration

To compare the effect of different support types on the beam vibrations, simulations are done for all three types: nonlinear viscoelastic supports, linear supports [7] (the nonlinear factor of the supports is removed) and rigid supports [6] with the same parameters of the beam and the loads.

Figure 5a shows the dynamic deflection of the entire beam at time $t = 3.1$ s. The graphs show that the displacement at the mid-beam linear support is much larger than the displacement at the two ends of the beam. The displacements at the rigid supports are zero as expected. The deflection of the beam on the nonlinear supports is larger than that on rigid supports, and smaller than that on linear supports. It is because the stiffness coefficients of the nonlinear supports increase with their deformations so that they are as soft as the linear supports to allow the beam to start deflecting, but quickly becomes stiff to limit the deflection of the beam.

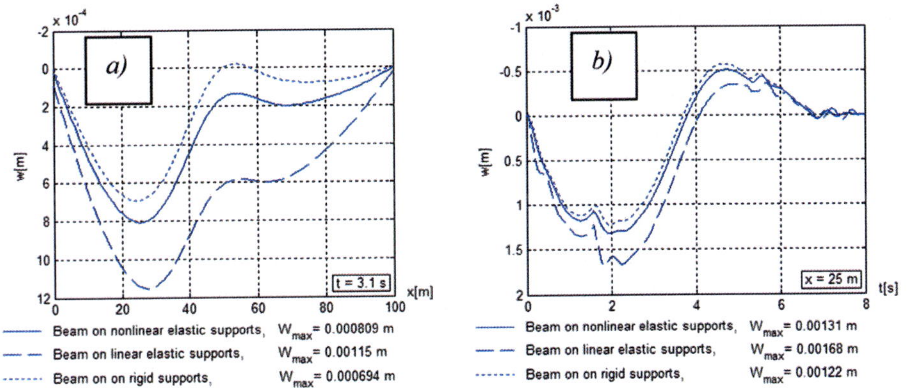

Fig. 5. Dynamic deflection of the beam

Figure 5b shows the dynamic deflection at the cross section x = 25 m. In most of the survey time, the deflection of the beam on nonlinear supports lies between the values of the other cases. When the loads enter or leave the beam, the survey section vibrates strongly, so that the graphs may intersect. When all loads run off the beam (t > 6.69 s), the survey section witnesses decaying vibration.

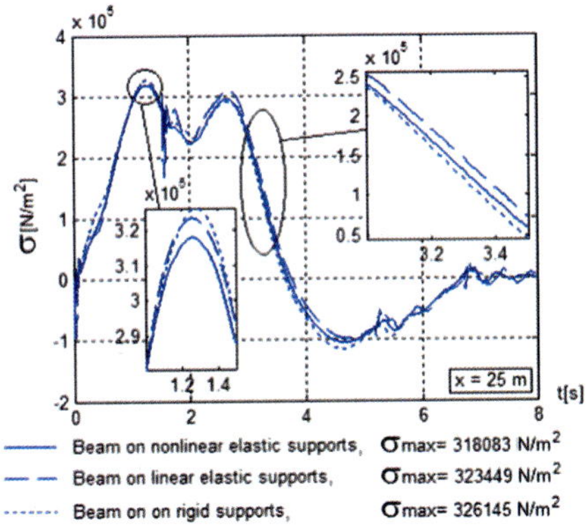

Fig. 6. Dynamic stress of beam

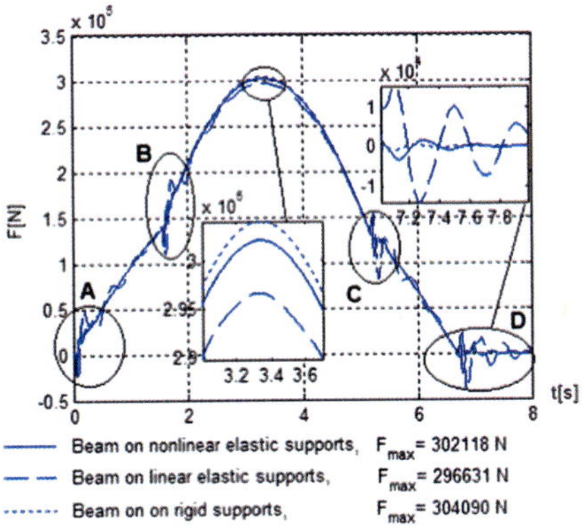

Fig. 7. Dynamic force acting on the mid-beam support

Figure 6 shows the simulation results of the dynamic stress at the cross section x = 25 m. The graphs of the 3 models are very close together. The stress amplitude oscillates strongly when any of the loads enter or leave the beam. For most of the survey time, the stress in the rigid case is the lowest and that in the linear supports case is the highest. However, the extreme value of stress in the nonlinear supports case is even lower than that in the rigid supports case.

Figure 7 shows the dynamic force acting on the mid-beam support. The dynamic force also oscillates strongly when any of the loads enter or leave the beam (areas A, B, C and D). The maximum value of the force is highest in the rigid supports case and lowest in the linear supports case.

5 Conclusions

This paper contributes an efficient approach to analyze the nonlinear vibration of a continuous beam using the method of substructures and mode shapes of the free-free beam. The proposed approach is successfully applied to simulate and compare the dynamic behaviors of a continuous beam under the action of moving loads in three cases: the beam rests on rigid supports, on linear supports, and on nonlinear supports.

Future works may apply the proposed method to the expanded study that considers the nonlinearities in the vehicles and in the contacts between the vehicles and the beam.

Acknowledgements. This research is funded by the Hanoi University of Science and Technology (HUST) under project number T2018-PC-211.

References

1. Clough, R.W., Penzien, J.: Dynamics of Structures. McGraw-Hill, New York (1986)
2. Bratosin, D.: On dynamic behaviour of the antivibratory materials. Proc. Roman. Acad. **4**(3), 205–210 (2003)
3. Ding, H., Dowell, E.H., Chen, L.-Q.: Transmissibility of bending vibration of an elastic beam. J. Vibr. Acoust. **140**(3), 031007 (2018). https://doi.org/10.1115/1.4038733
4. Yang, Y.B., Lin, C.L., Yau, J.D., Chang, D.W.: Mechanism of resonance and cancellation for train-induced vibrations on bridges with elastic bearings. J. Sound Vibr. **269**(1–2), 345–360 (2004). https://doi.org/10.1016/S0022-460X(03)00123-8
5. Yau, J.D., Wu, Y.S., Yang, Y.B.: Impact response of bridges with elastic bearings to moving loads. J. Sound Vib. **248**(1), 9–30 (2001)
6. Van Khang, N., Phuong, N.M.: Transverse vibrations of a continuous beam on rigid and elastic supports under the action of moving bodies, Technische Mechanik, Band 22. Heft **4**, 300–316 (2002)
7. Nguyen, M.P.: Calculating transverse vibration of a continuous beam on elastic supports under the action of moving loads (in Vietnamese). In: Proceedings of the National Conference on Engineering Mechanics, Vol. 1, Hanoi, pp. 569–574 (2014)
8. Qian, C.Z., Chen, C.P., Zhou, G.W., Dai, L.M.: Dynamic response analysis for elastic bearing beam under moving load. Nonlinear Eng. **1**, 109–114 (2012)
9. Thomson, W.T.: Vibration of continuous systems. In: Theory of Vibration with Applications, pp. 268–300. Springer, Boston, MA (1993). https://doi.org/10.1007/978-1-4899-6872-2_10
10. Yang, Y.B., Yau, J.D., Wu, Y.S.: Vehicle-Bridge Interaction Dynamics - With Applications to High-Speed Railways. World Scientific Publishing Co. Pte. Ltd., Singapore (2004). https://doi.org/10.1142/9789812567178

Exact Analytical Periodic Solutions in Special Cases and Numerical Analysis of a Half-Undamped 1-DOF Piecewise-Linear Vibratory System

Minh-Tuan Nguyen-Thai(✉)

Hanoi University of Science and Technology, Hanoi, Vietnam
tuan.nguyenthaiminh@hust.edu.vn

Abstract. A lot of realistic systems are better described by piecewise-linear models than by continuous models, including systems with dry friction or impact as well as some controlled systems. Such models are usually strongly nonlinear and do not have general exact analytical solutions. This paper analyzes a harmonically excited 1-DOF piecewise-linear vibratory system that has a damped domain and an undamped domain. The purposes are to find some exact analytical solutions of the system and find the connection between the analytical solutions and the numerical ones. Though the system is linear and can be analytically analyzed in each domain, no general closed-form solution for its motion has been found because the switching times between the two domains are solutions of transcendental equations. To solve the problem, a special method is proposed: the initial conditions and the parameters are adjusted so that the homogeneous solution in the damped domain is cancelled, and the excitation is subharmonic in the undamped domain. As a result, the transcendental equations are simplified and exact analytical expressions for periodic motions are found for special cases of the parameters. The obtained motions are single-penetration multi-periodic, which is a typical behavior of the considered bilinear system as seen in the bifurcation diagram. Thus, the proposed method may be used to investigate dynamic behaviors of more complicated piecewise-linear systems.

Keywords: Piecewise-linear · Exact analytical solution · Coexisting motions

1 Introduction

Piecewise ordinary differential equations (ODEs) are found useful to model real systems where non-smooth phenomena, e.g., dry friction and impact, may occur [1–3]. With the development of control methods [4–6], the controllers can also produce non-smoothness. The solutions of piecewise ODEs, which reflex the dynamics of the modelled systems, are the topic of a lot of studies, and they do show abundant of phenomena including periodic and chaotic vibrations [7–10]. Some solutions may coexist, leading to difficulties of global dynamics analysis. For instance, examining the same piecewise-linear system

N. V. Khang et al. (Eds.): ASIAN MMS 2021, MMS 113, pp. 783–792, 2022.
https://doi.org/10.1007/978-3-030-91892-7_75

with asymmetric piecewise stiffness and damping, two coexisting periodic solutions are reported in [9] while it is shown later that there are more coexisting solutions including a chaotic solution [10], and there is still no guarantee that all solutions have been found. Furthermore, the non-smooth nature of the system may make common methods for smooth system become less effective. Thus, specialized numerical algorithms [7–10] and semi-analytical method [2] have been developed to study this type of systems.

Even in the case of piecewise-linear, i.e., the ODEs are linear in each domain, it is well-known that there is usually no analytical solution if harmonic excitation presents [2]. It is because the difficulties in determining the switching times between the domains require solutions using numerical methods. This approach is adopted by all the mentioned studies on the topic.

A curios question arises as to find piecewise-linear ODEs that have exact analytical periodic solutions, and if they exist, the next question would be to find the connection between the obtained results and the solutions of the non-analytical-solution cases. To answer the former question, this paper presents an original method: simplifying the equations for switching times by adjusting the system parameters.

This paper focuses on a harmonically excited 1-DOF system with a gap-activate gap-deactivate (GAGD) linear viscous-elastic support and a GAGD spring. Firstly, the model and its equations are presented. Secondly, analytical analysis is performed where possible. Then, the parameters are adjusted to get special cases that have exact analytical solutions. Lastly, numerical analysis is performed to investigate the global dynamics of the system.

2 The Model Problem

The system depicted in Fig. 1 has two different domains of motion caused by the GAGD supports [10]: the first is the damped domain in which the displacement of the body is positive (Fig. 1a); the second is the undamped domain in which the displacement of the body is negative, and no damping is active (Fig. 1b). The governing equations are

$$\begin{cases} m\ddot{u} + d\dot{u} + k_1 u = F\cos(\Omega t) \text{ if } u \geq 0 \\ m\ddot{u} + k_2 u = F\cos(\Omega t) \text{ if } u < 0 \end{cases} \tag{1}$$

The parameters m, d, k_1, k_2, and Ω are positive.

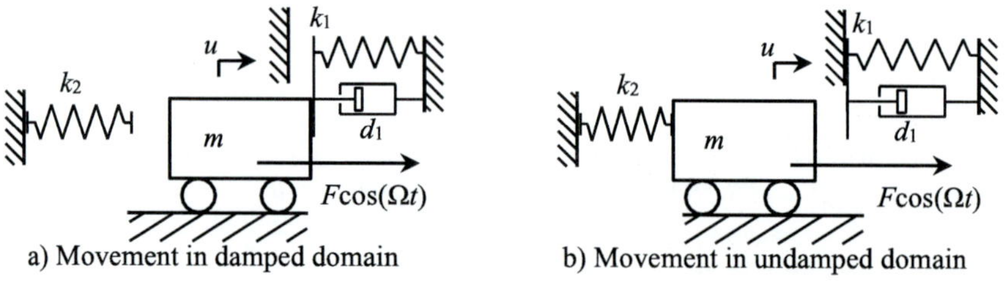

Fig. 1. A half-undamped 1-DOF piecewise-linear vibratory system.

The dimensionless forms read

$$\begin{cases} \ddot{x} + 2\gamma\dot{x} + \omega_1^2 x = \cos\tau \text{ if } x \geq 0 \\ \ddot{x} + \omega_2^2 x = \cos\tau \text{ if } x < 0 \end{cases} \tag{2}$$

in which $\tau = \Omega t$, $x = \frac{m\Omega^2}{F}u$, $\omega_1 = \frac{1}{\Omega}\sqrt{\frac{k_1}{m}}$, $\omega_2 = \frac{1}{\Omega}\sqrt{\frac{k_2}{m}}$, $\gamma = \frac{d}{2m\Omega}$, and the over dot now denotes $\frac{\mathrm{d}}{\mathrm{d}\tau}$ for the sake of simplicity. Consider the initial conditions

$$\begin{cases} x(\tau_0) = 0 \\ \dot{x}(\tau_0) = v_0 > 0 \end{cases} \tag{3}$$

in which $0 \leq \tau_0 < 2\pi$. The body initially moves in the damped domain and until a time point τ_1 that

$$\begin{cases} x(\tau_1) = 0 \\ \dot{x}(\tau_1) = v_1 < 0 \end{cases}, \tag{4}$$

the body moves into the undamped domain until a time point τ_2 that the displacement reaches zero and the velocity is positive, and so on. One can write the body's motion from τ_0 and τ_2 as follows

$$\begin{cases} x_I(\tau) = x_{Ih}(\tau) + x_{Ip}(\tau)\ \tau_0 \leq \tau \leq \tau_1 \\ x_{II}(\tau) = x_{IIh}(\tau) + x_{IIp}(\tau)\ \tau_1 \leq \tau \leq \tau_2 \end{cases}, \tag{5}$$

Where the subscription I is for the damped domain and II for the undamped domain.

3 Analytical Analysis

3.1 Damped Domain

The particular solution is

$$x_{Ip}(\tau) = A_I \cos\tau + B_I \sin\tau \tag{6}$$

with the coefficients

$$A_I = \frac{\omega_1^2 - 1}{(\omega_1^2 - 1)^2 + 4\gamma^2}, \quad B_I = \frac{2\gamma}{(\omega_1^2 - 1)^2 + 4\gamma^2} \tag{7}$$

The homogeneous solution in the damped domain is

$$\begin{cases} x_{Ih}(\tau) = e^{-\gamma(\tau-\tau_0)}\left(C_I \cos\left(\sqrt{\omega_1^2 - \gamma^2}(\tau - \tau_0)\right) + D_I \sin\left(\sqrt{\omega_1^2 - \gamma^2}(\tau - \tau_0)\right)\right) \text{ if } \gamma < \omega_1 \\ x_{Ih}(\tau) = C_I e^{-\gamma(\tau-\tau_0)} + D_I(\tau - \tau_0)e^{-\gamma(\tau-\tau_0)} \text{ if } \gamma = \omega_1 \\ x_{Ih}(\tau) = C_I e^{-(\gamma+\sqrt{\gamma^2-\omega_1^2})(\tau-\tau_0)} + D_I e^{-(\gamma-\sqrt{\gamma^2-\omega_1^2})(\tau-\tau_0)} \text{ if } \gamma > \omega_1 \end{cases} \tag{8}$$

The constants of integral C_{I} and D_{I} are obtained from the initial conditions (3): in underdamped case

$$\begin{cases} C_I = -A_I \cos\tau_0 - B_I \sin\tau_0 \\ D_I = \dfrac{v_0 + A_I \sin\tau_0 - B_I \cos\tau_0 - \gamma(A_I \cos\tau_0 + B_I \sin\tau_0)}{\sqrt{\omega_1^2 - \gamma^2}} \end{cases}, \tag{9}$$

in critical damped case

$$\begin{cases} C_I = -A_I \cos\tau_0 - B_I \sin\tau_0 \\ D_I = v_0 + A_I \sin\tau_0 - B_I \cos\tau_0 - \gamma(A_I \cos\tau_0 + B_I \sin\tau_0) \end{cases}, \tag{10}$$

and in overdamped case

$$\begin{cases} C_I = \dfrac{-v_0 - A_I \sin\tau_0 + B_I \cos\tau_0 + (\gamma - \sqrt{\gamma^2 - \omega_1^2})(A_I \cos\tau_0 + B_I \sin\tau_0)}{2\sqrt{\gamma^2 - \omega_1^2}} \\ D_I = \dfrac{v_0 + A_I \sin\tau_0 - B_I \cos\tau_0 - (\gamma + \sqrt{\gamma^2 - \omega_1^2})(A_I \cos\tau_0 + B_I \sin\tau_0)}{2\sqrt{\gamma^2 - \omega_1^2}} \end{cases}. \tag{11}$$

3.2 Undamped Domain

In the non-resonant case $\omega_2 \neq 1$, the particular solution is

$$x_{IIp}(\tau) = A_{II} \cos\tau + B_{II} \sin\tau \tag{12}$$

with the coefficients.

$$A_{II} = \frac{1}{\omega_2^2 - 1}, \quad B_{II} = 0 \tag{13}$$

In the resonant case $\omega_2 = 1$, the particular solution is

$$x_{IIp}(\tau) = A_{II}\tau \cos\tau + B_{II}\tau \sin\tau \tag{14}$$

with the coefficients

$$A_{II} = 0, \quad B_{II} = \frac{1}{2} \tag{15}$$

The homogeneous solution in the undamped domain is

$$x_{IIh}(\tau) = C_{II} \cos(\omega_2(\tau - \tau_1)) + D_{II} \sin(\omega_2(\tau - \tau_1)). \tag{16}$$

The constants of integral are

$$\begin{cases} C_{II} = -A_{II} \cos \tau_1 - B_{II} \sin \tau_1 \\ D_{II} = \dfrac{v_1 + A_{II} \sin \tau_1 - B_{II} \cos \tau_1}{\omega_2} \end{cases} \text{if } \omega_2 \neq 1, \tag{17}$$

$$\begin{cases} C_{II} = -A_{II}\tau_1 \cos \tau_1 - B_{II}\tau_1 \sin \tau_1 \\ D_{II} = \dfrac{v_1 + A_{II}\tau_1 \sin \tau_1 - A_{II} \cos \tau_1 - B_{II}\tau_1 \cos \tau_1 - B_{II} \sin \tau_1}{\omega_2} \end{cases} \text{if } \omega_2 = 1 \tag{18}$$

Since the body is back to the first domain at τ_2, and generally at τ_{2k} for any non-negative integer k, its motion in the time interval $[\tau_{2k}, \tau_{2k+2}]$ can be obtained in a similar way. The equations in this section, however, do not give the analytical solution for the initial problem because the switching times τ_i $(i = 1, 2, \ldots)$ is not known yet. Generally, each of them is a solution of one of the transcendental equations

$$A \cos x + B \sin x + C \cos(\omega x) + D \sin(\omega x) = 0, \tag{19}$$

$$Ax \cos x + Bx \sin x + C \cos x + D \sin x = 0, \tag{20}$$

$$A \cos x + B \sin x + e^{-\beta x}(C \cos(\omega x) + D \sin(\omega x)) = 0, \tag{21}$$

$$A \cos x + B \sin x + e^{-\beta x}(C + xD) = 0, \tag{22}$$

$$A \cos x + B \sin x + Ce^{-\alpha x} + De^{-\beta x} = 0, \tag{23}$$

where A, B, C and D are real constants, and α, β and ω are positive constants.

Such transcendental equations do not have general closed-form solutions. Thus, the switching times and the solutions of the ODEs do not have general analytical expressions.

4 Special Cases with Analytical Periodic Solutions

On the one hand, the transcendental equations to solve for switching times depend on five parameters: ω_1, ω_2, γ, τ_0 and v_0. On the other hand, if

$$\begin{cases} \tau_2 - \tau_0 = 2n\pi \\ v_2 = v_0 \end{cases} \quad n = 1, 2, \ldots \tag{24}$$

a period-n motion will be found. The critical idea is to make the transcendental equations simpler. In the damped domain, simplification can be done by eliminating the homogeneous solution, which means

$$C_I = D_I = 0, \tag{25}$$

The following initial conditions and terminal conditions must hold

$$\begin{cases} x_I(\tau_0) = \dfrac{\omega_1^2 - 1}{(\omega_1^2 - 1)^2 + 4\gamma^2} \cos\tau_0 + \dfrac{2\gamma}{(\omega_1^2 - 1)^2 + 4\gamma^2} \sin\tau_0 = 0 \\ v_0 = \dot{x}_I(\tau_0) = -\dfrac{\omega_1^2 - 1}{(\omega_1^2 - 1)^2 + 4\gamma^2} \sin\tau_0 + \dfrac{2\gamma}{(\omega_1^2 - 1)^2 + 4\gamma^2} \cos\tau_0 > 0 \\ x_I(\tau_1) = \dfrac{\omega_1^2 - 1}{(\omega_1^2 - 1)^2 + 4\gamma^2} \cos\tau_1 + \dfrac{2\gamma}{(\omega_1^2 - 1)^2 + 4\gamma^2} \sin\tau_1 = 0 \\ v_1 = \dot{x}_I(\tau_1) = -\dfrac{\omega_1^2 - 1}{(\omega_1^2 - 1)^2 + 4\gamma^2} \sin\tau_1 + \dfrac{2\gamma}{(\omega_1^2 - 1)^2 + 4\gamma^2} \cos\tau_1 > 0 \end{cases} \quad 0 \le \tau_0 < 2\pi. \tag{26}$$

The solution for the first two lines of (26) are

$$\tau_0 = \text{atan2}(1 - \omega_1^2, 2\gamma), \;\; v_0 = \frac{1}{\sqrt{(\omega_1^2 - 1)^2 + 4\gamma^2}} \tag{27}$$

The motion of the body in the damped domain is harmonic, so the displacement reaches zero after a half period; hence

$$\tau_0 = \tau_1 + \pi, \;\; v_1 = -v_0. \tag{28}$$

The motion of the body in the undamped domain (non-resonant) is deduced from Subsect. 3.2

$$x_{II}(\tau) = \frac{1}{\omega_2^2 - 1} \cos\tau - \frac{1}{\omega_2^2 - 1} \cos\tau_1 \cos(\omega_2(\tau - \tau_1)) + \frac{-v_0 + \frac{1}{\omega_2^2 - 1} \sin\tau_1}{\omega_2} \sin(\omega_2(\tau - \tau_1)). \tag{29}$$

Equation (29) already satisfies initial conditions for the motion in the undamped domain. It is desired to also satisfy the terminal conditions according to (24):

$$\begin{cases} x_{II}(\tau_1 + (2n - 1)\pi) = 0 \\ \dot{x}_{II}(\tau_1 + (2n - 1)\pi) = v_0 \end{cases} \quad n = 1, 2, ... \tag{30}$$

It is easily proved that if

$$\omega_2 = \frac{1}{2n - 1} \quad n = 2, \; 3, \; \dots \;, \tag{31}$$

Equation (30) is satisfied. The case $n = 1$ is excluded since it leads to resonance and invalidates (29).

The motion of the body in the time interval from τ_0 to τ_2 is then expected to be

$$\begin{cases} x(\tau) = \dfrac{\omega_1^2 - 1}{(\omega_1^2 - 1)^2 + 4\gamma^2} \cos\tau + \dfrac{2\gamma}{(\omega_1^2 - 1)^2 + 4\gamma^2} \sin\tau \;\; \tau_0 = \mathrm{atan2}(1 - \omega_1^2, 2\gamma) \le \tau < \tau_0 + \pi \\ x(\tau) = -\dfrac{(2n-1)^2}{4n(n-1)} \cos\tau - \dfrac{(2n-1)^2}{4n(n-1)} \cos\tau_0 \cos\left(\dfrac{\tau - \tau_0 - \pi}{2n-1}\right) \\ -\left(\dfrac{2n-1}{\sqrt{(\omega_1^2 - 1)^2 + 4\gamma^2}} - \dfrac{(2n-1)^3}{4n(n-1)} \sin\tau_0\right) \sin\left(\dfrac{\tau - \tau_0 - \pi}{2n-1}\right) \;\; \tau_0 + \pi \le \tau \le \tau_0 + 2n\pi \end{cases} \tag{32}$$

It should be noted that (32) only becomes a valid solution of the original system if

$$x(\tau) \le 0 \;\; \tau_0 + \pi \le \tau < \tau_0 + 2n\pi \tag{33}$$

Inequality (33) is transcendental, so it is not easy to solve. However, its exact solution is not necessary here.

To conclude, the following system

$$\begin{cases} \ddot{x} + 2\gamma\dot{x} + \omega_1^2 x = \cos\tau \text{ if } x \ge 0 \\ \ddot{x} + \dfrac{1}{(2n-1)^2} x = \cos\tau \;\; n = 2, 3, \ldots \text{ if } x < 0 \end{cases} \tag{34}$$

has a period-n solution of the form (32) if (33) is satisfied.

A question arises, as to whether there exists a valid solution for any value of n. It has not been strictly proved, but it can be illustrated by considering an even more specific case of $\omega_1 = 1$

$$\begin{cases} \ddot{x} + 2\gamma\dot{x} + x = \cos\tau \text{ if } x \ge 0 \\ \ddot{x} + \frac{1}{(2n-1)^2} x = \cos\tau \, n = 2, 3, \ldots \text{ if } x < 0. \end{cases} \tag{35}$$

In this case, (32) becomes

$$\begin{cases} x(\tau) = \dfrac{1}{2\gamma} \sin\tau \;\; 0 \le \tau < \pi \\ x(\tau) = -\dfrac{(2n-1)^2}{4n(n-1)} \cos\tau - \dfrac{(2n-1)^2}{4n(n-1)} \cos\left(\dfrac{\tau - \pi}{2n-1}\right) - \dfrac{2n-1}{2\gamma} \sin\left(\dfrac{\tau - \pi}{2n-1}\right) \;\; \pi \le \tau \le 2n\pi \end{cases} \tag{36}$$

In the second line of (36), only the last term can be varied by changing positive parameter γ. This term is negative in the open interval $(\pi, 2n\pi)$ and gets smaller as γ is reduced. In other words, one may probably reduce γ to a sufficiently small value so that (33) is satisfied.

For illustration, the parameters are chosen as follows:

$$\gamma = 0.5, \;\; \omega_1 = 1, \;\; \omega_2 = 1/(2n-1), \tag{37}$$

where n is the control parameter that is not restricted to be integer.

Some periodic solutions obtained with the exact analytical expressions are shown in Fig. 2.

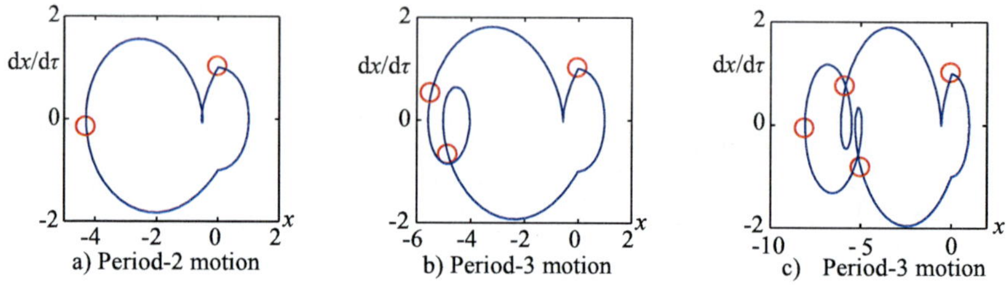

Fig. 2. Phase portraits and Poincaré points of periodic motions with analytical expressions.

5 Numerical Results and Discussion

Simulation where exact analytical expressions are not applicable is done by a specialized method for piecewise-linear systems based on matrix exponential [10]. The bifurcation diagram (Fig. 3) is obtained by integrating with initial conditions $(x(0), \dot{x}(0)) = (0,1)$ and ignoring the first 6000 periods. It bridges the gaps between the exact analytical solutions. Chaos may occur between two consecutive integer values of n as the lower order solution bifurcates while the higher order does not exist yet.

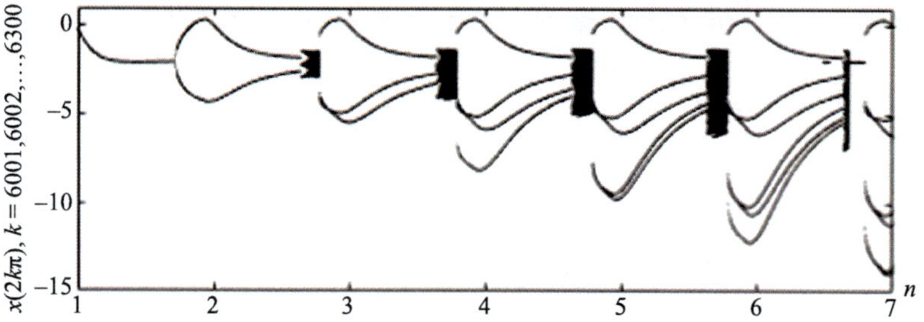

Fig. 3. Bifurcation diagram obtained by integrating with initial conditions $(x(0), \dot{x}(0)) = (0,1)$.

It should be noted that the bifurcation diagram obtained by integrating with fixed initial conditions does not show coexisting solutions. Hence, interpolated cell-mapping method is used to plot the basins of attraction [11]. Figure 4 shows that at $n = 6.5$, there are two coexisting stable periodic motions with fractal basins of attraction. At $n = 3$, a chaotic motion coexists with a stable periodic motion that has exact analytic expressions (Fig. 5).

Though the exact analytical periodic solutions are only valid for very specific parameters and do not fully characterize the global dynamics, they describe a type of single-penetration multi-periodic motion that is typical for the bilinear system as seen in the bifurcation diagram. The exact analytical solutions can be used as a tool to validate numerical methods.

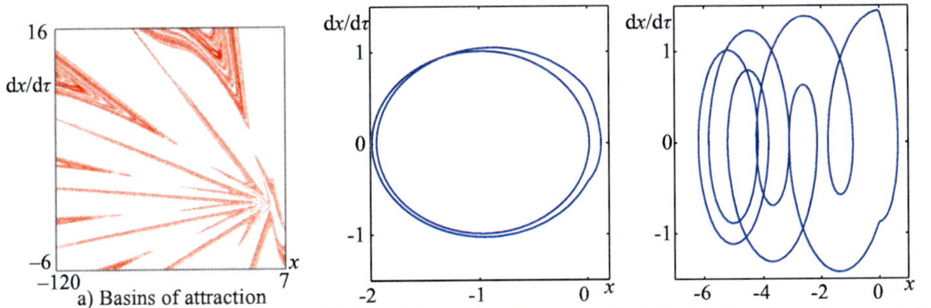

a) Basins of attraction
Red: Period-2, White: Period-6 b) Period-2 motion: phase portrait c) Period-6 motion: phase portrait

Fig. 4. Global dynamics at $n = 6.5$.

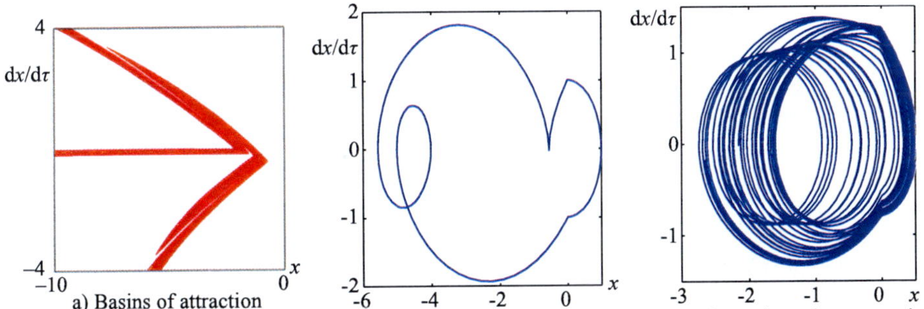

a) Basins of attraction
Red: Chaotic, White: Period-3 b) Period-3 motion: phase portrait c) Chaotic motion: phase portrait

Fig. 5. Global dynamics at $n = 3$.

6 Conclusions and Outlook

A 1-DOF piecewise-linear system that has a damped domain and an undamped domain excited by a harmonic force is investigated. The analytical analysis has been performed for each domain; however, the piecewise motion does not have a general closed-form solution because the switching times between domains are solutions of transcendental equations.

The exact analytical periodic motions are found for special cases of parameters: the frequency of the homogenous solution of the undamped domain is a whole number n (higher than one) multiple of the frequency of the excitation, and parameters in the damped domain are suitable, e.g., natural frequency equal to excitation frequency and low damping. Each of the motion has a period of n times the period of excitation but only moves from undamped domain to damped domain one time during its period. Thus, the obtained motions are single-penetration multi-periodic.

The exact analytical solutions can be used to partly predict the dynamics of the system. It can also be used as an effective tool to validate numerical methods.

Future work will apply the presented method to investigate more complicated systems, e.g., multi-DOF systems.

Acknowledgements. This research is funded by Hanoi University of Science and Technology (HUST) under project number T2021-PC-037.

References

1. Wiercigroch, M., Budak, E.: Sources of nonlinearities, chatter generation and suppression in metal cutting. Philos. Trans. R. Soc. London. Ser. A. Math. Phys. Eng. Sci. **359**(1781), 663–693 (2001)
2. Karpenko, E.V., Wiercigroch, M., Pavlovskaia, E.E., Cartmell, M.P.: Piecewise approximate analytical solutions for a Jeffcott rotor with a snubber ring. Int. J. Mech. Sci. **44**(3), 475–488 (2002)
3. Pavlovskaia, E., Wiercigroch, M., Grebogi, C.: Two-dimensional map for impact oscillator with drift. Phys. Rev. E. **70**(3), 036201 (2004)
4. Martínez-Miranda, M.A., San Miguel, C.R.T., Flores-Campos, J.A., Ceccarelli, M.: Numerical simulation of a 2D harmonic oscillator as coupling system for child restraint systems (CRS). In: The International Conference of IFToMM ITALY, 492–502. Springer, Cham (2020)
5. Perrusquía, A., Flores-Campos, J.A., Torres-San-Miguel, C.R.: A novel tuning method of PD with gravity compensation controller for robot manipulators. IEEE Access **8**, 114773–114783 (2020)
6. Perrusquía, A., Flores-Campos, J.A., Torres-San-Miguel, C.R., González, N.: Task space position control of slider-crank mechanisms using simple tuning techniques without linearization methods. IEEE Access **8**, 58435–58442 (2020)
7. Yu, S.D.: An efficient computational method for vibration analysis of unsymmetric piecewise-linear dynamical systems with multiple degrees of freedom. Nonlinear Dyn. **71**(3), 493–504 (2013)
8. He, D., Gao, Q., Zhong, W.: An efficient method for simulating the dynamic behavior of periodic structures with piecewise linearity. Nonlinear Dyn. **94**(3), 2059–2075 (2018). https://doi.org/10.1007/s11071-018-4475-8
9. Xu, L., Lu, M.W., Cao, Q.: Bifurcation and chaos of a harmonically excited oscillator with both stiffness and viscous damping piecewise linearities by incremental harmonic balance method. J. Sound Vib. **264**(4), 873–882 (2003)
10. Minh-Tuan, N.T., Khang, N.V.: Calculating periodic and chaotic vibrations of piecewise-linear systems using matrix exponential function. In: The 19th Asia Pacific Vibration Conference. Qingdao China (2021). (accepted)
11. Tongue, B.H.: On obtaining global nonlinear system characteristics through interpolated cell mapping. Phys. D Nonlinear Phen. **28**(3), 401–408 (1987)

Vibration Analysis of Laminated Composite Beams Using a Novel Two-Variable Model with Various Boundary Conditions

Quoc-Cuong Le[1](✉), Trung-Kien Nguyen[2], and Ba-Duy Nguyen[3]

[1] Institute of Engineering and Technology, Thu Dau Mot University, 06 Tran Van On Street, Thu Dau Mot City, Vietnam
cuonglq@tdmu.edu.vn

[2] Faculty of Civil Engineering, Ho Chi Minh City University of Technology and Education, 1 Vo Van Ngan Street, Thu Duc City, Vietnam

[3] Faculty of Architecture, Thu Dau Mot University, 06 Tran Van On Street, Thu Dau Mot City, Vietnam

Abstract. The kinematics of the beam having only two variables are increased in a hybrid form under polynomial, and trigonometric series in thickness and axial directions, respectively. Lagrange's equations are then used to derive characteristic equations of the beams. Numerical results for laminated composite beams are equalled with previous studies and are used to investigate the effects of length-to-depth ratio, fibre angles and material anisotropy on the vibration of laminated composite beams.

Keywords: Laminated composite beams · Vibration · Elasticity solution

1 Introduction

Laminated composite materials are created by assembling multiple layers of fibrous materials to achieve the superior engineering properties such as bending stiffness, strength to weight ratio and thermal performance.

As a result, laminate composite has been widely applied in aerospace engineering, mechanical engineering as well as construction technology. In order to maximise the potential advantage of this multilayered material, numerous studies and computation modelling have been conducted to fine-tune the static and dynamic behaviours of laminated composite beams. Various beam theories have been developed in order to predict accurately their structural responses and capture anisotropy of laminated composite materials.

Classical beam theory (CBT) is the simplest one in analysing responses of laminated composite beams. Nonetheless, this theory underestimates deflections and overestimates natural frequencies of the beams due to neglecting effects of transverse shear deformation.

In order to account for this effect, thanks to its simplicity in formulation and programming, the first-order shear deformation beam theory (FSBT) is commonly used by

N. V. Khang et al. (Eds.): ASIAN MMS 2021, MMS 113, pp. 793–804, 2022.
https://doi.org/10.1007/978-3-030-91892-7_76

researchers and commercial soft wares for the analysis of laminated composite beams [1, 2]. However, in this theory, the inadequate distribution of transverse shear stress in the beam thickness requires a shear correction factor to calculate the shear force.

This adverse in practice could be overcome by using higher-order deformation beam theory (HSBT) [2, 3] or Quasi-3D beam theory (Quasi-3D) [4, 5] owing to the higher-order variation of axial displacement or both axial and transverse displacements, respectively. In such an approach, stresses of the beam can be directly computed from constitutive equations without shear coefficient requirement.

Many higher-order shear deformation theories have been developed with different approaches in which its kinematics could be expressed in terms of polynomial [6, 7], trigonometric [8, 9], exponential ones [10], hyperbolic [11, 12] and hybrid higher-order shear functions [13].

A literature review shows that a vast number of researches on development HSBT and Quasi-3D have been developed, however the accuracy of these theories strictly depends on the choice of shear functions and number of variables defining the problem. The development of new beam theories as well as suitable solution methods is a complicated problem and needs to be studied further.

The purpose of this paper is to develop a two-directional elasticity solution for vibration analysis of laminated composite beams. Based on the elasticity equations, the proposed theory only requires two unknowns in which the axial and transverse displacements are approximated in series terms in its two in-plane directions for different boundary conditions and Lagrange's equations are used to derive characteristic equations.

Numerical results are presented to investigate the effects of length-to-depth ratio, material anisotropy, Poisson's ratio and fiber angles of laminated composite beams.

2 Theoretical Formulation

Considering a laminated composite beam with rectangular section b x h and length L as shown in Fig. 1, the beam is composed of n layers of orthotropic materials.

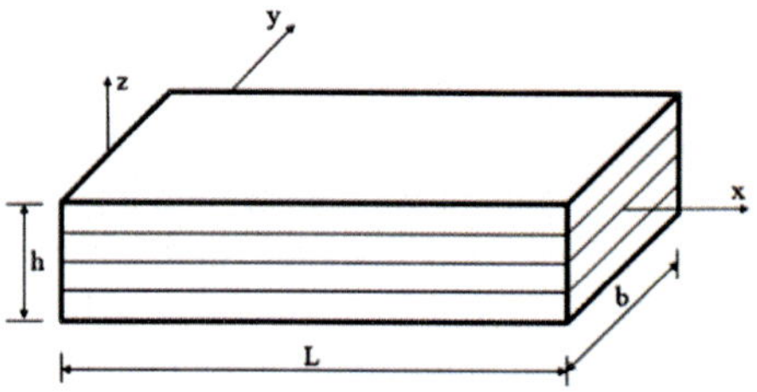

Fig. 1. Geometry of laminated composite beams.

2.1 Kinematic, Strain and Stress Relations

Denoting u and w are axial and transverse displacements at location (x, z) of the beam. The linear displacement-strain relations of the beam are given by:

$$\varepsilon_x = u_{,x} \tag{1}$$

$$\varepsilon_z = w_{,z} \tag{2}$$

$$\gamma_{xz} = u_{,z} + w_{,x} \tag{3}$$

where the comma indicates partial differentiation with respect to the coordinate subscript that follows. Based on an assumption of the plan stress in the plane (x, z) of the beam, i.e. $\sigma_y = \sigma_{yz} = \sigma_{xy} = 0$, the elastic constitutive equation in the global coordinate system is expressed by:

$$\begin{Bmatrix} \sigma_x \\ \sigma_z \\ \sigma_{xz} \end{Bmatrix} = \begin{bmatrix} \overline{\overline{C}}_{11} & \overline{\overline{C}}_{13} & 0 \\ \overline{\overline{C}}_{13} & \overline{\overline{C}}_{33} & 0 \\ 0 & 0 & \overline{\overline{C}}_{55} \end{bmatrix} \begin{Bmatrix} \varepsilon_x \\ \varepsilon_y \\ \gamma_{xz} \end{Bmatrix} \tag{4}$$

where $\overline{\overline{C}}_{11}$, $\overline{\overline{C}}_{13}$ and $\overline{\overline{C}}_{55}$ are the reduced in-plane and out of plane elastic stiffness coefficients of the laminated composite beam in the global coordinates (see [5] for more details).

2.2 Variational Formulation

Lagrangian function is used to derive the equations of motion:

$$\Pi = U - K \tag{5}$$

where U, and K denote the strain energy, and kinetic energy, respectively.

The strain energy U of a system is given by:

$$\begin{aligned} U &= \frac{1}{2}\int_V (\sigma_x\varepsilon_x + \sigma_z\varepsilon_z + \sigma_{xz}\gamma_{xz})dV \\ &= \frac{1}{2}\int_V \left[\overline{\overline{C}}_{11}u_{,x}^2 + 2\overline{\overline{C}}_{13}u_{,x}w_{,z} + \overline{\overline{C}}_{33}w_{,x}^2 + \overline{\overline{C}}_{55}\left(u_{,z}^2 + 2u_{,z}w_{,x} + w_{,x}^2\right)\right]dV \end{aligned} \tag{6}$$

The kinetic energy K is obtained as:

$$K = \frac{1}{2}\int_V \rho\left(\dot{u}^2 + \dot{w}^2\right)dV \tag{7}$$

where the differentiation with respect to the time t is denoted by dot-superscript convention; $\rho(z)$ is the mass density of each layer.

By substituting Eq. (6) and (7) into Eq. (5), Lagrangian function is explicitly expressed as:

$$\begin{aligned} \Pi = &\frac{1}{2}\int_V \left[\overline{\overline{C}}_{11}u_{,x}^2 + 2\overline{\overline{C}}_{13}u_{,x}w_{,z} + \overline{\overline{C}}_{33}w_{,x}^2 + \overline{\overline{C}}_{55}\left(u_{,z}^2 + 2u_{,z}w_{,x} + w_{,x}^2\right)\right]dV \\ &-\frac{1}{2}\int_V \rho\left(\dot{u}^2 + \dot{w}^2\right)dV \end{aligned} \tag{8}$$

2.3 Two-Directional Ritz Solution

Based on the Ritz method, the axial and transverse displacements at location (x, z) of the beam can be generally approximated in the following forms:

$$u(x, z, t) = \sum_{r=1}^{R} \sum_{s=1}^{S} \psi_{rs}(x, z) u_{rs} \tag{9}$$

$$w(x, z, t) = \sum_{r=1}^{R} \sum_{s=1}^{S} \varphi_{rs}(x, z) w_{rs} \tag{10}$$

where u_{rs},w_{rs} are unknown displacement values to be determined; $\psi_{rs}(x, z)$, $\varphi_{rs}(x, z)$ are the two-directional shape functions which are composed of admissible hybrid exponential-trigonometric function in the x-axis and polynomial function in the z-axis are given in Table 1.

Table 1. Shape functions and essentials BCs Beams.

BCs	$\psi_{rs}(x, z)$	$\varphi_{rs}(x, z)$	$x = 0$	$x = L$
S-S	$\cos\left(\frac{\pi x}{L}\right)e^{-rx/L}z^{s-1}$	$\sin\left(\frac{\pi x}{L}\right)e^{-rx/L}z^{s-1}$	$w = 0$	$w = 0$
C-F	$\sin\left(\frac{\pi x}{2L}\right)e^{-rx/L}z^{s-1}$	$\left(1 - \cos\left(\frac{\pi x}{2L}\right)\right)e^{-rx/L}z^{s-1}$	$u = 0$, $w = 0$, $w_{,x} = 0$	
C-C	$\sin\left(\frac{\pi x}{L}\right)e^{-rx/L}z^{s-1}$	$\sin^2\left(\frac{\pi x}{L}\right)e^{-rx/L}z^{s-1}$	$u = 0$, $w = 0$, $w_{,x} = 0$	$u = 0$, $w = 0$, $w_{,x} = 0$

The governing equations of motion can be obtained by substituting Eq. (9), (10) into Eq. (8) and using Lagrange's equations:

$$\frac{\partial \Pi}{\partial q_{rs}} - \frac{d}{dt}\frac{\partial \Pi}{\partial \dot{q}_{rs}} = 0 \tag{11}$$

with q_{rs} representing the values of (u_{rs}, w_{rs}), that leads to:

$$\left(\begin{bmatrix} \boldsymbol{K}^{11} & \boldsymbol{K}^{12} \\ {}^{T}\boldsymbol{K}^{12} & \boldsymbol{K}^{22} \end{bmatrix} - \omega^2 \begin{bmatrix} \mathbf{M}^{11} & 0 \\ 0 & \mathbf{M}^{22} \end{bmatrix}\right)\begin{Bmatrix} \boldsymbol{u} \\ \boldsymbol{w} \end{Bmatrix} = \begin{Bmatrix} 0 \\ 0 \end{Bmatrix} \tag{12}$$

where the components of the stiffness matrix **K** and the mass matrix **M** are given by:

$$\begin{aligned}
K_{rspq}^{11} &= \int_0^L \int_{-h/2}^{h/2} \overline{\overline{C}}_{11} \psi_{rs,x} \psi_{pq,x} bdxdz + \int_0^L \int_{-h/2}^{h/2} \overline{\overline{C}}_{55} \psi_{rs,z} \psi_{pq,z} bdxdz \\
K_{rspq}^{12} &= \int_0^L \int_{-h/2}^{h/2} \overline{\overline{C}}_{13} \psi_{rs,x} \varphi_{pq,z} bdxdz + \int_0^L \int_{-h/2}^{h/2} \overline{\overline{C}}_{55} \psi_{rs,z} \varphi_{pq,x} bdxdz, \\
K_{rspq}^{22} &= \int_0^L \int_{-h/2}^{h/2} \overline{\overline{C}}_{33} \varphi_{rs,z} \varphi_{pq,z} bdxdz + \int_0^L \int_{-h/2}^{h/2} \overline{\overline{C}}_{55} \varphi_{rs,x} \varphi_{pq,x} bdxdz, \\
M_{rspq}^{11} &= \int_0^L \int_{-h/2}^{h/2} \rho \psi_{rs} \psi_{pq} bdxdz, \; M_{rspq}^{22} = \int_0^L \int_{-h/2}^{h/2} \rho \varphi_{rs} \varphi_{pq} bdxdz
\end{aligned} \tag{13}$$

Finally, the vibration responses of the laminated composite beams can be determined by solving Eq. (12). It should be noted that the Eq. (12) does not consider damping materials, the investigation of vibration frequencies of damping materials [14] is also a very interesting problem and will be the future development of this paper.

3 Numerical Results

In this section, convergence and verification studies are carried out to demonstrate the accuracy of the present study. For vibration analysis, laminates are assumed to be equal thicknesses and made of the same orthotropic materials whose properties are given in Table 2. For convenience, the following non-dimensional terms are used:

$$\overline{\omega} = \frac{\omega L^2}{h}\sqrt{\frac{\rho}{E_2}} \text{ for Material I (MAT I) and } \overline{\omega} = \frac{\omega L^2}{h}\sqrt{\frac{\rho}{E_1}} \text{ for MAT II} \tag{14}$$

Table 2. Material properties of composite beams.

Material properties	MAT I [15]	MAT II [16]
E_1 (GPa)	E_1/E_2 = open	144.8
$E_2 = E_3$(GPa)	-	9.65
$G_{12} = G_{13}$ (GPa)	0.6$E2$	4.14
G_{23} (GPa)	0.5E_2	3.45
$\nu_{12} = \nu_{13} = \nu_{23}$	0.25	0.3
ρ (kg/m^3)	-	-
L (m)	L/h = open	$L/h = 15$
h (m)	-	-
b (m)	-	-

The composite beams (MAT I, $0^0/90^0/0^0$, $L/h = 5$, $E_1/E_2 = 40$) with different BCs are considered to evaluate the convergence. The non-dimensional fundamental frequencies with respect to the number of series in $x-$ direction (R) and $z-$ direction are given in Table 3. It can be seen that the responses converge quickly in x-direction and number of series in this direction $R = 10$ can be the point of convergence of the fundamental frequencies for the boundary conditions, whereas the frequency of vibration tends to decrease with increasing number of series in z-direction, and the beam tends to become softer. As an example for further verification, $R = 10$ and $S = 4$ will be chosen in the following examples.

Table 3. Convergence studies for normalized fundamental frequencies of $0^0/90^0/0^0$ laminated composite beams (MAT I, $L/h = 5$, $E_1/E_2 = 40$).

BC	Number of series (S)	Number of series (R)					
		2	4	6	8	10	12
S-S	1	11.8886	11.8251	11.8245	11.8245	11.8245	11.8245
	2	9.9289	9.8192	9.8178	9.8178	9.8178	9.8178
	3	9.9258	9.8160	9.8146	9.8146	9.8146	9.8146
	4	9.3105	9.2046	9.2033	9.2033	9.2033	9.2033
	5	9.3095	9.2036	9.2023	9.2022	9.2022	9.2022
	6	9.3088	9.2032	9.2019	9.2019	9.2019	9.2019
	7	9.3088	9.2032	9.2019	9.2019	9.2019	9.2019
C-F	1	6.2477	6.0473	5.9839	5.9562	5.9420	5.9355
	2	4.5409	4.4005	4.3737	4.3638	4.3589	4.3572
	3	4.5401	4.3993	4.3723	4.3624	4.3572	4.3547
	4	4.3294	4.2047	4.1808	4.1719	4.1676	4.1656
	5	4.3292	4.2045	4.1804	4.1716	4.1672	4.1655
	6	4.3273	4.2011	4.1770	4.1681	4.1636	4.1625
	7	4.3273	4.2011	4.1770	4.1682	4.1639	4.1612
C-C	1	13.8049	12.6527	12.2761	12.1087	12.0330	12.0311
	2	13.2193	12.0773	11.7457	11.5999	11.5358	11.5326
	3	13.2185	12.0765	11.7449	11.5990	11.5338	11.5287
	4	12.5178	11.4420	11.1473	11.0216	10.9655	10.9639
	5	12.5168	11.4406	11.1455	11.0195	10.9625	10.9604
	6	12.4768	11.4031	11.1073	10.9811	10.9258	10.9243
	7	12.4767	11.4030	11.1072	10.9809	10.9252	10.9243

Vibration behaviors of cross-ply laminated composite beams are investigated in Table 4 which presents changes of the non-dimensional fundamental frequencies with S-S, C-F and C-C boundary conditions, span-to-thickness ratio $L/h = 5,\ 10,\ 50$ of the 0°/90°/0° and 0°/90° laminated composite beams. The solutions are computed with MAT I and $E1/E2 = 40$. The accuracy of the solutions is tested by verification with those derived from HSBTs (Nguyen et al. [3], Nguyen et al. [5], Khdeir et al. [17], Vo, T.P et al. [18], Murthy et al. [19]) and Quasi-3Ds (Nguyen et al. [5], Mantari et al. [20], Matsunaga [21]). It can be seen that the present solutions comply with those from the Quasi-3Ds, however there are slight deviations between them for the thick beams ($L/h = 5$) and for C-F and C-C boundary conditions. The softer characteristic of the present beam model is again found for all solutions in comparison with the HSBTs and Quasi-3Ds.

Table 4. Non-dimensional fundamental frequencies of $0^0/90^0/0^0$ and $0^0/90^0$ laminated composite beams (MAT I,$E_1/E_2 = 40$).

BCs	Theory	$0^0/90^0/0^0$			$0^0/90^0$		
		$L/h = 5$	10	50	$L/h = 5$	10	50
S-S	HSBT [5]	9.206	13.607	17.449	6.125	6.940	7.297
	HSBT [3]	9.208	13.614	17.462	6.128	6.945	7.302
	HSBT [17]	9.208	13.614	-	6.128	6.945	-
	HSBT [18]	9.206	13.607	17.449	6.058	6.909	7.296
	HSBT [19]	9.207	13.611	-	6.045	6.908	-
	Quasi-3D [5]	9.208	13.610	17.449	6.140	6.948	7.297
	Quasi-3D [20]	9.208	13.610	-	6.109	6.913	-
	Quasi-3D [21]	9.200	13.608	-	5.662	6.756	-
	Present	9.203	13.610	17.449	5.831	6.833	7.292
C-F	HSBT [5]	4.230	5.490	6.262	2.381	2.541	2.603
	HSBT [3]	4.234	5.498	6.267	2.383	2.543	2.605
	HSBT [17]	4.234	5.495	-	2.386	2.544	-
	HSBT [19]	4.230	5.491	-	2.378	2.541	-
	Quasi-3D [5]	4.223	5.491	6.262	2.382	2.543	2.604
	Quasi-3D [20]	4.221	5.490	-	2.375	2.532	-
	Present	4.168	5.478	6.262	2.314	2.522	2.603
C-C	HSBT [5]	11.601	19.707	37.629	10.019	13.653	16.414
	HSBT [3]	11.607	19.728	37.679	10.027	13.670	16.429
	HSBT [17]	11.603	19.712	-	10.026	13.660	-
	HSBT [19]	11.602	19.719	-	10.011	13.657	-
	Quasi-3D [5]	11.499	19.672	37.633	9.944	13.664	16.432
	Quasi-3D [20]	11.486	19.652	-	9.974	13.628	-
	Present	10.966	19.454	37.652	8.8361	12.913	16.388

In order to verify the vibration behaviors of present theory further, TableS 5 and Table 6 report non-dimensional fundamental frequencies with respect to the fiber angle. The results are obtained with three boundary conditions, two lay-ups 0°/θ°/0° and 0°/θ°, material MAT I, $E1/E2 = 40$ and $L/h = 5$. It is worth noticing that the solutions are compared with those of Nguyen et al. [5] based on a Quasi-3D theory. It is observed that there are significant differences between two models for C-C boundary conditions, whereas the solution field of the theories is in agreement for S-S and C-F boundary conditions.

The unsymmetrical (45^0/-45^0/45^0/-45^0) and (30^0/-50^0/50^0/-30^0) composite beams (MAT II) with various BCs are considered. The results of fundamental frequencies are given in Fig. 2 and Fig. 3. A good agreement between the present solutions and previous studies (Nguyen et al. [5], Chen et al. [22], Chandrashekhara et al. [23]) is again found.

Table 5. Non-dimensional fundamental frequencies of $0^0/\theta$ $/0^0$ and $0^0/\theta$ laminated composite beams (MAT I, $E_1/E_2 = 40$, $L/h = 5$).

Lay-up	BCs	Theory	Fiber angle (θ)						
			00	15^0	30^0	45^0	60^0	75^0	90^0
$0^0/\theta$ $/0^0$	S-S	HSBT [5]	9.5498	9.5165	9.4487	9.3630	9.2831	9.2279	9.2083
		Present	9.5360	9.5008	9.4354	9.3531	9.2759	9.2223	9.2033
	C-F	HSBT [5]	4.3628	4.3307	4.3047	4.2754	4.2484	4.2297	4.2231
		Present	4.2988	4.2707	4.2464	4.2182	4.1919	4.1740	4.1676
	C-C	HSBT [5]	12.0240	11.9365	11.8341	11.7130	11.6020	11.5260	11.4992
		Present	11.3751	11.3341	11.2621	11.1521	11.0571	10.9886	10.9655
$0^0/\theta$	S-S	HSBT [5]	9.5498	7.9829	6.8336	6.3948	6.2215	6.1561	6.1400
		Present	9.5360	7.8827	6.6059	6.1104	5.9178	5.8478	5.8314
	C-F	HSBT [5]	4.3628	3.3266	2.7077	2.4964	2.4173	2.3887	2.3819
		Present	4.2987	3.2836	2.6501	2.4314	2.3498	2.3208	2.3140
	C-C	HSBT [5]	12.0240	11.0882	10.4823	10.1844	10.0347	9.9640	9.9435
		Present	11.3769	10.5547	9.6617	9.1718	8.9404	8.8557	8.8300

Finally, the symmetric $(\theta/-\theta)_s$ composite beams (MAT II) are considered. The effects of the fiber angle on the natural frequencies is illustrated in Table 7. It can be seen that the present natural frequencies are closer to those of HSBT (Nguyen et al. [5] and Aydogdu [24]) smaller than those of HSBT (Nguyen et al. [3]), which neglected the Poisson's effect, especially for $10^0 \leq \theta \leq 60^0$. This phenomenon can be explained by

Table 6. Non-dimensional fundamental frequencies of $0^0/\theta/0^0$ and $0^0/\theta$ laminated composite beams (MAT I, $E_1/E_2 = 40$, $L/h = 10$).

Lay-up	BCs	Theory	Fiber angle (θ)						
			00	15^0	30^0	45^0	60^0	75^0	90^0
$0^0/\theta/0^0$	S-S	HSBT [5]	13.9976	13.8822	13.8130	13.7400	13.6729	13.6264	13.6099
		Present	13.9968	13.8802	13.8118	13.7395	13.6728	13.6264	13.6098
	C-F	HSBT [5]	5.6259	5.5622	5.5403	5.5220	5.5059	5.4948	5.4909
		Present	5.6111	5.5484	5.5268	5.5087	5.4927	5.4816	5.4776
	C-C	HSBT [5]	20.4355	20.3428	20.1923	20.0062	19.8335	19.7144	19.6723
		Present	20.1538	20.0769	19.9452	19.7694	19.6065	19.4931	19.4541
$0^0/\theta$	S-S	HSBT [5]	13.9976	10.0656	7.9772	7.3028	7.0561	6.9682	6.9475
		Present	13.9968	10.0200	7.8862	7.1945	6.9425	6.8536	6.8330
	C-F	HSBT [5]	5.6259	3.7996	2.9428	2.6785	2.5837	2.5505	2.5428
		Present	5.6110	3.7877	2.9253	2.6588	2.5633	2.5300	2.5224
	C-C	HSBT [5]	20.4355	17.3592	15.0934	14.2004	13.8389	13.6989	13.6637
		Present	20.1626	17.0377	14.5214	13.4923	13.0949	12.9442	12.9133

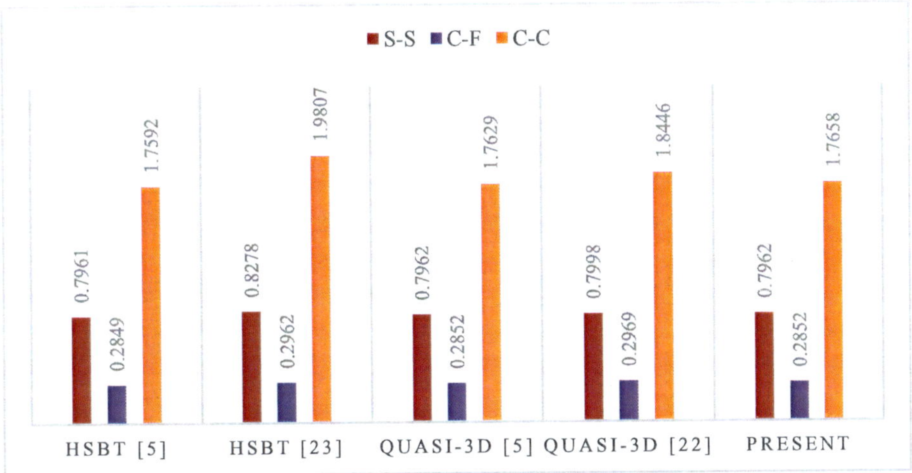

Fig. 2. Non-dimensional fundamental frequencies of $45^0/-45^0/45^0/-45^0$ laminated composite beams (MAT II).

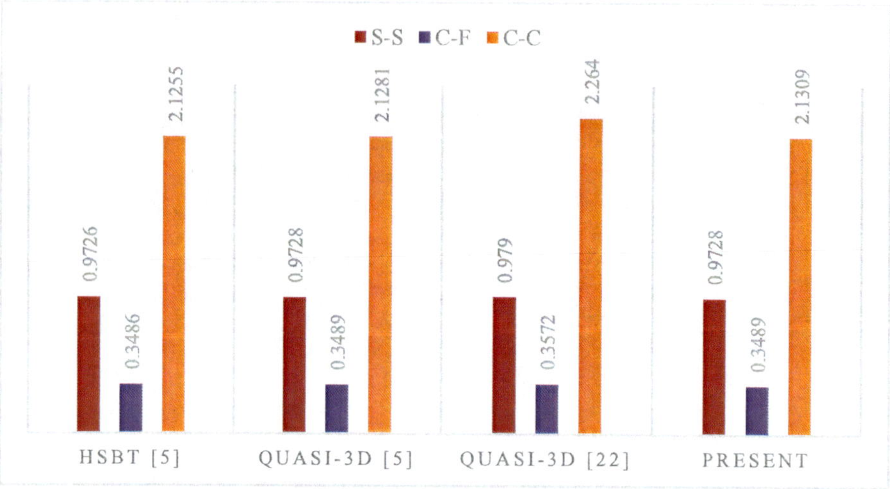

Fig. 3. Non-dimensional fundamental frequencies of $30^0/\text{-}50^0/50^0/\text{-}30^0$ laminated composite beams (MAT II).

Table 7. Non-dimensional fundamental frequencies of $(\theta/-\theta)_s$ laminated composite beams (MAT II).

BCs	Theory	Fiber angle (θ)						
		00	15^0	30^0	45^0	60^0	75^0	90^0
S-S	HSBT [5]	2.649	1.579	0.999	0.796	0.731	0.725	0.729
	HSBT [24]	2.651	1.896	1.141	0.804	0.736	0.725	0.729
	HSBT [3]	2.656	2.511	2.103	1.537	1.012	0.761	0.732
	Quasi-3D [5]	2.650	1.580	0.999	0.796	0.731	0.725	0.730
	Present	2.650	1.580	0.999	0.796	0.731	0.725	0.730
C-F	HSBT [5]	0.980	0.570	0.358	0.285	0.261	0.259	0.261
	HSBT [24]	0.981	0.676	0.414	0.288	0.262	0.258	0.260
	HSBT [3]	0.983	0.926	0.768	0.555	0.363	0.272	0.262
	Quasi-3D [5]	0.980	0.571	0.358	0.285	0.262	0.260	0.262
	Present	0.980	0.571	0.358	0.285	0.262	0.260	0.262
C-C	HSBT [5]	4.897	3.288	2.180	1.759	1.620	1.605	1.615
	HSBT [24]	4.973	4.294	2.195	1.929	1.669	1.612	1.619
	HSBT [3]	4.912	4.717	4.131	3.197	2.202	1.683	1.621
	Quasi-3D [5]	4.901	3.290	2.183	1.762	1.626	1.614	1.625
	Present	4.898	3.295	2.186	1.765	1.630	1.619	1.631

the fact that Poisson's effect is incorporated in the constitutive equations by assuming $\sigma_y = \sigma_{xy} = \sigma_{yz} = 0$. It means that the strains $(\varepsilon_y, \gamma_{yz}, \gamma_{xy})$ are nonzero and this causes the beams more flexible. This indicated that the Poisson's effect is quite significant to composite beams with arbitrary lay-ups, and neglecting this effect is only suitable for cross-ply composite beams.

4 Conclusions

The authors proposed a new two-unknown model for vibration analysis of laminated composite beams. The axial and transverse displacements of the beam are expanded in a hybrid form under polynomial, and trigonometric series. Lagrange's equations are used to derive characteristic equations of the beams. Numerical results for laminated composite beams with different boundary conditions are compared with previous studies and investigate effects of length-to-depth ratio, material anisotropy, Poisson's ratio and fiber angles on the natural frequencies of laminated composite beams. The obtained results show that the normal strain effects are significant for un-symmetric and thick beams. The Poisson's effect is also important for composite beams with arbitrary lay-ups, and thus omitting this effects is only suitable for the cross-ply ones. The present model is found to be appropriate for vibration analyses of laminated composite beams.

Acknowledgments. This research is funded by Thu Dau Mot University (TDMU).

References

1. Komijani, M., et al.: An analytical approach for thermal stability analysis of two-layer Timoshenko beams. Int. J. Struct. Stab. Dyn. **13**(08), 1350036 (2013)
2. Nguyen, T.-K., et al.: A novel unified model for laminated composite beams. Compos. Struct. **238**, 111943 (2020)
3. Nguyen, T.-K., et al.: Trigonometric-series solution for analysis of laminated composite beams. Compos. Struct. **160**, 142–151 (2017)
4. Mantari, J., Canales, F.: Finite element formulation of laminated beams with capability to model the thickness expansion. Compos. B Eng. **101**, 107–115 (2016)
5. Nguyen, N.-D., et al.: Ritz-based analytical solutions for bending, buckling and vibration behavior of laminated composite beams. Int. J. Struct. Stab. Dyn. **18**(11), 1850130 (2018)
6. Murin, J., et al.: Modal analysis of the FGM beams with effect of the shear correction function. Compos. B Eng. **45**(1), 1575–1582 (2013)
7. Nguyen, T.N., Thai, C.H., Nguyen-Xuan, H.: On the general framework of high order shear deformation theories for laminated composite plate structures: a novel unified approach. Int. J. Mech. Sci. **110**, 242–255 (2016)
8. Thai, C.H., et al.: Isogeometric analysis of laminated composite and sandwich plates using a new inverse trigonometric shear deformation theory. Eur. J. Mech. A/Solids **43**, 89–108 (2014)
9. Mantari, J., Oktem, A., Soares, C.G.: A new trigonometric shear deformation theory for isotropic, laminated composite and sandwich plates. Int. J. Solids Struct. **49**(1), 43–53 (2012)
10. Aydoğdu, M.: A new shear deformation theory for laminated composite plates. Compos. Struct. **89**, 94–101 (2009)

11. El Meiche, N., et al.: A new hyperbolic shear deformation theory for buckling and vibration of functionally graded sandwich plate. Int. J. Mech. Sci. **53**(4), 237–247 (2011)
12. Akavci, S., Tanrikulu, A.: Buckling and free vibration analyses of laminated composite plates by using two new hyperbolic shear-deformation theories. Mech. Compos. Mater. **44**(2), 145 (2008)
13. Mantari, J., Oktem, A., Soares, C.G.: A new higher order shear deformation theory for sandwich and composite laminated plates. Compos. B Eng. **43**(3), 1489–1499 (2012)
14. Mohammadian, M.: Nonlinear free vibration of damped and undamped bi-directional functionally graded beams using a cubic-quintic nonlinear model. Compos. Struct. **255**, 112866 (2021). https://doi.org/10.1016/j.compstruct.2020.112866
15. Aydogdu, M.: Vibration analysis of cross-ply laminated beams with general boundary conditions by Ritz method. Int. J. Mech. Sci. **47**(11), 1740–1755 (2005)
16. Chandrashekhara, K., Krishnamurthy, K., Roy, S.: Free vibration of composite beams including rotary inertia and shear deformation. Compos. Struct. **14**(4), 269–279 (1990)
17. Khdeir, A., Reddy, J.: Free vibration of cross-ply laminated beams with arbitrary boundary conditions. Int. J. Eng. Sci. **32**(12), 1971–1980 (1994)
18. Vo, T.P., Thai, H.-T.: Vibration and buckling of composite beams using refined shear deformation theory. Int. J. Mech. Sci. **62**(1), 67–76 (2012)
19. Murthy, M., et al.: A refined higher order finite element for asymmetric composite beams. Compos. Struct. **67**(1), 27–35 (2005)
20. Mantari, J., Canales, F.: Free vibration and buckling of laminated beams via hybrid Ritz solution for various penalized boundary conditions. Compos. Struct. **152**, 306–315 (2016)
21. Matsunaga, H.: Vibration and buckling of multilayered composite beams according to higher order deformation theories. J. Sound Vib. **246**(1), 47–62 (2001)
22. Chen, W., Lv, C., Bian, Z.: Free vibration analysis of generally laminated beams via state-space-based differential quadrature. Compos. Struct. **63**(3–4), 417–425 (2004)
23. Chandrashekhara, K., Bangera, K.M.: Free vibration of composite beams using a refined shear flexible beam element. Comput. Struct. **43**(4), 719–727 (1992)
24. Aydogdu, M.: Free vibration analysis of angle-ply laminated beams with general boundary conditions. J. Reinf. Plast. Compos. **25**(15), 1571–1583 (2006)

Investigation of Random Vibrations of a Rigid Body on Vibration Dampers with Straightened Surfaces

K. Bissembayev[1](✉), Zh. Iskakov[2], and A. Sagadinova[3]

[1] Institute of Mechanics and Machine Science named after the Academician U.A. Dzholdasbekov, Abai Kazakh National Pedagogical University, Almaty, Kazakhstan
[2] Institute of Mechanics and Machine Science named after the Academician U.A. Dzholdasbekov, Almaty, Kazakhstan
[3] Abai Kazakh National Pedagogical University, Almaty, Kazakhstan

Abstract. Creation of vibration protection devices using rolling bearings is now widely used in transport equipment and seismic protection of structures.

Design of economical and reliable engineering construction of buildings and structures requires abandoning current deterministic approach to the calculation of structures and creating new methods based on statistical approach.

This work investigates random vibrations of a vibration-proof body on vibration dampers with straightened surfaces. Random vibrations of a rigid body on rolling bearings bounded by high-order surfaces of revolution with allowance made to rolling friction on relaxing soils were investigated using the apparatus for cumulant analysis of random non-Gaussian processes. The law of establishing the mean and second-order cumulants is determined for arbitrary initial conditions. The stability of stochastic vibration isolation systems with rolling bearings in terms of second-order moments - the so-called root-mean-square stability - has been investigated.

Keywords: Vibration protection · Vibration damper · Cumulative analysis · Random vibrations · Seismic isolation

1 Introduction

Rolling bearings, bounded by various surfaces of rotation as the main element, are used in many earthquake and vibration protection devices.

The idea of seismic isolation of buildings during earthquakes using a rolling bearing is one of the simplest and at the same time the most fruitful in the history of the development of earthquake-resistant construction.

The principle of operation of seismic protection devices on rolling bearings is to reduce the reaction force acting on the building, creating its movable base.

The works [1, 2] are devoted to the study of vibrations of a rigid body on rolling bearings bounded by high-order surfaces of rotation with allowance made to rolling friction on relaxing soils.

N. V. Khang et al. (Eds.): ASIAN MMS 2021, MMS 113, pp. 805–813, 2022.
https://doi.org/10.1007/978-3-030-91892-7_77

Seismic effects that occur during an earthquake are classified as random effects, though in practice, simplified design models are usually considered as deterministic. The magnitude and the nature of seismic impacts cannot be accurately predicted in advance. Instrumental records characterizing the consistent patterns of seismic impacts change in time, inherent in individual earthquakes, never repeat each other, even when they occur in the same place.

Most of the dynamic disturbances acting in buildings and structures are of a very complex and irregular nature. An adequate representation of such perturbations – and, consequently, of the vibrations of mechanical systems caused by them – is possible only within the framework of the theory of random processes.

A number of theoretical and experimental studies have been devoted to the design of structures for dynamic actions taking into account random parameters, but up to now there are many unresolved issues in this area.

The study of nonlinear systems based on the Fokker-Plank equation as applied to problems of statistical dynamics is considered in a number of monographs [3, 4]. They provide some exact solutions and set out a number of approximate approaches to the formation and solution of this equation.

Many approximate analytical methods are aimed at simplifying the Fokker-Plank equation. Here, approaches based on a combination of the theory of Markov processes with asymptotic methods of nonlinear mechanics prevail. Thus, on the basis of a combination of the theory of Markov processes with the method of stochastic averaging in [5–7], a number of problems on stability and oscillations of linear and nonlinear systems under the combined action of external and parametric broadband random influences were solved.

Some authors consider it more useful to obtain approximate solutions of the Fokker-Plank equation than to develop approximate methods for solving problems of statistical dynamics. The Fokker-Plank equation is mainly solved by approximate variational or numerical methods [8, 9]. In [10] it is proposed to replace the Fokker-Plank equation for a nonlinear system with an "equivalent" equation for a linear system excited by additive white noise. It is proposed to find the unknown coefficients of the equivalent system using the weighted residuals scheme.

In [11], equations have been obtained for the correlation function of a random process, which describes the motion of a vibration-protected body on rolling bearings with straightened surfaces. The obtained equations have been used to determine the probabilistic characteristics that describe the random vibrations of the vibration-proof body. An assessment of the effectiveness of vibration protection by rolling bearings is given. Calculation formulas have been determined for the impact represented by a random function of the "white noise" type. The probability density is found by the direct approximation method.

The main way to represent random variables and processes, in addition to distribution functions and characteristic functions, is to describe them by various statistical averages, including moments and moment functions. When analyzing transformations of random variables, especially nonlinear transformations, the method of transformation of probability distributions, moments and moment functions are of primary interest;

equations are written for them and with their help one or another statistical regularities of transformations are analyzed.

At the same time, there is another approach to the study of random variables and processes – their description using cumulants and cumulant functions, which are nonlinear combinations of statistical means.

Distribution cumulants in many respects are much more convenient distribution parameters than moments (including central ones). Among other reasons, this is due to the fact that in many practical important cases, the higher cumulants of distributions, in contrast to the moments, can be neglected. On the other hand, it is possible to consider such distributions, the cumulants of which, starting from a certain order, all vanish, while the moments are not equal to zero.

This paper studies nonlinear vibrations of vibration-protective systems with rolling bearings, limited by high-order surfaces of rotation, taking into account rolling friction on relaxing soils under random disturbances. Using the hypothesis of "quasi-Gaussian", the equation of evolution and the law of establishment of cumulants for nonlinear oscillations of vibration protection systems are obtained.

Statistical analysis is performed using the apparatus for cumulative analysis of random non-Gaussian processes.

2 Statement of the Problem

Consider a rolling support bounded from below and from above by surfaces given by equations $y_1 = a_1 x_1^n$ and $y_2 = a_2 x_2^m$ (Fig. 1).

When $n = m$ the equation of motion of a body on rolling bearings with straightened surfaces, taking into account rolling friction on relaxing bases, has the form

$$\ddot{X} + \varepsilon\dot{\Phi}(X) + \Phi(X) - \omega_0^2 X = -\ddot{x}_0(t) \tag{1}$$

where

$$X = x - x_0,\ \Phi(X) = \omega_0^2 N_n X^{\frac{1}{n-1}},\ N_n = \frac{1}{(nH)^{\frac{1}{n-1}}}\left[\frac{1}{a_1^{\frac{1}{n-1}}} + \frac{1}{a_2^{\frac{1}{n-1}}}\right],\ \omega_0^2 = \frac{g}{H} \tag{2}$$

$x_0(t)$ and $x(t)$– horizontal displacement of the bases and the upper body bearing on the rolling bearing, respectively. movement X– of the body on the rolling bearings relative to the base (see Fig. 1, Fig. 2). ε– damping coefficient (soil relaxation period), g– free-fall acceleration, H– support height, $\ddot{x}_0(t)$– stationary random functions of time with known static characteristics and mean values equal to zero.

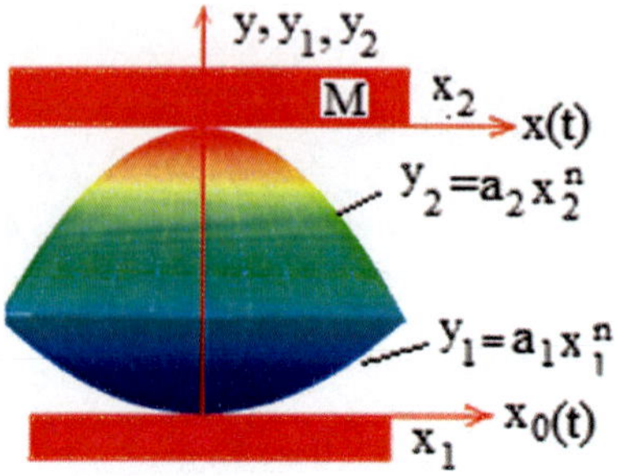

Fig. 1. Scheme of rolling bearers with bearing.

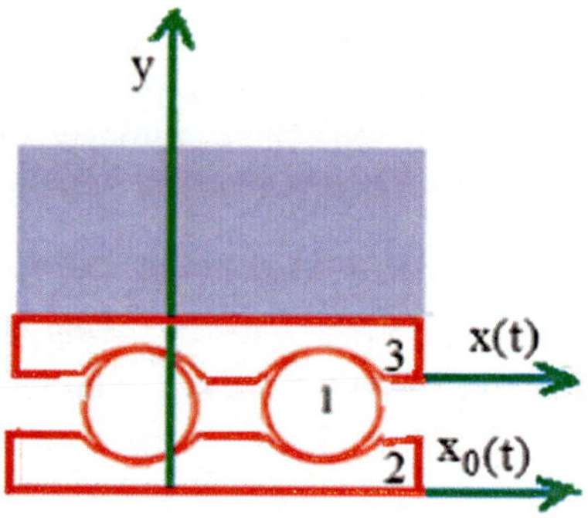

Fig. 2. Scheme of vibro-protective device on a movable base.

3 Analysis of Stochastic Vibrations of a Vibration-Proof Body

The difficulty of analysis of random oscillations of the system is associated with the presence of fractional degrees in the expressions of the restoring force and the rolling friction force. Therefore, in order to avoid these difficulties and to use the apparatus of cumulant analysis of random non-Gaussian processes for the analysis of random oscillations, transform Eq. (1), replacing the variables by setting

$$X = u^{n-1}, u = u_1, \frac{du_1}{dt} = u_2, u_3 = (n-1)u_1^{n-2}u_2$$

then instead of Eq. (1) the following system will be obtained

$$\begin{aligned} \dot{u}_1 &= u_2; \\ \dot{u}_3 &= -\varepsilon\omega_0^2 N_n u_2 - \omega_0^2 N_n u_1 + \omega_0^2 u_1^{n-1} - \ddot{x}_0(t); \\ u_3 &= (n-1)u_1^{n-2}u_2. \end{aligned} \tag{3}$$

Take the random process as stationary Gaussian white noise:

$$< x_0(t) >= 0, < x_0(t), x_0(t+\tau) >= D\delta(\tau)$$

Where, the parentheses $<>$ are statistical, and the parentheses with a comma separating the random variables $<\ ,\ >$ are the cumulative parentheses [3].

The evolution equations for the cumulants of the considered Markov process have the for

$$\begin{aligned} &\frac{d}{dt} < u_1 >=< u_2 > \\ &\frac{d}{dt} < u_1, u_1 >= 2 < u_1, u_2 > \\ &\frac{d}{dt} < u_3 >= -\varepsilon\omega_0^2 N_n < u_2 > -\omega_0^2 N_n < u_1 > +\omega_0^2 < u_1^{n-1} > \\ &\frac{d}{dt} < u_3, u_3 >= -2\varepsilon\omega_0^2 N_n < u_3, u_2 > -2\omega_0^2 N_n < u_3, u_1 > +\omega_0^2 < u_3, u_1^{n-1} > +D \end{aligned}$$

$$\frac{d}{dt} < u_1, u_3 >= -\varepsilon\omega_0^2 N_n < u_1, u_2 > -\omega_0^2 N_n < u_1, u_1 > +\omega_0^2 < u_1, u_1^{n-1} > + < u_3, u_2 >$$

....

$$< u_3 >= (n-1) < u_1^{n-2} u_2 >$$
$$< u_1, u_3 >= (n-1) < u_1, u_1^{n-2} u_2 >$$
$$< u_2, u_3 >= (n-1) < u_2, u_1^{n-2} u_2 >$$

.... (4)

The system of Eqs. (4) determines the law of setting of mean and second-order cumulants for arbitrary initial conditions:

$$m_1(0) = m_{10}, m_3(0) = m_{30}, \sigma_1^2(0) = \sigma_{10}^2, \sigma_3^2(0) = \sigma_{30}^2, \sigma_{13}^2(0) = \sigma_{13}^2,$$

Now, to solve the problem in the first approximation, the "quasi-Gaussian" hypothesis shall be used, according to which the unknown process has the following property: its odd moments are equal to zero, and the higher-order moments of even order are expressed through the second order moments.

If the average values $m_{10} = 0, m_{30} = 0$ are then $m_1(t) \equiv 0$ $m_3(t) \equiv 0$ for all $t > 0$. Setting of second-order cumulants in this case for all $n \geq 2$, according to (4) is described by the equation

$$\frac{d\sigma_1^2}{dt} = \frac{2\sigma_{13}^2}{(n-1)(n-1)!!\sigma_1^{n-2}},$$
$$\frac{d\sigma_3^2}{dt} = -\frac{2(n-1)!!\varepsilon\omega_0^2 N_n}{(n-1)^2(2n-5)!!}\frac{\sigma_3^2}{\sigma_1^{n-2}} + \frac{2(n-2)\varepsilon\omega_0^2 N_n}{(n-1)^2(n-1)!!}\frac{\sigma_{13}^4}{\sigma_1^n} - 2\omega_0^2 N_n \sigma_{13}^2$$
$$+ 2(n-1)!!\omega_0^2\sigma_1^{n-2}\sigma_{13}^2 + D,$$
$$\frac{d\sigma_{13}^2}{dt} = -\frac{\varepsilon\omega_0^2 N_n}{(n-1)(n-1)!!}\frac{\sigma_{13}^2}{\sigma_1^{n-2}} - \omega_0^2 N_n \sigma_1^2 + (n-1)!!\omega_0^2\sigma_1^n$$
$$+ \frac{(n-1)!!}{(n-1)^2(2n-5)!!}\frac{\sigma_3^2}{\sigma_1^{n-2}} - \frac{(n-2)}{(n-1)^2(n-1)!!}\frac{\sigma_{13}^4}{\sigma_1^n}. \quad (5)$$

It follows from (5) that the steady-state values of the second-order cumulants are determined by the equations

$$\sigma_{13}^2 = 0,\ \sigma_3^2 = \sigma_{\dot{X}}^2 = \frac{(n-1)^2(2n-5)!!\sigma_1^{n-2} D}{2(n-1)!!\varepsilon\omega_0^2 N_n},$$
$$\sigma_1^n - \frac{N_n}{(n-1)!!}\sigma_1^2 + \frac{D}{2(n-1)!!\varepsilon\omega_0^4 N_n} = 0. \quad (6)$$

To study the transient process of oscillatory systems, consider a special case, at $n = 4$ and restrict to the Gaussian approximation.

In the Gaussian approximation, Eqs. (4) for the first and second order cumulants take the form

$$\frac{dm_1}{dt} = m_2$$

$$\frac{d\sigma_1^2}{dt} = 2\sigma_{12}^2$$
$$\frac{dm_3}{dt} = -\varepsilon\omega_0^2 N_n m_2 - \omega_0^2 N_n m_1 + 3\omega_0^2 m_1 \sigma_1^2 + \omega_0^2 m_1^3$$
$$\frac{d\sigma_3^2}{dt} = -2\varepsilon\omega_0^2 N_n \sigma_{23}^2 - 2\omega_0^2 N_n \sigma_{13}^2 + 6\omega_0^2 \sigma_1^2 \sigma_{13}^2 + 6\omega_0^2 m_1^2 \sigma_{13}^2 + D$$
$$\frac{d\sigma_{13}^2}{dt} = -\varepsilon\omega_0^2 N_n \sigma_{12}^2 - \omega_0^2 N_n \sigma_1^2 + 3\omega_0^2 \sigma_1^4 + 3\omega_0^2 m_1^2 \sigma_1^2 + \sigma_{23}^2 \quad (7)$$

$$m_3 = 3m_2\sigma_1^2 + 3m_1^2 m_2 + 6m_1\sigma_{12}^2$$
$$\sigma_{13}^2 = 9\sigma_{12}^2\sigma_1^2 + 6\sigma_1^2 m_1 m_2 + 3m_1^2\sigma_{12}^2$$
$$\sigma_{23}^2 = 6\sigma_{12}^4 + 6\sigma_{12}^2 m_1 m_2 + 3\sigma_1^2\sigma_2^2 + 3m_1^2\sigma_2^2$$
$$\sigma_3^2 = 27\sigma_1^4\sigma_2^2 + 54m_1^2\sigma_1^2\sigma_2^2 + 9m_1^4\sigma_2^2 + 18\sigma_1^4 m_2^2 + 36m_1^2\sigma_1^2 m_2^2$$
$$+ 180m_1\sigma_1^2 m_2\sigma_{12}^2 + 36m_1^3 m_2\sigma_{12}^2 + 108\sigma_1^2\sigma_{12}^4 + 108m_1^2\sigma_{12}^4$$

The steady-state values of the cumulants can be found from the system of equations

$$m_2 = 0, m_3 = 0, \sigma_{12}^2 = 0, \sigma_{13}^2 = 0, m_1\left(-\omega_0^2 N_n + 3\omega_0^2\sigma_1^2 + \omega_0^2 m_1^2\right) = 0,$$
$$3\omega_0^2\sigma_1^4 - \omega_0^2 N_n\sigma_1^2 + 3\omega_0^2 m_1^2\sigma_1^2 + \frac{D}{2\varepsilon\omega_0^2 N_n} = 0,$$
$$\sigma_2^2 = \frac{D}{6\varepsilon\omega_0^2 N_n\left(\sigma_1^2 + m_1^2\right)}, \sigma_3^2 = \frac{\left(27\sigma_1^4 + 54m_1^2\sigma_1^2 + 9m_1^4\right)D}{6\varepsilon\omega_0^2 N_n\left(\sigma_1^2 + m_1^2\right)}. \quad (8)$$

For the steady-state value of the mean $m_1 = 0$, the steady-state values of variance correspond

$$\sigma_1^2 = \frac{N_n}{6} \pm \sqrt{\frac{N_n^2}{36} - \frac{D}{6\varepsilon\omega_0^4 N_n}}, \sigma_3^2 = \frac{9\sigma_1^2 D}{2\varepsilon\omega_0^2 N_n} \quad (9)$$

if $m_1 = \sqrt{N_n - 3\sigma_1^2}$ then

$$\sigma_1^2 = \frac{N_n}{6} + \sqrt{\frac{N_n^2}{36} + \frac{D}{12\varepsilon\omega_0^4 N_n}}, \sigma_3^2 = 9\left[12\sigma_1^4 - 18\sigma_1^2 + N_n^2\right]\frac{D}{6\varepsilon\omega_0^2 N_n\left(N_n - 3\sigma_1^2\right)}. \quad (10)$$

The statistical characteristics of the process $X(t)$ are determined by a non-linear transformation $X(t) = u_1^{n-1}(t)$. With a Gaussian input $u_1(t)$, the output process $X(t) = u_1^{(n-1)}(t)$ has the following mean value (see [3]):

$$m_x(t) = m_1^{n-1}(t) + (n-1)(n-2)m_1^{n-3}(t)\frac{\sigma_1^2(t)}{2} + \ldots + \frac{2(n-1)!}{(n-1)!}\frac{\sigma_1^{2(n-1)}(t)}{2^{n-1}}$$

and mean-square value

$$< X^2(t) >= m_x^{2(n-1)}(t) + \sum_{s=1}^{n-1} \frac{1}{S!}\left[\frac{d^{2S}}{du_1^{2S}} u_1^{2(n-1)}(t)\right]_{u_1=m_1(t)} \left(\frac{\sigma_1^2(t)}{2}\right)^S$$

if $m_{\dot{x}}(t) \equiv 0$, then

$$\sigma_x^2(t) = \frac{2(n-1)!}{(n-1)!} \frac{\sigma_1^{2(n-1)}(t)}{2^{n-1}} = (2n-3)!!\sigma_1^{2(n-1)}(t) \tag{11}$$

4 Results and Discussion

From relations (6) and (11), it is possible to determine the average value of the frequency of the vibration-protected object.

$$\omega_{cp}^2 = \frac{\sigma_{\dot{x}}^2}{\sigma_x^2} = \frac{(n-1)^2}{(2n-3)}\omega_0^2\left(\frac{\alpha_n N_n}{\sigma_x^{\frac{n-2}{n-1}}} - 1\right)$$

where $\alpha_n = \frac{[(2n-3)!!]^{\frac{1}{2(n-1)}}}{(n-1)!!}$.

The dependences of the average value of the frequency of the vibration-protected object on the dispersion coincide in nature with the frequency-amplitude characteristics of a determined nonlinear system with free oscillations.

The dispersions of displacement σ_X^2, velocity $\sigma_{\dot{X}}^2$ and the average value ω_{av} of vibration frequency of the body, on the rolling bearings, determined by formulas (6) and (11) coincide with the results obtained in [11] by the correlation method.

Transient modes of establishing the mean value and variance under arbitrary initial conditions m_{i0} and σ_{i0}^2 are described by Eqs. (9). The system of Eqs. (9) is numerically solved by the Runge-Kutta method for the following values of the rolling support parameters $n = 4$; $a_1 = 6,25 \cdot 10^{-8}\,\text{cm}^{-3}$; $a_2 = 15 \cdot 10^{-8}\,\text{cm}^{-3}$; $H = 300\,\text{cm}$; $\omega_0^2 = 3,261/c^2$.

Figure 3, 4, 5 shows settings of the mean value and variance of the variables u_1 for $D_1 = 60$, $D_2 = 80$, $D_3 = 120$ and $\varepsilon_1 = 0,03$; $\varepsilon = 0,04$; $\varepsilon_2 = 0,4$ under the initial conditions $m_{10} = 0,03$; $m_{130} = 0$; $m_{30} = 0$; $\sigma_{10}^2 = 0,053$;$\sigma_{30}^2 = 0$; $\sigma_{130}^2 = 0$.. The mean value of the variables u_1 for all D tends to zero according to the law of damped oscillations. The setting process $\sigma_1^2(t)$ will proceed in a more complex manner.

Figure 5 shows a graph of dependence of the variance on u_1 time under the action of a non-stationary kinematic excitation of the type of white noise, the intensity of which changes according to the law $D(t) = a + b\sin pt$; $D(t) = a_1 - b_1 t$, where $a = 80$; $b = 20$; $a_1 = 40$; $b_1 = 6,06$.

Using formula (10), it is possible to construct the dependence of the dispersion on σ_1^2 the intensity of kinematic excitation for the steady-state condition with the value $n = 4$; 6; 8 (Fig. 6).

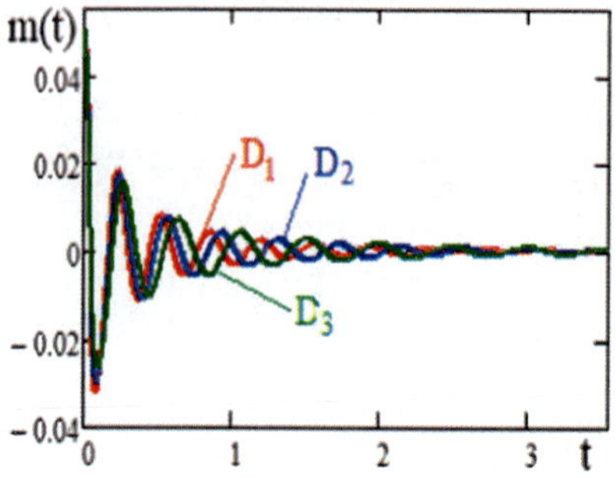

Fig. 3. Time dependence of the first-order cumulant for different values of the intensity of disturbance

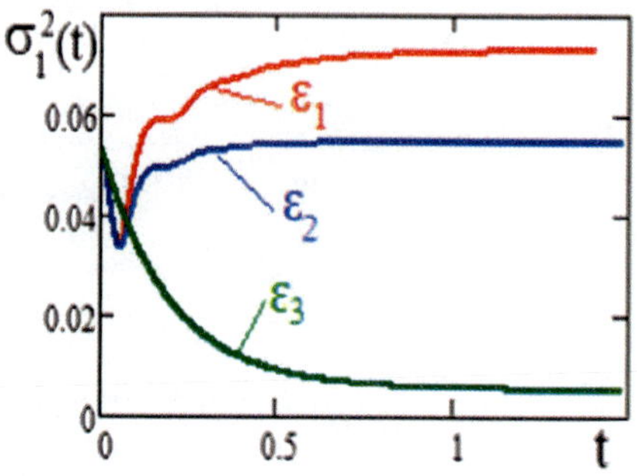

Fig. 4. Time dependence of the second-order cumulant for different values of the rolling friction coefficient

In Fig. 5, the dotted lines show the boundaries of the stability region.

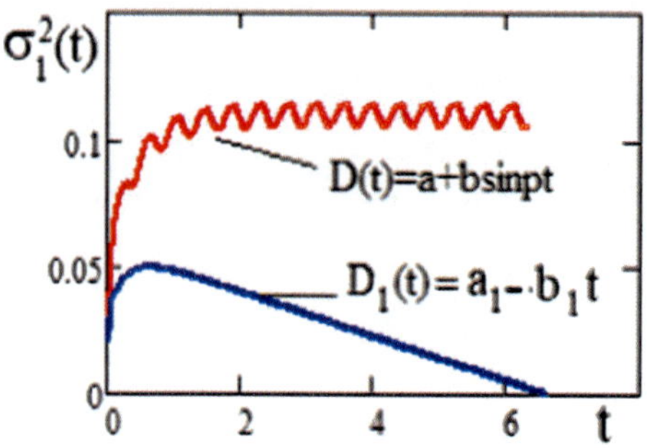

Fig. 5. Time dependence of the second-order cumulant with variable intensity of disturbance

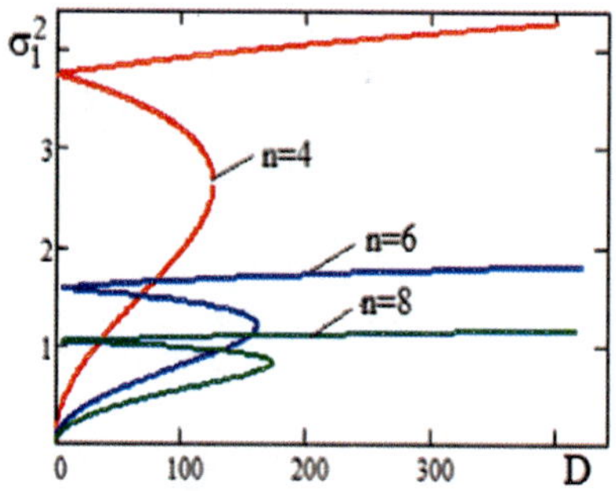

Fig. 6. Dependence of the second-order cumulant on the intensity for different value of the order of the rolling bearing surface

5 Conclusions

An analytical technique for studying the stationary and non-stationary process of random vibrations of a rigid body on rolling bearings with straightened surfaces has been developed. Statistical analysis of random vibrations of a vibration-protected body is performed using the apparatus for cumulative analysis of random non-Gaussian processes.

The average value of the frequency of the vibration-proof object has been determined. It has been established that the dependences of the average value of the frequency of the vibration-protected object on the dispersion coincide in nature with the frequency-amplitude characteristics of a determined nonlinear system with free oscillations.

Transient modes of establishment of the mean value and variance for arbitrary initial conditions have been investigated.

Investigated for stability of stochastic vibration protection devices in the second-order moments – the so-called stability in the mean-square value. It has been found that the steady motion of stochastic systems is asymptotically stable in the mean-square value under the following conditions $\sigma_{1c}^{n-2} \leq \frac{2N_n}{n(n-1)!!}$.

Funding. This research is funded by the Science Committee of the Ministry of Education and Science of the Republic of Kazakhstan (Grants No. AP08856744).

References

1. Bissembayev, K., Jomartov, A., Tuleshov, A., Dikambay, T.: Analysis of the oscillating motion of a solid body on vibrating bearers. Machines **7**(58), 1–22 (2019). https://doi.org/10.3390/machines7030058
2. Bissembayev, K., Smanov, A.: Stochastic oscillations of a solid body with a kinematic system of vibration isolation. In: Uhl, T. (eds.) Advances in Mechanism and Machine Science. IFToMM WC 2019. Mechanisms and Machine Science, vol. 73, pp.4205–4215. Springer, Cham (2019). https://doi.org/10.1007/978-3-030-20131-9_419
3. Malakhov, A.N.: Cumulant analysis of random non-Gaussian processes and their transformations, p. 376. Sovetskoe Radio (1978)
4. Makarov, B.P.: Nonlinear problems of statistical dynamics of machines and devices, p. 262. Mechanical engineering (1983)
5. Dimentberg, M.F.: Nonlinear stochastic problems of mechanical vibrations, p. 368. Nauka, Moscow (1980)
6. Dimentberg, M.F.: Random processes in dynamic systems with variable parameters, p. 176. Nauka (1989)
7. Nikolayenko, N.A., Uliyanov, S.V.: Statistical dynamics of engineering structures, p. 368. Mechanical Engineering (1977)
8. Sapsis, T.P., Vakakis, A.F., Bergman, L.A.: Effect of stochasticity on targeted energy transfer from a linear medium to a strongly nonlinear attachment. Probab. Eng. Mech. **26**(2), 119–133 (2011). https://doi.org/10.1016/j.probengmech.2010.11.006
9. Cross, E.J., Worden, K.: Approximation of the Duffing oscillator frequency response function using the FPK equation. J. Sound Vib. **330**(4), 743–756 (2011). https://doi.org/10.1016/j.jsv.2010.08.034
10. Lacquanti, S., Riccardi, G.: A probabilistic linearization method for nonlinear systems subjected to additive an multiplicative excitations. Int. J. Non-Linear Mech. **41**(10), 1191–1205 (2006). https://doi.org/10.1016/j.ijnonlinmec.2006.12.002
11. Bissembayev, K.B.: Statistical studies of the oscillatory motion of a rigid body on rolling bearings with straightened surfaces in the presence of rolling friction on relaxing soils. Bull. Natl. Eng. Acad. Repub. Kazakhstan Almaty **3**(37), 60–68 (2010)

Application of Finite Element Method to Analyze the Vibration and Dynamic Impact Factor of Displacement in I-Girder Bridge with Link Slab Due to Random Vehicle Load

Toan Nguyen-Xuan, Thuat Dang-Cong, Loan Nguyen-Thi-Kim(✉), and Thao Nguyen-Duy

The University of Danang – University of Science and Technology, Danang, Vietnam
ntkloan@dut.udn.vn

Abstract. I-girder bridge with link slab is the solution to reduce the number of expansion joints and improve the exploitability of the bridge. However, studies on the vibration of this type of structure under the effect of moving vehicles are limited. This paper presents some analysis results of vibration and dynamic impact factor (DIF) of displacement in I-girder bridge with link slab subjected to random vehicle load. The structural bridge consisting of three spans of I-girder with link slab is simulated by the finite element method. The moving load has three axes and is idealized by 6 degrees of freedom. The study is applied to Song Dinh bridge in Phan Thiet - Dau Giay highway construction Project, Vietnam. The obtained results of the vibration and DIF analysis are quite large compared to the values in current bridge design codes of Vietnam and the United States. The ruler of probability distribution of random variables about the DIF of displacement is not the same as the one of the input load. This proves that the mapping relationship between the DIF and the vehicle load has nonlinear signs.

Keywords: Vibration · Dynamic impact factor (DIF) · Finite element method · Random vehicle load · I-girder bridge with link slab

1 Introduction

The problem of bridge vibration subjected to moving vehicles has been studied since the 50s of the 19th century, Willlis [1]. Substantial dynamic effects can be triggered by the increasing proportion of heavy and high-speed vehicles in highway traffic [2, 3]. The vibration of the bridge structure due to vehicle load must include the influence of factors such as vehicle speed, vehicle weight, road unevenness, vehicle braking force, etc.... Law and Zhu [4] studied the dynamic behavior of continuous three spans under moving vehicle taking into account the braking force and the roughness of the road surface. X.T. Nguyen et al. [5, 6] showed that the dynamic vehicle – bridge interaction considering the braking force, the dynamic impact factor (DIF) increased significantly. Deng [7], X.T. Nguyen et al. [8] provided that the random roughness of the road affects the vibration of the bridge structure.

N. V. Khang et al. (Eds.): ASIAN MMS 2021, MMS 113, pp. 814–824, 2022.
https://doi.org/10.1007/978-3-030-91892-7_78

The DIF plays a vital role in the practice of bridge design and condition assessment. Frýba [9] showed that dynamic vehicle-bridge interaction results in an increase or decrease of the bridge deformation, which is described by the DIF that reflects how many times the constant load must be multiplied to cover additional dynamic effects. There is a large number of studies on the DIF including experimental impact factors, analytical methods and code specifications [7, 10–12]. The research indicated that the DIF of different bridge responses is different.

In general, most of the publications studied the dynamic vehicle – bridge interaction for simple or continuous girder bridges subjected to the predetermined loads. The most common model applied to beams or plates, traversed by either concentrated [13, 14] or distributed loads [15]. The bridge structures with continuous deck link slab are created by series of girder spans or simple slab girders connected bridge decks together, an effective alternative of replacing the traditional expansion joint with a link slab. However, due to the complexity of the structure, the problem of studying the vibration of the multi–span slab beam bridge with link deck considering the actual vehicle load is still very limited.

In the initial study, the authors present the model of vehicle-bridge interaction by using numerical methods to analyze the vibration and DIF of displacement in three-span bridge structure with link slab under the influence of random vehicle load. Actual data on vehicle loads crossing the bridge is collected at Dau Giay weighing station. Numerical simulations were performed for Song Dinh bridge of Phan Thiet - Dau Giay highway construction Project, Vietnam.

2 The Model of Vehicle – Bridge Interaction and the Differential Equations of Motion

2.1 Computational Models and the Differential Equations of Motion

The reinforced concrete I-girder bridge structure consisting of three spans with link slab subjected to the vehicle loads is described in Fig. 1. The vehicle load is modeled with three axles. Assume the mass of the vehicle including the good conveyed to the ith axle is m_{1i}, the mass of ith axle is m_{2i}. The dynamic interaction model between a three-axle vehicle and a girder element is described in Fig. 2.

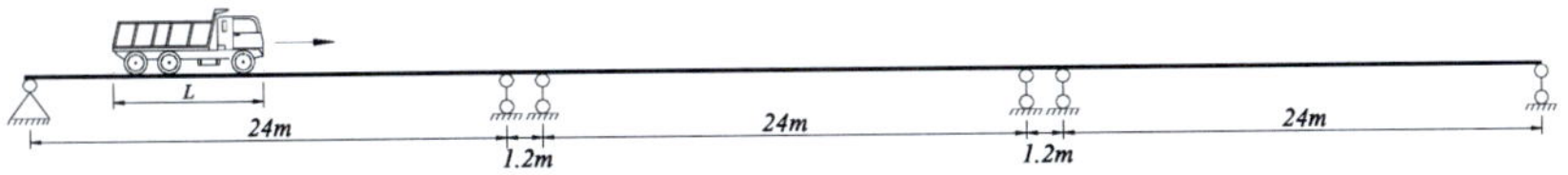

Fig. 1. Schematic of vehicle moving on girder bridge

In Fig. 2, $G_i.\sin\psi_i$ is the engine excitation force at i^{th} axle, with an assumption of a harmonic function; k_{1i}, d_{1i} are the spring and daspot of suspension at i^{th} axle, respectively; k_{2i}, d_{2i} are the spring and damping of tire at i^{th} axle, respectively; L is the length of the girder elements; $w_i(x_i, t)$ is the vertical displacement of girder element at i^{th} axle of vehicle; z_{1i} is the vertical displacement of chassis at i^{th} axle of vehicle; z_{2i} is the vertical displacement of ith axle of vehicle; y_{1i} is the relative displacement between the chassis and i[th] axle; y_{2i} is the relative displacement between i[th] axle and girder element; x_i is the coordinate of the ith axle of the vehicle at time t (i = 1, 2, 3).

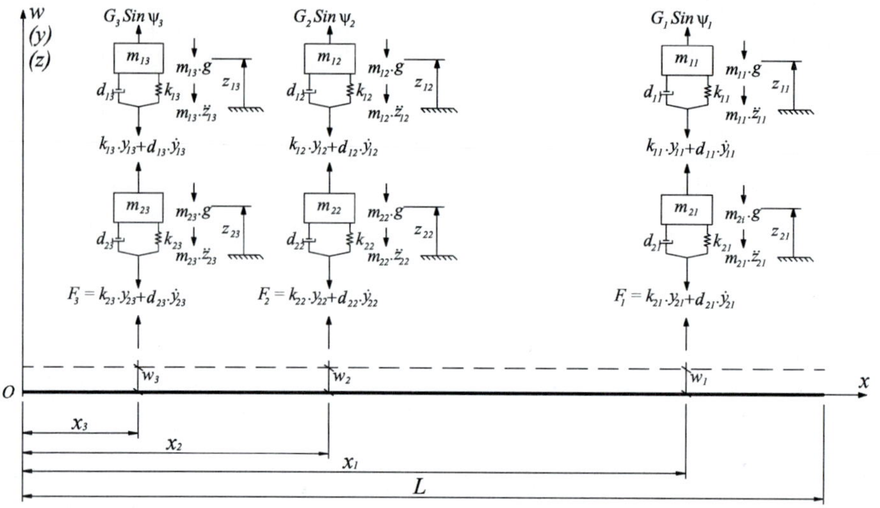

Fig. 2. Schematic of vehicle-bridge interaction

According to [5] and [16], the differential equations of girder element due to a distributed load p (x, z, t) taking into account the influence of internal and external friction, is presented as follows:

$$\frac{\partial^2 w}{\partial x^2}\left[EJ_d \cdot \left(\frac{\partial^2 w}{\partial x^2} + \theta \cdot \frac{\partial^3 w}{\partial x^2 . \partial t}\right)\right] + \rho F_d \cdot \frac{\partial^2 w}{\partial t^2} + \beta . \frac{\partial w}{\partial t} = p(x, z, t) \quad (1)$$

where EJ_d is the bending stiffness of the girder element; ρF_d is the mass per unit length of girder; θ and β are the coefficients of internal friction and external friction of the girder element, respectively; *p (x, z, t)* is the uniform loading over the girder element horizontally.

Using d'Alembert's principle, the dynamic equilibrium of each mass m_{1i}, m_{2i} on the vertical axis can be written as follows:

$$m_{1i}.\ddot{z}_{1i} + d_{1i}.\dot{z}_{1i} + k_{1i}.z_{1i} - d_{1i}.\dot{z}_{2i} - k_{1i}.z_{2i} = G_i. \sin \Psi_i - m_{1i}.g$$
$$m_{2i}.\ddot{z}_{2i} + (d_{1i} + d_{2i}).\dot{z}_{2i} + (k_{1i} + k_{2i}).z_{2i} - d_{1i}.\dot{z}_{1i} - k_{1i}.z_{1i} = -m_{2i}.g + d_{2i}\dot{w}_i + k_{2i}w_i \quad (2)$$

Adding on the logic control function $\xi_i(t)$ [17], Eq. (2) can be written as follows:

$$\xi_i(t).[m_{1i}.\ddot{z}_{1i} + d_{1i}.\dot{z}_{1i} - d_{1i}.\dot{z}_{2i} + k_{1i}.z_{1i} - k_{1i}.z_{2i}] = \xi_i(t).\left[G_i. \sin \psi_i - m_{1i}.g\right] \quad (3)$$

$$\xi_i(t)[m_{2i}.\ddot{z}_{2i} - d_{1i}.\dot{z}_{1i} + (d_{1i} + d_{2i})\dot{z}_{2i} - k_{1i}.z_{1i} + (k_{1i} + k_{2i})z_{2i}] = \xi_i(t)\left[k_{2i}w_i + d_{2i}\dot{w}_i - m_{2i}.g\right]$$

$$\xi_i(t) = \begin{cases} 1 \;\; khi \;\; t_i \le t \le t_i + T_i \\ 0 \;\; khi \;\; t < t_i \;\; and \;\; t > t_i + T_i \end{cases} ; \; T_i = \frac{L}{v_i} \quad (4)$$

From Fig. 2, the contact force between the ith axle and girder element is described by:

$$F_i = k_{2i}.y_{2i} + d_{2i}.\dot{y}_{2i} \tag{5}$$

The combined Eq. (1), (3) and (5), the differential equations of girder element due to a distributed load p(x, z, t) can be rewritten as follows:

$$EJ_d.\left(\frac{\partial^4 w}{\partial x^4} + \theta.\frac{\partial^5 w}{\partial x^4.\partial t}\right) + \rho F_d.\frac{\partial^2 w}{\partial t^2} + \beta.\frac{\partial w}{\partial t} = p(x, z, t)$$

$$p(x, z, t) = \sum_{i=1}^{N} \xi_i(t).[G_i \sin \Psi_i - (m_{1i} + m_{2i}).g - m_{1i}.\ddot{z}_{1i} - m_{2i}.\ddot{z}_{2i}].\delta(x - a_i) \tag{6}$$

where $\delta(x - a_i)$ is the Dirac delta function, i = 1 to N (N is number of axles, N = 3).

The Galerkin method and Green theory are applied to Eq. (1) to transform these equations into matrix forms, and the differential equations of girder element and vehicle can be written as follows:

$$M_e.\ddot{q} + C_e.\dot{q} + K_e.q = f_e \tag{7}$$

where M_e, C_e, K_e are the mass matrix, damping matrix, and stiffness matrix of the system including girder and vehicle, respectively:

$$M_e = \begin{bmatrix} M_{ww} & M_{wz1} & M_{wz2} \\ 0 & M_{z1z1} & 0 \\ 0 & 0 & M_{z2z2} \end{bmatrix};\ C_e = \begin{bmatrix} C_{ww} & 0 & 0 \\ 0 & C_{z1z1} & C_{z1z2} \\ C_{z2w} & C_{z2z1} & C_{z2z2} \end{bmatrix};\ K_e = \begin{bmatrix} K_{ww} & 0 & 0 \\ 0 & K_{z1z1} & K_{z1z2} \\ K_{z2w} & K_{z2z1} & K_{z2z2} \end{bmatrix} \tag{8}$$

$\ddot{q}, \dot{q}, q, f_e$ are the complex acceleration vector, complex velocity vector, complex displacement vector, complex forces vector of the system including girder and vehicle, respectively;

$$\{\ddot{q}\} = \begin{Bmatrix} \ddot{W} \\ \ddot{Z}_1 \\ \ddot{Z}_2 \end{Bmatrix};\ \{\dot{q}\} = \begin{Bmatrix} \dot{W} \\ \dot{Z}_1 \\ \dot{Z}_2 \end{Bmatrix};\ \{q\} = \begin{Bmatrix} W \\ Z_1 \\ Z_2 \end{Bmatrix};\ \{f_e\} = \begin{Bmatrix} F_w \\ F_{z1} \\ F_{z2} \end{Bmatrix} \tag{9}$$

$W = \left[u_1\ \varphi_1\ u_2\ \varphi_2\right]^T$ is the node displacement vector of the beam element in the local coordinate system. u_1, φ_1, u_2, φ_2 are the vertical displacement, rotation displacement on the left and right node of the beam element, respectively.

M_{ww}, C_{ww} and K_{ww} are mass, damping and stiffness matrices of the girder elements, respectively. They can be found in Zienkiewicz [18].

$$M_{z1z1} = \lceil m_{11} \quad m_{12} \ldots m_{1i} \ldots \quad m_{1N} \rfloor;\ M_{z2z2} = \lceil m_{21} \quad m_{22} \ldots m_{2i} \ldots \quad m_{2N} \rfloor; \tag{10}$$

$$M_{wz1} = P.M_{z1z1};\ M_{wz2} = P.M_{z2z2} \tag{11}$$

$$P = \begin{bmatrix} P_{11} & P_{12} & \cdots & P_{1i} & \cdots & P_{1n} \\ P_{21} & P_{22} & \cdots & P_{2i} & \cdots & P_{2n} \\ P_{31} & P_{32} & \cdots & P_{3i} & \cdots & P_{3n} \\ P_{41} & P_{42} & \cdots & P_{4i} & \cdots & P_{4n} \end{bmatrix}_{(4\times n)} \quad P_i = \begin{Bmatrix} p_{1i} \\ p_{2i} \\ p_{3i} \\ p_{4i} \end{Bmatrix} = \frac{\xi_i(t)}{L^3} \cdot \begin{Bmatrix} (L+2a_i)(L-a_i)^2 \\ L.a_i(L-a_i)^2 \\ a_i^2(3L-2a_i) \\ -L.a_i^2(L-a_i) \end{Bmatrix} \tag{12}$$

with P_i is the ith column of the matrix

$$C_{z1z1} = \left\lceil d_{11} \quad d_{12} \ldots d_{1i} \ldots \quad d_{1N} \right\rfloor; \; C_{z2} = \left\lceil d_{21} \quad d_{22} \ldots d_{2i} \ldots \quad d_{2N} \right\rfloor \tag{13}$$

$$C_{z1z2} = C_{z2z1} = -C_{z1z1}; \; C_{z2z2} = C_{z1z1} + C_{z2}; \; C_{z2w} = (N_a.C_{z2})^T; \tag{14}$$

$$K_{z1z1} = \left\lceil k_{11} \quad k_{12} \ldots k_{1i} \ldots \quad k_{1N} \right\rfloor; \; K_{z2} = \left\lceil k_{21} \quad k_{22} \ldots k_{2i} \ldots \quad k_{2N} \right\rfloor \tag{15}$$

$$K_{z1z2} = K_{z2z1} = -K_{z1z1}; \; K_{z2z2} = K_{z1z1} + K_{z2}; \; K_{z2w} = (N_a.K_{z2})^T + (\dot{N}_a.C_{z2})^T; \tag{16}$$

$$N_a = \begin{bmatrix} N_{11} \, N_{12} \ldots N_{1i} \ldots N_{1N} \\ N_{21} \, N_{22} \ldots N_{2i} \ldots N_{2N} \\ N_{31} \, N_{32} \ldots N_{3i} \ldots N_{3N} \\ N_{41} \, N_{42} \ldots N_{4i} \ldots N_{4N} \end{bmatrix}_{(4xN)} \text{ where } \left. \begin{aligned} N_{1i} &= \frac{1}{L^3}.(L^3 - 3.L.a_i^2 + 2.a_i^3) \\ N_{2i} &= \frac{1}{L^2}.(L^2.a_i - 2.L.a_i^2 + a_i^3) \\ N_{3i} &= \frac{1}{L^3}.(3.L.a_i^2 - 2.a_i^3) \\ N_{4i} &= \frac{1}{L^2}.(a_i^3 - L.a_i^2) \end{aligned} \right\} \tag{17}$$

$$F_w = \left[F_1 \ldots F_i \ldots F_4 \right]^T; \; F_i = \sum_{i=1}^{N} [G_i \sin \psi_i - (m_{1i} + m_{2i}).g].P_i; \tag{18}$$

$$F_{z1} = \begin{Bmatrix} G_1 \sin \Psi_1 - m_{11} g \\ \vdots \\ G_i \sin \Psi_i - m_{1i} g \\ \vdots \\ G_N \sin \Psi_N - m_{1N} g \end{Bmatrix}; \; F_{z2} = \begin{Bmatrix} -m_{21}.g \\ \vdots \\ -m_{2i}.g \\ \vdots \\ -m_{2N}.g \end{Bmatrix}; \tag{19}$$

The case that the axle load $m_{axle\ i}$ is a random quantity taken from the vehicle weighing stations. Therefore, the quantity $m_{1i} = m_{axle\ i} - m_{2i}$ in the Eqs. (2), (3), (6), (10), (18) and (19) is random, and other factors are considered as deterministic.

3 Application to Analyze Vibration of Song Dinh Bridge

3.1 Properties of Structure Bridge and Vehicle

Song Dinh Bridge belongs the Phan Thiet - Dau Giay highway construction project, which is a three-span I girder bridge with link slab. The length of each span is 24 m. The

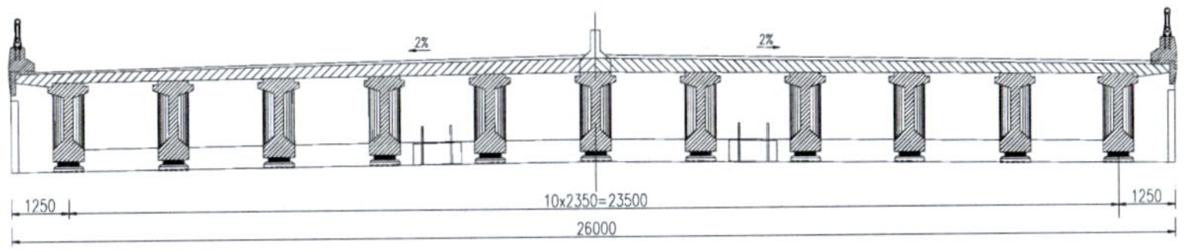

Fig. 3. Cross section of Song Dinh Bridge

deck slab is continuously reinforced concrete, with a flexible joint of 1.2 m in length. The cross-section of the bridge consists of 11 prestressed reinforced concrete I girder as shown in Fig. 3.

Using the finite element method, the bridge structure was discrete as Fig. 4.

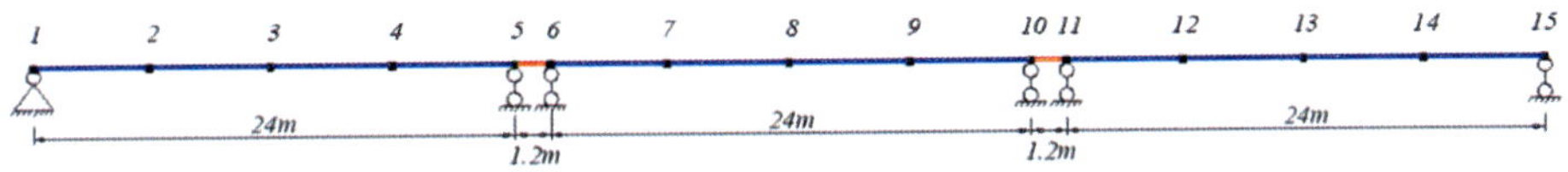

Fig. 4. Schematic of discrete bridge structure

The properties of slab beam are extracted from design documents. The properties of three-axle Foton are given by the manufacturing company. The parameters of the slab beam and vehicle used in the vibration analysis are listed in Table 1.

Table 1. Properties of slab beam and vehicle Foton

Item	Notation	Unit	Value	Item	Notation	Unit	Value
Slab beam concrete				***Dumper truck vehicle***			
Lenght	L	m	24	Mass m_{11}	m_{11}	T	5.2
Young's modulus	E	T/m^2	3230769.2	Mass m_{21}	m_{21}	T	0.26
Density	ρ	kg/m^3	2500	Mass m_{12},m_{13}	m_{12},m_{13}	T	8.9
Coefficient of internal friction	θ	-	0.027	Mass m_{22},m_{23}	m_{22},m_{23}	T	0.87
Coefficient of external friction	β	-	0.01	Suspensions spring k_{11}	k_{11}	T/m	120
Cross sectional área	A	m^2	0.9522	Suspensions spring k_{12};k_{13}		T/m	260
Second moment of area	I	m^4	0.30921	Tires spring k_{21}	k_{21}	T/m	240
				Tires spring k_{21};k_{22};k_{24}		T/m	380
Link slab (deck)				Suspensions dashpot d_{11}	d_{11}	Ts/m	0.7344
Height	h	m	0.02	Suspensions dashpot d_{12};d_{13}		Ts/m	0.3672
Cross sectional area	A	m^2	0.47	Tires dashpot d_{21}	d_{21}	Ts/m	0.4
Second moment of area	I	m^4	0.001567	Tires dashpot d_{22};d_{23}		Ts/m	0.8

3.2 Vibration Analysis of the Bridge

Vibration analysis of a three-span I-girder bridge with link slab due to vehicle load are performed with speed of 10 m/s. The input data is as in Table 1. The displacements at each node were investigated with 2 components: vertical displacement (U_y) and rotational

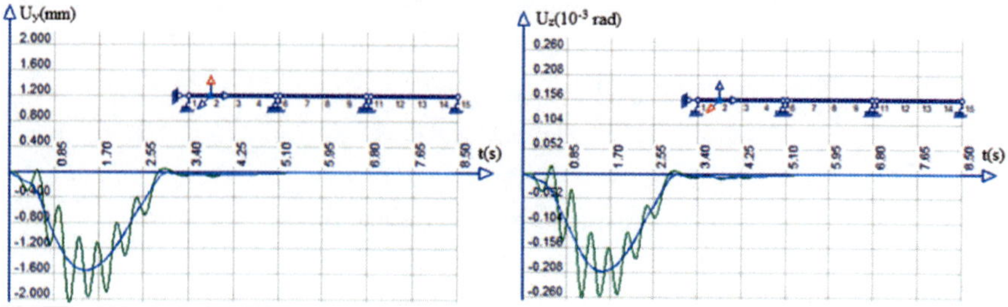

a. Vertical displacement b. Rotational displacement

Fig. 5. Vibration of U_y and U_z displacement at node 2

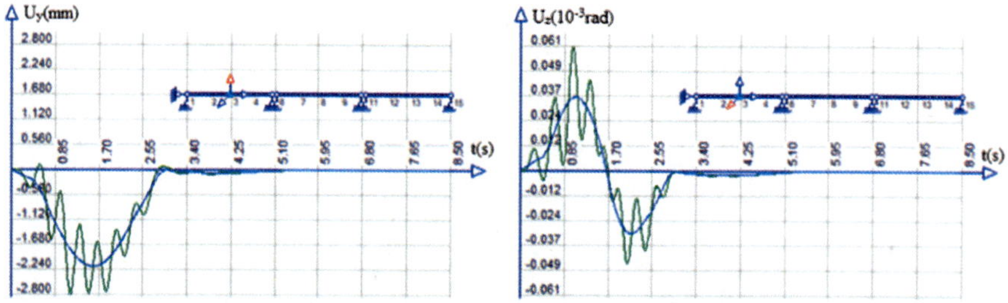

a. Vertical displacement b. Rotational displacement

Fig. 6. Vibration of U_y and U_z displacement at node 3

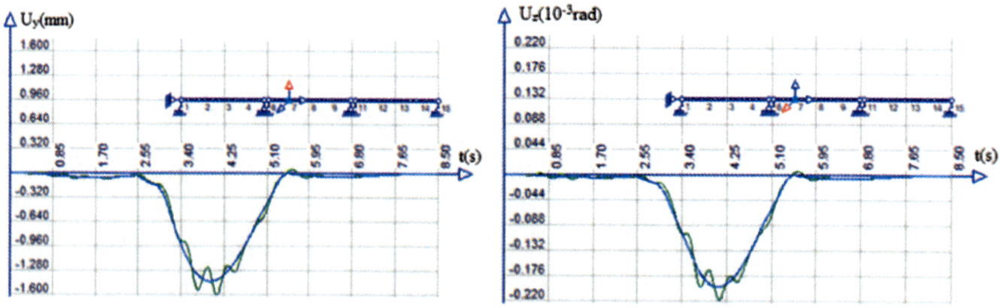

a. Vertical displacement b. Rotational displacement

Fig. 7. Vibration of U_y and U_z displacement at node 7

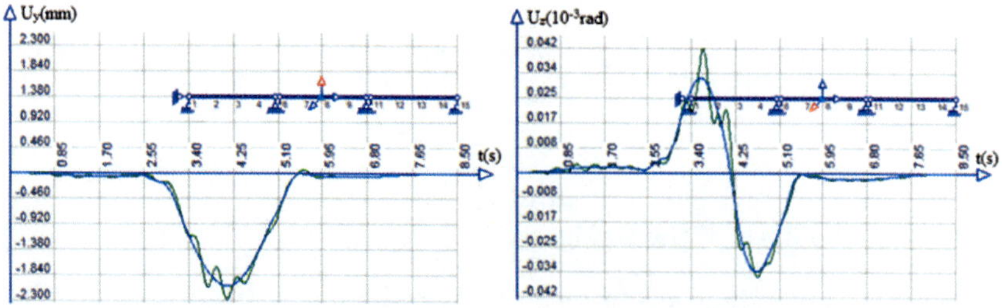

a. Vertical displacement b. Rotational displacement

Fig. 8. Vibration of U_y and U_z displacement at node 8

displacement (U_z). The results of results on the vibration of displacement at nodes 2 and 3 of the first span, nodes 7 and 8 of the second span are shown in Figs. 5, 6, 7 and 8.

From the Figures, it can be seen that the influence of vehicle load on each span is quite clear. When the vehicle moves on the investigated span, the displacement is much larger than that in the case of the vehicle passing through adjacent spans. This shows that the influence of the load on the surveyed span is very small when the vehicle is running at other spans. Although the bridge deck is continuous, the span structure still works quite independently of each other.

The results show that the dynamic displacements are harmonically distributed around the static analysis results. After a period of time, the system's vibration decreased rapidly due to the influence of the damping force, the analysis results of dynamic displacement converge on the static analysis results. That is entirely consistent with the theory of analysis vibration of the structure taking into account damping force.

3.3 Dynamic Impact Factor Analysis of the Bridge

The dynamic impact factor (DIF) or (1 + IM) of three spans of I girder with link slab is determined according to Eq. (20):

$$(1 + IM) = DIF = \frac{S_{D\,max}}{S_{S\,max}} \tag{20}$$

where S_{Dmax}, S_{Smax} - the absolute maximum displacements of dynamic and the static at the same point, respectively.

In this study, the vibration and DIF of Song Dinh bridge under the effect of three-axle vehicle load were analyzed, in which, the vehicle load and the axles are random variables. Vehicle loads and axles are enriched based on actual vehicle weighing data at Dau Giay weighing station.

On the basis of vehicle weighing data at Dau Giay weighing station, Nguyen Lan [19] estimated the probability law for the load of the three-axle vehicle and the load of the axles as shown in Fig. 9.

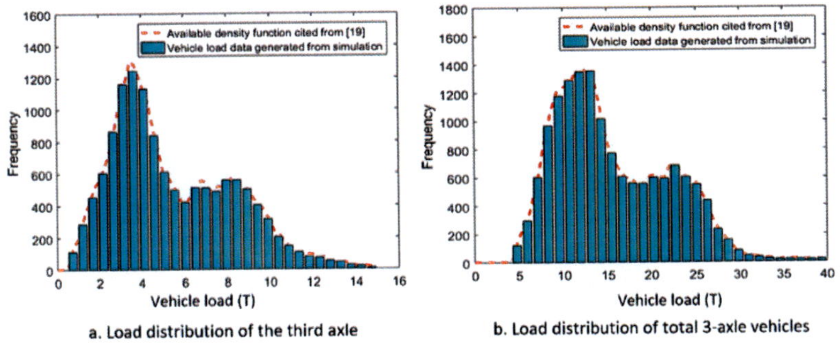

Fig. 9. Simulations of random vehicle load data

After analyzing the input random data set with N = 20000, the results obtained the DIF of vertical and rotation displacement at nodes 2, 3 of the first span and nodes 7, 8 of the second span are shown in Figs. 10, 11, 12 and 13:

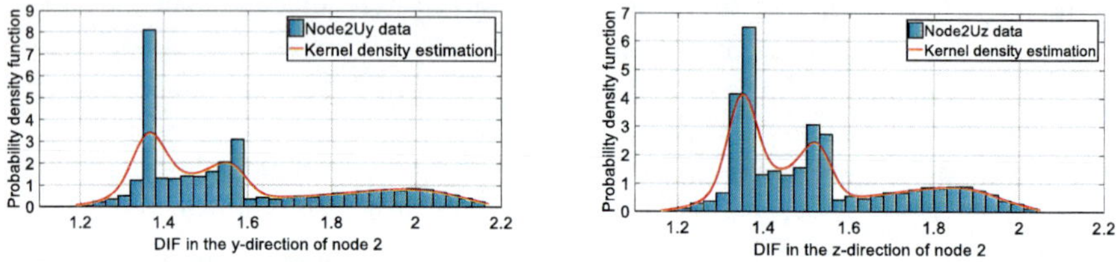

Fig. 10. DIF of Uy and Uz displacement at node 2

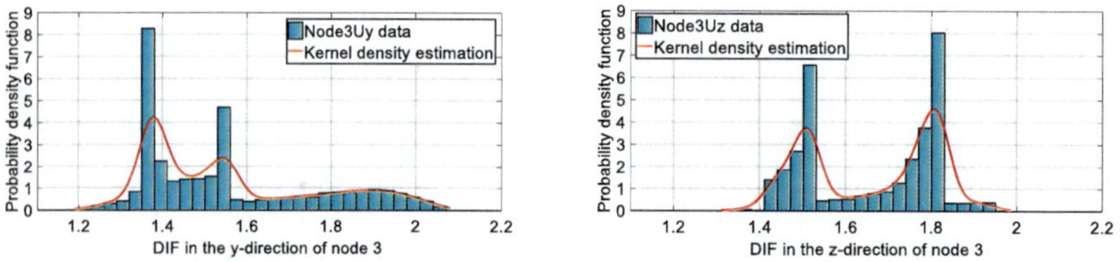

Fig. 11. DIF of Uy and Uz displacement at node 3

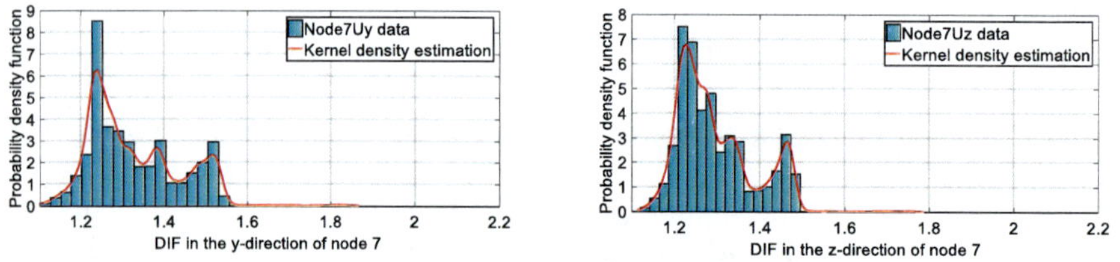

Fig. 12. DIF of Uy and Uz displacement at node 7

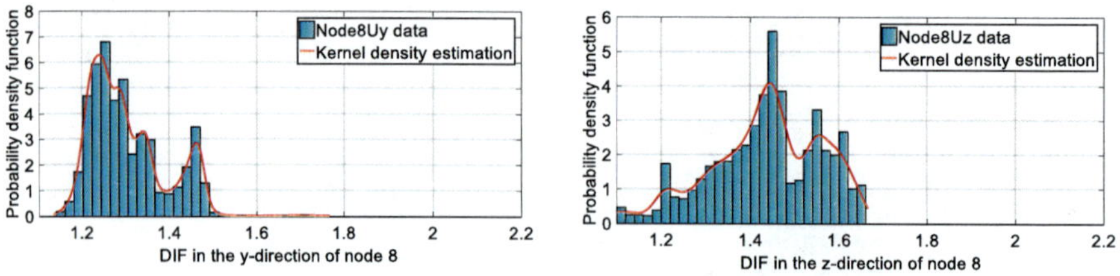

Fig. 13. DIF of Uy and Uz displacement at node 8

The results from Figs. 10, 11, 12 and 13 show that the ruler of probability distribution of random variables about the DIF of displacement is not the same as the one of the input load (Fig. 9). This proves that the mapping relationship between the DIF and the vehicle load has nonlinear signs.

Table 2. Statistical characteristics of the DIF of the displacement at nodes

Statistical	DIF of the vertical displacement				DIF of the rotational displacement			
	Node 2	Node 3	Node 7	Node 8	Node 2	Node 3	Node 7	Node 8
Mean value	1.466	1.461	1.274	1.272	1.438	1.632	1.267	1.381
Maximum	1.562	1.546	1.318	1.328	1.524	1.822	1.320	1.559
Minimum	1.370	1.376	1.229	1.216	1.352	1.443	1.213	1.203
Standard deviation	0.135	0.120	0.063	0.079	0.122	0.268	0.076	0.252

The mean value, maximum, minimum and standard deviation of the DIF of displacement at nodes 2, 3, 7 and 8 are shown in Table 2.

The results show that the DIF of displacement in the first span has a larger value than the DIF in the second span of three-span bridges. This is reasonable because when the vehicle just enters the bridge, it will create a larger vibration when the vehicle goes past the first span. The mean value of DIF in Table 2 shows that the increase in the DIF of the bridge structure subjected to random vehicle loads is mostly larger than the requirements for bridge design according to AASHTO [20] and Vietnamese Design of Road Bridges [21], where the DIF is specified as 1.33.

4 Conclusion

The article has analyzed some results on vibration and DIF of displacement in I-girder bridge with link slab due to random vehicle load by using the finite element method. Actual data on vehicle loads crossing the bridge was collected at Dau Giay weighing station and applied to the analysis for Song Dinh bridge. The obtained results of the vibration and DIF analysis are quite large compared to the values specified in current bridge design codes of Vietnam and the United States. The ruler of probability distribution of random variables about the DIF of displacement is not the same as the one of the input load. This proves that the mapping relationship between the DIF and the vehicle load has nonlinear signs.

References

1. Willis, R.: The effect produced by causing weights to travel over elastic bars. Report of the commissioners appointed to inquire into the application of iron to railway structures, Appendix B, Stationery office, London, England (1849)
2. Tan, G.H., Brameld, G.H., Thambiratnam, D.P.: Development of an analytical model for treating vehicle–bridge interaction. Eng. Struct. **20**, 54–61 (1998)
3. Fafard, M., Bennur, M.: A general multi-axle vehicle model to study the bridge–vehicle interaction. Eng. Comput. **14**(5), 491–508 (1997)
4. Law, S.S., Zhu, X.Q.: Bridge dynamic responses due to road surface roughness and braking of vehicle. J. Sound Vibr. **282**(3), 805–830 (2005)

5. Nguyen, X.T., Tran, V.D., Hoang, N.D.: A study on the dynamic interaction between three-axle vehicle and continuous girder bridge with consideration of braking effects. J. Constr. Eng. 1–12 (2017). https://doi.org/10.1155/2017/9293239
6. Toan, N.X.: Dynamic interaction between the two-axle vehicle and continuous girder bridge with considering vehicle braking force.Vietnam J. Mech. **36**(1), 49–60 (2014)
7. Deng, L., Cai, C.S.: Development of dynamic impact factor for performance evaluation of existing multi-girder concrete bridges. Eng. Struct. **32**(1), 21–31 (2010)
8. Nguyen-Xuan, T., Kuriyama, Y., Nguyen-Duy, T.: Analysis of dynamic impact factors of bridge due to moving vehicles using finite element method. In: Nguyen-Xuan, H., Phung-Van, P., Rabczuk, T. (eds.) ACOME 2017. LNME, pp. 1105–1119. Springer, Singapore (2018). https://doi.org/10.1007/978-981-10-7149-2_77
9. Frýba, L.: Dynamics of Railway Bridges. Thomas Telford, London (1996)
10. Deng, L., He, W., Shao, Y.: Dynamic impact factors for shear and bending moment of simply supported and continuous concrete girder bridges. J. Bridg. Eng. **20**(11), 04015005 (2015). https://doi.org/10.1061/(ASCE)BE.1943-5592.0000744
11. Toan, N.X., Van Duc, T.: Determination of dynamic impact factor for continuous girder bridge due to vehicle braking force by finite element method and experimental. Vietnam J. Mech. **39**(2) (2017). https://doi.org/10.15625/0866-7136/8745
12. Chang, D., Lee, H.: Impact factors for simple-span highway girder bridges. J. Struct. Eng. **120**(3), 704–715 (1994)
13. Sniady, P.: Vibration of a beam due to a random stream of moving forces with random velocity. J. Sound Vib. **97**(1), 23–33 (1984)
14. Turner, J.D., Pretlove, A.J.: A study of the spectrum of traffic-induced bridge vibration. J. Sound Vib. **122**(1), 31–42 (1988)
15. Sorrentino, S., Catania, G.: Dynamic analysis of rectangular plates crossed by distributed moving loads. Math. Mech. Solids **23**(9), 1291–1302 (2018)
16. Clough, R.W.: Dynamics of Structrures. McGraw-Hill, Inc., Singapore (1993)
17. Toan, N.X.: Vibration analysis of cable-stayed bridges subjected to moving vehicles. Thesis of Doctor, University of Danang, Vietnam (2007)
18. Zienkiewicz, O.C., Taylor, R.L., Zhu, J.Z.: The Finite Element Method: Its Basis and Fundamentals. 7th edn. Butterworth-Heinemann (2013)
19. Lan, N.: Research, evaluate and determine the allowable load to cross the bridge on the basis of the bridge test results. Thesis of Doctor, University of Transport and Communications, Vietnam (2015)
20. AASHTO: LRFD bridge design specifications, Washington, DC (2012)
21. Vietnamese National Standards: TCVN11823-13:2017 Design of Road Bridges (2017)

Application of High Order Averaging Method to Van Der Pol Oscillator

N. D. Anh[1,2], Nguyen Ngoc Linh[3], Nguyen Nhu Hieu[4], Nguyen Van Manh[5](✉), and Anh Tay Nguyen[6]

[1] Institute of Mechanics, VAST, Hanoi, Vietnam
[2] Faculty of Mechanical Engineering and Automation, UET, VNU Hanoi, Hanoi, Vietnam
[3] Faculty of Mechanical Engineering, Thuyloi University, Hanoi, Vietnam
[4] University Phenikaa, Hanoi, Vietnam
[5] Faculty of Mechanical Engineering, National University of Civil Engineering, Hanoi, Vietnam
manhnv@nuce.edu.vn
[6] Department of Mechanical Engineering, SUNY Korea, Incheon, South Korea

Abstract. In the paper, the procedure of the high order averaging method is applied to Van der pol system subject to white noise excitation, and the third order stationary joint probability density function of amplitude and phase is given. The accuracy of the approximate solution is confirmed by the Monte-Carlo method. The effects of the system parameters on the response are analyzed in detail. It is shown that the higher order averaging solution is more accurate than the one obtained by the traditional first order stochastic averaging method.

Keywords: High order averaging method · Random excitation · Van der Pol oscillator · Probability density function

1 Introduction

There has been a high interest in the effects of random noises on dynamical systems, due to various problems encountered in engineering applications [1, 2]. Non-linear systems subjected to random excitations have been investigated for many models, in particular for Van der Pol oscillator which was originally proposed by Balthasar Van der Pol [3]. This equation has been used in the design of various systems, for example in the theory of mechanisms and machines, and especially in the field of robotics [4–7] and studied with many interesting effects [8–12]. Among the approximate analytical methods, the well-known averaging method originally given by Krylov and Bogoliubov and then developed by Mitropolskii [13–16] has proved to be a very powerful approximate tool for investigating deterministic weakly nonlinear vibration problems. The stochastic averaging method was extended by Stratonovich [17] into the field of random vibrations and has mathematically rigorous proof by Khasminskii [18]. The interest and use of the averaging method have increased significantly recently because many authors have applied this method to study various nonlinear phenomena [19–22].

In the field of random oscillations, the averaging method, in most cases, is restricted to the first approximation. It is known that the accuracy of the first approximation may be

N. V. Khang et al. (Eds.): ASIAN MMS 2021, MMS 113, pp. 825–834, 2022.
https://doi.org/10.1007/978-3-030-91892-7_79

insufficient for many nonlinear random systems. Thus, different averaging procedures for obtaining more accurate approximate solutions have been developed [23–25]. A higher order stochastic averaging procedure using Fokker-Planck (FP) equation was developed by Anh [26, 27], and then applied to Van der Pol oscillator under white noise excitation [28]. However, in [28], only high order approximate solutions have been obtained, but the effects of the system's parameters on the response have not been analyzed yet.

In the paper this procedure is further applied to Van der pol systems subject to white noise excitation and the third order stationary joint probability density function (PDF) of amplitude and phase is given. The accuracy of the approximate solution is confirmed by the solution of Monte-Carlo method. The effects of the system parameters on the response are analyzed in detail.

2 Third Order Stochastic Averaging Procedure

Consider the equation of Van der Pol oscillator subject to random excitation

$$\ddot{x} + \omega^2 x = \varepsilon(1 - \gamma x^2)\dot{x} + \sqrt{\varepsilon}\sigma\xi(t) \tag{1}$$

where the symbols have their customary meaning, ω, γ, σ are positive constants. The excitation $\xi(t)$ is a zero mean Gaussian white noise process with unit intensity

$$E(\xi(t)\xi(t+\tau)) = \delta(\tau) \tag{2}$$

The operator E denotes the mathematical expectation, $\delta(\tau)$ is Dirac-Delta function. Introducing new variables amplitude and full phase, $a(t)$, $\phi(t)$ by the transformation

$$x(t) = a(t)\cos\phi(t),\ \dot{x}(t) = -\omega a(t)\sin\phi(t) \tag{3}$$

We obtain from Eq. (1) the system of Ito stochastic differential equations for amplitude and full phase [15, 27]

$$\begin{aligned} da(t) &= \varepsilon K_1(a,\phi)dt - \sqrt{\varepsilon}\frac{\sigma}{\omega}\sin\phi dB(t) \\ d\phi(t) &= (\omega + \varepsilon K_2(a,\phi))dt - \sqrt{\varepsilon}\frac{\sigma}{a\omega}\cos\phi dB(t) \end{aligned} \tag{4}$$

where $B(t)$ is a Wiener process and

$$\begin{aligned} K_1 &= \left[a/2 - \gamma a^3/8 + \sigma^2/(4a\omega^2)\right] - \left[a/2 - \sigma^2/(4a\omega^2)\right]\cos 2\phi + \left(\gamma a^3/8\right)\cos 4\phi \\ K_2 &= \left[1/2 - \gamma a^2/4 - \sigma^2/(2a^2\omega^2)\right]\sin 2\phi - (\gamma a^2/8)\sin 4\phi \end{aligned} \tag{5}$$

The Fokker-Planck equation for the stationary joint probability density function (PDF) of amplitude and phase $W(a,\phi)$ related to the system (4) takes the form [1, 2]

$$\omega\frac{\partial W}{\partial\phi} = -\varepsilon\frac{\partial}{\partial a}(K_1 W) - \varepsilon\frac{\partial}{\partial\phi}(K_2 W)$$

$$+\frac{\varepsilon}{2}\left[\frac{\partial^2}{\partial^2 a}(K_{11}W)+2\frac{\partial^2}{\partial a\partial\phi}(K_{12}W)+\frac{\partial^2}{\partial^2\phi}(K_{22}W)\right] \tag{6}$$

where K_1, K_2 are given by (5) and

$$K_{11}(a,\phi)=\frac{\sigma^2}{2\omega^2}-\frac{\sigma^2}{2\omega^2}\cos 2\phi;\ K_{12}(a,\phi)=\frac{\sigma^2}{2a\omega^2}\sin 2\phi$$
$$K_{22}(a,\phi)=\frac{\sigma^2}{2a^2\omega^2}+\frac{\sigma^2}{2a^2\omega^2}\cos 2\phi \tag{7}$$

A higher-order stochastic averaging procedure was proposed by Anh in [26, 27] to solve the Fokker-Planck Eq. (6). Here the third order procedure is implemented and for that purpose, the solution of Eq. (6) is determined in a form of a series with respect to the small parameter as follows

$$W(a,\phi)=W_0(a)+\varepsilon W_1(a,\phi)+\varepsilon^2 W_2(a,\phi)+\ldots \tag{8}$$

where $W_0(a)$ is determined from the first order averaging method, i.e. one has the averaged Fokker-Planck equation for Van der Pol system (1) [26, 27]

$$\frac{\partial}{\partial a}\left[\left(\frac{a}{2}-\frac{\gamma a^3}{8}+\frac{\sigma^2}{4a\omega^2}\right)W_0(a)\right]-\frac{1}{2}\frac{\partial^2}{\partial a^2}\left[\frac{\sigma^2}{2\omega^2}W_0(a)\right]=0 \tag{9}$$

which gives the first order stationary probability density function for amplitude

$$W_0(a)=Ca\exp\left(\frac{\omega^2}{\sigma^2}a^2-\frac{\gamma\omega^2}{8\sigma^2}a^4\right) \tag{10}$$

Substituting (8) into (6) yields

$$\omega\frac{\partial(W_0(a)+\varepsilon W_1(a,\phi)+\varepsilon^2W_2(a,\phi)+\ldots)}{\partial\phi}=-\varepsilon\frac{\partial}{\partial a}\Big(K_1(W_0(a)+\varepsilon W_1(a,\phi)+\varepsilon^2W_2(a,\phi)+\ldots)\Big)$$
$$-\varepsilon\frac{\partial}{\partial\phi}\Big(K_2(W_0(a)+\varepsilon W_1(a,\phi)+\varepsilon^2W_2(a,\phi)+\ldots)\Big)+\frac{\varepsilon}{2}\left[\frac{\partial^2}{\partial^2 a}\Big(K_{11}(W_0(a)+\varepsilon W_1(a,\phi)+\varepsilon^2W_2(a,\phi)+\ldots)\Big)\right.$$
$$\left.+2\frac{\partial^2}{\partial a\partial\phi}\Big(K_{12}(W_0(a)+\varepsilon W_1(a,\phi)+\varepsilon^2W_2(a,\phi)+\ldots)\Big)+\frac{\partial^2}{\partial^2\phi}\Big(K_{22}(W_0(a)+\varepsilon W_1(a,\phi)+\varepsilon^2W_2(a,\phi)+\ldots)\Big)\right] \tag{11}$$

Comparing the coefficients of ε, ε^2 in both sides of (11) gives

$$\omega\frac{\partial W_1}{\partial\phi}=-\frac{\partial}{\partial a}(K_1W_0)+\frac{\partial}{\partial\phi}(K_2W_0)$$
$$+\frac{1}{2}\left[\frac{\partial^2}{\partial^2 a}(K_{11}W_0)+2\frac{\partial^2}{\partial a\partial\phi}(K_{12}W_0)+\frac{\partial^2}{\partial^2\phi}(K_{22}W_0)\right] \tag{12}$$
$$\omega\frac{\partial W_2}{\partial\phi}=-\frac{\partial}{\partial a}(K_1W_1)+\frac{\partial}{\partial\phi}(K_2W_1)$$
$$+\frac{1}{2}\left[\frac{\partial^2}{\partial^2 a}(K_{11}W_1)+2\frac{\partial^2}{\partial a\partial\phi}(K_{12}W_1)+\frac{\partial^2}{\partial^2\phi}(K_{22}W_1)\right] \tag{13}$$

Substituting (5), (7) and (10) into (12) and integrating this equation with respect to the variable ϕ leads to

$$W_1(a,\phi) = W_0(a)\left[\frac{\gamma a^2}{32\sigma^2\omega}\left(12\sigma^2 + 4\omega^2 a^2 - \gamma\omega^2 a^4\right)\sin 2\phi + \frac{\omega\gamma a^4}{64\sigma^2}\left(\gamma a^2 - 4\right)\sin 4\phi\right] \quad (14)$$

Substituting (5), (7) and (14) into (13) and integrating this equation with respect to the variable ϕ leads to

$$W_2(a,\phi) = W_0(a)[W_{20}(a) + W_{22}(a,\phi)] \quad (15)$$

where

$$\begin{aligned} W_{20}(a) = &-\frac{9\gamma}{16\omega^2}a^2 + \left(\frac{39\gamma^2}{256\omega^2} - \frac{9\gamma}{64\sigma^2}\right)a^4 + \frac{29\gamma^2}{384\sigma^2}a^6 \\ &+ \left(\frac{5\gamma^2\omega^2}{1024\sigma^4} - \frac{109\gamma^3}{8192\sigma^4}\right)a^8 - \frac{5\gamma^2\omega^2}{2048\sigma^4}a^{10} + \frac{5\gamma^4\omega^2}{16384\sigma^4}a^{12} \end{aligned} \quad (16)$$

$$W_{22}(a,\phi) = E_1(a)\cos 2\phi + E_2(a)\cos 4\phi + E_3(a)\cos 6\phi + E_4(a)\cos 8\phi$$

$$\begin{aligned} E_1(a) &= -\frac{9\gamma}{16\omega^2}a^2 + \left(\frac{13\gamma^2}{64\omega^2} - \frac{3\gamma}{16\sigma^2}\right)a^4 + \frac{5\gamma^2}{64\sigma^2}a^6 - \left(\frac{7\gamma^3}{1024\sigma^4} + \frac{\gamma^2\omega^2}{256\sigma^4}\right)a^8 + \frac{\gamma^3\omega^2}{512\sigma^4}a^{10} - \frac{\gamma^4\omega^2}{4096\sigma^4}a^{12} \\ E_2(a) &= \left(\frac{3\gamma}{64\sigma^2} - \frac{11\gamma^2}{128\omega^2}\right)a^4 - \frac{17\gamma^2}{256\sigma^2}a^6 + \left(\frac{15\gamma^3}{1024\sigma^2} - \frac{\gamma^2\omega^2}{256\sigma^4}\right)a^8 + \frac{\gamma^3\omega^2}{512\sigma^4}a^{10} - \frac{\gamma^4\omega^2}{4096\sigma^4}a^{12} \\ E_3(a) &= \frac{\gamma^2}{48\sigma^2}a^6 + \left(\frac{\gamma^2\omega^2}{256\sigma^4} - \frac{19\gamma^2}{3072\sigma^2}\right)a^8 - \frac{\gamma^3\omega^2}{512\sigma^4}a^{10} + \frac{\gamma^4\omega^2}{4096\sigma^4}a^{12} \\ E_4(a) &= \left(\frac{\gamma^3}{4096\sigma^2} - \frac{\gamma^2\omega^2}{1024\sigma^4}\right)a^8 + \frac{\gamma^2\omega^2}{2048\sigma^4}a^{10} - \frac{\gamma^4\omega^2}{16384\sigma^4}a^{12} \end{aligned} \quad (17)$$

Substituting (10), (14) and (15) into (8) one obtains the third order PDF of amplitude and full phase $W(a,\phi)$ for Van der Pol oscillator (1)

$$\begin{aligned} W(a,\phi) = Ca\exp\left(\frac{\omega^2}{\sigma^2}a^2 - \frac{\gamma\omega^2}{8\sigma^2}a^4\right)\Big[1 + \frac{\varepsilon\gamma}{32\omega\sigma^2}\left(12\sigma^2 a^2 + 4\omega^2 a^4 - \gamma\omega^2 a^6\right)\sin 2\phi \\ + \frac{\varepsilon\gamma\omega a^2}{64\sigma^2}\left(\gamma a^4 - 4a^2\right)\sin 4\phi + \varepsilon^2 W_{20}(a) + \varepsilon^2 W_{22}(a,\phi)\Big] \end{aligned} \quad (18)$$

where $W_{20}(a)$, $W_{22}(a,\phi)$, C are given in (16), (17) and C is the normalization coefficient determined from the condition

$$\int_0^\infty \int_0^{2\pi} W(a,\phi)\,da\,d\phi = 1 \quad (19)$$

Using (18) the second moment of the displacement response of Van der Pol system (1) can be calculated numerically by

$$\left\langle x^2\right\rangle = \int_0^\infty \int_0^{2\pi} a^2\cos^2\phi\, W(a,\phi)\,da\,d\phi \quad (20)$$

In the next section, the third order stationary joint PDF will be investigated and the accuracy of the corresponding approximate solution will be verified.

3 Numerical Results

In this section, numerical simulations for Van der Pol system (1) are implemented where the third order stationary joint probability density function $W(a, \phi)$ given by (18) will be analyzed, and then the accuracy of the corresponding approximate solution will be compared with the ones obtained by the Monte Carlo method [29–31] and by the first order stationary joint probability density function. For the present investigation it was taken $\omega = 1$ and various values of γ, σ and ε are taken.

Figure 1 and Fig. 2 show the joint probability density function $W(a, \phi)$ in two cases, $(\gamma, \sigma^2) = (0.1, 0.1)$ and $(\gamma, \sigma^2) = (1, 1)$ with $\varepsilon = 0.02$, respectively. It is observed that $W(a, \phi)$ are multi-peak surfaces along with the *ϕ-axis* in both cases, there are eight peaks in the first case and three peaks in the latter case. This is a main difference in comparison with the first order PDF, $W_0(a)$ which excludes the phase ϕ and parameter ε as shown in (10). Besides, the shape of $W(a, \phi)$ along with the *a-axis* has the form of normal distribution in the first case with $\gamma < 1$, and Rayleigh distribution in the latter case with $\gamma = 1$.

In more detail, the contours of $W(a, \phi)$ on the *aW-plane* are illustrated in Fig. 3, 4, 5, 6, 7 and 8. As we can see, the peaks and valleys of $W(a, \phi)$ have large differences in amplitude for case $(\gamma, \sigma^2) = (0.1, 0.1)$ (Fig. 3, magenta line) but smaller differences for case $(\gamma, \sigma^2) = (1, 1)$ (Fig. 4, blue line). The maximum PDF in the first case is almost four times larger than that in the latter case, i.e. 0.14 and 0.497, and the corresponding mode is about 2.1 and 6.4, respectively. For other values of γ, it is seen that the larger value of σ^2, the larger variance will be. On the contrary, the variance seems to decrease as γ increases. The first order PDF $W_0(a)$ is also highlighted in Fig. 3 and Fig. 4 with dotted lines (only see when zooming in). The maximum PDF of $W_0(a)$ in both cases is 0.4 and 0.13, and the corresponding mode is 6.33 and 2.1. Clearly, the curve $W_0(a)$ is only fixed with a contour of $W(a, \phi)$ on the *aW-pla*.

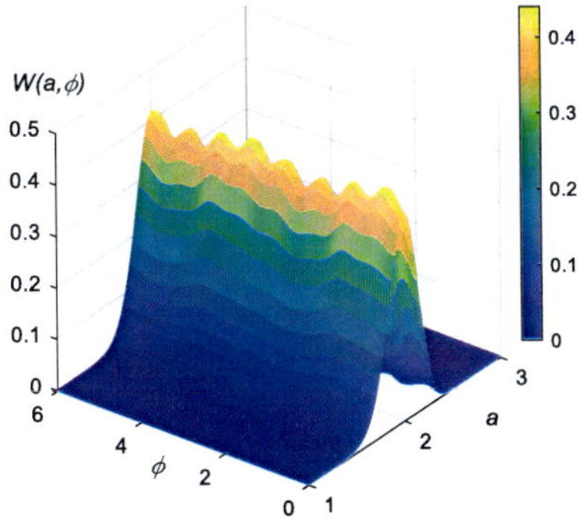

Fig. 1. $W(a, \phi)$ with $\gamma = 0.1, \sigma^2 = 0.1$

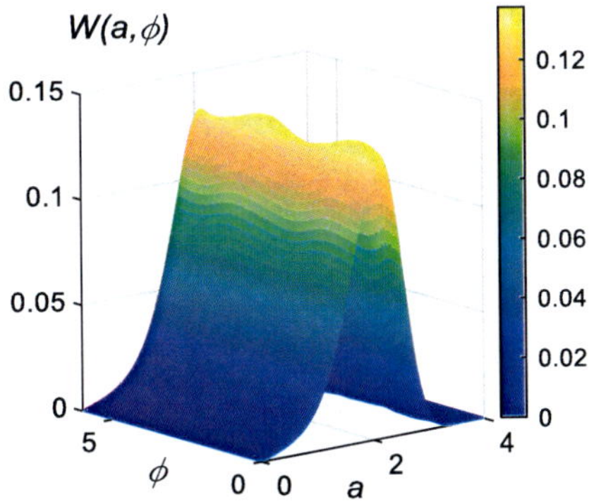

Fig. 2. $W(a, \phi)$ with $\gamma = 1, \sigma^2 = 1$

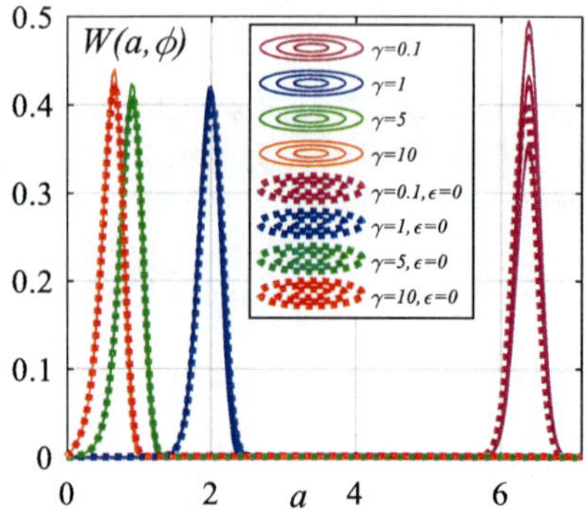

Fig. 3. $W(a, \phi)$ contours on the *aW-plane* with $\sigma^2 = 0.1$ and γ varies

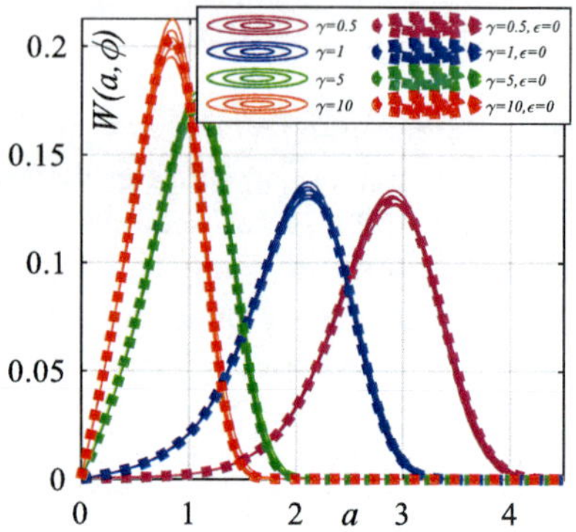

Fig. 4. $W(a, \phi)$ contours on the *aW-plane* with $\sigma^2 = 1$ and γ varies

Figures 5, 6, 7 and 8 present the contours of $W(a, \phi)$ on the *aW-plane* with four cases $\gamma = 0.1, 1, 5, 10$ with a variety of white noise intensity σ^2. Similar to Rayleigh distribution σ^2 takes the role of the shape parameter. In general, as the shape parameter increases, the distribution gets wider and flatter, and the peaks of $W(a, \phi)$ move forward slightly along the *a-axis*. On the contrary, for a specific value of σ^2, the value of mode decreases as γ gets larger. Namely, the relative mode and maximum PDF for $\sigma^2 = 0.1, 0.5$ respectively are presented in Table 1, respectively.

Table 1. Mode and Maximum PDF of $W(a, \phi)$

γ	σ² = 0.1		σ² = 0.5	
	Mode	Max. PDF	Mode	Max. PDF
0.1	6.4	0.497	6.6	0.258
1	1.97	0.42	2.07	0.19
5	0.93	0.42	1	0.22
10	0.67	0.44	0.77	0.25

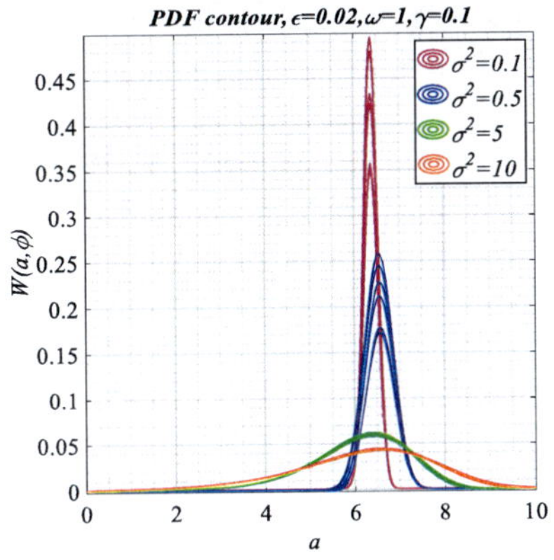

Fig. 5. $W(a, \phi)$ contours on the *aW-plane* with $\gamma = 0.1$ and σ varies

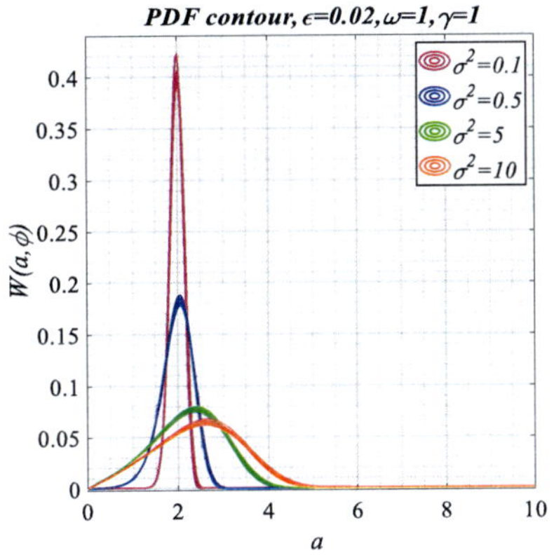

Fig. 6. $W(a, \phi)$ contours on the *aW-plane* with $\gamma = 1$ and σ varies

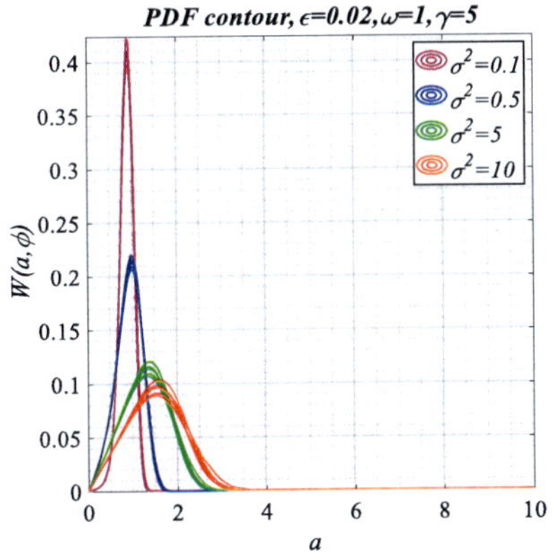

Fig. 7. $W(a, \phi)$ contours on the *aW-plane* with $\gamma = 5$ and σ varies

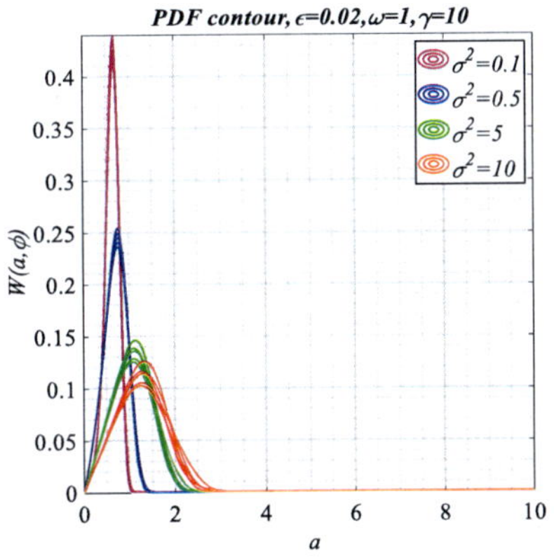

Fig. 8. $W(a, \phi)$ contours on the *aW-plane* $\gamma = 10$ and σ varies

Table 2 shows that as the white noise intensity increases from 0.1 to 6.0 the percentage of relative error between the third order approximation with the Monte-Carlo simulation is significantly smaller than the one of the first order approximation. Their maximum errors are 5.55% and 8.44%, respectively. However, when the white noise intensity increases above 8.0 the advantage of the third order approximation decreases slightly. This is explained by the reason that the third order approximation has the higher PDF than that of the first order one as σ^2 is small, especially the multi-peak PDF provides the distribution in the full phase from 0 to 2π. As σ^2 becomes very large, the third order PDF is widely scattered and its peaks get flatter, which has the same effect as that of the first order PDF.

Table 2. The mean-square responses obtained by Monte-Carlo simulation and by first- and third order stochastic averaging methods with $\gamma = 0.5$, $\varepsilon = 0.01$

$\gamma = 0.5$ $\varepsilon = 0.01$	Monte-Carlo simulation	1st approximation		3rd approximation	
σ^2	$\left\langle x^2 \right\rangle_{MC}$	$\left\langle x^2 \right\rangle_1$	Error (%)	$\left\langle x^2 \right\rangle_3$	Error (%)
0.1	3.6887	4.0000	8.44	3.8651	4.78
0.5	3.7797	4.0001	5.83	3.9892	5.55
1	3.8381	4.0104	4.49	4.0059	4.37
1.5	3.9837	4.0485	1.63	4.0461	1.57
2	4.0869	4.1105	0.57	4.1092	0.55
2.5	4.1601	4.1870	0.64	4.1865	0.64
3	4.2986	4.2715	0.63	4.2717	0.63
5	4.6519	4.6319	0.43	4.6340	0.38
6	4.8350	4.8101	0.52	4.8130	0.45
8	5.0921	5.1504	1.15	5.1549	1.23
10	5.4058	5.4684	1.15	5.4746	1.27

4 Conclusion

The stochastic averaging method has been a very useful tool for investigating non-linear random vibration systems. Thus, the further development of this method remains attractive for researchers. In the paper, the third order stochastic averaging procedure is applied to obtain the first and third order stationary joint probability density functions of amplitude and phase for Van der Pol system to white noise process. The accuracy of the approximate solutions is confirmed by the solution of the Monte-Carlo method. Using the third order PDF the probabilistic characters of the response and the effects of system parameters are analyzed in detail. It is shown that the accuracy of the third order approximate solution is significantly improved than the one of the traditional first order approximate solution.

Acknowledgments. This study is funded by the National University of Civil Engineering (NUCE) under Grant No. 27-2021/KHXD-TĐ.

References

1. Roberts, J.B., Spanos, P.D.: Random Vibration and Statistical Linearization. Dover Publications Inc., New York (2003)
2. Lutes, L.D., Sarkani, S.: Random Vibration: Analysis of Structural and Mechanical Systems. Elsevier, Amsterdam (2004)

3. Van der Pol, B.: On relaxation-oscillations, The London, Edinburgh and Dublin Phil. Mag. J. Sci. **2**(7), 978–992 (1926)
4. Veskos, P., Demiris, Y.: Robot swinging using van der Pol nonlinear oscillators. In: Third International Symposium on Adaptive Motion of Animals and Machines (2005)
5. Veskos, P., Demiris, Y.: Experimental comparison of the van der Pol and Rayleigh nonlinear oscillators for a robotic swinging task. In: Proceedings of the AISB 2006 Conference, Adaptation in Artificial and Biological Systems, Bristol, pp. 197–202 (2006)
6. Kuwata, N., Hoshi, Y., Nohara, B.T.: Analysis of coupled van der Pol oscillators and implementation to a myriapod robot. In: Proceedings of the 17th World Congress the International Federation of Automatic Control Seoul, Korea, 6–11 July 2008
7. Yu, H., Guo, W., Deng, J., Li, M., Cai, H.: A CPG-based locomotion control architecture for hexapod robot. In: 2013 IEEE/RSJ International Conference on Intelligent Robots and Systems (IROS) 3–7 November 2013, Tokyo, Japan (2013)
8. Mickens, R.E.: Fractional Van der Pol equations. J. Sound Vib. **259**(2), 457–460 (2003)
9. Elabbasy, E.M., El-Dessoky, M.M.: Synchronization of van der Pol oscillator and Chen chaotic dynamical system. Chaos Solit. Fractals **36**(5), 1425–1435 (2008)
10. Pinto, C.M.A., Machado, J.A.T.: Complex order van der Pol oscillator. Nonlinear Dyn. **65**(3), 247–254 (2010)
11. Anh, N.D., Zakovorotny, V.L., Hao, D.N.: Response analysis of Van der Pol oscillator subjected to harmonic and random excitations. Probab. Eng. Mech. **37**, 51–59 (2014)
12. Klimina, L.A.: Method for finding periodic trajectories of centrally symmetric dynamical systems on the plane. Differ. Eqn. **55**(2), 159–168 (2019). https://doi.org/10.1134/S0012266119020022
13. Bogoliubov, N.N., Mitropolskii, Y.: Asymptotic Methods in the Theory of Nonlinear Oscillations. Gordon & Breach, New York (1961)
14. Mitropolskii, Yu.A.: Averaging Method in Nonlinear Mechanics. Naukova-Dumka, Kiev (1971). (In Russian)
15. Mitropolskii, Yu.A., Dao, N.V., Anh, N.D.: Nonlinear Oscillations in Systems of Arbitrary Order. Naukova-Dumka, Kiev (1992). (In Russian)
16. Sanders, J.A., Verhulst, F., Murdock, J.: Averaging Methods in Nonlinear Dynamical Systems. Applied Mathematical Sciences, vol. 59, 2nd edn. Springer, New York (2007). https://doi.org/10.1007/978-0-387-48918-6
17. Stratonovich, S.: Topics in the Theory of Random Noise. Gordon & Breach v. 1, New York (1963)
18. Khasminskii, R.: Principle of averaging for parabolic and eliptic differential equations and for Markov processes with small diffusion. Theor. Prob. Appl. **8**, 1–21 (1963). (In Russian)
19. Tang, Y.Q., Chen, L.Q., Lim, C.W.: Dynamic stability of axially accelerating Timoshenko beam: averaging method. Eur. J. Mech. A. Solids **29**(1), 81–90 (2010)
20. Kovacic, I., Zukovic, M.: Oscillators with a power-form restoring force and fractional derivative damping: application of averaging. Mech. Res. Commun. **41**, 37–43 (2012)
21. Gu, X., Zhu, W.: A stochastic averaging method for analyzing vibro-impact systems under Gaussian white noise excitations. J. Sound Vib. **333**(9), 2632–2642 (2014)
22. Jiang, W.A., Chen, L.Q.: Stochastic averaging of energy harvesting systems. Int. J. Non-Linear Mech. **85**, 174–187 (2016)
23. Sri Namachchivaya, N., Lin, Y.K.: Application of stochastic averaging for nonlinear systems with high damping. Probab. Eng. Mech. **3**(3), 159–167 (1988)
24. Red-Horse, J.R., Spanos, P.D.: A generalization to stochastic averaging in random vibration. Int. J. Nonlinear Mech. **27**, 85–101 (1992)
25. Zhu, W.Q., Yu, M.Q., Lin, Y.K.: On improved stochastic averaging procedure. Probab. Eng. Mech. **9**, 203–212 (1994)

26. Anh, N.D.: Higher-order approximate solutions in stochastic averaging method. In: Proceedings of NCSR of VN, vol. 5, pp. 19–26 (1993)
27. Anh, N.D.: Higher-order averaging method of coefficients in Fokker-Planck equation. Indian J. SADHANA **20**(Parts 2–4), 373–387 (1995)
28. Mitropolskii, Yu.A., Anh, N.D., Tinh, N.D.: Random oscillations in the Van der Pol system under influence of wide-band stochastic process. Ucrainian Math. J. **50**(11), 1517–1521 (1998). (In Russian)
29. Debrabant, K., Rößler, A.: Families of efficient second order Runge-Kutta methods for the weak approximation of Itô stochastic differential equations. Appl. Numer. Math. **59**, 582–594 (2009)
30. Roy, R.V., Spanos, P.D.: Pade-type approach to nonlinear random vibration analysis. In: Lin, Y.K., Elishakoff, I. (eds.) Stochastic Structural Dynamics 1: New Theoretical Developments, pp. 155–172. Springer, Heidenberg (1991). https://doi.org/10.1007/978-3-642-84531-4_9
31. Spanos, P.-T.D.: Numerical simulations of a Van der Pol oscillator. Comput. Math. Appl. **6**(1), 135–145 (1980)

Prediction of Open-Ended Pile Driving Performance Under Dynamic and Static Driving Forces

Nguyen Anh Ngoc[1(✉)], Tong Duc Nang[2], Nguyen Binh[1], Do Van Nhat[2], Nguyen Van Kuu[3], and Nguyen Ngoc Linh[3]

[1] Faculty of Mechanical Engineering, University of Transport and Communications, Hanoi, Vietnam
nguyenanhngocmxd@utc.edu.vn
[2] Faculty of Mechanical Engineering, National University of Civil Engineering, Hanoi, Vietnam
[3] Faculty of Mechanical Engineering, Thuyloi University, Hanoi, Vietnam

Abstract. This paper deals with the pile driving problem of open-ended piles. An averaging model of the driver-pile-soil system is developed, in which a force equilibrium of the system can be obtained in time average. The averaging model is then performed for a hammer-pile-soil system supplemented by an extra static driving force. The main characteristic quantities of the pile driving performance are obtained by explicit formulations. A case study examination is carried out to verify the effectiveness of the proposed model.

Keywords: Open-ended pile · Case model · Dynamic pile driving

1 Introduction

In recent years, open-ended piles have played an increasingly important role in many fields of engineering in both onshore and offshore applications. The plugging effect of open-ended piles, including unplugged, partially plugged, and fully plugged state, becomes an issue of concern because of the large influence on the pile driving performance. In some practical cases, it is necessary to supplement additional force to the normal open-ended pile driving process. There are several studies and design guide methods on this issue primarily based on observations of model tests or in situ tests, such as [1–5]. Among test methods, the widely used Case method [6], named after Case Western Reserve University, consists of data measured from pile top forces and accelerations in pile load tests or laboratories, then provides predictions of pile bearing capacity, pile stresses, and hammer energies. Besides, the dynamic model (the Case model) of this method can be applied to theoretical models due to its simplicity. A recent remarkable investigation on open-ended piles using the Finite-element method based on the Case method has been carried out in [7]. Detailed analysis of pile capacity achieved from dynamic measurements through analytical methods, such as the Case method, was

N. V. Khang et al. (Eds.): ASIAN MMS 2021, MMS 113, pp. 835–844, 2022.
https://doi.org/10.1007/978-3-030-91892-7_80

introduced in [8], in which it was emphasized that "*the accurate prediction of pile capacity remains a challenge due to the complex response of piles during driving*". In the case of open-ended piles, it is also concerned with plugging effect.

In this paper, we develop a new averaging model derived from the Case model for analyzing open-ended pile driving performance under dynamic and additional static driving forces. The remainder of the paper is structured as follows: The averaging model of the driver-pile-soil system is introduced in Sect. 2, in which hammer-pile and pile-soil models are used in considering the soil plugging effect. A case study examination is carried out in Sect. 3. Conclusions are presented in Sect. 4.

2 Averaging Model of a Driver-Pile-Soil System for Open-Ended Pile Driving Under Dynamic and Static Driving Forces

2.1 Averaging Model of the Driver-Pile-Soil System

Following the Case method [6], the driver-pile-soil system as illustrated in Fig. 1 can be presented by a simple dynamic model (also called the Case model)

$$R = F - ma \tag{1}$$

where F is the force exerted by the driver on the pile, m is the mass, a is the acceleration of the moving mass, R is the total dynamic resistance of the soil. Normally R is the sum of the static component, R^S, and the dynamic one, R^D, which respectively represent total friction and damping effects of the pile-soil interaction at the pile wall (skin) and the pile tip [8]. Hence, we have

$$R = R^S + R^D \tag{2}$$

For the open-ended pile, the total resistance R can be formulated as below

$$R = R^S + R^D = R^S_{s,e} + R^S_{s,i} + R^S_t + R^D_{s,e} + R^D_{s,i} + R^D_t \tag{3}$$

where $R^S_{s,e}$ and $R^S_{s,i}$ are external and internal skin frictions, $R^D_{s,e}$ and $R^D_{s,i}$ are external and internal skin dampings.

It is noted that R^S and R^D are commonly estimated by functions of the pile penetration z and independent of time [6, 8] due to very short duration of impaction. Otherwise, F and a are functions of time. To overcome this obstacle, the Case model [6] proposed to replace F and a by time-independent quantities

$$R = \frac{1}{2}(F(t_1) - F(t_2)) + \frac{mc_p}{2L}(v(t_1) - v(t_2)) \tag{4}$$

where $t_2 = t_1 + 2L/c_p$, and t_1 is a selected time during the blow, L is the pile length, c_p is the wave transmission speed on the pile material, v is the velocity of the pile top. Obviously, the values of $F(t_1)$, $F(t_2)$ and $v(t_1)$, $v(t_2)$ are measured at two specific times. As a consequence, they may not provide adequate contributions of exerted forces as well as pile top velocities during the time interval $[t_1, t_2]$, as shown in Fig. 2.

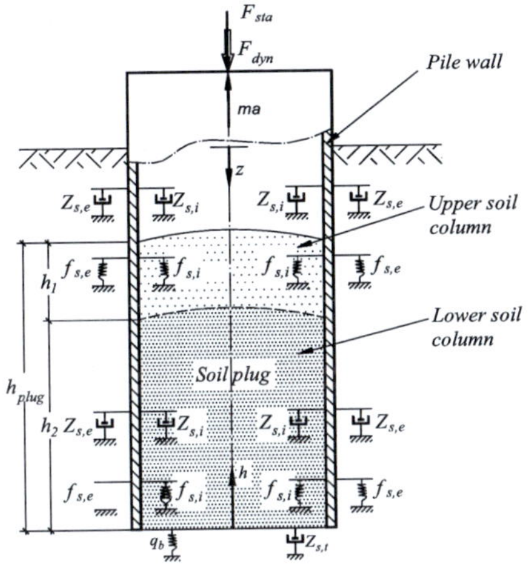

Fig. 1. Driver-pile-soil model

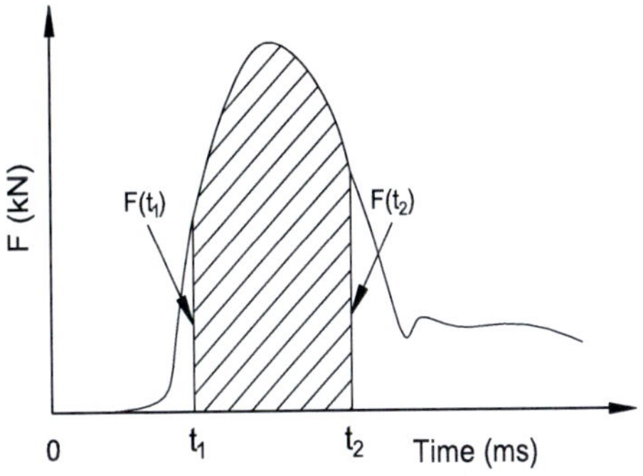

Fig. 2. Selected values of exerted forces (Case method)

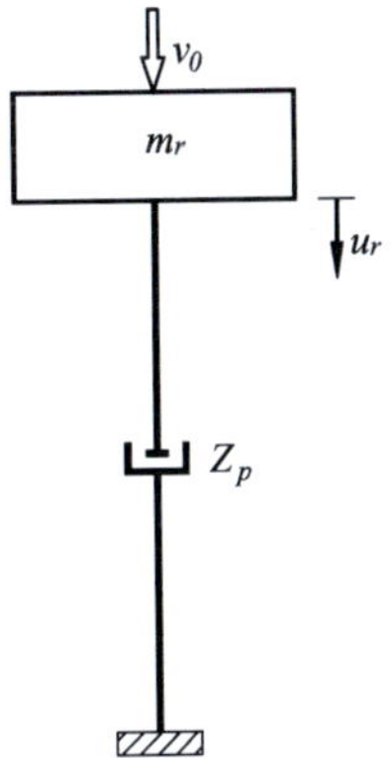

Fig. 3. Dashpot hammer-pile model

Fig. 4. MHP-01 pile driver. 1- hammer, 2- ram, 3- open-end pile

Therefore, we consider here an averaging model of the driver-pile-soil system in concern with time interval $[t_1, t_2]$ during driving

$$R = \langle F - ma \rangle \tag{5}$$

where F and a are time-dependent functions, $\langle \cdot \rangle$ is time averaging operator

$$\langle \cdot \rangle = \frac{1}{t_2 - t_1} \int_{t_1}^{t_2} (\cdot) dt \tag{6}$$

In case of open-ended pile driving under both dynamic and static driving forces, F_{dyn} (time-dependent) and F_{sta} (time-independent), noting Eq. (2) then Eq. (5) becomes

$$R^S_{s,e} + R^S_{s,i} + R^S_t + R^D_{s,e} + R^D_{s,i} + R^D_t = F_{sta} + \langle F_{dyn} - ma \rangle \tag{7}$$

By using averaging operator, the contribution of time-dependent force is considered in time interval $[t_1, t_2]$ in average meaning (crossed area in Fig. 2), which may help to improve the accuracy of force analysis as well as required energy calculation in the next section. The averaging model (5)–(7) can describe simultaneous interaction between pile driver, pile, and soil with a variety of driving forces. To concretize the problem, next we will concentrate on the hammer-pile-soil system in which an extra static driving force is added along with the dynamic force. The main problem now is to determine the force components based on hammer-pile interaction and pile-soil interaction.

2.2 Dynamic Driving Force from Hammer-Pile Model

Consider a hammer-pile model as illustrated in Fig. 3 in which the response of the pile to the impacting mass can be modeled by replacing the pile with a dashpot of equal impedance to the pile [10]. The dynamic force exerted by the hammer on the pile can be given by multiplying the pile impedance Z_p by the ram velocity $\dot{u}_r$

$$F_{dyn} = Z_p \dot{u}_r; \, Z_p = \frac{E_p A_p}{c_p} \tag{8}$$

where E_p is Young's modulus of the pile, A_p is the cross-section area of the pile, c_P is the speed of stress wave in pile. The equation of motion for the ram is

$$m_r \ddot{u}_r + Z_p \dot{u}_r = 0 \tag{9}$$

The initial conditions supplemented to Eq. (9) is derived from the instant the hammer strikes the pile

$$u_r(t_0) = 0, \; \dot{u}_r(t_0) = v_0 \tag{10}$$

Solving Eq. (9) with initial conditions (10) gives the displacement, velocity, and acceleration responses of the ram after impact

$$u_r = v_o(m_r/Z_p) - v_o(m_r/Z_p)e^{-Z_p t/m_r}; \, \dot{u}_r = v_o e^{-Z_p t/m_r}; \, \ddot{u}_r = -v_o(Z_p/m_r)e^{-Z_p t/m_r} \tag{11}$$

Assuming that the velocity of the ram coincides with that of the pile top due to soft impact, and the elastic force of the pile cushion is neglected. Consequently, the moving mass m equals the pile weight m_p plus the ram weight m_r, the acceleration of the moving mass a is represented by the hammer acceleration $\ddot{u}_r$ after impact. Applying to Eqs. (8) and using (11), yields averaging values of the force exerted on the pile and the initial force of the moving mass after impact, respectively

$$\langle F_{dyn} \rangle = \langle Z^p \dot{u}_r \rangle = \frac{Z_p}{t_2 - t_1} \int_{t_1}^{t_2} v_o e^{-Z_p t/m_r} dt = \frac{m_r v_o}{t_2 - t_1} \left(e^{-Z_p t_1/m_r} - e^{-Z_p t_2/m_r} \right) \tag{12}$$

$$\langle m \ddot{u}_r \rangle = -\frac{m_r + m_p}{t_2 - t_1} \int_{t_1}^{t_2} v_o (Z_p/m_r) e^{-Z_p t/m_r} dt = -\frac{(m_r + m_p) v_o}{t_2 - t_1} \left(e^{-Z_p t_1/m_r} - e^{-Z_p t_2/m_r} \right) \tag{13}$$

2.3 Damping Resistance

During pile penetration, soil damping resistance is generated along the pile shaft as well as at the pile tip at the same time. As formulated in Eq. (2), the total damping resistance of an open-ended pile is the sum $R^D = R^D_{s,e} + R^D_{s,i} + R^D_t$. The damping resistance at the pile tip, R^D_t, can be estimated by [11]

$$R^D_t = Z_{s,t} v_0 \tag{14}$$

In (14), v_0 is particle velocity of the ram at impact, $Z_{s,t}$ is soil impedance given by

$$Z_{s,t} = A_p c_s \gamma_t \tag{15}$$

where c_s is wave transmission speed on the soil at the pile-soil interface, γ_t is the unit weight of soil under the pile tip.

To determine the skin damping resistance of both outside and inside of the pile, we first consider the case of closed-ended piles in which the skin damping along the pile shaft, $R^D_{s,closed}$, can be estimated by [11]

$$R^D_{s,closed} = 0.5\pi v_0 R_c R_R D_e c_s L^w \gamma_s \tag{16}$$

where γ_s is the unit weight of surrounding soil, R_c is reduction factor for strain-softening, R_R is reduction factor for remolding/disturbance, D_e is pile external diameter, L^w is the length of the stress wave propagating down the pile that is calculated by

$$L^w = 2L_r c_p / c_r \tag{17}$$

where L_r is the length of the ram, c_r is the speed of stress wave in the ram. It is seen from (16) and that R_c, R_R, D_e, L_r, c_s, c_r, c_p, v_0 and γ_s can be treated as constants. In the case of open-ended pile driving, the unit weight of soil inside the pile will vary with the penetration depth z depended on the formation of soil plugging states. Hence, we may use the skin damping model (16) for the case of the open-ended pile in considering γ_s by the mean of external and internal soil unit weights, γ_e and γ_i

$$\gamma_s = (\gamma_e + \gamma_i)/2 \tag{18}$$

Consequently, the total skin damping of the open-ended pile can be calculated by

$$R^D = R^D_{s,e} + R^D_{s,i} + R^D_t = \left(\frac{\pi}{2} R_c R_R D_e L_r \frac{c_s c_p}{c_r} (\gamma_e + \gamma_i) + A_p c_s \gamma_t\right) v_0 \tag{19}$$

2.4 Friction Resistance

As formulated in Eq. (2), the total friction resistance of an open-ended pile is

$$R^S = R^S_{s,e} + R^S_{s,i} + R^S_t = \sum A_{s,e} f_{s,e} + \sum A_{s,i} f_{s,i} + R^S_t \tag{20}$$

where $f_{s,e}$ and $f_{s,i}$ are unit external and internal skin frictions at the pile shaft, $A_{s,e}$ and $A_{s,i}$ are outer and inner surface areas of the pile. A full description of skin frictions and tip friction has been recently investigated in considering the soil plugging effect [9]. Following this, the formulation of $f_{s,e}$ is of the form

$$f_{s,e} = K_e \sigma'_{v,e} \tan \delta_e \tag{21}$$

where δ_e is the pile-soil friction angle at the external interface, K_e is the lateral earth pressure coefficient, $\sigma'_{v,e}$ is effective vertical soil stress

$$\sigma'_{v,e} = \gamma_e z \tag{22}$$

The formulation of $f_{s,i}$ is also considered to be the same as that of $f_{s,e}$

$$f_{s,i} = K_i \sigma'_{v,i} \tan \delta_i \tag{23}$$

where K_i, $\sigma'_{v,i}$, δ_i have similar definitions as K_e, $\sigma'_{v,e}$, δ_e but for the internal pile-soil interaction. Due to the plugging effect, i.e. unplugged, partially plugged, and fully plugged states during pile driving, the values of K_i and $\sigma'_{v,i}$ vary with the pile penetration z. Detail analysis and explicit formulations of those quantities are available in [9] but are not presented again in this paper due to their long expressions.

Also depending on plugged states, the tip friction can be calculated by [9]

$$R_t^S(z) = A_{tip} q_b(z) = \begin{cases} \pi \left(r_e^2 - r_i^2\right) q_b(z), & \textit{unplugged \& partially plugged} \\ \pi r_e^2 q_b(z), & \textit{fully plugged} \end{cases} \tag{24}$$

where r_e, r_i are outer and inner pile radius, $q_b(z)$ is the stress under the pile tip or unit ultimate end-bearing

$$q_b(z) = N_q \sigma'_v(z) \tag{25}$$

and N_q is the bearing capacity factor [5], and $\sigma'_{v,t}(z) = \gamma_t z$ is the effective vertical stress of soil beneath the pile tip.

Substituting (21), (23), (24) into (20) we have

$$R^S = \sum A_{s,e} K_e \sigma'_{v,e} \tan \delta_e + \sum A_{s,i} K_i \sigma'_{v,i} \tan \delta_i + A_{tip} q_b \tag{26}$$

2.5 Performance Analysis of Pile Driving with the Averaging Model

Substituting obtained results of the force exerted on the pile by Eq. (12), the initial force by Eq. (13), the damping resistance by Eq. (19), and the damping resistance by Eq. (26) into Eq. (6), we get the averaging model

$$\begin{aligned} &\sum A_{s,e} K_e \sigma'_{v,e} \tan \delta_e + \sum A_{s,i} K_i \sigma'_{v,i} \tan \delta_i + A_{tip} q_b + \left(\frac{\pi}{2} R_c R_R D_e L_r \frac{c_s c_p}{c_r}(\gamma_e + \gamma_i) + A_p c_s \gamma_t\right) v_0 \\ &= F_{sta} + \frac{2m_r + m_p}{t_2 - t_1}\left(e^{-Z_p t_1/m_r} - e^{-Z_p t_2/m_r}\right) v_o \end{aligned} \tag{27}$$

From Eq. (27), the velocity of the ram when it strikes the pile can be determined

$$v_0 = \frac{\sum A_{s,e}K_e\sigma'_{v,e}\tan\delta_e + \sum A_{s,i}K_i\sigma'_{v,i}\tan\delta_i + A_{tip}q_b - F_{sta}}{\frac{2m_r+m_p}{t_2-t_1}\left(e^{-Z_pt_1/m_r} - e^{-Z_pt_2/m_r}\right) - \frac{\pi R_cR_RD_eL_r(\gamma_e+\gamma_i)c_sc_p}{2c_r} - A_pc_s\gamma_t} \quad (28)$$

Using (28), we can determine the kinetic energy at impact, the potential energy, and the required stroke of the ram by the following equations

$$E_{impact} = m_rv_0^2/2 \quad (29)$$

$$E_{potential} = \frac{E_{impact}}{\eta} = \frac{1}{2}\frac{m_rv_0^2}{\eta} \quad (30)$$

$$h = \frac{1}{2}\frac{v_0^2}{g\eta} \quad (31)$$

where η is hammer efficiency.

It is noted from the formulation of v_0 in (28) that $v_0 = 0$ as $F_{sta} = R^S$, this will give an upper bound of F_{sta}. Physically, the extra static force takes the role of reducing friction resistance. Otherwise, $v_0 \to \infty$ as the denominator of (28) tends to zero, that $\langle F_{dyn}\rangle = R^D$, this will give a lower bound of the average dynamic force $\langle F_{dyn}\rangle$.

3 Case Study Examination

In this case study, performance analysis of a mini pile driver commonly used in piling road safety barrier post or solar farm array construction is carried out for the design process. The pile driver (Fig. 4), which has the model number MHP-01, is a hammer-equipped machine combined with a hydraulic press to provide dynamic and static driving forces. This device has been successfully developed by Experimental Research Center of Construction Machines of the University of Transport and Communications. The input parameters of pipe pile, soil, and pile driver are given as follows

- Pile: pile diameter $D_e = 141$ mm, $D_i = 129$ mm, pile weight $m_p = 15$ kg, pile elastic modulus $E_p = 210$ GPa, the embed length of the pile $L = 2.6$ m
- Soil: clay with soft, stiff, and hard types according to API clay descriptions [5] (surrounding soil and soil at the pile tip are considered to be homogeneous)
- Pile driver: ram weight $m_r = 280$ kg, ram length $L_r = 0.5$ m, the maximum velocity at impact $v_{0\,\max} = 2.4$ m/s, the maximum impact force $F_{\max} = 6.5$ T, the maximum impact energy $E_{impact\,\max} = 800$ Nm.

Based on the pile-soil interaction model in [9], the plugging states formed during driving are considered to depend on the penetration depth z as follows: an unplugged state within $z/D_i = [0, 5]$, partially plugged state within $z/D_i = [5, 15]$, fully plugged state within $z/D_i = [15, 20]$. Then the length of upper and lower soil columns are $h_1 = 5D_i$ and $h_2 = 10D_i$, respectively. Based on empirical investigations in [7, 8, 9, 10, 11], the values of R_c, R_R, c_s, c_r, c_p in Eq. (19) are taken by $R_R = 0.01, R_c = 0.01, c_s = 668$ m/s, $c_r = 5700$ m/s, $c_p = 5000$ m/s, $t_1 = 0, t_2 = 0.005$ s.

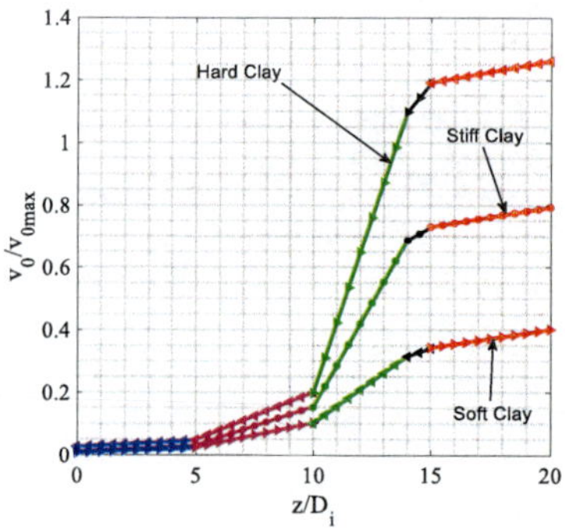

Fig. 5. $v_0/v_{0\,\max}$ versus z/D_i with various clay types

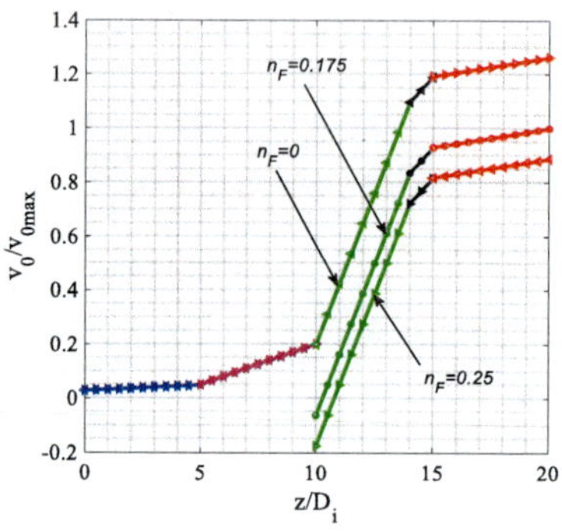

Fig. 6. v_0 versus z/D_i with n_F varies (hard clay)

The simulation of pile driving for this open-ended pile is implemented in two scenarios

- Scenario 1: impact driving ($F_{sta} = 0$).
- Scenario 2: impact driving with an extra static driving force ($F_{sta} = n_F F_{\max}$, where n_F is the ratio factor).

In general, the largest soil resistance is caused by hard clay, then by stiff clay, then soft clay. As we can see in Fig. 5, the maximum ratio $v_0/v_{0\,\max}$ for hard clay, stiff clay, and soft clay respectively are 1.26, 0.79, and 0.4. This means the impact driving needs to supplement extra driving force in the case of hard clay. Obviously, the deeper the penetration, the larger the value of v_0 is required. In detail, for hard clay, the required value of v_0 within the unplugged state and the early stage of the partially unplugged state is small according to the small external and internal skin frictions, i.e. $v_0/v_{0\,\max} < 0.2$; then v_0 needs a high rise in value within the late stage of the partially unplugged state ($z/D_i = [10; 15]$), i.e. $v_0 = v_{0\,\max}$ at $z = 13.5D_i$ and up to $v_0 = 1.2v_{0\,\max}$ at $z = 15D_i$, since the full formation of the soil plug; v_0 needs the largest value within the fully plugged state ($z/D_i = [15, 20]$), up to $v_0 = 1.26v_{0\,\max}$ at $z = 15D_i$, since the open-ended pile behaves like a close-ended one. The influence of extra static force F_{sta} on v_0 is shown in Fig. 5 where F_{sta} is supplemented within the late stage of the partially unplugged state ($z/D_i = [10; 15]$) and fully plugged state ($z/D_i = [15, 20]$) with various values of the ratio factor n_F. It is seen that $n_F = 0.175$ is the minimum value for reducing the value of required impact velocity to $v_{0\,\max}$ for hard clay. It is noted that with $n_F = 0.175$, the extra static force should be applied after the pile penetration reaches to $z = 13.5D_i$ when the applied force balance the friction force, $F_{sta} = R^S$, as above mentioned in analyzing Eq. (28).

Similar to the analysis procedure for v_0, the expressions of kinetic energy are presented in Fig. 7 and Fig. 8. Since the effect of squared value in velocity, the change in slope of the kinetic energy ratio slightly differs from that of the velocity ratio for plugging states. As we can see in Fig. 6, at the end of the driving process when the fully plugged has occurred, the maximum kinetic energy E_{impact} for hard clay requires 2.53

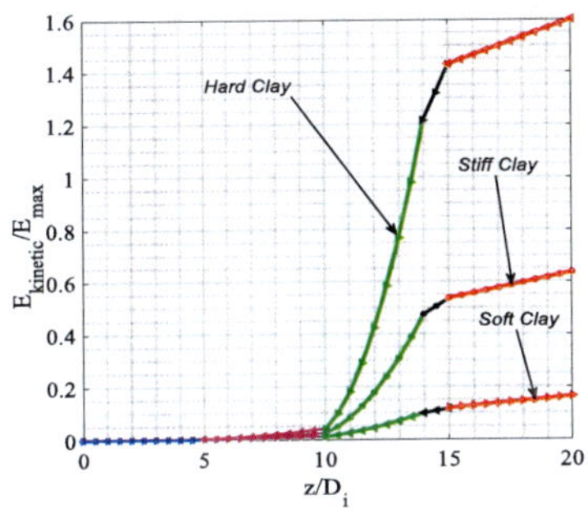

Fig. 7. $E_{impact}/E_{impact\,\max}$ versus z/D_i with various clay types

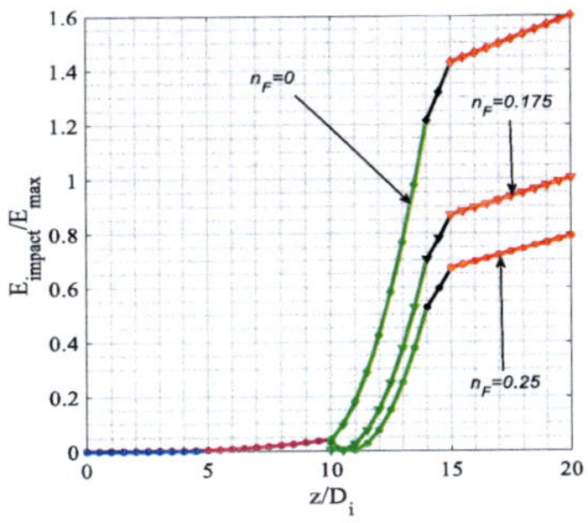

Fig. 8. $E_{impact}/E_{impact\,\max}$ versus z/D_i with n_F varies (hard clay)

times that of stiff clay and 10 times that of soft clay. Figure 8 shows the kinetic energy ratio $E_{impact}/E_{impact}^{\max}$ versus z/D_i with various values of n_F. When applying extra static force $F_{sta} = 0.175F_{\max}$, the required impact velocity reduces from $v_0 = 1.26v_{0\,\max}$ to $v_0 = v_{0\,\max}$, then the corresponding kinetic energy reduces from $1.26^2E_{impact}^{\max}$ to $E_{impact}^{\max}$.

4 Conclusions

From the achievements of the paper, the following conclusions could be drawn

a) An averaging model of the driver-pile-soil system for analyzing open-ended pile driving is developed in which the forces exerted on the pile and the inertia of moving mass are considered in time average. The averaging model is then performed for a hammer-pile-soil system supplemented by a static driving force. The dynamic force exerted on the pile is derived from a dashpot hammer-pile model, the damping resistance is defined by soil impedance based on the wave transmission, while friction resistance is obtained in considering the soil plugging effect during pile driving.
b) As a consequence, the main characteristic quantities of pile driving, such as the velocity of the ram, kinetic and potential energies, are explicitly obtained.
c) An examination of the proposed model is carried out on a Barrier post pile driver. It shows a good prediction of the performance analysis of the pile driver.

References

1. De Nicola, A., Randolph, M.F.: The plugging behavior of driven and jacked piles in sand. Geotechnique **47**(4), 841–856 (1997)
2. Paik, K., Salgado, R., Lee, J., Kim, B.: Behavior of open-and closed-ended piles driven into sands. J. Geotech. Geoenv. Eng. **129**(4), 296–306 (2003)
3. Liu, J., Zhang, Z., Yu, F., Xie, Z.: Case history of installing instrumented jacked open-ended piles. J. Geotech. Geoenv. Eng. **138**(7), 810–820 (2012)
4. Han, F., Ganju, E., Salgado, R., Prezzi, M.: Comparison of the load response of closed-ended and open-ended pipe piles driven in gravelly sand. Acta Geotech. **14**(6), 1785–1803 (2019). https://doi.org/10.1007/s11440-019-00863-1

5. API (American Petroleum Institute): Geotechnical and foundation design considerations. API, Washington DC, USA (2011)
6. Gravare, C.J., Goble, G.G., Rausche, F., Likins, G.: Pile driving construction control by the case method. F Ground Eng. **13**(2), 20–25 (1980)
7. Aghayarzadeh, M., Khabbaz, H., Fatahi, B., Terzaghi, S.: Interpretation of dynamic pile load testing for open-ended tubular piles using finite-element method. Int. J. Geomech. **20**(2), 04019169 (2020)
8. Salgado, R., Bisht, V., Prezzi, M.: Pile driving analysis for pile design and quality assurance (Joint Transportation Research Program Publication No. FHWA/IN/JTRP-2017/15). Purdue University, West Lafayette (2017)
9. Linh, N.N, Ngoc, N.A., Kuu, N.V., Diem, N.D.: Weighted dual approach to an equivalent stiffness-based load transfer model for jacked open-ended pile. J. Appl. Comput. Mech. **7**(3), 1751–1763 (2021). https://doi.org/10.22055/JACM.2021.37430.3013
10. Deeks, A.J., Randolph, M.F.: Analytical modeling of hammer impact for pile driving. Int. J. Numer. Anal. Methods Geom. **17**(5), 279–302 (1993)
11. Massarsch, K.M, Fellenius, B.H.: Prediction of ground vibrations induced by impact pile driving. In: The Sixth International Conference on Case Histories in Geotechnical Engineering, Missouri University of Science and Technology, 12–16 August 2008, Virginia, 38 p. (2008)

Shaping the Controller Gain of a Velocity Feedback Active Isolation with Time Delay

La Duc Viet[1](✉), Nguyen Van Hai[1], and Nguyen Tuan Ngoc[2]

[1] Institute of Mechanics, Vietnam Academy of Science and Technology, 264 Doi Can, Hanoi, Vietnam
ldviet@imech.vast.vn

[2] Freyssenet Vietnam, 11 Tran Hung Dao, Hanoi, Vietnam

Abstract. In a velocity feedback active vibration isolation, the time delay limits the controller gain, which should be shaped to be frequency-dependent. This paper applies a frequency range selector to shape the controller gain. The selector is a tuned combination of the isolated mass's acceleration and velocity. This approach has simple static switching law, suppresses high frequency control and does not add additional phase difference in control loop. Numerical simulations are performed to illustrate the tuning effect.

Keywords: Velocity feedback · Active isolation · Time delay · Phase-lag

1 Introduction

A passive isolator works without external power but still has some limited performance. Improved isolator using active control becomes more popular because of the rapid development of smart sensors, smart actuators, and microprocessors [1–5]. While various control strategies were discussed [1, 4], only the absolute velocity feedback control has been widely used in commercial active vibration isolation [4, 6]. Although the velocity feedback strategy is unconditionally stable in principle, the phase-lag due to time delay can destabilize the system [7, 8]. It causes the high-frequency instability if a high-gain feedback is used.

Some approaches can be used such as hybrid control [3, 9] or velocity feedforward of base platform [10]. However, those require modifying the isolator's mechanism or increasing the number of sensors. To keep the simplicity, the feedback gain should be frequency-dependent, which is low at high frequency to avoid the instabilities. The introduced conventional low-pass filter in the feedback loop can not solve the problem because of the additional phase-lag. Some recent approaches [11, 12] have used the switching type filter. This paper applies the frequency range selector presented in [13, 14] to the active isolation problem to shape the controller gain. To illustrate the shaping effect, active control of a single-degree-of-freedom system is studied. The proposed approach does not require more sensor than the conventional velocity feedback controllers. Some numerical simulations are presented to show the effectiveness of the frequency range selector.

N. V. Khang et al. (Eds.): ASIAN MMS 2021, MMS 113, pp. 845–851, 2022.
https://doi.org/10.1007/978-3-030-91892-7_81

2 Instability Due to Time Delay

Consider a velocity feedback isolation of a single degree-of-freedom (SDOF) base excited system (Fig. 1).

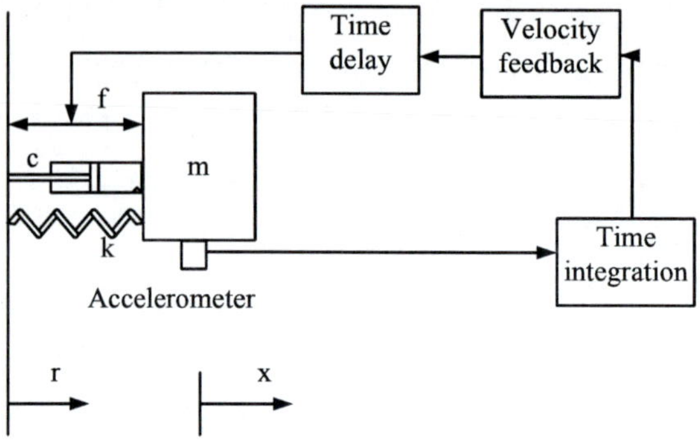

Fig. 1. Active vibration isolator with time delay

The isolator has stiffness k, damping c and control force f to support a rigid mass m. The foundation motion r results in the absolute displacement x of the isolated mass. The acceleration is integrated to supply to the velocity feedback control. A time delay is introduced in the feedback loop.

The equation of motion has form [1]:

$$m\ddot{x} + c\dot{x} + kx = c\dot{r} + kr + f \tag{1}$$

Consider the harmonic responses:

$$x = Xe^{j\omega t}, r = Re^{j\omega t} \tag{2}$$

where X and R are two complex amplitudes, ω is the vibration frequency and j is the imaginary unit. The control scheme in Fig. 1 gives:

$$f(t) = -g\dot{x}(t - \Delta) = -jg\omega Xe^{j\omega(t-\Delta)} \tag{3}$$

where Δ is the physical time delay. Substituting (2) and (3) into (1) give the transmissibility:

$$T = \frac{X}{R} = \frac{j\omega c + k}{\left(k - m\omega^2\right) + j\omega c + jg\omega e^{-j\omega\Delta}} \tag{4}$$

Denote $\omega_n = \sqrt{k/m}$ as the natural frequency, $\Omega = \omega/\omega_n$ as the normalized forcing frequency, $\zeta = c/\left(2\sqrt{km}\right)$ as the damping ratio, $\tau = \omega_n\Delta/(2\pi)$ as the normalized delay. We reduce the transmissibility (4) to:

$$T = \frac{1 + j2\zeta\Omega}{1 - \Omega^2 + j2\zeta\Omega + \frac{g}{m\omega_n}j\Omega e^{-j2\pi\Omega\tau}} \tag{5}$$

The critical frequency and the critical gain at which the system becomes unstable are determined by setting the denominator of (5) to zero. Expanding the denominator of (5) into real and imaginary parts and setting them to zeros give:

$$\begin{cases} 1 - \Omega^2 + \dfrac{g}{m\omega_n}\Omega \sin(2\pi\Omega\tau) = 0 \\ 2\zeta\Omega + \dfrac{g}{m\omega_n}\Omega \cos(2\pi\Omega\tau) = 0 \end{cases} \tag{6}$$

The critical frequency Ω_c and the critical gain g_c then are determined from:

$$\tan(2\pi\Omega_c\tau) = \frac{1 - \Omega_c^2}{2\zeta\Omega_c} \tag{7}$$

$$\frac{g_c}{c_{crit}} = \frac{-\zeta}{\cos(2\pi\Omega_c\tau)} \tag{8}$$

Where $c_{crit} = 2m\omega_n$ is the critical damping. These results were presented in [7]. If the delay is small, the critical frequency and critical gain are approximated as $1/(4\tau)$ and $1/(8\tau)$, respectively [7]. This instability occurs at high frequency and due to the phase-lag of time delay.

3 Shaping Control Gain

Because the instability occurs at high frequency, it is desired to shape the control gain such as it decrease with frequency. The conventional low pass filters can provide the desired frequency-dependent gain but the additional phase-lag (due to low-pass filter) can result in more instability. To avoid adding phase-lag, this paper applies an effective frequency range selector, which was presented in [13, 14].

The active control force is shaped as:

$$f_m(t) = \begin{bmatrix} -g\dot{x}(t) & \ddot{x}^2(t) < \omega_s^2\dot{x}^2(t) \\ 0 & \text{otherwise} \end{bmatrix} \tag{9}$$

in which ω_s is a selector frequency. The effect of the selector is explained as follows. If the forcing frequency ω is larger than ω_s, over a period, the interval over which $\ddot{x}^2(t) > \omega_s^2\dot{x}^2(t)$ is larger than that when $\ddot{x}^2(t) > \omega_s^2\dot{x}^2(t)$ (Fig. 2), and conversely (Fig. 3). Therefore, at high frequency, the shaped control force (9) is mostly zero. The selector, therefore, can provide the desired shaped gain. More important, when $\dot{x}(t)$ reaches its extreme values, as seen from (9), the modified control force $f_m(t)$ is in phase with the original one $f(t)$ $(= -g\dot{x}(t))$ for any excitation frequency. This means there is no more phase-lag is added to the loop.

In brief, the effects of the selector are drawn as:

- At high frequency region, the gain is mostly zero that reduces the high-frequency destabilization.
- The tuning selector frequency shapes the gain. If $\omega_s = 0$, the control force returns to the passive one ($f_m = 0$). Conversely, if $\omega_s = \infty$, the control force changes to the full active one $f_m(t) = -g\dot{x}(t)$.
- The selector works like a low-pass filter but adds no more phase-lag, which can destabilize the system.

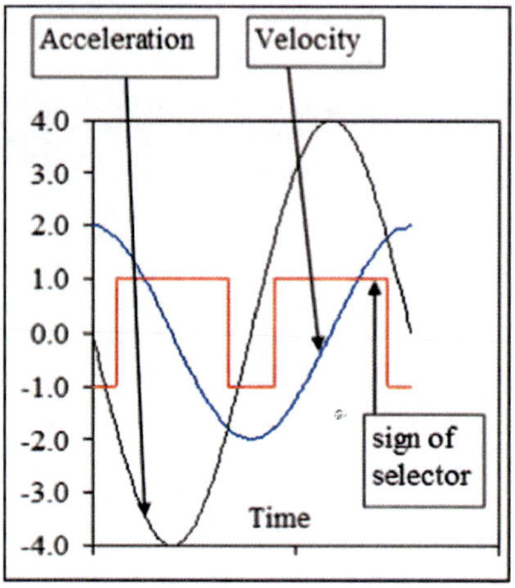

Fig. 2. Sign of selector over time when $\omega > \omega_s$

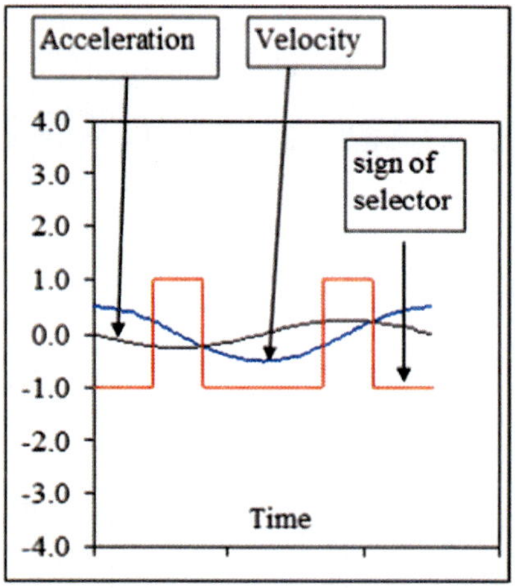

Fig. 3. Sign of selector over time when $\omega < \omega_s$

4 Numerical Verification

The differential Eq. (1) with the control force (9) is solved numerically. Because we only consider the dimensionless responses, m is normalized as unit, k is varied by varying the natural frequency ω_n, c is determined from the critical damping. The following procedure [13] is carried out to obtain the frequency response.

- The sinusoidal disturbance r with a predefined frequency and with unit amplitude is applied to the equation.
- For each predefined frequency, the simulation time is taken large enough (50 natural periods) to eliminate the transient responses.

– The discrete Fourier Transform is used to compute the amplitude of the output at the corresponding frequency.

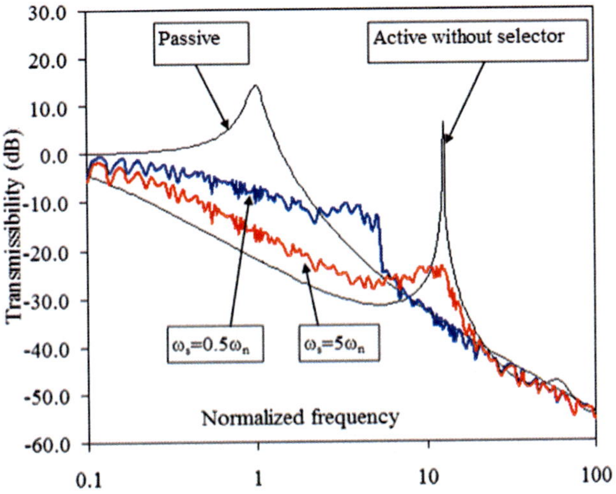

Fig. 4. Frequency response of transmissibility, $\tau = 0.02$

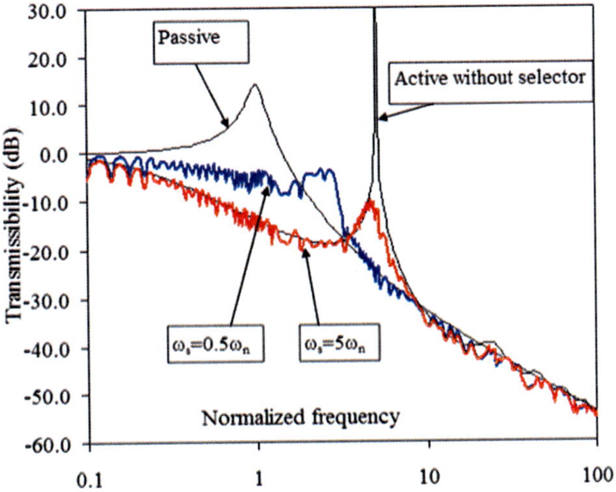

Fig. 5. Frequency response of transmissibility, $\tau = 0.05$

Some cases of delay are investigated in the simulations. To see the unstable behavior, the normalized gains in the calculation are taken from (8). The frequency responses are plotted in Figs. 4, 5 and 6 for various values of normalized time delay τ and selector frequency ω_s.

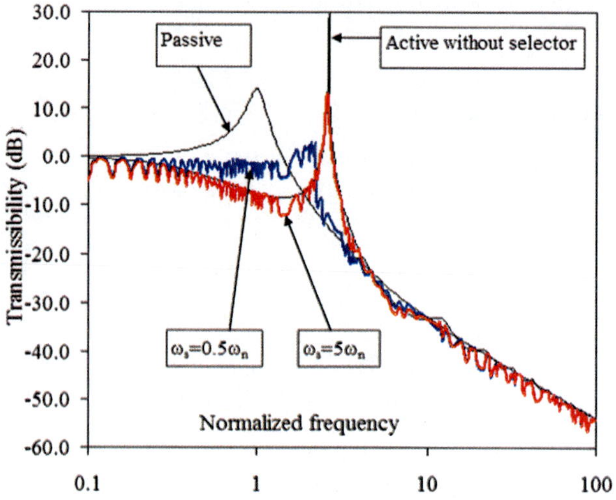

Fig. 6. Frequency response of transmissibility, $\tau = 0.1$

In all the figures, the influences of the selector frequency ω_S are clear as discussed above. The smaller ω_S moves the curve closer to that of the passive case, and also reduces the unstable behaviors in the high frequency range. The important point here is that the modified control force (9) has the desired frequency-dependent gain but does not introduce the phase-lag.

5 Conclusion

To reduce the unstable behavior due to high frequency phase-lag, this paper applies a frequency-range-selector to shape the controller gain of an active isolation system using velocity feedback. The feedback gain is simply shaped by a selector frequency. The desired frequency-dependent gain is produced without adding the phase-lag. Numerical simulations showed the expected effect of the frequency selector.

Acknowledgement. This paper is funded by Vietnam Academy of Science and Technology under grant number "VAST01.04/21-22" and "NVCC03.08/21-21".

References

1. Preumont, A.: Vibration Control of Active Structures. Springer-Verlag, Heidelberg (2011)
2. Liu, C., Jing, X., Daley, S., Li, F.: Recent advances in micro-vibration isolation. Mech. Syst. Sig. Process. **56–57**, 55–80 (2015)
3. Sun, L.L., Hansen, C.H., Doolan, C.: Evaluation of the performance of a passive–active vibration isolation system. Mech. Syst. Sig. Process. **50–51**, 480–497 (2015)
4. Kerber, F., Hurlebaus, S., Beadle, B.M., Stobener, U.: Control concepts for an active vibration isolation system. Mech. Syst. Sig. Process. **21**, 3042–3059 (2007)

5. Hurlebaus, S., Gaul, L.: Smart structure dynamics. Mech. Syst. Sig. Process. **20**(2), 255–281 (2006)
6. Balas, M.J.: Direct velocity feedback control of large space structures. J. Guidance Control **2**(3), 252–253 (1979)
7. Brennan, M.J., Ananthaganeshan, K.A., Elliott, S.J.: Instabilities due to instrumentation phase-lead and phase-lag in the feedback control of a simple vibrating system. J. Sound Vibr. **304**, 466–478 (2007)
8. Agrawal, A.K., Yang, J.N.: Effect of fixed time delay on stability and performance of actively controlled civil engineering structures. Earthquake Eng. Struct. Dyn. **26**, 1169–1185 (1997)
9. Wang, M., Li, X., Chen, X.: Active hybrid control algorithm with sky-hook damping and lead-lag phase compensation for Multi-DOFs ultra-low frequency active vibration isolation system. Shock Vibr. **2017** (2017). Article ID 1861809
10. Yan, T.H., Pu, H.Y., Chen, X.D., Li, Q., Xu, C.: Integrated hybrid vibration isolator with feedforward compensation for fast high-precision positioning X/Y tables. Meas. Sci. Technol. **21**(6) (2010). Article ID 065901
11. Saikumar, N., Sinha, R.K., Hossein Nia Kani, H.: Constant in gain lead in phase element - application in precision motion control. IEEE/ASME Trans. Mechatron. **24**(3), 1176–1185 (2019)
12. van den Eijnden, S.J.A.M., Knops, Y., Heertjes, M.F.: A hybrid integrator-gain based low-pass filter for nonlinear motion control. In: 2018 IEEE Conference on Control Technology and Applications, Copenhagen, Denmark, 21–24 August 2018 (2018)
13. Savaresi, S.M., Poussot-Vassal, C., Spelta, C., Sename, O., Dugard, L.: Semi-Active Suspension Control Design for Vehicles. Butterworth-Heinemann (2010)
14. Savaresi, S., Spelta, C.: A single sensor control strategy for semi-active suspension. IEEE Trans. Control Syst. Technol. **17**(1), 143–152 (2009)

Investigation Nonlinear Dynamic Behavior of the Compliant Tristable Mechanism

Van Chinh Truong, Cong Hoc Hoang, and Ngoc Dang Khoa Tran(✉)

Faculty of Mechanical Engineering, Industrial University of Ho Chi Minh City, Ho Chi Minh City, Vietnam
trannngocdangkhoa@iuh.edu.vn

Abstract. Compliant tristable mechanisms have three stable equilibrium positions in the linear motion as well as energy expenditure. In this paper, a research of the dynamic response of the compliant mechanism, which comprises the two compliant bistable mechanisms, is presented. The kinetostatic behavior of the mechanism is analyzed in the motion. Two operation stages are defined, the first stage is the bending of a bistable mechanism and the second stage is the movement of the tristable mechanism. The dynamic modeling of the tristable mechanism created for each operation stage is archived. The finite element method also supports investigating the natural frequency modes. A prototype is fabricated to investigate the characteristic of the mechanism. The experiment from the prototype defines the damping ratio. An experiment for the dynamic response was set up and implemented to verify the results from the modeling model.

Keywords: Bistable mechanims · Nonlinear behavior · Stable position · Frequency

1 Introduction

Complaint multistable mechanisms keep three distinct positions or more in their range motion. There are operations based on the deflections of the flexible elements. The compliant tristable mechanism is one of the kinds of this mechanism; it consists of three separate positions in operation. Many researchers are attracted to studying this mechanism due to the advantage in self-locking positions, no need for lubrication, and saving energy. They have many potential applications in valves [1], sensors [2], microsystems [3].

Due to the complexity of the elastic behavior of the compliant mechanism, almost all studies research the kinetostactic and dynamic characteristics of the compliant bistable mechanisms [4–6], which have two stable equilibrium positions in the motion. Compliant tristable mechanisms receive less attention than the bistable mechanism in the researches. A few reports present the design of the compliant tristable mechanism, such as a compliant mechanism constructed based on the four linkage bar model, which is achieved three stable positions when two flexible beam bends [7]. Oberhammer et al. [8]

N. V. Khang et al. (Eds.): ASIAN MMS 2021, MMS 113, pp. 852–861, 2022.
https://doi.org/10.1007/978-3-030-91892-7_82

introduce a micro tristable structure that effortlessly holds the position based on the latching mechanism. A bi-directional tristable mechanism device work on the tension flexible beams and rigid bodies that operate in two-directional in translate motion are demonstrated by Chen et al. [9]. Chen and Yu [10] synthesis two Young bistable mechanisms that generate a tristable mechanism. Several configurations of tristable mechanisms from the bistable mechanisms were additionally studied, Chen et al. [11] employ two bistable mechanisms and place them in two directions to create an orthogonally oriented compliant mechanism. The connection of a series of bistable mechanisms to design tristable mechanics is observed and investigated [12, 13].

The development of the energy harvester devices impulses the researchers to investigate the dynamic response of the bistable mechanism. The dynamic response is analyzed theoretically and experimentally, which shows that the bistable structure has a good frequency distinguishing capacity for mechanical shock pulses [14]. Ando et al. [15] apply the buckled beam bistable mechanism with the piezoelectrics to investigate the nonlinear behavior of the mechanism. The dynamic response of the nonlinear behavior of the bistable mechanism is also surveyed through the snap-through movement [16]. Our initial prototype energy harvester demonstrates that the compliant bistable mechanism featuring negative stiffness outperforms the conventional vibration energy harvester in the infra-low frequency range [17, 18]. The subharmonic motion of the bistable mechanism explored in the frequency sweep investigation assists in predicting the precious nonlinear harmonic behavior [19]. Moreover, most of the study of dynamic behavior involves the bistable mechanism. The prediction of the harmonic behavior of the compliant tristable mechanism has many challenges due to the complexity of the steady-state of flexible elements in the mechanism.

In the article, the design of the compliant tristable mechanism is demonstrated and analyzed the nonlinear behavior in Sect. 2. Section 3 investigates the dynamic model of the mechanism. Section 4 introduces the setup experiment and discusses the results.

2 Complaint Tristable Mechanism

The design of the compliant tristable mechanism is employed by Tran et al. [20]. The mechanism includes two masses. They connect through to two springs. One end of spring 1 is fixed to the anchor, the other end of spring 1 links to mass 1 and mass 1 also connect with spring 2 in one end. The other end of spring 2 connects to mass 2. Sping 1 and spring 2 are two compliant bistable mechanisms. Figure 1a demonstrates the compliant tristable mechanism and the components of the mechanism. The coordinate system is similarly performed in this figure. During operation, the external force F is applied to mass 2 of the mechanism. The spring 2 compress and releases, it causes mass 2 to move in the y-direction. Because spring 1 is constructed by a compliant bistable mechanism with the nonlinear characteristic, the jump phenomenon applies to the mechanism that induces mass 1 to move quickly from the first stable position P_1 to the second stable position P_2 with a distance of y_1. Mass 2 keeps the position without any external energy implemented. Figure 1b displays mass 2 in the second stable position and the spring 2 deformed. When mass 2 continually impinged an external force that is large enough, spring 1 also compressed and released, making mass 1 move down the position in y-direction. It also brings the mass translate from the second stable position P_2 to the third stable position P_3 with a distance of y_2. Spring 1 is also constructed by a compliant mechanism, so the jump phenomenon also utilizes mass 1. Figure 1c displays the compliant tristable mechanism in the third stable position and the deflection of spring 1 and spring 2.

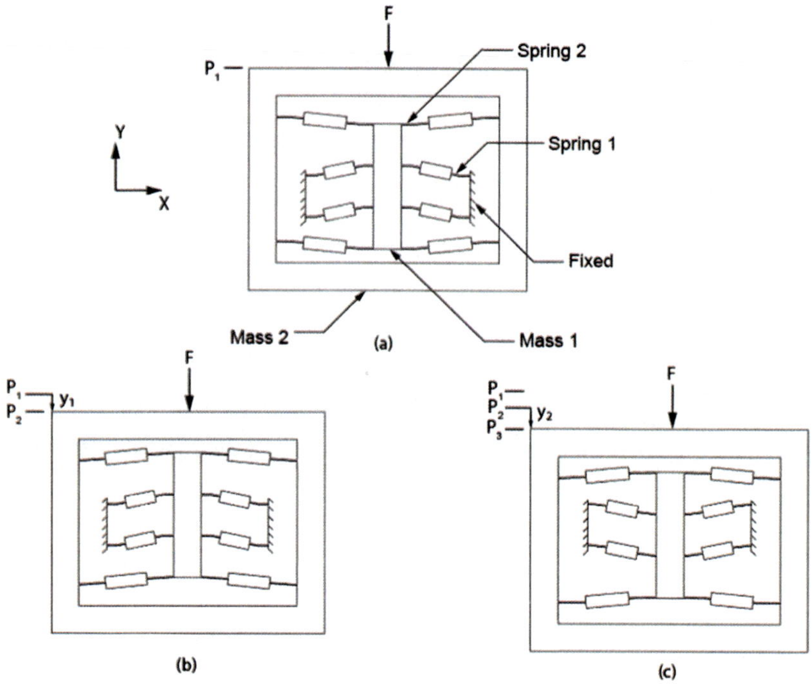

Fig. 1. Design and operation of the compliant tristable mechanism, a) the first stable position, b) the second stable positon, c) the third stable position of the mechanism.

The elastic characteristic of the springs in the tristable compliant mechanism is analyzed with the assistance of the finite element method. The material of the mechanism is Polyoxymethylene (POM), with Young's modulus is 2150 Mpa. The structure is homogenous in thickness. Fixed end of spring 1 attaches to the anchors, the holes in the masses support the fabrication process.

In this investigation, the operation of the mechanism includes two stages. In the first stage, the mechanism move up and down between the first stable position and the second stable position. In this stage, only the spring 2 is active. The mass 1 is almost consistent. Figure 2a illustrates the nonlinear characteristic of spring 2 in the first stage operation. The force increases from zero to the maximum force in the forward motion, $F_{max1} = 1.344$ (N). This is the compress action in spring 2. After the force reaches the maximum value, the force decreases until zero, the mechanism archive the unstable position and the spring 2 release the energy make the mass 2 jump instantly from the unstable position to the second stable position P_2, display in the black arrow in the figure while the force move to negative force. In the reverse motion, mass 2 returns from the second stable position to the first stable position, the force move from the zero value at P_2 and reaches the $F_{min1} = 0.978$ (N), follow the direction of the red arrow in this figure. The curve of backward and forward motion is coincident. The distance y_2 of between the first stable position and the second stable position is 6.71 mm.

The second stage operation of the mechanism is the deflection of spring 1. The performance of the nonlinear behavior of spring 2 is shown in Fig. 2b. When the mass reaches the second stable position, the force increase to the second maximum value, $F_{max2} = 5.2$ (N), then the spring 2 release the energy causes the force decrease quickly to zero value, at this unstable position. Due to the plastic deformation of spring 1, the mass 1 move down and bring the mass 2 jump to the third stable position P3, the black arrow in this figure indicates the direction of the forward movement of the mechanism. In the backward motion, the mass 2 move from the third stable position P_3 to the second stable position P_2 and stand in the first stable position P_1. The red arrow displays the direction of backward motion. The mechanism archive the third stable position at 13.8 mm. This action also determines two minimum values are F_{min1} and F_{min2}, are 0.978 (N) and 5.96 (N), respectively. However, the curves of backward and forward motion are different.

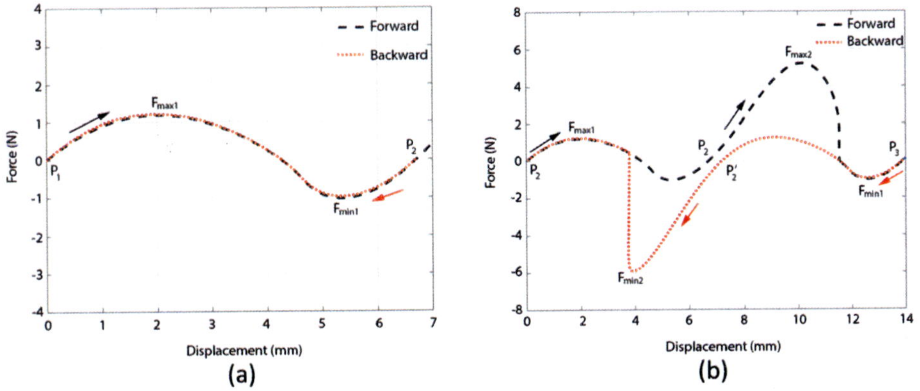

Fig. 2. Nonlinear behavior of the springs a) first stage, b) second stage.

3 Dynamic Modelling

As a result of the symmetric of the compliant mechanism, a quarter model is considered to analyze the harmonic vibration. The 2 degree-of-freedom model spring-mass vibration system is employ to predict the dynamic response of the mechanism. Here, m_1 and m_2 are the masses of the mechanism. k_1 and k_2 refer to the spring stiffness of spring 1 and spring 3. Due to the spring stiffness being nonlinear, the determination of the equation for these springs of such kind mechanisms has complicated challenges in researches [12, 20]. Therefore the interpolation method is employed to define the spring stiffness in the vibration analysis, which k_1 value is estimated based on the curve of Fig. 3a and k_2 value is computed with the curve of Fig. 3b. c_1 and c_2 are the damping ratio of each mass. Sinusoidal vibration is the excitation vibration input in the system. G is the acceleration of the excitation signal. The governing equation of the system is:

$$m_2\ddot{x}_2 + c_2(\dot{x}_2 - \dot{x}_1) + k_2(x_2 - x_1) = 0 \quad (1)$$

$$m_1\ddot{x}_1 + c_1\dot{x}_1 + k_1x_1 + c_2(\dot{x}_1 - \dot{x}_2) + k_2(x_1 - x_2) = Asin(\omega t) \quad (2)$$

$$A = G/\omega^2 \quad (3)$$

In order to define the damping ratio of the system, the experimental technique is employed to predict the value. A prototype of the compliant tristable mechanism is fabricated. The laser displacement sensor implemented to measure the displacement of mass 2 in the free oscillations of the mechanism is applied to archived the damping ratio. The mass 2 is pushed to a small distance and release to create the vibration. The assumed values of the damping coefficient are specified to simulate and archive the results, which are evaluated with the experiment data to predict the accuracy value. Figure 3 brings into comparison results of the finite element method and experimental method. The damping value is determined as 0.009 for the ς_1 and 0.002 for ς_2. The damping ratio can be described as

$$c_1 = 2\varsigma_1\sqrt{k_1(m_1 + m_2)} \quad (4)$$

$$c_2 = 2\varsigma_2\sqrt{k_2m_2} \quad (5)$$

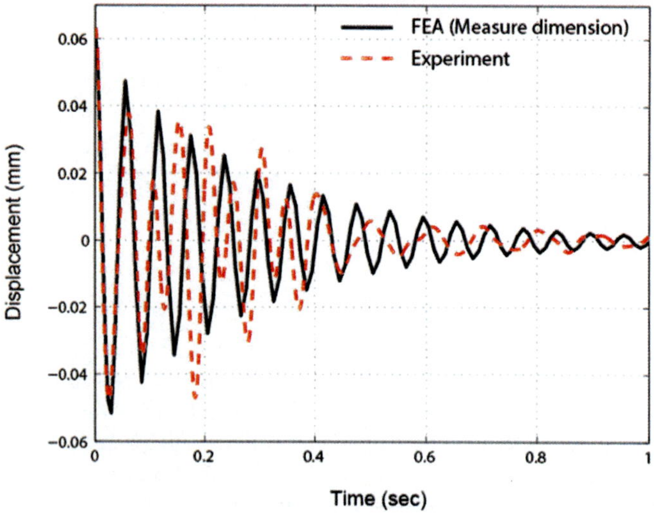

Fig. 3. Damping coefficient of the tristable mechanism.

The investigation of the frequency mode shapes is examined with the finite element method. The first frequency mode shape is defined as the bending of the springs makes that the masses translate in the Y direction, with the frequency around 17 Hz. Figure 4a is the mechanism in the original stage and Fig. 4b performs the mechanism's motion in the first frequency mode of the mechanism. In the second mode, the springs' torsion occurs and makes the mass twist around the anchor due to the unbalance of mass 1 and mass 2 in the movement between stable positions. The value of the second mode is near 25 Hz. Figures 4c and 4d depict the action of the frequency second mode, which are the ordinary stage and the twisting stage of the mechanism.

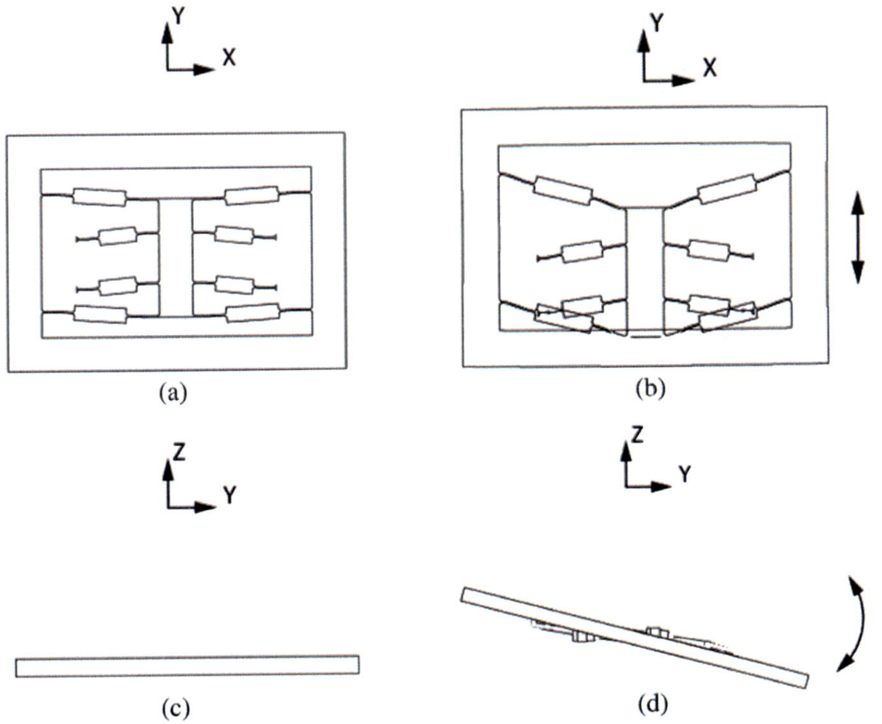

Fig. 4. Mode shapes of the mechanism. (a), (b) The first frequency mode shape. (c), (d) The second frequency mode.

4 Experiment Results

A prototype of the compliant tristable mechanism is set up and implements experiment. Figure 5 demonstrates the system of the frequency response experiment. The model is attached to the shaker. The computer controls the function generator, which creates the excitation sinusoidal signal input. The signal from the function generator transfer to the power amplifier to increase the signal for the shaker. Two laser displacement sensors are the feedback system, one of them measure the displacement of the shaker, another measure the movement of the mass 2 of the model. Both sensors acquire the data and send it to the data acquisition card to transfer to the computer.

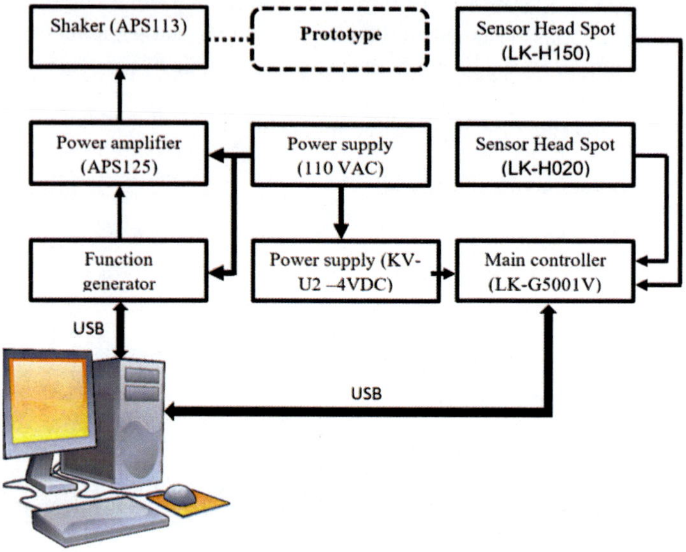

Fig. 5. Function block schematic for frequency harmonic experiment of the tristable model.

Figure 6 describes the flowchart for control the experiment system. In the upsweep experiment, the displacement of the shaker is kept constant at a value is 0.04 mm with many frequencies. When the frequency changed, the sensors wait and obtain the steady stage value of the shaker and compare it with the fixed value. If the shaker's value is not satisfied with the setup value, the amplitude of the function generator increase or decreases belongs to the data acquired. After the displacement of the shaker is recorded, the sensor realizes the vibration of mass 2 active and the data transfer to the computer for processing. The process of the down sweep experiment is likewise to the upsweep frequency experiment.

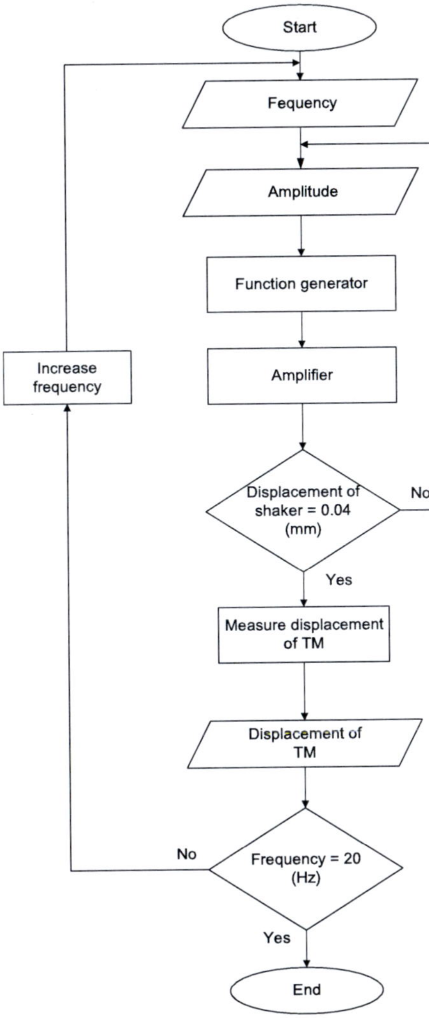

Fig. 6. Control system flowchart for the harmonic testing.

Figure 7 exposes the frequency sweep of the compliant tristable mechanism. Figure 7a illustrate the upward and downward frequency response from Eq. (1) and (2). The frequency of excitation sinusoidal signal is in the range 10–30 Hz. The mechanism's nonlinear behavior demonstrates through the springs' jump phenomenon, which values are around 17.8 Hz for the down sweep and 15.8 Hz for the up sweep. Figure 7b reveals the bandwidth of the mechanism from the experiment. The jump phenomenon appears at 17.4 Hz for upward and 16 Hz for downward frequency response. However, the mechanism also reaches the second frequency mode at frequency 25.5 Hz, which is demonstrated in Figs. 4c and 4d. Due to the nonlinear characteristic of the springs, the mechanism has a narrow bandwidth compare with the linear spring.

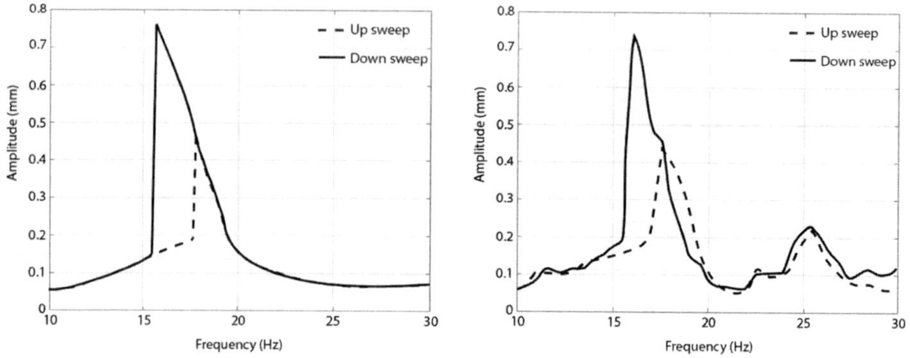

(a) (b)

Fig. 7. Frequency response of the tristable mechanism. a) analytical method, b) experiment.

5 Conclusion

This paper investigates the jump phenomenon characteristic in the harmonic response of a compliant tristable mechanism constructed by two compliant bistable mechanisms. The nonlinear behavior of the mechanism is analyzed by the finite element method and explains the movement during the operation process. A 2 degree of freedom system is employed to drive the mechanism. Two frequency modes are specified based on the finite element method. POM materials fabricate a prototype and the damping ratio of the mechanism is defined by the experimental method. The experimental system for frequency sweep is described. The narrow bandwidth is acquired from 15 to 18 Hz in the softening springs of the tristable mechanism. This mechanism can be assisted in designing the devices which work on the long-range frequency.

References

1. Wang, D.A., Pham, H.T.: Dynamical switching of an electromagnetically driven compliant bistable mechanism. Sens. Actuators A **149**(1), 143–151 (2009)
2. Hansen, B.J., Carron, C.J., Jensen, B.D., Hawkins, A.R., Schultz, S.M.: Plastic latching accelerometer based on bistable compliant mechanisms. Smart Mater. Struct. **16**, 1967–1972 (2007)
3. Huang, S.W., Lin, F.C., Yang, Y.J.: A novel single-actuator bistable microdevice with a moment-driven mechanism. Sens. Actuators A: Phys. **310**, 111934 (2020)
4. Le, G., Chen, G.: A function for characterizing complete kinetostatic behaviors of compliant bistable mechanisms. Mech. Sci. **5**(2), 67–78 (2014)
5. Liu, P., Yan, P.: A modified pseudo-rigid-body modeling approach for compliant mechanisms with fixed-guided beam flexures. Mech. Sci. **8**, 359–368 (2017)
6. Tran, N.D.K., Wang, D.A.: Design of a crab-like bistable mechanism for nearly equal switching forces in forward and backward directions. Mech. Mach. Theory **115**, 114–129 (2017)
7. Pendleton, T., Jensen, B.: Development of a tristable compliant mechanism. In: Proceedings of the 12th IFToMM World Congress (2007)

8. Oberhammer, J., Tang, M., Liu, A.Q., Stemme, G.: Mechanically tri-stable, true single-pole-double-throw SPDT switches. J. Micromech. Microeng. **16**(11), 2251–2258 (2007)
9. Chen, G., Wilcox, D.L., Howell, L.L.: Fully compliant double tensural tristable micromechanisms (DTTM). J. Micromech. Microeng. **19**(2), 025011 (2009)
10. Chen, G., Du, Y.: Double-young tristable mechanisms. J. Mech. Robot. **5**(1), 7 (2013)
11. Chen, G., Aten, Q.T., Zirbel, S., Jensen, B., Howell, L.L.: A tristable mechanism configuration employing orthogonal compliant mechanisms. J. Mech. Robot. **2**, 6 (2010)
12. Wang, D.A., Chen, J.H., Pham, H.T.: A tristable compliant micromechanism with two serially connected bistable mechanisms. Mech. Mach. Theory **71**, 27–39 (2014)
13. Tran, N.D.K.: Configuration compliant tristable mechanism based on characteristics of differential compliant bistable mechanisms. J. Mech. Eng. Res. Dev. **44**(1) (2021)
14. Zhao, J., Huang, Y., Gao, R.J., Wang, H.X.: Dynamics of a bistable mechanism with parallel beams and permanent magnets. Adv. Mater. Res. **308** (2011)
15. Andò, B., Baglio, S., Bulsara, A.R., Marletta, V.: A bistable buckled beam based approach for vibrational energy harvesting. Sens. Actuators A **211**, 153–161 (2014)
16. Du, H., Chau, F.S., Zhou, G.: Harmonically-driven snapping of a micromachined bistable mechanism with ultra-small actuation stroke. J. Microelectromech. Syst. **27**, 34–39 (2018)
17. Casals-Terré, J., Shkel, A.: Snap-action bistable micromechanism actuated by nonlinear resonance. IEEE Sens. (2005)
18. Kim, G.-W., Kim, J.: Compliant bistable mechanism for low frequency vibration energy harvester inspired by auditory hair bundle structures. Smart Mater. Struct. **22**(1) (2013)
19. Huguet, T., Badel, A., Druet, O., Lallart, M.: Drastic bandwidth enhancement of bistable energy harvesters: Study of subharmonic behaviors and their stability robustness. Appl. Energy **226**, 607–617 (2018)
20. Tran, H.V., Ngo, T.H., Tran, N.D.K., Dang, T.N., Dao, T.P., Wang, D.A.: A threshold accelerometer based on a tristable mechanism. Mechatronics **53**, 39–55 (2018)

Vehicle Dynamics and Control

Dynamic Analysis of Continuous Beams with Tuned Mass Damper Under the Moving Vehicles

Trong Phuoc Nguyen[1(✉)], Van Duc Dang[2], and The Tuan Nguyen[1]

[1] Ho Chi Minh City Open University, Ho Chi Minh City, Vietnam
phuoc.nguyen@ou.edu.vn
[2] KumKang Kind Vietnam Ltd. Company, Ho Chi Minh City, Vietnam

Abstract. The dynamic analysis of continuous beams mounted Tuned Mass Damper (TMD) under the moving vehicles is presented in this paper. The model of the moving vehicle is the sprung mass to consist of a mass of vehicle and wheel. The finite element method and dynamic balance are applied to discrete the beam structure and establish the governing equation of motion of the system. The property matrices of the system are established to describe the interaction between the beam, moving vehicle, and TMD with all components of the vehicle inertia forces. The equation of motion of the vehicle - beam - TMD system is solved by using Newmark's algorithm. The numerical results show that the behavior of beams is significantly dependent on the speed of vehicles. In addition, the influence of the parameters as frequency and mass of TMD on the dynamic response of the beam is also investigated.

Keywords: Continuous beam · Dynamic analysis · Finite element method · Moving vehicles · Tuned Mass Damper

1 Introduction

Nowadays, with the fast development of modern construction science and technology; high-rise projects, large span bridges crossing rivers and seas have been built. Along with that, there are urgent issues that need to be solved related to the safety level and longevity of that work under the effects of natural disasters and traffic. The study and understanding of the dynamic behavior of the bridge structure system under the influence of external forces are an issue of urgent concern, which is significant in both theory and practice. The problem of using TMD attached to the structure has been studied and applied in practice: large span bridges, continuous girder structures, runways, railway tracks, and moving trams…

Quite a lot of theoretical and experimental research was carried out, starting with the first TMD proposal by Hermann Frahm (1909) which aimed to reduce the swaying vibrations of the ship under the influence of the actuator. The selection of optimal parameters for TMD was first proposed by Den Hartog (1985), since then the optimal parameters for the TMD mass damping system have been widely researched, developed, and applied.

N. V. Khang et al. (Eds.): ASIAN MMS 2021, MMS 113, pp. 865–875, 2022.
https://doi.org/10.1007/978-3-030-91892-7_83

In 1989, Akin and Mofid [1] proposed a numerical analysis method that could be used to determine the dynamic behavior of beams, with different boundary conditions, subjected to moving mass. In 1992, Das and Dey [2] analyzed the damping effect of multiple tuned mass damper systems in reducing the stochastic response of simple bridge girders by finite element method. In 2005, Lin et al. [3] analyzed the trainload frequency to apply MTMD mass damping systems to prevent trains from causing vibration on the bridge, when the resonance effect will occur when the frequency of the bridge is close to the frequency of the impact of the trainload acting on the bridge. In 2014, Wang et al. [4] implemented a Simulation Study on train-induced vibration control of a long-span steel truss girder bridge by tuned mass dampers. In 2016, Kahya and Araz [5] published series of multiple tuned mass dampers for vibration control of high-speed railway bridges. In 2017, Mokrani et al. [6] mentioned the Passive damping of suspension bridges using multi-degree of freedom tuned mass dampers. In 2019, Fisli [7] studied the dynamic response of a multi-span, orthotropic bridge deck under moving truck loading with tandem axles. Also in this year, Lee and Eun [8] studied lever-type tuned mass damper for alleviating dynamic responses. In 2020, Alhassan et al. [9] studied control of vibrations of common pedestrian bridges in Jordan using tuned mass dampers. In 2021, Miguel and Santos [10] implemented optimization of multiple tuned mass dampers for road bridges taking into account bridge-vehicle interaction.

Through the above studies, it was found that the effectiveness of using the TMD system in damping the construction projects subject to all kinds of loads, especially many significant influencing factors can be mentioned such as 1) Vehicle speed; 2) Characteristics of the vehicle; 3) Number of vehicles; 4) Characteristics of the structure; 5) The number of parameter characteristics of the TMD system. Since then, this article proposes to research and apply the TMD system for the bridge girders that are continuously subjected to vehicle loads (sprung mass). This article analyzes the influence of all the inertial components of the vehicle - girder - TMD and important influencing parameters on the response of the continuous multi-span bridge girder using numerical methods. The problem of vehicle - girder - TMD system interaction is developed with a model that is relatively consistent with actual structures and with satisfactory accuracy.

2 Formulation

The vehicle - girder - TMD system problem model is shown in Fig. 1.

In which L is the length of a girder; E is the elastic modulus of the girder; I is the moment of inertia of the girder cross-section. Respectively, the masses m_1, m_2 are wheel mass and body mass; k_s, c_s are spring stiffness and viscosity damping coefficient of the vehicle; m_d, w_d are mass and vertical displacement of the TMD; k_d, c_d are the stiffness and viscosity damping coefficient of the TMD. The position of the TMD is fixed in the middle of the span of the girder during vehicle movement and the TMD is always firmly attached to the girder during the girder oscillation. The beam is modeled as Euler - Bernoulli beams, uniform cross-section, and single support.

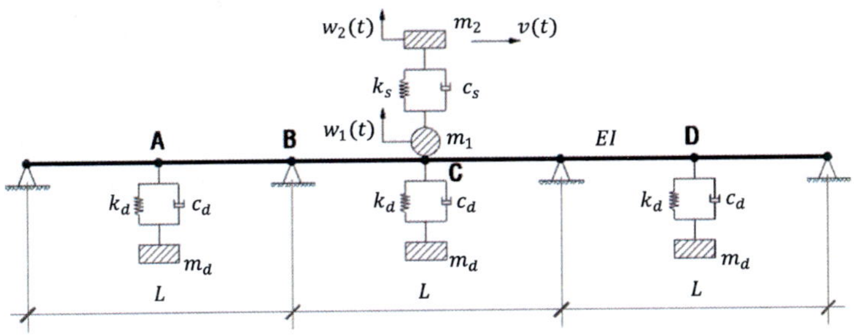

Fig. 1. The vehicle - girder - TMD system model

For the equation of the TMD system, denote the interaction force at the contact position between the bridge and TMD as $F'(t)$ as

$$F'(t) = -m_d g + m_d \ddot{w}_{di} \tag{1}$$

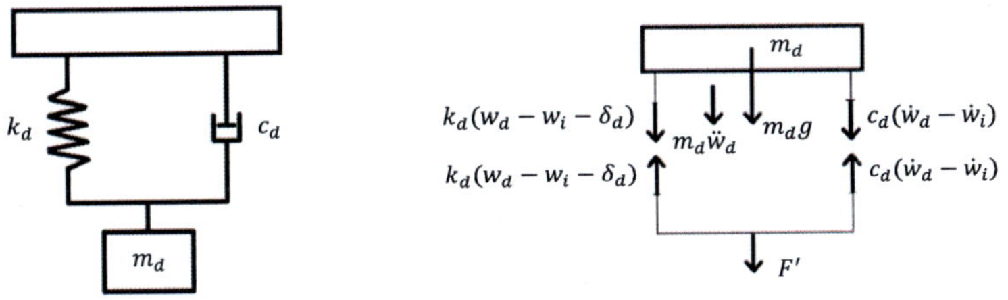

Fig. 2. Model and force equilibrium diagram of the TMD

Where w_i is the corresponding displacement at the position of the TMD in the middle of the span of the girder, and g is the gravity as shown in Fig. 2.

For the equation of vehicle oscillation, the model and force equilibrium diagram are shown in Fig. 3. Respectively, where m_1, w_1 are the mass and vertical displacement of the wheel and m_2, w_2 are the ones of the body of the vehicle; k_s is the spring stiffness and c_s is the viscosity damping coefficient of the vehicle. In real systems, the interaction force between the vehicle and the bridge is normally derived from the degree of roughness of the bridge surface. In this paper, it can be modeled by the wheel is attached to the bridge surface in the vertical direction using the equal degree of freedoms with the bridge is the master node and the vehicle is the slave node. Ignoring the frictional force between the wheel and the bridge surface (horizontal degree of freedoms), and the deformation between the wheel axle and the bridge profile, we have the force interaction $F(t)$ between the vehicle and the bridge as

$$F = -(m_1 + m_2)g + m_1\ddot{w}_1 + m_2\ddot{w}_2 \tag{2}$$

The general oscillation equation of the system is established as

$$[M]\{\ddot{\overline{q}}\} + [C]\{\dot{\overline{q}}\} + [K]\{\overline{q}\} = \{P\} \tag{3}$$

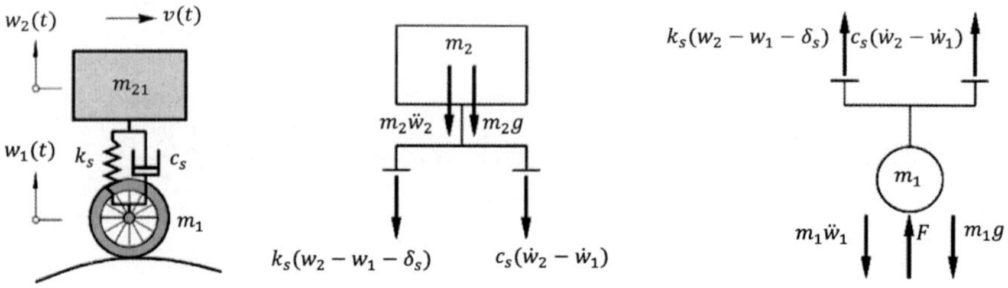

Fig. 3. Model and force equilibrium diagram of vehicle

Where:

$$[M] = \begin{bmatrix} [M_b] + m_1[N]^T[N] & m_2[N]^T & m_d[N_0]^T & m_d[N_0]^T & m_d[N_0]^T \\ 0 & m_2 & 0 & 0 & 0 \\ 0 & 0 & m_d & 0 & 0 \\ 0 & 0 & 0 & m_d & 0 \\ 0 & 0 & 0 & 0 & m_d \end{bmatrix} \quad (4)$$

$$[C] = \begin{bmatrix} [C_b] + 2m_1 v_m[N]^T[N]_x & 0 & 0 & 0 & 0 \\ -c_s[N] & c_s & 0 & 0 & 0 \\ -c_d[N_0] & 0 & c_d & 0 & 0 \\ -c_d[N_0] & 0 & 0 & c_d & 0 \\ -c_d[N_0] & 0 & 0 & 0 & c_d \end{bmatrix} \quad (5)$$

$$[K] = \begin{bmatrix} [K_b] + m_1 v_m^2[N]^T[N]_{xx} + m_1 a_m[N]^T[N]_x & 0 & 0 & 0 & 0 \\ -k_s[N] - c_s v_m[N]_x & k_s & 0 & 0 & 0 \\ -k_d[N_0] & 0 & k_d & 0 & 0 \\ -k_d[N_0] & 0 & 0 & k_d & 0 \\ -k_d[N_0] & 0 & 0 & 0 & k_d \end{bmatrix} \quad (6)$$

$$\{P\} = \begin{Bmatrix} -[N]^T(m_1 + m_2)g - [N_0]^T m_d g - [N_0]^T m_d g - [N_0]^T m_d g \\ 0 \\ 0 \\ 0 \\ 0 \end{Bmatrix} \quad (7)$$

$$\{\ddot{q}\} = \begin{Bmatrix} \ddot{q} \\ \ddot{w}_2 \\ \ddot{w}_{dA} \\ \ddot{w}_{dC} \\ \ddot{w}_{dD} \end{Bmatrix}, \ \{\dot{q}\} = \begin{Bmatrix} \dot{q} \\ \dot{w}_2 \\ \dot{w}_{dA} \\ \dot{w}_{dC} \\ \dot{w}_{dD} \end{Bmatrix}, \ \{q\} = \begin{Bmatrix} q \\ w \\ w_{dA} \\ w_{dC} \\ w_{dD} \end{Bmatrix} \quad (8)$$

$[M_b]$, $[C_b]$, and $[K_b]$ are the mass matrix, damping matrix and stiffness matrix of the bridge girder, respectively. $[N]_x$ and $[N]_{xx}$ are respectively the first and second derivatives of the matrix of shape functions $[N]$ with respect to x; v_m and a_m are the velocity and acceleration of the vehicle along the bridge. The $[N_0]$ is the matrix of shape functions at *A, B, C* as in Fig. 3. In order to analyze this problem, a computer program written by MATLAB language has been implemented to present typically the effectiveness of using the TMD system in the bridge structure.

3 Numerical Investigations

3.1 The Verification of the Computer Program

The program which is applied for numerical analysis is verified and the results then reflect the program's reliability with previous studies. The dynamic responses of the three spans beam subjected to moving load are shown in Figs. 4 and 5. Through the above examples, the numerical results from the computer program based on the MATLAB language with the formulation of this paper show quite good agreement with those in the previous studies. Therefore, this program can be used to analyze the dynamic responses of the beams with the TMD subjected to a moving vehicle in the next part.

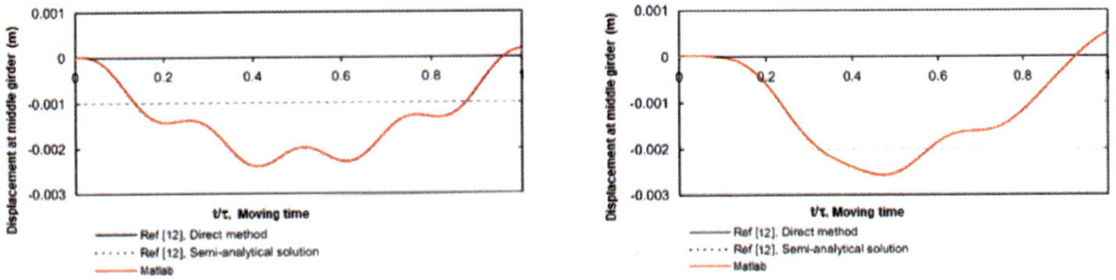

Fig. 4. Comparison of the results of displacements between the program and [12]

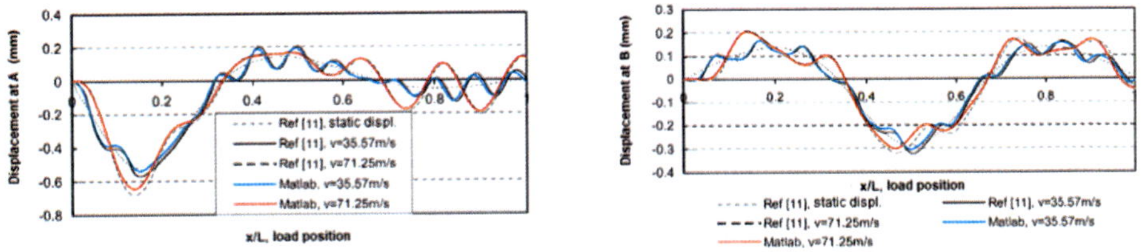

Fig. 5. Comparison of the results of displacements between the program and [11]

3.2 Numerical Analysis

Analysis of continuous Euler – Bernoulli girder with multiple spans and TMD, described in Fig. 6 as

For the parameters of the bridge girder, let length of a girder span $L = 30$ m, cross-sectional area $F = 0.6\ \text{m}^2$, moment of inertia $I = 0.05\ \text{m}^4$, elastic modulus $E = 2 \times 10^{10}\ \text{N/m}^2$, specific-weight $\rho = 2500\ \text{kg/m}^3$, viscous damping ratio $\xi_b = 2\%$. For the

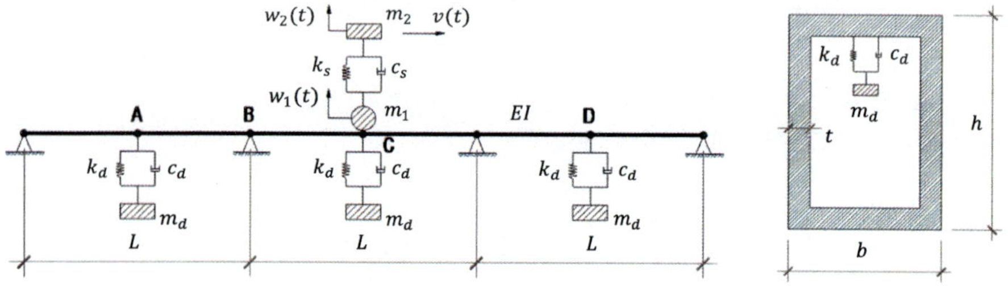

Fig. 6. Diagram of continuous girder of the TMD

parameters of the vehicle, let wheel mass $m_1 = 2500$ kg, vehicle mass $m_2 = 30000$ kg, spring stiffness $k_s = 8.63 \times 10^6$ N/m, damping ratio $c_s = 8.14 \times 10^4$ Ns/m, viscous damping ratio $\xi_v = c_s/2m_2\omega_v$. For the parameters of the TMD, let natural frequency $\omega_d = \sqrt{\frac{k_d}{m_d}}$, viscous damping ratio $\xi_d = \frac{c_d}{2\omega_d m_d}$, mass ratio $\mu = \frac{m_d}{m}$, frequency ratio $\gamma_{opt} = \frac{\omega_d}{\omega}$, where m, ω are the mass of a span girder and the first natural frequency of the girder, respectively. The mass ratio μ preselected from 0–5% for analysis.

Firstly, the effects of the mass ratio μ on response shown in Fig. 7, Fig. 8 and Fig. 9, describe when the mass ratio μ increases from 0% to 5%, the oscillation coefficients of displacement at A, C, and D also decrease respectively.

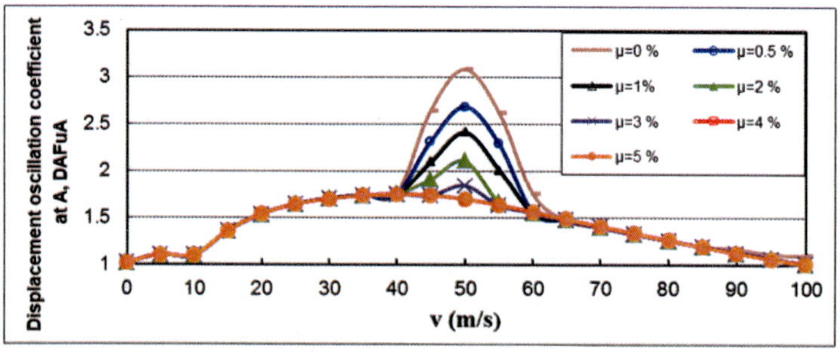

Fig. 7. Oscillation coefficient of displacement at point A when TMD at A

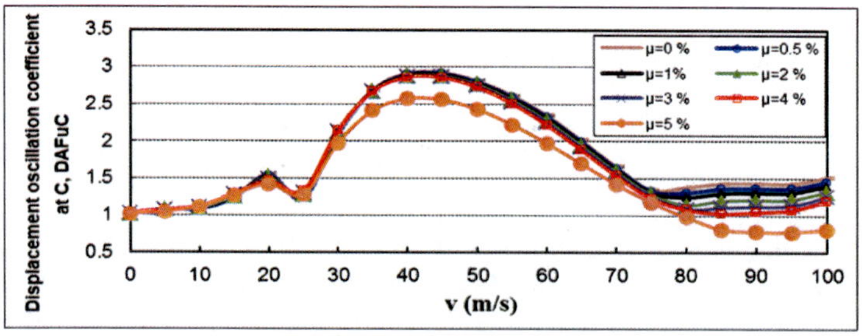

Fig. 8. Oscillation coefficient of displacement at point C when TMD at C

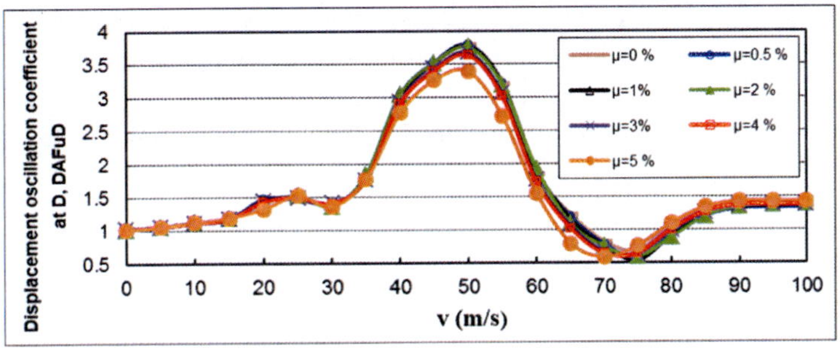

Fig. 9. Oscillation coefficient of displacement at point D when TMD at D

Secondly, the effect of velocity on dynamic responses is derived. Figures 9, 10, 11 and 12 describe when the mass ratio μ increases from 0% to 5%, the oscillation coefficients of the moment at A, C, and D also decrease respectively. And corresponding to the value of the TMD mass ratio is $\mu = 5\%$, the system works relatively efficiently when the velocity reaches a value of 50 m/s. So, the mass ratio of $\mu = 5\%$ and various velocities are chosen to analyze the next problems.

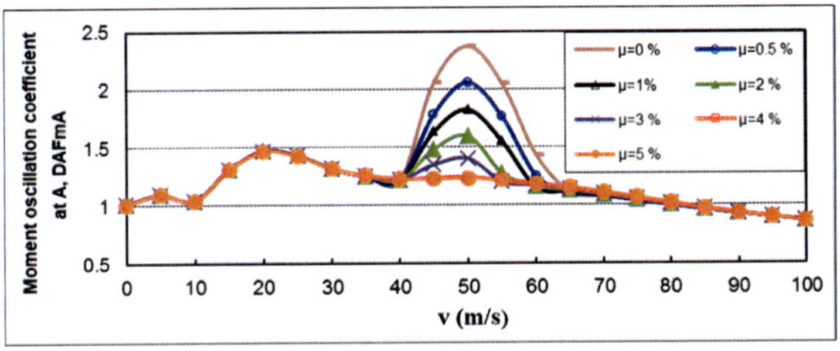

Fig. 10. Oscillation coefficient of moment at point A when TMD at A

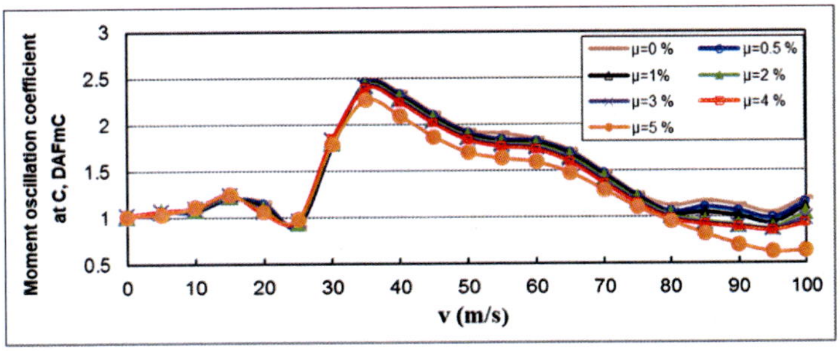

Fig. 11. Oscillation coefficient of moment at point C when TMD at C

Next, from Fig. 13, Fig. 14 and Fig. 15 describe vertical displacements of points A, C, D over time corresponding to fixed value $\mu = 5\%$ and variable velocity from 20 m/s,

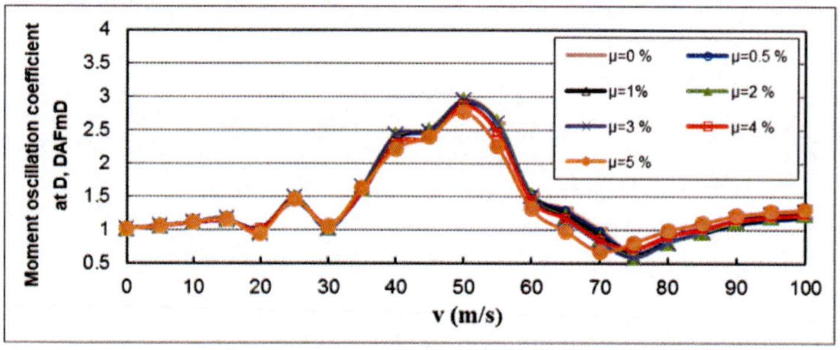

Fig. 12. Oscillation coefficient of moment at point D when TMD at D

40 m/s, 50 m/s, 60 m/s, 80 m/s. From the plots, it can be seen that the displacement of the girder trends to decrease. This is due to the effect of the damper caused by the forced displacement of the TMD against the direction of displacement of the girder.

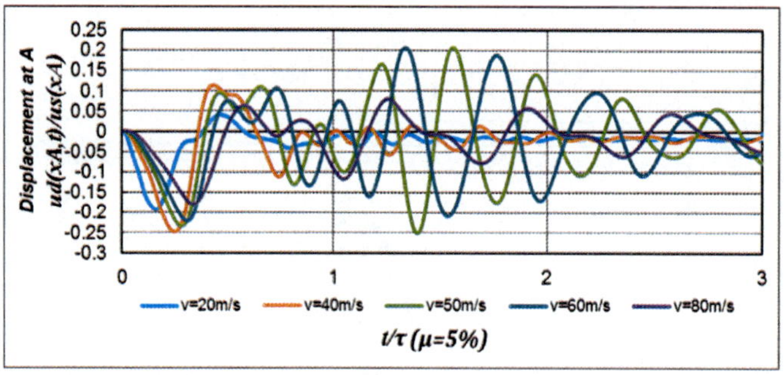

Fig. 13. Displacements at point A over time with TMD at A

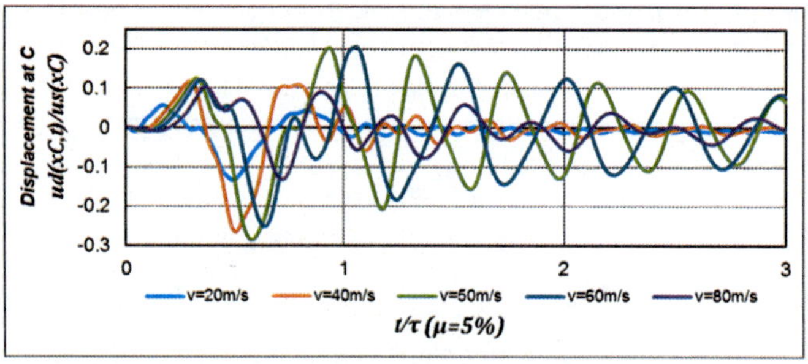

Fig. 14. Displacements at point C over time with TMD at C

Finally, the effect of the number of TMD on responses is given. From Fig. 16, Fig. 17 and Fig. 18 describe the displacements of the girder at A, C, D for the cases where there is no TMD, the case where there is one TMD located at either A, or C, or D; two TMDs located at A, C; and three TMDs located at A, C, D. When the girder

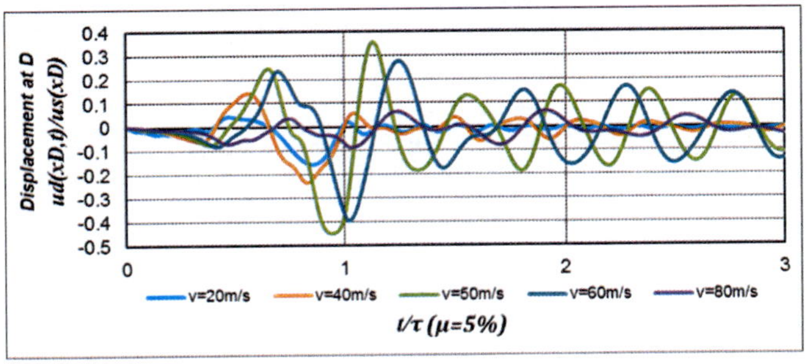

Fig. 15. Displacements at point D over time with TMD at D

is mounted TMDs, the amplitude of the girder's oscillation is decreased obviously and quickly. When increasing the number of TMDs, the oscillation amplitude of the girder is further decreased and the most desirable results when mounting three TMDs at A, C, D at the same time.

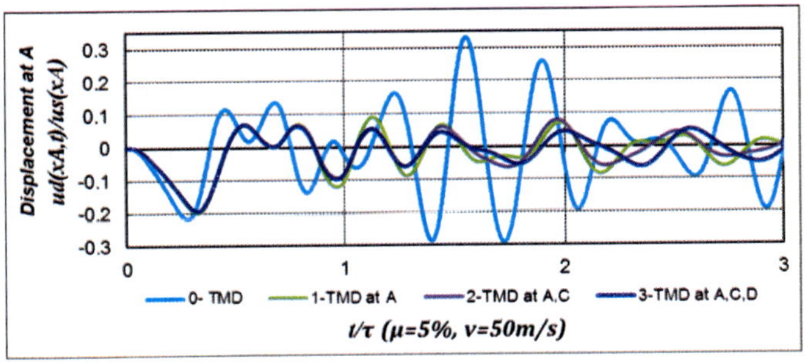

Fig. 16. Displacement of point A over time

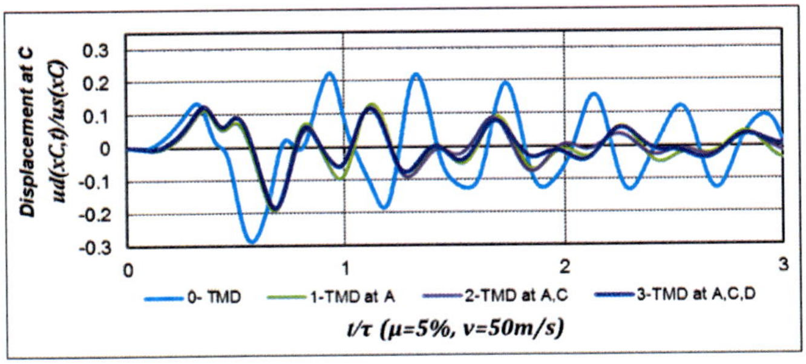

Fig. 17. Displacement of point C over time

In summary, the factors of the mass, number, and position of TMDs are significantly influenced by the dynamic response of multi-span continuous girder.

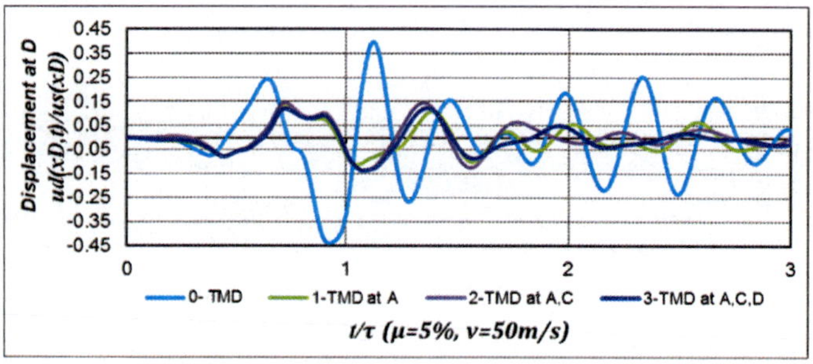

Fig. 18. Displacement of point D over time

4 Conclusion

From the above results of this study, the numerical analyses showed that the response of the girder with the TMDs is significantly reduced. When the number of TMDs increases, the effect of the reductional vibration is better under the moving vehicle. The TMD system expresses positive results and has practical significance in reducing the oscillations of continuous multi-span girder when subjected to vehicle loads. The TMD system expresses positive results and has practical significance in reducing the oscillations of continuous multi-span girder when subjected to moving vehicles.

The results obtained are still limited. The various recommendations need to be researched and developed for the continuous girder mounted TMDs under vehicle load as the influence of uneven bridge surface, displacement analysis with many vehicles, or optimizing the numerical method because the Finite Element Method updated matrices continuously in each time step leading to more difficult if many moving vehicles.

References

1. Akin, J.E., Mofid, M.: Numerical solution for response of beams with moving mass. J. Struct. Eng. **115**(1), 120–131 (1989)
2. Das, A.K., Dey, S.S.: Effects of tuned mass dampers on random response of bridges. J. Comput. Struct. **43**(4), 745–750 (1992)
3. Lin, C.C., Wang, J.F., Chen, B.L.: Train-induced vibration control of high-speed railway bridges equipped with multiple tuned mass dampers. J. Bridg. Eng. **10**(4), 398–414 (2005)
4. Wang, H., Tao, T., Cheng, H, He X.: Simulation study on train-induced vibration control of a long-span steel truss girder bridge by tuned mass dampers. Math. Probl. Eng. (2014). Article ID 506578
5. Kahya, V., Araz, O.: Series multiple tuned mass dampers for vibration control of high-speed railway bridges. Taylor & Francis Group, London. ISBN 978-1-138-02927-9 (2016)
6. Mokrani, B., Tian, Z., Alaluf, D., Meng, F., Preumont, A.: Passive damping of suspension bridges using multi-degree of freedom tuned mass dampers. Eng. Struct. **153**, 749–756 (2017)
7. Fisli, Y., Rezaiguia, A., Guenfoud, S., Laefer, D.F.: Dynamic response of a multi-span, orthotropic bridge deck under moving truck loading with tandem axles. DIAGNOSTYKA **20**(4) (2019)
8. Lee, E.T., Eun, H.C.: Lever-type tuned mass damper for alleviating dynamic responses. Adv. Civi. Eng. (2019). Article ID 5824972

9. Alhassan, M.A., Al-Rousan, R.Z., Al-Khasawneh, S.I.: Control of vibrations of common pedestrian bridges in Jordan using tuned mass dampers. Proc. Manuf. **44**, 36–43 (2020)
10. Miguel, L.F.F., Santos, G.P.: Optimization of multiple tuned mass dampers for road bridges taking into account bridge-vehicle interaction. Random Pavement Roughness, and Uncertainties (2021)
11. Henchi, K., Fafard, M., Dhatt, G., Talbot, M.: Dynamic behaviour of multi-span beams under moving loads. J. Sound Vibr. **199**(1), 33–50 (1997)
12. Neves, S.G.M., Azevedo, A.F.M., Calçada, R.: A direct method for analyzing the vertical vehicle -structure interaction. Eng. Struct. **34**, 414–420 (2012)

Multi-objective Optimization of Hedge-Algebra-Based Controllers for Quarter-Car Active Suspension Models

Hai-Le Bui(✉)

Hanoi University of Science and Technology, Hanoi, Vietnam
le.buihai@hust.edu.vn

Abstract. This work deals with multi-objective optimization of controllers based on the hedge-algebras (HA) theory for active control of quarter-car suspension models. Essential objectives include: minimizing body-car acceleration to ensure occupant comfort and minimizing suspension deflection to ensure system safety. Necessary constraints are: the vehicle's road-holding is guaranteed, the settling time is short, and the maximum control force is within an allowable range. The simulation results show that the HA-theory-based controllers have high efficiency, fast computation time, and short settling time. Moreover, from the Pareto front, established when performing multi-objective optimization, choosing a configuration for the HA-theory-based controller suitable for the controlled model is possible.

Keywords: Quarter-car active suspension · Hedge-algebras · Optimization

1 Introduction

Based on Zadeh's fuzzy set theory, fuzzy controllers have many advantages, and they have been widely applied in various industrial fields. Research on fuzzy control is still a field that attracts much attention from researchers in the world [1, 2].

Fuzzy logic, where the truth value of a variable is a real number between 0 and 1, can be considered an extension of Boolean logic. With a different approach, HA theory studies the real-world-semantics-based interpretability of fuzzy systems and the semantic order of linguistic values [3]. Following HA theory, based on an algebraic structure of a linguistic variable, the fuzziness measure of its linguistic values can also be quantified in the range from 0 to 1. The physical semantic order of these linguistic terms is always preserved [4, 5]. Hence, linguistic values can be digitized instead of using fuzzy sets as in fuzzy set theory. This is an advantage of HA theory because computational operations on real numbers are more convenient than fuzzy sets. Therefore, HA theory has been effectively applied in the problems of fuzzy database, logic programming, classification and regression, forecasting with time series, and linguistic summarization [6].

In particular, it is necessary to emphasize the role of HA theory when applied to the control field. The preliminary study of control based on HA theory was published in

N. V. Khang et al. (Eds.): ASIAN MMS 2021, MMS 113, pp. 876–885, 2022.
https://doi.org/10.1007/978-3-030-91892-7_84

2008 for a simplified pendulum model [7]. Then, the application of HA theory in controlling mechanical models has obtained encouraging results, including swing-up control problem for an inverted pendulum model [8] and vibration control of high-rise buildings subjected to seismic loads [6, 9–15]. Among these publications, the optimization problem of HA-theory-based controllers is also interested in improving the efficiency of the controllers [6, 8, 11, 14].

In the present study, the optimization problem of HA-theory-based controllers is extended in active control of quarter-car models with goals and constraints appropriate for the controlled object.

2 Controlled Model

Consider a quarter-car model with an active suspension, as shown in Fig. 1. In which m_1, m_2, z_1, and z_2 respectively are masses and displacements of the car body and wheel. The suspension and tire stiffness and damping coefficients are k_1, k_2, c_1, and c_2. The bump road profile z_r is determined through its length L and height A, and the car velocity V.

The control force u, placed between masses m_1 and m_2, will be determined through the state variables $x_1 = (z_1\text{-}z_2)$ and $x_2 = (\dot{z}_1 - \dot{z}_2)$, and the controllers represented in Sect. 3.

The state equation for the system is given in Eq. (1).

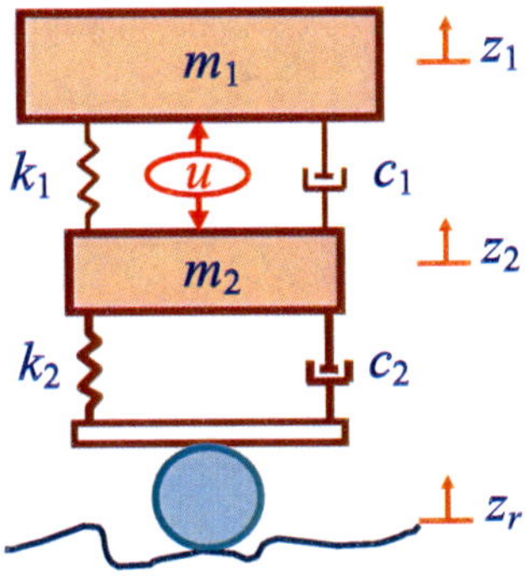

Fig. 1. The quarter-car model with active suspension.

$$\begin{bmatrix} m_1 & 0 \\ 0 & m_2 \end{bmatrix}\begin{Bmatrix} \ddot{z}_1 \\ \ddot{z}_2 \end{Bmatrix} + \begin{bmatrix} c_1 & -c_1 \\ -c_1 & c_1 + c_2 \end{bmatrix}\begin{Bmatrix} \dot{z}_1 \\ \dot{z}_2 \end{Bmatrix} + \begin{bmatrix} k_1 & -k_1 \\ -k_1 & k_1 + k_2 \end{bmatrix}\begin{Bmatrix} z_1 \\ z_2 \end{Bmatrix} = \begin{Bmatrix} 0 \\ c_2\dot{z}_r + k_2 z_r \end{Bmatrix} + \begin{Bmatrix} u \\ -u \end{Bmatrix} \quad (1)$$

The necessary conditions for the parameters in this model include [16]:

(1) The peak value of the car body acceleration max $|\ddot{z}_1|$, this parameter should be minimized to ensure passenger comfort;
(2) The suspension deflection max $|z_1 - z_2|$, this parameter also needs to be minimized for the system lifetime preservation;

(3) The relative tire force $x_3 = k_2(z_2 - z_r)/(g(m_1 - m_2))$, this parameter needs to be less than 1 to ensure the car road holding, where g is the gravity acceleration;
(4) The actuator limitation max $|u| \le u_{max}$, where u_{max} the allowable control force of the actuator; (5) The system settling time should be minimized.

3 Controller Design

Consider a linguistic variable Z with its linguistic values established by the primary terms Negative and Positive (denoted by N and P), and the hedges Little and Very (represented by L and V). Following HA theory, The fuzziness measure of all linguistic values of Z can be quantified between 0 and 1 through their quantitative semantic mapping (SQM). These SQM values faithfully reflect the inherent semantic order of these linguistic values. Figure 2 represents the SQM values of several typical linguistic terms of Z when these linguistic values are represented symmetrically, where W is the neutral element. The operating principle of HA-based controllers is shown in Fig. 3.

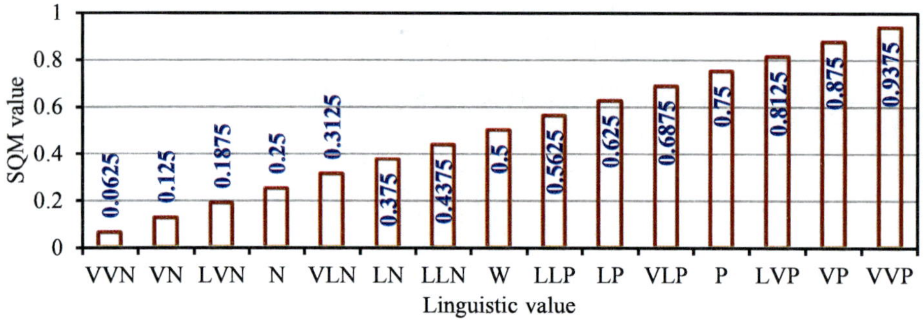

Fig. 2. SQM values of typical linguistic terms.

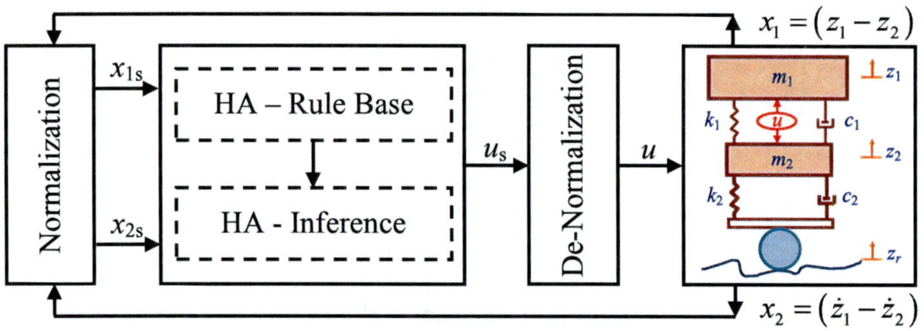

Fig. 3. Operating principle of HA-theory-based controllers.

Where x_{1s}, x_{2s}, and u_s are state and control variables in the semantic domain. Reference ranges and linguistic terms with SQM values of the variables are:

– The state variable x_1: $[-a, a]$; [N(0.25) LN(0.375) W(0.5) LP(0.625) P(0.75)].
– The state variable x_2: $[-b, b]$; [LN(0.375) W(0.5) LP(0.625)].
– The control variable u: [-c, c]; [VN(0.125) N(0.25) LN(0.375) W(0.5) LP(0.625) P(0.75) VP(0.875)].

The Normalization step to convert state variables from the real number domain to the semantic one is shown in Fig. 4a and 4b. The control rule base of the controllers is shown in Fig. 5a. SQM values of the terms in this rule base allow presenting the rule base in terms of a rule-base surface, as plotted in Fig. 5b. When the linguistic terms for variables and the rule base are determined above, the rule-base surface is a plane. The De-Normalization step to convert the control variable from the semantic domain to the real number one is shown in Fig. 4c.

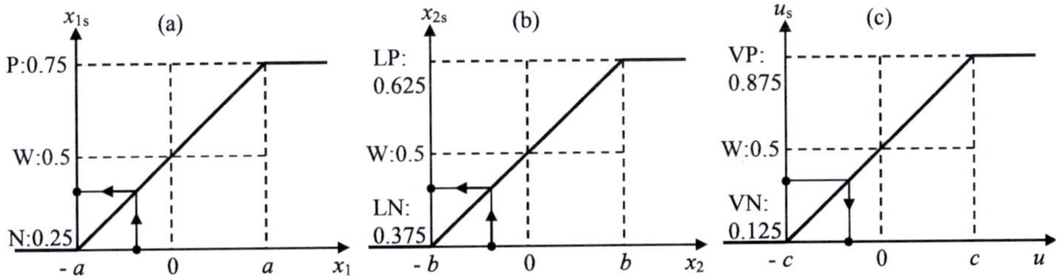

Fig. 4. Normalization and De-Normalization steps.

(a)

x_{1s}	x_{2s}		
	LN	W	LP
N	VP	P	LP
LN	P	LP	W
W	LP	W	LN
LP	W	LN	N
P	LN	N	VN

(b)

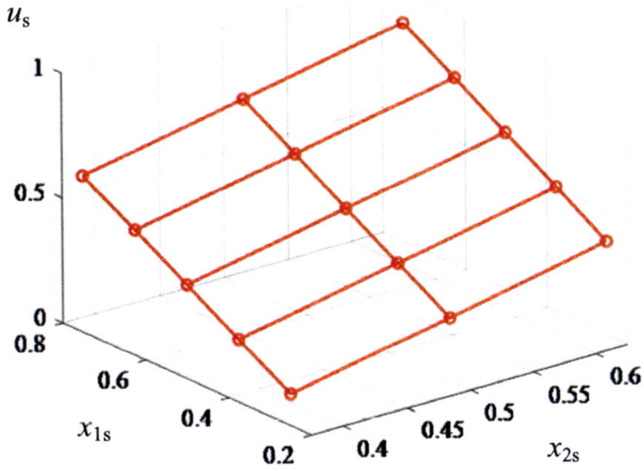

Fig. 5. Control rule base (a) and rule-base surface (b).

Hence, through Figs. 4 and 5, the relationship between the control and state variables can be explicitly expressed as follows [13]:

$$u = \frac{2c}{3a}x_1 + \frac{c}{3b}x_2 \tag{2}$$

It can be seen from Eq. (2) that the reference range of the variables significantly affects the control variable. Normally, these ranges are determined by the designer's experience or by trial and error steps. Therefore, optimizing these reference ranges is necessary to improve the controllers' efficiency and make the controllers more suitable for the investigated model. For the above quarter-car model, the optimization problem is stated as below:

Objective functions:

$$\max|\ddot{z}_1| \rightarrow \min \tag{3a}$$

$$\max|z_1 - z_2| \rightarrow \min \tag{3b}$$

Design variables:

$$a \in [a_1 \div a_2];\ b \in [b_1 \div b_2];\ c \in [c_1 \div c_2] \tag{4}$$

Essential constraints:

$$\max|u| \leq u_{\max} \tag{5a}$$

$$x_3 < 1 \tag{5b}$$

$$\max|u(t_1 \rightarrow t_2)| \leq u* \tag{5c}$$

The constraint (5c) is to reduce the vibration time of the system because after being excited from the bump road profile at the initial time, the controller needs to give a sufficiently small control force (u*) to obtain a short settling time.

4 Simulation Results

This section presents the simulation results for the quarter-car model, as shown in Sect. 2. The parameters of the system are given as: $m_1 = 300$ kg; $m_2 = 40$ kg; $c_1 = 1000$ Ns/m; $c_2 = 10$ Ns/m; $k_1 = 20000$ N/m; $k_2 = 200000$ N/m; $u_{\max} = 1000$ N. The bump road profile z_r is:

$$z_r = \begin{cases} \frac{1}{2}A\left(1 - \cos\left(2\pi t\frac{V}{L}\right)\right) & \text{if } 0 \leq t \leq \frac{L}{V} \\ 0 & \text{if } \quad t > \frac{L}{V} \end{cases} \tag{6}$$

where $A = 0.1$ m, $L = 5$ m; $V = 12.5$ m/s (45 km/h).

The initial design of the HA-based controller (HAC) is selected as follows:

- The state variables' reference ranges (a and b) are determined from the peak values of x_1 and x_2 in the open-loop (OL) case.
- The control variable's reference range (c) is calculated to satisfy the condition: the control force's peak value is equal to $u_{\max}$ (1000 N) (see Fig. 6).

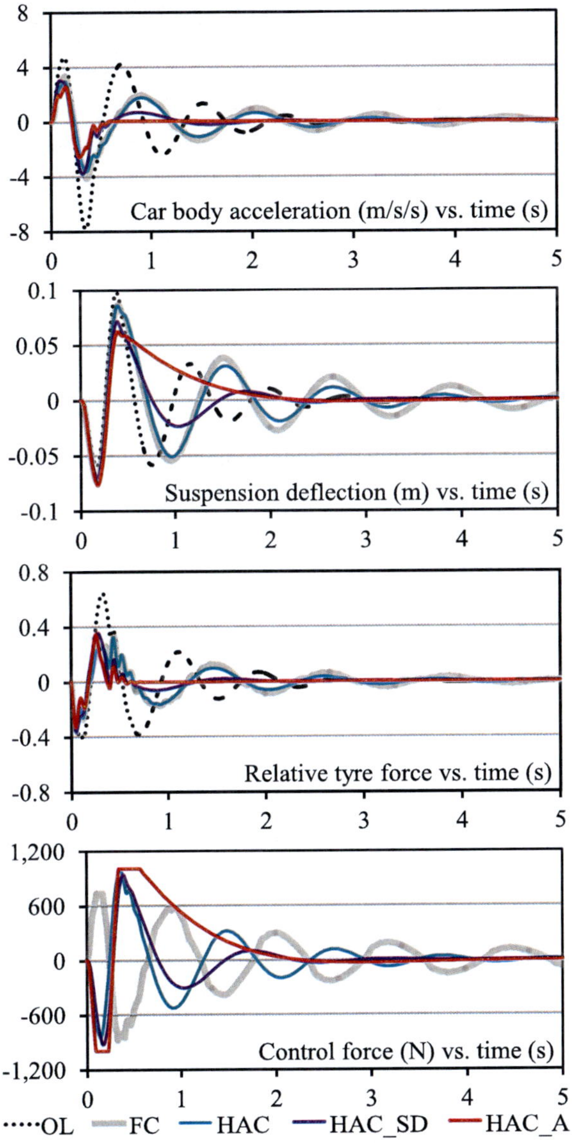

Fig. 6. Time responses of parameters.

- Time responses of car-body acceleration, suspension deflection, relative tire force, and control force are shown in Fig. 6.

In which, fuzzy-set-based controller (FC) is also included for comparison. The variables' reference ranges and control rule-base (Fig. 5a) of FC and HAC are similar. In FC, triangular membership functions are used for the fuzzification step, Mamdani's method is utilized for the inference step, and the defuzzification step is performed by the centroid method.

The notations HAC_SD and HAC_A are explained below.

To improve the efficiency of HAC, the optimization problem as described in Sect. 3 is performed. The optimal controllers using objective functions (3a) and (3b) are denoted HAC_A and HAC_SD, respectively. The variation range of the design variables is $a_1 = 0.1a$; $a_2 = 10a$; $b_1 = 0.1b$; $b_2 = 10b$; $c_1 = 0.1c$; $c_2 = 10c$. The time parameter in the constraint (5c) is $t_1 = 2$s and $t_2 = 5$s. The timed responses of HAC_A and HAC_SD are also shown in Fig. 6. The variation (%) of car-body acceleration (A_1), suspension deflection (SD), relative tire force (RTF), and control force (u) are shown in Fig. 7. The variation of A_1, SD, and RTF is compared with the OL case, and the change of u is compared with u_{max}.

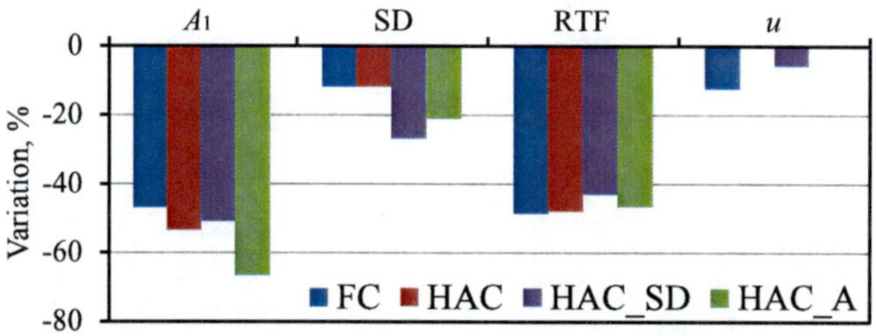

Fig. 7. Variations (%) of parameters.

Results in Figs. 6 and 7 show that the reduction of the car-body acceleration of HAC is higher than that of FC (~54% vs. ~47%). The decrease of SD and RTF criteria of HAC and FC are similar. The maximum control force of FC is about 12% smaller than that of HAC. HAC gives a shorter settling time compared to FC.

For the optimal controller HAC_SD, the reduction of the SD criterion is the best, and the deduction for the A_1 and RTF criteria is approximately equal to that of the HAC and FC controllers. It can be considered that the HAC_A controller gives the highest control efficiency when the car-body acceleration is reduced by ~69%, the SD parameter is diminished by ~21%, and the RTF parameter is varied by ~−47%. The optimal controllers' settling time is short (about 2.3 s), while this parameter of the remaining controllers is greater than 5 s.

From Fig. 7, it can be seen that the criteria A_1 and SD have a trade-off with each other, and therefore, the multi-objective optimization problem is performed to investigate this trade-off with the combined formula of these criteria as follows:

$$k\left(\frac{\max|A_1|}{A_{1OL}}\right) + (1-k)\left(\frac{\max|SD|}{SD_{OL}}\right) \to \min \tag{7}$$

In which A_{1OL} and SD_{OL} are peak values of car-body acceleration and suspension deflection in the OL case. The weight k can be varied between 0 and 1. The Pareto front representing the trade-off level between objectives A_1 and SD is shown in Fig. 8. An optimal controller with a balanced configuration between objectives A_1 and SD can be determined from the Pareto front in Fig. 8.

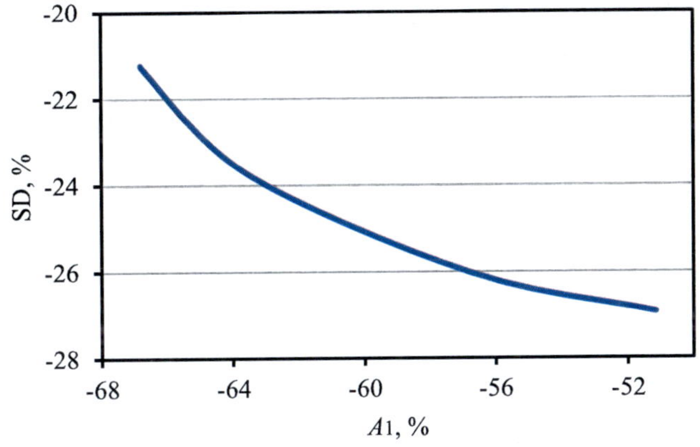

Fig. 8. Pareto front for criteria A_1 and SD.

The explicit formula (2) allows significantly reducing the control force calculation time (CPU time) for HA-based controllers compared with FC. This reduction is clearly shown in Table 1.

Table 1. Computational time (CPU time, s) of controllers.

Controller	FC	HA-based controllers
CPU time, s	~414.081	~0.408

In addition, the controllers presented in this study only use two state variables ($z_1 - z_2, \dot{z}_1 - \dot{z}_2$) to determine the control variable. In comparison, several controllers in other published investigations need up to 4 state variables ($z_1 - z_2$, $z_2 - z_r$, $\dot{z}_1$, $\dot{z}_2$) to calculate the value of control force for the above quarter-car model [16, 17].

5 Conclusion

In this study, the multi-objective optimization problem of HA-based controllers for quarter-car active suspension models is performed. The main results are summarized as follows:

- Using suitable objective functions and constraints, the optimal HA-based controllers provide high control efficiency and short system settling time. The Pareto front between the essential objectives allows selecting an optimal controller that best appropriates the controlled model.
- The explicit formula between state and control variables allows to significantly reduce CPU time of HA-based controllers compared with fuzzy-set-theory-based controllers.

The controllers investigated in this study use a small number of state variables to determine the control force. This is an advantage of these controllers when applied in industrial practice.

Active and semi-active controls for half-car and full-car models and taking into account the nonlinear factors of the models will be interesting problems to expand the research results of this investigation.

Acknowledgement. This work was supported by the Autonomous Higher Education Project (SAHEP) grant funded under number T2021-SAHEP-009.

References

1. Precup, R.-E., Hellendoorn, H.: A survey on industrial applications of fuzzy control. Comput. Ind. **62**(3), 213–226 (2011)
2. Nguyen, A.-T., Taniguchi, T., Eciolaza, L., Campos, V., Palhares, R., Sugeno, M.: Fuzzy control systems: Past, present and future. IEEE Comput. Intell. Mag. **14**(1), 56–68 (2019)
3. Nguyen, C.H., Alonso, J.M.: Looking for a real-world-semantics-based approach to the interpretability of fuzzy systems. In: 2017 IEEE International Conference on Fuzzy Systems (FUZZ-IEEE), Naples, Italy, pp. 1–6. IEEE (2017)
4. Ho, N., Nam, H.: Towards an algebraic foundation for a zadeh fuzzy logic. Fuzzy Set Syst. **129**, 229–254 (2002)
5. Ho, N.C., Van Long, N.: Fuzziness measure on complete hedge algebras and quantifying semantics of terms in linear hedge algebras. Fuzzy Sets Syst. **158**(4), 452–471 (2007)
6. Bui, H.-L., Tran, Q.-C.: A new approach for tuning control rule based on hedge algebras theory and application in structural vibration control. J. Vibr. Control **27**(23–24), 2686–2700 (2021)
7. Ho, N.C., Lan, V.N.: Optimal hedge-algebras-based controller: design and application. Fuzzy Sets Syst. **159**(8), 968–989 (2008)
8. Bui, H.-L., Tran, D.-T., Vu, N.-L.: Optimal fuzzy control of an inverted pendulum. J. Vibr. Control **18**(14), 2097–2110 (2012)
9. Duc, N.D., Vu, N.-L., Tran, D.-T., Bui, H.-L.: A study on the application of hedge algebras to active fuzzy control of a seism-excited structure. J. Vibr. Control **18**(14), 2186–2200 (2012)
10. Anh, N.D., Bui, H.L., Vu, N.L., Tran, D.T.: Application of hedge algebra-based fuzzy controller to active control of a structure against earthquake. Struct. Control. Health Monit. **20**(4), 483–495 (2013)
11. Bui, H.-L., Nguyen, C.-H., Vu, N.-L., Nguyen, C.-H.: General design method of hedge-algebras-based fuzzy controllers and an application for structural active control. Appl. Intell. **43**(2), 251–275 (2015). https://doi.org/10.1007/s10489-014-0638-6
12. Bui, H.-L., Nguyen, C.-H., Bui, V.-B., Le, K.-N., Tran, H.-Q.: Vibration control of uncertain structures with actuator saturation using hedge-algebras-based fuzzy controller. J. Vibr. Control **23**(12), 1984–2002 (2017)
13. Bui, H.-L., Le, T.-A., Bui, V.-B.: Explicit formula of hedge-algebras-based fuzzy controller and applications in structural vibration control. Appl. Soft Comput. **60**, 150–166 (2017)
14. Bui, V.-B., Tran, Q.-C., Bui, H.-L.: Multi-objective optimal design of fuzzy controller for structural vibration control using Hedge-algebras approach. Artif. Intell. Rev. **50**(4), 569–595 (2017). https://doi.org/10.1007/s10462-017-9549-3
15. Tran, D.-T., Bui, V.-B., Le, T.-A., Bui, H.-L.: Vibration control of a structure using sliding-mode hedge-algebras-based controller. Soft. Comput. **23**(6), 2047–2059 (2017). https://doi.org/10.1007/s00500-017-2919-6

16. Gao, H., Sun, W., Shi, P.: Robust sampled-data H-infinity control for vehicle active suspension systems. IEEE Trans. Control Syst. Technol. **18**(1), 238–245 (2009)
17. Sun, W., Gao, H., Shi, P.: Advanced Control for Vehicle Active Suspension Systems. Springer, Switzerland (2020)

Mechatronics

A Novel Bidirectional MRF Based Actuator: Configuration, Optimal Design and Experimental Validation

Hiep Dai Le[1], Qui Duyen Do[1], Khai Vo[1], Bao Tri Diep[2], Van Bo Vu[2], Van Dang Chuong Le[2], Xuan Hung Nguyen[3], and Quoc Hung Nguyen[1(✉)]

[1] Vietnamese German University, Thủ Dầu Một, Binh Duong, Vietnam
hung.nq@vgu.edu.vn
[2] Industrial University of Ho Chi Minh City, Ho Chi Minh City, Vietnam
[3] CERTech Insitute, Ho Chi Minh University of Technology, Ho Chi Minh City, Vietnam

Abstract. In this research, development of a new bidirectional MRF based actuator (BMRA) is presented. In this configuration, the housing and the input shaft of BMRA rotate in opposite direction at the same speed via bevel gear system. On the input shaft, a disc made of magnetic steel is fastened. Another disc, also made of magnetic steel, is fastened to the output shaft of the BMRA. The two discs are embedded inside housing. The gap between two discs and housing is filled by magnetorheological fluid (MRF). Two exciting coils are placed on second disc to generate the required magnetic field to transfer a required torque from the first disc and from the housing to the second disc. When exciting first coil, the output shaft rotates in clockwise direction (the input shaft rotation direction) whereas exciting second coil makes the output shaft rotates in counter-clockwise (the housing rotation direction). The optimal design of BMRA is conducted considering maximum braking torque and minimum the mass of the structure. The balancing composition motion optimization algorithm (BCMO) is employed to solve the optimization problem. A prototype of the optimized BMRA is manufactured to validate the optimal results. Besides, experimental work is performed to evaluate the performance of the prototype. Torque response in clockwise and counter-clockwise at different value of step applied current is measured and compared with the calculated ones. The results show that the novel proposed configuration of BMRA is more compact than other previous configurations in the same transmitting torque. Moreover, the bottle neck magnetic phenomenon of conventional configurations is eliminated.

Keywords: Magnetorheological fluid (MRF) · Magnetorheological brake (MRB) · Bidirectional MRF based actuator (BMRA) · Optimal design

1 Introduction

Haptic devices gain more interest from numerous researchers. Its applications are employed in a wide range of engineering, including medical, aerospace, military, virtual reality devices [1, 2]. In recent years, studies on smart materials and their applications

N. V. Khang et al. (Eds.): ASIAN MMS 2021, MMS 113, pp. 889–898, 2022.
https://doi.org/10.1007/978-3-030-91892-7_85

attract more attention. Magnetorheological fluid (MRF) is one of smart materials, which their rheological properties depend on excitation magnetic flux. The merits of its properties such as fast response time, high controllability, and lower power consumption are deserved for consideration. For this reason, MRF can be applied in various engineering fields such as damper, brake, clutch [5–10]. One of interesting application of MRF is the haptic system using bidirectional actuator (BMRA). Bidirectional actuator (BMRA) using MRF is first proposed by P B Nguyen [11, 12], its output torque can be controlled varying from negative to positive depending on excitation scheme of BMRA. However, this configuration is limited in magnetic flux with higher required torque of BMRA since bottle-neck of magnetic flux. Therefore, in [13, 14], Q H Nguyen et al. proposed a new BMRA configuration, in which the coils are placed on each side of the housing. This configuration solved the bottle-neck magnetic problem in BMRA [11]. Although, the bottle-neck of magnetic flux was solved, the mass burden is still an open problem.

Consequently, the main contribution of this work is to focus on developing the novel configuration BMRA employing in haptic system. This configuration eliminates the bottle-neck of magnetic flux in the configuration of PB Nguyen [11] and could gain lighter mass than the BMRA proposed by Q H Nguyen [13]. BCMO algorithm is used to optimize the proposal configuration with a induced toque is greater than 5Nm in both directions. Then, prototype is fabricated and tested to assess its performance.

2 Configuration

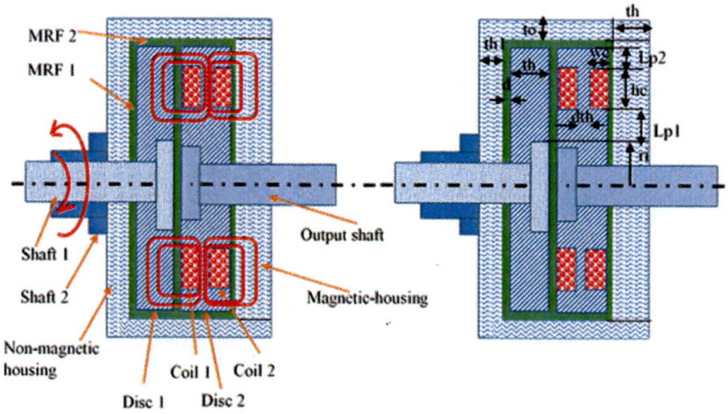

Fig. 1. The proposed BMRA and significant dimensions

Proposed configuration is given in Fig. 1. In this work, the housing is connected with shaft 2 and disc 1 connected with shaft 1. Two input shafts rotate with the same angular velocity and counter rotation. The coils are located on both sides of disc 2 which is connected with the output shaft. Because the structure is not symmetry, distributed magnetic flux in both sides when active each coil is not similar. In this work, the housing comprises non-magnetic and magnetic parts. When coil 1 is excited, the MRF between disc 1 and disc 2 becomes solid-like instantaneously. Therefore, this results in a controllable torque transmitting torque from right side of disc 1 to output shaft connected with disc 2. Then, the output shaft rotates in the same direction as shaft 1. When coil 2 is triggered, the MRF between right side of disc 2 and housing becomes solid-like instantaneously. Therefore, this results in a controllable torque transmitting torque from right side of disc 2 to output shaft connected with disc 2. Then, the output shaft rotates in the same direction as shaft 2.

The braking torque of BMRA comprises mainly end-face torque and annual duct. The frictional moment acting on a single end-face of the disc can be defined as follows [15]

$$T_e = \frac{\pi \mu R_0^4}{2d}\left[1 - \left(\frac{R_i}{R_0}\right)^4\right]\Omega + \frac{2\pi \tau_y}{3}\left(R_0^3 - R_i^3\right) \tag{1}$$

Where R_i, R_o are the inner and outer radius of disc; d is the gap of MRF, Ω is the angular velocity of disc. μ, τ_y are known as post-yield viscosity and yield stress of MRF in gap, respectively. It should be noted that these values depend on exerted magnetic flux across MR gap and calculated by [16]:

$$\mu = \mu_\infty + (\mu_0 - \mu_\infty)(2e^{-B\alpha_\mu} - e^{-2B\alpha_\mu}) \tag{2}$$

$$\tau_y = \tau_{y\infty} + (\tau_{y0} - \tau_{y\infty})(2e^{-B\alpha_{\tau y}} - e^{-2B\alpha_{\tau y}}) \tag{3}$$

where τ_{y0} and μ_0 are the yield stress and the post-yield viscosity of the MRF at zero applied field respectively, $\tau_{y\infty}$ and μ_∞ are respectively the yield stress and the post-yield viscosity of the MRF and at the saturated applied field, $\alpha_{\tau y}$ and α_μ are the saturation moment index of the yield stress and the viscosity respectively, B is the applied magnetic density. For the MRF132-DG produced by Lord corporation, which is employed in this research, the curve fitting parameters determined from experiment are: $\mu_0 = 0.1 pa \cdot s$; $\mu_\infty = 4 pa \cdot s$; $\alpha_\mu = 4.5T^{-1}$; $\tau_{y0} = 15 pa$; $\tau_{y\infty} = 40000 pa$; $\alpha_{t_y} = 2.9T^{-1}$.

Similarly, the friction torque of MRF in an annual duct acting on a disc can be determined as follows:

$$T_a = 2\pi R^2 L\tau = 2\pi R^2 L\left(\tau_y + \frac{\mu \Omega R}{d}\right) \tag{4}$$

Where L is the effective length of MR annual gap, R is the outer radius of disc.

The resultant torque of BMRA comprises two friction torque components whose directions are opposite. Therefore, the direction torque of BMRA depends on the dominance of one of these torque components. It can be determined as follows:

$$T = |T_1 - T_2| + T_{or} \tag{5}$$

where T_1, T_2 are induced torque of disc 1 and disc 2, respectively. T_{or} is friction torque of lip seal determined from [17]. By implementing Eq. (1) and Eq. (2), the induced torque T_1, T_2 can be evaluated by

$$T_1 = \frac{\pi \mu_{d11} R_d^4}{2d}\left[1 - \left(\frac{R_i}{R_d}\right)^4\right]\Omega + \frac{2\pi \tau_{yd11}(R_d^3 - R_i^3)}{3} + 2\pi R_d^2 t_d\left(\tau_{yd12} + \mu_{d12}\frac{\Omega R_d}{d_0}\right) \tag{6}$$

$$T_2 = \frac{\pi \mu_{d21} R_d^4}{2d}\left[1 - \left(\frac{R_i}{R_d}\right)^4\right]\Omega + \frac{2\pi \tau_{yd21}(R_d^3 - R_i^3)}{3} + 2\pi R_d^2 t_d \left(\tau_{yd22} + \mu_{d22}\frac{\Omega R_d}{d_0}\right) \tag{7}$$

Where R_i, R_d are, respectively, the inner and outer radius of discs, t_d is the width of discs, Ω is the angular velocity of the disc, μ_{d11}, μ_{d12} are respectively the average post-yield viscosity of MRF denoted by MRF1, MRF2 relating to disk 1 and τ_{yd11}, τ_{yd12} are the corresponding yield stress, μ_{d21}, μ_{d22} are the average post-yield viscosity relating to disk 2 and τ_{yd21}, τ_{yd22} are the corresponding yield stress.

3 Optimal Design

In this section, the optimal design of BMRA is considered. The optimization problem in this study is to find the optimal value of geometric dimensions to minimize the mass of BMRA while BMRA reaches the required transmitting torque. The mass of BMRA can be approximately determined by

$$m_B = V_{d1}\rho_d + V_{d2}\rho_d + V_h\rho_h + V_{s1}\rho_s + V_{s2}\rho_s + V_{MR}\rho_{MR} + V_C\rho_c \tag{8}$$

where V_{d1}, V_{d2}, V_h, V_{s1}, V_{s2}, V_{MR}, V_c are the geometry volume of disc 1, disc 2, housing, shaft 1, shaft 2, MRF and coils. Their parameters are the functions of geometry dimensions of BMRA structures changing in the optimization process. ρ_{d1}, ρ_{d2}, ρ_h, ρ_{s1}, ρ_{s2}, ρ_{MR}, ρ_c are the mass density of disc 1, disc 2, housing, shaft 1, shaft 2, MRF and coils. Each material has a different mass density deciding the mass of parts in BMRA. In this work, low carbon steel (C45 steel) is used as magnetic components since high magnetic field density at saturation along with permeability, cost and availability whereas MRF-132 DG is employed since average viscosity, yield stress and broader temperature range. The shaft of BMRA should be non-magnetic to keep magnetic flux far away the seals (to avoid the solidification of MR Fluid in the vicinity of seals). The coil wire sized as 24- gauge (diameter = 0.511 mm) is chosen and the maximum working current is 2.5 A. It is noted that small MRF gap size results in high induced torque, however, causes cost and difficulty in manufacturing. In this research work, the MRF gap size is set as 0.8 mm considering manufacturing convenience and stiffness housing of structure. The angular velocity of input shaft is set by 300 rpm. In this work, it is assumed that the shape of BMRA coil is rectangular and the filling ratio of the coil is set by 80% based on empirical experience.

The aforementioned explanation about optimal design is to finding optimal dimension while reaching required transmitting torque. To determine transmitting torque using Eq. (5), the post-yield viscosity and yield stress of MRF must be calculated. Of course, these values are depended on magnetic flux through the MRF ducts. In this work, finite element method (FEM) is employed to solve the magnetic flux across MRF ducts. The FE model using 2D axisymmetric element (PLANE 13) of ANSYS APDL, shown in Fig. 2, is employed for solving magnetic circuit in BMRA. The outer lines of the BMRA are set as boundary condition with parallel magnetic flux. Thus, it is assumed that there is no leaking magnetic flux at boundary of the BMRA. The excitation current through the coils is set as 2.5 A (maximum working current) to ensure the safe working of coils. It is noted that the meshing size is defined by the number element of each line to ensure the number of elements is unchanged during the optimization process.

In this work, BCMO proposed in [18] is employed to optimize the BMRA structure. The significant advantages of BCMO are reducing computational resources and memory in terms of large scale and high dimensional optimization problems compared with PSO. Moreover, it is more dynamic and effective than DE optimization since balancing exploration and exploitation. Therefore, BCMO is employed in optimizing the proposed BMRA. Firstly, BCMO generates the modeling data for BMRA. Then, these data are used to create CAD model in ANSYS for magnetic analysis. Then transmitting torque and mass of BMRA are determined and input into for evaluation. The constraints, objective and convergence criteria are evaluated. If the convergence criteria are satisfied, the optimization is completed. In this work, maximum iteration is convergence criteria and set as 300. The design variables of the optimization are: the inner radius of the disc (R_i), the inner pole length (l_{p1}), the outer pole length (l_{p2}), the disc thickness (d_{th}), the housing thickness (t_h), the number of coil wire layers in the width (n_{wc}) and in the height (n_{hc}).

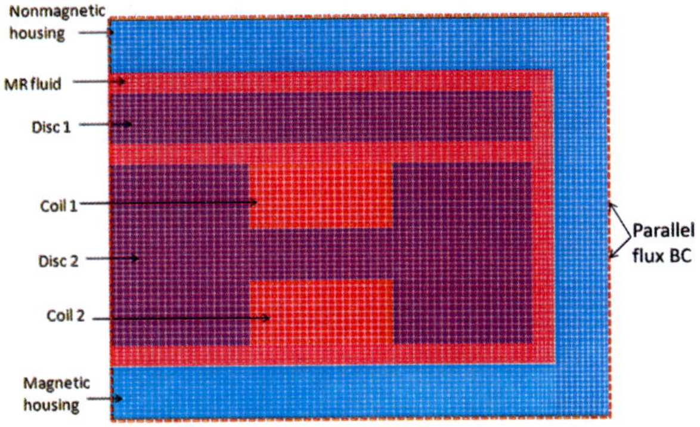

Fig. 2. Finite element model to solve magnetic circuit of BMRA

The optimal dimensions of BMRA are given in Table 1. The outer radius and length of BMRA are 25.57 mm and 47.91 mm, respectively. The applied current for coils is 2.5 A considering the safety and power consumption. Therefore, the maximum power of BMRA is 13.3 W with resistance in each coil is 2.1 Ω. The optimal mass of BMRA with required torque 5 Nm is 1.53 kg. Optimal processes are explained in Fig. 3 and Fig. 4. It can be seen that after 150 iterations, the convergence of optimization is reached. At the optimum, the values of R_i, l_{P1}, d_{th}, h_c, w_c, l_{P2}, R_o and t_h are 13.6 mm, 14 mm, 5.5 mm, 8.91 mm, 3.64 mm, 8.6 mm and 4.2 mm respectively.

Figure 5 shows the magnetic flux density of BMRA obtained from optimization in ANSYS. The magnetic flux density of proposed configuration at magnet housing (For exciting second coil) reaches saturated magnet 1.67 T. It can be seen that the magnetic flux almost cannot through annual duct because of non-magnetic housing. Therefore, induced moment is generated mainly at single end-face. In comparison with BMRA

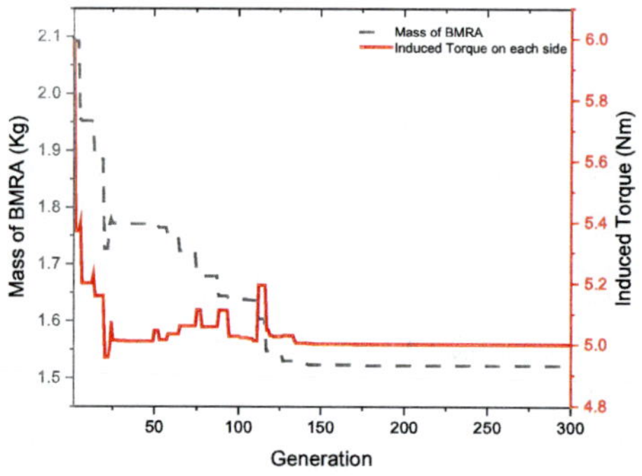

Fig. 3. Optimization solution of BMRA with required braking torque 5 Nm

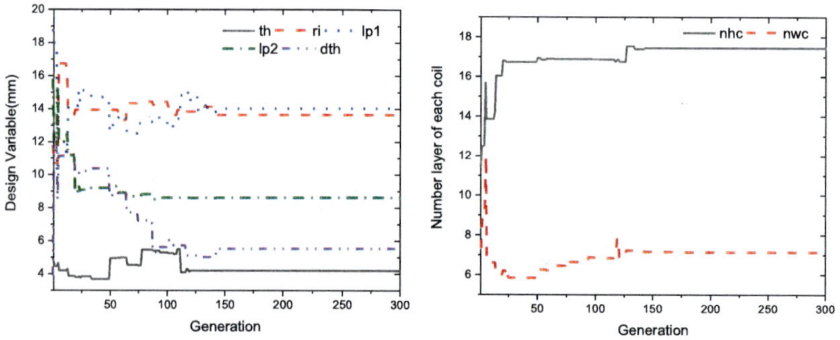

Fig. 4. Convergence of design variable of each configuration

Table 1. Optimal solution of BMRA when required braking torque is 5 Nm

Design Parameters (mm)	Characteristics
Coil: width $w_c = 3.64$; height $h_c = 8.91$; Radius $R_{w1} = 27.6$; No. of turn: 124 Disc: inner radius $R_i = 13.6$, outer radius $R_o = 45.11$; thickness $d_{th} = 4.2$ MRF gap: 0.8 Housing: Outer radius: 25.57; overall length: 47.91	Maximum torque: 5 Nm Mass: 1.53 kg Power consumption: 13.3 W Coil resistance: Rc = 2.1 Ω

[13], both coils in this work are located in the same disc. It can reduce the length of BMRA result in reducing mass.

4 Experiment

To validate optimal results, the prototype for proposed BMRA with required torque 5Nm is designed and evaluated actual performance. 2D detailed drawing of BMRA is given in Fig. 6. Firstly, the winding is wrapped in plastic bobbin. After that, the winding is

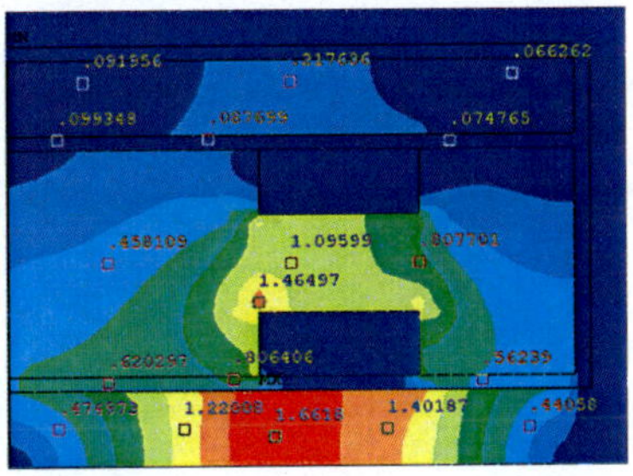

Fig. 5. Magnetic density distribution of the BMRA at the optimum

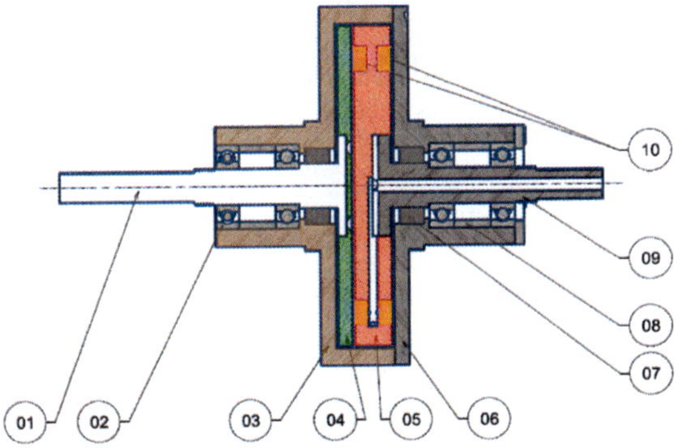

Fig. 6. Drawing of BMRA (1) First input (shaft), (2) Double row bearing, (3) Second input (housing), (4) First disc of first input, (5) Second disc of second input, (6) Housing, (7) Lip seal, (7) Coils, (8) Double row bearing, (9) Output shaft, (10) Coils

detached from bobbin and placed directly in second disc. The wire of coil then passes thoroughly the output shaft via small slot in second disc and shaft. Ball-bearings are employed to coincide the housing axis with shaft axis. Lip-seals are installed between shafts and housings to prevent leaking MRF. CNC machines are used to manufacture all components of prototype with high accuracy.

The test-bed for BMRA is given in Fig. 7. The AC servo motor (MSMD022S1T) with gearbox is used to drive two shafts of BMRA via bevel gear. The programming power (PPW-8011) is employed to supply power for each side of BMRA. It can communicate with DAQ to adjust desired current. The output shaft of BMRA is connected with torque sensor (TS20) to measure the BMRA's torque. The signal of torque sensor is transmitted to amplifier (LCVU-10) before sending to DAQ (NI-Myrio 1900). LABVIEW software is used to read data of DAQ via COM port.

The step currents are provided separately for each coil of BMRA from 0.5 A to 2.5 A with increment step of 0.5 A. The transmitting torque when triggering each side of BMRA shows in Fig. 8. Transmitting torque when exciting coil 1 at other applied currents of 0.5A, 1.0A, 1.5A, 2.0A, 2.5 A are −1.63 Nm, −2.7 Nm, −3.9 Nm, −4.7 Nm, −5.5 Nm, respectively whereas the transmitting torque in exciting coil 2 at each step currents are respectively 0.7 Nm, 1.8 Nm, 2.8 Nm, 3.6 Nm, 4.5 Nm. It can be seen

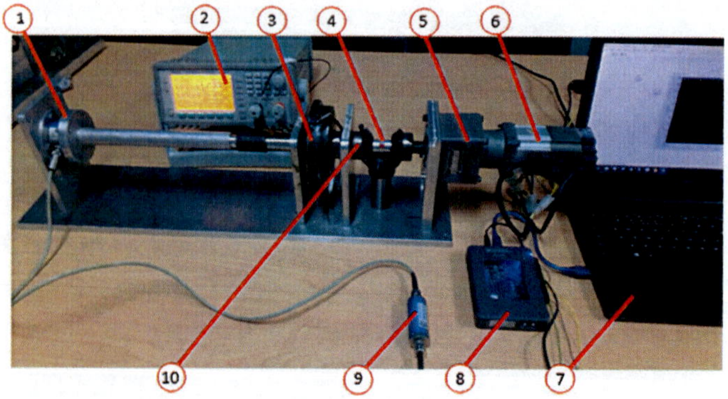

Fig. 7. Schematic of the braking measurement system (1) Torque sensor, (2) Programming Power, (3) BMRA, (4) First shaft, (5) Gear box, (6) Servo motor, (7) Computer, (8) DAQ, (9) Amplifier, (10) Second shaft.

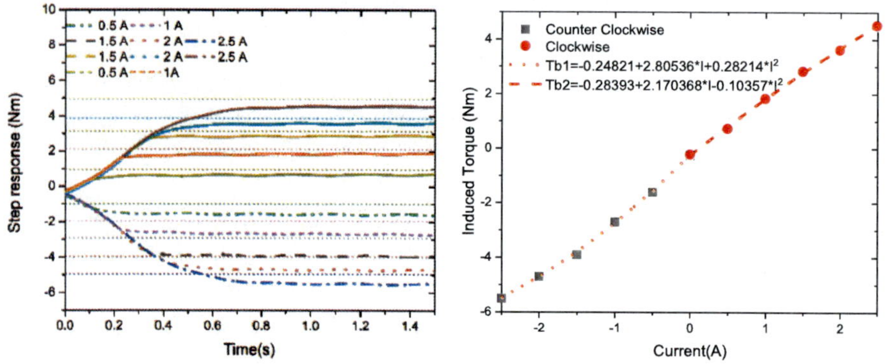

Fig. 8. Experimental results of clockwise and counterclockwise torque

that the measured torque of exciting coil 1 is slightly higher than simulation results, whereas, the measured torque of exciting coil 2 is a bit smaller than simulation results. It can be explained that the initial torque of both sides is not zero due to different friction and inaccuracy of fabrication. Curve fitting is employed to approximate the relationship between input current and transmitting torque given in Fig. 8. The difference between transmitting torque in exciting coil 1 and coil 2 may come from the assembly tolerance which makes the size of MRF gaps is not equal. The smaller MRF gap generates higher induced torque than the larger MR gap. Besides, the inaccuracy of magnetic loss also causes this error. The time response of induced torque is around 0.75 s which comprises both mechanical and induced output torque response time. In comparison with required maximum torque of each side, the error between measured value and desired value is around 10%. In addition, it can be seen that the transmitting torque of BMRA can be adjusted to zero by supplying small current in the second coil around 0.3 A. If the output torque at off-state positive, it is necessary to supply small current in the first coil to have a zero output torque.

5 Conclusion

In this work, a new configuration of bidirectional magneto-rheological actuator (BMRA) is proposed, optimally designed, fabricated and experimentally tested. A first magnetic disc is fastened with input shaft. Another magnetic disc is fastened with output shaft. The two discs are embedded inside housing. The gap between two discs and housing is filled by magnetorheological fluid (MRF). Two exciting coils are placed on second disc to generate the required magnetic field to transfer a required torque from the first disc and from the housing to the second disc. When exciting first coil, the output shaft rotates in clockwise direction whereas exciting second coil makes the output shaft rotates in counter-clockwise. BCMO is employed to optimize the mass of BMRA with required transmitting torque. A prototype of the optimized BMRA was manufactured and tested. It was found that there is difference between the measured torques in clockwise and counter-clockwise. The different may come from fabrication inaccuracy, different friction and magnetic loss. The induced torque difference is around 1Nm and dominated torque in counter-clockwise rotation. The results also showed a reasonable agreement between the simulated results and the measured ones. In the future works, this configuration is compared with previously developed BMRA for performance evaluation. In addition, a controller is employed to control the output torque and the BMRA is implemented in several applications.

Acknowledgement. This work was supported by research fund from Ministry of Education and Training (MOET) of Vietnam.

References

1. Anupam, A., Pratik, S., Aditya, J.: Haptic technology: a comprehensive review of its applications and future prospects. Int. J. Comput. Sci. Inf. Technol. **5**, 6039–6043 (2014)
2. Aude, B., Stéphane, R.: A review of haptic feedback teleoperation systems for micromanipulation and microassembly. IEEE Trans. Autom. Sci. Eng. **3**, 496–502 (2013)
3. Shah, K., Phu, D.X., Choi, S.-B.: Rheological properties of bi-dispersed magnetorheological fluids based on plate-like iron particles with application to a small-sized damper. J. Appl. Phys. **115**, 203907 (2014)
4. Park, B.J., Song, K.H., Choi, H.J.: Magnetic carbonyl iron nanoparticle based magnetorheological suspension and its characteristics. Mater. Lett. **63**, 1350 (2009)
5. Nguyen, Q.H., Choi, S.B.: Optimal design of a vehicle magnetorheological damper considering the damping force and dynamic range. Smart Mater. Struct. **18**(1), 015013 (2009)
6. Nguyen, Q.-C., Choi, S.-B.: Optimal design of a novel hybrid MR brake for motorcycles considering axial and radial magnetic flux. Smart Mater. Struct. **21**, 055003 (2012)
7. Nguyen, Q.H., Choi, S.B.: Selection of magnetorheological brake types via optimal design considering maximum torque and constrained volume. Smart Mater. Struct. **21**(1), 015012 (2011)
8. Nguyen, Q.H., Lang, V.T., Nguyen, N.D., et al.: Geometric optimal design of a magneto-rheological brake considering different shapes for the brake envelope. Smart Mater. Struct. **23**(1), 015020 (2014)

9. Nguyen, Q.H., Nguyen, N.D., Choi, S.B.: Design and evaluation of a novel MR brake with coils placed on side housings. Smart Mater. Struct. **24**(4), 90590I (2015)
10. Lee, U., Kim, D., Hur, N., Jeon, D.: Design analysis and experimental evaluation of an MR fluid clutch. J. Intell. Mater. Syst. Struct. **10**, 701–707 (1999)
11. Nguyen, P.B., Choi, S.B.: A bi-directional magneto-rheological brake for medical haptic system: optimal design and experimental investigation. Adv. Sci. Lett. **13**(1), 165–172 (2012)
12. Nguyen, P.B., Choi, S.B.: Accurate torque control of a bi-directional magneto-rheological actuator considering hysteresis and friction effects. Smart Mater. Struct. **22**(5), 055002 (2013)
13. Nguyen, Q.H., Diep, B.T., Le, D.H.: Design and evaluation of a bidirectional magnetorheological actuator for haptic application. J. Vietnam Mech. Eng.
14. Diep, T.B., Dai Le, H., Van Vo, C., Nguyen, H.Q.: Performance evaluation of a 2D-haptic joystick featuring bidirectional magneto-rheological actuators. In: International Conference on Advances in Computational Mechanics, pp. 1051–1059
15. Nguyen, Q.H., Choi, S.B.: Optimal design of an automotive magnetorheological brake considering geometric dimensions and zero-field friction heat. Smart Mater. Struct. **19**(11), 115024 (2010)
16. Zubieta, M., Eceolaza, S., Elejabarrieta, M.J., Bou-Ali, M.M.: Magnetorheological fluids: characterization and modeling of magnetization. Smart Mater. Struct. **18**, 95019 (2009)
17. P. S. Division: Rotary Seal Design Guide (Parker Hannifin Corporation) Catalog EPS 5350 (2006)
18. Le-Duc, T., Nguyen, Q.-H., Nguyen-Xuan, H.: Balancing composite motion optimization. Inf. Sci. **520**, 250–270 (2020)

Multi Objective Optimization of Dimension Accuracy and Surface Roughness in High-Speed Finishing Milling for the Hard Steel Alloy After Heat-Treatment

Duong Xuan Bien[1(✉)], Dao Van Duong[2], Do Tien Lap[1], Le Cong Doan[1], Ngo Tan Loc[1], and Duong Van Nguy[1]

[1] Le Quy Don Technical University, Hanoi, Vietnam
duongxuanbien@lqdtu.edu.vn
[2] Dong Nai Technical College, Biên Hòa, Vietnam

Abstract. This paper presents a multi-objective optimization (MO) problem including dimensional accuracy (DA) or dimensional error (DE) and surface roughness (SR) in high-speed finishing milling (HSFM) the SKD11 alloy steel after heat-treatment with hardness above 55HRC. Regression equations describing the relationship between cutting parameters (CP) and DE and SR are built on the basis of the specific experiments results. The Central Composite Design model (CCD) in the Response Surface Methodology (RSM) method is used to construct the experimental matrices. The ANOVA method in DESIGN EXPERT software was used to analyze experimental data. The results of the MO problem are found along with the suitable CP values based on the NSGA-II algorithm. The distinct property of the objective function DE is also considered based on the division of the MO problem solving cases. The research results have important value in finding suitable sets of cutting parameters and meeting multiple objectives simultaneously when machining HSM for hard materials.

Keywords: HSM · Multi-objective optimization · Dimension accuracy · Hard steel alloy

1 Introduction

Today, industries are increasingly demanding on time progress and economic efficiency in production in order to improve competitiveness in the fierce market, especially in the era of industrial revolution 4.0. In the field of mechanical processing, the HSM method is one of the solutions chosen first because of its great applications and efficiency [1]. HSM method is mainly used in machining low hardness materials such as aluminum alloy, alloy steel with hardness below 40 HRC [2, 3]. The application of HSM method to materials with high hardness (over 50 HRC) still faces many challenges in terms of machines, cutting tools and technological modes [4].

Common problems in the HSM method are cutting force [5, 6], cutting heat [7], vibration [8], tool wear [9], surface roughness [10],... However, these problems become

N. V. Khang et al. (Eds.): ASIAN MMS 2021, MMS 113, pp. 899–909, 2022.
https://doi.org/10.1007/978-3-030-91892-7_86

more difficult in high-speed machining for hard materials. On the other hand, surface quality and dimensional accuracy are always one of the leading criteria to evaluate machining efficiency and product quality. In particular, DA is an extremely complicated issue because it is governed by many factors. Many studies have shown that the change of technological parameters during the cutting process such as cutting speed, feed rate, cutting depth, cutting width has a great influence on the DA. Therefore, building a mathematical model describing the relationship of DA and SR with technological parameters is extremely necessary to ensure the required DA in machining in general and in HSM of hard alloys in particular. There have been many studies on surface roughness in HSM of hard alloys such as [6, 10, 11] but there are almost no specific publications on DA in HSM. Some publications mentioning this issue, although not HSM, but it can be considered such as [12–18]. The influence of CP on DA when machining turning aluminum alloys experimentally is studied in [12]. The issue of tool wear, SR and DA when machining AISI 4140 alloy using low temperature coolant is considered in [13, 16]. The method of measuring and analyzing the cutting force and vibration during the steel alloy turning process to determine the SR and DE based on the neural algorithm is described in [14]. The causes of DE including errors in machine geometry, cutting heat, control system and cutting force are presented in [15]. The prediction model of SR and DA through imaging recognition technology during machining is described in [17]. The optimal toolpath strategy to ensure DA when HSM of micro parts is presented in [18].

Optimizing the machining process is always a complex problem when building mathematical models that show the relationship between technological parameters and factors appearing in the machining process such as cutting force, tool wear, cutting heat, vibration and factors to evaluate product quality directly such as SR and DA. Although there have been many published works on the optimal problem of cutting force [19], wear of cutting tools [20], SR [21] with milling and turning methods but most are not high-speed machining. There are only few papers presenting on optimization problems in HSM such as [22, 23]. The MO problem of cutting force and SR is considered in [22] when high-speed turning of Inconel 718 alloy using TGRA theory. The SR is optimized for HSM of aluminum alloys based on the TAGUCHI experimental method and the ANOVA method described in [23].

This paper presents the results of HSFM experiments for alloy steel SKD11 after heat-treatment with hardness above 55HRC in order to study the influence of CP on DA and SR. The regression equations are built based on CCD method and ANOVA method on DESIGN EXPERT software. Based on these regression equations, the MO problem is built to find the technology modes that match the optimal requirements for DA and SR. The NSGA-II algorithm is used to solve this MO problem. The research results have important implications in providing appropriate sets of technological parameters to meet multiple objectives simultaneously when high-speed machining of hardened alloy steels, and contribute to improving product quality in industrial production.

2 Materials and Methods

2.1 Experimental Design

The experimental objective of HSFM of alloy steel SKD11 after heat treatment (55HRC) is to determine the relationships between cutting parameters (depth of cutting, feed rate, and spindle speed) in HSFM with dimension accuracy and surface roughness through the regression equations. These equations are built based on the CCD method. The workpiece material is SKD11 (Japanese Standard JIS G4404) with the composition described in Table 1.

Table 1. Chemical composition of SKD11 alloy steel workpiece

Ingredient	C (%)	Si (%)	Mn (%)	Cr (%)	Mo (%)	V (%)	P (%)
SKD11	1.4–1.6	≤ 0.6	≤0.6	11–13	0.7–1.2	≤1.1	≤0.03

SKD11 workpiece was heat-treated in the Turbo-IPSEN furnace and hardness measured on an FR-1E/Hard machine (Fig. 1a). Hardness measurement results with the average value of SKD11 alloy steel workpiece after heat treatment reached 55 HRC (Fig. 1b).

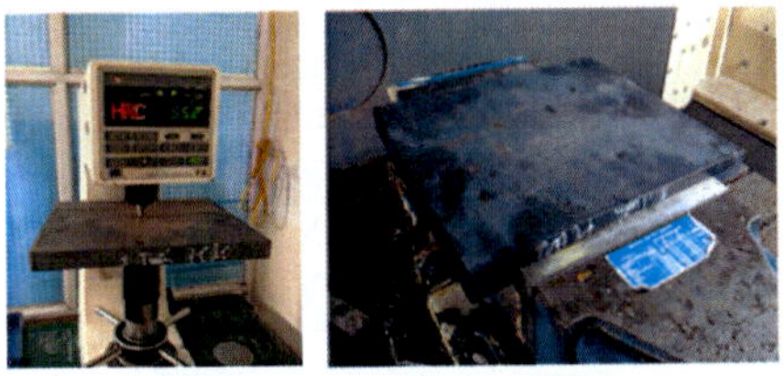
(a) (b)

Fig. 1. FR-1E/Hard hardness tester machine and SKD11 55HRC workpiece

(a) (b)

Fig. 2. YG SGN09080H endmill and experimental model

The cutting tool used for the experiment is a specialized high-speed carbide Endmill tool YG SGN09080H with cutting diameter is 8 mm, shank diameter is 8 mm, cutting length is 20 mm, total tool length is 65 mm with 4 cutting teeth (Fig. 2a). Regarding machining machines, experiments were carried out on HSM machines DMC1450V (Fig. 2b). Roughness measuring device is MITUTOYO SJ-201 (Fig. 3). The measuring device is a CMM CONTURA G2 (Fig. 4).

The CCD method is a nonlinear planning method, the structure has a center [24, 25]. The number of influencing factors is 3 factors which are cutting depth (mm), feed rate (mm/min), and spindle speed (rev/min) or cutting speed (m/min). The stepover parameter is fixed. The total number of experiments to be performed is 15. The α quantity is the distance from the center of the experimental structure to the position of points on the coordinate axis of the structure. Conventional influencing factors according to the

Fig. 3. MITUTOYO SJ-201 machine

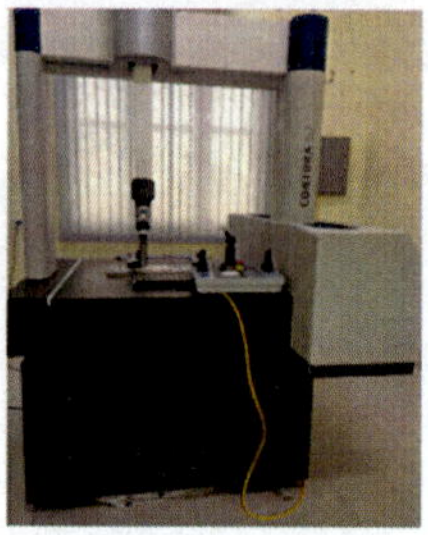

Fig. 4. CMM CONTURA G2 machine

variables are as follows, cutting depth t(mm) is x_1, the feed rate F(mm/min) is x_2, spindle speed n(rev/min) is x_3. It should be noted that accuracy of dimensions (DA) is expressed through error of dimensions (DE). The smaller the error, the higher the accuracy. In this case, the DE will be used to represent the DA. Therefore, the regression equation built is the equation of DE. Based on the study of previous publications on HSM of different materials, different hardness, assessment of the machining ability of the CNC machine for SKD11 steel alloy with HRC55, cutting parameters domain recommended of cutting tool manufacturer and test results, the value range of experimental cutting parameters are specifically selected as shown in Table 2.

Table 2. Limited cutting parameters

Cutting parameter	Limit lower ($-\alpha$)	Low (-1)	Basic (0)	High (+1)	Limit upper (α)	Variables
$t(mm)$	2.25	3	6	9	9.75	x_1
$F(\frac{mm}{\min})$	575	600	700	800	825	x_2
$n(\frac{rev}{\min})$	3800	4000	4800	5600	5800	x_3
$v_c(\frac{m}{\min})$	95.5	100.5	120.6	140.7	145.7	x_3

The strategy for finish milling is the Profile cycle. The diameter of the circular pocket before finishing is 21.8 mm. The required pocket diameter is 22 mm. Pocket depth reaches 11 mm.

2.2 ANOVA/DESIGN EXPERT Analyzing

The experimental matrices are coded and the experimental results are presented in Fig. 5 and Fig. 6.

Std	Factor 1 A:x1	Factor 2 B:x2	Factor 3 C:x3	Response 1 DA (mm)
1	-1	-1	-1	-0.0498749
2	1	-1	-1	0.04825
3	-1	1	-1	0.047239
4	1	1	-1	0.0342537
5	-1	-1	1	-0.0460665
6	1	-1	1	0.0375305
7	-1	1	1	-0.047062
8	1	1	1	-3.7E-06
9	-1.215	0	0	-0.048373
10	1.215	0	0	0.0408903
11	0	-1.215	0	0.0204121
12	0	1.215	0	0.0186679
13	0	0	-1.215	0.025781
14	0	0	1.215	0.0260444
15	0	0	0	0.0292724

Fig. 5. The Box-Wilson experimental matrix (CCD) and experimental results DE

Std	Factor 1 A:x1 mm	Factor 2 B:x2 mm/min	Factor 3 C:x3	Response 1 SR micromet
1	-1	-1	-1	0.9
2	1	-1	-1	0.845
3	-1	1	-1	0.77
4	1	1	-1	0.7
5	-1	-1	1	0.6
6	1	-1	1	0.48
7	-1	1	1	0.47
8	1	1	1	0.54
9	-1.215	0	0	0.83
10	1.215	0	0	0.735
11	0	-1.215	0	0.495
12	0	1.215	0	0.395
13	0	0	-1.215	0.645
14	0	0	1.215	0.575
15	0	0	0	0.64

Fig. 6. The Box-Wilson experimental matrix (CCD) and experimental results SR

The analyzing results of ANOVA method on DESIGN EXPERT software for DE are shown in Fig. 7a. The coefficients of the regression equation are shown in Fig. 7b with the excluded coefficients of expressions x_2 and x_1x_3, x_2^2, x_3^2.

Transform | Fit Summary | Model | ANOVA | Diagnostics

Analysis of Variance

ANOVA for Quadratic model

Response 1: Yss

Source	Sum of Squares	df	Mean Square	F-value	p-value	
Model	0.0181	9	0.0020	6.17	0.0295	significant
A-A	0.0096	1	0.0096	29.53	0.0029	
B-B	0.0002	1	0.0002	0.5065	0.5085	
C-C	0.0017	1	0.0017	5.13	0.0729	
AB	0.0027	1	0.0027	8.38	0.0340	
AC	0.0003	1	0.0003	0.7966	0.4130	
BC	0.0018	1	0.0018	5.69	0.0627	
A²	0.0017	1	0.0017	5.31	0.0694	
B²	0.0001	1	0.0001	0.2282	0.6530	
C²	1.611E-07	1	1.611E-07	0.0005	0.9831	
Residual	0.0016	5	0.0003			
Cor Total	0.0197	14				

a

Coefficients | Coded Equation | Actual Equation

Final Equation in Terms of Actual Factors

R1	=
+0.026529	
+0.029605	* x1
+0.003878	* x2
-0.012340	* x3
-0.018456	* x1 * x2
+0.005689	* x1 * x3
-0.015206	* x2 * x3
-0.019896	* x1²
-0.004125	* x2²
+0.000192	* x3²

b

Fig. 7. a. The analyzing results of ANOVA for *DE*. b.*C*oefficients of the regression equation *DE*

The regression equation for DE is established as follows

$$Y_{DE} = 0.02365 + 0.02961x_1 - 0.01234x_3 - 0.018456x_1x_2 \\ -0.01521x_2x_3 - 0.01989x_1^2 \quad (1)$$

The analyzing results of ANOVA method on DESIGN EXPERT software for SR are shown in Fig. 8a. The coefficients of the regression equation are shown in Fig. 8b. Although the analytical results show that the coefficients are acceptable in the regression equation, because the SR value is quite small, the coefficients of x_1, x_2 (Fig. 8b) can be ignored.

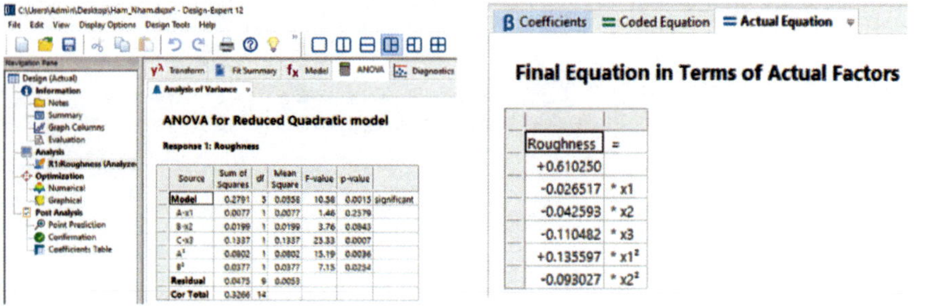

a b

Fig. 8. a. The analyzing results of ANOVA for *SR*. b. Coefficients of the regression equation SR

Therefore, the regression equation for SR is written as follows

$$Y_{SR} = 0.6097 - 0.11048x_3 + 0.1359x_1^2 - 0.0926x_2^2 \tag{2}$$

Equation (1) and Eq. (2) describe the relationship of DE and SR with cutting parameters, respectively. Boundary conditions of the equations are described as (3) below

$$- 1.215 \le x_1,\ x_2,\ x_3 \le 1.215 \text{ or } 2.25 \le \text{t} \le 9.75(\text{mm}),\ 575 \le \text{F} \le 825(\text{mm/min}),$$
$$3800 \le n \le 5800(rev/\text{min});\ es \le Y_{DE} \le ES(mm);\ Y_{SR} > 0(\mu m) \tag{3}$$

Where *es* and *ES* are the upper and lower limits of the diameter tolerance, respectively.

2.3 Multi-objective Optimization Problem

Multi-objective optimization means finding the best solution in a certain sense to achieve (maximum or minimum) multiple objectives at the same time. However, it is very difficult and rare to find the optimal value that satisfies the requirements of the objective functions. The best way is to find a solution that satisfies all the goals set out within an acceptable level. That is the Pareto solution [26]. This paper uses improved genetic algorithm (NSGA-II) [26] to find those solutions. In this case, it is necessary to find a suitable set of technological parameters in HSFM ensures the minimum DE and SR values. The objective functions DE and SR are described as in Eq. (1) and Eq. (2) with the boundary condition (3). Due to the different nature of the DE objective function (the smallest value is the value to be found but not necessarily the best, because the error value approaches 0 is the best value). It is necessary to divide the case for intervals different tolerances for more comprehensive and realistic evaluation data. The DE objective function is divided into two cases corresponding to two different limit domains.

Case 1 (C1): $es = -0.025(mm)$, $ES = 0$. This tolerance range value is used for tight assemblies.

Case 2 (C2): $es = 0$, $ES = 0.02(mm)$. This tolerance range value is used for assemblies with gap.

The basic parameters of the NSGA-II algorithm for the MO problem are described in Table 3. The results of the MO problem for two cases mentioned above are described in Fig. 9 and Fig. 10, respectively.

Table 3. Parameters of the NSGA-II

No	Parameters	Symbols	Values
1	Number of generations	MaxGen	50
2	Initial population size	npop	100
3	Random selection rate	Se	0.5
4	Mutation probability	Mu	0.05

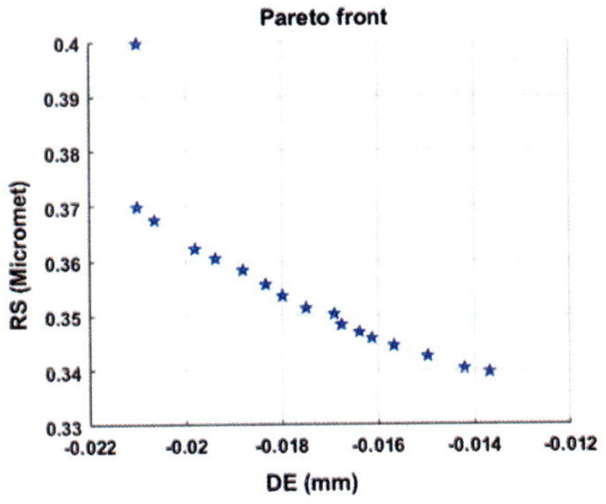

Fig. 9. MO results in C1

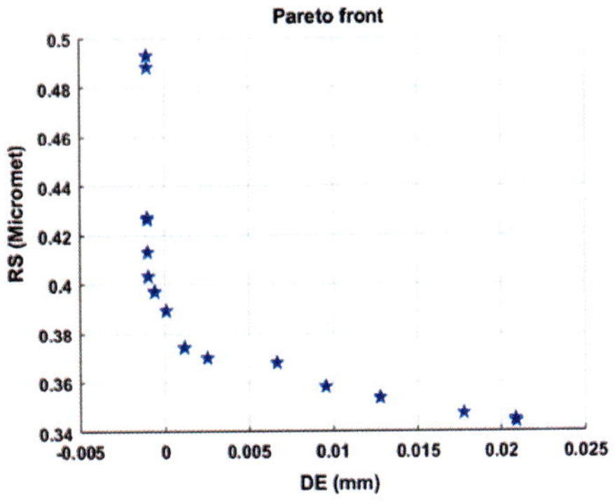

Fig. 10. MO results in C2

According to the results in Fig. 8 of the C1, the best solution of MO problem is location with the smallest SR and the DE value is located near the value 0, corresponding to position $y_{SR} = 0.34(\mu m)$ and $y_{DE} = -0.014(mm)$. These values achieved when $t = 6(mm), F = 821(mm/\min)$ and $n = 5771(rev/\min)$. Table 4 describes some other multi-objective optimal locations on the Pareto boundary. Notably, the Pareto (acceptable) values of the objective functions are found in the range of values of specific technological parameters $4.4 \leq t \leq 6.1(mm)$, $812 \leq F \leq 822(mm/\min)$ and $5740 \leq n \leq 5771(rev/\min)$. Values of the objective functions are determined in the range $-0.021 \leq y_{DE} \leq -0.014(mm)$ and $0.34 \leq y_{SR} \leq 0.4(\mu m)$.

Same evaluating as C1. The results of the C2 are shown in Fig. 9 and Table 5. The Pareto boundary is found with the range of technological parameters such as $4.5 \leq t \leq 6.7(mm)$, $578 \leq F \leq 793(mm/\min)$, $5598 \leq n \leq 5770(rev/\min)$. Values of the objective functions were obtained in the range $0 \leq y_{DE} \leq 0.02(mm)$ and $0.34 \leq y_{SR} \leq 0.5(\mu m)$.

Table 4. MO solutions for C1

No	x_1	x_2	x_3	$y_{DE}(mm)$	$y_{SR}(\mu m)$	$t(mm)$	$F(mm/\text{min})$	$n(rev/\text{min})$
1	−0.460	1.209	1.206	– 0.020	0.370	4.62	820.9	5765.2
2	−0.213	1.212	1.213	– 0.016	0.346	5.36	821.2	5770.6
3	−0.522	1.125	1.175	– 0.021	0.400	4.43	812.5	5740.2
4	−0.404	1.211	1.212	– 0.020	0.362	4.79	821.1	5769.5
5	−0.136	1.212	1.211	– 0.015	0.342	5.59	821.2	5769.2
6	−0.269	1.211	1.208	– 0.017	0.350	5.19	821.1	5766.4
7	– 0.363	1.211	1.209	– 0.019	0.358	4.91	821.1	5767.4
8	– 0.229	1.212	1.213	– 0.016	0.347	5.31	821.2	5770.7
9	– 0.522	1.125	1.175	– 0.021	0.400	4.43	812.5	5740.2
10	– 0.338	1.211	1.211	– 0.018	0.356	4.99	821.1	5768.8
11	– 0.318	1.211	1.213	– 0.018	0.354	5.05	821.1	5770.5
12	– 0.444	1.210	1.208	– 0.021	0.367	4.67	821.0	5766.3
13	0.003	1.212	1.214	– 0.014	0.340	6.01	821.2	5771.1
14	– 0.062	1.212	1.214	– 0.014	0.340	5.81	821.2	5770.8
15	– 0.183	1.211	1.213	– 0.016	0.344	5.45	821.1	5770.5
16	– 0.387	1.211	1.211	– 0.019	0.360	4.84	821.1	5768.7
17	– 0.252	1.212	1.213	– 0.017	0.348	5.25	821.2	5770.6
18	– 0.292	1.212	1.213	– 0.017	0.351	5.12	821.2	5770.7

Table 5. MO solutions for C2

No	x_1	x_2	x_3	$y_{DE}(mm)$	$y_{SR}(\mu m)$	$t(mm)$	$F(mm/\text{min})$	$n(rev/\text{min})$
1	– 0.301	– 0.040	1.168	– 0.001	0.493	5.1	696.0	5734.7
2	0.227	0.93	0.998	– 0.001	0.426	6.7	793.0	5598.7
3	– 0.283	– 0.009	1.199	– 0.001	0.488	5.2	699.1	5759.5
4	– 0.496	– 1.091	1.174	– 0.001	0.403	4.5	590.9	5739.0
5	– 0.489	– 1.032	1.182	– 0.001	0.413	4.5	596.8	5745.7
6	– 0.183	– 1.213	1.212	0.021	0.344	5.5	578.7	5769.6
7	– 0.313	– 1.212	1.207	0.013	0.354	5.1	578.8	5765.3
8	0.238	0.928	1.001	– 0.001	0.427	6.7	792.8	5601.1
9	– 0.489	– 1.138	1.204	0.000	0.389	4.5	586.2	5763.2
10	– 0.465	– 1.212	1.206	0.003	0.370	4.6	578.9	5764.8

(*continued*)

Table 5. *(continued)*

No	x_1	x_2	x_3	$y_{DE}(mm)$	$y_{SR}(\mu m)$	$t(mm)$	$F(mm/\min)$	$n(rev/\min)$
11	– 0.235	– 1.212	1.212	0.018	0.347	5.3	578.8	5769.2
12	– 0.363	– 1.212	1.209	0.010	0.358	4.9	578.8	5767.3
13	– 0.183	– 1.213	1.204	0.021	0.345	5.5	578.7	5763.4
14	– 0.301	– 0.040	1.168	– 0.001	0.493	5.1	696.0	5734.7
15	– 0.495	– 1.111	1.192	– 0.001	0.397	4.5	588.9	5753.7
16	– 0.483	– 1.203	1.206	0.001	0.374	4.6	579.7	5765.0
17	– 0.401	– 1.189	1.204	0.007	0.368	4.8	581.1	5763.4
18	– 0.183	– 1.213	1.212	0.020	0.344	5.5	578.7	5769.6

3 Conclusion

For SKD11 alloy steels after heat-treatment reaching high hardness above 55HRC in particular and hard materials in general, the problem of machining becomes extremely difficult due to the limitations of technological systems such as CNC machines, cutting tools, technology modes and physical phenomena occurring during machining such as cutting force, vibration, cutting heat, tool wear. Moreover, the requirements for dimensional accuracy and surface quality such as surface roughness when high-speed machining for these materials become even more difficult. This paper has built and solved the MO problem for DA and SR in HSFM for hard alloy steel SKD11 with specific requirements for the value of DA. Based on the NSGA-II algorithm, suitable sets of cutting parameters were found and the optimal objectives were met. To get these results, regression equations describing the relationship between cutting parameters with DE and SR were built based on ANOVA module in RSM method with experimental model CCD. The research results have significate implications in applied HSM method in machining hard and super-hard materials with dimensional accuracy and surface quality as required. The problems solved in this article have practical significance and high applicability in modern industrial production.

References

1. Schulz, H.: The history of high-speed machining. Revista de Ciencia and Tecnologia **13**, 9–18 (1999)
2. Evans, R., Platt, E.: Aluminum High Speed Machining Metalworking Fluid Performance in Aluminum High Speed Machining. Department of Industrial & Manufacturing Engineering, The Pennsylvania State University (2009)
3. Wang, Z., Ranman, M.: High speed machining. Compr. Mater. Process. **11**, 221–252 (2014)
4. Zhao, Z., Xiao, Y., Zhu, Y., Liu, B.: Influence of cutting speed on cutting force in high-speed milling. Adv. Mater. Res. **139–141**, 835–838 (2010)
5. Fang, N., Pai, P.S., Edwards, N.: A comparative study of high-speed machining of Ti-6Al-4V and Inconel 718 - part I: effect of dynamic tool edge wear on cutting forces. Int. J. Adv. Manuf. Technol. **68**(5–8), 1839–1849 (2013). https://doi.org/10.1007/s00170-013-4981-2

6. Tian, X., Zhao, J., Iqbal, J.M, Konneh, M., Bin, M.H., Abdallah, K.A., Binting, M.F.B.: Cutting temperature in high-speed milling of silicon carbide using diamond coated tool. Int. J. Mech. Prod. Eng. 62–66 (2015)
7. Zhao, Z., Gong, Y., Dong: Effect of cutting speed on cutting forces and wear mechanisms in high-speed face milling of Inconel 718 with Sialon ceramic tools. Int. J. Adv. Manuf. Technol. (2013)
8. Krishnakumara, P., Rameshkumarb, K., Ramachandranc, K.I.: Tool wear condition prediction using vibration signals in high-speed machining (HSM) of titanium (Ti-6Al-4V) alloy. Proc. Comput. Sci. **50**, 270–275 (2015)
9. Silva, R.B., Machado, A.R., Ezugwu, E.O., Bonney, J., Sales, W.F.: Tool life and wear mechanisms in high-speed machining of Ti-6Al-4V alloy with PCD tools under vaious coolant pressure. J. Mater. Process. Technol. **213**, 1459–1464 (2013)
10. Cui, X., Zhao, J.: Cutting performance of coated carbide tools in high-speed face milling of AISI H13 hardened steel. Int. J. Adv. Manuf. Technol. **71**(9–12), 1811–1824 (2014). https://doi.org/10.1007/s00170-014-5611-3
11. Honghua, S., Peng, L., Yucan, F., Jiuhua, X.: Tool life and surface integrity in high-speed milling of titanium alloy TA15 with PCD/PCBN tools. Chin. J. Aeronaut. **25**, 784–790 (2011)
12. Shouckry, A.S.: The effect of cutting conditions on dimensional accuracy. Wear **80**, 197–205 (1982)
13. Dhar, N.R.: Machining of AISI 4140 steel under cryogenic cooling-tool wear, surface roughness and dimensional deviation. J. Mater. Process. Technol. **123**, 483–489 (2002)
14. Risbood, K.A.: Prediction of surface roughness and dimensional deviation by measuring cutting forces and vibrations in turning process. J. Mater. Process. Technol. **132**, 203–214 (2003)
15. Schmitz, T.L.: Part Accuracy in High-Speed Machining: Preliminary Results. MSEC2006-2110
16. Dhar, N.R.: Cutting temperature, tool wear, surface roughness and dimensional deviation in turning AISI-4037 steel under cryogenic condition. Int. J. Mach. Tools Manuf **47**, 754–759 (2007)
17. Shahabi, H.H., Ratnam, M.M.: Prediction of surface roughness and dimensional deviation of workpiece in turning: a machine vision approach. Int. J. Adv. Manuf. Technol. **27**, 213–226 (2010)
18. Dadgari, A., Huo, D., Swailes, D.: The effect of machining toolpath on surface roughness and dimensinal accuracy for high-speed micro milling. Solid State Phenom. **261**, 69–76 (2017)
19. Yildiz, A.R.: Cuckoo search algorithm for the selection of optimal machining parameters in milling operations. Int. J. Adv. Manuf. Technol. 55–61 (2013)
20. Bhushan, R.A.: Optimization of cutting parameters for minimizing power consumption and maximizing tool life during machining of Al alloy SiC particle composites. J. Clean. Prod. **39**, 242–254 (2013)
21. Sathish, T., Tamizharasan, N., Siva, S.T.B., Varu, K.E., Radhakrishman, B.: Optimization of surface roughness in CNC milling process using RMS. Int. J. Mech. Eng. 5–12 (2017)
22. Pawade, R.S., Joshi, S.S.: Multi-objective optimization of surface roughness and cutting forces in high-speed turning of Inconel 718 using Taguchi grey relational analysis (TGRA). Int. J. Adv. Manuf. Technol. **56**, 47–62 (2011)
23. Reddy, S.M.: Optimization of surface roughness in high-speed end milling operation using Taguchi's method. J. Impact Factor **4**, 249–258 (2013)
24. Arslan, F.N., Kara, H.: Central composite design and response surface methodology for the optimization of Ag+-HPLC/ELSD method for triglyceride profiling. J. Food Meas. Characteriz. **11**(2), 1–11 (2017)

25. Danbaba, N., Nkama, I., Badau, M.H.: Application of response surface methodology (RSM) and central composite design (CCD) to optimize minerals composition of rice-cowpea composite blends during extrusion cooking. Int. J. Food Sci. Nutr. Eng. **5**(1), 40–52 (2015)
26. Alawode, K.O., Adegboyega, G.A., Muhideen, J.A.: NSGA-II/EDA hybrid evolutionary algorithm for solving multi-objective economic/emission dispatch problem. Electr. Power Components Syst. (2018)

Investigation the Influence of the Welding Seams Trajectory on Driving Energy for Industrial Robots

Duong Xuan Bien(✉)

Le Quy Don Technical University, Hanoi, Vietnam
duongxuanbien@lqdtu.edu.vn

Abstract. This paper focuses on investigating the effect of the welding seams in the working space on the driving energy for the six degrees of freedom (6DOF) FD-V8 industrial welding robot. The system of kinematics equations is built on the basis of multi-bodies mechanics theory and industrial robotics theory. The AGV algorithm is applied to solve the inverse kinematics problem (IKP) for each specific welding seam. The Lagrange - Euler method is used to build the differential equations of the robot motion. The robot driving energy is determined on the basis of solving the inverse dynamics problem (IDP). The results of this study have important implications in designing an appropriate welding seams trajectory to reduce energy consumption, as a basis for step by step building the optimizing energy consumption and production costs problem meeting the challenge of the current energy shortage in modern industrial production.

Keywords: Industrial robots · Welding seam · Driving energy

1 Introduction

Smart manufacturing is the core of industrial revolution 4.0. In particular, industrial robots in general and welding robots, in particular, are one of the main subjects that are interested in research and application more and more. The need to improve the performance of robots is always an urgent issue. In the current and near future, the number of robots will be used large, with continuous operation frequency and long working cycles in smart factories. Therefore, they are subject to high energy consumption in production. Finding solutions to reduce energy consumption is an urgent matter in order to minimize costs and enhance production efficiency.

The problem of reducing costs and improving machining productivity of robots is solved in many directions such as optimization of machining trajectory [1–3], feed rate [4–6], technology parameters [7–9] and reduction of machining time leads to reduced energy consumption during machining [10–12].

The optimization direction of production planning, operation based on optimization of production processes, management of delay time, waiting time, and operation time is an important and key direction studied by most enterprises. However, this solution

N. V. Khang et al. (Eds.): ASIAN MMS 2021, MMS 113, pp. 910–920, 2022.
https://doi.org/10.1007/978-3-030-91892-7_87

depends heavily on the specific conditions of the management level, the external organization but has not yet entered into the operational nature of each production equipment. The optimal research direction of the technology parameters depends a lot on the object of processing. The technological parameters change continuously when changing materials, textures of objects, and machining methods [7–9]. In order to find the optimal parameters, it is necessary to carry out many tests, measurements, and statistics, which consumes time and research costs. The research direction to optimize the energy consumption based on the optimal trajectory, jerk, and feed rate goes into the nature of the kinematics and dynamics of the robot, which is feasible and highly cost-effective. These methods are implemented based on basic researches right in robot structure design, trajectory design, and optimization algorithms.

The issue of energy consumption in production is directly mentioned in [13–20]. Experimental studies on the effects of robot operating factors on energy consumption are discussed in [10]. The algorithm that minimizes the energy required to move in a point-to-point trajectory is proposed by [11] and applies the simulation for a 2DOF planar robot. The issue of energy consumption for a system of many robots is examined in [12] with a 3DOF robot used to illustrate the algorithm. The technique for adjusting the parameters of the dynamics model and recognition technology is used in [13] to plan the trajectory of industrial robots with the minimum energy consumption. The problem of evaluating the effects of feed rate and load on energy consumption is mentioned in [14] through the construction of an industrial robot simulation model. Determining the optimal positions in the robot workspace to minimize the driven energy is considered in [15]. Similarly, using energy efficiency is considered for the 4-DOF parallel robot in [16] based on the position optimization in the workspace. The energy consumption optimization solution through the structural optimization design study considered in [17], the 5-bar mechanism model and the SCARA robot are the illustrated objects. The redundant properties of the robot in [18] are exploited to improve the efficiency of driving energy. In essence, this proposal is also based on adjusting the machining trajectory of the robot. The proposed model for calculating the energy consumption of the robot with the effect of temperature at the robot's driving joints is presented in [19]. The solution to reduce energy consumption for the pick-and-drop robot based on the optimal working trajectory is also presented in [20]. Energy consumption in the production process has been studied for a long time, but mainly focused on the cutting process on CNC machines, but mainly through experimental research. For industrial robots, this issue has only been interested in recent years due to the development and application of more and more robots in industrial production, the explosion of the industrial revolution 4.0, and the challenge of the global energy shortage.

This paper focuses on surveying the effect of the welding seams in the workspace on the driving energy of the robot joints. This driving energy is described in terms of the torque value of each joint. The robot mathematical model is built using the multi-bodies mechanics theory and industrial robot theory. The welding seams in the workspace are used as an input for the IKP to determine the values of the joint variables. The AVG method [21] is effectively used to solve the IKP. The dynamics equations are established based on the energy differential equations Lagrange - Euler. The driving torques of joints are determined by solving the IDP.

2 Materials and Methods

2.1 The Kinematics and Dynamics Modeling

Consider the kinematics model of industrial welding robot FD-V8 with 6DOF as shown in Fig. 1. The fixed coordinates system is $(OXYZ)_0$ located at point O_0 and $(OXYZ)_i$, $(i = 1 \div 6)$ are the local coordinate systems attached link i. Table 1 describes the kinematic parameters according to the D-H rule [23]. Accordingly, the transformation homogeneous matrices $\mathbf{H}_i$, $(i = 1 \div 6)$ are determined.

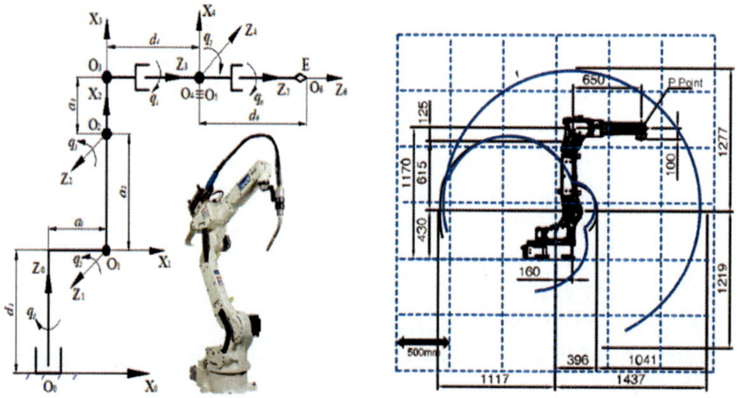

Fig. 1. Kinematic model of the FD-V8 robot and the EEP working range [22]

Table 1. DH parameters

Links	DH parameters			
	$\theta_i = q_i$	d_i	a_i	α_i
1	q_1	d_1	a_1	$\pi/2$
2	q_2	0	a_2	0
3	q_3	0	a_3	$\pi/2$
4	q_4	d_4	0	$\pi/2$
5	q_5	0	0	$-\pi/2$
6	q_6	d_6	0	0

The position and direction of the end-effector point (EEP) from the $\mathbf{D_6}$ matrix following the fixed coordinate system are determined as follows [23]. In this paper, the tip point of the welding torch is the end-effector point.

$$\mathbf{D}_6 = \mathbf{H}_1\mathbf{H}_2\mathbf{H}_3\mathbf{H}_4\mathbf{H}_5\mathbf{H}_6 \tag{1}$$

Define the generalized vector of robot is $\mathbf{q} = [q_1\ q_2\ q_3\ q_4\ q_5\ q_6]^T$ and $\mathbf{x}_{EEP}(t) = \left[x_E\ y_E\ z_E\right]^T$ is the coordinate vector of end-effector point following fixed coordinate system. The forward kinematics equations can be written as

$$\mathbf{x}_{EEP} = f(\mathbf{q}) \quad (2)$$

Where, f is a vector function representing the robot forward kinematics. Derivative (2) with respect to time, the relation between generalized velocities is obtained as

$$\dot{\mathbf{x}}_{EEP} = \mathbf{J}(\mathbf{q})\dot{\mathbf{q}} \quad (3)$$

Where, $\mathbf{J}(\mathbf{q})$ is the Jacobian matrix with size 3×6. The acceleration of the end-effector point can be given by derivation (3)

$$\ddot{\mathbf{x}}_{EEP} = \dot{\mathbf{J}}\dot{\mathbf{q}} + \mathbf{J}\ddot{\mathbf{q}} \quad (4)$$

The IKP equations of robots are written as

$$\mathbf{q} = f^{-1}(\mathbf{x}_{EEP}) \quad (5)$$

The values of vector $\mathbf{q}$ have been determined from (5), the joints velocity is determined as

$$\dot{\mathbf{q}} = \mathbf{J}^{+}(\mathbf{q})\dot{\mathbf{x}}_{EEP} \quad (6)$$

Where, $\mathbf{J}^{+}(\mathbf{q})$ is the pseudo-inverse matrix of $\mathbf{J}(\mathbf{q})$ matrix and is defined as [23]

$$\mathbf{J}^{+}(\mathbf{q}) = \mathbf{J}^{T}(\mathbf{q})\left[\mathbf{J}(\mathbf{q})\mathbf{J}^{T}(\mathbf{q})\right]^{-1} \quad (7)$$

The joints acceleration is calculated from (6)

$$\ddot{q} = \mathbf{J}^{+}(\mathbf{q})(\ddot{x}_{EEP} - \dot{J}\dot{q}) \quad (8)$$

For the given $\mathbf{x}$, $\dot{\mathbf{x}}$, $\ddot{x}$ vectors and using the algorithms for adjusting the increments of generalized vector which was proposed in [21], the approximately joint variables value can be determined exactly. The dynamics equations show the relationship between forces and torques with the motion characteristics of robots such as joint position $\mathbf{q}$, velocity $\dot{\boldsymbol{q}}$, joint acceleration $\ddot{q}$. The dynamics equations of the robot are described as follows [24]

$$\mathbf{M}(\mathbf{q})\ddot{q} + \mathbf{C}(\mathbf{q}, \dot{\mathbf{q}})\dot{q} + \mathbf{g}(\mathbf{q}) = \tau \quad (9)$$

Where, $\mathbf{M}(\mathbf{q})$is the mass matrix, $\mathbf{C}(\mathbf{q}, \dot{\mathbf{q}})$ is Coriolis matrix, $\mathbf{g}(\mathbf{q})$is the gravity vector, τ is the joints torque vector. The components of (9) are determined similarly in [24]. The generalized vectors $\mathbf{q}$, $\dot{\mathbf{q}}$, $\ddot{q}$ and $\dddot{q}$ are calculated from solving the IKP.

2.2 The IKP and IDP

Due to the robot is a redundant system solving the system of (5) will give countless answers. Choosing the most suitable answer is a quite difficult problem. Therefore, building an effective algorithm to solve the problem of inverse kinetics is always interested in. Apply the AGV algorithm [21] to find $\mathbf{q}(t)$ value with given rules $\mathbf{X}_{EEP}(t)$, $\dot{X}_{EEP}(t)$, $\ddot{X}_{EEP}(t)$. The position error $\mathbf{e_x}$ of the EEP can be determined as follows

$$\mathbf{e_x} = \mathbf{x}_{EEP} - f(\mathbf{q}) \tag{10}$$

The dynamic problem includes the forward and inverse dynamics. The forward dynamic problem has input data that are driving torques or forces and outputs are the dynamic behaviors such as the position, velocity, and acceleration of the joints in the joint space or the EEP in the workspace. The inverse dynamics allows determining the value of torques or driving forces required to ensure that the motion of the system according to the given path $\mathbf{x}_{EEP}(t)$ in the workspace or $\mathbf{q}(t) = \begin{bmatrix} q_1 \; q_2 ... \; q_6 \end{bmatrix}^T$ in the joint space. Solving the inverse dynamic problem in the workspace is truly complicated because it is necessary to solve before the already complicated IKP.

The driven torque vector can be found as follows

$$\boldsymbol{\tau}(t) = \mathbf{M(q)\ddot{q}} + \mathbf{C(q,\dot{q})\,\dot{q}} + \mathbf{g(q)} \tag{11}$$

The calculational diagram for solving the IDP is described as Fig. 2.

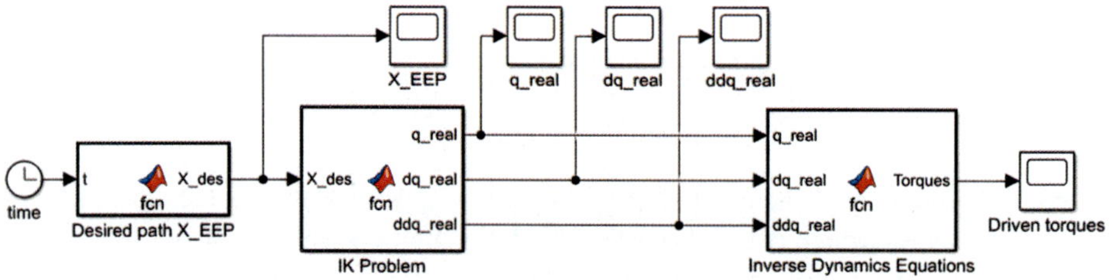

Fig. 2. The calculation diagram for the inverse dynamic problem in MATLAB/SIMULINK

2.3 Numerical Simulation Results and Discussions

This section presents the numerical simulation results for welding robot FD-V8 with three weld seams. Some dynamics parameters of the system can be showed as Table 2.

Table 2. Dynamics parameters of the system

No	Link 1	Link 2	Link 3	Link 4	Link 5	Link 6
Length of links (m)	0.42	0.15	0.56	0.13	0.6	0.325
Mass of links (kg)	127.9	37.4	79.9	19.2	3.8	3.7
Inertial moment of links (I_{xx}) (kg.m^2)	2.26	0.064	0.76	0.81	0.022	0.04
Inertial moment of links (I_{yy}) (kg.m^2)	3.04	1.25	0.88	0.78	0.0045	0.031
Inertial moment of links (I_{zz}) (kg.m^2)	2.49	1.281	0.95	0.075	0.021	0.011

Given the welding seams of the EEP in three cases are shown in Table 3.

Table 3. Trajectories in the workspace

Trajectory	$x_E(m)$	$y_E(m)$	$z_E(m)$
Case 1	$0.7 + 0.3\sin(2t)$	$0.3\cos(2t)$	0.45
Case 2	0.7	$0.3\cos(2t)$	$0.65 + 0.3\sin(2t)$
Case 3	$1.075 + 0.3\sin(2t)$	0	$0.81 + 0.3\cos(2t)$

Three welding seams in these basic planes are used to determine the torques of a redundant manipulator with 6DOF fixed in a vertical plane. There are many different types of trajectories in space depending on the specific task and these are all based on the three basic planes. The position, velocity, and acceleration of the joints are the results of this problem. The IKP solving algorithm takes the given error of the joints variables and limits the joints as the conditions for performing the calculation. There are several reasons for choosing 3 basic planes for planning the EEP trajectory in the workspace. Firstly, this study is only at the beginning of research on the driven torques of robots with simple and basic trajectories. Secondly, parameters of the EEP trajectories on these 3 planes can ensure to bring the EEP of the robot to the positions that need to be investigated such as the position outstretched, close to the robot's body, the position of rising or falling low close to the base. Thirdly, the EEP trajectories are easily built on these basic planes and easily verify reliability in both theoretical and experimental geometry calculations. On the other hand, easily fabricate auxiliary equipment such as jigs for experimenting, measuring, and verifying calculation results. Next, the analysis results of the problem in this paper can be used immediately because in reality most welded structures are mainly machined on these planes. Finally, the generalized EEP trajectory in the workspace can completely be built and investigated, but verifying the reliability and accuracy of the calculations will be a huge challenge, especially experimentally verified.

The numerical simulation results of case 1 with the trajectory in the workspace, the value of joint variables and simulation model in MATLAB are described respectively in Figs. 3, 4, and 5. Similar, simulation results of case 2 are shown in Figs. 6, 7, and 8.

Case 1 (C1): The trajectory on OXY plane.

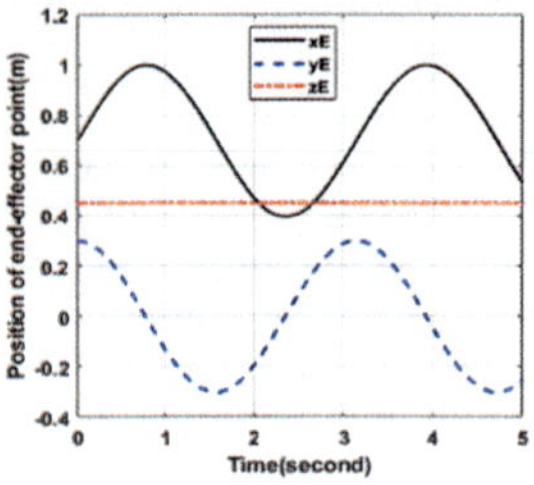

Fig. 3. Trajectory in C1

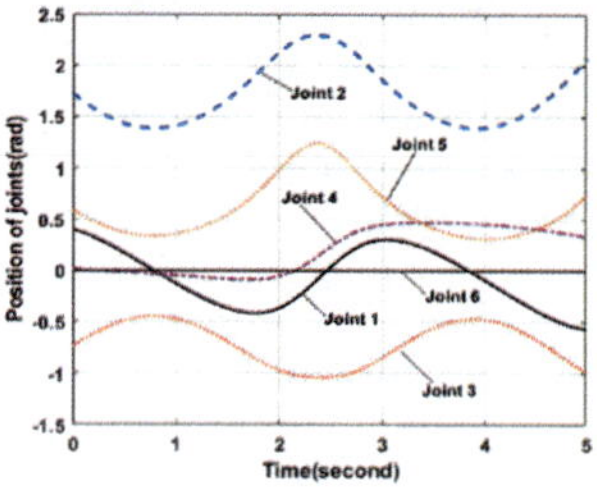

Fig. 4. Joint position in C1

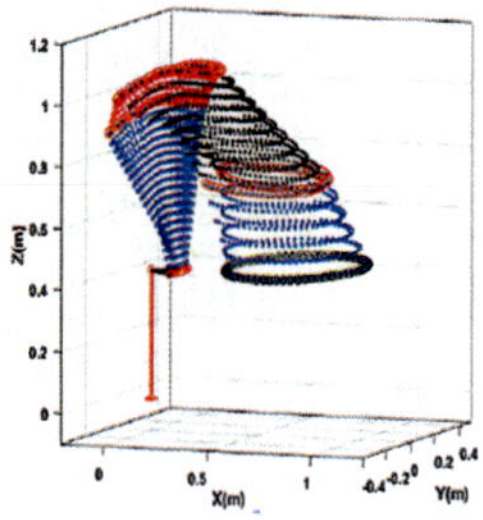

Fig. 5. Case 1

Case 2 (C2): The trajectory on OYZ plane.

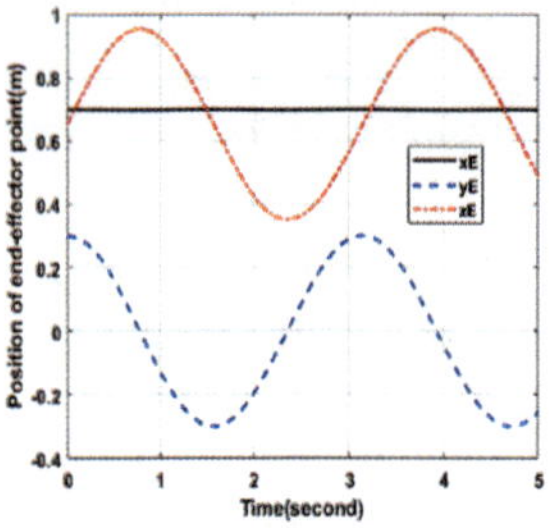

Fig. 6. Trajectory in C2

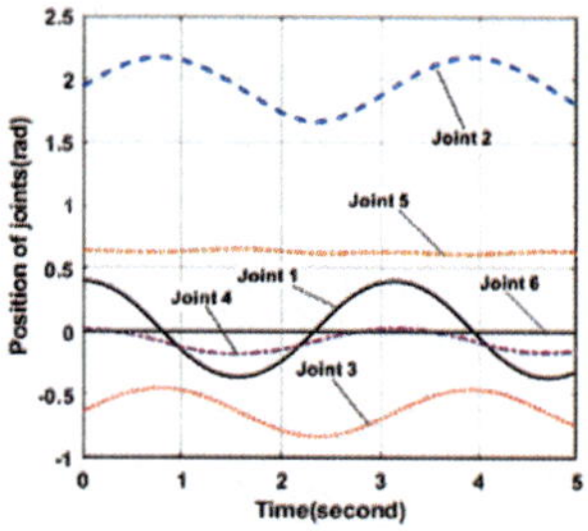

Fig. 7. Joint position in C2

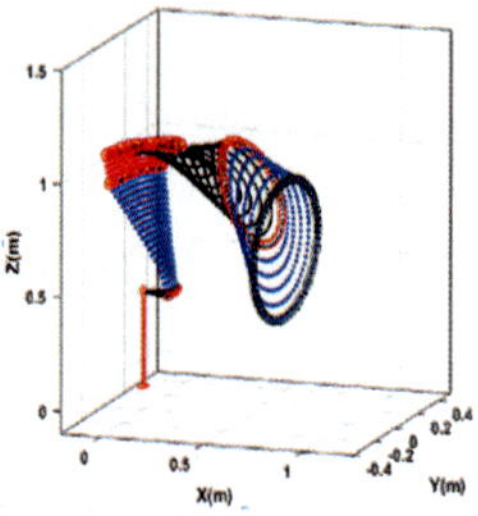

Fig. 8. Case 2

Case 3 (C3): The trajectory on OXZ plane.

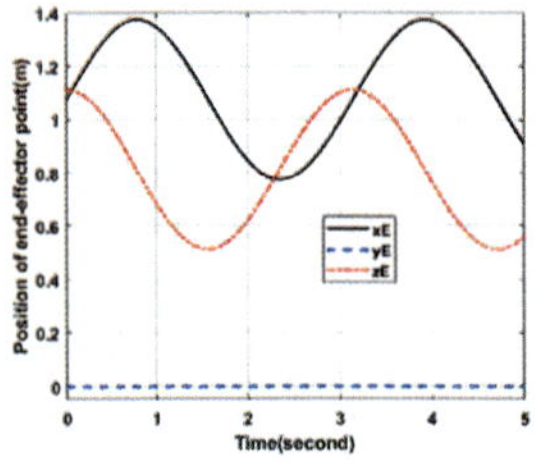

Fig. 9. Trajectory in C3

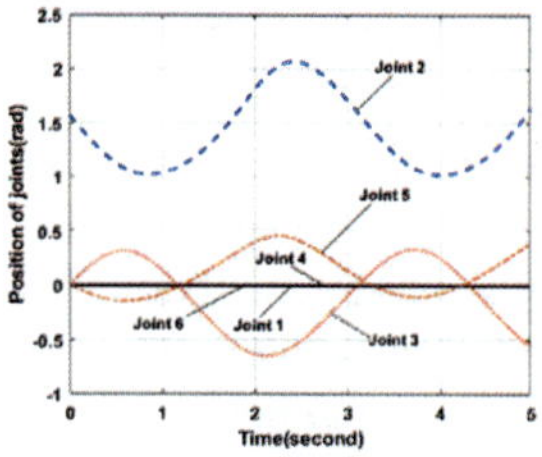

Fig. 10. Joint position in C3

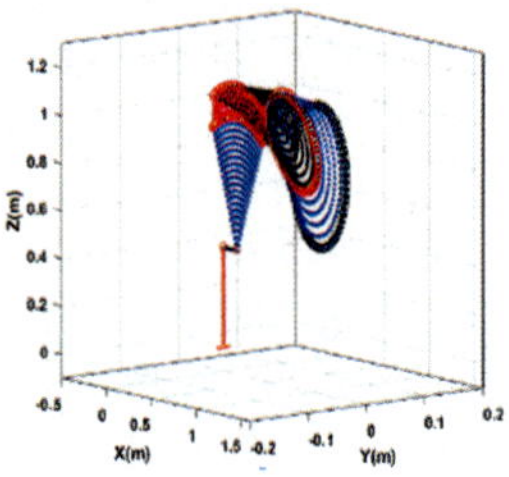

Fig. 11. Case 3

Case 3 is determined in Figs. 9, 10, and 11, respectively. It is easy to see that the values of the joint variables change continuously and there is no singularity point. The 3D model in MATLAB of the robot in Figs. 5, 8 and 11 is created from joint variable values obtained in the IKP solving.

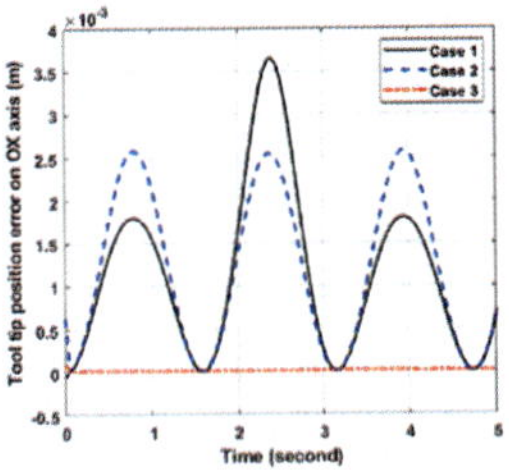

Fig. 12. Error position OX

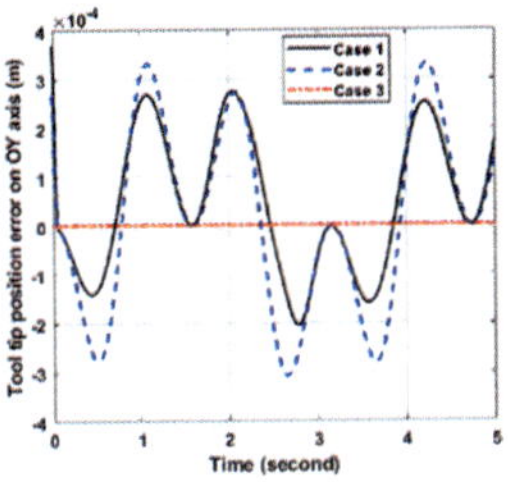

Fig. 13. Error position OY

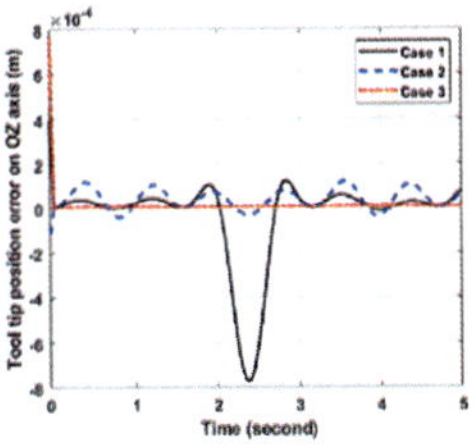

Fig. 14. Error position OZ

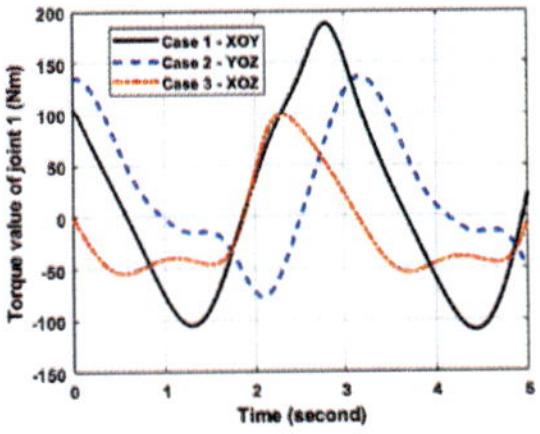

Fig. 15. Torque of q_1

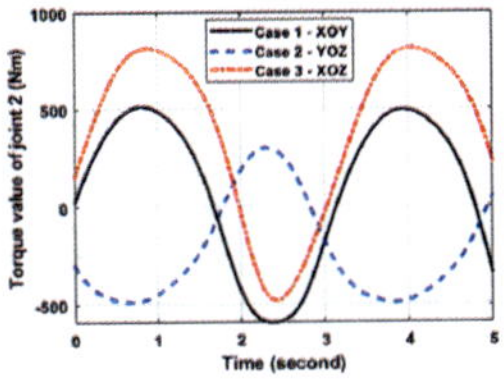

Fig. 16. Torque of q_2

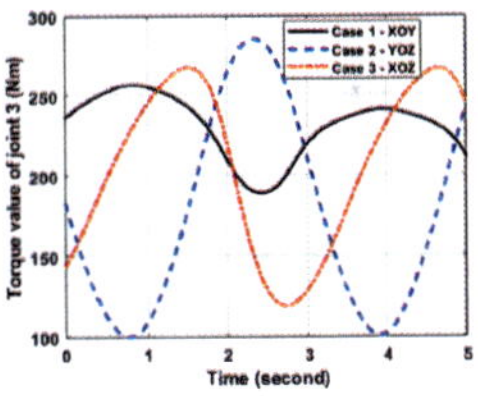

Fig. 17. Torque of q_3

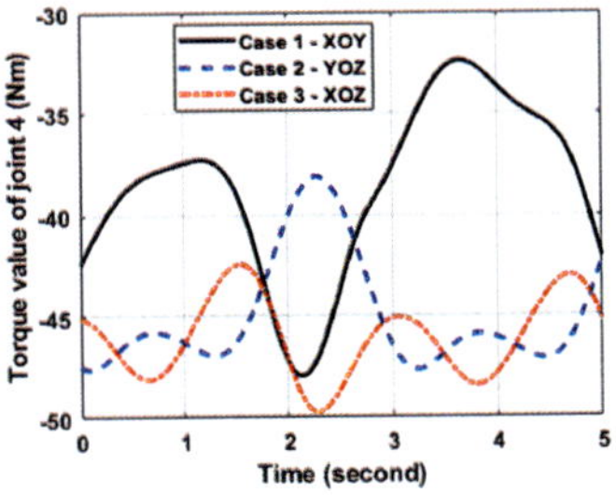

Fig. 18. Torque of q_4

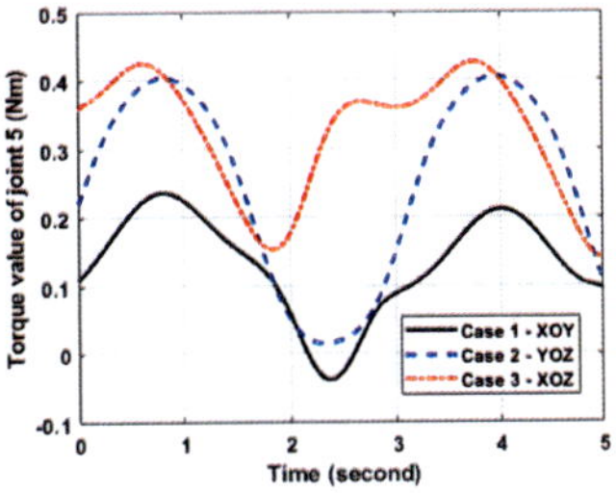

Fig. 19. Torque of q_5

The errors position of the EEP in the workspace are presented in Figs. 12, 13 and 14. The values of errors position are small. Those results prove the high reliability and efficiency of the AGV algorithm. The results of the IDP are the driving torques of joints values that are described from Figs. 15, 16, 17, 18 and 19. For C1-XOY, the maximum driving torque value is 500 (Nm) and is located at joint 2. Likewise, the torque at joint 2 in C2-YOZ and C3-XOZ also reaches the maximum values are 806.8 (Nm) and 297.6 (Nm).

The maximum driving torque value in all cases belongs to joint 2 in C3-XOZ with 806.8 (Nm). This means that the drive motor of joint 2 needs the most drive power. The reason C3 reaches the maximum value is because the robot links must reach the farthest

Table 4. The maximum torque values of joints (Nm)

No	Joint 1	Joint 2	Joint 3	Joint 4	Joint 5	Maximum
Case 1 - XOY	188	500	256.8	48	0.24	500
Case 2 - YOZ	135.5	297.6	285.1	47.7	0.4	297.6
Case 3 - XOZ	97.6	806.8	267	50	0.42	806.8
Maximum	188	806.8	285.1	50	0.42	

from the fixed origin position. The C2-YOZ gives a much smaller torque value than the other two cases. Table 4 presents the maximum torque value of each joint in three cases.

For each specific joint, the maximum driving torque value in joint 1 reaches 188 (Nm) and belongs to C1–XOY. Joint 2 requires the maximum drive torque as described above. Joint 3 gives the same driving torque value in all 3 cases. Joint 4 and joint 5 require a much smaller value of driving torque than joint 1. Joint 5 has the smallest torque value because it only carries the link 6.

In summary, the order of the links with the torque value from the largest to the smallest is joint 2, joint 3, joint 1, joint 4 and finally joint 5, respectively. The torque at the joints is greater when joint the further the operation moves away from the fixed origin. The C2-YOZ gives the driving torque value at the joints is the similarity, does not make a big difference between the joints. Thus, in order to design the welding trajectory to ensure that the driving energy is not too large, the trajectories should be designed according to the C2-YOZ. All cases where there is a change of the OX axis coordinates lead to a high demand for driving power.

3 Conclusion

In general, the effect of the welding path trajectories on the basic planes in the workspace on the driving energy of 6DOF industrial robot has been specific considered and evaluated. The results show that welding trajectories that change in the direction far away from the fixed coordinate system origin require large driving moments. On the other hand, the welding trajectory designed on the YOZ plane (Case 2) requires the smallest driving torque compared to the other cases. The value of the driving torque at joint 2 is always the largest requirement in all cases. This is the basis for calculating and selecting the driving motor power when designing the transmission, calculating the structural strength and cost expected when investing in manufacturing.

References

1. Liu, Y., Tian, X.: Robot path planning with two-axis positioner for non-ideal sphere-pipe joint welding based on laser scanning. Int. J. Adv. Manuf. Technol. **105**(1–4), 1295–1310 (2019). https://doi.org/10.1007/s00170-019-04344-3
2. Liu, F., Lin, L.: Time-jerk optimal planning of industrial robot trajectories. Int. J. Robot. Autom. **31**, 1–7 (2016)

3. Dai, C., Lefebvre, S., Yu, K.M., Geraedts, J.M.P., Wang, C.C.L.: Planning jerk-optimized trajectory with discrete time constraints for redundant robots. IEEE Trans. Autom. Sci. Eng. **17**, 1711–1724 (2020)
4. Liang, F., Zhao, J., Ji, S.: An iterative feed rate scheduling method with confined high-order constraints in parametric interpolation. Int. J. Adv. Manuf. Technol. **92**(5–8), 2001–2015 (2017). https://doi.org/10.1007/s00170-017-0249-6
5. Mansour, S.Z., Seethaler, R.: Feedrate optimization for computer numerically controlled machine tools using modeled and measured process constraints. J. Manuf. Sci. Eng. Trans. ASME **139**, 1–9 (2017)
6. My, C.A., Bien, D.X., Tung, B.H., Cong, N.V., Hieu, L.C.: New Feed Rate Optimization Formulation in a Parametric Domain for 5-Axis Milling Robots. Springer International Publishing, New York, 1121 (2020)
7. Sterling, T., Chen, H.: Robotic welding parameter optimization based on weld quality evaluation. In: 6th Annual IEEE International Conference Cyber Technology Automatic Control Intelligent System IEEE-CYBER 2016, pp. 216–221 (2016)
8. Yan, S.J., Ong, S.K., Nee, A.Y.C.: Optimal pass planning for robotic welding of large-dimension joints with deep grooves. Proc. CIRP **56**, 188–192 (2016)
9. Kumaran, K.S., Raj, S.O.N.: Optimization of parameters involved in robotic MIG welding process based on quality responses. IOP. Conf. Ser. Mater. Sci. Eng **402**, 021016 (2018)
10. Garcia, R.R., Bittencourt, A.C., Villani, E.: Relevant factors for the energy consumption of industrial robots. J. Braz. Soc. Mech. Sci. Eng. **40**(9), 1–15 (2018). https://doi.org/10.1007/s40430-018-1376-1
11. Shugen, M.A.: Real-time algorithm for quasi-minimum energy control of robotic manipulators. In: The 21st International Conference on Industrial Electronics, Control, and Instrumentation (IECON 1995), Orlando, pp. 1324–1329 (1995)
12. Wigstrom, O., Lennartson, B.: Towards integrated OR/CP energy optimization for robot cells. In: IEEE International Conference on Robotics and Automation, Hong-Kong (2014)
13. Paes, K., Dewulf, W., Elst, K.V., Kellens, K., Slaets, P.: Energy efficient trajectories for an industrial ABB robot. Proc. CIRP **15**, 105–110 (2014)
14. Paryanto, Brossog, M., Bornschlegl, M., Franke, J.: Reducing the energy consumption of industrial robots in manufacturing systems. Int. J. Adv. Manuf. Technol. **78**, 1315–1328 (2015)
15. Wolniakowski, A., Valsamos, C., Miatliuk, K., Moulianitis, V., Aspragathos, N.: Optimization of dynamic task location within a manipulator's workspace for the utilization of the minimum required joint torques. Electronics **10**, 1–18 (2021)
16. Scalera, L., Boscariol, P., Carabin, G., Vidoni, R., Gasparetto, A.: Enhancing energy efficiency of a 4-DOF parallel robot through task-related analysis. Machines **1**, 1–14 (2020). https://doi.org/10.3390/machines8010010
17. Palomba, E., Wehrle, G.C., Vidoni, R.: Minimization of the energy consumption in industrial robots through regenerative drives and optimally designed compliant elements. Appl. Sci. **10**(21), 1–18 (2020). https://doi.org/10.3390/app10217475
18. Boscariol, P., Caracciolo, R., Richiedei, D., Trevisani, A.: Energy optimization of functionally redundant robots through motion design. Appl. Sci. **10**(9), 3022 (2020). https://doi.org/10.3390/app10093022
19. Eggers, K., Knochelmann, E., Tappe, S., Ortmaier, T.: Modeling and experimental validation of the influence of robot temperature on its energy consumption. In: Proceedings of IEEE International Conference Industrial Technology, pp. 239–243 (2018)
20. Pellicciari, M., Berselli, G., Leali, F., Vergnano, A.: A method for reducing the energy consumption of pick-and-place industrial robots. Mechatronics **23**, 326–334 (2013)

21. Khang, N.V., Dien, N.P., Vinh, N.V., Nam, T.H.: Inverse kinematic and dynamic analysis of redundant measuring manipulator BKHN-MCX-04. Vietnam J. Mech. VAST **32**, 15–26 (2010)
22. FD-V8 robot. https://www.daihen-usa.com/product/fd-v8-robot-6kg-payload-1-4m-reach/. Accessed 21 June 2021
23. Spong, M.W., Hutchinson, S., Vidyasagar, M.: Robot Modeling and Control, 1st edn. Wiley, New York (2001)
24. Khang, N.V: Dynamics of Multi-bodies. Hanoi Science and Technology Publishing House (2007)

Optimizing a Design of Constant Volume Combustion Chamber for Outwardly Propagating Spherical Flame

M. T. Nguyen(✉), M. T. Phung, H. P. Ho, and V. Đ. Nguyen

The University of Danang – University of Technology and Education, Danang City, Vietnam
nmtien@ute.udn.vn

Abstract. Constant Volume Combustion Chamber (CVCC) is valid for fundamental studies under internal combustion engine-like conditions. Their configuration depends on the engine types and research purposes. Hence, considering many parameters such as the chamber shape, component materials, dimensions, cost, and operating safety plays a vital role in designing a proper vessel. This work aims to optimize the structure of a cylindrical CVCC and its optical quartz windows using finite element analysis of Solid Simulation. The chamber, which can withstand the maximum pressure of 10 MPa, was designed for outwardly propagating spherical flame of gaseous and liquid fuels under engine-like conditions. By setting the boundary conditions and other constraints, the optimized CVCC satisfies the material's criterion and reduces approximately 20% in its mass compared to the preliminary design. The preliminary test results suggest that the finite element method could successfully be applied to design such a CVCC system.

Keywords: Constant volume combustion chamber · Optical quartz window · Finite element · Outwardly propagating spherical flame

1 Introduction

Improving vehicle fuel economy and emission has drawn extensive effort to achieve a sustainable society by which the enhancement of engine thermal efficiency offers an effective way. Many studies on practical engines have indicated the thermal efficiency could achieve up to 50% [1]. However, the complex operating conditions (e.g., in-chamber tumble flow, high-pressure/temperature, mixture preparing) alongside many unexpected phenomena (misfire, cycle-to-cycle variation, knocking, etc.) may cause research difficulties and inadequate understanding of the aforesaid achievement in the practical engine. An alternative-simplified apparatus that can operate under engine-like conditions is thus essential for a better understanding of the fundamental internal combustion engines (ICE) by which higher thermal efficiency and lower emission could be achieved.

Constant volume combustion chamber becomes the most common apparatus [2–6] for fundamental studies relating to ICEs due to its convenience to control and study the

N. V. Khang et al. (Eds.): ASIAN MMS 2021, MMS 113, pp. 921–931, 2022.
https://doi.org/10.1007/978-3-030-91892-7_88

effect of different parameters on the combustion process. For instance, Nguyen et al. [5–9] studied turbulent ignition and turbulent flame speed of different gaseous fuel and liquid iso-octane in a 3D-cruciform burner under high pressure and high temperature. Using a spherical combustion chamber, Zhao et al. [3] explored the effect of multi-channel spark on the ignition of lean n-pentane/air mixture. Chen et al. [10] investigated the ignition and combustion of blended ethanol-gasoline spray in DISI engine-like conditions in the cylindrical chamber. Regarding the experimental purposes, CVCC can be designed in various configurations, i.e., cuboid, cylindrical, and spherical shapes with the optical quartz windows. The in-chamber combustion process can be seen more clearly and captured for other analyses (e.g., shockwave development, flame propagation speed) using the optical windows alongside the schlieren system. Furthermore, it is easy to control engine-like conditions such as fuel/air ratio, turbulent flow characteristics, ignition energy source, spark gap, pressure, and temperature. Therefore, researchers could have a better understanding of the internal relationship of these parameters.

The constant volume combustion chamber operates under such engine-like conditions, no matter what configurations; CVCC must withstand the high-pressure and high temperature. Therefore, the selected optimal dimensions and materials of CVCC's parts, especially the optical quartz window, are essential and can reduce the manufacturing cost. For these purposes, several works applied the finite element method (FEM) for strength analysis and structure optimization of CVCC. Tian et al. [11] employed FEM to the cuboid CVCC with the rectangular quartz window and then reached a 50%-decrease in volume. Similarly, Haji et al. [12] designed and optimized a cylindrical CVCC for high-pressure variations and thermal shocks by FEM. The satisfied practical results on their optimized CVCC system [11, 12] proved that FEM could successfully be employed to design such a CVCC system. In the present work, we design and optimize the cylindrical CVCC and its quartz windows for the outwardly propagating spherical flame of gaseous and liquid fuels used in ICEs under maximum in-chamber pressure of 10 MPa using FEM in Solid Simulation.

2 Method

2.1 Preliminary Design of CVCC

Figure 1 shows the schematic view of the cylindrical CVCC, including three main parts: a central cylindrical cavity, two-end flange covers, and two flat quartz windows that can be view in the axial direction. A pair of pin-to-pin electrodes is mounted on the central cavity surface to ignite the combustible mixtures. Pressure transducers are used to detect partial pressure of fuel/air sequentially during the mixture preparation and the in-chamber pressure rise during the combustion process. Two-end quartz windows as the flat cylinders with the dimension of (Φ200 mm $\times$ L40 mm) is convenient to see the in-chamber combustion process. Dimension and material for the CVCC parts are presented in Fig. 2 and Table 1.

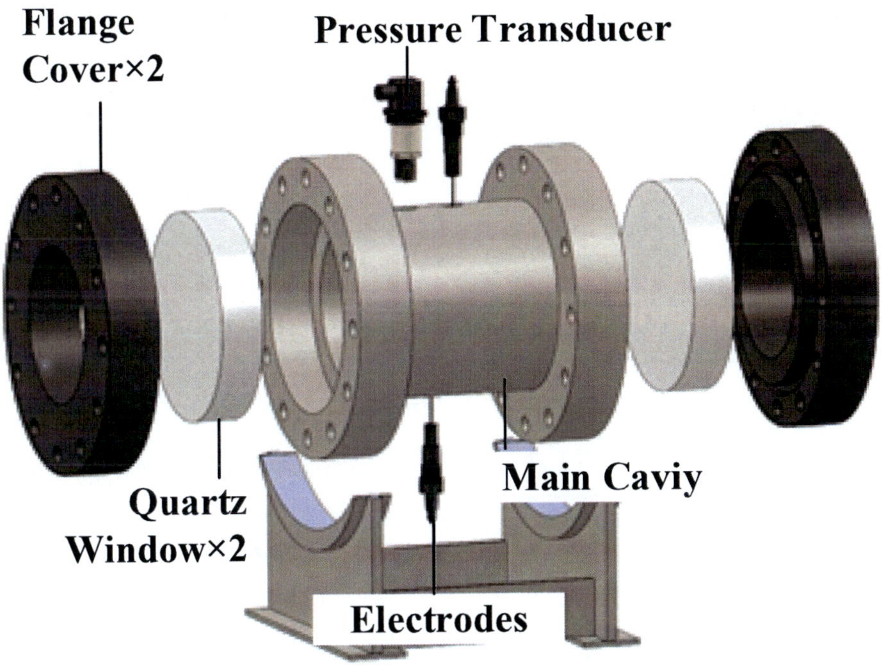

Fig. 1. Schematic view of the cylindrical constant volume combustion chamber

2.2 Finite Element Model and Boundary Conditions

The solid model of CVCC is automatically meshed using curvature-based mesh to build the finite element model, as shown in Fig. 3. The minimum and maximum element sizes are 4.3 mm and 21.6 mm, respectively. The total elements number of the meshed model is 89480, and the nodes number is 135988. The holes are fully constrained, and the other components (i.e., bolts, pressure transducers, valves) are fixed before analyzing. The preload of bolts is set to 150 Nm. The uniform gas pressure of 10 MPa is applied to the in-chamber surface. The static analysis alongside the FFEPlus solver is then employed to analyze the CVCC solid meshed model.

2.3 Structure Analysis

By fixing the boundary conditions and statically analyzing the solid-meshed model in Fig. 3, we obtain the Von Mises stress distribution on the whole-CVCC and its components, as indicated in Fig. 4. The stress mainly affects the central cavity and the quartz windows because the gas pressure directly contacts these component surfaces, where the maximum stress value is 90 MPa. It is noted that the maximum stress at the bolts is probably due to the high preload of 150 Nm. Based on these preliminary results, the maximum stress is approximately a haft of the material's yield strength (i.e., 206.8 MPa for stainless steel 304 and 155 MPa for glass SiO_2). The maximum stress on the quartz

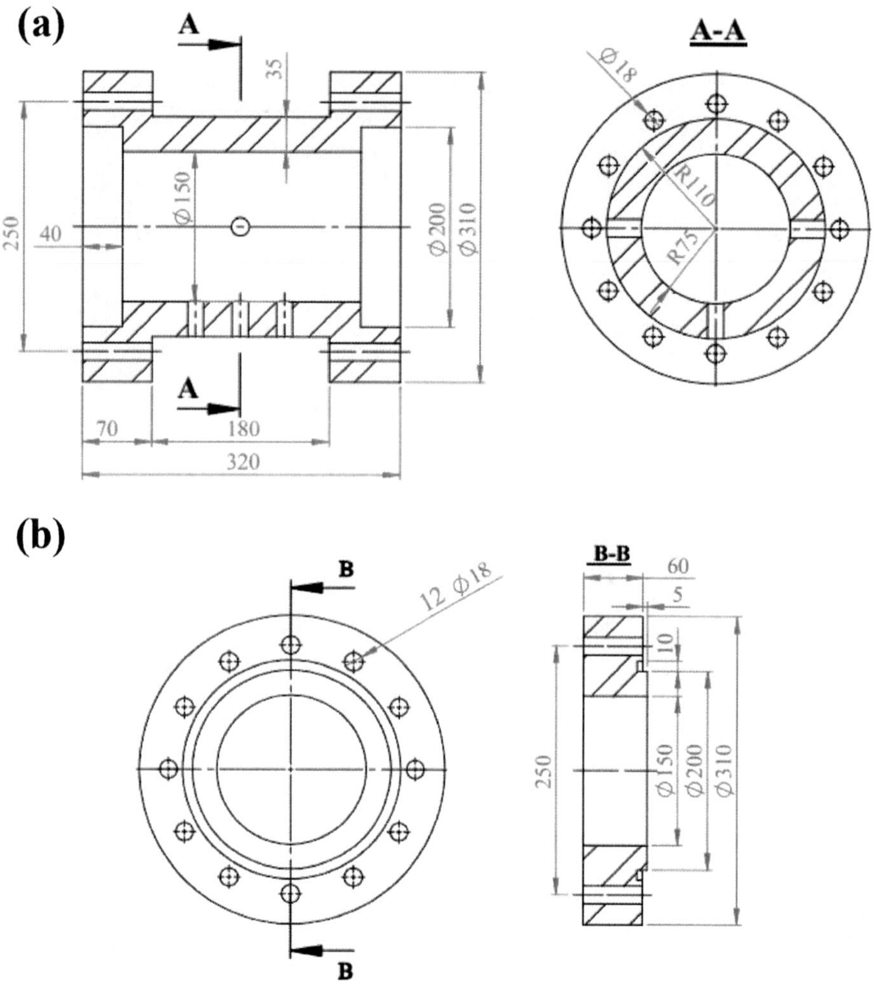

Fig. 2. Two-dimension map. (a) Main cylindrical cavity and (b) Flange cover

Table 1. Material property.

Component	Material	Yield strength (MPa)	Poison's Ratio	Density (Kg/m^3)
Quartz window	SiO_2	155.0	0.17	2210
Main cavity	Stainless steel 304	206.8	0.29	8000
Flange cover	Stainless steel 304	206.8	0.29	8000

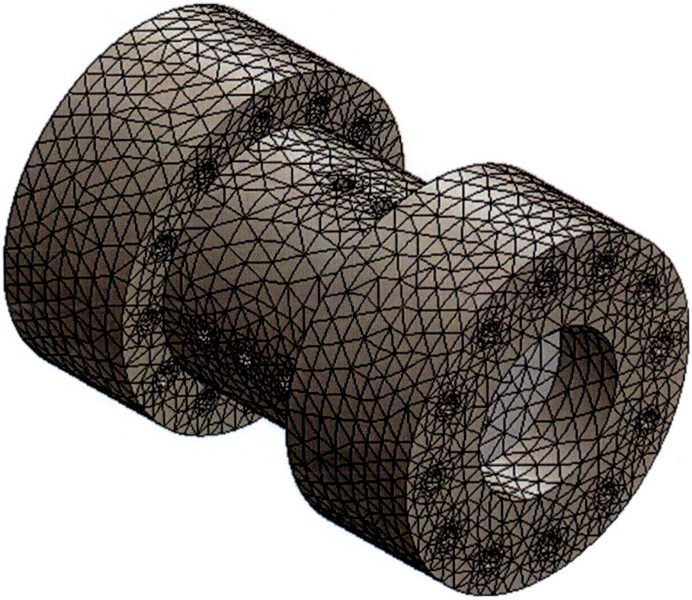

Fig. 3. Finite element model of CVCC. The total element number is 89480, and the nodes number is 135988.

window is also smaller than the critical Modulus of Rupture (200 MPa [13]), an additional criterion for SiO_2 material [12]. Therefore, structure optimization is necessary for saving materials and manufacturing costs.

3 Structure Optimization

3.1 Main Cavity

In order to optimize the structure of the central cavity, it is necessary to set the initial parameter and the constraints by which the maximum stress of optimized variables is less than 206 MPa and the mass is minimum, as indicated in Fig. 5, and Table 2.

According to the setup range in Table 2, after 27 interactions (the results are not cpresented here due to the limitation of space), we obtain an optimal result (please see the 3rd column in Table 3) that indicates approximately 20%-decrease of mass and/or volume as compared to the preliminary design. The highest stress distribution in Fig. 6 suggests that it should be better to make weld reinforcement between the connecting valves and the chamber, although the maximum stress (92.9 MPa) is much smaller than the critical value (206 MPa).

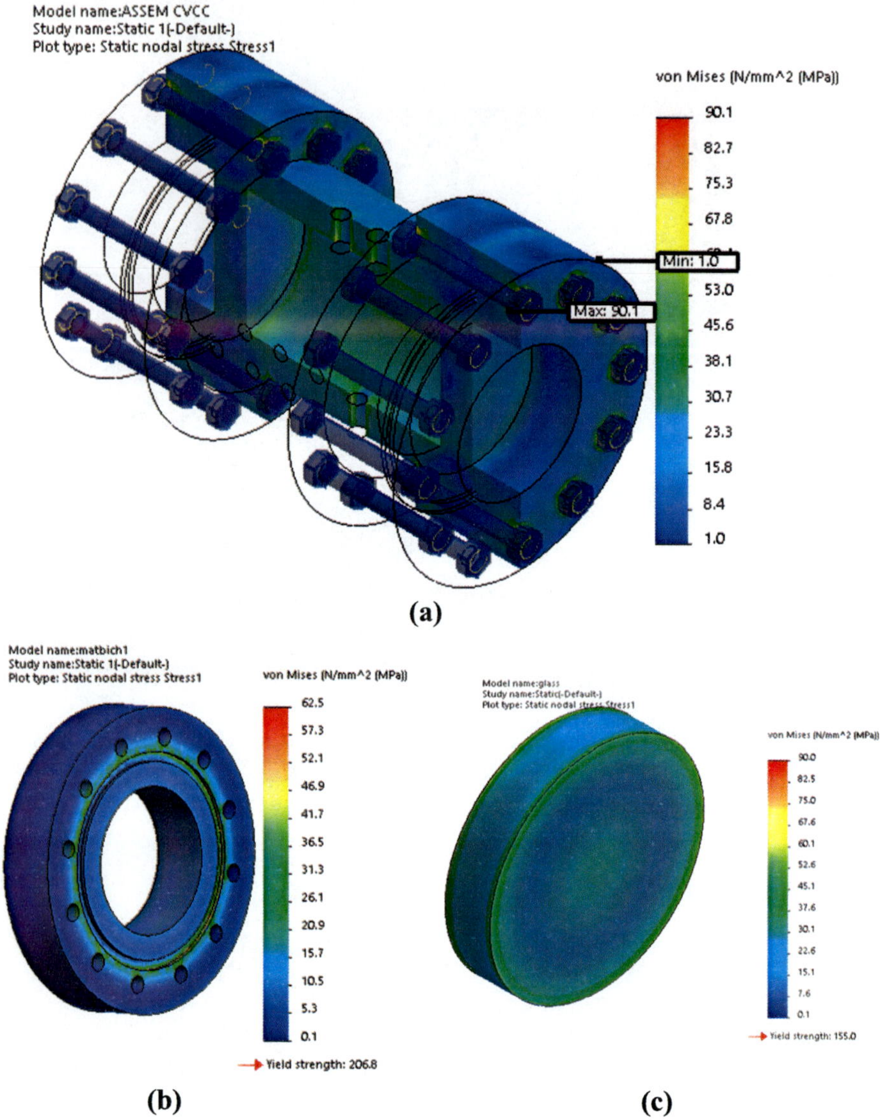

Fig. 4. Von Mises stress distribution. (a) The CVCC, (b) Flange cover, (c) Quartz window

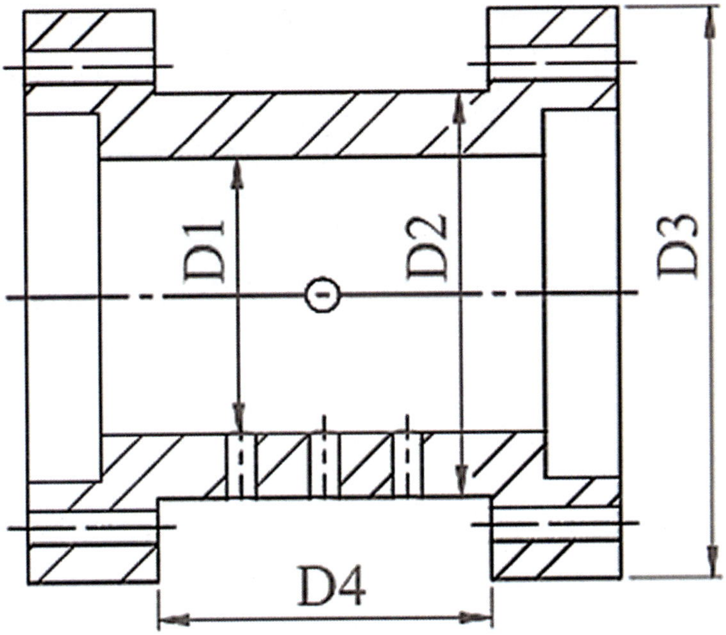

Fig. 5. Design variables

Table 2. Initial design variables

State variables: the maximum stress is less than 206.8 MPa			
Optimized target: Minimum of mass			
Name	Type	Value	Units
D1	Range	Min:140 Max:150	mm
D2	Range	Min:210 Max:220	mm
D3	Range	Min:280 Max:310	mm
D4	Range	Min:150 Max:180	mm

3.2 Quartz Windows

In the results above, we used the quartz windows with a dimension of Φ200 mm × L40 mm to analyze the CVCC structure, indicating that the maximum Von Mises stress of 90 MPa (please see Fig. 4c) mainly concentrates in the edge of the quartz window. The distributed area of Von Mises is as same as the other cases having a smaller thickness of 30 mm (Fig. 7a) or 20 mm (Fig. 7b); however, the maximum Von Mises stress increases with decreasing the thickness. The result in Fig. 7(a) suggests that the quartz window with 30-mm thickness satisfies the material criteria in which the maximum Von Mises stress of 123.5 MPa is less than Yield strength (155 MPa) or even Modulus of

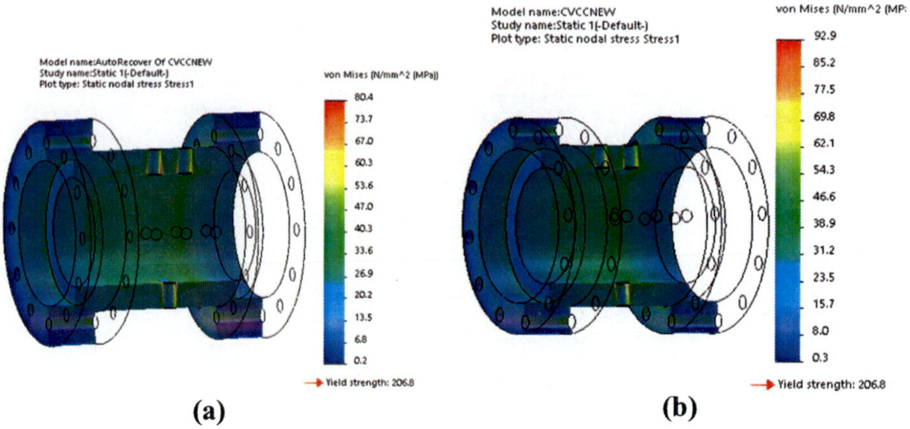

Fig. 6. Von Mises stress distribution on the main cavity. (a) Before optimization and (b) After optimization

Table 3. Optimized result for the main cavity

Name	Initial	Optimal	Units
D1	150	150	mm
D2	220	210	mm
D3	310	281	mm
D4	180	150	mm
Stress	80.4	92.9	(MPa)
Mass	81.5	65	Kg
Volume	0.0102	0.0082	m^3

Rupture (200 MPa). Hence, we selected the quartz windows (Φ200 mm × L30 mm) to re-analyze the optimized-assembling CVCC structure (please see Table 4 for more detail). Furthermore, the silicone rubber gasket is used between the quartz window and the chamber to prevent the gases leakage and dampen the shock due to the in-chamber pressure rise.

3.3 The Optimized-Assembling CVCC Structure Analysis

After analyzing the structure of each component, we re-analyze the optimized-assembling structure. Figure 8(a) indicates that the maximum Von Mises stress increases to 154.1 MPa, which still satisfies the material criteria. In addition, the factor of safety (FOS) as a ratio of the critical Yield strength and the maximum Von Mises stress is greater than unity and in the range of 1.34 to 10, as indicated in Fig. 8(b).

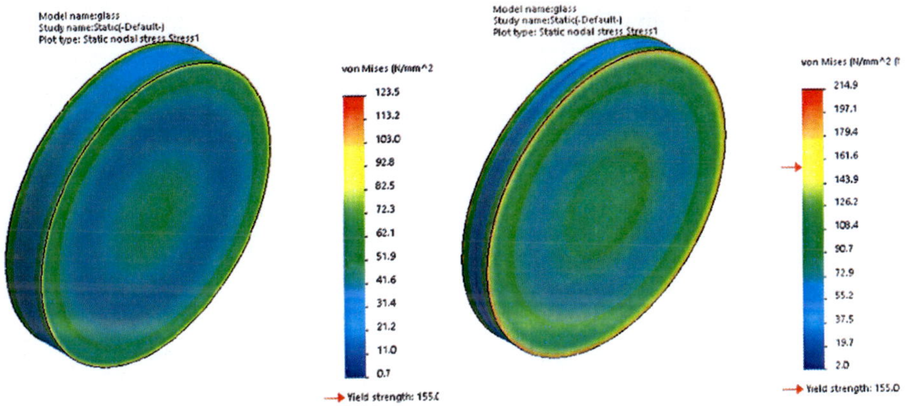

(a) **(b)**

Fig. 7. Von Mises stress distribution on the quartz window at different thicknesses. (a) 30 mm and (b) 20 mm. Yield strength is 155 MPa, and the Modulus of Rupture is 200 MPa [13].

Table 4. Maximum stress and displacement magnitudes of quartz window based on Von Mises

Thickness (mm)	Von mises stress (MPa)	Displacement (mm)
20	214.9	0.289
30	123.5	0.1012
40	90	0.05236

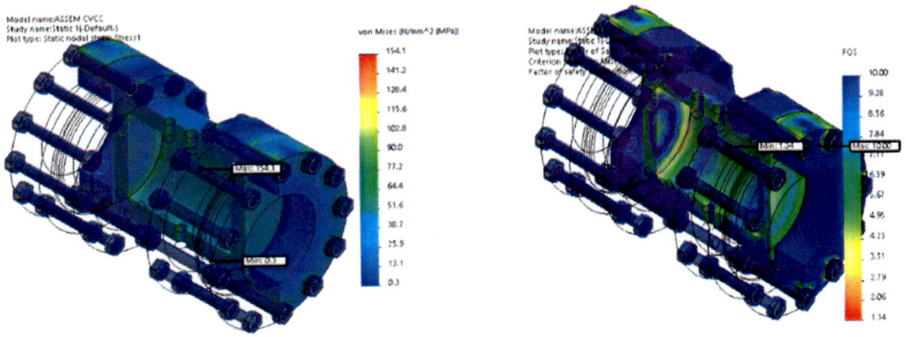

(a) **(b)**

Fig. 8. (a) Von Mises stress distribution on the optimized CVCC. (b) The factor of safety (FOS)

3.4 Practical Tests in the Chamber

After manufacturing, the chamber's performance was investigated by conducting lean CH4/air combustion [14].

4 Conclusion

In order to research ignition phenomena and flame propagation under engine-like conditions (e.g., high temperature, high pressure, and turbulence), we have designed a cylindrical constant volume combustion chamber for simulating such similar events in the practical engines when the piston is near at top dead center. The preliminary design includes three major parts: a cylindrical cavity, two-end flange covers, and two quartz windows. The preliminary design of CVCC is statically analyzed and optimized under 10 MPa of in-chamber pressure using the FFEPlus solver in the finite element analysis. The optimized result shows that the total mass decreases approximately 20%, and the stresses can be in their critical values. After manufacturing, the CVCC chamber is tested by conducting CH_4/air combustion [14]. The preliminary test results in Ref. [14] suggest that the finite element method could successfully be applied to design such a CVCC system.

Acknowledgments. The financial support from the Ministry of Education and Training, Viet Nam, under grant B2021-DNA-02 is greatly appreciated.

References

1. Nakata, K., Nogawa, S., Takahashi, D., Yoshihara, Y., Kumagai, A., Suzuki, T.: Engine technologies for achieving 45% thermal efficiency of S.I. engine. SAE. Int. J. Engines **9**(1), 179–192 (2015)
2. Wang, Y., Zhang, X., Li, Y., Qin, S.: Experimental study on premixed turbulent flame of methane/air in a spherical closed chamber. Energy Convers. Manage. **222**, 113219 (2020). https://doi.org/10.1016/j.enconman.2020.113219
3. Zhao, H., et al.: Studies of multi-channel spark ignition of lean n-pentane/air mixtures in a spherical chamber. Combust. Flame **212**, 337–344 (2020)
4. Hayashi, N., Sugiura, A., Abe, Y., Suzuki, K.: Development of ignition technology for dilute combustion engines. SAE Int. J. Engines **10**(3), 984–995 (2017)
5. Nguyen, M.T., Yu, D., Chen, C., Shy, S.: General correlations of iso-octane turbulent burning velocities relevant to spark ignition engines. Energies **12**(10), 1848 (2019)
6. Nguyen, M.T., Shy, S.S., Chen, Y.R., Lin, B.L., Huang, S.Y., Liu, C.C.: Conventional spark versus nanosecond repetitively pulsed discharge for a turbulence facilitated ignition phenomenon. Proc. Combust. Inst. **38**(2), 2801–2808 (2021)
7. Nguyen, M.T., Yu, D.W., Shy, S.S.: General correlations of high pressure turbulent burning velocities with the consideration of Lewis number effect. Proc. Combust. Inst. **37**(2), 2391–2398 (2019)
8. Shy, S.S., Nguyen, M.T., Huang, S.Y.: Effects of electrode spark gap, differential diffusion, and turbulent dissipation on two distinct phenomena: turbulent facilitated ignition versus minimum ignition energy transition. Combust. Flame **205**, 371–377 (2019)

9. Nguyen, M.T., Phung, M.T.: 'Effect of nanosecond repetitively pulsed frequency on laminar flame propagation of lean n-butane/air mixture with Le >> 1'. In: Editor (Ed.): 'Book Effect of Nanosecond Repetitively Pulsed Frequency on Laminar Flame Propagation of Lean n-Butane/Air Mixture with Le >> 1', pp. 698–704. Springer, Cham (2020)
10. Chen, R., Okazumi, R., Nishida, K., Ogata, Y.: Effect of ethanol ratio on ignition and combustion of ethanol-gasoline blend spray in disi engine-like condition. SAE Int. J. Fuels Lubr. **8**(2), 264–276 (2015)
11. Tian, X.F., Zhang, H.G., Bai, X.L.: Strength analysis and structure optimization of constant volume combustion bomb. Adv. Mater. Res. **233–235**, 821–826 (2011)
12. Hajialimohammadi, A., Ahmadisoleymani, S., Abdullah, A., Asgari, O., Rezai, F.: Design and manufacturing of a constant volume test combustion chamber for jet and flame visualization of CNG direct injection. Appl. Mech. Mater. **217–219**, 2539–2545 (2012)
13. AZO materials. https://www.azom.com/article.aspx?ArticleID=1114. Accessed July 29 2021
14. Tien, N.M., Thanh, N.L.C., Dong, N.V., Phi, H.H.: A study on the influence of ignition energy on ignition delay time and laminar burning velocity of lean methane/air mixture in a constant volume combustion chamber. J. Sci. Technol. Accepted for Publishing, 18 August 2021

Thermo-mechanical Modeling and Analysis of Adaptive NiTi Shape Memory Alloy Plates

Rashmi Jadhav[1](✉) and Rajan L. Wankhade[2]

[1] Dr. D.Y. Patil Pratishthan's College of Engineering, Salokhenagar, Kolhapur, Maharashtra, India

[2] Government College of Engineering, Nagpur, Maharashtra, India

Abstract. It is observed that structure cab me made adaptive so as to have better performance. The structure can behave efficiently by having adaptive and responsive properties. SMA-Shape Memory Alloy belongs to a smart materials family which can sense the change and activate its special properties. These SMA-shape memory alloys have very high recovery capacity which in term needed for active/passive vibration control. In this study, the response of plate with SMA is observed under loading, unloading cycling. Hence, the steel and NiTi plate analysis is performed to compare the residual deflection after loading and unloading. The reference maximum deflection is categorized depending upon dimension of steel plate. Incremental load is applied to attain the maximum deflection. Further loading and unloading is done with same incremental and decrease in loads. The maximum and residual deflection of both steel and NiTi plates is compared and results are presented through graphs and tables.

Keywords: Stress induced · NiTi · Residual deflection

1 Introduction

Smart SMA-NiTi plates are now days widely used at a variety of places and in many structural applications. Such plates are used as dampers, at the beam-column connections and base plates in civil engineering applications. It is observed that during dynamic loading greatly affects the yield strength if we vary the aspect ratio of such plated structure. Material behavior thus influences the yield strength in case of NiTi plates due to its ability to recover the shape and size. While changing the thickness of the plate via plate aspect ratio, it results in different load carrying capacity and vice-versa. Thinner the plate experiences low load carrying capacity. As we increase the thickness of the plate, load carrying capacity also increases. Plates can also be made more stiff by changing its material property or buy making them efficient, adaptive and responsive. Thus SMA is one of the material in smart materials category which can be used to make a plated structure efficient, adaptive and responsive. Also fatigue life of these NiTi composite plates is four times than that of steel plates. As higher pre-strain provides more fatigue life SMA tends to be application in structural vibration control. Behavior of NiTi plate is compared with that of steel plate and observed that it yields 25% more fatigue life than that of steel plate with the considered aspect ratio.

N. V. Khang et al. (Eds.): ASIAN MMS 2021, MMS 113, pp. 932–943, 2022.
https://doi.org/10.1007/978-3-030-91892-7_89

Amini et al. (2010) proposed a technique to increase the resistance of steel plate under dynamic blast loading. The experimental study suggest that the polyuria layer on steel plate increases the resistance to dynamic loading and this technique can be used for retrofitting of existing plates. Yen and Chien (2010) worked on study of vertical cyclic loads exerted on steel plates for rehabilitating RC beam–column joints. Huawen et al. (2010) studied fatigue performance of tension steel plates strengthened with prestressed CFRP laminates. Bajoria and Wankhade (2012) studied vibration characteristics for SSSS piezolaminated composites via fem. Wankhade and Bajoria (2013a, 2013b, 2015) also showed buckling for stability of SSSS piezolaminated composites. Shariat et al (2013) performed analytical modelling of functionally graded NiTi shape memory alloy plates under tensile loading and recovery of deformation upon heating. Wankhade and Bajoria (2015; 2016; 2017; 2019; 2021) further provided HOST model for shape, vibration and buckling control for smart piezolaminated beams/plates under static and dynamic excitations. Rasid et al. (2014) developed FEM models for analysis of SMA symmetric cross-ply and angle-ply composites using FEM.

Samadpour et al. (2015) Provided vibration analysis of thermally buckled SMA hybrid composite sandwich plates. Bendine and Wankhade (2016; 2017; 2019) examined vibration control of FGM piezoelectric plate based on LQR genetic search. Dehkordi and Khalili (2015) performed frequency analysis of sandwich plate with active SMA hybrid composite face-sheets and temperature dependent flexible core. Shahverdi et al. (2016; 2018) studied iron-based shape memory alloys for prestressed near-surface mounted strengthening of reinforced concrete beams. Mahabadi et al. (2016) performed free vibration of laminated composite plate with shape memory alloy fibers. Izadi et al. (2018) developed an iron-based shape memory alloy (Fe-SMA) strengthening system for steel plates. Yang et al. (2019) did non linear finite element analysis of simply supported rectangular plate under velocity impact. During the dynamic analysis it is observed that the ultimate dynamic compressive strength depends on aspect ratio and impact duration, but the plate thickness and material yield strength have negligible effect on dynamic compressive strength. They proposed impirical formula to predict dynamic compressive strength of plate gives results matching to experimental results.

Chang et al. (2019) investigated the behavior of SMA reinforced composite plate under impact load. The effect of high velocity by applying impact with different velocities is studied using finite element analysis. It is observed that energy absorption increases as impact energy increases for SMA reinforced composite plate. Wankhade and Niyogi (2020) performed buckling analysis of symmetric laminated composite plates for various thickness ratios and modes. Joaquín et al. (2020) demonstrated a FE modeling of RC beams externally strengthened with iron based shape memory alloy (Fe-SMA) strips, including analytical stress-strain curves for Fe-SMA. Bendine and Wankhade (2021) also developed piezoelectric energy harvesting system from a curved plate subjected to time-dependent loads. The Quasi-static load was applied to those plates failure, the stress recovery, strength and yield strength is high for the FeSMA Anchored steel plates. The main important advantage of this system is, and strengthening using those plates is hassle-free.

The manuscript contains the introduction of material and previous studies of plate analysis in the initial part. The NiTi plate formulation in finite element method and the simulation techniques in ANSYS is explained in the later part. The manuscript is concluded by mentioning the analysis results of Steel and NiTi finite element analysis under loading unloading cycle.

2 Methodology Developed for Study

2.1 Convergence Study

The results obtained in finite element analysis are highly dependent on the meshing or element size and the connection between elements. A convergence study is performed initially to obtain the meshing size which will provide the results converging with the exact solution. The maximum mid span deflection of a fixed plate is calculated using a formula available in literature. The maximum mid span deflection obtained in analytical solution for different mesh sizes is compared with the result obtained by the actual calculation. The maximum mid span deflection for 10mm and 5mm mesh sizes show a difference of 2% with the actual results obtained. The actual maximum mid span deflection is 6.9 mm, the mid span deflection obtained when a coarse mesh of 25 mm is used is 4.3 mm, and it is 6.1 mm and 6.7 mm respectively when 10 mm and 5 mm mesh sizes are used. As the 5 mm mesh size deflection converges with the actual value, the 5 mm mesh is used to study the combined effect of steel SMA in composite plates and to investigate the stress strain behavior of different materials. The plate size is 100 mm × 100 mm × 10 mm. The analysis of plate is done for element size 20 mm, 10 mm and 5 mm and 2 mm. The optimum results obtained, converging with actual results are for 5 mm and 2 mm mesh element size. To reduce the computation time 5 mm mesh is used. Figure 1 shows a typical bending of plate under uniform load under coarse/fine mesh.

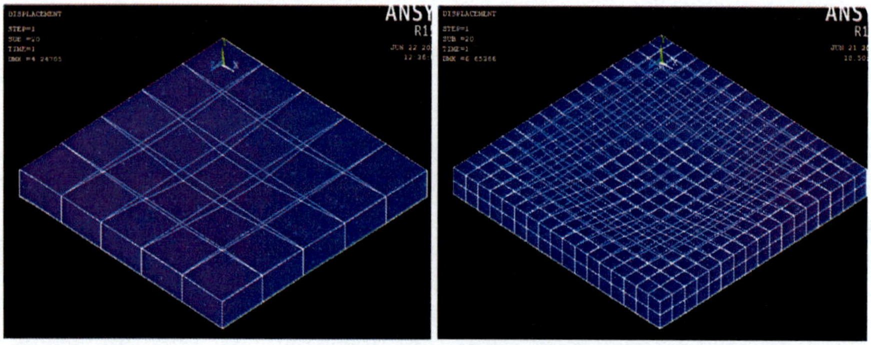

Fig. 1. Bending of plate under uniform load a) Coarse mesh b) Fine mesh

$$\textit{Maximum Deflection} = C_1 \frac{P * \min(Lx, Ly)^4}{Eh^3}$$

Where, P = Uniform load = 1000 N
Lx, Ly = Dimension of plate = 100 mm*100 mm
E = Youngs Modulus
H = Thickness of plate = 10 mm
C_1 – coefficient based on Ly/Lx ratio – 0.0138 (Fig. 2).

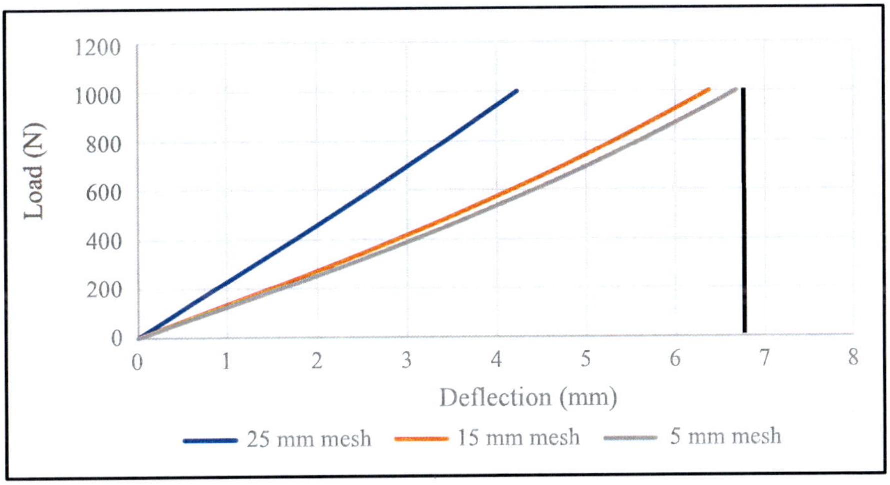

Fig. 2. Load and deflection for plates meshed with different sizes (deflection calculated analytically is 6.9 mm)

2.2 Material Modelling

The special properties of SMA can be simulated in very few softwares. To study the SMA material in detail, finite element analysis software where the special stress strain curve of SMA can be modelled was identified. Generally, OPENSEES, ABAQUS, SEISMOSTRUCT and ANSYS softwares are used for simulation of SMA as the superelastic and shape memory effect properties can be modelled in this. For this case study the analysis is done using ANSYS software version 15.

As the analysis done in this software is based on finite element analysis, the results obtained in the evaluation depend on element type, mesh size, material model and the connection between the elements.

A variety of elements, about 200 are available in ANSYS, supporting different materials. After studying the different elements, SOLID185 has mixed formulation capability for simulating deformations of nearly incompressible elastoplastic materials, and fully incompressible hyperelastic materials, also it supports (BISO) and (SUPE) material models hence this element is considered for the present study.

The behaviour of steel and SMA is reviewed to know the possibility of using ANSYS software. To study the elasto-plastic behaviour of steel, a bilinear material model is used. The stress strain curve plotted for the steel plate shows an elastic curve till the stress reaches 250 N/mm^2 and the plastic behaviour is observed as it exceeds the elastic range. Superelastic material behaviour is also obtained for the SMA plate. The material option for superelasticity is based on Auricchio model, in which the material undergoes large-deformation without showing permanent deformation under isothermal conditions, as shown in Figs. 3–4.

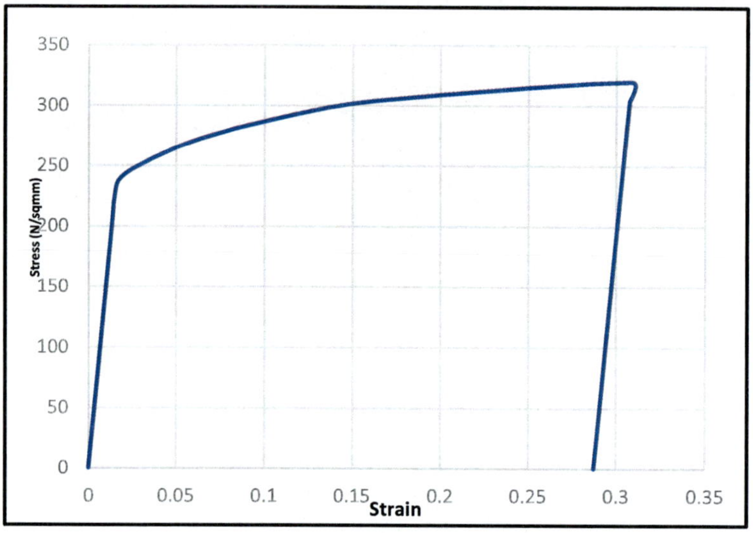

Fig. 3. Material model for Bilinear model (BISO) for steel

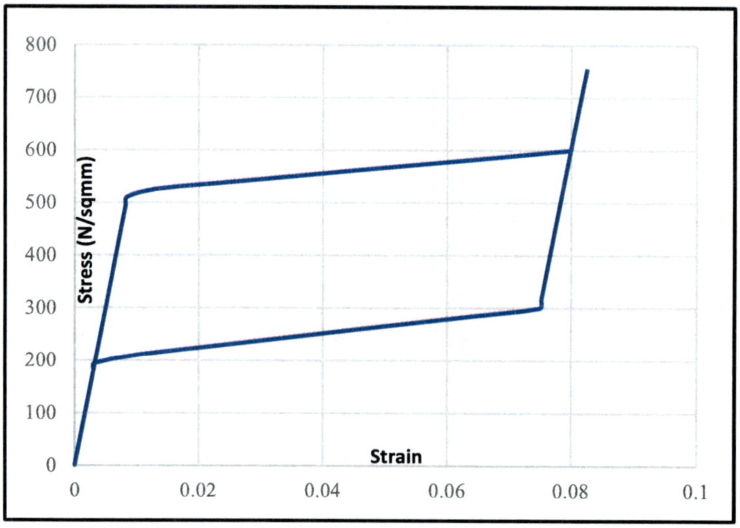

Fig. 4. Material model for Superelastic (SUPE) for SMA

The model for Superelastic (SUPE) for SMA considers reversible phase transformation between the austenite and single martensite phase.

2.3 Finite Element Model of Plate

The plate model considered has the dimensions 100 mm × 100 mm, with a thickness of 10 mm. Since previous research has suggested the use of solid element which provides better results for the investigation of stress location and stress pattern, a solid 185 element is used in this model. The boundary conditions are fixed, the nodes at all boundaries are restrained for translation and rotation in all directions as shown in Fig. (5-a). The uniformly distributed load is applied on plate as shown in Fig. (5-b).

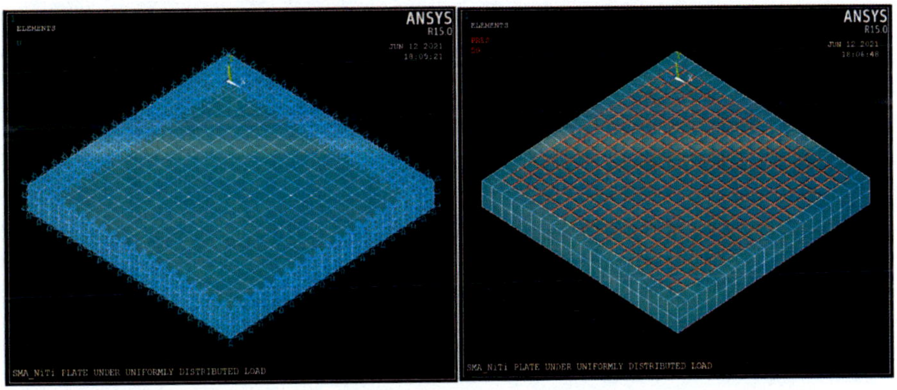

Fig. 5. Ansys Model a) Boundry condition- Fixed Boundry b) Loading

2.4 Material and Structural Properties

The load is applied in increments, then it is unloaded totally the load for which, 1% residual deflection (1 mm–10% of 10 mm). The same load is applied to NiTi plate for comparative study. Hence Table 1 shows material and structural properties of steel and NiTi SMA.

3 Results and Discussions

3.1 Analysis of Steel Plates

The performance of steel plate for steel is investigated applying incremental load upto 10% deflection of thickness which is 50 kN. The deflection is liner in elastic zone till the load reaches the yielding load and then the deflection increases rapidly in plastic zone. The maximum deflection reached is 1.4 mm. While unloading the 4–6% deflection is recovered and 1mm deflection is remained in the material as a residue.

3.2 Analysis of NiTi SMA Plate

The 50kN load same load which is applied to steel plate is applied to compare the response in the similar pattern. As the load increases the deflection increases, the maximum deflection of NiTi is 1.56% which is 11% more than steel. 100% Recovery of maximum deflection is achieved for this load. The plate comes to its original position after unloading and no residual deflection is observed. To understand the load carrying capacity, the load

Table 1. Material and Structural Properties

Property	Structural steel	Units
Young's Modulus (E)	2.1E11	Pascal
Yield Strength (Fy)	355	N/ mm^2
Poissons Ratio	0.28	
Density	7.8e−6	Kg/mm^2
Property	Nitinol – NiTi	Units
Young's Modulus (E)	6.7E10	Pascal
Poisons Ratio (υ)	0.33	
SIG-SAS (σsAS)	3.656E8	Pascal
SIG-FAS(σfAS)	4.22E8	Pascal
SIG-SSA (σsSA)	2.11E8	Pascal
SIG-FSA (σfSA)	1.41E8	Pascal
EPSILON	0.07	m/m
ALPHA	0.0	

developing 8% strain is applied gradually. For 8% deflection 90 kN maximum load is applied. Complete recovery is observed even after increasing the load by 80% compared to steel plate. The deflection increases gradually till the forward phase transformation stress value is reached and then the deflection increases rapidly till it reaches the finish stress of forward phase transformation. During unloading exactly the forward phase transformation curve is achieved.

3.3 Comparative Performance Assessment

Here the comparative study of the behavior of NiTi and Steel plates under loading unloading cycle is studied. This study aims to use the stress induced properties of SMA only temperature induced SME is not considered in the proposed study. Stress induced super elastic SMA are used which are temperature independent and recover their shape when loading is removed. Due to this special property, the residual deflection of the structure is recovered and thus the damage is minimized. To understand the recovery capacity of SMA its behavior under loading unloading cycle is compared with steel plate analysis. For 50 kN load the maximum deflection of steel plate is 1.4 mm and residual deflection after removal of load is 1.04 mm. 4% recovery of deflection is attained for steel. For the same load 50 kN, the maximum deflection observed in NiTi plate is 1.56 mm, 11% more than maximum deflection in steel plate. The residual deflection in steel plate is 1.04 mm and NiTi is 0% which is 96% more than steel plate recovery. Figures 7, 8, 9 and 10 shows load-deformation curves under study. The load carrying capacity of NiTi plate for its maximum strain rate is 90 N which regains 100% recovery. It is 80% more than the load carrying capacity of steel with 4% recovery of deflection (Fig. 6) (Tables 2 and 3).

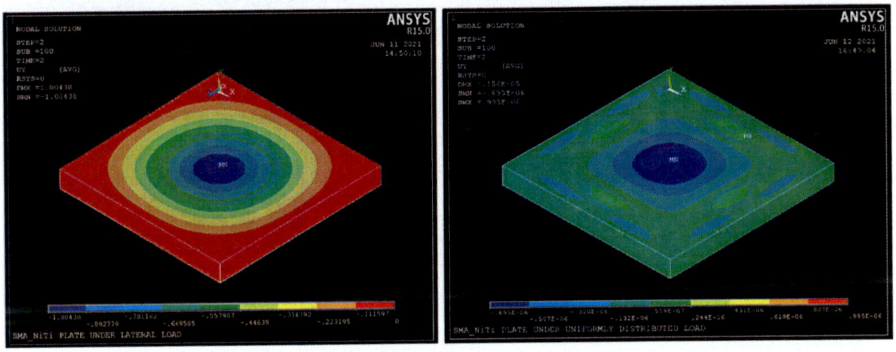

Fig. 6. Deflection of plates a) Steel b) NiTi

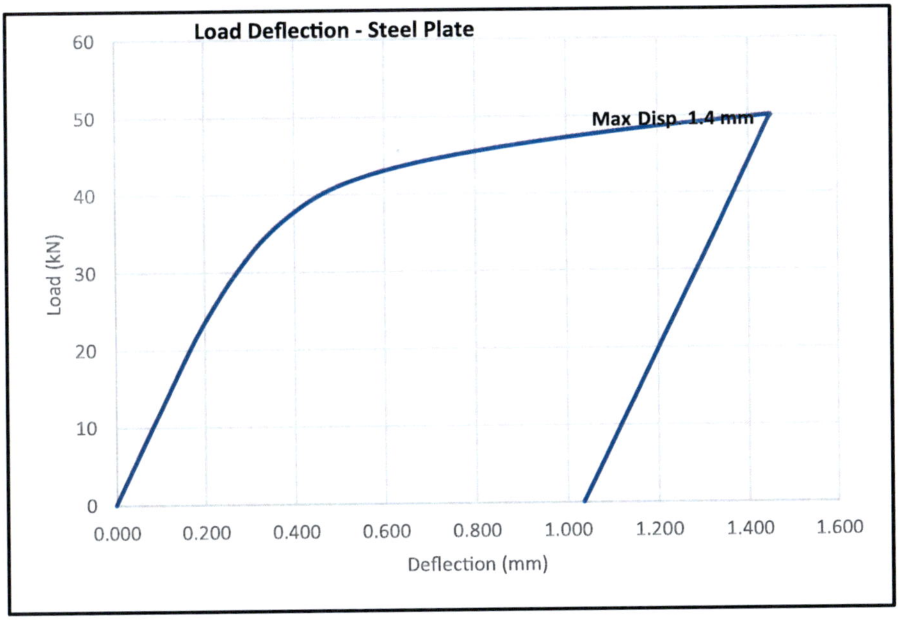

Fig. 7. Load deflection curve for loading unloading cycle a) Steel b) NiTi

Table 2. Comparison for residual deflection and % recovery

Material	Load	Max deflection	Residual deflection	Recovery
Steel	50 kN	1.4 mm	1.04 mm	4%
NiTi	50 kN	1.56 mm	0 mm	100%
NiTi	90 kN	7.96 mm	0 mm	100%
NiTi	100 kN	8.12 mm	0 mm	100%
NiTi	110 kN	9.08 mm	0 mm	100%

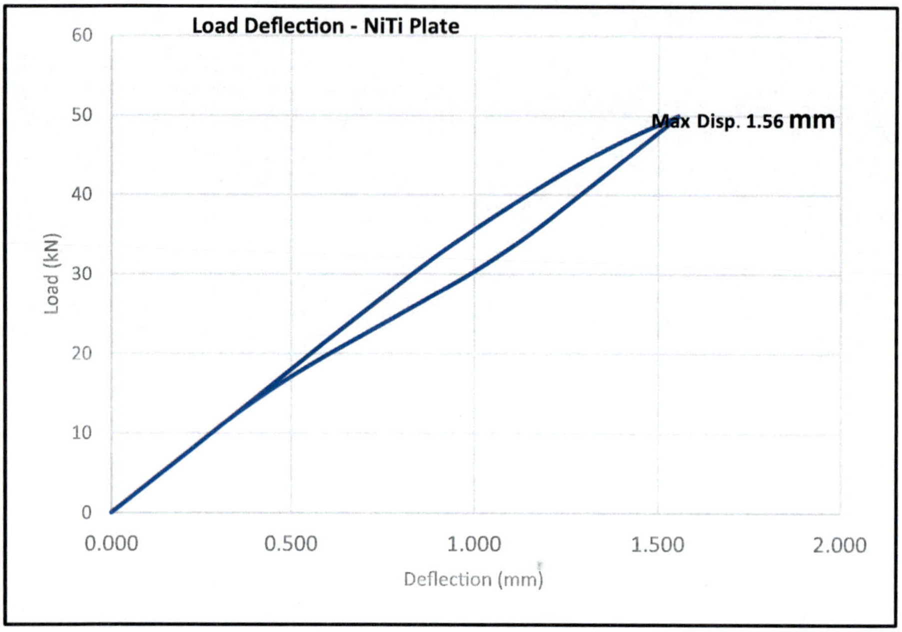

Fig. 8. Load deflection curve for loading unloading cycle a) Steel b) NiTi

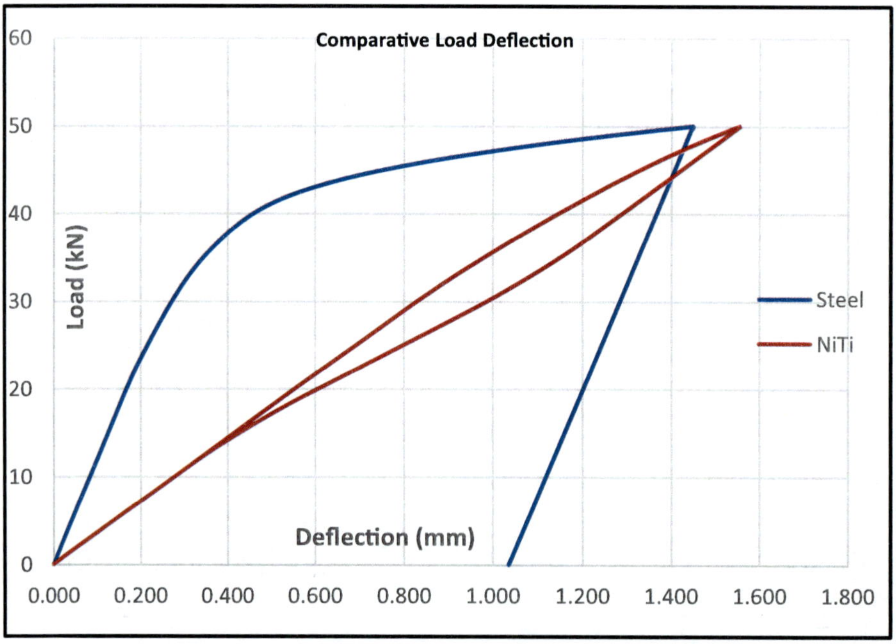

Fig. 9. Comparative load deflection for steel and NiTi

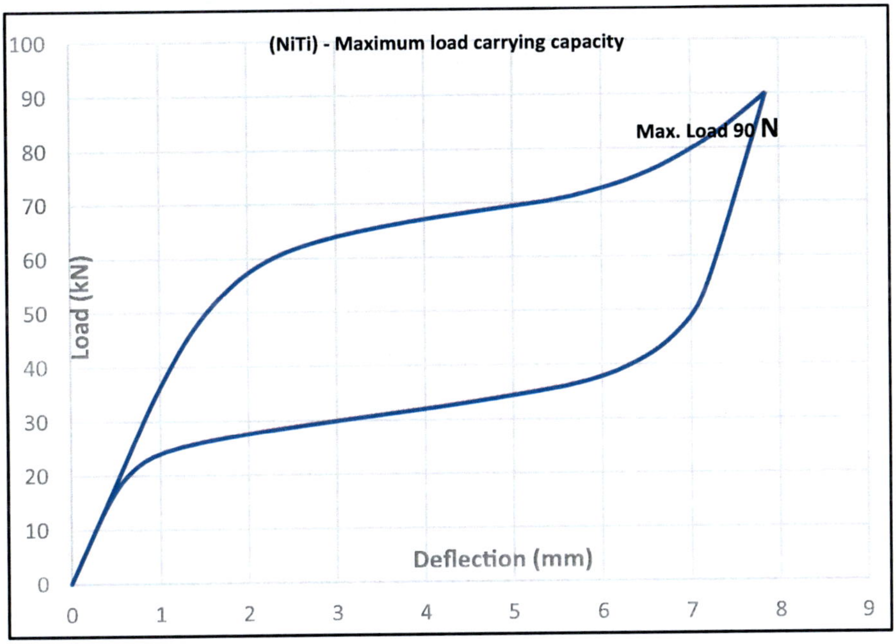

Fig. 10. Load carrying capacity of NiTi- Max load at 8% strain

Table 3. Comparison for increase in load carrying capacity (%) and recovery

Material	Steel	NiTi (A)	NiTi (B)	NiTi (C)
Load (kN)	50	90	110	120
Increase in load carrying capacity (%)	-	80	84	87
Increase in recovery	-	96%	93%	90%

From Table 2 and 3 it is observed that deflection increases as we increase the loads but residual deflection decreases for NiTi plate than that of steel plates. It is because of the residual strain energy stored in NiTi via phase transformation. Increase in load carrying capacity and recovery are in proportion.

4 Conclusion

From the study it is observed that load carrying capacity can be greatly increased in case of NiTi plates as compared to that of steel plates. These NiTi plates recover almost 100% deflection to regain its original shape and size. Loading and unloading via incremental load is given to these NiTi plates to observe its performance. Flexibility of the structure can also be monitored for such NiTi plates to sustain the loads. About 96% recoveries is observed in case of NiTi plates as compared to steel plates. Increase in load carrying capacity and recovery are in proportion with respect to aspect ratio of the plate. Hence structures undergoing reversal of stresses like beam column connections, bracing members, base plates can be made with composite NiTi plates. In vibration control such NiTi plates plays vital role in suppressing the response. During earthquake excitation's also base plate made up of NiTi can be effective.

References

Tanaka, K., Nagaki, S.: A thermomechanical description of materials with internal variables in the process of phase transitions. Archive App. Mech. **51**(5), 287–299 (1982)

Tanaka, K.: A thermomechanical sketch of shape memory effect: one dimensional tensile behaviour. Mater. Sci. Res. Int. **18**, 251–263 (1985)

Liang, C., Rogers, C.A.: One-dimensional thermomechanical constitutive relations for shape memory materials. J. Intell. Mater. Syst. Struct. **1**(2), 207–234 (1990). https://doi.org/10.1177/1045389X9000100205

Brinson, L.: One-dimensional constitutive behaviour of shape memory alloys: thermomechanical derivation with non-constant material functions and redefined martensite internal variable. J. Intell. Mater. Syst. Struct. **4**, 229–242 (1993)

Auricchio, F.: A robust integration-algorithm for a finite-strain shape-memory-alloy. Int. J. Plast. **17**(7), 971–990 (2001)

Murasawa, G., Yoneyama, S., Sakuma, T., Takashi, M.: Influence of cyclic loading on inhomogeneous deformation behavior arising in NiTi shape memory alloy plate. Mater. Trans. **47**(3), 780–786 (2006). https://doi.org/10.2320/matertrans.47.780

Fawzia, A., Riadh, A., Zhao, X.: Experimental and finite element analysis of a double strap joint between steel plates and normal modulus CFRP. Sabrina Compos. Struct. **75**, 156–162 (2006)

Amini, M., Isaacs, J., Nemat, S.: Experimental investigation of response of monolithic and bilayer plates to impulsive loads. Int. J. Impact. Eng. **3**(7), 82–89 (2010)

Yen, J., Chien, H.: Steel plates rehabilitated RC beam–column joints subjected to vertical cyclic loads. Constr. Build. Mater. **24**, 332–339 (2010)

Ye, H., Christian, K., Ummenhofer, T., Shizhong, Q., Plum, R.: Fatigue performance of tension steel plates strengthened with prestressed CFRP laminates. J. Compos. Constr. **14**(5), 609–615 (2010). https://doi.org/10.1061/(ASCE)CC.1943-5614.0000111

Bajoria, K.M., Wankhade, R.L.: Free vibration of simply supported piezolaminated composite plates using finite element method. Adv. Mater. Res. **587**, 52–56 (2012)

Wankhade, R.L., Bajoria, K.M.: Stability of simply supported smart piezolaminated composite plates using finite element method. In: Proceedings of International Conference Advance Aeronautical Mechanical Engineering AME, pp. 14–19 (2012)

Shariat, B., Liu, Y., Meng, Q., Rio, G.: 'Analytical modelling of functionally graded NiTi shape memory alloy plates under tensile loading and recovery of deformation upon heating. Acta Mater. **61**, 3411–3421 (2013)

Wankhade, R., Bajoria, K.M.: Free vibration and stability analysis of piezolaminated plates using finite element method Smart. Mater. Struct. **22**(125040), 1–10 (2013)

Wankhade, R.L., Bajoria, K.M.: Buckling analysis of piezolaminated plates using higher order shear deformation theory. Int. J. Compos. Mater. **3**, 92–99 (2013)

Wankhade, R.L., Bajoria, K.M.: Vibration of cantilever piezolaminated beam with extension and shear mode piezo actuators. Active. Passive. Smart. Struct. Integr. Syst. Proc. SPIE. **9431**, 943122–943131 (2015)

Rasid, Z.A., Zahari, R., Ayob, A.: The instability improvement of the symmetric angle-ply and cross-ply composite plates with shape memory alloy using finite element method. Adv. Mech. Eng. **6**, 632825 (2014). https://doi.org/10.1155/2014/632825

Samadpour, M., Sadighi, M., Shakeri, M., Zamani, H.: Vibration analysis of thermally buckled SMA hybrid composite sandwich plate. Compos. Struct. **119**, 251–263 (2015)

Wankhade, R.L., Bajoria, K.M.: Shape control and vibration analysis of piezolaminated plates subjected to electro-mechanical loading. Open J. Civ. Eng. **6**, 335–345 (2016)

Bajoria, K.M., Wankhade, R.L.: Vibration of cantilever piezolaminated beam with extension and shear mode piezo actuators. Proc. SPIE. **9431**(943122), 1–6 (2015)

Dehkordi, B., Khalili, S.: Frequency analysis of sandwich plate with active SMA hybrid composite face-sheets and temperature dependent flexible core. Compos. Struct. **123**, 408–419 (2015)

Shahverdi, M., Czaderski, C., Motavalli, M.: Iron-based shape memory alloys for prestressed near-surface mounted strengthening of reinforced concrete beams. Constr. Build. Mater. **112**, 28–38 (2016)

Bendine, K., Wankhade, R.L.: Vibration control of FGM piezoelectric plate based on LQR genetic search. Open J. Civ. Eng. **6**, 1–7 (2016)

Wankhade, R.L., Bajoria, K.M.: Shape control and vibration analysis of piezolaminated plates subjected to electro-mechanical loading, open. Open J. Civ. Eng. **6**, 335–345 (2016)

Mahabadi, K., Shakeri, M., Pazhooh, D.: Free vibration of laminated composite plate with shape memory alloy fibers. Latin Am. J. Solids Struct. **13**(2), 314–330 (2016). https://doi.org/10.1590/1679-78252162

Wankhade, R.L., Bajoria, K.M.: Numerical optimization of piezolaminated beams under static and dynamic excitations. J. Sci. Adv. Mater. Dev. **2**(2), 255–262 (2017)

Bendine, K., Wankhade, R.L.: Optimal shape control of piezolaminated beams with different boundary condition and loading using genetic algorithm. Int. J. Adv. Struct. Eng. **9**(4), 375–384 (2017). https://doi.org/10.1007/s40091-017-0173-x

Izadi, M.R., Ghafoori, E., Shahverdi, M., Motavalli, M., Maalek, S.: Development of an iron-based shape memory alloy (Fe-SMA) strengthening system for steel plates. Eng. Struct. **174**, 433–446 (2018). https://doi.org/10.1016/j.engstruct.2018.07.073

Shahverdi, M., Michels, J., Czaderski, C., Motavalli, M.: Iron-based shape memory alloy strips for strengthening RC members: material behavior and characterization. Constr. Build. Mater. **173**, 586–599 (2018)

Wankhade, R.L., Bajoria, K.M.: Vibration analysis of piezolaminated plates for sensing and actuating applications under dynamic excitation. Int. J. Struct. Stab. Dyn. **19**(10), 1950121 (2019)

Yang, B., Soares, G., Wang, D.: Dynamic ultimate compressive strength of simply supported rectangular plates under impact loading. Mar. Struct. **66**, 258–271 (2019)

Chang, M., Wang, Z., Liang, W., Sun, M.: A novel failure analysis of SMA reinforced composite plate based on a strain-rate-dependent model: low-high velocity impact. J. Mater. Res. Technol. **8**(1), 812–826 (2019). https://doi.org/10.1016/j.jmrt.2018.06.012

Wankhade, R.L., Niyogi, S.B.: Buckling analysis of symmetric laminated composite plates for various thickness ratios and modes. Innov. Infrastruct. Solut. **5**(3), 1–12 (2020). https://doi.org/10.1007/s41062-020-00317-8

Joaquin, G., Pinilla, R., Montoya-Coronado, L.A., Ribas, C., Cladera, A.: Finite element modeling of RC beams externally strengthened with iron-based shape memory alloy (Fe-SMA) strips, including analytical stress-strain curves for Fe-SMA. Eng. Struct. **223**, 111152 (2020). https://doi.org/10.1016/j.engstruct.2020.111152

Wankhade, R.L., Bajoria, K.M.: Vibration attenuation and dynamic control of piezolaminated plates for sensing and actuating applications. Arch. Appl. Mech. **91**(1), 411–426 (2021)

Wankhade, R.L., Niyogi, S., Gajbhiye, P.: Buckling analysis of laminated composites considering the effect of orthotropic material. IOP Sci. J. Phys. Conf. Ser. **1706**(1), 012188 (2021)

Bendine, K., Wankhade, R.L.: Piezoelectric energy harvesting from a curved plate subjected to time-dependent loads using finite elements. IOP Sci. J. Phys. Conf. Ser. **1706**(1), 012008 (2021)

2D Lidar Data Matching

Using Simulated Annealing on Point-Based Method

Linh Tao[1](✉), Tinh Nguyen[1], Trung Nguyen[1], Toshio Ito[2], and Tam Bui[1,2]

[1] Hanoi University of Science and Technology, Hanoi, Vietnam
linh.taongoc@hust.edu.vn
[2] Shibaura Institute of Technology, Tokyo, Japan

Abstract. The paper proposes a novel, simple but effective method to align 2D LiDAR laser data. The method uses point-based approach with a simulated annealing searching algorithm. Iterative Closest Point (ICP) is a common method used to solve 2D Lidar alignment problem and widely used to solve Simultaneous localization and mapping (SLAM) problem. The local minima problem allows ICP can be applied only in final aligning steps where data are roughly aligned. The proposed method solves this problem by using simulated annealing (SA) searching algorithm to align two data from distance. SA works on point medium on a point-based approach to reduce the searching dimensions and enhance convergence rate. The method has proved its robustness and efficiency in aligning 2D LiDAR laser data.

Keywords: Point-based method · 2D scan matching · 2D LiDAR laser · Simulated annealing

1 Introduction

Point-clouds registration is an attempt to search for the movement between two point-clouds and align overlapped regions. The reference is fixed, while the source is transformed to align with the reference at similar regions. Point-cloud registration is the key algorithm in computer graphics, computer vision, reserve technology and many tasks including surface reconstruction, scenario reconstruction, mobile robot localization, etc. The alignment movement usually includes two steps. The first step is raw alignment, which roughly moves the source close to the reference as possible. Then coming to the fine step, ICP [1] have been used for this step as the default option. The closeness of two point-clouds is a loose definition. How close is enough for ICP, that is an unanswered question. The obvious is that the closer they are they higher chance ICP converge into global minima with smallest error. KinectFusion [2] is a famous algorithm for 3D scene reconstruction. It applies ICP for consecutive frames in real-time with helps from GPU units. Consecutive means frames are quite close to each other. Because of that reason, ICP can work well in doing the alignment task. However, if the camera moving fast, the consecutive frames do not fulfill the close enough requirement, ICP could result in wrong alignment.

N. V. Khang et al. (Eds.): ASIAN MMS 2021, MMS 113, pp. 944–949, 2022.
https://doi.org/10.1007/978-3-030-91892-7_90

In recent works, we tried to work around closeness problem by using heuristic searching algorithms. Searching algorithms work as a global searching tool while ICP is a local tool. Global searching algorithm scramble through searching spaces for the most suitable rough movement. ICP, as local search, simplifies searching spaces for more efficient global searches [3]. Another method directly uses global heuristic named ISADE [4], an adaptive differential evolution algorithm. Without ICP, this method proved its ability in registering 3D point-clouds without closeness requirement. They prove the potential of applying heuristic search into alignment problem.

In autonomous vehicles research areas, vehicles must locate its current position in the map. A correct of start position helps them plan a good moving path such as the smoothest path, the shortest path, or the path with smallest energy consumption. Many researchers have proposed good solutions to tackle this problem. Sensor fusion is one of new trend for solving the problem, different sensors are installed in vehicles such as GPS, IMU, 2D laser [5], 3D laser [6], etc. Like alignment problem, the localization problem is usually divided in two main steps which are raw and fine searches. From the global scale data of GPS or IMU, their current position can be recorded with some error. Those errors can be large or small depends on sensors' accuracy with GPS or sliding between the road and wheels with IMU. The current position and sensors' accuracy boundary create a searching region, fine searching algorithms, such as ICP, come into use. However, with unavailable or unavailable or having IMU without knowing the original position, ICP and its variants are not able to solve autonomous vehicles localization problem.

To solve the localization for autonomous vehicle with 2D LiDAR without other sensor, we propose a novel, simple but effective method to apply ICP on 2D LiDAR laser data. The method uses point-based approach integrated with simulated annealing searching algorithm. Experiments shows robustness and effectiveness of the new method in comparison with original ICP.

2 Autonomous Senior Car and 2d Laser

2.1 Senior Car Model

In developed countries, population aging has become an urging problem. More senior citizens need caring from society. It takes labors and resources for caring services. Helping senior citizens can relies on themselves on moving around in daily basic tasks is a purpose of our senior car system.

We aim to design an autonomous navigation system for this autonomous senior car and other vehicles in general. In senior car 2D LiDAR laser is installed as shown in Fig. 1.

2.2 LiDAR Laser Scanner

A LiDAR camera like SICK LMS200 as in Fig. 2 is used for front range data capturing. The resolution of this sensor varies from 0.25, 0,5 and 1°. The choosing a right resolution can affects the computation cost. To lessen computation burden, the algorithm takes half of data at resolution of 1°.

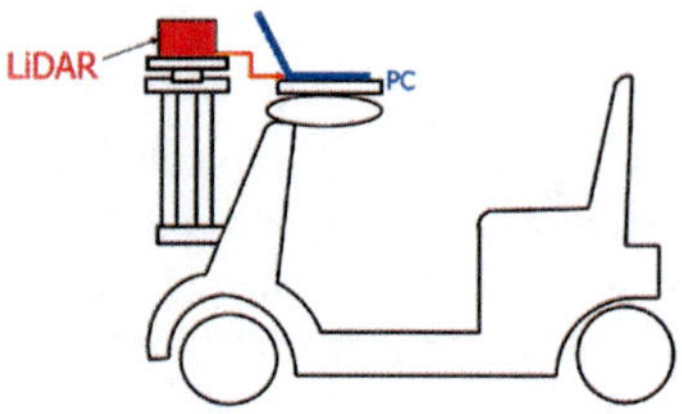

Fig. 1. Senior car model

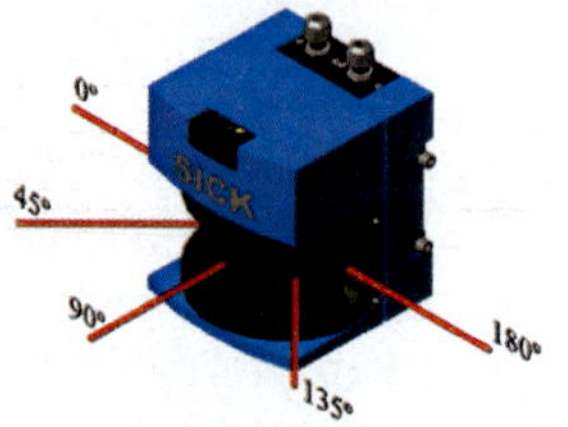

Fig. 2. SICK 2D Lidar sensor

3 Iterative Closet Point

3.1 Iterative Closest Point

Given two point-clouds: $X = \{x_1, \ldots, x_n\}$ as reference and $P = \{p_1, \ldots, p_n\}$ as source one.

The alignment have to find the translation t and rotation R that minimizes the sum of squared error:

$$E(R, t) = \frac{1}{N_p} \sum_{i=1}^{N_p} \|x_i - Rp_j - t\|^2 \tag{1}$$

where x_i and p_j are corresponding points.

Essentially, there are four steps in ICP:

1. For each point in the source point-cloud, search for the closest corresponding point in the reference point-cloud.
2. Estimate the combination of rotation and translation using a root mean square point to point distance metric minimization technique which will best align each source point to its match found in the previous step.
3. Transform the source points using the obtained transformation.
4. Iterate (re-associate the points and so on) (Fig. 3).

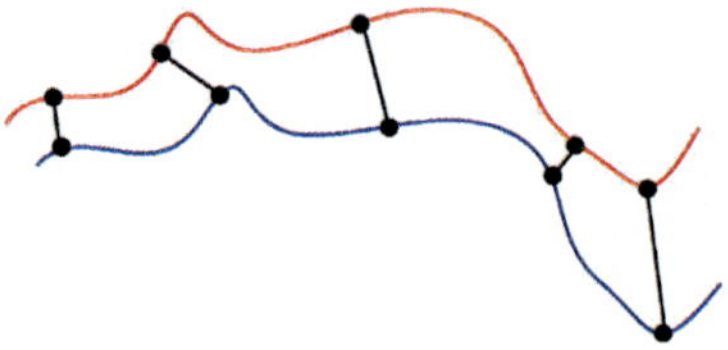

Fig. 3. Point to Point ICP

KD-tree algorithm and its variants is an efficient method for determining corresponding pair points.

3.2 Two Steps Alignment Movement ICP

If we assign one point of the reference point-cloud as the based point and knowing its corresponding on the source point-cloud, registration perform two steps which are shown in Fig. 1.

- Step 1: The first movement is to make two chosen corresponding points aligned is performed.
- Step 2: To rotate the source point-cloud about the corresponding point in step 1. The angle is calculated by using root mean square method as in original ICP method (Fig. 4).

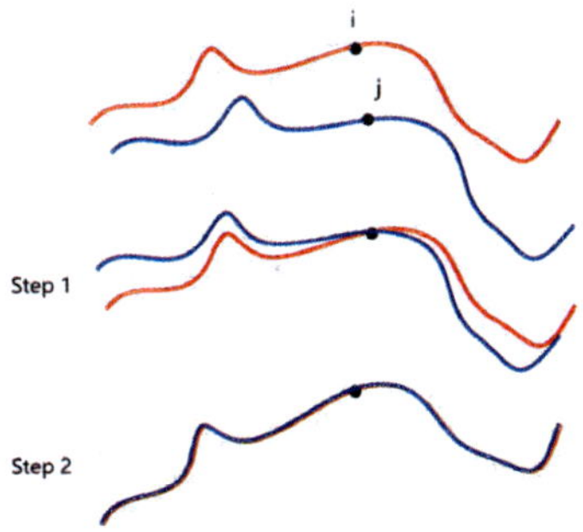

Fig. 4. Two step alignment

Choosing the correct corresponding point in the first step decides the correction of this method. Simulated annealing method is employed for this task (Fig. 5).

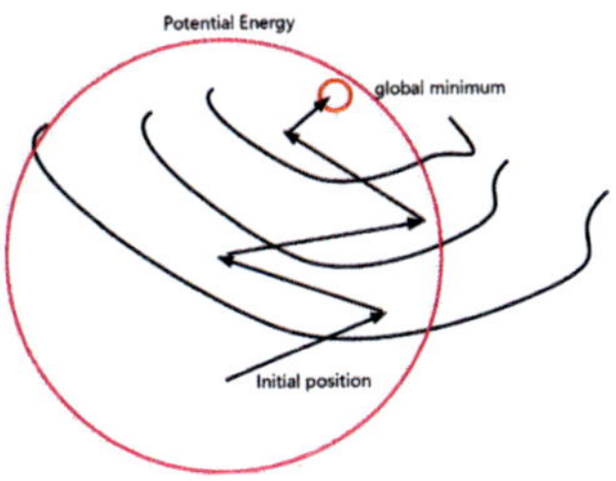

Fig. 5. Simulated annealing for global minimum error search

Simulated annealing is a metaheuristic to approximate global optimization in a large search space for an optimization problem. Instead of being trapped in small area search, searching region of SA goes from large to small as the energy ball is getting smaller. This allows SA to have chance searching for global optima instead of local optima.

4 Experiment and Results

4.1 Experiment and Setup

The map is downloaded from MRPT (Mobile Robot Programing Toolkit) library sharing data website, a map of Intel Research Lab. From this map, we collect simulated laser scans in different position. Those scans are used as an input for matching algorithm (Fig. 6).

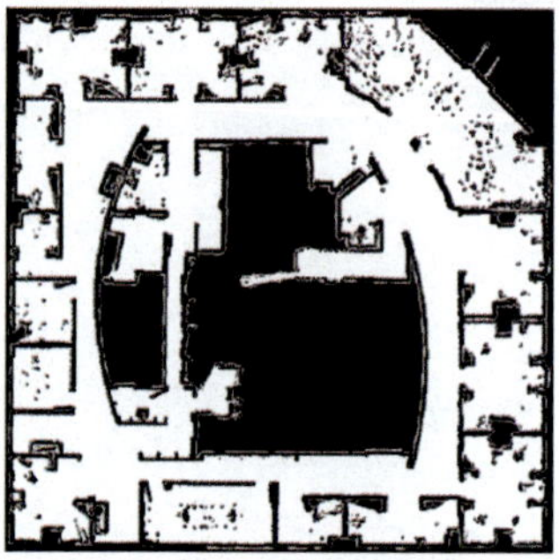

Fig. 6. Intel office map

The code is written in python language with Jupyter Notebook as editor and run on an Intel core I5–4690 3.5 GHz with 16 GB memory.

4.2 Results

The red points in Figs. 7 and 8 are reference point-clouds. Current point-clouds are taken from a different position and presented in blue. Next, using our algorithm, source point-clouds are aligned with reference. Table 1 shows success rate, and errors of three different cases. Figures 7and 8 show the first case alignment result.

Table 1. Alignment results of three different cases

Cases	Rate	RMSE mean	RMSE Std
[140,400,60] and [140,420,65]	100%	15.334	0.82
[140,120,60] and [130,110,55]	100%	2.697	0.37
[600,140,60] and [610,130,55]	100%	0.948	0.19

5 Discussion and Conclusion

The paper proposed the point-based method which integrates point-based method with simulated annealing searching algorithm. Experiments results show the feasibility of applying method for 2D LiDAR laser data with promising results.

Currently the point-to-point error is use which make the error is inaccurate since the distance affects the gap between two consecutive data points. In the future work we would like to use point-to-plane method to gain a more accurate error.

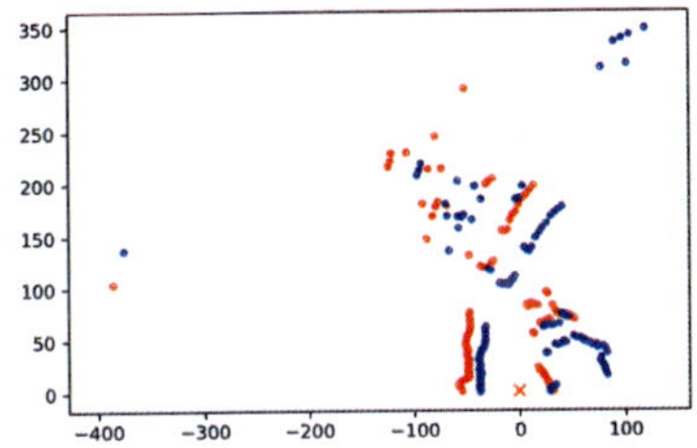

Fig. 7. Initial 2D lidar data

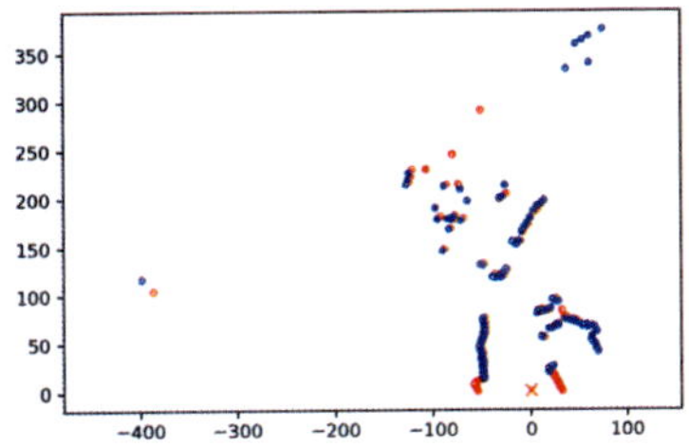

Fig. 8. Aligned data

Ackknowledgement. This work is supported by School of Mechanical Engineering, Hanoi University of Science and Technology under a project named T2020-PC-010.

References

1. Besl, P.J., McKay, N.D.: A method for registration of 3-D shapes. IEEE Trans. Pattern Anal. Mach. Intell. **14**(2), 239–256 (1992). https://doi.org/10.1109/34.121791
2. Newcombe, R.A., et al.: KinectFusion: real-time dense surface mapping and tracking. In: 10th IEEE International Symposium on Mixed and Augmented Reality Basel, pp. 127–136 (2011). https://doi.org/10.1109/ISMAR.2011.6092378
3. Tao, L., Nguyen, T., Hasegawa, H.: Global hybrid registration for 3D constructed surfaces using ray-casting and improved self adaptive differential evolution algorithm. Int. J. Mach. Learn. Comput. **6**(3), 160–166 (2016). https://doi.org/10.18178/ijmlc.2016.6.3.592
4. Tao, L., Bui, T., Hasegawa, H.: Global ray-casting range image registration. IPSJ. Trans. Comput. Vision App. **9**(1), 1–15 (2017). https://doi.org/10.1186/s41074-017-0025-4
5. Wen, J., et al.: 2D LiDAR SLAM back-end optimization with control network constraint for mobile mapping. Sensors **18**(11), 36–68 (2018). https://doi.org/10.3390/s18113668
6. Hening, S., et al.: 3D LiDAR SLAM Integration with GPS/INS for UAVs in Urban GPS-Degraded Environments (2017)

Development of Smart Irrigation & Sunshade Systems for Chrysanthemum Cultivation in Greenhouse Based on the Internet of Things

Ngoc Tran Le(✉)

Industrial University of Ho Chi Minh City, Ho Chi Minh City, Vietnam
lengoctran@iuh.edu.vn

Abstract. To be favored by nature in terms of climatic conditions, Dalat - Lam Dong of Vietnam has the leading agriculture in the country in terms of vegetable and flower products. One of the economically valuable flowers for Dalat agriculture is the chrysanthemum plant. For chrysanthemum strongly growth, it is necessary for monitoring and adjustment two important factors such as moisture and light. This study presents the design and implementation of smart irrigation and sunshade systems that have the ability to automatic sprinkler irrigation all over the surface of the chrysanthemum plant based on the sensor signals. Meanwhile, the light in the greenhouse is optimal based on the sunshade nets that are adjusted automatically for preventing the sunlight shines directly on the chrysanthemum plants. The control system uses micro-control STM32F103C8T6 as the central control to collect measured data from the sensors and to control the irrigation and sunshade systems. The measured data are sent to the Webserver by Lora device. The webserver interface is created by the IoT MyDvices Cayenne software where can be monitored all climate parameters in the greenhouse through web browser applications. This method can help the farmers to manage their flower cultivation better, remove human labor and the optimal time to harvest, increasing the flower quality and productivity compare with the traditional methods.

Keywords: Chrysanthemum cultivation · IoT technology · Lora gateway · Microprocessor · Irrigation & sunshade systems · Humidity and sunlight measurement

1 Introduction

In a previous, despite having advantages in terms of climate, soil, and standing floriculture history, in general, the flower agriculture of Dalat city, Lam Dong province is still not better than other sections due to outdated farming techniques, fragmented planning, and small-scale cultivation mainly in households. After 1994, the floriculture of Dalat city is more strongly developed when there was the appearance of foreign companies investing strongly with scientific farming techniques, the yield, and quality of flowers markedly increased. This has changed the traditional farming views of thousands of flower farmers in Dalat city [1]. Today, there are more and more individuals and

N. V. Khang et al. (Eds.): ASIAN MMS 2021, MMS 113, pp. 950–960, 2022.
https://doi.org/10.1007/978-3-030-91892-7_91

businesses domestically and abroad investing in high-tech floriculture with revenue of many VND billions per year per hectare. This land is quickly becoming the agricultural capital of Vietnam's high-tech vegetables and flower farming, aiming to Dalat's agricultural products is reaching new heights to conquer the domestic market and export to foreign countries. Currently, flower farming in Dalat city has grown tremendously in both quantity and quality by applying advanced technology in flower farming. However, the application of automation technology which was sensor technique not concerned due to high investment costs, the only foreign companies dare to make an investment, most individual flower growers still use human labor, farming efficiency is not high and still depends on nature. Nowadays, IoT technology has been widely used in agriculture cultivation for the measurement, monitoring and controlling of climate parameters in greenhouses [2–10]. By using the IoT, the farmers can monitor the climate parameters in their garden for 24 h non-stop even when they left and control garden appliances from outside. The watering and light control also is executed automatically based on measured data of the sensors. Data from sensors installed in the garden can help the farmers plan the optimal time from cultivating to harvest or ensure that harvest time is ready with the maximum results [10].

Chrysanthemum is one of the flowers that have high economic value in Dalat city, this plant is sensitive to the climate parameters such as temperature, moisture, humidity and light. The conditions for chrysanthemum flowers growing will be optimized fulfilled if they are cultivated in a greenhouse where the climate parameters can be adjusted easily based on IoT technology. Two important parameters that influence on the growth of chrysanthemum in the greenhouse are light and moisture. To improve the chrysanthemum growth and its flower quality, it is necessary to monitor and control two parameters all day and night through watering and lighting control regularly. There are several kinds of research to focus these works based on the IoT technology [8–10].

For the development of a monitoring system in the chrysanthemum greenhouse, Riskiawan et al. [8] proposed a design of an online greenhouse monitoring system based on the Raspberry Pi. In this research, the sensors installed in the greenhouse include temperature and humidity, soil moisture, light intensity, and an RGB camera. Measured data from sensors is sent to the cloud database via raspberry pi. Data is sent online to the server through the internet network connectivity and displayed in graphical on the website page in real-time. By utilizing this method, technicians can find out the environmental condition in the greenhouse without staying there. This can help farmers to manage their greenhouse effectively as well as analysis of crop yields based on greenhouse conditions during cultivation.

For the development of a control system for watering and lighting in the chrysanthemum in greenhouse based on the Internet of Things. Fajrin et al. [9] introduced a control system aims to control the light of lamps and the watering based on controlling of water pump through a website. This study uses an ESP8266 module as a main controller and makes a local web-server without an additional ethernet shield. The remote control is executed on the actuators and sensor monitoring through a web server-based that can be accessed by smartphones and computers. Meanwhile, Adam et al. [10] proposed a design of monitor and control system for the climate parameters in greenhouse based on sensors and controller. In this research, they use the sensors such as temperature and humidity sensor (DHT-22), soil moisture (SEN0057 sensor) for measuring the climate parameters in greenhouse, and a Real Time Clock (RTC) to provide timing input as an extra light intensity timer. Measured data of sensors are sent to a controller (Arduino Mega 2560) and to give output instructions to condition the temperature and humidity for controlling the devices such as fan, lamp, heating lamp and water pump.

Approach from a different aspect of research for chrysanthemum cultivation in the greenhouse. In this research, we developed smart irrigation and sunshade mechanism that have the ability to automatic sprinkler irrigation all over the surface of the chrysanthemum plant based on the sensor signal such as temperature and humidity, soil moisture. Meanwhile, the sunshade net is adjusted automatically based on the comparison between sunlight outside and inside of the greenhouse. Sensor data is sent to the webserver through a gateway made by the Lora device. The monitoring of climate parameters as well as control of the remote devices can be executed through web browser applications. This method can help the farmers to manage their flower cultivation better, remove human labor and optimal time to harvest, increasing the flower quality and productivity compare with traditional method.

2 Material and Method

2.1 Systematic Procedure for Developing a Proposed System

To develop the proposed system, the chrysanthemum biology is first analyzed to find the optimal climate parameters in the greenhouse for strong chrysanthemum growth, the greenhouse structure is then designed, and the irrigation and sunshade mechanisms are also designed. Based on the analysis of the system operation mechanisms, a control system is designed for controlling the sprinkle irrigation and sunshade operations which to adjust the humidity and light parameters to reach the desired values. After that, the wireless communication is designed using the Lora device, the climate parameters in the greenhouse can be monitored by mobile devices through a web browser. The webserver interface is developed using IoT Cayenne software. A systematic procedure for developing a proposed system is shown in Fig. 1.

2.2 Chrysanthemum Biology Analysis

Chrysanthemum grows optimally at an altitude of 700-1200m above sea level. It is the short-day flower plants, the growth period of the plant is from 10–12 weeks, including three stages: the root development phase (1–2 weeks), the plant growth phase (from 6–8 weeks), and the period from bud break to harvest (2–3 weeks). During the growth of chrysanthemums, there are harsh requirements on climate conditions, on the one hand, the plant needs water but cannot withstand rain, so it is necessary to design an appropriate irrigation system. During the flowering period, chrysanthemum plants are also not tolerant of direct sunlight and need to be treated with light to shorten the day to promote the differentiation of flower buds and adjust the flowering time appropriately. Based on the chrysanthemum technology guide of Lam Dong agriculture extension center [11], suitable climate parameters for chrysanthemum cultivation in periods are recommended in Table 1.

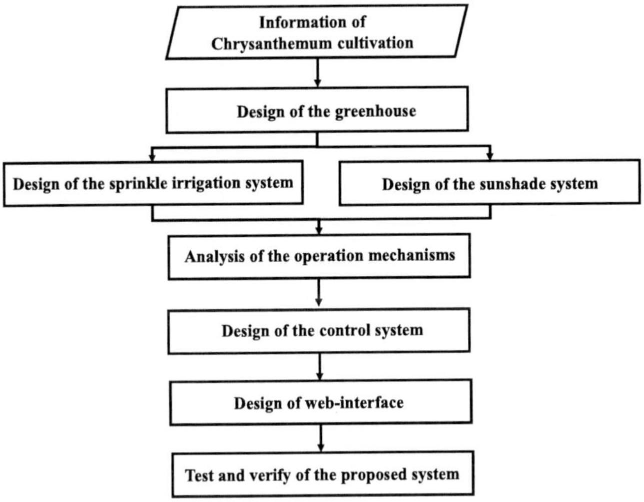

Fig. 1. Systematic procedure for developing a proposed system.

Table 1. Recommended climate parameters for chrysanthemum growth in greenhouse.

Chrysanthemum growth stages	Time	Parameters	Units	Accepted thresholds
Roots	1–2 weeks	Temperature	°C	22 ÷ 26 °C
		Humidity	%	90 ÷ 95%
		Soil moisture	%	70 ÷ 80%
		Light intensity	h (hour)	-
Trunks	6–8 weeks	Temperature	°C	15 ÷ 20 °C
		Humidity	%	65 ÷ 70%
		Soil moisture	%	70 ÷ 80%
		Light intensity (32–108 lx)	h (hour)	6 ÷ 8 h
Blooming	2–3 weeks	Temperature		16 ÷ 18 °C
		Humidity		65 ÷ 75%
		Soil moisture		70 ÷ 80%
		Light intensity (32–108 lx)		7 ÷ 9 h

2.3 Mechanical System Design

Design of the Sprinkle Irrigation System

In order to adjust the climate parameters entering the selected ranges, it is necessary to choose the types of irrigation systems accordingly. There are three famous irrigation system types such as localized irrigation, drip irrigation, and sprinkler irrigation [12]. In this research, the sprinkler irrigation mechanism is chosen due to conformity for the shrub plant as the chrysanthemum by forming the dewdrops spray onto the surface of the plants. The sprinkler irrigation system was designed using the water hose and sprinklers which are hinged on the high and horizontal movements to provide enough water to plants in a large area (Fig. 2). The sprinklers are mounted on the bracket which is fixed in the trolley and movement along with the trolley. The water hose which supplies water to sprinklers is hanged on cable and it is extended or contracted when the trolley moves forward or reverse. The trolley is driven by a DC motor through the belt transmission, it moves on the rails based on four wheels, four guide wheels are mounted to contact the inside of the rails to protect the trolley is not derailed out of the rails. The sprinkler irrigation system operation is automatically controlled by the main controller when the temperature and humidity or soil moisture is out of their thresholds. Two the ends of the rail system are installed two limit switches and the dampers aiming to reverse the trolley direction.

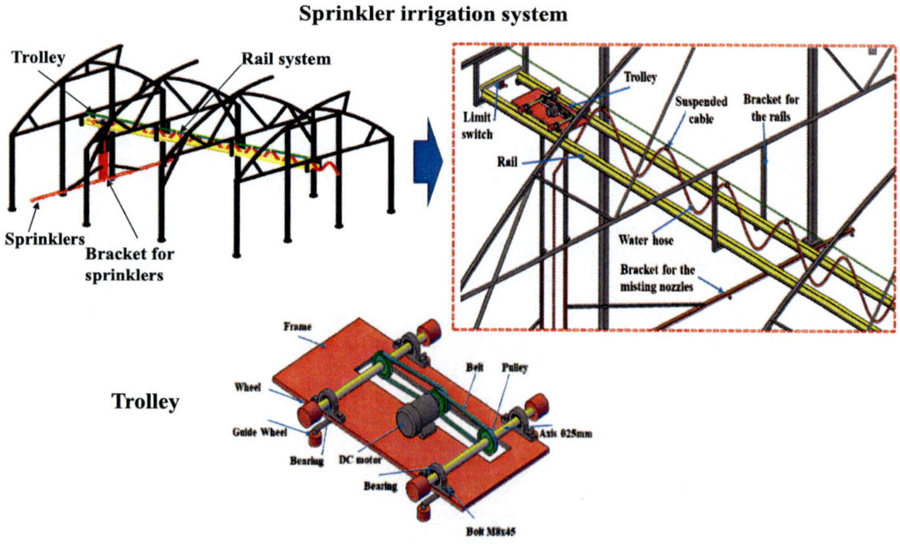

Fig. 2. The sprinkler irrigation mechanism.

The Sunshade System Design

The sunshade nets are used to block a part of the sunlight to reach the ground and creates shade for plants as well as to prevent the harmful ultraviolet rays by UV coating on the

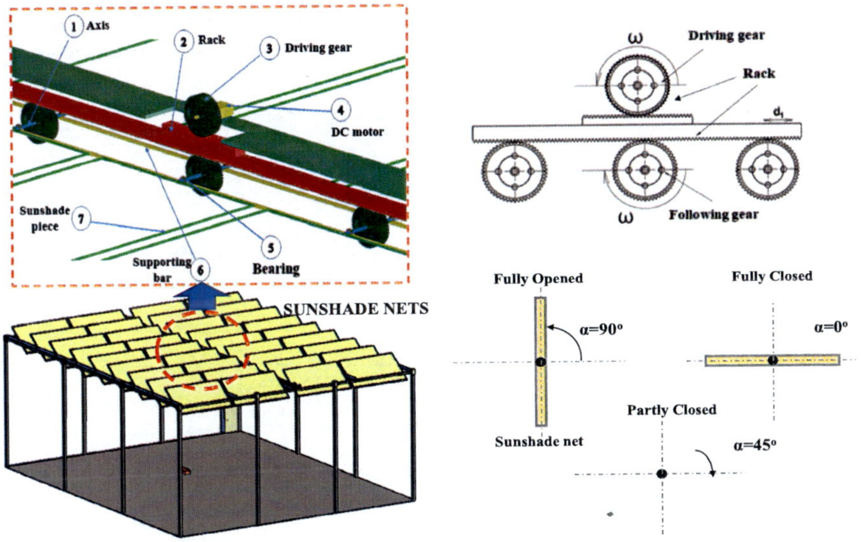

Fig. 3. The sunshade mechanism.

surface of the material. In addition, the sunshade nets also help to limit strong winds that can affect weak trees and reduce the force of raindrops on the ground to avoid scouring and soil erosion during heavy rain. The sunshade mechanism is designed as shown in Fig. 3. It consists of the multi-sunshade pieces which have the dimension are 2000x1000 and are installed above the irrigation system. The sunshade pieces are fixed in the gears and they are driven by the DC motor (4) through the rack-and-gear transmission (2,3). The adjustment of sunlight in the greenhouse is executed by turning the gear (3).

2.4 Control System Design

Hardware Design

The control system consists of five main units such as the main controller, the sensors, the actuators, the wireless communication, and the webserver interface. The block diagram of the proposed control system for the chrysanthemum cultivation in the greenhouse is shown in Fig. 4.

Software Design

In this project, three types of programming are used, namely: programing in a micro-controller, programming in a gateway, and programming in a web interface. The main program is developed in main controller for data collection of sensors, controlling the irrigation and sunshade operations, and send measured data to the gateway through the Lora SX1278 module. The gateway program is developed for receiving data packages from the controller (Node), storage to the database, and send to the webserver, conversely, receiving commands from the web server, comparison measured values with

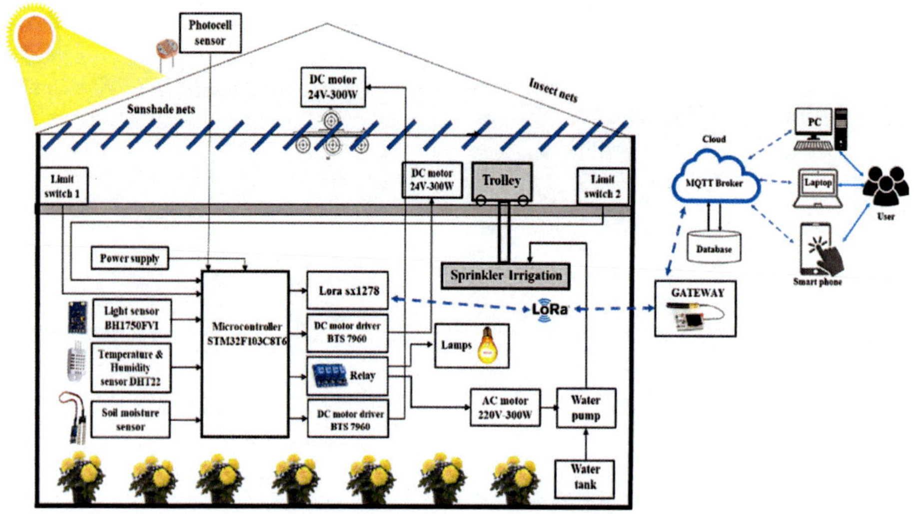

Fig. 4. The block diagram of the proposed control system.

their thresholds, creating the data packages, and sending the packages to Node through Lora protocol. The web-server interface program is created using the Cayenne myDevices software. The Cayenne is an IoT platform with superior features such as data visualization, easy to use by drag-and-drop, the dashboard is created to visualize data coming from remotely connected devices or control them, using widgets to visualize information, supporting several devices (Arduino, Raspberry, ESP,...). The flowcharts of Node and gateway are shown in Fig. 5.

3 Results and Conclusion

3.1 Implementation

Before applying the proposed solution for the reality of the chrysanthemum greenhouse, the proposed method will be implemented and verification on the testbed. The testbed was built and shown in Fig. 6. The power supply of the devices including a 12VDC for two DC motors, and a voltage 5VDC for others such as sensors, microcontroller, ESP32 Lora kit, bridge circuit BTS7960. Pin mapping between microcontroller and devices is described in Table 2.

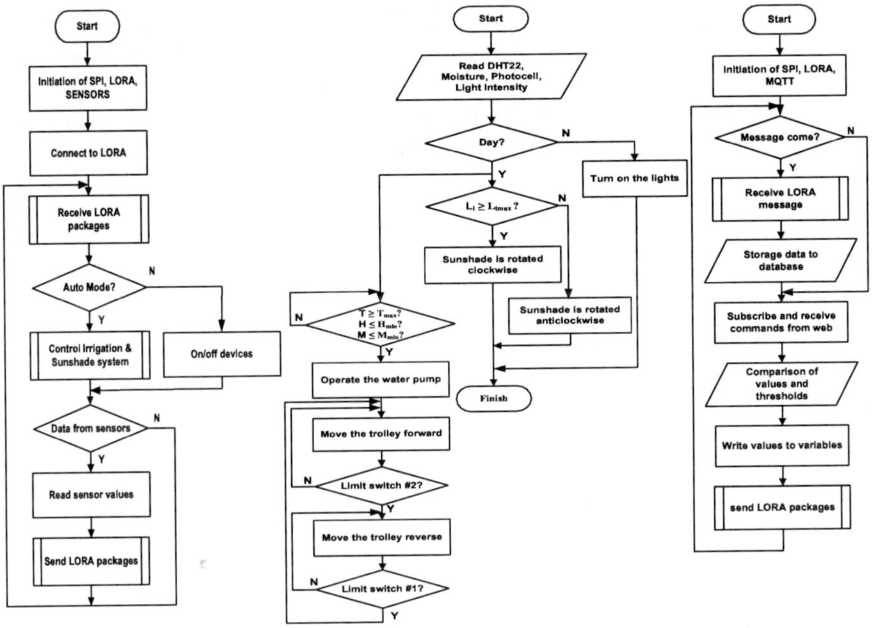

(a) Flow chart Node (b) Flow chart control of Node (c) Flow chart gateway

Fig. 5. Flow chart of node and gateway.

Table 2. Pin mapping.

Controller	PIN	Devices
STM32F103C8T6 BLUE PILL	PB_8 (I2C) - 2(Data)	DHT22
	PB_3 - analog in	Soil moisture
	PA_0, PA_1 - SDA, SCL	Light intensity
	PB_4 – analog in	Photocell
	B1, B0, B11 – R_PWM, L_PWM, R_EN&L_EN	BTS7960_1
	B6, B7, B10 – R_PWM, L_PWM, R_EN&L_EN	BTS7960_2
	A5, A7, A6, A12, A11, B15, B14-SCK, MISO, MOSI, RESET, DI00, DI01, DI02	Lora SX1778

3.2 Testing Results

After the devices are mounted in the testbed, wiring connection and parameters setting are then executed. The system testing is executed with the distance of the gateway over 1km from the testbed. System testing aims to evaluate the accuracy of measured data

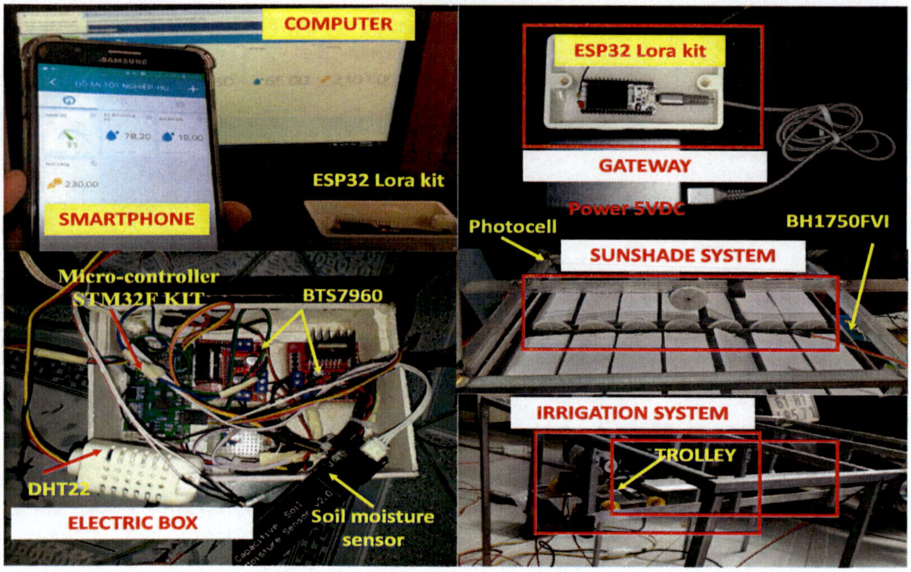

Fig. 6. Hardware implementation.

from sensors and the reaction of actuators such as the water pump, the trolley, and sunshade net when the measured data is out of the selected thresholds.

Testing and Evaluation of the Sprinkler Irrigation Mechanism:
The operation of the irrigation system is tested based on climate various conditions. Initial condition, the water pump, and trolley are not operating. The reference values are adjusted out of the accepted thresholds (Table 1), the water pump and the trolley are immediately started and the operations are maintained until the measured values are to reach the accepted thresholds. The results of the irrigation control are shown in Table 3.

Testing and Evaluation of the Sunshade Mechanism: At night, the shade nets are fully opened for ventilation and they are fully closed when it's heavy rain. During the day, the sunshade nets are turned to reach suitable angular based on the light intensity in the greenhouse. The control results are shown in Table 4.

Table 3. The results of the irrigation control based on various conditions.

Air temperature (°C)	Air humidity (%)	Soil moisture (%)	On/Off condition & Response time (s) Water pump Trolley operation	
30.0	81	64	ON (5s)	ON (5s)
29.5	78	68	-	-
28.3	85	70	-	-
27.2	87	72	-	-
26.1	90	74	-	-
25.8	89	74	-	-
24.3	92	76	-	-
23.1	92	78	-	-
22.8	94	80	-	-
21.5	97	85	OFF(5 s)	OFF(5 s)

Table 4. The results of the sunshade control based on various conditions.

Time	Light Intensity (Lux)	Shade net angle (Deg)	Shade net condition
Night	00.0	90	Fully Opened (5 s)
Day	45.0	80	Partly Closed (5 s)
	50.0	70	-
	65.0	60	-
	72.0	50	-
	78.0	40	-
	84.0	30	-
	96.0	20	-
	100.0	10	-
	110.0	0	Fully Closed (5 s)
Heavy rain	0	0	Fully Closed (5 s)

3.3 Conclusion

Based on the testing results of the proposed system for chrysanthemum cultivation in greenhouse. It can be concluded that the proposed method has solved the problems in which the traditional cultivation for chrysanthemum was never implemented:

- Real-time measurement and monitor of the climate parameters in the greenhouse with high accuracy. Measured data are stored in the database up to one year.
- Automatic adjustment of the climate parameters in the greenhouse through the sprinkler irrigation and sunshade mechanisms that were implemented based on the changes in the measured data compared with their threshold values. Timely response to control task below 5 s.
- Remotely monitor and control of the chrysanthemum greenhouse can be accessed by various types of smartphones, as well as browsers.

The achieved experiment results on the testbed demonstrated that the proposed solution can able to apply the real chrysanthemum cultivation in the greenhouse in Dalat city.

References

1. http://cand.com.vn/Kinh-te/Ky-1-Luong-sinh-khi-moi-436106/
2. Mustafa, A.A., Radosveta, S.: An IoT-based greenhouse monitoring system with Micaz motes. In: Procedia Computer Science, vol.13, pp. 603–608. Elsevier (2017)
3. Sonali, M., Jayashri, S., Ankita, K., Kajal, L.: Smart farming using IOT. Int. Res. J. Eng. Technol. **06**, 2363–2365 (2019)
4. Kranthi Kumar, M., Srenivasa Ravi, K.: Automation of irrigation system based on Wi-Fi technology and IOT. Indian. J. Sci. Technol. **9**(17), 1–5 (2016)
5. Ashwini, B.V.: A study on smart irrigation system using IoT for surveillance of crop-field. Int. J. Eng. Technol. **7**(45), 370–373 (2018)
6. Pavankumar, N., Arun, K., Kirthishree, K., Nagaraj, T.: Automation of irrigation system using iot. Int. J. Eng. Manuf. Sci. **8**(1), 77–88 (2018)
7. Suganya, K., Preethi, S., Pavithra, M., Sathishkumar, M.: Improved irrigation system for agricultural environmnts using sensors GSM and water meter. Int. J. Eng. Res. Technol. **8**(12), 53–56 (2020)
8. Riskiawan, H.Y., Anwar, S., Kautsar, S., Setyohadi, D.P.S., Arifin, S.: On-line monitoring system in greenhouse area for chrysanthemum cultivation based on raspberry PI and IOT. IOP Conf. Ser. Earth Environ. Sci. **672**(1), 012084 (2021). https://doi.org/10.1088/1755-1315/672/1/012084
9. Fajrin, N., Taufik, I., Ismail, N., Kamelia, L., Ramdhani, M.A.: On the design of watering and lighting control systems for chrysanthemum cultivation in greenhouse based on internet of things. IOP Conf. Ser. Mater. Sci. Eng. **288**, 012105 (2018). https://doi.org/10.1088/1757-899X/288/1/012105
10. Adam, F., Muhammad, A.R., Muhammad, F.A., Lia, K., Eneng, N.: Light control and watering system in greenhouse for the cultivation of chrysanthemum SP. Indonesian J. Elect. Eng. Comput. Sci. **12**(3), 950–957 (2018)
11. Chrysanthemum cultivation. http://khuyennong.lamdong.gov.vn/
12. Dong, T.T., Zhao, X.S., Ma, Z.K., Yu, H.L., Liang, Y.: Design of automatically sprinkling irrigation system to water saving in greenhouse. In: International Conference on Computer Information Systems and Industrial Applications, pp. 270–272 (2015)

Experimental Study of a Small-Capacity Wind-Powered Generator Based on Aeroelasticity Phenomenon

Vu Dinh Quy, Le Thi Tuyet Nhung(✉), and Luu Thanh Trung

Hanoi University of Science and Technology, No. 1, Dai Co Viet, Hai Ba Trung, Hanoi, Vietnam
nhung.lethituyet@hust.edu.vn

Abstract. Nowadays, the development of wind power industry is going up and being applied in many fields. Along with this development are the inventions of wind generator models. Among them, Windbelt is a small-capacity wind-powered generator introduced by Humdinger in 2007. Based on two main phenomena, aeroelasticity and electromagnetic induction, the windbelt model can generate electric current with low power. In the past, there have been many studies to investigate the design models to the output power of the windbelt. However, the theoretical model is not convincing enough, and some parameters have not been studied. In this study, parameters will be investigated such as the size of the belt film (length, width), the size of the magnet (thickness). Besides, the theoretical model for calculating the output voltage of the windbelt is presented as the basis for comparing and rating the parameters affecting the model's performance. The results show that the voltage and the power generated by the windbelt will increase when we increase the length of the belt and the thickness of the cap magnet, whereas the result will decrease when the width of the belt is increased. In addition, at the wind speed of 2.5 m/s, windbelt can achieve a capacity of 1 mW.

Keywords: Windbelt · Aeroelasticity · Small-capacity wind-powered generator

1 Introduction

Today, the role of electricity is increasingly important because of the need to provide uninterrupted electricity for all manufacturing industries and social life. Moreover, the production of electricity today is especially focused on the environment. Wind energy is not only an inexhaustible source of natural energy but also a renewable and environmentally friendly energy source. The most common way to take advantage of wind energy is to spin a turbine generator. However, wind turbines only work well for large-scale applications, with large power and voltage. In fact, in the market, there is still a need for a small, simple, and cheap milliwatt wind generator such as power supply for traffic light sensors, wifi base stations… Windbelt is a small power generator that works on two main phenomena: aerodynamic elasticity and electromagnetic induction. In the condition of wind speed from 2.5 m/s, this model can produce power of 1 mW.

N. V. Khang et al. (Eds.): ASIAN MMS 2021, MMS 113, pp. 961–970, 2022.
https://doi.org/10.1007/978-3-030-91892-7_92

Nowadays, several windbelt models have been developed. Allen and Smits (2001) [1] studied the capacity of a piezoelectric membrane which is placed in the wake of a bluff body and using the induced Von Karman vortices formed behind the bluff body to generate an output power by oscillations of the membrane. A novel piezoelectric energy-harvesting was investigated by Bryant and Garcia (2011) [2] that worked by aeroelastic flutter vibrations of a simple pin-connected flap and beam. In the condition of the flow velocity of 8m/s, maximum power output can achieve 2.2 mW. Sirohi and Mahadik (2009) reported a tested device that uses a galloping D-shaped beam exciting a PZT piezoelectric beam [3]. The power output can achieve 1.14 mW. In 2011, Li et al. [4] published an investigated device that converts wind energy into electrical energy by wind-induced fluttering motion by a bioinspired piezo-leaf architecture. The research was reported that maximum output power of 0.6 mW and a maximum power density can achieve 2 mW/cm^3 from a single leaf. Other research on micro wind energy harvesters was published by Tang et al. (2009) [5] Erturk et al. (2010) [6].

In 2016, the design parameters affecting the windbelt's capacity such as wire tension, magnet position, wind speed, magnet diameter, angle of attack, direction of windbelt were investigated by Vu et al. [7]. It was shown that, at the wind speed from 3–12 m/s, the voltage and the power will increase when increasing parameters such as wind speed, wire tension. The calculation of the voltage was presented by Prajyot Satish Tathode et al. [8]. The results showed that the voltage of the windbelt depends on factors such as cross-sectional area of fabric, wind speed, distance between cap magnet to the coil, wire displacement, cap magnet size, cap magnet material... In 2019, the influences of the width of the membrane on the voltage and the capacity were investigated by V. A. Vinayan [8]. It was shown that the voltage and the power decrease as the width of the membrane increases.

In this paper, experimental testing is carried out to investigate the effects of the size of the membrane and the effects of cap magnet size (magnet thickness) on the windbelt's power.

2 Theory

To serve the theoretical calculation, the parameters affecting the output voltage of the windbelt such as the number of turns of conducting coil, size of the magnet, size of the membrane... are denoted as shown in Table 1.

The electrical power that the windbelt can produce depends on the voltage and amperage:

$$P = U \cdot I \tag{1}$$

The power is usually calculated after a resistor is applied to the source. For the windbelt model, the voltage is calculated without resistance, and it depends on the parameters of the windbelt model. The no-load voltage generated by the windbelt is shown in Eq. (2) [8].

$$U = 2\pi N f A_c B \tag{2}$$

In which

$$f = v/d \tag{3}$$

The cross-sectional area of the membrane length is calculated according to Eq. (4):

$$Ac = L \cdot b \tag{4}$$

In this paper, the N35 cap magnet is used with a relative magnetic field (depending on the N35 material) from 1.17–1.21 T [10]. The actual magnetic field strength Br which depends on the material and size of the cap magnet (thickness, diameter) is presented in Eq. (5) [8].

$$B = \frac{B_r}{2}\left(\frac{D+z}{(R^2 + (D+z)^2)} - \frac{z}{R^2 + z^2}\right) \tag{5}$$

Table 1. Nomenclature

Name	Nomenclature
Voltage generated (mV)	U
Number of turns of conducting coil (cycle)	N
Area of the membrane (m)	Ac
Magnetic field intensity (Tesla)	B
Frequency of vibration of the membrane (Hz)	f
Wind speed (m/s)	v
Length of the membrane (m)	L
Thickness of the membrane (m)	b
Relative magnetic field strength for given magnet material (Tesla)	Br
Thickness of the cap magnet (m)	D
Vibration amplitude of the membrane during fluttering (m)	d
Radius of the cap magnet (m)	R
Distance between the pole and nearest conducting surface (m)	z

In general, the design parameters that affect the windbelt's output voltage specifically are:

- The voltage increases with increasing parameters such as: the number of turns of the conducting coil, the length of the membrane, the thickness of the membrane, wind speed, the thickness of the cap magnet.
- The voltage decreases as design parameters increase such as: distance between the pole and nearest conducting surface, displacement of magnet …

Based on these equations, experimental methods to investigate the influences of winbelt design parameters on the voltage and the power can be developed. In addition, theoretical results can also be pre-calculated to evaluate experimental results.

3 Experiment

3.1 Experimental Model

Based on humdinger's original model [11], the Windbelt kit is preliminarily designed 3D through CATIA V5R21 software (Fig. 1). 2D drawings are exported to support the model manufacturing using CNC cutting machines. Plywood material is applied to models because of its durability, lightness, and reasonable price. Membrane belt uses ripstop fabric material and is CNC laser cut to the exact size. The core of Teflon tape (20 × 10 mm) is used as a copper coil core. Parts of the model can be easily disassembled by the connection through the bolts. The position of the magnet, the position of the copper coil and the length of the model can be flexibly changed following the requirements of the research.

The operating principle of the winbelt is based on the oscillation of the belt membrane. The vibration of the belt film depends on many factors such as belt size, magnet size, angle of incidence, wire tension. An experimental model was built to study the influence of the following parameters: Dimensions of the belt (length and width of the belt) and Cap magnet size (thickness of the magnet).

To investigate the influences of the above parameters on the generator performance, an experimental model was designed and fabricated as shown in Fig. 1. The main characteristics of the model are presented in Table 2. The experimental model consists of four main components including:

- Coil: a coil of 2500 rounds of 0.18 mm diameter copper.
- Cap magnet: a cap magnet with a diameter of 20 mm, made of N35 material
- Membrane: ripstop nylon fabric (a kite fabric), with a thickness of about 0.25 mm
- Frame: made of plywood. The belt base is drilled with pairs of holes 20 mm apart so that the position of the belt mount can be easily changed. This helps to change the length of the belt for the experimental process.
- Features of the model:
- Pre-applied tension load is 15 N tension; The windbelt is placed in the horizontal direction; the cap magnet is positioned in the center of the belt; the angle of attack is 0 (rad).
- The length of the belt can be changed from 420 mm to 620 mm.
- Distance from the cap magnet to the nearest copper coil plane can be varied
- The position of the copper coil can be changed along the windbelt
- The model works well in front of the wind tunnel propeller (Fig. 2).

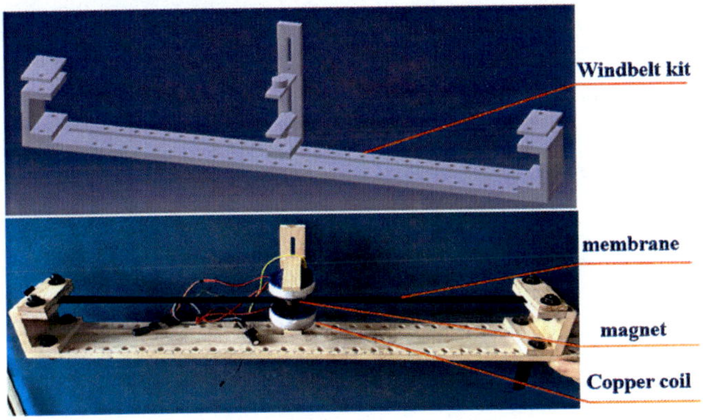

Fig. 1. An experimental model of the windbelt

Table 2. The main characteristics of the model.

Parameters	Value	Unit
Pre-applied tension	15	N
Angle of attack	0	Rad
Length of the membrane	420–620	mm
Width of the membrane	15, 18, 20, 22, 25	mm
Distance between the pole and nearest conducting surface	15	mm
Wind speed	2.5	m/s
Thickness of the cap magnet (same diameter of 20 mm)	1.5, 2, 3	mm

3.2 Experimental Setup

Figure 2 shows the experimental setup. The windbelt model is placed in front of the blower end of the wind tunnel. The Agilent data converter measures the output voltage of the windbelt and the results are saved and processed directly in the computer. Each measurement case will give voltage data. This value is measured once per second and lasts for about 5 min. Then the final voltage result will be the average of the above values.

During the operation of the windbelt model, the wind speed is measured at any 5 positions along the length of the belt membrane (Fig. 3) by the UNI-UT363 wind anemometer as shown in Table 3. All measurements are carried at a wind speed of about 2.5 m/s.

Fig. 2. The experimental setup

Table 3. The estimated wind speed in the experiment

Position	1	2	3	4	5	Average
Wind speed (m/s)	2.5	2.8	2.2	2.3	2.7	2.5

Fig. 3. Five positions along the length of the membrane

4 Results and Discussions

4.1 Effects of the Membrane Dimensions

Effects of the Length of the Membrane

In this investigation, the width of the membrane is kept fixed at 15 mm. Figure 4a shows the evolution of the voltage generated in the function of the length of the membrane in the case there is no electrical load. We can observe that the voltage increases as the length of the membrane increases. The highest achievable no-load voltage value is 8.7 V When the membrane length is 620 mm.

To investigate the power generated by this model, a 1 kΩ resistor is used. Figure 5a Fig. 6a show the evolution of the power and the voltage generated when a 1 kΩ resistors are installed at both ends of the power supply. The voltage drop is obvious and significant from 8.7 V down to about 1 V. In general, the voltage and the power tend to increase as we increase the length of the membrane. The maximum power the can be achieved is about 1 mW when the membrane is 620 mm long.

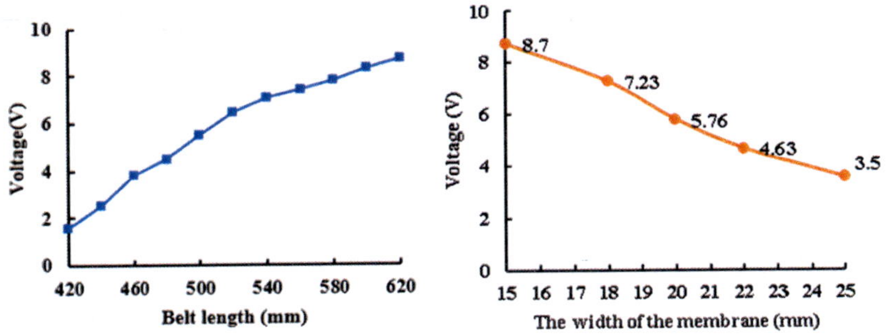

Fig. 4. The voltage in function of the length (a) and width (b) of the membrane (no resistor)

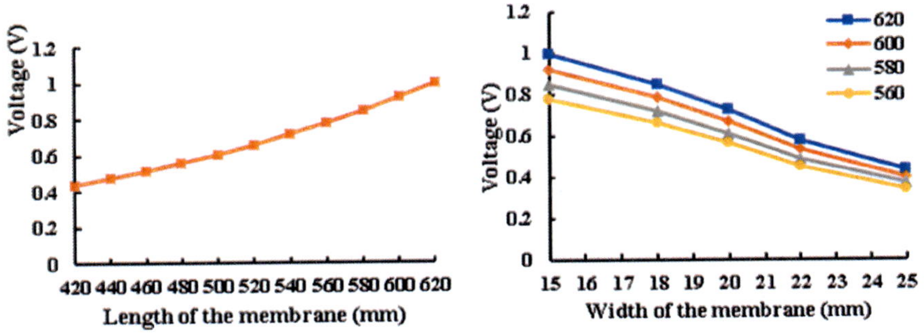

Fig. 5. The voltage in function of the length (a) and width (b) of the membrane (1 kΩ resistor)

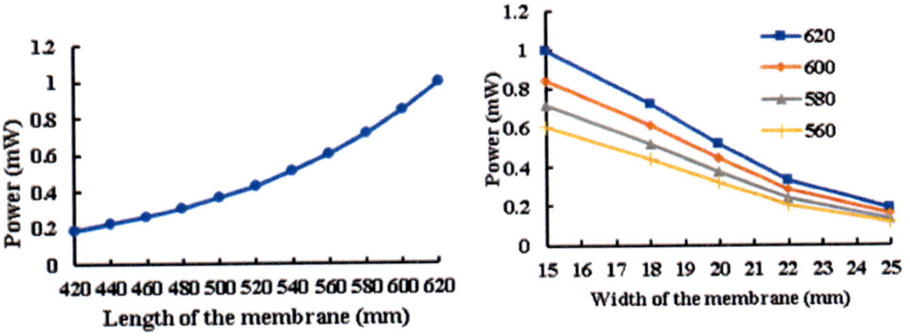

Fig. 6. The Power in function of the length (a) and width (b) of the membrane (1 kΩ resistor)

According to the experimental results, the voltage and the power generated by the windbelt is directly proportional to the length of the membrane. As the membrane length increases, the cross-sectional area of the membrane increases, the plane exposed to the wind increases, helping the belt to oscillate easily even though the wind speed is not high.

Based on the Eqs. (2) and (4), the length of the membrane is proportional to the voltage generated by the windbelt. The results tend to be consistent with the experimental ones.

Effects of the Width of the Membrane
In this testing, the width of the membrane will be changed from 15 mm, 18 mm, 20 mm, 22 mm, 25 mm accordingly Figure 4b shows the dependence of the no-load voltage generated by the windbelt on the change of the width of the membrane with the same length of 620 mm. Specifically, the voltage decreases as the width of the membrane increases. Although the membrane width change is not mentioned in the theory, it does contribute to the voltage change generated by the windbelt. Moreover, through experimentation, the amplitude of angular oscillation of the cap magnet decreases as we increase the width of the membrane. This is because the torsional stiffness of the membrane decreases as the width of the membrane increases.

Figures 5b and 6b show the voltage and the power generated by the windbelt when a 1 kΩ resistor is applied. The voltage drop occurs quite clearly when the voltage drops from 8.7 V to about 1 V. To be more specific, the cases of different lengths of the membrane (600 mm, 580 mm, 560 mm) are added to evaluate the variation of the voltage and the power depending on the width of the membrane. At the wind speed of 2.5 m/s, the maximum power that can be generated in this model is about 1mW with a voltage of about 1 V.

Effects of the Thickness of the Cap Magnet
The phenomenon of aeroelasticity occurs with a combination of aerodynamic forces, elastic forces, and inertial forces. The variation of thickness of the cap magnet not only leads to the change of the inertia force which affects the vibration of the cap magnet, but also affects the magnetic induction of the magnet.

Figure 7 shows the no-load voltage generated by the windbelt using three different cap magnets of the same 20mm diameter with different thicknesses varies from 1.5 mm, 2 mm, 3 mm. We can observe that as the thickness of the cap magnet increases, the output voltage of the windbelt increases.

Figures 8a and b show the voltage and the power generated by the windbelt after a 1 kΩ resistor is applied. In general, the tendency of the voltage and the power increases with the increasing thickness of the cap magnet. At the wind speed of 2.5 m/s, the maximum voltage is 0.848 V and the maximum power is 1,095 mW when the thickness of the cap magnet is 3 mm.

The explanation for this is that as the thickness of the cap magnet increases, the magnetic field B (Eq. 5) - is proportional to the output voltage of the windbelt. Moreover, increasing the thickness of the cap magnet is equivalent to increasing its weight. This makes the cap magnet's ability to vibrate better, the amplitude of the magnet's oscillation will be larger.

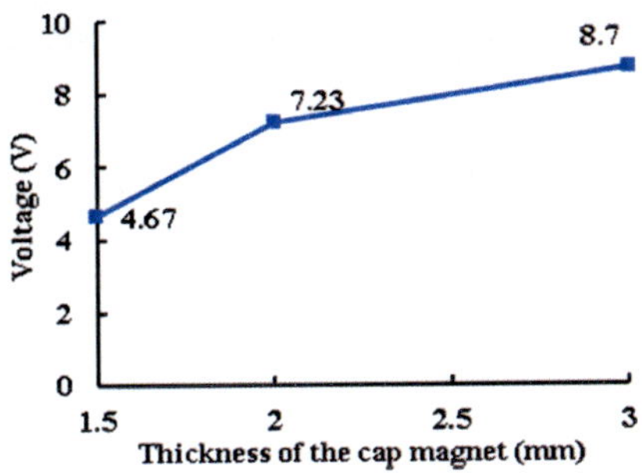

Fig. 7. The Voltage in function of the cap magnet's thickness (no resistor)

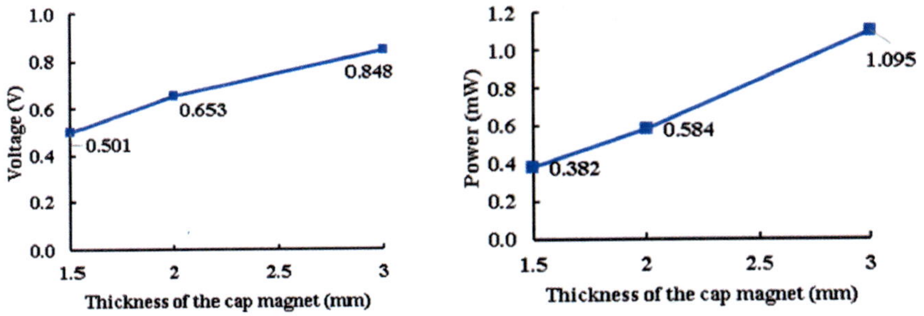

Fig. 8. The Voltage (a) and Power (b) in function of the cap magnet 's thickness (1 kΩ resistor)

5 Conclusion

In this paper, the windbelt no-load voltage calculation theory is demonstrated. From there, we can determine the parameters that directly affect the windbelt's ability to generate voltage such as: the dimension of the membrane (length, width), the thickness of the cap magnet, wind speed, number of turns of copper coil, magnetic induction of the cap magnet (magnet material), the amplitude of oscillation of the cap magnet.

The influences of the design parameters of the windbelt model such as belt size, magnet size on the parameters of no-load voltage, load voltage, and output power were investigated. At a wind speed of 2.5 m/s, the output power can reach 1mW. From the experimental process, some conclusions about the influences of design parameters on the windbelt's performance was presented. The voltage and the power increase with the increase of the length of the membrane. The voltage and the power decrease also with the increase of the width of the membrane. The voltage and the power increase as the thickness of the cap magnet increases.

There will be a difference between theoretical and experimental results. This can be explained in detail as two reasons. Firstly, some windbelt design parameters have not been included in the theoretical model for example the width of the windbelt, the magnetic forces that interfere with the vibration from the copper coil. Secondly, it comes from the accuracy of measuring instruments (ruler, voltage meter, wind speed meter). Finally, other reasons may cause the errors: flutter of the membrane is not stable, the current generated there is oscillation and instability, conductor with small resistance causes a voltage drop.

References

1. Allen, J.J., Smits, A.J.: Energy harvesting EEL. J. Fluids Struct. **15**, 1–12 (2001)
2. Bryant, M., Garcia, E.: Modeling and testing of a novel aeroelastic flutter energy harvester. J. Vibr. Acoust. **133**(011010), 1–12 (2011)
3. Sirohi, J., Mahadik, R.: Harvesting wind energy using a galloping piezoelectric beam. J. Vibr. Acoust. **134**(011009), 1–8 (2012)
4. Li, S., Yuan, J., Lipson, H.: Ambient wind energy harvesting using crossflow fluttering. J. Appl. Phys. **109**(2), 1–3 (2011)
5. Tang, L., Païdoussis, M.P., Jiang, J.: Cantilevered flexible plates in axial flow: energy transfer and the concept of flutter-mill. J. Sound Vib. **326**, 263–276 (2009)
6. Erturk, A., Vieira, W.G.R., De Marqui, C., Inman, D.J.: On the energy harvesting potential of piezoaeroelastic systems. Appl. Phys. Lett. **96**(184103), 1–3 (2010)
7. Quy, V.D., Sy, N.V.: Wind tunnel and initial field tests of a micro generator powerd by fluid-induced flutter. Energy Sustain. Dev. **33**, 75–83 (2016)
8. Tathode, P.S., Phapale, M.S., Teli, P.B., Lomate, P.J.: Generation of clean energy using concept of wind belt. Int. J. Eng. Res. Technol. (IJERT) **6**, 121–124 (2017)
9. Vinayan, V.A., Yap, T.C., Go, Y.I.: Design of aeroelastic wind belt for low-energy wind harvesting. IOP Conf. Ser. Earth Environ. Sci. **268**(1), 012069 (2019). https://doi.org/10.1088/1755-1315/268/1/012069
10. Humdinger: Wind Belt Kit, Kid Wind Project. Paul, Minnesota, USA. https://fliphtml5.com/mjoa/nclm/basic?fbclid=lwAROrxnLxwwCiKT5rZtPfzH8ko7N9VNEuxDNMLesi6dXtSZNq0v4p0LIA_ZU

Dynamics of the Micromechanical Gyroscope with the Elastic Disk Resonator

Vu The Trung Giap(✉)

Le Quy Don Technical University, Hanoi, Vietnam
vuthetrunggiap@lqdtu.edu.vn

Abstract. In the present paper, we consider a model of a thin elastic disc vibrating in its plane and subjected to inertial rotation. We obtained a new mathematical model of small longitudinal oscillations of the elastic resonator. Using the Lagrange formalism, the equations of low-frequency longitudinal oscillations of the resonator are obtained. Expressions for the natural frequencies and forms of oscillations of the disk resonator are obtained. It is shown that the oscillations of the resonator are a standing wave processing with an angular velocity proportional to the angular velocity of the base.

Keywords: Micromechanical gyroscope · The resonator · Elastic system

1 Introduction

In recent years, one of the most intensively and dynamically developing areas is microsystem technology, which includes miniature sensors of inertial and external information. Applications of new technologies of microelectromechanical systems has allowed reducing considerably weight and geometrical characteristics, consumption of energy and cost of sensors of the inertial information that has allowed expanding sphere of their application. To increase the sensitivity of the gyroscope to the angular velocity of the base, the materials with low internal losses, such as quartz and silicon, are used as well as the evacuated housing to reduce friction.

It is shown in paper [1, 2] that all the fundamental questions of the micromechanical gyroscope theory can be considered as part of the same mathematical model similar to the classical Foucault pendulum. Dynamics of sensitive elements of micromechanical gyroscopes is investigated in [3, 4]. The impact on the accuracy of the gyroscope splitting frequency oscillations arising due to instrumental errors of manufacturing is investigated in [5]. The purpose of this work is construction of new mathematical model of small longitudinal oscillations of the disk resonator made of a monocrystal with hexagonal symmetry towards determining the error of the gyroscope resonator causing caused by the material anisotropy of the resonator and manufacturing defects.

N. V. Khang et al. (Eds.): ASIAN MMS 2021, MMS 113, pp. 971–976, 2022.
https://doi.org/10.1007/978-3-030-91892-7_93

2 Dynamics of the Micromechanical Gyroscope with the Elastic Disk Resonator

Let us consider a resonator of a gyroscope as an elastic disk of the radius R and the thickness h, rigidly fixed to the basis on the radius R_0 (Fig. 1). Let $Ox_1x_2x_3$ be the right Cartesian system of coordinates connected to the basis of a gyroscope and a plane Ox_1x_2, containing an elastic disk; r, φ - polar coordinates in a plane Ox_1x_2. Let $v = v(t, \varphi, r)$ and $w = w(t, \varphi, r)$ are displacements of the points of the resonator along axes y_2 and y_3 accordingly.

The longitudinal displacements of a disk are raised by electrostatic electrodes which are located along a free edge of the resonator. Oscillations of a disk are registered by measuring gauges of capacitor type [4]. For the description of deformations of the resonator we shall enter laboratory orthogonal system of coordinate's $y_1y_2y_3$, connected with meridians and parallels of its median surface. Assume that $\mathbf{u} = (0, v, w)^T$ is the vector of linear displacements of an arbitrary point of a disc in the axes $y_1y_2y_3$.

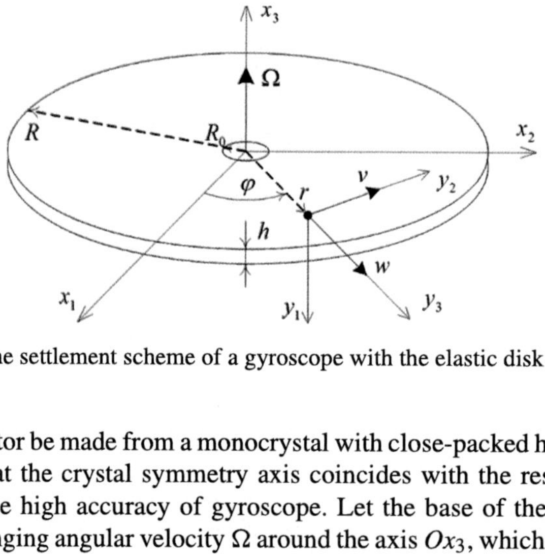

Fig. 1. The settlement scheme of a gyroscope with the elastic disk resonator.

Let the resonator be made from a monocrystal with close-packed hexagonal latitude. It is necessary that the crystal symmetry axis coincides with the resonator symmetry axis to provide the high accuracy of gyroscope. Let the base of the gyroscope rotate with a slowly changing angular velocity Ω around the axis Ox_3, which are the gyroscope sensitivity axis.

At a conclusion of the equations of the movement of the resonator we will use Lagrange's formalism. The density of kinetic energy of the disk resonator is:

$$T = \frac{1}{2}\rho h\Big((\dot{v} + \Omega(r + w))^2 + (\dot{w} - \Omega v)^2\Big) \tag{1}$$

where $\rho-$ resonator material density, $h-$ resonator thickness, the dot denotes the derivative with respect to time t.

The density of potential energy of a disk is:

$$\Pi = \frac{1}{2}h\big(\sigma_{rr}e_{rr} + 2\sigma_{r\varphi}e_{r\varphi} + \sigma_{\varphi\varphi}e_{\varphi\varphi}\big) \tag{2}$$

where $\sigma_{rr}, \sigma_{r\varphi}, \sigma_{\varphi\varphi}$ are the strains of an element of the resonator, determined by generalized law [6]:

$$\sigma_{rr} = \frac{E}{1-\mu^2}(e_{rr} + \mu e_{\varphi\varphi}), \quad \sigma_{r\varphi} = (1-\mu)\frac{E}{1-\mu^2}e_{r\varphi}, \quad \sigma_{\varphi\varphi} = \frac{E}{1-\mu^2}(e_{\varphi\varphi} + \mu e_{rr}), \tag{3}$$

here E– Young's module, μ– Poisson's coefficient, $e_{rr}, e_{r\varphi}, e_{\varphi\varphi}$ are components of deformation of a median surface of the resonator:

$$e_{rr} = w'_r, \quad e_{r\varphi} = \frac{1}{2}\left(\frac{w'_\varphi}{r} + v'_r - \frac{v}{r}\right), \quad e_{\varphi\varphi} = \frac{1}{r}\left(v'_\varphi + w\right), \tag{4}$$

hereafter the comma denotes the partial derivative of the coordinate specified by the bottom index, the dot denotes the derivative with respect to time t.

The density of the Lagrangian $L = T - \Pi$, taking into account (1), (2), (3) and (4), is as follows:

$$L = \frac{\rho h}{2}\Big[(\dot v + \Omega(r+w))^2 + (\dot w - \Omega v)^2 - \\ -\frac{\kappa}{r^2}\Big(\left(v'_\varphi + rw'_r + w\right)^2 + \frac{1-\mu}{2}\Big(\left(w'_\varphi + rv'_r - v\right)^2 - 4rw'_r\left(v'_\varphi + w\right)\Big)\Big)\Big], \tag{5}$$

where $\kappa = \frac{E}{(1-\mu^2)\rho}$.

To obtain the equations of motion of an elastic disk we use the variation principle of Hamilton-Ostrogradsky. The equations of vibrations of an elastic disk on a movable base are:

$$\begin{cases} \ddot v + 2\Omega\dot w + \dot\Omega(r+w) - \Omega^2 v - \dfrac{(1-\mu)\kappa}{2}v''_{rr} - \dfrac{(1+\mu)\kappa}{2}\dfrac{w''_{r\varphi}}{r} - \dfrac{(1-\mu)\kappa}{2}\dfrac{v'_r}{r} - \\ -\dfrac{\kappa}{r^2}v''_{\varphi\varphi} - \dfrac{(3-\mu)\kappa}{2}\dfrac{w'_\varphi}{r^2} + \dfrac{(1-\mu)\kappa}{2}\dfrac{v}{r^2} + \lambda'_\varphi = 0, \\ \ddot w - 2\Omega\dot v - \dot\Omega v - \Omega^2 w - \kappa w''_{rr} - \kappa\dfrac{w'_r}{r} - \dfrac{(1+\mu)\kappa}{2}\dfrac{v''_{r\varphi}}{r} - \\ -\dfrac{(1-\mu)\kappa}{2}\dfrac{w''_{\varphi\varphi}}{r^2} + \dfrac{(3-\mu)\kappa}{2}\dfrac{v'_\varphi}{r^2} + \kappa\dfrac{w}{r^2} - \lambda = 0. \end{cases} \tag{6}$$

Differentiate the second equation of the system (6) by φ and subtract from the first. This difference is differentiated φ once again and take into account the condition of non-extensibility. As a result, we have a differential equation:

$$\ddot w''_{\varphi\varphi} - \ddot w + 4\Omega\dot w'_\varphi + 2\dot\Omega w'_\varphi - \kappa w^{IV}_{rr\varphi\varphi} + \frac{(1-\mu)\kappa}{2}w''_{rr} + \frac{(1-\mu)\kappa}{2}\frac{w''_r}{r} - \kappa\frac{w'''_{r\varphi\varphi}}{r} \\ -\frac{(1-\mu)\kappa}{2}\frac{w^{IV}_{\varphi\varphi\varphi\varphi}}{r^2} - (1-\mu)\kappa\frac{w''_{\varphi\varphi}}{r^2} - -\frac{(1-\mu)\kappa}{2}\frac{w}{r^2} - \Omega^2\left(w''_{\varphi\varphi} - w\right) = 0, \tag{7}$$

in which Ω is considered a given time function.

Fixed Base: Consider the case where the gyroscope is on a fixed base ($\Omega = 0$). In this case, the Eq. (7) is transformed as follows:

$$\ddot{w}''_{\varphi\varphi} - \ddot{w} - \kappa w^{IV}_{rr\varphi\varphi} + \frac{(1-\mu)\kappa}{2} w''_{rr} + \frac{(1-\mu)\kappa}{2} \frac{w'_r}{r} - \kappa \frac{w'''_{r\varphi\varphi}}{r} - \frac{(1-\mu)\kappa}{2} \frac{w^{IV}_{\varphi\varphi\varphi\varphi}}{r^2} - (1-\mu)\kappa \frac{w''_{\varphi\varphi}}{r^2} - \frac{(1-\mu)\kappa}{2} \frac{w}{r^2} = 0. \quad (8)$$

Eventually find the own form of oscillations of the resonator:

$$W_{nk} = A_1 \left[B_J\left(\sqrt{a_n}, r\omega_{nk}\sqrt{b_n}\right) + \frac{B_J\left(\sqrt{a_n}, R_0\omega_{nk}\sqrt{b_n}\right)}{B_Y\left(\sqrt{a_n}, R_0\omega_{nk}\sqrt{b_n}\right)} B_Y\left(\sqrt{a_n}, r\omega_{nk}\sqrt{b_n}\right) \right], \quad (9)$$

where A_1– arbitrary constant numbers defined by boundary conditions; $B_J(a, x)$ and $B_Y(a, x)$ Bessel functions of the first and second kinds, respectively, which are solutions to the Bessel equation of the form $- x^2 y'' + xy' + \left(x^2 - a^2\right)y = 0$. In the future, k it is supposed to be equal to 1.

The final solution of the Eq. (8) is as follows:

$$w(t, \varphi, r) = W_n(r)\cos(n\varphi)\cos(\omega_n t), \quad (10)$$

where $W(r)$– is the function of the eigenform of oscillations with the number n, ω_n– the frequency of natural oscillations.

A Numerical Example: Let the resonator with a radius of $R = 20$ mm fixed on the rod of radius $R_0 = 2$ mm. The material of the resonator is fused quartz density $\rho = 2210\,\text{kg/m}^3$, modulus of elasticity $E = 7.36 \cdot 10^{10}$ Pa, Poisson coefficient $\mu = 0.17$.

Using the procedure for numerical solution of the transcendental Eq. (9) for given numerical values, we obtain a second own frequency:

$$\omega_1 = 2.716 \cdot 10^5\ \text{c}^{-1}\ (43.2\ \text{kHz}), \quad \omega_2 = 4.918 \cdot 10^5\ \text{c}^{-1}\ (78.3\ \text{kHz})$$

An example of calculating the forms of elastic oscillations of the disc resonator is presented in Fig. 2.

Mobile Base: Let the gyroscope be on a uniformly rotating base ($\Omega \neq 0$).

The solution of the Eq. (7) will be sought in the form of:

$$w(t, \varphi, r) = W_n(r)\cos(n\varphi + \lambda t), \quad (11)$$

where W_n and λ – the desired natural form and frequency of oscillations of the resonator on a mobile base.

After substituting the function (11) into the Eq. (7), we get:

$$W''_n + \frac{W'_n}{r} - a_n \frac{W_n}{r^2} + c_n W_n = 0 \quad (12)$$

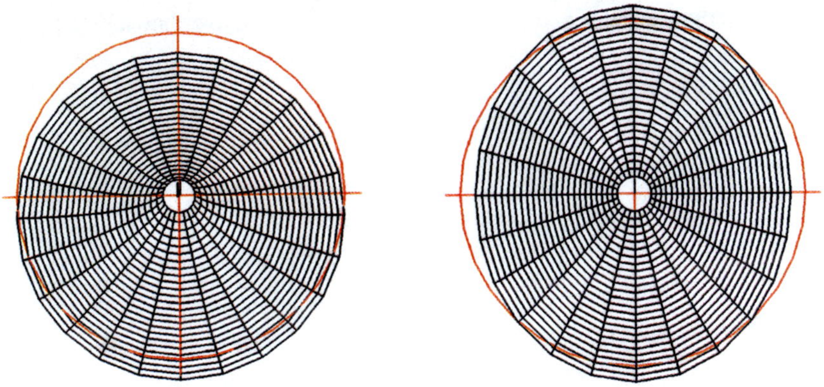

a) b)

Fig. 2. Forms of longitudinal own oscillations of the disk resonator on a fixed base:

where

$$c_n = \frac{2(n^2+1)(\lambda^2+\Omega^2) - 8n\Omega\lambda}{\kappa(2n^2+1-\mu)}.$$

The solution of the Eq. (7) is written as the sum of two traveling waves:

$$w(t, \varphi, r) = C_1 W_n \cos(n\varphi + \lambda_1 t) + C_2 W_n \cos(n\varphi - \lambda_2 t), \tag{13}$$

where

$$\lambda_{1,2} = \frac{2n\Omega \pm \sqrt{(n^2+1)^2\omega_n^2 - (n^2-1)^2\Omega^2}}{n^2+1}.$$

To create a standing wave of oscillations of the resonator, set the amplitudes of the oscillations of the traveling waves equal to $C_1 = C_2 = C$. In this case, convert (13) to the form:

$$w(t, \varphi, r) = CW_n \cos\left(n\varphi + \frac{\lambda_1 - \lambda_2}{2}t\right)\cos\left(\frac{\lambda_1 + \lambda_2}{2}t\right). \tag{14}$$

Note that:

$$\frac{\lambda_1 + \lambda_2}{2} = \frac{\sqrt{(n^2+1)^2\omega_n^2 - (n^2-1)^2\Omega^2}}{n^2+1} \approx \omega_n\,(at\,\Omega << \omega_n), \tag{15}$$

$$\frac{\lambda_1 - \lambda_2}{2} = -\frac{2n}{n^2+1}\Omega. \tag{16}$$

Thus, the angular velocity of the base causes the splitting of the own frequencies of the resonator oscillations into two close frequencies. The wave pattern of oscillations of the resonator is determined from the expressions (14), (16):

$$w(t, \varphi, r) = CW_n \cos(n\varphi + n\dot{\theta}t)\cos(\omega_n t), \tag{17}$$

where $\dot{\theta}$ is the angular rate of precession of the wave of oscillations relative to the resonator:

$$\dot{\theta} = -\frac{2}{n^2 + 1}\Omega. \tag{18}$$

Note that the scale coefficient in (18), also called the Brian coefficient, coincides with the scale coefficient of a thin elastic ring and a long thin cylinder. For the second basic form of oscillation, the scale coefficient is 2/5. Thus, the own form of oscillation of the elastic disk resonator on the mobile base is a standing wave processing with an angular velocity proportional to $\dot{\theta}$ the angular velocity of the base.

3 Conclusion

Low-frequency longitudinal oscillations of elastic disk resonator for a new type of micromechanical gyroscope are investigated. With using the Lagrange formalism, the equations of low-frequency longitudinal oscillations of the resonator are obtained. Formulas for calculating scale coefficients, natural frequencies and forms of oscillations of the resonator are obtained. Expressions for the natural frequencies and forms of oscillations of the disk resonator are obtained. It is shown that the oscillations of the resonator are a standing wave processing with an angular velocity proportional to the angular velocity of the base. A set of programs for mathematical modeling of the functioning of micromechanical gyroscopes has been created, which allows to study the influence of various design parameters of sensitive elements and base vibrations, introduced into the educational process of training specialists in the direction of "Applied Mechanics".

References

1. Zhuravlev, V.P., Klimov, D.M.: Hemispherical Resonator Gyro. Nauka, Moscow (1985). [in Russian]
2. Zhuravlev, V.P.: Theoretical foundations of solid-state wave gyroscopes. Theoretical Foundations of Wave Solid Gyroscope (WSG). Izv. Akad. Nauk. Mekh. Tverd. Tela, **3**, 15–26 (1993). (Mech. Solids (Engl. Transl.) **28**(3), 3–15 (1993))
3. Merkuryev, I.V., Podalkov, V.V.: Dynamics wave solid-state and micromechanical gyroscopes. Fizmatlit, 228 p. (2009). (in Russian)
4. Challoner, A.: Systematic disk resonator gyroscope tuning. US Pat. 2009/0301194. December 2009. Int.Cl.G01C 19/56
5. DeNatale, J.F., Stupar, P.A. Disc resonator gyroscope with improved frequency coincidence and method of manufacture. U.S. Patent 2009/0095077, April 2009. Int.Cl. G01C19/56
6. Sirotin, Y., Shaskolskaya, M.P.: Theoretical Bases of Crystal Physics. Nauka, Moscow (1975). [in Russian]

Author Index

N. V. Khang et al. (Eds.): ASIAN MMS 2021, MMS 113, pp. 977–980, 2022.
https://doi.org/10.1007/978-3-030-91892-7